Molecular Embryology of Flowering Plants

Molecular Embryology of Flowering Plants

V. Raghavan
The Ohio State University

CAMBRIDGE UNIVERSITY PRESS

PUBLISHED BY THE PRESS SYNDICATE OF THE UNIVERSITY OF CAMBRIDGE
The Pitt Building, Trumpington Street, Cambridge CB2 1RP, United Kingdom

CAMBRIDGE UNIVERSITY PRESS
The Edinburgh Building, Cambridge CB2 2RU, United Kingdom
40 West 20th Street, New York, NY 10011-4211, USA
10 Stamford Road, Oakleigh, Melbourne 3166, Australia

First published 1997

Printed in the United States of America

Typeset in Palatino

Library of Congress Cataloging-in-Publication Data
Raghavan, V. (Valayamghat), 1931–
Molecular embryology of flowering plants / V. Raghavan.
p. cm.
Includes index.
ISBN 0-521-55246-X (hardcover)
1. Botany – Embryology. 2. Angiosperms – Embryology I. Title.
QK665.R34 1997 96-44410
571.8'62–dc21 CIP

A catalog record for this book is available from the British Library

ISBN 0 521 55246 X hardback

This work is dedicated to the memory of my parents.

Contents

Preface

This book is derived from my long-standing interest in the reproductive biology of vascular plants, in particular of the ferns and flowering plants. During the initial planning stages of this venture, the book was intended as a revision of my *Experimental Embryogenesis in Vascular Plants* (Academic Press) published in 1976, with the specific aim of describing how our present ideas in molecular and genetic biology apply to embryo development in vascular plants. After I wrote two or three chapters on angiosperm embryogenesis, I began to view the project essentially as a revision of my 1986 book, *Embryogenesis in Angiosperms: A Developmental and Experimental Study* (Cambridge University Press), and aborted the idea of covering embryogenesis of ferns and gymnosperms. Reflecting the perspective of current research, it began to dawn on me that molecular and genetic principles underlying embryo development in angiosperms have much in common with principles regulating the whole gamut of reproductive processes in flowering plants. This periodic reshaping and refining of the various drafts during the past five years has resulted in the present volume providing a synthetic review of molecular and cellular aspects of the familiar sequence in the reproductive biology of flowering plants, beginning with the flower and terminating with the embryo. It is thus a wholly new book and has very little to share with its predecessors.

My objective in writing this book is to explain the progress achieved toward a molecular understanding of the reproductive processes in angiosperms with particular emphasis on embryology. In reviewing the materials for each chapter, my prime consideration has been to suppress the enthusiasm to cover strictly molecular work, but to provide the necessary structural and developmental framework into which the molecular data are inserted. The view of angiosperm embryology that this book presents, therefore, owes as much to research done during these past few decades as to the research performed during the early part of this century. I have also made an attempt to show how, by a shift in focus and by the application of new and sophisticated techniques, the study of angiosperm embryology has become a dynamic and exciting discipline. The book is written as a scholarly work of reference for use by researchers interested in the field of angiosperm embryology and as a reference text in graduate level courses on the subject.

I have attempted to provide coverage of materials that were published up to mid-1996, when the final draft of the manuscript was made press-ready. I am aware of the fact that some of the topics reviewed in this book have continued to be enlivened by a steady stream of high-quality publications reporting interesting new observations. To incorporate this information in the book at advanced stages of production would have required additions and revisions of a magnitude beyond what could be deemed practical.

The book has benefited tremendously from the generosity, cooperation, and help of many professional colleagues and others. I am deeply appreciative to many colleagues who have provided illustrative materials in the form of original pictures or photographs, prints, negatives, or slides from their publications for inclusion in the book. These scientists, listed individually, along with others have provided the necessary stimulus for this book through their devotion to the study of plant embryology. The final draft of the book was read by Professor Abraham D. Krikorian (State University of New York at Stony Brook) and Professor Joseph P. Mascarenhas (State University of New York at Albany). I am thankful to them for their comments and for convincing Dr. Robin Smith, Life Sciences Editor, Cambridge University Press, New York, that the length of the book is one of its strengths. The credit for providing consistency in style and punctuation and for converting my manuscript into the finished product goes to Ms. Dorothy Duncan, the copyeditor at Bookworks. I appreciate her professional expertise and the informed intuitions that she brought to the job. Ms. Elizabeth A. (Bette) Hellinger, administrative secretary of my department, aided me immensely in various ways and spent considerable time to rectify what appeared to be never-ending problems in my word-processing operations. Ms Vicki A. Payne, secretary of the department, provided extensive and dependable help in typing and printing. I was really fortunate to have had this kind of assistance in the "front office." At this time, I should also like to acknowledge with gratitude the role played by my teachers, Professor William P. Jacobs (Princeton University) and the late Professor John G. Torrey (Harvard University), in leading me to the study of problems in plant development, of which embryology is just one.

During the gestation period of this book, I was helped by my wife, Lakshmi, who dispensed wisdom with understanding, and by my daughter, Anita, by her sense of humor. Their love and devotion enabled me to complete this work sooner than I expected.

V. Raghavan
April 1, 1997

Abbreviations

ABA	=	abscisic acid
ACC	=	1-aminocyclopropane-1-carboxylic acid
ACP	=	acyl carrier protein
ADF	=	actin-depolymerizing factor
ADH	=	alcohol dehydrogenase
ADP	=	adenosine diphosphate
AMP	=	adenosine monophosphate
ATP	=	adenosine triphosphate
ATPase	=	adenosine triphosphatase or synthase
cAMP	=	cyclic adenosine monophosphate
CaMV	=	cauliflower mosaic virus
CAT	=	chloramphenicol acetyl transferase
CCC	=	2-chloroethyl-trimethyl ammonium chloride
cDNA	=	complementary DNA
cms	=	cytoplasmic male sterility; cytoplasmic male steriles
coA	=	coenzyme-A
con-A	=	concanavalin-A
2,4-D	=	2,4-dichlorophenoxyacetic acid
DAPI	=	4'6-diamidino-2-phenylindole
DCCD	=	dicyclohexylcarbodiimide
DMS	=	dimethylsulfate
DNA	=	deoxyribonucleic acid
DNase	=	deoxyrixbonuclease
ds-DNA	=	double-stranded DNA
dUTP	=	deoxyuridine triphosphate
EBN	=	endosperm balance number
EMS	=	ethyl methanesulfonate
ER	=	endoplasmic reticulum
ethrel	=	2-chlorethylphosphonic acid, ethylene-releasing agent
FeEDTA	=	ferric ethylenediamine-tetraacetic acid
FITC	=	fluorescein isothiocyanate
G1, G2	=	growth phases
GA	=	gibberellic acid
GTP	=	guanosine triphosphate
GTPase	=	guanosine triphosphatase
GUS	=	β-glucuronidase
H2B histone	=	lysine-rich histone
H3 histone	=	arginine-rich histone
HSP	=	heat-shock protein
IAA	=	indoleacetic acid
IBA	=	indolebutyric acid
ICL	=	isocitrate lyase
2iP	=	$N^6(\Delta^2$-isopentenyl) adenine
L-PPT	=	L-phosphinothricin
L-protein	=	leptotene protein
morphactin	=	methyl-2-chloro-9-hydroxy-fluorene-(9)carboxylate
mRNA	=	messenger RNA
MS	=	malate synthase
NAA	=	naphthaleneacetic acid
NAD	=	nicotinamide adenine dinucleotide
NAD-3	=	nicotinamide adenine dinucleotide dehydrogenase subunit-3
NADH	=	nicotinamide adenine dinucleotide, reduced
NADPH	=	nicotinamide adenine dinucleotide phosphate
NMR	=	nuclear magnetic resonance
oligo-dT	=	oligo-deoxythymidylic acid
PAS	=	periodic acid-Schiff reagent
PAT	=	phosphinothricin acetyltransferase
PCR	=	polymerase chain reaction
P-DNA	=	pachytene DNA
P-grains	=	premitotic pollen grains
PHA	=	phytohaemagglutinin
PHA-E	=	erythrocyte-agglutinating phytohaemagglutinin
PHA-L	=	lymphocyte-agglutinating phytohaemagglutinin
poly(A)-RNA	=	polyadenylic acid–containing RNA
poly(U)	=	polyuridylic acid
P-particles	=	polysaccharide-rich particles
PPDK	=	pyruvate orthophosphate dikinase
psn-DNA	=	pachytene small-nuclear DNA sequences
psn-RNA	=	pachytene small-nuclear RNA
rDNA	=	ribosomal DNA
recA-protein	=	DNA-binding or DNA-reassociation protein

RFLP = restriction fragment length polymorphism
RH-0007 = fenridazon
RH-531 = sodium-1(*p*-chlorophenyl)-1,2-dihydro-4,5 dimethyl-2-oxonicotinate
RNA = ribonucleic acid
RNase = ribonuclease
RNase T1 = RNase from *Aspergillus oryzae*
rRNA = ribosomal RNA
Rubisco = ribulose-1,5-bisphosphate carboxylase
SC-1058, SC-1271 = phenylcinnoline carboxylate compounds
SDS = sodium dodecyl sulfate
SDS-PAGE = sodium dodecyl sulfate polyacrylamide gel electrophoresis
S-period (phase) = DNA synthetic period
ss-DNA = single-stranded DNA
2,4,6-T = 2,4,6-trichlorophenoxyacetic acid
T-DNA = transferred DNA
tRNA = transfer RNA
UDP = uridine diphosphate
U-protein = DNA-unwinding protein
WGA = wheat germ agglutinin
Z-layer = Zwischenkörper or oncus
zyg-DNA = zygotene DNA
zyg-RNA = zygotene DNA transcript

List of cDNA Clones, Genes, Protein Products, and Mutants

(Abbreviations and names of cDNA clones and genes are presented here and in the text in italicized capital letters; mutants are indicated in italicized lowercase letters. Protein products are given in capital letters. In a few cases, the convention used by the authors is followed. Abbreviations for the same gene, protein product, and mutation are not listed separately.)

A-3, A-6, A-8, A-9 = tapetum-specific cDNA clones from *Brassica napus*

aba = ABA-deficient mutant seed of *Arabidopsis thaliana*

abi = ABA-insensitive mutant seed of *Arabidopsis thaliana*

ABR = protein induced by ABA in embryos of *Pisum sativum*

ac = *activator*, mobile mutator element of *Zea mays*

ACT-1 = actin gene of *Oryza sativa*

ae = *amylose extender* mutant of *Zea mays*

ag = *agamous* mutant of *Arabidopsis thaliana* flower

AG-1 = protein that binds to AT-rich sequences of phaseolin gene promoter

agl = *agamous-like* mutant of *Arabidopsis thaliana* flower

al = viviparous mutant of *Zea mays*

AMBA-1 = pollen allergen of *Ambrosia artemisiifolia*

AMBtV = gene for allergen from *Ambrosia trifida*

ant = *aintegumenta*, ovule mutation of *Arabidopsis thaliana* flower

ap = *apetala* mutant of *Arabidopsis thaliana* flower

APG = anther-specific cDNA clone from *Arabidopsis thaliana*

ARK-1 = receptor protein kinase gene from *Arabidopsis thaliana*

AT2S = albumin gene from *Arabidopsis thaliana*

ATEM-1 = *LEA* gene of *Arabidopsis thaliana*

ATP-6 = mitochondrial subunit-6 of F_1-ATP-synthase

ATP-9 = mitochondrial subunit-9 of F_1-ATP synthase

ATP-A = α-subunit of F_1-ATP synthase

ats = *aberrant testa shape* mutant of *Arabidopsis thaliana* ovule

AX-92 = axis-abundant gene of *Brassica napus* embryos

AX-110 = cDNA clone from somatic embryos of *Daucus carota*

B11E = aleurone-specific cDNA clone from *Hordeum vulgare*

B-19 = *LEA* gene family of *Hordeum vulgare*

B22E = cDNA clone from the embryo and endosperm of *Hordeum vulgare*

B-32, B-70 = proteins from *Zea mays* endosperm

BA-112, BA-118 = tapetum-specific cDNA clones from *Brassica napus*

BAG-1 = *MADS*-box gene from *Brassica napus*

BAR = gene encoding phosphinothricin acetyltransferase

BARNASE = gene for RNase from *Bacillus amyloliquefaciens*

BARNSTAR = gene that counteracts *BARNASE*

BCP-1 = pollen-specific cDNA clone from *Brassica campestris*
bel-1 = *bell* mutant of *Arabidopsis thaliana* ovule
BETV-I = pollen allergen of *Betula verrucosa*
BGP-1 = genomic clone of *BCP-1* from *Brassica campestris*
bHLH = basic region/helix-loop-helix protein
bio-1 = biotin auxotrophic mutant of *Arabidopsis thaliana* embryo
BiP = immunoglobulin heavy-chain–binding protein
BmT or HmT = toxin produced by the fungus *Bipolaris (Helminthosporium) maydis*
bno = defective kernel (embryo) mutant of *Zea mays*
BP-4, BP-10, BP-19 = pollen-specific genomic-clones from *Brassica napus* corresponding to cDNA clones *BP-401, BP-408*, and *BP-405*, respectively
BP-4A, BP-4B, BP-4C = pollen-specific cDNA clones of *BP-4* family from *Brassica napus*
BP-401, BP-405, BP-408 = pollen-specific cDNA clones from *B. napus*
BSSS-53 = inbred line of *Zea mays*
bt = *brittle* endosperm mutant of *Zea mays*
bZIP = domain of a leucine zipper DNA-binding protein
C-1 = gene in the anthocyanin pathway of *Zea mays*
C-1, C-2 = callus-specific proteins of somatic cells of *Daucus carota*
3C-12 = pollen-specific cDNA clone from *Zea mays*
C-98 = cDNA clone from anthers of *Brassica napus*
CA-1 = DNA-binding protein of phaseolin gene
CAB = chlorophyll a/b–binding protein
CAN = protein that binds to motifs in the phaseolin gene
CAT = *CATALASE* gene from *Zea mays*
CDC-2 = cell cycle control protein kinase gene from *Arabidopsis thaliana* and *Zea mays*
cDNA clone *17* = pollen-specific cDNA clone from *Brassica napus*
cDNA clone *9612* = cDNA clone expressed in the style of *Lycopersicon esculentum*
cDNA clones *92-b, 108, 127-A* = tapetum-specific cDNA clones from *Lycopersicon esculentum*
CDPK = gene for calmodulin-dependent protein kinase from pollen of *Zea mays*
CELP = gene for cysteine-rich, extensinlike proteins from flowers of *Nicotiana tabacum*
CEM-1, CEM-6 = cDNA clones from embryogenic cells of *Daucus carota*
CHI = gene for chitinase
CHI-A = gene encoding chalcone flavanone isomerase
4-CL = gene that encodes parsley 4-coumarate:coenzyme A ligase
CLV-3 = *CLAVATA3*, gene for shoot and floral meristem development in *Arabidopsis thaliana*
CMS-3 = cytoplasmic male-sterile line of *Helianthus annuus*
cms-C = cytoplasmic male-sterile Charrua line of *Zea mays*
cms-S = cytoplasmic male-sterile U.S. Department of Agriculture line of *Zea mays*
cms-T = cytoplasmic male-sterile Texas line of *Zea mays*
COB = apocytochrome-6
COR-6.6 = *COLD-REGULATED* gene induced by cold temperature in *Arabidopsis thaliana*
COX-I, COX-II = cytochrome oxidase-*c* subunits

cp	=	defective kernel (embryo) mutant of *Zea mays*
CRU	=	gene for the storage protein cruciferin
CRY-1A	=	gene encoding a protein derived from *Bacillus thuringiensis*
CVC-A	=	gene for the storage protein convicilin
CYC1-At	=	cell cycle control cyclin gene from *Arabidopsis thaliana*
cyd	=	*cytokinesis-defective* mutant of *Pisum sativum*
D-7, D-11, D-19	=	LEA proteins of *Gossypium hirsutum*
DC-2, DC-3, DC-5, DC-7, DC-8, DC-9, DC-13, DC-59	=	cDNA clones from embryogenic cells of *Daucus carota*
de-17	=	*defective seed* mutant of *Zea mays*
de-B30*	=	*defective endosperm-B30* mutant of *Zea mays*
def	=	*deficiens* mutant of *Antirrhinum majus* flower
dek	=	*defective kernel* mutant of *Zea mays*
DLEC	=	gene encoding lectin in *Phaseolus vulgaris*
ds	=	*dissociation* transposable element
DT-A	=	diphtheria toxin chain-A
E-1, E-2	=	embryonic proteins of somatic cells of *Daucus carota*
E-2	=	cDNA clone from anthers of *Brassica napus*
ECP-31, ECP-40	=	cDNA clones from embryogenic cells of *Daucus carota*
EF-1	=	elongation factor-1α
EM	=	*EARLY METHIONINE*–labeled gene from *Triticum aestivum* embryos
emb	=	*embryo-specific* mutant of *Zea mays*
EMB-1	=	cDNA clone from embryogenic cells of *Daucus carota*
EMB-30	=	embryo-lethal gene *(112-A)* of *Arabidopsis thaliana*
EMB-564	=	*LEA* gene from *Zea mays* embryos
eml-1	=	*embryoless* mutant of *Oryza sativa*
EP-1, EP-2, EP-4	=	extracellular proteins secreted by embryogenic or nonembryogenic cells of *Daucus carota*
ett	=	*ettin* mutation of the gynoecium of *Arabidopsis thaliana*
EXOPG	=	gene for exopolygalacturonase from *Zea mays*
F2S	=	pollen-specific cDNA clone from *Brassica napus*
FBP	=	*FLORAL BINDING PROTEIN* gene of *Petunia hybrida* ovule
fdh	=	*fiddlehead* mutant of *Arabidopsis thaliana*
fie	=	*fertilization-independent endosperm* mutant of *Arabidopsis thaliana*
fk	=	*fackel* pattern mutant of *Arabidopsis thaliana* embryo
fl	=	homeotic floral mutant of *Arabidopsis thaliana*; also *floury* endosperm or embryo mutant of *Zea mays*
flo	=	*floral* mutant of *Arabidopsis thaliana* flower
FR, FR-2	=	nuclear restorer genes for restoring fertility in *Phaseolus vulgaris*
FS	=	*FASS* gene that affects embryo shape in *Arabidopsis thaliana*
fus	=	*fusca* seedling mutant of *Arabidopsis thaliana*
G-1	=	gene for glycinin 1
G-10	=	genomic clone of the pollen-specific cDNA clone *TP-10* from *Nicotiana tabacum*
G-box	=	CACGTG motif of the phaseolin gene
G-box/C-box	=	GCCACGTCAG or CACACGTCAA motifs of the phaseolin gene

GCN4 = (for General Control Nonderepressible); a class of mammalian and yeast regulatory proteins
gH2B = variant lysine-rich histone of *Lilium longiflorum* pollen nuclei
gH3 = variant arginine-rich histone of *Lilium longiflorum* pollen nuclei
gib-1 = gibberellin-deficient mutant of *Lycopersicon esculentum*
GLB = gene for globulin storage protein
glo = *globosa* mutant of *Antirrhinum majus* flower
GLTP = gene for nonspecific lipid transfer protein from ray florets of *Gerbera hybrida*
gn = *gnom* pattern mutant of *Arabidopsis thaliana* embryo
GOS-2 = *Oryza sativa* genomic clone involved in the initiation of translation
G-protein = GTP-binding protein
GT = glutelin gene from endosperm of *Oryza sativa*
gu = *gurke* pattern mutant of *Arabidopsis thaliana* embryo
GY = gene for glycinin storage protein of *Glycine max* embryo
H3-1 = histone H3 gene from *Medicago sativa*
HA = parent cell line of *Daucus carota*
HAG = gene for helianthinin of *Helianthus annuus* embryos
HBP-1 = *Triticum aestivum* histone DNA-binding protein
HMG-2 = gene encoding 3-hydroxy-3-methylglutaryl coenzyme A reductase
HmT or BmT = toxin produced by the fungus *Bipolaris* (*Helminthosporium*) *maydis*
HSP-17, HSP-18.2, HSP-70, HSP-81, HSP-82, HSP-90 = heat-shock protein genes
I-3 = pollen-specific cDNA clone from *Brassica napus*
I-49 = protein involved in the resistance of pea pods to plant pathogens
ig = *indeterminate gametophyte* mutant of *Zea mays*
Iwt = conserved region in the promoter of an ABA-responsive gene
JIM-5 = monoclonal antibody for unesterified pectin
JIM-7 = monoclonal antibody for esterified pectin
JIM-8 = monoclonal antibody for arabinogalactans
JUN = JUNNA, a class of yeast regulatory proteins
KBG = cDNA clone of allergenic protein of *Poa pratensis* pollen
KEU = *KEULE* gene that affects embryo shape in *Arabidopsis thaliana*
KIN-1 = gene induced by cold temperature in *Arabidopsis thaliana*
KN = *KNOLLE* gene for pattern formation in *Arabidopsis thaliana* embryo
KN-1 = *KNOTTED* gene of *Zea mays*
KNF = *KNOPF* gene that affects embryo shape in *Arabidopsis thaliana*
KTI = gene for Kunitz trypsin inhibitor
L-3 = gene for 16-kDa oleosin isoform of *Zea mays*
LAT-51, LAT-52, LAT-56, LAT-58, LAT-59 = pollen-specific cDNA clones from *Lycopersicon esculentum*
LE = lectin gene from embryos of *Glycine max*
LEA = late embryogenesis abundant gene
LEB = gene for the storage protein legumin, B-type

lec = *leafy cotyledon* seedling mutant of *Arabidopsis thaliana*

LEG = genes for the storage protein legumin, A- and B-types

LELB = gene for the storage protein legumin, B-type

lfy = *leafy* mutant of *Arabidopsis thaliana* flower

LIM = meiotic-specific gene from microsporocytes of *Lilium longiflorum*

LIM = (for *LIN-11, ISL-1*, and *MEC-3* genes) a conserved motif of metal-binding, cysteine-rich proteins of animals

LIM-15 = bacterial reassociation protein

LMP-131A, LMP-134 = cDNA clones expressed in mature pollen grains of *Lilium longiflorum*

LOLP = cDNA clone of allergenic protein of *Lolium perenne* pollen

LTP = for *LIPID TRANSFER PROTEIN*, aleurone-specific genomic DNA clone from *Hordeum vulgare*

lug = *leunig* mutant of *Arabidopsis thaliana* flower

MA = a myrosinase gene from embryos of *Sinapis alba*

MA-12 = *LEA* gene from *Zea mays* embryos, now referred to as *RAB-17*

MAC-207 = monoclonal antibody for arabinogalactans

MADS-box = functional motif shared by *MCM-1, AG, DEF,* and *SRF* genes

MB = myrosinase gene from embryos of *Sinapis alba*

mc = *mucronate* endosperm mutant of *Zea mays*

MCM = gene that causes defects in the maintenance of minichromosomes of yeast

MHSP18-1 = heat-shock protein gene from *Zea mays*

MIC = *MICKEY* gene that affects embryo shape in *Arabidopsis thaliana*

MLG-3 = LEA protein of *Zea mays* embryos

mn = *miniature seed* endosperm mutant of *Zea mays*

Mo-17 = inbred line of *Zea mays*

mp = *monopteros* pattern mutant of *Arabidopsis thaliana* embryo

ms-4 = male-sterile mutant of *Glycine max*

ms-H = male-sterile mutant of *Arabidopsis thaliana*

mu = *mutator* transposon of *Zea mays*

MYB = recognition site in the genome, identified with myeloblastosis-associated viruses

MYC = recognition site in the genome, associated with myelocytomatosis-associated viruses

NAG-1 = *MADS*-box gene from *Nicotiana tabacum*

NAM = *NO APICAL MERISTEM* gene required for shoot meristem formation in *Petunia* sp. embryo

NEIF-4A8 = pollen-specific cDNA clone from *Nicotiana tabacum*

NOD = nodulation gene of *Rhizobium leguminosarum*

NOS = gene for nopaline synthase

NPG-1 = cDNA clone highly expressed in *Nicotiana tabacum* pollen

NPTII = gene encoding neomycin phosphotransferase

NTGLO = *MADS*-box gene from *Nicotiana tabacum*

NTP-303 = pollen-specific cDNA clone from *Nicotiana tabacum*

ogu = *ogura* cms line of *Brassica napus* and *Raphanus sativus*

OHP-1 = protein product of a *Zea mays* cDNA clone

OM-1 = *MADS*-box gene from the orchid, *Aranda* sp.
op = *opaque* endosperm mutant of *Zea mays*
OP-2 MODIFIER = gene that modifies *op-2* mutation
ORF-13, ORF-25 = open reading frames of *TURF-13* gene associated with cms-T line of *Zea mays*
ORF-138 = open reading frame associated with cms in *Brassica napus* and *Raphanus sativus*
ORF-522 = open reading frame associated with cms in *Helianthus annuus*
OSG-6B = Tapetum-specific genomic clone from *Oryza sativa* anthers
OSMADS = *MADS*-box gene from *Oryza sativa*
ovu = *ovulata* mutant of *Antirrhinum majus* flower
P-2 = pollen-specific cDNA clone from *Oenothera organensis*
$p34^{cdc2}$ = yeast cell division cycle phosphoprotein
PA = conserved sequence in the promoter of an ABA-responsive gene
PA-1, PA-2 = promoters of a gene for chalcone synthase isomerase from *Petunia hybrida*
PAL-2 = gene for phenylalanine ammonia lyase from *Phaseolus vulgaris*
PCBC-3 = monoclonal antibody for arabinofuranosyl residues
PCF = *Petunia* fusion gene, mitochondrial fusion gene associated with cms
PDLEC = phytohaemagglutinin gene of a lectin-nonproducing mutant of *Phaseolus vulgaris*
pEMB-4, pEMB-94 = embryoid-specific cDNA clones from cultured anthers of *Triticum aestivum*
PEX-1 = pollen-specific cDNA from *Zea mays*
pi = *pistillata* mutant of *Arabidopsis thaliana* flower
pin-1 = *pin-formed* floral mutant of *Arabidopsis thaliana*
ple = *plena* mutant of *Antirrhinum majus* flower
Pm = toxin produced by the fungus *Phyllosticta maydis* (teleomorph *Mycosphaerella zeae-maydis*)
PMA-2005 = *LEA* gene from embryos of *Triticum aestivum*
PMAH-9 = *LEA* gene from embryos of *Zea mays*
po = a recessive mutation of *Petunia hybrida*
pop = mutation in *Arabidopsis thaliana* defective in pollen–pistil interactions
PPE-1 = pollen-specific cDNA clone from *Petunia inflata*
PREM = pollen retroelement of *Zea mays*
PRK-1 = pollen-specific cDNA clone from *Petunia inflata*
prl = *prolifera* mutant of *Arabidopsis thaliana* affecting the embryo sac
PS-1 = gene highly expressed in *Oryza sativa* pollen
PS-1B = cDNA clone encoding *S*-associated proteins from *S-1* allele of *Petunia hybrida*
PSB-A = 32-kDa quinone-binding protein associated with photosystem II
ptd = defective kernel (embryo) mutant of *Zea mays*
PVALF = gene for *Phaseolus vulgaris* ABAI-3–like factor from embryos of *Phaseolus vulgaris*
PVS = mitochondrial DNA sequence associated with cms in *Phaseolus vulgaris*
pZE-40 = embryo-specific cDNA clone from *Hordeum vulgare*

r, rb = *rugosus*, genetic loci that control wrinkled condition in *Pisum sativum* seeds
RAB = RESPONSIVE TO ABA protein synthesized in response to ABA application
RAD-51 = bacterial reassociation protein
RBC-L = large subunit of Rubisco
RBC-S = small subunit of Rubisco
re = *reduced endosperm* mutant of *Zea mays*
RF-1, RF-2, RF-3, RF-4 = nuclear restorer genes for fertility restoration in cms *Zea mays*
rgh = defective kernel (embryo) mutant of *Zea mays*
Rho = protein factor from *Escherichia coli* that causes the termination and release of RNA molecules during transcription in vitro
Ri = root-inducing plasmid of *Agrobacterium rhizogenes*
Risø-56, Risø-1508 = mutants of *Hordeum vulgare* defective in hordein accumulation
RITA-1 = transcription factor from *Oryza sativa* endosperm
ROL-B, ROL-C = *ROOT LOCI* genes for the hairy root of *Agrobacterium rhizogenes*
RPS-12 = gene for small ribosomal subunit protein-12
RRN-26 = gene encoding 26S rRNA
RTBV = gene of *RICE TUNGRO BACILLIFORM VIRUS*
RTL = *ROOTLESS* gene of *Arabidopsis thaliana* embryo
RY-repeats = 5'-CATGCAT-3', 5'-CATGCAC-3', and 5'-CATGCATG-3' motifs found in the glycinin gene
S1, S2 = linear episomal DNAs of cms-S line of *Zea mays*
S-1, S-2, S-3, S-5, S-6, S-7, S-8, S-9, S-12, S-13, S-14, S-22, S-29, S-63 = *S*-alleles identified in *Brassica, Nicotiana*, and *Petunia*
Sa, Sf-11, Sz = *S*-alleles identified in *Nicotiana alata*
SBH-1 = *Glycine max* homeobox-containing gene
scf-1 = mutation from self-incompatibility to compatibility in *Brassica campestris*
se, seg = *shrunken endosperm* mutants of *Hordeum vulgare*
SEC-7 = secretory protein of yeast
SEF = soybean embryo factor, a protein that binds to the conglycinin gene of *Glycine max*
sep = *sepaloidea* mutant of *Antirrhinum majus* flower
sex = *shrunken endosperm expressing xenia* mutant of *Zea mays*
SF-3, SF-16, SF-17 = pollen-specific cDNA clones from *Helianthus annuus*
S-gene = sterility gene
sh = *shrunken endosperm* mutant of *Zea mays*
sin-1 = *short integuments* mutant of *Arabidopsis thaliana*
sl-2 = *stamenless-2* mutant of *Lycopersicon esculentum* flower
SLA = *S-LOCUS ANTHER* gene
SLG = *S-LOCUS GLYCOPROTEIN* gene
SLR = *S-LOCUS–RELATED* gene
sp-1 = *small pollen* mutant of *Zea mays*
spm = *suppressor mutator*, mobile mutator of *Zea mays*
SRF = serum response factor from vertebrates
SRK = *S-LOCUS RECEPTOR KINASE* gene

S-RNase = STYLAR RNase, a product of the *S*-gene

STA39-3, STA39-4 = late pollen genes from *Brassica napus*

STA44-4 = gene highly expressed in *Brassica napus* pollen

STIG-1 = stigma-specific gene from *Nicotiana tabacum* pistil

STM = *SHOOT MERISTEMLESS* gene required for shoot meristem formation in *Arabidopsis thaliana* embryo

STS-14, STS-15 = pistil-specific genes from *Solanum tuberosum*

su = mutation from self-incompatibility to compatibility in *Brassica oleracea*; also *sugary* endosperm mutant of *Zea mays*

sup = *superman* mutant of *Arabidopsis thaliana* flower

sus = *suspensor* embryo-lethal mutant of *Arabidopsis thaliana*

SUS = *SUCROSE SYNTHASE* gene of *Zea mays*

t-4 = male-fertile mutant of *Zea mays*

TA-13, TA-26, TA-29, TA-32, TA-36 = tapetum-specific cDNA clones from *Nicotiana tabacum*

TA-20, TA-56 = cDNA clones from *Nicotiana tabacum* anthers

Ti plasmid = tumor-inducing plasmid of *Agrobacterium tumefaciens*

TM-6 = *MADS*-box gene from *Lycopersicon esculentum*

TP-10 = pollen-specific cDNA clone from *Nicotiana tabacum*

TPC-44, TPC-70 = pollen-specific cDNA clones from *Tradescantia paludosa*

TR-2′ = nannopine synthase gene from *Agrobacterium tumefaciens*

TRP-E = gene for tryptophan biosynthesis from *Escherichia coli*

ts = *tassel-seed* mutant of *Zea mays*

ts-11c, ts-59, ts-85 = temperature-sensitive mutant cell lines of *Daucus carota* impaired in somatic embryogenesis

tu = *tunicate* mutant of *Zea mays*

TUA-1 = α-tubulin gene from *Arabidopsis thaliana*

TUBα = α-tubulin gene from *Zea mays*

TUB-3, TUB-4, TUB-5 = β-tubulin genes from *Zea mays*

TURF-13 = gene associated with male sterility in cms-T line of *Zea mays*

twn = *twin* embryo-lethal mutant of *Arabidopsis thaliana*

ufo = *universal floral organs* mutant of *Arabidopsis thaliana* flower

URF-13 = 13-kDa protein associated with cms-T line of *Zea mays*

URF-RMC = chimeric gene associated with cms in *Oryza sativa*

VIR = *VIRULENCE* gene of *Agrobacterium tumefaciens*

vp = *viviparous* mutant

w-3 = viviparous mutant of *Zea mays*

WGA-B = gene encoding wheat germ agglutinin isolectin B

WUN = gene encoding wound-inducible protein in potato tuber

WUS = *WUSCHEL* gene required for shoot meristem formation in *Arabidopsis thaliana* embryo

wx = *waxy* mutant

y-9 = viviparous mutant of *Zea mays*

YEC-2 = gene from yeast

ZAG = *MADS*-box gene from *Zea mays*

ZAP-1 = *MADS*-box gene from *Zea mays*

Z-locus = locus unlinked to *S*-locus that controls self-incompatibility in members of Poaceae

ZM-58.1, *ZM-58.2* = pollen-specific cDNA clones from *Zea mays*

ZMABP-1, *ZMABP-2*, *ZMABP-3* = genes from *Zea mays* pollen that encode actin-depolymerizing factors

ZMC-13, *ZMC-26*, *ZMC-30* = pollen-specific cDNA clones from *Zea mays*

ZMEMPR-9' = heat-shock protein gene from *Zea mays*

ZMG-13 = genomic clone of pollen-specific cDNA clone *ZMC-13* from *Zea mays*

ZMPK-1 = receptor protein kinase gene from *Zea mays*

Chapter 1

Reproductive Biology of Angiosperms: Retrospect and Prospect

Recent advances in molecular biology and genetic engineering have brought new and powerful methodologies to bear upon investigations into the reproductive biology of flowering plants or Angiospermae (angiosperms). These studies, which have been undertaken using a few model systems, have provided novel insights into the role of genomic information during a dynamic phase in the life of higher plants. Naturally, the modern approaches owe their origin to foundations laid in the past; taken together, the old and the new studies have led to the conclusion that the steps in the reproductive biology of plants are integrated systems of changes that occur at levels ranging from morphological to molecular. The purpose of this introductory chapter is to highlight the impact of novel conceptual approaches as well as emerging technologies in unraveling the complex mechanisms that underlie the various phases of sexual and asexual reproduction in angiosperms. Since most of our food and many other natural commodities are generated directly or indirectly from angiosperms, these studies collectively hold great promise of opening new frontiers to improve our commercial exploitation of flowering plants. Thus, the story to be told in the following chapters is as intellectually exciting as it is economically important, with many ramifications of a practical nature.

For purposes of orientation, we will begin with a brief account of the flower and the processes leading to gamete, zygote, seed, and fruit formation. The detailed structural changes that occur during these processes and the underlying mechanisms will be considered later at appropriate places in the book. Because the topics covered in this chapter are treated in a general way, no attempt has been made to key specific statements to books or to journal articles.

1. AN OVERVIEW OF ANGIOSPERM REPRODUCTION

Angiosperms constitute the division Anthophyta, which is separated into two large classes, Dicotyledonae (dicotyledons or dicots) and Monocotyledonae (monocotyledons or monocots). As the names imply, dicots have two cotyledons in their embryos, whereas just one cotyledon is characteristic of monocot embryos. Reproductive potential of almost all angiosperms is due to sexual or asexual processes or a mixture of both. The future of research in angiosperm reproduction has never been brighter, and the prospect of introduction of useful foreign genes into flowering plants to subvert the reproductive processes and modify the progeny to our advantage appears a distinct reality.

(i) Sexual Reproduction

Angiosperms are by far the most diverse and widespread of all plants and constitute the dominant vegetation of the land. Among the features attributed to the ecological spread of angiosperms are their unique sexual reproductive structure, the flower, and the fruit that is born out of it. The conventional flower is an abbreviated shoot with four whorls of modified leaves constituting the sterile and the fertile parts. The sterile parts are an outer whorl of sepals, which are usually green and enclose the rest of the flower before it opens, and an inner whorl of brightly colored petals, which aid in attracting insects and other pollinators. The fertile organs of the flower directly concerned with sexual reproduction are the stamens, representing the male units, and the carpels (or pistil, consisting of one carpel or of several), representing the female units. These four whorls produced by the floral meristem or the receptacle are precisely determined according to a genetic blueprint characteristic of each species.

The stamen consists of a threadlike stalk known as the filament, which bears at its distal end four sporangia held together in a swollen portion, the anther. While the flower is developing in the bud, microspores or pollen grains are formed in the sporangia by meiotic division of diploid microsporocytes (microspore mother cells). A subsequent step in the differentiation of the pollen grain gives rise to a large vegetative cell and a small generative cell. While these changes occur within the pollen grain, a two-layered wall, the outer layer of which (exine) is often finely sculptured, develops around each pollen grain. The final step in the differentiation of the pollen grain is the division of the generative cell to form two sperm cells. The mature pollen grain enclosing the two sperm constitutes the male gametophyte; the complex of the vegetative cell and the two sperm is christened the male germ unit.

The essential parts of the carpel are an enlarged basal portion known as the ovary and a distal receptive surface known as the stigma; these two parts are often separated by a robust stalk known as the style. The ovary contains ovules, which enclose a mass of undifferentiated cells or the nucellus. While the male gametophyte is being fabricated, a parallel process leading to the formation of the female gametophyte is orchestrated in the ovule. Within each ovule, the megasporocyte or megaspore mother cell, which is differentiated by the prior division of a nucellar cell, undergoes meiosis to form four haploid megaspores. Not long after they are carved out, three of these cells disintegrate, and the fortunate one that survives enlarges to form the embryo sac. As a result of three sequential mitotic divisions of the nucleus of the surviving megaspore followed by wall formation, the mature embryo sac comes to enclose six haploid cells: an egg, two synergids, and three antipodals. There are also two free haploid nuclei within the embryo sac that fuse to form a diploid polar fusion nucleus, now confined to the remaining part of the embryo sac known as the central cell. The mature embryo sac, consisting of seven cells, represents the female gametophyte. Of these cells, the egg and the polar fusion nucleus persist in the embryo sac, whereas the synergids and antipodals have a transitory existence. In analogy to the male germ unit, the egg, synergids, and the central cell are collectively designated as the female germ unit, but the term has not been really caught on.

Fertilization of the egg is preceded by the transfer of pollen grains from the anthers of one flower to the stigmatic surface of another flower on the same or a different plant and by the subsequent germination of pollen grains. Upon germination, a pollen grain develops a tube that pushes through the style and the tissues of the ovary to reach the vicinity of the embryo sac, where the sperm cells are discharged. "Double fertilization" in angiosperms is a unique process that has no match in other divisions of the plant kingdom, although some genera in the gymnosperms are believed to come close. Double fertilization occurs when one sperm fuses with the egg and a second sperm fuses with the polar fusion nucleus. These two fusion events result in the formation of a diploid zygote and a triploid primary endosperm nucleus, respectively.

The zygote is the first cell of the sporophytic generation. It displays a rapid burst of mitotic activity to develop into the embryo, whereas the primary endosperm nucleus differentiates into a nutritive tissue known as the endosperm. Externally, the striking changes in the embryo during this time are caused by the formation of embryonic organs like the root, hypocotyl, cotyledons, and shoot. At the cellular level, during their growth both the endosperm and parts of the embryo begin to store massive amounts of nutrients such as carbohydrates, proteins, and fats. The ovule enclosing the developing embryo and endosperm is transformed into a seed contained within the ovary. A ripened ovary becomes the fruit, which protects the seeds and aids in their dispersal.

The mature, full-grown embryo enclosed within the seed is the structural link between the completed gametophytic generation and the future sporophytic generation of the plant. Two fundamentally distinct, but interdependent, phenomena characterize the fate of the embryo in the seed. One is the capacity of the embryo to remain metabolically inactive in a quiescent or dormant state spanning a sizable segment of an unfavorable season or until such time that an essential environmental condition is met. Second is the reactivation of growth during germination leading to sweeping structural changes that result in a seedling plant. The seedling matures into the adult, which eventually flowers and sets seeds for the next generation to continue the life cycle.

(ii) Asexual Reproduction

Propagation of plants by cuttings of stems and leaves and by underground organs like roots, tubers, and rhizomes is an age-old horticultural practice. When nurtured under appropriate conditions of moisture and humidity, a favorable temperature, and the normal composition of the atmosphere, meristematic regions of the transplanted organs regenerate new plants, which continue the life cycle exactly the same way as did the original plant. A well-known natural phenomenon in angiosperms leading to the generation of new plants is adventive or asexual embryogenesis, that is, the development of embryo-like units that are not products of sexual recombination. Obviously, when plants reproduce in the absence of sex, there is no sorting of genes. Although many plants frequently reproduce asexually from vegetative organs or, rarely, reproduce from a variety of cells and tissues like the nucellus, synergids, and antipodals of the embryo sac under natural conditions, it was the development of tissue culture methods that brought to light the startling regenerative powers of plant cells, tissues, and organs. Early investigations showed that when tissues such as the pith of *Nicotiana tabacum* (tobacco; Solanaceae)[1] stem were cultured aseptically in a defined medium, they grew as disorganized masses of cells known as callus. By subtle manipulations of growth hormones, such as auxins and cytokinins, supplied to the medium, it was possible to induce a pervasive state of growth of the callus, resulting in the regeneration of shoot buds. At this stage it was necessary only to remove the buds from the original medium and transplant them onto a maintenance medium of a different composition or to vermiculite or soil to induce rooting and formation of normal plants in full multicellularity, sexuality, and structure. In an extension of this work, it was found that when single cells and clusters of cells sloughed off from a rapidly proliferating callus generated from the secondary phloem of *Daucus carota* (carrot; Apiaceae) were nurtured in a nutrient broth, they gave rise to organized growth through units that resembled various stages of embryogenic development of the zygote. Because the embryolike structures (embryoids) have their origin in somatic cells, as opposed to germ cells, the term "somatic embryogenesis" has been frequently applied to this phenomenon. Normal plants are easily regenerated from embryoids by simple manipulative methods, such as transferring them to a medium of a different composition or nurturing them under high humidity in soil or in vermiculite. A simmering controversy about the precise origin of embryoids – namely, whether they are from single cells or from cell clumps – was laid to rest with the unequivocal demonstration that in different plants, single callus cells carefully nurtured in isolation give rise to complete mature plants by organogenesis or embryogenesis. Thus, activation of a single cell to choose the embryoid pathway may significantly resemble the triggering of the egg to embryogenic divisions by fertilization.

The challenge is to apply these technologies adequately and economically to crop improvement. Methods are at hand for the long-term cryopreservation of nonembryogenic and embryogenic callus cells and for retrieving them later to induce plant regeneration. Storage of embryoids under ordinary conditions may be achieved by encapsulating them in synthetic seed coats that allow embryoids to be handled more or less like conventional seeds. Many attempts have been made to achieve this, but the results have for the most part been equivocal due to limitations in the development of appropriate coating systems and the low quality of embryoids and seedlings produced. Overcoming or bypassing the technical problems is central to the further use of this method for rapid, large-scale vegetative propagation of plants from embryoids.

By employing a mixture of enzymes that hydrolyze cell wall components, it is possible to isolate protoplasts from cells; these naked cells reform walls and attain respectability, divide to form calluses, and eventually regenerate plantlets by organogenesis or by embryogenic episodes. In extensions of this work along several lines, progress has been made that permits production of triploid hybrids by the fusion of gametic and somatic protoplasts, isolation of generative cells and sperm cells from pollen grains, microinjection of sperm cells into embryo sac, fusion of isolated

[1] When first mentioned in the book, the binomials are followed in parentheses by the common name and/or the family name of the plant.

sperm and egg cells, and isolation of viable embryo sacs and their component cells and protoplasts. Protoplasts are the choice systems for fusion experiments, because cells that retain their walls would not be able to fuse.

In yet another unexpected discovery, it was shown that when anthers of certain plants excised at an appropriate stage of development are cultured in a medium supplemented with growth hormones and other additives, a small number of the enclosed pollen grains dedifferentiate and develop into embryoids and plantlets with the haploid or gametic number of chromosomes. Confirmation of the ploidy of the regenerants is an important control in these experiments, since it provides direct support for the origin of plants from pollen grains. Embryogenic development of pollen grains has proved to be a boon for geneticists and plant breeders, because haploid plants can be obtained in quantity by anther culture techniques and transformed into true-breeding isogenic diploid lines by simple procedures. From anther culture has developed the free pollen culture system designed to control the development of pollen grains along either the gametophytic or the sporophytic pathways. Although the pollen grain is normally programmed for terminal differentiation, under culture conditions it follows a sporophytic type of program, demonstrating that the nucleus retains a developmental potential that vastly exceeds its normal fate. In a general sense, these observations and those demonstrating the formation of plants from single somatic cells carry implications far beyond asexual reproduction; they support the concept of totipotency, that is, all plant cells, except those that are committed to irreversible differentiation, possess the ability to reprogram their genome and regenerate a normal plant.

Finally, in the context of direct programming of specific cells into new developmental pathways, the induction of flower buds from identifiable cells of cultured explants is of great interest. When surface layer explants consisting of only the epidermal and subepidermal cells derived from the inflorescence stalk of tobacco are cultured, floral meristems and fertile flowers are produced by divisions of the subepidermal cells without intermediate callus formation, and they appear outside on the surface of the explant. It is appropriate to emphasize here that the nonmeristematic cells of the explant reconstitute the shape and the cell, tissue, and organ specificities that we recognize in a flower without the benefit of internal control mechanisms that operate in the shoot apex where flowers normally appear. These nonmeristematic cells of the explant should obviously contain the genetic information for the initiation and formation of floral meristems and their differentiation into flowers. The diverse ways of regenerating plants without involving sexuality and of regenerating flowers capable of sexuality from differentiated cells indicate that cells of the angiosperm plant body are in an unstable state and do not retain their existing state of differentiation under experimental conditions.

The reproductive biology of flowering plants described in this section involves several independent developmental systems suitable for analysis at various levels. As we shall see in the chapters that follow, each phase of angiosperm reproduction is a clearly ordered process whereby the structural and functional organization of the participating cells becomes expressed and integrated. This provides the necessary continuum for the biological processes of reproduction.

2. HOW REPRODUCTIVE BIOLOGY IS STUDIED

Much of our basic knowledge of reproduction of angiosperms has been garnered from studies at the morphological, histological, cytological, and physiological levels. For the most part, achievements made thus far in the study of floral morphology and pollination biology have not required an array of sophisticated instruments and equipment or specialized facilities. Especially during the early part of this century, plant scientists acquired new and unprecedented opportunities to explore the flora in the different parts of the world, conduct field observations, collect plants, and even cultivate them in new habitats. These studies, which were directed to other objectives, have proved to be invaluable to the basic science of plant reproductive biology. Investigations of this nature are being continued even to this day. Not only have these endeavors allowed researchers to probe the mysteries of sex in diverse groups of plants, but they have also opened up new approaches for developing economically important strains of some of our cultivated plants and for preserving their wild relatives.

(i) Developmental Cytology

It is light microscopy, involving examination of stained serial sections of fixed materials, that has contributed the basic framework for our attempts to compile a complete structural description of the remarkable processes of anther and ovule development, microsporogenesis, and megasporogenesis in angiosperms. This approach has also helped to define the developmental changes that occur at the level of individual cells, an accomplishment that is not possible by biochemical, molecular, or physiological methods. Initiation of floral primordia is reflected in a series of morphological changes not only in the shoot apex of the vegetative plant but also in its individual cells. The chronological sequence of changes in the distribution and positional appearance of the different parts of the flower as seen in sections in the light microscope or in the scanning electron microscope has provided a means of correlating the results of more sophisticated studies. Unfortunately, both superficial and sophisticated analyses frequently involve only a limited number of experimentally amenable species. Comparative studies of the origin of floral parts in a wide range of plants is needed to consider the currently available data in a broad biological context and to focus attention on features that are common to flowers of many species and those that are unique to flowers of just one or two species.

Some of the early studies on anther development were bedeviled by controversy as to whether the tissues of the anther arose from a single hypodermal cell (archesporial initial; archesporium) or from a group of cells several layers beneath the epidermis; it is now clearly established that a hypodermal origin of the wall layers and sporogenous cells is the mode in most, if not all plants, examined. Present interest in the tapetum – a layer of cells immediately surrounding the sporogenous cells of the anther – has been stimulated by divergent views on its origin, as deduced from light microscopic observations. Because of the dense cytoplasmic nature of the tapetal cells and the frequent divisions they undergo, some investigators erroneously thought that the tapetum is a modified outer layer of cells of the sporogenous tissue; however, based on later studies, the view on the origin of the tapetum from the descendants of the parietal cell, which itself is born out of a periclinal division of the archesporial initial, is now uncontested.

More than ever, the coming of the transmission electron microscope has enriched research in the reproductive biology of angiosperms. During the past three decades, the electron microscope has portrayed the structure of the cells of the anther and of the pollen grain as amazingly complex packets of organelles, differing from cell type to cell type, yet fundamentally the same in all. We also owe to light and electron microscopic analyses our current level of knowledge of the structure and origin of the exine of the pollen wall. Here the focus has been on the tapetal cells, which apparently synthesize precursors of sporopollenin, the primary component of the polymer that forms the exine. Sophisticated electron microscopic protocols such as rapid fixation and freeze-substitution, immunological methods, and biophysical approaches are beginning to unravel the structure of the pollen tube and the mechanism governing the ebb and flow of its cytoplasm and growth of the tube.

The identification of deoxyribonucleic acid (DNA) as the genetic material was preceded by the formulation of an elegant cytochemical procedure, the Feulgen reaction, that provided the basis for the quantitative estimation of this material. A colored complex is generated when DNA interacts with Schiff's reagent under acidic conditions that hydrolyze the purine–deoxyribose linkages and expose the free aldehyde groups. The color produced is proportional to the amount of DNA present and the quantity of DNA can be measured by densitometry. The application of this method combined with microspectrophotometry led to the determination of DNA amounts in pollen nuclei; it was confirmed that doubling of DNA occurs only in the nucleus of the generative cell, which gives rise to the sperm, whereas the vegetative cell nucleus, which does not divide, retains the expected haploid level of DNA.

A multidisciplinary approach involving histological, immunochemical, and cytochemical methods and electron microscopy has also been employed to study the female reproductive units of angiosperms. In the majority of angiosperms examined, the basic structure of the embryo sac appears to be conserved, although multiple pathways are followed to evolve this structure. At the ultrastructural level, the angiosperm egg provides a fine example of a cell whose metabolic activity is at a low ebb. In a similar way, the structure of synergids with internal projections known as the filiform apparatus, can be rationalized in terms of their function in the transport of metabolites from the nucellus to the developing egg cell.

Histochemical changes are valuable for monitoring the chemical state of individual cells, especially for localizing specific substances or sites of chemical activity in the cells. Relative levels of histochemically detectable proteins, nucleic acids, and carbohydrates have been used to define the component cells of the embryo sac and of the developing anther, embryo, and endosperm. The presence of ribonucleic acid (RNA) and proteins in particular cell types, such as the egg and the synergids, has in a limited way alerted workers to the possibility that gene activation is involved in the differentiation of these cells. This is not surprising, since a cell which is programmed to behave in a certain way in response to an internal signal will inevitably show changes when the signal is received; what is surprising is that such changes can be detected in the light microscope.

Although the histochemical approach presents a picture of cells that accumulate macromolecules at a given point in time, it does not demonstrate the unique ability of cells to synthesize any of the substances nor does it identify the sites of their synthesis. Synthesis of macromolecules by specific cells is monitored by autoradiography using radioactive precursors. Autoradiography of incorporation of precursors of DNA, RNA, and proteins into developing anthers has reinforced the image of differential nucleic acid and protein synthetic activity of the vegetative and generative cells of the pollen grain. Because autoradiographic methods are generally applied at the cellular level, they are particularly useful when heterogeneity of the tissues precludes the use of conventional biochemical techniques.

Interest in the process of fertilization has been revived by computer simulations of the three-dimensional structure of sperm cells based on electron microscopic analysis of serial sections. In attempts to understand the essentially uncontrolled way in which one sperm is presumed to fuse with the egg and the other with the diploid fusion nucleus in the act of double fertilization, a hopeful note was introduced with the discovery that in some plants the two sperm cells are structurally and functionally different. In the analysis of the beginnings of the sporophytic development of the plant, the embryo and endosperm have figured in several different levels of investigations. One of the important concepts to emerge from light microscopic analysis of embryo development in plants is that the early division sequences of the zygote are carried forward according to a blueprint characteristic of each species. In this scenario, the first division of the zygote is almost invariably asymmetric and transverse, cutting off a large vacuolate basal cell attached to the embryo sac wall (micropylar end) and a small, densely cytoplasmic terminal cell toward the embryo sac cavity (chalazal end). The organogenetic part of the embryo is derived from the terminal cell with little or no contribution from the basal cell; on its way to form the embryo, the terminal cell may divide transversely, longitudinally, or obliquely. Divisions of the basal cell create a small subtending structure of the embryo, known as the suspensor. The advantage of identifying these precise division steps is that it enables one to trace the adult organs of the embryo back to the cells cut off during early embryogenesis. Embryo development has been studied so exhaustively from a descriptive point of view that it is easily one the best understood areas of angiosperm embryology. Although fusion events that give rise to the zygote and the endosperm occur almost synchronously, examination of postfertilization embryo sacs of a number of plants has clearly shown that the primary endosperm nucleus begins to divide ahead of the zygote. The development of a free-nuclear or cellular endosperm in the embryo sac before the division of the zygote has an adaptive value in ensuring the availability of an adequate food supply to the developing embryo.

There is a further type of behavior of the embryo and endosperm, which was uncovered by a combination of light and electron microscopy, microspectrophotometry, and biochemical analysis and which mirrors their distinctive functions. This is the synthesis and accumulation of an acervate complex of food reserves, mainly storage proteins that are used by the embryo during its development and germination. In the embryo itself, storage proteins accumulate almost exclusively in the cells of the cotyledons, which undergo senescence after germination. In contrast, in the endosperm, all of its cells dedicate their synthetic machinery to the production and accumulation of food reserves. The study of storage protein synthesis has done much to stimulate interest in the dynamics of DNA changes in the embryo and endosperm cells when major increases in their fresh and dry weights occur. It has been found that the cells of developing cotyledons and endosperm continue to synthesize DNA after cell divisions cease, leading to the production of cells with disproportionately high DNA values. This abbreviated version of the mitotic cycle, known as endoreduplication, is fairly far removed from normal cellular activity, but it does offer some explanation for the continued accumulation of food reserves in the cells. It can be envisaged that an increase in gene copies might affect the amplification of those sequences concerned with the transcription of storage protein messenger RNAs (mRNAs). This is of some advantage to the plant since it allows rapid utilization of amino acids and other metabolites that are attracted to the developing fruits and seeds.

(ii) Genetics

We now consider the role of genetic studies in unraveling the reproductive biology of angiosperms. In many flowering plants, self-fertilization is prevented by a genetically controlled mechanism known as self-incompatibility. Self-incompatibility is manifest in the failure of the pollen grains to germinate on the stigma or of the pollen tube to grow through the style of the same flower. One of the outstanding contributions of genetic studies to angiosperm reproductive biology is the formulation of the view that incompatibility reactions are controlled by a single gene (sterility or *S*-gene) with multiple alleles. The result is that a pollen grain bearing an *S*-allele identical to one of the two alleles carried by the pistil suffers growth inhibition, whereas pollen bearing alleles different from both pistil alleles is able to grow.

There does not exist very much published material commenting on genetic control in other aspects of angiosperm reproduction because, until recently, research emphasis was on the comparative and descriptive aspects of the processes and not on their genetics. As alluded to earlier, this approach has resulted in a wealth of information on floral morphology, anther and ovule development, micro- and megagametogenesis, embryogenesis, and endosperm formation in a large number of plants. At the biochemical level, chemical components of the various organs and tissues concerned with reproduction have been analyzed in a covert attempt to find a chemical basis for their function. As a result, morphologists, cytologists, physiologists, and biochemists working with different aspects of plant reproduction considered genetic control to be a sideshow to the development of the structure or organ modulated by other mechanisms. Although *Drosophila melanogaster* (fruit fly) was mainly worshiped by geneticists on the zoological side, and a small core of animal cytologists and embryologists took to the liking of the fly to champion the role of genes in embryogenesis, no comparable organism was available to plant biologists.

This was the story until about 1970. During the last 30 years there have been several breakthroughs that have laid the foundation for the genetic analysis of flower pattern formation and floral morphogenesis, isolation of incompatibility genes, isolation and characterization of genes that are expressed during male gametogenesis and embryogenesis, isolation of seed protein genes, gene transfer, and transposon mutagenesis. These investigations, coupled with the development of *Arabidopsis thaliana* (Brassicaceae; hereafter referred to as *Arabidopsis*) as a model system comparable in many respects to *D. melanogaster*, promise to provide new insights into the genetic control of angiosperm reproductive biology. A brief background of this research is provided in the following section.

3. GENETIC AND MOLECULAR PERSPECTIVES

The morphological changes that occur during the reproductive development of plants are correlated with modifications in the biochemical properties of the participating cells. The central dogma of developmental biology and genetics is that these modifications are due to differential gene action. Assuming that all cells contain an identical complement of genes, the hypothesis implies that the primary cells involved in each differentiation event utilize different genes from their inherited gene pool. The major question in the analysis of the reproductive biology of plants then becomes, How is the genetic information regulated at different stages of the life cycle of the plant? When we recall that the reproductive life of angiosperms consists of a series of landmark stages, it becomes apparent that different cell types should be expressing different genes. Principles governing the genetic and molecular analysis of reproduction in angiosperms then involve the differential expression of specific genes in specific cells at specific times.

(i) Genes and Flowering

Because of the ease of generating mutants, flower development has proved to be simpler and more amenable to genetic analysis than other aspects of plant reproduction. Genetic studies of flower development have concentrated on gene systems that regulate the transformation of the vegetative shoot apex into a floral meristem and on those that cause the meristem to differentiate into cell types characteristic of each floral organ. Although the transformation of a vegetative shoot apex into a floral meristem is clearly distinct from the determination of floral organs, genes of both classes are considered to be homeotic, as mutations in them cause certain organs to be replaced by other organs not normally found in that position. In *Arabidopsis*, a broad range of mutants have been described, including plants with petals instead of stamens in the wild-type flowers, or sepals developing in positions occupied by petals in the wild type, or extra flowers in place of the ovary. Cloning and sequencing of representative organ identity genes have led to the conclusion that the wild-type products of the genes involved in certain mutations act by allowing cells to determine their positions within the floral meristem and thus to specify the fate of the organ. Were this not the case, the entire structural organization of the flower would be in jeopardy.

Special DNA sequences known as transposons possess the ability to move about in the genome. Because the insertion of a transposon into the coding sequence of a gene will prevent the expression of that gene, transposon insertion has been employed to produce homeotic mutants for structural abnormalities and pigmentation pattern in flowers. A series of mutations similar to those found in *Arabidopsis* that alter the identities of floral organs have been characterized by this strategy in *Antirrhinum majus* (snapdragon; Scrophulariaceae), which belongs to a taxonomically distant family. This suggests that the genetic mechanisms controlling floral organ identity are very ancient and are therefore highly conserved in the evolution of the dicotyledons. Transposon mutagenesis, pursued in *Petunia hybrida* (Solanaceae), promises to provide a useful molecular probe to identify genes involved in flower development. Using a complementary DNA (cDNA) library made to mRNA isolated from thin-layer explants of tobacco induced to flower, it has been possible to isolate recombinants corresponding to messages present during the early stages of differentiation of the floral meristem and to follow their expression at the cell and tissue levels. What is evident from a consideration of these results is that the morphological pattern of the flower and its organs is determined by a multitude of influences, of which genes are paramount.

Clearly, if floral differentiation involves the activation of genes, then the stability of the differentiated state within the flower also requires the involvement of genes. Support for this notion comes from the observation that each floral and vegetative organ of the tobacco plant has only a small number of common mRNAs, but has a large number of mRNAs specific for the particular organ. Cells of the floral organs are capable of a degree of organization that is matched by the presence of discernible mRNA subsets.

(ii) Gene Expression during Anther and Pollen Development

Several independent gene expression programs that correlate with the differentiated state of the specific cell types have been established in developing anthers. These include mainly mRNAs that are specific for the wall layers and tapetum of the anther and for the premeiotic pollen precursor cells and mature pollen grains. A delayed gene expression pattern apparently accounts for the elaboration of the exine of the pollen; here, transcription that determines the exine pattern is initiated during the premeiotic phase, but the exine itself is not formed until after meiosis. At least two sets of genes have been implicated in the development of the pollen grain. Genes of the first set, which are not specific to the pollen, appear to be active soon after meiosis and reach a peak during pollen development but diminish in the mature pollen. Genes of the second group, comprising pollen-specific genes, are activated at the bicellular pollen stage and their transcripts are found maximally in the mature pollen. Clearly, other gene sets, having different patterns of expression, must also be involved in pollen development; on the whole, data thus far available present compelling evidence for differential gene expression during male gametogenesis in angiosperms.

Male sterility resulting in pollen abortion in the anther has been previously shown to be caused by mitochondrial genes. At the molecular level, in comparison to corresponding DNA segments of the male-fertile plant, a highly ordered rearrangement of portions of the mitochondrial genome appears to be characteristic of the male-sterile plant. From this, one can predict that the rearranged gene might cause mitochondrial dysfunction by its inability to code for functional enzymes. Since many male-sterile mutations interfere with tapetal development, genetic engineering methods have opened up the way to introduce into plants chimeric genes that selectively destroy the tapetum during anther development, prevent pollen formation, and lead to the production of male-sterile plants.

In the area of self-incompatibility, impressive progress has been made in identifying products of the *S*-allele in members of Solanaceae and Brassicaceae and in determining how they might cause the inhibition both of pollen germination and of pollen tube growth following an incompatible pollination. This study has been made possible by immunological localization of antigens specific to various *S*-locus alleles in stigma homogenates, identification of stylar glycoproteins with individual *S*-alleles, and by the cloning and sequencing of the *S*-genes and their protein products. Suggestive of a functional involvement of S-proteins in the self-incompatibility interactions are the correlations of the temporal expression of these proteins with the onset of self-incompatibility and the spatial expression at the sites of pollen tube rejection in the stigma and style and the inhibition of self-pollen tube growth in vitro by purified S-proteins. The most interesting current question is how the S-proteins might mediate recognition between the pollen grain and the stigma. There is considerable focus now on arabinogalactan as a protein mediating in pollen recognition on the stigma and in pollen tube growth in the style independent of incompatibility reactions. Gene expression during female gametogenesis in angiosperms is an area that has not lent itself to molecular analysis. This has been attributed to the fact that the formation of the egg cell occurs within the embryo sac, which itself is buried within several layers of somatic cells of the ovule. Protocols for the enzymatic isolation of living embryo sacs from ovules and of egg from the embryo sacs of certain plants have introduced a new and hopeful note that before long it will be possible to identify genes that are required to establish and maintain polarity of the egg and to specify maternal factors for embryogenic development. A strategy most appropriate for this would be to select egg-specific recombinants from a cDNA library made to mRNAs of isolated embryo sacs or egg cells by subtraction hybridization or polymerase chain reaction (PCR) amplification.

(iii) Embryo-specific Genes

The developing embryo has already been recognized as the major organ of the plant where cell-specific gene expression programs operate. Genetics has been a powerful tool in the early stages of this work. Analysis of embryo- and seedling-defective mutants has led to the identification of genes responsible for pattern formation and morphogenesis in early-stage embryos. Growth arrest at a particular stage by a mutation implies that, due to a defective gene, the embryo is unable to complete a metabolic reaction or to synthesize a specific nutrient substance required for development. But this type of information is no substitute for hard-core molecular studies of the processes that direct polarity and specify cell lineages in the embryo; unfortunately, very little work has been done to address these problems, which first appear in the zygote and in the cells immediately derived from it. Identifying the protein products of some genes affecting divisions of early-stage embryos as ubiquitous regulators of cell wall synthesis has only reinforced the importance of cell wall formation in embryogenesis. To a large extent, genes expressed during later stages of embryogenesis, especially during cotyledon initiation and growth, persist throughout maturation and are stored in the embryo of the mature seed; many are present in both the seedling and the adult plant. Another set of genes, designated as late embryogenesis abundant (*LEA*), whose mRNAs begin to accumulate at still later stages of embryogenesis, are thought to be involved in seed desiccation when the embryo lapses into a state of quiescence or dormancy. Seed protein genes that encode the abundant storage proteins, however, represent the most striking example of a highly regulated embryo-specific gene set. The general features of a seed protein gene expression program include accumulation and decay at precise stages of embryogenesis, almost exclusive expression in embryos in contrast to other parts of the plant, and spatial expression within specific parts of the embryo. There is also evidence to indicate that seed protein gene expression is regulated by both transcriptional and posttranscriptional events. Finally, the expression of different seed protein genes in transgenic plants suggests that DNA sequences of seed proteins are highly conserved and that the regulatory factors that control seed protein synthesis are common to seeds of different plants. Although a great deal has been learned about the expression of storage protein genes and their structure, our understanding of the mechanisms controlling their temporal, spatial, and cell-specific expression still needs to be fine tuned.

The endosperm of cereal grains has also proved to be valuable for basic studies of storage protein gene expression. With the benefit of existing background knowledge on the chemistry of endosperm proteins of our common cereals, research on the isolation, characterization, and expression of endosperm storage protein genes has taken off dramatically. However, analysis of the gene activation program in the endosperm concerned with expressional divergence from the embryo developmental program has proved to be difficult in both monocots and dicots.

Progressive embryogenesis depends upon the control of the gene expression program by the environment, both in its hormonal and, to some extent, in its physical aspects also. For example, the expression of some classes of embryo-specific genes, especially those encoding storage proteins and proteins that confer desiccation tolerance in embryos, appears to be under the control of the hormone abscisic acid (ABA). This reliance of embryo gene expression on ABA is based on the fact that mRNAs for storage protein genes accumulate precociously in immature embryos in response to ABA application. Current knowledge indicates that in *Triticum aestivum* (wheat; Poaceae) embryos, an 8-bp region of the gene is the conserved core in the ABA-responsive element and that a *trans*-acting factor is functionally important in the ABA response of the gene.

Gene transfer is a powerful tool for tailoring flowering plants to perform functions far beyond the scope of their intended role and for the demonstration of DNA regions important for seed protein gene expression. This is an area of research that has made rapid strides since the 1980s. The highly conserved nature of storage protein genes of the embryo and endosperm and the similarity of the regulatory factors involved in their expression in different plants require that the input of chimeric seed protein gene promoters into the cellular machinery of heterologous hosts be subject to correct molecular processing. This has been found to be the case when embryo and endosperm storage protein genes of several plants introduced into tobacco or *Petunia* using *Agrobacterium tumefaciens*–mediated plasmid vectors became incorporated into the genome of the recipient's cells and were found to be expressed in a controlled, spatial, and temporal fashion. Significant as this discovery was, the amount of protein made by the gene in the alien environment was not sufficient to match its action in a homologous milieu. As storage protein transcriptional units are identified, they can be attached to regulatory sequences that allow their expression in plants, and the constructs can be used for transformation experiments. This will also simplify the methods to assess the stability of the genetic material in the engineered plant and to increase the yield of the protein product.

For a number of years, monocots have remained highly recalcitrant in their regenerative ability in tissue culture. This roadblock has now been largely cleared, and methods are at hand to regenerate many of the common cereals by organogenesis or by somatic embryogenesis. Unfortunately, until recently, monocots have not been readily infected by *A. tumefaciens*, although they are amenable to direct gene transfer into protoplasts or by microprojectile bombardment.

Because of their versatility and ease of handling, embryogenic cells growing in suspension cultures and embryogenic pollen grains might be exploited for a functional characterization of the gene expression program of cells embarking on the embryogenic pathway. In view of the fact that embryogenesis is induced in carrot cell suspension by the simple expedient of removing auxin from the medium, not surprisingly, only minor qualitative changes occur in the protein profile during transition from a nonembryogenic to an embryogenic state. The benefits to the embryogenic cell of using more or less the same genetic program as its nonembryogenic counterpart for differentiation are not fully understood. Genes identified during somatic embryogenesis are also found to be expressed during zygotic embryogenesis. In addition, analysis of proteins secreted into the medium by embryogenic cells has revealed the presence of a plant lipid transfer protein not found in the medium of nonembryogenic cells; expression of the gene for this protein in the early-stage somatic embryos of carrot has suggested a role for lipid transfer proteins in cutin biosynthesis on the epidermal surface of developing embryoids. The similarity in the gene expression programs of embryogenic cells of carrot and developing zygotic embryos may be more than a coincidence; it might reflect the universality of the two phenomena, resulting in identical end products.

The evidence for gene activation during pollen embryogenesis has been hard to come by. Differences have been observed in the regenerative ability of pollen grains of cultured anthers of different genotypes of the same species. This might suggest that the process is largely determined by a genetic component of the individual plant that at some point impinges on the process of pollen dedifferentiation. The clearest model of gene action during pollen dedifferentiation is based on cytological studies. The results suggest that in contrast to normal pollen development, embryogenic development is due to the transcription of new mRNA in the uninucleate pollen grains of cultured anthers. It appears that as much as possible of the gametophytic program is masked and that a sporophytic program is unfolded in but a few pollen grains. At the molecular level, some specific phosphoproteins have been associated with the induction of embryogenic divisions in pollen grains, whereas other phosphoproteins seem to be characteristic of pollen grains that continue the gametophytic program.

(iv) Gene Activity during Fruit Development and Ripening

Fruit development and ripening involve a complex series of physiological processes whose molecular biology is still incompletely understood. A brief look at the gene activation program during these events is justified not only because of their biological importance, but also because genes concerned with ripening are those that cause cellular degradation. One group of genes that has been monitored is the photosynthetic genes whose expression patterns vary at different times during fruit development. During ripening, a decline in the transcript levels of a number of plastid-encoded proteins is found to correlate with the dedifferentiation of chloroplasts into chromoplasts. In addition, genes for polygalacturonase, cellulase, and other ripening-related proteins have been cloned and their expression followed. These studies, along with the analysis of extant and in vitro synthesized proteins of fruits of various ripening stages, have shown that genes for both synthesis and degradation of proteins are activated during the ripening process. Since the hormone ethylene controls the ripening of many fruits, antisense RNA strategy has been used to inactivate the synthesis of a rate-limiting enzyme in the biosynthetic pathway of ethylene and to delay fruit ripening. This discovery has raised the prospects of extending the life span of fruits, thereby preventing spoilage. Indeed, genetically engineered tomatoes with a longer shelf-life than normal ones are already being sold in the supermarkets.

4. MOLECULAR AND GENETIC ENGINEERING TECHNIQUES

The achievements in the realm of angiosperm reproductive biology described in the previous section have been made possible by refinements in the existing methods in biochemistry and cell and molecular biology and by the introduction of the new disciplines of biotechnology and genetic engineering. These methods are not only numerous, but they are also very complex. Since it is assumed that readers will be familiar with these methods, no attempt is made here to describe them. However, the principles underlying some of these techniques to which references are made in this book without further explanation will be described.

Sodium dodecyl sulfate polyacrylamide gel electrophoresis (SDS-PAGE): This method is used to fractionate nucleic acids and proteins using polyacrylamide as a support medium. When electrophoresis is performed in the presence of the ionic detergent sodium dodecyl sulfate (SDS), the method can be used to determine the molecular mass of migrating proteins. After electrophoresis, the gel is stained to locate the bands or is autoradiographed against an X-ray film to locate radioactively labeled macromolecules.

Two-dimensional electrophoresis involves the fractionation of proteins in two dimensions based on independent properties such as the isoelectric point and the ability to complex with SDS. A staggering number of proteins can be separated by this method.

Cell-free translation: This method employing preparations from wheat germ or rabbit reticulocytes as the source of cytoplasmic ingredients necessary for protein synthesis is used to confirm the functional competence of mRNA or polyadenylic acid–containing RNA [poly(A)-RNA] isolated from cells. A labeled amino acid is added to the reaction mixture to determine the amount and nature of proteins synthesized. For the latter purpose, the proteins are separated by SDS-PAGE.

Molecular hybridization: It is possible to assay for the immediate products of gene expression by RNA–DNA hybridization and to determine gene sequence complexity by DNA–DNA hybridization. In practice, one population of the single-stranded nucleic acid, present in low concentrations, is radioactively labeled; the other population, present in excess, is unlabeled. Separation of the hybrids is accomplished by fractionation on hydroxyapatite column or by S1-nuclease digestion. The former is used to bind selectively double-stranded molecules, whereas the latter selectively degrades single-stranded molecules. By either method, the percentage of hybrids formed is obtained by determining the radioactivity in the fractions.

In situ hybridization: This method is ideal for localizing gene sequences on chromosomes at the light microscope level by DNA–DNA hybridization using labeled DNA probes and in cells by RNA–RNA hybridization using labeled antisense RNA probes. Total RNA is localized in cells and tissues by annealing with labeled polyuridylic acid [poly(U)] as a probe. Sites of annealing of labeled probes are detected in both cases by autoradiography. Nonradioactive digoxygenin-labeled oligodeoxythymidylic acid (oligo-dT) probes are also widely used to localize specific transcripts and total RNA in cells. To localize proteins, antibodies that are radioactively labeled, or coupled with an enzyme, or linked to an electron-dense heavy metal like gold, are used and the sections are viewed in a light, fluorescence, or electron microscope.

cDNA library: A cDNA library or a cDNA clone bank is a population of DNA copies of homogeneous mRNA sequences coding for a specific protein, or DNA copies of representatives of most, if not all, mRNAs of a cell, tissue, organ, or organism, inserted into a plasmid or phage vector and cloned into bacterial cells. Cloning is initiated by the isolation of mRNAs transcribed by genes involved in the synthesis of a particular protein or active in a specific organism or its constituent parts.

Differential screening is employed to isolate genes specifically expressed in a cell or organ or genes involved in physiological processes whose protein products are unknown. For example, many genes specifically involved in pollen development as well as others common to the pollen, stem, and leaf of the plant are likely to be present in a cDNA library made to poly(A)-RNA from mature pollen grains. To eliminate recombinants common to all systems and to enrich the library for pollen-specific clones, the cDNA library is replica-plated and probed with labeled cDNAs made to mRNAs from pollen, stem, and leaf. In this method, bacterial colonies with inserts that are pollen specific will give positive signals only to the probe made from pollen mRNA, whereas colonies with inserts containing sequences similar to those present in the pollen, stem, and leaf will react positively to all three cDNA probes.

If the protein product of the gene is known, one of the simplest procedures for screening the cDNA library is the use of a protein detection method based on immunological techniques. Alternatively, oligonucleotides synthesized to the amino acid sequence of the protein of interest are used as probes to pick specific clones from the library.

Genomic library: In contrast to the cDNA library, which is made from mRNA, the genomic library has its origin in the total DNA of the organism. Genomic clones are useful for revealing the structure of the gene in the chromosome and the structure of the sequences flanking it. They may also provide information about the presence or absence of introns that are not found in the mRNA. Genomic cloning is generally accomplished using bacteriophage lambda as a vector, and hybridization techniques similar to those described for cDNA clones are applied to identify individual genomic clones.

Filter hybridization: This constitutes a family of methods in which nucleic acids or proteins are initially adsorbed to a nitrocellulose filter before they are hybridized with labeled probes or reacted with antibodies to assess their size, tissue distribution, and quantitative expression. Southern blot is employed to identify electrophoretically separated DNA fragments transferred from agarose or polyacrylamide gels to membrane filters. When RNA fragments separated on agarose gels under denaturing conditions are transferred to a filter and hybridized with complementary single-stranded DNA (ss-DNA), the method is known as Northern blot. For some experiments it is preferable to spot nucleic acids directly on the filter membrane using a specially designed apparatus to produce uniform spots, rather than transfer them after electrophoretic separation. This is the basis for slot–blot or dot–blot hybridization; after hybridization, the spots are excised and the hybridized radioactivity is counted. Another filter hybridization technique, known as Western blot, is used for the identification and quantification of specific proteins in complex mixtures by transferring them from a SDS-PAGE gel to a membrane and probing with antibodies that react with target proteins.

DNA sequencing: Determination of the arrangement of the particular sequences of base pairs in a DNA molecule is the basis of DNA sequencing and constitutes the method of choice to determine the differences between individual molecules of DNA. The Maxam–Gilbert chemical degradation method and the Sanger dideoxy-mediated chain termination method are currently employed to sequence DNA.

Polymerase chain reaction: This new technique is extremely efficient for exponential amplification of DNA sequences. Using two oligonucleotides as primers and repeated cycles of denaturation, annealing, and DNA synthesis, substantial amounts of double-stranded DNA (ds-DNA) terminated by the primers is obtained. The method is especially useful for amplifying small pieces of target sequences from a pool to generate probes to screen a cDNA library.

Antisense RNA strategy: This is used to down-regulate the expression of specific genes of interest in cells and tissues. When antisense genes are introduced into plants, they promote the synthesis of noncomplementary RNAs, which subsequently hybridize with target RNAs and prevent their expression. Cells that express antisense genes are thus starved for proteins normally coded by complementary RNAs and eventually they disintegrate.

Construction of chimeric genes: Here, the regulatory elements from one gene are fused to the coding sequence of a prokaryotic reporter gene whose product can be easily detected by simple staining reactions or which produces discernible effects, such as cytotoxicity, on target cells. Among the reporter genes that can be readily detected by means of sensitive histochemical tests are those encoding for β-glucuronidase (GUS) and chloramphenicol acetyltransferase (CAT). Diphtheria toxin is a member of a group of protein toxins that inhibit protein synthesis, and expression of diphtheria toxin chain-A gene (*DT-A*) driven by a promoter causes cell death as a visible phenotypic expression.

Gene transfer: Gene transfer methods involve the introduction of DNA into plant cells, integration of the foreign DNA into the host plant's genome, regeneration of transgenic plants, and selection of transformants. Techniques used to deliver foreign DNA into plant cells include particle bombardment, microinjection, electroporation, and *Agrobacterium*-mediated transformation. Particle bombardment capitalizes on the fact that metal microprojectiles (particles) carrying DNA fragments of appropriate size, if propelled by an acceleration device to high velocities, penetrate the cell walls and cell membranes of plants without damaging them. For bombardment, gold and tungsten particles coated with precipitated DNA or frozen micropellets of DNA acting as microprojectiles are used. In the microinjection method, glass micropipettes with openings of less than 1 μm in diameter are used to deliver DNA into targets such as protoplasts, isolated pollen grains, and single cells or cells in small clusters. Electroporation is based on the application of a pulse of high-voltage electric field to targets suspended in a buffer, which results in a reversible permeabilization of the plasma membrane to macromolecules.

Because of the ease with which different strains of *A. tumefaciens* deliver foreign DNA into the host genome, transformations with high efficiencies have been achieved in many species by the use of this vector. As is well known, *A. tumefaciens* readily infects various species of dicots to produce tumorous outgrowths known as crown-gall tumors. Tumor induction is the result of transfer and integration of a specific segment of the tumor-inducing (Ti) plasmid DNA from the bacterium into the nuclear genome of the susceptible plant cells. The inserted DNA segment of the Ti plasmid is known as the transferred DNA (T-DNA); the other critical region of this plasmid is a sequence encompassing the virulence (*VIR*) genes encoding proteins required for T-DNA transfer.

To serve as an effective vehicle for gene transfer, the bacterium is first disarmed by deleting the T-DNA genes and then is equipped with a marker gene for antibiotic resistance to give a selectable advantage to the transformants. The desired foreign gene is then ligated between the T-DNA borders of the vector, and bacteria with the genetically engineered plasmid are used to infect cultured plant materials, which typically include leaf discs, hypocotyls, stems, roots, embryos, and embryogenic callus. Depending upon the plant material, the type of explant, and the composition of the regeneration medium, regenerants may appear as callus, shoot buds, or somatic embryos. These are further subjected to appropriate tissue culture protocols to obtain normal plants. Among the methods used to screen regenerants and identify transformants, perhaps the easiest are assays for *GUS* or *CAT* genes.

At a more sophisticated level, DNA is isolated from the putative transformants and subjected to Southern blot hybridization to identify the inserted genes, or the inserts are separated from restriction endonuclease–digested DNA by electrophoresis and amplified by PCR.

5. SCOPE OF MOLECULAR EMBRYOLOGY

Against the background of review of the reproductive biology of angiosperms presented in this chapter, it is now necessary to describe the scope of molecular embryology, the subject of this book. Whereas embryogenesis as applied to angiosperms is concerned with the study of zygotic, somatic, and pollen-derived embryos and of the endosperm, most plant biologists include under embryology the study not only of the embryo and endosperm, but also of the ensemble of morphological, cytological, and anatomical changes concerned with sporogenesis, gametogenesis, and fertilization. As conceived in this book, embryology involves both premeiotic and postmeiotic events in the anther and ovule, encompassing both diploid and haploid phases in the life cycle of the plant; in short, embryology is the plant's own strategy to perpetuate the progeny. Thus, the topics that are germane for discussion under the rubric of embryology include anther and ovule development; microsporogenesis and megasporogenesis; prefertilization events; fertilization; zygotic, somatic and pollen embryogenesis; and endosperm formation. These past studies in embryology, described in some classical books and scores of scientific papers, have been carried forward by many investigators in laboratories around the world. They provided a useful framework for an appreciation of the range of variations in the development and organization of the male and female reproductive units of angiosperms and the process of embryogenesis. Subsequent evaluation of the data led to a diversification in the outlook in research in plant embryology. The utilization of embryological information in determining the phylogenetic affinities of certain controversial families, genera, and species of flowering plants was the immediate outcome of such diversification. Embryological data have also been used to find new affinities for certain species and genera of plants that were rather uncomfortable in their old alliances.

Since the 1930s, advances made in the fields of plant physiology, biochemistry, and genetics, and refinements in the culture of plant organs, tissues, cells, and protoplasts under aseptic conditions have had much influence on the direction of embryological research. The outcome of research in these fields introduced the discipline of experimental embryology, involving the control of pollination and fertilization and manipulations of the anther, pollen grains, ovary, ovule, and embryo by excision and culture, by chemical, hormonal, and surgical treatments, and by exposure to selected day-lengths and temperature conditions, as a way to study the controlling mechanisms that affect the form and structure of plant reproductive organs. Studies in experimental embryology were enshrined in the hope that as we gain familiarity with the laws that control the reproductive processes in plants, we will get new clues about how to control them to our advantage. Thus, experimental embryology transformed an era of descriptive studies into one of experiments and deductions, designed to discover the physiological basis of the reproductive processes in angiosperms.

From experimental embryology, it is perhaps one further step of complexity to molecular embryology. In recent years, the most profound impact on the study of angiosperm embryology has come from advances in molecular biology, recombinant DNA technology, and gene cloning. The ability to isolate specific genes that modulate developmental processes in living organisms opened the way for dramatic new approaches in the study of angiosperm embryology. The insights obtained in our understanding of the molecular and genetic basis for the different embryological processes in angiosperms are considered here as molecular embryology. In general, these studies are concerned with the spatial and temporal activation of genes involved in embryological processes, isolation and cloning of genes, sequencing the genes, determination of the amino acid sequence of the proteins encoded by the genes, and identification of the protein products. The use of embryos and embryogenic cells for gene transfer is also considered to be a topic in molecular embryology. The developments that were to herald molecular embryology may be said to have had their origin in the 1970s; although somewhat late blooming, research in this field has taken off dramatically, providing an increasing awareness of the reality of genetic engineering for improvement of our agricultural crops. Therein lies the importance of molecular embryology in the reproductive biology of angiosperms.

6. ORGANIZATION OF THE TEXT

Following is an abbreviated survey of the contents and organization of this book. Based on the wider comprehension of embryology, the book is divided into four major sections, entitled "Gametogenesis" (five chapters), "Pollination and Fertilization" (five chapters), "Zygotic Embryogenesis" (four chapters), and "Adventive Embryogenesis" (two chapters). Included in the section entitled "Applications" is a concluding chapter on the genetic engineering of embryos. The first section of the book closely examines the morphological, cytological, biochemical, and molecular aspects of the development of the anther and the ovule that result in the formation of functional male and female gametophytes. The background of events thereby established serves as a basis for the second section, dealing with pollination and fertilization. The problems of pollination and fertilization are approached from the points of view of pollen deposition on the stigma and the resulting recognition reactions, in vivo and in vitro growth of pollen tubes, phenomena of self- and cross-incompatibility, and, finally, sexual recombination heralding the sporophytic growth. The third section is concerned with the growth of the embryo and endosperm and with gene expression in cells with different genetic backgrounds. Two chapters on adventive embryogenesis, dealing with somatic embryogenesis and pollen embryogenesis, form the focus of the fourth section. Despite the fact that work on the transformation of embryos has not made a significant dent in practical or theoretical considerations, a short account of the current work in this area rounds off the descriptive part of the book. Although the framework of discussion is provided mainly by angiosperms, some contributions from the gymnosperms are also stressed, especially when they illustrate principles that are generally applicable. According to one viewpoint, the life of the embryo is not considered complete until the seed germinates; unfortunately, the many fascinating and intriguing biochemical and molecular aspects of germination of the seed (which by themselves would have formed another bulky section of the book) have not been included here. Notwithstanding this deliberate omission, it is hoped that the book will be of use in bringing together the diverse and scattered literature on the developmental and molecular aspects of plant embryology and will give a general idea of the current and future perspectives of this exciting field.

SECTION I

Gametogenesis

Chapter 2

Anther Developmental Biology

The anther is a morphologically simple organ of the flower concerned with some unique functions, such as microsporogenesis and the production of pollen grains. The size, shape, and orientation of the anther on the stamen and the diversity of the cells and tissues of the anther make it a morphogenetic system of great interest. A typical anther is a two-lobed organ with two locules or microsporangia in each lobe and, so, the anther is functionally a group of four microsporangia. A single vascular strand is embedded in the connective, which is the tissue found in the central region of the anther between the two lobes. In most plants, anthers are readily accessible to observations and manipulations; for these reasons, and in recognition of their role in the sexual reproduction of plants, anther developmental biology is one of the most extensively studied topics in plant embryology. Over the years, the goal of these studies, aptly termed "classical," has been to gain insight into the mechanisms by which the anther reacts to developmental information communicated to it. As a result, there has been an accumulation of considerable data on the histology underlying the differentiation of specialized cells and tissues of the anther and the physiology of anther growth, as well on the isolation and characterization of genes preferentially expressed during anther development.

In this chapter, most attention will be devoted to the molecular aspects of anther development in angiosperms leading to meiosis in the microsporocytes and the production of haploid microspores. However, for a proper appreciation of these studies, it is necessary to preface the accounts with brief surveys of a cluster of related problems that include anther determination and the ontogeny and physiology of the anther. These topics have been reviewed periodically during the past 30 years (Vasil 1967; J. Heslop-Harrison 1972; Mascarenhas 1975; Bhandari 1984; Shivanna and Johri 1985).

1. ANTHER DETERMINATION

In the analysis of the factors concerned with stamen or anther determination in a flower, various experimental approaches coalesce. As in the case of leaves, the filament is the first-formed part of the stamen and the anther is a later-derived part. The stamens begin their life on the floral meristem, which initially is a mass of homogeneous cells but later becomes a mosaic of determined cell types. Following the initiation of the sepals and petals, the stamens arise as small, radially symmetrical outgrowths of cells with potentially organogenetic properties. These outgrowths presage determination, after which their developmental pathway becomes fixed as stamens. With further growth, the protuberances become cylindrical at the base and bilaterally symmetrical at the top. Once initiated, the stamen primordium completes its development rapidly, within a few days, to satisfy the structural and functional requirements of the organ. During development, the position of the primordia on the floral apex apparently determines their identity as stamens and causes them to follow the characteristic genetically determined program. In this scenario, determined cells of each of the incipient stamens must assess or know their relative positions within the primordium and differentiate into the cell types appropriate for their parts, such as the filament, connective, and anther. However, the factors responsible for the way in which cells destined to become the anther determine their position on the stamen primordium, establish priority, and differentiate are not fully known. This is attributed partly to the microscopic size of the floral meristem and partly to the lack of clues about the diversity of causal factors. For example, a cell's knowledge of its position in the floral meristem preparing to form an organ could be imparted by the concentration gradient of a hormone in the meristem.

(i) Timing of Anther Determination

In an early study, stamen development was monitored in flower buds of *Aquilegia formosa* (Ranunculaceae), in various stages of development, cultured in media supplemented with plant hormones such as indoleacetic acid (IAA), 2,4-dichlorophenoxyacetic acid (2,4-D), gibberellic acid (GA), and kinetin. Even the very early stage flower buds cultured in a favorable medium developed stamens with normal anthers, indicating that determination has already occurred in the floral primordium (Tepfer et al 1963). Similar results were obtained when excised flower primordia of *Nicotiana tabacum* were cultured in a medium with or without kinetin (Hicks and Sussex 1970; McHughen 1980). These observations do not necessarily identify a role for hormones in fixing the developmental fate of cells in the differentiative pathway, nor do they give any insight into the process itself.

Other studies have shown that under certain conditions it is even possible to alter the developmental pathway embarked upon by cells of the stamen primordium. For example, Hicks (1975) found that potential stamens of *N. tabacum* are partially transformed into carpeloid and stigmatoid organs when the primordia are cultured in a medium containing kinetin. This perplexing finding presumably indicates that the stamen is a developmentally flexible organ of the flower. The reversion of stamens with twisted, distorted anthers bearing naked external ovules found in the *stamenless-2* (*sl-2*) mutant of *Lycopersicon esculentum* (tomato; Solanaceae) to wild-type stamens has yielded further evidence of developmental flexibility of the stamen. Despite the striking phenotypic expression of the mutation, the primordia of normal and mutant stamens resemble each other. Whereas treatment of mutant plants at the stamen primordial stage with GA leads to the development of stamens with wild-type features, later application of the hormone results in the appearance of abnormal mutant stamens (Sawhney and Greyson 1973a, 1979). It is logical to interpret these data as indicating that the primordia that respond to GA are undetermined and are not affected by the mutation. In contrast, older primordia are presumably determined and become sensitive to the genetic lesion.

Insofar as overt morphological differentiation of the stamen is concerned, there is some evidence suggesting that cues from other floral organs on the meristem may influence the process. For example, in cultured floral meristems of tobacco, excision of the territory of the sepals and petals or suppression of growth of the sepals and petals interferes with the formation of stamens (McHughen 1980). It also appears that developmental interactions between floral organs affect the initiation of stamens. Tepfer et al (1963) found that in cultured flower buds of *A. formosa*, the sepals had to be removed if the remainder of the floral organs were to faithfully recapitulate their normal mode of differentiation. The course of development pursued by the stamens is also markedly affected by the presence of GA in the medium, because the hormone stimulates the growth of all floral organs except that of the stamens. This observation is of interest in indicating that the altered potential for organ initiation on the floral meristem might result from its changing hormonal status.

It is also possible that interactions between other floral organs are involved in the formation of stamens in unisexual flowers. In *Cucumis sativus* (cucumber; Cucurbitaceae), there are separate genetic lines of plants, which produce male, female, and bisexual flowers. Although the kind of organ destined to develop may be inherent in the bud stage, there is compelling evidence from the culture of flower buds to show that the fate of the stamen is originally undetermined. Thus, very young potential male buds tend to form carpels when cultured in the basal medium; it is also possible to increase the prevalence of femaleness in slightly older buds with visible stamen primordia by the inclusion of IAA in the medium. Simultaneous addition of IAA and GA negates the effect of auxin in promoting carpel development. Suggestive of an antagonism between the development of stamens and of carpels is the observation that in no case do the flowers become bisexual; no hormone treatment is also found to transform the potential female flowers into males (Galun, Jung and Lang 1963). These experiments provide tangible evidence for the operation of subtle and complex influences even as early as the determinative phase of anther development in the flower. Some observations on the role of kinetin in the selective stimulation of stamen development and of kinetin and GA in the development of the pistil and abortion of stamens in cultured ear shoots of *Zea mays* (maize; Poaceae) also seem to bear on the undetermined nature of the stamen (Bommineni and Greyson 1987, 1990).

(ii) Genetics and Molecular Biology of Anther Determination

The fundamental influences that determine whether and where a stamen with a normal anther will differentiate on the floral meristem are ultimately attributed to genetic, biochemical, and molecular factors. In *Nicotiana langsdorfii* × *N. alata*[1] hybrids, divergence of the stamen primordium from the normal developmental pattern, giving rise to a stigmatoid organ from the anther, is reported to be a heritable trait involving nuclear genes (White 1914). In *N. debneyi* × *N. tabacum* hybrids, development of stigmatic characters on the anthers has been traced to cytoplasmic elements of the female parent (Sand and Christoff 1973). In this case, determination of the fate of the organ appears to have been made early in ontogeny, since the phenotypic expression is not altered by in vitro culture of undifferentiated, prestigmatic stamen primordia (Hicks and Sand 1977). By observing changes in the phenotype of the stamen following crosses, these experiments have revealed something of the origin of the genetic factors controlling determination of this floral organ. In line with the genetically based biochemical differentiation of floral organs on the receptacle, polypeptides unique to stamens and carpels have been detected in one-dimensional and two-dimensional electrophoresis of soluble and in vitro–translated proteins of floral primordia of certain plants, but the available data do not permit informed speculation about the nature of the proteins triggering sex expression (Sawhney, Chen and Sussex 1985; Bassett, Mothershed and Galau 1988; Galli et al 1988; Bhadula and Sawhney 1989a; Bommineni et al 1990; Rembur et al 1992; Thompson-Coffe et al 1992).

[1] In designating crosses, the first-named is the female parent.

Maize is a particularly suitable object for genetic studies because many mutations that affect sex determination of the normally unisexual flowers are known (Veit et al 1993). Some mutations affect sex determination of the florets, permitting the development of pistils in the staminate floret. *Tassel seed* (*ts*) and *tunicate* (*tu*) are two mutants in this category that develop pistillate florets instead of staminate florets in the tassel; genetic analysis has shown that the genes concerned normally function in pistil abortion in the male florets (Irish and Nelson 1993; Irish, Langdale and Nelson 1994; Langdale, Irish and Nelson 1994). The molecular and biochemical basis for these mutations remains largely unknown. However, the sequence similarity between a TS protein and short-chain alcohol dehydrogenases (ADH) has fueled the speculation that the mutation may be brought about by a series of sex determination genes some of which act on gibberellins or steroidlike molecules of the floral meristem (DeLong, Calderon-Urrea and Dellaporta, 1993).

Homeotic Mutants in *Arabidopsis*. A major contribution to our understanding of the mechanism of anther determination has come from an impressive series of studies on *Arabidopsis* and *Antirrhinum majus*; much of this work has been reviewed (Coen and Meyerowitz 1991; Meyerowitz et al 1991; Sommer et al 1991; Okamuro, den Boer and Jofuku 1993; Ma 1994; Haughn, Schultz and Martinez-Zapater 1995). Flowers of *Arabidopsis* are favorable for research because of their bilateral symmetry consisting of only one whorl each of the calyx with four sepals, corolla with four petals, androecium with six stamens, and gynoecium of two fused carpels. Other fortuitous biological features, such as the small size of the plant, its abbreviated life cycle, and the small size of its genome, as well as the availability of a good background of classical genetics, have also made this species a model system for research in plant molecular biology and genetics. In the approach used to generate mutants, seeds are chemically mutagenized and phenotypes that interfere with the orderly development of floral parts are selected. A general characteristic of the mutations is that they cause havoc on the floral meristem as a whole, rather than specific defects on the individual floral organs. Although abnormalities in floral structure are usually held in abeyance in the wild type, mutations affect the differentiation of organs in particular positions on the meristem and produce abnormal flowers. Mutant flowers therefore reveal the operation of genetic influences and indeed provide what is in many respects a convenient system for the molecular investigation of the determination of floral organs. Especially useful in such studies are genetic lesions that produce floral organs in alien locations on the receptacle, such as petals in the place of stamens in the third whorl and a new flower primordium instead of carpels in the fourth whorl (*agamous* [*ag*]) (Bowman, Smyth and Meyerowitz 1989, 1991), facsimiles of carpels and stamens in the place of sepals and petals, respectively, (*apetala* [*ap-2*], *fl-40, fl-48,* and *leunig* [*lug*]) (Komaki et al 1988; Kunst et al 1989; Bowman, Smyth and Meyerowitz, 1989, 1991; Liu and Meyerowitz 1995), carpels where stamens are normally present (*pistillata* [*pi*], *ap-3,* and *universal floral organs* [*ufo*]) (Hill and Lord 1989; Bowman, Smyth and Meyerowitz 1989, 1991; Levin and Meyerowitz 1995), stamens that lack anther locules (*fl-54*) (Komaki et al 1988), stamens in the place of carpels (*floral* [*flo-10*], and *superman* [*sup*]) (Schultz, Pickett and Haughn 1991; Bowman et al 1992), carpels in the place either of sepals or of the entire perianth (*flo-2, flo-3,* and *flo-4*) (Haughn and Somerville 1988), and secondary flowers in the axils of leaflike first whorl organs of the primary flower (*ap-1*) (Irish and Sussex 1990). The phenotypes of these mutants, and the many ancillary observations about them that are described in the references cited, leave little doubt that determination of the floral organs in the wild-type plant is dictated by the products of the corresponding genes. Further analysis of the genetic control of organ formation in the flowers of *Arabidopsis* has involved cloning and sequencing of the genes and identification of their protein products. Cloning of the *AG* gene was facilitated by using a mutant line to tag genes using the T-DNA of the Ti plasmid of *Agrobacterium tumefaciens* as the insertional mutagen. Transformation was accomplished by infecting germinating seeds of *Arabidopsis* with the *Agrobacterium* construct. One of the transformed lines selected by this method was found to display stable phenotypic alterations of the *ag* mutant. Based on this observation and genetic analysis, it has become clear that T-DNA can be used as a molecular probe to isolate the *AG* gene. This has opened the way for the use of restriction fragments spanning the T-DNA insertion site to probe a cDNA library made to poly(A)-RNA of wild-type *Arabidopsis* flowers, to select sequences complementary to the *AG* gene, to analyze their structure and regulatory function, and to identify the protein products they encode. The amino acid sequence comparison of the *AG* gene product based on DNA sequence analysis showed similarities in the amino terminal portion to the DNA-binding regions of transcription factors from a yeast gene that causes defects in the maintenance of minichromosome (*MCM-1*) and to the vertebrate serum response factor genes (*SRF*). Based on this relationship, the apparent uniqueness of the *AG* gene is that it probably encodes a transcription factor that regulates genes determining stamen and carpel function in wild-type flowers (Yanofsky et al 1990). The transcription factor in all likelihood binds to key elements of the *AG* gene, thereby triggering a cascade of regulatory events whose ultimate consequence is the activation of stamen- and carpel-determining genes (H. Huang et al 1993). The distinctive feature of the mutants defective at the *AG* locus is that they fail to produce the transcription factor, thereby causing biochemical dysfunction of the stamen- and carpel-determining pathways. Additional *AG-LIKE* genes (*AGL-1* to *AGL-6, AGL-8*) that share this conserved sequence motif have been isolated from flowers of *Arabidopsis* (Ma, Yanofsky and Meyerowitz 1991; Mandel and Yanofsky 1995).

The similarity between the phenotypes of *ap-3* mutant of *Arabidopsis* and *deficiens* (*def*) mutant of *Antirrhinum majus*, to be described later, offered an experimental opportunity to use a *DEF* cDNA clone to probe an *Arabidopsis* flower cDNA library and isolate *AP-3* gene. The surprising observation from a comparison of nucleotide and amino acid sequence data is that like DEF and AG, AP-3 protein contains a conserved motif similar to the transcription factors identified in other systems (Jack, Brockman and Meyerowitz 1992). Molecular cloning of *AP-1* (Mandel et al 1992b) and *PI* (Goto and Meyerowitz 1994) genes by other standard methods and determination of their amino acid sequences have also shown the protein products to belong to the same family of transcription factors as the *AG* gene product.

Since the small size of the floral organ primordia precludes RNA extraction and Northern blot analysis, gene expression in the floral meristem has been visualized by in situ hybridization of the complementary mRNA with labeled probes (Bowman, Drews and Meyerowitz 1991; Drews, Bowman and Meyerowitz 1991). When sections of floral meristems of wild-type *Arabidopsis* plants of different stages of development are hybridized with antisense ^{35}S-RNA probe derived from sequences within an *AG* cDNA clone, the label is detected first in the floral meristem as the sepal primordia are initiated. Subsequently, *AG* RNA is expressed almost exclusively in the stamen and carpel primordia with barely detectable label in the sepals and petals.

These data are consistent with the view that the *AG* gene plays a role in the early determination of the stamen and carpel. In contrast, *AG* RNA is present in the primordia of all four sets of floral organs of an *ap-2* mutant that causes carpels and stamens to appear in whorls reserved for sepals and petals in the wild type. The paradox arising from these observations has been resolved on the basis of genetic evidence and data on the expression of *AP-1* gene in transgenic *Arabidopsis* flowers carrying an antisense *AG* gene construct; the results corroborate the view that the function of the *AP-2* gene product is to negatively regulate *AG* gene activity, rather than to act on its own (Bowman, Smyth and Meyerowitz 1991; Mizukami and Ma 1995). The product of the *LUG* gene appears to be earmarked for a similar function (Liu and Meyerowitz 1995).

That the *AG* gene interferes with the expression of *AP-1* activity has been revealed by in situ hybridization of *AP-1* probe to sections of floral meristems of wild type and *ag* mutants. Although *AP-1* transcripts are uniformly distributed during the early stage of development of the floral meristem of wild type *Arabidopsis,* they disappear from the potential sites of stamen and carpel primordia at later stages. Since this coincides with the expression of *AG* gene transcripts in the same sites of the floral meristem, a possible interaction between *AG* and *AP-1* genes was followed by monitoring the expression of *AP-1* in the floral meristem of an *ag* mutant. The surprising finding was that *AP-1* RNA levels were quite high in the inner whorls of the flower when *ag* was in the mutant form. This gives strong support to the idea that *AP-1* expression is negatively regulated by *AG* in the domains of the stamen and carpel primordia (Mandel et al 1992b; Gustafson-Brown, Savidge and Yanofsky 1994). Based on these and other results it has been proposed that organ specification in flowers depends upon regulatory interactions between different homeotic genes in the different parts of the meristem; this is no small feat considering all the network of peripheral reactions that cannot be left to chance (Bowman et al 1993).

The confluence of genetic and molecular studies has led to a working model to explain the function of mutant genes in the three extensively characterized homeotic mutations, *ag, ap-2,* and *ap-3* or *pi* (Figure 2.1). The model envisages that a different genetic activity necessary for normal flower pattern formation is missing in each of these mutations. Designating these activities as A, B, and C, *ap-2* mutant is assumed to lack A, *ap-3* and *pi* mutants lack B, and C is missing in *ag*. Later analyses, which showed that *AP-1* and *LUG* are expressed in the sites of potential sepal and petal primordia for most of floral development, has led to the addition of *ap-1* and *lug* to the list of mutants lacking A activity. From the functional perspective, the gene products are assumed to act in three overlapping fields such that field A controls the sepals and petals, B is responsible for petals and stamens, and C modulates stamens and carpels. By assigning genes *AP-1/AP-2, AP-3/PI,* and *AG* to function in fields A, B, and C, respectively, a picture of a unique pattern of expression of floral organs in each of the four whorls controlled by these gene products, acting singly or in combination, is apparent. Thus, when these conditions are fulfilled, cells in which *AP-1/AP-2/LUG* and *AG* alone are expressed might be expected to form sepals and carpels, respectively, whereas the combined effects of *AP-1/AP-2/LUG* and *AP-3/PI* will lead to petal formation. Similarly, both *AP-3/PI* and *AG* gene products are necessary to create conditions for the formation of stamens (Bowman et al 1993). Provided the correct gene product is diverted to the floral meristem, any type of organ can indeed regenerate in any whorl of the flower, as shown by the observation that ectopic expression of *AP-3* and *PI* genes gives rise to flowers with petals in the two outer whorls and stamens in the inner. The uniqueness of this ectopic expression is underscored by the fact that the same regions of the functional protein define the specificity of the two genes (Krizek and Meyerowitz 1996a, b).

Some of the results from in situ hybridization studies just considered are explained by yet another feature of the model, which proposes the existence of a mutual antagonism between A and C functions. Because this proposal implies that A function represses the action of C in whorls 1 and 2, and that C function prevents the action of A in whorls 3 and 4, it comes as no surprise that *AG* RNA is present in all four whorls of an *ap-2* mutant. Direct evidence that the product of the *AG* gene prevents *AP-2* function has come from the demonstration that chimeric *AG* gene introduced into transgenic *Arabidopsis* under the control of cauliflower mosaic virus (CaMV) 35S promoter induces homeotic transformations of sepals into carpels, and petals into stamens, very much like *ap-2* mutant phenotype (Mizukami and Ma 1992). Similar results were obtained when *AG* gene cloned from *Brassica napus* (rapeseed; Brassicaceae) was expressed in the flowers of transgenic tobacco (Mandel et al 1992a). These results clearly indicate that homologous genes control the development of

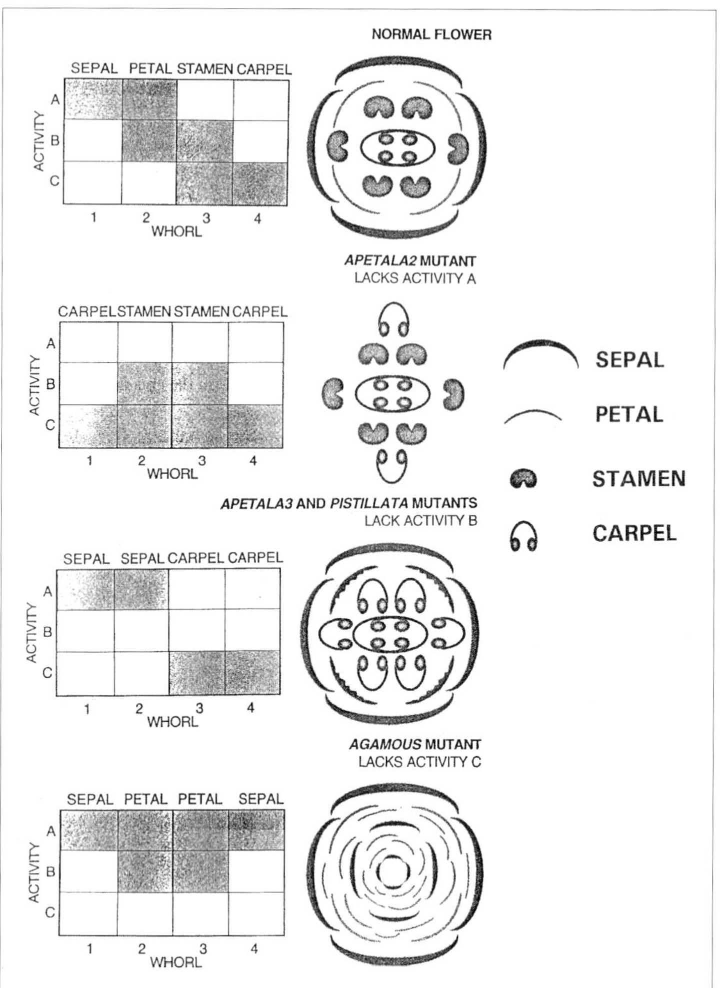

Figure 2.1. A diagram showing the arrangement of floral parts in the wild-type and three groups of homeotic mutants of *Arabidopsis*. The genetic activities in each whorl are shown in the block diagrams. (From E. M. Meyerowitz 1994. The genetics of flower development. *Sci. Am.* 271(5):40–47. © 1994 by Scientific American Inc. All rights reserved.)

the stamens and carpels even in distantly related plants and that it is feasible to manipulate in a predictable manner the production of these floral organs on the receptacle by altering the expression of a single regulatory gene.

AP-2 gene has been cloned and found to encode a putative nuclear protein that bears no significant similarity to any known plant, fungal, or animal regulatory protein. However, the gene is expressed at the RNA level in all organs of wild-type *Arabidopsis* flowers and is not restricted to sepals and petals as predicted by the model; this has raised some questions as to whether *AG* and *AP-2* genes are mutually antagonistic (Jofuku et al 1994).

AGL-1 to *AGL-6* and *AGL-8* genes isolated from flowers of *Arabidopsis* have provided somewhat different expression patterns when sections of the floral meristem are hybridized in situ with ^{35}S-labeled RNA probes complementary to *AGL* mRNA. For example, in contrast to *AG*, expression of *AGL-1* and *AGL-2* begins at a much later stage in floral ontogeny. Whereas *AGL-1* is expressed preferentially in the carpels, *AGL-2* is detected additionally in the sepals, petals, and stamens. Furthermore, *AGL-2* is also expressed during early development of all four whorls of flowers defective in the organ identity genes *AG, AP-2,* and *AP-3*; the function of these genes is therefore not essential for *AGL-2* expression per se. One possible implication of these results is that *AGL-1* and *AGL-2* may regulate a battery of genes required for the establishment of floral organs different from organs that are under the influence of *AG* and other genes (Ma, Yanofsky and Meyerowitz 1991; Flanagan and Ma 1994). Three other *AGL* genes studied in great detail are *AGL-4*, *AGL-5*, and *AGL-8*. The remarkable interaction of *AGL-4* and *AGL-5* genes with the *AP-2* gene is seen in a significant increase in their transcript levels in *ap-2* mutant flowers, suggesting that the *AP-2* gene might negatively regulate the expression of *AGL-4* and *AGL-5* genes (Savidge, Rounsley and Yanofsky 1995). Stamens in the third whorl of wild-type flowers normally lack *AGL-8* mRNA, but the gene is expressed in the carpels that replace stamens in *ap-3* mutants. This assigns an undetermined role for the *AP-3* gene product in blocking *AGL-8* expression in the third whorl organs (Mandel and Yanofsky 1995). A general role has been ascribed to the *AGL-3* gene because it is somewhat unique; unlike other *AGL* genes, its expression is global and is confined to all above-ground parts of the plant (Huang et al 1995). On the whole, the choice of whether a floral organ is established under the influence of one or the other set of organ identity genes probably depends upon the interaction of genes active in the meristem at the time of floral induction.

How the different genes interact at the floral apex is a conceptually puzzling problem, but some pointers have come from studies on the expression pattern of the *AP-3* gene in the wild-type and mutant flowers of *Arabidopsis*. Transcripts of *AP-3* are detected at the same stage of development in the floral meristem of wild-type flowers as those of the *AG* gene, but the subsequent expression of the gene is restricted to progenitors of the petals and stamens. The coincident expression of *AG* and *AP-3* in the floral meristem specifying the petal, stamen, and carpel primordia is just what one would expect if initiation of the stamen primordium does not depend upon the differentiation of other floral organs on the apex. The expression pattern of the *AP-3* gene in the floral meristem of *ag* mutant also offers a clue about the relationship between *AG* and *AP-3* genes. In *ag* mutant, *AP-3* gene expression is restricted to the second and third whorl petals. This indicates that the *AG* gene product normally does not regulate the function of the *AP-3* gene. What is more appealing for speculative purposes about gene interactions in floral organ specification is the observation that although the *AP-3* gene is normally expressed in the progenitors of both petals and stamens in the receptacle of *pi* mutant flowers, its later expression is restricted to the petals. Despite the lack of sufficiently detailed evidence to account for the failure of *AP-3* expression in the carpels formed in place of stamens in the *pi* mutant, there is compelling reason to believe that to form petals and stamens in the correct orientation on the meristem, interaction between *AP-3*, *AG*, and *PI* gene products is necessary (Jack, Brockman and Meyerowitz 1992).

This general picture of interaction of *AG*, *AP-3*, and *PI* genes is further complicated by the spatial and temporal expression patterns of other genes that have been cloned subsequently. For example, although the phenotypic effects of *ap-3* and *pi* mutations are nearly identical, the pattern of accumulation of their respective transcripts in the receptacle of wild-type flowers is strikingly variable. In contrast to *AP-3* expression confined to the petal and stamen primordia of wild-type flowers, *PI* transcripts, at a similar stage of development, accumulate in the three inner whorls of the floral meristem. The mystery is deepened by the fact that this is a transient stage in the expression of *PI* transcripts because by the time stamens begin to differentiate, these transcripts are no longer expressed in the carpel territory; at this stage, the patterns of expression of *PI* and *AP-3* genes become similar (Goto and Meyerowitz 1994). Interestingly, it has been shown that when *AP-3* is expressed in the fourth whorl of flowers of a transgenic line of *Arabidopsis*, high levels of transcripts of *PI* accumulate in all three inner whorls throughout flower development. This provides good evidence that the disappearance of *PI* transcripts from the carpel territory is due to the absence of simultaneous *AP-3* expression (Jack, Fox and Meyerowitz 1994). Associated with the homeotic transformation of stamens to carpels in the floral meristem of *ufo* mutant, there is a decrease in *AP-3* and *PI* RNA expression, suggesting that the *UFO* gene holds a sway over the activities of B-field genes (Levin and Meyerowitz 1995). Based on *PI* RNA expression levels in the receptacles of *sup* and *ap-2* mutant flowers, *SUP* and *AP-2* genes have also been identified as controlling *PI* gene expression (Goto and Meyerowitz 1994). The present view of the *SUP* gene is that it is a transcriptional regulator involved in regulating the boundary between the stamen and carpel whorls on the receptacle (Sakai, Medrano and Meyerowitz 1995).

In contrast to the genes promoting floral organ initiation, meristem identity genes control floral meristem initiation and size, but there are indications that they also control the initial activation of the homeotic genes. The candidate genes that control such activation are *LEAFY* (*LFY*), *AP-1*, *AP-2*, and *UFO*. Like *ap-1* mutant, *lfy* mutation causes the partial transformation of flowers into abnormal structures, shoots, or inflorescences (Schultz and Haughn 1991, 1993; Huala and Sussex 1992). Because the determination of the floral meristem precedes the determination of floral organs, there is justification to consider that *LFY* and *AP-1* genes are expressed before homeotic genes are activated; indeed, transcripts of cloned *LFY* gene are expressed strongly in wild-type flower primordia (Weigel et al 1992). In situ hybridizations performed with *AP-3, PI,* and *AG* gene probes on sections of floral meristems of wild type and single or double *lfy* and *ap-1* mutants have indicated that the initial activation of the homeotic genes is accomplished by synergistic interaction with products of *LFY* and *AP-1* genes (Weigel and Meyerowitz 1993). Apparently, the gap between the developmental pathway from a determined floral meristem to a determined flower primordium is bridged in a specific way by the products of *LFY* and *AP-1* genes.

Homeotic Mutants in *Antirrhinum* and Other Plants. Following the discovery of homeotic mutants in *Antirrhinum majus* in the early 1920s, there has been a recent revival of interest in the molecular analysis of these mutants. This has been aided by the isolation and characterization of transposable elements that can be used for transposon tagging and mutagenesis. Additional advantages of *A. majus* for molecular studies are the large size of the flowers, availability of a good genetic map with many well-characterized mutations affecting flower development, and the ease of vegetative propagation by cuttings. However, unlike *Arabidopsis, Antirrhinum* flowers are zygomorphic, having only one plane of mirror-image symmetry, with a pentamerous arrangement of sepals, petals, and stamens, and two united carpels. In the approach followed to generate homeotic mutants, plants known to carry active transposons are self-pollinated. M1 plants that are raised are also self-pollinated, allowing the isolation – in the M2 generation – of homozygous recessive mutations displaying the same characteristic misrepresentation of positions of the floral organs as in *Arabidopsis*. Among such mutations, those that are of interest to us at this stage are *def, globosa* (*glo*), and *sepaloidea* (*sep*), in which the first three whorls of the flower are altered to yield sepal, sepal, and carpel instead of sepal, petal, and stamen; *plena* (*ple*), with homeotic conversion of sex organs to sterile whorls; and *ovulata* (*ovu*), in which carpels replace sepals, and stamens replace petals in the two outer whorls (Carpenter and Coen 1990). The *def, glo,* and *sep* mutants have phenotypes similar to the *ap-3* and *pi* mutants of *Arabidopsis* and thus affect the B region on the receptacle. The *ple* mutant is similar to *Arabidopsis ag* mutant and affects the C region, whereas *ovu*, like *ap-2*, controls the A region. The *DEF* gene, which has been cloned, is also found to encode a protein with the same type of DNA-binding domain as *AG* and *AP-3* genes from *Arabidopsis*. The functional motif shared by these diverse genes has been given the name *MADS*-box (for *MCM-1, AG, DEF,* and *SRF*) (Sommer et al 1990; Ma, Yanofsky and Meyerowitz 1991). In situ hybridization of sections of flower buds of wild-type *A. majus* has shown that *DEF* gene is very strongly expressed in the petal and stamen primordia (Schwarz-Sommer et al 1990); this is not surprising, since these organs are most sensitive to the effects of the mutation. Moreover, a homologue of the *DEF* gene isolated from wild-type *Arabidopsis* complements the *ap-3* mutation in transgenic plants of the same species and produces flowers with petals and stamens of normal size, shape, and number (Okamoto et al 1994). As *DEF* and *AP-3* have very similar protein products, mutant phenotypes, functions, and spatial expression patterns, we cannot fail to be impressed by the argument that the same molecular mechanisms probably operate in organ specification in the flowers of the two genera.

To gain further insight into the molecular basis for the interaction of the three sets of genes in the floral meristem of *A. majus*, both *PLE* and *GLO* genes have been cloned and characterized. The *PLE* locus was identified and isolated using *AG* from *Arabidopsis* as a probe. The functional similarity of *PLE* to the *AG* gene was shown by the restricted expression of *PLE* transcripts in the stamens and carpels of wild-type *Antirrhinum* flowers. The complementary nature of the phenotypes of *ple* (normally restricted to the two inner whorls) and *ovu* (expressed in the two outer whorls) has been attributed to the opposite orientations of a transposon in the intron of the *PLE* gene (Bradley et al 1993). The spatial and temporal expression patterns of *DEF* and *GLO* genes in the second and third whorls of the floral receptacles of wild-type and various *DEF* and *GLO* alleles have implied that the two genes are transcribed independently. From other studies it appears that, whereas GLO or DEF proteins do not bind DNA, in combination they bind to a motif found in the promoters of *DEF* or *GLO* genes (Schwarz-Sommer et al 1992; Tröbner et al 1992; Zachgo et al 1995). These results favor a cross-regulatory role for the two gene products in specifying organ identity at the floral meristem; if this is correct, mutations could cause alterations in the transcription of genes normally controlled by MADS proteins and lead to homeotic changes in the floral meristem.

These pioneering studies on *Arabidopsis* and *Antirrhinum* have not only resulted in a common genetic model of flower development but have also led to further research on the principles of organ identity in flowers of several other plants. *MADS*-box genes associated with floral organ development have now been described in tomato (Pnueli et al 1991, 1994a, b), *Brassica napus* (Mandel et al 1992a), *Petunia hybrida* (Angenent et al 1992, 1993, 1995a; Kush et al 1993; Tsuchimoto, van der Krol and Chua 1993; van der Krol et al 1993; Cañas et al 1994), tobacco (Hansen et al 1993; Kempin, Mandel and Yanofsky 1993), *Solanum tuberosum* (potato; Solanaceae) (Garcia-Maroto, Salamini and Rohde 1993), *Aranda* sp. (Orchidaceae) (Lu et al 1993), maize (Schmidt et al 1993; Mena et al 1995), and *Sinapis alba* (Brassicaceae) (Menzel, Apel and Melzer 1995). Close homology has been shown to exist between the homeotic genes *BAG-1* of *B. napus* (Mandel et al 1992a), *ZAG-1* of maize (Schmidt et al 1993), *OM-1* of *Aranda* (Lu et al 1993), *NAG-1* of tobacco (Kempin, Mandel and Yanofsky 1993), *OSMADS-3* of *Oryza sativa* (rice; Poaceae) (Kang et al 1995) and the *AG* gene of *Arabidopsis*; between *ZAP-1* of maize and *AP-1* of *Arabidopsis* (Mena et al 1995); between *ZAG-3*, *ZAG-5* of maize and *AGL-6* of *Arabidopsis* (Mena et al 1995); between *TM-6* of tomato and *DEF* of *Antirrhinum* (Pnueli et al 1991); between *NTGLO* of tobacco and *GLO* of *Antirrhinum* (Hansen et al 1993); and between *OSMADS-2*, *OSMADS-4* of rice, *GLO* of *Antirrhinum* and *PI* of *Arabidopsis* (Chung et al 1995). *MADS*-box genes also appear to play a role in sex differentiation in the flowers of dioecious plants, *Silene latifolia* (Caryophyllaceae) (Hardenack et al 1994) and *Rumex acetosa* (Polygonaceae) (Ainsworth et al 1995). Using cDNA clones with *MADS*-box homology isolated from a male-flower cDNA library of *S. latifolia*, Hardenack et al (1994) have shown that each gene is expressed in the same floral whorl of the male and female flowers as is the homologous gene from snapdragon or *Arabidopsis*, such as *AG*, *AP-3*, *DEF*, and *PI*. The *MADS*-box gene is presumed to function here by reducing the size of the fourth whorl in the male flower. A recessive flower mutant described in tomato, which shows transformation of petals to sepals in whorl two, and of stamens to carpels in whorl three, also supports in principle the mechanism underlying the phenomenon of homeosis in floral architecture (Rasmussen and Green 1993).

In summary, tissue culture, genetic, and molecular investigations have led to some generalizations regarding determination of the floral organs, especially of the stamens and carpels. One is that floral organs are largely undetermined as they are initiated on the meristem, but the primordia soon acquire features of the organs they are destined to represent. Mutations that provide access to detailed molecular processes underlying determination of stamens indicate that specific genes direct stamen position on the floral meristem. However, sufficient data are not at hand to understand the role of genes in determining how the stamen primordium attains its mature form, consisting of the anther, filament, and connective, after its identity has been specified. Since the basic mechanisms that specify stamen determination are the same in widely distant species, the genetic organization for the functioning of genes and for their expression appears to have a long evolutionary history.

2. ANTHER DEVELOPMENT

Study of anther development in angiosperms has spanned more than a century and has provided a stable foundation for later work in the areas of physiology and, more recently, in molecular biology. Much of the early work is compiled by Maheshwari (1950); the later studies have been surveyed by Bhandari (1984). For physiological and molecular investigations, a nondestructive method of predicting the stage of microsporogenesis in the anther while it is still enclosed in the flower bud, is necessary. For several plants, this has resulted in the establishment of allometric relationships between the growth of the flower bud, the length of the anther, and the stage of microsporogenesis. Erickson (1948) showed for the first time that in the flower buds of *Lilium longiflorum* (lily; Liliaceae), from an initial length of 10 mm to a final length of 150 mm, cytological events of microsporogenesis occur at predictable anther lengths. This does not necessarily imply a steady distribution of activity along the entire length of the anther, as marking experiments have shown the existence of a rather basipetal wavelike distribution pattern of growth in length and cell division activity (Gould and Lord 1988). Mature florets and anthers of *Oryza sativa* attain only about 5% of the length of the flower buds and anthers of lily; however, the correlations observed between the growth of florets and anthers and stages of microsporogenesis in rice are similar

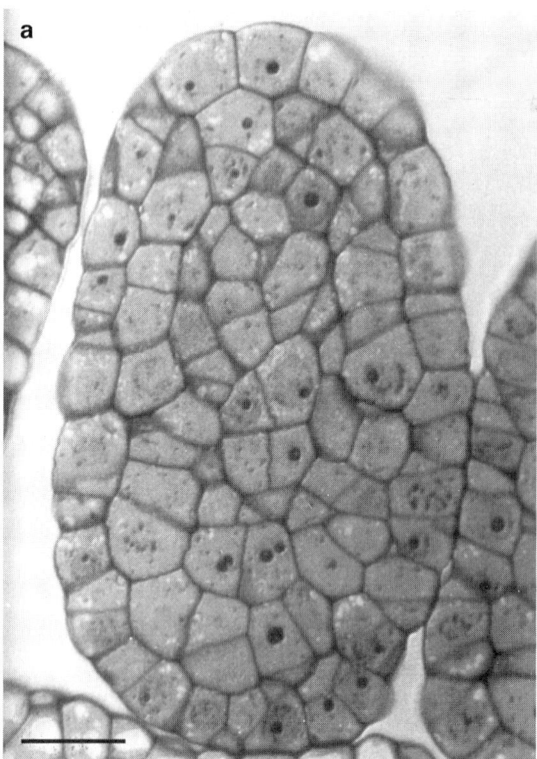

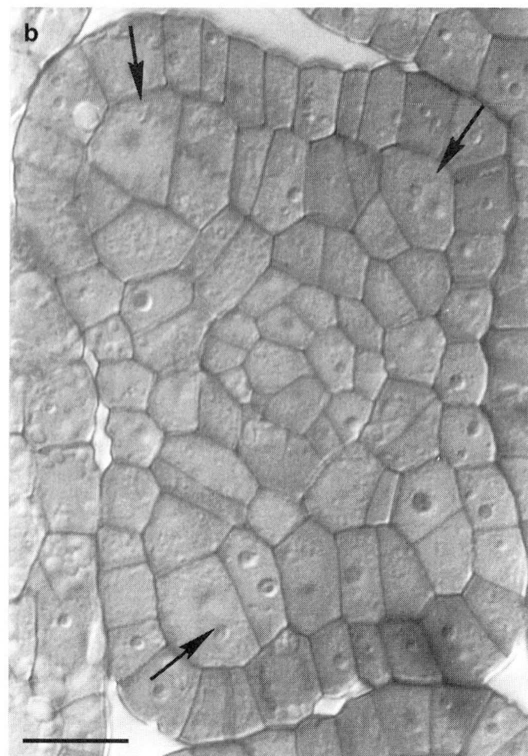

Figure 2.2. Anther development in rice. (**a**) Cross section of an anther primordium showing the mass of meristematic cells. (**b**) Differentiation of archesporial initials (arrows). Both photographed with Nomarski optics. Scale bars = 10 μm. (From Raghavan 1988.)

to those found in lily (Raghavan 1988). Significant correlations have been established between the cytological stages of microsporogenesis and the morphological features of anther development in *Triticum aestivum* (Bennett et al 1973b) and *Zea mays* (Chang and Neuffer 1989).

Initially, the anther consists of a mass of meristematic cells that do not display any differentiated characteristics, bounded on the outside by an epidermis. Subsequently, cell differentiation proceeds in such a manner that the anther becomes a four-lobed organ. At this stage, each lobe becomes populated along its entire length by rows of large cells with dense cytoplasm and deeply staining nuclei. There is virtually universal agreement that these cells, which constitute the archesporial initials, are hypodermal in origin (Figure 2.2). Although multiple rows of archesporial initials are the norm in anthers of many plants studied, in others the archesporium may consist of only a single longitudinal file of cells. Indicative of its morphogenetic potential, each archesporial initial divides periclinally (parallel to the outer wall of the anther lobe) to yield a primary parietal cell toward the periphery and a primary sporogenous cell toward the inside. Although these two cells are dedicated to produce different cell types, their origin from the same initial cell indicates that they are ontogenetically related. When the distribution of mRNA [poly(A)-RNA] in the anther primordium of rice is assayed by in situ hybridization using ^{3}H-poly(U) as a probe, poly(A)-RNA is found to be uniformly distributed in all cells. During the formation of the archesporial initial and the primary parietal and primary sporogenous cells, there is no differential accumulation of poly(A)-RNA in the progenitor cells (Raghavan 1989).

Later divisions of the primary parietal cell and the primary sporogenous cell can be summarized as follows. The former undergoes repeated anticlinal (perpendicular to the outer wall of the anther lobe) and periclinal divisions to form two to five concentric layers of cells constituting the anther wall and tapetum. The primary sporogenous cell functions directly as the microsporocyte or resumes mitotic divisions to generate a mass of secondary sporogenous cells that function as microsporocytes (Figure 2.3). Four microspores, with the haploid or

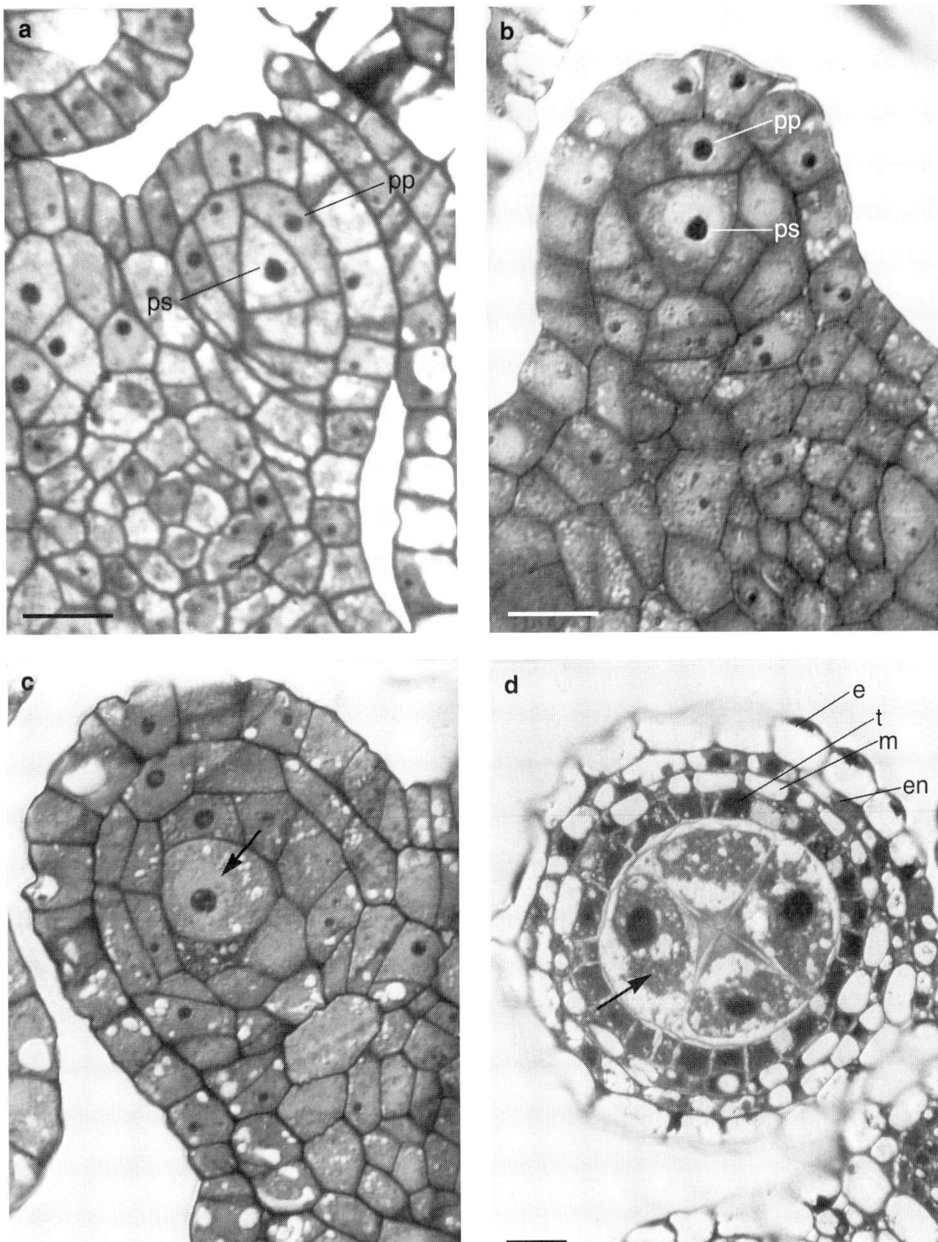

Figure 2.3. Differentiation of wall layers and microsporogenesis in the anther of rice. **(a, b)** Cross sections of anther lobes showing two successive stages in the formation of the primary parietal cell (pp) and the primary sporogenous cell (ps). **(c)** An anther lobe showing the primary sporogenous cell (arrow) surrounded by a three-layered wall. **(d)** An anther lobe showing the wall constituted of the epidermis (e), endothecium (en), and middle layers (m) and tapetum (t). The primary sporogenous cell has undergone cross-wise vertical divisions to form secondary sporogenous cells or microsporocytes (arrow). Sections were photographed with Nomarski optics. Scale bars = 10 μm. (From Raghavan 1988.)

gametic number of chromosomes, are born out of each microsporocyte by a meiotic division. This development, which in its extreme form may involve a single diploid cell, presents in a microcosm an excellent example of a meiotic pathway ending with the formation of specialized cells.

There is a large body of anatomical evidence showing that the anther is derived either exclusively from the subepidermal layer of cells or from both subepidermal and epidermal cell lineages of the floral meristem. Utilizing transposable elements and X-rays to generate clonal sectors in the tassels, Dawe and Freeling (1990) showed that either the epidermal or subepidermal cell layer in maize can independently form the anther wall, but cell lineages from only the subepidermal layer generate the sporogenous cells. The total number of cells in each of the two layers participating in anther development has been estimated to be 12 or less (Dawe and Freeling 1992).

(i) The Anther Wall

It has long been tacitly assumed that the function of the anther wall is to protect the sporogenous cells and developing microspores and to facilitate discharge of the baggage of enclosed pollen grains. What are the unique features of the anther wall that serve this function? The wall of the mature anther consists of an epidermis, followed to the inside by a layer of cells of the endothecium and cells comprising two or three middle layers. In some plants, cells of the anther wall and connective are traversed by plasmodesmata that allow assimilates to pass to the locule (Clément and Audran 1995). The most striking changes in the anther wall are observed in the endothecial cells. As perceived in *Lens culinare* (Fabaceae), endothecial cells develop wall thickenings at the vacuolate pollen grain stage; at the mature pollen grain stage, they harbor long strands of rough endoplasmic reticulum (ER), polysomes, and plastids with and without starch. Later, the cytoplasm becomes diffuse, accompanied by the appearance of fibrillar thickenings (Biddle 1979). These thickenings constitute the beginnings of fibrous bands of the endothecial cells of mature anthers. The bands, which arise chiefly along the inner tangential wall, assume a U-shaped pattern, with the gap directed toward the epidermis. During dehydration of the anther preparatory to dehiscence, the endothecial cells lose water and begin to shrink in an uneven manner, resulting in the opening of the anther. Opening occurs through longitudinal slits, known as stomium, located between the two locules of each anther lobe. As regards the middle layers, anthers inherit a fixed number, which is virtually invariant from species to species within a genus. Thus, the genome of each species is already programmed for determination of cell layers in the anther wall in a specific lineage and for the fate of the cells. However, after the middle layers are carved out, they become compressed, crushed, or obliterated by the burgeoning cells of the endothecium to the outside and the tapetum to the inside.

The classic question as to how descendants of the same progenitor cell differentiate into structurally and functionally different cell types can be posed in the analysis of the anther wall. Although the role of the anther wall previously alluded to makes understanding of the physiology and molecular biology of its cells clearly important, little is known of the gene activation program in these cells. The problem here is compounded by the limited number of cells present in the anther wall and the difficulty of separating them from the sporogenous cells and pollen grains. Among examples of gene expression in the wall layers of the anther that have been investigated is the localization of poly(A)-RNA in rice anthers by in situ hybridization with ^{3}H-poly(U). Preparatory to meiosis in the microsporocytes, there is a sharp decrease in the poly(A)-RNA concentration in the epidermis and middle layers of the anther wall, although the label is predominant in the endothecium. Poly(A)-RNA concentration attains low levels in the persistent endothecium of the postmeiotic anther. An interpretation consistent with these observations is that transcriptional activity is required for the differentiation of the endothecium, perhaps in the fabrication of the radial thickening bands (Raghavan 1989). In anther sections hybridized with a 1.3-kb-long H3 (arginine-rich) histone gene probe isolated from a genomic library of 10-day-old rice seedlings, transcripts are restricted to the endothecium (Figure 2.4); in line with this observation, the promoter of an H3 histone gene from wheat is found to confer expression of the *GUS* gene in the endothecium of anthers of transgenic rice (Raghavan 1989; Terada et al 1993). In attempts to delimit the nucleotide sequences in the 5'-region of wheat H3 histone gene responsible for its expression, Terada et al (1995) demonstrated that replication-independent expression of the gene in the anther wall of transgenic rice is not affected by mutations in the highly conserved hexamer and octamer motifs; in contrast, the mutated gene is expressed at an appreciably reduced intensity in the rapidly dividing cells of the root meristem of the transgenic

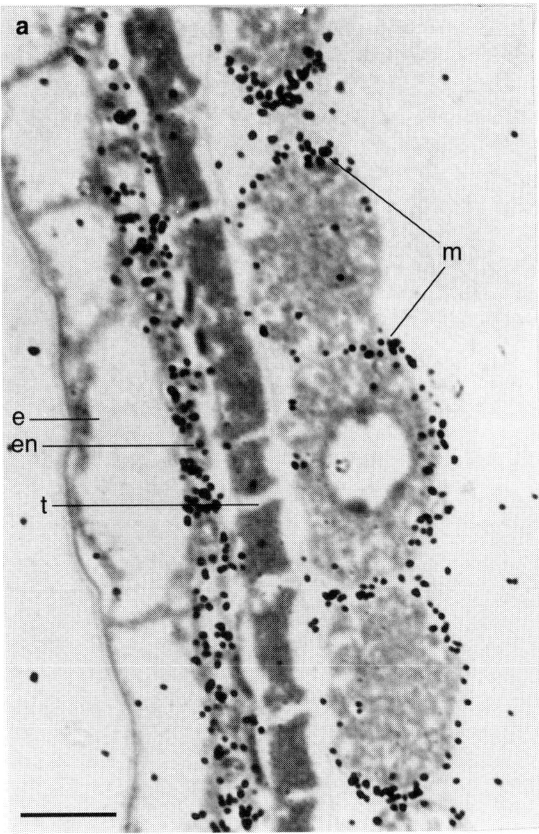

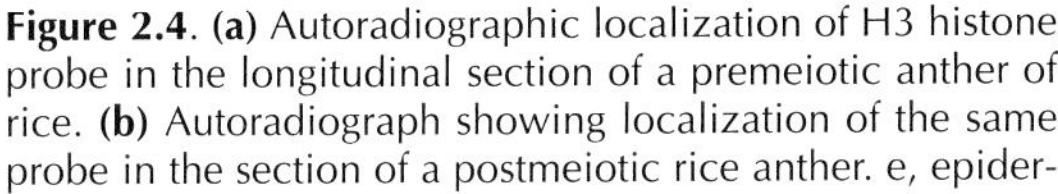

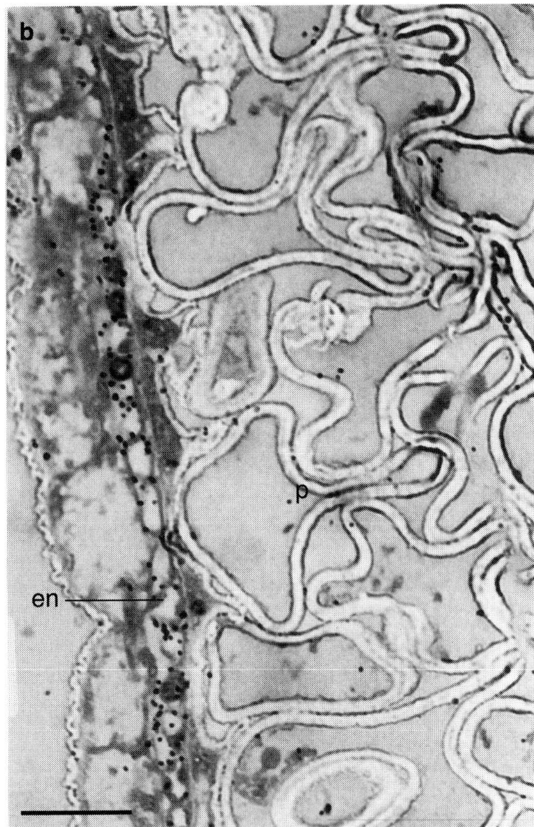

Figure 2.4. **(a)** Autoradiographic localization of H3 histone probe in the longitudinal section of a premeiotic anther of rice. **(b)** Autoradiograph showing localization of the same probe in the section of a postmeiotic rice anther. e, epidermis; en, endothecium; m, microsporocyte; p, uninucleate pollen grains; t, tapetum. Scale bars = 10 μm. (From Raghavan 1989. *J. Cell Sci.* 92:217–229. © Company of Biologists Ltd.)

plant. Rice histone H3 gene probe also hybridizes to sections of anthers of a dicot, *Hyoscyamus niger* (henbane; Solanaceae), indicating that the fidelity of the gene is highly conserved between the two major classes of angiosperms. In *H. niger*, autoradiographic silver grains generated by in situ hybridization are found to be present in more or less the same density in the cells of the anther primordium and, later, in the epidermis and endothecium (Raghavan, Jiang and Bimal 1992). Unexpected as these findings are, they serve to indicate that histone transcripts are not necessarily confined to actively dividing cells as one would expect if their protein products are used to complex with DNA. Specific hybridization of probes made to the gene for 4-coumarate:coenzyme A ligase from *Petroselinum crispum* (parsley; Apiaceae) and *Solanum tuberosum* to the cells of the endothecium is also observed in anthers of tobacco transformed with this gene (Reinold, Hauffe and Douglas 1993). Since the enzyme plays a key role in phenylpropanoid metabolism, presence of transcripts of the gene in the lignified cells of the endothecium is significant.

Among anther-specific cDNA clones identified in a cDNA library made to poly(A)-RNA of tobacco anthers are those encoding a thiol endopeptidase (*TA-56*) and an unknown protein (*TA-20*) (Koltunow et al 1990). When RNA extracted from anthers and other floral organs and vegetative parts of tobacco are hybridized with the cDNA clones in a Northern blot, they produce strong hybridization signals with anther RNA only; this is considered as good evidence for anther specificity of the two cDNA clones and for their relative prevalence in the anther. Dot–blot hybridization of RNA isolated from anthers of different stages of development showed that although mRNA encoding thiol endopeptidase accumulates from the stage of microsporocyte meiosis to the stage of the

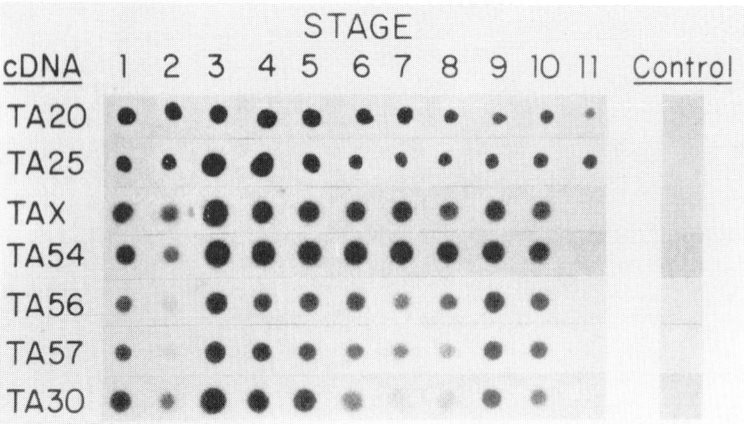

Figure 2.5. Dot–blot hybridization of tobacco anther cDNA clones with anther mRNAs at different stages of development. *TA-20* and *TA-25* dots contained 0.25 μg mRNA; the other dots contained 0.05 μg mRNA. Control dots contained an equivalent amount of soybean embryo polysomal mRNA. Stages of anther development are described in Koltunow et al (1990). (From Koltunow et al 1990. *Plant Cell* 2:1201–1224. © American Society of Plant Physiologists.)

first haploid mitosis in the pollen grain, mRNA encoded by *TA-20* becomes detectable long in advance and is present at the same level in anthers of all stages. Although the concentration of *TA-56* mRNA remains relatively constant until the disruption of the connective, it declines to low levels thereafter (Figure 2.5). As predicted from this observation, in situ hybridization shows that at the earliest stage of anther development, transcripts for *TA-56* are localized in the anther wall in the region that corresponds to the stomium. As the anther matures, the intensity of hybridization signals in the stomium decreases. The outcome was different in similar experiments using *TA-20* mRNA as a probe; here the gene transcripts are expressed initially in all layers of the anther wall, although in older anthers they become concentrated in the connective adjacent to the vascular bundle. From the functional point of view, the early and from-the-outset-localized appearance of *TA-56* transcripts at the stomium can be reasonably related to the fact that the gene belongs to a class whose primary function is to encode enzymes involved in the degeneration of cells in this part of the anther. Both *TA-56* and *TA-20* genes apparently function by responding to regulatory signals from the anther wall and are expressed during anther ontogeny when enzymes necessary for the differentiation of the anther wall are required. A similar function can be ascribed to genes encoding proline- and glycine-rich proteins, which are expressed in the epidermis of *Helianthus annuus* (sunflower; Asteraceae) anthers before pollen maturation (Domon et al 1990, 1991; Evrard et al 1991; Domon and Steinmetz 1994), and chalcone synthase and thionin, which are expressed throughout the wall layers of cells of tobacco anther (Drews et al 1992; Gu et al 1992). The expression of *Phaseolus vulgaris* (bean; Fabaceae) phenylalanine ammonia lyase (*PAL-2*)–*GUS* gene construct and of the gene for 4-coumarate:coenzyme A ligase from parsley and potato in the potential cells of the stomium of transgenic tobacco plants is thought to be critically related to the biosynthesis of lignin (Reinold, Hauffe and Douglas 1993; Shufflebottom et al 1993).

Anther specificity has been established for several cDNA clones isolated from *Lycopersicon esculentum* anthers. Transcripts of these recombinants are also expressed in the somatic cells of the anther, especially in the epidermis and endothecium (Figure 2.6). The predicted protein sequence of the *LAT-52* gene showed some homologies to the protein encoded by a pollen-specific gene from maize and to trypsin inhibitors (Twell et al 1989b; Ursin, Yamaguchi and McCormick 1989; McCormick 1991). Collectively interpreted, these data show that there are a few known molecular markers for the tissues of the anther wall. It should be noted that the evidence to date concerns genes that are selected from cDNA libraries made to poly(A)-RNA of the entire anther. The appearance of low-abundance mRNAs during the development of the anther wall would escape detection by this method. Such sequences could be crucially significant for the analysis of gene expression in the cells of the anther wall.

A series of antibodies with a high degree of specificity to the cells of the stomium have been raised in

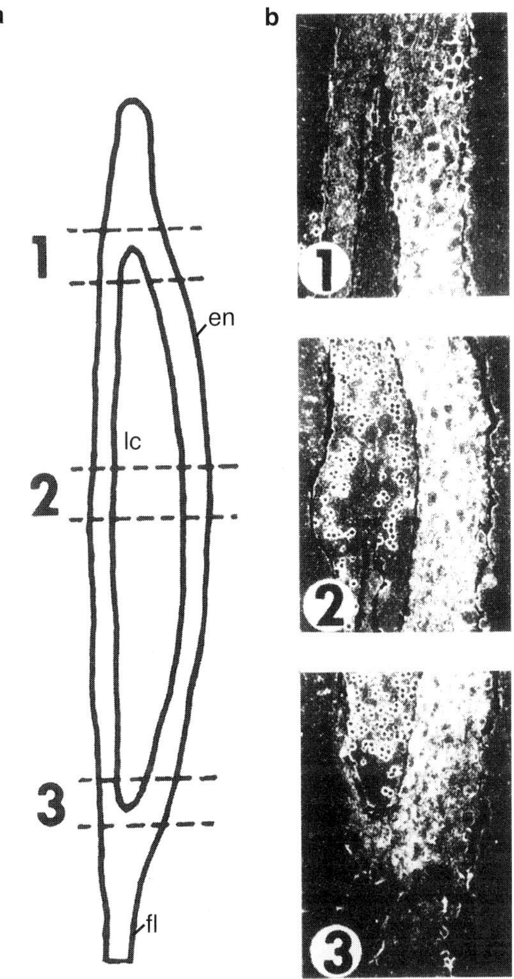

Figure 2.6. In situ localization of a tomato anther-specific cDNA clone (*LAT-58*) to longitudinal sections of tomato anther. **(a)** Drawing of the section of an anther to identify endothecium (en), filament (fl), and locule (lc). Regions 1, 2, and 3 correspond to tip, middle, and base of the anther as shown by dashed lines. **(b)** Dark-field autoradiographs of longitudinal sections from regions 1, 2, and 3 hybridized with ^{35}S-labeled RNA probe. (From Ursin, Yamaguchi and McCormick 1989. *Plant Cell* 1:727–736. © American Society of Plant Physiologists.)

tobacco flower extracts; this observation further supports the view that at the protein level also, the stomium is biochemically different from the rest of the cells of the anther wall (Trull et al 1991; Cañas and Malmberg 1992). Immunocytochemical analysis has shown that patatin, the major glycoprotein of the potato tuber, is present in the epidermal cells of anthers of *Solanum tuberosum* and *Capsicum annuum* (pepper; Solanaceae) (Vancanneyt et al 1989). Strengthening the protective function of the anther epidermis in some unknown ways is envisaged as the role for patatin in the cells.

(ii) The Tapetum

As the limiting tissue between the somatic cells and germ cells of the anther, the tapetum is in immediate contact with the sporogenous cells to the inside and with the middle layers to the outside. Compared to the cells of the anther wall, the tapetal cells are in a dynamic state during their short life period; they constantly undergo divisions; there is a substantial increase in the total DNA content of the cells, mostly by endoreduplication or polyteny (presence of bundles of interphase chromosomes composed of many parallel fibrils resulting from repeated rounds of DNA replication without separation of the daughter chromatids); cytoplasmic upheavals, such as accumulation of diverse kinds of complex macromolecules, are common in the tapetal cells; and the differentiated state of the tapetum is associated with the expression of unique genes. By virtue of its control of the passage of materials into and out of the anther locule, the tapetum has an important role in the nutrition of the sporogenous cells and microspores. The morphological, structural and cytological features of the tapetum that facilitate this function, as well as the other postulated role of the tapetum in the formation of the pollen exine, will be first considered here. This will be followed by an analysis of the cytology and gene expression patterns in the tapetum. For reviews of the evolution, structure, and function of the tapetum and its changing physiological relationship with the sporogenous cells, see Pacini, Franchi and Hesse (1985), Albertini, Souvré and Audran (1987), and Chapman (1987). Papers presented at a symposium on the tapetum collected together in the book edited by Hesse, Pacini and Willemse (1993) are also germane to our discussion.

Structure–Function Relationships. Following the genesis of the tapetal cells, their subsequent fate might take one of two routes. In most plants, the tapetum remains as a single layer of cells characterized by a densely staining cytoplasm and prominent nuclei; however, in a few plants it becomes biseriate or multiseriate throughout the anther (Bhandari 1984). The tapetal cells attain their maximum development at the tetrad stage of microsporogenesis, after which degradative changes set in, resulting in the collapse of cells. Beginning with a study by Echlin and Godwin

(1968a), there have been several investigations on the ultrastructure of development of the tapetum in both dicotyledons and monocotyledons. These studies have clearly established that the tapetal cell is quite different from the cells of the wall layer to which it is ontogenetically related. However, the differences in the structure of the tapetum among various species are mainly concerned with the relative timing of developmental episodes compared with the stage of microsporogenesis, whereas generalizations mostly involve the type and distribution of the different organelles. In *Avena sativa* (oat; Poaceae), the newly formed tapetal cell walls are perforated by plasmodesmata that connect the cells to each other, to the pollen mother cells, and to the cells of the middle layer. A mechanism for the possible coordination of various functions of the tapetum emerges from the observation that during microsporocyte meiosis in *Zea mays*, the plasmodesmata between tapetal cells are replaced by cytoplasmic channels (Figure 2.7) that allow for free exchange of cytoplasm and organelles (Perdue, Loukides and Bedinger 1992). Microtubules, which show preferential orientations during the division of the tapetal cells of *A. sativa*, are also found in the peripheral cytoplasm of the nondividing cells along the tangential and radial walls (Steer 1977a). The rest of the cytoplasm of the cells is highly organized in spatial terms with a generous supply of mitochondria, plastids, dictyosomes, ribosomes, and smooth and rough ER. An interesting feature of the tapetal cells of *A. sativa* is the marked association among the smooth ER, plastids, and mitochondria. The majority of these organelles are completely encircled by the distended cisternae of the ER, but a few are ensheathed only partially; in either case, there are no direct connections between the organelles and the ER. It has been suggested that the ER–organelle association represents a route for the transfer of metabolites from the organelles to the anther locule. ER has also been implicated in the movement of dictyosome products to the surface of tapetal cells of *A. sativa* (Steer 1977b). In many plants, a distinct phase in the tapetal cell ontogeny that can be easily observed in the electron microscope at the time of microsporocyte meiosis is characterized by the amplification of ER, dictyosomes, mitochondria, and ribosomes (Echlin and Godwin 1968a, b; Christensen, Horner and Lersten 1972; Carraro and Lombardo 1976; Lombardo and Carraro 1976a; Pacini and Cresti 1978; Pacini and Juniper 1979b; Stevens and Murray 1981; Misset and Gourret 1984; Tiwari and Gunning 1986b; Keijzer and Willemse 1988; Audran and Dicko-Zafimahova 1992; Perdue, Loukides and Bedinger 1992; Brighigna and Papini 1993; Fernando and Cass 1994; Hess and Hesse 1994; Owen and Makaroff 1995). In *Olea europaea* (olive; Oleaceae), cytoplasmic vesicles found to peak in the tapetal cells during the meiotic phase of microsporogenesis are believed to function in the

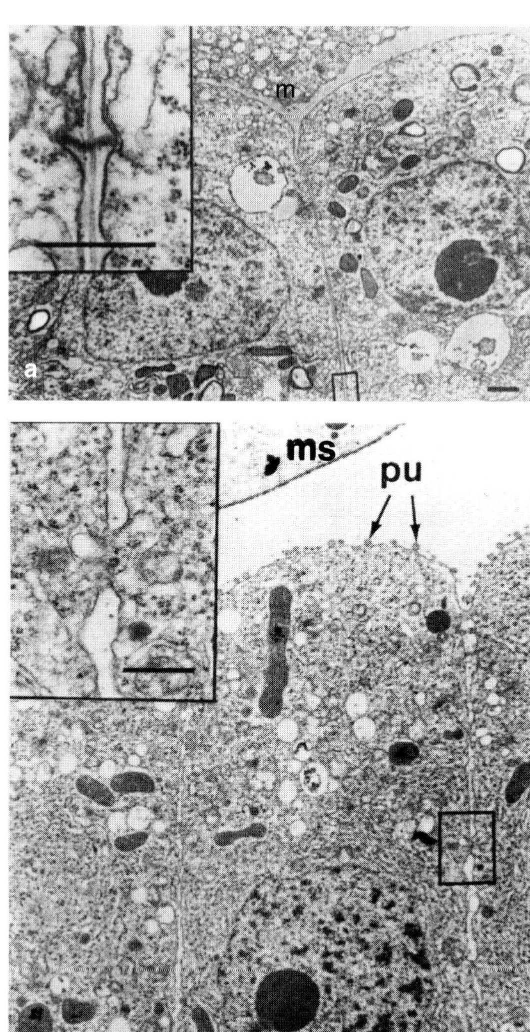

Figure 2.7. Development of the wall of the tapetum in *Zea mays*. **(a)** Premeiotic tapetal cells and microsporocyte (m). Inset shows typical plasmodesmata between cells. **(b)** Postmeiotic tapetal cells and microspore (ms) soon after release from the tetrad. Pro-Ubisch bodies (pu) line the tapetal wall. Inset shows cytoplasmic channel. Scale bars = 1 μm; inset scale bars = 0.5 μm. (From Perdue, Loukides and Bedinger 1992.)

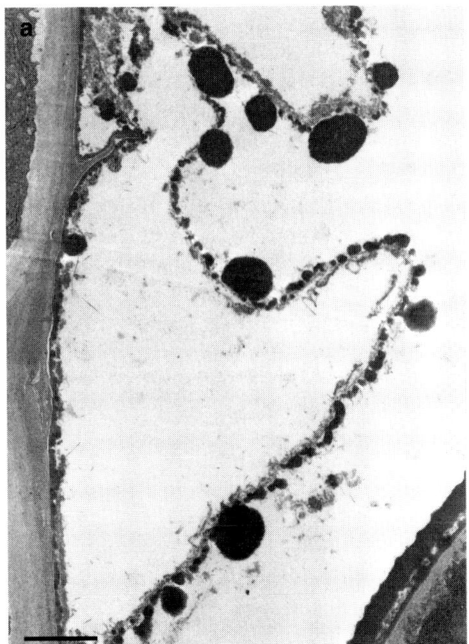

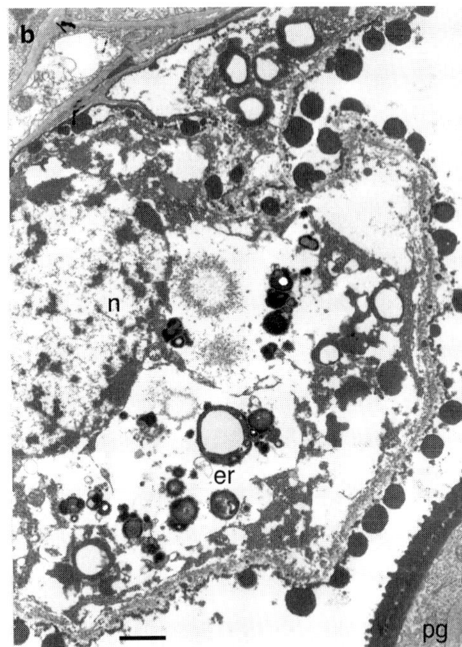

Figure 2.8. Electron micrographs of tapetal cells of *Lycopersicon esculentum* at the binucleate pollen grain stage. **(a)** A tapetal cell devoid of cytoplasm, but with orbicules lining the cell boundary. **(b)** Another tapetal cell with degenerated cytoplasm. Organelles are reduced to the nucleus (n) and isolated fragments of endoplasmic reticulum (er); part of the pollen grain (pg) is also seen. Orbicules are seen within the tapetum as well, lining the outside wall. Scale bars = 1 μm. (From P. L. Polowick and V. K. Sawhney 1993a. Differentiation of the tapetum during microsporogenesis in tomato [*Lycopersicon esculentum* Mill.] with special reference to the tapetal cell wall. *Ann. Bot.* 72:595–605, by permission of Academic Press Ltd., London. Photographs supplied by Dr. V. K. Sawhney.)

transport of substrates mobilized from the anther wall (Pacini and Juniper 1979a). The roles attributed to the ER–organelle complex and dictyosomes in the tapetal cells of *A. sativa* and to the cytoplasmic vesicles in *O. europaea* appear to be attractive from the structural point of view, but as yet there is no clear evidence in support of the notions. This uncertainty reflects a general difficulty in assigning a role for the tapetum based on static electron micrographs alone.

Of interest from the functional perspective is the presence of membrane-lined lipoidal bodies in the tapetal cells (Figure 2.8). These units are currently known as orbicules in preference to older terms such as Ubisch bodies or plaques (Echlin and Godwin 1968a; J. Heslop-Harrison 1968f; J. Heslop-Harrison and Dickinson 1969; Risueño et al 1969; Horner and Lersten 1971; Christensen, Horner and Lersten 1972; Horner and Pearson 1978; Stevens and Murray 1981; El-Ghazaly and Nilsson 1991; Polowick and Sawhney 1993a). As shown in *Helleborus foetidus* (Ranunculaceae), progenitors of the orbicules (pro-orbicules) appear at the sporogenous cell stage in the tapetal cytoplasm as medium electron-dense vesicles in close association with ribosomes and ER. Coincident with the completion of microsporocyte meiosis, the tapetal cell walls lyse, extruding the pro-orbicules through the cell membrane into the anther locule. After their release into the locule, there is some mechanism, yet to be identified, that causes these bodies to become rapidly impregnated with sporopollenin and transformed into orbicules. From these observations it is possible to consider the orbicules as fragments of tapetal cytoplasm sufficiently organized to contribute to the formation of the pollen exine (Echlin and Godwin 1968a). A wall fraction isolated from the tapetal cells of anthers of *Tulipa* sp. (Liliaceae) by acetolysis was found to show an infrared spectrum similar to that of acetolyzed pollen grain from the same species, asserting the chemical identity of the two materials (Rittscher and Wiermann 1983). Materials originating from the tapetum of certain members of Asteraceae (J. Heslop-Harrison 1969b) and the gymnosperm *Pinus banksiana* (Pinaceae) (Dickinson 1970b) form a snugly fitting membrane around the microspore tetrads. In *Gasteria verrucosa* (Liliaceae) and *Petunia hybrida*, there is an excellent correlation between the timing of disappearance of Ca^{2+} from the tapetum and its presence on the pollen surface, suggesting that the degenerating tapetum is the source of Ca^{2+} that mediates in the pollen recognition reaction on the stigma (Tirlapur and Willemse 1992; Bednarska and Butowt 1994). These observations tell us that orbicule formation is not always the final fate of the material released by the tapetal cells.

Electron microscopic observations indicate that the proteins produced by tapetal cells aggregate in the outer cavities of the exine. Perhaps this is not surprising, given the coincidence between the cytochemical reactions of the proteins in the tapetum and on the exine. These proteins, together with pigments, lipids, and monosaccharides, apparently function as recognition substances during pollen–stigma interactions in plants. Another group of proteins generally found in the intine is enzymatic and antigenic in nature and is probably produced in the pollen grain. In several members of Malvaceae (J. Heslop-Harrison et al 1973) and in *Iberis amara* (Brassicaceae) (J. Heslop-Harrison, Knox and Heslop-Harrison 1974), the exine proteins are believed to have their origin in the tapetal cells, specifically in the membrane-bound cisternae of the ER, with a large number of attached ribosomes and with inclusions of a fibrillar material. Esterase is a marker enzyme for exine proteins that shows a dramatic increase in activity in the tapetal cells of *Brassica oleracea* (cabbage) beginning with the microspore to the stage of mature pollen grain. In contrast, acid phosphatase, a marker enzyme for the intine proteins, shows barely detectable activity in the tapetum (Vithanage and Knox 1976). The difficulty in assigning specific roles to the exine and intine proteins is underscored by the fact that both apparently act together in the recognition reaction on the stigmatic surface.

Claims made to the effect that DNA from the tapetum, especially the free nucleic acid bases and nucleotides, is transferred to the microsporocytes during meiosis have proven to be spurious. By monitoring the fate of labeled tapetal nuclei during microsporogenesis in *Lilium longiflorum*, it has been shown that materials from the disintegrating tapetal nuclei are not incorporated into either the sporogenous cells or the microsporocytes (Takats 1959, 1962).

What happens to the protoplasm of the collapsed tapetal cells is just as important as how it functions in the intact cell. From this perspective, two principal types of tapetum are recognized. One is the glandular or the secretory type, in which the cell wall lyses, but the cell contents remain stationary and progressively become disorganized and undergo autolysis; the other is the amoeboid or the periplasmodial type, in which the cell walls break down and the protoplasts fuse to form a periplasmodium. This syncytium later invades the anther locule and envelops and nourishes the developing pollen grains. Rarely, as in *Canna* sp. (Cannaceae), the tapetal protoplasts invade the locule individually without forming a periplasmodium (Tiwari and Gunning 1986d). At the cellular level, the activities that culminate in the glandular mode of tapetum are analogous to those described in other cell types that senesce under natural or experimental conditions. In *Helleborus foetidus*, mitochondria and plastids are the first organelles that succumb as the tapetum follows the degradative pathway. Later, similar considerations apply to the dictyosomes and ribosomes, which show a decrease in number. Shortly thereafter, the tapetum resembles a ghost cell consisting of only the plasma membrane together with the nucleus and ER (Echlin and Godwin 1968a). A different mode of cell breakdown is seen in *Citrus limon* (lemon; Rutaceae), in which the tapetal cells intrude among the microspores at the stage of starch accumulation; later, the plasma membrane ruptures, releasing the protoplasm, which consists of more or less intact mitochondria, plastids, and nucleus (Horner and Lersten 1971). In *Pisum sativum* (pea; Fabaceae), cytoplasmic breakdown of tapetal cells is reflected in their extreme vacuolation and is accompanied by the disintegration of the nucleus (Biddle 1979).

A major difference between the glandular tapetum and amoeboid tapetum is that the cells of the latter possess an organized, functional ultrastructure even as they go into a decline. This is evident from the detailed account of development of the tapetum in *Tradescantia bracteata* (Commelinaceae) (Mepham and Lane 1969) and *T. virginiana* (Tiwari and Gunning 1986a, b). In addition to the presence of abundant rough ER, plastids, mitochondria, and dictyosomes, the tapetal cells accumulate raphides at the microsporocyte stage. Another feature of this phase of tapetal development is the presence of small vesicles and their coalescence, followed by the convolution of the plasma membrane and lysis of the cell wall. During meiosis, the tapetal cytoplasm, now bereft of its plasma membrane, ER, dictyosomes, and microtubules, streams into the anther locule, where it engulfs the microsporocytes. The reorganization of the naked protoplasm that occurs at this stage involves the reappearance of rough ER, dictyosomes, and microtubules, the appearance of electron-lucent vesicles, and the synthesis of enzymes that degrade callose (a polymer of (1,3)-β-glucan). The periplasmodium that surrounds the tetrads develops long amoeboid processes, the callose walls of the tetrad being progressively degraded in front of them. Other signs of tapetal activity evident at this and later stages are a decrease in the number of raphides, accumulation of plastoglobuli in the plastids and their release into the cytoplasm, proliferation of both rough and smooth ER, accumulation of granular material in the smooth ER, and increase in polysomes and dictyosomes. No signs of disintegration are seen in the tapetal nuclei. At the binucleate pollen grain stage, the tapetal plastids enter a new phase of polysaccharide synthesis. These changes in tapetal cells are analogous to reorganization rather than degeneration; this view is supported by the work of Souvré and Albertini (1982) on tapetal development in *Rhoeo spathacea* (Commelinaceae) and Owens and Dickinson (1983) in *Gibasis karwinskyana* and *G. venustula* (Commelinaceae). In the tapetal periplasmodium of *Arum italicum* (Araceae), cytoplasmic reorganization mainly centers around a cluster of microtubules that acquire various orientations in the vicinity of the microsporocytes and later surround the microspores (Pacini and Juniper 1983).

Some of the aforementioned facts suggest that the amoeboid tapetum is not only the site of the packaging of metabolites, but also of their synthesis. Whether these are exclusive roles performed by the amoeboid tapetum is not certain. Cytochemical and autoradiographic experiments on the amoeboid tapetum of *R. spathacea* have shown increases in the amount of polysaccharides and basic and acidic proteins of the tapetal cells and in the incorporation of ^{3}H-acetate and ^{3}H-choline into the lipids (Albertini, Grenet-Auberger and Souvré 1981) and of ^{3}H-tryptophan into proteins (Albertini 1975) of the cells during the tetrad stage of microsporocyte meiosis. Demonstration of synthetic activity in the tapetal cells is an example of the type of evidence that is necessary to establish a correlation between the structure and function of the tapetum.

As electron microscopic observations show, wall dissolution and invasion of the locule by the tapetal contents occur in both glandular and amoeboid types of tapetal cells. Presence or absence of orbicules was previously considered as a striking difference between the two types of tapetal cells; however, the presence of bodies having similar electron density and structure as orbicules in plasmodial tapetal cells of *Gentiana acaulis* (Gentianaceae) (Lombardo and Carraro 1976b), *Tradescantia virginiana* (Tiwari and Gunning 1986b, c), and *Butomus umbellatus* (Butomaceae) (Fernando and Cass 1994) has led to some questions about the validity of orbicules as a distinguishing character. This should perhaps alert us to focus on the formation of the periplasmodium as the most stable criterion to differentiate between glandular and amoeboid types of tapetum.

Among the substances released from the degenerating tapetum is pollenkitt, a generic term applied to compounds imparting stickiness to the pollen. Pollenkitt is composed of lipoidal materials, flavonoids, carotenoids, and degeneration products of tapetal proteins. Based on an assay using isolated cell layers of *Tulipa* sp. anthers, the complete sequence of enzymes involved in flavonoid metabolism has been localized predominantly in the tapetal cells (Herdt, Sütfeld and Wiermann 1978; Rittscher and Wiermann 1983; Kehrel and Wiermann 1985; Beerhues et al 1989). In *Lilium longiflorum* (J. Heslop-Harrison 1968f; Dickinson 1973), *L. hybrida* (Reznickova and Dickinson 1982; Keijzer 1987), *Olea europaea* (Pacini and Casadoro 1981), *Raphanus sativus* (radish; Brassicaceae) (Dickinson 1973), *Gasteria verrucosa* (Keijzer 1987), *Anemarrhena asphodeloides* (Liliaceae) (Chen, Wang and Zhou 1988), *Brassica oleracea*, *B. napus* (Murgia et al 1991a; Evans et al 1992), and *Apium nodiflorum* (Apiaceae) (Weber 1992), coalesced lipids or their precursors, proteins, and carotenoids of pollenkitt are synthesized in the plastids, which degenerate prior to tapetal collapse. At this stage the contents of plastids are seen as osmiophilic globules freely distributed in the cytoplasm of the tapetal cells. Tapetal activity is terminated by the collapse of cells and the release of the globules. It is not unusual for the tapetal cytoplasm also to participate in the production of pollenkitt (Pacini and Casadoro 1981; Pacini and Keijzer 1989). Osmiophilic bodies harboring pollenkitt precursors are associated with the ER during tapetal development in *Ledebouria socialis* (Hyacinthaceae) anthers; however, it is not established whether these bodies are derived from the ER (Hess and Hesse 1994). A variety of roles, such as insect attractant, agent for pollen dispersal, and an antidote against the damaging effects of UV rays, have been attributed to the pollenkitt. Lipoidal components of the pollenkitt have an important role in controlling water loss from the germ pores of the exine following pollen dispersal. The pollenkitt is also probably involved in controlling the recognition reactions on the stigmatic domain. Tryphine is often treated as different from pollenkitt. It is a mixture of hydrophilic substances derived from the breakdown of tapetal cells, whereas pollenkitt is principally made up of hydrophobic lipids containing species-specific carotenoids. In *R. sativus*, the main bulk of the tryphine is synthesized in the tapetal cells after they have been stripped off their walls; this coincides with the stage of the first haploid pollen mitosis. Tryphine apparently consists of proteinaceous materials secreted by the ER and degenerate lipid-rich cytoplasm, along with lipid-filled plastids. It is believed that besides pollenkitt, tryphine is the main encrustation of the pollen exine and aids in the adhesion of pollen grains to the stigma and also possibly regulates recognition reactions on the stigma (Dickinson 1973; Dickinson and Lewis 1973b).

Cytology and Gene Expression. Few areas of investigations on the anther demonstrate more forcefully the interplay of old-fashioned cytology and modern molecular biology and genetic engineering than does the study of the tapetum. Tapetal cells may be uninucleate, binucleate, or multinucleate. Nuclear divisions are initiated in the cells during meiotic division of the microsporocytes and may continue through the active life of the tapetum. According to Wunderlich (1954), who has reviewed the cytology of the tapetum of angiosperms, the multinucleate tapetum is more common than the uninucleate type and has consequently received much experimental attention. Since the early 1900s, several reports of abnormal divisions of tapetal nuclei resulting in increased DNA contents, polyploidy, and polyteny have also appeared to suggest that nuclear aberrations are an intrinsic feature of tapetal cytology. Although these abnormal divisions have been precisely documented, their impact on the function of the tapetum, for the most part, is largely unknown, and none have yet been studied at the molecular level. For these reasons, only a summary of the cytological abnormalities is given here. Because disturbances in the mitotic spindle result in a slow movement of individual chromosomes and due to their stickiness, polyploid restitution nuclei are a frequent occurrence in both uninucleate and multinucleate tapetal cells. A variation of this theme, resulting in the same end product, namely, polyploidy, is endomitosis, which causes a doubling of chromosome number without the formation of a mitotic spindle. Endoreduplication leads to modal increase in DNA values of the nuclei of tapetal cells in the absence of chromosome duplication. In yet other cases, the chromatids formed in each chromosome by DNA replication remain aligned with one another, forming polytene chromosomes. In some species with multinucleate tapetal cells, cytologically complex cells are formed when restitution nuclei undergo endoreduplication, resulting in nuclei with high chromosome numbers and high DNA content (D'Amato 1984). Ploidy levels in the uninucleate tapetal cells range from 4*N* in *Helleborus foetidus* (Carniel 1952) to 16*N* in *Cucurbita pepo* (Cucurbitaceae) (Turala 1958). In plants with multinucleate tapetum, 4*N* nuclei formed by restitutional mitosis are common in *Zea mays* (Carniel 1961), whereas 32*N* nuclei by endomitosis and restitutional mitosis are found in *Lycopersicon esculentum* (Brown 1949).

The timing of synthesis and accumulation of DNA in tapetal nuclei was demonstrated in the early applications of autoradiography and microdensitometry to plant tissues and later by histochemical methods. The first autoradiographic studies carried out using $^{32}PO_4$, ^{3}H-cytidine, and ^{14}C-thymidine showed that tapetal cells of *Tulbaghia violacea* (Liliaceae) (Taylor 1958), *Lilium longiflorum* (Taylor 1959), and *L. henryi* (J. Heslop-Harrison and Mackenzie 1967), respectively, begin to synthesize DNA after the microsporocytes enter meiosis. Earlier, microdensitometry had shown that the DNA content of tapetal nuclei of *L. longiflorum* increases during microsporocyte meiosis, followed by a decrease and then an increase even up to 8*C* level (Taylor and McMaster 1954). In maize, the tapetum reveals a unique pattern of DNA accumulation. Although the early cell divisions are not accompanied by a modal increase in DNA values, synthesis does ensue during the later times and attains 4*C* values (Moss and Heslop-Harrison 1967). According to Albertini (1971), tapetal cells of *Rhoeo spathacea* begin to incorporate ^{3}H-thymidine as the microsporocytes enter the pachytene stage of meiosis. In *Nigella damascena* (Ranunculaceae), the intensity of Feulgen staining of the tapetal nuclei appears to increase during the entire period of meiosis of the microsporocytes (Bhandari, Kishori and Natesh 1976). The range of *C* values observed in the tapetal nuclei of *Petunia hybrida* varies from 2*C* to 4*C* during meiotic prophase in the microsporocytes, advancing to 7*C* to 8*C* at the early pollen grain stage (Liu, Jones and Dickinson 1987). Thus, the timing mechanism responsible for DNA synthesis in the tapetal nuclei appears to be linked to the entry of the sporogenous cells into the meiotic cycle.

Because of the increase in genome size of the

tapetal cell nuclei as revealed by observations just described, it is clear that they engage in widespread and intense transcriptional activity. Evidence for transcription has come from the autoradiographic pattern of incorporation of precursors of RNA synthesis into tapetal cells. As shown by ^{3}H-uridine labeling, in *R. spathacea* and *L. longiflorum* (Albertini 1965, 1971; Williams and Heslop-Harrison 1979), marked RNA synthesis occurs in the tapetal cells at all stages of meiosis, with a peak during diplotene. Somewhat similar results were obtained by Sauter (1969a) on the incorporation of ^{3}H-cytidine into the tapetum of *Paeonia tenuifolia* (Ranunculaceae). Although *L. longiflorum* has the glandular type and *R. spathacea* has the amoeboid type of tapetum, it is interesting to note that RNA synthetic activity is related to the stage of microsporogenesis and not to the ultimate fate of the tapetum.

Transcription of mRNA in the tapetal cells has been visualized by in situ hybridization with ^{3}H-poly(U) as a probe. As seen in Figure 2.9, in anther sections of *Hyoscyamus niger*, the tapetal cells act as a focal point for binding of the label from the time of completion of meiosis in the microsporocytes to the stage when the tapetum begins to disintegrate (Raghavan 1981a). A similar pattern of poly(A)-RNA distribution has been observed in the anthers of rice (Raghavan 1989). The absence of poly(A)-RNA accumulation in the premeiotic tapetum argues against any active role for this tissue in the nutrition of the microsporocytes or in inducing meiosis in them. In contrast, the tapetal tissue of the anthers of *Lilium* sp. is found to anneal ^{3}H-poly(U) throughout meiosis (Porter, Parry and Dickinson 1983). Intense annealing of a cloned rice histone H3 gene to tapetal cells of *H. niger* presumably indicates that histones are made in the tapetal cells (Raghavan, Jiang and Bimal 1992). The overall impact of these studies is to confirm that long before the tapetum begins to disintegrate, it acquires a complex set of specifications in the form of mRNAs.

Tapetal cells also actively synthesize proteins, although there is some variation between species in the periods of peak synthetic activity. At the same time, there are similarities in the timing of accumulation and synthesis of RNA and proteins in the tapetum. As noted by Taylor (1959), intense incorporation of ^{14}C-glycine into proteins of tapetal cells occurs during the later stages of meiosis in the microsporocytes of *L. longiflorum*. In contrast, in *P. tenuifolia*, compared to tapetal cells at the early stage of microsporocyte meiosis, less than 30% of ^{3}H-leucine incorporation is detected at later stages (Sauter 1969a). As shown by ^{3}H-arginine and ^{3}H-tryptophan incorporation, tapetal cells of *Tradescantia paludosa* synthesize histone and nonhistone nuclear proteins coincident with microsporocyte meiosis (De 1961). Moss and Heslop-Harrison (1967) showed that like RNA, proteins are present in the premeiotic tapetum of maize anthers, but increase to attain a peak at the tetrad stage. By immunological methods, two specific proteins have been shown to accumulate exclusively in the tapetum of lily anthers coincident with the peak of tapetal secretory functions (Wang et al 1992a, b, 1993; Balsamo et al 1995).

Although protein synthetic activity of the tapetal cells might reflect changes in enzyme titer, there is little experimental support for this belief. The only published study that addresses this question is that of Linskens (1966), who found marked fluctuations in the activity of lactic dehydrogenase, thiamine pyrophosphatase, and acid phosphatase in isolated tapetal cells of *L. henryi*; these fluctuations were related to the stages of microsporocyte meiosis. It seems probable that tapetal metabolism is the outcome of a stage-specific gene expression program.

Tapetum-specific Genes. Recent investigations have correlated the expression of certain unique classes of mRNAs with the differentiated state of the tapetum (see Schrauwen et al 1996 for review). From a cDNA library made to poly(A)-RNA of preanthesis stamens of tomato, Smith et al (1990) isolated three tapetum-specific clones that are stringently regulated in this tissue over time. As shown by Northern blot, clones *92-B*, *108*, and *127-A* hybridize to RNAs of 600, 700, and 750 bp, respectively, and are detected only in RNA isolated from stamens prior to anthesis. In situ hybridization showed that their expression is limited to the tapetum spanning a narrow window from late microsporocyte meiosis to immature pollen grain stage; this provides a convincing correlative interpretation of the data from Northern blot (Figure 2.10). The deduced proteins of tapetum-specific clones have the characteristics of glycine-rich or cysteine-rich secretory proteins (Aguirre and Smith 1993; Chen and Smith 1993; Chen, Aguirre and Smith 1994). Tapetum-specific genes that encode small secretory proteins have also been isolated from flower buds of *Antirrhinum majus* (Nacken et al 1991a, b), flowering apices of *Sinapis alba* (Staiger and Apel 1993; Staiger et al 1994), tassels of maize (Wright et al 1993), and microsporocytes of *Lilium henryi* (Crossley, Greenland and Dickinson

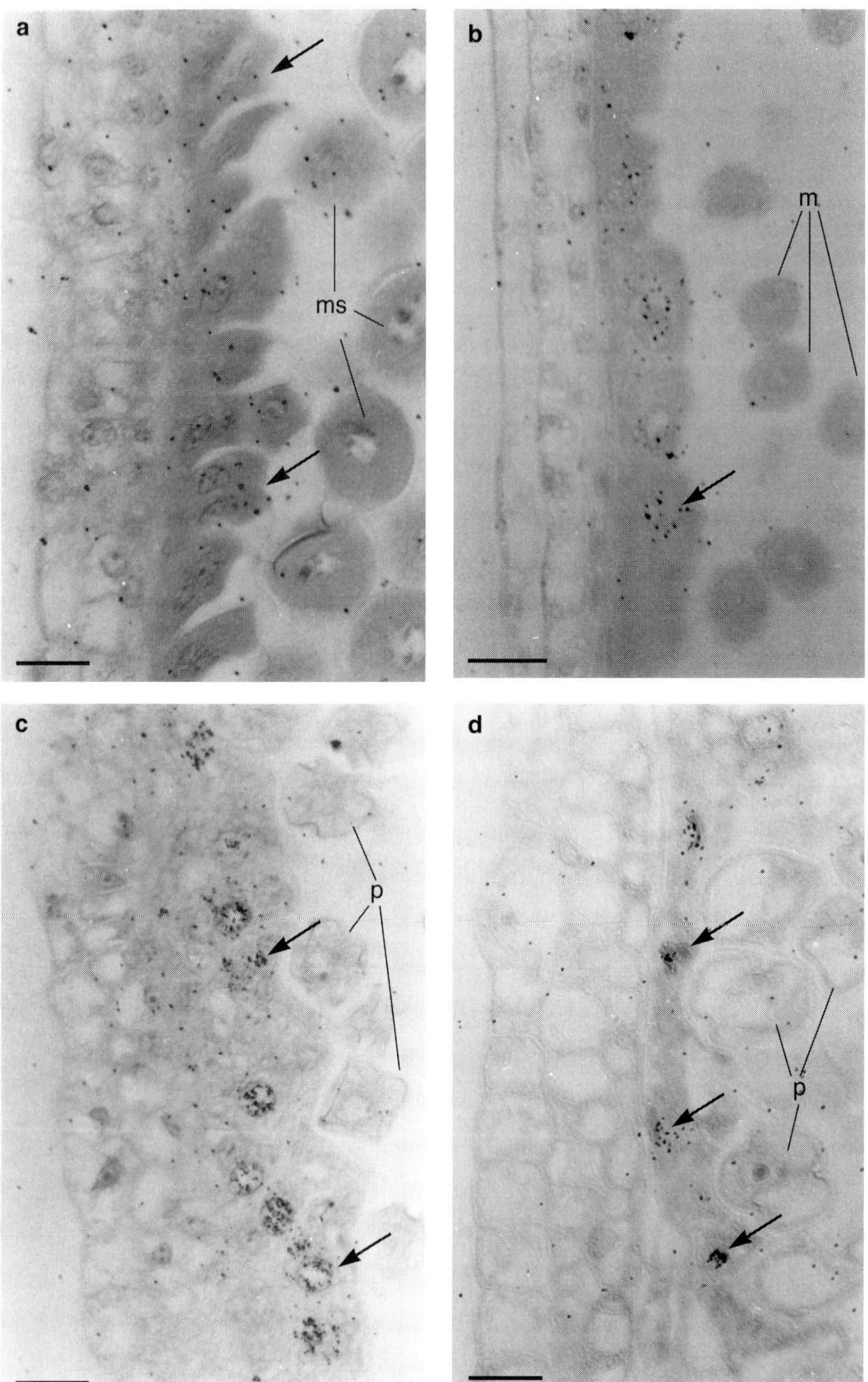

Figure 2.9. Autoradiographs showing the binding of ^{3}H-poly(U) to tapetal cells of anthers of *Hyoscyamus niger* at different stages of development. **(a)** Anther with microsporocytes (ms) at an early stage of meiosis. **(b)** Anther with microspores (m) after release from tetrads. **(c)** Early unicellular pollen grain stage of the anther (p). **(d)** Unicellular, partially vacuolate pollen grain stage of the anther (p). In all cases, arrows point to tapetal cells. Scale bars = 20 μm. (From Raghavan 1981a.)

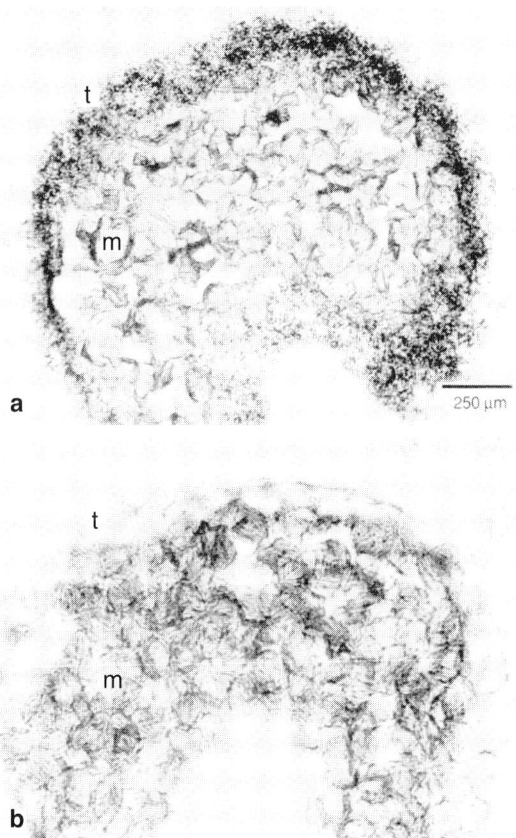

Figure 2.10. Localization of ^{35}S-labeled antisense and sense probes of tapetum-specific cDNA clone *108* in sections of tomato anthers. **(a)** Section hybridized with the antisense probe showing intense localization in the tapetal cells (t). **(b)** Section hybridized with the sense probe showing lack of hybridization in the tapetum. Little or no hybridization is seen in the microspores (m) of both sections. (From Smith et al 1990. *Mol. Gen. Genet.* 222:9–16. © Springer-Verlag, Berlin.)

1995). By immuno–electron microscopic methods, the secreted proteins have been localized in the boundary between the tapetum and the middle layers of the anther (peritapetal membrane) and in the exine of early vacuolate microspores of *S. alba* (Staiger et al 1994). The identification of a protein of tapetal origin in the microspore exine by a high-resolution method is significant as support of the view that some of the exine materials are synthesized in the tapetum.

Several cDNA clones (*TA-13, TA-26, TA-29, TA-32,* and *TA-36*) that represent transcripts expressed exclusively in the tapetum have been described from the tobacco anther cDNA library considered earlier (Koltunow et al 1990). These transcripts encode proteins that are functionally important in the physiology of the tapetum. For example, *TA-13/TA-29* and *TA-32/TA-36* mRNAs encode glycine-rich cell wall proteins and lipid transfer proteins, respectively, that are present in the tapetum and are probably transferred to the exine of the pollen grains. Southern blot and DNA sequencing studies have indicated that each of these mRNAs is encoded by a small gene family (Seurinck, Truettner and Goldberg 1990); mRNAs for these genes accumulate coordinately in the tapetum following microsporocyte meiosis and decay before the first haploid pollen mitosis. Because positive signals are obtained by hybridizing blots containing *TA-29* gene sequences with run-off transcripts made by isolated anther nuclei, it has been concluded that tapetal-specific expression of this gene is controlled at the transcriptional level. This is based on the assumption that run-off transcripts reflect in vivo transcription, since the intact nuclei should preserve the native state of chromatin and its regulatory proteins. To localize the tapetum-specific *TA-29* transcriptional domain, chimeric *TA-29/GUS* gene or *TA-29/GUS* gene containing a minimal CaMV 35S promoter (–52 to +8 bp) and different amounts of the 5'-region of *TA-29* gene were constructed. The chimeric genes were next transferred to tobacco plants using the disarmed *Agrobacterium tumefaciens* Ti plasmid vector in the tobacco leaf disc assay. This assay (Figure 2.11) showed that a 122-bp region (–207 to –85 bp) of *TA-29* gene contains all the information required to program tapetal-specific expression in anthers of transgenic plants (Koltunow et al 1990). In another work, Mariani et al (1990) demonstrated that, like the hybridization profile of

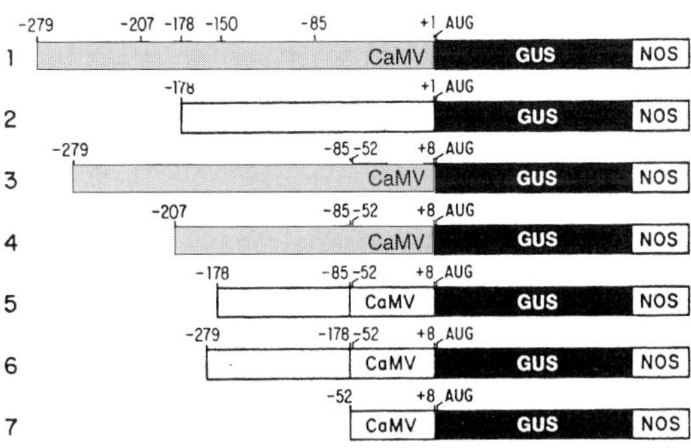

Figure 2.11. Deletion analysis of tobacco *TA-29* gene. Schematic representation of the –279 bp 5'-region of chimeric *TA-29/GUS* constructs with or without a minimal CaMV promoter. Shaded 5'-regions represent chimeric genes that produced the tapetal-specific *GUS* expression. *NOS*, nopaline synthase gene, used as a selectable marker. (From Koltunow et al 1990. *Plant Cell* 2:1201–1224. © American Society of Plant Physiologists.)

TA-29 antisense RNA probe in the normal tobacco plant, *TA-29/GUS* antisense RNA probe produces intense hybridization signals in the tapetum of the transformed tobacco plant. In sum, it appears that sequences within a restricted domain of the *TA-29* gene are required to activate transcription factors in the anther tapetum during specific times in the life cycle of the tobacco plant.

Several genes (*A-3, A-8, A-9, BA-112*, and *BA-118*) isolated from a cDNA library made to poly(A)-RNA of anthers of *Brassica napus* at the sporogenous cell stage have been found to be tapetum specific. Characterization of the *A-9* gene has shown that its predicted protein sequence has homology to a family of seed proteins. It is difficult to assign a critical role for seed proteins in the development of the tapetum; in line with this fact, it was found that transgenic *B. napus* plants containing *A-9* antisense gene are phenotypically normal and produce viable pollen grains (Scott et al 1991; Paul et al 1992; Shen and Hsu 1992; Turgut et al 1994). In a similar way, the tapetum specificity of a storage protein gene from the endosperm of maize in transgenic *Petunia hybrida* is puzzling (Quattrocchio et al 1990). Another cDNA clone from *B. napus* (*A-6*) and its corresponding *Arabidopsis* genomic clone encode a protein with some similarity to callase (1,3)-β-glucanase. Because the *A-6* gene is tapetum specific and is expressed in the tapetum of transgenic plants with a peak in activity coincident with callase activity in the anther, its protein is considered to be a component of the callase enzyme complex (Hird et al 1993). A gene for callase isolated from the anthers of tobacco is expressed in the tapetum just prior to tetrad dissolution (Bucciaglia and Smith 1994). Clones from a cDNA library made from anthers of *B. napus* at the microspore stage appear to be expressed in the tapetum as well as in the microspores. A protein encoded by a cDNA clone (*E-2*) from this group is found to be homologous to a lipid transfer protein, possibly functioning in the formation of both membranes and storage lipids; another cDNA clone encodes a novel proline-rich protein with no homology to any other known protein (Scott et al 1991; Foster et al 1992; Roberts et al 1993a). Acetolactate synthase is an enzyme in branched-chain amino acid biosynthesis, and the expression of transcripts of the gene encoding this enzyme in the tapetal cells of tobacco anthers suggests a requirement for branched-chain amino acids for tapetal function (Keeler et al 1993).

The importance of the tapetum of *B. napus* as a source of lipid materials has been brought into sharp focus by the characterization of genes related to oleosins from a flower cDNA library. Oleosins are specialized proteins of the electron-dense proteinaceous membrane enclosing the naked oil droplets of storage triglycerides; they are present in embryos and endosperms of various plants, as well as in pollen grains of *B. napus*. The transcripts of these genes are specifically expressed in the tapetum during a narrow window in its active life when lipid accumulation is also believed to occur (Robert et al 1994c; Ross and Murphy 1996). In this context, the demonstration that the promoter of a stearoyl–acyl carrier protein (ACP) desaturase gene isolated from embryos of *B. napus* is active in the tapetum of anthers of transgenic tobacco, is noteworthy (Slocombe et al 1994).

By differential screening of a mature *B. campestris* pollen cDNA library, Theerakulpisut et al (1991) have identified a clone, *BCP-1*, whose unique feature is its occurrence in the tapetum and in pollen grains. In situ hybridization with probes labeled with digoxygenin–deoxyuridine triphosphate (dUTP) reporter molecule and an alkaline phosphatase detection system showed that the gene is activated in the tapetal cells beginning at the bicellular pollen grain stage and that its expression continues until tapetal dissolution. The molecular mass of the putative protein encoded by *BCP-1* mRNA is approximately 12 kDa. Although the identity of the protein has not been established, it appears to be a major pollen protein.

Transcripts of two clones isolated from a cDNA library made from rice anthers at the microspore stage are localized exclusively in the tapetum. Further analysis of the genomic clone of one cDNA revealed that its upstream region can regulate *GUS* gene expression in the tapetum of transgenic tobacco. A set of 5′-deletions generated from this clone (Figure 2.12) showed that tissue regulatory elements reside between the –1,273 and –1,095 bp regions of the gene for promoter activation in transgenic tobacco (Tsuchiya et al 1994).

Mitochondrial replication occurs very actively in the developing tapetum to meet the changing energy demands of the cell. The dispersion of genes in the mitochondria suggests that the developmental control of organelle biogenesis may be regulated at the level of mitochondrial gene expression. In attempts to equate increasing mitochondrial number to changes in mitochondrial gene expression, it has been shown by tissue imprinting that both nuclear-encoded alternative peroxidase protein and mitochondrially encoded proteins of α-subunit of

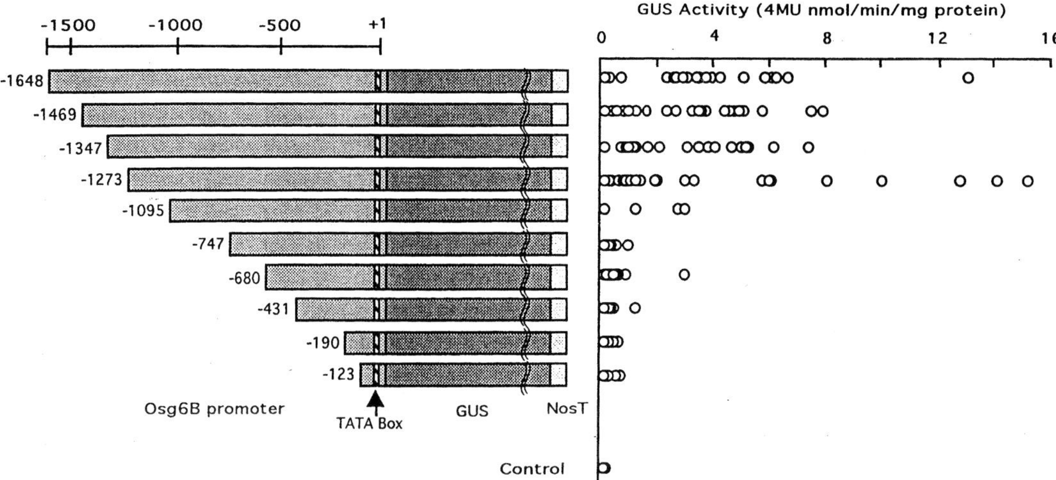

Figure 2.12. Deletion constructs of a tapetum-specific rice genomic clone (*OSG-6B*) and their expression in transgenic tobacco. The deletion constructs are indicated on the left and GUS activity in individual plants in each group and in an untransformed control plant are plotted on the right. (From Tsuchiya et al 1994.)

F_1-adenosine triphosphate (ATP) synthase (ATP-A) and cytochrome oxidase-*c* subunit-II (COX-II) accumulate in the tapetal cells of *Petunia hybrida* at different stages of microsporocyte meiosis (Conley and Hanson 1994). The accumulation of transcripts of *ATP-A*, mitochondrial subunit-9 of F_1-ATP synthase (*ATP-9*), and mitochondrial 26S ribosomal RNA (rRNA) (*RRN-26*) genes in the tapetum of sunflower anthers is also indicative of a cell-specific regulation of gene expression (Smart, Monéger and Leaver 1994).

The mounting evidence in favor of a role for tapetal cells in sporopollenin and pollenkitt synthesis has led to a renewed interest in their flavonoid metabolism. Tapetal-specific expression of genes encoding two key enzymes of flavonoid pathways in *P. hybrida* is an interesting case study. When promoter regions of chalcone synthase and chalcone flavanone isomerase genes are fused with *GUS* reporter gene, expression of the genes in the anthers of transgenic plants is confined mostly to the tapetum (Koes et al 1990; van Tunen et al 1990). Transcripts of a cDNA clone isolated from *Brassica napus* anthers sharing reasonably good amino acid homology with chalcone synthase are also expressed in the tapetum (Shen and Hsu 1992).

The work on the tapetum of the angiosperm anther has now come full circle to provide a conceptual framework for its probable function. The most influential recent approach in regard to the functional aspect of the tapetum is to introduce chimeric ribonuclease (RNase) genes *TA-29/RNase-T1* (regulatory fragment of tobacco *TA-29* fused to a gene encoding RNase-T1 from the fungus *Aspergillus oryzae*) and *TA-29/BARNASE* (*TA-29* fused to RNase gene *BARNASE* from *Bacillus amyloliquefaciens*) into tobacco and *B. napus* plants and show that their expression selectively destroys the tapetum, prevents pollen formation, and leads to male sterility in the transformed plants (Mariani et al 1990; de Block and Debrouwer 1993; Denis et al 1993). Similar results were obtained (Figure 2.13) when promoter fragments from *B. napus A-9* and *A-6* genes fused to *BARNASE* were expressed in transgenic tobacco (Paul et al 1992; Hird et al 1993) or when an antisense construct using the regulatory region of *BCP-1* gene from *B. campestris* was introduced into *Arabidopsis* (Xu et al 1995a). Premature tapetal degeneration was also observed when *Arabidopsis* anther-specific gene *APG-BARNASE* construct was introduced into transgenic tobacco, although tapetal ablation after the late uninucleate stage of microspore development was no longer effective in causing pollen sterility (Roberts, Boyes and Scott 1995). In this respect, the onset of tapetal independence of the microspores offers a clear contrast with results from the other studies. In any event, the tapetum appears to be involved in pollen formation; the challenge now is to provide a complete formal description of how the tapetum accomplishes this function.

3. MICROSPOROGENESIS

Investigation of the structure, physiology, and cytology of the microsporocytes as they are transformed into microspores is considered in this sec-

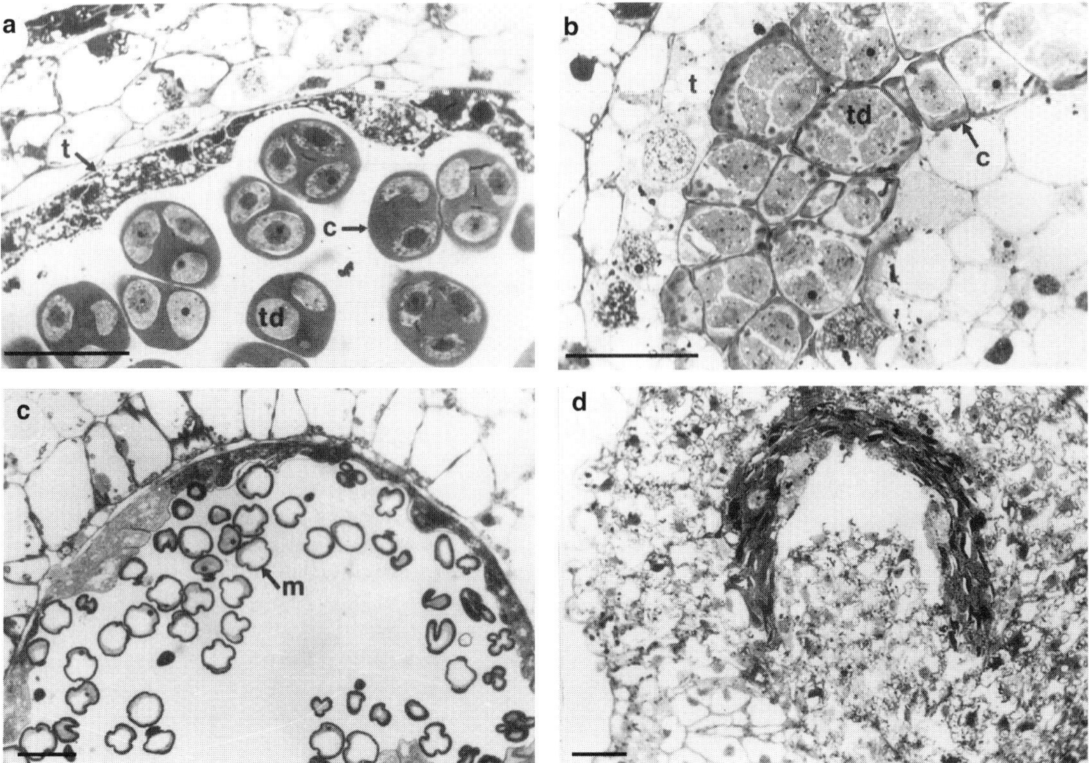

Figure 2.13. Transverse sections of wild-type and transgenic tobacco anthers, transformed with *A-6-/BARNASE* construct. **(a)** Wild-type anther (3 mm) at the tetrad stage. **(b)** Transgenic anther (2.5 mm) with tetrads. **(c)** Wild-type anther (4 mm) at the microspore stage. **(d)** Transgenic anther (3.5 mm) showing the complete degeneration of the tapetum. c, callose wall; m, microspore; t, tapetum; td, tetrad. Scale bars = 100 µm. (From Hird et al 1993; photographs supplied by Dr. R. J. Scott.)

tion under microsporogenesis. Research in microsporogenesis in flowering plants has, as its major goals, an understanding of the causes and consequences of the origin of cells with a balanced but reduced number of chromosomes, and a comprehension of the molecular mechanisms involved in meiosis of the microsporocytes. In a sense, by a series of events that flow as a continuum in time, microsporogenesis bridges the gap between the diploid microsporocytes, on the one hand, and the haploid microspores, on the other. Cytological and biochemical discoveries have become an indispensable information package to define landmarks in the study of microsporogenesis in angiosperms.

(i) Ultrastructural Changes

We start our account of microsporogenesis with the microsporocyte. Processes concerned with cell growth and development converge on the microsporocyte and events of meiosis diverge from it. Soon after the microsporocyte is cut off, it is covered by a primary cellulose cell wall. However, shortly before meiosis, the primary wall gives way to another distinctive wall made of callose; in occasional species, the cellulose wall persists around the microsporocyte and callose deposition occurs within the primary wall (Bhandari, Bhargava and Geier 1981). At the same time as callose appears as a sheath around each microsporocyte, massive protoplasmic strands connect the microsporocytes with one another. Attention has been called to the fact that the presence of cytoplasmic channels between microsporocytes makes the contents of an anther locule a single massive coenocyte and thus probably facilitates synchrony of meiosis (J. Heslop-Harrison 1966b).

At the ultrastructural level, the cytoplasm of the microsporocyte presents the profile of a metabolically active cell enriched with ribosomes, mitochondria, and plastids. Among the distinctive properties of the latter two organelles, one is that they show little internal differentiation.

Dictyosomes and fine strands of rough ER dot the rest of the cytoplasm (van Went and Cresti 1989). The cytoplasmic contour of the microsporocytes of *Canna generalis* is ascribed to the presence of stacked annulate lamellae, which merge into the rough ER. Pores, which suggest similarity to similar openings in the nuclear membrane, are a regular feature of the architectural design of the lamella (Scheer and Franke 1972). In the microsporocytes of *Lycopersicon esculentum*, major changes are restricted to the nucleus, nucleolus, and mitochondria. Formation of vacuoles by the invagination of the inner membrane of the nuclear envelope during the early prophase period is the change observed in the nucleus. The nucleolus is transformed from a granular and fibrillar condition during the preprophase stage to a fibrillar state during late prophase; mitochondria are simple, with lightly stained matrix and a few cristae to begin with, but they acquire a densely stained matrix with dilated cristae later (Sheffield et al 1979; Polowick and Sawhney 1992). Transformation of the ER into double-membraned islands enclosing blobs of cytoplasm enmeshed in a stroma of the regular cytoplasm has been found to be an important cytological feature of the microsporocytes of *Lilium longiflorum* as they enter the meiotic prophase (Dickinson and Andrews 1977). Although these cytoplasmic enclaves are detected readily in the microsporocyte cytoplasm of other plants also, their functions are not clearly defined (Echlin and Godwin 1968a; Biddle 1979). Perhaps they represent a temporary storage facility for the cytoplasm for the duration of meiosis or serve to support and maintain the integrity of the cell during meiosis. More will be said about this later.

As is well known, meiosis consists of two successive nuclear divisions, but chromosome replication occurs only once. Most authors recognize five distinct stages of meiosis, namely, leptotene, zygotene, pachytene, diplotene, and diakinesis, and our account of microsporogenesis here will also be keyed to these stages. DNA synthesis is initiated shortly before leptotene and continues into leptotene. Leptotene lapses into zygotene, the visible evidence of which is apposition or pairing of homologous chromosomes (synapsis), leading to the formation of bivalents. What constitutes the basic mechanism that facilitates homology recognition once the homologues are side by side is yet to be determined, but it appears that chromosomes undergo several dramatic morphological changes, such as partial separation of sister chromatids, increase in the surface complexity of the chromatid fibers, and elongation of the heterochromatic knobs that make this possible (Dawe et al 1994). In *L. longiflorum* and other plants, pairing – and probably the physical basis of it – is the formation of a tripartite structure, known as the synaptonemal complex, between meiotic homologues (Moens 1968). Indicative of the proteinaceous nature of this structure, more than 20 proteins were identified in SDS-PAGE as components of the synaptonemal complex of *L. longiflorum* (Ohyama et al 1992). Zygotene is followed by pachytene, when a shortening and thickening of the bivalents takes place. In each bivalent, the homologous chromatids break apart and exchange segments, thus re-forming complete chromatids but with a reassortment of genetic units. The region of contact between homologous chromatids where crossing-over occurs is known as the chiasma. The events that highlight pachytene are followed by diplotene, when the bivalents begin to contract. A major feature of diplotene is the separation of the two paired chromosomes by splitting along the synaptonemal complex. However, maximum contraction of the bivalents is attained at diakinesis, which signals the end of prophase.

The conventional view of chromosome pairing at zygotene and crossing-over at pachytene has not been universally accepted. Although data on the timing of DNA synthesis and chromosome replication in the premeiotic interphase period challenged the view of pachytene crossing-over, later cytological and biochemical studies seem to reinforce the thesis that synapsis takes place during zygotene, followed by crossing-over at pachytene (Walters 1970). Considerable uncertainty also prevails about the precise time at which crossing-over occurs in relation to the assembly of the synaptonemal complex. It has been suggested that some undefined cytological and biochemical changes occur in the microsporocytes prior to synapsis, entailing an arbitrary selection of potential sites for crossing-over (Stern, Westergaard and von Wettstein 1975).

Prophase is followed by the other regular stages of mitosis (meiosis I, heterotypic division), resulting in the formation of two daughter nuclei with half the number of chromosomes of the mother cell. These nuclei promptly complete a second mitotic division (meiosis II, homotypic division), generating four cells with the haploid number of chromosomes. If we view meiosis in its totality, the dismantling of the paired chromosomes and their reconstruction stand in sharp contrast to other events. It is necessary to point out here that this bare-bones account of meiosis is intended to guide the reader through the biochemical and molecular events that are considered later and does not do justice to the many processes that together make up this division.

With keen foresight, some investigators working at the light microscopic level suggested that meiosis in the microsporocytes is accompanied by some striking changes in cytoplasm and organelle complement. The consequences of these changes, later confirmed in the electron microscope, are most easily seen in the ribosome population, nucleoli, plastids, and mitochondria. In *L. longiflorum*, for which extensive data are available, by diakinesis the microsporocytes are almost completely denuded of ribosomes; this is followed by a repopulation of the cytoplasm with ribosomes beginning about the metaphase–anaphase stages of meiosis I (see Dickinson and Heslop-Harrison 1977, Porter et al 1984, for reviews). When changes in the RNA content of microsporocytes of *L. longiflorum, Trillium erectum* (Liliaceae) (Mackenzie, Heslop-Harrison and Dickinson 1967), and *Cosmos bipinnatus* (Asteraceae) (Knox, Dickinson and Heslop-Harrison 1970) are determined, substantial reductions during the zygotene–pachytene interval and an increase toward the end of the division are observed. This supports connections between the elimination of ribosomes and a decline in the RNA content of the microsporocytes, on the one hand, and the increase in RNA content and the flooding of cells with ribosomes, on the other hand. In *C. bipinnatus* (Knox, Dickinson and Heslop-Harrison 1970) and *Lilium* sp. (Bird, Porter and Dickinson 1983), the low RNA content of microsporocytes also follows a period of lytic enzyme activity, mainly of acid phosphatase. According to Dickinson and Heslop-Harrison (1970b), replenishment of the ribosome population in this RNA cycle is related to the formation of nucleolus-like bodies from nucleolus-organizing regions of the chromosomes; these supernumerary nucleoli, termed nucleoloids, are later set free in the cytoplasm of the microsporocyte at anaphase of meiosis I. Additional evidence from cytochemistry, autoradiography of ^{3}H-uridine incorporation, and in situ hybridization using cloned rRNA probes has shown that the increase in RNA content is due to a significant synthesis of new rRNA in the accessory nucleoli and in the nucleolus-organizing regions, as well as in a small number of nucleolus-like inclusions that appear associated with the chromosomes (Williams, Heslop-Harrison and Dickinson 1973; Dickinson and Willson 1985; Sato, Willson and Dickinson 1989; S. Sato et al 1991; Majewska-Sawka and Rodríguez-García 1996). Apparently, the ribosomal DNA (rDNA) formed by chromosomal amplification is packaged in the extruded nucleoli. Although not widespread, chromosome-associated nucleolus-like bodies and cytoplasmic nucleoloids have been noted during microsporocyte meiosis in *Rhoeo spathacea* (Williams and Heslop-Harrison 1975) and *Olea europaea*, respectively (Rodríguez-García and Fernández 1987; Alché, Fernández and Rodríguez-García 1994).

Changes seen in the structure of plastids and mitochondria during microsporocyte meiosis in *L. longiflorum* are the very epitome of new developmental potencies in the organelles. To begin with, normal plastids present in the microsporocyte undergo a cycle of dedifferentiation and redifferentiation, characterized by the loss of starch, changes in shape, and erosion of internal structure. These events, which are completed by late zygotene, relegate plastids to the status of hollow double-membrane organelles consisting solely of osmiophilic droplets, small vesicles, and membranous tubes. The redifferentiation of the plastids is initiated during metaphase of meiosis I with the transient appearance of granule–double membrane associations in the stroma. Following the release of microspores from the tetrad, the membrane–particle associations disappear and the plastids reconstitute their regular structure. Based on enzymatic digestion and cytochemical tests, there is some evidence to show that RNA-, protein-, and carbohydrate-containing structures are involved in the dedifferentiation of plastids (Dickinson and Heslop-Harrison 1970a; Dickinson 1981; Dickinson and Willson 1983). Since these plastids develop into amyloplasts in the microspore, the possibility that the dedifferentiated plastid is the progenitor of the amyloplast deserves serious consideration (Dickinson and Willson 1983).

In lily microsporocytes, the mitochondria also show a cycle of changes similar to that of the plastids, such as structural simplification by the zygotene stage and recovery of the normal structure in the microspore. In certain plants, there is a coincident accumulation of transcripts of mitochondrial genes and their proteins during the early stages of microsporocyte meiosis (Conley and Hanson 1994; Smart, Monéger and Leaver 1994). It was mentioned earlier that a considerable portion of the microsporocyte cytoplasm is engulfed by unit membranes. Interestingly enough, the ribosome population in the enclosed cytoplasm undergoes much less degradation than the ribosomes in the rest of the cytoplasm. Coincident with the redifferentiation of plastids and mitochondria in the microspores, the investing membranes tear apart, allowing diffusion of the contents into the microsporocyte cytoplasm, while segments of the membranes themselves move to the vicinity of the plasma membrane. It is believed that these membrane fragments play a major role in the formation of the pollen wall layer (Dickinson and Andrews 1977). Spherosomes also show some changes correlated with meiosis; these organelles display a cycle of aggregation and dispersion without undergoing any internal gyrations (J. Heslop-Harrison and Dickinson 1967). On the whole, these observations leave little doubt that transition of the microsporocyte to the gametophytic phase involves a radical reorganization of the cytoplasm, resulting in the elimination of much of the sporophytic program and in the installation of a program for gametophytic functions. Although the ultrastructural changes during microsporogenesis have been followed in *Tradescantia paludosa* (Bal and De 1961; Maruyama 1968), *Hyacinthoides non-scripta* (Liliaceae) (Luck and Jordan 1980), *Nymphaea alba* (Nymphaeaceae) (Rodkiewicz, Duda and Bednara 1989), *Triticum aestivum* (Mizelle et al 1989), and *Gossypium hirsutum* (cotton; Malvaceae) (Hu, Wang and Yuan 1993), the data are neither complete nor as comprehensive as those provided for *L. longiflorum*.

The basic features of the cytokinetic apparatus and the factors that control the division planes of the nuclei during microspore meiosis have hardly been investigated. As is well known, the cytoskeletal network of plant cells is constituted of microtubules and microfilaments. Microtubules are easily identified in the electron microscope; they are also detected by immunocytochemical methods using an antibody for tubulin. During cell division, microtubules are thought to play an important role in the formation of the preprophase band that predicts the division plane and helps to guide the cytokinetic apparatus along a predetermined path. However, electron microscopic and/or immunofluorescence analyses of microtubular patterns during microspore meiosis in *Tradescantia paludosa* (Clapham and Östergren 1984), *Gasteria verrucosa* (van Lammeren et al 1985), *Lilium henryi* (Sheldon and Dickinson 1986), *L. longiflorum* (Tanaka 1991), *Lycopersicon esculentum, Ornithogalum virens* (Liliaceae) (Hogan 1987), *Impatiens sultani* (Balsaminaceae), *Lonicera japonica* (Caprifoliaceae) (Brown and Lemmon 1988; van Went and Cresti 1988b), *Solanum melongena* (eggplant) (Traas, Burgain and de Vaulx 1989), *Doritis pulcherrima* (Orchidaceae) (Brown and Lemmon 1989), *Zea mays* (Staiger and Cande 1990), *Phalaenopsis* sp. (Orchidaceae) (Brown and Lemmon 1991a), *Magnolia denudata, M. tripetala* (Magnoliaceae) (Brown and Lemmon 1992a), and *Cypripedium californicum* (Orchidaceae) (Brown and Lemmon 1996) have not revealed a configuration of microtubules in the form of a preprophase band. In *G. verrucosa*, microtubules are randomly oriented during meiosis, suggesting that their main function is to ensure the integrity of the cytoplasm during division. As shown in Figure 2.14, following meiosis II, arrays of microtubules radiate from the nucleus and thus apparently define the cytoplasmic domains of the future microspores (van Lammeren et al 1985). A similar role has been ascribed to a postmeiotic system of microtubules in establishing the division plane for quadripartitioning of microsporocytes in most of the other species just listed. Thus, the disposition of the four haploid nuclei with their own microtubule cytoskeleton seems to determine the future cleavage pattern of the tetrad. The radiating microtubules are found to exclude plastids from around the microsporocyte nuclei of *L. longiflorum*; a wildly speculative view is that this paves the way for the subsequent formation of plastid-free generative cells from microspores (Tanaka 1991).

Microfilaments are best detected by immunofluorescence using antibodies for actin (the main chemical constituent of the microfilament) or by treating unfixed, permeabilized cells with fluorescent-dye–labeled phallotoxins (rhodamine-phalloidin staining). According to Sheldon and Hawes (1988), actin cytoskeleton appears to be independent of the microtubule system both within the mitotic spindle and in the cytoplasm during microspore meiosis in *Lilium hybrida*; these authors suggest that here actin may function with the microtubules in maintaining cytoplasmic integrity. However, actin filaments accompany microtubules in phragmoplast formation during microspore meiosis in *L. hybrida* (Sheldon and Hawes 1988), *G. verrucosa* (van Lammeren, Bednara and Willemse 1989), *Z. mays* (Staiger and Cande 1991), and *Phlaenopsis* sp. (Brown and Lemmon 1991a, b). In *S. melongena*, a dense web of randomly oriented actin filaments is thought to mark the future division plane after meiosis I, very much like the preprophase band of microtubules, but such a configuration of actin is not detected during meiosis II (Traas, Burgain and de Vaulx 1989). Bands of actin filaments are also seen at the level of cell division planes constituting cleavage furrows following both meiosis I and II in the microsporocytes of *Magnolia soulangeana* (Dinis and Mesquita 1993). In the most detailed studies on *G. verrucosa*, microfila-

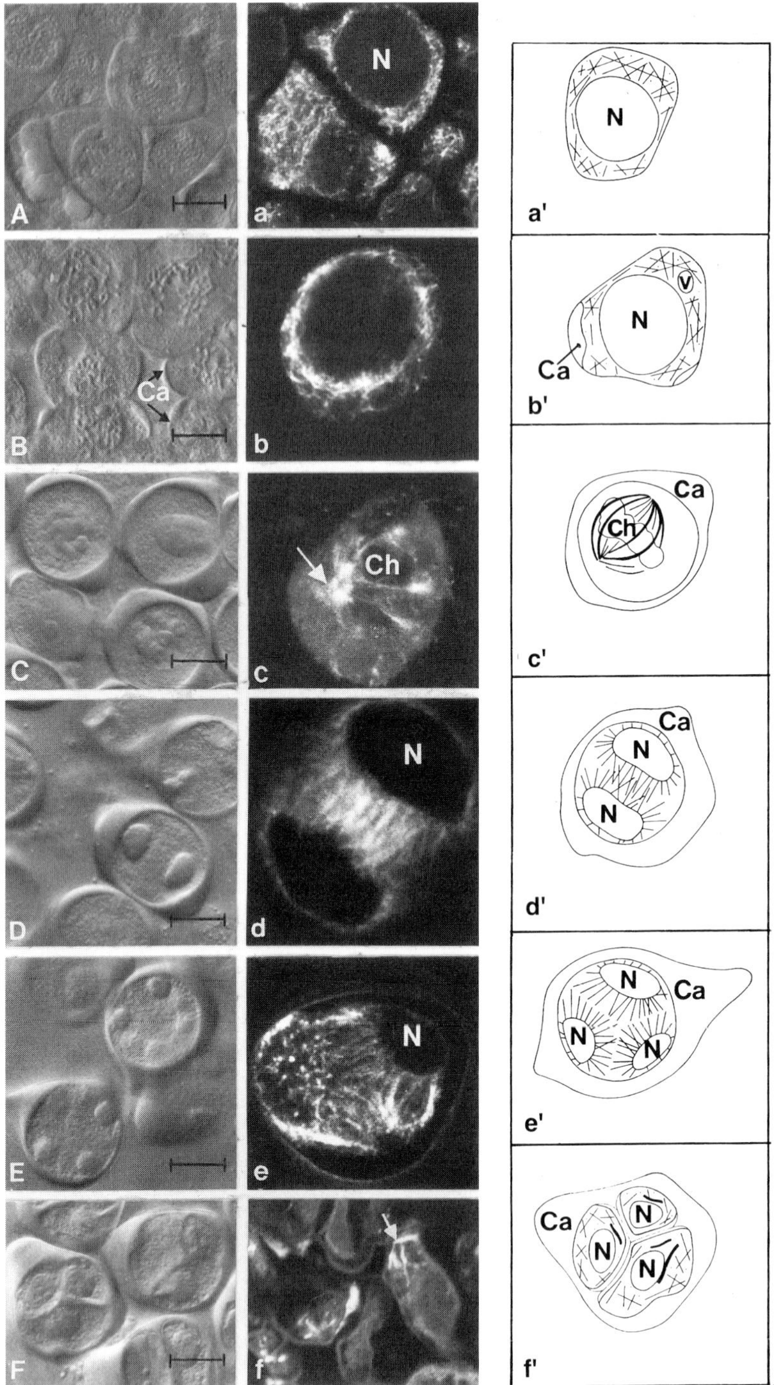

Figure 2.14. Microsporogenesis in *Gasteria verrucosa.* **(A–F)** Stages in microsporogenesis as visualized by Nomarski optics. **(a–f)** Fluorescently labeled microtubules in the corresponding stages. **(a'–f')** Diagrammatic representations of microtubule configurations in the corresponding stages. (A, a, a') Microsporocytes at leptotene with microtubules arranged in a crisscross pattern. (B, b, b') Microsporocytes at pachytene; microtubules are arranged in a crisscross pattern and are also concentrated near the nucleus. (C, c, c') Microsporocytes at diakinesis with microtubules at the polar center (arrow), and the bundles at the equatorial poles at metaphase. (D, d, d') Dyads, showing microtubules radiating from the nuclei and interdigitating between the nuclei. (E, e, e') Tetrads of nuclei, with microtubules radiating from the nuclei. (F, f, f') Tetrads after wall formation, with thick bundles of microtubules (arrow). Ca, callose wall; Ch, chromosome; N, nucleus. Scale bars = 10 μm. (From van Lammeren et al 1985. *Planta* 165:1–11. © Springer-Verlag, Berlin. Photographs and line drawing supplied by Dr. A. A. M. van Lammeren.)

ments and microtubules appear to interact during the different stages of microsporocyte meiosis (van Lammeren, Bednara and Willemse 1989). Since no site-specific actin was localized during meiosis, it is difficult to assign a function for actin in the division process. Perhaps there is sufficient reason to consider a spatial control of meiosis in terms of the dynamics of the entire cytoskeleton, rather than of the microtubules or microfilaments alone. Moreover, it seems that key processes of meiosis involving microtubules and microfilaments are controlled by genes. Recent studies on certain meiotic mutants of maize have shown that they exhibit distinct alterations in cytoskeletal behavior, especially of the microtubules, associated with aberrations in the meiotic pathway (Staiger and Cande 1990, 1991; Liu, Golubovskaya and Cande 1993). Identification of the genes regulating entry into, and exit from, the meiotic processes and of the proteins responsible for the phenotype will provide further insight into cytoskeletal functions during microsporocyte meiosis.

Considerable variation exists in the formation of the phragmoplast and cell plate during microsporocyte meiosis. When meiotic cytokinesis occurs successively after each nuclear division, cell plates are laid down typically in association with centrifugally expanding phragmoplasts. An usual feature of cell plate formation in the microsporocytes of *Lilium longiflorum* is the presence of numerous Golgi-derived, coated vesicles with fuzzy hairlike spikes that contribute to the formation of the phragmoplast (Nakamura and Miki-Hirosige 1982). Absence of phragmoplast after meiosis I is a general feature of simultaneous cytokinesis, in which the four microspores are cleaved from a common cytoplasm. A complex of phragmoplasts is, however, generated after meiosis II, and wall formation occurs by centripetally growing furrows, by cell plate formation, or by a combination of both processes with the help of microtubules that determine spore domains. With simultaneous cytokinesis in *Doritis pulcherrima* (Brown and Lemmon 1989), *Phalaenopsis* sp. (Brown and Lemmon 1991b), *Magnolia denudata,* and *M. tripetala* (Brown and Lemmon 1992a), typical phragmoplasts and associated cell plates are formed after meiosis I, but they are short-lived and are routinely resorbed before meiosis II.

(ii) Molecular Biology of Meiosis

The sense and purpose of microsporogenesis attain their climax at meiosis. Tangible evidence of this is a series of biochemical and molecular activities concerned with pairing, crossing-over, and a reduction in chromosome number in a small population of cells over a short period of time. Two aspects of the molecular biology of meiosis that will be addressed here are those concerned with the transition of microsporocytes from a state of mitosis to meiosis and the completion of meiosis itself. A question of primary importance concerns the nature of the trigger that prompts sporogenous cells to cease mitotic divisions and enter the meiotic cycle. It has been suggested that the synthesis of some factors in tissues other than the sporogenous cells is involved in triggering the cells to enter meiosis. These substances have been given such trivial names as "meiosis-inducing substance" or "meiosis determinants," but their chemical nature remains obscure (Walters 1985).

It would seem that some activities connected with meiosis, such as the synthesis of the spindle material, are initiated during the premeiotic period and might even be a prerequisite to division, but we do not yet have a catalogue of these events. Among other biochemical events of the immediate premeiotic period is the apparent accumulation of nucleic acid precursors in the microsporocytes. According to Foster and Stern (1959), the premeiotic interval in the microsporocytes of *Lilium longiflorum* is associated with a marked accumulation of soluble deoxyribosidic compounds. Other studies have shown the presence of enzymes associated with the production of deoxynucleosides and their phosphorylation during the interval adjacent to DNA synthesis, with the caveat that some enzymes that mediate in the nucleoside transformations are not directly tied to DNA synthesis (Hotta and Stern 1961a, b). The timing of appearance of the nucleosides may be variable, as in *L. henryi,* in which nucleosides, nucleotides, and free bases are detected in the midmeiotic prophase; this observation might influence the judgment as to whether they really accumulate before meiosis (Linskens 1958).

An important group of compounds that are thought to have a structural involvement in cell division is the –SH group, whose presence is related to the formation and functioning of the mitotic spindle. Although the role of –SH groups in the bonding of macromolecules of the mitotic apparatus is not well understood, fluctuations have been noted in the relative levels of free and bound –SH groups associated with microsporocyte meiosis in *L. longiflorum* and *Trillium erectum.* In both species, particularly striking is a rise in the concentration of soluble sulfhydryls preceding meiosis and a decline before the fabrication of the mitotic apparatus (Stern 1958). This observation does not vitiate the suggestion that sulfhydryl compounds are used prior to prophase; however, it appears that we have a long way to go before the preparative molecular changes as they relate to discrete events in the premeiotic microsporocytes involving the utilization of –SH groups are understood. A promotive effect of glutathione on meiosis in cultured anthers of *L. henryi* has been linked to a requirement for –SH compounds in cell wall formation during meiosis (Pereira and Linskens 1963).

The Timing of DNA Synthesis. To understand the concepts of pairing of homologous chromosomes and recombination during microsporocyte meiosis, we must first examine whether there are any unique patterns of DNA synthesis associated with the meiotic cell cycle. The problem here is infinitely more complex than DNA synthesis during mitosis, since it is conceivable that the freedom of the homologous chromosomes for recombination depends upon a round of DNA synthesis for the repair of breaks formed during crossing-over. During the past several decades, the study of DNA synthesis during microsporocyte meiosis in angiosperms has proceeded from microspectrophotometry through autoradiography to biochemical and molecular approaches. A pioneering investigation of Swift (1950), which showed an increase in the DNA content of microsporocytes of *Tradescantia paludosa* during the early period of meiosis, suggested that the conventional synthetic period (S-period) is at the onset of prophase. In a later work, Taylor and McMaster (1954) substantiated this finding from their work with anthers of *L. longiflorum*. Particularly striking was their observation that as the microsporocytes underwent synchronous DNA replication in anticipation of meiosis, they settled at a DNA content corresponding to the 4*C* amount. Subsequently, they attained a transient phase with 2*C* DNA and ended up as microspores containing the haploid amount of DNA. Following this study, the pattern of autoradiographically or microspectrophotometrically measurable DNA synthesis during microsporocyte meiosis has been followed in other species of *Lilium* (Plaut 1953; Taylor 1953, 1959), *T. paludosa* (Taylor 1953; Moses and Taylor 1955; De 1961), *Tulbaghia violacea* (Taylor 1958), and *Secale cereale* (rye; Poaceae) (Bhaskaran and Swaminathan 1959). These studies have generally established that DNA synthesis in the microsporocytes occurs at late premeiotic interphase or at the onset of prophase and that there is no further synthesis until after the completion of meiosis. The premeiotic DNA synthetic period was referred to as the preleptotene period (Plaut 1953; Taylor and McMaster 1954). Although label incorporation in these investigations is found almost exclusively in the nucleus, Holm (1977) has described ^{3}H-thymidine label in the cytoplasm of premeiotic microsporocytes of *L. longiflorum*; more than half of the incorporation was resistant to digestion by deoxyribonuclease (DNase).

Two early reports indicated that a reduced level of DNA synthesis occurred during prophase after completion of the main S-period. This might appear necessary for chromosome pairing at zygotene and crossing-over at pachytene. According to Sparrow, Moses and Steele (1952), in the microsporocytes of *Trillium erectum* a small increase in DNA content occurs between pachytene and diplotene, but it is insufficient to account for an expected doubling if the entire premeiotic DNA is synthesized during this interval. In anthers of *L. henryi*, there is a burst of DNA increase in the microsporocytes during prophase after the S-period (Linskens and Schrauwen 1968b). The reality of a midprophase DNA synthesis during microsporocyte meiosis in the liliaceous plants *Lilium* and *Trillium* has now been established by the detailed investigations of Hotta and Stern (see Stern and Hotta 1977, 1984, for reviews). The microsporocytes of several hybrids of *L. longiflorum*, *L. henryi*, *L. speciosum*, *L. tigrinum*, and *T. erectum* have proved to be favorable for this work because of their abundance, natural synchrony, and ability to undergo part of the meiotic division under in vitro conditions. Moreover, compared to other plants, meiotic division in the microsporocytes of liliaceous plants occurs at a leisurely pace and is spread over a period of 9 to 12 days (Ito and Stern 1967). In much of the work on liliaceous plants, meiotic cells at different stages were labeled by exposing flower buds to solutions containing the isotope or by suspending extruded microsporocytes directly in the isotope; both $^{32}PO_4$ and ^{3}H-thymidine were used to label the cells. DNA isolated by a standard method was next subjected to centrifugation to equilibrium in cesium chloride, and the positions of the labeled peaks in the gradient were determined by analyzing fractions (Figure 2.15). The most interesting outcome of these studies is the identification and characterization of two groups of DNA sequences that replicate during the prophase interval. This was accomplished by DNA–DNA hybridization involving DNA synthe-

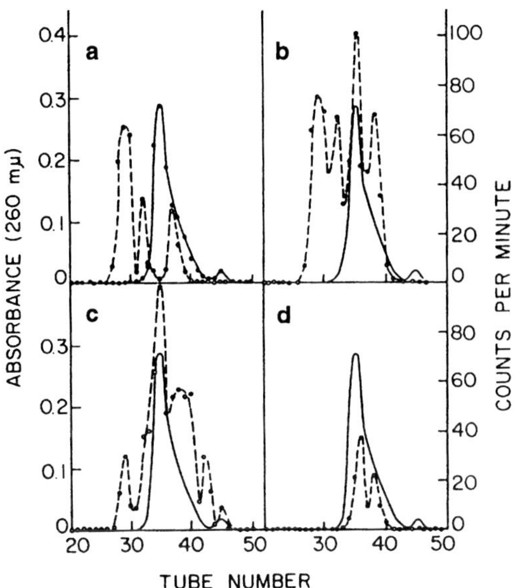

Figure 2.15. DNA labeling during microsporocyte meiosis in lily anthers. Microsporocytes cultured for 24 hours were exposed to $^{32}PO_4$ for 20 hours. DNA extracted was centrifuged in cesium chloride, fractions (indicated by tube number) were collected, and radioactivity (broken lines) and absorbance (solid lines) were determined. The stages during which microsporocytes were exposed to the isotope are **(a)** zygotene, **(b)** late zygotene, early pachytene, **(c)** late pachytene, and **(d)** late pachytene, early diplotene. (From Hotta, Ito and Stern 1966.)

sized exclusively during the premeiotic S-phase and zygotene and pachytene intervals and DNA prepared from somatic cells. The synthesis of the first group of DNA sequences coincides with the beginning of chromosome pairing at zygotene and continues to the end of pachytene. This sequence is designated as zygotene DNA (zyg-DNA) and constitutes about 0.1% to 0.2% of the genome whose replication is delayed until halfway through the meiotic prophase (Hotta, Ito and Stern 1966; Hotta and Stern 1971b; Hotta et al 1985b). Despite its presence in limited amounts, zyg-DNA is remarkable for the complex it forms with proteins and phospholipids (Hecht and Stern 1971). Another interesting aspect of this DNA synthesis is that the newly replicated strands are not immediately ligated to the body of the nuclear DNA, thus causing gaps between zyg-DNA and adjacent nuclear DNA. This was demonstrated in sedimentation profiles of ^{14}C-labeled S-phase DNA and ^{3}H-labeled zyg-DNA in sucrose or glycerol gradients, where the zyg-DNA sediments toward the top of the gradient while the S-phase DNA remains at the bottom (Hotta and Stern 1976). The characteristic buoyant density of zyg-DNA is probably due to its protein–phospholipid association.

In an attempt to account for the selective suppression of zyg-DNA synthesis until zygotene and the delayed completion of its replication, Hotta, Tabata and Stern (1984) have invoked a specific protein named leptotene protein (L-protein). This protein has been purified from the nuclear membranes of meiotic cells of *L. speciosum* and has a molecular mass of 73 kDa. It has been found to be highly specific in inhibiting zyg-DNA synthesis in isolated prezygotene nuclei, whereas nuclei from premeiotic anthers, from postzygotene microsporocytes, and from microsporocytes before or during S-phase are insensitive to the protein inhibitor. The rationale for a protein inhibitor in suppressing zyg-DNA synthesis during the premeiotic S-phase and through the prezygotene interval is the need for zyg-DNA synthesis to occur coincident with the pairing of homologous chromosomes.

Synthesis of zyg-DNA has proved to be a useful molecular marker for the onset of meiosis in liliaceous plants. Hotta et al (1985b) utilized poly(A)-RNA prepared from microsporocytes of *L. speciosum* to test for its ability to hybridize with fractions of lily DNA centrifuged to equilibrium in a cesium chloride gradient. A major peak of hybridization coinciding with zyg-DNA has pointed to the poly(A)-RNAs, being a zyg-DNA transcript (zyg-RNA). The accumulation of this mRNA occurs only after leptotene and does not extend beyond midpachytene (Figure 2.16). Since zyg-RNA was not detected in nonmeiotic tissues such as the anther wall, young anthers, and callus, it is believed that its function is essentially meiotic. A speculative view is that zyg-RNA is translated into proteins that are involved in the formation of the synaptonemal complex.

It has been argued that a critical role must be envisaged for zyg-DNA in meiosis, perhaps in the actual process of chromosome pairing itself. The effects of inhibitors of DNA synthesis (for example, deoxyadenosine) on meiosis in cultured microsporocytes of *L. longiflorum* have added substance and support to this argument (Ito, Hotta and Stern 1967; Sakaguchi et al 1980). If the inhibitors are applied at late leptotene or very early zygotene stages, microsporocytes are totally arrested in their development and end up with fragmented chromosomes. Anarchic chromosome shattering is reduced when microsporocytes are exposed to the inhibitors at the midzygotene stage; these cells also go through an abortive meiosis I. Effects of inhibition of DNA synthesis during late zygotene or early

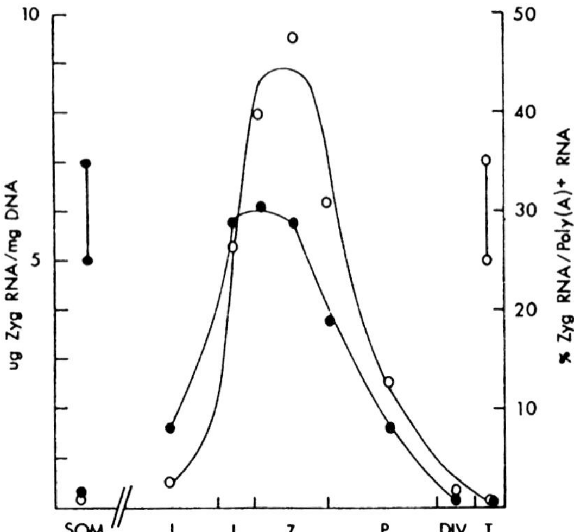

Figure 2.16. Changes in the level of zygotene DNA transcript (zyg-RNA) during microsporocyte meiosis in lily anthers. The amount of zyg-RNA was determined from the amount of RNA adsorbed from poly(A)-RNA in the zyg-DNA column. Closed circles are values expressed relative to DNA; open circles are values expressed as percentage of total poly(A)-RNA. Values for somatic tissues are also shown. SOM, somatic tissues; I, interphase; L, leptotene; Z, zygotene; P, pachytene; DIV, end of meiosis; T, tetrads. (From Hotta et al 1985b. © Cell Press.)

pachytene stages are generally reflected in chromosome fragmentation at meiosis II. The best pointer to the effect of inhibitors of DNA synthesis is the electron microscopic demonstration that deoxyadenosine apparently interferes with the initiation and formation of the synaptonemal complex with subsidiary effects on the disjunction of the paired homologues at diplotene stage and the separation of sister chromatids at anaphase (Roth and Ito 1967). Mitomycin-C is another inhibitor of DNA synthesis; it interferes with the formation of the synaptonemal complex if it is applied to microspores at the leptotene–zygotene stage (Sen 1969). It is difficult to judge from the electron micrographs alone whether inhibitors prevent the organization of the axial core or the interdigitation of the transverse filaments of the synaptonemal complex. In an attempt to define the relationship of zyg-DNA synthesis to chromosome pairing, it has been proposed that synthesis of DNA during the zygotene–pachytene stage requires the simultaneous synthesis of certain nuclear proteins. Selective application of cycloheximide, an inhibitor of protein synthesis, not only arrests protein synthesis during the zygotene interval, but also results in the inhibition of DNA synthesis and in the failure of chromosome pairing and fabrication of the chiasma (Hotta, Parchman and Stern 1968; Parchman and Stern 1969). This has fueled some speculation that the protein is the component of an axial element involved in chromosome pairing and chiasma formation and that its function is related to simultaneous DNA replication.

Various types of evidence have been adduced to support the view that a second group of DNA sequences that replicate during the preprophase interval of microsporocyte meiosis afford an efficient mechanism for genetic recombination, especially in the limited repair activities at the break points of DNA molecules involved in crossing-over at pachytene. Synthesis of this DNA is of a much smaller magnitude than that occurring at zygotene and, since pachytene replication does not cause a net increase in DNA, it has been concluded that the regions of the DNA strand undergoing replication are simply replacing preexisting ones (Hotta and Stern 1971b). The first major indication that breakage and restoration of continuity by repair are largely confined to pachytene came from the demonstration that activities of a microsporocyte-specific endonuclease and of polynucleotide kinase and polynucleotide ligase, which oversee chromosomal DNA breakage and reunion events, peak at the end of the zygotene phase and decline sharply during late pachytene (Howell and Stern 1971). Evidence for a programmed breakage of chromosomal DNA during pachytene has been provided by zonal sedimentation profile of denatured, ss-DNA prepared from microsporocytes of *Lilium* sp. at different stages of the meiotic cycle. The striking result of this experiment reflecting the introduction of nicks at pachytene is that the denatured DNA displayed a bimodal profile during pachytene, with two principal peaks corresponding to sedimentation values of 104S and 62S, and a more or less unimodal profile at all other stages of meiosis. No change was seen in the sedimentation profile of native DNA, which showed a major peak in the 250S region of the gradient (Figure 2.17). In addition, it was found that sedimentation profiles of DNA prepared from microsporocytes of hybrids of *L. speciosum* in which chiasma formation was suppressed or very much reduced by colchicine treatment were distinguished by the absence of a 62S peak in the chiasmatic forms (Hotta and Shepard 1973; Hotta and Stern 1974). In other studies, the seemingly elaborate organization of pachytene DNA (P-DNA) has been unraveled by reannealing kinetics and thermal stability analysis. It has been deduced from these experiments that families of

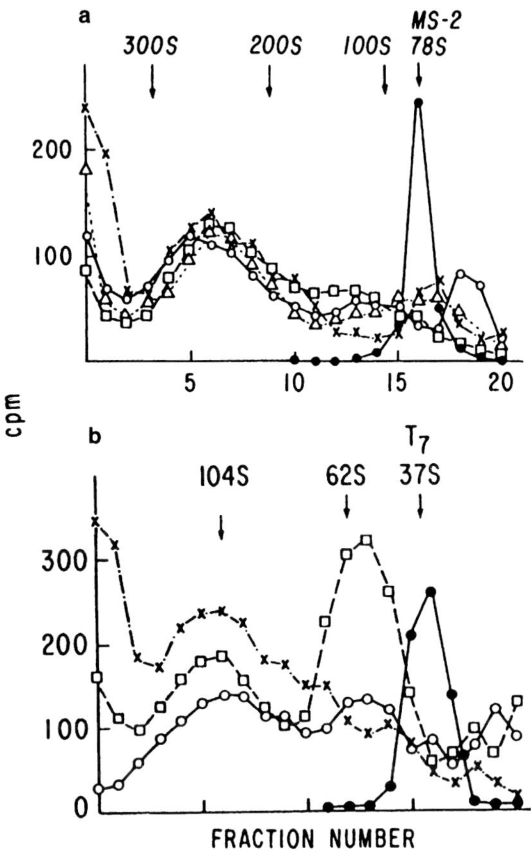

Figure 2.17. Sedimentation profiles of native and alkali-denatured DNA extracted from microsporocytes of lily at meiotic stages. **(a)** Sedimentation of native DNA. **(b)** Sedimentation of denatured DNA, both in a 10%–30% glycerol gradient. ○, premeiotic interphase to early leptotene; △, zygotene; □, early to midpachytene; ×, diplotene to anaphase I. Zygotene DNA was not measured in (b). Sedimentation values are given at the top of each figure. MS-2 and T_7 are marker DNAs. (From Hotta and Stern 1974. *Chromosoma* 46:279–296. © Springer-Verlag, Berlin.)

moderately repeated sequences, of modal length of about 1,500–2,000 bp, scattered throughout the chromosomes may be the sites of specific nicks and repair synthesis. Moreover, there is a strong preference for nicking and rejoining functions to a particular subset of repeats as opposed to a weak preference for repetitive sequences in general (Smyth and Stern 1973; Bouchard and Stern 1980). Perhaps the most remarkable complexity associated with the functioning of P-DNA results from factors that render these DNA sequences susceptible to endonuclease nicking. Hotta and Stern (1981) found that when pachytene cells of *L. speciosum* are cultured in the presence of ^{3}H-uridine, P-DNA appears to be labeled. By RNase digestion, the label was identified in RNA bound to P-DNA. These RNA molecules, referred to as pachytene-small-nuclear RNA (psn-RNA), revealed a single size class, with a sedimentation value in the 5S to 6S range, and were found to be synthesized during meiotic prophase coincident with chromosome pairing. Since P-DNA sequences rendered accessible to endonuclease nicking are partially complementary to psn-RNA, the specificity of chromatin sites selected by psn-RNA is attributed to sequence complementarity. By modifying chromatin organization in the regions housing P-DNA sequences, it has been claimed that psn-RNA is a major factor in rendering P-DNA sequences open to endonuclease nicking.

Based on the work just reviewed, a tentative model of the organization of pachytene nick-repair sites in the genome of *L. speciosum* has assigned each P-DNA three different regions. According to this model, each end of the P-DNA consists of pachytene-small-nuclear DNA sequences (psn-DNA) ranging in length from about 150 to 300 bp and a middle region that does not share homology with psn-DNA (Hotta and Stern 1984). A discordant note in this otherwise convincing account of P-DNA is the concern raised by the work of Smyth and Shaw (1979) showing that a lot more DNA synthesis, monitored by ^{3}H-thymidine incorporation, occurs in the cytoplasm than in the pachytene nuclei of microsporocytes of *L. henryi*. A later work involving electron microscope autoradiography showed that about 75% of pachytene synthesis might represent plastid rather than nuclear DNA synthesis, the latter accounting for only about 10% (Smyth 1982; see also Bird, Porter and Dickinson 1983). In addition to nuclear DNA synthesis, varying degrees of cytoplasmic DNA synthesis have also been described during microsporocyte meiosis in *Agapanthus umbellatus* (Liliaceae) (Lima-de-Faria 1965).

In conformity with biochemical studies, DNA synthesis during early meiotic prophase has been demonstrated by autoradiographic methods, although an unambiguous separation of the synthesis of zyg-DNA and P-DNA was not possible. In the microsporocytes of cultured anthers of *Triticum aestivum*, where meiosis is completed in about 24 hours, incorporation of ^{3}H-thymidine is found to occur at all stages of meiosis, although a possible effect of wounding on the observed response is not eliminated (Riley and Bennett 1971). Ito and Hotta (1973) provided evidence to show that in the microsporocytes of *L. longiflorum*, zyg-DNA is synthesized by all the chromosomes without any obvious localization, although in long-term ^{3}H-thymidine labeling experiments, the sites of DNA synthesis are obscured by nonspecific labeling by products of thymidine catabolism. The reality of DNA synthesis during zygotene and pachytene episodes in the microsporocytes of *L. longiflorum* has been confirmed by high-speed scintillation autoradiography of DNA fibers (Sen, Kundu and Gaddipati 1977). In the microsporocytes of this same species, electron microscope autoradiography has revealed preferential association of label with the synaptonemal complex (Kurata and Ito 1978) and with the condensed or condensing chromatin (Porter, Bird and Dickinson 1982), supporting the speculation that DNA synthesis is involved in some aspect of chromosome pairing.

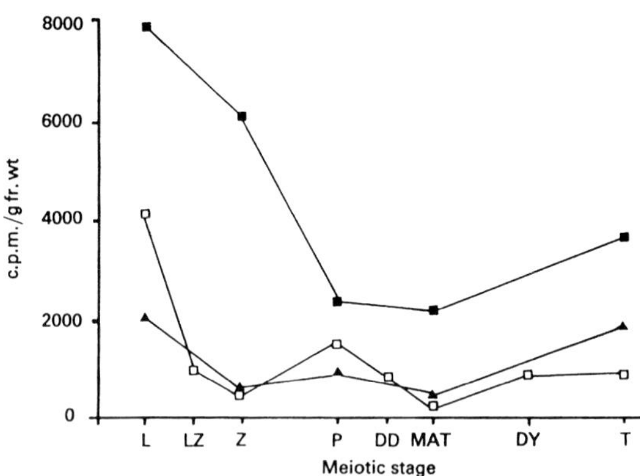

Figure 2.18. Changes in the specific activity of poly(A)-RNA isolated from microsporocytes of lily at different stages of meiosis. The three lines indicate three separate experiments. L, leptotene; LZ, leptotene–zygotene; Z, zygotene; P, pachytene; DD, diplotene–diakinesis; MAT, metaphase, anaphase, telophase; DY, dyad; T, tetrad. (From Porter, Parry and Dickinson 1983. *J. Cell Sci.* 62:177–186. © Company of Biologists Ltd.)

RNA Synthesis and Metabolism. The characteristics of RNA synthesis and metabolism during microsporocyte meiosis correlate in most general respects with what has already been described at the ultrastructural level in *Lilium longiflorum* (Mackenzie, Heslop-Harrison and Dickinson 1967). At the heart of these changes is the simultaneous decrease in the RNA content and in the ribosome population of microsporocytes during meiosis; this result was foreshadowed by previous works from other laboratories. Using $^{32}PO_4$ as an indicator of autoradiographically detectable RNA synthesis, Taylor (1958) identified a period of active synthesis during premeiotic interphase and one of declining synthesis during zygotene pairing in the microsporocytes of *Tulbaghia violacea*. Somewhat similar results were obtained in *L. longiflorum* using ^{14}C-orotic acid as a precursor of RNA synthesis; the comparatively high resolution of the autoradiographs afforded by this isotope compared to $^{32}PO_4$ also made it possible to follow cytoplasmic and nuclear incorporation of the label. Generally, no cytoplasmic labeling of microsporocytes is observed during the period of decreasing RNA synthesis beginning at leptotene (Taylor 1959). *Rhoeo discolor* is another monocot in which a decreasing pattern of RNA synthesis is noted in the microsporocytes beginning with pachytene and extending into metaphase of meiosis I (Albertini 1965, 1971). The strategy utilized for RNA synthesis during meiosis by microsporocytes of the dicot *Paeonia tenuifolia* is strikingly similar to that of the monocots (Sauter and Marquardt 1967b; Sauter 1969a). Evidently, even in taxonomically distinct classes of plants, there is a regulated pattern of RNA synthesis during meiotic intervals when chromosomes undergo extensive alterations. The role played by specific RNA species during meiosis is not apparent from these studies, although the possibility of a selective inhibition of rRNA and mRNA synthesis has been implied. For example, Das (1965) showed that in the microsporocytes of *Zea mays*, the nucleolus rapidly loses its ability to incorporate precursors of RNA synthesis beginning with leptotene and extending into diakinesis. In the microsporocytes of *L. longiflorum*, there is not only a decrease in the rate of rRNA transcription during meiosis, but the processing of nucleolar RNA into mature ribosomes is delayed as well (Parchman and Lin 1972). Indicative of the inhibition of mRNA synthesis, dramatic decreases have been found in the levels of poly(A)-RNA during prophase of meiosis in the microsporocytes of *Lilium* sp. (Figure 2.18) and *Nicotiana tabacum*, with a low level being registered at the pachytene stage (Porter, Parry and Dickinson 1983; Chandra Sekhar and Williams 1992).

Cytochemical analyses of the changes in RNA content of microsporocytes of *Ribes rubrum* (Saxifragaceae) (Genevès 1966), *Z. mays* (Moss and Heslop-Harrison 1967), *P. tenuifolia* (Sauter and Marquardt 1967a, b), and *L. longiflorum* (Linskens and Schrauwen 1968a) at different stages of meiosis are consistent with the autoradiographic data. These studies have also shown that later stages of meiosis of the microsporocytes are invariably accompanied by a definitive increase in RNA contents. According to Linskens and Schrauwen (1968b), there is a considerable increase in the polysome content of the microsporocytes of *L.*

longiflorum during metaphase, indicating the engagement of mRNAs.

In contrast to the autoradiographic and cytochemical data, biochemical analyses of RNA synthesis during microsporocyte meiosis have been bedeviled by a set of conflicting data. A decrease in RNA synthesis monitored by ^{14}C-uracil incorporation is noted between leptotene and zygotene stages in the microsporocytes of *Tulipa gesneriana* (tulip), but this is not associated with corresponding changes in RNA polymerase activity (Hotta and Stern 1965b). However, a precipitous decrease in RNA synthesis coincident with the zygotene stage is not always the case. For example, Hotta and Stern (1963a) determined the synthesis of RNA by continuous labeling of microsporocytes of intact anthers of *Trillium erectum* with ^{14}C-cytosine and found two short peaks at leptotene–zygotene and a substantial peak at the tetrad stage. In this experiment, in contrast to a short pulse of radioactivity, the net incorporation reflects the difference between synthesis and turnover. Similar results were also obtained using isolated microsporocytes and a short pulse with ^{14}C-uracil. Based on the data relating to inhibition of RNA synthesis by actinomycin-D and nucleotide analysis of pulse-labeled RNA, it has been deduced that part of the RNA synthesized during meiotic prophase in *T. erectum* has messenger properties (Hotta and Stern 1963b; Kemp 1964).

Protein Synthesis and Metabolism. Patterns of protein synthesis and metabolism during microsporocyte meiosis are in many respects similar to those described for RNA. Perhaps this is not surprising, because variations in RNA metabolism might be expected to reflect variations in enzyme activity. Several investigators have exploited the possibilities offered by autoradiographic localization of labeled amino acids to follow protein synthesis during microsporocyte meiosis, with the same end result in the diverse examples. This is illustrated by the results from Taylor's (1959) work on *Lilium longiflorum*, which showed intense nuclear and cytoplasmic incorporation of ^{14}C-glycine into the microsporocytes through leptotene and a reduced labeling during the rest of the meiotic prophase. In the microsporocytes of *Rhoeo discolor* (Albertini 1967, 1971) and *Paeonia tenuifolia* (Sauter 1968), there is very little protein synthesis from leptotene until pachytene; these data are also consistent with a decline in the accumulation of cytophotometrically detectable proteins during meiosis in the microsporocytes of *Zea mays* (Moss and Heslop-Harrison 1967). Based on SDS-PAGE of proteins of anthers of *L. longiflorum*, there is some indication that specific polypeptides may be involved in the events of meiosis (Wang et al 1992a).

A uniform pattern is also encountered in considering the synthesis of histones and nonhistone nuclear proteins during microsporocyte meiosis. By using ^{3}H-arginine as a marker for histone synthesis and ^{3}H-tryptophan for nonhistone proteins, combined with ^{3}H-thymidine incorporation, De (1961) showed that microsporocytes of *Tradescantia paludosa* synthesize histones at the early leptotene stage synchronously with DNA synthesis. Considerable nonhistone nuclear protein synthesis occurs transiently at late leptotene when no further DNA is synthesized. Similar results have been reported for the incorporation of ^{3}H-arginine and ^{3}H-tryptophan during meiosis in the microsporocytes of *R. spathacea* (Albertini 1971, 1975). According to Sauter and Marquardt (1967b), there is an inverse correlation between the nucleohistone content of the microsporocytes of *P. tenuifolia* and their RNA and protein contents: Meiotic cells characterized by high RNA and protein contents have little or no nucleohistones, whereas those with reduced RNA and proteins have invariably high nucleohistone content. In *Hippeastrum belladonna* (Amaryllidaceae), the lysine-rich histone profile of the sporogenous cells appears to be complex and different from that of the tapetum, suggesting that it may be associated with changes in the metabolic activity of the cells (Pipkin and Larson 1972).

An interesting observation about the microsporocytes of *L. longiflorum* and *Tulipa gesneriana* is that they contain a unique histone that is easily separated by acrylamide gel electrophoresis. This histone, designated as the meiotic histone, is virtually absent in the somatic tissues of the plants but is prevalent throughout microsporocyte meiosis (Sheridan and Stern 1967). A meiotic-specific histone has also been detected in microsporocytes of *L. candidum* (Bogdanov, Strokov and Reznickova 1973; Strokov, Bogdanov and Reznickova 1973) and *L. speciosum* (Sasaki et al 1990; Sasaki and Harada 1991). The meiotic histone is probably identical, or related, to a chromatin-associated protein designated as meiotin-1 (Figure 2.19), found in microsporocytes of *L. longiflorum* immediately preceding and during meiosis (Riggs and Hasenkampf 1991; Hasenkampf et al 1992). The origin, nature, and molecular mode of action of this protein remain elusive; one does not even know whether it affects directly or indirectly the meiotic cycle of the microsporocyte.

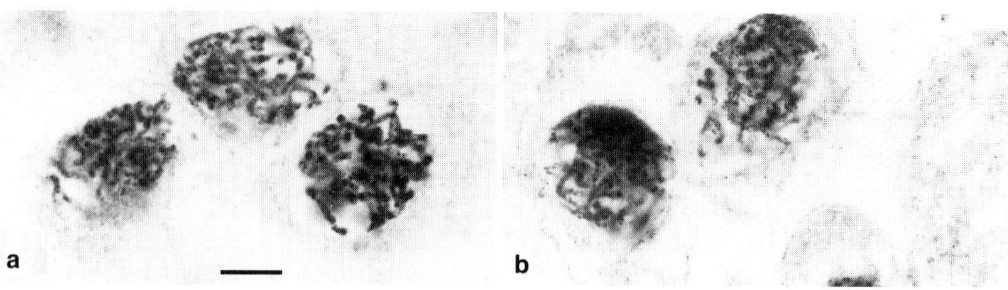

Figure 2.19. Immunostaining of meiotic microsporocytes of *Lilium longiflorum.* **(a)** Microsporocytes at pachytene stage immunostained with lily histone H1 immune serum. **(b)** Same, immunostained with lily meiotin-1 immune serum. Scale bar = 10 μm. (From C. Hasenkampf et al 1992. Temporal and spatial distribution of meiotin-1 in anthers of *Lilium longiflorum. Devel. Genet.* 13:425–434. © 1992. Reprinted by permission of Wiley-Liss Inc., a subsidiary of John Wiley & Sons Inc. Photographs supplied by Dr. C. Hasenkampf.)

Current thinking about the role of protein synthesis during microsporocyte meiosis has been largely influenced by a series of studies by Hotta, Stern, and associates. This has led to the view that one aspect of the molecular regulation of meiosis in liliaceous plants lies in the controlled synthesis of structural proteins required for the formation of the synaptonemal complex and of catalytic proteins involved in the recombination events. An early work (Hotta and Stern 1963b) established that in isolated microsporocytes of *Trillium erectum,* there is a surge in protein synthesis at the pachytene stage. Although the four labeled precursors used, namely, ^{14}C-arginine, ^{14}C-glycine, ^{14}C-leucine, and ^{3}H-tryptophan, showed some fluctuations in the rate of incorporation, in other respects the four patterns were similar (Figure 2.20). Part of the proteins synthesized during pachytene could be inhibited by actinomycin-D; seen in the light of the claimed specificity of this drug as an inhibitor mRNA synthesis, this result lends itself to the conclusion that some of the proteins are encoded by newly transcribed mRNA. Reference was made earlier to the relationship between the synthesis of these proteins, synthesis of DNA, and the key events of meiosis like synapsis and crossing-over (Hotta, Parchman and Stern 1968; Parchman and Stern 1969). Other proteins that might play a role in initiating meiotic recombination are the endonucleases (Howell and Stern 1971).

Among the catalytic proteins, a DNA-unwinding protein ("U-protein") has been found to increase in the microsporocytes of *Lilium* sp. coincident with the appearance of chromosomes in homologous alignment and subsequent crossing-over during meiosis. This protein, whose properties are similar to those of the prokaryotic U-proteins, apparently unwinds DNA duplexes from ends or nicks and thus provides single-strand pieces for heteroduplex formation (Hotta and Stern 1978). U-protein from lily microsporocytes has not yet been examined in sufficient detail to specify its role. There is nonetheless the intriguing possibility that it promotes recombination between DNA duplexes by producing single-strand tails of DNA.

The finding of a specific DNA-binding or reassociation protein ("recA-protein") in the nuclei of microsporocytes of *L. speciosum* has made it highly probable that this protein has a special meiotic function, namely, that of catalyzing the reassociation of ss-DNA. Its transient appearance at prophase and absence at all other intervals of meiosis suggest a role for this protein in chromosome pairing and/or recombination (Hotta and Stern 1971a). Two homologues of the bacterial recA-protein (RAD-51 and LIM-15) are found abundantly in the leptotene and zygotene chromosomes and decrease significantly at pachytene (Terasawa et al 1995). Moreover, certain primary properties of recA-protein are modified by its state of phosphorylation. This is illustrated by the observation that dephosphorylation with phosphatase increases the affinity of the protein for ss-DNA and bestows the ability to bind ds-DNA also. Although cyclic adenosine monophosphate (cAMP)-dependent protein kinase cannot reverse the effects of dephosphorylation, a protein kinase present in the microsporocytes is able to reactivate the dephosphorylated form (Hotta and Stern 1979). According to Hotta et al (1979), the activity of this protein is regulated in some manner by homologous chromosome pairing at zygotene. This is based on the evidence that microsporocytes of a variety of lily, "Black Beauty," which is characterized by the absence or near absence of synapsis, shows a reduced level of recA-protein during zygotene and pachytene. Induction of tetraploidy in the microspore mother cells by colchicine has, however, allowed extensive chromo-

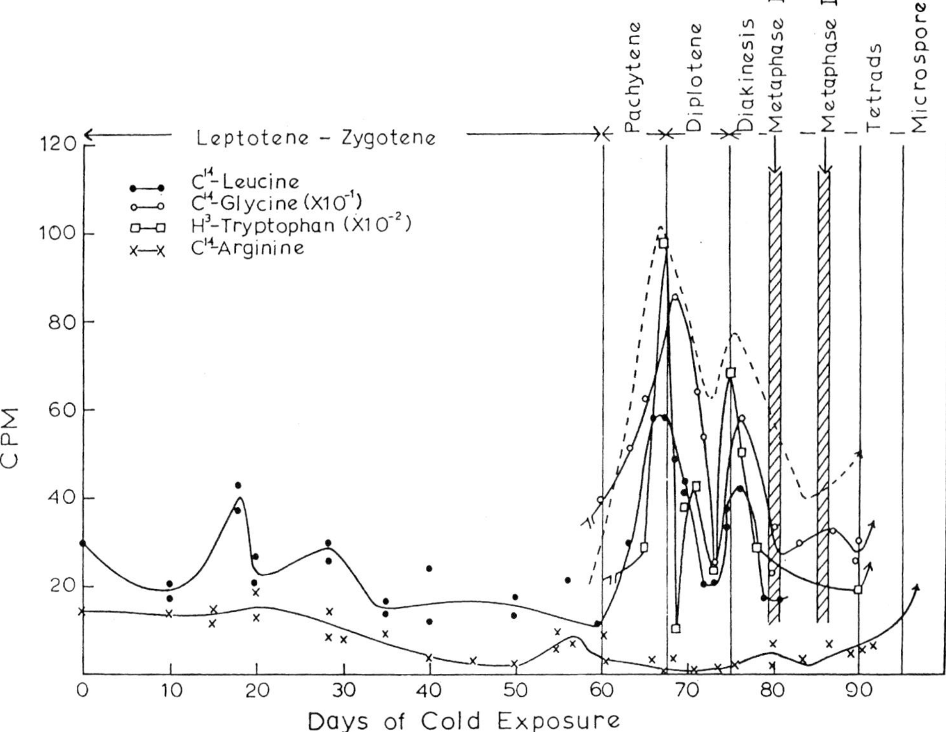

Figure 2.20. Incorporation of ^{14}C-arginine, ^{14}C-glycine, ^{14}C-leucine, and ^{3}H-tryptophan into proteins of isolated microsporocytes of lily at different stages of meiosis. The dotted line represents the pattern of ^{14}C-leucine incorporation by microsporocytes in situ. Plants were kept in storage at 1 °C to induce microsporocyte meiosis. (Reproduced from Hotta and Stern 1963a. *J. Cell Biol.* 19:45–58 by copyright permission of the Rockefeller University Press.)

some pairing as well as restoration of the characteristically high level of recA-protein activity during the zygotene–pachytene stages. Another clue to the relationship of this protein to chromosome pairing is the observation that disruption of homologous pairing affects the levels only of recA-protein, but not of others. How this protein might function in the process of recombination is speculative, although it is likely that it facilitates pairing between complementary DNA strands at the zygotene and pachytene stages. Whether phosphorylation is a mechanism for modifying the DNA-binding properties of the protein remains to be determined.

Hotta et al (1985a) have also purified a recA-like protein from meiotic microsporocytes of a *Lilium* hybrid. Central to the concept of homologous recombination, recA-protein from *Escherichia coli* is known to rapidly promote the pairing of homologous DNA strands and to more slowly modulate the exchange of one strand in a linear duplex DNA for a strand from a closed circular ss-DNA molecule, thereby creating heteroduplex DNA (Radding 1982). The tests carried out with recA-protein isolated from lily microsporocytes have revealed its great similarity to the protein from *E. coli*. From the standpoint of meiotic physiology, an important observation made in this work is that there is a major increase in recA-protein activity in the microsporocytes at meiosis with a peak activity at early pachytene (Figure 2.21). In another experiment, recombination between two mutant plasmids (recA$^-$, HB101) that restores tetracycline resistance in transformed *E. coli* was used in an assay to measure the activity of recA-protein in extracts of meiotic cells. This work showed that, paralleling the increase in protein titer, the recombination frequency begins to increase as the microsporocytes enter zygotene, reaching a peak at pachytene. These observations point strongly to the conclusion that the enzymatic steps leading to recombination are consummated at the pachytene interval (Hotta et al 1985a).

Meiosis-specific Genes. In an attempt to clone the genes that transcribe meiosis-related mRNAs, a cDNA library was constructed to poly(A)-RNA

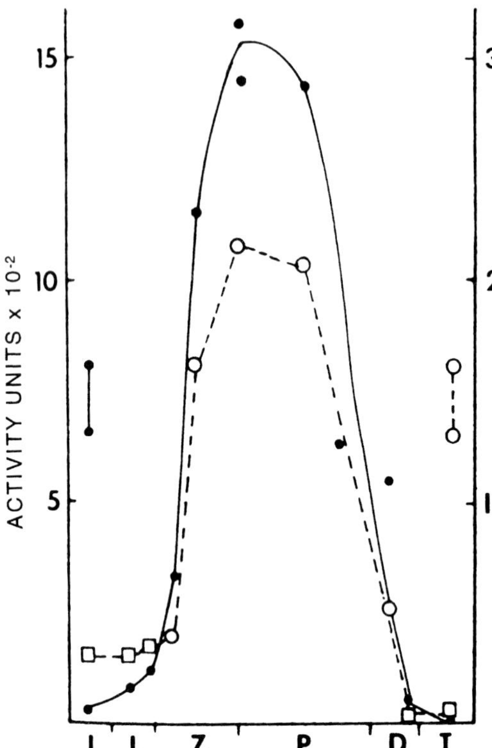

Figure 2.21. Changes in the levels of recA-like proteins during microsporocyte meiosis in lily. The protein was purified from meiotic cells at different stages and the DNA-dependent adenosine triphosphatase (ATPase) activity was assayed as an indicator of recA-like protein activity. The open squares (bottom right) probably do not represent recA-like protein because of its very low DNA-dependent ATPase activity. I, premeiotic interphase; L, leptotene; Z, zygotene; P, pachytene; D, diplotene–diakinesis; T, tetrads. (From Hotta et al 1985a. *Chromosoma* 93:140–151. © Springer-Verlag, Berlin.)

from *Lilium* sp. microsporocytes spanning the entire premeiotic interphase through pachytene. Of the several meiotic-specific clones identified by differential screening of the library with cDNA from mitotic cells, one has been inferred from its DNA sequence to code for a protein homologous to the soybean (*Glycine max;* Fabaceae) heat-shock protein HSP-17 (Appels, Bouchard and Stern 1982; Bouchard 1990).

Subtraction screening of a cDNA library made to zygotene-specific mRNAs of *L. longiflorum* microsporocytes with a probe made to premeiotic microsporocytes has led to the identification of a large number of meiotic-specific genes. As seen in Figure 2.22, most of the recombinants continue to be expressed throughout meiosis beginning at the zygotene stage. Included in this group are also clones that are expressed at low levels before microsporocytes enter the meiotic cycle (*LIM-17, LIM-18*) and decrease subsequently; one clone (*LIM-15*) appears transiently during early prophase. Heat-shock proteins (HSP), serine proteases, and proteins involved in DNA repair are some of the translated amino acid sequences of these gene products (Kobayashi et al 1994). Although it seems that further analysis of these recombinants will help to elucidate the molecular biology of meiosis, it is possible that most of the proteins identified will be more homologous to those that are generally found in other meiotic cells than to those specific to lily.

An early meiosis cDNA clone from wheat florets identified by indirect screening using a lily meiosis-specific clone encodes a leucine-rich protein with a structure that resembles leucine zipper proteins. This DNA-binding motif generally found in eukaryotic genomes consists of a heptameric repeat of leucines that serves as the "zipper" in dimer formation. The dimers are located adjacent to a cluster of positively charged amino acids ("basic motif") that recognize sequence-specific target DNA. The gene is expressed only after leptotene, predominantly at the zygotene and pachytene stages, and the protein probably plays a structural role in the fabrication of the synaptonemal complex (Ji and Langridge 1994). Enhanced *GUS* gene expression during the leptotene, zygotene, and pachytene stages of meiosis in lily microsporocytes by activation of yeast-derived meiotic promoter elements has been demonstrated; this is a strong indication of the ubiquitous nature of the *cis*-acting sequences required for transcription of meiosis-specific genes (Tabata et al 1993).

In summary, studies reviewed here, elaborate though they may seem, are only the beginnings for a general molecular analysis of meiosis in the microsporocytes of angiosperms. The impression that one gains at first glance is that the entry of the microsporocyte into meiosis is a specialty far removed from the events of mitosis. This is probably true because of the unique process of genetic recombination that occurs during the meiotic episode. The molecular studies have extended our knowledge of recombinational activities to a point where we can talk in terms of a vulnerable period when the activity of a vast array of enzymes for initiating recombination is controlled in a sensitive way. The key to this activity involves not only the availability of genes, but also the frequency with which available genes are transcribed. Obviously, any insight into the mechanisms whereby the activities of meiotic genes are regulated would be valuable.

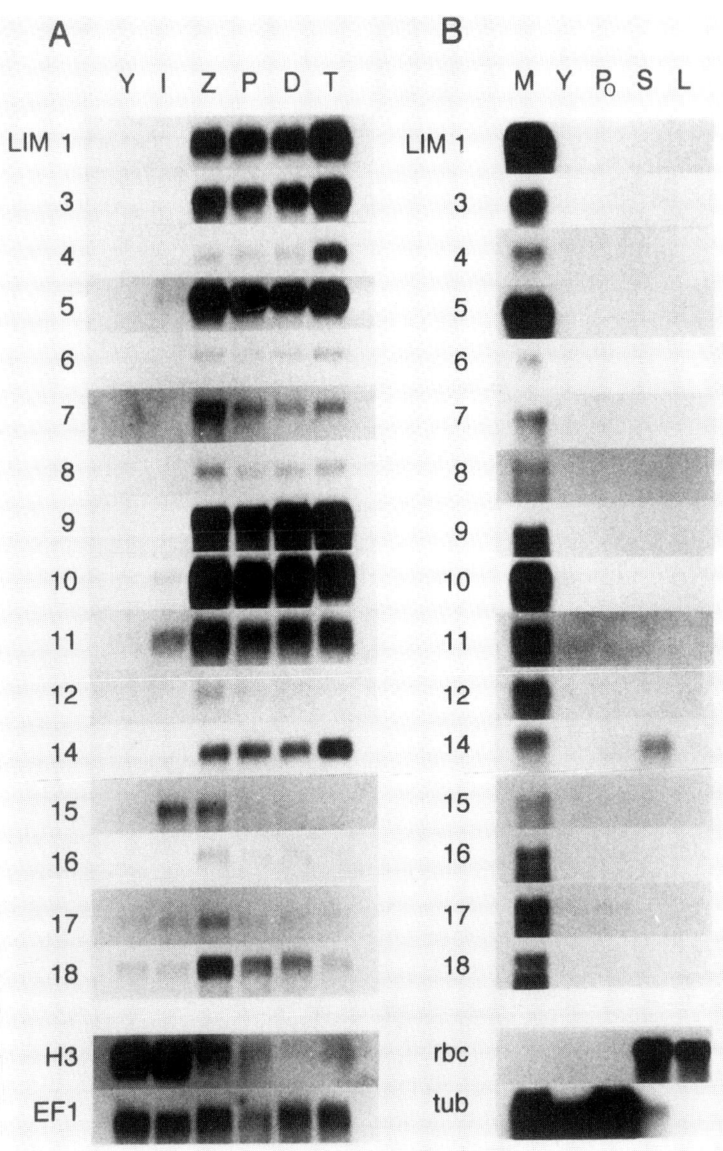

Figure 2.22. Characterization of meiotic-specific genes from microsporocytes of *Lilium longiflorum*. Poly(A)-RNA from microsporocytes (A) and total RNA from other tissues (B) were subjected to Northern blot hybridization using probes made to premeiotic microsporocytes. Controls consisted of probes containing wheat histone H3 gene (H3), lily elongation factor 1 (EF1), tobacco large subunit of ribulose-1,5-bisphosphate carboxylase (rbc), and lily α-tubulin (tub). Y, young anthers; I, interphase and leptotene stages; Z, zygotene; P, pachytene; D, diplotene–diakinesis; T, tetrads; M, total microsporocytes; P_0, mature pollen grains; S, stem; L, leaves. The cDNA clones are numbered from LIM-1 to LIM-18. (From Kobayashi et al 1994; photograph supplied by Dr. S. Tabata.)

4. ANTHER CULTURE IN THE STUDY OF MEIOSIS

Some of our modern understanding of the molecular biology of meiosis is derived from experiments in which a part, or the whole, of the syndrome of meiotic events is followed in microsporocytes grown under controlled conditions. This technique involves excision of flower buds and anthers or extrusion of the microsporocytes and their subsequent culture in a defined medium. The conditions of an ideal culture system would be such that the microsporocytes undergo synchronous meiotic development in culture when they are grown in the absence of the somatic tissues of the anther, but this aim has not been achieved.

Early experiments demonstrated that microsporocytes need to go through part of the meiotic division cycle on the mother plant before anthers can be cultured and microsporocytes allowed to advance through the rest of the division. Generally, success has been sporadic if cultures are initiated at stages earlier than pachytene. Moreover, other conditions being equal, anthers attached to the flower bud are more prone to complete meiosis in culture than anthers separated from the bud and cultured. For a period of time, these observations led to a lively debate on the possible deficiencies of the nutrient medium and on the absence of certain essential substances required for completion of meiosis in the anther at the time of culture. See Vasil (1967) for a detailed discussion of these investigations; a few representative cases will be described.

Tradescantia paludosa illustrates a case in which the sporogenous cells of anthers excised and cultured 5–8 days before leptotene continue mitotic divisions and end up as abnormal cells showing no parallel to meiotic stages. Microsporocytes of anthers excised during pachytene and later stages nearly always complete the meiotic division in culture (Taylor 1950). In an attempt to induce microsporogenesis in excised anthers of *Trillium erectum*, Sparrow, Pond and Kojan (1955) used a variety of media and found that the best development was achieved by the incorporation of 25%–50% coconut milk in the medium. A small percentage of anthers excised even as early as the pachytene stage completed meiosis in culture, although the survival rate increased in anthers cultured at later stages of the meiotic prophase. Vasil (1957, 1959) has reported experiments that show that with the addition of kinetin, GA, and autoclaved nucleic acids to the medium, meiosis proceeds to completion in the microsporocytes of anthers of *Allium cepa* (onion; Liliaceae) cultured as early as the leptotene stage. In cultured anthers of *Secale cereale*, naphthaleneacetic acid (NAA) and kinetin trigger microsporocyte meiosis without necessarily completing the process and thus have a rather limited role (Rueda and Vázquez 1985). Survival and limits of meiotic development of microsporocytes of *Lilium candidum* in culture have much in common with anthers of other species. Completion of tapetal development is also correlated with the stage of microsporocyte meiosis at the time of culture of the anthers, collapse of this tissue being almost complete in anthers cultured at the leptotene stage and less so at later stages (Reznickova and Bogdanov 1972).

A variation of the anther culture method was developed by Hotta and Stern (1963a). In this method, the leaves and perianth from flowers of *Trillium erectum* were excised, and the intact group of anthers attached to the receptacle along with a segment of the pedicel was transferred to the medium. This protocol served as the cornerstone for several investigations on mitotic and meiotic regulation in the anthers of liliaceous plants (Hotta and Stern 1963b, c, d; 1974, 1976; Shepard, Boothroyd and Stern 1974). In all of these different methods of anther culture, the problem of explaining the role of the somatic tissues of the anther, especially of the tapetum, in meiotic development of the microsporocytes remains difficult. From this perspective, a method developed for the in vitro culture of microsporocytes of liliaceous plants is significant (Ito and Stern 1967). The basic protocol has centered around the feasibility of extruding microsporocytes from the cut end of an anther into a nutrient medium as a coherent filament by gentle pressure from the opposite end. Interestingly, even in the most complex medium tested, only a small percentage of the microsporocytes cultured at the leptotene stage completes meiosis without causing aberrant segregation and failure of the spindle organization at meiosis I. As the extruded microsporocytes approach a higher stage limit, such as late zygotene or pachytene, meiotic division is nearly normal. Microsporocyte culture has been employed in several cytological and molecular studies of meiosis in liliaceous plants (Hotta, Ito and Stern 1966; Ito, Hotta and Stern 1967; Hotta, Parchman and Stern 1968; Parchman and Stern 1969; Hotta and Hecht 1971; Hotta and Shepard 1973; Ito and Hotta 1973; Sakaguchi et al 1980; Appels, Bouchard and Stern 1982). What has emerged from subsequent investigations is that the commitment of isolated microsporocytes to enter the meiotic cycle is a function of the stage of the premeiotic cell cycle (Parchman and Roth 1971; Ninnemann and Epel 1973; Takegami et al 1981; Ito and Takegami 1982). It has been shown that microsporocytes transplanted at growth phase 1 (G1), S, or early G2 phases divide mitotically, whereas those explanted at the late G2 phase proceed to divide meiotically (Figure 2.23). This has led to the view that commitment of microsporocytes to meiosis is made during the G2 phase of premeiosis (Takegami et al 1981; Ito and Takegami

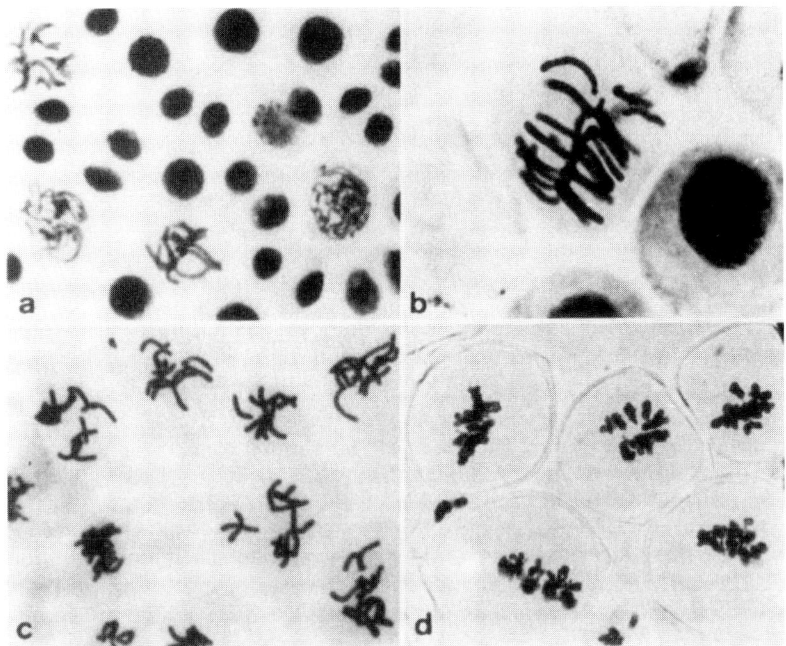

Figure 2.23. Meiosis in cultured microsporocytes of *Trillium* sp. **(a, c)** and *Lilium longiflorum* **(b, d)**. Microsporocytes cultured at the completion of premeiotic DNA synthesis reverted to mitotic division (a, b), whereas those cultured at the late G2 phase completed normal meiosis (c, d). (From Takegami et al 1981.)

1982). Although high yields of protoplasts of microsporocytes at meiotic stages can be obtained routinely by enzyme treatments, success in inducing them to form walls and complete the cell division cycle has been sporadic (Ito 1973; Ito and Maeda 1973; Bajaj 1974a, b; Takegami and Ito 1975; Maeda, Yoshioka and Ito 1979).

A simplification of the culture medium and completion of meiosis in the microsporocytes of anthers cultured at the primordial stage of development have been achieved when flower buds or inflorescences are cultured. Using this approach, it was possible to follow by ^{3}H-thymidine labeling the entire meiotic cycle of the microsporocytes in hermaphrodite flower buds of *Cucumis melo* (melon). The only difference between in situ and in vitro development is the lack of synchrony between individual anthers of a flower bud in the latter condition (Porath and Galun 1967). Izhar and Frankel (1973b) found that culture of flower buds of *Petunia hybrida* and *P. axillaris* in a simple mineral-salt–sucrose medium supports the complete cycle of microsporocyte meiosis in a small number of anthers. The spartan requirement for anthers of cultured flower buds to complete normal development is underscored by the fact that the addition of casein hydrolyzate or coconut milk does not improve the development of anthers in culture. Meiotic division appeared to proceed with minimal disorder in isolated microsporocytes of *Secale cereale* cultured in a mineral-salt medium containing amino acids and sucrose, although microspore development was arrested at the uninucleate stage (de la Peña 1986).

Zea mays is the only example in which microsporocyte meiosis has been induced in inflorescence cultures. The explants were generally 1-cm-long pieces of the tassel bearing approximately six hundred spikelet primordia; the most mature spikelets in the explant consisted of undifferentiated anther primordia enclosing essentially nonvacuolate meristematic cells. Although microsporogenesis is found to proceed to the stage of uninucleate microspores in anthers of tassels cultured in a mineral-salt–sucrose medium, addition of kinetin is found to enhance the frequency of responding florets (Polowick and Greyson 1982, 1984; Pareddy and Greyson 1985). When spikelets are cultured, microsporocytes proceed through meiosis and produce normal pollen grains even in the absence of kinetin in the medium (Stapleton and Bedinger 1992). From these results one can appreciate why in most cases excised anthers are unable to complete meiotic development in vitro. It is not because the effects of the culture medium are trivial; it is because the culture medium lacks some essential ingredients that are provided by the flower or by the inflorescence.

5. GENERAL COMMENTS

This chapter has assembled a considerable body of information on the development of the anther from its primordial beginnings and on the fate of each of the cell types formed. Although somewhat outside the scope of this account, the genetic and epigenetic factors concerned with the determination of the anther on the floral meristem have also been considered. The results show the power of mutational approaches in unraveling the complex mechanisms involved in the determination and differentiation of the floral parts. The investigations reviewed in the first part of this chapter underlie the dominant theme that will be evident throughout this book, namely, the study of plant reproductive biology has undergone a quiet revolution from a descriptive science to one emphasizing molecular and genetic concepts.

The account of anther development has been built upon the traditional view that the anther as a whole consists of two distinct cell types programmed for different functions. The anther wall and the tapetum have taken on added significance with the demonstration that specific genes are expressed in these cells to define their functions. In this context the endothecium and stomium no longer appear to be passive cell types, but their assigned role in the dehiscence of the anther seems to be modulated by the expression of specific genes. Although the tapetum remains an enigmatic tissue, evading a thorough examination, the ability to manipulate this tissue offers new opportunities to genetically engineer male-sterile plants.

The picture of meiosis in the microsporocytes appears increasingly complex as more and more data have become available. Considering the fact that microsporocyte meiosis, like its counterpart in the megasporocyte, is a critical phase in the life cycle of the plant, it is surprising that what we know at the molecular level is restricted to studies on a model system. Moreover, in comparison to the work done on meiosis in animal cells and yeast, our knowledge of the molecular biology of meiosis in the microsporocytes is very limited. No meiotic genes from plants have yet been cloned and their protein products identified.

The premeiotic anther enclosing diploid microsporocytes proceeds through an orderly series of changes to produce haploid microspores. Refinements in tissue culture have made it possible to study meiosis under controlled conditions in cultured microsporocytes, flower buds, and inflorescences or tassels. Future advances in the study of anther development will certainly involve the isolation and characterization of additional anther-specific genes, especially those activated during meiosis. The following chapter will return to the products of meiotic division of the microsporocyte to complete the discussion of the development of the angiosperm male gametophyte.

Chapter 3

Pollen Development and Maturation

The work reviewed in the previous chapter has established that meiotic division of the microsporocyte results in the formation of four haploid microspore nuclei. They remain encased in the original callose wall of the microsporocyte to form a tetrad – the four-celled stage at the end of meiosis. There are two basic patterns of wall formation followed by these nuclei before they attain the status of cells. In most monocotyledons, immediately after each meiotic division of the microsporocyte, cell plate formation occurs in concert with a centrifugally expanding phragmoplast to produce the tetrad (successive cytokinesis). Alternatively, the norm in dicots is a type of division in which the four nuclei are walled off at the end of meiosis II (simultaneous cytokinesis). In either case, the first wall delimiting the microspore nuclei from each other is constituted of callose and not of cellulose. Later, after its release from the tetrad, each microspore forms its own wall comprising the exine and intine. According to an informal morphological concept, the microspore represents the beginning of the male gametophytic generation, with the term "pollen grain" being reserved for the older microspore, particularly after its release from the tetrad. In some accounts, the first haploid mitosis is considered to terminate the life of the microspore and usher in the reign of the pollen grain. However, the terms "microspore" and "pollen grain" continue to be used interchangeably and synonymously in standard embryology literature to refer to the first cell of the male gametophytic generation of angiosperms.

At a gross level, the pollen grain represents the structural unit that is programmed for terminal differentiation into sperm cells. For descriptive purposes, it is convenient to consider pollen development in terms of a series of changes beginning with the microspore after its release from the tetrad. These changes are very diverse and are geared to the ultimate function of the pollen grain as the transport vector for the sperm cells or their progenitor cell. Outside the microspore wall, sporopollenin is deposited in a diagnostic way to form the characteristic architectural pattern of the exine. Within the cytoplasm, vacuolation occurs, accompanied by a rearrangement of the organelle population of the microspore and leading to the first haploid pollen mitosis. The result is the formation of a large vegetative cell and a small generative cell. In some plants, the generative cell divides to form two sperm cells while enclosed in the pollen grain, but in the large majority of angiosperms, this division is postponed until the pollen tube is formed (Figure 3.1). It is believed that the bicellular condition is primitive and that the tricellular state originated indepen-

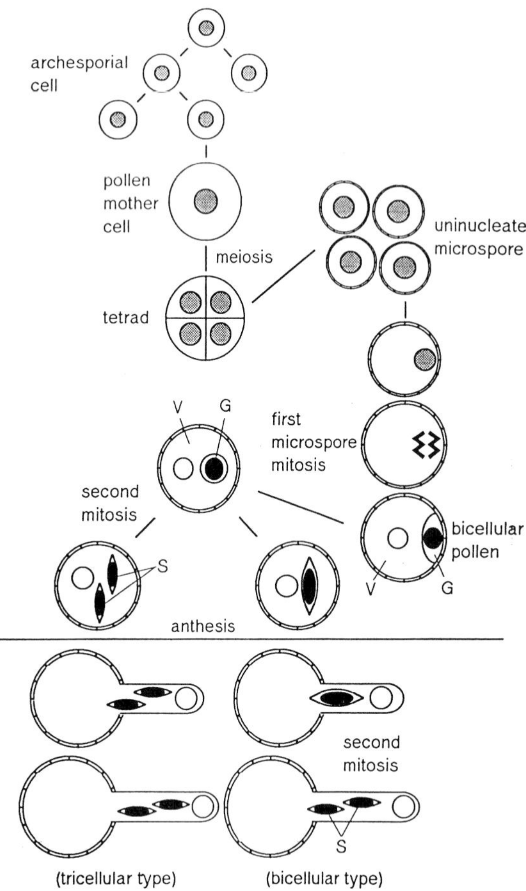

Figure 3.1. Diagrammatic representation of microspore development and male gametogenesis beginning with the archesporial cell. G, generative cell; S, sperm cell; V, vegetative cell. (From Tanaka 1993.)

dently many times during angiosperm evolution (Brewbaker 1967).

In order to cover in a logical manner the many diverse and seemingly unrelated events of pollen development, this chapter will begin with a review of the structure and morphogenesis of the pollen wall. Working from the outside to inside, the chapter will then examine the cytology of pollen development, which culminates in the formation of the vegetative cell and generative cell. Integration of the biochemical and molecular work into a review of the state of our present knowledge on gene expression during pollen development will be taken up in the next chapter. For an appraisal of the early literature on pollen development, the book by Maheshwari (1950) is still unsurpassed; the later work has been reviewed by Steffen (1963), Mascarenhas (1975), and Knox (1984a).

1. MORPHOGENESIS OF POLLEN WALLS

There are several important questions concerning pollen wall morphogenesis that reflect, on the whole, the process of maturation of the microspore from the tetrad stage to the point when sperm cells are generated. What is the significance of the impervious callose wall around the microspores? What is the source of the exine materials and what are the level and limits of morphogenetic control of exine fabrication? What specific processes concerned with wall differentiation are orchestrated in the pollen cytoplasm and in the tapetal cytoplasm? As will be seen, what emerges from a study of the diversity of processes is that pollen wall assembly is the result of a series of coordinated changes both outside and inside the cell.

(i) The Callose Wall

The deposition of the callose wall begins centripetally on either side of the newly formed cell plate immediately after completion of microsporocyte meiosis. The daughter cells born out of meiotic division are devoid of plasmodesmatal connections, so that each cell carved out of the microsporocyte floats in the incipient tetrad, isolated from its neighbors but ensheathed in its own callose wall (J. Heslop-Harrison 1966a). Because of the nature of callose as a sealing agent, it is apparent that the walls of the tetrad and of the microspore present effective barriers to the penetration of large molecules from the anther locule and to the traffic of materials between the microspores. From this perspective, the reason why ^{14}C-thymidine and ^{3}H-phenylalanine do not penetrate the callose-invested microspores of cultured anthers or flower buds is understandable (J. Heslop-Harrison and Mackenzie 1967; Southworth 1971). These same precursors are freely incorporated into microsporocytes or into microspores that are at later stages of development and are denuded of the callose sheath. At the cytochemical level, fluorescein diacetate is easily detected by fluorescence microscopy in microspores of various angiosperms mechanically released from their callose walls, but intact dyads and tetrads fail to show fluorochromasia. This could not happen if the callose wall had allowed free access to fluorogenic substrate into the microspore (Knox and Heslop-Harrison 1970b). J. Heslop-Harrison (1964, 1968b) has made a case that the significance of insulating the newly formed microspores from the external milieu is to allow them to assert their genetic independence and establish cellular autonomy as the first cells of a new generation. As one can see, meiosis ensures the production of genetically different cells; the callose wall eliminates any possible effects of the informational pool of the sibs on the cytoplasm bequeathed by the microsporocyte. Another view of the significance of the callose wall is that it acts as a template for the development of the exine under the control of the microspore cytoplasm (Waterkeyn and Bienfait 1970). This remains a moot point; although the absence of callose in the tetrad is associated with poor exine deposition on the microspores (Vijayaraghavan and Shukla 1977; Pettitt 1981; Takahashi 1987; Worrall et al 1992; Fitzgerald et al 1994), tetrads without callose sheath also generate microspores with perfect exines (Periasamy and Amalathas 1991). In mutant *Arabidopsis* plants that release pollen as permanent tetrads, a callose wall may be absent but the exine development is no less complete as compared to wild-type pollen (Preuss, Rhee and Davis 1994).

Despite its postulated importance, the callose wall is only transiently associated with the tetrad as it is degraded at the termination of meiosis. This results in the liberation of microspores with partially constituted walls from the sectored microsporocyte; the timing of callose breakdown is correlated with the activity of callase (Eschrich 1961; Echlin and Godwin 1968b; Mepham and Lane 1970; Frankel, Izhar and Nitsan 1969; Izhar and Frankel 1971; Stieglitz and Stern 1973; Stieglitz 1977). Analysis of callase activity separately in the microsporocytes and anther tissues of *Lilium* sp. has disclosed the interesting fact that the enzyme is concentrated mostly in the somatic cells of the anther and that the tapetum is probably the source (Stieglitz and Stern 1973; Stieglitz 1977). Exclusive localization of transcripts of an anther cDNA clone from *Nicotiana tabacum* encoding callase in the tapetal cells of tobacco anthers provides strong support for this view (Bucciaglia and Smith 1994). These observations underscore the fact that all component tissues of the developing anther are responsive to one another through mechanisms whereby individual cells in a tissue can assess the developmental state of cells in a neighboring tissue.

(ii) Assembly of Pollen Walls

Wall development involves the synthesis of materials in which the pollen grain is wrapped and is part of the preparation toward the final journey of the pollen to the stigmatic surface of another flower. The familiar differentiation of the pollen wall into an outer exine and an inner intine is now widely accepted. In members of Poaceae, a granular, nonstructured middle layer termed Zwischenkörper (also known as Z-layer or oncus) is recognized; this layer is particularly prominent beneath the germination pores on the exine (Rowley 1964; Christensen and Horner 1974; J. Heslop-Harrison 1979a; El-Ghazaly and Jensen 1986a). However, other investigations have tended to view the Z-layer, which overlies the main intine layer, essentially as a part of the intine. It has a slightly higher electron density than the main intine material, but lacks the cytoplasmic tubules that traverse the latter (J. Heslop-Harrison and Heslop-Harrison 1980a; Kress and Stone 1982).

In this chapter, pollen wall morphogenesis will be broadly categorized as the deposition of the outer patterned layers of the exine and the inner nonpatterned layers of the intine. One might say that morphological, developmental, chemical, and genetic differences between the exine and intine could not be more distinct. It is now known that the pollen walls function not only as protective envelopes for the cytoplasm but also as storage sites for a variety of proteins and other physiologically active substances concerned in the interactions of the pollen grain with the stigma and in some critical early functions in pollen germination. Electron microscopy has played a major role in uncovering the timing and pattern of the formation of the pollen walls. The development of the exine will be considered first.

The Exine. As alluded to in the previous chapter, the exine consists of sporopollenin, a wall substance known for its resistance to chemicals and biodegradation. The exact chemical composition of sporopollenin is not known; one view is that it is a polymer of carotenoids and carotenoid esters (Brooks and Shaw 1978). Another view of more recent vintage, based on solid state nuclear magnetic resonance (NMR) spectroscopy, is that it consists mainly of aliphatic domains in addition to carbohydrates and aromatic compounds (Guilford et al 1988; Espelie et al 1989). Negative evidence against the presence of carotenoids as a major constituent of sporopollenin has come from the observation that norflurazon, an inhibitor of carotenoid synthesis, does not significantly inhibit the accumulation of sporopollenin in the anthers of *Cucurbita pepo* (Prahl et al 1985). Of potential value for gaining a complete understanding of the chemical composition of the exine is the development of methods for the isolation of purified exine fractions

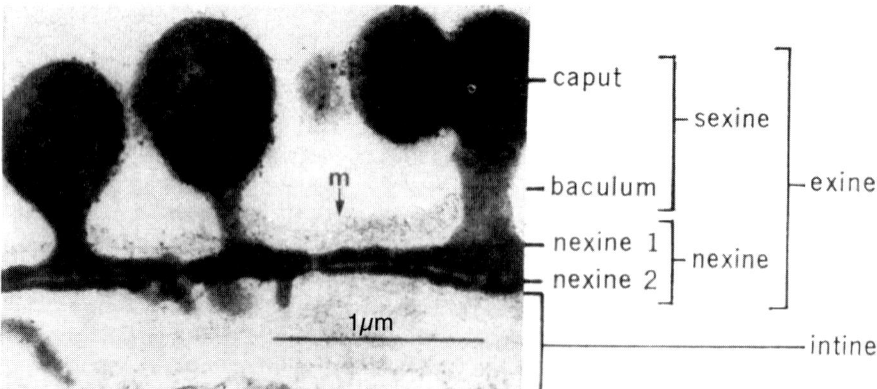

Figure 3.2. Section of the wall of an untreated pollen grain of *Lilium* sp. showing the stratification. m, fibrillar matrix material. (Reprinted with permission from J. Heslop-Harrison 1968b. Pollen wall development. *Science* 161:230–237. ©1963. American Association for the Advancement of Science.)

from pollen grains and the production of antibodies to exine proteins (Southworth 1988; Southworth et al 1988; Chay et al 1992).

Based on early light microscopic images, a complex terminology is now in use to describe exine stratigraphy. Briefly, this subdivides the exine into an outer, intricately sculptured portion, the sexine, and an inner, more or less continuous floor layer, the nexine. Rarely, as described in the pollen grains of *Brassica oleracea* and *B. napus*, the exine is surrounded by a coating with properties of a biological membrane for possible involvement in the recognition reaction on the stigma (Gaude and Dumas 1984; Elleman and Dickinson 1986). The sculpturing of the sexine is attributed to the presence of radially directed free-standing rods, or baculae, with enlarged heads (caputs). It should also be noted that heads of the baculae are linked at the top to produce a reticulate pattern or are sealed to create the facsimile of a roof or tectum. The tectum and the nexine are, however, perforated by micropores to maintain communication with the outside. The sexine architecture dominates the entire pollen grain surface except at the site of the germination apertures, where it is absent or attains a low point. The nexine is subdivided into an outer nexine-1 and an inner nexine-2, separated by an electron-transparent discontinuous region (Figure 3.2). One recurrent feature of exine development in the microspores of various angiosperms is the cytochemical variability between species and in the same species during a developmental time span (J. Heslop-Harrison 1968b, 1975a).

Pioneering work by J. Heslop-Harrison (1963a, b) on pollen wall ontogeny in *Silene pendula* and *Saponaria officinalis* (both Caryophyllaceae) showed that the signal for formation of the sexine is the completion of microsporocyte meiosis. At this stage, when the microspores are still in the tetrad configuration, the first elements of the sexine, referred to as primexine, appear as a layer of cellulose microfibrils interspersed between the plasma membrane of the microspore and its callose wall. The basic outline of the final sexine pattern is determined shortly thereafter by the formation of radially oriented columns of electron-opaque material in the cellulose matrix of the primexine. These, referred to as probaculae, represent the progenitors of the baculae. Observations on microspore ontogeny in *Lilium longiflorum* have shown a dynamic picture of pleated and folded lamellae that eventually compact into the baculae. The lamellae are probably of plasmalemmal origin, organized in an assembly of membranelike structures to form the baculae (Dickinson and Heslop-Harrison 1968). As described in *S. pendula* and *S. officinalis,* in the final phase of microspore wall morphogenesis within the tetrad, the radial columns extend and link together over the outer face of the primexine to assemble the tectum; at the same time, they become confluent below the radial columns to form the beginnings of nexine-1. With some variations, the essential details of the presence of isolating callose walls around microspores and of the primexine template of fibrillar material have been confirmed as common denominators in the formation of the pollen wall sexine in several other plants (Banerjee, Rowley and Alessio 1965; Skvarla and Larson 1966; Angold 1967; Godwin, Echlin and Chapman 1967; Echlin and Godwin 1968b; J. Heslop-Harrison 1969a; Dickinson 1970a; Horner and Lersten 1971; Christensen, Horner and

Lersten, 1972; Pettitt, 1976; Horner and Pearson 1978; Nakamura 1979; Shoup, Overton and Ruddat 1980; Owens and Dickinson 1983; Southworth 1983; El-Ghazaly and Jensen 1986b; Takahashi and Kouchi 1988; Takahashi 1989a, 1994; Jordaan and Kruger 1993; Pérez-Muñoz, Jernstedt and Webster 1993a; F. Xu et al 1993; Hess and Frosch 1994; Simpson and Levin 1994; Nepi, Ciampolini and Pacini 1995). An extreme case, where a virtually mature sexine is assembled around microspores while they are still protected within the callose sheath, has been described in *Tradescantia bracteata* (Mepham and Lane 1970).

The chemical nature of primexine remains unknown, but from the work of J. Heslop-Harrison (1968g) on microspore development in *L. longiflorum,* we have some pointers about its closeness to sporopollenin. The strategy used in this investigation was to follow by a combination of ultrastructural and cytochemical methods the characteristics of the wall during the tetrad period and also after the release of microspores from the tetrad. These experiments showed that at the time of its first appearance, the material of the probaculae is not resistant to acetolysis and apparently does not contain sporopollenin; however, like sporopollenin, the material later deposited within the tetrad wall displays some physical and chemical properties that confer a capacity to survive acetolysis. In other words, the microspore protoplast holds the key to the synthesis of the real sporopolleninlike materials; what form this key might take, and whether it is present in the microsporocyte as a long-lived molecule or is a transient product of meiosis, requires further work.

From a developmental perspective, two other views on the pattern determination of the exine are worthy of consideration. One invokes a role for a plasma membrane–associated compound rich in polysaccharides and proteins, known as glycocalyx, in the process. According to Rowley and Skvarla (1975), spinules of the exine of *Canna generalis* pollen are initiated within a three-dimensional network constituted of a compound resembling glycocalyx. On this basis, it has been suggested that genetic expression of the form of the exine prior to primexine formation is mediated by glycocalyx, perhaps in association with the plasma membrane. The other view, with a different perspective, is that the initial exine pattern is determined by the plasma membrane of the microspores while they are still enclosed within the callose sheath of the tetrad. Attempts to identify explicit early ultrastructural indications of exine development in *Caesalpinia japonica* (Fabaceae) and *Bougainvillea spectabilis* (Nyctaginaceae) have shown that prior to exine formation, the plasma membrane invaginates and forms a reticulate pattern corresponding to the mature exine pattern. The plasma membrane, rather than the primexine, it is argued, is the prime mover in the determination of the basic exine pattern (Takahashi 1989b, 1995; Takahashi and Skvarla 1991). According to this view, a protectum, which appears as irregularly oriented threads, is the first exine layer to be deposited on the reticulate plasma membrane. The threads probably serve as the scaffolding upon which the exine is built (Takahashi 1993). An examination of primexine formation in pollen grains of *Brassica campestris* by rapid freeze-substitution technology has added another dimension to the controversy. Although it has confirmed the appearance of the primexine matrix as the first event followed by an undulation of the plasma membrane, it has also shown that fibrous materials inserted into the invaginations of the plasma membrane contribute to the beginning of probacula formation (Fitzgerald and Knox 1995). That there is no universal exine initiation template is not surprising in view of the bewildering array of mature exine patterns in angiosperm pollen grains.

A foreshadowing of the germination pores on the exine is first evidenced by the discontinuity of primexine over those areas of the microspore wall destined to become pores. An important question concerning pores is what cytoplasmic or genetic factors determine their location. If a single cytoplasmic process does exist in the control of sexine morphogenesis, this process is the association of elements of ER with certain areas of the plasma membrane, marking out the sites of the pores. This relationship, first described in a study of pollen wall ontogeny in *Silene pendula* (J. Heslop-Harrison 1963a), has since then been visualized in microspores as diverse as those of *Zea mays* (Skvarla and Larson 1966), *Helleborus foetidus* (Echlin and Godwin 1968b), *Lilium longiflorum* (Dickinson 1970a), *Sorghum bicolor* (sorghum; Poaceae) (Christensen and Horner 1974), *Cosmos bipinnatus* (Dickinson and Potter 1976), *Helianthus annuus* (Horner and Pearson 1978), *Gibasis venustula* (Owens and Dickinson 1983), *Triticum aestivum* (El-Ghazaly and Jensen 1986a), and *Vigna vexillata* (Fabaceae) (Pérez-Muñoz, Jernstedt and Webster 1993b). A possible explanation for a role for ER in defining the location of the germination aperture on the exine is that it acts as a shield, preventing access to precursors of cellulose biosynthesis outside the plasma membrane (J. Heslop-Harrison

1971b); unfortunately, we have very little experimental evidence to support this view. Other observations indicate that the location of pores is programmed into microspores by virtue of their geometrical relationship in the formative tetrad stage. Attention was first drawn to this relationship by Wodehouse (1935), whose extensive light microscopic studies revealed that pollen grains have an inherent polarity because of the constant position of germination pores; in later years, it was shown that the siting of pores in the microspores is determined by the polarization imposed by the mitotic spindle and the plane of division during meiosis II (El-Ghazaly and Jensen 1986a). A causal relationship between the orientation of the mitotic spindle and the location of pores on the exine has also been established in experiments using centrifugation to displace cytoplasmic components, or colchicine to disturb spindle formation during microsporocyte meiosis. For example, if microsporocytes of *L. henryi* are centrifuged before the assembly of the mitotic spindle, pores are found to be wholly lacking in the microspores formed. In contrast, displacement of the spindle by centrifugation at later stages of microsporogenesis results in microspores with pores mostly in the normal configuration (J. Heslop-Harrison 1971b; Sheldon and Dickinson 1983). When microsporocytes are treated with colchicine, primary effects of the drug are evident in the failure of the surviving microspores to become polarized due to spindle disassembly; in such cells, pore sites assume a random distribution or appear ill defined (J. Heslop-Harrison 1971b). Colchicine treatment has also been shown to induce anarchic divisions in the microsporocytes of *T. aestivum*, with the location of pores remaining constant in relation to the spindle axes of the surviving microspores (Dover 1972). Taken together, these observations show that the spindle axes specifying the cleavage planes of the microsporocyte tetrad also identify the pore sites in the sexine of the microspores. Whether centrifugation and colchicine effects on spindle formation during meiotic division of the microsporocytes are reflected in the assembly of cytoplasmic components in the region of the microspores destined to form germination pores, remains to be seen.

A regular participation of membranous structures is possibly the most distinctive feature of the disposition of baculae on the sexine. Just as there is a correlation between the position of the ER and sites of the germination pores on the exine, so too there is an association between the formation of primexine and a random orientation of elements of the ER close to the plasma membrane. It has been observed that in *Silene pendula*, although there is no increase in the elements of ER in the microspore cytoplasm immediately after meiosis, this situation is quickly followed by the presence of ER at intervals in ordered arrays closely apposed to the plasma membrane. This precedes the appearance of the cellulose matrix outside the plasma membrane, as described earlier (J. Heslop-Harrison 1963a). The involvement of ER as a membranous template for primexine formation has also been observed in the microspores of *Zea mays* (Skvarla and Larson 1966), *Berberis vulgaris* (Berberidaceae) (Gabara 1974), *Cosmos bipinnatus* (Dickinson and Potter 1976), and *Vigna vexillata* (Pérez-Muñoz, Jernstedt and Webster 1993b), whereas in other plants, the primexine-forming potential is associated with cytoplasmic organelles resembling vesicles or mitochondria. In *Pinus banksiana*, the vesicles appear to be derived from stacks of membranous cisternae somewhat similar to dictyosomes in their structure, and they have as their destination the periphery of the microspore cytoplasm beneath the plasma membrane (Dickinson 1971a). Although the disposition of dictyosomes and membrane proliferations seems to presage the arrangement of the baculae on the sexine of *Lilium longiflorum* (Dickinson 1970a), *Gasteria verrucosa* (Willemse 1972), and *Helianthus annuus* (Horner and Pearson 1978), mitochondria are reported to be associated with the organization of the sexine in *Linum usitatissimum* (flax; Linaceae) (Vazart 1970). The presence of microtubules in the peripheral cytoplasm in the vicinity of the plasma membrane of microspores of *P. banksiana*, *C. bipinnatus*, and *L. longiflorum* (Dickinson 1976), and as elements radiating from the nucleus to the plasma membrane in *L. henryi* (Dickinson and Sheldon 1984), *V. vexillata* (Muñoz, Webster and Jernstedt 1995), and *V. unguiculata* (Southworth and Jernsted 1995), reinforces the view that the cytoskeleton is responsible for the movement of the vesicles and membrane elements involved in sexine pattern formation on the cell surface. However, based on experiments involving colchicine treatment of microsporocytes of *L. henryi*, microtubules do not appear to be responsible for determining the sexine pattern. These experiments showed that despite the obliteration of microtubules by the drug and the chaotic nature of meiosis that ensued, pattern generation on the sexine was unaffected (Sheldon and Dickinson 1986). Somewhat similar results were obtained by treating microspores of *Tradescantia virginiana* with colchicine during the period of wall development (Tiwari 1989). These observations relegate the cytoskeleton to a secondary role in sexine formation, perhaps molding the materials into the pattern already established by some other agents.

A particularly fascinating aspect of these observations is the attention they have directed to the genetic control of sexine pattern formation. To what extent does the activity of the haploid genome determine the species-specific sexine architecture on the microspores? Since microscopically detectable elements of the sexine first appear on microspores encased in the tetrad, it is logical to assume that the process is under control of the haploid genome. In a thoughtful essay, Godwin (1968) has discussed some of the older evidence from the literature in support of this view, largely borne out by electron microscopy. Later work has, however, shown that the deposition of the basic patterned component of the sexine is associated with cytoplasmic control extending back to the diploid microsporocyte. Particularly instructive are the results of experiments using colchicine or centrifugation stress to disrupt meiosis in the microsporocytes of *L. henryi*. An extreme situation is seen where anucleate cell fragments produced by these treatments bear the typical sexine pattern approaching that of unstressed cells (J. Heslop-Harrison 1971b; Sheldon and Dickinson 1983). Similarly, the development of the normal sexine pattern in microspores with incomplete chromosome complements in sterile hybrids arising out of crosses between the tetraploid *Linum cateri* and various related diploids also indicates a cytoplasmic, rather than a nuclear, control of sexine pattern determination (Rogers and Harris 1969). These results excited quite a lot of interest at that time, for they seemed to argue against the possibility that the control of sexine differentiation depends on gene action in the microspore. For how could the sexine pattern be formed so precisely in such genetically deficient, enucleate cytoplasmic enclaves if the microspore nucleus was the controlling agent? The mechanism by which the enucleate microspore fragments assemble the sexine is not understood, but it is hard to imagine how this could happen otherwise than by transcription of the genes in the microsporocyte and transmission of information from the diploid parent to haploid daughter cells by "delayed action" messages.

As described in *Lilium longiflorum*, nexine-1 is invariably organized after the early definition of the probaculae, but the precise events are subtle and often difficult to detect. Accompanying the advance of the matrix material at the base of the probaculae and fully integrated with them are lamellae of plasma membrane origin. Nexine-1 is formed by the apposition of the lamellae, a process essential for the structural reinforcement of this layer (Dickinson and Heslop-Harrison 1968). Once the microspore has thus assembled the primexine and nexine-1, dissolution of the callose wall of the tetrad is the next order of business. This ensures the release of microspores, growth of nexine-2, maturation of the sexine, and assembly of the intine, all drawing upon the metabolites of the anther locule. Rowley and Southworth (1967) first showed that, during the formation of the inner exine layer (presumably nexine-2) around germination pores of the microspore of *Anthurium* sp. (Araceae), layers of lamellae consisting of unit membranes are routinely formed and that sporopollenin is subsequently deposited on them to form a homogeneous layer. Similar observations have been made on the mode of origin of nexine-2 in other plants, such as *Hippuris vulgaris* (Hippuridaceae), *Populus tremula* (Salicaceae) (Rowley and Dunbar 1967), *Ipomoea purpurea* (Convolvulaceae) (Godwin, Echlin and Chapman 1967), *Lilium longiflorum*, *Lilium* sp. (Dickinson and Heslop-Harrison 1971; Willemse and Reznickova 1980), and *Silene alba* (Shoup, Overton and Ruddat 1980, 1981). Of interest in terms of the origin of the lamella, in *L. longiflorum* the initiating structure has been designated as the parent lamella, which is presumed to be formed on the outer surface of the plasma membrane (Dickinson and Heslop-Harrison 1971). Structures strongly suggestive of lamellar organization of the lower layer of the sexine have also been described in earlier studies on the ontogeny of nexine-2 (Larson, Skvarla and Lewis 1962; Larson and Lewis 1963), so that the basis for the origin of this layer can be considered as a secure generalization.

Sporopollenin Synthesis. It is clear from the foregoing that the actual patterning of the exine is imprinted while the young microspores are protected within the callose sheath. Subsequent work has left little doubt that the molding of the structure of the mature exine by sporopollenin accumulation occurs after the release of microspores from the tetrad. Research on the origin and accumulation of sporopollenin on the pollen exine has a long history dating back to the mid-19th century. Early work was, necessarily, a domain in which the light microscope was the preferred tool. By the 1930s, several cytologists had documented the accumulation of bodies with the chemical properties of sporopollenin on the locular faces of tapetal cells. The specific location of these bodies or orbicules immediately suggested a connection between the tapetum and sporopollenin synthesis. A development that followed from this in later years has been the electron microscopic examination of orbicule development in pollen grains of a number of plants, most notably, *Poa annua* (Poaceae) (Rowley 1963), *Cannabis sativa* (hemp; Moraceae), *Silene pendula* (J. Heslop-Harrison 1962, 1963b), *Zea mays* (Skvarla and Larson 1966), certain species of *Oxalis* (Oxalidaceae) (Carniel 1967b), *Salix humilis* (Salicaceae), *Populus tremula* (Rowley and Erdtman 1967), *Helleborus foetidus* (Echlin and Godwin 1968a), *Lilium longiflorum*, *Lilium* sp. (J. Heslop-Harrison and Dickinson 1969; Reznickova and Willemse 1980), *Citrus limon* (Horner and Lersten 1971), *Sorghum bicolor* (Christensen, Horner and Lersten 1972), *Pinus banksiana* (Dickinson and Bell 1972a, 1976), *Helianthus annuus* (Horner and Pearson 1978), *Triticum aestivum* (El-Ghazaly and Jensen 1986b), and *Ledebouria socialis* (Hess and Frosch 1994). In *H. foetidus* (Echlin and Godwin 1968a), *L. longiflorum* (J. Heslop-Harrison and Dickinson 1969), and *T. aestivum* (El-Ghazaly and Jensen 1986b) in which the ontogeny of the tapetal cell has been closely followed, progenitors of orbicules that have a different electron density than the rest of the tapetal cytoplasm have been identified. Subsequently, these cytoplasmic territories accumulate in the vicinity of the plasma membrane, from which they are extruded, followed by their acquisition of a sporopollenin coating in the anther locule. In the final lap of their journey from the tapetum, the orbicules align themselves on the microspore wall and are integrated into the developing exine (see Figure 2.8). These studies have enormously augmented the light microscopic observations and reasserted the chemical identity of tapetal orbicules with sporopollenin of the exine. At the same time, some conflicting suggestions regarding the origin of substrates for sporopollenin synthesis have appeared in the literature. Among these suggestions are the possible formation of substrates in the mitochondrial population of tapetal cells (J. Heslop-Harrison 1962), their elaboration in both the tapetum and the anther locule (Rowley, Mühlethaler and Frey-Wyssling 1959; Rowley 1964), and their elaboration in the microspore cytoplasm through dictyosome vesicles (Larson and Lewis 1963). These ideas either have been withdrawn or remain unconfirmed, and so the origin of sporopollenin substrates remains a mystery. What is now established is that following the early period of sexine assembly within the tetrad, continued thickening of the wall occurs through the polymerization of sporopollenin precursors derived mostly from the tapetum and developing microspores. The goal of a diagnostically patterned exine is attained when the precursors are laid down upon the scaffolding of special initiating sites outside the plasma membrane such as the probaculae or the lamellae. Considered in its entirety, the dramatic aspect of exine pattern formation on the pollen wall is the precise timing of its initiation, the apparently simple beginning, the speed with which it goes forward, the complexity of the final product, and the gaps in our knowledge of the critical steps.

Exine Inclusions. Intuitively, one might assume that the function of the exine with its load of sporopollenin is to provide mechanical protection to the developing male gametophyte. Though this is undoubtedly true, it also goes without saying that several structural modifications of the exine have adaptive significance for the role of the pollen in gamete transport. For example, it is a matter of common observation that exine elaboration attains its highest level of complexity in the pollen grains of insect-pollinated flowers, whereas pollen of wind-pollinated species have relatively smooth exine. In recent years considerable attention has been focused on another facet of exine function, namely, in the storage and release of lipids, carotenoids, and proteins during pollen–stigma interaction. Reference was made in the previous chapter to the fact that the presence of precursors of pollenkitt and tryphine is a typical feature of degenerating tapetal cells of certain plants. As to the origin of pollenkitt, it is now known that osmiophilic globules of precursor materials discharged from the tapetal cells migrate between microspores and are eventually trapped in the cavities of the exine, where they polymerize into pollenkitt. Some globules are even lodged in the interprobacular cavities (Pankow 1957; J. Heslop-Harrison 1968e, f; J. Heslop-Harrison et al 1973).

The only detailed treatment of tryphine deposition on the exine is that described in *Raphanus sativus* (Dickinson 1973; Dickinson and Lewis 1973b). Of the two components of tryphine in this species, the proteinaceous material is applied first, followed by a thick layer of the lipoidal mass. During the final stages of pollen maturation, these components dry down, forming a thick encrustation on the exine. The possible role of tryphine in pollen–stigma interaction will be considered later in this section.

A significant component of the exine inclusion of pollen grains of *Brassica napus* consists of lipids. These are presumed to be the products of the tapetum and apparently transferred to the pollen surface late in development (Evans et al 1987). Various analyses have demonstrated interesting biochemical differences between the internal cytoplasmic lipids of pollen grains, presumed to be the products of the haploid genome and the external exine-associated lipids (Evans et al 1987, 1990, 1991).

The Intine. Developmentally, structurally, and physiologically, the intine is quite different from the exine. This layer is hidden within the exine, although it begins to develop only after the dissolution of the tetrad. Thus, in contrast to the exine, intine organization is programmed entirely by the haploid genome of the microspore. Unlike the exine, the intine is nonsculptured and is therefore structurally a simple layer of fibrillar material whose components merge with those of the exine. The presence of cellulose microfibrils embedded in a matrix of pectin and hemicellulose also defines the intine and gives it an identity different from that of the exine, but redolent of the primary wall of a somatic cell. By progressive chemical extraction of the exine and the enclosed cytoplasmic constituents, the microfibrillar skeleton of the intine can be isolated as an intact component (intine "ghost") for analysis of its structural features. As shown in representative species, the orientation of microfibril aggregates in the intine necessarily tends to conform to the major topographical features of the exine and to the shape of the pollen grain as a whole (J. Heslop-Harrison 1979a; Y. Heslop-Harrison and Heslop-Harrison 1982b). In some plants, especially in members of Cannaceae (Skvarla and Rowley 1970), Heliconiaceae (Kress, Stone and Sellers 1978), Cymodoceaceae (McConchie, Knox and Ducker 1982), and Asclepiadaceae (Vijayaraghavan and Shukla 1977), the pollen grains display extreme reduction in wall structure, with the exine very much reduced and the intine considerably thickened. In these plants and in members of Poaceae, it is not difficult to identify by cytochemical methods an outer layer of the intine, presumably pectic polysaccharide in nature, and an inner layer primarily of cellulose (Kress and Stone 1982); rarely, as in *Lilium longiflorum*, a third, fibrous layer of intine has been identified (Nakamura 1979). Even in other plants, in which a definitive separation of two intine layers is not possible, the outer stratum is universally thought to be pectic, and the inner cellulosic, in nature. Although the intine appears rapidly as a continuous layer at the inner edge of the incipient exine of the microspore, often it does not attain its final thickness until late in pollen ontogeny. Intine specialization subtending the germination pores is somewhat more complex than in the nonapertural areas and involves the incorporation of additional cellulosic strata, as in *Linum grandiflorum* (Dulberger 1989) and *Cobaea scandens* (Polemoniaceae) (J. Heslop-Harrison and Heslop-Harrison 1991b), the formation of transfer–cell type wall ingrowths (Charzyńska, Murgia and Cresti 1990), or the incorporation of pectocellulosic thickenings encrusted with proteins (Nepi, Ciampolini and Pacini 1995). In terms of function, the intine layers aid in the interaction of the pollen grains with the stigma by serving as a repository of enzymatic and other proteins and by facilitating pollen germination and pollen tube growth (J. Heslop-Harrison and Heslop-Harrison 1991b).

Since intine assembly is a rapid, multistep process, it is not easy to identify a single, truly initial step. However, in *Olea europaea* there is little doubt that the early reactions occur at the plasma membrane, which separates irregularly from the inner layer of the exine, leaving an undulating area beneath. Soon this area becomes filled with fibrillar precursors of pectocellulosic wall materials; attainment of uniform thickness follows (Pacini and Juniper 1979a). Dictyosomes are the prime movers in the formation of the intine and apparently supply wall precursors for its microfibrillar organization. Suggestive of the contribution of dictyosomes to the growing intine layer is the finding that coated vesicles are dispersed near the plasma membrane of freshly released microspores of *Lilium longiflorum* (J. Heslop-Harrison 1968c; Dickinson and Heslop-Harrison 1971; Nakamura 1979) and *O. europaea* (Pacini and Juniper 1979a). Attempts to correlate these observations with a general mechanism for the transfer of Golgi-derived materials to the intine layer have not yet produced unequivocal results; one idea is that intine matrix substances leave the Golgi vesicles by exocytosis (Hess 1993). According to Mepham and Lane (1970), in addition to dictyosomes, ER also appears to be involved in intine secretion in the microspores of *Tradescantia bracteata*. The basis for this assertion is the frequent occurrence of profiles of ER-producing terminal vesicles that fuse with the plasma membrane, and the correspondence between the materials of the tubular elements of ER and the developing intine, with electron microscopically detected microtubule disposition beneath the intine layer at this time. It has been suggested that ER plays a role by absorbing soluble precursors and transferring them to the dictyosomes for incorporation into the growing intine skeleton. A similar role has been envisaged for ER in intine function in the microspores of *L. longiflorum* (Nakamura 1979). This is not surprising in view of the fact that dictyosomes are an obligate intermediary between ER, on the one hand, and secretory granules and plasma membrane, on the other. The ability of isolated pollen protoplasts to regenerate a wall in culture does not seem to result in the formation of an intine; we can assume that besides the cytoplasmic organelles, there must be some other precursors in limited supply necessary for intine assembly (Miki-Hirosige, Nakamura and Tanaka 1988; Y. Wu and Zhou 1990).

Summarizing, we see that formation of the pollen wall is a complex process for which information is derived from both sporophytic and gametophytic genomes. What are the forces that drive this dynamic morphogenetic process? Although this question will remain unanswered for some time, the temporal and spatial separation of the actions of the two genomes suggests that the primary events rely on information transfer and the synthesis of specific proteins.

(iii) Pollen Wall Proteins

There is general agreement that pollen wall development is a period of accumulation of great many proteins of sporophytic origin. The proteins accumulate in both the exine and intine layers; deposits in the exine are confined to the outer sexine, especially to the interbacular spaces, where they are found in association with pollenkitt and tryphine. Although proteins are initially deposited in the pectic part of the intine, in the mature pollen grain they form a conspicuous zone between the pectic and cellulosic strata that can be considered as a separate layer by itself (J. Heslop-Harrison and Heslop-Harrison 1991b). Tsinger and Petrovskaya-Baranova (1961) are generally credited with the first demonstration of the presence of included proteins in the pollen walls. By staining tests, these investigators noted high concentrations of proteins in the walls of pollen grains of *Paeonia* sp. and *Amaryllis* sp. (Amaryllidaceae) and showed that some of the proteins possess enzymatic properties. As later research showed, their conclusion that the pollen "wall . . . appears to be a living, physiologically active structure, playing a very responsible role in the processes of interchange between the pollen grain and its substrate," has proved to be prophetic. This conclusion also provided the starting point for a chain of interactive investigations, utilizing electron microscopy, cytochemistry, and immunology to gain a deep insight into the nature and function of the wall proteins. Outstanding among the many contributions of more recent years has been the work of Knox and Heslop-Harrison (1969, 1970a), who showed by high resolution cytochemistry the presence of hydrolytic enzymes like acid phosphatase, esterase, RNase, amylase, and protease in the inner cellulosic intine layer of pollen walls of about 50 angiosperms. In this survey, which included all major pollen types, it is interesting that the enzyme activity is strikingly localized in the vicinity of the germination pores, without entirely bypassing the rest of the intine. Upon hydration of the pollen grain, the proteins diffuse through the exine, particularly at the germination pores, and are released into the medium. SDS-PAGE electrophoresis of diffusates from pollen grains of *Cosmos bipinnatus* and *Ambrosia elatior* (Asteraceae) has shown that the wall proteins are heterogeneous in nature. The diffusate from *C. bipinnatus* revealed more than 15 bands, including two glycoproteins and three or four major fractions (Howlett et al 1975). Using antisera prepared in rabbits against diffusates from pollen walls of *Gladiolus gandavensis* (Iridaceae), *Phalaris tuberosa* (canary grass; Poaceae), *Populus alba, P. deltoides* (poplars), *A. trifida*, and *C. bipinnatus*, the intine was shown by immunofluorescence to be the principal site of protein accumulation (Knox, Heslop-Harrison and Reed 1970; Knox and Heslop-Harrison 1971c; Knox, Willing and Ashford 1972; Howlett, Knox and Heslop-Harrison 1973). The antiserum from *Ambrosia trifida* pollen was found to cross-react with pollen from four other species of the genus, namely, *A. artemisiifolia* (ragweed), *A. elatior, A. psilostachya*, and *A. tenuifolia*, and in each case the association was restricted to the intine and, to some extent, to the exine; however, there was no evidence for the presence of antigen in the microspore cytoplasm or in the vegetative cell of the mature pollen grain (Knox and Heslop-Harrison 1971a; Howlett, Knox and Heslop-Harrison 1973). The characteristic parts of the intine where enzymes predominate are the ribbonlike tubules within the cellulosic layer (Figure 3.3). This was established by ultracytochemical localization of acid phosphatase activity in the intine of *Crocus vernus* (Iridaceae) (Knox and Heslop-Harrison 1971b). Similar tubules, presumed to harbor RNase, have been described in the intine of *Malvaviscus arboreus* (Malvaceae) (J. Heslop-Harrison et al 1973). Another fine-structural study has called attention to the presence of stratified ER encrusted with ribosomes and of cytochemically detectable proteins encased in the cellu-

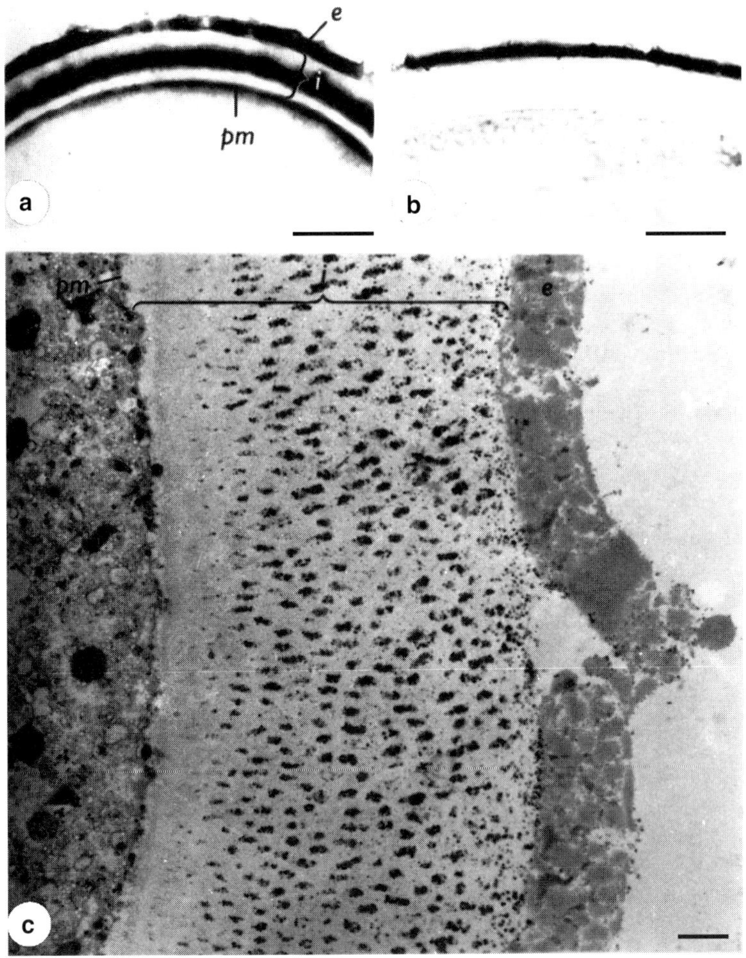

Figure 3.3. (**a**) Naphthol AS-B1 phosphate pararosaniline reaction for localization of acid phosphatase in the intine of pollen grains of *Crocus vernus*. (**b**) Control: reaction without substrate. (**c**) Electron micrograph of the pollen wall showing localization of acid phosphatase in the ribbonlike tubules of the intine. e, exine; i, intine; pm, plasma membrane. Scale bars = 10 μm (a, b) and 1 μm (c). (From Knox and Heslop-Harrison 1971b. *J. Cell Sci.* 8:727–733. © Company of Biologists Ltd.)

lose matrix of the microspores of *Cosmos bipinnatus* during the period of intine deposition (Knox and Heslop-Harrison 1970a). As shown in Figure 3.4, immunolocalization of cutinase (the enzyme that breaks down cutin) in the pollen grains of *Brassica napus* by a polyclonal antiserum to *Fusarium* cutinase has revealed that the enzyme is confined predominantly to the intine, with a faint presence in discrete regions of the pollen cytoplasm (Hiscock et al 1994).

In extending this work in a developmental context, both cytochemical and immunofluorescence methods have shown that although microspores freshly released from the tetrad lack wall-associated enzyme activity, they acquire it soon after, during the period of expansion and vacuolation, when the intine growth is most active (Knox and Heslop-Harrison 1970a; Knox 1971). When the activity of acid phosphatase as a marker enzyme for intine proteins of pollen grains of *B. oleracea* was followed by quantitative cytochemical methods, two periods of enzyme accumulation were observed. The first coincided with the late vacuolation phase of the microspore, and the second with the period of pollen maturation (Vithanage and Knox 1976). A similar pattern of acid phosphatase activity has been verified by qualitative observations and quantitative analyses in the intine of *Lolium perenne* (ryegrass; Poaceae) pollen (Vithanage and Knox 1980). In *Silene alba*, acid

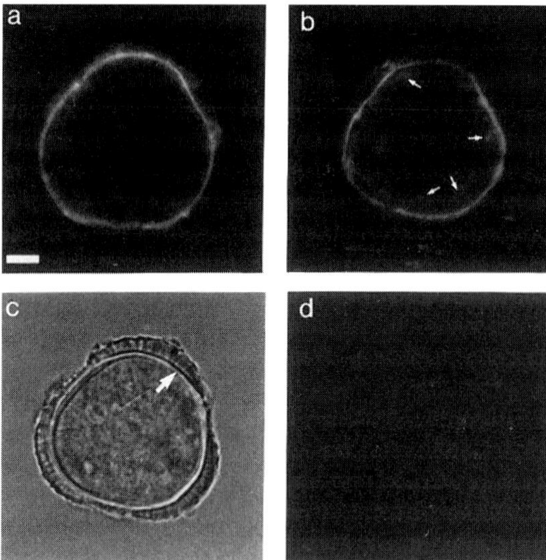

Figure 3.4. Immunolocalization of cutinase in the intine of pollen grains of *Brassica napus.* **(a, b)** Fluorescence micrographs of sections of pollen grains treated with cutinase polyclonal antiserum followed by goat antimouse fluorescein isothiocyanate (FITC)-conjugated secondary antibody showing fluorescence in the intine and in discrete regions of the cytoplasm (arrows). **(c, d)** Controls showing light micrograph (c) and fluorescence micrograph (d) of sections of pollen grains treated with mouse pre-immune serum followed by FITC-conjugated secondary antibody. Arrow in (c) indicates the intine. Scale bar = 8 μm. (From Hiscock et al 1994. *Planta* 193:377–384. © Springer-Verlag, Berlin. Photographs supplied by Dr. H. G. Dickinson.)

phosphatase is initially detected in the peripheral pollen cytoplasm prior to its exclusive localization in the domain of the intine (Shoup, Overton and Ruddat 1981). It would seem from these and the foregoing observations that intine proteins are synthesized by the microspore cytoplasm and are therefore gametophytic in origin. This makes sense since, unlike the exine proteins, the intine proteins do not show homology to proteins released by the tapetum.

Because of its morphological nature with free-standing baculae or the reticulate pattern with open cavities, the exine provides large volumes of more or less enclosed space for the storage of proteins. An unequivocal demonstration of the deposition of proteins in the exine was possible using the large pollen grains of *Malvaviscus arboreus, Hibiscus rosa-sinensis, Abutilon megapotamicum,* and *Anoda crispa* (all of Malvaceae) (J. Heslop-Harrison et al 1973), *Cosmos bipinnatus* (Howlett, Knox and Heslop-Harrison 1973), *Iberis semperflorens,* and *I. sempervirens* (J.Heslop-Harrison, Knox and Heslop-Harrison 1974). In contrast to the intine proteins, the exine materials are derived from the tapetum, which during dissolution releases its melange of materials, including membrane-bound cisternae, into the anther locule, where they lodge in the exine cavities. The cisternae containing granular materials are thought to be precursors of the proteins. In the tapetal cells of *B. oleracea,* esterase, a marker enzyme of the exine, attains its peak activity before the cells begin to disintegrate. Indicative of the transfer of the enzyme from the tapetum to the exine, an increase in the activity of esterase in the pollen wall exine is found to coincide with the dissolution of the tapetum (Vithanage and Knox 1976). The pattern is comparable to that seen for esterase activity in anthers of *Helianthus annuus* (Vithanage and Knox 1979) and *Lolium perenne* (Vithanage and Knox 1980). Although acid phosphatase is of gametophytic origin and is present mainly in the intine in the pollen of *B. oleracea* and *L. perenne,* in *H. annuus* it is present in both exine and intine sites. Moreover, coincident with tapetal dissolution, enzyme activity in *H. annuus* is localized in the sexine; this is indicative of a sporophytic origin of the exine-bound enzyme (Vithanage and Knox 1979). It should be pointed out that pollen grains of the species just listed also harbor intine proteins, which are found mainly, but not exclusively, in the vicinity of the germination pores.

Among other proteins accumulating in the pollen walls are the allergens that cause the seasonal disease known as hayfever. In *Ambrosia,* a genus that includes the ragweeds, the most active allergens are two globular proteins, antigen-E and antigen-K, with the former being somewhat more active than the latter. By immunofluorescence method, Knox and Heslop-Harrison (1971a) showed that antigen-E is localized in the intine of the pollen grain of *A. artemisiifolia* and that its appearance coincides with the earliest indication of antigen reaction with the whole pollen extract antiserum. The ragweed pollen, with the structural specialization of its intine as a repository for allergens, provides a striking contrast to the ryegrass pollen, where the active allergens are found in the exine. The major allergen of ryegrass pollen is an acidic glycoprotein, and immunocytochemical methods have permitted detection of the allergen not only in the exine but also in the pollen cytoplasm (Vithanage et al 1982; Staff et al 1990). The allergic protein of *Cryptomeria japonica* (Japanese cedar; Taxodiaceae) is localized all over, and between, the exine and intine layers, as well as in the generative cell wall and in organelles like the

Golgi vesicles (Miki-Hirosige et al 1994). Recent developments in the characterization of the allergens of pollen grains have included isolation of cDNA clones encoding the allergens and determination of their nucleotide sequences and the deduced amino acid sequences; these investigations are described in Chapter 4. The discovery of allergens in the exine and intine layers of pollen grains and their characterization hold considerable promise for the future in terms of a better understanding of the allergic response, which could help alleviate the suffering due to hayfever.

(iv) Adaptive Significance of Pollen Walls

An appropriate way to end the discussion on pollen walls is to consider their functional significance in a broad biological context. Pollen grains of angiosperms are astonishingly diverse in the details of their exine ornamentation. Palynology literature is replete with examples of the species-specific architecture of the pollen exine, which has permitted investigation of the systematic affinities of certain groups of plants; in some groups, additional features related to pollination ecology are associated with the exine. Although the pollen wall structure is now recognized as an adaptation of great importance in the evolution of angiosperms, the concern here is with the significance of pollen walls in the overall reproductive strategy of this group of plants. From this perspective, interest in pollen walls arises due to their roles such as protection against excessive desiccation, facilitation of efficient germination of the pollen grain, and promotion of interaction of the pollen grain with the stigma. In periodic reviews, J. Heslop-Harrison (1971a, 1975a, 1976), J. Heslop-Harrison and Heslop-Harrison (1991b), and J. Heslop-Harrison et al (1975) have extensively commented on these aspects of the function of pollen walls and the following account has been drawn largely from these articles.

For a long time, researchers have been preoccupied with the protective function of the pollen walls. The presence of a thick sporoderm material is clearly not the only mechanism employed in nature to prevent the loss of water from cells, for both single-celled and multicellular plants achieve the same result with the cuticle. The development of the concept of pollen walls nevertheless provides an interesting example of the upper limit of structural specialization that is achieved by plant cells in response to selective pressures. The reason why pollen grains need to have a complex wall for protection is clearly related to the unpredictable conditions of their life on the dry land after they leave the anther. There are difficulties and uncertainties associated with this interpretation, but it provides a framework for understanding how different cell types respond to the same environmental stress.

A problem of enduring interest relates to the role of pollen walls during the critical period when a dehydrated pollen grain lands on the receptive stigma and begins to germinate. In recent years it has been possible to correlate the development of the germination pores of the pollen grain with physiological changes that follow upon hydration and germination. This has made it possible to demonstrate that besides its traditional role as a preferred path of emergence of the pollen tube, the aperture controls the water balance during germination. The ingress of water into the dehydrated pollen due to exposure of the intine surface at the pore site, in contrast to the exclusion of water due to occlusion of the pores by sporopollenin deposit, has revealed a positive feedback system during hydration; this results in an increase in the surface of the intine available for water uptake with each increase in volume. The exine also functions in a similar way in the control of water uptake in non-aperturate pollen grains (J. Heslop-Harrison 1971a). Examples of specialization of the intine for a critical early function in pollen germination are described by J. Heslop-Harrison and Heslop-Harrison (1991b).

Investigations of the movement of included proteins from the exine and intine domains are also changing our concepts of their possible role in the physiology of the pollen grain. Indications are that following pollination, proteins leach out of the pollen walls and take up temporary residence on the surface of stigmatic cells. Some of these proteins, especially those held in the exine, include enzymes that play important roles in the pollen–stigma interaction, possibly in the germination of the pollen grain and early growth of the pollen tube on the stigma. Intine-borne proteins, on the other hand, appear responsible for the penetration of the stigma by the emerging pollen tube, as shown in studies in certain Brassicaceae. In *Raphanus sativus* and other members of Brassicaceae, and in members of Caryophyllaceae, Asteraceae, and Malvaceae, the surface of the stigma papillae bears an external proteinaceous coating overlying the cutinized outer layer of the cell wall (Mattsson et al 1974; J. Heslop-Harrison, Heslop-Harrison and Barber 1975). This layer, known as the pellicle, is the first barrier encountered by pollen grains and serves to attach them to the stigma. Dickinson and Lewis (1973a) showed that penetration of the pollen tube through the cuticle proceeds without hindrance in both compatible and incompatible pollinations and that incompatibility reactions set in as the pollen tube begins to grow. It appears that pollen tube penetration of the cuticle depends on the presence of a precursor of cutinase that is carried in the intine and is activated when the pollen tube comes in contact with the pellicle. The basis for this interpretation is the observation that although enzymatic degradation of the pellicle in *Agrostemma githago* (Caryophyllaceae) does not impair pollen germination on the stigma, it prevents the penetration of the pollen tube, possibly due to the failure of cutinase activation (J. Heslop-Harrison and Heslop-Harrison 1975). The interpretation also receives support from similar experiments done in *Crocus chrysanthus*. In this species, in which the stigmatic papillae do not bear a pellicle, the germinating pollen grain directly encounters the cuticle, whose enzymatic erosion permits the penetration of the pollen tube. Protein secretions from the underlying cell, stored in the lacunae or cavities of the cuticle, apparently contain the factor that activates intine-held cutinase; here also, as in *A. githago*, pollen tube fails to penetrate the stigmatic papillae from which the surface secretions have been removed by enzyme treatment (Y. Heslop-Harrison 1977).

A conceptual reorientation in the field of plant breeding was initiated by studies that showed that exine-held proteins released by pollen grains during their early period of attachment to the stigma play a role in determining compatibility relationships. Although this remarkable fact was suspected toward the end of the last century, the impetus for much of the later experimental studies was the postulation of J. Heslop-Harrison (1968a) that implicated exine-based tapetum-derived proteins in sporophytic incompatibility systems. A series of experiments involving germination of pollen grains and growth of pollen tubes in compatible and incompatible pollinations in a wide range of plants, and the effects of proteins isolated from pollen walls of compatible genotypes in overcoming the inhibition of germination of incompatible pollen, have demonstrated the interference by proteinaceous factors of the pollen wall as a basic physiological characteristic of the incompatibility system. These proteins are designated as "recognition substances" that control pollen germination and pollen tube growth during both compatibility and incompatibility interactions.

The role of exine proteins in compatibility control was first clarified in experiments involving members of Brassicaceae and Asteraceae. In *Iberis sempervirens*, which has an intraspecific incompatibility system, the early events following self- and cross-pollinations are more or less similar. These include hydration of the pollen grain and release of exine-bound proteins onto the stigmatic surface. In selfed plants, the rejection reaction develops soon after pollen tube formation and is accompanied by callose occlusion of the stigmatic papilla adjacent to the pollen grain. The rejection reaction is also induced in excised stigmas by agar blocks enriched with exine proteins, as well as by isolated tapetal fragments as a substitute for the exine proteins (J. Heslop-Harrison, Knox and Heslop-Harrison 1974). A similar syndrome of changes at the stigmatic surface accompanies compatible and incompatible pollinations in *Raphanus sativus*. A chemical fraction isolated from the exine of *R. sativus*, consisting mainly of proteins and tryphine, elicited the production of a callosic reaction on the stigmatic papillae in the appropriate genotype in a way similar to an incompatible pollination. This is what one would expect if the incompatibility reaction is due to exine-held materials (Dickinson and Lewis 1973a, b). Callose occlusion of the stigmatic papillae following incompatible pollination has also been described in *Cosmos bipinnatus*. Genetic evidence for the role of exine-borne proteins in intraspecific incompatibility in this species was provided by offsetting self-incompatibility by pollination with pollen mixes of killed compatible pollen and fresh self-pollen or by self-pollination following application of compatible pollen wall diffusate to the stigma (Knox 1973; Howlett et al 1975). Although the percentage of successful illegitimate crosses was not overwhelming, these observations indicate that exine proteins are factors to be reckoned with in controlling the breeding behavior of plants.

A function for the intine proteins in the control of interspecific incompatibility has been demonstrated in poplars. Failure of pollen grains to germinate on the foreign stigma is the usual cause of hybrid failure in *Populus deltoides* × *P. alba* crosses. However, this incompatibility barrier can be breached by disguising pollen grains from the foreign donor by mixing them with acceptable pollen or by substituting for the recognition pollen a protein preparation leached from the pollen intine. By this method, *P. deltoides* × *P. alba* hybrids were obtained, using *P. deltoides* as recognition pollen (Knox, Willing and Ashford 1972; Knox, Willing and Pryor 1972).

Studies reviewed in this section are perhaps only the beginnings of future research geared to apply the new knowledge about pollen wall structure toward ensuring efficient pollination and fertilization in flowering plants. There are several questions at the molecular level that remain to be answered before our understanding of the functions of the pollen wall can be considered complete. They range from the reactions triggered by the recognition materials to the secondary signals elicited in the participating cells. Research addressed to these questions as they apply to the genetics of self-incompatibility will be considered in Chapter 9 of this book.

2. THE FIRST HAPLOID MITOSIS

In the ontogeny of the pollen grain, the most studied and the most impressive division is the first haploid mitosis, heralding the formation of the vegetative and generative cells. The elaborate network of preparatory steps that are completed by the microspore before this division suggests that they may represent a set of common denominators for an asymmetric division. Although microspores of different stages of development collected from intact anthers continue to provide excellent materials for cytological and biochemical studies, successful attempts have been made to induce division in microspores of cultured immature inflorescences (Barnabás and Kovács 1992; Pareddy and Petolino 1992) and cultured anthers (Sparrow, Pond and Kojan 1955), in microspores cultured at the uninucleate stage of development (Tanaka and Ito 1980, 1981b; Tanaka, Taguchi and Ito 1980; Benito Moreno et al 1988; Stauffer, Benito Moreno and Heberle-Bors 1991; Tupý, Řihová and Žárský,1991; Touraev et al 1995b) and in microspore and pollen protoplasts divested of their walls by enzymatic digestion (Bhojwani and Cocking 1972; Rajasekhar 1973; Bajaj 1974a; Pirrie and Power 1986; Baldi, Franceschi and Loewus 1987; Tanaka, Kitazume and Ito 1987; Lee and Power 1988; Miki-Hirosige, Nakamura and Tanaka 1988; Pental et al 1988; Zhou 1989a, b; Y. Wu and Zhou 1990, 1992). Apparently, the microspore genome contains all the information needed to complete its maturation, which thus appears to be independent of the somatic tissues of the anther. Results of investigations on the maturation of microspores under in vitro conditions have opened up questions relating to the need for the mitotic division of the microspore nucleus and its asymmetrical disposition for continuation of the gametophytic program. The work of Touraev et al (1995b) showed that the gametophytic program resulting in pollen tube growth is faithfully continued by cultured microspores of *Nicotiana tabacum* even when they divide symmetrically into two cells. Pollen tubes are also formed in microspores of a transgenic tobacco cultured in the presence of high concentrations of colchicine, which effectively blocks the first haploid mitosis (Eady, Lindsey and Twell 1995). The in vitro methods are obviously of great advantage in the manipulation of pollen cells and in the analysis of cell cycle parameters of microspores unimpeded by the influence of the parent plant and in the absence of the tapetum. In addition, a high probability of success in the gametophytic selection of desirable genes for transmission to the progeny has been indicated using a transgenic approach with in vitro cultured microspores (Touraev et al 1995a).

A striking and consistent feature of the newly released microspore is the presence of a dense, granular, nonvacuolate cytoplasm poised for resumption of metabolic activity. This cytoplasm is in part originally bequeathed from the microsporocyte and is in part synthesized by the microspore immediately after its liberation from the tetrad. It is, of course, not unreasonable to expect nucleo–cytoplasmic interactions to be important determinants of gametophytic expression in the microspore at this stage. Ultrastructural features suggestive of such interactions are nuclear invaginations in the microspores of *Podocarpus macrophyllus* (Podocarpaceae) (Aldrich and Vasil 1970) and *Pinus banksiana*, invaginations in the latter probably enclosing RNA (Dickinson and Bell 1970, 1972b; Dickinson and Potter 1975), and the association of the nuclear envelope with membrane-bound inclusions in the microspores of *Lilium longiflorum* (Dickinson 1971b) and *Cosmos bipinnatus* (Dickinson and Potter 1979). Unfortunately, the control exerted by these interactions in the subsequent reprogramming of the microspore remains a mystery. Increase in the microspore cytoplasm inevitably leads to an increase in the volume of the microspore, which is accentuated by vacuolation. The process of vacuolation is truly remarkable, because it presents an entirely new structural profile of the microspore and is accomplished by the coalescence of a large number of tiny vacuoles or by the formation of a single large vacuole. In *Phalaris tuberosa*, vacuolation occurs in a predictable time interval and occupies a major part of the developmental time span of the microspore, when its diameter more than doubles to about three-quarters of the final size. Toward the end of maturation of the microspore, the vacuole is resorbed and the resulting void is filled with new cytoplasm (Vithanage and Knox 1980). A phenomenal increase in size, up to about 12 times the volume of the postmeiotic products, occurs during vacuolation of microspores of *Triticum aestivum* (Mizelle et al 1989). The other event in the microspore that is almost invariably linked to vacuolation is the displacement of the nucleus from its central position to the periphery of the cell, where mitosis occurs. It is remarkable that in some plants nuclear migration is not a random event, but is site specific as the organelle is propelled to a site opposite the germination pore (Sanger and Jackson 1971a; Christensen and Horner 1974). The location of the pore thus appears to dictate where the nucleus will migrate and where the wall partitioning the cell will be formed.

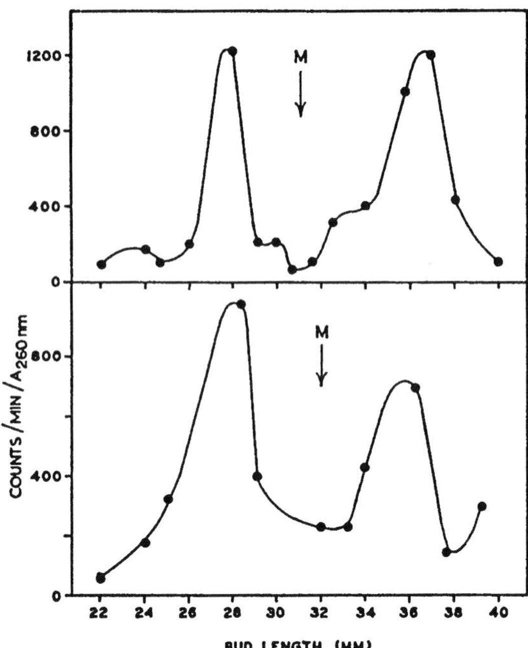

Figure 3.5. Incorporation of ^{3}H-thymidine into microspores of lily anthers. Bud lengths on the abscissa are closely correlated with stages of the mitotic cycle. The first haploid mitosis is indicated by M. Meiotic division of the microsporocytes is completed in 22-mm-long buds. Each graph is derived from separate experiments using different groups of buds. (Reprinted from S. B. Howell and N. B. Hecht 1971. The appearance of polynucleotide ligase and DNA polymerase during the synchronous mitotic cycle in *Lilium* microspores. *Biochim. Biophys. Acta* 240: 343–352, with kind permission of Elsevier Sciences-NL, Sara Burgerhartstraat 25, 1055 KV Amsterdam, The Netherlands.)

(i) Biochemical Changes

Since the standard plan of preparation for the first haploid mitosis includes the synthesis of DNA, the many cytological and molecular studies on DNA synthesis in the microspore nucleus have had a profound influence in crystallizing our ideas on the functional divergence of the division products. These studies have been aided in part by the relatively long duration of the cell cycle in developing microspores. For example, the approximate durations of the G1, S, and G2 + M phases of mitosis in the microspores of *Lilium longiflorum* are estimated to be 12, 2, and 1 day(s), respectively (Tanaka, Taguchi and Ito 1979). A series of microspectrophotometric measurements of DNA amounts in the microspores of *L. longiflorum* showed an increase in the 1C amount of DNA to the expected 2C level just before nuclear division, and a decrease to the 1C class in the nuclei of the daughter cells. The time course of DNA synthesis established that this is correlated with autoradiographically detectable incorporation of precursors of DNA synthesis into the nucleus before mitosis (Taylor and McMaster 1954; Tanaka, Taguchi and Ito 1979). As shown in Figure 3.5, chromatographic separation of labeled DNA revealing a steep increase in the amount of label during the S-period of the microspore nuclei also clearly supports the simple assumption of DNA doubling before division (Steffensen 1966; Howell and Hecht 1971). By microspectrophotometry alone or combined with autoradiography, a period of DNA doubling in anticipation of division has been observed in the microspore nuclei of *Tradescantia paludosa* (Moses and Taylor 1955), *Tulbaghia violacea* (Taylor 1958), and maize (Moss and Heslop-Harrison 1967). It is interesting to note that whereas DNA synthesis occurs after a long interruption in the interphase in *L. longiflorum* and *T. paludosa* microspores, in *T. violacea* it is confined to a period immediately after meiosis.

Although we do not have convincing cytological guideposts, we can associate the period of DNA synthesis in the microspores with certain biochemical activities of replication. For example, in *L. longiflorum* (Foster and Stern 1959) and *Trillium erectum* (Hotta and Stern 1961b) the S-period of microspore mitosis is generally preceded by a build-up of soluble deoxyribosidic precursors in the anther. The appearance of DNase, thymidine kinase, and deaminases – enzymes associated with the production of deoxyribosides, their phosphorylation, and their deamination, respectively – at brief intervals in propinquity to DNA synthesis during microspore mitosis is compatible with their involvement in the impending synthetic phase. Whereas there is no doubt about the presence of deoxyribosides and the relevant enzymes for the synthesis and conversion of nucleosides in the anther locule, the question relates to their origin. The rise in the activity of DNase has been taken to indicate that deoxyriboside precursors are the products of DNA breakdown in the tapetum (Hotta and Stern 1961b; Stern 1961). Thymidine kinase is thought to be formed in the microspores by de novo synthesis. The use of inhibitors of RNA and protein synthesis led to the conclusion that enzyme synthesis is dependent upon the simultaneous synthesis of RNA. However, the conditions that lead to the cessation of enzyme synthesis following S-phase are not fully understood (Hotta and Stern 1963a, b). Another study showed that thymidine kinase is susceptible to induction by exogenous thymidine as in microorganisms; the surprising finding was that the inducer stimulates thymidine kinase activity only during the period when the enzyme begins to appear in the microspores (Hotta and Stern 1965a). Evidently, a network of intercellular control mechanisms functions in regulating the timing of DNA synthesis during the first haploid mitosis; of these mechanisms, substrate availability for enzyme induction is but one. In *L. longiflorum*, enzymes that deaminate dephosphorylated derivatives of deoxycytidine and 5-methyldeoxycytidine are more abundant and active in the somatic tissues of the anther than in the microspores. It does not seem likely that the enzyme levels are linked to DNA synthesis in the microspores, since they do not match the sharp periodicity of appearance of deoxyriboside pools (Hotta and Stern 1961a). Nevertheless, deaminases could perhaps deaminate phosphorylated substrates for DNA synthesis in the tapetum, but this remains to be established.

Two enzymes concerned with synthesis of the polynucleotide chains of DNA during microspore mitosis in *Lilium* sp. show contrasting patterns of activity. Although DNA polymerase activity is generally high during S-phase, enzyme appearance is not strictly correlated with the onset of DNA synthesis. On the other hand, polynucleotide ligase activity fluctuates during the mitotic cycle, registering a peak prior to S-phase and declining markedly through the interval leading to the next cell cycle (Howell and Hecht 1971).

Analyses of other cellular activities in preparation for microspore mitosis have generated data that may even serve as a frame of reference for mitosis in general. One study relates to the energy requirements for division. As Nasatir and Stern (1959) have shown, activities of aldolase and glyceraldehyde-3-phosphate dehydrogenase are low during the period of microspore mitosis in anthers of *L. longiflorum*, although both premitotic and mitotic phases are associated with high enzyme titer. These data parallel the overall decrease in oxygen consumption noted during mitosis in the microspores of the related *Trillium erectum* (Stern and Kirk 1948). Clearly, neither is an increased glycolytic activity the source of energy for mitotic divisions, nor do intrinsic changes in respiratory metabolism account for the entry of microspores into mitosis. The finding that microspore mitosis in *L. longiflorum* is preceded by a rise in soluble –SH supports the idea that high sulfhydryl concentration is important to the phase of chromosome condensation and separation during cell division. Since a major component of the soluble sulfhydryls is probably glutathione, this observation assigns a role for glutathione not only in the physical aspects of mitosis, but also in the activity of the –SH enzymes involved in the division process (Stern 1956, 1958). Ascorbic acid is another metabolite that shows a striking increase in concentration during microspore mitosis in *L. longiflorum* (Stern and Timonen 1954). However, this finding remains isolated, and much work is needed before we can truly assess its significance in the division of the microspore.

There is some evidence for the view that microspore mitosis is associated with an asymmetrical partitioning of membrane-associated Ca^{2+}. An interesting finding suggestive of a role for Ca^{2+} in the alignment of the cell plate during mitosis is that lithium, which blocks the asymmetric distribution of Ca^{2+} in the microspores of tobacco, induces a symmetric division in these microspores (Zonia and Tupý, 1995). Sensitive immunological and cytochemical methods have revealed that a common stepping stone in the maturation and division of the microspore involves changes in the distribution of specific nucleolar proteins and various species of RNA (Testillano et al 1995).

The foregoing survey demonstrates the diversity of preparative steps that occur in the microspore before the first haploid mitosis. This is still only a partial list, as there is little doubt that other steps will be added in the future. Considerable work has also been done on the timing and pattern of RNA and protein synthesis associated with this division in microspores of various plants. This will form the subject matter of the next chapter and so will not be taken up here.

(ii) Ultrastructural Changes

Despite some anomalies, on the whole one gains the impression that in many ways the microspore is an ordinary cell containing typical organelles distributed randomly in the cytoplasm. Although light microscopical aspects of the first haploid pollen mitosis have been described in many species, electron microscopy provides a striking visual portrait of the structure of the vegetative and generative cells born out of this division. Over the years, it has become clear that these cells are not simply a stable reflection of the nucleus and cytoplasm they contain, but that they undergo important physiological and biochemical changes as well.

The cytoplasm of the microspore of *Helleborus foetidus* freshly released from the callose sheath of the tetrad is notable for its ultrastructural simplicity, since it contains few readily recognizable organelles other than the nucleus and nucleolus. At the vacuolate stage it has electron-dense mitochondria and large, starch-filled plastids, but they are not numerous. The scarcity of dictyosomes and ER is also consistent with the low level of synthetic activity of the microspore (Echlin 1972). The newly formed microspore of *Tradescantia bracteata* has a profile strikingly similar to that of *H. foetidus*; however, as vacuoles appear, the parts of cytoplasm containing mitochondria and plastids are engulfed in the vacuoles. It is believed that this cytoplasm is degraded and gives way to the formation of an organelle-rich, stable cytoplasm before division (Mepham and Lane 1970).

Role of the Cytoskeleton. An important aspect of the dynamics of the first haploid mitosis is the molecular mechanism that polarizes the nucleus within the microspore. It was noted earlier that in preparation for this division, the microspore undergoes a cytoplasmic reorganization that includes displacement of the nucleus to the cell periphery. It is reasonable to suppose that nuclear polarity in the microspore is mediated by the presence of microtubules and microfilaments; the question is how. It is easy to conceive how a preprophase band of microtubules girdling the nucleus could encompass part of the microspore cytoplasm where the future cell plate is formed; however, in spite of careful searches, no one has documented the presence of a preprophase band of microtubules during pollen mitosis (J. Heslop-Harrison 1968d; Burgess 1970c; van Lammeren et al 1985; Terasaka and Niitsu 1990; Hause, Hause and van Lammeren 1992). In studies of microspore mitosis in the orchid, *Phalaenopsis equestris*, Brown and Lemmon (1991c, 1992b) have identified by immunofluorescence technique the reorganization of microtubules, from a nuclear-based radial array to their concentration at the distal pole of the microspore, as the first cytological event associated with nuclear polarity. These microtubules, designated as the generative pole microtubule system, extend between the plasma membrane and the nuclear envelope and apparently guide the nucleus to its new position near the cell surface prior to mitosis. During mitosis, the microtubules disappear, although their site marks the pole of the future mitotic spindle, which delimits the generative cell. Another indication of the unusual nature of the cytoskeleton of the dividing microspore is the presence of a multipolar spindle, characterized by the presence of several distinct cables of microtubules (Figure 3.6). Although the pattern of actin distribution remains unchanged during the premitotic rearrangement of microtubules, actin filaments are co-aligned with the spindle microtubules during division, with a substantial array present between the spindle and the cytoplasm. It has been suggested that this array of microfilaments could function in maintaining the acentric position of the spindle. However, the first haploid pollen mitosis in another orchid, *Cypripedium fasciculatum*, is not accompanied by the formation of a microtubule system at the future generative pole (Brown and Lemmon 1994). A network of intermediate filaments, a rarity in plant cells, has been demonstrated in protoplasts of uninucleate and binucleate pollen grains of tobacco; whether they function alone or in concert with the microtubules in the division of the pollen nucleus is not known (Yang, Xing and Zhai 1992).

The role of microtubules in nuclear maneuvering during pollen mitosis has been studied in some experiments using colcemid (colchicine). The results are instructive. Tanaka and Ito (1981b) found that although the microspore nucleus of *Tulipa gesneriana* initially migrates to one side of the cell in the presence of the drug, it returns to its original position in the center of the cell before mitosis. It seems likely, therefore, that nuclear migration is not governed by microtubules, but

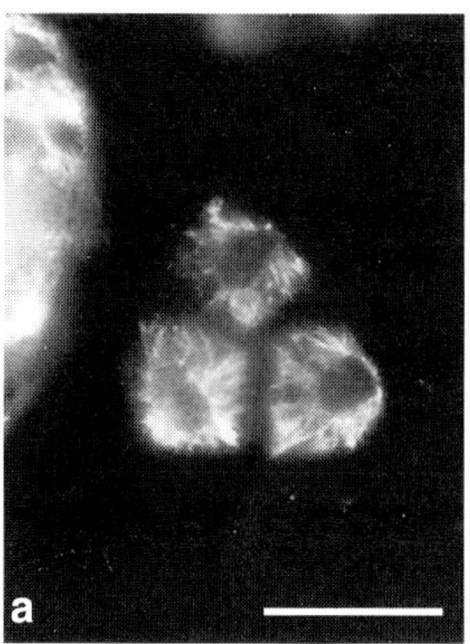
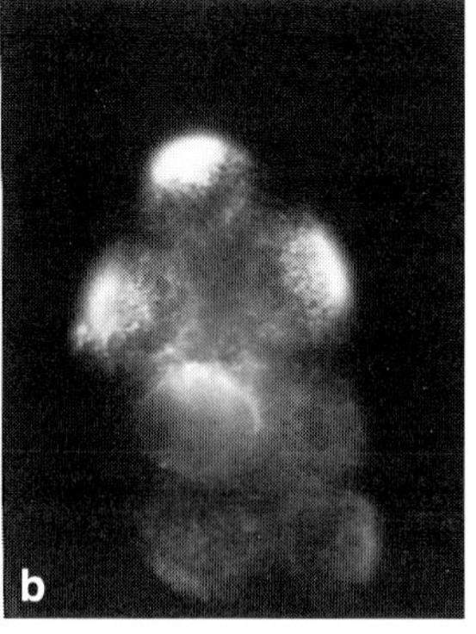
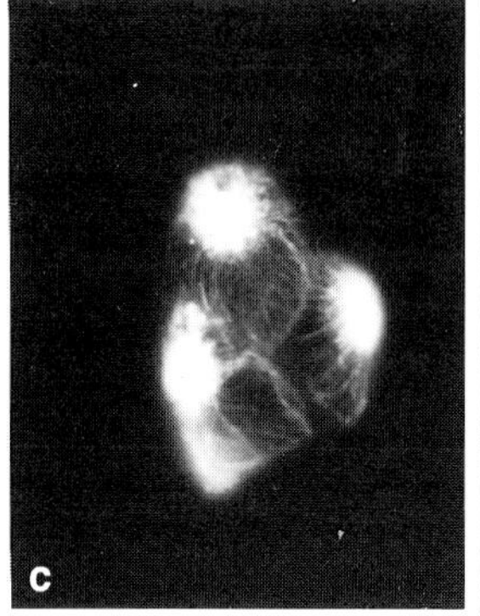
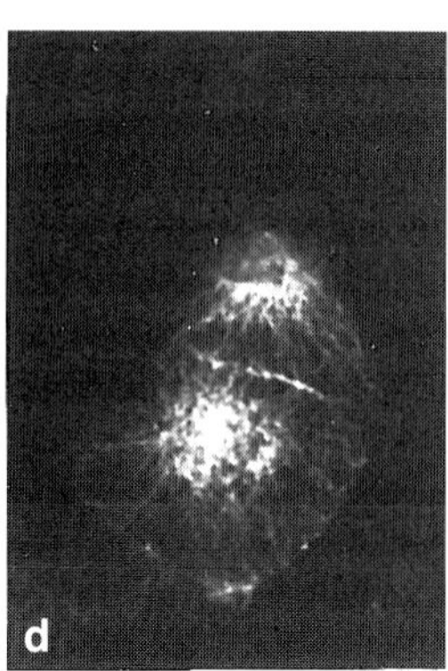

Figure 3.6. Fluorescence micrographs showing the pattern of microtubule changes prior to microspore mitosis in *Phalaenopsis equestris.* **(a)** Newly formed microspores showing microtubules radiating from the nuclei. **(b)** Organization of microtubules into the generative pole microtubule system. **(c)** Radiation of peripheral microtubules from the generative pole microtubule system. **(d)** Confocal laser scanning microscope image showing multiple foci of microtubules and radiating cortical microtubules in the generative pole microtubule system. Scale bar = 10 µm, except in (d), bar = 8.6 µm. (From Brown and Lemmon 1991c. *J. Cell Sci.* 99:273–281. © Company of Biologists Ltd. Photographs supplied by Dr. R. C. Brown.)

keeping the nucleus at its new address depends in part on the microtubules. The observation that in the microspores of *Brassica napus*, following nuclear migration, microtubules are found connecting the nucleus with the plasma membrane, supports this contention (Hause, Hause and van Lammeren 1992). Terasaka and Niitsu (1990) have provided another clear example of the limited role of microtubules in nuclear migration. In the microspores of *Tradescantia paludosa*, there are two separate predivision nuclear movements: an initial displacement from the center to one side, and a second one back to the center, where division occurs. Microtubules appear to have little to do with the first migration since it is insensitive to colchicine. The second nuclear migration, which is signaled by a shifting of the position of microtubules, is, however, colchicine sensitive, and hence microtubule dependent.

Wall Formation. Following the first haploid mitosis, appearance of a cell plate is presaged by the fabrication of a phragmoplast perpendicular to the axis of the mitotic spindle. As the cell plate follows the contour of the spreading phragmoplast, it slowly bisects the microspore into a large vegetative cell and a small generative cell. In *Brassica napus* microspores, microfilaments are found co-aligned with microtubules in the phragmoplast, suggesting that the cytoskeletal framework causes the centripetal movement of the phragmoplast (Hause, Hause and van Lammeren 1992). From the point of view of students of mitosis, perhaps the most salient feature of microspore division is that the new cell plate grows around the generative cell rather than fuses with the parent wall, resulting in a generative cell that is more or less surrounded by the vegetative cell. In the absence of a preprophase band of microtubules to guide the cell plate, the cytoplasm controlled by a nucleus is considered to be the determinant of the division plane. According to Brown and Lemmon (1991d), a possible factor at play in the formation of the cell plate in the microspores of *Phalaenopsis equestris* is a radial system of microtubules emanating from the nucleus of the future generative cell. The extent of these microtubules appears to outline the cytoplasm surrounding the generative cell and guide the expanding margin of the phragmoplast in its unusual curved path. In addition, the fact that microtubules extend from the margins of the phragmoplast to the nucleus of the generative cell, but not to that of the vegetative cell, might also be important in giving directionality to the phragmoplast to follow the outline of the generative cell domain.

The existence of a cell wall separating the generative cell from the vegetative cell has been controversial for some time. Although many early electron microscopic investigations led to the belief that the two cells are separated by a double membrane and not by a true wall, later studies showed the presence of a clear wall between the cells at least for a period of time (Sassen 1964; Maruyama, Gay and Kaufmann 1965; Angold 1967, 1968; J. Heslop-Harrison 1968d; Jensen, Fisher and Ashton

1968; Lombardo and Gerola 1968b; Hoefert 1969a; Burgess 1970c; Sanger and Jackson 1971a; Dupuis 1972; Echlin 1972; Dunbar 1973; Lutz and Sjolund 1973; Hu et al 1979; Brighigna, Fiordi and Palandri 1981; Nakamura and Miki-Hirosige 1985; Murgia et al 1986). It has become clear that in various species, the substance of the cell wall is contributed by different sets of organelles. In *Tradescantia paludosa*, fusion of pectic vesicles and formation of a phragmoplast that curves around the nucleus of the generative cell are the primary means by which the wall is assembled. Later, the wall grows around the nucleus and establishes contact with the intine of the pollen grain (Maruyama, Gay and Kaufmann 1965; Terasaka and Niitsu 1995). An intense activity of rough ER in the vicinity of the intine and the generative cell wall of the pollen grain of *Monotropa uniflora* (Pyrolaceae) has suggested the involvement of ER in cell wall formation (Lutz and Sjolund 1973). The division details of the pollen grain of *Lilium longiflorum* are consistent with the view that the wall of the generative cell is formed by fusion of polysaccharide vesicles derived from lipid bodies, coated vesicles, and numerous fragments of rough ER (Nakamura and Miki-Hirosige 1985). Cell plate microtubules that probably direct the initial deposition of vesicles at the site of the future cell wall have been identified in dividing microspores of some plants (Angold 1968; J. Heslop-Harrison 1968c; Burgess 1970c; Echlin 1972; Nakamura and Miki-Hirosige 1985).

Organelle Inheritance. A feature of the microspore cytoplasm that may serve as a basis for the maternal inheritance of certain organelles is the polarity of their distribution prior to the first haploid mitosis. In *Haemanthus katherinae* (African blood lily; Amaryllidaceae), as the chromosomes begin to condense, most of the plastids, mitochondria, and lipid bodies appear to concentrate toward the pore side of the microspore, away from the nucleus (Sanger and Jackson 1971a). A similar one-way traffic of plastids away from the immediate vicinity of the nucleus has been described in the microspores of *Hyoscyamus niger* (Reynolds 1984b), *Impatiens glandulifera, I. walleriana* (van Went 1984), and *Gasteria verrucosa* (Schröder 1985b) on the verge of the first haploid mitosis. This ensures that following division, the vegetative cell is cut off with most of the plastids, and sometimes with most of the mitochondria. In other plants studied, such as *Hippeastrum belladonna, Hymenocallis occidentalis* (Amaryllidacae), *Parkinsonia aculeata* (Fabaceae), *Ranunculus macranthus* (Ranunculaceae) (Larson 1963, 1965), *Tradescantia paludosa* (Maruyama, Gay and Kaufmann 1965), orchids (Chardard 1958, 1962; J. Heslop-Harrison 1968d; Cocucci and Jensen 1969b; Brown and Lemmon 1991c, d, 1992b), *Beta vulgaris* (beet root; Chenopodiaceae) (Hoefert 1969a), *Helleborus foetidus* (Echlin 1972), *Tillandsia caput-medusae* (Bromeliaceae) (Brighigna, Fiordi and Palandri 1981), *Chlorophytum comosum* (Liliaceae) (Schröder 1986b), *Aloe secundiflora, A. jucunda* (Liliaceae) (Schröder and Hagemann 1986), and *Tulbaghia violacea* (Schröder and Oldenburg 1990), a distinct polarization of the cytoplasmic organelles occurs during microspore division with most of the plastids, mitochondria, and lipid bodies being displaced toward the side of the pollen grain that becomes the vegetative cell, whereas the generative cell inherits some mitochondria, a normal ribosome population, and a few profiles of ER, but very few or no plastids at all (Figure 3.7). A variable number of plastids are, however, reported to be present in the generative cell or sperm cells of *Lilium candidum* (Bopp-Hassenkamp 1960), *L. martagon* (Schröder 1984), *Oenothera hookeri* (Onagraceae) (Diers 1963), *Pelargonium zonale* (Geraniaceae) (Lombardo and Gerola 1968a), *P. hortorum* (Guo and Hu 1995), *Plumbago zeylanica* (Plumbaginaceae) (Russell 1984), *Prunus avium* (sweet cherry; Rosaceae) (Pacini, Bellani and Lozzi 1986), *Pisum sativum* (Hause 1986), *Rhododendron* sp. (Ericaceae) (Taylor et al 1989), *Medicago sativa* (alfalfa; Fabaceae) (Zhu, Mogensen and Smith 1990), and *Nicotiana tabacum* (Yu and Russell 1994a). The last-mentioned example is illustrated in Figure 3.8. A rather unusual situation exists in *Solanum chacoense, S. tuberosum* × *S. chacoense* (Clauhs and Grun 1977), *Hosta* sp. (Liliaceae) (Vaughn et al 1980), *Fritillaria imperialis* (Liliaceae) (Schröder 1985a), *Convallaria majalis* (Liliaceae) (Schröder 1986a), *Epilobium* sp. (Onagraceae) (Schmitz and Kowallik 1987), *Triticum aestivum, Lilium longiflorum* (Miyamura, Kuroiwa and Nagata 1987), and *Lolium perenne* (Pacini et al 1992), in which the generative cell is cut off with one to many plastids. However, during the maturation of this cell, a mechanism resulting in the degeneration of the plastids seems to be operating. In *C. majalis*, signs of the degeneration of plastids are the disappearance of the starch grain and its replacement by electron-dense bodies and myelinlike structures (Schröder 1986a). In *T. aestivum*, there is a decrease in the number of plastid DNA aggregates (plastid nucleoids) from the generative cell and a loss of DNA (Miyamura, Kuroiwa and Nagata 1987). According to Pacini et al (1992), prior to their dis-

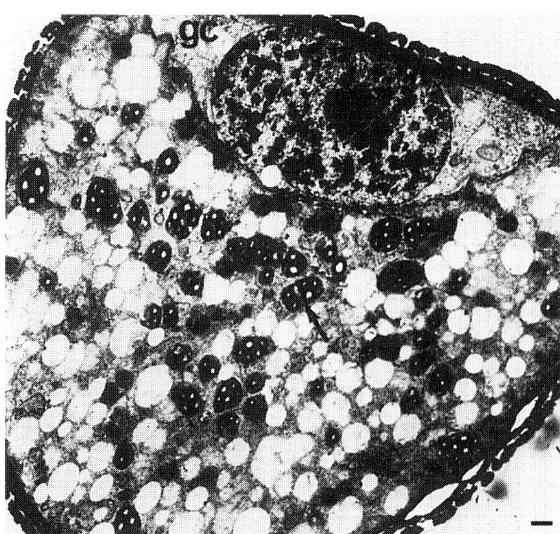

Figure 3.7. The first haploid mitosis in the microspore of *Tulbaghia violacea*. The generative cell (gc) is attached to the wall of the vegetative cell. Plastids (arrow) are present in the vegetative cell cytoplasm, but the generative cell does not contain any plastids. Scale bar = 1 μm. (From Schröder and Oldenburg 1990.)

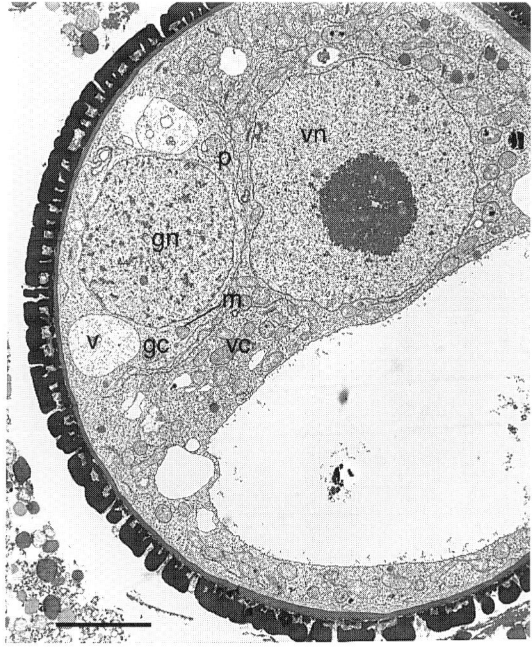

Figure 3.8. Section of part of a pollen grain of *Nicotiana tabacum* showing the generative cell. gc, generative cell; gn, nucleus of the generative cell; m, mitochondrion; p, plastid; v, vacuole; vc, vegetative cell; vn, nucleus of the vegetative cell. Scale bar = 0.5 μm. (Photograph supplied by Dr. S. D. Russell.)

integration, plastids of the generative cell of *L. perenne* are enveloped by small vacuoles that are autophagic in function. Analysis of nucleolytic activities of certain plants that inherit plastids maternally has revealed the presence of high nuclease-C (a Ca^{2+}-dependent nuclease) activity, suggesting its involvement in the digestion of chloroplast DNA derived from the generative cell, but the enzyme activity is greatly reduced or hardly discernible in plants showing biparental plastid inheritance (Nakamura et al 1992; Sodmergen et al 1992; Li and Sodmergen 1995). A generally high concentration of mitochondria and plastids in the vegetative cells, as well as their lack or their scanty occurrence in the generative cells, has been reported in *Petunia* sp. (Sassen 1964), *Lobelia erinus* (Campanulaceae) (Dexheimer 1965), cotton (Jensen, Fisher and Ashton 1968), *Mirabilis jalapa* (Nyctaginaceae) (Lombardo and Gerola 1968a), certain species of *Castilleia, Cordylanthus, Orthocarpus, Ophiocephalus* (all Scrophulariaceae) (Jensen, Ashton and Heckard 1974), *Endymion non-scriptus* (Campanulaceae), *Scilla peruviana* (Liliaceae), *Allium cepa* (Rodríguez-García and García 1978), *Nicotiana alata* (Cresti et al 1985), *Euphorbia dulcis* (Euphorbiaceae) (Murgia et al 1986), and *Brassica napus* (Murgia et al 1991b), but it is not established whether these organelles are left out of the generative cell during the asymmetric mitosis or whether the organelles included in the generative cell are subsequently lost from it.

Genetic evidence for the pattern of organelle transmission has established that a majority of the angiosperms are characterized by maternal inheritance of plastids (Kirk and Tilney-Bassett 1978). In most cases studied, ultrastructural evidence has corroborated genetic data for the inheritance of plastids. However, because of the difficulty of distinguishing undifferentiated proplastids from mitochondria in the electron microscope and the need to establish unequivocally the presence of DNA in the putative plastids, an alternative diagnostic method to detect plastid DNA in the generative cell or sperm cell has been developed (Corriveau and Coleman 1988). The method, based on the use of a DNA fluorochrome, 4'6-diamidino-2-phenylindole (DAPI), in conjunction with fluorescence microscopy, has proved to be relatively rapid and simple. In a survey of plastid inheritance using this method in 235 species representing 80 families, 81% of the families were found to follow uniparental (from the egg only) plastid DNA transmission mode; 18% were potentially capable of biparental (from both egg and sperm) transmission. As a caveat to the uncritical use of this technique in determining male plastid transmission, in some cases, a majority of plastids of the generative cell of a plant having genetically established biparental inheritance have been shown to lack DNA detectable by fluorescence (Shi et al 1991). However, it is indisputable that in all plants, maternal plastids are expressed in the zygote and in the sporophyte derived from it; the question is with regard to the extent of involvement in the next generation of genetic information from the plastids of the male germ line. Based on electron microscopic evidence, Hagemann and Schröder (1989) have recognized four modal patterns of plastid inheritance in angiosperms, in which different cytological and physiological mechanisms operate. These are (a) *Lycopersicon* type, in which the newly formed generative cell does not contain any plastids; (b) *Solanum* type, in which the few plastids present in the generative cell disappear during maturation of the cell; (c) *Triticum* type, in which the plastids present in the generative cell are eliminated from the sperm cell during fertilization; and (d) *Pelargonium* type, in which plastids contained in the generative cell are transmitted to the zygote. The distinctiveness of the *Lycopersicon, Solanum,* and *Triticum* types is the different mechanisms resulting in uniparental plastid inheritance; that of the *Pelargonium* type lies in the unambiguity by which biparental transmission occurs. The fact of plastid transmission from the sperm in *P. zonale* is also emphasized by the finding that the plastids cut off in the generative cell enter a round of dedifferentiation as proplastids, whereas those confined to the vegetative cell are doomed to store starch as amyloplasts (Sodmergen et al 1994).

3. VEGETATIVE AND GENERATIVE CELLS

There are several investigations on the ultrastructure and cytochemistry of the vegetative and generative cells as components of the pollen grain and of the pollen tube. A survey of these papers (Sassen 1964; Kroh 1967a; Jensen, Fisher and Ashton 1968; Crang and Miles 1969; Hoefert 1969a; Mepham and Lane 1970; Sanger and Jackson 1971b, c; Echlin 1972; Jensen, Ashton and Heckard 1974; Kozar 1974; Pacini and Cresti 1976; Hu et al 1979; Brighigna, Fiordi and Palandri 1981; Clarke and Steer 1983; Cresti, Ciampolini and Kapil 1983; Cresti et al 1985, 1988, 1990; Dickinson and Elleman 1985; Cresti, Lancelle and Hepler 1987; Charzyńska, Murgia and Cresti 1989; Mizelle et al 1989; van Went and Gori 1989; Noguchi and Ueda 1990; Tarasenko and Bannikova 1991; van Aelst and van Went 1991; Lancelle and Hepler 1992; Caiola, Banas and Canini 1993; Polowick and Sawhney 1993b, c; van Aelst et al 1993; Yu and Russell 1994a) shows that organelles such as ER, dictyosomes, ribosomes, mitochondria, lipid bodies, and vacuoles are distributed in different configurations in the two pollen cells of the different species. However, as discussed earlier, plastid distribution is asymmetric, with most of the plastids being confined to the vegetative cell and none, or very few, ending up in the generative cell. Although occasional references have been made to the presence of microbodies in the pollen grain, only lately have they been examined in detail from a functional point of view. This is reflected in their association with lipid bodies, as observed in the pollen grains of *Ophrys lutea* (Orchidaceae) (Pais and Feijo 1987) and *Brassica napus* (Charzyńska, Murgia and Cresti 1989). In *O. lutea*, there is a concurrent proliferation of microbodies and a decrease in lipid bodies; this probably suggests an involvement of microbodies in the mobilization of lipids. The decrease in the number of lipid bodies in the pollen grains of *B. napus* is paralleled by an increase in the population of polysaccharide-containing vesicles. Accordingly, it appears that the conversion of lipids to polysaccharides is mediated by enzymes conserved in the microbodies.

(i) The Vegetative Cell

After it is cut off, the vegetative cell goes through a period growth, characterized by the synthesis of new cytoplasm and an increase in the number of organelles and in the elaboration of the structure of individual organelles. In the vegetative cell of the pollen grain of *Tillandsia caput-medusae,* signs of accelerated biosynthetic activity are proliferation of mitochondria and modifications in their ultrastructure, increase in the amount of rough ER, clustering of Golgi bodies and a simultaneous reduction of starch reserves, increased ribosomal content, increase in the exchange surface between the nucleus and cytoplasm, and consolidation of vacuolate lipid droplets into osmiophilic bodies (Brighigna, Fiordi and Palandri 1981). A unique feature of the structure of the vegetative cell of the pollen grain of *Gossypium hirsutum* is a distended form of ER that is frequently folded to form pockets containing lipid bodies and small vesicles, presumably derived from dictyosomes. In this configuration, the ER foldings function as remark-

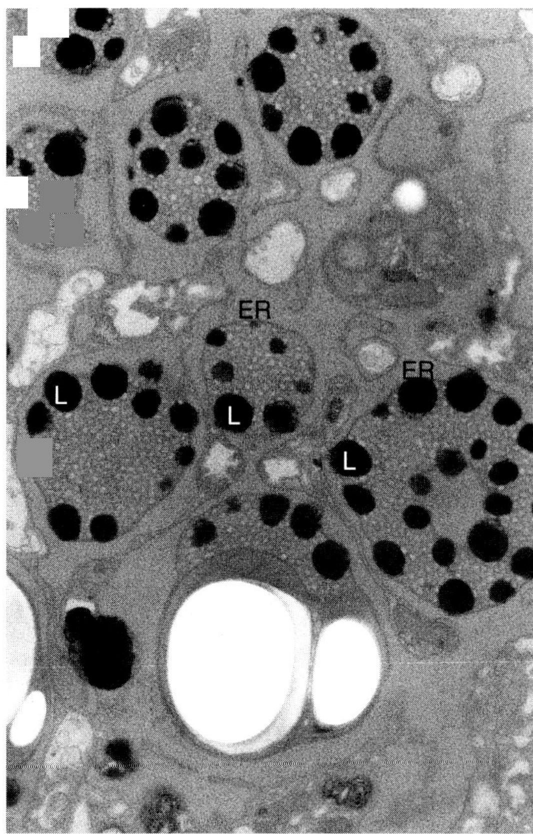

Figure 3.9. A view of the pollen cytoplasm of *Gossypium hirsutum* showing the ER pockets containing lipids (L) and small vesicles. Scale bar = 0.5 μm. (From Fisher, Jensen and Ashton 1968. *Histochemie* 13:169–182. © Springer-Verlag, Berlin. Photograph supplied by Dr. W. A. Jensen.)

able storage enclaves, and as they unfold during germination, the contents mix with the rest of the cytoplasm (Figure 3.9). In the mature pollen grain, the central core region filled with ER pockets is surrounded by a peripheral cytoplasm devoid of these pockets (Fisher, Jensen and Ashton 1968; Jensen, Fisher and Ashton 1968; Wetzel and Jensen 1992). According to Jensen, Ashton and Heckard (1974), vegetative cells of pollen grains of certain species of *Castilleia, Cordylanthus, Orthocarpus*, and *Ophiocephalus* contain pronounced stacks of ER, with each genus displaying a characteristic stack dimension and intracisternal width. In *Apium nodiflorum*, the rough ER of the vegetative cell shows a progressive transformation from a highly dilated state to a vesicular form, and it ends up in stacks or in a narrow cisternal configuration in the mature pollen grain (Weber 1989). Immunocytochemical studies have shown that in several members of Oleaceae, allergenic proteins are synthesized and stored in the rough ER of the vegetative cell, prior to their transmittal to the pollen walls (Rodríguez-García, Fernández and Alché 1995; Rodríguez-García et al 1995).

An important component of the maturation process of the vegetative cell of the pollen grain of *Parietaria officinalis* (Urticaceae) is the development of a series of wall projections from the intine near the pore region. These outgrowths, which are akin to transfer cells in their morphology, probably facilitate the traffic of sporophytic metabolites from the anther locule during pollen maturity and of cell wall proteins from the pollen grain during germination (Franchi and Pacini 1980). Strip-shaped projections, attached to the cytoplasmic face of the inner plasma membrane of the vegetative cell bordering the generative cell, described in the pollen grains of *Amaryllis belladonna* (Cresti, Ciampolini and van Went 1991; Southworth, Salvatici and Cresti 1994), *Capparis spinosa* (Capparidaceae) (van Went and Gori 1989), and *Convallaria majalis* (Bohdanowicz, Ciampolini and Cresti 1995) constitute yet another structural modification of the vegetative cell. Although the function of these cytoplasmic structures has not been defined, it appears that they may be related to the movement of the generative cell in the pollen tube. This view is based on the fact that the inner membrane of the vegetative cell is in reality the outer membrane of the generative cell and, at least in *C. majalis*, the projections have been shown to contain myosinlike proteins. Since not many studies have focused on the cytoskeletal elements of the vegetative cell, a composite impression obtained from mature pollen grains and germinating pollen grains is that both microtubules and microfilaments are distributed in perinuclear locations in the vegetative cell cytoplasm (Tiwari 1989). The comparatively well developed state of the organelles in the vegetative cell, their modifications, and the presence of wall ingrowths convey the tang and flavor of a cell that is programmed to resume immediate metabolic activity.

The nucleus of the vegetative cell generally maintains an irregular shape with dispersed chromatin and appears to maraud the cytoplasm. The plasma membrane of the vegetative cell of pollen grains of several plants has been shown, through the use of monoclonal antibody, to express an arabinogalactan protein epitope, although there is little evidence for its participation in any cell-inductive processes (Pennell et al 1989; Pennell and Roberts 1990; van Aelst and van Went 1992).

(ii) The Generative Cell

Except for the plastids, the cytoplasm of the generative cell contains the same organelles found in the vegetative cell, although some variability is seen in the density of organelle distribution in pollen grains of different plants. With the exception of microtubules, there are no changes in the number or appearance of cytoplasmic organelles during maturation of the generative cell. The nucleus of the generative cell, compared to that of the vegetative cell, is highly condensed; a small nucleolus and a membrane with pores complete the general description of the nucleus (Jensen, Fisher and Ashton 1968; Burgess 1970b; Jensen, Ashton and Heckard 1974).

A great deal of interest was generated by the report of the presence of a confining callose wall around newly formed generative cells of pollen grains of *Chlorophytum comosum, C. elatum, Hyacinthus orientalis* (Liliaceae), *Tradescantia bracteata,* and *Impatiens balsamina* (Górska-Brylass 1967a, b). By resorcinol blue staining and fluorescence reaction with aniline blue, this work showed that callose first appears in the region of the wall between the two daughter nuclei of the dividing pollen grain and then progresses around the generative cell, completely enveloping it. The transitory nature of this wall layer is underscored by the fact that it disappears with the acquisition of a spindle shape by the generative cell. Later studies have shown that callose alone, or along with cellulose, is a component of the early formed wall around the generative cell of pollen grains of certain other plants (Angold 1968; J. Heslop-Harrison 1968d; Górska-Brylass 1970; Dunbar 1973; J. Heslop-Harrison and Heslop-Harrison 1988b; Keijzer and Willemse 1988; Zee and Siu 1990; Schlag and Hesse 1992). Whether there is a morphogenetic significance for the presence of callose or whether this is a wasteful process remains to be determined.

When newly formed, the generative cell is appressed to the pollen wall, but it gradually detaches therefrom by an inward extension of the wall at the point of contact with the intine. As the generative cell moves inward, it becomes entirely engulfed by the vegetative cell. Ablation of the vegetative cell of tobacco pollen by the expression of a cytotoxic gene (*DT-A*) in transgenic plants is found to be coupled with the failure of the generative cell to migrate from the pollen wall. A plurality of mechanisms must be envisioned for generative cell migration, including the possibility that it is fostered by metabolites produced by the vegetative cell (Twell 1995). Occasionally, plasmodesmata have been observed connecting the generative cell to the vegetative cell (Brighigna, Fiordi and Palandri 1981; Nakamura and Miki-Hirosige 1985; Murgia et al 1986; Lancelle and Hepler 1992). In *Tradescantia reflexa* (Noguchi and Ueda 1990) and *Cyrtandra pendula* (Gesneriaceae) (Luegmayr 1993), cisternae of the ER of the vegetative cell are in close contact with the generative cell throughout its development and contribute to the surface undulations of the cell. It is a curious fact that although a cell wall is a prominent feature of the early ontogeny of the generative cell, with further development, the wall assumes the contour of two membranes (Angold 1968; J. Heslop-Harrison 1968d; Mepham and Lane 1970; Sanger and Jackson 1971a; Echlin 1972). Through improved staining and fixation techniques, the paired plasma membranes of the generative cells of *Rhododendron laetum* and *R. macgregoriae* have been shown to enclose matrix components suggestive of the presence of wall materials (Kaul et al 1987).

Two significant departures from the classical model of the generative cell have been described. One is the presence of wall ingrowths; these appear from the intine that links the generative cell to the vegetative cell in *Acacia retinodes* (Fabaceae) (McCoy and Knox 1988). In *Pyrus communis* (pear; Rosaceae), the wall processes arise from all faces of the generative cell and protrude into the cytoplasm of the vegetative cell (Tiwari 1994). The other modification is found in *R. laetum,* in which the ends of the generative cell narrow into long evaginations that coil around and embrace the cell. One of the tails is connected, like a life-support system, to wall ingrowths from the intine (Theunis, McConchie and Knox 1985). Because of the transfer cell–like nature of the wall labyrinths, we can assume that, in both cases, transport of metabolites from other parts of the pollen grain is necessary for development of the generative cell.

After the generative cell is cut off, its further differentiation marks the beginning of a developmental process that produces the male gametes. A critical aspect of this differentiation is the change in the shape of the cell from a spherical to a spindle shape, accompanied by a corresponding change in the shape of the nucleus. Change in shape, attributed to microtubules, helps to streamline the cell and facilitate its passage through the narrow confines of the pollen tube. In electron microscopic studies, microtubules appear to float randomly throughout the cytoplasm of the spherical generative cell, but they take up positions in longitudinal parallel bundles in the vicinity of the plasma membrane in the elongated stage (Burgess 1970a, b;

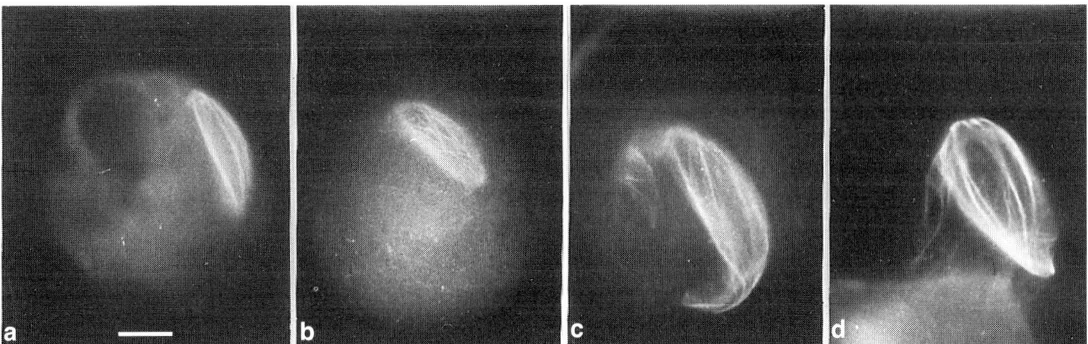

Figure 3.10. Immunofluorescent localization of microtubules in the generative cell of pollen grains of *Zephyranthes grandiflora.* **(a, b)** Views of pollen protoplasts after removal of the exine by enzyme treatments, showing spindle-shaped generative cells with longitudinal (a) and cross orientation (b) of microtubule bundles. **(c, d)** Longitudinally oriented microtubule bundles in spindle-shaped generative cells. Scale bar = 10 μm. (From Zhou, Zee and Yang 1990. *Sex. Plant Reprod.* 3:213–218. © Springer-Verlag, Berlin. Photographs supplied by Dr C. Zhou.)

Sanger and Jackson 1971b; Jensen, Ashton and Heckard 1974; Brighigna, Flordi and Palandri 1981; Cresti, Ciampolini and Kapil 1984; Murgia et al 1986; Del Casino et al 1992). Parallel bundles of microtubules seem to arise as a result of cross-bridge interactions (Burgess 1970a, b). When the elongate generative cell of *Haemanthus katherinae* pollen is treated with colchicine or isopropyl-N-phenylcarbamate, microtubules are found to perish, with a concomitant reversion of the cell to a spherical shape (Sanger and Jackson 1971b). Since these inhibitors are known to interfere with the polymerization of microtubule precursors or the patterned assembly of the finished products, the best interpretation of the results is that microtubules in the configuration observed in the elongate state of the generative cell are primarily structural, assisting in maintaining the asymmetric shape.

The microtubular network of the generative cell has also been characterized by antitubulin immunofluorescence staining. Although immunofluorescence microscopy has been employed mostly to investigate generative cells within pollen tubes, in some studies the arrangement of microtubules in generative cells inside mature pollen grains, or in cells isolated from pollen grains, has also been followed. In pollen grains of *Zephyranthes grandiflora* (Amaryllidaceae), the cytoskeleton of the spindle-shaped generative cell consists mainly of longitudinally oriented microtubule bundles (Figure 3.10). The generative cell changes its shape upon isolation from the pollen grain by osmotic shock or by grinding, and this shape change is found to depend upon the orientation of the bundles of microtubules. Transition of the spindle-shaped cell to ellipsoidal form is associated with a mixture of microtubule bundles and meshes, whereas in the spherical cells, the mesh structure appears to predominate (Zhou, Zee and Yang 1990). A reverse-order transition from mesh structure to axially oriented bundles through an intermediate form is observed in the three-dimensional organization of microtubules in developing generative cells of pollen grains of *Hippeastrum vittatum* (Zhou and Yang 1991). A fine reticulate network of microtubules in the cytoplasm also plays a key role in maintaining the rounded shape of isolated generative cells of *Allamanda neriifolia* (Apocynaceae) pollen (Zee and Aziz-Un-Nisa 1991). However, in other plants it has been shown that the loss of the spindle shape of isolated generative cells and their reversion to spherical shape could be due to a complete elimination of the microtubule cytoskeleton, rather than due to its rearrangement (Theunis and van Went 1989; Theunis, Pierson and Cresti 1992).

The microtubular cytoskeleton of the generative cell in the pollen tube is organized more or less like that described in isolated generative cells. As examined both by electron microscopy and by immunofluorescence microscopy, the cytoskeleton in the spindle-shaped cell consists of basketlike arrays of microtubules aligned helically or longitudinally relative to the long axis of the cell. The microtubules occur singly or as aggregates in bundles of up to 35 members (Derksen, Pierson and Traas 1985; Pierson, Derksen and Traas 1986; Raudaskoski et al 1987; Zhou 1987b; J. Heslop-Harrison et al 1988; Palevitz and Cresti 1989; Taylor et al 1989; Noguchi and Ueda 1990; Xu, Zhu and Hu 1990; Yu and Russell 1993). Disagreement has arisen over the presence of microfilaments in the generative cell. Although Taylor et al (1989) were able to observe antiactin immunofluorescence patterns indicative of axially oriented microfilaments in the generative cell of germinated pollen grains of *Rhododendron laetum*, a subsequent work by Palevitz and Liu (1992) on the generative cells in pollen tubes of *Tradescantia virginiana, Nicotiana tabacum*, and *R. laetum* using three antiactins, rhodamine phalloidin, and antimyosin has failed to reveal a microfilament-based cytoskeletal system. This observation relegates the generative cell to the status of perhaps the only plant cell so far examined that is devoid of microfilaments.

(iii) Biochemical Cytology of Vegetative and Generative Cells

Cytologists have been fascinated by the vegetative and generative cells ever since the widespread occurrence of the bicellular condition in angiosperm pollen grains was demonstrated in the late 19th century. Clearly, the asymmetric division destabilizes the genetic and epigenetic status of the microspore and induces new developmental potencies that enable the daughter cells to follow divergent pathways of differentiation. What causes deviation in the developmental pathways of daughter cells born out of a mitotic division of the microspore is a difficult question to ponder. It has been suggested that the presence of a callose wall separating the two cells is instrumental in determining their fates (Górska-Brylass 1967b). Support for this view comes from the observation that elimination of the callose wall in pollen grains of *Tradescantia bracteata* by caffeine treatment results in the formation of binucleate pollen grains (Charzyńska and Pannenko 1976). This occurs irrespective of whether the nucleus was in the center of the microspore or was poised at the proximal end in anticipation of the asymmetric division. According to Nitsch (1974a), administration of a cold shock to anthers of *Datura innoxia* (Solanaceae) leads to an increase in the number of pollen grains with two identical cells. Apparently, the cold shock immobilizes the nucleus in its original position in the center of the microspore, thus forcing a symmetrical division when the thermal stress is removed. This indicates that the fate of the daughter cells in the angiosperm pollen grain is determined largely by the placement of the nucleus and the orientation of the mitotic spindle.

It has been known for some time that pollen differentiation is accompanied by qualitative and quantitative changes in the nucleic acid and protein metabolism of the vegetative and generative cells. Here, the changes in DNA metabolism of the pollen nuclei will be considered; details of the changes in RNA and proteins are presented in the next chapter. After the first haploid mitosis in microspores of *T. paludosa*, the nucleus of the vegetative cell enlarges, although its DNA content remains unchanged; on the other hand, the nucleus of the generative cell undergoes DNA synthesis to attain a 2*C* DNA level (Woodard 1956, 1958; Takats 1965). Cytophotometry, flow cytometry, and autoradiography of incorporation of labeled precursors have also failed to reveal DNA synthesis in the nucleus of the vegetative cell of pollen grains of other plants (Taylor 1958; Charzyńska and Maleszka 1978; Thiébaud and Ruch 1978; Górska-Brylass, Bednarska and Świerzowicz 1986; Bino, van Tuyl and de Vries 1990). Some cytophotometric and autoradiographic investigations have revealed variability in the extent of DNA replication in the nucleus of the vegetative cell. The amount of DNA is known to be completely doubled in the vegetative cell nucleus of pollen grains of tobacco, barley (*Hordeum vulgare*; Poaceae) (D'Amato, Devreux and Scarascia Mugnozza 1965), and *Allium chamaemoly* (Corsi and Renzoni 1972), whereas in others, including *T. bracteata, Hyacinthus orientalis, D. innoxia*, and *Muscari racemosum* (Liliaceae), the vegetative cell nucleus shows some increase in DNA content or synthesizes some DNA (Moses and Taylor 1955; Rodkiewicz 1960; Sangwan-Norreel 1979; Bednarska 1981; Górska-Brylass, Bednarska and Świerzowicz 1986). Even in the pollen grains of *Tradescantia* sp., the potential for DNA synthesis appears to be present in the nucleus of the vegetative cell because it shows a slightly higher level of DNA polymerase activity than does the generative cell nucleus (Takats and Wever 1971). Thus, whatever may be the true function of the vegetative cell, it does not appear to involve processes that require extensive genome replication. As will be seen in Chapter 4, a contrasting situation is seen with regard to the RNA and protein synthetic activities of the vegetative and generative cells.

Attention has been called to the fact that there are some fundamental differences between the structure of the nuclei of the generative and vegetative cells of the pollen of *T. paludosa* (LaFountain and LaFountain 1973) and *Phoenix dactylifera* (date palm; Arecaceae) (DeMason and Chandra Sekhar 1988; Southworth et al 1989). The generative cell nucleus is spherical and contains more heterochromatin than the nucleus of the vegetative cell, which is elongate. The differences also extend to the nuclear envelopes, with the generative cell having fewer nuclear pores than the vegetative cell. In the mature pollen grains of *Papaver dubium* (Papaveraceae) (van Aelst et al 1989), *Nicotiana tabacum* (Wagner et al 1990), and *Medicago sativa* (Shi, Mogensen and Zhu 1991) in which the nucleus of the vegetative cell forms a close physical association with the generative cell, pore density on the former is considerably higher than on the generative cell nucleus. Undoubtedly, these differences foreshadow the divergent fates and functions of these cells.

(iv) Division of the Generative Cell

With the organization of the generative cell in mind, the discussion now returns to the events of spermiogenesis, or the division of the generative cell. In species that shed their pollen at the bicellular stage, this division takes place in the pollen tube. This is indeed the common situation encountered in the vast majority of angiosperms; in the minority of species that shed their pollen at the trinucleate stage, the generative cell divides in the pollen grain (Figure 3.11). Regardless of whether the division occurs in the pollen grain or in the pollen tube, a preprophase band of microtubules does not form prior to division.

In the first ultrastructural characterizations of the division of the generative cell in the three-celled pollen grains of barley and rye, it was shown that cytokinesis is initiated and progresses to a moderately advanced stage while the cell is still attached to the pollen wall. Although a conventional cell plate is not formed after division of the generative cell in barley pollen grains, an incipient wall appears between the two sperm cells. After the generative cell is detached from the pollen wall, final separation of the sperm cells is thought to result by dissolution of the partitioning and surrounding wall (Cass and Karas 1975; Karas and Cass 1976). Other studies on barley (Charzyńska, Ciampolini and Cresti 1988), *Galium mollugo* (Rubiaceae), *Trichodiadema setuliferum* (Aizoaceae), *Avena sativa* (Weber 1988), *Brassica napus* (Charzyńska et al 1989; Murgia et al 1991b), and *Sambucus nigra* (Caprifoliaceae) (Charzyńska and Lewandowska 1990) have assertively shown that the generative cell is detached from the pollen wall before it divides and that formation of a regular cell plate by the fusion of vesicles associated with microtubules is a common feature during the division of the generative cell. In many plants, after the division of the generative cell the sperm cells do not separate but remain associated with each other and with the vegetative cell nucleus in a complex way to form the male germ unit (Chapter 11).

Interest in the division of the generative cell in the pollen tube has centered around the establish-

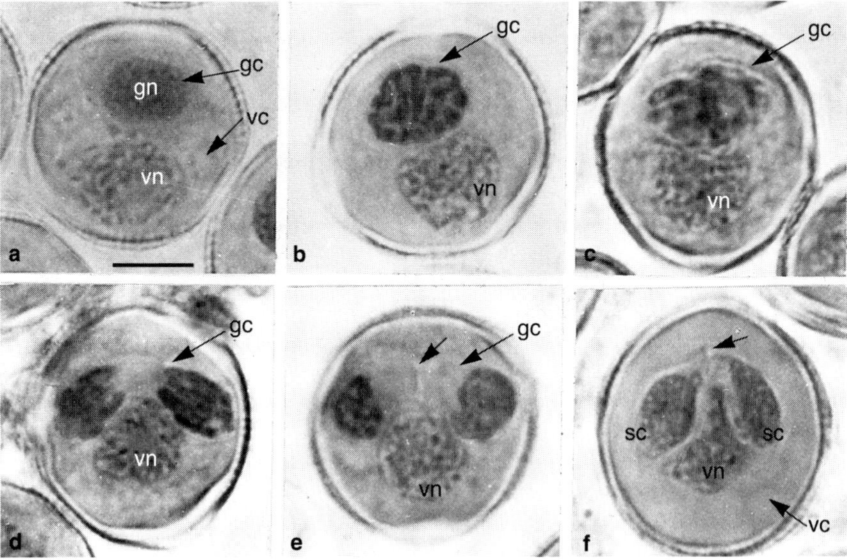

Figure 3.11. Division of the generative cell and formation of sperm cells in the pollen grain of *Sambucus nigra.* **(a)** Generative cell before division. **(b)** Prophase. **(c)** Metaphase. **(d)** Anaphase; the generative cell is elongated and curves around the vegetative cell. **(e)** Late anaphase, showing cell plate formation (arrows). **(f)** Tricellular pollen grain; the two sperm cells appear to be connected (arrow). gc, generative cell; gn, generative cell nucleus; sc, sperm cell; vc, vegetative cell; vn, vegetative cell nucleus. Scale bar = 10 μm. (From M. Charzyńska and E. Lewandowska 1990. Generative cell division and sperm cell association in the pollen grain of *Sambucus nigra. Ann. Bot.* 65:685–689, by permission of Academic Press Ltd., London. Photographs supplied by Dr. M. Charzyńska.)

ment of the mitotic spindle and metaphase plate and the mode of cytokinesis. Because of the spatial restraints within the narrow pollen tube, a clear picture of these events has been hard to come by and some of the published light microscopic observations of division in the same species have been contradictory. An account of these conflicting views is given by Palevitz and Tiezzi (1992). Modern cytologists have taken up the task of describing the steps in generative cell division in the pollen tube that are different from conventional mitosis and have provided information that might form the foundation of future studies on spermiogenesis in angiosperms. As one might expect, the use of electron microscopy and immunocytochemistry for the analysis of cell division in plants has reached its most sophisticated level in studies of the division of the generative cell in the pollen tube.

An ultrastructural account of the division of the generative cell in the pollen tube of *Endymion non-scriptus* characterized the process as a normal type of mitosis. Illustrating the dynamic nature of the division is a decline in the number of cytoplasmic microtubules present at the early prophase in favor of mitotic spindle fibers (Burgess 1970b). Microtubules are also programmed to disappear from the cytoplasm of generative cells of *Nicotiana tabacum* (Raudaskoski et al 1987; Yu and Russell 1993), *Allamanda neriifolia* (Zee 1992), *Ornithogalum virens* (Charzyńska and Cresti 1993), and *Zantedeschia aethiopica* (Araceae) (Xu, Zhu and Hu 1993) as the mitotic spindle is fabricated. On the other hand, a wholesale breakdown of the cortical microtubular network does not signal the formation of the mitotic spindle in dividing generative cells of *Tradescantia virginiana* (Palevitz and Cresti 1989), *N. tabacum* (Bartalesi et al 1991; Palevitz 1993), *Hyacinthus orientalis* (Del Casino et al 1992), *Gagea lutea* (Liliaceae) (H.-Q. Zhang et al 1995), and *O. virens* (Banaś et al 1996).

It has been known for a number of years that the traditional, radially oriented mitotic spindle of a typical somatic cell gives way in the pollen tube to a spindle oriented obliquely. Associated changes include the scattering and the generally frantic movements of the chromosomes, resulting in atypical alignment on the mitotic spindle. These observations have been confirmed repeatedly in more recent years (Lewandowska and Charzyńska 1977; Palevitz and Cresti 1989; Taylor et al 1989; Terasaka and Niitsu 1989; Palevitz 1990; Yu and Russell 1993) and only rarely has the occurrence of conventional spindles and metaphase plates been reported.

The origin of the kinetochore fibers that connect the condensed chromosomes to the poles of the mitotic spindle during generative cell mitosis has been a contentious issue. One view is that microtubules in and around the generative cell disassemble at the beginning of division and that the spindle fibers are subsequently synthesized from this pool of tubulin subunits (Raudaskoski et al 1987). Alternatively, it has been argued that kinetochore fibers arise from, and remain connected to, the interphase array of microtubules. Appropriately stained preparations of the generative cell in division in the pollen tube vividly show kinetochores strung out at intervals along the length of the pollen tube with connecting fibers linking successive kinetochores to each other and to the surrounding interphase cytoskeleton (Palevitz and Cresti 1989; Palevitz 1990; Liu and Palevitz 1991). During nuclear division in the generative cell of tobacco, kinetochore fibers display an unusual intimacy with ER lamellae, and it has been suggested that parts of the dispersed mitotic spindle are anchored by this means (Yu and Russell 1993).

A satisfactory framework for the orderly chromosome movement toward opposite poles of the atypical spindle at anaphase is yet to be formulated. Generative cells that divide within the confines of very narrow pollen tubes and produce extremely abnormal mitotic plates, as in *Tradescantia virginiana*, offer a good system for answering questions about chromosome movement. According to Liu and Palevitz (1992), in the generative cell of this species on the verge of anaphase, there is a reorganization of the axial system of microtubules and kinetochore fibers into two thick bundles (superbundles). As a result, chromatids of each pair are segregated to different superbundles bound to opposite ends of the cell. The shortening of the kinetochore fibers observed at this stage of division of the generative cell provides compelling evidence that the chromatids become directly linked to the superbundles at the kinetochore. This sets the stage for the separation of the superbundles, thus moving the kinetochores to opposite ends of the cell. During the anaphase–telophase interval in the generative cells of *Lilium* sp. (Xu, Zhu and Hu 1990), *Amaryllis vittata* (Zhu and Liu 1990), *Alocasia macrorrhiza* (Araceae) (Xu 1991), *Allamanda neriifolia* (Zee and Aziz-Un-Nisa 1991; Zee 1992), and *Zantedeschia aethiopica* (Xu, Zhu and Hu 1993), a new generation of microtubules is formed between the poles and it is believed that chromatids are moved further to the poles by the elongation of these fibers.

Two different modes of cytokinesis have been described that account for the partition of the cytoplasm between the two daughter nuclei in the final phase of division of the generative cell. In most species investigated, the appearance of a phragmoplast derived from microtubules and disconnected vesicles and the formation of an associated cell plate have been described, reinforcing the idea that the process is not different from a typical somatic cell mitosis (Raudaskoski et al 1987; Taylor et al 1989; Terasaka and Niitsu 1989; Charzyńska and Lewanowska 1990; Zee and Aziz-Un-Nisa 1991; Zee 1992; Palevitz 1993; Xu, Zhu and Hu 1993; Yu and Russell 1993; H.-Q. Zhang et al 1995; Banaś et al 1996). However, no phragmoplast has been observed in the generative cell of *Tradescantia virginiana*, in which cytokinesis is believed to occur in an apparently simple and direct manner by constriction of the microtubules (Palevitz and Cresti 1989).

From this review it can be concluded that, despite certain unusual features, the basic cytology of division of the generative cell in the pollen tube remains essentially close to that of a normal mitotic division. This is precisely what is expected of a cell that is the immediate progenitor of sperm cells that participate in genetic recombination.

(v) Isolation of Pollen Cells and Their Nuclei

To investigate nuclear and cellular differentiation during the ontogeny of the generative and vegetative cells, efforts have been made to isolate these cells, their nuclei, and their protoplasts from bicellular pollen grains and pollen tubes (Wever and Takats 1971; LaFountain and Mascarenhas 1972; Pipkin and Larson 1973; Zhou 1987b, 1988; C. Zhou et al 1986a, 1988; Tanaka 1988; Tanaka, Nakamura and Miki-Hirosige 1989; X.-L. Wu and Zhou 1990, 1991a; Zhou and Wu 1990; Southworth and Morningstar 1992; Ueda and Tanaka 1994). A procedure for the isolation of large quantities of protoplasts (gametoplasts) from generative cells of *Lilium longiflorum* has also been described (Tanaka 1988). Further developments along these lines include induction of divisions in generative cells cultured in vitro (X.-L. Wu and Zhou 1990), fusion between generative cells isolated from various species (X.-L. Wu and Zhou 1991b) and between pollen protoplasts and gametoplasts, and the division of the generative cell in the fusion product (Ueda, Miyamoto and Tanaka 1990). In a few cases, fusions between microspore protoplasts and somatic cell protoplasts seem to offer promising alternative methods for the synthesis of hybrids, bypassing pre- and postzygotic sexual incompatibility barriers (Pirrie and Power 1986; Lee and Power 1988; Pental et al 1988; Desprez et al 1995). The results of these investigations are significant, as they indicate possible ways of manipulating the male gametic cells in future investigations into angiosperm reproductive biology.

4. GENERAL COMMENTS

The view presented in this chapter is that the microspore released from the extracellular matrix of the callose wall of the tetrad proceeds through a series of distinctive changes to achieve its developmental potential as the male gametophyte. Although we have a reasonably clear picture of the assembly of the pollen wall layers, it does not necessarily delve into such questions as how the pattern is established, the nature of the sporopollenin precursors, and the mechanism of their polymerization into the exine template. The evidence currently available also does not give us a clue to the nature of the critical proteins that might be involved in the patterning of the exine. Yet, the contributions of both gametophytic and sporophytic genes to the final architecture of the exine is established and is intriguing.

The study of the pollen grain can be credited with documenting the considerable cytological changes associated with the development of a cell whose division yields two cells with entirely different structural and functional attributes. Although there is a high probability that specific cytological changes herald cell differentiation, such changes are often limited to the elimination of plastids in the generative cell and to their abundance in the vegetative cell during the first haploid pollen mitosis. The failure to identify more specific cellular changes associated with differentiation of the vegetative and generative cells perhaps indicates that the process involves subtle molecular and biochemical changes. This is to some extent borne out by cytochemical analyses of the cells of the pollen grain.

The intent of this chapter has been to emphasize the importance of the coordinated events of male gametogenesis, most importantly, the formation of the vegetative and generative cells and the genesis of the sperm cells. These and other events have been documented with a number of examples. When there are no significant differences between the accounts given for different species, references to additional studies of a similar nature are provided to underscore the continuing interest in the study of the angiosperm male gametophyte. The next chapter will consider the central aspect of this topic – namely, what are the genes that are expressed during male gametogenesis?

Chapter 4

Gene Expression during Pollen Development

The account presented in the previous two chapters has established that the differentiation of the diploid microsporocyte proceeds in several discrete stages to produce a two- or three-celled haploid pollen grain with a complex structural and functional organization. Included in the progression of events in this multistep process are meiotic division and the evolution of cells endowed with different characteristics, form, and function. The existence of a tightly controlled series of cytological and biochemical changes associated with pollen development and maturation underlies a coordinated gene expression pattern, involving genes from both gametophytic and sporophytic generations. Although we do not yet fully understand how the genetic program operates during pollen ontogeny, we have some notion of its machinery and how the machinery might be used. This understanding has come from three independent approaches – genetic, molecular–biochemical, and gene cloning – that have been pursued from time to time during the past 50 years or so. At the genetic level, examination of the segregation of pollen characters and of spontaneous and induced mutations has indicated that a very large set of genes is expressed during male gametogenesis. Molecular–biochemical studies have included analyses of the accumulation and synthesis of nucleic acids and proteins during anther and pollen development. The main thrust of the research in recent years has been on the isolation and characterization of anther- and pollen-specific genes by differential screening of cDNA libraries from whole anthers and pollen grains. The developmental expression of these genes and the sequence analysis of their promoters and coding regions have provided new information about the control of gene expression during male gametogenesis.

What follows in this chapter is a review of the research findings relevant to gene activity during pollen development in angiosperms based on these approaches, coupled with a synthesis of our current understanding in this area. The account will be limited to the more widely investigated systems; however, such a limitation is only slightly restrictive, because most of the significant work has been done with a few model systems.

1. GENETIC BASIS FOR POLLEN GENE EXPRESSION

Part of a review by Ottaviano and Mulcahy (1989) provides an excellent account of the results obtained, and progress made, by a host of investigators in analyzing the genetic control of development of the angiosperm pollen, based on segregation of pollen characters; the following is a distillate from this review. Many studies referred to in this article were conducted during the first half of this century, when classical genetics flourished; others have followed upon them, using modern biochemical tools.

(i) Analysis of Pollen Characters

An instructive comparison of gene actions that pervade pollen ontogeny shows that certain pollen characters are influenced exclusively by the sporophytic genotype, others by both sporophytic and gametophytic genotypes, and still others exclusively by the gametophytic genotype. Though it is of great developmental significance, sporophytic control is exerted only on a relatively small number of pollen characters so far identified. One of the best analyzed is the sporophytic control of pollen exine sculpturing. In particular, the formation of the basic patterned component of the sexine is now believed to be under control of the diploid microsporocyte (Chapter 3).

In many plants, pollen size is determined by either the sporophytic or the gametophytic genome or by the combined action of both. Several instances of pollen dimorphism or polymorphism are known to improve the chances of cross-pollination in heterostylous plants (Chapter 9). Pollen size differences in plants associated with differences in the length of the stamen and style can reasonably be attributed to stable genomic changes prior to microsporocyte meiosis and hence to transcription and translation of the sporophytic genome. Gametophytic determination of the pollen size is illustrated by the gene *SMALL POLLEN* (*SP-1*) in maize, whose obvious effect is to cause a reduction in pollen size. However, gene action is manifest only late in pollen development, because pollen grains harboring the *SP-1* gene appear slightly smaller than the normal grains just before the division of the generative cell (Singleton and Mangelsdorf 1940). The concept of combined gametophytic and sporophytic influences on pollen grain size accompanying inbreeding has received strong support from another study on maize (Johnson, Mulcahy and Galinat 1976). As expected, within a series of progressively inbred lines, F1 through F7, a significant decrease in mean pollen diameter, as well as a decrease in variation in pollen sizes within individual plants, is observed. Certainly, a sporophytic effect is implied in the decrease in pollen mean diameter accompanying selfing. However, since there is an appreciable reduction in pollen size within individual plants, a pervasive effect of the gametophytic genotype is also evident here.

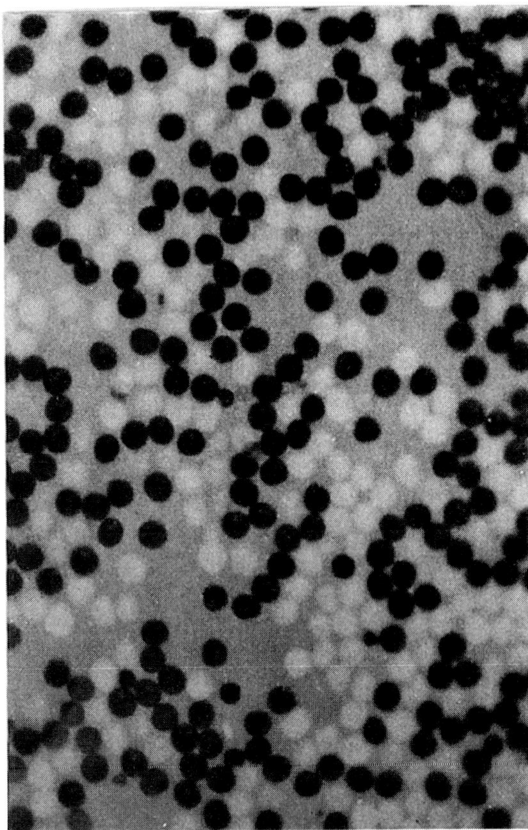

Figure 4.1. Pollen grains of maize segregating for ADH^+ and ADH^- after staining specifically for ADH enzyme activity. The darkly stained pollen grains are ADH^+. (From Freeling 1976.)

Some of the best evidence for the occurrence of postmeiotic gametophytic gene expression has come from the analysis of pollen grains of plants heterozygous for genes encoding specific enzymes or other characters, segregating according to Mendelian expectations. The phenotypic effects are easily revealed by staining the pollen grains, whereas electrophoretic separation of enzymes provides evidence for genotypic effects. The process of allelic segregation of genes for pollen characters during meiosis yields two groups of pollen grains, each possessing only one of the two alleles coding for the character. By examining the anthers of heterozygous plants, it is possible to detect approximately equal numbers of pollen phenotypes with contrasting characters in one and the same anther (Figure 4.1). Based on this rationale, gametophytic gene expression has been detected in pollen grains of endosperm mutants *waxy* (*wx*) (Brink and MacGillivray 1924; Demerec 1924) and *amylose extender* (*ae*) (Moore and Creech 1972) of maize, as well as for ADH (Freeling 1976) and β-galactosidase activities (Singh and Knox 1985) in pollen grains of maize and *Brassica campestris*, respectively. Another line of evidence for gametophytic expression of the *ADH-1* gene has come from a mutational analysis in maize. It has been found that in certain *adh-1* mutations of maize generated by transposons, excision of the transposable element generates new alleles that overexpress or underexpress *ADH-1* gene in the pollen grains (Dawe, Lachmansingh and Freeling 1993).

(ii) Enzyme Electrophoresis

Electrophoretic analysis of dimeric or multimeric enzymes to discriminate between sporophytic and gametophytic gene expression relies on the banding patterns and mobilities of polypeptides of heterozygous plants. Since the haploid pollen grains contain only one allele of each locus, pollen grains of heterozygous plants can produce either subunit, but not both subunits, of an enzyme. Consequently, electrophoresis of pollen grain extracts from heterozygotes will reveal only the two homodimeric enzymes, but no heterodimers. Heterodimers are seen in extracts of somatic tissues, and thus comparison of specific enzyme profiles of pollen grains and somatic tissues makes it possible to determine whether enzymes in the pollen grains are synthesized before or after meiosis. Use of this method to generate isozyme profiles for sporophytic and gametophytic tissues of *Clarkia dudleyana* (Onagraceae) (Weeden and Gottlieb 1979), *Lycopersicon esculentum* (Tanksley, Zamir and Rick 1981), *Zea mays* (Frova et al 1983; Sari Gorla et al 1986; Acevedo and Scandalios 1990; Frova 1990), *Populus deltoides, P. nigra, P. maximowiczii* (Rajora and Zsuffa 1986), and *Hordeum vulgare* (Pedersen, Simonsen and Loeschcke 1987) has established that genes for a variety of enzymes operating in the pollen are gametophytically controlled. As shown in Table 4.1, genes for seven of the dimeric isozymes of acid phosphatase, glutamic-oxaloacetic transaminase, phosphoglucoisomerase, ADH, and triosephosphate isomerase are expressed postmeiotically in the pollen grains of tomato. Evidence for postmeiotic genetic activity in pollen grains of *Cucurbita* species hybrids has been provided by an ingenious microelectrophoretic separation of proteins and enzymes from single pollen grains (Mulcahy, Mulcahy and Robinson 1979; Mulcahy et al 1981).

(iii) Chromosome Deficiencies

Isozyme changes are only one of a series of genetic mechanisms that operate on the haploid pollen genome. Variant pollen grains arising with known chromosomal deficiencies might be expected to provide information on gene action affecting pollen development at the cellular level. A–B translocations, which are reciprocal interchanges between the normal chromosome, A, and a supernumerary element, called B chromosome, are well known in maize. "Hypoploid" is the preferred term used to designate plants carrying segmental deficiencies in chromosome arms due to B–A translocations. The clearest indications of alteration in cellular properties of pollen grains as a direct expression of disparity in chromosome structure are seen in the work of Kindiger, Beckett and Coe (1991) on hypoploids generated in maize. Although meiosis is uneventful in the hypoploids, there is good reason to believe that a normal chromosome structure is indispensable for the production of a full complement of viable pollen grains in them. This is strikingly illustrated by the observation that although one-half of microspores in the anther locule develop normally, the other half have predictable defects leading to the loss of viability.

When the cellular phenotype is related to chromosome deletions, a general pattern of pollen developmental arrest beginning with the completion of microsporocyte meiosis is observed, with the point of arrest depending upon which chromosome arm is deleted. Typically, deficiencies in parts of chromosomes 1, 3, 5, 7, and 9 affect materially the development of microspores beyond the uninucleate stage, and only occasional pollen grains complete the first haploid mitosis according to routine. Changes affecting the second pollen mitosis are associated with deletions in the long arm of chromosomes 6 and 10 (Figure 4.2). Both groups of hypoploids complete the first haploid mitosis normally, but the subsequent behavior of the nucleus of the vegetative cell is different and abnormal. In hypoploids with a deficiency in chromosome 6, the vegetative nucleus hypertrophies; in hypoploids for chromosome 10, the vegetative nucleus behaves as if it is entering the mitotic cycle. Because the anomalous behavior of the deficient pollen grains may be due to the loss of genes that code for some essential proteins, it is obvious that genetic infor-

Table 4.1. Isozymes in various sporophytic and gametophytic tissues of tomato. (+, present; N.D., not detected; –, data inconclusive). ADH, alcohol dehydrogenase; APS, acid phosphatase; EST, esterase; GOT, glutamic-oxaloacetic transaminase; PGI, phosphoglucoisomerase; PGM, phosphoglucomutase; PRX, peroxidase; SKDH, shikimate dehydrogenase; TPI, triosephosphate isomerase. (Reprinted with permission from S. D. Tanksley, D. Zamir and C. M. Rick 1981. Evidence for extensive overlap of sporophytic and gametophytic gene expression in *Lycopersicon esculentum. Science* 213:453–455. © 1981. American Association for the Advancement of Science.)

	SPOROPHYTE			GAMETOPHYTE		POSTMEIOTIC EXPRESSION	
Iso-zyme	Roots	Leaves	Developing and mature seeds	Pollen	Germ-inated pollen	Tested	Positive
APS-1	+	+	+	+	+	+	+
APS-2	+	+	+*	+	+		
APS-A	N.D.	+*	N.D.	+*	+*		
GOT-1	+	+	+*	+*	+*		
GOT-2	+	+	+	+	+	+	+
GOT-3	+	+	+	+	+	+	+
GOT-4	+	+	+	+	+	+	+
PRX-1	+	+	N.D.	N.D.	N.D.		
PRX-2	+	+	+	+*	N.D.		
PRX-3	+	+	+	N.D.	N.D.		
PRX-4	+	N.D.	N.D.	N.D.	N.D.		
PRX-5	+	N.D.	N.D.	+*	N.D.		
PRX-6	+	N.D.	+*	+*	N.D.		
PRX-7	+	N.D.	+*	N.D.	N.D.		
PRX-A	N.D.	N.D.	+*	N.D.	N.D.		
EST-1	+	N.D.	N.D.	N.D.	N.D.		
EST-3	+*	+	+	+	+		
EST-4	+	N.D.	N.D.	N.D.	N.D.		
EST-5	+	N.D.	N.D.	N.D.	N.D.		
EST-6	+	N.D.	N.D.	N.D.	N.D.		
EST-7	+	+*	N.D.	N.D.	N.D.		
EST-A	N.D.	–	N.D.	+†	+†		
EST-B	N.D.	+	N.D.	–	–		
PGI-1	+	+	+	+	+	+	+
PGM-1‡	N.D.	+	+*	+*	+*		
PGM-2	+	+	+	+	+		
ADH-1	N.D.	N.D.	+	+	+	+	+
ADH-2	+	+	+	N.D.	N.D.		
SKDH-1	+	+†	+	+	+		
TPI-1‡	N.D.	+	N.D.	+*	+*		
TPI-2	+	+	+	+	+	+	+

*Weekly expressed. †Double-banded. ‡Enzyme localized in chloroplast.

mation concerned with the differentiation of the pollen grain is controlled by the haploid genome.

Exploitation of the properties of pollen walls of hypoploids as an indicator of the gametophytic or sporophytic genome has also proved to be fruitful. The presence of fragile walls in all pollen grains produced by hypoploids for the long arm of chromosome 6 provides one setting in which sporophytically acting genes necessary for pollen wall development can be mapped to this chromosome. Since only the hypoploid pollen grains produced on plants deficient in the short arm of chromosome 10 have fragile walls, the factors involved in the production of normal pollen wall in this situation must accordingly be deemed to be regulated by the gametophytic genome (Kindiger, Beckett and Coe 1991).

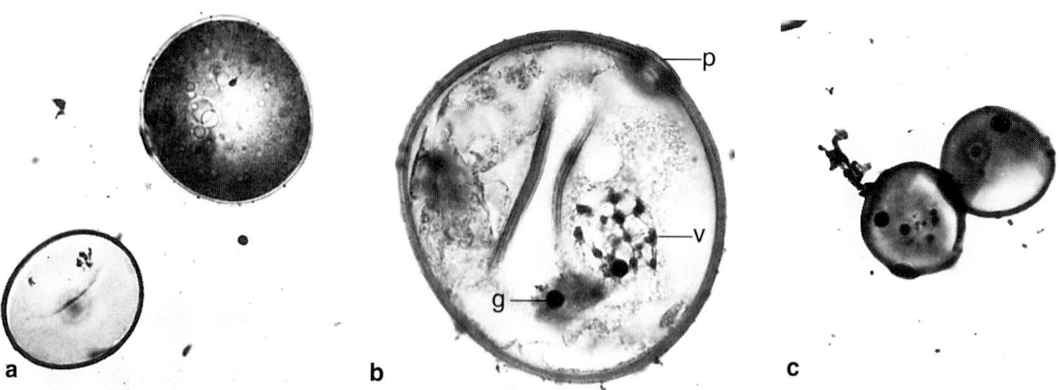

Figure 4.2. Effects of chromosomal deletions on pollen grains of maize. **(a)** A hypoploid microspore (bottom) from a plant deficient in part of chromosome 5, showing a knotted appearance of the chromosomes during the first haploid mitosis. A normal pollen grain is at the top. **(b)** A hypoploid binucleate pollen from a plant deficient in part of chromosome 6, showing the abnormal vegetative cell nucleus (v) and a normal generative cell nucleus (g). The pollen aperture (p) is also seen. **(c)** Presence of several nucleolus-like bodies in the nucleus of the vegetative cell of the pollen grain (left) of a plant hypoploid for part of chromosome 10. (From Kindiger, Beckett and Coe 1991; negatives supplied by Dr. B. Kindiger.)

2. NUCLEIC ACID AND PROTEIN METABOLISM

The historical evolution of our knowledge of gene expression during the ontogeny of pollen grains is almost inextricably linked to investigations of their changing patterns of nucleic acid and protein metabolism. A general account of the biochemical changes during pollen development as they relate to DNA metabolism was given in Chapter 3, and so this will not be taken up here again. As the following account shows, other investigations have yielded data on RNA and protein synthesis that are highly relevant to the evaluation of the control of gene activity during pollen development and maturation. Much of this work is reviewed by Mascarenhas (1975).

(i) RNA Metabolism

A good part of the early and subsequent work on RNA metabolism during pollen ontogeny consists primarily of cytochemical and autoradiographic analyses. However, research in this area is of necessity a modern subject, in the sense that most efforts have occurred since the discovery in the 1940s of DNA as the genetic material. The first and the most notable was the work of La Cour (1949), who showed by staining reactions that after the first haploid mitosis there is a rapid increase in the RNA content of the vegetative cell; the nucleus of this cell also enlarges, but its DNA content remains unchanged. On the other hand, there is no appreciable increase in the RNA content of the generative cell, although its nucleus undergoes DNA replication preparatory to division. Development of the pollen grain thus involves switching off DNA synthesis in one cell without interfering with transcription; in the other cell, transcription is slowed down without impairing DNA synthesis.

Following this work, an interesting assortment of methods have been employed to follow the timing of accumulation and synthesis of RNA and to study the nature of RNA synthesized during pollen development and maturation. In one of the first experiments using autoradiography to follow RNA synthesis in pollen grains, anthers of *Lilium longiflorum* and *Tradescantia paludosa* were incubated in $^{32}PO_4$ for 8–48 hours. Although the resolution of the autoradiographs was not impressive by current standards, the results nonetheless showed that periods of RNA synthesis occur in the pollen grains during and after the first haploid mitosis (Taylor 1953). In another work, autoradiography was combined with silver grain counts to show that in anthers of *Tulbaghia violacea*, $^{32}PO_4$ is incorporated into RNA of pollen grains of all stages of development except the very mature stage (Taylor 1958). When RNA synthesis was monitored in the microspores of *Trillium erectum* by electron microscopic autoradiography of ^{3}H-uridine incorporation, it was found that a change in the organizational state of the DNA-containing regions of the nucleus from a condensed to a relaxed state favors a high RNA synthetic activity, whereas a condensed nucleus supports a low level of RNA synthesis (Kemp 1966). Somewhat similar observations have also been made in a study of RNA synthesis during pollen ontogeny in *Tradescantia paludosa* (Dryanovska 1981). These results, however, do not tell us whether RNA synthetic activity is indicative of changes that are characteristic of landmark stages in pollen differentiation.

As the scope of these investigation was extended, it gradually became clear that changes in RNA synthesis and accumulation can serve as markers of the metabolic state of the pollen grain and its constituent cells. A cytophotometric study of the changes in RNA contents of postmitotic pollen grains of *T. paludosa* showed that independent nuclear and cytoplasmic RNA-synthesizing systems operate in the vegetative cell. For example, an almost continuous increase in cytoplasmic RNA appears to underlie the differentiation of the vegetative cell. The increase in RNA content of the nucleolus and chromosomes of the vegetative cell is of a smaller order than that of the cytoplasm and continues only up to about 40 hours after division; a steep decrease then follows (Figure 4.3). No RNA is detected in the nucleolus or chromosomes of the generative cell, and probably the amount of cytoplasm in this cell is too small to be separately measured (Woodard 1958). A later, much less detailed cytochemical study also led to the impression that RNA concentration in the vegetative cell of *Crocus longiflorus* is very high (Jalouzot 1969b).

Similar conclusions about the relative RNA synthetic activities in the vegetative and generative cells have been drawn from various autoradiographic investigations using ^{3}H-labeled precursors. In *Paeonia tenuifolia*, ^{3}H-cytidine is incorporated into RNA of the microspore nucleus at a modest level, but intense incorporation is noted in the vegetative cell (Sauter 1969a). Autoradiography of ^{3}H-uridine incorporation into developing pollen grains of *Lilium candidum* (Jalouzot 1969a), *Hyoscyamus niger* (Reynolds and Raghavan 1982), and *Hyacinthus orientalis* (Bednarska 1984) has also provided us with confirmatory data on RNA synthetic activity that is more intense in the vegetative cell than in the generative cell (Figure 4.4).

In autoradiographic studies using precursors of RNA synthesis, most of the incorporation seen as silver grains over sections is due to the synthesis of rRNA almost to the exclusion of mRNA and transfer RNA (tRNA). The cellular localization of mRNA in sections has been facilitated by in situ hybridization using radioactive or nonradioactive labeling methods. Experiment with ^{3}H-poly(U) as a probe to monitor mRNA has shown that although the generative cell of *Hyoscyamus niger* does not synthesize any appreciable appreciable rRNA, it nonetheless accumulates mRNA in tandem with the vegetative

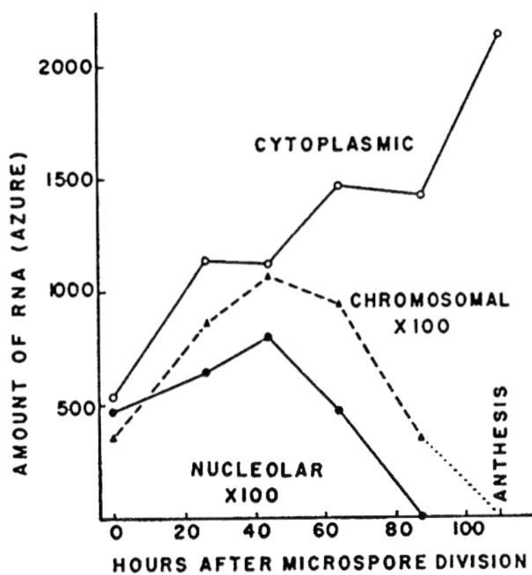

Figure 4.3. Changes in the RNA contents of chromosomal, cytoplasmic, and nucleolar fractions of the vegetative cell of postmitotic pollen grains of *Tradescantia paludosa*. The dotted part of the curve between 87 and 110 hours is based on estimated values. (Reproduced from Woodard 1958. *J. Cell Biol.* 4:383–390 by copyright permission of the Rockefeller University Press.)

cell. With the approaching maturity of the pollen grain, there is a steep decrease in the mRNA concentrations of both cells (Raghavan 1981b). Changes in the pattern of mRNA accumulation during pollen development in rice include an increase in the concentration of mRNA in the generative and vegetative cells after the first haploid mitosis, and a decrease as the pollen grains become completely starch filled (Raghavan 1989). Use of immunological detection methods to monitor mRNA distribution during pollen ontogeny in *Nicotiana tabacum* has shown that the most readily detectable change is an increase in labeling beginning with the uninucleate microspore, with an intense labeling in the nucleus of the vegetative cell of the nearly mature pollen grain (Chandra Sekhar and Williams 1992). It has been suggested that the increase in mRNA is correlated with the reprogramming of the pollen nuclei, which leads to a postmeiotic production of gametophytic information.

Some pertinent information about the timing and synthesis of stable RNA species has been derived from biochemical analyses of pollen grains undergoing synchronous development within the anther. An early work on *Lilium longiflorum* showed a linear rise in the RNA content of microspores, with a sharp increase at mitosis continuing until pollen maturity (Ogur et al 1951). A later study was much more explicit in showing that the synthesis of high-specific-activity RNA is confined to a period just before the first haploid mitosis; a second peak of synthesis of low-specific-

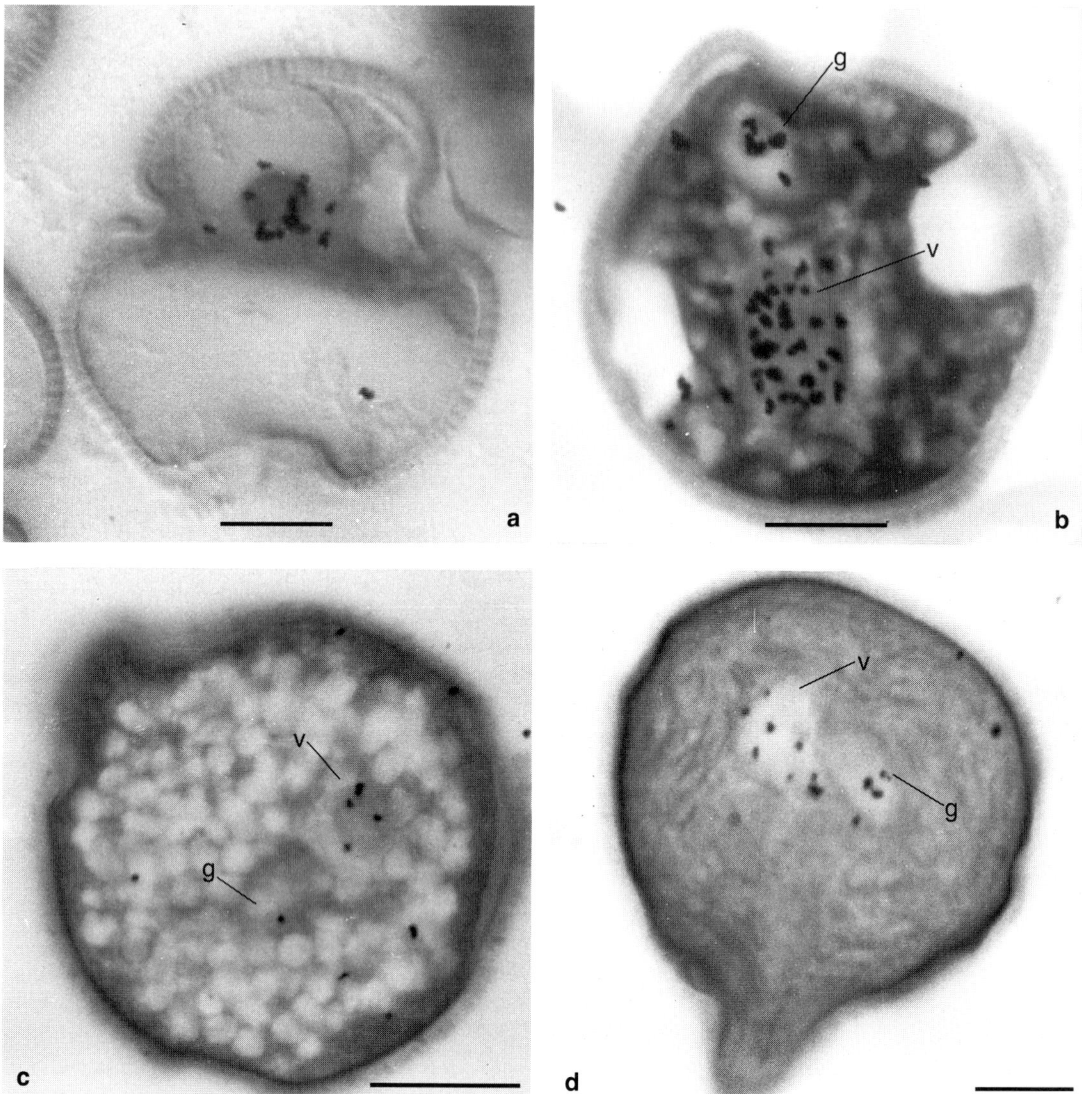

Figure 4.4. Autoradiographs showing incorporation of ^{3}H-uridine into developing pollen grains of *Hyoscyamus niger*. **(a)** Unicellular, vacuolate microspore. **(b)** Bicellular pollen grain; the generative cell is still attached to the intine. **(c)** Mature, bicellular pollen; the generative cell is detached from the intine. **(d)** Mature pollen grain from an open flower. Incorporation in all cases is into the nuclei. g, generative cell; v, vegetative cell. Scale bars = 10 μm. (From Reynolds and Raghavan 1982.)

activity RNA occurs in the bicellular pollen grains. Chromatographic fractionation of RNA showed that rRNA contributes the major fraction to the total RNA synthesized by the pollen grains. The periods of RNA synthesis were also correlated with nucleolar activity and occurred when the nucleoli were large (Steffensen 1966). The trend is similar with regard to the accumulation of total RNA in pollen grains of *L. henryi* (Linskens and Schrauwen 1968b). Measurements on whole anthers of *Agave americana* (Amaryllidaceae) showed a steady decrease in RNA content coinciding with the maturation of microspores (Maheshwari and Prakash 1965).

The observations of Mascarenhas and Bell (1970) on the timing of rRNA synthesis in the pollen grains of *Tradescantia paludosa* provide interesting support to the data from *L. longiflorum*. As

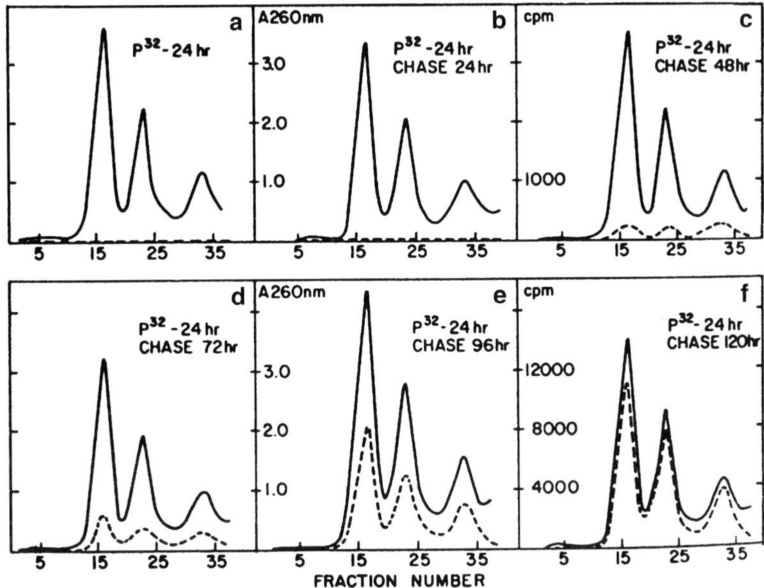

Figure 4.5. Synthesis of ribosomal RNA during pollen development in *Tradescantia paludosa*. Inflorescence cuttings were initially incubated for 24 hours in a solution containing $^{32}PO_4$ followed by chase in an unlabeled medium. Pollen grains were collected at 24-hour intervals for RNA extraction. RNA was centrifuged in a sucrose gradient and fractions were collected to determine absorbance and radioactivity. **(a)** RNA from pollen grains after incubation in the isotope for 24 hours. **(b)** Same as (a), followed by 24 hours' chase. **(c)** Same as (a), followed by 48 hours' chase. **(d)** Same as (a), followed by 72 hours' chase. **(e)** Same as (a), followed by 96 hours' chase. **(f)** Same as (a), followed by 120 hours' chase. The RNA peaks have sedimentation values (from left to right) of 25S, 16S, and 5S. Solid line, absorbance at 260 nm; dotted line, counts per minute. (From Mascarenhas and Bell 1970.)

characterized by sucrose density gradient centrifugation, not only is most synthesized RNA ribosomal, but synthetic activity is also restricted to a window between 96 hours and 144 hours before anthesis, corresponding to the interval of the first haploid mitosis (Figure 4.5). The synthesis of 5S RNA follows the same pattern as rRNA. Although small amounts of 5S RNA are synthesized before or during microsporocyte meiosis, the bulk of it is made before microspore mitosis. Subsequently, there is a sharp decrease in the synthesis of 5S RNA, with a cutoff point 96 hours later. This has been attributed to the existence of multiple copies of 5S RNA genes, some of which are turned off rapidly following the peak synthesis, although a few persist for an additional period in development (Peddada and Mascarenhas 1975). Despite the wealth of detail provided in these works, information available does not permit a distinction to be made between the role of RNA synthesis in the differentiation of the vegetative and generative cells.

In *Nicotiana tabacum*, the postmitotic increase in RNA synthesis is related to the differentiation of the vegetative cell and is probably due to the activation of its rRNA genes (Tupý et al 1983). Here the transition of the microspore to a mature pollen grain is a critically important phase in which mRNA synthesis outstrips turnover. During this period, the accelerated synthesis of poly(A)-RNA is accompanied by a shift in its size toward a high molecular mass type (Tupý 1982). Qualitative variations in mRNA populations have been noted in in vitro translation profiles of developing pollen grains of lily, tobacco (Schrauwen et al 1990), and maize (Bedinger and Edgerton 1990). These changes include the presence of specific mRNAs related to pre- and postmitotic stages and the stable accumulation of some mRNAs during the final stages of pollen maturation (Figure 4.6). These observations provide a basis for the conclusion that mRNA produced is probably stored in mature pollen grains to provide templates for the first proteins of germination.

A major conceptual issue relating to the presence of stored mRNA in mature pollen grains concerns its sequence complexity. Although sequence complexity of a population of nucleic acid molecules is evaluated in terms of the total length of the diverse sequences represented, it is used here in a broad sense to include information about the dif-

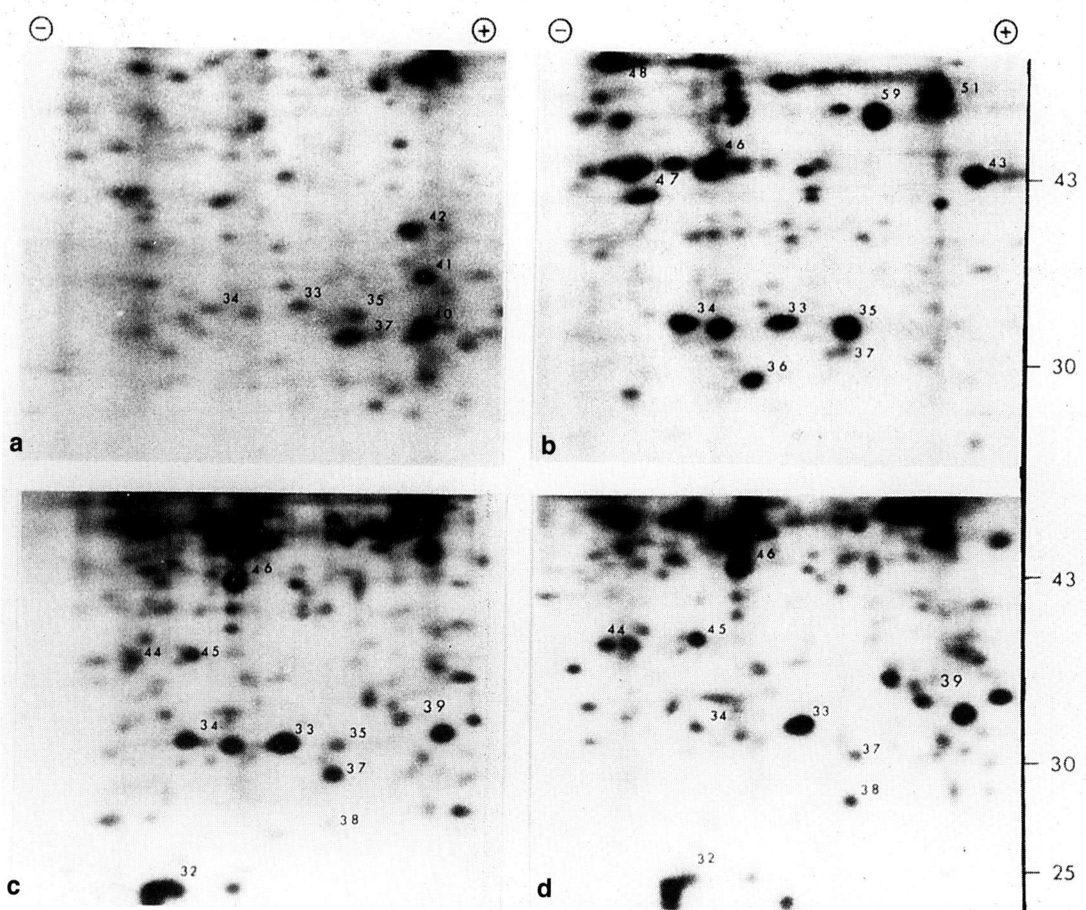

Figure 4.6. In vitro translation of mRNA isolated from pollen grains of *Lilium longiflorum* at various stages of development. ^{35}S-methionine-labeled proteins synthesized were separated electrophoretically and gels were subjected to autoradiography. Only the parts of gels with protein spots numbered arbitrarily are shown. **(a)** Early unicellular pollen. **(b)** Late unicellular pollen. **(c)** Mid-bicellular pollen. **(d)** Pollen grains at anthesis. (From Schrauwen et al 1990. *Planta* 182:298–304. © Springer-Verlag, Berlin. Photograph supplied by Dr. J. A. M. Schrauwen.)

ferent kinds of mRNAs and their abundant classes. Altogether these data provide an estimate of the number of different genes active in the pollen grain. Most measurements of sequence complexity of primary transcripts are obtained from analyses of the kinetics of hybridization of labeled cDNA with an excess of poly(A)-RNA, with particular reference to the reassociation kinetics and the percentage of hybrids formed. Based on these measurements, the total complexity of mature pollen mRNA of *Tradescantia paludosa* has been estimated to be 2.3×10^7 nucleotides and is equivalent to about 20,000 different genes. Three abundance classes with complexities of 5.2×10^4, 1.6×10^6, and 2.1×10^7 nucleotides comprise the mRNA. In terms of the number of copies per sequence per pollen grain, about 15% of the mRNAs are included in the very abundant category, which comprises about 40 different sequences, with each sequence present in 26,000 copies in a pollen grain. About 60% of the mRNA consists of 1,400 different sequences, each present, on average, in about 3,400 copies per pollen grain. About 24% of the mRNAs are considered to be the least abundant, because the 18,000 different sequences they comprise are present in only about 100 copies per cell (Willing and Mascarenhas 1984). A similar analysis of maize pollen mRNA showed that it has a total complexity of 2.4×10^7 nucleotides, corresponding to about 24,000 different genes. As in the case of *T. paludosa*, the mRNAs present in maize pollen have been separated into three abundance classes with different complexities. In both *T. paludosa* and maize, the spectrum of mRNAs is much more abundant in the pollen grains than in the shoot, although there is a substantial overlap between genes active in gametophytic and sporophytic tissues (Willing, Bashe and Mascarenhas 1988). Of the many interesting molecular implications of these data, perhaps the most important one is that a very extensive range of transcriptional activity is encountered in the development of the pollen grain. This is likely to make identification of a broad selection of genes that are expressed in the pollen a formidable task. Apparently, the limited cytoplasmic domain of the pollen grain appears highly stable, at least with respect to the messages that are transcribed.

(ii) Protein Metabolism

As expected, cytochemical, autoradiographic, and biochemical studies of protein metabolism of developing pollen grains show similarities with, and differences from, the trend described for RNA. The characteristic pattern noted in postmitotic pollen grains of *Tradescantia paludosa* is an increase in cytoplasmic proteins, accompanied by a decrease in chromosomal and nucleolar proteins (Woodard 1958). In the pollen grains of *Zea mays*, naphthol-yellow staining showed that proteins begin to accumulate at a modest level from postmeiosis until maturity; whether these proteins are synthesized by the pollen grains or transferred from the tapetum, as suggested, remains to be established (Moss and Heslop-Harrison 1967). The idea that some of these proteins may not fit into the mold of typical housekeeping proteins has come from reports of the predominant occurrence of a specific β-tubulin isotype in pollen grains of carrot (Hussey, Lloyd and Gull 1988) and of specific mitochondrial proteins in pollen grains of *Nicotiana sylvestris* (de Paepe et al 1993). Despite the simplicity of the pollen structure, a number of additional polypeptides not found in leaves have been identified in two-dimensional SDS-PAGE of mitochondrial proteins. The identification of one of these proteins as an additional ATP synthase β-subunit may reflect the general importance of respiratory proteins in pollen development.

When protein contents of the vegetative and generative cells are analyzed separately, there is an unequal partitioning of the proteins between the two cells, with more proteins being synthesized and accumulated in the former than in the latter. Coincidentally, this was also first demonstrated in the pollen grains of *T. paludosa*, in which both total and acid-insoluble proteins of the nucleus of the vegetative cell are found to be twice as plentiful as in the generative cell nucleus (Bryan 1951). Apart from the large size of the vegetative cell compared to that of the generative cell, how and why proteins selectively accumulate in the vegetative cell and its nucleus is not known. In the bicellular pollen grains of *Z. mays* and *Agave attenuata*, staining reactions have shown the presence of a higher concentration of basic proteins in the nucleolus of the vegetative cell than in the corresponding organelle of the generative cell. Since the total concentration of proteins in the postmitotic nucleoli does not differ from the amount in the premitotic nucleolus, the logical assumption is that the basic proteins of the microspore nucleolus are partitioned unequally between the nucleoli of the vegetative and generative cells (Martin 1960).

The mechanisms that implement protein accumulation in the developing pollen grains must ultimately be expressed in terms of altered synthesis. There are several reports of autoradiographic incorporation of labeled amino acids that indicate a pattern of differential protein synthesis associated with pollen differentiation. The first effort, which still retains some historical interest, was that of Taylor (1958), who followed protein synthetic activity in pollen grains of *Tulbaghia violacea* anthers incubated in $^{32}PO_4$. Silver grain counts on autoradiographs appeared to remain stable throughout the unicellular and bicellular stages of pollen development but decreased in the mature pollen. Other autoradiographic studies with anthers of *Paeonia tenuifolia* (Sauter 1969a) and *Tradescantia paludosa* (Dryanovska 1981) using ^{3}H-leucine have shown low to intense incorporation in the unicellular pollen grains and intense incorporation in the vegetative cell of bicellular pollen grains. There is good evidence that in the pollen grains of *Hyacinthus orientalis*, the generative cell is endowed with the ability to synthesize some proteins, although the activity never reaches the level of the vegetative cell (Bednarska 1984).

From these results, it is clear that a principal problem in analyzing the synthesis of proteins during pollen development is the difficulty of distinguishing the various types of proteins by the use of a single radioactive amino acid precursor. Part of this limitation was overcome in a study of protein synthetic activity during pollen development in *Hyoscyamus niger* by using ^{3}H-leucine, ^{3}H-lysine, ^{3}H-arginine, and ^{3}H-tryptophan as precursors. An interesting difference in protein synthetic activity between the unicellular and bicellular pollen grains

Table 4.2. Average number of autoradiographic silver grains per pollen grain (± standard error) of *Hyoscyamus niger* of various stages of development after incubation of flower buds in ^{3}H-arginine, ^{3}H-leucine, ^{3}H-lysine, and ^{3}H-tryptophan (10 µC/ml each) for 4 hours. Stages of pollen development: I, soon after release from the tetrad; II, pollen grain with the nucleus displaced to one side; III, vacuolate pollen grain; IV, immediately after the first haploid mitosis; V, beginning of starch accumulation in the pollen grain; VI, pollen grain with a protoplasmic protrusion at the pore; VII, pollen grain from an open flower. (From Raghavan 1984.)

Stage of pollen development	^{3}H-arginine	^{3}H-leucine	^{3}H-lysine	^{3}H-tryptophan
I	9.0 ± 0.5 0	0	0	
II	34.0 ± 3.0	0	18.0 ± 1.1	0
III	17.8 ± 0.8	0	0	6.1 ± 0.5
IV	[a]	12.3 ± 0.8	10.4 ± 0.3	9.1 ± 0.5
V	46.5 ± 3.2	33.5 ± 2.2	26.5 ± 2.2	13.6 ± 1.2
VI	8.5 ± 0.6	14.2 ± 0.8	10.3 ± 0.6	10.6 ± 0.8
VII	14.0 ± 0.8	10.4 ± 0.5	10.6 ± 1.1	9.5 ± 0.6

[a] Pollen grains of this developmental stage were not found in the sampling.

relates to the type of proteins synthesized. Silver grain counts showed that unicellular pollen grains incorporate ^{3}H-arginine and ^{3}H-lysine almost to the total exclusion of ^{3}H-leucine and ^{3}H-tryptophan. A different situation prevailed in the bicellular pollen grains, which incorporated all four amino acids into proteins (Table 4.2). Incorporation generally occurred into the vegetative cell and only rarely were silver grains found over the generative cell. There was also a propensity for pronounced stage-specific synthesis of basic proteins in the nucleus and for a tardy synthesis of cytoplasmic proteins during the development of the unicellular pollen grain (Raghavan 1984). In the absence of any discernible protein synthesis, how the generative cell continues to function is a point about which, at present, it is not possible to form a conclusion.

Although histones are now accepted essentially as structural constituents of chromatin, there was a time when they were believed to play a role in the regulation of gene activity. In that context, using cytochemical (Sauter and Marquardt 1967b; Sauter 1969b), microspectrophotometric (Rasch and Woodard 1959; Sangwan-Norreel 1978; Thiébaud and Ruch 1978), and autoradiographic (Bednarska 1981) methods, several investigators have shown that histone accumulation and synthesis in the generative cell of the angiosperm pollen grain is a fairly common phenomenon. Refined methods for the separation of nuclei of the vegetative and generative cells and for histone analysis have shown that different types of proteins are complexed with these nuclei, but the evidence on the whole is far from compelling. In *Hippeastrum belladonna*, no differences are seen in the number or types of chromatographically separable components of the "very" lysine-rich histones of the vegetative and generative cell nuclei. However, the acidic proteins of the two nuclei show some specificity in electrophoretic banding patterns, suggesting that the acidic nuclear proteins are more sensitive to metabolic alterations than are the basic proteins (Pipkin and Larson 1973). Another study has shown that vegetative nuclei isolated from mature pollen grains of *Lilium longiflorum* contain only arginine-rich histones whereas the generative nuclei have, in addition, lysine-rich histones and meiotic histones (Sheridan 1973). Recent investigations using two-dimensional SDS-PAGE of basic proteins of isolated pollen nuclei of *L. longiflorum* have identified two variants of lysine-rich H2B and arginine-rich H3 histones, designated as gH2B and gH3, respectively, which begin to accumulate in the nucleus of the generative cell. The specificity of association of these histones with the generative cell nucleus and sperm nuclei was confirmed by immunocytochemistry using antisera raised against gH2B and gH3. As seen in Figure 4.7, these antisera recognize only the generative cell nucleus of pollen protoplasts (Ueda and Tanaka 1994, 1995a, b). Apparently, these and other proteins are present in the nucleus of the generative cell in a highly concentrated form for possible use as key factors in regulating differential gene activity during male gametogenesis. A

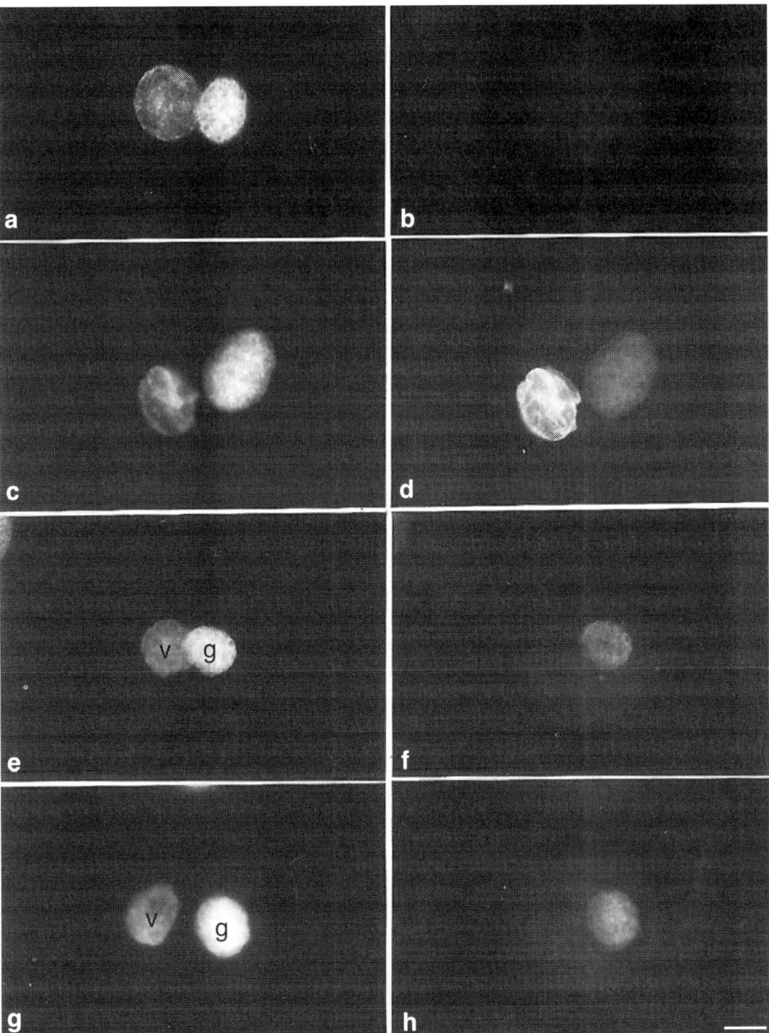

Figure 4.7. Immunofluorescence staining of gH2B and gH3 histones in isolated pollen protoplasts of *Lilium longiflorum*. **(a, c, e, g)** Stained with DAPI. **(b, d, f, h)** The same protoplasts were stained with primary antibody followed by FITC-labeled secondary antibody. Each protoplast contained a generative cell (g) and vegetative cell (v) nucleus. Primary antibodies were serum of nonimmunized rat (b), antihistone monoclonal antibody (d), antiserum against gH2B (f), and antiserum against gH3 (h). (From Ueda and Tanaka 1995b. *Planta* 197:289–295. © Springer-Verlag, Berlin. Photographs supplied by Dr. I. Tanaka.)

hypothesis to explain these observations might consider a preferential segregation of specific histones and other proteins in the generative cell nucleus during the first haploid pollen mitosis. Verification of the model in other systems might provide a firm basis to explain nuclear differentiation in the pollen grain.

Currently, few researchers are interested in the analysis of proteins of pollen nuclei. This is unfortunate, because direct determination of nuclear and cytoplasmic constituents of differentiating cells is the cornerstone of developmental biology and a gap exists in our knowledge of proteins that seem to have important consequences in male gametogenesis.

Changes in the amounts and types of proteins synthesized during pollen development involve the increase or decrease in the concentrations, as well as in the appearance or disappearance, of specific polypeptides and enzymes. An electrophoretic analysis of anthers of *Lilium henryi* showed clear stage-specific differences in the protein band patterns and in the concentrations of certain enzymes of the pollen grains covering a wide developmental spectrum (Linskens 1966). Other experiments using carefully staged whole anthers of *Datura metel* (Sangwan 1978), *Petunia hybrida* (Nave and Sawhney 1986), *Zea mays* (Delvallée and Dumas 1988), *Nicotiana tabacum* (Villanueva, Mathivet and Sangwan 1985; Chibi et al 1993), *Brassica oleracea* (Detchepare, Heizmann and Dumas 1989), and *L. longiflorum* (Wang et al 1992a) have also augmented our understanding of the changes in protein patterns during successive stages of pollen development. In bicellular pollen grains of *L. longiflorum*, appearance or disappearance of new proteins in two-dimensional SDS-PAGE coincides roughly with the first haploid mitosis, whereas in tricellular pollen grains of *Z. mays* and *B. oleracea*, similar changes are associated with the first haploid mitosis and the mature stage. These protein alterations seem to be linked to transcription in the haploid pollen genome. However, the contributions of the proteins of the anther wall and tapetum to the observed patterns of changes were not completely eliminated in these experiments.

That these conclusions are probably valid can be judged from electrophoresis of proteins of pure fractions of pollen grains of different ages isolated by gradient centrifugation. Changes in the concentrations of certain protein bands occur during the bicellular pollen stage in wheat and, in particular, new polypeptide bands are associated with the formation of the tricellular pollen grain (Vergne and Dumas 1988). In the pollen grains of *Z. mays*, major changes in protein populations and enzymes accompany the first and second haploid mitotic divisions (Figure 4.8). Consistent with this observation, in vivo labeling of microspores followed by two-dimensional SDS-PAGE has led to the identification of significant changes in protein synthesis during the transition from the tetrad stage to the vacuolate microspore stage, and again just prior to anthesis (Bedinger and Edgerton 1990; Mandaron et al 1990). Another programmed change in protein metabolism associated with pollen differentiation is seen in *N. tabacum*, in which a rise in the amount of polysomes, and soluble and insoluble proteins, and an increase in protein synthetic activity essentially coincide with the growth of the vegetative cell (Tupý et al 1983; Žárský et al 1985). Whether these changes are due to new transcription or due to alterations in the translational capacity of the preexisting messages, is not known.

Interesting questions on the reprogramming of pollen grains for rapid germination and tube emergence have surfaced from investigations on the stage-specific changes in hydroxyproline and ubiquitin contents of developing pollen grains of *Z. mays*. There is a progressive increase in the hydroxyproline content beginning with the microspore stage, reaching a maximum in the mature pollen. Since hydroxyproline-containing proteins, such as extensins and arabinogalactans, have important roles in the formation of the structural elements of the pollen tube wall and in the pollen recognition reaction on the stigma, it is possible that high concentrations of hydroxyproline in the mature pollen grain might facilitate the production of these proteins (Rubinstein, Prata and Bedinger 1995). In contrast, ubiquitins, which target cellular proteins for degradation by covalent attachment, display a developmentally regulated loss, attaining a very low level in the mature pollen (Callis and Bedinger 1994; Kulikauskas et al 1995). This may have consequences for preventing the degradation of proteins from the mature pollen grains. The success of empirical methods to identify functionally significant proteins is also attested by the cloning and characterization of profilin from pollen grains of tobacco (Mittermann et al 1995). This protein is believed to regulate actin polymerization and signal transduction involving the phosphoinositide pathway. Since actin is a component of pollen grains and pollen tubes, various explanations can be advanced for the role of profilin during pollen maturation and pollen tube growth.

An experimental strategy that facilitates the analysis of gene action in terms of protein synthesis is the administration of heat stress. It is well known that living organisms respond to a permissive elevated temperature or heat shock by synthesizing a specialized set of mRNAs and HSPs at the same time that synthesis of proteins currently being made is inhibited. The new proteins are believed to be the products of heat-shock genes. That developing pollen grains synthesize HSPs was demonstrated in an experiment in which *Z. mays* anthers of different stages of development were incubated at 38 °C in the presence of ^{35}S-methionine. Analysis of the newly synthesized proteins showed that unicellular microspores are most sensitive to the heat stress, synthesizing at least seven HSPs. There is a general decline in the synthesis of HSPs after the first haploid pollen mitosis, and only two HSPs are synthesized by the bicellular and tricellular pollen grains. Some of the HSPs synthesized by the unicellular pollen grains appear to be characteristic of the gametophytic phase, because they are not found in heat-shocked sporophytic tissues such as root tips (Frova, Taramino and Binelli 1989). Immature pollen grains of

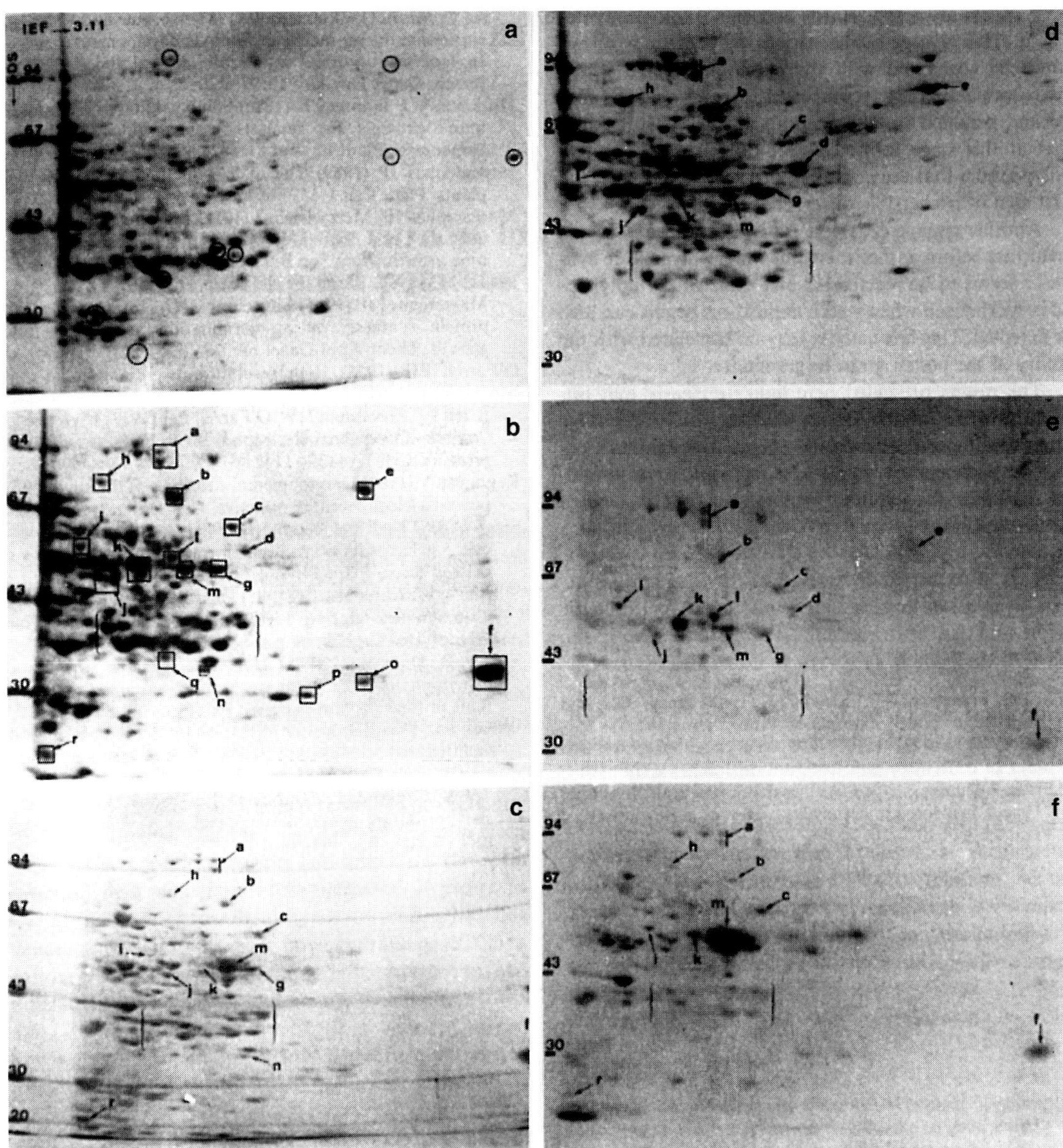

Figure 4.8. Two-dimensional SDS-PAGE of extant and in vivo synthesized proteins of pollen grains of *Zea mays* of three different stages of development. **(a–c)** Stained gels of extant proteins. **(d–f)** Autoradiograms of the same gels. (a, d) Vacuolate microspore stage. (b, e) Starch accumulation stage. (c, f) Pollen grains at anthesis. Molecular mass markers are indicated in kDa. Most polypeptides are indicated by letters and arrows. Others are indicated as (○), polypeptides present at the vacuolate microspore stage disappearing later; (□), newly appearing polypeptide or one whose synthesis dramatically increases at starch accumulation stage. A group of polypeptides present at all stages of pollen development, but whose synthesis varies greatly, is indicated by brackets. (From Mandaron et al 1990. *Theor. Appl. Genet.* 80:134–158. © Springer-Verlag, Berlin. Slide supplied by Dr. F. Monéger.)

Sorghum bicolor also synthesize a set of HSPs that do not have a counterpart in the sporophytic tissues (Frova, Taramino and Ottaviano 1991). It is clear from these results that the regulation of heat-shock response in developing pollen grains is due to the synthesis of new proteins.

As a key process typifying the living state, cells protect against heat stress by the synthesis of HSPs, yet mature pollen grains of many plants lack the ability to synthesize HSPs in response to heat stress. In mature tomato pollen grains, which do not respond to heat shock by the synthesis of new HSPs, a cognate HSP-70 (HSP expressed in the absence of heat stress) is found to be present throughout microspore development (Duck and Folk 1994). The cognate proteins stored in the mature pollen grains may anticipate the heat shock and function as HSPs.

As seen from the foregoing account, a good part of our understanding of gene action during microsporogenesis has come from developmental and biochemical studies. If instances of gene segregation for specific pollen-based enzymes are detected and their validity established by genetic crosses, these markers can be considered to be products of postmeiotic transcription of the haploid genome. To date, the only cases of biochemical segregation of enzyme markers detected in pollen grains are ADH activity in *Zea mays* (Freeling 1976) and β-galactosidase activity in *Brassica campestris* (Singh and Knox 1985). Quantitative cytochemical evidence has shown that these enzymes are absent in the microsporocytes and tetrads but begin to appear first in the young microspores. Thereafter, the enzyme titers continue to increase up to the time of the generative cell division. In both species, plants that are heterozygous for the enzyme locus produce pollen, 50% of which stains positive for the enzyme and 50% of which stains negative (Singh, O'Neill and Knox 1985; Stinson and Mascarenhas 1985). The synthesis of enzymes by metabolic reactions modulated by the expression of genes in the gametophytic genome is logically consistent with the knowledge derived from the segregation of genes.

Lipids are the major components of pollen grains of certain plants, and thus the activity of enzymes involved in lipid biosynthesis would be expected to change during pollen development. The fact that ACP is involved in lipid synthesis as an acyl carrier for the fatty acid synthase enzyme complex has led to the use of ACP activity as an index of lipid accumulation in pollen grains of *B. napus*. ACP appears first in the postmeiotic pollen and is found maximally during the bicellular stage (Evans et al 1992). Like the other proteins considered earlier, the pattern of ACP accumulation could fit with the model of transcription from the haploid genome.

3. ISOLATION OF GENES CONTROLLING POLLEN DEVELOPMENT

From the work reviewed thus far, a case can be made that successful completion of the male gametophytic program depends upon the sequential expression of different sets of genes at unique developmental stages for the control of pollen functions. Isolation of genes from pollen grains of several plants and their characterization have of late become major areas of investigations of angiosperm reproduction in many laboratories in the world. The initial impetus for this effort stemmed from RNA–excess/single copy DNA hybridization analyses of Kamalay and Goldberg (1980, 1984), which showed that at least 25,000 diverse genes are active in the tobacco anther during the advanced stage of microspore development. Comparisons of the sequence contents of anther mRNAs with mRNAs from the leaf, stem, root, petal, and ovary of tobacco showed that a majority (about 15,000) of the diverse mRNAs are shared between the different organs, whereas on a conservative estimate, about 10,000 mRNAs are considered to be anther specific. In principle, these well-documented results reaffirm the tenets that differentiation of the anther is dependent upon the differential regulation of an unusually large number of genes and that their isolation and characterization could fruitfully provide molecular markers for the analysis of pollen development. Molecular studies of pollen differentiation that lie within the scope of this section are reviewed by Mascarenhas (1989, 1990, 1992), McCormick (1991), and Twell (1994), and by several authors in a recent book edited by Mohapatra and Knox (1996).

(i) Characterization of Pollen-specific Genes

The large size of anthers of certain plants and the ease of collecting them from flowers at anthesis ensured that transcripts from pollen grains at advanced stages of development were the first to be isolated. In a pioneering investigation, Stinson et al (1987) made cDNA libraries to mRNA of mature pollen grains of *Zea mays* and *Tradescantia paludosa* and isolated pollen-specific clones by differential screening against sporophytic tissues. The pollen specificity of the clones was confirmed by Northern blot hybridization to RNAs from pollen grains and other tissues and organs of the plant such as shoot, root, endosperm, ovule, and stigma-style. Additional examples were soon forthcoming and included *Lycopersicon esculentum* (Twell et al 1989b), *Oenothera organensis* (Brown and Crouch 1990), *Brassica napus* (Albani et al 1990; Scott et al 1991; Shen and Hsu 1992), *B. oleracea* (Theerakulpisut et al 1991), *Helianthus annuus* (Baltz et al 1992a; Dudareva et al 1993; Reddy et al 1995), *Oryza sativa* (Tsuchiya et al 1992; Zou et al 1994; Xu et al 1995b), *Nicotiana tabacum* (Rogers, Harvey and Lonsdale 1992; Weterings et al 1992; Brander and Kuhlemeier 1995), *Lilium longiflorum* (Kim, Kim and An 1993),

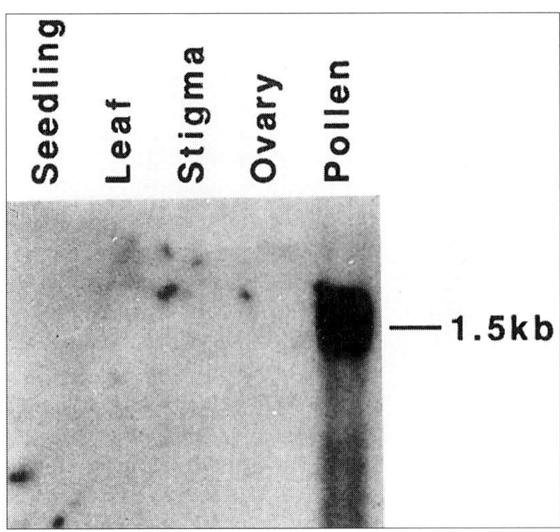

Figure 4.9. Pollen specificity of transcripts of a cDNA clone from *Oenothera organensis*. Northern blot hybridization of labeled *P-2* cDNA to RNAs from pollen grains and from different tissues of the plant. The length of the RNA is indicated on the right. (From Brown and Crouch 1990. *Plant Cell* 2:263–274. © American Society of Plant Physiologists.)

Gossypium hirsutum (John and Petersen 1994), and *Petunia inflata* (Mu, Stains and Kao 1994; Mu, Lee and Kao 1994). These studies have also identified cDNA clones whose transcripts are present exclusively, or at elevated levels, in the pollen grains, as compared to the vegetative parts of the plant, floral organs, or somatic tissues of the anther (Figure 4.9). As the following account shows, pollen-expressed genes have received much attention in recent years and there is a large and rapidly expanding literature on their molecular features.

In considering the properties of pollen-specific cDNA clones, the question arises whether the genes represented by the clones are members of large or small gene families. Southern blot hybridization showed that pollen-specific genes *ZMC-13* and *ZMC-26* from maize are present in one or a few copies in the genome (Stinson et al 1987). Such a conclusion also appears valid for *STA44-4, NPG-1,* and *PS-1* genes that are highly expressed in *B. napus, N. tabacum,* and rice pollen, respectively (Robert et al 1993; Tebbutt, Rogers and Lonsdale 1994; Zou et al 1994). Genes *LAT-51, LAT-52,* and *LAT-58* that are expressed in abundance in pollen grains of tomato are present in single copies (Twell et al 1989b; Ursin, Yamaguchi and McCormick 1989). In contrast, a gene expressed in the mature pollen grains of *O. organensis* (*P-2*) is present in 6–8 copies per genome and is thus a member of a relatively small family (Brown and Crouch 1990). Genomic clones *BP-4* and *BP-10*, isolated from *B. napus* pollen, are composed of 8–15 closely related genes constituting small multigene families (Albani et al 1990, 1992). The existence of a small multigene family has also been confirmed for the genomic clone of *SF-3* gene from sunflower pollen (Baltz et al 1992a). A maize pollen cDNA clone, *3C-12*, hybridizes strongly to five or six fragments and weakly to several others, indicating that it is a member of a medium-sized gene family (Allen and Lonsdale 1993). On the whole, it appears that the different pollen-expressed genes can at best be characterized as not belonging to large gene families.

It will be recalled that in accordance with genetic and biochemical studies, the concept of pollen development requiring the timely expression of genes in both gametophytic and sporophytic tissues was suggested. This postulate, which has formed the cornerstone of many discussions on pollen gene expression, has gained support from the work using cloned genes. The main outlines of the time frame for the expression of new genes during pollen development were established by Stinson et al (1987) in the work previously referred to. This study showed that the first detectable expression of pollen-specific genes *TPC-44* and *TPC-70* from *T. paludosa* and *ZMC-30* from maize occurs in the young pollen grains emerging from the first haploid mitosis. Evidently, the genome is programmed during division to express the new transcripts. Following their initial appearance, these genes continue to accumulate in the bicellular pollen grains, reaching their maximum concentration in the mature pollen (Figure 4.10). Variations are, however, noted in the pattern of accumulation of other genes. For example, transcripts of an actin gene are detected soon after meiosis and, after a transient peak at late pollen interphase, they decrease substantially in the mature pollen grain. A case can be made from these results that two distinct sets of genes are activated in a temporal fashion during pollen development. Transcripts of one set, christened as "early genes," which includes actin, become active soon after meiosis, but have a short half-life. On the other hand, the "late genes," represented by the pollen-specific clones, are expressed after pollen mitosis and continue to accumulate as the pollen grain matures. Circumstantial evidence points to the early genes' coding for cytoskeletal proteins, whereas the late genes encode proteins necessary for pollen maturation and early pollen tube growth.

The expression of the following pollen-specific cDNA clones displays a pattern that is consistent with that of late genes: *LAT-51, LAT-52, LAT-58,* and

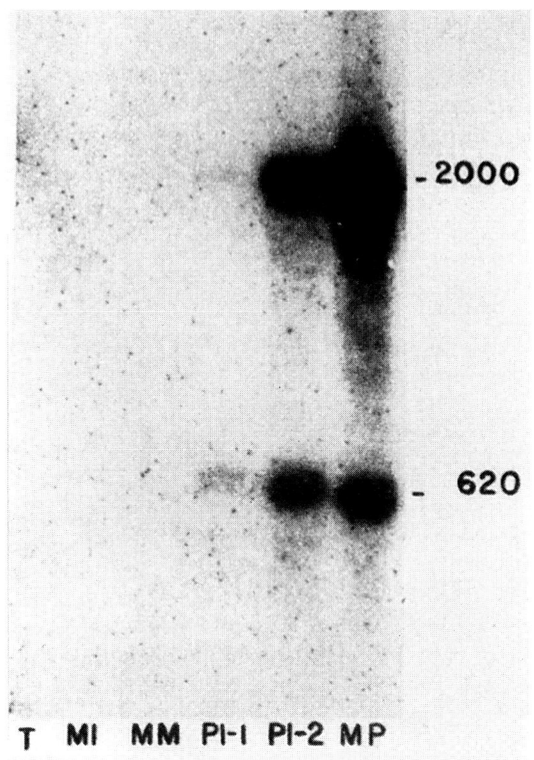

Figure 4.10. Developmental expression of transcripts of pollen-specific cDNA clones of *Tradescantia paludosa*. Northern blot hybridization of total RNA from microspores of different stages of development to both *TPC-44* and *TPC-70* cDNA clones, which hybridize to microspore mRNAs of 2,000 and 600 bp, respectively. T, tetrads; MI, microspores during interphase; MM, microspores in mitosis; PT-1, pollen at early interphase; PT-2, pollen at late interphase; MP, mature pollen. (From Stinson et al 1987. *Plant Physiol.* 83:442–447. © American Society of Plant Physiologists.)

LAT-59 from tomato (Twell et al 1989b; Ursin, Yamaguchi and McCormick 1989; Wing et al 1989); *P-2* from *O. organensis* (Brown and Crouch 1990); *BCP-1* from *B. campestris* (Theerakulpisut et al 1991); *BP-10, STA39-3, STA39-4,* and *STA44-4* from *B. napus* (Albani et al 1992; Robert et al 1993; Gerster, Allard and Robert 1996); *TP-10, NTP-303,* and *NEIF-4A8* from tobacco (Rogers, Harvey and Lonsdale 1992; Weterings et al 1992; Brander and Kuhlemeier 1995); *PPE-1* and *PRK-1* from *Petunia inflata* (Mu, Lee and Kao 1994; Mu, Stains and Kao 1994); members of the profilin gene family (Staiger et al 1993); *CDPK* (Estruch et al 1994) and *PEX-1* (Rubinstein et al 1995a) from maize; *BETV-1* from *Betula verrucosa* (birch; Betulaceae), encoding a major allergen (Swoboda et al 1995a); and *SF-17* from sunflower (Reddy et al 1995). The characteristic late-gene-expression pattern of *NTP-303* prevails without change even in immature pollen grains of tobacco allowed to complete the full developmental program in culture (van Herpen et al 1992). A gene (*CHI-A*) encoding the flavonoid biosynthetic enzyme, chalcone flavanone isomerase from *Petunia hybrida*, is expressed in floral tissues like petals, and can be included in the list of late genes by its abundant expression in nearly mature pollen grains (van Tunen et al 1989). Final proof that messages encoded by these genes are functionally active during pollen development has come from the demonstration that downgrading the activity of the gene by an antisense approach effectively arrests pollen development and causes pollen abortion (Xu et al 1995a).

The importance of the class of early genes in pollen development is underscored in some reports that have described cDNA clones that are expressed in microspores soon after microsporocyte meiosis, peak in pollen grains about the time of the first haploid mitosis, and then decline as pollen grains mature. These reports have all come out of work done on pollen-specific clones of *B. napus*. Because the clones studied have a unique temporal pattern of expression during pollen development, they will be described briefly. One of them is a genomic clone, *BP-4*, which is typically expressed beginning around the tetrad stage of microspores and increases in concentration until after the division of the generative cell, but the transcripts do not accumulate in the mature pollen grain (Albani et al 1990). The pattern of expression of another genomic clone, *BP-19*, is somewhat similar to the pattern of *BP-4*, but transcripts are not expressed at early stages of pollen development at as high levels as *BP-4* (Albani et al 1991). The time of the first appearance of transcripts is by no means uniform nor is it predictable in all cases, as shown by pollen-specific cDNA clones *I-3, 17, E-2,* and *F2S* isolated from anthers of *B. napus*. Messages corresponding to clone *I-3* suddenly appear at a high level around microspore mitosis and decline dramatically in the mature pollen grain in which the generative cell is poised to divide (Roberts et al 1991). The expression patterns of cDNA clones *17* and *E-2* are more or less similar to one another and begin during a discrete developmental window slightly earlier than the expression of clone *I-3*, but transcripts of both maintain their high levels until the division of the generative cell. The mRNA of clone *F2S* continues to accumulate from the time it first appears in the microspores prior to the first haploid mitosis until after generative cell division, and then it disappears in mature pollen grains

(Scott et al 1991). Although the possibility of a selective overlap of expression with pollen of other developmental stages cannot be rigorously excluded in these experiments, on the whole it is clear that, during pollen ontogeny, many genes are put to work at periods centering around the first haploid mitosis and the division of the generative cell. This strategy is necessary because the development of the microspore is a rapid process and a whole new program of the gametophyte needs to be elaborated in this relatively short time span.

(ii) Expression of Other Genes during Pollen Development

Expression of several genes isolated from other parts of plants are developmentally regulated during microsporogenesis. As the principal component of microtubules, tubulin genes are ubiquitous in plant cells. Tubulin is a heterodimer of two subunits, namely, α- and β-tubulin; γ-tubulin has also been identified as a third class, but it appears to be a rare protein and its participation in microtubule assembly is not established (Fosket and Morejohn 1992). The expression of α- and β-tubulin in pollen is highly dependent upon the stage of development, because mature pollen grains appear to be the most vulnerable stage for transcript expression. This was shown by using probes for α-tubulin (*TUA-1*) from *Arabidopsis* (Carpenter et al 1992) and for β-tubulin (*TUB-3, TUB-4,* and *TUB-5*) from *Zea mays* (Rogers, Greenland and Hussey 1993) in Northern blots of pollen RNA from the respective species. Like tubulin, actin subunits polymerize to form microfilaments, and expression of members of the actin gene family in pollen grains is developmentally regulated (Thangavelu et al 1993).

In view of the importance of mitochondria and chloroplasts during microsporogenesis, especially in their pattern of inheritance, it was of interest to follow the expression of these organelle genes. In maize, transcripts of three mitochondrial genes, apocytochrome-6 (*COB*), ATP synthase subunit-6 (*ATP-6*), and subunit-9 (*ATP-9*), are found to peak at the mid-term of pollen development, coincident with the rapid replication of mitochondria, and then to decrease to undetectable levels in the mature pollen. On the other hand, suggestive of maternal inheritance of plastids, transcripts of plastid-encoded genes for the 32-kDa quinone-binding protein associated with photosystem II (*PSB-A*) and for *RBC-L*, the large subunit of ribulose-1,5-bisphosphate carboxylase (Rubisco), are not detected at any appreciable level during microsporogenesis (Lee and Warmke 1979; Monéger et al 1992). Other studies have shown that mature pollen grains of maize express transcripts of catalase and superoxide dismutase genes, whereas those of tobacco express transcripts of *ADH* and pyruvate decarboxylase (Acevedo and Scandalios 1990; Bucher et al 1995). Collectively, these four enzymes are believed to play a central role against oxidative stress in plants during photosynthesis, respiration, and photorespiration, and it is reassuring to know that pollen grains are ideally packaged to withstand the stress.

Consistent with the frequent accumulation of starch in mature pollen grains, transcripts of a sucrose synthase gene isolated from potato are expressed to an appreciable extent toward the late stages of development of tobacco pollen grains (Olmedilla, Schrauwen and Wullems 1991). By PCR amplification, Zhang et al (1994) have shown that there is a correlation between the appearance of mRNAs encoding glyoxylate-cycle enzymes, isocitrate lyase (ICL), and malate synthase (MS), and the metabolism of storage lipids in pollen grains of *Brassica napus*, because these enzymes are detected only during the period when storage lipid levels decrease.

Investigations on the accumulation of HSP genes during pollen development in maize illustrate a set of conclusions different from those drawn from the other systems considered. Marrs et al (1993) obtained from a maize leaf genomic library two clones (*HSP-81* and *HSP-82*) that carry sequences corresponding to the 83-kDa HSP gene of *Drosophila melanogaster* and showed that proteins coded by these sequences have 64%–88% amino acid homology to the *HSP-90* family of genes from a variety of organisms. Dot–blot and Northern blot analyses in which RNA isolated from normal and from heat-shocked tassels with microsporocytes and anthers enriched for microspores and pollen grains of different ages was probed with labeled *HSP-81* led to the conclusion that expression of this gene is developmentally regulated and occurs even in the absence of a heat stress during microsporogenesis. In contrast, expression of *HSP-82* appears to be regulated differently, since its transcripts are seen only after a heat shock. High levels of transcripts of both genes are detected in the premeiotic microsporocytes and in microsporocytes undergoing meiosis; this is followed by a decrease in signal intensity with pollen maturation, but the levels of *HSP82* transcripts are 4- to 5-fold lower than those of *HSP-81*. When these cells are heat shocked, *HSP-*

82 transcripts are expressed at about 40-fold higher levels than in uninduced cells. The generality of these effects and the relationship between the two genes remain to be investigated. Transcripts of two heat-shock protein genes from maize (*MHSP18-1* and *ZMEMPR-9'*), encoding proteins of low molecular mass (about 18 kDa), also accumulate during microsporogenesis in a stage-specific manner, independent of heat-shock response (Dietrich et al 1991; B. G. Atkinson et al 1993). The constitutive nature of *HSP-81*, *MHSP18-1*, and *ZMEMPR-9'* is probably of some importance in the defense of the microsporocytes and pollen against heat stress.

(iii) Structure of Pollen-specific Genes and Their Proteins

Functions of the pollen-specific genes are best understood from sequence data and from protein analyses, and so the perspective of this section now shifts to the structure of individual genes. Central to this discussion is the nature of the protein products of the genes and their relationship to the biosynthetic pathways unique to pollen development. Moreover, nucleotide sequence data of genomic clones provide a means of identifying the promoter sequences of genes.

Some Genes Involved in Pectin Metabolism. The comprehensive analysis of the structure of cDNA and genomic clones of *LAT-52*, *LAT-56*, and *LAT-59* from tomato anthers serves to project the protein products of pollen-specific genes into a general format. Comparison of the cDNA and genomic sequences showed the presence of a single intron of 458 bp in *LAT-52* and two small introns in *LAT-56* (71 and 86 bp) and *LAT-59* (73 and 102 bp). The putative proteins encoded by *LAT-52*, *LAT-56*, and *LAT-59* are of molecular masses 17.8 kDa, 40.6 kDa, and 50.9 kDa, respectively. The proteins possess hydrophobic domains at the amino termini with potential signal sequence cleavage sites, suggesting that they are secreted molecules. These general features predicted by the amino acid sequences of the three proteins should not be taken to exclude some of their specific features. For example, LAT-52 protein is characterized by the presence of nine cysteine residues, which are probably involved in intra- and intermolecular cross-linking through disulfide bridges. The protein is also distantly related to Kunitz trypsin inhibitor and is believed to play a role in pollen hydration or germination. A comparison of the protein sequences of *LAT-56* and *LAT-59* showed that these two proteins have a 55% amino acid identity. Particularly intriguing is the limited sequence homology of the predicted proteins of *LAT-56* and *LAT-59* to a pectinase group of enzymes, namely, the pectate lyases of tomato and of the bacterial plant pathogen *Erwinia chrysanthemi* (Twell et al 1989b; Wing et al 1989; Muschietti et al 1994). The amino terminal region of proteins encoded by *LAT-56* and *LAT-59* also shows sequence similarity to the tryptic peptide of a pollen allergen from *Cryptomeria japonica*. The sequence similarities of LAT-56 and LAT-59 proteins to the major pollen allergen AMBA-1 from *Ambrosia artemisiifolia* and to the protein encoded by *9612*, the cDNA clone of a gene that is expressed predominantly in the style of tomato, are also impressive. Some interpretations have been offered for the unusual homology observed between a protein found in tomato pollen and the proteins of other organisms as diverse as fungi and gymnosperms. It is conceivable, for example, that pollen tubes require pectin-degrading enzymes to make their way through the intracellular matrix of the style. A more convincing view is the possibility that, like many allergens, proteins encoded by *LAT-56* and *LAT-59* are typical pollen wall–held molecules. General features of LAT-56 and LAT-59 proteins are reminiscent of those of *S*-allele–expressed glycoproteins, suggesting an interaction between the two groups of proteins during compatible pollinations (McCormick 1991). There is an obvious need for considerable experimental support before these views can be further discussed. Sequence analysis of a tobacco pollen–specific cDNA clone (*TP-10*) and its corresponding genomic clone (*G-10*) has indicated that they encode proteins with pectate lyase activity (Rogers, Harvey and Lonsdale 1992).

The essential role of genes involved in pectin metabolism of pollen grains has also been documented from sequence analysis of the pollen-specific *P-2* gene family of *Oenothera organensis*. The principal finding of this work is the extensive protein sequence similarity of cDNA clones representing the *P-2* gene family to another pectinase group of enzymes, polygalacturonase [poly(1,4-α-D-galacturonide)glycanohydrolase] of tomato fruit. In addition to the presence of an identical number of glycosylation sites, the predicted amino acid sequences of pollen cDNA and tomato fruit enzyme exhibit homology in the range of 54% (Brown and Crouch 1990). The protein products of pollen-specific genes isolated from *Nicotiana tabacum* (Tebbutt, Rogers and Lonsdale

1994), *Gossypium hirsutum* (John and Petersen 1994), and *Medicago sativa* (Qiu and Erickson 1996) also show homology to polygalacturonase. The impact of these observations is to compel us to consider the possibility that polygalacturonase genes expressed in the pollen grains may function in depolymerizing pectin during pollen development, germination, and pollen tube growth. Indeed, there is a long list of plants in which pollen grains express polygalacturonase activity; all polygalacturonases so far isolated from pollen grains are exopolygalacturonases (Pressey and Reger 1989; Pressey 1991).

cDNA Clones from Pollen of Maize and Other Plants. A cDNA clone expressed in mature pollen grains of maize beginning with the stage of starch accumulation and the formation of sperm cells encodes a polypeptide that has significant homology to tomato polygalacturonase. Immunoblot analysis with tomato polygalacturonase antibodies has identified two polypeptides of molecular masses 49 kDa and 53 kDa in the active polygalacturonase of maize pollen. Based on the similarity of the N-terminal sequences of the two proteins, their apparent dualism can be considered to originate from a single precursor through posttranslational modifications (Niogret et al 1991). Transcripts of the polygalacturonase gene, as well as its protein product, are abundantly present in maize pollen grains, suggesting that at least some of the accumulating messages are concurrently translated for immediate use (Dubald et al 1993). cDNA clones of the polygalacturonase gene family have been isolated from maize pollen by other investigators (Rogers et al 1991; Allen and Lonsdale 1992, 1993).

A different category of pollen-specific gene is represented by cDNA clone *ZMC-13* from maize. As shown by primer extension analysis, *ZMC-13* turns out to be a full-length copy of the mRNA and is 976 nucleotides long, including a poly(A) tail of 47 nucleotides. The ZMC-13 protein (molecular mass 18.3 kDa) has about 170 amino acid residues, and from the tentative presence of a hydrophobic signal peptide, it can be considered to be a secreted protein. It shows substantial amino acid homology with LAT-52 protein, including the presence of a cluster of six conserved cysteine residues. Characterization of the genomic clone of *ZMC-13* shows that it is entirely colinear with that of the mRNA, but no introns are present (Hamilton et al 1989; Hanson et al 1989). A pollen-specific gene of rice (*PS-1*) and its deduced amino acid sequence show high degrees of homology with the *ZMC-13* gene and its deduced protein (Zou et al 1994). Like *LAT-56* and *LAT-59* genes from tomato, two other pollen-specific genes from maize, *ZM-58.1* and *ZM-58.2*, are noteworthy for their significant sequence homology to pectate lyases of *Erwinia* sp. and to pollen allergens from *Ambrosia artemisiifolia* (Turcich, Hamilton and Mascarenhas 1993). The presence of putative signal peptide sequences and of an unusual grouping of cysteine residues has suggested that a protein encoded by a pollen-specific gene (*TPC-70*) from *Tradescantia paludosa* is secreted for extracellular functions (Turcich et al 1994). Estruch et al (1994) have cloned from mature maize pollen a gene for calcium-dependent calmodulin-independent protein kinase (*CDPK*); because of its pollen-specific expression, it is one of the few protein kinases that are specific for a particular cell type. *PEX-1* is a pollen-specific gene of maize that encodes an unusual protein containing multiple repeats of an extensinlike domain and a putative globular domain at the N-terminus. Because of its localization in the pollen intine and in the pollen tube wall, this protein probably serves as a recognition molecule in pollen–stigma interaction or as a structural element for pollen tube growth (Rubinstein et al 1995a, b).

One of the keys to understanding how actin reorganization might control pollen maturation and germination resides in the activity of actin-depolymerizing factors (ADF). The emphasis on this group of proteins seems particularly well justified in the light of a recent report of isolation of a group of genes that encode ADF-like proteins from pollen grains of maize. An unusual finding from this study is that although two genes of this group (*ZMABP-1* and *ZMABP-2*) are specifically expressed in the quiescent and germinated pollen, a third gene (*ZMABP-3*) is excluded from these very cells but is expressed in all other tissues of the maize plant (Figure 4.11). Biochemical properties of the protein product of *ZMABP-3* are consistent with actin-binding and actin-depolymerizing activities commonly associated with ADF (Rozycka et al 1995; Lopez et al 1996). Similar functions can be envisaged for profilin-like proteins encoded by cDNA clones isolated from maize pollen (Staiger et al 1993).

Screening of a maize genomic library has resulted in the identification in maize pollen of a family of highly dispersed repetitive sequences, designated as *PREM* (for Pollen RetroElement Maize). The *PREM* nucleotide sequence has characteristics of plant retroactive elements, which are described as mobile DNA elements. The sequences are first transcribed into RNA and reverse-tran-

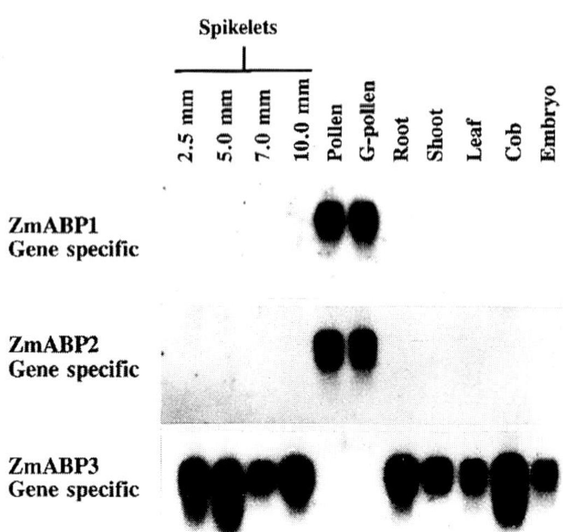

Figure 4.11. Differential expression of transcripts of maize *ZMABP-1, ZMABP-2,* and *ZMABP-3* cDNA clones seen in Northern blots of total RNA from spikelets, developing pollen grains, germinating pollen grains (G-pollen), and vegetative organs of maize. (From Lopez et al 1996; photograph supplied by Dr. P. J. Hussey.)

scribed into DNA before being inserted at a new site in the genome. Transcripts homologous to *PREM* elements are detected during pollen development, especially at the unicellular microspore stage. The general interpretation suggested for this anomalous finding is that transpositions occurring in the DNA of the male germ cells have a greater potential for being inherited than those in the somatic cells. This line of argument seems to be supported by the timing of transcription of *PREM* elements before the first haploid pollen mitosis (Turcich and Mascarenhas 1994; Turcich et al 1996).

cDNA Clones from Pollen of *Brassica napus.* Molecular characterization of a series of genomic clones (*BP-4, BP-19,* and *BP-10*) and their homologous cDNA clones (*BP-401, BP-405,* and *BP-408*) from pollen-specific gene families of *Brassica napus* has focused on the role of specific genes not only in the structural aspects of pollen grains but also in their metabolism. *BP-4* is a composite of three genes, *BP-4A, BP-4B,* and *BP-4C* of about 2,700, 3,160, and 2,720 bp, respectively. Whereas *BP-4B* seems to be nonfunctional because of critical sequence rearrangements, the first and third genes, besides having single introns of almost similar length, are also structurally nearly identical in other respects. Based on this observation, it comes as no surprise that the two functional genes and the homologous cDNA clones of the *BP-4* gene family encode proteins with similar amino acid sequences but with differing sizes. One also gains the impression that all three proteins share a common alternation of hydrophobic and hydrophilic regions and that they are enriched for lysine and cysteine residues. Left unresolved by these findings is the function of these proteins; a speculation based on their high lysine and cysteine contents is that they are associated with pollen wall biosynthesis (Albani et al 1990).

The proteins of pollen-specific clones from tomato and other plants that show identity to enzymes of the pectinase group have a counterpart in yet another pectinase, pectin esterase; this is deduced from the nucleotide sequence of a genomic clone and several homologous cDNA clones of the *BP-19* gene. Two specific features of the *BP-19* gene are that it has a single intron of 137 bp and that it encodes an mRNA of about 1.9 kb. The protein product of the gene has 584 amino acid residues (molecular mass 63 kDa) and reveals a hydrophobic amino terminal region with features of a signal peptide. The predicted pectin esterase identity and the presence of a putative signal peptide in the protein are considered to support the role of the gene in pollen wall reorganization, especially in intine deposition (Albani et al 1991). Robert et al (1993) have shown that the cDNA clone *STA44-4*, corresponding to an mRNA highly expressed in *B. napus* pollen grains, encodes a protein showing sequence homology to polygalacturonase from *Oenothera organensis* and maize; on the other hand, the predicted proteins of two related clones, *STA39-3* and *STA39-4*, resemble arabinogalactans (Gerster, Allard and Robert 1996).

Whereas the *BP-4, BP-19,* and *STA44-4* gene families are thus likely to be involved in the structural aspects of pollen development, members of the *BP-10* gene family appear to code for proteins that have regulatory or metabolic functions. As deduced from sequence analysis and comparison with cDNA clones, the *BP-10* gene contains two introns and codes for a protein of 555 amino acids (molecular mass 62 kDa). The protein has six N-linked potential glycosylation sites and a hydrophobic amino region representing a signal peptide. This protein is assigned an important enzymatic function in pollen development because of its considerable sequence identity to cucumber and pumpkin (*Cucurbita* sp.) ascorbate oxidase (Albani et al 1992). Ascorbate oxidase belongs to a group of widely distributed copper-containing enzymes that possess conserved clusters of histidines. The precise biological function of this group of enzymes in plants in not clearly defined; moreover, since histidine residues are not conserved in the BP-10 protein, it is likely that ascorbate oxidase encoded by the *BP-10* gene has a slightly different enzymatic function. Putative ascorbate oxidase genes that are preferentially expressed in pollen grains have also been identified in tobacco (Weterings et al 1992).

Two genes (*I-3* and *C-98*) that are expressed in abundance in the pollen grains and anthers of *B. napus* have been identified as members of a gene family that encodes an oleosin-like protein. The amino acids encoded by these genes display significant sequence similarity with oleosins characterized from embryos. The oleosin of pollen grains has a molecular mass of 14 kDa and its N-terminal amino acid sequence is homologous to the predicted sequence of the anther-specific cDNA, *I-3*. The deduced amino acid sequences of both *I-3* and *C-98* also possess a characteristic C-terminal domain, somewhat different from that found in oleosins characterized from embryos. These observations are of fundamental importance in implying that pollen-specific oleosin genes are genetically and biochemically different from the embryo-specific class (Roberts et al 1993b; Roberts, Hodge and Scott 1995).

Other Pollen-expressed Genes. Rounding out this account of the structure of pollen-specific genes will be a discussion of sequence data of other pollen-expressed clones that have provided information about the variety of proteins involved in the programmed development of pollen grains. The deduced amino acid sequence of a cDNA clone (*LMP-131A*) expressed in mature pollen grains of *Lilium longiflorum* is partially identical to that of an animal actin-depolymerizing factor, cofilin. The probe from lily pollen was also used to isolate a recombinant clone from a cDNA library of *B. napus* pollen, encoding a protein similar to that encoded by *LMP-131A* (Kim, Kim and An 1993). The deduced amino acid sequence of another pollen-preferential cDNA clone (*LMP-134*) from lily is closely related to a family of thioredoxins (Kim et al 1994).

A cDNA clone, *SF-3A*, isolated from pollen grains of *Helianthus annuus* encodes a 219-amino-acid-long polypeptide containing two potential zinc finger domains of 30 amino acids alternating with 10-amino-acid basic domains. This zinc finger class includes the conserved LIM motif, which is a DNA-binding cysteine-rich protein used in a variety of molecular contexts in animals. Other features of SF-3A protein that set it apart from the mainstream proteins are the absence of a hydrophobic signal peptide at the amino terminus, indicating that the protein is intracellular in nature, and the presence of a sixfold repeat of the pentapeptide sequence at the C-terminus, serving as a possible activator domain. From this it appears that SF-3A protein fits the bill as a transcription factor that regulates the expression of genes coding for late pollen functions (Baltz et al 1992a, b). In physical terms, zinc finger projections on the transcription factor grip specific sites on DNA that prepare genes for activation. Two additional genes expressed in mature pollen grains of sunflower are *SF-16* and *SF-17*. The latter encodes a protein containing blocks of leucine-rich repeat elements, suggesting that it might have a role in signal transduction during pollen development or during pollen–stigma interaction; much less is known about the functional domains of SF-16 protein (Dudareva et al 1994; Reddy et al 1995).

Two genes isolated from a *Petunia inflata* pollen tube cDNA library are abundantly expressed in the pollen grains. One of these (*PPE-1*) encodes a protein that shows sequence similarity to plant, fungal, and bacterial pectin esterases (Mu, Stains and Kao 1994), whereas the other (*PRK-1*) encodes a protein with structural features and biochemical properties of receptor kinases (Mu, Lee and Kao 1994). As a late pollen-expressed gene, the protein product of *PRK-1*, like that of the sunflower *SF-17* gene, may serve in signal transduction during postmeiotic pollen development. In support of this view, it was found that *P. inflata* plants, when transformed with an antisense construct containing the promoter of the *LAT-52* gene fused to *PRK-1* cDNA encoding part of the extracellular domain, produce a large number of aborted pollen grains in which the cytoplasm and nuclei disintegrate after the uninucleate stage (Lee et al 1996). The components of the signal transduction cascade regulated by the PRK-1 protein have not been elucidated.

A pollen-specific sequence (*NEIF-4A8*) and its corresponding genomic clone isolated from a cDNA library of mature tobacco pollen possess rather unusual properties. The genomic clone has a coding region interrupted by three introns and a 5'-untranslated region split by a fourth intron. The encoded protein exhibits a high degree of homology to the typical eukaryotic translation initiation factor with properties similar to RNA helicases. The inherent ability of *NEIF-4A8* to express during late stages of pollen development is duplicated in transgenic tobacco by a promoter construct fused to the *GUS* gene (Brander and Kuhlemeier 1995). Sustaining the high translational activity of general proteins or promoting the translation of pollen-specific mRNAs are two possible roles proposed for the enzymatic protein encoded by *NEIF-4A8* in pollen development, but there is no evidence to support either roles.

(iv) In Situ Localization of Gene Transcripts

It was stated earlier that pollen-specific genes are expressed in the pollen grains and not in the vegetative parts of the plant. Other studies have shown that several genes are expressed in the anther with the enclosed pollen grains. However, Northern blot hybridization as the basis for these conclusions does not provide the kind of resolution necessary to relate gene expression to the cellular level. Pollen specificity of cDNA clones *ZMC-13* and *3C-12* from maize and *STA44-4* from *Brassica napus* is underscored by the fact that when anther sections are subjected to in situ hybridization with single-stranded riboprobes, the marker label accumulates in the pollen grains and not in the anther wall (Hanson et al 1989; Allen and Lonsdale 1993; Robert et al 1993). A question of enduring interest is whether there is a differential pattern of gene expression in the vegetative and generative cells born out of the first haploid mitosis. Hanson et al (1989) showed that transcripts of *ZMC-13* are localized in the cytoplasm of the vegetative cell only, indicating a preferential activation of the gene in this cell. Similarly, transcripts of a pollen-specific gene from tobacco are segregated in the vegetative cell (Weterings et al 1995).

Localization of other pollen-specific genes by in situ hybridization has shown that transcripts cross gametophytic and sporophytic cell types within the anther. This is true of cDNA clones *LAT-52, LAT-56,* and *LAT-59* from tomato pollen, whose transcripts are localized not only in the pollen grains but also in the endothecium and epidermis of the anther wall (Ursin, Yamaguchi and McCormick 1989). In *Brassica campestris*, transcripts of *BCP-1* are expressed at a low level in the microspores and in the tapetum early during their ontogeny; gene expression is considerably enhanced in the mature pollen grains and continues in the tapetal cells until their dissolution (Theerakulpisut et al 1991).

4. PROMOTER ANALYSIS OF POLLEN-SPECIFIC GENES

Regulation of gene expression is under the control of promoter sequences; hence, dissecting the components of promoters that are functional in controlling the spatial and temporal patterns of expression of genes considered in this chapter has become an important part of ongoing investigations into the molecular biology of pollen development. These investigations have the goals of identifying the molecular events that confer differential patterns of gene expression during microsporogenesis and the factors involved in the activation of the promoters during gene expression, but we are obviously a great distance from achieving these goals. However, considerable progress has been made in identifying the promoter sequences of several pollen-specific genes that follow cell- and tissue-specific expression patterns. These studies have been made possible by the ease of regenerating fertile transgenic plants containing chimeric pollen-specific genes, the development of microprojectile systems to deliver chimeric genes into intact pollen grains (transient assay), and the formulation of histochemical assay systems to monitor reporter gene expression.

(i) Genes from Tomato

Deletion analyses with promoters of pollen-specific genes from tomato elegantly illustrate the level of information that can be generated about the sequence elements of genes required for regulated expression during pollen development; studies with tomato genes also appear to be the most comprehensive to date of the promoter regions of pollen-specific genes. In the initial investigations, it was shown that 0.6 kb of the *LAT-52* and 1.4 kb of the *LAT-59* 5'-flanking DNA regions fused to the *GUS* gene are sufficient to direct high levels of expression exclusively in the pollen grains of transgenic tomato, tobacco, and *Arabidopsis* plants. Chimeric genes containing the promoter segment from *LAT-52* are also found to be transiently expressed following their introduction into tobacco pollen by microprojectile bombardment. The expression of the two gene constructs is developmentally and coordinately regulated during microsporogenesis and is closely correlated with the first haploid pollen mitosis (Twell et al 1989a; Twell, Yamaguchi and McCormick 1990). The results described by Twell et al (1991) and, more recently, by Eyal, Curie and McCormick (1995) have further established the functional organization and properties of *cis*-acting regulatory elements in the promoter regions of *LAT-52* and *LAT-59* genes. This was done by analyses of the expression of 5'-deletion constructs and of internal deletion and linker substitution mutants of the promoters both by tran-

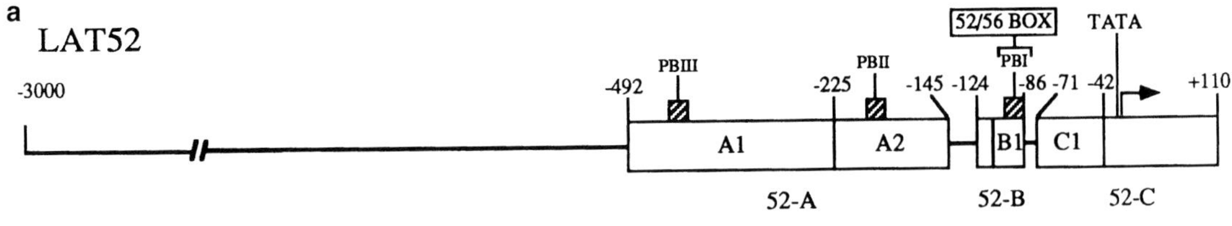

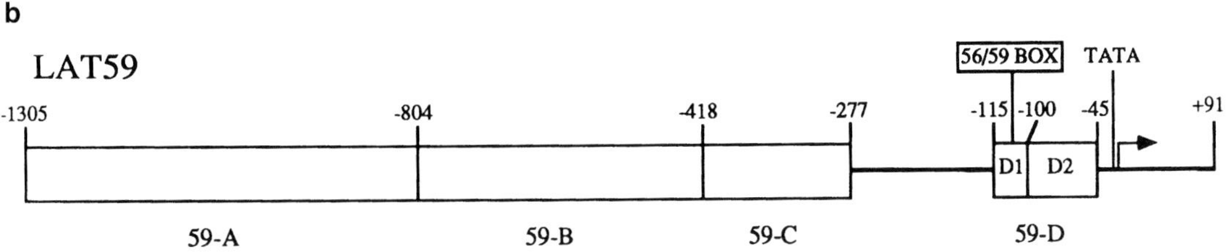

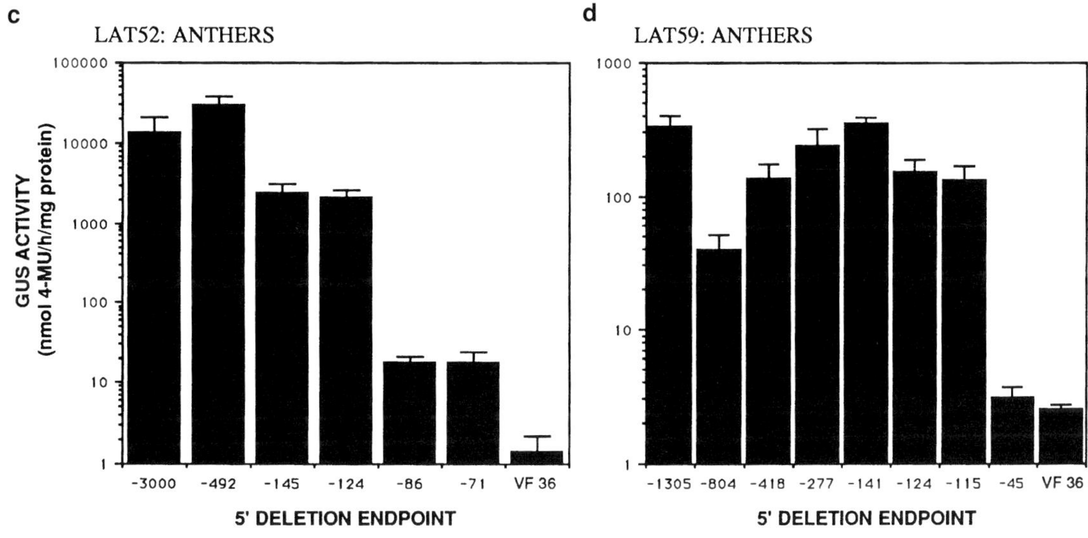

Figure 4.12. Promoter analysis of pollen-specific genes from tomato. **(a, b)** Functional maps of promoters of *LAT-52* (a) and *LAT-59* (b) genes as defined by transient assays and/or expression in transgenic plants. **(c, d)** GUS activity of 5' deletion mutants of promoters of *LAT-52* (c) and *LAT-59* (d) genes in anthers of transgenic tomato plants. (From Twell et al 1991; print supplied by Dr. S. McCormick.)

sient assay and in transgenic plants. It was found that only small stretches of the promoter sequences of *LAT-52* (–71 to +110 bp) and *LAT-59* (–115 to +91 bp) linked to the *GUS* gene are sufficient to drive their expression in the pollen grains of transgenic tomato plants, although the addition of small upstream segments (–492 to –145 bp and –124 to –86 bp for *LAT-52*, and –1,305 to –804 bp for *LAT-59*) appeared to enhance the activity of these minimal promoters. Similarly, regulatory elements located between –163 to –118 bp and –103 to –95 bp appear to be largely responsible for the activity of *LAT-56* promoter in pollen grains (Figure 4.12). Further dissection of the proximal promoters of the two genes showed that common 30-bp regions, defined by –84 to –55 bp in *LAT-52* and –98 to –69 bp in *LAT-59*, alone are sufficient to direct a high level of pollen-specific expression in transient assays or in transgenic plants. In sum, it has been reasoned from these and other analyses that *LAT-52* and *LAT-59* genes share a functionally similar promoter core that probably binds a common *trans*-acting factor, but several upstream regulatory elements apparently modulate the function of each gene in a specific manner in the pollen grains and in certain sporophytic tissues of the plant.

Using a construct in which a *LAT-52* promoter was fused to a modified version of the *GUS* gene containing a nuclear targeting signal, Twell (1992) followed the differential gene expression patterns in the vegetative cell and generative cell of transgenic tobacco pollen. The striking result of this work was the promoter-directed *GUS* gene activity in the nascent vegetative cell to the complete exclusion of the nucleus and cytoplasm of the generative cell. Similar observations were made with isolated microspores of transgenic tobacco allowed to undergo the first haploid mitosis in culture (Eady, Lindsey and Twell 1995) and with tricellular pollen grains of transgenic *Arabidopsis* harboring the *LAT-52* construct (Eady, Lindsey and Twell 1994). These results provide unambiguous evidence for differential gene expression during the first haploid pollen mitosis, a fact that was suspected from other studies. Additional investigations that have illustrated the great versatility of *LAT-52* promoter are the demonstrations of transient expressions of the *GUS* gene driven by this promoter in pollen grains of *Nicotiana glutinosa* (van der Leede-Plegt et al 1992), *N. rustica, N. tabacum, Lilium longiflorum,* and *Paeonia lactiflora* (Nishihara et al 1993) and in pollen protoplasts of *L. longiflorum* (Miyoshi, Usami and Tanaka 1995).

Expression of the *LAT-52* construct in colchicine-inhibited, in vitro cultured microspores of transgenic tobacco has yielded a set of insightful observations concerning the basis of differentiation of the vegetative and generative cells. One is that *LAT-52* promoter is activated even in microspores in which the first haploid mitosis is blocked by high concentrations of the drug; secondly, activation of the gene promoter occurs in both cells of microspores induced to divide symmetrically in a medium containing a low concentration of colchicine (Eady, Lindsey and Twell 1995). Evidently, a mitotic division is not a prerequisite for the expression of a vegetative cell–specific gene, but an asymmetrical alignment of this division silences gene expression in the generative cell.

(ii) Genes from *Brassica*

The promoter sequences of genes *BP-4, BP-10,* and *BP-19* from *Brassica napus* are remarkably varied and have provided new information about promoter functions of pollen-expressed genes. Promoters of the three clones in the *BP-4* gene family range from 235 bp (*BP-4A*) to 360 bp (*BP-4B* and *BP-4C*) in length. Although no enhancerlike elements are defined, all three promoters are punctuated by two 11-nucleotide direct repeats (TAAATTAGATT) approximately 20 and 60 nucleotides upstream of the putative TATA-box. Northern blot analysis of anther RNA from tobacco plants transformed with a segment of the genomic clone of *BP-4A* containing the promoter region indicated that expression of this gene is spatially and temporally regulated similarly to the pattern seen in *B. napus*. These results confirm that transcripts in the transgenic plant correspond to those of the *BP-4A* gene and that the promoter region controls the correct temporal and spatial expression of the gene in tobacco (Albani et al 1990).

The promoter sequence of the *BP-19* gene shows several putative regulatory regions containing repeated elements and short palindromes at the 5'-flanking region. Comparison between *BP-19* promoter and the promoter sequences of a few other pollen-expressed genes has provided the first indication of a potential mechanism that may confer pollen specificity to the promoters (Figure 4.13): the presence of several regions that are partially conserved in all promoters and that could correspond to pollen-specific regulatory elements (Albani et al 1991).

The *BP-10* gene has a promoter region containing two imperfect palindromes, which span positions –107 to –92 bp and –271 to –256 bp of the sequence. They show homology to DNA sequences conserved in various ABA-regulated promoters and other transcription factors. Another interesting feature of the *BP-10* promoter, extending from positions –364 to –338 bp, is an array of TAA trinucleotides tandemly repeated 12 times. This DNA sequence shows similarity to binding sites for homeodomain proteins involved in insect morphogenesis. Promoter fusion experiments have shown that a 396-bp region is sufficient to induce high levels of expression of the gene in the pollen grains of transgenic tobacco plants (Albani et al 1992).

(iii) Genes from Maize and Other Plants

Guerrero et al (1990) have shown that a 375-bp segment spanning the sequences within the first –314 bp upstream and +61 bp downstream relative to the transcription start site of the promoter region of a maize genomic clone (*ZMG-13*) is sufficient to drive pollen-specific expression of a reporter gene in transgenic tobacco plants. That the *cis*-acting elements of the *ZMG-13* gene of a monocot can be recognized correctly by the *trans*-acting factors of a dicot and can drive genetically stable, cell-specific activity was also confirmed by the temporal expression pattern of *ZMG-13* in the pollen grains of progenies of the transformed tobacco plants. Electroporated microspores of *Brassica napus* repre-

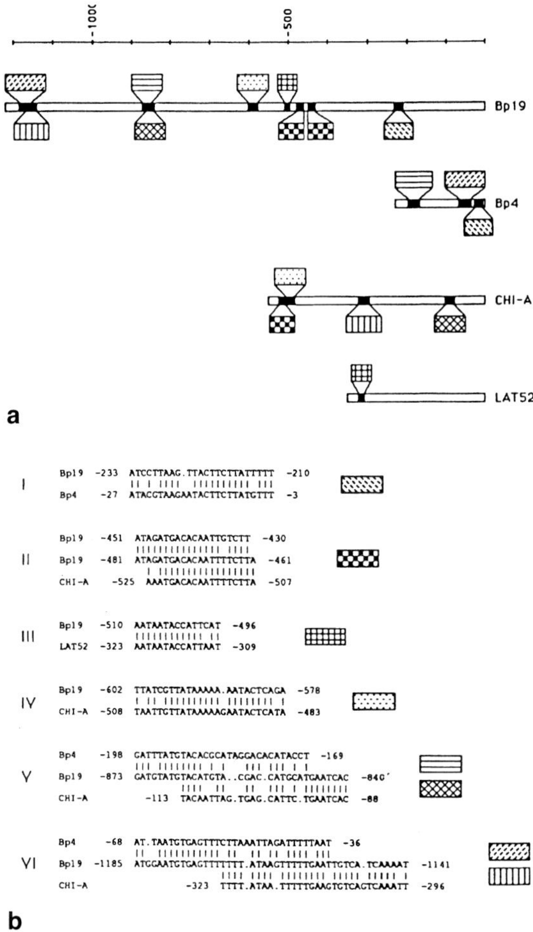

Figure 4.13. **(a)** Schematic representation of sequences conserved between the 5'-flanking regions of *BP-19, BP-4, CHI-A4,* and *LAT-52* genes. The positions are relative to the site of transcription. **(b)** Comparative alignment of the same sequences. (From Albani et al 1991.)

sent another example of a dicot expressing a reporter gene under the control of *ZMG-13* (Jardinaud, Souvré and Alibert 1993). Similarly, the promoter of the *PS-1* gene from rice mediates pollen-specific expression in rice and in transgenic tobacco (Zou et al 1994). A high level of reporter gene expression in postmeiotic microspores and mature pollen grains of transgenic tobacco is observed when promoter segments of maize, tobacco, and cotton pollen polygalacturonase genes are fused with the *GUS* gene (Allen and Lonsdale 1993; Dubald et al 1993; John and Petersen, 1994; Tebbutt, Rogers and Lonsdale 1994). Deletion analysis of the tobacco polygalacturonase gene has revealed the presence of several motifs related to the *cis*-acting sequences of promoters of pollen-specific genes of tomato and maize polygalacturonase (Tebbutt and Lonsdale 1995).

Use of a transient assay system in *Tradescantia paludosa* pollen to study the expression of constructs containing the *GUS* gene under the control of various fragments from the 5'-flanking region of maize *ZMG-13* gene has affirmed the complexity of the maize gene promoter and has identified additional regions that affect the expression of this gene. Deletion of all sequences upstream of –54 bp resulted in a very low to undetectable (1%) level of *GUS* gene activity. Therefore, although sequences both necessary and sufficient for full pollen-specific expression reside upstream of –54 bp, sequences between –100 and –54 bp are sufficient for expression, and those between –260 and –100 bp enhance the expression. The longest deletion construct tested (–1,001 to –260 bp) seems to generate only a 50% level of expression of the –260 bp construct, indicating the presence of a negative regulatory element in the promoter region (Hamilton et al 1992).

It was mentioned earlier that the *CHI-A* gene, which codes for chalcone flavanone isomerase, is expressed in the petals and pollen grains of *Petunia hybrida*. A feature of the *CHI-A* gene that is at the heart of this activity is the presence of two distinct, tandemly arranged promoters that function differently. A simplified interpretation, based on several experiments, is that a downstream promoter (*PA-1*) is active in the petals, whereas the upstream one (*PA-2*) is active in the pollen grains. By analyzing transgenic plants containing chimeric genes, it was possible to identify a 440-bp *PA-2* promoter fragment that is necessary for the pollen-specific gene expression (van Tunen et al 1989, 1990). The regulatory aspects of expression of this gene have been analyzed in a recessive mutation of *P. hybrida* (*po*) in which *CHI-A* expression is restricted to the petals and is absent in pollen grains. Sequences of *CHI-A* gene from *po* dominant and *po* recessive lines showed that they are generally comparable, with differences due to deletions, additions, and the presence of an inverted repeat in the promoter region of the mutant (van Tunen et al 1991). The apparent involvement of these changes as accessory elements contributing to the inactivation of the promoter is intriguing and deserves further investigations.

In some cases, analyses of promoter sequences of anther-expressed genes have opened up new insights into the regulation of gene expression during pollen development. These genes were originally identified as apparently pollen specific, but they have subsequently been found to be expressed in various combinations of somatic tissues of the anther. H. Xu et al (1993) used a series of 5'-deletion constructs made from the genomic clone *BGP-1*, homologous to cDNA clone *BCP-1*, from *Brassica oleracea* to transform *Arabidopsis* plants and showed that a construct containing the 5'-region up to position –168 bp is the minimal promoter sequence necessary for pollen expression in the transformants. This work also identified a negative *cis*-acting element between –322 and –580 bp that affects pollen expression. The distinction between pollen grains and tapetum offered little difficulty in establishing that different *cis*-acting DNA elements control *BGP-1* expression in the latter tissue. Anther-specific gene, *APG*, from *Arabidopsis* encodes a novel proline-rich protein, and its promoter region directs *GUS* activity in the anther tissues as well as in pollen grains (Roberts et al 1993a). The experiments of Twell et al (1993) offer an instructive example of reporter gene activity in pollen grains of transgenic tobacco driven by the *APG* gene promoter that first appears to be uniform throughout pollen development but eventually follows a biphasic pattern. Their data indicate that besides significant *GUS* activity in the unicellular microspores, there is a peak in the mid-bicellular pollen grain stage. This has been attributed to an age-dependent capacity of developing microspores to activate the promoter and is probably related to the role of transcription factors.

(iv) Promoters of Other Genes

The expression of various other genes has been localized in pollen grains by transient assay or by transformation technology. These genes were originally isolated from different parts of plants excluding anthers and pollen grains. In some of the examples investigated, pollen specificity of the promoters has been clearly established, whereas in others the promoters are active in pollen grains as well as in the vegetative tissues of the plant. The interest in these observations lies in the demonstration that pollen grains contain the transcription factors necessary to recognize heterologous promoters. Only future research will tell us whether these genes have any regulatory functions in the developing pollen grains.

The best-studied case of pollen specificity of the promoter of a nonpollen gene comes from the work on the expression of *Arabidopsis* α_1-tubulin promoter monitored by *GUS*, *CAT*, and *DT-A* reporter gene assays in transgenic *Arabidopsis* and tobacco plants (Kim and An 1992). *CAT* and *GUS* gene activities were detected exclusively in mature pollen grains of both species carrying the transgene. Because of the cytotoxic nature of the *DT-A* gene, plants carrying the chimeric gene showed defective pollen development or formed pollen grains unable to transmit the transgene to the progeny. Preferential expression of the *GUS* gene driven by *Arabidopsis* α_1-tubulin promoter in pollen grains of transformed *Arabidopsis* (Carpenter et al 1992) and by maize *TUBα-1* promoter in transgenic tobacco (Rigau et al 1993) has also been demonstrated.

Among other nonpollen genes, pollen localization has been shown in transgenic tobacco for promoters of the following genes encoding specific proteins: 5-enolpyruvylshikimate-3-phosphate synthase from *Petunia* sp., which catalyzes a step in the biosynthesis of aromatic amino acids (Benfey and Chua 1989); wound-inducible proteins encoded by *WUN-1*, derived from potato, which accumulate upon wounding (Siebertz et al 1989); *Arabidopsis* ubiquitin and *H^+-ATPase* (Callis, Raasch and Vierstra 1990; DeWitt, Harper and Sussman 1991); tobacco osmotin, which encodes a protein that is related to the osmotic stress of plants (Kononowicz et al 1992); a desiccation-related protein from *Craterostigma plantagineum* (Scrophulariaceae) (Michel et al 1993); pathogenesis-related proteins of unknown function induced in tobacco and *Asparagus officinalis* (Liliaceae) (Uknes et al 1993; Warner, Scott and Draper 1993); *Arabidopsis ACP*, which encodes cofactors in fatty acid biosynthesis (Baerson and Lamppa 1993; Baerson, Vander Heiden and Lamppa 1994); *Petroselinum crispum* 4-coumarate:coenzyme A ligase (*4-CL*), which plays a key role in phenylpropanoid metabolism (Reinold, Hauffe and Douglas 1993); rice endochitinase (Zhu, Doerner and Lamb 1993); maize *ADH-2* (Paul and Ferl 1994); chlorophyll a/b-binding protein (*CAB*) from *Pinus thunbergii* (Kojima, Sasaki and Yamamoto 1994); *Brassica napus* stearoyl-ACP desaturase, a fatty acid desaturation enzyme (Slocombe et al 1994); ICL and MS of *B. napus*, which encode substrates for glyoxysomal functions (Zhang et al 1994); *Nicotiana plumbaginifolia* superoxide dismutase, which plays a key role in cellular defense against reactive oxygen species (Hérouart, van Montagu and Inzé 1994); 3-hydroxy-3-methylglutaryl coenzyme A reductase (*HMG-2*) from *Arabidopsis* and potato, which catalyzes a key step in isoprenoid biosynthesis (Bhattacharyya et al 1995; Enjuto et al 1995); *Arabidopsis LEA* gene *ATEM-1* and genes *KIN-1* and *COR-6.6*, which are transcriptionally activated by low temperature, as well as by ABA and drought (Wang and Cutler 1995; Hull et al 1996); and the *Phaseolus vulgaris* storage protein, phaseolin (van der Geest et al 1995). Additional researchers have described or recorded, in different transgenic systems, expressions driven by promoters of genes encoding the following proteins: rice actin-1 gene (*ACT-1*) in pollen grains of transgenic rice and maize (Zhang, McElroy and Wu 1991; Zhong et al 1996); patatin and adenosine diphosphate (ADP)-glucose pyrophosphorylase genes, which encode in potato tubers the abundant polypeptides and a key regulatory enzyme in starch biosynthesis, respectively, in pollen grains of transgenic potato (X.-Y. Liu et al 1991; Müller-Röber et al 1994); soybean gene encoding root and root nodule cytosolic glutamine synthetase in pollen grains of transgenic *Lotus corniculatus* (Fabaceae) (Marsolier, Carrayol and Hirel 1993); bean and *Arabidopsis PAL* genes in pollen grains of transformed *Arabidopsis*, potato, and tobacco (Ohl et al 1990; Shufflebottom et al 1993; Hatton et al 1995); heat-inducible *HSP-81.1* from *Arabidopsis* in pollen grains at all stages of development of transgenic *Arabidopsis* (Yabe, Takahashi and Komeda 1994); rice *WX* gene in pollen grains of transgenic rice and *Petunia* sp. (Hirano et al 1995); *Arabidopsis* phytochrome and actin genes in pollen grains of transgenic *Arabidopsis* (Somers and Quail 1995; An et al 1996); and rice tungro bacilliform virus (*RTBV*) in pollen grains of transgenic rice (Yin and Beachy 1995). Deletion analysis of the maize *ADH-1* gene promoter has shown that pollen-specific expression in transgenic rice requires part of a leader sequence outside the promoter region (Kyozuka et al 1994). After particle bombardment–mediated transformation, localization has been demonstrated for promoters of the following genes: *Petunia hybrida CHI-A* gene in pollen grains of *Nicotiana glutinosa* and *N. tabacum*; mannopine synthase gene (*TR-2′*) of *Agrobacterium tumefaciens* in pollen grains of *N. glutinosa* and *Lilium longiflorum* (van der Leede-Plegt et al 1992; Stöger et al 1992); CaMV 35S promoter in pollen grains of *N. tabacum, N. rustica, L. longiflorum*, and *Paeonia lactiflora* (Nishihara et al 1993); promoter of a constitutively expressed rice gene (*GOS-2*) involved in the initiation of translation in pollen grains of barley (Hensgens et al 1993); and rice actin *ACT-1* and tomato *LAT-52* genes in pollen grains of maize and *Picea abies* (Pinaceae) (Martinussen, Junttila and Twell 1994; Jardinaud et al 1995). Pollen grains of certain plants also express in transient assays a *GUS*-RNA construct with an in vitro synthesized 5' cap structure that was introduced by particle bombardment. This observation reinforces the view that the pollen cytoplasm possesses factors for translation of foreign mRNA, although RNA-mediated gene expression is less efficient than that mediated by DNA (Tanaka et al 1995). Since there are several ways by which *GUS* gene activity is expressed in transgenic plants and in transient assays as a blue color, including spurious background color, a more explicit approach becomes necessary to follow the expression of these genes in developing pollen grains.

5. POLLEN ALLERGENS

As noted in Chapter 3, pollen walls accumulate several proteins that become allergens and contribute in a major way to hayfever. Although most allergenic proteins are of intine origin, the active allergens of ryegrass are found in the exine. Because intine proteins are synthesized late in pollen development, they are encoded by late pollen-expressed genes, indicating haploid gene expression. On the other hand, allergens of the exine are synthesized in the tapetum and their presence on the pollen wall more appropriately reflects gene action in the diploid tapetal cells. Although the following account deals with the characterization of genes encoding allergenic proteins of pollen, this distinction between the two types of allergens should be kept in mind.

From a practical standpoint, the most productive approach for isolating genes that code for allergenic proteins is to screen an expression cDNA library made to mRNA from pollen grains with serum from patients allergic to pollen of known species or with antibodies prepared specifically against allergenic proteins. The first complete amino acid sequence of an allergen (BETV-I) was reported from pollen of *Betula verrucosa* and was deduced from a cDNA clone selected from an expression library made to mature pollen mRNA and screened with the serum of a birch-pollen-allergic patient. Subsequently, as many as 13 different cDNA clones coding for BETV-1 isoforms with a basic primary structure have been characterized. Although the amino acid sequence of BETV-I shows some homology to other known allergens, a pronounced similarity is found to a protein (I-49) involved in the resistance response of pea pods to plant pathogens (Breiteneder et al 1989; Swoboda et al 1995b) and to RNase (Bufe et al 1996). The deduced amino acid sequence of another pollen allergen cDNA clone from birch is homologous to profilins of a variety of organisms, including humans (Valenta et al 1991). The presence of profilins as allergens poses a dilemma in interpretation because of the primary function of this group of proteins as actin-binding factors.

Pollen grains of *Poa pratensis* (Kentucky bluegrass) have provided another system for an incisive analysis of allergenic protein genes. In the work on *P. pratensis*, several cDNA clones encoding pollen allergens were obtained by immunologically screening a pollen mRNA expression library with human allergic serum. One of the clones (*KBG-7.2*) was found to hybridize in a specific manner to a 1.5-kb mRNA transcript from *P. pratensis* pollen, as well as to mRNA from pollen grains of eight other grasses. Comparisons of clones *KBG-31, KBG-41*, and *KBG-60* with respect to nucleotide sequence similarity, pollen specificity, transcript length, amino acid sequence similarity, and presence of leader peptides in the amino acid sequence have revealed an exceptionally high degree of homology to one another. However, when the derived amino acid sequences of the four cDNAs are analyzed for possible sequence homologies with known proteins, they do not match any other sequences and thus appear to be new proteins (Mohapatra et al 1990; Silvanovich et al 1991). Three clones from an anther cDNA library of *Dactylis glomerata* (orchard grass; Poaceae) encoding a pollen allergen have been characterized by screening an expression library with human allergic serum and rabbit polyclonal antiserum raised to a pollen extract (Walsh et al 1989). An expression library of anthers of *Brassica napus* and *B. rapa* has yielded a pollen allergen belonging to a new class of Ca^{2+}-binding proteins (Toriyama et al 1995).

By screening a cDNA library with appropriate antibodies, genes encoding AMBA-I, the allergen of ragweed pollen, have been characterized. cDNA clones divided into three groups according to their sequence similarity (*AMBA-I.1, AMBA-I.2*, and *AMBA-I.3*) share an astonishing degree (greater than 99%) of identity within the group at the nucleotide level and 85%–90% identity between groups. Typical of the close identity of the cDNA groups are also other features, such as hybridization to a 1.5-kb message, proteins containing 396–398 amino acids coded by the sequences, and the presence of a signal polypeptide. Because of the multiple isoelectric forms of proteins encoded by the three groups of cDNAs, AMBA-I is classified as a family of closely related proteins (Rafnar et al 1991). Using PCR, the gene for *AMBtV* from another ragweed, *Ambrosia trifida*, has been cloned and sequenced (Ghosh, Perry and Marsh 1991).

The central problem identified from the cloning and sequencing of genes encoding the allergenic protein (LOLP-I) of ryegrass also appears to be the presence of more than one isoform of the same allergen. By screening a pollen cDNA library with a synthetic oligonucleotide probe complementary to a segment of the amino acid sequence of LOLP-I protein, Perez et al (1990) isolated two cDNA clones, corresponding to two isoforms of the allergen. Comparison of the nucleotide and deduced amino acid sequences of the two clones showed more than 20 nucleotide differences, although only 4 resulted in amino acid changes. Two full-length cDNA clones subsequently isolated and characterized from an expression library by immunological screening with antibodies are *LOLP-IA* and *LOLP-IB*. Both clones are found to be pollen specific, hybridizing to 1.35-kb (*LOLP-IA*) and 1.2-kb (*LOLP-IB*) transcripts. The mature processed proteins encoded by *LOLP-IA* and *LOLP-IB* have molecular masses of 26.6 kDa and 31.3 kDa, respectively, and both possess signal peptide at the N-terminal sequence (Griffith et al 1991; Singh et al 1991). Consistent with the exine origin of the allergen, LOLP-IA is immunologically localized in the exine and its central chamber. The experimental picture is complicated to some extent by the apparent occurrence of LOLP-IA in the pollen cytoplasm, especially in the mitochondria (Staff et al 1990), and of LOLP-IB in the starch granules (Singh et al 1991). No direct observations have been made of the pathway of the accumulation of the allergen in the starch grains; the presence of the signal peptide in the protein supports the view that precursors synthesized in the cytoplasm are targeted initially into the amyloplast by the signal peptide that separates after transport into the organelle.

Although contributions from the characterization of genes for pollen allergens to the pollen gene expression pattern is marginal, this approach has paved the way for the synthesis of allergenic proteins in pure form in a less labor-intensive way than was possible before.

6. GENERAL COMMENTS

Hitherto, evidence for a coordinated program of gene expression during angiosperm pollen development has been advanced mostly from genetic and biochemical studies. With the cloning and isolation of pollen-specific genes, the study of the gene expression program during pollen development may be said to have come of age. Cloning of the genes, their characterization, and identification of their protein products have provided new information about the regulatory aspects of pollen gene expression. Assays that reconstitute the promoters of these genes in a transgenic background have led to the identification of sequences that mediate in pollen gene expression. The pollen specificity of many of the genes investigated suggests that they can serve as useful target genes for the development of genetically engineered plants. Information presently available indicates that genes that code for proteins necessary for pollen germination and pollen tube growth are activated during pollen development. Yet, much remains to be learned about the function of the proteins encoded by the pollen-specific genes and about the genes that determine the fate and function of the vegetative and generative cells. Taking into account the diversity of pollenspecific genes cloned, the complete story of pollen gene expression is likely to become very complicated.

The role of the pollen grain as the carrier of sperm cells lends great importance to the goal of isolating and characterizing the genes and proteins that are essential at critical stages of male gametogenesis. Complementary approaches, such as cell ablation techniques combined with the use of marker genes, could provide insightful information about gene expression in the pollen grain in the absence of the vegetative cell or the generative cell. Isolation of mutants that affect the development of these pollen cells would be useful in elucidating the importance of cellular interactions before the final phase of male gametogenesis. These studies are expected to usher in a new level of understanding of the molecular biology of the angiosperm pollen grain and the mechanisms controlling the spatial and temporal expression of pollen-specific genes.

orously established. Besides, progenies of many interspecific and intergeneric crosses have also been reported to display genic male sterility. In about 35 species, genic male sterility has been induced by mutagens, mainly by gamma rays. On the whole, genic male sterility is probably of wide occurrence in angiosperms, and a number of male sterile lines have been identified in crop plants such as maize, barley, pea, tomato, and pepper. Cytoplasmic male sterility arising spontaneously and in hybrids is known to occur in at least 153 species, including representatives of both wild and cultivated plants. Outstanding among the crops known for spontaneous cms is *Zea mays*, which possesses three major types of male-sterile lines. These lines, designated as cms-T (Texas), cms-S (U.S. Department of Agriculture), and cms-C (Charrua), are distinguished according to the nuclear-encoded restorer genes that compensate for incompatibilities that are phenotypically expressed during anther and pollen development. The cms-S and cms-C lines each require a single dominant restorer gene, *RF-3* and *RF-4*, respectively, for fertility restoration whereas two restorer genes, *RF-1* and *RF-2*, override sterility in the cms-T line. Spontaneous occurrence of gene–cytoplasmic male sterility has been reported in only about 40 species, although a large number of gene–cytoplasmic male-sterile lines have arisen from interspecific and intergeneric hybridization programs. Mutation-induced cases of gene–cytoplasmic male sterility are rare, since simultaneous or successive mutations at the nuclear and cytoplasmic levels are required for its expression (Edwardson 1970; Kaul 1988).

In recent years, molecular and genetic tools have been employed for the generation of male-sterile plants. These include the use of antisense RNA to demolish tapetal activity and pollen development (Mariani et al 1990; van der Meer et al 1992; de Block and Debrouwer 1993; Denis et al 1993; Xu et al 1995a; Matsuda et al 1996; Zhan, Wu and Cheung 1996); the expression of the exotoxin-A gene from *Pseudomonas aeruginosa* fused to regulatory sequences of the storage protein napin gene (Koning et al 1992); the expression of the *DT-A* gene fused to promoters of the *S*-locus glycoprotein and storage protein phaseolin genes (Kandasamy et al 1993; Thorsness et al 1993; van der Geest et al 1995); the expression of the root loci (*ROL-C*) gene of the T-DNA of *Agrobacterium rhizogenes* under the control of a promoter (Schmülling, Schell and Spena 1988; Schmülling et al 1993); tapetum-specific genes for overproduction or underproduction of specific metabolites for disruption of microsporogenesis (Worrall et al 1992; Spena et al 1992); overexpression of a mitochondrial gene that disrupts anther development (Hernould et al 1993); and transposon tagging (Aarts et al 1993). A promising new approach to induce conditional male sterility in plants is by engineering tapetal expression of chimeric genes that selectively convert exogenously applied nontoxic compounds to toxic compounds and cause pollen abortion (O'Keefe et al 1994; Kriete et al 1996).

2. STRUCTURAL, DEVELOPMENTAL, AND FUNCTIONAL ASPECTS OF MALE STERILITY

Mutations affecting male fertility are so very diverse that it is impossible here to describe the variations in individual species or indeed the diversity that occurs in a single widely investigated species such as *Lycopersicon esculentum*. To this can be added a general statement that mutations range from the production of flowers completely lacking stamens to those in which the stamens are perverted into other floral organs, such as sepals, petals, carpels, or phyllodes. In male-sterile mutants of *Zea mays* (Beadle 1932), *L. esculentum* (Rick 1945), *Nicotiana tabacum* (Raeber and Bolton 1955), cultivated members of Cucurbitaceae (Shifriss 1945), *Gossypium hirsutum* (Allison and Fisher 1964), and *Pisum sativum* (Klein and Milutinović 1971), microsporogenesis is a wasteful process because there is only a vestigial development of stamens that under most conditions, produce no anthers or pollen; in *Antirrhinum majus* (Schick and Stubbe 1932), *Sorghum bicolor* (Karper and Stephens 1936), and *Tagetes erecta* (Asteraceae) (Bolz 1961), stamens are completely absent. In a genic mutant of *L. esculentum* described by Sawhney and Greyson (1973a) stamen development is normal until the primordium is about 100 μm long; thereafter, development of the stamen of the wild type outstrips that of the mutant. In their final form, mutant stamens are dwarf and pale and they bear naked ovules at the base of the anther. Differentiation of the sporogenous tissue is confined to the adaxial surface of the anther, and sporangia – ranging in number from zero to three in each anther – generally enclose shriveled, or occasionally normal, pollen grains. Potentialities for sex reversal that results in the formation of external ovules on staminal tubes have been noted in an interspecific hybrid from the cross (*Gossypium anomalum* × *G. thurberi*) × *G. hirsutum* (Meyer and Buffet 1962). Application of morphactin [methyl-2-chloro-9-hydroxy-fluorene-(9)-carboxylate] interferes with stamen development in certain plants and induces feminization symptoms ranging from a reduction in the number of stamens to the transformation of stamens to carpellate structures or to the complete suppression of stamen function (Mohan Ram and Jaiswal 1971; Jayakaran 1972). In *Arabidopsis*, there have been reported some homeotic mutants in which male fertility is diminished due to a lack of stamens, others in which anthers are absent or infrequently converted to sepals (Bowman, Smyth and Meyerowitz 1989; Hill and Lord 1989; Chaudhury et al 1992), and still others in which anthers fail to elongate (Estelle and Somerville 1987). All these mutations result in varying degrees of male sterility.

Various operational criteria are used to assess gene action at the structural level when the transformation of stamens into foliar or other floral organs occurs. These include the formation of different types of floral organs in place of stamens, and whether the substitute organs contain pollen grains or not. The most frequent form of transformation appears to be carpeloidy of stamens, although occasional cases of petaloidy and of the simultaneous absence of petals and stamens are also on record (Kaul 1988). Mutations known as *pi* in *Arabidopsis* and *def*, *glo*, and *sep* in *Antirrhinum majus* have a general tendency for enhanced feminization and are characterized by the presence of carpels in the place of stamens (see Chapter 2). Different degrees of stamen malformation have been observed in a spontaneous mutation in soybean that generates male-sterile flowers (Skorupska, Desamero and Palmer 1993). Evidently, what appears to result in sterility of the male reproductive system in many angiosperms involves, in fact, several clearly defined structural intermediates representing the action of a network of interacting genes.

(i) Cytological Basis of Male Dysfunction

A complete account of the changes associated with microsporogenesis in male steriles is clearly an unrealistic goal for this section, in view of the task of summarizing more than one hundred publications in the field. The simpler task undertaken here is to focus on the changes in a few representative types with the goal of identifying the mode of action of the different male-sterile genes.

Most of the cytological studies are based on simultaneous light or electron microscopic observations of the development of anthers in fertile and sterile lines of the species. From these accounts, a general sequence of events causing male sterility can be considered to begin with tapetal abnormalities and to end with the failure of pollen formation. During microsporogenesis, the effects of the sterility gene can be identified before microsporocyte meiosis, at various stages of meiosis, at the time of release of microspores from the tetrad, and as microspores mature into pollen grains. Indeed, in some plants, like tomato (Rick 1948), pea (Gottschalk and Klein 1976), *Cirsium arvense* (Asteraceae) (Delannay 1979), maize (Albertsen and Phillips 1981), rice (Lu and Rutger 1984), and *Arabidopsis* (Dawson et al 1993; Chaudhury et al 1994) among others, mutations covering the entire developmental sequence of microsporogenesis have been identified. Especially in maize, a staggering number of mutations affecting the cytogenetic events during meiosis have been generated and described (Golubovskaya 1989).

Gene Action in Premeiotic Anthers. Based on the final form of the anther in the male-sterile lines, assumptions about the time of gene action during the premeiotic stages are entirely arbitrary, although in many cases cytological observations have provided some clarifications. If gene action occurs at the primordial stage of the anther before the differentiation of sporogenous tissue, a program of continued cell divisions ensues without the formation of the archesporium. Accordingly, there is an accumulation of undifferentiated parenchymatous cells in the potential anther locule and the real effect of the gene is concealed as the anther continues to elongate. A case of male sterility resulting in anthers lacking sporogenous tissue has been described in *Pennisetum americanum* (pearl millet; Poaceae). Lack of wall layers in sections of defective anthers is often a tangible indication of the failure of sporogenous cells to differentiate (Rao and Devi 1983). Even when the sporogenous tissue escapes the action of the mutant gene, the phenotype of the anther produced is generally characterized by morphological simplicity and limited internal differentiation. The absence of microsporocytes, or their degeneration after they are formed, is often diagnostic of the timing of gene action. Whereas male sterility resulting in the degeneration of microsporocytes has been described in several plants, the formation of sporogenous cells that are uncommitted to develop into microsporocytes has been found in only a few cases (Kaul 1988). Included in the latter category is

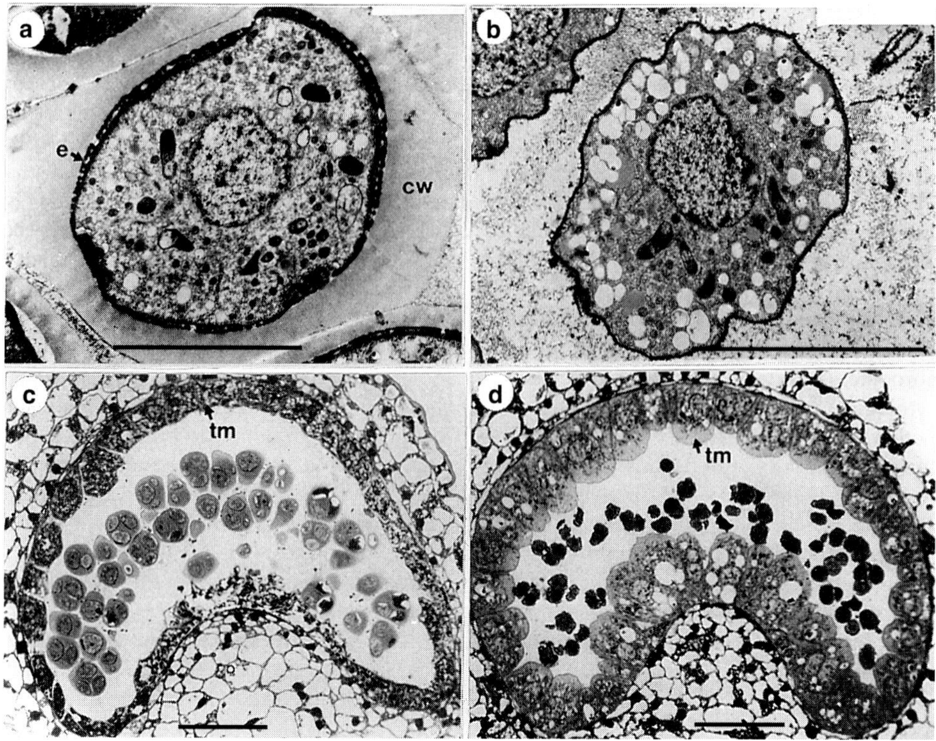

Figure 5.2. Premature callose dissolution and aberrant microspore and tapetum development in transgenic tobacco. **(a)** Microspore of a tetrad from wild-type tobacco with well-developed exine within a callose sheath. **(b)** Microspore from the transformant, showing the absence of the callose wall and exine patterning. **(c)** Section of the wild-type anther. **(d)** Section of the transformed anther showing the hypertrophied tapetum and sterile pollen grains. cw, callose wall; e, exine; tm, tapetum. Scale bars = 10 μm (a, b) and 45 μm (c, d). (From Worrall et al 1992. *Plant Cell* 4:759–771. © American Society of Plant Physiologists. Photographs supplied by Dr. R. J. Scott.)

(Dundas, Saxena and Byth 1982); and cytomixis or the movement of pachytene chromatin from one microsporocyte to another (Stelly and Palmer 1982) (Figure 5.3). Meiotic abnormalities observed in chemically induced mutants of rice and pea are not fundamentally different from those seen in spontaneously occurring mutants and include desynapsis, asynapsis, vacuolation of microsporocyte cytoplasm, and fusion of microsporocytes (Kitada et al 1983; Kaul and Nirmala 1993).

Despite the fact that only a relatively few genes strike during meiosis I and II, similarities in the abnormal syndromes observed due to gene action during these periods are impressive. As seen in male-sterile mutants of *Solanum tuberosum* (Fukuda 1927), *Hevea brasiliensis* (rubber; Euphorbiaceae) (Ramaer 1935), *Lathyrus odoratus* (Fabergé 1937), *Hebe townsoni* (Scrophulariaceae) (Frankel 1940), *Pennisetum americanum* (Rao and Koduru 1978), and *Secale cereale* (Cebrat and Zadęcka 1978), the mutant gene impairs the normally error-free segregation of chromosomes during metaphase of meiosis I. Irregular spindle formation and its disharmonious orientation during meiosis I or II, chromosome bridge formation, random distribution of chromosomes, and incomplete separation of dyads and tetrads have been commonly observed in mutants of *S. tuberosum* (Fukuda 1927), hemp (Breslavetz 1935), maize (Clark 1940), *Saccharum officinarum* (sugarcane; Poaceae) (Janaki-Ammal 1941), *Tripsacum laxum* (Poaceae) (Dodds and Simmonds 1946), *Cucumis melo* (Bohn and Principe 1964), carrot (Zenkteler 1962), and barley (Herd and Steer 1984). A range of abnormalities, including chromosome stickiness and hypercontraction, and premature centromere division, often accompanies varying degrees of male sterility in inbred populations of *Alopecurus myosuroides* (Poaceae) (Johnsson 1944). A particularly good example of late meiotic irregularities is provided by *Impatiens sultani*, in which male sterility is preceded by the precocious disjunction and tripolar separation of chromosomes, the formation of chromosome bridges, various patterns of atypical cytokinesis,

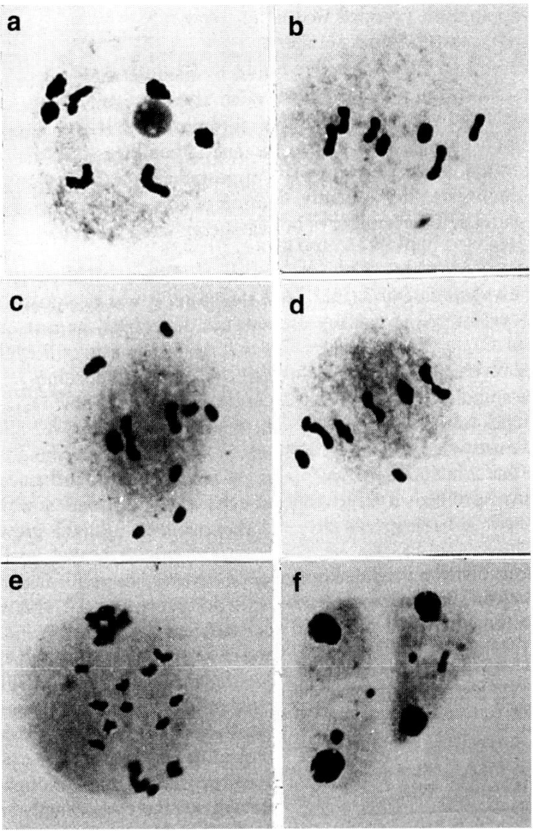

Figure 5.3. Meiosis in normal (a, b) and cms lines (c–f) of *Phleum nodosum* (Poaceae). **(a)** A normal diakinesis showing seven bivalents. **(b)** A normal metaphase. **(c)** Metaphase I with three bivalents and eight univalents. **(d)** Metaphase I with six bivalents and two univalents. **(e)** Late anaphase I, showing the division of five univalents. **(f)** Telophase II, showing micronuclei that have been excluded from the nuclei. (From May and Kasha 1980.)

supernumerary division of meiotic products, and irregular division of microspores (Tara and Namboodiri 1974). How do these irregularities affect the quadripartitioning of the microsporocyte and the formation of microspores? Recurrent features of the action of male-sterile genes that cause breakdown of late meiotic processes are incomplete cytokinesis and failure of normal tetrad formation. In *C. melo*, an incomplete division apparently promotes the formation of triads with one large nucleus and two small nuclei (Bohn and Principe 1964). Degeneration of tetrads before microspores become disjoined occurs in *Hebe salicifolia* (Frankel 1940), *Brassica oleracea* (Cole 1959), *B. campestris* (Chowdhury and Das 1968), *Manihot esculenta* (cassava; Euphorbiaceae) (Magoon, Jos and Vasudevan 1968), *Lupinus mutabilis* (Fabaceae) (Pakendorf 1970), *Cirsium spinosissimum, C. palustre* (Delannay 1979), *Cajanus cajan* (pigeon pea; Fabaceae) (Reddy, Green and Bisen 1978; Dundas, Saxena and Byth 1981), and maize (Albertsen and Phillips 1981). In certain male-sterile lines of *Glycine max*, cytokinetic failure results in as many as four nuclei becoming enclosed within a single wall to form coenocytic microspores (Albertsen and Palmer 1979; Delannay and Palmer 1982). In other male-sterile lines, microsporocytes that survive up to the tetrad stage display characteristic degenerative symptoms such as vacuolation and formation of darkly staining crescent-shaped structures; a proclivity to disintegrate in situ in the tetrad without the dissolution of the callose wall is intrinsic to these microspores (Stelly and Palmer 1982; Graybosch et al 1984; Graybosch and Palmer 1985a). Defective cytokinesis as a mechanism of male sterility leading to the formation of coenocytic microspores has also been described in barley (Herd and Steer 1984). In an X-ray–induced male-sterile mutant of *Pisum sativum* (Klein 1969) and in cytoplasmic male-sterile *Phaseolus vulgaris* (Johns et al 1992), incomplete quadripartitioning results in the formation of tetrads in which sister microspores are joined by conspicuous cytoplasmic strands. The range of meiotic abnormalities described in male-sterile mutant lines of *Arabidopsis* generated by T-DNA tagging is almost kaleidoscopic, covering many stages of meiosis and producing tetrads containing as many as eight microspores (He et al 1996; Peirson et al 1996). Irrespective of the stage of meiosis when mutant genes act, the final product in all cases is an anther with nonfunctional pollen grains in the locule.

As is well known, the cytoskeletal elements (microtubules and microfilaments) are deeply involved in the events of mitosis and cytokinesis, but little is known about their rearrangements during abortive pollen development in male steriles. Staiger and Cande (1990, 1991) have shown that utilization and reorganization of microtubules and microfilaments correlate well with abnormalities observed during microsporocyte meiosis in mutants of maize. A typical lesion is found to disrupt microtubule arrangement during the transition from a prometaphase to a metaphase spindle during meiosis I, resulting in diffuse and divergent metaphase spindle poles; other lesions cause an abundance of microtubules in the cortical cytoplasm during diplotene and diakinesis, the formation of a spindle with multiple poles, and the appearance of extra spindles. Mutations that alter

In several types of cms, the first detectable divergence from normal development frequently involves changes in the tapetum, with one of the consequences of tapetal dysfunction being the premature disintegration of the tapetum. Since tapetal activity must synchronize with the development of the microspores for efficient transfer of nutrients, this early disintegration deprives the microspores of certain unique substances at a time when they are most needed. Several investigators have described in exquisite detail tapetal cytology in cytoplasmic male steriles and their fertile counterparts and have invariably shown that male sterility is temporally linked with tapetal abnormalities. A few examples will be described here. In a male-sterile line of *Sorghum bicolor*, tapetal dysfunction is evident at about the time of completion of microsporocyte meiosis. Several basic changes seen in the tapetum during the whole cycle of cell demolition serve as an instructive paradigm of progressive gene action: (a) thickening and vacuolation of cells; (b) cell enlargement and disorganization; (c) cytoplasmic disintegration and rupture of radial walls; and (d) migration of tapetal contents into the locule as a syncytium. These changes contrast with the more regular developmental repertoire of the tapetum of the fertile anther, in which the cell contents remain fully functional until late in microspore development (Overman and Warmke 1972). The ontogeny of the tapetum in male-sterile *Secale cereale* illustrates a different level of interaction with the contents of the anther locule. Here, following the breakup of the tetrad, the tapetum becomes coenocytic and invades the locule; this is followed by the formation of an aggregate between the tapetal mass and degenerating microspores. Still later, the aggregate becomes compressed as a thin layer constituting a lining over what is left of the anther wall (Scoles and Evans 1979). According to Horner and Rogers (1974), a period of cell enlargement and vacuolation in the tapetal cells precedes microsporocyte meiosis in male-sterile *Capsicum annuum*, and the tapetum remains in this state even as the tetrads collapse. The basic plan of tapetal disintegration in cytoplasmic male-sterile *Helianthus annuus* is similar to that seen in *C. annuum*, although the tapetum does not persist here beyond the disintegration of the tetrads (Horner 1977). There is a considerable increase in the size of the tapetal cells of anthers of male-sterile *Daucus carota* and *H. annuus*; this is also associated with an increase in the size or number of the nuclei (Zenkteler 1962; Kini, Seetharam and Joshi 1994). Although the degenerating tapetum wanders into the anther locule as a periplasmodium in many cms lines, this cannot be generalized. Pollen sterility in anthers of *Beta vulgaris* is associated with either a periplasmodial or a cellular tapetum and, outwardly, the signs of pollen degeneration under both conditions remain the same (Artschwager 1947).

Tapetal cells of some cms are more sensitive than the sporogenous cells to gene action and start along the degenerative pathway with conspicuous vacuolation of the contents (Warmke and Lee 1977; Lee, Gracen and Earle 1979; van Went 1981; Bino 1985a, b; Conley, Parthasarathy and Hanson 1994). Profiles of tapetal behavior described in cms lines of *Sorghum vulgare* (Brooks, Brooks and Chien 1966), *Phaseolus atropurpureus* (Pritchard and Hutton 1972), *Iris pallida* (Iridaceae) (Lippi et al 1994), and *Lycopersicon esculentum* (Polowick and Sawhney 1995) differ from one another only in minor details and have essentially confirmed that this monolayer of cells is ahead of the pack in leading the charge for abortive processes in the anther. In a cms line of *Petunia hybrida*, the most striking changes observed in the tapetum even before the onset of ultrastructural upheavals are inhibition of DNA synthesis and inhibition of nuclear division (Liu, Jones and Dickinson 1987). Close correlation between tapetal abnormalities and pollen abortion has been reported in both T- and C-cms lines of *Zea mays* (Warmke and Lee 1977; Lee, Gracen and Earle 1979); however, the tapetal cells of the S-version behave similarly to those of fertile anthers (Lee, Earle and Gracen 1980). These differences in tapetal function between what appears to be phenotypically similar cases of male sterility in the same species illustrate the striking diversity of this phenomenon at the ultrastructural, and perhaps at the molecular, level.

Three types of abnormal tapetal development in relation to pollen abortion have been documented in a cms line of wheat. In the first type, disintegration of the tapetum begins early in the anther ontogeny, and the disorganized tissue carries along with it the sporogenous cells to the same fate; in another type, the tapetum forms a periplasmodium that engulfs the microsporocytes and destroys them prior to meiosis. In the third type, tapetal degeneration, marked by tangential elongation of cells and decrease in size of the nuclei, is delayed until tetrads have been formed (Chauhan and Singh 1966). In another male-sterile line of wheat, the tapetum persists up to the mature pollen grain stage, even though the tissue has disappeared from the male-fertile anther (Joppa, McNeal and Welsh 1966).

What is the level of integrity maintained by the tapetum in anthers of genic male steriles? Present evidence indicates that there are striking similarities between cytoplasmic and genic male steriles in the pattern of ultrastructural changes in the tapetal cells; obviously, the mechanism underlying male sterility in the two groups of plants must share some properties in common. A study by Cheng, Greyson and Walden (1979) illustrates the fate of the tapetum in genic male-sterile maize. The first changes are detected in the tapetum at the young microspore stage and include the appearance of large vacuoles around the nucleus. This is followed by degeneration of the microspores, and by the young pollen grain stage, only remnants of the tapetum remain (see Figure 5.1). Degeneration of tapetal cells in male-sterile lines of *Cajanus cajan* (Dundas, Saxena and Byth 1981) and soybean (Graybosch et al 1984; Graybosch and Palmer 1985a), also associated with intense vacuolation beginning with meiosis in the microsporocytes, is complete at the time of the breakdown of tetrads. Tapetal dysfunction in male-sterile *Medicago sativa* is initiated just before the onset of meiosis in the microsporocytes and encompasses swelling and overgrowth of the cells and acquisition of a dense cytoplasm. The complete disintegration of the tapetum does not occur until after the demise of the microsporocytes and is characterized by the presence of large oil-like bodies in the locule (Childers 1952).

One unsettled aspect of tapetal breakdown relates to the function of inclusions like oil bodies and other materials. Some of these inclusions form discrete deposits on the outer surface of the tapetal cells of a male-sterile *Oenothera hookeri* hybrid, and it has been suggested that they may be sporopollenin precursors (Noher de Halac, Cismondi and Harte 1990). An overproduction of sporopollenin, identified as globules in the locule of a cms line of barley, illustrates a further link between tapetal function and male sterility (Ahokas 1978). An extreme case of failure of the secondary parietal cells to divide and form the tapetum has been reported in another male-sterile line of barley. The abortion of microsporocytes while they are still enclosed in a callose sheath can be traced to the absence of callase due to the failure of tapetal function (Herd and Steer 1984). In several mutants of *Arabidopsis*, tapetal cells degenerate in the young anthers and this probably causes abortion of microsporocytes before they enter meiosis or of pollen grains shortly after they escape from the tetrads (Aarts et al 1993; Dawson et al 1993; Chaudhury et al 1994).

To the extent that a perceived link between the premature breakdown of the tapetum and the incidence of male sterility tells us something about the nutritive role of this tissue, a similar generalization emerges when we relate pollen abortion to the constant presence of an intact tapetum. Apparently, the diversity of interactions between the tapetum and sporogenous tissues predicts that failure of tapetal disintegration deprives the microspores of nutrient materials and starves them to death. Moreover, a persistent tapetum would divert a great deal of metabolites for its own survival, at the sacrifice of microspore development. In male-fertile *Sorghum vulgare*, the tapetal cells become binucleate and start degenerating as meiosis is completed in the microsporocytes; in contrast, in the cms line, the cells become multinucleate and persist long after the few surviving pollen have matured (Singh and Hadley 1961). Rapid degeneration of the tapetum at the tetrad stage of microspores is the rule in fertile lines of *Linum usitatissimum*, whereas in sterile anthers the tapetal cells remain intact even at the pollen grain stage (Dubey and Singh 1965). In some genic male-sterile lines of barley in which free microspores begin to degenerate, the tapetum remains as a healthy tissue and undergoes changes similar to those of fertile anthers (Kaul and Singh 1966; Mian et al 1974). One aspect of tapetal cell behavior that can be correlated with the onset of sterility in the anthers is the proliferation of the tissue into several layers of cells that eventually fill the anther locule. This change, observed in male-sterile *Cucurbita maxima* (Francis and Bemis 1970) and *Brassica napus* (Grant, Beversdorf and Peterson 1986), suggests that the morphogenetic switch of the tapetum into a multilayered tissue may be a causal factor in the collapse of the microsporocytes and microspores. Degenerative changes in the tapetum of the *sl-2* mutant of tomato appear to be permissive rather than directive and, despite the early vacuolation of cells indicative of the onset of degeneration, the process is delayed until after the microspores have aborted (Sawhney and Bhadula 1988).

One of the most important and neglected aspects of the relationship between the behavior of the tapetum and pollen abortion concerns the nature of tapetal metabolites whose absence contributes to male sterility. In the classical literature, these substances are generally envisioned as enzymes, hormones, and general food materials for the nurture and nutrition of the developing microsporocytes and microspores, and as sporopollenin precursors for the pollen wall. Although no other somatic tissue of the anther is as well understood at the cytological level as the tapetum, it seems clear that new approaches will be required to associate specific substances of the tapetum with pollen sterility. The failure of microspores to differentiate a well-defined wall following gametocide treatment of florets has led to the suggestion that inadequacies in the supply of sporopollenin precursors from the tapetum and interference in their cross-linking and polymerization contribute to pollen abortion (Mizelle et al 1989; Schulz, Cross and Almeida 1993). Regardless of the validity of this suggestion, it is frustrating that these important consequences of tapetal activity cannot be subject to direct experimentation.

The whole question of the role of the tapetum in male sterility is a subject of intense current investigation using genetic engineering methods. It was mentioned in Chapter 2 (pp. 39–41) that the *TA-29* gene isolated from tobacco anthers is transcribed specifically in the tapetum and that RNase-degrading fusion genes, *TA-29/RNase-T1* and *TA-29/BARNASE*, selectively destroy the tapetum, thereby leading to the formation of pollenless male-sterile plants, and induce male-sterility in transgenic plants. This could happen only if male sterility in the anthers were linked to a high RNase activity, resulting in a decrease in the level of RNA in the tapetum. This was found to be the case by both in situ hybridization and RNA dot–blot hybridization. As shown in a later work, it was possible to restore male fertility in genetically engineered male-sterile *B. napus* by introducing the gene *BARSTAR* for counteracting *BARNASE* action. Crossing male-sterile plants harboring chimeric *BARNASE* gene with male-fertile plants containing the *BARSTAR* gene was found to produce male-fertile plants. Although both genes are coexpressed in anthers of the male fertiles, *BARNASE* action is apparently suppressed by *BARSTAR* through the formation of complex proteins in the tapetum (Mariani et al 1992).

Since the tapetum is the principal site of flavonoid biosynthesis in the anther, down-regulation of its biosynthetic pathway might be expected to affect tapetal function. Van der Meer et al (1992) were able to induce male sterility in transgenic *Petunia hybrida* by expressing an antisense flavonoid-specific gene of immature *P. hybrida* anthers under the control of the CaMV 35S promoter. The role of the tapetum in inducing male sterility was inferred from the observation that transgenic plants expressing the *GUS* gene driven by the chimeric CaMV 35S promoter showed staining in the tapetal cells. Anthers of transgenic plants displayed severe reduction in flavonoid content and in the level of chalcone synthase mRNA. In yet another strategy, a chimeric gene consisting of a tapetum-specific promoter from *Antirrhinum majus* and the *ROL-B* gene of *Agrobacterium rhizogenes* was expressed in transgenic tobacco plants. The anthers of transgenic plants displayed symptoms reminiscent of male sterility, such as shriveled appearance and decreased pollen output. Since *ROL-B* gene product causes increased auxin activity, the phenotypic alterations in the anther have been attributed to increased auxin production, possibly in the tapetum (Spena et al 1992). *BCP-1*, a gene isolated from pollen grains of *Brassica campestris*, is also expressed in the tapetum; interference with the expression of this gene either in the tapetum or in pollen grains thwarts the production of fertile pollen grains in transgenic *Arabidopsis* (Xu et al 1995a). These techniques find practical application in the production of male-sterile plants and in the genetic engineering of novel hybrids.

3. PHYSIOLOGICAL AND BIOCHEMICAL ASPECTS OF MALE STERILITY

Cytological investigations reviewed in the previous section have helped to elucidate the sequence of events leading to male sterility in plants. Unfortunately, the mechanism by which genetic information is translated into altered anther phenotype is poorly understood. From previous accounts we conceive of the anther primordium of male-sterile mutants as being similar in structure to that of the male-fertile lines; yet, at defined stages during anther ontogeny, pollen development is arrested. What are the physiological and biochemical differences between anthers of male-sterile and male-fertile plants that can be construed as impinging on microsporogenesis? Do male-sterile anthers synthesize a particular set of enzymes connected with the abolition of male function? What is the nature of the male-sterility gene identified by mutations? Answers to these questions have not been obtained, but there is considerable amount of background information that argues for environmental and hormonal involvement and for a role for mitochondrial plasmid in the process.

(i) Environmental and Hormonal Control

Most studies of environmental control of male sterility have emphasized the effects of temperature and photoperiod; the responses can be broadly classified as causing the transformation of male steriles into fertiles at a particular environmental condition and of male fertiles into steriles at another condition. Moreover, because identical conditions might induce one set of changes in some species and different changes in others, it is difficult to arrive at an operational generalization. Careful investigations have revealed that temperatures above optimum generally accelerate the progress toward sterility in male steriles and are inimical to the production of viable pollen grains in male fertiles (Meyer 1969; Welsh and Klatt 1971; Izhar 1975; Polowick and Sawhney 1988). There are also examples such as *Allium cepa* (van der Meer and van Bennekom 1969), *Sorghum bicolor* (Brooking 1979), and *Capsicum annuum* (Polowick and Sawhney 1985) in which low temperatures are known to cause pollen infertility. The effect of temperature on stamen determination at the primordial stage is strikingly illustrated in the *sl-2* mutant of tomato, which produces external ovules on the anther. Flowers, complete with stamens displaying surface features of the wild type (pattern of hair formation, shape of epidermal cells, and cuticular thickenings) and normal-looking anthers enclosing viable pollen grains, are formed on mutant plants grown at a low temperature regime; on the other hand, a high temperature regime favors the production of normal carpels in place of stamens (Sawhney 1983; Sawhney and Polowick 1986).

Phenotypic form of the stamen and progress of microsporogenesis in *ogura* (*ogu*) cms line of *Brassica napus* are dependent upon narrow limits of temperature. Stamens typical of the male-sterile line, with a reduced output of pollen, are produced in plants grown at a high 28–23 °C day–night temperature regime. At a low temperature regime (18–15 °C day–night) the stamens resemble carpels, whereas at an intermediate temperature regime (23–18 °C day–night), stamens that also bear characteristics of carpels are produced. Microsporogenesis in the stamens produced at the intermediate temperature cycle generally stops at the tetrad stage. The value of this system for analyzing the environmental control of male sterility lies in the correlated changes observed in the tapetum during temperature-induced anther development. These range from degenerating tapetal cells with sparse deposits of sporopollenin in anthers reared at high temperature, highly vacuolated cells at intermediate temperature, and the complete absence of the tapetum in the low temperature cycle (Polowick and Sawhney 1987, 1990, 1991).

There is a long list of plants in which pollen fertility is affected by the photoperiodic conditions of their growth (J. Heslop-Harrison 1957). In maize, it has been known for a long time that, compared to plants grown in summer, plants grown in autumn exhibit reduced pollen fertility. Moss and Heslop-Harrison (1968) brought these observations into focus by showing that short photoperiods promote male sterility; analogous to the classical dark-interruptions in short-day plants, male fertility is partially restored by interrupting the long dark period with low intensity light. Moreover, the arrest of microsporocyte development during meiosis and the visually detectable tapetal malfunction in the anthers provide a structural basis for relating the photoperiodic effect to gene action. Close parallels to the effects of photoperiod on male sterility in maize are also found in other members of Poaceae, such as *Rottboellia exaltata* (J. Heslop-Harrison 1959) and *Dichanthium aristatum* (Knox and Heslop-Harrison 1966); in the latter, pollen failure follows as a consequence of tapetal malfunction. A spontaneous mutant of rice that is male sterile under long-day conditions and that becomes fertile under short days, has been described; consistent with the operational criterion for the involvement of phytochrome, male sterility in this mutant is induced by red light administered during the long dark period of short days, and the effect of red light is reversed by far-red light (Tong, Wang and Xu 1990).

To understand the effects of environmental factors on male sterility it is necessary to delve into the action of hormones, as in many instances extracellular control of plant growth and development is intertwined with changes in endogenous hormone levels. A review of the literature shows that all major groups of hormones are implicated in the male sterility of plants (Sawhney and Shukla 1994). In several plants, auxin treatment promotes male sterility and shifts the balance to increased femaleness (Rehm 1952; Sawhney and Greyson 1973b). In the *sl-2* mutant of tomato in which auxin effects parallel the effects of high temperature, not only do mutant stamens contain higher endogenous IAA than stamens of normal plants, but an increase in IAA concentration is noted in stamens of mutants exposed to high temperature (Singh, Sawhney and Pearce 1992). These results suggest that temperature effects are mediated through changes in endogenous IAA levels.

A totally different effect seen in the *sl-2* mutant treated with GA is the reversion of mutant stamens to the wild type enclosing viable pollen (Sawhney and Greyson 1973b). Corroborative evidence that the development of wild-type stamens is dependent on the biosynthesis or metabolism of gibberellins has come from some related observations. One is that flowers of the mutant contain a lower level of extractable gibberellin-like substances than those of the wild type (Sawhney 1974). Secondly, the requirements for growth in culture of floral buds of the mutant are different from those for the wild type and include the addition of GA to the medium (Rastogi and Sawhney 1988a). Supplementation of the medium with 2-chloroethyl-trimethyl ammonium chloride (CCC), an inhibitor of GA biosynthesis, or with ABA, known to counteract GA action, inhibits stamen growth of the mutant flower buds without affecting female fertility (Rastogi and Sawhney 1988b). Reports of complete or partial reversal of male sterility by GA in other tomato mutants (Phatak et al 1966; Schmidt and Schmidt 1981; Jacobsen et al 1994) and in mutants of barley (Kasembe 1967) on the one hand, and of suppression of maleness or induction of male sterility by GA in maize (Nelson and Rossman 1958; Hansen, Bellman and Sacher 1976), sunflower (Schuster 1961), *Allium cepa* (van der Meer and van Bennekom 1973), *Brassica oleracea* (van der Meer and van Dam 1979), and *Capsicum annuum* (Sawhney 1981), on the other hand, attest to the continuing interest in the role of this hormone in stamen development. In a gibberellin-deficient mutant of tomato (*gib-1*), microsporocytes are arrested at the premeiotic G1 stage of the cell cycle, and the application of GA allows the completion of meiosis and formation of viable pollen grains (Jacobsen and Olszewski 1991).

Cytokinins are another group of hormones involved in the development of sex organs and of male sterility in plants. Although it has not been possible to induce male sterility or fertility in plants in a major way by cytokinin application, bioassays have revealed differences in endogenous cytokinin levels in mutant and wild-type plants. Analysis of the endogenous cytokinins of plants is extremely complex, since there are many different cytokinins, not all of which respond to a particular bioassay or show up in an analytical procedure. This fact should be taken into account in reviewing the work on the role of cytokinins in the expression of male sterility. In root exudates of a cms line of barley, Ahokas (1982b) found a lower level of cytokinins compared to the normal and fertility-restored lines. The suggestion that an increase in cytokinin activity is associated with anther fertility in barley is compatible with the results. Comparable observations have been made in genic male-sterile and *ogu* cms lines of *Brassica napus* (Shukla and Sawhney 1992; Singh and Sawhney 1992). Results of cytokinin changes in male-sterile and bisexual plants of *Plantago lanceolata* (Plantaginaceae) indicate that roots of the former have a higher amount of zeatin riboside than those of bisexual plants (Olff et al 1989). Here, male sterility is associated with cytokinin changes in a different way, presumably reflecting a different underlying mechanism. Culture of young floral meristems of a male-sterile tobacco hybrid in a medium containing a specific concentration of kinetin led to differentiation of flowers showing features of male sterility (Hicks, Browne and Sand 1981). Since kinetin is necessary for the expression of male-sterile phenotype beyond the stage of initial differentiation of floral organs, one could speculate that under normal growth conditions, the cytokinin required for expression of male sterility is provided by the parent plant.

Investigations on the dioecious plant *Mercurialis annua* (Euphorbiaceae) have played a major part in relating male sterility to cytokinin metabolism. At least some degree of change in the phenotype of the flower can be induced by exogenous application of cytokinins to floral primordia, with a trend toward feminization of genetic males being most prominent (Durand 1966). With regard to the cytokinin contents of the floral apices, the free base zeatin is associated with femaleness; zeatin riboside, rather than zeatin, is found in the male apex (Dauphin-Guerin, Teller and Durand 1980). Further work has identified two separate cytokinin pathways during the differentiation of fertile and sterile flowers: a *trans*-cytokinin pathway linked to fertile lines, and a *cis*-cytokinin pathway active in the male-sterile lines (Louis, Augur and Teller 1990). The precise regulation of each pathway is attributed to the action of genes controlling the production of specific metabolic intermediates in cytokinin biosynthesis.

How changes in cytokinin metabolism cause male sterility in plants sharing identical genomes is illustrated in a male-sterile mutant of *Arabidopsis* deficient in the purine salvage pathway enzyme adenine phosphoribosyl transferase, which converts adenine to adenylic acid. At the cytological level, the genetic lesion affects the development of microspores after their release from the tetrads; at the biochemical level, high concentrations of adenine or its intermediates probably inhibit in a nonspecific way continued pollen development in the mutant (Moffatt and Somerville 1988; Moffatt, Pethe and Laloue 1991).

It has already been noted that the ethylene-releasing agent, ethrel, acts as a gametocide in barley. ABA has also been related to male sterility in certain plants in an indirect way, such as the application of jasmonate (or jasmonic acid, a compound with chemical and biological similarities to ABA) to restore fertility in male-sterile *Arabidopsis* mutants that contain negligible amounts of certain fatty acids (McConn and Browse 1996). In wheat plants, water stress leads to the formation of small, shriveled anthers containing nonviable pollen grains and causes male sterility and low seed set (Morgan 1980; Saini and Aspinall 1981). Not only were wilted plants found to have an increased level of ABA, but it was also possible to mimic the effect of water stress on anther and pollen development by exogenous application of the hormone to healthy plants (Morgan 1980; Saini, Sedgley and Aspinall 1984). Since the sequence of anatomical events leading to male sterility by water stress is different from that induced by ABA, it is probably misleading to suggest that ABA is the sole regulatory agent that induces male sterility in water-stressed plants. Molecular biologists have largely overlooked the impact of adverse environmental conditions on the reproductive biology of plants, but the relationships between water stress, ABA, and pollen infertility may well provide a classic example of the importance of interacting environmental and hormonal conditions on reproductive development. The possible involvement of ABA in the expression of male sterility in a genic male-sterile line of *B. napus* is noted in the higher endogenous level of ABA in the mutant stamens in comparison to the wild type and in the impairment of stamen development and pollen viability in wild-type flower buds cultured in the presence of ABA in the medium (Shukla and Sawhney 1994).

(ii) Biochemical Control

Though the visible effects of male-sterility genes are manifest during the chaotic microsporocyte meiosis and other subsequent events, there are some subtle biochemical changes that precede or follow gene action. Many of the causal changes are probably cryptic and are poorly understood, whereas common metabolic alterations are unlikely to provide the type of insight into the mechanism of male sterility.

Our present knowledge of the respiratory characteristics of anthers of fertile and male-sterile lines remains scanty, but there is little doubt that our awareness of the involvement of the mitochondrial genome in male sterility established in some plants will accelerate research in this area. Although the participation of a cyanide-resistant, alternative respiration pathway in some male steriles and their fertile counterparts has remained doubtful or has been deemed insignificant (Musgrave, Antonovics and Siedow 1986; van der Plas et al 1987; Håkansson, Glimelius and Bonnett 1990; Johns et al 1993), it has been established to be the case in cultured cells and immature anthers of *Petunia hybrida* (Connett and Hanson 1990). Irrespective of the electron transport pathway chosen by the mitochondria, one factor that influences mitochondrial activity in the cells is ATP production and its transport across mitochondrial membranes by ADP–ATP translocator. A finding of some interest in this context is that mitochondria isolated from leaves of a cms line of *P. hybrida* export less ATP than mitochondria from leaves of the fertile line (Liu, Jones and Dickinson 1988). A subsequent work reported that ATP–ADP ratios are lower in cms anthers than in anthers of male fertiles. This is also correlated with corresponding changes in the ratios of the oxidized and reduced forms of nicotinamide adenine dinucleotide phosphate (NADPH–NADH) in anthers of the two lines of plants. Since major reserves of NADPH are found almost exclusively in the tapetal cells, pollen abortion can be traced to the failure of the tapetum to generate high levels of NADPH (Liu and Dickinson 1989).

Changes in the amino acid composition of anthers of fertile and male-sterile lines have been studied as a factor in male sterility. In keeping with this role, the accumulation of certain specific amino acids in the anther has been implicated as the stimulus for initiation of gene action. Chromatographic analysis of anthers of male fertiles and their sterile counterparts has revealed that during early stages of development, free amino acids are distributed in roughly equal proportions in both types; at later stages of development, an excess of certain amino acids or their amides, specific for each species, begins to accumulate in the anthers of the male steriles. Thus, asparagine accumulates in anthers of wheat (Fukasawa 1954), asparagine and alanine in anthers of maize (Fukasawa 1954; Khoo and Stinson 1957), and glycine in anthers of *Sorghum vulgare* (Brooks 1962). From where do these amino acids flow into the anther, and how does the accumulation of one or two amino acids in the anther cause disintegration of the sporogenous cells? One view is that the amino acids may represent breakdown products of the sporogenous cells and tapetum and their presence in excess may accelerate the disintegration process already under way.

The importance of amino acids in the anthers of fertile and sterile lines of plants goes beyond accumulation to transient stage-specific changes. This is shown in a comparative study of the quantitative differences in the free amino acid contents of anthers of male-fertile, cms, and genic male-sterile lines of *P. hybrida*. Although breakdown of sporogenous cells occurs at different times during microsporocyte meiosis in the cms and genic male-sterile anthers, the event is linked to an increase in asparagine content in both sterile types (Izhar and Frankel 1973a).

Differences in metabolism experienced by anthers of normal and male-sterile lines are also revealed by one- and two-dimensional electrophoretic profiles of soluble proteins. In anthers of male-sterile *Sorghum vulgare*, the soluble protein profile shows fewer bands in comparison to two fertile lines (Alam and Sandal 1969). This may be due to proteolysis, as shown in *P. parodii*, although protein degradation is more drastic in extracts of fertile anthers than in sterile anthers (Wu and Murry 1985b). Moreover, fertile anthers of *P. parodii* show enhanced levels of certain high molecular mass polypeptides as well as an accumulation of proline (Wu and Murry 1985a). Although there are no differences in the protein bands between normal and male-sterile anthers of barley at early stages of development, coincident with the onset of degenerative changes in the sterile anther, some short-chain polypeptides arising out of proteolysis are found in the electrophoretic profile (Ahokas 1980). Cytochemical studies have shown that there is a gross underrepresentation of DNA and proteins in the microspores and tapetum of male-sterile lines of several species (Chauhan and Kinoshita 1979; Banga, Labana and Banga 1984). Decreases in the levels of these macromolecules are detected after meiosis and it is more than likely that in some species the primary effect of the male-sterility gene is on the tapetum, whose hypertrophy affects the microspores.

Male sterility in the *sl-2* mutant of tomato is associated with an enrichment in the anther of a 23-kDa and a 31-kDa protein (Sawhney and Bhadula 1987). These proteins are also detected in anthers of plants grown at a high temperature that induces sterility; conversely, they are diffuse or absent in anthers of the wild type and of mutants grown at a low temperature. The relationship of these proteins to pollen degeneration in the mutant is, however, problematical, since they are not synthesized in anthers incubated in ^{35}S-methionine. Instead, a 53-kDa protein is synthesized in the mutant anthers, coincident with pollen degeneration (Bhadula and Sawhney 1991). Current speculation on the source of this protein centers on the tapetum.

Our understanding of the regulation of male sterility has been greatly increased by other biochemical studies on anthers of the *sl-2* mutant of tomato. A comparative study of polyamines in the wild-type and mutant anthers has disclosed the unsuspected fact that the latter contain significantly higher levels of putrescine, spermidine, and spermine, and of their biosynthetic enzymes, ornithine decarboxylase and *S*-adenosylmethionine decarboxylase, than the wild type anthers. Another observation relevant to the possible relationship between stamen development and changes in polyamine levels is that the concentration of polyamines in stamens of mutant plants grown at low temperatures decreases to a level comparable to that of the wild type. That high levels of polyamines contribute to male sterility is also supported by the observation that supplementation of the medium with polyamines induces symptoms reminiscent of sterility in the anthers of cultured flower buds of the wild type, and that inhibitors of polyamine biosynthesis induce the development of the normal type of pollen in anthers of cultured mutant flower buds (Rastogi and Sawhney 1990a, b). In anthers of certain cms-T lines of maize, the presence of low concentrations of hydroxycinnamic acid amide constituting polyamine conjugates is associated with sterility, whereas normal anthers contain high levels of the conjugates. Restoration of fertility reinstates the synthesis of these substances (Martin-Tanguy et al 1982).

It is generally believed that esterases hydrolyze sporopollenin and thus their activities are intrinsic to the formation of the pollen wall. Work on the subject using wild-type and *sl-2* mutant of tomato has disclosed an increased specific activity of the enzyme in the male-fertile anthers, as well as differences between the two genotypes in the timing of appearance of certain isozymes. This is particularly true of a major band, presumably related to tapetum development, that persists for a longer time in the mutant than in the wild type; appearance of another band, attributed to wall formation, is somewhat delayed in the mutant and coincides with microspore formation (Bhadula and Sawhney 1987). Differences in the esterase isozyme pattern also distinguish the development of anthers of fertile and male-sterile mutants of maize (Abbott, Ainsworth and Flavell 1984), wheat (Höhler and Börner 1980), and *P. hybrida* (van Marrewijk, Bino and Suurs 1986; Karim, Mehta and Singh 1984). In anthers of *P. hybrida*, the enzyme is mostly localized in the tapetum and, in male-sterile lines, its accumulation stops with the breakdown of this tissue. Other enzymes that have been implicated in cms are acid phosphatase, cytochrome oxidase, and succinic dehydrogenase in maize (Watson, Nath and Nanda 1977; Nath and Watson 1980), lipoxygenase in barley (Ahokas 1982a) and PAL in *Brassica oleracea* (Kishitani et al 1993).

Some evidence points to a decreased amylolytic activity as contributing to male sterility in the *sl-2* mutant of tomato. Stamens of the wild type appear to be unusual in having higher concentrations of amylase and sugar than the mutant stamens, which are enriched for starch. The reason for the preponderance of starch in the mutant is the low level of endogenous gibberellins; this reduces amylase activity, a change that may cause starch accumulation in the stamens (Bhadula and Sawhney 1989b).

4. ROLE OF THE MITOCHONDRIAL GENOME IN MALE STERILITY

Male sterility elicits far-reaching changes in the organelles of the microspores and tapetum. Most changes impinge on the mitochondria of plants exhibiting cms. As the involvement of mitochondria in cms became known from ultrastructural studies, substantial evidence was adduced from molecular investigations to indicate that this trait is encoded by the mitochondrial genome.

In the accounts of male sterility related to organelle morphology of anther cells, one feature seems common to all. There is a variable lag period in the development of the anther primordium before visible differences between cms and their fertile counterparts are seen in the organelle complement of the microspores and tapetum. This alone alerts us to the fact that in spite of their common differentiative pathways, there are differences in the degree and timing of organelle changes in the anther cells of the fertile and sterile plants. An ultrastructural study of wheat anthers has shown that there are fewer organelles in the pollen grains of male steriles than in those of their fertile counterparts and that the organelles in the former appear physiologically inactive and subject to rapid degeneration (de Vries and Ie 1970). The idea that sterility in anthers is due to a loss of mitochondrial structure came from the work of Warmke and Lee (1977), who reported mitochondrial degeneration in the tapetum and middle layer as the earliest detectable sign of pollen abortion in cms-T maize. Signs of degeneration are the loss of cristae, lack of internal structure, and swelling. A quantitative study revealed a rapid replication of mitochondria in the tapetum of both sterile and fertile anthers resulting in a 20- to 40-fold increase in mitochondrial number during early ontogeny between the precallose and tetrad stages of the microsporocyte; in sterile anthers, the burst of mitochondrial activity precedes tapetal breakdown (Warmke and Lee 1978; Lee and Warmke 1979). An important question relating to mitochondrial influence on anther development is concerned with differences in the ultrastructure and replication rate of the organelle beginning at the primordial stage of the anther. A study by Pollak (1992) showed that mitochondria of cells of the anther primordium of a cybrid (somatic hybrid or cytoplasmic hybrid produced by fusion of protoplasts) of tobacco have a smaller volume density than those of cells of the fertile counterpart, although the organelles are structurally similar in both. This is consistent with the notion that a low rate of mitochondrial replication could limit ATP production in the anther primordium, leading to sterility. A few cases are known in which breakdown of mitochondria represents only a part of the degeneration syndrome associated with male sterility, because other organelles also respond in strange ways to gene action. These include fragmentation of the ER and lack of dictyosomes in the tapetum of sunflower (Horner 1977), presence of an array of vacuoles containing membranous inclusions and osmiophilic material in the tapetum of *Petunia hybrida* (Bino 1985a), occurrence of plastids enriched with amyloplasts but with poorly developed grana in the parenchymatous cells of the anther primordium of a tobacco cybrid (Pollak 1992), and appearance of ER in concentric configurations in the tapetum and microsporocytes of sugar beet (Majewska-Sawka et al 1993), preceding or accompanying mitochondrial changes in cms lines. In other male-sterile plants, mitochondria function as well-ordered organelles up to the time of pollen abortion; organelle degeneration closely follows pollen breakdown, as if the signal for the former comes from the latter. Both cms-C and cms-S lines of maize provide examples of this group (Lee, Gracen and Earle 1979; Lee, Earle and Gracen 1980); there is even some question whether the cms-T line might also belong to this fold (Colhoun and Steer 1981).

The issue of mitochondrial function during male sterility has been considered from other angles, too. In maize, the introduction of nuclear fertility restorer genes into cms lines partially reinstates the normal structure of mitochondria, with the degree of structural restoration depending upon the degree of fertility restoration (Turbin et al 1968). The identity of one of these genes (*RF-2*) to a class of proteins similar to that of mammalian mitochondrial aldehyde dehydrogenases also emphasizes the adaptive adjustment of mitochondrial function as a basis of fertility restoration (Cui, Wise and Schnable 1996). Mitochondria isolated from male-sterile wheat seedlings are found to have a higher oxidative phosphorylation efficiency than those from male-fertile seedlings (Srivastava, Sarkissian and Shands 1969). Similar is the case with regard to cytochrome oxidase activity in seedlings of fertile and sterile lines of maize; however, only fertile anthers retain the biochemical activity for this enzyme, which increases from the premeiotic anther to the mature stage (Watson, Nath and Nanda 1977).

There are practical difficulties in relating, at the cellular level, mitochondrial lesions to male sterility of anthers, but numerous biochemical and molecular studies have placed the alleged connection on a firm footing. These investigations will now be examined in some detail to trace the evolution of the problem in specific cases, with a view to providing a sense of the accomplishments of the past and of the outstanding questions of the future.

(i) Maize

Ironically, the impetus for biochemical and molecular studies on male sterility in maize had its origin under unusual circumstances in research conducted in the wake of the destruction of the crop in the United States in the 1970s by the southern corn leaf blight disease caused by *Bipolaris* (*Helminthosporium*) *maydis* (teleomorph *Cochliobolus heterostrophus*) race-T and by the yellow corn leaf blight caused by *Phyllosticta maydis* (teleomorph *Mycosphaerella zeae-maydis*). These pathogens produce host-specific toxins called BmT (or HmT) and Pm, respectively. One intriguing outcome of the research associating mitochondria with the cms-T line was the discovery that under both in vitro and in vivo conditions, mitochondria from cms-T plants are more susceptible to the toxin from *B. maydis* than those from plants carrying normal cytoplasm. The toxin causes a swelling of the mitochondria that leads to fragmentation of the cristae and vesiculation of the inner membrane. Secondarily, the toxin affects the permeability of cms-T mitochondria and promotes the uncoupling of oxidative phosphorylation and the leakage of small molecules like nicotinamide adenine dinucleotide (NAD^+) and Ca^{2+} (Miller and Koeppe 1971; Gengenbach et al 1973; Peterson, Flavell and Barratt 1975; Payne, Kono and Daly 1980; Holden and Sze 1984). Similar effects on the structure and function of mitochondria from the cms-T line of maize are caused by Pm toxin (Comstock, Martinson and Gengenbach 1973). An additional line of evidence for the involvement of altered mitochondria in cms is the observation that the presence of male restorer genes modifies the sensitivity of the organelle to the toxin from *B. maydis* (Barratt and Flavell 1975). Effects on membrane structure and function must be intrinsic to mitochondria of cms-T line, because digitonin, a detergent that disrupts biological membranes, is more effective in inhibiting malate and succinate oxidation in cms-T mitochondria than in cms-C or normal mitochondria (Gregory et al 1977). Effects somewhat similar to those caused by fungal toxins are seen when cms-T mitochondria are challenged with methomyl, an insecticide structurally unrelated to the toxin (Koeppe, Cox and Malone 1978). Although mitochondria mediate in a host of energy-related functions, such as electron transport and ATP formation, response of mitochondria from cms-T plants to fungal toxin has a significance extending beyond energy transfer.

Levings and Pring (1976) first employed restriction mapping to show that several fragments present in the mitochondrial DNA of the normal maize line are absent in the cms-T line. The distinction in mitochondrial restriction mapping patterns extended even to stocks of normal and cms-T cytoplasms with different genetic backgrounds. Analysis of mitochondrial DNA from different stocks of normal and cms lines of maize, however, revealed considerable heterogeneity among the different cytoplasms, with each maize line having a unique mitochondrial DNA that is not affected by the nuclear genotypes (Figure 5.5). Another perspective on these observations is provided by the fact that with the exception of cms-S line, the restriction mapping patterns of respective chloroplast DNAs are indistinguishable in nature (Pring and Levings 1978). Altogether these results show that, within the resolution afforded by the methods employed, there is provision for involving mitochondrial DNA in the male sterility and disease susceptibility traits in maize.

Later studies have revealed a rather complex picture of distribution of low-molecular-mass DNAs in mitochondria of a wide range of normal and cms-T, cms-C, and cms-S lines of maize

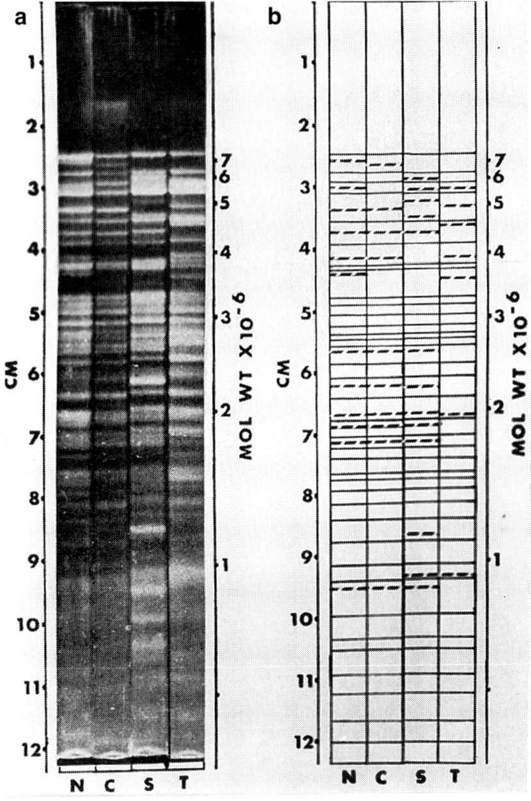

Figure 5.5. Agarose gel electrophoretic patterns **(a)** and schematic drawing **(b)** of *Hind*III digests of mitochondrial DNA from normal (N), cms-C (C), cms-S (S), and cms-T (T) lines of maize. Dashed lines in (b) indicate fragments that are not common to all lines. (From Pring and Levings 1978.)

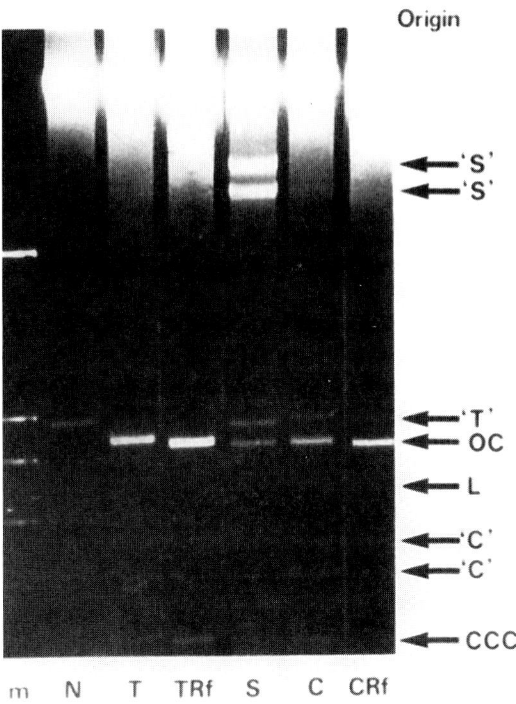

Figure 5.6. Electrophoresis of mitochondrial DNA preparations from normal (N), cms-T (T), cms-T restored (TRf), cms-S (S), cms-C (C), and cms-C restored (CRf) lines of maize. Sizes of molecular mass markers (m) on left are (from top to bottom) 3.9, 2.4, 2.1, 1.7, and 1.19 kb. 'S', 'T', and 'C' indicate band differences pertinent to cms-S, cms-T, and cms-C lines, respectively. CCC, OC, and L indicate supercoil, open circle, and linear bands, respectively. (From Kemble and Bedbrook 1980. Reprinted with permission from *Nature* 284:565–566. © Macmillan Magazines Ltd.)

(Kemble and Bedbrook 1980; Kemble, Gunn and Flavell 1980; Pring, Conde and Levings 1980). A common denominator for all four cytoplasms is a supercoiled DNA molecule of about 1,940 bp. In the normal, cms-C, and cms-S lines, the supercoiled DNA coexists with another DNA species of about 2,350 bp, which is not present in the cms-T cytoplasm. The presence of two additional unique DNA minicircles of about 1,570 bp and 1,420 bp in the cms-C cytoplasm completes the picture of mitochondrial DNA configuration in the cms lines (Figure 5.6). These low-molecular-mass DNAs have not been characterized to such a degree that a particular function in the specific cms line can be associated with a known DNA species.

That the mitochondrially inherited male-sterility trait occurs at the protein level is supported by the observation that certain discrete polypeptides synthesized by isolated mitochondria are uniquely associated with a particular cms line. Especially suggestive is the fact that a 13-kDa protein (URF-13) observed in cms-T mitochondria is absent in the fertile line and that a 21-kDa protein found in the mitochondria of the fertile line is lacking in the cms-T line. The 13-kDa protein product of cms-T cytoplasm is apparently under nuclear control, since its synthesis is significantly reduced in mitochondria of plants restored to fertility by nuclear genes (Forde, Oliver and Leaver 1978; Forde and Leaver 1980). As shown by Forde et al (1980), cms maize cytoplasm from several sources can be distinguished by their mitochondrial polypeptides and assigned to one of the three groups (Figure 5.7).

Role of Mitochondrial Genes. At this stage, the characterization of mitochondrial genes might be considered pivotal to the analysis of cms in maize. Toward this end, a 3,547-bp mitochondrial DNA fragment associated with male sterility in the cms-

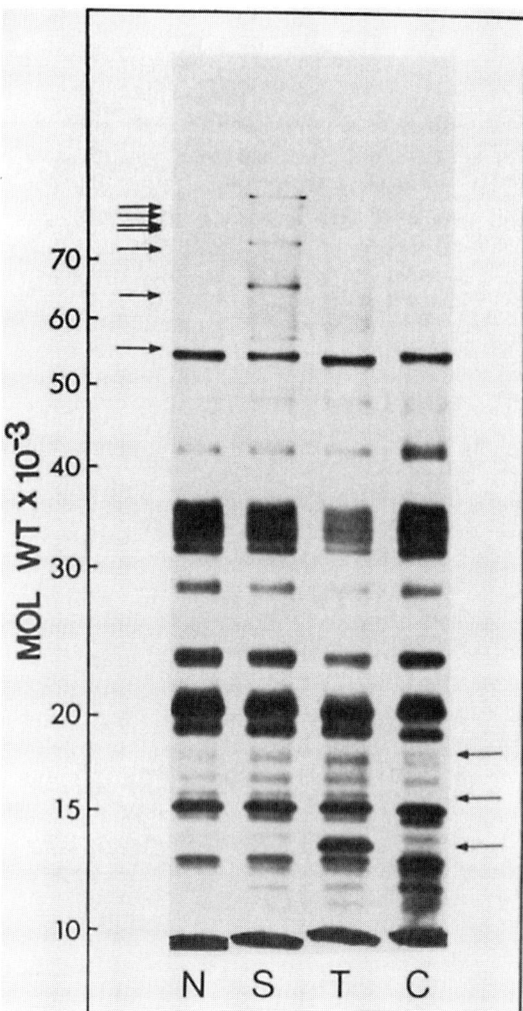

Figure 5.7. Polyacrylamide gel electrophoresis of proteins synthesized by mitochondria from normal (N), cms-S (S), cms-T (T), and cms-C (C) lines of maize. These four distinct protein profiles were obtained from mitochondria isolated from 32 male-sterile maize lines. The arrows indicate the polypeptides that distinguish the four cytoplasmic types. (From Forde et al 1980.)

T line of maize was selected from a mitochondrial cDNA library made to cms-T cytoplasm. This gene, designated as *TURF-13*, is located on a 6.6-kb *XhoI* fragment and has two long open reading frames that code for proteins of predicted molecular masses of 12.9 kDa (*ORF-13*) and 24.6 kDa (*ORF-25*). That *TURF-13* is an active mitochondrial gene responsible for the cms trait is inferred from Northern blot analysis of mitochondrial transcripts from fertile and cms maize cytoplasms probed with DNA sequences homologous to *ORF-13* and *ORF-25*. The latter are detected in the fertile, as well as in all three sterile, maize cytoplasms, whereas *ORF-13* sequences are unique to the cms-T line. Moreover, modifications occur in the mitochondrial gene transcripts of cms-T plants in which the male-sterile phenotype is masked by nuclear fertility restorer genes (Dewey, Levings and Timothy 1986).

Appropriate recognition of the importance of mitochondrial dysfunction in cytoplasmic male sterility requires the demonstration that the *TURF-13* gene or its protein product interferes in some way with normal pollen development. A 13-kDa protein in particular seems to be implicated in the sterility trait of the cms-T line. Dewey, Timothy and Levings (1987) showed that labeled mitochondrial translation products from normal and cms-T cytoplasms can be immunoprecipitated using antiserum specific to ORF-13 sequence, separated by SDS-PAGE, and identified by fluorography; a signal is found against a 13-kDa polypeptide in cms-T mitochondria but is absent in mitochondrial translation products of the fertile line. It is believed that the antiserum that binds to a 13-kDa polypeptide in blots of total mitochondrial proteins from cms-T plants is associated with mitochondrial membranes. Although cms-T plants restored by the nuclear restorer genes *RF-1* and *RF-2* show a reduced abundance of the 13-kDa protein, protein change can also be accomplished by the *RF-1* locus alone. The biochemical malfunction that causes a decrease in 13-kDa gene product in the restored lines is not yet fully defined; speculation is that it might be either transcriptional – due to a nuclear restorer gene–directed differential RNA processing event, or translational – resulting in a premature termination of translation of the transcripts (Kennell, Wise and Pring 1987; Kennell and Pring 1989).

Another approach to elucidating the relationship between cms and mitochondrial gene *TURF-13* has involved the development of tissue culture methods that can regenerate revertant plants insensitive to BmT and Pm toxins. Conditions for the regeneration of male-fertile, toxin-resistant cms-T mutants include culture of immature cms-T embryos in a medium containing 2,4-D to induce callus formation, selection of callus in the presence of increasing concentration of the fungal toxin, and regeneration of plants from the callus by subculture in a medium lacking 2,4-D (Gengenbach, Green and Donovan 1977); fortuitously, revertants appear even in the absence of the selective agent (Brettell, Thomas and Ingram 1980). In a similar way, tissue culture methods have been used to obtain male-fertile

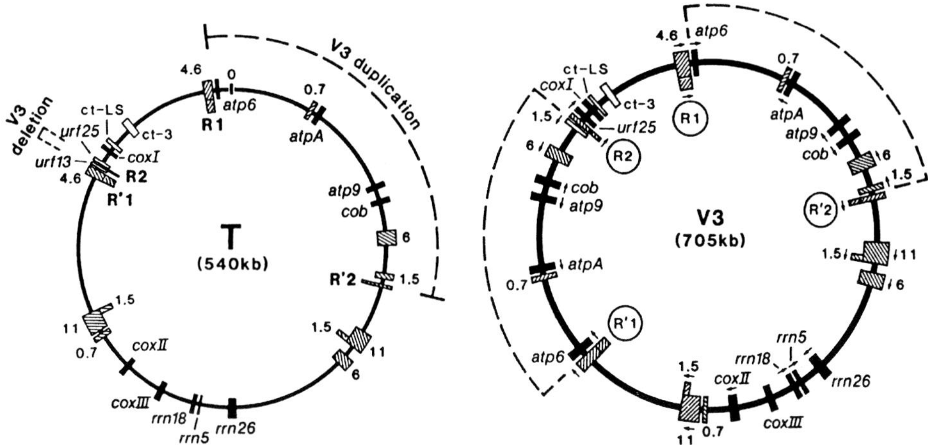

Figure 5.8. Comparison of the master mitochondrial genomes of cms-T (T) and a revertant mutant (V3) of maize. Located on the cms-T genome are the 165-kb region found duplicated in the V3 genome and the 0.423-kb region deleted from the V3 genome. The duplicated region is also marked on the V3 genome. (From Fauron, Havlik and Brettell 1990.)

methomyl-resistant plants by selection in the presence and absence of the insecticide (Kuehnle and Earle 1989). Early analyses of mitochondrial DNA restriction patterns showed a distinct change in the mitochondrial genome organization of the toxin-resistant revertants due to deletions of specific fragments (Gengenbach et al 1981; Kemble, Flavell and Brettell 1982; Umbeck and Gengenbach 1983). The mitochondrial genome of many male-fertile, toxin-insensitive regenerants analyzed suffers a deletion of a 6.7-kb *Xho*I restriction fragment.

Deletions of fragments of the genome are probably not in themselves concerned with phenotypic alterations but may be gross manifestations of genome reorganization. Further study of the mitochondrial genome organization of the revertant mutants showed that the majority of them fall into the general class in which the *TURF-13* locus has been eliminated by multi-recombinational events. In general, the recombination involves two sets of repeated sequences (Figure 5.8), although other genomic arrangements, such as intermolecular recombinations, also occur and produce a master circular mitochondrial DNA (Rottmann et al 1987; Fauron, Havlik and Brettell 1990; Fauron et al 1992). Mitochondria isolated from regenerated male-fertile phenotypes also fail to synthesize the 13-kDa polypeptide; thus, the regenerated plants maintain the correlation between male fertility, resistance to fungal toxin, and suppression of the synthesis of a specific polypeptide (Dixon et al 1982). An unusual mutant, *t-4*, was found to be male fertile and insensitive to the toxin, yet it retained *TURF-13* sequences. Molecular analysis revealed that the mutant is characterized by a 5-bp insertion in the coding region that places a premature stop codon in the correct reading frame. Although the *t-4* mutant, like other tissue culture–derived mutants, does not synthesize a 13-kDa protein, it makes a unique polypeptide of approximately 8 kDa (Wise, Pring and Gengenbach 1987; Wise et al 1987).

These observations evidently implicate the *TURF-13* gene in conferring fungal toxin sensitivity to mitochondria of cms-T maize. There is some evidence that the *TURF-13* gene is responsible for the deleterious effects of methomyl on cms-T mitochondria. This evidence is provided by the observation that when *TURF-13* gene is expressed in *Escherichia coli*, inhibition of whole-cell respiration and mitochondrial swelling seem to proceed along analogous lines after the addition of fungal toxins or methomyl (Dewey et al 1988). Expression of *TURF-13* in yeast also imparts sensitivity to methomyl and fungal toxins by targeting to the mitochondria a protein similar to the 13-kDa polypeptide from the cms-T line (URF-13) of maize (Glab et al 1990; J. Huang et al 1990). In contrast, when URF-13 protein is expressed in tobacco and insect cells, it confers toxin or methomyl sensitivity without being targeted to mitochondria (von Allmen et al 1991; Korth and Levings 1993). These observations are considered to provide direct evidence that a protein encoded by the *TURF-13* gene accounts for the susceptibility of cms-T maize to the fungal toxins. However, the question as to how

URF-13 protein confers toxin sensitivity and causes male sterility remains unanswered; the problem is compounded by another observation: Even targeting of the protein to mitochondria does not induce male sterility in transgenic tobacco (Chaumont et al 1995).

Mode of Action of the Gene Product. Although biochemical studies have indicated that the URF-13 protein is associated with the membrane fraction of maize mitochondria, immunocytochemical studies have extended this relationship to a high level of precision to show that this protein is localized on the inner membrane of mitochondria of roots, coleoptiles, and anther tapetum of cms-T maize (Dewey, Timothy and Levings 1987; Hack et al 1991). To hone in on the question as to how *TURF-13* gene causes male sterility, we must consider other properties of the gene product and its effects on mitochondrial membranes. The demonstration that labeled Pm toxin binds to the URF-13 protein in maize mitochondria and in *E. coli* expressing this protein suggested that toxin binding may be used to determine the relationship between the toxin and URF-13 protein. The binding is reversible, and the toxin and methomyl compete for the same overlapping binding sites. Studies performed with *E. coli* mutants expressing a toxin-insensitive version of the *TURF-13* gene have identified three specific regions of the gene required to maintain toxin sensitivity. Both a site-directed substitutional mutation at amino acid 39 and a mutant with a deletion of 33 amino acids from the carboxyl end show reduced toxin binding, whereas an internal deletion mutant missing amino acids 2–11 binds the toxin significantly (Braun, Siedow and Levings 1990). This leaves the possibility, however, that although the mutation does not diminish toxin insensitivity even with toxin binding, amino acids 2–11 may be redirected to activities connected with membrane modification following toxin–URF-13 protein interaction. Other investigations have shown that the effect of the fungal toxin can be abolished by pretreatment of mitochondria with dicyclohexylcarbodiimide (DCCD), an inhibitor of membrane-bound ATPase; a similar protection is afforded by DCCD to *E. coli* expressing URF-13 protein. One explanation of this effect is that it must be due to the interaction of DCCD with a membrane-bound protein binding site. This is supported by the observation that in site-directed mutations of *E. coli* expressing URF-13 protein, DCCD binds to two aspartate residues at positions 12 and 39. This of course happens because the amino acids are membrane bound (Bouthyette, Spitsberg and Gregory 1985; Braun et al 1989). How does the toxin impair mitochondrial function by binding with a toxic protein such as that encoded by the *TURF-13* gene? Ion efflux has been used to distinguish between *E. coli* expressing the standard genome (without *TURF-13* insert) and the *TURF-13* gene; results have shown that in contrast to bacterial cells without *TURF-13* insert, those expressing the *TURF-13* gene suffer massive ion loss when exposed to BmT toxin (Braun et al 1989). Similar was the pattern of ion traffic in mitochondria of cms-T maize treated with fungal toxins (Holden and Sze 1987). These results have attributed toxin–URF-3 protein interaction to the formation of a network of pores in the bacterial plasma membrane or in the inner mitochondrial membrane. Analysis of the configuration of URF-13 protein molecules on the plasma membrane of *E. coli* has provided some insights into how membrane pores are formed during toxin–URF-13 protein interaction (Korth et al 1991).

In summary, then, it is clear that although data at hand have not permitted a complete resolution of the question of how mitochondrial dysfunction impacts adversely on pollen development and leads to male sterility in maize, a number of useful leads have been generated to bring about an understanding of the basic mechanism. Any hypothesis formulated to explain male sterility in molecular terms will have to confront the fact that the protein that has been identified in the mitochondrial membrane of male-sterile lines is toxic only to the cells of the anther and to no other cells of the plant. One view to accommodate the recent molecular studies is that the toxic protein might interfere with mitochondrial replication in the tapetum by affecting an essential function of the organelle (Levings 1993). A hypothesis that invokes the action of a substance produced in the anther with properties similar to the fungal toxin has been in the center stage for some time now as an explanation of the mechanism of cms (Flavell 1974). It is thought that this substance interacts with cms-T cytoplasm to cause male sterility by inactivating some vital mitochondrial functions. Fertility restoration, on the other hand, probably occurs when nuclear fertility restorer genes correct the altered organelle structure, making the organelles insensitive to the anther-specific substance. If an anther-specific substance is identified, this hypothesis accords with the recent molecular data, but none has been found.

Male Sterility in cms-S Line. The distinguishing characteristic of mitochondrial DNA from cms-S cytoplasm of maize is the presence in high copy

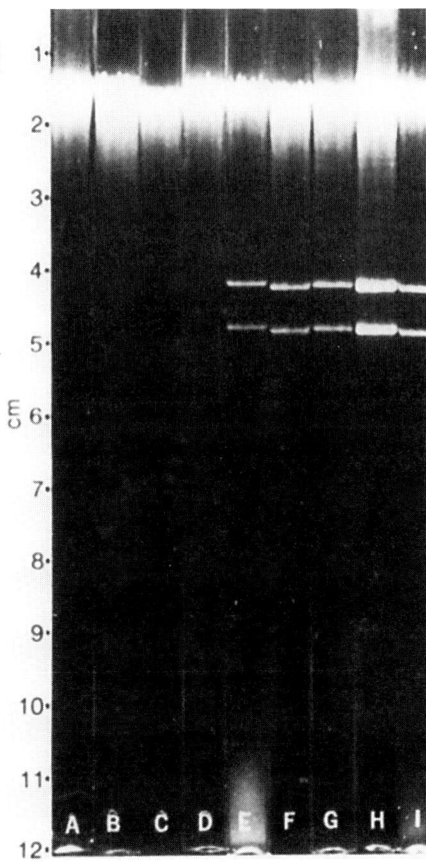

Figure 5.9. Agarose gel electrophoretic patterns of mitochondrial DNA from normal line (A), cms-T line (B), cms-C line (C), male-sterile line EP (D), and cms-S lines S (E), H (F), EK (G), SD (H), and CA (I) of maize. A high-molecular-mass DNA is common to all lines, but the cms-S lines yield two additional bands. (From Pring et al 1977.)

numbers of two low-molecular-mass linear episomal DNAs, S-1 (6.4 kb) and S-2 (5.4 kb) (Figure 5.9). These plasmidlike DNAs are present in different cms-S maize stocks regardless of the stocks' nuclear background and geographical location and are absent in mitochondrial DNA preparations of normal, cms-T, and cms-C lines (Pring et al 1977; Weissinger et al 1982, 1983). An independent approach that implicates mitochondrial genes in the cytoplasmic inheritance of male fertility in cms-S maize has followed the fate of S-1 and S-2 DNA in lines in which pollen fertility is restored by nuclear restorer genes. This work showed that restoration to fertility is correlated with the disappearance of S-1 and S-2 DNA and an ordered rearrangement of the structure of mitochondrial DNA of the restored stock by incorporation of S-1 and S-2 molecules. Because DNA of fertile restored strains revealed several regions of homology to S-1 and S-2 DNA, it was proposed that nonhomologous recombination is probably involved in the restoration of male sterility to male fertility (Levings et al 1980; Thompson, Kemble and Flavell 1980). Some evidence has also suggested that spontaneous reversion to fertility is coincident with the transposition of the S-2 sequences from one region of the mitochondrial genome to another (Kemble and Mans 1983). Other studies have defined the specific structural changes in the mitochondrial DNA during spontaneous reversion of cms-S cytoplasm to fertile type. One profound effect of the presence of episomes in the cms-S cytoplasm is the linearization and fragmentation of a high proportion of the circular mitochondrial DNA with episomes covalently linked to one end. This is probably achieved by a recombination of inverted repeats of S-1 and S-2 DNA with homologous sequences present on the mitochondrial chromosomes to form new functional chromosomes consisting, in part, of sequences from each species. Spontaneous reversion of mutants to fertility is believed to go hand in hand with a deletion of S-1 and S-2 episomes and the associated linear molecules, which results in an overall change in the organization of the mitochondrial genome; episomal sequences integrated in the mitochondrial chromosomes, however, are retained (Schardl et al 1984; Schardl, Pring and Lonsdale 1985). It is as if the requirement of a cytoplasmic genotype for specific nuclear fertility alleles has been fulfilled by a rearranged mitochondrial genome during spontaneous reversion to fertility. Another approach to analyzing the causes of mitochondrial function in cms-S line of maize has focused on the fate of the gene for *COX-I*. In contrast to the normal fertile, cms-T, and cms-C mitochondria, which contain single copies of the gene, cms-S mitochondrial DNA contains several distinct copies of the *COX-I* gene. Figure 5.10 shows that the appearance of extra copies of *COX-I* seems to depend upon DNA rearrangements just 5'- (at –175 bp) to the gene and involves most of the inverted sequence of S-1 and S-2 DNAs (Isaac, Jones and Leaver 1985). These observation are open to two interpretations regarding the function of mitochondrial genes in cms-S type male sterility. One is that the reconstituted gene encodes a polypeptide that is deleterious to mitochondrial activity; alternatively, mitochondrial proliferation is stymied in the mutant due to the high proportion of episomes, linear fragments of mitochondrial chromosomes, and extra gene copies. The first alternative, of course, is similar to the situation in cms-T maize.

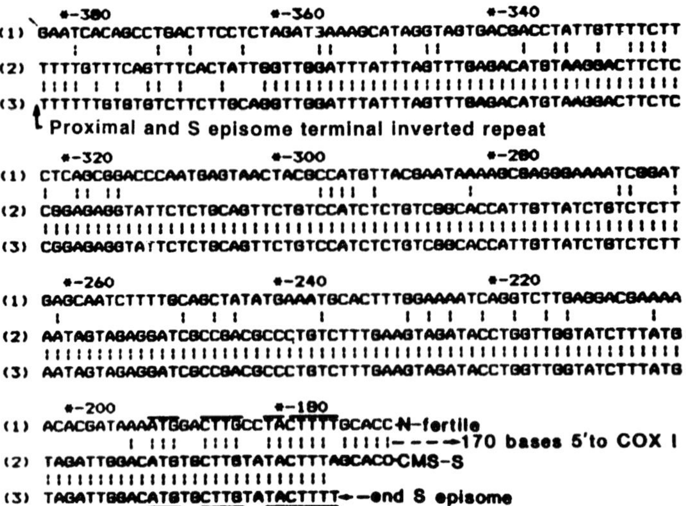

Figure 5.10. Comparison of DNA sequence homology 5'- to the *COX-I* gene around the point of divergence between mitochondrial DNAs from normal fertile (1) and cms-S (2) maize lines and the 208-bp tandem inverted repeats of S-1 and S-2 DNAs (3). The numbering refers to the number of bases 5'- to the ATG initiation codon of the *COX-I* gene in mitochondrial DNA from the normal maize line. Homology between a sequence in the mitochondrial DNA of the normal maize line and the distal end of the cms-S tandem inverted repeats around the point of divergence is underlined. (From Isaac, Jones and Leaver 1985. *EMBO J.* 4:1617–1623, by permission of Oxford University Press.)

(ii) *Petunia*

Among the variety of evidence indicating that some defect in mitochondrial functions results in cms in *Petunia* was the identification of a cms-associated mitochondrial DNA region in somatic hybrids. Unlike maize, *Petunia* has only one known cms cytoplasm, designated as S-cytoplasm, and fertility of male steriles is restored by a single dominant gene (Edwardson and Warmke 1967; Izhar 1978). Production of somatic hybrids is accomplished by fusion of male-fertile *P. hybrida* protoplasts with cms *P. parodii* protoplasts, followed by culture and regeneration of plants in a tissue culture medium. Most of the hybrids obtained are genetically stable male fertiles; a small number are stable male steriles; a few regenerants are genetically unstable for the cms trait (Izhar, Schlicter and Swartzberg 1983). When the chloroplast genomes of the cybrids are analyzed by restriction endonuclease digestion and by Southern blot hybridization with cloned chloroplast DNA probes, the cms phenotype is found to segregate independently of the chloroplast genome (Clark, Izhar and Hanson 1985). This shows that despite the dynamic nature of the chloroplast genome, this genome does not specify the cms trait.

The first evidence for the presence of a novel mitochondrial genome in the somatic hybrid was provided by restriction patterns of mitochondrial DNA from the hybrids as well as from the parents (Figure 5.11). This work showed that the somatic hybrid restriction pattern is unique in containing fragments common to both parents and some fragments characteristic of each parent (Boeshore et al 1983). If molecular recombination has occurred in the mitochondrial DNA of the somatic hybrids, the DNA region containing the recombined fragment (named the *PCF* locus) should segregate with the cms phenotype; this has been found to be true (Boeshore, Hanson and Izhar 1985). Experiments on the origin of the recombinant fragment that made use of restriction digest analysis and Southern hybridization confirmed a general presumption that in addition to parental DNA fragments, the novel mitochondrial genome is composed of nonparental and recombinant fragments. One of the nonparental fragments contains *ATP-9* gene generated by intergenomic recombination between *ATP-9* genes from the two parental lines. The recombinant *ATP-9* gene has the 5'-transcribed region of one parent and the 3'-transcribed region of the other parent (Rothenberg et al 1985; Rothenberg and Hanson 1987, 1988). Sequences that are involved in controlling the synthesis of the primary transcript or in the processing of the flanking regions have been sought as a means of gaining insight into the structure of the gene that generates the cms phenotype. Sequence data revealed that the *PCF* gene is a fusion product fab-

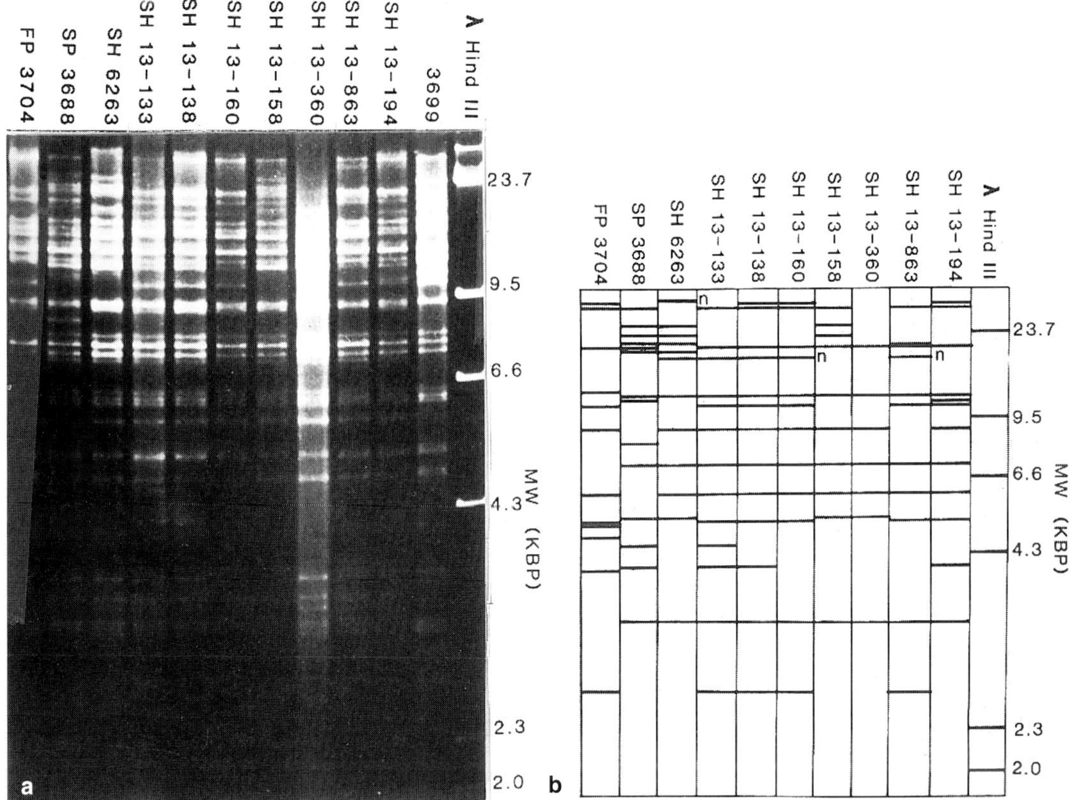

Figure 5.11. Agarose gel electrophoresis of mitochondrial DNA digested with *Bgl* I from parents and somatic hybrids of *Petunia* **(a)** and a schematic diagram of the gel **(b)**. Novel bands are indicated by n in the diagram. FP 3704, fertile parental line 3704 of *P. hybrida*; SP 3688, cms line 3688 with *P. parodii* nuclear background; 3699, *P. parodii* fertile line. SH followed by a number indicates the particular somatic hybrid analyzed. λ*Hind*III indicates molecular mass marker DNA fragments. (From Boeshore et al 1983. *Mol. Gen. Genet.* 190:459–467. © Springer-Verlag, Berlin.)

ricated from three cotranscribed genes, namely, an atypical open reading frame harboring part of the coding region of *ATP-9*, sequences with partial identity to the *COX-II* gene, and an unidentified open reading frame; located closely downstream of *PCF* are genes for NADH dehydrogenase subunit-3 (*NAD-3*) and for small ribosomal subunit protein-12 (*RPS-12*) (Young and Hanson 1987; Pruitt and Hanson 1989; Rasmussen and Hanson 1989). To determine whether the *PCF* gene is transcribed only in anthers of cms lines of *Petunia*, a *PCF*-specific DNA probe was directed against mRNA in an S1-nuclease protection study; the results showed that the transcripts are four to five times more abundant in anthers than in leaves (Young and Hanson 1987). Reversion of cms lines to fertility is correlated with a loss of most of the *PCF* sequences and transcripts (Clark, Gafni and Izhar 1988; Pruitt and Hanson 1991).

The best evidence for a role for the *PCF* locus in causing cms has come from the demonstration of a correspondence between the *PCF* gene and a 25-kDa protein (Figure 5.12). This protein, identified by antibodies prepared to the unidentified reading frame from synthetic polypeptides, is found only in cms lines of *Petunia*, is synthesized by their isolated mitochondria, and is derived by processing of a 43-kDa precursor protein representing the entire *PCF* gene product. The argument that the *PCF* gene is correlated with cms is also compelling in view of the fact that the 25-kDa protein is completely absent in male-fertile lines and is detected at a reduced amplitude in fertility-restored lines (Nivison and Hanson 1989; Nivison et al 1994). However, expression of the 25-kDa protein targeted to mitochondria was not sufficient to alter the fertility of transgenic *Petunia* and tobacco (Wintz et al 1995). The reasons for this are at present unknown.

The properties of genes in the *PCF* locus appear

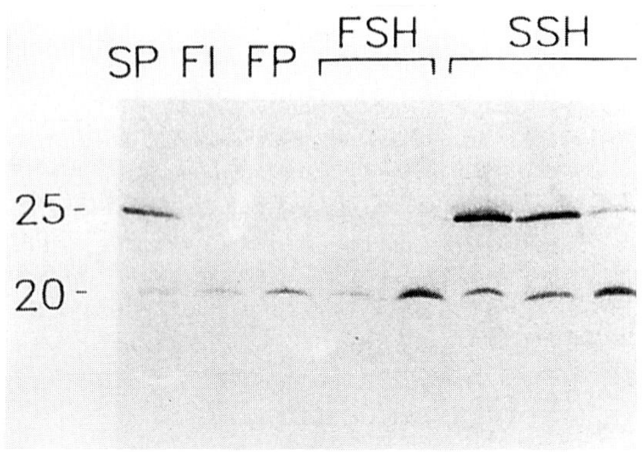

Figure 5.12. Immunoblot of mitochondrial proteins from sterile (SP, SSH) and fertile (FI, FP, and FSH) lines of *Petunia* probed with antibodies prepared to the unidentified reading frame of *PCF* gene. Positions of the 20-kDa and 25-kDa proteins are indicated. (From Nivison and Hanson 1989. *Plant Cell* 1:1121–1130. © American Society of Plant Physiologists.)

explicable largely in terms of their role in encoding respiratory enzymes. One of the criteria that indicates a close association between *PCF* and respiratory activity is the difference in utilization of a cyanide-insensitive, alternative oxidase pathway of electron transport by suspension cultures and by immature anthers of fertile, cms, and fertility-restored lines of *Petunia*. Although reduced alternative oxidase activity is consistently associated with cms lines, in plants that show normal pollen development due to a nuclear restorer gene, the activity appears comparable to lines lacking the *PCF* locus (Connett and Hanson 1990). The general conclusion suggested by this study is that the expression of novel polypeptides in the male-sterile plants interferes with the functioning of genes essential for normal mitochondrial activity. How this might cause such a drastic effect as pollen sterility remains puzzling.

(iii) *Phaseolus*

Investigations on cms in *Phaseolus vulgaris* with a focus on fertility restoration support the idea that mitochondrial DNA rearrangements occur in plants restored by the introduction of nuclear restorer genes or by spontaneous reversion. In investigations using the nuclear restorer gene *FR*, recovery of pollen fertility is found to be accompanied by the loss of a 25-kb mitochondrial DNA fragment, just as in the spontaneously reverted plants. This large DNA segment contains a 3.7-kb putative mitochondrial sequence uniquely associated with cms and designated as *PVS*. The *PVS* sequence is flanked by two different repeated sequences, one containing a complete coding region of the mitochondrial gene *ATP-A* and the other containing a coding segment from the mitochondrial *COB* gene (Chase and Ortega 1992). Sequence analysis has identified two long, and a number of small, open reading frames within *PVS*. The loss of mitochondrial sequence from fertility-restored plants was inferred from genetic analysis of F2 progenies that produced both male fertiles and steriles. In addition, in situ hybridization analysis with sterility-associated mitochondrial sequence as a probe showed that, whereas pollen grains of male-sterile plants segregating in the F2 population retain the mitochondrial genome configuration in the *PVS* sequence, pollen grains of fertile F2 plants show no hybridization signals (Mackenzie and Chase 1990; Johns et al 1992; Janska and Mackenzie 1993). A developmental analysis of the restoration process by the *FR* gene has suggested that an interaction of the FR product with the mitochondrial population, resulting in the selective elimination of mitochondria harboring the *PVS* sequence, accounts for the loss of mitochondrial DNA (He, Lyznik and Mackenzie 1995). In restriction analysis of mitochondrial DNA from other restored lines, a 6-kb fragment can be clearly identified as having been lost. When an important check was made by comparing this with the restriction pattern of mitochondrial DNA from spontaneously reverted plants, loss of the same segment of mitochondrial DNA is also apparent (Mackenzie et al 1988).

Somewhat different results are achieved when pollen fertility is restored by the nuclear gene *FR-2*, as this does not involve the loss of *PVS* sequence from the mitochondrial genome. In addition, *PVS* transcript patterns are identical in seedlings and floral buds of both cms and fertility-restored lines (Mackenzie 1991; Chase 1994). A revealing comparison involves the localization of mitochondrial peptides associated with pollen sterility in the anthers of cms lines and revertant lines. Antibodies made to the putative protein encoded by an open reading frame of the *PVS* sequence are localized in the pollen mother cells and microspores of anthers of the cms line of bean, but none are detected in the same cells of a fertile line containing *FR-2* genome or in the revertant (Abad, Mehrtens and Mackenzie 1995). These observations have essentially focused attention on the possible translational or posttranslational influence of *FR-2* on the expression of the *PVS* gene product.

(iv) *Sorghum*

The conventional approach to determining the molecular basis for cms in *Sorghum bicolor* has involved restriction endonuclease analysis of chloroplast and mitochondrial DNA from several male-fertile and male-sterile cytoplasms. In analogy with investigations in maize, these studies showed persistent differences in mitochondrial DNA arrangements among male-sterile sources that can be used to differentiate between several cytoplasmic groups; chloroplast DNA showed a lesser degree of variation (Conde et al 1982; Pring, Conde and Schertz 1982; Bailey-Serres et al 1986a). A measure of the relationship between mitochondrial DNA and genetic behavior of male-sterile cytoplasm is evident in the discovery of linear, plasmidlike DNAs in mitochondrial DNA preparations from a line of cms sorghum and by the observation that isolated mitochondria from various cms lines synthesize a unique polypeptide not found in the fertile line (Pring et al 1982). This polypeptide, which is a variant form of COX-I, has an apparent molecular mass of 42 kDa and replaces a 38-kDa polypeptide synthesized by the normal fertile line of sorghum (Dixon and Leaver 1982). It has been shown that the synthesis of the variant enzyme is due to genome rearrangements in the mitochondrial DNA both 100 bp 5'- to the presumed initiator methionine and within the 3'-coding sequence, which results in the formation of novel mitochondrial open reading frames (Bailey-Serres et al 1986b). Possibly, mitochondrial biogenesis or function impaired by the presence of the variant enzyme could interfere with pollen fertility.

(v) *Raphanus* and *Brassica*

The existence of a highly rearranged mitochondrial genome in male-sterile *Raphanus sativus* was implied from the alignment of a restriction map for the male-sterile *ogu* cytoplasm with that of the normal line. A minimum of 10 inversion events have been postulated to account for DNA alterations. In addition, altered transcript patterns were identified for the mitochondrial genes *ATP-A*, *ATP-6*, and *COX-I*, which coincidentally map near the rearranged segments. Since *ATP-6* and *COX-I* transcripts are not affected in fertility-restored *ogu* lines, differences in the transcriptional patterns of these genes are most likely due to DNA alterations (Makaroff and Palmer 1988). Extensive analysis of the *ATP-6* locus from both normal and *ogu* lines of *R. sativus* has shown an anarchic rearrangement of the locus in the cms line that results in the separation of the sequences of the normal *ATP-6* locus into three widely separated regions of the *ogu* mitochondrial genome; the generation of an *ogu*-specific *ATP-6* transcriptional unit containing both a disrupted *ATP-6* coding region and a novel 105-bp codon open reading frame; and elimination of the normal translation start site, due to multiple nucleotide differences at the 5'-end. Although these observations were initially marshalled to formally propose that the parameter determining the expression of male sterility in *R. sativus* is the synthesis of a new ATP-6 protein detrimental to mitochondrial function in the cms line, the proposed role of the protein appears be based on false premises (Makaroff, Apel and Palmer 1989; Krishnasamy, Grant and Makaroff 1994). Molecular analysis of an open reading frame (*ORF-138*) that has been implicated in male sterility in *Brassica napus* cybrids (to be considered next) suggested that this region might be involved in male sterility in *R. sativus*. The relationship of *ORF-138* to cms in *R. sativus* appears to be that a 20-kDa protein corresponding to this open reading frame is produced in the mitochondria of the *ogu* line (Figure 5.13). The presence of fertility restorer genes causes a major reduction in this protein in the leaves and flowers

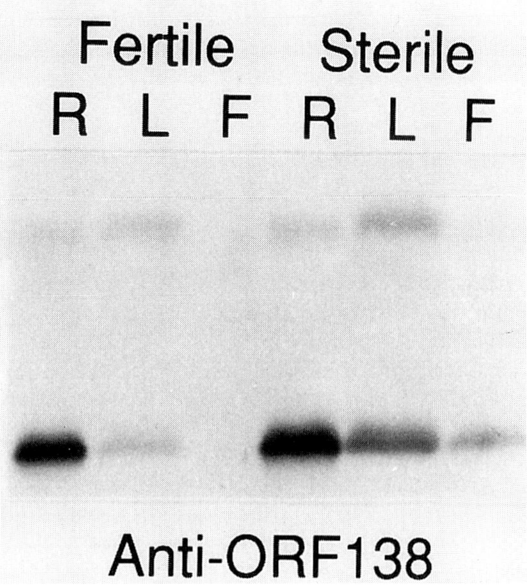

Figure 5.13. Immunoblot of mitochondrial proteins isolated from roots (R), leaves (L), and immature flower buds (F) of fertile and sterile *ogu Raphanus sativus* probed with antibodies to *ORF-138*. (From Krishnasamy and Makaroff 1994; photograph supplied by Dr. C. A. Makaroff.)

(Krishnasamy and Makaroff 1993, 1994). The results of this analysis introduce a note of caution: The regulation of mitochondrial function established in one cms system may not be wholly applicable to others.

Less complete information is available on the molecular biology of cytoplasmic male sterility in *Brassica napus*. Sterile lines of *B. napus* have their origin from intergeneric crosses with a cms line of *R. sativus*. Differences have been documented between fertile and cms lines with respect to mitochondrial DNA restriction patterns, mitochondrial translation products (Vedel et al 1982; Erickson, Grant and Beversdorf 1986a; Singh and Brown 1991), chloroplast DNA restriction patterns, location of chloroplast genes (Vedel et al 1982; Vedel and Mathieu 1983), and thylakoid protein profiles (Remy and Ambard-Bretteville 1983); however, other than providing useful biochemical markers for further studies, these investigations do not support a role for cytoplasmic DNA in male sterility. Some interest has focused on the presence of a linear 11.3-kb plasmid in mitochondria purified from several species of *Brassica* and *Raphanus*. There is an abundance of this plasmid in cms lines of *B. campestris* and *B. napus* but it is not detected, or is present at a low level, in the corresponding fertile lines (Palmer et al 1983). A correlation found in this study between the presence of the plasmid and the cms trait has been questioned because the plasmid has also been identified in both fertile and cms cytoplasms of certain accessions of *B. napus*, in maintainer cytoplasms, and in cytoplasms that do not contribute to sterility in rapeseed (Erickson, Grant and Beversdorf 1986b).

For a long time, studies on male sterility in *B. napus* were bedeviled by the unstable nature of male fertiles produced by conventional breeding. Analysis of mitochondria of cybrids of *B. napus* has shown that a specific mitochondrial DNA fragment consistently segregates with male sterility and is lost or rearranged in fertile revertants (Bonhomme et al 1991; Temple et al 1992). Sequence and transcript analyses subsequently identified *ORF-138*, whose transcripts are specific to sterile cybrids (Bonhomme et al 1992). The transcripts are translated into a mitochondrial membrane protein in the male-sterile cybrids, but the protein is absent in the fertile revertants (Grelon et al 1994).

(vi) Rice and Wheat

Fertile and sterile cytoplasms of rice can be distinguished by their mitochondrial DNA restriction patterns and by the presence of additional mitochondrial DNA molecules in the sterile cytoplasm (Yamaguchi and Kakiuchi 1983; Mignouna, Virmani and Briquet 1987). One approach to defining the role of mitochondrial genes in cms in rice has focused on the *ATP-6* gene. The mark of divergence of the sterile cytoplasm from the fertile one is the presence of an extra *ATP-6* gene in the former. The suggestion that this is a chimeric gene (*URF-RMC*) is borne out by cloning, sequencing, and hybridization data that showed that it is generated by a recombination event involving part of the *ATP-6* gene and an unidentified sequence. The need for *URF-RMC* to induce male sterility is confirmed by the demonstration that fertility restoration alters the transcription of *URF-RMC* but not that of *ATP-6* (Kadowaki, Suzuki and Kazama 1990). In rice generated by protoplast fusion, a second *ATP-6* gene is found to be uniquely associated with cms. It appears that the RNA transcribed by this gene is not processed and is inefficiently edited, resulting in a grossly altered polypeptide as the translation product (Figure 5.14). Competition between the normal and altered polypeptides is believed to tip the balance in favor of the latter, thus causing male sterility. The presence of restorer genes presumably influences the normal process-

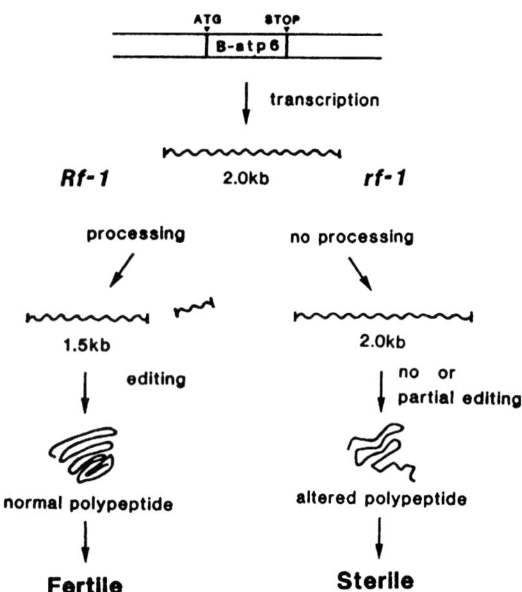

Figure 5.14. A model proposed for the regulation of the *ATP-6* gene in cms rice. The gene is indicated as B-atp6. Transcription, processing, and editing of the gene are shown separately for the fertile and sterile lines of rice. (From Iwabuchi, Kyozuka and Shimamoto 1993. *EMBO J.* 12:1437–1446, by permission of Oxford University Press.)

ing and editing of the second *ATP-6* RNA (Iwabuchi, Kyozuka and Shimamoto 1993).

Comparable results have been obtained with cms *Triticum timopheevi* and fertile *T. aestivum* mitochondrial DNAs. The *ATP-6* region of the male-sterile wheat is remarkable in that it has an organization different from that found in the fertile line due to a reduction in the copy number of the genes from 2 to 1 and the rearrangement of the sequences in the 5'-flanking regions (Mohr et al 1993). There do not appear to be any DNA rearrangements in other mitochondrial genes, such as *COB, COX-I*, or *COX-II*, associated with *T. timopheevi*–derived male sterility (Rathburn, Song and Hedgcoth 1993). Song and Hedgcoth (1994) have identified a 7-kDa protein in the cms lines as an integral part of the inner mitochondrial membrane. This protein is absent in cms plants restored to fertility by introduction of nuclear genes. A reasonable interpretation of the role of this protein in cms and in cms restoration might revolve around its toxic action on mitochondrial function in cms lines and the blocking of its synthesis in fertility-restored lines. The occurrence of high mitochondrial transcription is characteristic of a male-sterile alloplasmic hybrid of *T. aestivum* with *Agropyron trichophorum*; how mitochondrial overexpression causes male sterility here remains to be investigated (T. Suzuki et al 1995).

(vii) Sunflower

Sunflower has joined the ranks of plants in which the cms phenotype is known to be associated with a mutation in the mitochondrial genome. The mutation events involved are a 12-kb inversion and a 5-kb insertion/deletion in the vicinity of the ATP locus, resulting in the creation of a novel open reading frame (*ORF-522*) located 3'- to the α-subunit of *ATP-A* gene in the sterile line (Siculella and Palmer 1988; Köhler et al 1991; Laver et al 1991; Horn et al 1996). Translation experiments with mitochondria purified from seedlings of fertile and sterile lines showed that a novel polypeptide of about 15 kDa is synthesized by mitochondria of the sterile line but is not detectable in the fertile line (Horn, Köhler and Zetsche 1991; Laver et al 1991). Although the presence of *ORF-522* transcripts and the 15-kDA protein in the fertility-restored lines of sunflower raised some question as to whether the protein is functionally related to the cms phenotype, a subsequent work showed that in the presence of a restorer gene, the steady-state level of this protein is significantly reduced in the male florets. This is also associated with a male floret–specific reduction in the level of *ORF-522* transcripts (Monéger, Smart and Leaver 1994). These results implicate the nuclear restorer gene in a posttranscriptional reaction that destabilizes the novel mitochondrial transcript in a tissue-specific manner to restore fertility.

Molecular characterization of another line of cms in sunflower (CMS-3) has shown that genome sequences of at least five mitochondrial loci are altered. However, the typical 15-kDa protein is absent in the new cms line. Some evidence has been provided for potential recombination events in *COX-II* and *ATP-6* genes in the cms phenotype (Spassova et al 1994).

(viii) Other Plants

The pioneering studies described here ushered in a radical change in the perspective from which the mechanism of male sterility was investigated in later years. The involvement of mitochondrial rearrangements in cms of several other plants has been demonstrated, although these studies have not been carried forward to the level of sophistication needed for unequivocal conclusions. Moreover, such a broad range of mitochondrial alterations have been found, that it is safe to predict the occurrence of a unique set of changes for each system.

Given the ease of protoplast fusion and plant regeneration in tobacco, organelle DNA restriction patterns have been studied in cytoplasmic hybrids. Here also, the main detectable change relates to mitochondrial DNA, suggesting possible mitochondrial recombination events in cms lines (Belliard, Vedel and Pelletier 1979; Asahi, Kumashiro and Kubo 1988).

Many fertile and cms lines of *Beta vulgaris* show variations in mitochondrial DNA restriction fragment patterns, although restriction patterns of chloroplast DNA are identical (Powling and Ellis 1983; Mikami, Sugiura and Kinoshita 1984; Mikami et al 1985; Mann et al 1989). An interesting feature with implications on mitochondrial DNA organization in *B. vulgaris* is that mitochondria from cms lines lack two small, supercoiled DNA species that are present in the male-fertile lines (Powling 1981, 1982). Although these results are consistent with the assumption that mitochondrial DNA encodes the cms trait, alternative explanations have also been proposed (Duchenne et al 1989). A renewed strategy for differential screening of mitochondrial cDNA libraries of fertile and cms lines and for characterization of the genomic organization, DNA sequence, and transcription of *ATP-6, ATP-A, COX-I*, and *COX-II* loci has revealed some genomic rearrangements that can serve as a basis for studying the differences between the two types (Xue et al 1994; Senda et al 1991; Senda, Mikami and Kinoshita 1993).

Studies on cms in *Vicia faba* (broad bean; Fabaceae) have indicated that hereditary information for pollen abortion can be carried in forms other than mitochondrial genes coding for a unique protein. In an electron microscopic study of male-sterile, maintainer, and restored male-sterile *V. faba* lines, Edwardson, Bond and Christie (1976) found that the presence of cytoplasmic spherical bodies about 70 nm in diameter distinguishes the male-sterile plants from the rest. Subsequent work indicated that the spherical bodies are made up of unit membranes enclosing single- or double-stranded high-molecular-mass RNA (Grill and Garger 1981; Scalla et al 1981). The fact that it is possible to cause sterility symptoms in a fertile *V. faba* host by transmission of this RNA through a graft union is probably suggestive of the involvement of a viruslike particle in cms in this plant (Grill and Garger 1981).

Reliance upon the analysis of mitochondrial DNA has opened up the question of mitochondrial inheritance of male sterility in *V. faba*. Compared to normal fertile lines, sterile lines display an altered mitochondrial DNA restriction pattern and carry in their cytoplasm small supercoiled mitochondrial DNA molecules (Boutry and Briquet 1982; Negruk et al 1982; Goblet et al 1983). However, the nucleotide sequence of plasmids isolated from the sterile lines is analogous to that of plasmids isolated from the fertile lines (Goblet, Flamand and Briquet 1985; Flamand et al 1992). Here there is no escape from the conclusion that a mitochondrial plasmid is unlikely to be associated with cytoplasmic male sterility.

Specific transcripts identified in two male-sterile cytoplasms of *Plantago lanceolata* have been associated with the presence of a unique 1.1-kb DNA fragment of mitochondrial origin (Rouwendal, van Damme and Wessels 1987). Although this finding may provide a marker to screen populations for male sterility, its implication on the role of mitochondrial DNA in male sterility should not be overlooked. Since restriction fragment patterns of mitochondrial DNA in male-sterile interspecific crosses in *Epilobium* appear to be identical to those of the female parents, the basis for male sterility has been sought in the expression of mitochondrial genes. It turns out that in the male-sterile hybrid of *E. hirsutum* × *E. montanum*, the transcription of the *COX-II* gene is significantly altered by the presence of a major transcript of 2 kb and two minor fragments of 1.2 kb and 0.8 kb in place of a 1.7-kb transcript found in the *E. hirsutum* maternal parent (Schmitz and Michaelis 1988). Unique patterns of mitochondrial transcripts or proteins have been described in male-sterile cytoplasms of *Pennisetum glaucum* (Smith, Chowdhury and Pring 1987; Smith and Chowdhury 1991), *P. americanum* (Munjal and Narayan 1995), *Daucus carota* (Scheike et al 1992), *Allium schoenoprasum* (Potz and Tatlioglu 1993), and *Lolium perenne* (Kiang and Kavanagh 1996). In *D. carota*, restoration of fertility by nuclear genes does not affect the mitochondrial DNA structure or the synthesis of a 17-kDa polypeptide found exclusively in the anthers of cms lines.

A few other studies that indirectly implicate mitochondrial DNA in male sterility are also worth mentioning. Reference was made in a previous section to the isolation of a male-sterile mutant of *Arabidopsis* by transposon tagging. Characterization of the corresponding gene has revealed a short region of homology with an open reading frame in wheat mitochondrial genome (Aarts et al 1993). Male-sterile mutants have been generated by fusion of protoplasts of *Lycopersicon esculentum* with those of *Solanum acaule* and *S. tuberosum* after treatments that damage the mitochondria or the nucleus. The molecular event identified in the hybrids is the loss of some elements of the mitochondrial DNA of both parents, as well as the appearance of recombinant mitochondrial fragments not present in either parent (Melchers et al 1992).

This section will conclude with a brief account of studies that have related cms to alterations in chloroplast DNA. In the investigations described earlier in this section, when chloroplast DNA was analyzed simultaneously with mitochondrial DNA, little or no differences were found in the chloroplast DNA restriction patterns between male-sterile and fertile plants. This, coupled with the mounting molecular evidence for alterations in mitochondrial DNA in male-sterile lines of several species, led to the near universal acceptance of the view that chloroplast genes are unlikely to be involved in male sterility. However, some claims have been made for an interaction of altered chloroplast DNA with nuclear DNA in the induction of male sterility, perhaps at an early stage in floral development before mitochondrial genes strike. Studies were performed using approaches that mainly involved restriction fragment analysis of chloroplast DNA and SDS-PAGE of chloroplast proteins to gauge the degrees of affinity and difference between fertile and sterile plants. In seven male-sterile cultivars of tobacco, all of which have abnormal stamens that fail to produce pollen, the appearance and the isoelectric points of the large subunit polypeptides of Rubisco are different from those of male-fertile cultivars (Chen, Johal and Wildman 1977). In most of the male-sterile stocks studied, the polypeptide pattern can be traced to an original F1 hybrid with a maternal parent possessing chloroplast DNA that codes for the altered polypeptide. An analogous correlation exists between a male-sterile line of cotton possessing petaloid stamens and Rubisco large subunit structure like that of the source of cytoplasm (Chen and Meyer 1979). The observation that chloroplast DNA isolated from certain isonuclear male-sterile lines of tobacco has a restriction endonuclease digestion pattern identical with that of the maternal parents is also compelling in its implications to male sterility (Frankel, Scowcroft and Whitfeld 1979). Li and Liu (1983) have noted differences in the restriction fragment pattern of chloroplast DNA from cms lines of maize, wheat, and rapeseed and from their fertile counterparts. The differences are most visible in two-dimensional electrophoresis of restriction fragments using partial denaturation of DNA for fragment differentiation. When this method yields the kind of resolution necessary to analyze changes in the composition of DNA fragments, the results might indicate that some nucleotide sequence differences in the chloroplast DNA are associated with male sterility. A possible relationship has been suggested between a 165-bp deletion in the plastid gene that encodes RNA polymerase β''-subunit and male sterility in certain lines of sorghum that produces small anthers (Chen et al 1995). However, in none of the cases described here is the evidence overwhelming for the contention that male sterility is coded by chloroplast genes.

5. GENERAL COMMENTS

It has been well documented that genetic defects in pollen development in angiosperms may occur at any stage of microsporogenesis or microgametogenesis and result in male sterility. Thus, diploid cells of the anther primordium earmarked to produce pollen grains, as well as the mature haploid pollen grains programmed to produce gametes, are prime targets of action of male-sterility genes. Given the complex development of the anther, it is not surprising that pollen sterility may be secondarily caused by disturbances in the development of the somatic tissues of the anther, such as the tapetum. In spite of the array of defects that are associated with male sterility in different plants, gene action does not interfere with the development of the female part of the flower. From a practical point of view, one of the most significant aspects of the study of male sterility in plants is the bearing it has upon our means of increasing the cytoplasmic genetic diversity in crops by identifying or creating new sources of male sterility.

Starting off at a modest level from genetic and cytological concepts, the subject of male sterility has matured to take a molecular outlook. At the molecular level, one of the particularly striking features of male sterility is its relationship to the rearrangement of the mitochondrial genome. It is important to note that male-sterility genes do not act by causing the loss of segments of the gene but rather act by introducing new recombinant molecules that produce the translation of abnormal mitochondrial proteins. Recombination thus plays an important role in introducing new genetic diversity by generating mutants with new mitochondrial genomes.

Now that the role of the mitochondrial genome in male sterility has been established, a more speculative topic can be addressed, namely, the manner in which the altered mitochondria are able to influence events specifically in the haploid cells of the anther. Very little attention has been focused on this question; however, the many parallels between cms in such widely studied systems as maize, *Petunia*, and bean suggest that the toxic protein produced acts like any toxic compound, namely, it causes the death of the cells in its immediate vicinity. Moreover, why the cells of the anther are most vulnerable to the action of the sterility gene is important, because an answer to this question may help elucidate the most fundamental character of the gene product. The mechanism by which cms interrupts normal pollen development may reside not only in the toxic protein produced, but also in the specialized cells of the anther, especially the sporogenous cells and their division products, and the tapetum. Interestingly enough, the isolation of male-sterility genes remains at the top of our wish list. The lack of progress seems to indicate that new and unorthodox screening procedures with model systems are probably required to achieve a breakthrough in this field.

Chapter 6

Megasporogenesis and Megagametogenesis

The carpel is the gross morphological part of the flower concerned with female sporogenesis and gametogenesis and is the homologue of the stamen. In its simplest form, the carpel is a leaflike organ composed of the ovary, style, and stigma, although various kinds of fusions between carpels have produced a complex organ in the modern flower. As is well known, along its margins the ovary encloses ovules, which become seeds in the mature fruit. In its role of sheltering the progenitors of seeds, the carpel determines the extent of physical and chemical influences that regulate the transformation of ovules into seeds. Tissues constituting the wall of the carpel serve as supporting structures when the ovary becomes the fruit.

The ovule is functionally a megasporangium because it defines the structural unit of the carpel in which sporogenesis and subsequent differentiation of the female gamete, or the egg, take place. In the context of the reproductive biology of angiosperms, the functional unit of the plant that produces the egg for fusion with the sperm is the megagametophyte or the female gametophyte. The megagametophyte is a highly reduced group of cells dependent upon the sporophytic plant for its nurture and nutrition; it is generally equated with the embryo sac. The central position of the embryo sac in the reproduction of flowering plants has been recognized for well over a century. The organization of the mature embryo sac in a wide variety of angiosperms is now fairly well established as a result of light microscopic histological studies. It has been one of the conclusions from comparative studies that the different types of embryo sacs are fabricated by a series of divisions, although the final organizations of the embryo sacs might differ. Understanding of the detailed structure of the cells of the embryo sac has been given a sharp definition during the past 30 years by the ability to examine the cells in the electron microscope. This new wealth of information has provided insights into the function of the cells of the embryo sac. The emergence of light microscopic histochemical and molecular techniques has shown that the behavior and functional competence of the embryo sac are reflected in the integrated action of its constituent cells. If the full structural and functional nature of the female gametophyte is to be fully unraveled, it is clearly necessary to isolate the embryo sac from its environment and study it uninfluenced by the surrounding cells. Research undertaken during the past 10 years has made a remarkable beginning toward this goal.

This chapter will first examine the development of the carpel and the evolution of a group of cells on the floral receptacle into the carpel. The latter parts of the chapter will describe the development and organization of the megagametophyte with an emphasis on both the old and the new information available. No attempt is made to present a comprehensive treatment, and specific examples are chosen because of their priority in discovery or for the insights they provide.

1. CARPEL DETERMINATION

As a rule, the carpels closely follow the stamens on the floral meristem and are the final appendages to be carved out of the meristem during the ontogeny of the flower. The method of initiation and early differentiation of the carpel closely agrees with the early ontogeny of the stamens considered in Chapter 2. The development of the carpel described in *Ranunculus repens, Aquilegia formosa,* and other members of Ranunculaceae is typical of that in many other plants and can be subdivided into several phases (Tepfer 1953; van Heel 1981). In the first phase, specific groups of cells are determined as carpels and are set apart at precise intervals on the receptacle. In the second phase, cell differentiation occurs in such a way that the carpel primordium assumes the shape of a shallow cup. The third phase is concerned with the formation of two longitudinal ridges or flanges that give the carpel the shape of a pitcher with an open adaxial side. The formation of the locule usually begins with the growth of margins toward one another and with their subsequent fusion. Following closure of the margins, ovules are initiated from the placenta, which is the enlarged tissue between the fused edges of the carpel and the inner face of the locule. To what extent differential mitotic activity is involved in the attainment of form by the carpel is not known.

The question now concerns the events that determine the establishment of prospective carpel primordia. Similar to the situation described in stamens, culture of early stage flower buds of *A. formosa* (Tepfer et al 1963) and *Nicotiana tabacum* (Hicks and Sussex 1970) has shown that irrespective of the stage of the flower bud at the time of culture, part of the floral meristem is already determined as carpels. Other observations, such as the partial transformation of stamen primordia of *N. tabacum* into carpels in a medium containing kinetin (Hicks 1975), the induction of ovules on stamens of *sl-2* mutant of tomato (Sawhney and Greyson 1973a, 1979), and the propensity of cultured potential male buds of cucumber to form carpels (Galun, Jung and Lang 1963), show that there is a great deal of developmental flexibility in the stamen and carpel primordia even after they are determined on the floral receptacle. The existence of a long-lived determinant of carpel primordium that spreads to other parts of the flower, implied in studies of tobacco and tomato, is supported by the results of culture of excised placentae of tobacco. The significant observation is that, at an initial stage in the ontogeny of the ovule on the placenta, the regenerate formed very much resembles a gynoecium, although this is replaced by an ovule at a later developmental stage (Evans and Malmberg 1989). In tobacco flowers, carpels rarely appear if the stamen primordia are surgically removed; this is in keeping with the thesis that carpel determination depends not only on the influence of the meristem but also of the stamen primordia (McHughen 1980). On the whole, like the stamens, the carpels seem to fit well into our line of thought about determination of floral organs and their interrelationships.

Chapter 2 described a genetic approach using homeotic mutations to study the factors controlling stamen determination in the flowers of *Arabidopsis* and *Antirrhinum majus*. It was emphasized therein that interactions between floral organs are common in the floral meristem during their determinative phase and that the same genetic program may control both stamen and carpel determination. To avoid duplication, reference is made to the section "Anther Determination," pp. 17–25, for a full discussion of these experiments. In summary, determination of the carpel in *Arabidopsis* is under the control of *AG* and *SUP* genes with intervention in the fringe by other genes. Expression of the *AG* gene is antagonized by action of the *AP-2* gene in the whorls reserved for sepals and petals, and by the combined action of the *PI* and *AP-3* genes in the whorls where stamens and carpels appear. On the other hand, the *SUP* gene normally prevents the activity of the *PI* and *AP-3* genes; consistent with this is the observation that transcripts of the latter two genes are localized in the fourth whorl of *sup* mutants where stamens instead of carpels appear. The gene in *A. majus* that comes close to the *Arabidopsis AG* gene is *PLE*. A common genetic pathway underlying stamen and carpel determination represents one of possibly several means by which homologous organs are derived, but it does not answer the question of what separates one kind of organ from the other during progressive ontogeny. This consideration should be kept in mind in an examination of the molecular processes underlying carpel determination.

Evidence from periclinal chimeras of tomato points to a possible role of cells occupying the third inner layer of the floral meristem in controlling carpel formation. In chimeras produced between a wild-type *Lycopersicon esculentum* that forms two carpels per flower and the mutant *fasciated,* which contains an increased number of carpels, and between *L. esculentum* and *L. peruvianum* with more than two carpels, the genotype of cells in the third inner layer contributed by either the *fasciated* mutant or *L. peruvianum* causes an increase in the number of carpels initiated in the chimeric flower (Szymkowiak and Sussex 1992). The authors consider the results as evidence for the transmission of inductive information from the third inner layer to the outer layers, where carpel primordia appear.

2. MEGASPOROGENESIS

In Chapter 2 it was shown that in many plants the development of the anther occurs in a predictable time interval, thus making it possible to relate stages in anther ontogeny to certain aspects of floral morphology as markers. In contrast, developmental stages of the ovule have not been used as morphological indicators that can be easily correlated with changes in floral morphology. Although the development of the flower and of the ovule cannot be dissociated from each other, the relative inaccessibility of the ovule has hampered attempts at systematic developmental analyses of this organ. To fill this void, a developmental timetable for maize ovules has been constructed by relating stepwise changes in megasporogenesis and megagametogenesis to the length of the silk (Huang and Sheridan 1994). Similarly, the distinguishing features seen in cleared and stained whole mounts of ovules of *Arabidopsis* have been incorporated into a comprehensive classification of ovule development in this plant (Schneitz, Hülskamp and Pruitt 1995). Although it is appropriate that maize and *Arabidopsis* ovules have figured in these pioneering morphological studies, developmental timetables for other widely studied ovules would be highly desirable.

In a discussion of megasporogenesis and the subsequent development of the mature megagametophyte, a description of the ovule merits some attention. This is necessary for the simple reason that the single most important cell from which the megagametophyte evolves is housed in the ovule. Nutrition of the developing megagametophyte must perforce take place by cell-to-cell interactions and by mechanisms that allow the flow of nutrients into the ovule. In consequence, much of the developmental dynamics of the megagametophyte must ultimately be interpreted and explained in terms of the fate and functions of the tissues of the ovule. Two reviews have covered the developmental biology of sporogenesis and female gametogenesis in angiosperms fairly extensively and from different perspectives (Kapil and Bhatnagar 1981; Willemse and van Went 1984).

(i) The Ovule

Figure 6.1 provides a diagrammatic overview of the development of the ovule with special reference to the events of megasporogenesis and megagametogenesis described in this section. The ovule has its origin by periclinal divisions of cells lying below the epidermal layer of the placenta. Later, by repeated anticlinal and periclinal divisions, an undistinguished mass of cells of the ovule primordium, known as the nucellus, is formed; early during its ontogeny, the primordium differentiates from its base two multilayered coverings, or integuments, at the same time as it is raised on a stalk known as the funiculus. In the final stage of ovule maturation, the integuments grow around the nucellus in a kind of intimacy that leaves a small opening at the free end of the ovule; this opening is termed the micropyle. The opposite end of the ovule, where the integuments fuse with the funiculus, is known as the chalaza. The terms "micropylar end" and "chalazal end" are in common usage to designate these respective extremities of the ovule. A remarkable feature of the histogenesis of the integuments is the differentiation of the inner layer of the inner integument into a radially extended layer of cells known as the endothelium or integumentary tapetum. Occasionally, as in *Arabidopsis*, a certain degree of asymmetry prevails in the growth of the outer integument, which consequently arches over the inner integument on one side of the ovule.

The discovery of *MADS*-box genes that encode putative transcription factors in the floral meristem has led to a study that shows that a similar class of genes accounts for ovule identity in *Petunia hybrida*. These genes, designated as *FLORAL BINDING PROTEIN* (*FBP*), are expressed exclusively in the developing ovule primordia. Inhibition of the expression of the genes in transgenic *Petunia* by cosuppression leads to the development of abnormal ovaries packed with spaghetti-shaped structures that are obviously products of the remodeling of normal ovules. In the ovaries of transgenes, expression of *FBP* transcripts is dramatically reduced, indicating that they are probably involved in the formation of normal ovules (Angenent et al 1995b). Finally, there is evidence that overexpression of the gene in transgenic plants leads to the development of ovulelike structures on the adaxial side of the sepals and on the abaxial side of the petals (Colombo et al 1995). This evidence provides an entering wedge for a new role for *MADS*-box genes in the reproductive biology of angiosperms,

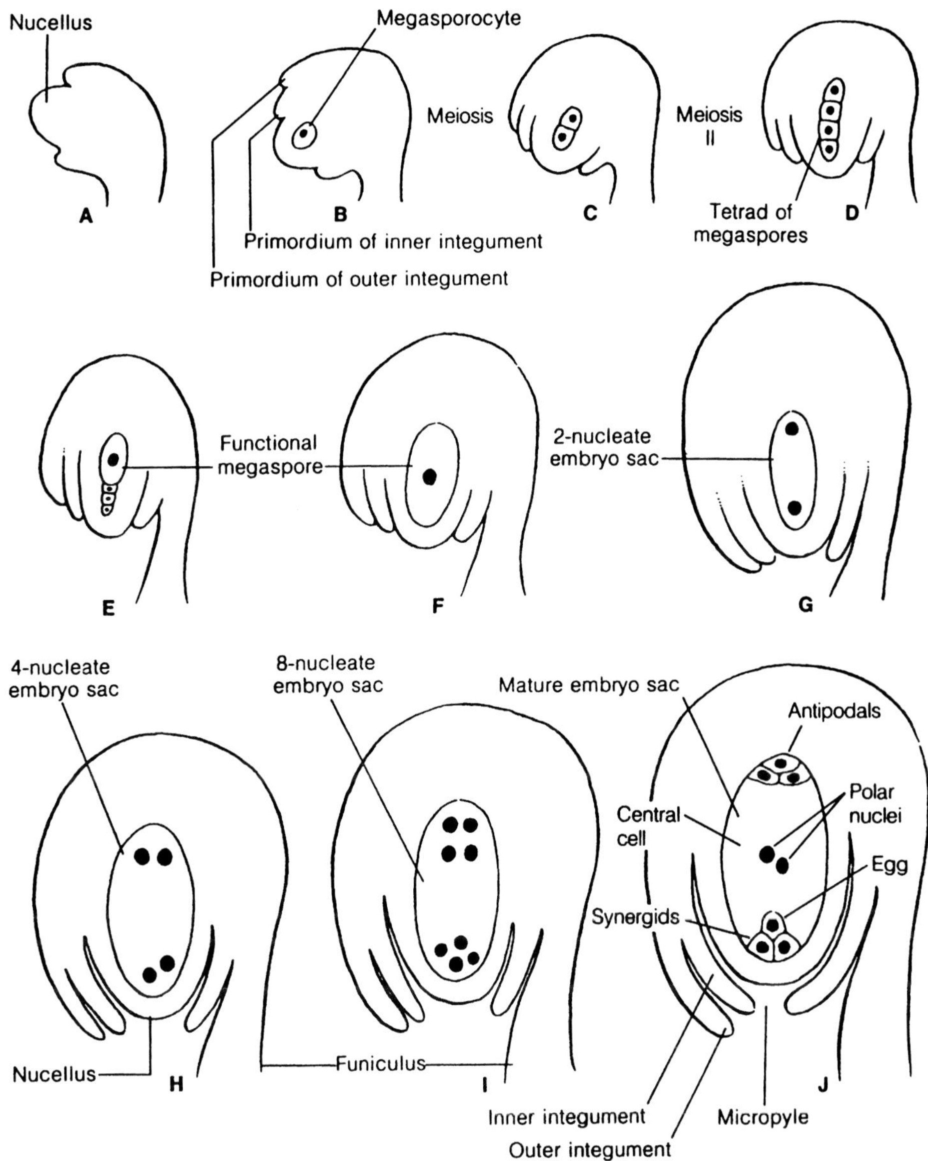

Figure 6.1. A diagrammatic representation of ovule development with reference to megasporogenesis and megagametogenesis. **(A, B)** Formation of integuments and megasporocyte. **(C, D)** Meiosis and formation of the linear tetrad. **(E, F)** The functional megaspore. **(G–I)** Two-, four-, and eight-nucleate embryo sacs, respectively. **(J)** Organization of the typical, seven-celled mature embryo sac. (From *Morphology and Evolution of Vascular Plants* 3d ed., by Gifford and Foster. Copyright © 1989 by W. H. Freeman and Company. Used with permission.)

perhaps as a master regulator of ovule development, but the sheer diversity of these genes and their interactions rules out any simple or universal pattern of transcriptional regulation during ovule determination. Nadeau et al (1996) have isolated several cDNA clones representing genes expressed in a stage-specific manner during ovule development in *Phalaenopsis* sp. Diverse proteins, such as a homeobox transcription factor, cytochrome P450 monooxygenase, a glycine-rich protein, and cysteine proteinase, are encoded by these genes. This reinforces the possibility that the molecular mechanisms during ovule development may differ among species or that multiple mechanisms may come into play.

Some *Arabidopsis* mutants that are important in

the genetic analysis of the ovule are those that cause discrete defects in the normal development of the integuments, such as *bell* (*bel-1*), in which the outer integument develops into a collarlike structure; *short integuments* (*sin-1*), in which growth of both integuments is disrupted (Robinson-Beers, Pruitt and Gasser 1992; Modrusan et al 1994); *aberrant testa shape* (*ats*), with a reduced number of cell layers in the integument (Léon-Kloosterziel, Keijzer and Koornneef 1994); *sup*, which induces a symmetric rather than an asymmetric growth of the outer integument (Gaiser, Robinson-Beers and Gasser 1995); and *aintegumenta* (*ant*), characterized by the lack of integuments (Klucher et al 1996; Elliott et al 1996). In several alleles of *bel-1*, the collar proceeds to form a carpel complete with stigmatic papillae and second-order ovules tucked away on the margins. *ANT* and *BEL-1* loci both encode DNA-binding transcription factors. The protein product of the *ANT* gene is related to that of the homeotic gene *AP-2* and contains sequences homologous to the DNA-binding sequences of ethylene-response–binding proteins, whereas the *BEL-1* gene encodes a homeodomain transcription factor. Suggestive of the involvement of these genes in integument formation, expression of their transcripts in the developing ovule infallibly predicts the position where the integuments arise (Reiser et al 1995; Elliott et al 1996; Klucher et al 1996).

That the abnormal development of the carpel phenotype on the *bel-1* mutants is due to the presence of the *AG* gene was suggested by the unrestricted expression of *AG* transcripts in the developing collar tissue of *bel-1* ovules. Confirmation of this causal relationship came from an experiment in which *AG* gene linked to a viral promoter was overexpressed in transgenic *Arabidopsis* and caused the development of phenocopies of *bel-1* mutant ovules (Ray et al 1994). Based on these observations, the protein product of the *BEL-1* gene can reasonably be regarded as a regulator of normal integument formation, functioning perhaps by suppressing *AG* activity in this structure. However, the substantial overlap of expression of *BEL-1* and *AG* transcripts seen in the wild-type *Arabidopsis* ovules rules out a *BEL-1*–*AG* interaction at the RNA level as a working explanation (Reiser et al 1995).

The cell that initially specifies the pathway to embryo sac formation in the ovule belongs to the nucellus. This cell, designated as the archesporial cell (initial), is distinguished from the other cells of the nucellus in two respects; first, it is always larger than the rest of the pack, and second, it is hypodermal in position at the micropylar end of the ovule. Both these remain as relatively stable properties and are transmitted faithfully from one generation to the other. Most commonly, the archesporial cell divides transversely to form a primary parietal cell to the outside and a primary sporogenous cell to the inside, the latter functioning as the megasporocyte. As to the fate of the primary parietal cell, it may either remain undivided or undergo a series of anticlinal and periclinal divisions to form a parietal tissue. In another scenario, the archesporial cell directly functions as the megasporocyte. In either case, the megasporocyte comes to occupy the tip of the nucellus (Figure 6.2).

Irrespective of the route by which the megasporocyte is formed, major interest in this cell has been its ultrastructural or biochemical identity from the rest of the cells of the nucellus. As seen in *Dendrobium* sp. (Orchidaceae) (Israel and Sagawa 1964), *Lilium candidum* (Rodkiewicz and Mikulska 1963, 1965b), *L. longiflorum* (Dickinson and Potter 1978), *Helianthus annuus* (Newcomb 1973a), *Cytinus hypocistis* (Rafflesiaceae) (Ponzi and Pizzolongo 1976), *Zea mays* (Russell 1979), *Capsella bursa-pastoris* (Brassicaceae) (Schulz and Jensen 1981), *Glycine max* (Folsom and Cass 1989), and *Viscum minimum* (Loranthaceae) (Zaki and Kuijt 1995), the primary components of the megasporocyte are the same typically found in nucellar cells, namely, a centrally placed nucleus surrounded by a blend of organelles in a nonvacuolate cytoplasm. Except for their terminal position in the nucellar row, the potential megasporocytes of *Dendrobium* sp. and *C. bursa-pastoris* do not show any differences from the cells of the nucellus in the complement and disposition of the organelles. The presence of plasmodesmata on the lateral cell walls or end walls of the megasporocyte indicates that this cell, in all likelihood, communicates with the cells of the nucellus for the transfer of metabolites (Israel and Sagawa 1964; Ponzi and Pizzolongo 1976; Dickinson and Potter 1978; Russell 1979; Schulz and Jensen 1981). Microtubule distribution in the megasporocytes of *Gasteria verrucosa*, *Chamaenerion angustifolium* (Onagraceae) (Bednara, van Lammeren and Willemse 1988; Willemse and van Lammeren 1988), *Arabidopsis* (Webb and Gunning 1990), and *Cymbidium sinense* (Orchidaceae) (Zee and Ye 1995) has been the object of investigations using immunofluorescence labeling of the tubulin components. The main import of these studies is that the generally random orientation of microtubules in the megasporocyte, or their occasional concentration in the perinuclear region radiating from the megasporocyte nucleus, is suggestive of the role of this cytoskeletal system in maintaining the

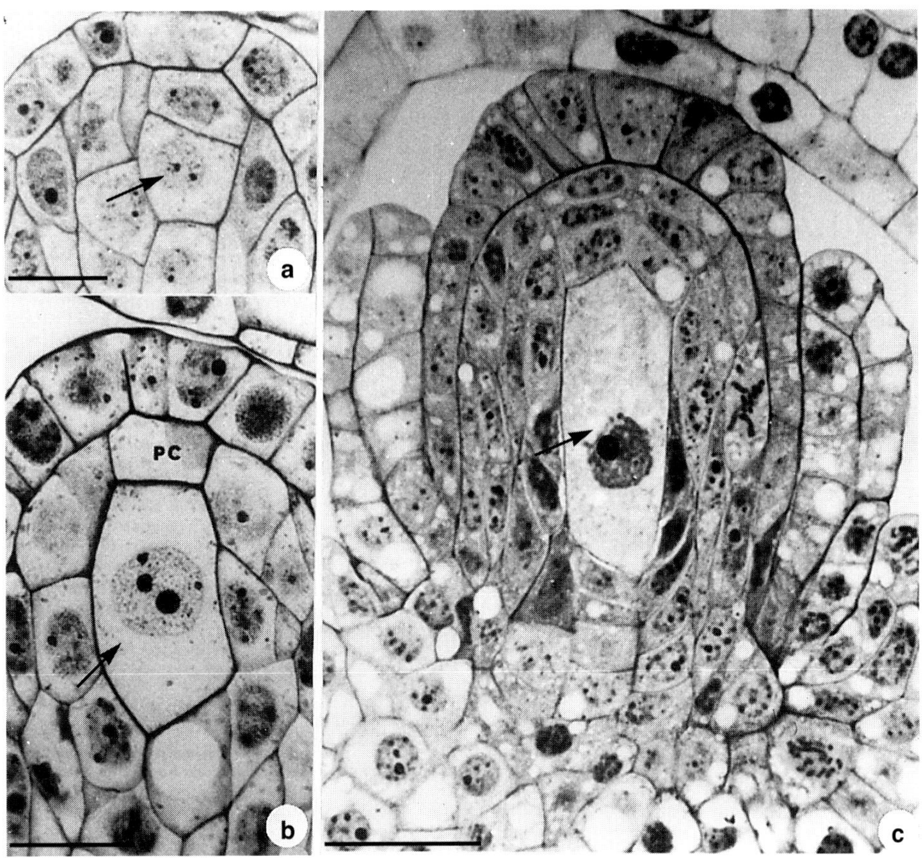

Figure 6.2. Megasporogenesis in *Ornithogalum caudatum.* **(a)** Nucellus with the archesporial initial just divided and the primary sporogenous cell (arrow) beginning to enlarge. **(b)** Nucellus showing the primary sporogenous cell (arrow) and the primary parietal cell (PC). **(c)** Nucellus with the megasporocyte (arrow) and three layers of parietal tissue. Scale bars = 10 μm. (From Tilton 1981.)

integrity of the cell and the position of the nucleus. In the megasporocyte of *G. verrucosa,* microfilaments invariably share a distribution pattern with microtubules, but postmeiotically they are concentrated in the functional megaspore (Bednara, Willemse and van Lammeren 1990).

Since the archesporial cell is small relative to its final size as the megasporocyte, it probably releases a biosynthetic program of some magnitude to undergo cell expansion. In ultrastructural terms, as described in *Dendrobium* sp. and *Lilium candidum,* there is a marked multiplication of organelles that causes the increase in cytoplasmic growth. In *L. candidum,* this is vividly seen with regard to the ER. Although the newly formed megasporocyte is devoid of ER, growth of this cell is associated with a progressive development of the organelle as parallel strands around the nucleus, followed by the appearance of the strands in the cytoplasm in spirals like a watchspring (Israel and Sagawa 1964; Rodkiewicz and Mikulska 1965b). Although megasporocytes have no preformed developmental axis, the megasporocyte of *Z. mays* displays a polarized distribution of cytoplasmic components with an abundance of rough ER toward the micropylar half of the cell and of mitochondria and plastids toward the chalazal part (Russell 1979).

Variable results have been obtained with regard to the changes in RNA and protein contents of the megasporocyte. Whereas histochemical methods have revealed moderate to high concentrations of RNA and proteins in the archesporial cells or megasporocytes of *Stellaria media* (Caryophyllaceae) (Pritchard 1964a), *Dipcadi montanum* (Liliaceae) (Panchaksharappa and Syamasundar 1975), *Zephyranthes rosea, Lagenaria vulgaris* (Cucurbitaceae) (Malik and Vermani 1975), *Farsetia hamiltonii, Eruca sativa* (both of Brassicaceae) (Prasad 1977), and *Argemone mexicana* (Papaveraceae) (Bhandari, Soman and Bhargava 1980), in situ hybridization of

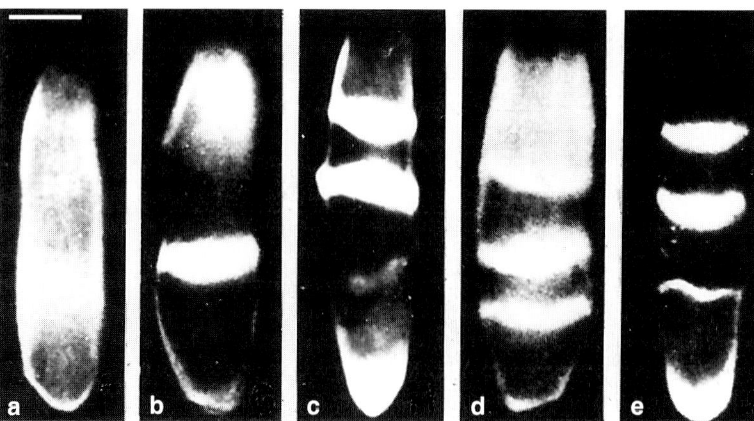

Figure 6.3. Megasporogenesis in *Oenothera biennis,* showing callose fluorescence of cell walls. **(a)** Megasporocyte in late meiotic prophase. **(b)** Dyad. **(c)** Triad. **(d)** Tetrad. **(e)** Tetrad at a later stage without fluorescence in the side walls. Scale bar = 10 µm. (From Rodkiewicz 1970. *Planta* 93:39–47. © Springer-Verlag, Berlin. Photographs supplied by Dr. B. Rodkiewicz.)

ovules of *Capsella bursa-pastoris* with ^{3}H-poly(U) showed the absence of poly(A)-RNA in the megasporocyte as compared to the rest of the nucellar cells (Raghavan 1990b). Use of a nonradioactive probe to detect poly(A)-RNA has disclosed that both the megasporocyte and the functional megaspore of *Medicago sativa* accumulate poly(A)-RNA during major periods of their existence (Bimal et al 1995). An important question concerning the differentiation of the archesporial cell and the megasporocyte relates to the factors that cause a particular cell of the nucellus to be recruited as a specialized cell. Data currently available on the ultrastructure and biochemical cytology of these cells do not yet give us a clue.

(ii) Meiosis

Once the megasporocyte is established, it is virtually at the end of its life as a diploid cell, since it immediately prepares to undergo meiosis. Meiotic division leads to the formation of four haploid megaspores. A characteristic feature of the heterotypic division is that it is always transverse to the long axis of the cell and generates two dyad cells. This is followed by the homotypic division, which in the majority of angiosperms studied is also transverse, and results in the formation of a linear tetrad. Although this represents the norm, there is also a bewildering array of variations in the orientation of the division wall following the homotypic division; Fabaceae, in which 12 different tetrad patterns have been described, is a paradigm of this diversity (Rembert 1971). In all cases the total number of megaspores or nuclei formed remains the same, although, as will be seen later, there is a distinction between divisions that form megaspores and those that generate the nuclei of the megagametophyte. If the orientation of the division wall can be changed without a change in the number of nuclei formed, there is a reasonable ground for the conclusion that production of nuclei is the primary event and that planes of division are specified under its control. In a few instances that have been examined, there is a conspicuous lack of a preprophase band of microtubules in the megasporocytes as they divide meiotically (Bednara, van Lammeren and Willemse 1988; Willemse and van Lammeren 1988; Webb and Gunning 1990; Huang and Sheridan 1994).

A notable change in the megasporocyte as it enters the meiotic cycle is callose impregnation in the wall. This was first described by Rodkiewicz and Górska-Brylass (1967, 1968), who found a distinct aniline blue fluorescence, indicative of callose deposition, in the chalazal end of the megasporocyte of *Orchis maculata* (Orchidaceae) beginning at diakinesis. Because of the simplicity of its use and reproducibility of the results, this has been a most useful assay, widely employed (Figure 6.3). Consequently, we have now an impressive list of plants in which callose formation during megasporogenesis has been demonstrated (Rodkiewicz 1970; Jalouzot 1971; Schwab 1971; Kuran 1972; Maze and Bohm 1974, 1977; Rodkiewicz and Bednara 1976; Kapil and Tiwari 1978a; Russell 1979; Willemse and Bednara 1979; Noher de Halac 1980b; Schulz and Jensen 1981, 1986; Kennell and Horner 1985; Folsom and Cass 1989; Webb and Gunning 1990). Although it was thought for some time that the initial appearance of callose at the chalazal end of the megasporocyte was related to the fate of the megaspore formed at this pole, the evidence has turned out to be misleading. It is now clear that irrespective of the fate of the chalazal megaspore after meiosis, the first signs of callose deposition can appear at either end of the megasporocyte. Later, as the megasporocyte enters the meiotic cycle, the entire cell wall is impregnated with callose and then callose-rich walls are formed in the dyad and tetrad. Callose deposition is transient, since it disappears as the three megaspores begin to degenerate and as a functional megaspore is established after meiosis. An analogous situation occurs during microsporogenesis when the microspores are temporarily ensheathed in a callose wall immediately after meiosis (Chapter 3). As in microsporogenesis, the callose walls during megasporogenesis have been conjured up to justify the need for isolating the newly formed haploid megaspores from the influence of the surrounding diploid cells, thus allowing the megaspores to assert their genetic individuality. It should be noted incidentally that callose deposition around megaspores is frequently observed in the monosporic and bisporic embryo sac types, but it has not been documented in the few tetrasporic types examined (Rodkiewicz 1970).

Although it is possible to isolate by enzymatic digestion of ovules certain stages of megasporogenesis that include megasporocytes (Yang and Zhou 1984; Golubovskaya, Avalkina and Sheridan 1992; Huang and Sheridan 1994; Mouritzen and Holm 1995), the technical difficulties of isolating these cells in enriched fractions remain at present insurmountable, raising the prospects that the biochemical and molecular details of megasporocyte meiosis may never be known. It is for this reason that thus far only a limited number of cytological studies of female meiosis in angiosperms have been undertaken (Bennett et al 1973a; Bennett and Stern 1975; Mogensen 1977). However, as in microsporocyte meiosis, the essential validity of the five distinct stages (leptotene, zygotene, pachytene, diplotene, and diakinesis) in megasporocyte meiosis has long been established; at least in the case of *Zea mays*, based upon reconstruction from electron micrographs, the pachytene stages of megasporocyte and microsporocyte meiosis appear to be similar with regard to the makeup and dimensions of the synaptonemal complex (Mogensen 1977). As reviewed in Chapter 2, molecular studies on microsporocyte meiosis have verified beyond doubt the occurrence of midprophase DNA synthesis, genetic recombination, DNA repair activity, and RNA and protein synthesis; similar molecular changes probably should be commonplace during meiosis in the megasporocyte.

Study of meiotic mutants of maize and rye has provided some insight, if not an explanation, into the genetic events underlying the formation of megaspores. Critical stages in megasporocyte meiosis that are controlled by distinct genes are the switch of the hypodermal cell into the archesporial initial, entry of megaspores into meiosis, early prophase of meiosis I, typical prophase stages, synapsis at diakinesis, spindle orientation at meiosis II, and divisions of the functional megaspore. Mutations of the genes involved highlight how relatively simple genetic changes alter the progress of meiosis, lead to abnormal cellular configurations in the female gametophyte, and eventually cause female sterility (Golubovskaya, Avalkina and Sheridan 1992; Sosnikhina et al 1992; Liu, Golubovskaya and Cande 1993; Sheridan et al 1996).

To the extent that meiosis bridges the transition from the sporophytic to the gametophytic phase of the life cycle, it is reasonable to expect some striking and probably significant ebb and flow of cytoplasmic activities associated with meiosis. Since specific gene activation may be involved in the transition, cell organelles that harbor genes and that are involved in nucleic acid and protein synthesis are the prime candidates for initiating the kind of stable changes concerned with repression of genetic information or synthesis of new transcripts. The nucleolus has figured in some studies that showed that at the pachytene stage of megasporocyte meiosis in *Pisum sativum*, this organelle segregates in a caplike fashion against the nuclear membrane until it disperses as dense fibrillar components at later stages of meiosis (Galán-Cano, Risueño and Giménez-Martín 1975; Medina and Risueño 1981; Medina et al 1983). This is reminiscent of the synthetic activity of the nucleolus associated with male meiosis, but it occurs at a reduced level and is unlikely to be the same at the molecular level (Chapter 2). In *Lilium longiflorum*, a particularly arresting feature is that plastids and mitochondria cycle through a phase of dedifferentiation beginning at the early prophase of megasporocyte meiosis and a phase of redifferentiation at the tetrad stage. Typical symptoms of dedifferentiation are the loss of contents, including starch and ribosomes from the plastids, and the spherical or rodlike conformation of the mitochondria, accompanied by the appearance of electron-dense moi-

disintegration and survival of megaspores following meiosis are reflections of the possible nutritional influence of the sporophytic tissues, especially of the nucellus. The presence of plasmodesmata, initially in the cell wall at the chalazal end of the megasporocyte and subsequently at the chalazal end of the dyad and the tetrad, makes these cells so porous as to provide the functional megaspore with a nutritional edge over the others and to favor its survival (Russell 1979; Willemse and Bednara 1979; Schulz and Jensen 1986; Folsom and Cass 1989). Because the degenerating megaspores show an increased titer of proteins and nucleic acids, they have attracted some attention as a possible source, along with the products of cell degeneration, of metabolites to stabilize the functional megaspore (Noher de Halac and Harte 1977). Since there is absolutely no indication of the nature of the materials coming from the aborting cells, this idea must rest now where it is, somewhere aside of the mainstream.

3. ORGANIZATION OF THE MEGAGAMETOPHYTE

The study of megagametophyte development in angiosperms has been pursued with great fervor for more than a century, but it will be treated here in a superficial way. For finer details of the ontogeny of the different embryo sac types, the book by Maheshwari (1950) and the reviews by Willemse and van Went (1984) and Haig (1990) and the papers cited therein are recommended. A two-volume treatise on the comparative embryology of angiosperms has also recently been published (Johri, Ambegaokar and Srivastava 1992).

The organization of the megagametophyte begins with the completion of megasporocyte meiosis, which has as its requirement a reduction in chromosome number from the diploid to the haploid value. Up to this point the discussion has focused almost entirely on a pattern of meiosis that gives rise to a linear tetrad of megaspores, abortion of three megaspores toward the micropylar pole, and survival of the remaining, chalazal one as the functional megaspore. This was done with good reason – the normal course of megasporogenesis in the majority of angiosperms follows this pattern, designated as the monosporic type. Comparative studies of megasporogenesis in a wide variety of angiosperms have disclosed unsuspected and very interesting deviations: For instance, in several species, after a dyad is formed, degeneration of one of the cells is the norm and only the remaining cell undergoes the second meiotic division. Since wall formation does not occur after the second meiotic division, both megaspore nuclei contribute to the formation of the female gametophyte, which is of the bisporic type. Still more surprising is the fact in some species, meiosis I and meiosis II of the megasporocyte are unaccompanied by wall formation, so that all four megaspore nuclei take part in the construction of the gametophyte. This is designated as the tetrasporic type.

This general pattern of megasporogenesis is further complicated by the existence of various types of gametogenesis, characterized by a variable number of nuclear divisions. In angiosperms, the goal of megagametogenesis is to produce the egg and the polar fusion nucleus, whose further development can be triggered by fertilization. In the monosporic type, the development of the female gametophyte begins with the elongation of the functional megaspore along the micropylar–chalazal axis. At the same time the nucleus of the megaspore is partitioned into two nuclei, one of which is pushed to the micropylar pole and the other to the chalazal pole of the cell. Later, both the nuclei divide twice, so that eight haploid nuclei, four at each pole, come to cohabit in a common cytoplasm. After the final nuclear division, the cell undergoes appreciable growth to form a saclike supercell, the embryo sac. This cell, embedded in the nucellus, apparently supports and binds together the various nuclei and the cells derived from them, and thus serves to maintain the integrity of the female gametophyte. The existence of the embryo sac as a multinucleate cell is very transient as the nuclei imprisoned at the two poles organize as cells in a characteristic pattern. Of the four nuclei at the micropylar pole, three organize as the egg apparatus, consisting of an egg cell flanked on either side by a synergid. Similarly, three nuclei at the chalazal pole become distinct membrane-enclosed cells anointed as the antipodals. The main body of the embryo sac remaining after the definition of the egg apparatus and antipodals is occupied by the central cell, which contains two polar nuclei. Initially, the polar nuclei come to lie side by side in the benign environment of the central cell; later, they become invested by a single membrane to form a diploid polar fusion nucleus. The widely accepted view about the origin and disposition of the four nuclei at the micropylar end is that the two synergids are derived from one of the two nuclei of a pair in the four-nucleate embryo sac, whereas the other nucleus of the pair gives rise to the egg and a polar nucleus. This typical monosporic, eight-nucleate embryo sac was first described in *Polygonum divaricatum* (Polygonaceae) and, according to convention, is known as the Polygonum type. A variation of the Polygonum type is the Oenothera type, where the functional megaspore is positioned at the micropylar end of the ovule. Moreover, the megaspore nucleus divides only twice to constitute an embryo sac consisting of an egg apparatus and a single haploid polar nucleus, antipodals being completely absent. Incidentally, the Polygonum mode of embryo sac development occurs with high frequency among angiosperms, although it represents only one example of a process that has many variations.

Because the development of the mature embryo sac just described depends critically on the number of participating megaspore nuclei and the number of intervening divisions, deviations in these parameters might be expected to result in changes in embryo sac structure. This does indeed appear to be true, and variations have been noted in the composition of the egg apparatus, ploidy level of the polar fusion nucleus, and the number of antipodals cut off in embryo sacs (Figure 6.5). In the bisporic type, the two surviving nuclei undergo only two successive mitotic divisions to yield an 8-nucleate embryo sac. Representatives are the Allium type, in which the chalazal dyad is functional, and the Endymion type, where the micropylar dyad is functional. In the tetrasporic type, some very bizarre embryo sacs are produced by the participation of all four megaspore nuclei in gametogenesis. The four megaspore nuclei may not divide further but instead may form a 4-nucleate embryo sac (Plumbagella type), or may undergo a single postmeiotic division to produce an 8-nucleate embryo sac (Adoxa type, Plumbago type, Fritillaria type) or may go through two rounds of divisions to form a 16-nucleate embryo sac (Penaea type, Peperomia type, Drusa type); after they are formed, the four nuclei may fuse in a complex manner before dividing further (Fritillaria type, Plumbagella type). Regarding the organization of the egg apparatus in the tetrasporic type of embryo sacs, in one variation (Peperomia type), there is only one synergid, whereas in the Plumbago and Plumbagella types the egg appears orphaned without flanking synergids. The central cell may harbor four haploid nuclei (Plumbago and Penaea types), eight haploid nuclei (Peperomia type), or one haploid and one triploid nucleus (Fritillaria and Plumbagella types). These nuclei apparently fuse before fertilization to form fusion nuclei with more than the usual diploid chromosome complement. The different types of embryo sacs also show a trend toward reduction in the number of antipodals, ranging from 11 in the Drusa type to 1 in the Plumbagella and Plumbago types, and none, as indicated earlier, in the Oenothera type. This finding has led to the suggestion that the major embryo sac types in angiosperms have probably evolved from a gymnosperm ancestry by the progressive specialization of a structure resembling an archegonium (Favre-Duchartre 1978). Although the final organization of the bisporic Oenothera type and some of the tetrasporic types of embryo sacs is identical to that of the Polygonum type, it is clear, because of the participation of more than one meiotically segregating megaspore nuclei, that genetic heterogeneity has been introduced.

(i) Structure of the Embryo Sac

In an ovule, the embryo sac is limited on the outside by the remaining cells of the nucellus, or where the nucellus has disintegrated, by the endothelium. As is well known, during fertilization one sperm fuses with the egg to generate the zygote, whereas the second sperm fuses with the polar fusion nucleus to generate a nutritive tissue known as the endosperm. To accomplish these fusion events, the embryo sac has evolved into a highly specialized structure enclosing equally specialized cells with a surprising degree of structural similarity. To get a sense of this specialization, the ultrastructural features of the constituent cells of the embryo sac, beginning with the egg cell, will be examined.

The Egg. The egg of angiosperms comes in various sizes and shapes, but it is generally a pear-shaped cell attached to the micropylar pole of the embryo sac by its pointed end (Figure 6.6). A prominent nucleus and a mixture of cytoplasmic organelles, including ER, dictyosomes, mitochondria, and ribosomes, are distributed in the egg in myriad different ways. As a result, although there are common principles of organization, it is unwise to attempt to generalize a structural base for widely separate species. Our knowledge of the structure of the egg has been greatly enhanced since the initial studies of Jensen (1963) and van der Pluijm (1964) on *Gossypium hirsutum* and *Torenia fournieri* (Scrophulariaceae), respectively. In later years, a succession of papers has appeared, describing the ultrastructural appearance of eggs of *Zea mays* (Diboll and Larson 1966; van Lammeren 1986; Faure et al 1992), *Crepis tectorum* (Asteraceae)

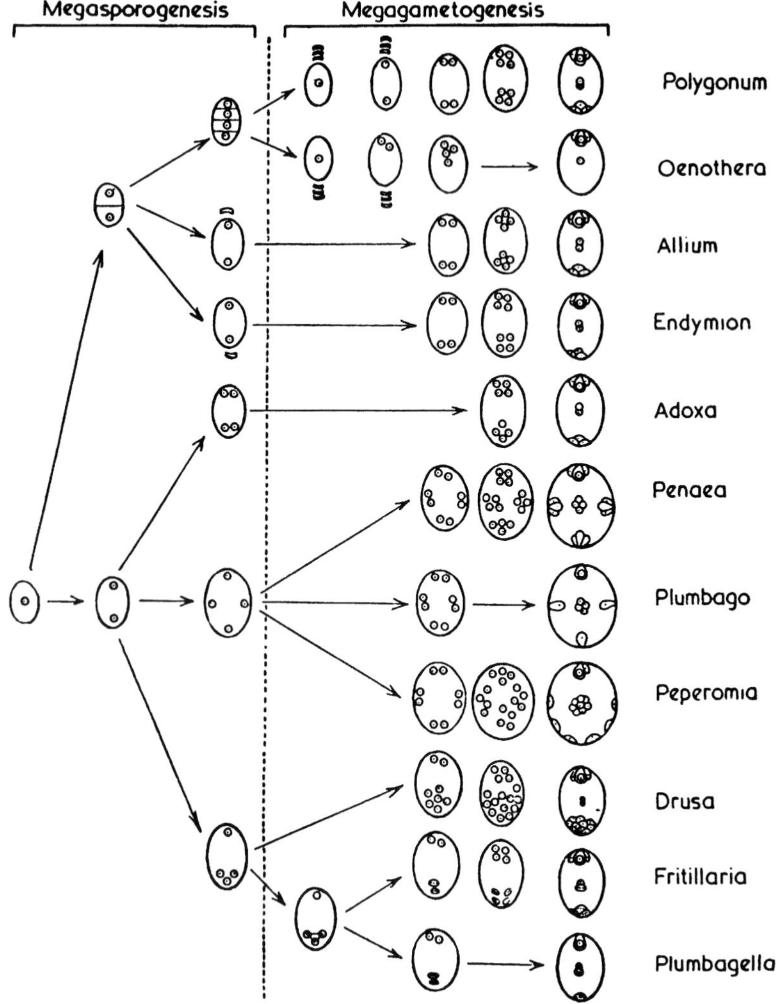

Figure 6.5. Diagrammatic representation of the different monosporic, bisporic, and tetrasporic types of embryo sacs. Micropylar pole is toward the top of the page. (From Johri 1964.)

(Godineau 1966), *Linum usitatissimum* (Vazart 1969; Secor and Russell 1988), *L. catharticum* (D'Alascio Deschamps 1973), *Capsella bursa-pastoris* (Schulz and Jensen 1968c), *Myosurus minimus* (Ranunculaceae) (Woodcock and Bell 1968b), *Epidendrum scutella* (Orchidaceae) (Cocucci and Jensen 1969a), *Petunia hybrida* (van Went, 1970b), *Quercus gambelii* (Fagaceae) (Mogensen 1972), *Helianthus annuus* (Newcomb 1973a; Yan, Yang and Jensen 1989, 1991b), *Plumbago zeylanica* (Cass and Karas 1974; Russell 1982, 1987), *Oenothera lamarckiana* (Jalouzot 1975), *Stipa elmeri* (Poaceae) (Maze and Lin 1975), *Nicotiana tabacum* (Mogensen and Suthar 1979), *Agave parryi* (Tilton and Mogensen 1979), *Spinacia oleracea* (spinach; Chenopodiaceae) (Wilms 1981a), *Glycine max* (Folsom and Peterson 1984; Folsom and Cass 1990, 1992), *Triticum aestivum* (You and Jensen 1985), *Hordeum vulgare* (Cass, Peteya and Robertson 1986), *Beta vulgaris* (Bruun 1987), *Brassica campestris* (Sumner and van Caeseele 1989), *Oryza sativa* (Maeda and Maeda 1990), *Arabidopsis* (Mansfield, Briarty and Erni 1991), *Daucus carota, D. aureus, D. muricatus* (Hause 1991), *Allium tuberosum* (Tian and Yang 1991), *Solanum nigrum* (Briggs 1992), and *Viscum minimum* (Zaki and Kuijt 1994). In the most detailed of these studies on *G. hirsutum* (Jensen 1963, 1965c), *C. bursa-pastoris* (Schulz and Jensen 1968c), and *Arabidopsis* (Mansfield, Briarty and Erni 1991), the eggs are described as strongly polarized cells with

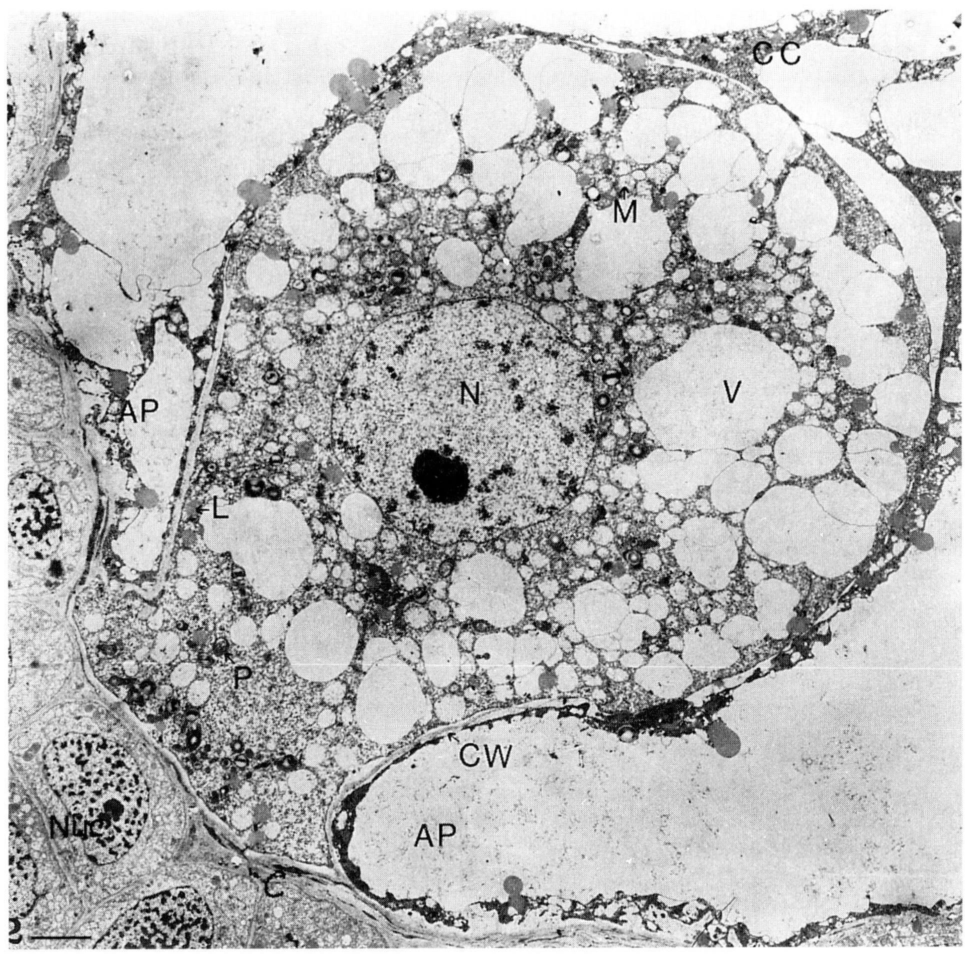

Figure 6.6. Longitudinal section of the egg cell of *Triticum aestivum* before pollination. Micropyle is toward the bottom of the page. AP, apical pocket; CC, central cell; CW, cell wall; L, lipid bodies; M, mitochondria; N, nucleus; V, vacuole. Scale bar = 10 μm. (From You and Jensen 1985; photograph supplied by Dr. R. You and Dr. W. A. Jensen.)

a large vacuole toward the micropylar end and an aggregation of cytoplasmic organelles and the nucleus toward the chalazal end. The egg has a limited amount of cytoplasm, which is spread in a thin layer surrounding the vacuole except near the nucleus. Plastids, mitochondria, and dictyosomes are randomly and parsimoniously distributed in the cytoplasm. Strands of ER are relatively abundant in the egg of cotton, where they seem to partially enclose other organelles. Occasional strands of ER also appear unique because of the presence of an internal network of tubes probably formed by invagination of the inner membrane of ER. By contrast, the egg of *C. bursa-pastoris* has very little ER, which occurs in the form of short, randomly oriented strands. Eggs of both species also contain generous supplies of ribosomes, which exist predominantly as monosomes. Two reports have identified unusual mitochondria in egg cells: a polymorphic type, forming a reticulation in *Zea mays* (Faure et al 1992), and a giant type, in the form of long, coiled, cup-shaped structures in *Pelargonium zonale* (Kuroiwa and Kuroiwa 1992). On the whole, from the point of view of functional significance, the ultrastructural simplicity of the mature egg, in particular the comparative poverty of its cytoplasmic organization, tends to suggest that it is a quiescent cell that has a limited morphogenetic repertoire.

The situation seems to be quite different, however, in the egg of *Plumbago zeylanica,* which appears as an orphaned cell in the micropylar end of the embryo sac without any flanking synergids. The egg has a munificent supply of organelle-enriched cytoplasm, which suggests that it is in a metabolically active state (Cass and Karas 1974; Russell 1987). As described later in this chapter, the egg of *P. zeylanica* has both egglike and synergidlike features.

One of the most profound questions is how, following fertilization, the complex structure of an embryo and, subsequently, of the adult sporophyte develops from what appears to be a structurally simple egg cell. At this stage, we do not have an answer to this question.

A common feature of the angiosperm egg is the attenuation of the cell wall toward the chalazal end. Generally, the wall appears thickest at the micropylar part and gradually thins out toward the chalazal end, as if the signal for wall organization operates in a gradient. As seen in *Torenia fournieri* (van der Pluijm 1964), *Gossypium hirsutum* (Jensen 1965c), *Crepis tectorum* (Godineau 1966), *Zea mays* (Diboll and Larson 1966; van Lammeren 1986), *Linum usitatissimum* (Vazart 1969), *Petunia hybrida* (van Went 1970b), *Quercus gambelii* (Mogensen 1972), *Cytinus hypocistis* (Ponzi and Pizzolongo 1976), *Nicotiana tabacum* (Mogensen and Suthar 1979), *Spinacia oleracea* (Wilms 1981a), *Beta vulgaris* (Bruun 1987), *Brassica campestris* (Sumner and van Caeseele 1989), *Oryza sativa* (Maeda and Maeda 1990), and *Solanum nigrum* (Briggs 1992), a cell wall is present only around the micropylar half of the cell, the chalazal portion being covered by just the plasma membrane. An interesting situation is found in *Capsella bursa-pastoris* (Schulz and Jensen 1968c), *Epidendrum scutella* (Cocucci and Jensen 1969a), *Plumbago zeylanica* (Cass and Karas 1974), *Agave parryi* (Tilton and Mogensen 1979), *Ornithogalum caudatum* (Tilton 1981b), and *Glycine max* (Folsom and Peterson 1984), in which isolated deposits of wall material dot the chalazal part of the wall of the egg cell. Finally, in *Helianthus annuus*, the juvenile egg is covered with a complete wall over its entire surface, but the wall gradually becomes thin and disappears before anthesis of the flower (Yan, Yang and Jensen 1991b). Viewed in the context of fertilization, the naked or partially naked chalazal part of the egg is of considerable significance. It means that the "business" part of the egg – its first line of contact with the outside milieu – is already adapted for the facilitation of the entry of the sperm, as well as for the absorption of food materials from the central cell. Be that as it may, the problem of the external architectural pattern of the chalazal part of the angiosperm egg becomes one of defining the conditions that control differential accumulation of cell wall precursors.

Synergids. The association of the egg with one or two synergids in the majority of embryo sac types investigated is a matter of great interest to plant embryologists. This stems from the fact that although the existence of synergids has been known since the first embryo sacs were described, their possible function has been mired in controversy. As will be seen, electron microscopic profiles of synergids of several plants have provided new clues about their probable function. Some aspects of the structure and function of synergids have been reviewed by Vijayaraghavan and Bhat (1983).

The synergid, like the egg, is a pear-shaped cell, vacuolate at the chalazal end and lined by the nucleus at the micropylar end. At the micropylar end of the synergid there are also elaborate proliferations of the wall material, known as the filiform apparatus, which extend as small tubular projections into the cytoplasm. These ingrowths are not spatially ordered, but are labyrinthine and random in form (Figure 6.7). After the synergids are formed, their subsequent fate is highly variable. In general, they have a limited life span; in many plants, they do not survive beyond fertilization, but in others, one or both synergids persist as haustorial cells for a period of time after fertilization. Most synergids appear to have a basic ultrastructure and organelle complement with a typically dense cytoplasm enriched with ER, ribosomes, dictyosomes, mitochondria, plastids, and, occasionally, microbodies. Some synergids are also endowed with variable amounts of starch or lipid as storage products. This structural profile of the synergid arose from research carried out on a wide variety of plants, notably *Gossypium hirsutum* (Jensen 1965b), *Zea mays* (Diboll and Larson 1966; van Lammeren 1986), *Capsella bursa-pastoris* (Schulz and Jensen 1968b), *Crepis tectorum, Picris echioides, Cichorium intybus* (chicory), *Calendula officinalis* (all of Asteraceae) (Godineau 1969), *Petunia hybrida* (van Went 1970a), *Quercus gambelii* (Mogensen 1972), *Aquilegia formosa* (Vijayaraghavan, Jensen and Ashton 1972), *A. vulgaris* (Fougère-Rifot 1975), *Helianthus annuus* (Newcomb 1973a; Yan, Yang and Jensen 1991b), *Stipa elmeri* (Maze and Lin 1975), *Cytinus hypocistis* (Ponzi and Pizzolongo 1976), *Proboscidea louisianica* (Martyniaceae) (Mogensen 1978a), *Nicotiana tabacum* (Mogensen and Suthar 1979), *Agave parryi* (Tilton and Mogensen 1979), *Ornithogalum caudatum* (Tilton 1981b), *Spinacia oleracea* (Wilms 1981a),

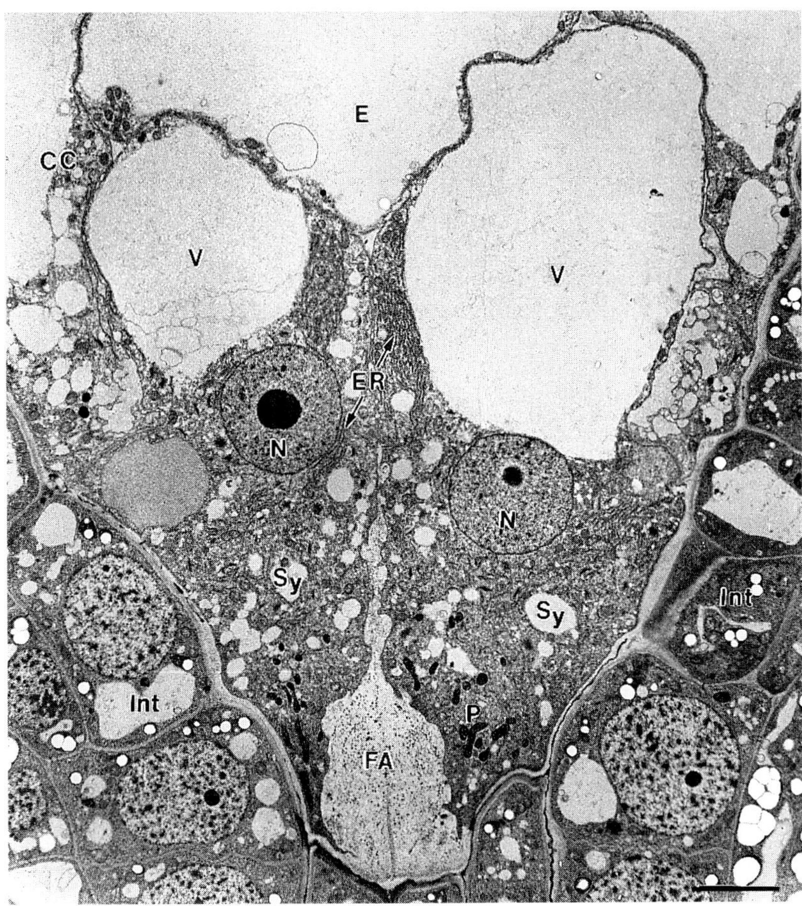

Figure 6.7. Longitudinal section of the embryo sac of *Nicotiana tabacum* showing the synergids (Sy) before fertilization. CC, central cell; E, egg; ER, rough endoplasmic reticulum; FA, filiform apparatus; Int, integument; N, nucleus; P, plastid; V, vacuole. Scale bar = 5 µm. (From Huang and Russell 1994. *Planta* 194:200–214. © Springer-Verlag, Berlin. Photograph supplied by Dr. S. D. Russell.)

Torenia fournieri (Tiwari 1982), *Glycine max* (Folsom and Peterson 1984), *Triticum aestivum* (You and Jensen 1985), *Beta vulgaris* (Bruun 1987), *Brassica campestris* (Sumner and van Caeseele 1989), *Crepis capillaris* (Kuroiwa 1989), *Arabidopsis* (Mansfield, Briarty and Erni 1991), *Solanum nigrum* (Briggs 1992), *Pennisetum glaucum* (Chaubal and Reger 1993), and *Viscum minimum* (Zaki and Kuijt 1994). The extensive range of the similarities and differences in the ultrastructure of synergids of various plants is consistent with the interpretation that, compared to the egg cell, the synergids are held in a very active state of metabolism.

This view of the active metabolic state of the synergids is supported by some very detailed studies of the structure of the filiform apparatus. The filiform apparatus first drew the attention of electron microscopists because of its refractive nature and high density compared to the rest of the synergid. Basically the filiform apparatus represents an example of the specialization of the inner wall of the synergid at the micropylar part for the absorption and transport of metabolites; these functions are aided by the wall material's being thrown into a series of fingerlike projections extending deep into the cytoplasm. A consequence of the wall infoldings is, of course, a great increase in the area of the plasma membrane around the synergid cell wall; in this sense, the synergid develops in a manner not unlike the transfer cells described in other parts of the plant. The matrix of the projections is cellulosic in nature and consists of polysaccharides and proteins. Unlike cellulose, which has a physical form and a chemical uniformity, the physical and chemical relationships of the polysaccharides of the filiform apparatus are not well understood. Despite this uncertainty about the nature of the matrix polysaccharides, their distribution within the wall projections is quite predictable and organized. In most plants examined, the wall projection is made up of a central, electron-dense core surrounded by an electron-translucent layer. A further point of structural interest is that in *Capsella bursa-pastoris*, microfibrils make up the bulk of the core and the peripheral layer of the filiform apparatus, with the difference being that the microfibrils in the core are more compactly packed than those in the surrounding sheath. The distribution of organelles indicates that the functional center of the synergid may reside in the vicinity of the filiform apparatus. This is based on the observation that coursing through the cytoplasm of the synergid is a striking gradient of distribution of organelles, with most of them being found close to the filiform apparatus. The concentration of organelles may therefore cause the differentiation of the part of the synergid around the wall projections for the purposes of absorption and translocation of metabolites from the surrounding nucellar cells, although how this is mediated remains obscure (Jensen 1965b; Diboll and Larson 1966; Rodkiewicz and Mikulska 1967; Schulz and Jensen 1968b; Godineau, 1969; Vazart 1969; van Went 1970a; Mogensen, 1972; Vijayaraghavan, Jensen and Ashton 1972; D'Alascio-Deschamps 1973; Newcomb 1973a; Fougère-Rifot 1975; Jalouzot 1975; Ponzi and Pizzolongo 1976; Mogensen and Suthar 1979; Tilton and Mogensen 1979; Wilms 1981a; Folsom and Peterson 1984; Hause and Schröder 1986; van Lammeren 1986; Bruun 1987). The only published cases where electron microscopy has failed to reveal the presence of a filiform apparatus in the synergid are *Crepis capillaris* (Kuroiwa 1989) and *Lilium longiflorum* (Janson and Willemse 1995). In *Plumbago capensis* (Cass 1972), *P. zeylanica* (Cass and Karas 1974; B.-Q. Huang et al 1990), and *Plumbagella micrantha* (Plumbaginaceae) (Russell and Cass 1988), which lack synergids, the egg apparently doubles as the gamete and as the synergid; here, the micropylar wall of the egg generates a facsimile of the filiform apparatus that, in all likelihood, performs the same functions as its counterpart in the synergid. Thrusting aside details, what is important to note is that irrespective of the final configuration of the egg apparatus, the filiform apparatus is involved in the ultimate fate of the egg. What form might this role of the filiform apparatus take? Studies of the fertilization process in certain angiosperms have shown that the pollen tube, which effects sperm delivery, enters the embryo sac by growing directly through the filiform apparatus into the synergid (van der Pluijm 1964; van Went and Linskens 1967; Diboll 1968; Jensen and Fisher 1968a; Schulz and Jensen 1968b; Cocucci and Jensen 1969c; Cass and Jensen 1970; Mogensen and Suthar 1979; Wilms 1981b; Russell 1982; You and Jensen 1985; Yan, Yang and Jensen 1991b). On the basis of the presence of high concentrations of Ca^{2+} in one or both synergids, the directed movement of the pollen tube to the embryo sac may be suspected to be triggered by a Ca^{2+} gradient (Jensen 1965b; Chaubal and Reger 1990, 1992a, b, 1993, 1994; He and Yang 1992; Tirlapur, van Went and Cresti 1993).

Cytological studies have also revealed a limited range of nuclear changes, such as endoreduplication and polyteny, in the synergids of certain species; however, these changes largely affect the persistent synergid and only marginally affect the degenerating one. Synergid nuclei with polytene chromosomes have been described in *Allium nutans* (Håkansson 1957) and *A. cepa* (Syamasundar and Panchaksharappa 1975). Absence of data in these studies on the degree of polytenization is compensated by observations on other species of *Allium*, such as *A. ursinum*, *A. pulchellum*, *A. angulosum*, and *A. ammophilum*, wherein polyteny has been shown to be associated with three to five cycles of endoreduplication (Hasitschka-Jenschke 1957, 1958). It would be interesting to know the relationship between nuclear changes and the postulated role of the synergids.

The Central Cell. An explanation that is often proposed for the existence of the central cell in the megagametophyte is that it sustains the biochemical growth of the egg to the minimum developmental stage necessary for fertilization and that it serves as the mother cell of the endosperm. At least for a short period after fertilization, the central cell nurtures cells of two competing ploidy levels, the diploid cells of the embryo and the triploid cells of the endosperm. To whatever degree the zygote and the endosperm nucleus proceed to differentiate, differentiation occurs in the specialized milieu of the central cell by reciprocal interactions between the nucleus and the cytoplasm. With this general picture in mind, the next logical step is to turn to studies that have provided the sum and substance to the specialized nature of the central cell.

The maturation of the central cell is seemingly accomplished by its enlargement after completion of the last gametophytic division and by the flooding of the cytoplasm with an array of organelles. From its modest beginning as a minor component of the cell, the central vacuole balloons out and serves as a reservoir for carbohydrates, amino acids, and inorganic salts (Ryczkowski 1964). At this stage, the cytoplasm is generally found in the periphery of the cell or both in the periphery and as strands crisscrossing the vacuole. Two large polar nuclei, suspended by cytoplasmic strands or floating in the cytoplasm close to the egg apparatus, constitute the organizational center of the central cell. In most species examined, the distribution of organelles in the central cell has provided a striking visual portrait of a metabolically active cell. Based on observations made on *Gossypium hirsutum* (Jensen 1965c; Schulz and Jensen 1977), *Crepis tectorum* (Godineau 1966), *Zea mays* (Diboll and Larson 1966; Diboll 1968; van Lammeren 1986), *Lilium regale* (Mikulska and Rodkiewicz 1967a, b), *Epidendrum scutella* (Cocucci and Jensen 1969a), *Linum usitatissimum* (Vazart and Vazart 1966; Vazart 1969), *Helianthus annuus* (Newcomb 1973a; Yan, Yang and Jensen 1991b), *Capsella bursa-pastoris* (Schulz and Jensen 1973), *Plumbago zeylanica* (Cass and Karas 1974), *Stipa elmeri* (Maze and Lin 1975), *Spinacia oleracea* (Wilms 1981a), *Glycine max* (Folsom and Peterson 1984), *Triticum aestivum* (You and Jensen 1985), *Beta vulgaris* (Bruun 1987), *Solanum nigrum* (Briggs 1992), and *Viscum minimum* (Zaki and Kuijt 1994), one can summarize the fine structure of the central cell as consisting of an extensive network of ER, numerous and well-developed plastids and mitochondria, dictyosomes, and polysomes. In contrast, in *Petunia hybrida* (van Went 1970b) and *Linum catharticum* (D'Alascio-Deschamps 1973), the central cell has a relatively impoverished cytoplasm, with poorly developed ER and few mitochondria and plastids. These descriptions provide several criteria by which the central cell may be recognized. Another possible ultrastructural marker of the central cell is glyoxysome, in which enzymes of the glyoxylic acid cycle are localized (Newcomb 1973a; Schulz and Jensen 1973).

The notion of intense metabolism in the central cell is also made plausible by the presence of storage products such as starch, proteins, and lipids, which attain peak accumulation just before fertilization (Hu 1964; Jensen 1965c; Vazart and Vazart 1965; Diboll and Larson 1966; Diboll 1968; Newcomb 1973a; Schulz and Jensen 1973; Sehgal and Gifford 1979; Folsom and Peterson 1984; You and Jensen 1985; Yan, Yang and Jensen 1991b). The presence of elaborate transfer-cell–type wall ingrowths found in the central cell of several plants, including *L. usitatissimum* (Vazart and Vazart 1966), *H. annuus* (Newcomb and Steeves 1971; Newcomb 1973a; Yan, Yang and Jensen 1991b), *C. bursa-pastoris* (Schulz and Jensen 1973), *Stellaria media* (Newcomb and Fowke 1973), *Jasione montana* (Campanulaceae) (Berger and Erdelská 1973), *Euphorbia helioscopia* (Gori 1977), *Macadamia integrifolia* (Proteaceae) (Sedgley 1981c), *Spinacia oleracea* (Wilms 1981a), *Scilla sibirica* (Bhandari and Sachdeva 1983), *Glycine max* (Folsom and Peterson 1984; Tilton, Wilcox and Palmer 1984), *Beta vulgaris* (Bruun 1987), *Oryza sativa* (Jones and Rost 1989b), *Brassica campestris* (Sumner and van Caeseele 1990), *Solanum nigrum* (Briggs 1992, 1995), and a starchless mutant of *Arabidopsis* (Murgia et al 1993), suggests that in these species the central cell may be modified for large-scale absorption of nutrients from the neighboring cells (Figure 6.8). Although physiological evidence is lacking, these observations support the view that the metabolic activity of the central cell supplies energy for the absorption of nutrients from the surrounding cells, their conversion into stored reserves, and finally their hydrolysis into simple precursors.

Antipodals. Not much attention has been devoted to the structure–function relations of the

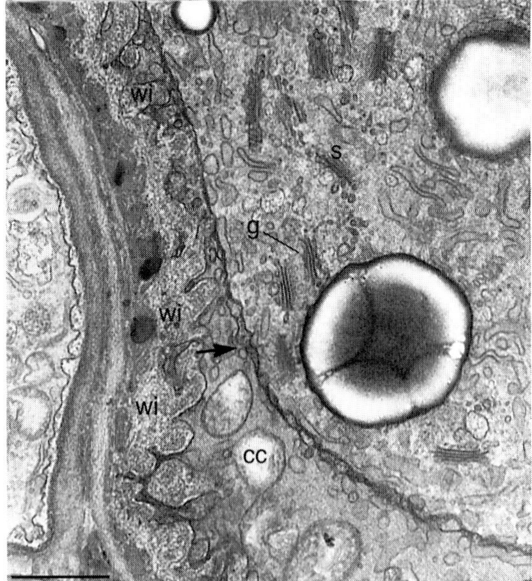

Figure 6.8. Section of the embryo sac of *Glycine max* showing wall ingrowths (wi) of the central cell (cc). Arrows point to the wall between the synergid (s) and the central cell. g, Golgi body. Scale bar = 1 μm. (From Folsom and Peterson 1984; photograph supplied by Dr. M. W. Folsom.)

antipodals because of the variability they display in the different embryo sac types. As described by Maheshwari (1950), embryo sacs come equipped with as many as 11 antipodals in the Drusa type, to none at all in the Oenothera type and everything in between in others. Since antipodal morphology has not figured as a basis for assignment of embryo sac types, variations in the number, life span, cytology, growth, and structure of the antipodals have been described in megagametophytes spanning the entire spectrum of mono-, bi- and tetrasporic embryo sac types. Perhaps the most unusual situation is that encountered in members of Poaceae, in which the antipodals proliferate into a tissue consisting of several hundred cells (Yamaura 1933). As to their life span, although usually short lived, antipodals in many plants persist for some time after fertilization; in *Allium ursinum* (Hasitschka-Jenschke 1957), *Aconitum vulparia* (Ranunculaceae) (Bohdanowicz and Turala-Szybowska 1987), and *Hordeum vulgare* (Engell 1994), antipodals retain their form up to the stage of free nuclear or cellular endosperm.

The only cells of the embryo sac in which nuclear cytology has been studied in some detail are those of the antipodals. According to Diboll and Larson (1966), division of antipodals in *Zea mays* is accompanied by incomplete cytokinesis that results in the formation of a multinucleate protoplast or syncytium. Besides this discrete change, the most frequent nuclear abnormalities in the antipodals are somewhat dramatic and involve polyteny and endoreduplication (Figure 6.9). As in the case of synergids, there are two lists: one in which polyteny has been documented without reference to the level of endoreduplication (Hasitschka-Jenschke 1962; Turala 1966), and the other in which the level of endoreduplication of the cells has been determined by DNA cytophotometry, chromosome counts, or measurement of nuclear volume. Based on these data, a range of ploidy levels (indicated as *N* or *C* values) has been documented in antipodal cells of various plants, such as 8 in *Allium pulchellum* (Hasitschka-Jenschke 1958); 16 in *Allium ursinum* (Hasitschka-Jenschke 1957), *Achillea millefolium* (Asteraceae) (Titz 1965), and *Caltha palustris* (Ranunculaceae) (Grafl 1941); 32 in *Arum maculatum* (Erbrich 1965), *Chrysanthemum alpinum* (Asteraceae) (Titz 1965), and *Clivia miniata* (Amaryllidaceae) (Tschermak-Woess 1957); 36 in *Pennisetum ciliare* (Sherwood 1995); 64 in *Corydalis cava, C. nobilis* (Papaveraceae) (Hasitschka-Jenschke 1959), and *Crocus suaveolens* (Turala 1966); 128 in several species of *Aconitum* (Tschermak-Woess 1956) and *Papaver rhoeas* (Hasitschka 1956); 196 in *Triticum aestivum* (Bennett et al 1973b); 256 in *Ammophila arenaria* (Poaceae) (Kubień 1968) and *Triticale* (Poaceae) (Kaltsikes 1973); 512 in *Hordeum distichum* (Erdelská 1966); and 1024 in *Scilla bifolia* (Nagl 1976a). In the last-mentioned species, autoradiography of ^{3}H-thymidine incorporation that indicates unusually heavy labeling over the nucleolus and the nucleolus organizers has led to the suggestion that some of the polytene nuclei undergo amplification of rDNA. It is known that polyploidy and other nuclear aberrations permit a certain increase in cell size beyond the normal limits for a haploid or diploid cell. By reducing mitotic activity, the polytene antipodal cells might be viewed as taking a shortcut to cell expansion and growth while conserving materials necessary for mitosis and cytokinesis.

In the electron microscopic examination of antipodals of a number of plants, there is a predictable relationship between their life span and the distribution of organelles. For example, in *Capsella bursa-pastoris* (Schulz and Jensen 1971), *Gasteria verrucosa* (Willemse and Kapil 1981a), and an *Arabidopsis* mutant (Murgia et al 1993), the short-lived antipodals are generally impoverished of organelles; those few organelles present show a minimum of internal organization. In other plants, such as *Zea mays* (Diboll and Larson 1966), *Epidendrum scutella* (Cocucci and Jensen 1969c), *Helianthus annuus* (Newcomb 1973a), *Aconitum napellus* (Zhukova and Sokolovskaya 1977), *A. vulparia* (Bohdanowicz and Turala-Szybowska 1985), *Triticum aestivum* (You and Jensen 1985), *Beta vulgaris* (Bruun 1987), *Ranunculus sceleratus* (Chitralekha and Bhandari 1991), and *Hordeum vulgare* (Engell 1994), the antipodals have dense cytoplasm with abundant organelles, including highly developed mitochondria, dictyosomes, and ER. The apparent correlation between the ultrastructure of antipodals and their differing life spans suggests a nutritive role for antipodals with long life spans. In this context it is interesting to note that the wall of the antipodal cell is by no means a static structure, and infoldings of the wall, as well as plasmodesmatal connections with the central cell and nucellar cells at the chalazal end, are common (Diboll 1968; Newcomb, 1973a; Rifot 1973; Maze and Lin 1975; Zhukova and Sokolovskaya 1977; Willemse and Kapil 1981a; Wilms 1981a; Bhandari and Bhargava 1983; Bhandari and Sachdeva, 1983; Bohdanowicz and Turala-Szybowska 1985, 1987; Chitralekha and Bhandari 1991; Mansfield, Briarty and Erni 1991;

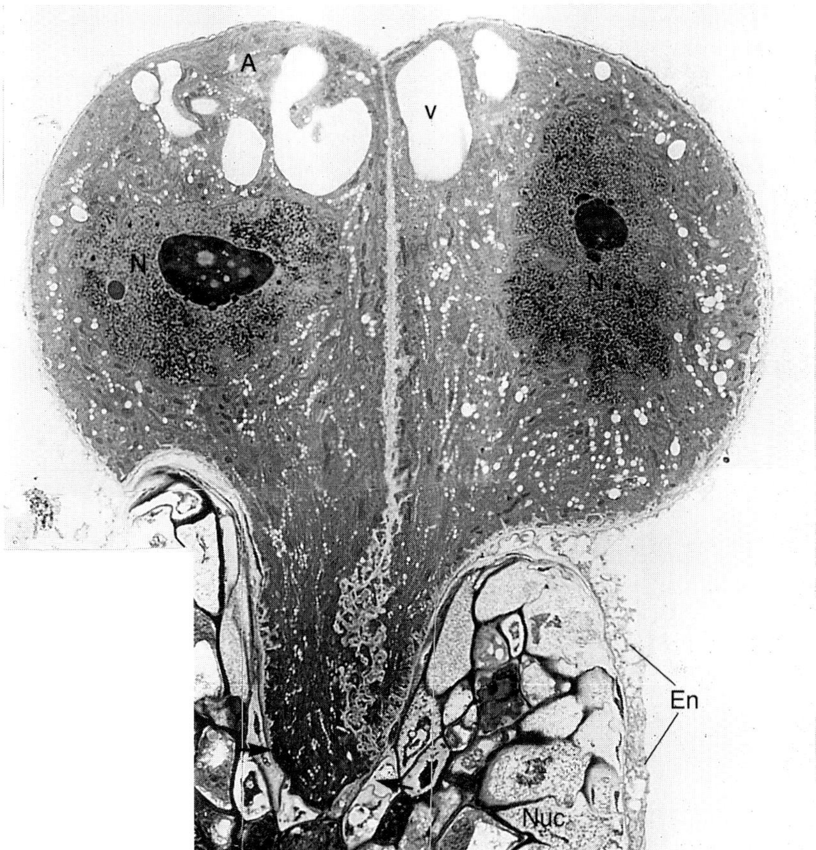

Figure 6.9. Section of the embryo sac of *Aconitum vulparia* showing two antipodals (A) with giant nuclei. Arrows indicate regions of the antipodal wall where ingrowths are not present. En, endosperm; N, nucleus; Nuc, nucellus; V, vacuole. Scale bar = 10 μm. (From Bohdanowicz and Turala-Szybowska 1987; photograph supplied by Dr. J. Bohdanowicz.)

Engell 1994). In *Aconitum vulparia*, the wall ingrowths increase in size and number, and eventually they decorate the entire antipodal wall, except for a small chalazal part, and function as an efficient system for the transport of metabolites (Bohdanowicz and Turala-Szybowska 1987). In addition, the nuclear behavior of persistent antipodals and the abundance of vesicles associated with the dictyosomes of these cells also suggest a secretory function. If the antipodals are that important, it is not understood how certain embryo sac types are able to dispense with them soon after they are formed and yet function normally.

The Cytoskeleton of the Embryo Sac. Electron microscopy and immunofluorescence labeling have shown that a complex system of microtubules, ranging from a few to an extensive array, appears in embryo sacs at various stages of development (Cass and Karas 1974; Cass, Peteya and Robertson 1985; Willemse and van Lammeren 1988; Franssen-Verheijen and Willemse 1990; Sumner and van Caeseele 1990; B.-Q. Huang et al 1993; Murgia et al 1993; Huang and Russell 1994; Huang and Sheridan 1994; Webb and Gunning, 1994). However, only in a few cases has it been possible to relate the presence of microtubules to the morphogenetic activities of cells, such as nuclear movement and cell shaping. Thus, microtubules present in the region between sister nuclei in the embryo sacs of *Hordeum vulgare* (Cass, Peteya and Robertson 1985), *Arabidopsis* (Webb and Gunning 1994), and maize (Huang and Sheridan 1994) might function in positioning and maintaining the nuclei at the two poles. The second role of microtubules comes after nuclear movement has ceased; the putative functions of cortical microtubules found in the egg, central cell, or antipodals of *Gasteria verrucosa* (Franssen-Verheijen and Willemse 1990), *Plumbago zeylanica* (B.-Q. Huang et al 1993),

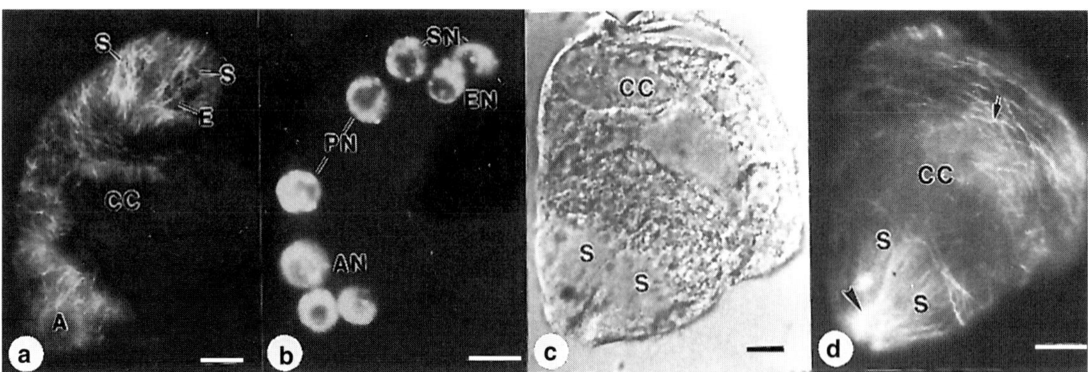

Figure 6.10. Organization of microtubules in the embryo sac of maize. **(a)** An isolated eight-nucleate embryo sac showing parallel microtubule bundles in the synergid cells, transverse microtubule arrays in the central cell, and random microtubules in the antipodals. **(b)** Corresponding nuclei in the embryo sac. **(c)** An isolated embryo sac showing two synergids, the egg cell (not in focus), and the central cell. **(d)** Distribution of microtubules in the synergids of the corresponding embryo sac; arrowhead points to the microtubules radiating from the filiform apparatus; arrow points to the transverse microtubules localized in the peripheral region of the central cell. A, antipodal; AN, antipodal nucleus; CC, central cell; E, egg cell; EN, egg nucleus; PN, polar nucleus; S, synergid; SN, synergid nucleus. Scale bars = 2 µm. (From Huang and Sheridan 1994. *Plant Cell* 6:845–861. © American Society of Plant Physiologists. Photographs supplied by Dr. W. F. Sheridan.)

Nicotiana tabacum (Huang and Russell 1994), and *Arabidopsis* (Webb and Gunning 1994) may be accepted as part of the mechanism to establish the shape and volume of these cells. Of more direct relevance in the functional context are the arrays of microtubules at the micropylar end of the synergids in the vicinity of the filiform apparatus in *G. verrucosa* (Willemse and van Lammeren 1988), *Arabidopsis* (Webb and Gunning 1994), maize (Huang and Sheridan 1994), *N. tabacum* (Huang and Russell 1994), and the synergidlike egg of *P. zeylanica* (B.-Q. Huang et al 1993). The activity in the synergids for which microtubules are responsible may be permissive, allowing the proper functioning of the filiform apparatus (Figure 6.10). Microfilaments colocalize with microtubules during critical stages of megagametogenesis, especially in the egg and central cell of *P. zeylanica* (B.-Q. Huang et al 1993), *Arabidopsis* (Webb and Gunning 1994), and *N. tabacum* (Huang and Russell 1994). We do not know in which, if any, of the morphogenetic activities of these cells microfilaments function alongside microtubules.

(ii) Enzymatic Isolation of Embryo Sacs

Within the last decade, an interesting method has developed for the study of megagametophytes of angiosperms under controlled conditions. It is essentially the use of pectolytic and cellulolytic enzymes to digest the integuments and nucellus, followed by mild centrifugation to isolate viable embryo sacs from ovules. Working on the observations of several previous investigators, Zhou and Yang (1982) isolated embryo sacs from fixed ovules of *Nicotiana tabacum, Vicia faba,* and *Brassica campestris.* Subsequently, this technique was used for the isolation of embryo sacs from fresh ovules of *Antirrhinum majus, Helianthus annuus,* and *N. tabacum.* Staining of isolated embryo sacs with auramine-O indicated the presence of cutin; this might explain why enzymes that macerate ovular tissues do not affect the embryo sac (Zhou and Yang 1985). There is no doubt that isolation of viable embryo sacs has made possible such diverse lines of investigations as cytochemical and cell–physiological analysis of embryo sacs of different ages and their constituent cells (Sidorova 1985; Wu and Zhou 1988; Wagner, Kardolus and van Went 1989; Wagner et al 1989; Huang and Russell 1990; B.-Q. Huang et al 1990, 1992; van Went and Kwee 1990; Wu, Zhou and Koop 1993; Ohshika and Ikeda 1994), isolation of viable egg cells (Kranz, Bautor and Lörz 1991a; Faure et al 1992; van der Maas et al 1993a; Kovács, Barnabás and Kranz 1994), in vitro culture of embryo sacs, and in vitro fertilization (Zhou 1987a; Kranz, Bautor and Lörz 1991a). A further step in this direction is the isolation of protoplasts from embryo sac cells for these same lines of research (Hu, Li and Zhou 1985; Mól 1986). The insight that such studies might provide is bound to give one an advantage in the experimental manipulation of fertilization and embryogenesis.

4. MOLECULAR BIOLOGY OF MEGAGAMETOGENESIS

As might be expected, a number of molecular-level questions on the megagametophyte of angiosperms beg resolution. A central question is concerned with the gene expression pattern that heralds the gametophytic program in the functional megaspore. Equally important is the basis for determining how cells born out of simple mitotic divisions – such as the cells of the micropylar and chalazal halves of the embryo sac, and the egg and the polar nucleus – show expressional divergence. A significant concern is the nature of the transcripts made by these different cells of the embryo sac as well as by the polar nuclei. Another molecular-level question about female gametogenesis relates to the synthesis and accumulation of mRNA in the egg, which may serve as a legacy of template information for the first proteins of the zygote. Finally, is there a molecular basis for polarity of the egg? Admittedly, these are important questions for which answers based on studies using modern molecular approaches are not currently available. It is no exaggeration to state that the microscopic size of the embryo sac, and the large volume of the ovule tissue it is embedded in, have proved to be effective bottlenecks for exploitation of the female gametophyte of angiosperms for molecular approaches. Consequently, information on macromolecule synthesis and gene expression pattern during female gametogenesis is largely derived from experiments based on cytophotometry, histochemistry, and in situ hybridization.

There has been considerable interest in the past in the DNA content and its quantitative stability in the individual cells of the embryo sac. In a classic study, Woodard (1956) showed by Feulgen microspectrophotometry that the egg, synergid, and antipodal nuclei of *Tradescantia paludosa* maintain the 1C DNA content to be expected of haploid cells and that not until after fertilization, when a new nucleus is reconstituted in the egg, does this cell attain the 2C DNA level. Results of a recent study using fluorimetry of DAPI-stained nuclei have also unambiguously affirmed a 1C level of DNA in isolated egg cells of maize (Mogensen et al 1995). A few other investigations have yielded inconsistent results; most often, Feulgen staining of the mature egg nucleus has been found to be anomalous and very much reduced, whereas that of the synergid and antipodal nuclei has been found to be normal (Klyuchareva 1960; Hu 1964; Hu and Chu 1964; Pritchard 1964a; Jensen 1965c; Sehgal and Gifford 1979). A note of caution about the interpretation of these results should be added here. The loss of DNA from the egg nucleus may be real and may be a prelude to its participation in the impending act of fertilization. Since the egg nucleus grows in volume before fertilization, it is possible that its DNA content has been diluted and is thus beyond the resolution of the Feulgen reaction. The disappearance of Feulgen staining is also attributable to chemical transformation of chromosomal DNA. A clue to the low Feulgen stainability of the egg nucleus has come from the work of Woodcock and Bell (1968a). These investigators showed that the egg and polar nuclei of *Myosurus minimus*, which have no detectable Feulgen-positive DNA, nevertheless contain a DNase-digestible fraction absorbing at 260 nm. DNA is also detected in these nuclei by fluorescence microscopy and microspectrography. Although egg maturity appears to be manifest by an unusual kind of nuclear DNA metabolism, it has to be kept in mind that DNA transformation is intertwined with processes that regulate enzyme synthesis, precursor levels, and the physiological state of the primers. It is also not clear at present whether organelle DNA has any role in the differentiation of the egg. With the advantage of enzymatically isolated embryo sacs and specific staining techniques, organelle DNA mostly of plastid origin has been shown to accumulate in the micropylar end of the embryo sac of *Plumbago zeylanica*; far from manifesting polarity of the embryo sac, this might represent the molecular basis of determination of cells in this region (Huang and Russell 1993).

Accumulation of RNA and proteins in the cells of embryo sacs of various plants has been monitored by histochemical methods. A point of great interest in these studies is the relatively intense accumulation of macromolecules in the egg cell, especially in the nucleus, nucleolus, and the immediately peripheral cytoplasm (Pritchard 1964a; Alvarez and Sagawa 1965a; Schulz and Jensen 1968c; Malik and Vermani 1975; Prasad 1977; Sehgal and Gifford 1979; Bhandari, Soman and Bhargava 1980). The RNA and protein contents of the synergids of *Stellaria media* are lower than those of the egg (Pritchard 1964a), whereas a reverse pattern holds true for *Argemone mexicana* (Bhandari, Soman and Bhargava 1980); these differences are sufficiently striking to suggest that the accumulation of macromolecules may be regulated by different mechanisms in these cells. The cytoplasm of the egg of *Vanda* sp. (Orchidaceae) stains less intensely for RNA and proteins than the nucleus, but the concentration of RNA- and protein-rich cytoplasm at the chalazal end of the egg reinforces the conclusion that polarity is already determined in the egg at this stage (Alvarez and Sagawa 1965a). The RNA and protein profiles of the egg of *Capsella bursa-pastoris* also show a dense accumulation at the chalazal end that similarly reflects a predetermined polarity (Schulz and Jensen 1968c). The force of these findings stems from their implication that at least part of the RNA detected in the mature egg may be used as templates for initiating sporophytic growth after fertilization.

Consideration of the histochemical localization of RNA and proteins leads to the broader question of the role of specific messages during megagametogenesis. One approach to elucidating this role has been to study the distribution of poly(A)-RNA by in situ hybridization of sections of ovules using appropriate probes. In *C. bursa-pastoris*, perhaps the most obvious pattern of annealing of the probe, beginning with the formation of the eight-nucleate embryo sac, is the concentration of label in the egg and synergids. With maturation of the egg, there is an increasing gradient in the accumulation of mRNA from the micropylar to the chalazal pole (Raghavan 1990a, b). An almost comparable type of annealing pattern is seen in the ovules of *Medicago sativa*, in which binding of the probe occurs in the central cell as well (Bimal et al 1995). From localization of total mRNA, it is one further step to follow the expression of specific genes during megagametogenesis. When a cloned rice H3 histone gene is used as a probe, no transcripts are detected in the cells of the embryo sac of *Oryza sativa*; surprisingly, annealing occurs in the cells of the integument and pericarp (Raghavan and Olmedilla 1989). In another work, Dow and Mascarenhas (1991a, b) have followed the pattern of synthesis of ribosomes during embryo sac development in *Zea mays* by in situ hybridization using a wheat rRNA probe combined with confocal microscopy to determine the changes in volume of the individual cells. It is estimated from these studies that the number of ribosomes present in the cells of the egg apparatus and in the central cell is several fold higher than that found in the antipodals and in the cells of the nucellus. There is also an increase in the quantity of ribosomes as development proceeds, with the rate of synthesis being high in the egg and central cell. The special requirement of these cells for a large endowment of ribosomes remains unexplained, although utilization of the ribosomes for increased protein synthesis following fertilization is a possibility.

Reference was made in the previous section ("Megasporogenesis," pp. 154–162) to the isolation of *Arabidopsis* mutants *bel-1*, *sin*, and *ant* with abnormalities in the form of the integuments. The ovules of these mutants are characterized by the failure of the megasporocyte to undergo meiosis and differentiate into the embryo sac (Robinson-Beers, Pruitt and Gasser 1992; Lang, Ray and Ray 1994; Elliott et al 1996; Klucher et al 1996). Since the same mutation induces defects in integument formation and megasporogenesis in the ovule, a common genetic pathway is probably operational in these events. Another *Arabidopsis* mutation, *prolifera* (*prl*), causes lesions at the four-nucleate stage of the embryo sac and, like *bell* and *sin* mutants, leads to partial female sterility. On the basis of homology of the PRL protein to a family of yeast proteins involved in DNA replication, the role of the gene seems to be in sustaining continued pre- and post-fertilization development in the embryo sac (Springer et al 1995).

Summarizing, it is clear that our understanding of the molecular aspects of female gametogenesis in angiosperms is shallow. What is presently known is limited to histochemical studies of nucleic acid and protein accumulation, and to the localization of bulk RNA and cloned transcripts by in situ hybridization. Procedures that allow enzymatic isolation of the embryo sac and its enclosed cells should permit construction of cDNA libraries and characterization of genes specific to the cells of the embryo sac. Indeed, a beginning has been made to generate a cDNA library to poly(A)-RNA from the egg cells of maize using PCR technology, although no egg-specific clones have been isolated (Dresselhaus, Lörz and Kranz 1994). Analysis of the expression of embryo sac–specific genes during megagametogenesis will lead to a greater understanding of the female gametophyte and the functions of its cells before, during, and after fertilization. This is one of the challenges that face molecular plant embryologists.

5. NUTRITION OF THE MEGAGAMETOPHYTE

Megagametophytes of angiosperms are astonishingly diverse in the details of their structure and organization. It is also remarkable that none of the different embryo sac types contain massive amounts of storage products to sustain their metabolism. A long-standing problem has been the need to understand how the nutritional demands of the gametophyte are met at various stages of its ontogeny. Although references to structural modifications of embryo sac cells that facilitate solute uptake and translocation have been made earlier in this chapter, it is the intention here to pull together the various facts into a unified theme. For a review of this subject, see Masand and Kapil (1966).

The entire surface of the embryo sac can be considered to perform an absorptive function, the tangible expression of this being the dissolution of the adjacent cells of the nucellus and even the cells of the integument. Support for this line of thought was provided by much of the early research on the haustorial nature of the embryo sac, which results in its growth beyond the ovular tissues. This has produced some very anomalous situations, such as the embryo sac's growing into the micropyle or into the style, its establishing contact with the placenta and funiculus, or digesting its way through the nucellus (Masand and Kapil 1966). That there are no mechanical restraints to the growth of synergids and antipodals as haustoria has also been deduced from certain observations. *Quinchamalium chilense* (Santalaceae) is an example in which the tips of the synergids become tubular and make their way into the tissues of the style along the vascular strand. In this species, the antipodal nuclei go through an absolutely unusual developmental phase during which they do not round off as cells. Instead, the nuclei are cut off by a wall at the chalazal end of the embryo sac and the tip of this multinucleate cell elongates as a tubular haustorium. The haustorium grows through the funiculus into the placenta, where it branches (Agarwal 1962). In contrast, the antipodal haustorium in *Rubia cordifolia* (Rubiaceae) is relatively simple and consists of a tubular extension that grows into the chalazal part of the ovule (Venkateswarlu and Rao 1958). In effect, the haustorial outgrowths may serve to extend the working area of the embryo sac and, because of the intimate association between the haustoria and ovular tissues, they may bring in metabolites from outside the boundary of the embryo sac. There are references in the literature to the following structures having roles seemingly connected with the translocation and supply of nutrient substances to the embryo sac: the vascular tissues of the integument; the group of nucellar cells at the base of the integument, known as the hypostase (Masand and Kapil 1966; Tiwari 1983); the persistent chalazal extension of residual nucellar cells into the embryo sac, known as the postament (Mogensen 1973); the endothelium (Kapil and Tiwari 1978b); and the later-formed nucellar cells surrounding the embryo sac (Folsom and Cass 1988). In many accounts, the entire nucellus is generally perceived as a source of nutrition for the embryo sac. In keeping with this role, storage materials, especially starch and periodic acid-Schiff (PAS)–positive substances (Noher de Halac 1980b; Bhandari and Sachdeva 1983; Bruun and Olesen 1989), proteins (Jensen 1965a; Gori 1976; Tilton and Lersten 1981; Bhandari and Sachdeva 1983), tannins (Noher de Halac 1980a; Olesen and Bruun 1990), ascorbic acid, and enzymes (Malik and Vermani 1975), localized in the nucellar cells of various plants might serve as the primary source of nutrients. However, no direct observations have been made of the transport of metabolites from the nucellar cells and haustoria into the embryo sac; as in all microscopic observations, corroborative physiological evidence is essential if we are to understand the significance of the structural studies.

Electron microscopic and other investigations are increasingly suggesting mechanisms for embryo sac nutrition that eluded earlier light microscopic approaches. The breakthrough in the electron microscopic era in support of the absorptive function of the embryo sac was the discovery of plasmodesmata and wall ingrowths in the central cell, synergids, and antipodals, as described earlier. It has been proposed that the embryo sac wall acts like a common apoplast hauling nutrients along its entire length. If the filiform apparatus is included as part of the apoplastic pathway, a typical scenario would have the nutrients that enter the common pool of the apoplast being transported through the filiform apparatus into the synergids. Further transport of the materials to the egg and the central cell is probably taken up by the transfer cells through the interconnecting symplast of the central cell and egg apparatus (Folsom and Cass 1990). From this perspective, the fact that the filiform apparatus attains its maximum development during the period of rapid expansion of the synergids and egg apparatus may be more than a coincidence.

However, some plants also have mechanisms for nutrient intake into the embryo sac that bypass the traditional routes. In this context, the presence of stored nutrients in the form of accumulated starch in the cells of the integument has attracted some attention (Mogensen 1973; Wilms 1980b; Olesen and Bruun 1990). Based on electron microscopic and histochemical analyses, the major food reserves of the ovule of *Quercus gambelii* are found to be confined to the outer integument. In the absence of plasmodesmata in the cells of the egg apparatus and the central cell, it has been proposed that food materials from the outer integument find their way to the embryo sac through the chalaza and the postament (Mogensen 1973). In a similar way, absence of ATPase activity in the embryo sac cells of *Saintpaulia ionantha* (African violet; Gesneriaceae) and tobacco and its concentration in the surrounding integumentary cells have indicated the involvement of the latter in the active transport of metabolites into the embryo sac (Mogensen 1981a, 1985). In the ovules of *Polygonum capitatum* and tobacco, the uptake of a fluorescent dye is consistent with the view that translocation proceeds through the funiculus into the micropylar and chalazal parts (Mogensen 1981b).

Because of the diversity of nutritional mechanisms involved, the embryo sac does not appear to be typical of many other cellular units of the plant. It is a specialized cell population that has to be effectively nurtured for a short period before the events of fertilization overtake the life of the cells. There is no compelling reason to suspect that a given embryo sac cannot utilize more than one nutritional pathway. Identification of the specific types of nutrients that are utilized by the female gametophyte during its ontogeny remains as a research priority in this field.

6. APOMIXIS

The many events discussed in this chapter have the common goal of preparing the egg for fertilization. Despite the well-known advantages of sexual reproduction for the transmission of hereditary characters, plants have also evolved various strategies for propagation of progeny without involving sex. These are collectively considered as asexual reproduction. A historical shadow lies over apomixis because the term was introduced early in this century to designate all forms of asexual reproduction in angiosperms; it is currently used in a more restricted sense to conceptualize asexual reproductive processes that occur in the ovule and lead to seed formation. The link between apomixis and the theme of this chapter is that apomixis generates unreduced embryo sacs from which embryos with maternal genotype (maternal offsprings) arise in the absence of fertilization. Production of unreduced embryo sacs unaccompanied by fertilization allows the continuation of alternation of generations without affecting the genotypes of the gametophyte and sporophyte.

Now that apomixis has been defined, the ramifications of the process and the parameters available to differentiate apomixis from regular sexual reproduction can be discussed. There are two main types of apomixis: obligate and facultative. Almost without exception, offspring resulting from the former harbor only the maternal genotype. Generally, most apomicts retain a propensity for sexual interludes, so that some offspring with genotypes of both parents (sexual offspring) usually occur in the progeny; this situation comes under the rubric of facultative apomixis. Based on the stability of apomixis, there are two subdivisions that merit mention here: nonrecurrent (unstable) apomixis and recurrent (stable) apomixis. The origin of embryos from reduced egg cells without fertilization, also known as haploid parthenogenesis, is the most prevalent form of nonrecurrent apomixis. On the other hand, diploid parthenogenesis, or embryogenesis by cells of an unreduced embryo sac unaccompanied by fertilization, is typically a recurrent type of apomixis. This classification thus shows that knowledge of the cytology of the origin of the embryo sac and of the chromosome constitution of the embryos formed is necessary to identify cases of apomixis.

This section has a limited objective, namely, to describe the salient features of the recurrent type of apomixis and to show how the unreduced embryo sacs formed are similar to, or different from, the reduced embryo sacs of the sexual types. Polyembryony and adventive embryogenesis, which are included under apomixis in the old definition, will not be taken up here. Several reviews and books that have considered apomixis from different perspectives are recommended for additional information and for numerous examples (Asker and Jerling 1992; Ramachandran and Raghavan 1992a; Naumova, 1993; Koltunow 1994; Koltunow, Bicknell and Chaudhury 1995).

Although the final configurations of embryo sacs in the sexually reproducing and apomictic species are identical, their origins can be traced to different cells. In contrast to the origin of the embryo sac in the reduced megaspores in the sexual types, the embryo sac in apomicts arises from an unreduced initial cell that can be either a germ cell (diplospory) or a somatic cell of the ovule (apospory). The germ cells from which unreduced embryo sacs are born in the diplosporous apomicts may be the female archesporial cell or the megasporocyte. Antennaria,

Taraxacum, Ixeris, and Allium are the four major types of diplosporous apomicts recognized. In the Antennaria type, the megaspore mother cell does not go through meiosis but, after a long interphase, it divides mitotically and becomes an unreduced functional megaspore. A typical eight-nucleate embryo sac is formed from the functional megaspore by two further mitotic divisions. The basic features of the Taraxacum and Ixeris types are the consistent derangement of meiosis and the resultant formation of restitution nuclei after meiosis I. In the Taraxacum type, the restitution nucleus divides mitotically to generate a dyad with somatic chromosome numbers, and the embryo sac is formed from one of the nuclei of the dyad. A variation of this theme is found in the Ixeris type, in which both nuclei of the dyad participate in embryo sac formation. The megaspore mother cell in the Allium type passes through a premeiotic endomitosis followed by normal meiosis in which the pairing of identical chromosomes generates bivalents. The result is the formation of a tetrad of unreduced nuclei from one of which an embryo sac is fabricated.

Somatic cells of the ovule from which aposporous embryo sacs are carved out generally belong to the nucellus or integuments. Contrary to the condition observed in sexual plants, in most aposporous apomicts, the development of several embryo sacs in a single ovule is not uncommon. A significant difference reported between megasporogenesis in the diplosporous apomict *Elymus rectisetus* (Poaceae) and its sexual relative, *E. scabrus*, is the absence of a regular callose wall around the megaspore mother cells in the apomict (Carman, Crane and Riera-Lizarazu 1991). In their final assembly, both diplosporous and aposporous embryo sacs have a typical egg apparatus consisting of an egg cell and two synergids, three antipodals, and two polar nuclei. Some comparative ultrastructural studies have shown that both sexual and aposporous types of embryo sacs are remarkably similar, confirming the view that the chief contrasts concern their origin (Chapman and Busri 1994; Naumova and Willemse 1995; Vielle et al 1995). Since diplospory and apospory produce diploid embryo sacs, the diploid egg develops into an embryo without fusion with a sperm. As the egg develops parthenogenetically, the polar nuclei, unencumbered by fusion with a sperm, also begin to divide to give rise to the endosperm. However, unlike the egg, the polar nuclei are not constrained to develop in the absence of fusion with a sperm, and indeed such fusions of the sperm with the polar nuclei are a way of life in some apomicts.

So, what does all this add up to concerning our understanding of the mechanism of apomixis in plants? The different types of apomixis, the varying routes to embryo sac development, the fusion of the sperm with the polar nuclei in some cases, and the retention of sexuality in certain apomicts, all make the phenomenon very complex. These are hallmarks of a developmental system that is regulated by a host of genes. The significance of these genes is that they direct the somatic cells of the ovule to bypass meiosis to form the embryo sac and direct the egg cell to bypass sexual recombination to form the embryo. The impasse in our knowledge of the molecular biology of apomixis points to the isolation and characterization of genes as the road to further understanding of this area of reproductive biology and the facilitation of the genetic transfer of apomixis to cultivated crops to produce uniform progeny independently of pollination and fertilization.

Our agriculture will stand on the threshold of a revolution if crop plants are induced to produce seeds without sex by apomixis. Some practical implications of the availability of crop plants engineered with genes for the apomictic trait are the production of new hybrid cultivars, enhancement of the stability of the hybrids, seed production by crops traditionally doomed for vegetative propagation, and independence of sexual reproductive processes in seed production. Unfortunately, research on the molecular biology of apomixis has been of peripheral interest in the past. Although attempts are being made to develop model systems such as *Hieracium* (Asteraceae) isolate apomictic genes, the work is in its infancy.

7. GENERAL COMMENTS

The megagametophyte of angiosperms is the end product of a series of reductions with the fundamental innovation of cells to move from the diploid to the haploid state. Although the megagametophyte is organized according to some common underlying patterns, the molecular basis for the pattern determination remains elusive. The progress made during the last two decades in our understanding of the structural organization of the component cells of the embryo sac has been impressive. Despite the descriptive nature of these data, they provide invaluable hints to elucidate much that is mysterious about megasporogenesis and megagametogenesis in angiosperms.

There are profound differences between the contributions of the different cells of the embryo sac to its ultimate function; perhaps the most interesting of these comes from a comparison of the fates of the egg cell and of the diploid polar nucleus. The former is the female sex cell, which fuses with the sperm to form the future sporophyte; the other provides the nurture for the developing sporophyte. Other contrasts of significance in the embryo sac from both structural and functional angles are those found between the egg and the synergids. Comparative studies of megasporogenesis and embryo sac development suggest that many of the developmental events are conserved throughout evolution.

Investigators interested in the genetic and molecular analysis of the angiosperm female gametophyte have acknowledged the intractability of the system. For this reason, compared to the male gametophyte, relatively little attention has been paid to the isolation and characterization of female gametophyte mutants or of genes expressed during megasporogenesis and megagametogenesis. It is still unclear whether genes expressed during megasporogenesis, megagametogenesis, and apomixis have any features in common. Thus, an integration of genetic, cellular, and molecular properties that underlie the development of the angiosperm megagametophyte into a unified theme appears to be a distant dream.

SECTION II

Pollination and Fertilization

Chapter 7

Stigma, Style, and Pollen–Pistil Interactions

In the majority of angiosperms the pollen grain matures at the two-celled stage, enclosing a vegetative cell and a generative cell. In some plants, pollen maturation occurs at the three-celled stage, when the generative cell divides to produce two sperm cells. Irrespective of the number of cells they enclose, mature pollen grains are released by the dehiscence of the anther and are passively carried to the receptive surface of the stigma of another flower in the act of pollination. This is the beginning of a cascade of events that ensure double fertilization in the embryo sac. This chapter will consider how the pollen grain makes its way through the stigma and style toward the ovary and ovule. Although our knowledge of the intimate details of individual events in the odyssey of the pollen grain is far from complete, there is a considerable body of descriptive information relating to these events.

The environment of the stigma and style where the events subsequent to pollination take place is so overwhelmingly complex that it almost defies analysis. Fortunately, recent advances in biochemical and cell biological methods have gone in tandem with exploitation by the electron microscope, with the result that a detailed account of the structure of most of the participating cells in the stigma and style has become available. Based on this knowledge, specific questions that are addressed here from developmental and molecular angles are the following: What are the specializations of cells of the stigma to trap pollen grains and to provide for their early nurture? What signals do the stigmatic cells release following a compatible pollination? What are the factors that control the interactions between pollen grains and stigmatic cells? What are the adaptations of the style to support the growth of pollen tubes? These questions are all interrelated because the events implied therein occur in fairly rapid succession and the consequences of each bear upon the others.

For purposes of discussion here, it is convenient to consider the questions just raised in the context of a compatible mating, in which both self- and cross-pollinations result in successful fertilization and seed set. This approach has been taken to emphasize the role of the cells of the stigma and style as they prepare to receive the pollen grain and pollen tube, respectively, and to stress the significance of the interactions between them as part of the basic system of reproduction. As will become evident, encounters between the pollen grain and stigma, on the one hand, and between the style and pollen tube, on the other hand, go a long way to define the principles of cell recognition and signaling and to establish their genetic control and physiological features.

1. THE STIGMA

Intensive research has been devoted to the study of pollination biology, particularly in terms of the morphology of the various flower types that facilitate pollination by the agencies of wind, water, and species of the animal kingdom. Among the animal species, insects are known to have been pollinators of the earliest angiosperms, whereas ancestors of the present-day angiosperms were pollinated by the action of wind. It has been claimed by evolutionary biologists that interactions between insects and flowers in the process of pollination have played a major role in the morphological specialization of flowers. The actual transfer of pollen grains from the anther to the stigma involving these external agents is not considered here. The various types of incompatibility reactions that enable the stigma and style to recognize and reject pollen grains from the same individual flower are also not discussed here; they form the subject of Chapter 9. Within this deliberately limited scope, this chapter will begin with an examination of the structure of the stigma, which is operationally defined as the part of the carpel or pistil that receives the pollen grains and whose cells first respond to the arrival of the pollen.

(i) Stigma Types and Their Structure

The study of the stigma of angiosperm flowers has assumed a challenging freshness in recent years, not only because it has been neglected through much of the present century, but also because it has become pertinent to our understanding of the control of breeding system in plants. The cells of the receptive surface of the stigma are more or less glandular in nature and are frequently elongated into unicellular or multicellular papillae. However, differences exist in the amount of glandular secretion covering the receptive cells at maturity and in the morphology of these cells. J. Heslop-Harrison (1975b) took advantage of this situation and broadly separated species in which the receptive surface is free of copious fluid secretion (designated as dry stigma) from those in which surface secretion decorates the receptive cells at maturity (designated as wet stigma). Based on a survey of stigmatic surfaces of nearly 1,000 species covering 900 genera distributed within some 250 families, the classification was further refined to accommodate stigma types with different degrees of morphological variation in the two basic groups (Y. Heslop-Harrison and Shivanna 1977; Y. Heslop-Harrison 1981). A summary of this classification, with some examples of families in each group, is given in Table 7.1. What is striking about this classification is that there is an impressive correlation between the surface morphology of stigmatic cells and the amount of secretion present during their receptive period. Thus, stigmas that are generally papillate and whose receptive surface is plumose or is concentrated in specific patterns are of the dry type, whereas those with little or no papillae belong to the wet type. Although only scattered examples have been examined, it is clear that there are some heterogeneous families, such as Amaryllidaceae, Commelinaceae, Liliaceae, Onagraceae and Rosaceae, that include genera with both wet and dry stigmas. Cases of extreme variability in the stigma surface morphology have been described in species of *Aneilema, Commelina,* and *Tradescantia* (all Commelinaceae), with representatives of three of the four groups in Table 7.1 present within each genus. Adding to the difficulty of assigning stigma types of these genera to a recognized system of classification is the fact that in some species the exudate may be absent or scanty when the flower opens, but it increases subsequently (Owens and Kimmins 1981). The limitations of this classification are further underscored by the fact that in heterostylous flowers (Chapter 9) of *Linum grandiflorum*, the stigma surface of the pin morph is of the dry type, whereas that of the thrum morph resembles the wet type (Ghosh and Shivanna 1980a). Types and proportions of surface exudates can also be useful in the classification of stigmas; however, one should characterize the exudate chemically to classify a given stigma type (Dumas 1978).

Anatomy. Besides light microscopic histology, extensive use has been made of transmission and scanning electron microscopy to study the surface features of stigmas (Figure 7.1). On a conservative estimate, approximately a thousand papers dealing with the microanatomy of stigmas have been published during the last quarter-century and the list continues to grow. Because of the sheer volume of this information, it is impossible to provide more than a cursory treatment of the topic here. A common feature of the anatomy of the stigma is that the epidermis that covers the style lower down differentiates into elongate, slightly thick-walled, unicellular or multicellular, uninucleate, glandular papillae on the stigma. An analysis of periclinal chimeras in *Datura stramonium* has shown that besides the surface layers of the stigma and style, the cells of the ovarian placental epidermis are also derived from the same epidermal layer (Satina 1944; Vithanage and Knox 1977). What causes simple epidermal cells to differentiate into glandular cells is tied to the onset of morphogenetic changes

Table 7.1. Classification of angiosperm stigma types based on the morphology of the receptive surface and the amount of secretion present during the receptive period. (Modified from Y. Heslop-Harrison and K. R. Shivanna 1977. The receptive surface of the angiosperm stigma. *Ann. Bot.* 41:1233–1258, by permission of Academic Press Ltd., London.)

Stigma type	Group	Description	Some families represented
Dry stigma	I	Plumose, with receptive cells dispersed on multiseriate branches	Poaceae (no dicot families)
	II	Receptive cells concentrated in distinct ridges, zones, or heads	
		A. Surface nonpapillate	Cyperaceae, Liliaceae (monocots); Acanthaceae, Ranunculaceae (dicots)
		B. Surface distinctly papillate	
		i. Papillae unicellular	Agavaceae, Alismataceae (monocots); Campanulaceae, Rutaceae (dicots)
		ii. Papillae multicellular	
		(a) Papillae uniseriate	Amaryllidaceae, Hypoxidaceae (monocots); Amaranthaceae, Onagraceae (dicots)
		(b) Papillae multiseriate	Bromeliaceae, Flagellariaceae (monocots); Cucurbitaceae, Oxalidaceae (dicots)
Wet stigma	III	Receptive surface with low to medium papillae; secretion fluid flooding interstices	Agavaceae, Orchidaceae (monocots); Apocyanaceae, Scrophulariaceae (dicots)
	IV	Receptive surface nonpapillate; cells often necrotic at maturity; usually with more surface fluid than Group III	Araceae, Zingiberaceae (monocots); Ericaceae, Rubiaceae (dicots)

in the developing stigma, so it will be best to consider both together in representative species.

In its early stages of development, the stigma of *Petunia hybrida* is covered with an epidermis, overlaid with a thin layer of cuticle, and is followed to the inside by several layers of densely meristematic subepidermal cells. Even at the primordial stage, the groundwork is laid for the division of occasional epidermal cells and for their eventual differentiation into two-celled papillae. Curiously, the metabolism of carbohydrates and lipids appears to be involved in this process, because cells in the differentiating region of the stigma become filled with starch and oil globules (Konar and Linskens 1966a). A slightly different organization of the stigma described in *Gossypium hirsutum* is characterized by the presence of a weft of unicellular hairs that cover the stigmatic lobes and extend a short distance along the style. Since these hairs are transitory and begin to degenerate before the stigma becomes receptive, the layer of cells immediately below the hair cells probably functions as the papillae (Jensen and Fisher 1969). A most unusual type of stigma is found in *Zea mays*. The terminal portion of the silk

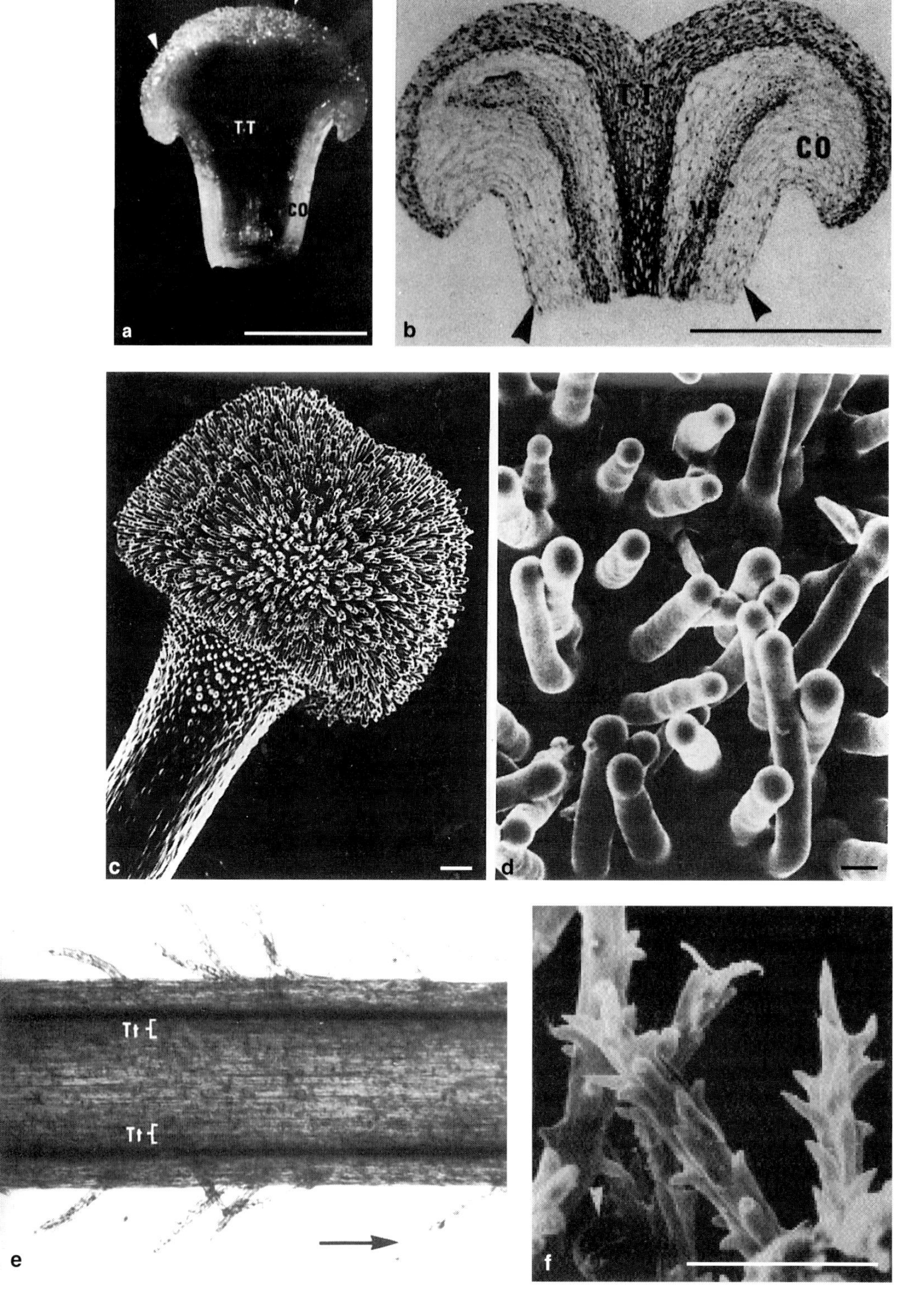
TT
CO
a
TT
CO
b
c
d
Tt
Tt
e
f

of common parlance is unevenly cleft, and in the older literature this branched portion is designated as the stigma (Miller 1919). In the present account, the whole length of the silk is considered as the stigma. It is certainly one of the longest stigmas (up to 75 cm) found in angiosperms and, throughout the entire length, it displays the full morphogenetic repertoire of typical stigmatic cells. The stigma is composed for most of its length of a single axis traversed by two vascular bundles. Associated with each vascular bundle is a specialized tissue touted as analogous to the transmitting tissue of the style through which pollen tubes grow. Lateral appendages known as trichomes arise from single cells of the epidermis of the axis and extend in 6–10 irregular ranks along the margin of the stigma. The cells of the trichome, often recurved at the tip, function as the papillae on which pollen grains land (Kroh, Gorissen and Pfahler 1979; Y. Heslop-Harrison, Reger and Heslop-Harrison 1984a).

The general ultrastructural features of a typical stigmatic papilla are shown in Figure 7.2. In most cases examined, these are found to include the usual complement of organelles, with a preponderance of mitochondria, plastids, ER, dictyosomes, ribosomes, microbodies, and vesicles (Konar and Linskens 1966a; Vasil'ev 1970; Dickinson and Lewis 1973a; Dumas et al 1978; Sedgley and Buttrose 1978; Herrero and Dickinson 1979; Clarke et al 1980; Herd and Beadle 1980; J. Heslop-Harrison and Heslop-Harrison 1980b; Tilton and Horner 1980; Wilms 1980a; Y. Heslop-Harrison, Heslop-Harrison and Shivanna 1981; Sedgley 1981a; Uwate and Lin 1981b; Dickinson, Moriarty and Lawson 1982; Owens and Horsfield 1982; Ciampolini, Cresti and Kapil 1983; Sedgley and Blesing 1983; Cresti et al 1986b; Kandasamy et al 1989; Bystedt 1990; Wróbel and Bednarska 1994). A distinctive ultrastructural feature of the stigmatic papillae of *Acacia retinodes* and certain other Fabaceae is the presence of plastids containing ferritin. This iron–protein complex generally associated with photosynthetically inactive chloroplasts is present in the stigmatic papillae of *A. retinodes* of all ages (Jobson et al 1983). However, the most conspicuous organelle of the stigmatic papilla is the nucleus, which acquires great dimensions in *Ornithogalum caudatum* (Tilton and Horner 1980) and *Brassica oleracea* (O'Neill, Singh and Knox 1988) or undergoes endomitotic divisions or endoreduplication in *Spironema fragrans* (Commelinaceae) (Tschermak-Woess 1959), *B. oleracea* (O'Neill, Singh and Knox 1988), and *Vicia faba* (Wróbel and Bednarska 1994). There are appreciable deposits of tannin in the vacuoles of the stigmatic papillae of *Lycopersicon esculentum* (Dumas et al 1978) and *Olea europaea* (Ciampolini, Cresti and Kapil 1983); the significance of these is not known.

In *Secale cereale*, the wall of the trichome is not homogeneous but is stratified into several layers. The external face of the cell is covered with an inert armor of a somewhat discontinuous layer of cuticle. Underlying the cuticle, the wall consists of a thinly dispersed microfibrillar layer, followed by a PAS-reactive electron-transparent layer. An additional layer of intermediate electron density is seen close to the plasma membrane. A conspicuous feature of the cytoplasm of the papillate cells is the presence of paramural bodies that possibly constitute a form of granulocrine secretory system, working in concert with vesicles to target proteins into the pathway of the pollen tube (J. Heslop-Harrison and Heslop-Harrison 1980b). The organization of the wall of individual receptive trichomes on the stigma of *Z. mays* is similar to that described in *S. cereale* (Y. HeslopHarrison, Reger and Heslop-Harrison 1984a). As seen in a freeze-fracture study, the principal elements of the cellulose wall of the stigmatic papillae of *Gladiolus gandavensis* are a series of channels that radiate from the plasma membrane and reach out to the cuticle. The channels probably allow the free flow of exudate to the cuticular surface (Clarke et al 1980). Similar, but smaller, channels have been observed traversing the papillar wall of the stigma of *B. oleracea* (Roberts, Harrod and Dickinson 1984a). The stigmatic papillae of *Lilium regale*, *L. davidii* (Vasil'ev 1970), *Ornithogalum caudatum* (Tilton and Horner 1980), *Citrullus lanatus* (water-

Figure 7.1. Morphology and anatomy of the stigma. **(a)** A cut view of the stigma and upper part of the style of *Nicotiana tabacum* showing the stylar transmitting tissue and the cortex. The stigmatic papillae are indicated by arrowheads. CO, cortex; TT, transmitting tissue. **(b)** The same in longitudinal section. Arrowheads point to the epidermis. VB, vascular tissue. Scale bars = 1 mm. (From Bell and Hicks 1976. *Planta* 131:187–200. © Springer-Verlag, Berlin.) **(c)** Scanning electron micrograph of the stigma head and upper part of the style of the pin morph of *Primula vulgaris*. Scale bar = 10 µm. **(d)** Detail of the papillae of the same type of stigma. Scale bar = 10 µm. (From Y. Heslop-Harrison, Heslop-Harrison and Shivanna 1981; photographs supplied by Dr. K. R. Shivanna.) **(e)** Central zone of the silk of *Zea mays* approximately 260 mm long. The arrow points toward the base of the silk. Tt, transmitting tissue. Scale bar = 100 µm. (From Y. Heslop-Harrison, Reger and Heslop-Harrison 1984a.) **(f)** Scanning electron micrograph showing the stigmatic hairs of *Hordeum vulgare* and a germinated pollen (arrowhead). Scale bar = 100 µm. (From Cass and Pateya 1979.)

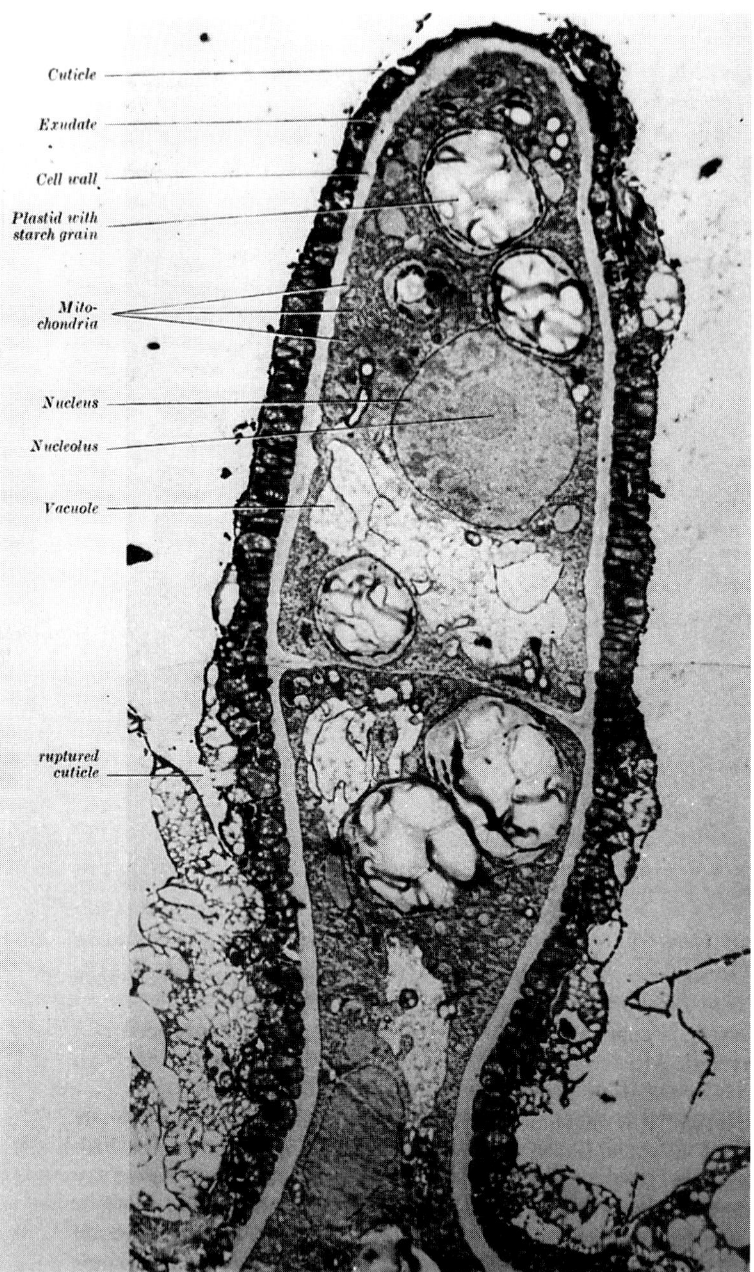

Figure 7.2. A stigmatic papilla of *Petunia hybrida*. The accumulated exudate is seen between the cellulose wall and the cuticle. Scale bar = 1 μm. (From Konar and Linskens 1966a. *Planta* 71:356–376. © Springer-Verlag, Berlin.)

melon; Cucurbitaceae) (Sedgley 1981a, 1982), and *Commelina erecta* (Owens and Horsfield 1982) have wall ingrowths similar to transfer cells. The increased absorbing surface provided by the ingrowths obviously facilitates absorption and transfer of nutrients between intercellular barriers.

Although the overlying cuticle is less obvious and less clearly defined than the cell wall, it presents a range of unsuspected complexity. In *Tropaeolum majus* (Tropaeolaceae) and several species of *Aneilema* and *Commelina*, the cuticle is more or less ridged or folded (Shayk, Kolattukudy and Davis 1977; Owens and Kimmins 1981). The cuticle over the receptive papillae of *Crocus chrysanthus* is characteristically chambered, and the stigmatic fluid accumulates in the chambers; however, toward the base of the cells, the cuticle extends as a simple overlying layer (Y. Heslop-Harrison 1977). In certain members of Caryophyllaceae, Brassicaceae, and Asteraceae, the cuticle at the tip of the stigmatic papillae is seen in the electron microscope as a discontinuous layer with numerous interruptions, possibly due to the presence of radially oriented cutinized rodlets separated by wall materials (J. Heslop-Harrison, Heslop-Harrison and Barber 1975). Investigations into the structure of the developing stigma of *Vicia faba* have shown that the cuticle is lifted from the epidermis and appears as a hovering membrane above the accumulated stigmatic exudate (Lord and Heslop-Harrison 1984). From the functional point of view, the significance of the cuticle in pollen–stigma interactions cannot be overemphasized. Firstly, the cuticle provides a surface for the deposition of secretory products from the cytoplasm of the stigmatic papillae and a route through which secretory products are transported. Secondly, it constitutes a barrier to water transport from the papillae during pollen hydration. Thirdly, pollen tubes need to penetrate the cuticle in order to grow into the style, so the cuticle thus provides the first structural impediment that helps prevent incompatible pollinations (Owens and Horsfield 1982).

The Pellicle. Stigmatic surfaces of certain plants are characterized by the presence of an extracellular, hydrated, proteinaceous–lipoidal layer termed the pellicle (Mattsson et al 1974). It has been proposed that as an extracellular matrix shared by several cells, the pellicle performs two functions on the stigmatic surface. One is a supportive role in facilitating the capture and hydration of pollen grains, and the other is a dynamic role in serving as a recognition site during pollen–stigma interactions. Although details of the origin of the pellicle are not fully understood, the cuticle has discontinuities through which the pellicle is presumed to be extruded to the surface of the papillae.

It was originally thought that a pellicle adorns the cuticle of only the dry type of stigma and that an exudate makes up the bulk of the materials found external to the cuticle of the wet type. Later research showed that the pellicle crosses these artificial boundaries and is present in both types of stigmas. In *Linum grandiflorum*, the papillae of the dry stigma of the pin morph are covered with a uniform cuticle–pellicle layer; in the wet thrum type, the cuticle–pellicle layer is interrupted at several places and embellished at the surface with secretion products (Ghosh and Shivanna 1980a). In *Petunia hybrida, Nicotiana tabacum, Crinum defixum* (Amaryllidaceae), and *Amaryllis vittata*, all characterized by wet stigma, a pellicle is present on the cells of the stigmatic papillae from a very early stage of development in a thin and continuous layer, as in the dry stigma. Whereas the pellicle is randomly interrupted over the papillae of mature stigmas of *P. hybrida* and *N. tabacum* by the newly secreted exudate, in *C. defixum* and *A. vittata*, stigmatic papillae retain the original contour of the pellicle even in the mature state (Shivanna and Sastri 1981). The pellicle that ensheathes the stigmatic papillae of *Glycine max*, another species with a wet stigma, is highly differentiated and gives the impression of being composed of three bands of different electron densities (Tilton et al 1984). The secretion of a pellicle seems to be one activity of the stigmatic papillae that transcends even ecological barriers, since a surface pellicle exhibiting esterase activity is present on the stigmatic papillae of certain marine angiosperms that live a wholly submerged existence (Ducker and Knox 1976; Pettitt 1980).

That the pellicle has surface molecules that match lectins has been shown by the binding of concanavalin-A (con-A), a lectin, to the pellicle (Pettitt 1980). Lectins represent a relatively new class of molecules that bind to sugars with a high degree of specificity. It is likely that in certain plants in which con-A binding occurs on the stigmatic surface, the pellicle is involved, since alteration of the surface properties of the stigma by con-A invariably leads to changes in the recognition reaction (Y. Heslop-Harrison 1976; Knox et al 1976; Vithanage and Knox 1977; Clarke et al 1979; Kerhoas, Knox and Dumas 1983; Gaude and Dumas 1986; O'Neill, Singh and Knox 1988).

Our understanding of the chemical nature of the pellicle is far from satisfactory. According to Clarke et al (1979), the surface of the stigmatic papillae of *Gladiolus gandavensis* which probably constitutes the pellicle, assembles and accumulates an array of proteins, glycoproteins, and glycolipids. The latter two are the major constituents of the surface matrix and contain the monosaccharides galactose, arabinose, glucose, mannose, and rhamnose. A compound isolated from the stigmatic surface has adhesive properties and is an arabinogalactan protein. The chemical organization of the pellicle thus seems to be primed to interact with complementary substances on the pollen grain in a recognition event. Cytochemical studies have established the presence of ATPase and possibly adenylate cyclase in the stigma pellicle of *Brassica oleracea* and *Populus lasiocarpa* (Gaude and Dumas 1986; Zhu, Ma and Li 1990) and of acetylcholinesterase in *Pharbitis nil* (Convolvulaceae) (Bednarska and Tretyn 1989). These compounds may be essential for the recognition reaction or they may modify the pellicle itself, permitting recognition events to occur.

In summary, the stigmatic cells constitute the specialized surface for the reception of pollen grains and for the facilitation of their germination. The nature of the stigmatic cells, which may be papillate or nonpapillate; the disposition of the papillae, which may be dispersed or concentrated; the characteristics of the papillae, which may be unicellular or multicellular; the wall texture of the stigmatic cells, including the cuticular pattern and the presence of pellicle – all contribute in various ways to trapping pollen grains and providing the milieu for their germination.

(ii) The Stigmatic Exudate

A liquid known as the exudate that appears over the stigmatic papillae forms the medium in which the pollen grains germinate. The exudate is secreted by modified subepidermal cells of the stigma. Secretion generally begins before anthesis, with a marked increase following pollination (Konar and Linskens 1966b; Sedgley and Scholefield 1980; Kandasamy and Kristen 1987a). The environmental factors that affect the secretion process and the molecular mechanisms involved in the regulation of transcription and translation of proteins of the exudate are not understood.

In *Petunia hybrida*, as the epidermal cells differentiate into the papillae, the subepidermal cells become committed to exudate production by their progressive transformation into a spongy zone of schizogenous cavities. Initially, the exudate accumulates in the intercellular spaces, which become transformed into large cavities as the exudation continues. Eventually, the exudate fills the space between the cuticle and the cell wall and overflows onto the receptive surface of the stigma through rupture of the cuticle (Figure 7.3). A second wave of exudation starts soon after anthesis of the flower, when the breakdown of the schizogenous cavities releases large volumes of the exudate between the epidermal cells (Konar and Linskens 1966a). The origin of the exudate on the stigmas of *Lycopersicon peruvianum* (Dumas et al 1978), *Prunus avium* (Raff, Pettitt and Knox 1981; Uwate and Lin 1981a), *Nicotiana sylvestris* (Kandasamy and Kristen 1987a), and *Solanum tuberosum* (MacKenzie, Yoo and Seabrook 1990) is not fundamentally different from that described in *P. hybrida*. In *L. peruvianum*, a column of densely cytoplasmic cells is the main secretory tissue of the stigma. The secretion products are initially deposited as small droplets in the intercellular spaces that form an efficient communication center between the stigmatic papillae to the outside and the transmitting tissue of the style to the inside (Dumas et al 1978). In *Vitis vinifera* (grape; Vitaceae) (Considine and Knox 1979), *Smyrnium perfoliatum* (Apiaceae) (Weber 1994), and *Hacquetia epipactis* (Apiaceae) (Weber and Frosch 1995), the stigmatic exudate appears to be secreted by the underlying cells of the transmitting tissue from whence it is directed to the papillae.

At the subcellular level, one can envision a stepwise process of secretion and targeting of the exudate outside the stigmatic papillae involving (a) a phase of synthesis of the substance by specific cell organelles, (b) its transport to the vicinity of the plasma membrane, and (c) the spilling of the finished product to the outside. Papilla of *Aptenia cordifolia* (Aizoaceae) is almost an ideal system for monitoring the secretion of the exudate. The cell starts off with a cytoplasm enriched in ribosomes, mitochondria, and ER. Soon after, the ER becomes active and develops numerous attached vesicles containing a fibrillar material. Most evidence suggests that the material is extruded by exocytosis. Although this function is traditionally assigned to the Golgi apparatus, the small number of dictyosomes present in the cell is believed to preclude their involvement in the secretion process. What distinguishes a young papilla from an old one is the role of vacuoles in the transfer of the exudate, as seen when some of the ER with attached vesicles is phagocytotically incorporated into the large cell vacuole. This is part of the holocrine secretion, which is attained by membrane dissolution, the mixing of the contents of the ER–vesicle complex with the vacuolar fluid, and the degeneration of the protoplast (Kristen 1977; Kristen et al 1979). In some species of *Aneilema* and *Commelina*, the exudate is generated in small, single-membrane-bound vesicles and appears in the subcellular space by exocytosis (Owens and Horsfield 1982).

According to Kroh (1967b), the exudate in the stigma of *Petunia hybrida* appears to be produced by the ER and transported through the cytoplasm in the form of droplets to the plasma membrane. A role is also envisaged for the plasma membrane in transferring the droplets to the cell wall. In a similar way, in the glandular cells of the stigma of *Forsythia intermedia* (Oleaceae), secretion vesicles connected with ER accumulate the exudate, which

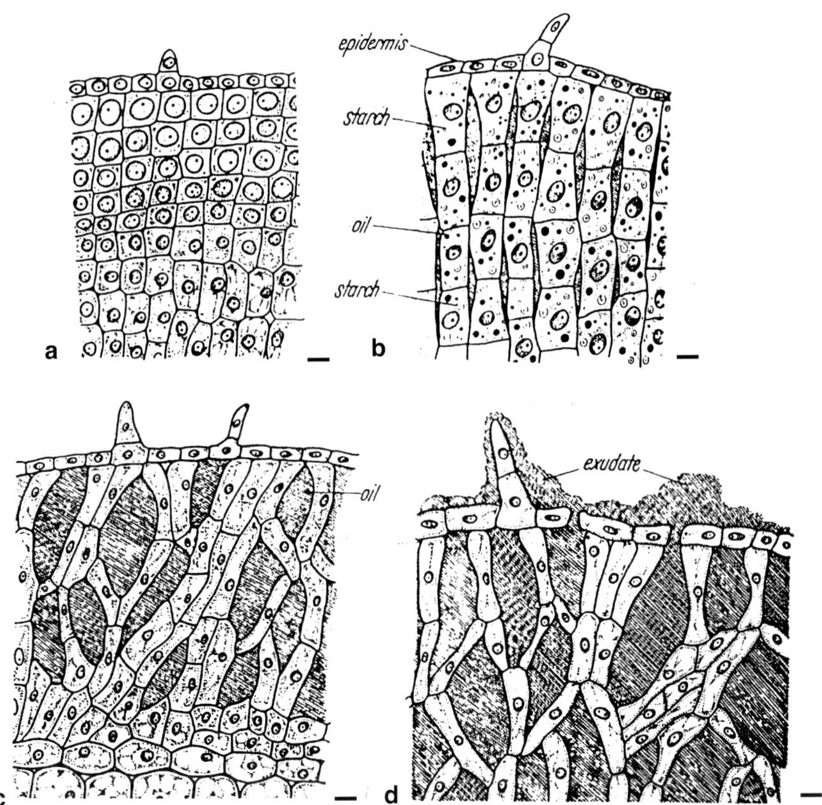

Figure 7.3. Development of the stigmatic papillae and formation of the exudate in *Petunia hybrida*. **(a)** Part of a young stigma showing cells with prominent nuclei. **(b)** Stigmatic cells showing accumulation of exudate in the intercellular spaces. **(c)** Schizogenous cavities in the secretory zone. **(d)** Separated epidermal cells and release of the exudate. Scale bars = 10 μm. (From Konar and Linskens 1966a. *Planta* 71:356–371. © Springer-Verlag, Berlin.)

is discharged under the cuticle. This general mechanism does not apply to lipids, which appear to pass through the wall by the eccrine mechanism due to a lack of bounding membranes (Dumas 1973, 1977b). Lipid bodies are produced in the stigmatic papillae of *Persea americana* (avocado; Lauraceae) and secreted outwardly by the eccrine mechanism; it has been suggested that lipids are synthesized in the plastids of the cell (Sedgley and Blesing 1983). The dynamics of secretion of the exudate by stigmatic papillae of *Lycopersicon peruvianum* rely on a cooperative interaction between the ER and dictyosomes. Before its release into the intercellular space, the secretory product resembles lipid droplets that are found in close proximity to the ER. A period of preponderance of dictyosome vesicles coincides with anthesis, when the exudate production attains its peak; this is probably suggestive of the operation of a basic plan involving dictyosomes to transport the droplets to the periphery of the cell from whence they are extruded by exocytosis (Dumas et al 1978). Although the stigmatic papillae of *Verbascum phlomoides* (Scrophulariaceae) (Dumas 1974a) and *Citrullus lanatus* (Sedgley 1981a) excrete carbohydrates, papillae of each species appear to be physiologically unique, exhibiting slightly different secretion pathways. In the latter, both ER and dictyosomes are associated with secretion; in the former, this function is under almost exclusive dictyosome control.

Despite the presence of an active Golgi system in the stigmatic papillae of grasses, dictyosomes do not seem to be involved in the secretion of surface components. In *Secale cereale*, it has been suggested that paramural bodies found in the papillae secrete the exudate, which finds its way through the vesicles to the outside of the cell (J. Heslop-Harrison and Heslop-Harrison 1980b). In a general way, differences noted in the origin of the exudate in the species investigated may reflect gaps in our knowledge of the ultrastructure of the cells rather than real differences, or may be due to the difficulty in deciding which organelles participate most directly in secretion. In some cases, more than one organelle may be involved and, obviously, many observations need to be synthesized into a coherent model of exudate origin and delivery in the stigma.

A model for the division of labor in exudate production is provided by the multicellular papillae of the stigma of *Citrus limon*. Of the principal components of the exudate, lipids are first secreted by the basal cells of the papillae, where they are trapped as droplets in the dilated parts of the smooth ER. Subsequently, files of these droplets accumulate at the plasma membrane for transfer to the intercellular spaces and the middle lamellar region. Tip cells of the papillae become active and begin to accumulate granular polysaccharides after lipid secretion by the basal cells ceases. This appears to be related to an increased function of the Golgi vesicles (Cresti et al 1982).

One of the metabolic changes observed in the stigmatic papillae of *Raphanus sativus* is the accumulation of conspicuous protein bodies. This begins with the appearance of large vesicles containing fibrillar materials. The vesicles, which appear to be derived from ER, finally discharge into the central vacuole, where the contents concentrate and assume a crystalline conformation (Dickinson and Lewis 1973a). Although as many as 10 such protein bodies have been identified in each cell, it is not clear whether they contribute in some way to the formation of the exudate.

The aforementioned studies refer to the origin of the exudate in solid-styled species; much less is known about the phenomenon in hollow-styled species. In *Crinum defixum* and *Amaryllis vittata*, the exudate that reaches the stigmatic surface is reported to emanate from the style (Shivanna and Sastri 1981). However, in *Lilium regale, L. davidii,* and *L. longiflorum*, the activity of a large number of vesicles that populate the vicinity of the plasma membrane of the stigmatic papillae has been associated with the secretion of the exudate. Inasmuch as these vesicles are the preferred sites of accumulation of the components of the exudate, it is likely that the material from the vesicles is discharged as the exudate (Vasil'ev 1970; Dickinson, Moriarty and Lawson 1982). We do not know whether the vesicles cooperate with some other organelles in the production of the exudate. Although the dry stigma of cotton does not form an exudate, cells of the stigmatic papillae go through phases of accumulation of membranous, lipoidal, and granular inclusions in the vacuoles, very similar to the sequence described in wet stigmas of some plants (Sethi and Jensen 1981).

The amount of the exudate and its chemical composition vary from species to species. Because of the difficulty of collecting the substance in quantity for biochemical analyses, much of what we know about the composition of the exudate has come from histochemical tests. Lipids, phenolic compounds, carbohydrates, amino acids, and proteins are generally present in the exudate. With this general statement as a backdrop, the composition of the stigmatic secretion in a few representative cases will be examined. In probably the first detailed chemical analysis of the stigma secretion, Konar and Linskens (1966b) showed that in *Petunia hybrida* it consists of an oil that is free of phospholipids, sterols, and fatty acids; sugars like sucrose, glucose, and fructose as major components and galactose as a minor component; and a range of about 19 amino acids. Depending upon the age of the flower, stigma cells contain the flavonol aglycone kaempferol or its sugar-conjugated form, kaempferol glycoside; the former is present in extracts of mature stigmas, whereas immature stigmas have the glycoside form (Mo, Nagel and Taylor 1992; Pollak et al 1993; Vogt et al 1994). The stigma exudate of *Lilium longiflorum* is composed mainly of a protein-containing polysaccharide rich in galacturonic acid, glucuronic acid, rhamnose, arabinose, and galactose; the very presence of these complex compounds engenders some functional versatility for the exudate. Cell wall polysaccharide precursors like *myo*-inositol, L-proline, and glucose are utilized by detached pistils for the synthesis of the stigmatic exudate. Especially in unpollinated pistils, production of stigmatic exudate is rather pronounced, rising slowly during the first two days after anthesis and then somewhat rapidly up to about seven days (Labarca, Kroh and Loewus 1970; Labarca and Loewus 1973). Stigmatic cells of legumes secrete scanty amounts of exudate, which includes lipids, amino acids, proteins, sugars, reducing acids, phenols, and polysaccharides, as well as esterases and acid phosphatases (Lord and Webster 1979; Ghosh and Shivanna 1982b; J. Heslop-Harrison and Heslop-Harrison 1982c). Other histochemical and biochemical characterizations of the stigmatic fluid have not uncovered any novel compounds; in some plants, chemically defined fractions of carbohydrates, proteins, and lipids have been identified (Kandasamy and Vivekanandan 1983; Cresti et al 1986b; Kandasamy and Kristen 1987a; Weber 1994; Weber and Frosch 1995), whereas in others, lipids (Vasil and Johri

1964; Dumas 1974b; Tilton 1984; Tilton et al 1984; Kadej, Wilms and Willemse 1985; MacKenzie, Yoo and Seabrook 1990), simple carbohydrates and sugars (Horovitz, Galil and Portnoy 1972; Portnoi and Horovitz 1977; Kristen et al 1979; Sedgley and Scholefield 1980; Kenrick and Knox 1981b), phenolic compounds and their derivatives (Martin 1969, 1970a, b, c; Martin and Telek 1971; Sedgley 1975), or arabinogalactans (Aspinall and Rosell 1978; Gleeson and Clarke 1980a, c) comprise major fractions of the exudate.

Buffer-soluble arabinogalactans represent about 65% of the high-molecular-mass carbohydrates of the stigma of *Nicotiana alata* and contain, for the most part, galactose and arabinose with small amounts of rhamnose, xylose, mannose, and glucose. During the ontogeny of the stigma, there is an increase in the total amount and cellular concentration of arabinogalactans until flower maturity. But the most prominently featured change in the stigma arabinogalactans is in their electrophoretic charge characteristics, which remain diffuse during early flower development but resolve into a major peak during later stages. These changes parallel the development of the capacity of the stigma for pollen reception and thus stand out as being functionally related (Gell, Bacic and Clarke 1986; Bacic, Gell and Clarke 1988).

The occurrence of Ca^{2+} on the stigmatic surface or in the stigma exudate of *Ruscus aculeatus* (Liliaceae), *Petunia hybrida*, and *Pharbitis nil* has been established by chlorotetracycline fluorescence and X-ray microanalysis. Because membrane-bound Ca^{2+} is found only in the cells of physiologically active stigmas, its presence is linked to the secretory activity of the stigma (Bednarska 1989a; Bednarska and Karbowska 1990; Bednarska and Butowt 1995a). In the stigmatic papillae of *Impatiens palmata*, both Ca^{2+} and calmodulin are present, although they are not colocalized (Tirlapur and Shiggaon 1988). These observations suggest that the Ca^{2+}-enriched medium is involved in promoting germination of pollen grains in vivo as it does in vitro.

As we learn more about the chemical constituents of the stigmatic exudate, some observations that had been difficult to interpret are beginning to make sense. Many of the substances referred to in this section are involved in recognition reactions with pollen grains. Other molecules that are synthesized, or that accumulate, in stigmatic cells play a role in protecting the stigma against potential pathogens and can thus be considered as defense-related molecules. Included in this group are proteinase inhibitors, chitinase, and certain phenolic compounds. In the stigma of *Petunia hybrida*, the activity of chitinase is developmentally regulated, the enzyme being several times more active in the mature than in the immature tissues (Leung 1992).

Compared to the stigma, the style has an unusually long life while it supports the growth of the pollen tube. How does the style remain as a healthy organ as its apical part atrophies? One view is that physiological demands for new substances to keep the style in prime condition are met by the stigma as it undergoes senescence. According to Vogt et al (1994), stigma cells of *P. hybrida* retain a high concentration of kaempferol even after pollen tubes are at least two-thirds of the way in the style. Because the levels of kaempferol present in the stigma at this stage are toxic, it has been proposed that the flavonol probably prevents colonization of the stigma and style by pathogens until the pollen tube has completed its journey to the ovule. This model is a useful working hypothesis to explain why stigmatic cells enriched with sugars and other substances that favor fungal and bacterial growth are seldom infected.

As will be seen later in this chapter, stigmatic exudate is essential for practically all pollination-related events. There is abundant evidence to indicate that after pollination, molecules of the exudate join forces with complementary molecules from the pollen grain in recognition reactions. Carbohydrates probably help in the germination of pollen grains. Lipoidal substances of the exudate facilitate the trapping of pollen grains, essentially functioning as the glue that keeps them attached to the stigma, and they protect the stigmatic surface from desiccation and wetting. The phenolic compounds are helpful in protecting the stigma from microbes and pests. Phenolics have also been implicated in pollen nutrition and in the selective promotion or inhibition of pollen germination on the stigma. This seems to raise the paradox of the same substance's playing essential roles in a variety of unrelated and occasionally opposing functions.

The stigma is a transient structure of the flower. Following anthesis, irrespective of whether the flower is pollinated or not, necrotic changes are initiated in the cells that eventually lead to the collapse of the stigma. In *P. hybrida*, many cytological changes contribute to restrictions in the developmental capacity of the stigmatic cells. For example, as degeneration commences, appearance of large drops of lipids in the cytoplasm is typical of most stigmatic cells. Directly related to the debilitating age-related changes is the rupture of the plasma membrane of cells; this releases the lipids, which mix with other extracellular droplets. Cell function ceases as the remaining cytoplasm is transformed into small islets of degenerate masses of no adaptive value to the cell (Herrero and Dickinson 1979). In *Lilium longiflorum*, older stigmatic cells accumulate electron-opaque, paracrystalline bodies in the vacuoles, although it is not clear whether they contribute to the senescence of these cells (Dickinson, Moriarty and Lawson 1982).

How the deterioration and senescence of the stigma are coordinated with anthesis is not known. Does the signal for initiating these changes originate in the stigma itself, or is it transmitted from another part of the carpel or of the flower? Do the appearance and disappearance of the signal coincide with the developmental state of the stigma? Here we are undoubtedly too handicapped by inadequate knowledge to formulate statements in molecular language linking anthesis of the flower with death of the stigma.

2. THE STYLE

The style connects the stigma to the ovary and is the floral part that facilitates the directed growth of the pollen tube toward the ovary and ovule. Programmed for a specialized function, the style derives its developmental potential from the presence of cells that provide nurture and nutrients to the pollen tube. Structural modifications during ontogeny have given rise to two types of styles, the solid (closed) and hollow (open) types. In the former, the essential part of the style through which pollen tubes grow is a central core of nutritive, glandular cells known as the transmitting tissue; in the open style, the transmitting tissue, known as canal cells, is confined to one or more cell layers lining the canal. The canal may be filled with mucilage, which also serves as a medium for pollen tube growth. Some authors recognize a third type of style, the semiclosed style, which has features of both the solid and the open types. The style grows by cell division during the primordial stage and subsequently by cell elongation. In *Petunia hybrida*, the early phase of growth of the style appears to dependent upon some stimulus from the anthers, because their excision inhibits stylar elongation (Linskens 1974).

Mutations in *Arabidopsis* designated as *ettin* (*ett*) cause structural alterations in the pistil that result in a reduction in the ovary and the appearance of stigmatic and stylar features in the basal regions of the pistil. The *ETT* gene is assumed to code for a pattern-forming factor that ensures the formation of the ovary, style, and stigma in their proper geometrical proportion along the longitudinal axis of the flower (Sessions and Zambryski 1995).

(i) The Transmitting Tissue

Topographically, the solid type of style has a tissue system comprising the epidermis, cortex, transmitting tissue, and vascular system. In some Fabaceae, the transmitting tissue breaks down lysogenously to different degrees, so that the style shows a transition from solid to hollow type, as in *Vigna unguiculata* (Ghosh and Shivanna 1982b) and *Phaseolus acutifolius* (Lord and Kohorn 1986), or becomes completely hollow as in *Trifolium pratense* (Y. Heslop-Harrison and Heslop-Harrison 1982a) and *Crotalaria retusa* (Malti and Shivanna 1984). Though a full complement of organelles is packaged into the cells of these four tissues, the transmitting tissue will be the focus of attention here. The cells of the transmitting tissue are generally elongate and fusiform in the axial direction and have modified pectocellulosic walls (Figure 7.4). Following the work of Sassen (1974) on *Petunia hybrida*, ultrastructural studies have been extended to transmitting tissues of styles of *Nicotiana tabacum* (Bell and Hicks 1976), *Lycopersicon peruvianum* (Cresti et al 1976), *N. alata* (Sedgley and Clarke 1986), *Gasteria verrucosa* (Willemse and Franssen-Verheijen 1986), *N. sylvestris* (Kandasamy and Kristen 1990), *Brugmansia* (*Datura*) *suaveolens* (Solanaceae) (Hudák, Walles and Vennigerholz 1993), and *Triticum aestivum* (Vishnyakova and Willemse 1994). In summary, it appears that the cells of the transmitting tissue have a dual property; they divide to form new cells and separate to form conspicuous intercellular spaces that later become filled with a secretion product. In terms of the ultrastructural profile, cells of stylar transmitting tissues of all species studied conform to metabolically active cells bountifully endowed with ribosomes, mitochondria, rough ER, dictyosomes in active configuration, and amyloplasts. Cellular communication occurs through plasmodesmata in the transmitting tissues of *P. hybrida* (Sassen 1974), *L. peruvianum* (Cresti et al 1976), and *N. tabacum* (Bell and Hicks 1976). Some cell populations in the transmitting tissue of *P. hybrida* possess wall ingrowths that snugly abut into neighboring cells in the file and apparently provide an efficient intercellular link as do plasmodesmata (Herrero and Dickinson 1979). The polar plasmalemma of the

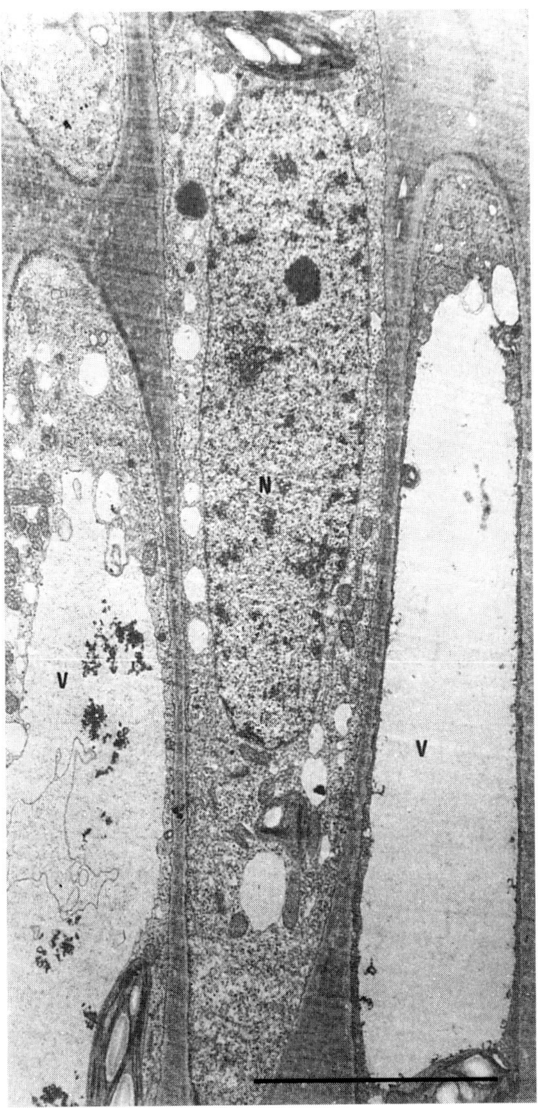

Figure 7.4. Cells of the transmitting tissue of the style of *Nicotiana tabacum* seen in longitudinal section. Approximately one-half of the middle cell is included. N, nucleus; V, vacuole. Scale bar = 5 µm. (From Bell and Hicks 1976. *Planta* 131:187–200. © Springer-Verlag, Berlin.)

transmitting cells of this species is also the site of free and loosely bound Ca^{2+} (Bednarska and Butowt 1995b). A noteworthy feature of the transmitting tissue of *N. tabacum* is the presence of crystal-containing bodies and myelinlike structures in the cells (Bell and Hicks 1976).

The general features of the ultrastructural organization of cells of the transmitting tissues of *Gossypium hirsutum* (Jensen and Fisher 1969), *Olea europaea* (Cresti et al 1978a; Ciampolini, Cresti and Kapil 1983), *Persea americana* (Sedgley and Buttrose 1978), *Malus communis* (Rosaceae) (Cresti, Ciampolini and Sansavini 1980), *Spinacia oleracea* (Wilms 1980a), *Primula vulgaris* (Primulaceae) (Y. Heslop-Harrison, Heslop-Harrison and Shivanna 1981), *Prunus avium* (Cresti et al 1978b; Uwate et al 1982), *Glycine max* (Tilton et al 1984), *Raphanus raphanistrum* (Hill and Lord 1987), and *Pyrus serotina* (Japanese pear) (Nakanishi et al 1991) have also been discussed and illustrated. Among modifications of the basic ultrastructural profile, the lateral walls of cells of the transmitting tissue of *G. hirsutum* have a complicated structure. From inside to outside, the wall consists of a uniformly textured, pecto-hemicellulosic wall close to the plasma membrane, a thin layer rich in hemicellulose, a loose-textured wall consisting of rings of fibrous material of pectin, and a prominent middle lamella (Jensen and Fisher 1969). The wall of the stylar transmitting tissue cells of soybean has three conspicuous layers, and the distinctive wall anatomy seems to distinguish these cells from others in the style (Tilton et al 1984).

Some important general concepts about stylar secretion products (or intercellular substances, as they are called), their origin, and their probable chemical nature have emerged in recent years. As in the case of the stigmatic exudate, the origin of the intercellular substance and its mode of delivery outside the cell remain in dispute, and both ER and Golgi bodies have been implicated in this function. Thus, the presence of active dictyosomes with their vesicles that fuse with the plasma membrane, as described in the cells of the transmitting tissue of cotton, tends to suggest that dictyosomes participate in the transport of the secretion product (Jensen and Fisher 1969). Secretion of the intercellular substance in the transmitting tissue of the style of *Lycopersicon esculentum* calls for two sets of instructions. When the style is about 3 mm long, dictyosomes dominate the production of intercellular substances, which consist mostly of polysaccharides. During the later phase of growth of the style, a second set of instructions are read and proteins are released into the intercellular matrix; this is associated with the presence of numerous rough ER and polyribosomes in the cells (Cresti et al 1976). The resemblance of the secretion process in the cells of the transmitting tissue of *Malus communis* to the situation in *L. esculentum* extends to differential activity of the rough ER and dictyosomes that results in the secretion of proteins and polysaccharides, respectively (Cresti, Ciampolini and

Sansavini 1980). In *Pyrus serotina*, an intercellular substance of undefined composition is produced in the vesicles originating from the ER. However, the lipid fraction of the matrix has a different origin. Lipid droplets are closely associated with Golgi bodies and seem to be incorporated into vacuoles. Later, the vacuolar substance apparently glides through the cell wall to form part of the intercellular substance (Nakanishi et al 1991). A variation of this theme is described in *Trifolium pratense,* in which secretion of proteins and a protein–carbohydrate mixture is associated with paramural bodies and Golgi-derived vesicles, respectively (Y. Heslop-Harrison and Heslop-Harrison 1982a).

Sedgley and Clarke (1986) have provided ultrastructural and immunocytochemical evidence that indicates a role for multivesicular bodies in the secretion of intercellular substance in the transmitting tissue of *Nicotiana alata*. Using monoclonal antibody to terminal α-L-arabinosyl residues, a highly specific labeling of the multivesicular bodies was detected. These bodies were closely associated with the Golgi apparatus and ER, which were also labeled to some extent. Furthermore, during the development of the style, there was an increase in the area of the intercellular matrix in the transmitting tissue and an increased labeling of the matrix relative to multicellular bodies. From these observations, it appears that the multivesicular bodies originating from the Golgi apparatus and ER secrete arabinogalactan proteins and facilitate their transfer from the intracellular to the intercellular matrix. Immunological methods have been used to show that in the cells of the transmitting tissue of *Brugmansia suaveolens*, unesterified pectins and polysaccharides, which are major constituents of the intercellular substance, are synthesized by the endomembrane system (Vennigerholz 1992; Hudák, Walles and Vennigerholz 1993). However, in these studies, we cannot assume without further evidence that a cell organelle that stains positively for a compound is the source of that compound, which also accumulates elsewhere in the tissue.

Often, the intercellular substance has a higher electron density than the walls of cells of the transmitting tissue in the immediate vicinity. This is due to the amorphous nature of the substance, which is generally a subtle blend of a mucilaginous base with other compounds. The chemical identity of secretion products from various plants has been defined by histochemical tests that have revealed the presence of carbohydrates, proteins, lipids, polysaccharides, pectins, phenolic compounds, and tannins, as well as enzymes like esterases, acid phosphatase, peroxidase, and glucose-6-phosphatase (Knox 1984b). Electron microscopic autoradiography of pistils of *Petunia hybrida* labeled with *myo*-inositol-^{3}H has shown the label predominantly in the intercellular substance, thus affirming its carbohydrate nature (Kroh and van Bakel 1973). Where both stigmatic and stylar exudates of the same flower are analyzed, broad similarity exists in the spectra of compounds found. If differences exist, they relate to the presence or absence of certain protein bands or isozymes of critical enzymes (J. Heslop-Harrison and Heslop-Harrison 1982c). Major carbohydrates identified in the stylar extracts of several dicotyledons and monocotyledons are arabinogalactans associated with hydroxyproline-rich proteins (Anderson, Sandrin and Clarke 1984; Hoggart and Clarke 1984).

By careful use of cytochemical and immunofluorescence methods, Sedgley et al (1985) found that arabinogalactan proteins are present in the extracellular matrix of the transmitting tissue of the style of *Nicotiana alata*. However, unlike in the stigma, there is no corresponding increase in the concentration of the proteins during development of the style. Chemical analysis showed that cell walls of isolated transmitting tissue of *N. alata* possess the structure of primary cell walls of dicotyledons, containing cellulose, xyloglucans, and pectic polysaccharides, but the presence of an unusually high proportion of arabinogalactans seems to set them apart from typical primary cell walls (Du et al 1994; Gane et al 1994). As seen in crossed electrophoresis, the charge nature of the arabinogalactans in the style is distinctive, is different from that in the stigma, and changes with maturity of the style (Gell, Bacic and Clarke 1986). This technique has also revealed a high concentration of arabinogalactans in the style of *Lycopersicon peruvianum* (van Holst and Clarke 1986). The one presumed function of arabinogalactans that may be important to emphasize here is the nutrition of the pollen tubes during their growth in the style.

Another major constituent of the extracellular matrix of the transmitting tissue of *N. alata* is a basic, 120-kDa, hydroxyproline-rich glycoprotein. Style specificity of this protein was established by the absence of any immunologically cross-reactive material in other reproductive or vegetative parts of the plant. This property, as well as the conservation of the glycoprotein in the styles of other members of Solanaceae and the presence of the protein in the extracellular matrix of the transmitting tissue and in the cytoplasm and cell walls of pollen tubes, suggests important stylar functions, such as pollen tube nutrition, elongation, and guidance, or defense against pathogens (Lind et al 1994, 1996). Ubiquitin is also a protein identified in the stylar transmitting tissue of *N. alata* that is correlated with the tissue's function of undergoing programmed disintegration (Li et al 1995b). Unfortunately, as in the case of stigmatic secretion, little information is available on the quantitative changes in these proteins during the ontogeny of the style. Immunolocalization has shown that antibodies specific to the style of *N. tabacum* are not distributed throughout the length of the style but are confined to a limited region (Evans, Holaway and Malmberg 1988; Cañas and Malmberg 1992).

Styles of several species possess RNase activity. The list includes species with solid and hollow styles, as well as those that are self-incompatible and self-compatible. Generally, species with solid styles have higher concentrations of the enzyme than do species with a stylar canal (Schrauwen and Linskens 1972). Clearly, in situations where complex catalytic enzymes such as RNase are present in the cells, it is very difficult to assign them a nutritive role. Fortuitously, in some plants that exhibit the gametophytic type of self-incompatibility, there is mounting evidence to indicate that the catalytic activity of RNase might explain the arrest of pollen tube growth in an incompatibility reaction. However, no relationship between the presence of RNase and stylar growth of pollen tubes has been established in self-compatible plants. Ironically, RNase inhibits in vitro germination of pollen grains and growth of pollen tubes in self-compatible species containing the enzyme in the stigma or style (Roiz, Goren and Shoseyov 1995; Roiz and Shoseyov 1995). The inherent conflict between the presence of RNase in self-compatible plants and its inhibitory effect on the pollen can be reconciled if it is assumed that the enzyme may have an indirect role in male gametophyte selection on the stigma.

(ii) The Canal Cells

The layer of cells lining the canal of hollow styles displays many cytological features of the transmitting cells of solid styles and stands clearly apart from other cells of the style. In a few species investigated, the ultrastructural ontogeny of these cells follows well-defined and related patterns. The most striking feature of the canal cells of the style of *Lilium longiflorum* is an elaborate wall on the side facing the canal. The wall is smooth toward the outside, but the inside is built around a framework of labyrinthine projections very much as a transfer cell is. The ingrowths possess a gap between the plasmalemma and the wall, and this region has an abundance of paramural bodies of undetermined function (Rosen and Thomas 1970; Dashek, Thomas and Rosen 1971; Mike-Hirosige, Hoek and Nakamura 1987). As shown in *L. regale* (Figure 7.5), canal cells with varying degrees of wall modifications for solute transport merge basipetally with the ovarian transmitting tract and form a secretory epithelium on the placenta (Singh and Walles 1992, 1995). Typical transfer cells are found to occur along the stylar canal of *Medicago sativa* (Johnson, Wilcoxson and Frosheiser 1975). In other cases, the wall modification may be less elaborate, consisting of localized expansions as in *Ornithogalum caudatum* (Tilton and Horner 1980) and *Citrus limon* (Ciampolini et al 1981).

The primary characteristic of the canal cells, as with cells of the transmitting tissue, is the presence of an extracellular matrix of exudate. It is mucilaginous in nature, is secreted by the canal cells, and probably provides rigidity and elasticity to these cells. Eventually the mucilage flows into the stylar canal, where it accumulates. Since hollow styles are found only in a few species, not much information is available about the secretory activities of the canal cells. The sequence of events during the secretory phase in the style of *O. caudatum* is divided into three periods. The first is marked by an abundance of ribosomes and rough ER and by the production of lipid and protein components of the exudate. During the second phase, the population of mitochondria and plastids increases and the ER becomes smooth; a different type of lipid, one having a dense core ensheathed by a less dense material, is secreted at this time. The third and probably the most dramatic phase is characterized by a decrease in the number of mitochondria and plastids and a concomitant increase in the population of microbodies, polysomes, and dictyosomes; carbohydrates constitute the predominant secretory component of this period (Tilton and Horner 1980).

From the studies discussed thus far, the secretion product of canal cells emerges as a mixture of lipids, proteins, and carbohydrates. Certainly it is much more than that, and indeed, a knowledge of the chemical composition of this substance is important for an understanding of the way in which pollen tubes are nurtured during their growth through the style. As in the stigmas and stylar transmitting tissues of several angiosperms, it appears that a critical component of exudates of hollow styles is arabinogalactan protein (Gleeson

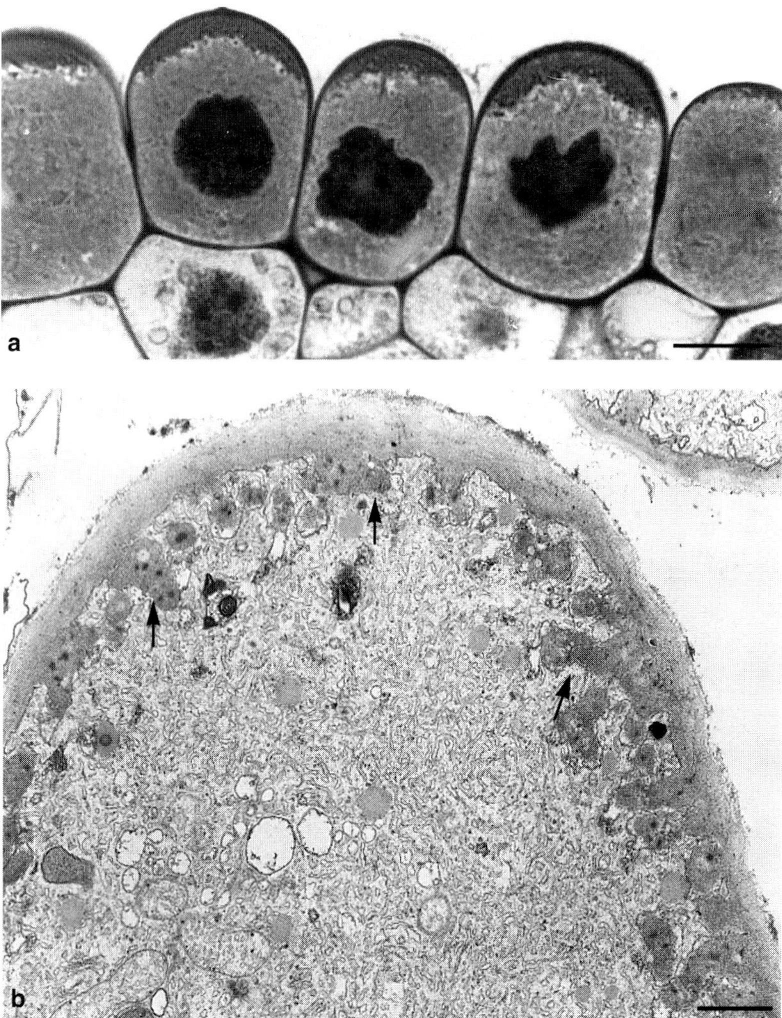

Figure 7.5. Section of the transmitting tissue of the style of *Lilium regale* cut at the transition region between the style and ovary. **(a)** Light microscopic image of epithelial cells of the stylar canal one day after anthesis. Scale bar = 10 μm. (From Singh and Walles 1992.) **(b)** Electron microscopic image of an epithelial cell two days after anthesis. Arrows point to the wall ingrowths. Scale bar = 1 μm. (From S. Singh and B. Walles 1995. Ultrastructural differentiation of the ovarian transmitting tissue in *Lilium regale*. *Ann. Bot.* 75:455–462, by permission of Academic Press Ltd., London. Both photographs supplied by Dr. S. Singh and Dr. B. Walles.)

and Clarke 1979, 1980b, c). In the style mucilage of *Gladiolus gandavensis*, the arabinogalactan accounts for 40% of the soluble stylar extract, which has a disproportionately high concentration of carbohydrates, mostly galactose and arabinose, compared to that of proteins, which have high contents of serine, glutamic acid, aspartic acid, glycine, and alanine (Gleeson and Clarke 1979).

The account of the style given here is justifiable largely because of prevalent ideas about the role of the stylar exudate in the nutrition of pollen tubes. In *L. longiflorum*, the secretory function is elicited by the canal cells only in response to pollination; this seems to suggest that secretion is not the terminal phase of a developmental pathway of the canal cells unless the flower is pollinated (Yamada 1965). In experiments with *L. longiflorum*, labeled *myo*-inositol and glucose supplied to detached styles were found to be utilized for the biosynthesis of pollen tube wall materials. It is believed that the label is transferred to the pollen tube by way of the stylar exudate (Kroh et al 1970; Labarca and

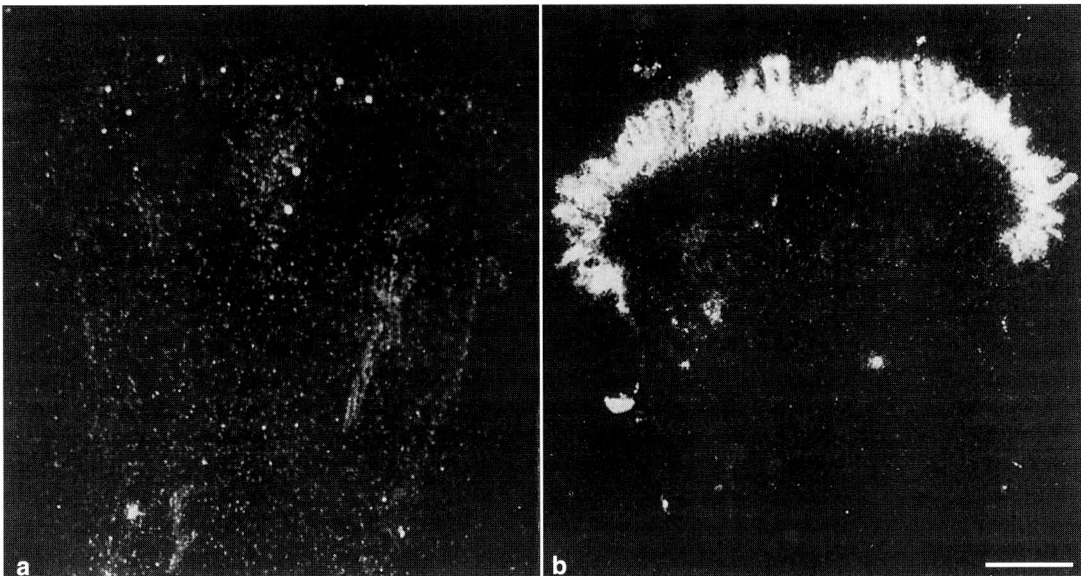

Figure 7.6. Localization of ^{35}S-labeled transcripts of a pistil-specific cDNA clone isolated from *Brassica napus* in the stigmatic papillae of the same plant. (**a**) Hybridization with the sense strand. (**b**) Hybridization with the antisense strand. Scale bar = 100 μm. (From Robert et al 1994b; photographs supplied by Dr. L. S. Robert.)

Loewus 1972). The correspondence between the presence of molecules in the stylar exudate and their utilization is shown by the incorporation of labeled substances by the style and by the appearance of the label in the pollen tube.

(iii) Genes Expressed in the Stigma and Style

The morphological differentiation of the stigma and style described earlier follows from a programmed, organ-specific control of gene expression. To further our understanding of the function of the stigma and style in pollination and fertilization, it is essential to identify the genes that regulate the development of these organs and the protein products of the genes. A. H. Atkinson et al (1993) screened a cDNA library made to poly(A)-RNA of the stigma of *Nicotiana alata* and isolated several highly abundant genes that were not associated with self-incompatibility. One of the cDNA clones was found to encode a protein with sequence homology to proteinase inhibitors from potato and tomato. In situ localization of the gene transcripts in the cells of both immature and mature stigmas of *N. alata* is consistent with the high levels of proteinase inhibitor activity in these tissues.

A stigma-specific gene (*STIG-1*), identified from a cDNA library of tobacco pistils, is expressed exclusively in the secretory zone of the stigma of transgenic tobacco. Transcripts of this gene are highly prevalent during the period of differentiation of the stigmatic papillae and then decrease during later periods. The deduced amino acid sequence shows that the gene encodes a 16-kDa, unidentified secretory protein (Goldman, Golberg and Mariani 1994). Stigma-specific expression, confined to the papillar cells and spanning the entire period of development of the flower, is characteristic of a gene isolated from *Brassica napus* (Figure 7.6). In this instance also, the identity of the secreted protein encoded by the gene is unknown (Robert et al 1994b). A gene encoding a lipid transfer protein isolated from *Arabidopsis* is expressed in the papillae of transgenic *Arabidopsis* stigma; this observation favors a role for the protein in the secretion or deposition of lipophilic substances in the stigmatic cells (Thoma et al 1994). Among genes that are not stigma specific, but whose promoters drive stigmatic expression, are *PAL-2* from bean (Liang et al 1989); 4-coumarate:coenzyme A ligase from tobacco and parsley (Hauffe et al 1991; Reinold, Hauffe and Douglas 1993); *ACP* and *HMG-2* from *Arabidopsis* (Baerson and Lamppa 1993; Baerson, Vander Heiden and Lamppa 1994; Enjuto et al 1995), all in transgenic tobacco; and *RTBV* in transgenic rice (Yin and Beachy 1995).

Attempts to identify additional biochemical components of the style have revealed two major protein groups: pathogenesis-related proteins and extensin-like proteins. Infection of plants by fungi, bacteria, and viruses causes the expression of pathogenesis-related proteins that function by degrading the main components of fungal and bacterial cell walls. The presence of such a protein in the extracellular matrix of the stylar transmitting tissue of tobacco was revealed by immunoblot using a specific antibody; subsequently, the protein was identified as a (1,3)-β–glucanase (Lotan, Ori and Fluhr 1989; Ori et al 1990). Further characterization of the protein has resulted in the isolation of corresponding cDNA clones and in their sequencing. The strong homology of the deduced amino acid sequence to acidic (1,3)-β-glucanase and the developmental regulation of expression of the protein and/or its mRNA in stylar transmitting tissue of wild-type and transgenic tobacco, beginning about eight days before anthesis, has led to the inference that the pathogenesis-related protein is a major stylar matrix factor concerned with some aspects of female reproduction (Ori et al 1990; Sessa and Fluhr 1995). Genes isolated from cDNA libraries made to poly(A)-RNA of tobacco flower buds and of pistils of *Petunia inflata* have been shown to encode another pathogenesis-related protein, thionin (Gu et al 1992; Karunanandaa, Singh and Kao 1994). In connection with specific cellular functions, it is noteworthy that gene transcripts are detected only in the cortex of the style, but not in the transmitting tissue or on the stigmatic surface of tobacco. This lends support for a role of the protein product of the gene in reactions unrelated to pollen tube growth or self-incompatibility.

Wemmer et al (1994) characterized an abundant protein from the pistil of *Solanum tuberosum* and identified it as an endochitinase (Figure 7.7). The pistil specificity of the gene for the protein was confirmed by Northern blot hybridization of RNA samples from pistil, leaf, and pollen grains with a cDNA probe. By immunochemical analysis, the protein has been localized in the secretory cells of the stigma and in the intercellular matrix of the stylar transmitting tissue (Li et al 1994b; Wemmer et al 1994). From its properties, the protein appears to provide both active and passive constraints to the ingress of pathogens or for pollen tube penetration. Based on their expression in the cortex of the style, other pistil-specific genes (*STS-14, STS-15*) isolated from *S. tuberosum* can also be considered to belong to the group encoding pathogenesis-related proteins (van Eldik et al 1995, 1996).

Identification of developmentally regulated genes in the style of tomato has been approached by differential screening of a pistil cDNA library with a seedling mRNA probe. One of the cDNA clones selected is expressed in the cells of the stylar transmitting tissue prior to fertilization (Gasser et al 1989). Two additional observations on the regulation of expression of this gene are of interest. One is that transcripts of the gene are present abundantly in a temporal fashion in the mature styles of tomato as well as of tobacco. Secondly, the promoter fragment of the gene is found to regulate the expression of a reporter gene in the style of transgenic tomato in a pattern similar to that observed with the cDNA probe in the wild-type style (Budelier, Smith and Gasser 1990). If the protein encoded by the gene is identified, it will be possible to determine whether it has any functional role in the style, perhaps in the differentiation of the transmitting tissue or in the growth of the pollen tube. The putative protein products encoded by other cDNA clones expressed predominantly or exclusively in the tomato pistil during early stages of its development include endo-β-1,4-glucanase, γ-thionin, leucine aminopeptidases, Fe_{2+}/ascorbate-dependent oxidases, and endochitinase (Milligan and Gasser 1995; Harikrishna et al 1996). Whether the transcripts of these genes or their protein products are expressed in all parts of the pistil or are confined to the style is far from clear. A large body of information has assigned roles connected with carbohydrate degradation and defense against pathogens to glucanases and thionins/endochitinases, respectively, whereas the precise functions of the other two proteins in plant tissues remain to be established. With these caveats, genes expressed in the tomato pistil may be considered to function in the development, maintenance, and defense of this floral organ, rather than in the primary events of pollination.

By a strange coincidence, genes identified in the styles of *Antirrhinum majus, Nicotiana alata,* and *N. tabacum* in different laboratories at about the same time have turned out to encode proline-hydroxyproline–rich proteins constituting common cell wall components. The protein products of genes isolated from *A. majus* (Baldwin, Coen and Dickinson 1992), *N. tabacum* (Goldman et al 1992), and *N. alata* (Chen, Cornish and Clarke 1992; Chen, Mau and Clarke 1993) are rich in proline and serine and contain several proline-rich repetitive amino acid motifs found in extensin. For the most part, transcripts of these genes begin to be expressed in the young flower buds but accumulate abundantly

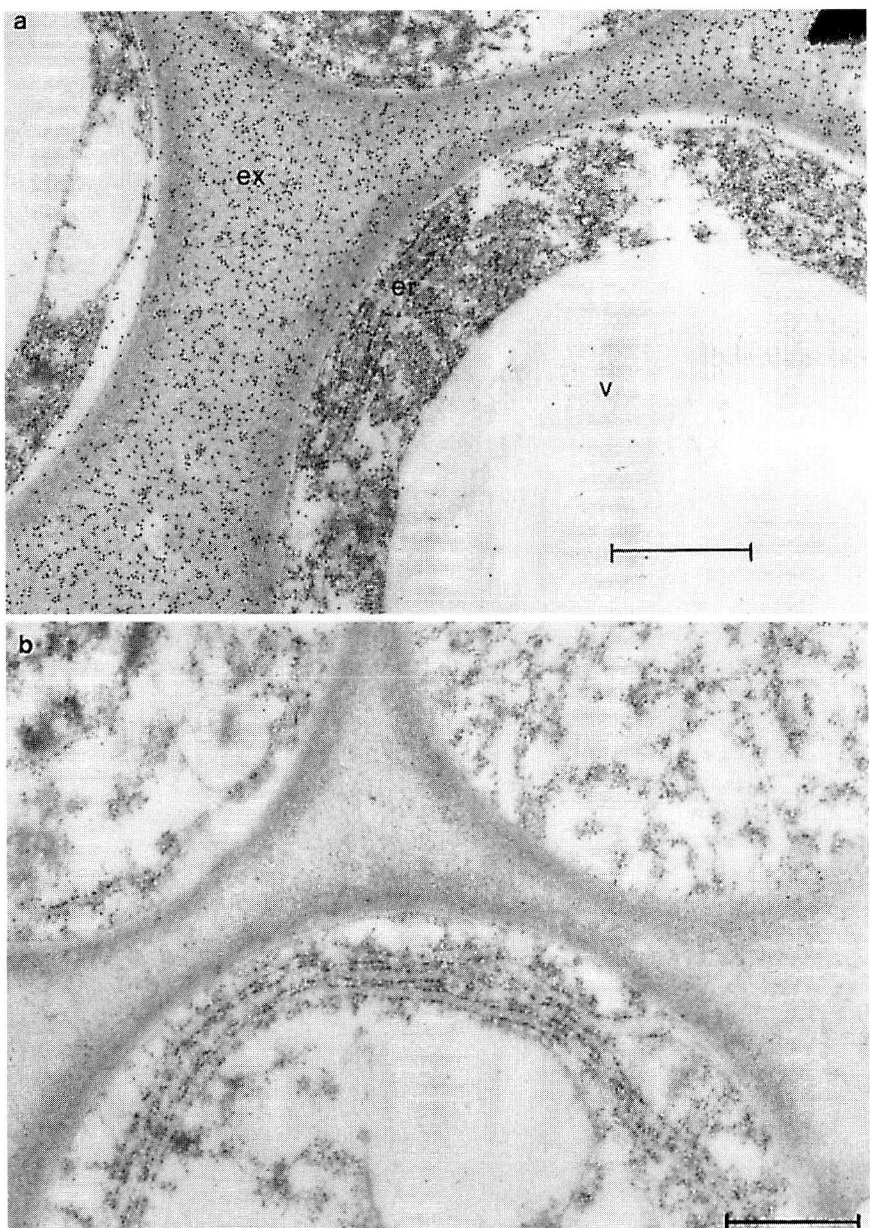

Figure 7.7. **(a)** Immunolocalization of an antiserum for a pistil-specific protein (chitinase) in the cells of the stylar canal of *Solanum tuberosum*. The protein is localized in the extracellular matrix (ex) between the cell walls. er, endoplasmic reticulum; v, vacuole. **(b)** A control section treated with preimmune serum. Scale bars = 500 μm. (From Wemmer et al 1994. *Planta* 194:264–273. © Springer-Verlag, Berlin. Photographs supplied by Dr. R. D. Thompson.)

in the mature buds. In respect to the spatial expression of the genes, their distribution is restricted to the transmitting tissues, and variations noted between the expression of cDNA probes from the three species are no greater than between different preparations of the material from the same plant. The style and/or stigma of *N. alata* also expresses transcripts corresponding to arabinogalactan proteins (Du et al 1994, 1996).

Based on the work done in another laboratory, two stylar transmitting tissue–specific proteins encoded by genes isolated from the style of *N.*

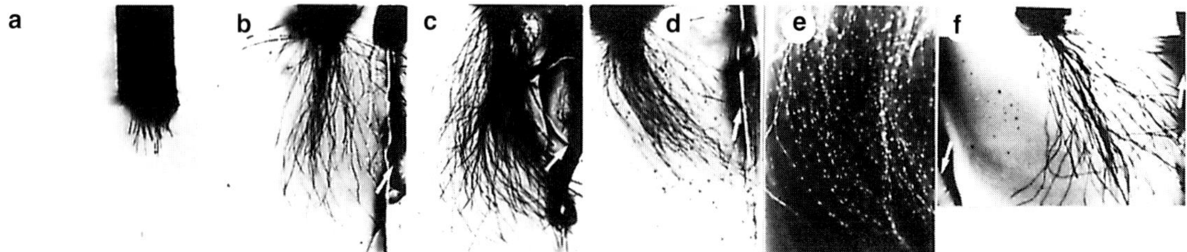

Figure 7.8. Semi–in vivo assay showing the attraction of pollen tubes of *Nicotiana tabacum* to purified transmitting tissue–specific proteins. In the assay, the pollinated pistil is cut close to the bottom of the style and cultured in a growth medium. Pollen tubes whose tips have not reached the base of the style will emerge from the cut end and grow into the medium. Test substances are applied in agarose plugs implanted a few millimeters to the left or right of the style. **(a)** Pollen tubes emerging from the cut style at 10 hours after culturing on a sugar-free medium. **(b)** Pollen tubes growing with sugar-free medium (arrow) implanted at 1.5 mm to the right of the style. **(c)** Pollen tubes growing with the test proteins at 0.2 μg/μl (arrow) implanted at 1.5 mm to the right of the style. **(d)** Pollen tubes growing with test proteins at 0.5 μg/μl (arrow) implanted at 3 mm to the right of the style. **(e)** Aniline blue–stained pollen tubes growing with test proteins at 0.2 μg/μl implanted at 2 mm to the right of the style. **(f)** Pollen tubes growing with test proteins at 0.2 μg/μl (arrow) implanted at 3 mm to the right of the style, and sugar-free medium (arrow) implanted at 3 mm to the left of the style. Scale bar = 1.6 mm. (From Cheung, Wang and Wu 1995; © Cell Press. Photographs supplied by Dr. A. Y. Cheung.)

tabacum have been found to possess much less homology to extensin. They are proline-rich, highly glycosylated proteins with molecular masses ranging between 50 and 100 kDa and belong to the arabinogalactan protein family. The localization of these proteins in the extracellular matrix of the stylar tissues, the temporal and spatial regulation of the proteins and of their corresponding mRNAs in the transmitting tissue of the style coincident with the growth of the pollen tubes, the pollination-stimulated increase in the mRNAs and proteins in the styles, and the restriction of the capacity to glycosylate the proteins to the transmitting tissue suggest that the proteins have an important function in the process of pollination and in the subsequent growth of the pollen tube. This view has been reinforced by other experiments, which show that the purified proteins stimulate the growth of pollen tubes in vitro and deflect their directional growth in a semi–in vivo assay (Figure 7.8). One of the most explicit functional manifestations of these proteins is seen in the inhibition of pollen tube growth in transgenic plants in which protein concentration is significantly reduced by antisense suppression (Cheung et al 1993, 1996; Wang, Wu and Cheung 1993; Cheung, Wang and Wu 1995). According to Wu et al (1993), members of a family of genes encoding a cysteine-rich, extensin-like protein (*CELP*), isolated from a tobacco floral cDNA library, are expressed in diverse cell types of the flower, including a thin layer of cells embedded between the transmitting tissue and the stylar cortex.

Stylar expression has been reported for genes encoding *HSP-70* from tomato (Duck, McCormick and Winter 1989), nonspecific lipid transfer protein (*GLTP-1*) isolated from ray florets of *Gerbera hybrida* (Asteraceae) (Kotilainen et al 1994), and a vitronectin-like adhesion protein from cultured tobacco cells (Zhu et al 1994); for promoters of a pea plastocyanin gene, the *Arabidopsis ACP* gene, and the alfalfa isoflavone reductase gene in transgenic tobacco (Baerson and Lamppa 1993; Pwee and Gray 1993; Oommen, Dixon and Paiva 1994); and for promoters of *Arabidopsis* low-molecular-mass *HSP-18.2* and anthranilate synthetase genes in transgenic *Arabidopsis* (Takahashi, Naito and Komeda 1992; Tsukaya et al 1993; Sessions and Zambryski 1995). Expression of the two HSP genes occurred under nonstressed conditions which suggests that their proteins may have a structural role in the style.

Extensins and related proteins are considered to have important functions in maintaining the structural integrity of the cells. For example, their presence in the extracellular matrix of the style might contribute to the strength and rigidity of the transmitting tissue. In turn, this might allow the unhindered growth of a large number of pollen tubes without inflicting localized deformation on the cells. Pathogenesis-related proteins might protect the style from environmental stresses or from attack by pathogens during the crucial periods of pollination and pollen tube growth. A general role might exist for arabinogalactans as a common glue to hold the cells of the transmitting tissue together. The compositional changes in the style, especially of those constituents that impart strength and rigidity to the cells and govern their defense-related needs, point to the development of functions that indirectly favor the normal growth of pollen tubes.

3. POLLEN–PISTIL INTERACTIONS

The structural features of the stigma and style have obvious implications in the biology of sexual reproduction. The major events that follow pollination are pollen recognition and the subsequent acceptance or rejection of the pollen grain by the stigma or of the pollen tube by the style. There are cellular mechanisms that operate to discriminate between the different types of pollen grains that dock on the stigmatic surface, germinate, and course through the style. These mechanisms ensure that only intraspecific pollinations are successful except when self-incompatibility genes act to prevent inbreeding. Intergeneric and interspecific crosses are frowned upon by the plant, which employs a different set of mechanism to avoid such crosses. As shown by nearly all of the published work in recent years, the emphasis in developmental and molecular studies of pollen-stigma interactions has been on the postpollination behavior of the self–pollen grains and pollen tubes and on the isolation and characterization of genes involved in self-incompatibility reactions. These aspects of interaction of the pollen grain with the stigma and style are considered in Chapters 9 and 10 and will not be taken up here.

The concern of this section is with pollen–stigma interactions during compatible pollinations. Following a compatible pollination, profound physiological changes occur in the pollen grain and in the stigmatic papilla and in signal transductions between these cell types in reciprocal acknowledgement of each other. The predominant changes reflect recognition between complementary molecules on the pollen grain and on the stigmatic papilla and the associated enzyme reactions. How these changes are translated into germination of the pollen grain will be explored now.

(i) Pollen Adhesion

Despite the fact that a variety of pollen grains are trapped on the stigmatic surface of a flower, the system favors the selection of only the compatible pollen. The changes that occur in the components of the pollen grain and of the stigmatic papilla after compatible pollination are now documented in a few cases, and the general model of pollen recognition is known in outline. In this context, the adhesion of all kinds of pollen grains to the stigmatic papillae is clearly of considerable interest and presents a challenge to explain the basis of one of the most important manifestations of recognition. Sufficient evidence has been adding up for some time now that one can assert that the idea of molecular recognition between pollen and stigma is not as unreasonable as was once thought, and several models are at hand to explain the phenomenon (Dumas and Gaude 1981). Lectins have been implicated as recognition molecules because of their wide distribution in the plant kingdom and their unique ability to bind saccharides and saccharide-containing proteins in a highly specific way. The possibility that differences in the physical binding of the pollen to accessible glycoprotein sites in the stigma may regulate pollen adhesion is supported by the discovery that con-A binds to the pellicle of the stigma of *Phalaris minor* (Y. Heslop-Harrison 1976). The surface determinants of the stigma of *Gladiolus gandavensis* contain a number of glycoprotein components and esterases, and the capacity to bind con-A is eliminated if these sites are occupied (Knox et al 1976; Clarke et al 1979). In *Brassica oleracea,* the initial contact of the pollen grain with the pellicle is through a superficial layer investing both the pollen and the exine coating (coating superficial layer). The pellicle appears to fuse with this layer as the first sign of adhesion (Elleman and Dickinson 1986). That some proteins responsible for pollen grain adhesion in *B. oleracea* are held in the pellicle is supported by the observation that treatment of the stigma with protease adversely affects adhesion of pollen grains. A further important general principle is that these proteins turn over rapidly, as evidenced by the full recovery by the stigma of its adhesive properties within a short time after protease treatment (Stead, Roberts and Dickinson 1980). The most likely component of the pollen grain to bind to the stigma in this manner is the tapetally derived pollen coating, tryphine. A comparison of the surface components with recognition potential present on the pollen and stigma of *G. gandavensis* has revealed the intriguing fact that both factors contain a complex mixture of proteins, glycoproteins, and glycolipids. Thus, the pollen surface molecules apparently complement the components of the stigma to produce an ideal adhesion (Clarke et al 1977b, 1979). These observations strongly point to the conclusion that interaction between molecules present on the pollen surface and the stigma papilla is the hallmark of adhesion of compatible pollen grains. Whether incompatible pollen grains adhere to the stigmatic papillae in a molecular sense or mechanically stick to the papillae has remained conjectural. Qualitative data collected on pollen grains landing on the stigmatic surface of *B. oleracea* after cross- and self-pollinations have revealed differences in adhesion due to physico-chemical reactions (Roggen 1972). From counts of pollen released from pollinated stigmas of *B. oleracea*, it appears that after an initial sluggish period of about two hours, during which time self-pollen grains bind less firmly than cross-pollen, the former adhere to an extent comparable to cross-pollen (Stead, Roberts and Dickinson 1980). These cases are exceptions rather than the rule.

Besides the stigma, other floral organs of *Arabidopsis*, such as the style, ovary, anther, petal, and sepal, are receptive to pollen grains and support their germination and pollen tube growth (Kandasamy, Nasrallah and Nasrallah 1994). In the *fiddlehead* (*fdh*) mutant of *Arabidopsis*, wild-type pollen grains germinate and grow on a floral part formed by the fusion of noncarpel organs of the shoot (Lolle and Cheung 1993). These observations suggest that components of the pollen recognition system are present in the shoot epidermis and in all floral organs during their ontogeny but are segregated to the surface cells of the stigma and other floral parts by the action of the *FDH* gene.

(ii) Pollen Hydration

Hydration by the uptake of water is the first change observed in compatible pollen grains after they make contact with the stigmatic surface; the best evidence establishing this as a recognition event is the fact that, in many plants, foreign pollen grains remain dry on the stigma and are not recognized. The recognition reaction apparently functions by regulating water traffic from the stigmatic cells to individual pollen grains; for example, when pollen grains are placed on the stigmatic papilla in a chain, but with only one pollen grain in contact with the surface of the papilla, the one grain is preferentially hydrated, but the others remain dry (Sarker, Elleman and Dickinson 1988; Hülskamp et al 1995). Apart from the sweeping changes in the cytoplasmic activities of the pollen grain, hydration plays an important role in restoring the structure of the plasma membrane of the vegetative cell. The exact sequence of changes that occur in the plasma membrane upon hydration is not known, but it appears that the partly dissociated membrane acquires normal osmotic properties during controlled hydration (Shivanna and Heslop-Harrison 1981). In most species, the pollen grain absorbs water rapidly from the stigmatic papilla, whereas in others it obtains water from the exudate on the stigma surface. The degree of hydration depends upon the state of the stigma, whether it is wet or dry, and the state of dryness in which the pollen grains are held prior to pollination (Gilissen 1977; Stead, Roberts and Dickinson 1980; Barnabás and Fridvalszky 1984; Zuberi and Dickinson 1985). We also need to know about the substances from the exudate taken up by the pollen grains. Autoradiographic studies have shown that immature pistils of *Primula officinalis* and *Ruscus aculeatus* cultured on an agar medium containing $^{45}Ca^{2+}$ transport the ion through the stigmatic papillae to germinating pollen grains; the implication of this observation is that any Ca^{2+} present in the stigmatic exudate might be similarly taken up by pollen grains germinating in vivo (Bednarska 1991). Little information is available on the dynamics of pollen hydration on wet stigmas as compared to that on dry stigmas. One reason for this limited knowledge is that the matrix of the exudate provides the pollen grains with an instant source of water at controlled osmotic pressure for hydration without recourse to other physical processes.

Hydration of pollen grains alighting on dry stigmas is apparently a complex process because dry stigmas offer a less hospitable surface than wet stigmas. A few cases will be considered in which events occurring on the stigmatic papillae and on the pollen grains have been followed after compatible pollinations of dry stigmas. The grasses are unusual among the species so far investigated in showing extremely rapid initial exchanges between pollen and stigma. A study of pollen hydration in *Secale cereale* has been particularly illuminating in showing by time-lapse microphotography three distinct phases in the hydrodynamics of the pollen grain immediately preceding germination. The first phase is the passage of water from the stigmatic papilla into the pollen grain. Although this influx greatly increases the water content primarily in the vegetative cell, some subtle changes are also initiated on the plasma membrane of the cell until it is re-formed. Continued uptake of water by the pollen leads to the second phase, which begins before the plasma membrane becomes an effective osmotic barrier. During this phase, rather than expanding to the point of bursting, the pollen grain begins to lose water and solutes by exudation from the germination pore and through the micropores

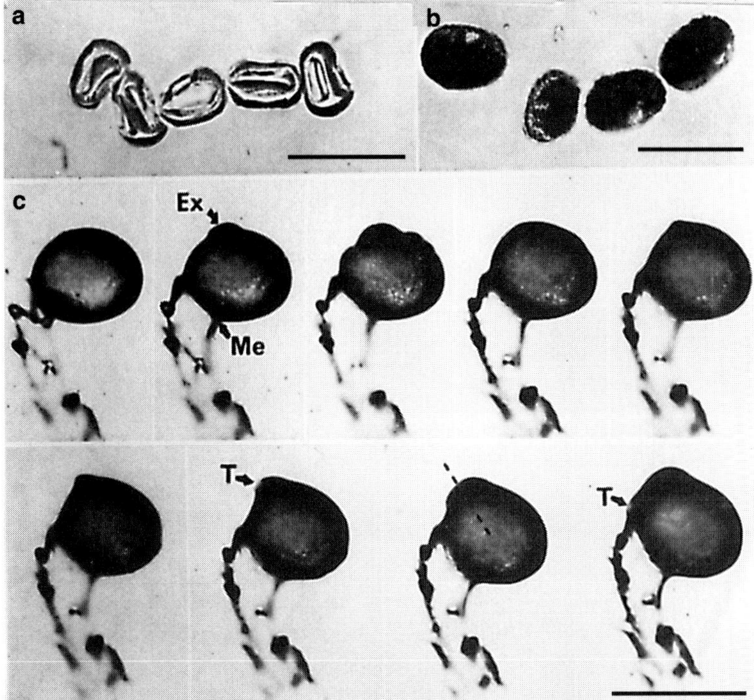

Figure 7.9. Germination of pollen grains of *Secale cereale* on the stigmatic trichome. **(a)** Pollen grains after natural drying. **(b)** Dried pollen grains 1 minute after hydration in 0.5 M sucrose. **(c)** Nine frames from a time-lapse sequence showing hydration, exudation, resorption, and germination of a single pollen grain on the trichome. The first frame is 20 seconds after application; time intervals are 30 seconds thereafter. A meniscus (Me) is formed in 20–25 seconds and is followed by the first apertural exudation (Ex). T, pollen tube tip. The dashed line in frame 8 is the plane of sectioning used in another study. Scale bars = 100 μm. (From J. Heslop-Harrison 1979a. Aspects of the structure, cytochemistry and germination of the pollen of rye (*Secale cereale* L.). *Ann. Bot.* 44: Suppl. 1, 1–47, by permission of Academic Press Ltd., London.)

of the exine. Proteins and other mobile constituents held in the pollen wall are also probably leached out from the surface of the pollen grain onto the stigmatic surface. The use of fluorescent-labeled pollen wall proteins seems to confirm that during pollination, exine-held proteins might indeed penetrate the cells of the stigma. The third phase, during which the plasma membrane is completely reconstituted, is the most critical, and further traffic of water molecules during this period is determined by the rules governing an osmotically active plant cell. With the completion of this sequence, the pollen grain has built up considerable hydrostatic pressure which will be relieved by the emergence of a pollen tube (Figure 7.9). By the most generous estimates, these events resulting in germination do not take more than two minutes from the time the pollen makes contact with the stigmatic papilla (J. Heslop-Harrison 1979a; Vithanage and Heslop-Harrison 1979). Although details of pollen–stigma interaction are scanty for other species of grasses studied, they also seem to follow the same general principles as in rye (J. Heslop-Harrison 1979b, c; J. Heslop-Harrison and Heslop-Harrison 1981). Special attention has been focused on the interpolated period of efflux from the pollen grain, which includes intine-held proteins in grasses like *Phalaris tuberosa* (Knox and Heslop-Harrison 1971c). A rapid release of pollen-held proteins such as antigen-E accounts for the presence of a fluid that coats the pollen grains of *Ambrosia trifida* and *Cosmos bipinnatus* after they land on their respective stigmatic papillae (Knox 1973).

It has proved difficult to observe pollination-related changes in pollen grains through the use of conventional fixatives without inducing artifactual hydration. Anhydrous fixation techniques have revealed that several structural changes in the exine coating accompany hydration of the pollen grain of *Brassica oleracea*. Along with water, some materials emitted from the stigma and loaded into the pollen grain account for these changes. The first event that is consequent upon the pollen grains making contact with the stigma is the movement of the exine coating to form an appresoria-like "foot" on the surface of the papilla. The control of this traffic is poorly understood, but there is no doubt that it is linked in some way to the occurrence of recognition reactions at the interface between the pollen grain and the stigma. Progressive hydration elicits structural changes in the coating, which becomes electron opaque due to densely packed membranous assemblies (Dickinson and Elleman 1985; Elleman and Dickinson 1986, 1990). Transfer of exine-held materials of the pollen onto the stigma surface has also been noted following pollination in other plants with dry stigmas (Elleman, Franklin-Tong and Dickinson 1992).

Although these examples serve to illustrate the physical and structural aspects of pollen hydration, they do not address the molecular nature of pollen–stigma communication. Contact with a compatible stigma triggers the synthesis of a particular set of proteins in the pollen grains of *Brassica napus*. Because some of these proteins are phosphorylated, a conceptual pattern of signaling mechanism during compatible pollinations that involves protein phosphorylation may be in prospect (Hiscock, Doughty and Dickinson 1995). To investigate the signaling mechanism during pollen–stigma interaction from a different angle, Preuss et al (1993) isolated mutants of *Arabidopsis* that affect pollen–stigma communication. Although wild-type pollen grains hydrate rapidly on the stigma and germinate, those of the mutant *defective in pollen–pistil interactions* (*pop-1*) do not absorb water from the stigma and consequently fail to germinate. A note of interest in linking the loss of germinability to the failure of pollen hydration is the observation that *pop-1* pollen grains are rescued in vivo on the stigma under humid growth conditions or in vitro by culture in a simple nutrient medium. A connection between pollen hydration and long-chain lipid molecules was indicated by the absence of wax on the stem of *pop-1* plants. Chemical analysis showed that wax deficiency on the mutant stem is correlated with deficiency in long-chain lipids on the mutant pollen grains, and electron microscopy confirmed that mutant pollen grains lack lipoidic tryphine on their exine coating. A model for pollen hydration in *Arabidopsis* suggested by these results is that during compatible pollination, the pollen grain signals the stigma surface and establishes communication with the help of lipid and tryphine molecules; this initial molecular interaction evidently sets the stage for the cascade of events beginning with the uptake of water. Analysis of other mutants that affect pollen hydration on the stigma of *Arabidopsis* has also focused on the lipid product of the pollen wall as a major player in the binding of the pollen grain to the stigmatic papilla (Aarts et al 1995; Hülskamp et al 1995). Work with mutant pollen grains of *Petunia* sp. deficient in flavonols has revealed a signaling role for kaempferol in pollen germination on the stigma. This is based on the finding that mutant pollen grains can be rescued by germinating on wild-type stigma or in micromolar quantities of kaempferol isolated from the stigma (Mo, Nagel and Taylor 1992). A different scenario is noted in transgenic plants of *P. hybrida* characterized by the absence of flavonols in the pollen grains. Plants are rendered self-sterile by this manipulation and, although pollen grains germinate normally in vitro, their growth is stymied after a short period (Ylstra et al 1994).

(iii) Stigmatic Response to Pollination

During pollen–stigma interactions, changes occurring on the stigmatic papillae are the very epitome of cell recognition events on the female side. Biological changes associated with these interactions are reflected in physical changes, such as voltage variations in the style specific for a compatible pollination (Spanjers 1978). As seen earlier, the morphogenesis of the papilla is the major event in the development of the stigma. Cell ablation techniques have been used to determine the extent to which full development of the papilla is necessary for normal interaction with compatible pollen grains. Wild-type pollen grains of *Arabidopsis* germinate and form tubes on the surface of stigmatic papillae that are stunted in growth by the introduction of *DT-A* toxic gene fusion. This shows that the pollen recognition mechanism functions even in flowers with stigmatic cells that, in contrast to cells in the normal stigma, are not biosynthetically fully active (Thorsness et al 1993). However, as shown in transgenic tobacco engineered by the introduction of the *BARNASE* gene, if the toxic gene wipes out the stigmatic papillae completely, pollen germination is reduced and the pollen tubes fail to penetrate the transmitting tissue of the style. The block to pollen germination and pollen tube growth is overcome by bathing the ablated surface in an exudate from the wild-type stigma (Goldman, Goldberg and Mariani 1994). Evidently, in ways unknown, the exudate provides conditions for pollen–stigma interactions even in the absence of the stigmatic cells.

The responses of the stigma to pollination have been examined by light and electron microscopy in representatives of plants with active wet and dry types of stigmas. In the wet type of stigma, a signal from pollination leads to a net increase in the amount of exudate, which in some cases may even be followed by the autolysis of papillar cells (Sedgley and Scholefield 1980; Kenrick and Knox 1981a; Sedgley 1982; Sedgley and Blesing 1982). It is clear that dry stigmas of some plants also secrete exudate after pollination (Owens and Kimmins 1981; Elleman, Franklin-Tong and Dickinson 1992). The chemical composition of postpollination exudate is as heterogeneous as that of the prefertilization product; in *Crocus chrysanthus* and *Pennisetum typhoides*, a dramatic increase in esterase activity on the surface of the stigmatic cuticle is noted after pollination (Y. Heslop-Harrison 1977; Y. Heslop-Harrison and Reger 1988). As to the fate of the stigmatic papillae after pollination, degeneration of the cytoplasm of cells immediately adjacent to the pollen grain and pollen tube, or necrosis of cells through which pollen tubes pass, are common occurrences in the stigmas of both wet and dry types studied (Jensen and Fisher 1969; Sedgley 1979, 1981b; J. Heslop-Harrison and Heslop-Harrison 1981; Elleman, Franklin-Tong and Dickinson 1992).

The fact that secretory products of the stigma contain enzymes calls for a molecular mechanism to explain stigmatic secretion in response to pollination. Such a mechanism could be provided for by the opening of channels through the cuticle and the passage of recognition molecules from the pollen surface to trigger enzyme secretion. We do not know the role played by pollination-induced additional secretion on the wet stigma or new secretion on the dry stigma; one possibility emphasized before in the general context of the role of exudate is that it allows hydration and germination of pollen grains.

(iv) Pollen Germination and Pollen Tube Growth

Following hydration on the stigma, some characteristic changes directly concerned with germination take place in the cytoplasmic domain of the pollen grain. Within a few minutes after hydration, the pollen cytoplasm appears quite different from the dehydrated pollen cytoplasm. In *Brassica oleracea*, the dry pollen grain is characterized by the presence of many spherical fibrillar bodies at the periphery of the protoplast, whereas the protoplast of the hydrated pollen is conspicuously stratified and contains a peripheral layer of membranous cisternae (Elleman and Dickinson 1986).

Growth of pollen tubes and their entry into the stigma after compatible pollination have been described in several plants. However, we face serious problems if our aim is to build a unified picture of the patterns observed. Pollen tube growth in *Crocus chrysanthus* fits in well with a model in which the tip of the tube enters the stigmatic papilla after lysis of the cuticle by cutinase. Thereafter, the pollen tube continues to cruise toward the ovary by growing beneath the cuticle close to the underlying pectocellulosic wall. From the fact that pollen tubes do not penetrate the cuticle when the proteins of the stigma exudate are degraded enzymatically, lysis of the cuticle by pollen-held cutinase without cooperation from the stigmatic exudate appears unlikely. Rather, it has been suggested that the pollen grain contributes a precursor of the enzyme, which is activated by a factor present on the stigmatic surface (J. Heslop-Harrison and Heslop-Harrison 1975, 1981; Y. Heslop-Harrison 1977). Once pollen germination has begun on the stigma of barley, penetration of the stigmatic hair by the pollen tube is sine qua non for its subsequent growth, as it is not unusual to see penetration of the hair by short tubes, random growth of tubes in the air before penetrating the hair, or pollen tubes bypassing one hair before penetrating another (Cass and Peteya 1979). Germination of pollen grains on the stigma of *Gladiolus gandavensis* is followed by pollen tube penetration of the cuticle and continued growth of the tube over the underlying pectocellulosic layer. Growth of the pollen tube into the style and through the stylar canal toward the ovary occurs in a matrix of mucilage that is secreted by the stigma and serves as a guide (Clarke et al 1977a). In *Lychnis alba* (Caryophyllaceae) (Crang 1966), in members of Asteraceae such as *Ambrosia tenuifolia* and *Helianthus annuus* (Knox 1973; Vithanage and Knox 1977), and in members of Brassicaceae such as *Arabidopsis, Brassica oleracea, Diplotaxis tenuifolia, Raphanus raphanistrum,* and *R. sativus* (Kroh and Munting 1967; Dickinson and Lewis 1973a; Hill and Lord 1987; Elleman et al 1988; Elleman, Franklin-Tong and Dickinson 1992), pollen tubes penetrate the cuticle and travel between the two wall layers of the stigma cells to the base of the papilla, where they enter the middle lamella of the transmitting tissue. Unlike in *C. chrysanthus*, in the other examples cited, a role for pollen-held cutinase in the penetration of the pollen tube appears likely, but has not been established. Pollen–stigma interaction in *Vicia faba* is somewhat unusual, since pollen tubes enter the vicinity of the stigmatic papillae through torn gaps in the thick layer of cuticle, thus circumventing the need for enzymatic disruption of cutin (Lord and Heslop-Harrison 1984). Cutinase action also appears unlikely in *Gossypium hirsutum* (Jensen and Fisher 1969) and *Oenothera organensis* (Dickinson and Lawson 1975). In the former, the pollen tube grows downward by riding on the surface of the papillae and penetrates neither the cuticle nor the cell wall until it reaches the transmitting tissue of the style. In *O. organensis*, the pollen tube grows through the middle lamella of adjacent cells of the stigmatic papillae and travels down the surface of the cells of the subjacent layer (alveolar tissue). Of course, the concept of pollen tube growth at this critical juncture in the absence of interaction with the cells of the stigma is a gross oversimplification in terms of recognition reactions.

In the stigmas of grasses, the trichomes are programmed for two purposes. First, they serve as the principal pollen-capturing surface and provide conditions for the germination of pollen grains. The pollen tube penetrates the trichome by digesting the cuticle. The second function of the trichome is to determine the pathway of entry and initial orientation of the pollen tube in the stigma. This is achieved by contact of the pollen tube tip with the trichome, with the base of the trichome serving as a guide for the entry of the pollen tube into the intercellular space in the central axis of the stigma (Y. Heslop-Harrison, Reger and Heslop-Harrison 1984b; Y. Heslop-Harrison and Reger 1988). The second function of the trichomes is also demonstrated by observations made on the fate of pollen tubes on a trichome-less mutant of *Pennisetum typhoides*. Here, without the cue given by the trichome structure, pollen tubes grow in the central axis of the stigma indifferently, rather than directionally toward the ovary (Y. Heslop-Harrison and Reger 1988). The results show that the development of trichomes is not an essential requirement for pollen germination. The directionality of the pollen tube in *Arabidopsis* is largely determined by the age of the style because pollen tubes growing through immature styles invade the epidermis and stylar cortex as much as they do the transmitting tissue (Kandasamy, Nasrallah and Nasrallah 1994).

In many plants, pollination syndromes extend into the style even before pollen tubes enter the transmitting tissue. Deterioration of cells of the transmitting tissue throughout the path of the pollen tube is a common postpollination phenomenon. At the molecular level, a reduction of the poly(A)-tail of transmitting tissue–specific mRNAs has been found to presage the deterioration of the pollinated transmitting tissue of tobacco. Interestingly, an mRNA that encodes glycoproteins essential for pollen tube growth in the style is the least affected (Wang, Wu and Cheung 1996). This degradation of the transmitting tissue as a programmed series of changes to facilitate passage of the pollen tube poses a novel situation for which no comparable models exist in the stigma. An important role of wall-degrading enzymes becomes manifest during the growth of pollen tubes through the transmitting tissue of the style. It has long been known that even though hundreds of pollen tubes grow through the style, there is no net increase in the diameter of the style, indicating that the cells of the transmitting tissue are actually chewed up by the pollen tubes. This simple idea is only one example of a process that has several variations. In most species examined, the pollen tube grows through the intercellular substance of the transmitting tissue (Crang 1966; van der Pluijm and Linskens 1966; Kroh and Munting 1967; Vithanage and Knox 1977; Hopping and Jerram 1979; Sedgley 1979; Hill and Lord 1987; O'Brien 1994). In cotton, the pollen tube bypasses the middle lamella of the cells of the transmitting tissue and travels through the peripheral layer of the cell wall (Jensen and Fisher 1969). In grasses, in which there is no defined transmitting tissue in the stigma, the route taken by the pollen tubes is through the intercellular spaces of a tissue of elongated cells between the epidermis and the vascular strands (Y. Heslop-Harrison and Reger 1988). Wall ingrowths in the form of tubular invaginations of the plasma membrane, known as plasmatubules, are found in abundance in the pollen tube of *Nicotiana sylvestris* during its growth in the style (Figure 7.10). The nature of the ingrowths and their appearance in the periplasmic space between the tube wall and the plasma membrane seem to reflect and accommodate the needs of the growing pollen tube for active solute transport (Kandasamy, Kappler and Kristen 1988). There is a further interesting observation that certain proteins of the pollen tube appear in the style of *Gasteria verrucosa*; this suggests a possible interaction between proteins excreted by the growing pollen tubes and those of the style (Willemse and Vletter 1995). A traffic in the reverse direction, identified in *Petunia hybrida* and having implications in the signaling phenomenon during pollen tube growth, is the uptake of Ca^{2+} from cells of the stylar transmitting

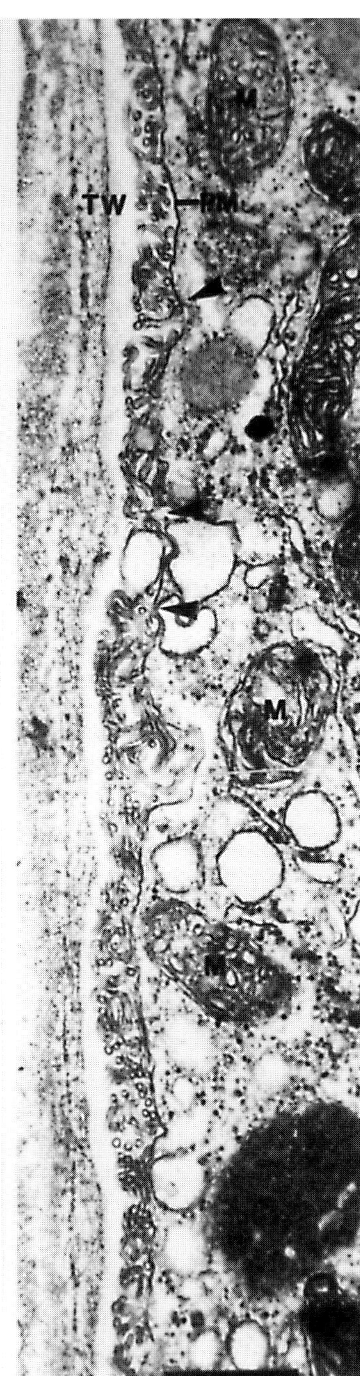

Figure 7.10. Section of a pollen tube of *Nicotiana sylvestris* grown in vivo for 24 hours, showing the twisted plasmatubules between the pollen tube wall (TW) and the plasma membrane (PM). Apparent continuities between plasma membrane and plasmatubules are indicated by arrowheads. M, mitochondria. Scale bar = 0.5 μm. (From Kandasamy, Kappler and Kristen 1988. *Planta* 173:35–41. © Springer-Verlag, Berlin.)

tissue and its concentration in the tip of the pollen tube (Bednarska and Butowt 1995b).

Even though the path of pollen tubes in the various species may appear seemingly different, in view of the pectin-rich nature of the cells they encounter along the way, we can expect that they arrive at their destination by using similar mechanisms. In a few examples studied, unusual dictyosome activity has been noted at the tips of pollen tubes growing in the stylar tissues, suggesting that dictyosome vesicles or products derived from them play a part in the formation of the pollen tube (van der Pluijm and Linskens 1966; Kroh 1967a; Jensen and Fisher 1970; Dickinson and Lawson 1975). The gynoecial pathway for pollen tubes of *Rhododendron fortunei* follows a series of exudate-filled channels from the stigma, through the style, all the way to the vicinity of the ovule (Palser, Rouse and Williams 1992). In *Monotropsis odorata* (Pyrolaceae), the exudate-coated stigma is connected with the exudate-filled stylar canal by discrete strands of transmitting tissue (Olson 1994). These modifications of the stylar canal ensure the growth of pollen tubes bathed in an unbroken milieu of nutrient medium leading from the receptive surface of the stigma.

This section will conclude with some comments about the changes in stylar metabolism following pollination. This information should provide a basis for an understanding of the effects of pollination on the physiology of the style and the interrelationships between pollen tubes and stylar tissues as both pass through the terminal stages of their differentiated state. Numerous observations made in orchid flowers have implicated both auxin and ethylene as signaling factors in the development of the ovary and ovule following pollination. The role of these hormones finds ready explanation in the hypothesis that auxin present in the pollen stimulates ethylene formation. In support of this view, it has been found that in the postpollination development of flowers of *Phalaenopsis* sp., inhibitors of ethylene biosynthesis inhibit some early pollination-induced morphological changes. Moreover, pollination, as well as the application of auxin or ethylene, permits the accumulation of transcripts of 1-aminocyclopropane-1-carboxylic acid (ACC) synthase and ACC oxidase in the gynoecium to similar levels. A model proposed envisages that pollination provides the signal that leads to the accumulation of ACC, which is converted to ethylene by ACC oxidase (O'Neill et al 1993; Zhang and O'Neill 1993). Other plants in which pollination is accompanied by a substantial increase in the ethylene contents of the style are *Petunia inflata* (Singh, Evensen and Kao 1992) and

Dianthus caryophyllus (carnation; Caryphyllaceae) (Larsen et al 1995). In tobacco, an increase in IAA content of the style is correlated with the growth of pollen tubes, suggesting that pollination might play a role in the synthesis of the hormone (Lund 1956). In the style of *Hemerocallis fulva* (Liliaceae), concentrations of wall-softening enzymes like cellulase and pectinase begin to decrease at anthesis irrespective of whether the flower is pollinated or not. This suggests that wall-softening processes in the style are independent of pollination and pollen germination, and it reinforces a possible wall-softening role for enzymes of pollen tube origin (Konar and Stanley 1969). If gene activation for the synthesis of specific wall-degrading enzymes is initiated in the style after pollination, this would be reflected in the synthesis or accumulation of RNA and proteins. However, although analytical studies have unearthed no significant differences in the total RNA contents or in the individual RNA species between pollinated and unpollinated styles, changes such as an increase in the polysome content of cells of the transmitting tissue close to the site of pollen tube growth have occasionally been noted (Godfrey and Linskens 1968; Tupý and Rangaswamy 1973; Herrero and Dickinson 1979). It seems that rather than passively growing through the style, the pollen tube influences the metabolism of the style in a significant way; comparative studies of pollinated and unpollinated styles should identify the signaling mechanisms that operate during pollen tube growth in the style. It is worth noting that the pollination signal might be propagated all the way down to the ovary and ovule and might influence their physiology in advance of the arriving pollen tube (Chandra Sekhar and Heij, 1995).

4. GENERAL COMMENTS

The stigma and style are the products of very complex developmental processes and have only a short-term implication for sexual reproduction. The evidence presented in this chapter strongly indicates that numerous factors, which vary in different plants, are involved in the proper functioning of the stigma and style and in the fulfillment of their respective roles. Stigma maturation includes anatomical remodeling of the cells and acquisition of biosynthetic capabilities for the production of substances for pollen recognition and germination. Since the stigma is the first female part of the flower that the pollen grain encounters, identification of the specific proteins of the stigma that function in pollen recognition would greatly increase our knowledge of the initial steps of double fertilization. As seen in Chapter 3, some information already exists about pollen wall proteins with recognition potential.

Apart from the primary role of the stigma in pollen recognition, its functions coalesce with those of the style in providing the cellular environment to foster the growth of pollen tubes. Formation of different organs with similar functions is somewhat rare in plants. A wide variety of substances, ranging from simple sugars and amino acids to complex molecules such as arabinogalactans and glycolipids, are found in the matrix of the stigma and style. This is suggestive of the functioning of an elaborate network of metabolic channels arising from gene activation. The identification and characterization of genes encoding proteins of the stigma and style will make a strong case for these organs as a nutrient support and as a directional guide for pollen tube growth.

The pollen–stigma interactions discussed in this chapter have been developed from the perspective of events following a compatible pollination. Changes observed in the pollen grain, stigmatic cells, and to some extent in the cells of the style following a compatible pollination pose a striking contrast to changes occurring in these cells following an incompatible encounter. Although it now appears that some sort of signal transduction is normal, if not universal, during pollen–stigma interaction, the signaling substances, as far as they have been identified, appear varied in different plants. The result of successful signal transduction is the germination of the pollen grains and the growth of the pollen tubes. As will be seen in the next chapter, in vitro conditions provide a more favorable setting for investigating the structure, physiology, and biochemistry of pollen germination and pollen tube growth than in vivo studies on the stigma and style.

Chapter 8

In Vitro Pollen Germination and Pollen Tube Growth

1. Pollen Germination
   (i) Pollen Hydration
   (ii) Pollen Activation
   (iii) Pollen Tube Emergence
2. Biochemistry of Pollen Germination
   (i) Respiratory Changes
   (ii) Evidence for Presynthesized mRNA
   (iii) RNA Metabolism during Pollen Germination and Pollen Tube Growth
   iv) Protein Metabolism
   (v) Genes Expressed during Pollen Germination and Pollen Tube Growth
3. Structure of the Pollen Tube
   (i) Dictyosomes, Vesicles, and Their Derivatives
   (ii) Pollen Tube Wall
   (iii) Pollen Tube Cytoskeleton
   (iv) Calcium Gradient in the Pollen Tube
4. Growth of the Pollen Tube
   (i) Nuclear Migration
   (ii) Thoughts on the Mechanism of Tip Growth
5. General Comments

Following pollination, the pollen grain absorbs water from the stigmatic exudate and germinates to produce a pollen tube. In the life cycle of flowering plants, the pollen tube serves as a transient structural link between the end of the male gametophytic phase and the beginning of the sporophytic phase. As is well known, the pollen tube elongates in a seemingly endless extension of itself as it travels through the style to the ovary, seeks out the ovule, grows into the embryo sac, and discharges the baggage of sperm. The proportional increase in wall area that occurs during elongation of the pollen tube is by tip growth. Much of the work on the germination of pollen grains on the stigma and growth of pollen tubes in the style was considered in Chapter 7; however, this work does not tell much about the requirements for pollen germination and pollen tube growth and about the structure, growth physiology, and metabolism of pollen tubes. Indeed, with the insight gained from the knowledge of the complexity and diversity of the components of the stigmatic exudate in general, one can only wonder about the specific molecules that promote pollen germination and pollen tube growth. Therefore, it suffices to remark here that the emerging model of pollen germination on the stigma and of pollen tube growth in the style hardly does justice to the molecular mechanisms of germination and the dynamic processes of pollen tube growth.

Because germination of pollen grains in vitro under controlled conditions is now possible, a great deal of effort has been devoted to elucidating the requirements for pollen germination and pollen tube growth in a wide variety of plants. One might naively assume that once the recognition reactions are completed on the stigmatic surface, metabolic precursors are synthesized in the pollen cytoplasm and wall materials are packaged and assembled into the pollen tube poised to emerge from the germinating pollen grain. However, as shown by in vitro studies, pollen germination and pollen tube growth are not neatly separated like this, but are integrated and interdependent processes.

The objective of this chapter is to reflect on the various facets of the germination of pollen grains and the growth of pollen tubes based on information gained from in vitro studies. In particular, the structural changes in the pollen grain and pollen tube, gene activity during germination, intracellular gradients in the germinating pollen grain, and tip growth of the pollen tube will be discussed on the basis of evidence from ultrastructural, physiological, and molecular studies. Because of the sheer volume of literature that has accumulated on each of these and other aspects of pollen germination and pollen tube growth to be described here, review articles will be heavily relied on for complete coverage and for access to older papers. A sampling of these reviews includes those on the viability, storage, and germination of pollen grains (Johri and Vasil 1961; Linskens 1964), the physiology of pollen tube growth (J. Heslop-Harrison 1987; Steer and Steer 1989; Derksen et al 1995), and the biochemistry and molecular biology of germination and tube growth (Mascarenhas 1975, 1993). For those interested in trying their own hand at the game of pollen germination for teaching or for research, the laboratory manual by Shivanna and Rangaswamy (1992) is recommended.

1. POLLEN GERMINATION

The starting point for a discussion of germination is the dehydrated pollen grain sown in a growth medium adjusted to a favorable pH and osmolarity under appropriate conditions of temperature and humidity. An ideal medium for pollen germination includes water in which are dissolved a carbohydrate and micromolar quantities of boron or Ca^{2+} or both. The most widely used carbohydrate is the disaccharide sucrose, and rarely has any other carbohydrate been as effective in promoting germination of pollen grains of a large number of species. Undoubtedly, the primary function of sugar is to provide a carbon energy source that can initiate the metabolic processes that trigger germination and support pollen tube growth. A secondary function, that of an osmoticum, has evolved for sucrose and other carbohydrates in pollen germination, because a high osmotic environment of the medium prevents the bursting and collapse of pollen grains immersed in a hypotonic medium. Among the roles suggested for boron in pollen germination and pollen tube growth are that it facilitates the uptake of sugar from the medium and that it is involved in the biosynthesis of cell wall precursors. In recent years, Ca^{2+} has come to center stage as one of the most important cations involved in many key metabolic reactions, especially signal transduction, in plants and animals. Although there are no firm ideas about the mechanism of Ca^{2+} action in pollen germination and pollen tube growth, it is believed that the effect is exercised through a large number of different cellular processes (Johri and Vasil 1961; Steer and Steer 1989).

The literature on pollen physiology is replete with studies in which a variety of substances, including various types of sugars, amino acids, nucleotides, plant hormones, and assorted chemicals, have been empirically added to the culture medium to promote germination of pollen grains and growth of pollen tubes. A number of analogues of nucleic acid bases, but 2-thiouracil in particular, have aided germination and tube growth in cultured pollen grains of *Nicotiana alata*; incorporation of labeled amino acids into proteins confirms the expectation that growth stimulation of the pollen tubes by 2-thiouracil is associated with increased protein synthesis (Tupý, Stanley and Linskens 1965; Tupý 1966). Polyamines (Linskens, Kochuyt and So 1968; Rajam 1989) and flavonols (Ylstra et al 1992) also stimulate germination and tube growth of pollen grains of certain plants. In a few instances, germinating pollen grains have been shown to possess active transport systems for utilization of exogenous substrates; these include systems for sugar uptake in lily pollen (Deshusses, Gumber and Loewus 1981), uridine uptake (Süss and Tupý 1982), and amino acid uptake (Čapková et al 1983) in *N. tabacum* pollen.

More than a century of observations with the light microscope, supplemented by other techniques including electron microscopy have led to a widely accepted stage-by-stage analysis of the process of pollen germination. This was developed primarily to rationalize observations made on pollen germination under in vitro conditions, although some of the changes observed in vitro may find parallels under in vivo conditions. In the context of alterations in external morphology, internal biochemical changes, and cell differentiation, pollen germination is considered to involve three stages, namely, hydration, activation, and pollen tube emergence; there are subsets of processes to be considered during each of these stages. Hydration and activation are interdependent, continue throughout the germination time course, and may even be initiated simultaneously; pollen tube emergence is the culmination of the hydration and activation processes.

(i) Pollen Hydration

The fact that pollen grains need to attain a certain degree of dehydration before they germinate is one of the important generalizations resulting from early classical physiological studies on the viability and germination of pollen grains. The consequences for the physiology and metabolism of pollen grains of this need are several, because they ensure that germination will not occur prematurely within the anther. For example, bicellular lily pollen grains harvested one or two days before anthesis give greatly improved germination after they have been dried to a moisture content of about 15%, whereas pollen grains from anthers earlier than two days before anthesis germinate poorly or not at all, before or after they are dried (Tanaka and Ito 1981a; Lin and Dickinson 1984). In contrast, tricellular pollen grains of grasses generally have a higher water content than the bicellular pollen of other groups, and grass pollen grains collected from freshly opened anthers germinate readily without any further drying (Lin and Dickinson 1984). It seems that premature germination of grass pollen is impaired by factors other than their water content.

Absorption of water by pollen grains is regulated by their walls. Among the most remarkable cases of pollen wall adaptations for osmoregulation during germination are those described in *Corylus avellana* (hazel; Betulaceae) (Y. Heslop-Harrison, Heslop-Harrison and Heslop-Harrison 1986) and *Eucalyptus rhodantha* (eucalypt; Myrtaceae) (J. Heslop-Harrison and Heslop-Harrison 1985). Volume changes of the pollen of hazel due to water uptake are attributed to the activity of the oncus. This thickened part of the intine along with the outer layer of sporopollenin forms a tight seal over the germination aperture; its unsealing is essential for germination. The key event is that during the early stage of hydration of the pollen grain, the oncus dissolves, emitting a cloud of gelatinous pectin and lifting the layer of sporopollenin as if it were a cap. This apparently facilitates the outgrowth of the pollen tube from the inner cellulosic layer of the intine. In eucalypt, the principal control over hydration of the pollen grain is assigned to a refractive layer of the oncus, which has no counterpart in hazel. This layer undergoes slow dissolution upon hydration of the pollen, and a result is the slow opening of the aperture to facilitate outgrowth of the pollen tube. It has been claimed that the behavior of the special layer of the oncus protects the pollen grain from the effects of short-term osmotic stress and guards against the destruction of the vegetative cell in a hypotonic medium.

Because absorption of water is a prerequisite for germination, changes that occur during pollen hydration have come under close scrutiny. This interest stems from the renewal of a long-standing view that apart from a simple increase in volume of the pollen grain, hydration initiates some critical changes that stir things up quite effectively and with amazing speed (Figure 8.1). As the regulation of water uptake must ultimately focus on the state of the plasma membrane of the cell, the changes in membrane architecture during pollen hydration will be examined here; other changes are considered under the section on pollen activation.

There are varying degrees of imperfections in the plasma membrane of the dry pollen grain that render the membrane ineffective as an osmotic barrier. Consequently, the ability of the plasma membrane to regain its unit membrane structure in a controlled fashion early during hydration has repeatedly been stressed as an essential requirement for successful germination. Based on a diagnostic fluorochromatic reaction, Shivanna and Heslop-Harrison (1981) concluded that in the dehydrated pollen grains of eight species examined, the plasma membrane of the vegetative cell is in a largely disorganized state and that upon hydration it re-forms the normal bilayer organization. More graphic evidence of the state of the membrane is provided by an electron microscopic examination of pollen grains of *Secale cereale*, in which the absence of a continuous membrane around the dehydrated protoplast is considered as the distinguishing feature. Additionally, the boundary of the protoplast is associated with electron-opaque bodies that stain for lipids (J. Heslop-Harrison 1979a). The best image of the plasma membrane of dry pollen grains of *Brassica oleracea* or of pollen grains at early stages of hydration is that of a discontinuous layer associated with infoldings of small vesiclelike inclusions into the cytoplasm. With the progress of hydration, the membrane becomes continuous and assumes the contour of a physiologically active bilayer (Elleman and Dickinson 1986; Elleman, Willson and Dickinson 1987). Studies using freeze-substitution show that multilamellate membrane profiles, vesicles, and densely osmiophilic bodies are in close proximity to the plasma membrane of the dry pollen grain of *Pyrus communis*; these organelles disappear upon hydration, and the plasma membrane assumes a clear bilayer organization throughout the pollen surface (Tiwari, Polito and Webster 1990). In light of these results, the common observation that the conditioning of pollen grains in a humid atmosphere improves their germination may reflect the fact that humidity leads to the reorganization of the membrane as an effective osmotic system (Hoekstra and Bruinsma 1975a; Gilissen 1977; Shivanna, Heslop-Harrison and Heslop-

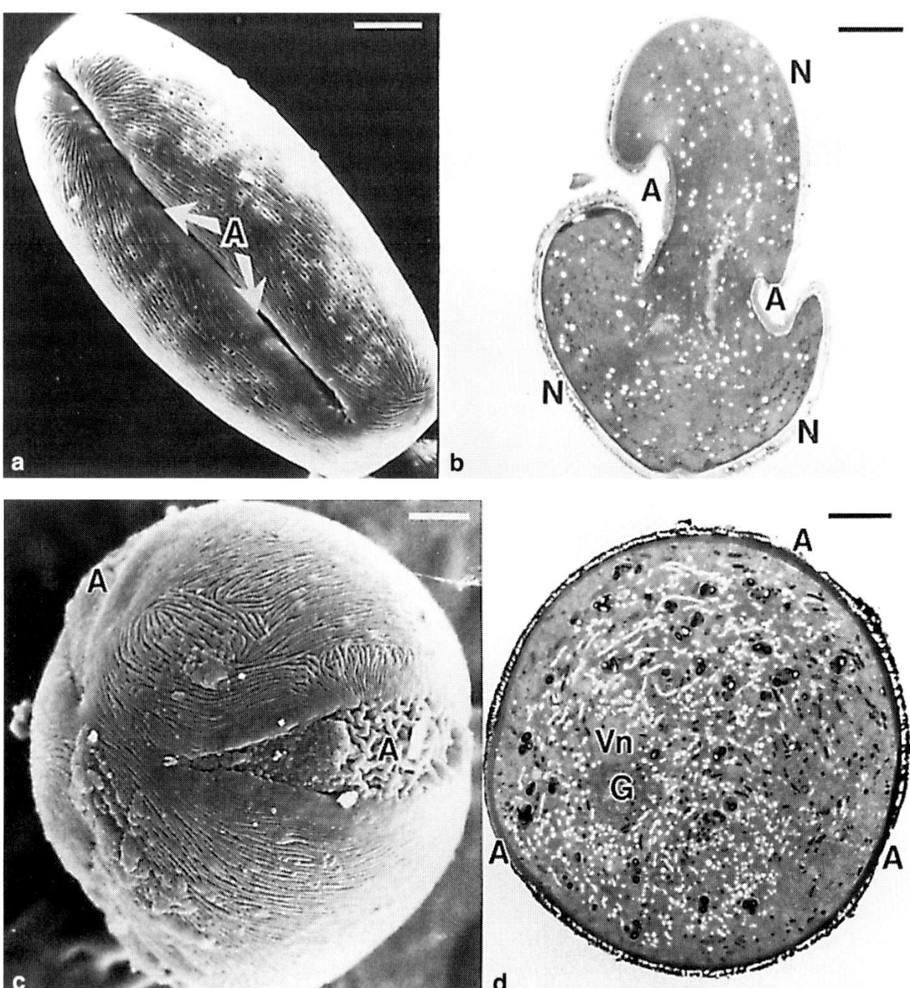

Figure 8.1. Dry and hydrated pollen grains of *Pyrus communis.* **(a)** Scanning electron micrograph of a dry pollen showing one of the three apertures (arrows). **(b)** An obliquely transverse section of a dry pollen showing two apertures. **(c)** Scanning electron micrograph of a hydrated pollen. **(d)** Transverse section of a hydrated pollen. A, apertural area; G, generative cell; N, nonapertural area; Vn, nucleus of the vegetative cell. Scale bars = 4 μm (a, b) and 5 μm (c, d). (From Tiwari, Polito and Webster 1990; photographs supplied by Dr. V. S. Polito.)

Harrison 1983). Because freeze-fracture electron microscopic images have shown that membranes in the dehydrated pollen grains of certain plants are stabilized with typical bilayer organization similar to that seen in the hydrated cells, there is some question about whether dehydration critically affects the membrane architecture (Platt-Aloia et al 1986; Kerhoas, Gay and Dumas 1987). Analysis of the proton-decoupled ^{31}P-nuclear magnetic resonance spectrum of pollen grains of *Typha latifolia* (Typhaceae) of low moisture content has indicated that membrane lipids are arranged in a bilayer (Priestley and de Kruijff 1982); this suggests that the plasma membrane of the dry pollen is capable of functioning as a selective permeability barrier. Continued improvements in techniques to visualize membranes of dry cells are needed before the question about the effects of dehydration on membrane architecture can be settled satisfactorily.

(ii) Pollen Activation

For purposes of discussion, pollen activation can be considered to occur during the lag period between hydration and pollen tube emergence. The activation phase of pollen germination has

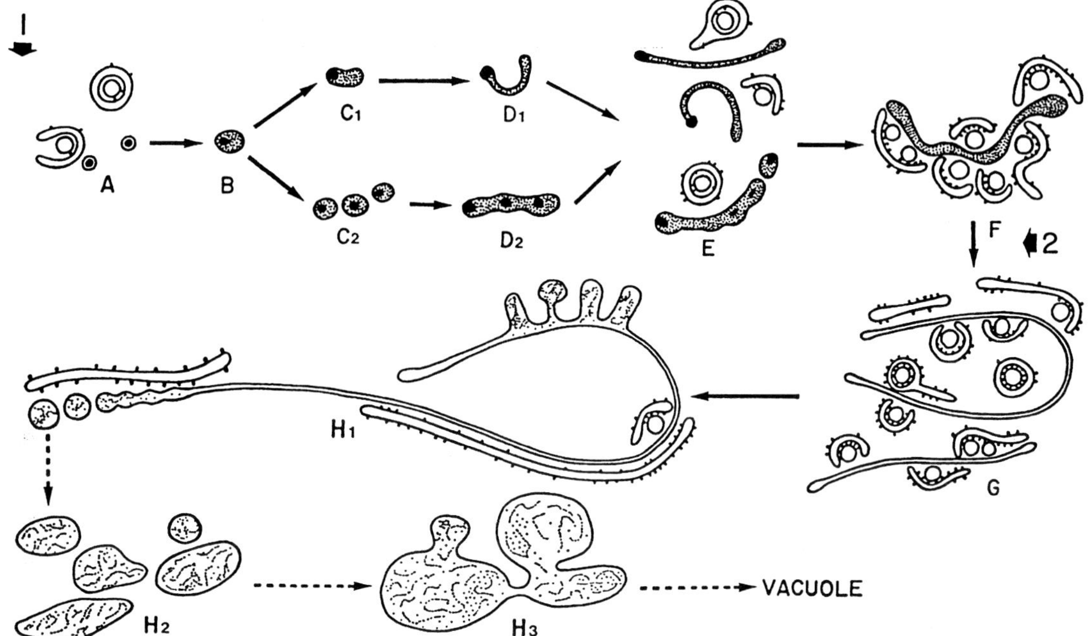

Figure 8.2. Diagrammatic representation of the formation of vesicles and vacuoles in relation to ER-associated lipid granules during activation of pollen grains of *Tradescantia reflexa*. Arrow 1, sowing of pollen grains; arrow 2, germination of pollen grains. (A) ER-associated lipid granules and vesicles containing electron-dense substances. (B–D) Enlargement of vesicles. (E, F) ER-associated lipid granules attached to electron-dense vesicles. (G) Bottle-shaped vesicle. (H) Formation of vacuoles. (From Noguchi 1990.)

been studied from a number of different angles. For many years, the respiratory metabolism of pollen grains during transition from the dormant to the active state was of prime concern to many students of pollen biology. During the last 30 years, a large part of research on the activation phase of pollen germination has been devoted to an understanding of the ultrastructural cytology, the changes in the cytoskeleton, and RNA and protein synthetic activities.

Ultrastructural Changes. The ultrastructure of normal, dehydrated pollen grains provides a standard of reference for assessing the unfolding sequence of changes that occurs during activation. The unique configuration of spherical ER pockets with lipid droplets and small vesicles described in the pollen grain of *Gossypium hirsutum* (Chapter 3) might well be regarded as an example of an organelle complex that responds rapidly to hydration. During germination, the ER pockets open and their load of lipid droplets and small vesicles loosens and mixes with plastids and mitochondria of the rest of the pollen cytoplasm. Simultaneously, the ER becomes encrusted with a large number of ribosomes and the cytoplasm is packed with mitochondria, lipid droplets, vesicles, and additional ER (Jensen, Fisher and Ashton 1968). A striking example of lipid degradation associated with germination is seen in pollen grains of *Tradescantia reflexa* (Figure 8.2). Early during germination, the ER-associated lipid granules become attached to electron-dense, thin vesicles and undergo degradation in this condition. Following the disintegration of lipids, the thin vesicles are reactivated and pinch off a generation of new vesicles that fuse together to form vacuoles in the emerging pollen tube (Noguchi 1990). The behavior of the ER–lipid complex in *G. hirsutum* and *T. reflexa* suggests the interesting possibility that the complex represents raw materials to provide energy for germination.

Intracellular rearrangements of ER have figured prominently during the activation of pollen grains of other plants. In *Lycopersicon peruvianum* pollen grains that contain stacks of rough ER, the first observable change upon hydration is the dissociation of the ER cisternae from the stacks. Other aspects of activation are polysome formation and increased dictyosome activity (Cresti et al 1977; Cresti, Ciampolini and Sarfatti 1980). Somewhat similar changes are associated with the activation of pollen grains of *T. virginiana* (Clarke and Steer 1983), *Nicotiana alata* (Cresti and Keijzer 1985; Cresti et al 1985), *Aloe ciliaris* (Ciampolini, Moscatelli and Cresti 1988), and *Agapanthus umbellatus* (Malhó and Pais 1992). Impressive documentation has been provided to show that in pollen grains of lily, ultrastructural rearrangements during activation are mainly concerned with the endomembrane system, such as dispersal of the ER, appearance of dictyosomes, decrease in the number of cisternae in the dictyosome, and appearance of secretory vesicles (Southworth and Dickinson 1981). Considering that hydration may impinge on the function of the pollen nuclei, changes observed during hydration and activation of pollen grains of *N. tabacum*, such as the expansion of the nuclei of the vegetative and generative cells and the reduction in the frequency of pores on the nuclear membrane of the vegetative cell, are significant (Wagner et al 1990).

Electron microscopic and biochemical studies have sought to determine the changes in mitochondrial structure and function during pollen activation. The impact of these studies has shifted the focus of discussion of mitochondrial properties for the most part to the function of the differences between bicellular and tricellular pollen grains. Bicellular pollen grains of *Typha latifolia*, for example, have structurally simple mitochondria in which numerous cristae appear during activation, whereas tricellular pollen grains of *Aster tripolium* (Asteraceae) have highly organized mitochondria that adapt to the activation phase without further development (Hoekstra and Bruinsma 1978). Compared to mitochondria isolated from tricellular pollen grains, those from bicellular pollen are slow to attain peak respiratory activity (Hoekstra 1979). These findings also reflect a difference in hydration metabolism between pollen grains of the primitive bicellular and advanced tricellular types.

The Pollen Cytoskeleton. Information concerning changes in the cytoskeletal elements of the pollen grain during germination is scanty. An early line of indirect evidence for the presence of microfilaments in pollen grains and for their possible role in pollen activation and germination came from the observation that cytochalasin, an inhibitor of actin polymerization, delayed germination of pollen grains of *Tradescantia paludosa* (Mascarenhas and Lafountain 1972). Two electron microscopic investigations of some importance in shaping our present concepts about the role of actin in pollen activation deserve mention. Cresti et al (1986a) described the occurrence of numerous crystalline fibrillar bodies in the vegetative cell of mature and activated pollen grains of *Nicotiana tabacum*. These bodies appeared to be bundles of fibrils measuring about 4–7 nm. It was also found that during activation prior to germination, the fibrils decreased in number and departed gradually from the aggregate to a dispersed state. The nature of the fibrillar materials and the reasons for their disaggregation remain unexplained, although their homology to a storage form of actin has been considered. A similar orientation of fibrillar elements, also identified as actin, was described in disrupted protoplasts derived from mature hydrated pollen grains of *Helleborus foetidus* (J. Heslop-Harrison, Heslop-Harrison and Heslop-Harrison 1986). At the same time, other works provided excellent demonstrations of actin, identified by actin-specific phalloidin binding in the pollen grains of several species of angiosperms (J. Heslop-Harrison et al 1986; Pierson 1988; J. Heslop-Harrison and Heslop-Harrison 1989b, 1992a, b; Tanaka and Wakabayashi 1992). Especially in hydrated but ungerminated monoporate pollen grains of *Endymion non-scriptus*, fusiform phalloidin-binding bodies are found scattered throughout the vegetative cell, soon to be replaced by an entangled web of filaments converging toward the aperture of pollen grains poised to germinate (J. Heslop-Harrison et al 1986). In the pollen grains of *Triticum aestivum*, actin cytoskeletal elements are already disposed toward the germination site at the time of dispersal (J. Heslop-Harrison and Heslop-Harrison 1992b). Although actin elements dot the plasma membrane, the generative cell, and the vegetative cell cytoplasm and its nucleus in the pollen grain of *Hyacinthus orientalis*, they all merge during the activation phase and concentrate at the potential germination site (J. Heslop-Harrison and Heslop-Harrison 1992a). Actin distribution during wall formation and germination of pollen protoplasts of *Lilium longiflorum* is basically similar to that described in intact pollen grains of other species (Figure 8.3). A selective accumulation of actin fibrils toward a preferred site of tube emergence in pollen with more than one such site has been dramatically demonstrated in pollen grains of *Pyrus communis* (Tiwari and Polito 1988a), *Narcissus pseudonarcissus* (Amaryllidaceae) (Y. Heslop-Harrison and Heslop-Harrison 1992), and *Amaryllis vittata* (Cai, Dong and Sodmergen 1995). In *P. communis*, in which the pollen grain is triaperturate, an initial polarization of actin fibrils toward each of

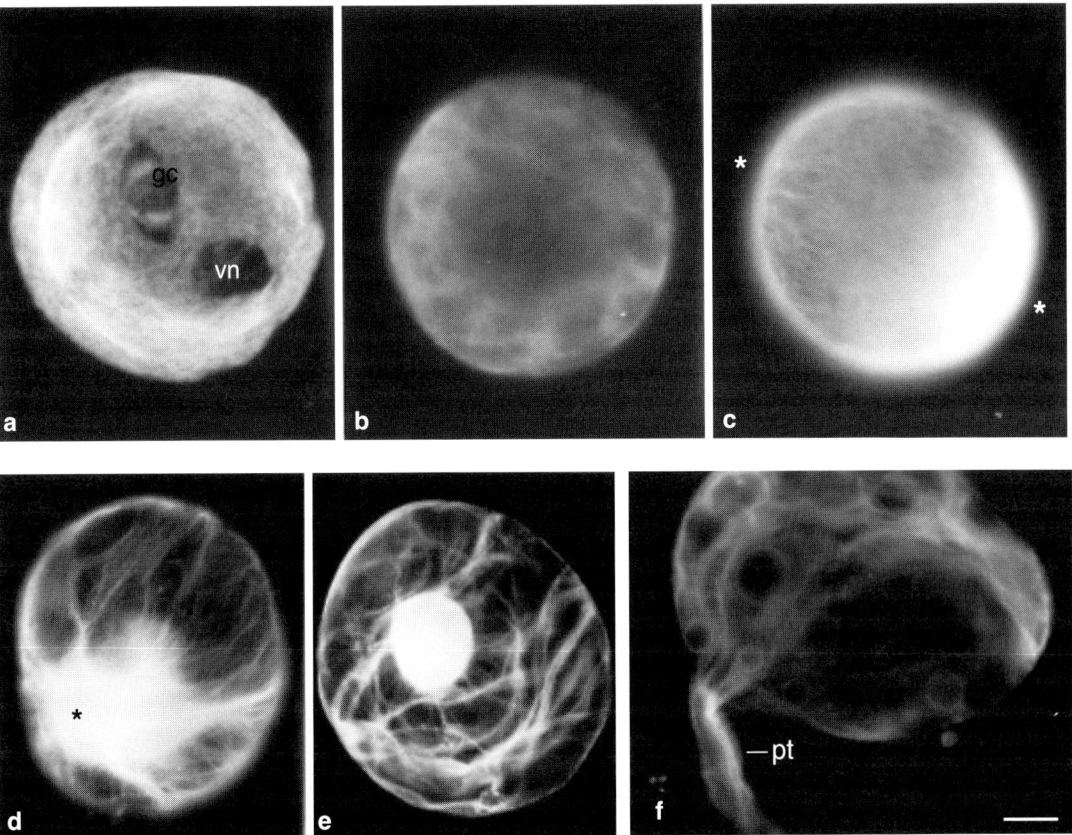

Figure 8.3. Immunofluorescence micrographs showing the organization of actin filaments in cultured pollen protoplasts of *Lilium longiflorum*. **(a)** A freshly isolated pollen protoplast showing the network of fine actin filaments throughout the cytoplasm of the vegetative cell; the generative cell (gc) and the nucleus of the vegetative cell (vn) do not have actin filaments. **(b)** A three-day culture showing the organization of cortical actin filaments. **(c)** A five-day culture showing the extension of actin filaments to opposite foci (asterisks). **(d)** A six-day culture showing actin filaments radiating from one focus (asterisk). **(e)** An eight-day culture showing the alignment of actin filaments transverse to the long axis of the cell. **(f)** Organization of actin filaments in an eight-day culture of germinated pollen; pt, pollen tube. Scale bar = 50 μm. (From Tanaka and Wakabayashi 1992. *Planta* 186:473–482. © Springer-Verlag, Berlin. Photographs supplied by Dr. I. Tanaka.)

the three apertures is changed to a final pattern in which the fibrils are concentrated around the preferred site of tube emergence (Figure 8.4). This behavior finds its counterpart in pollen grains of *N. pseudonarcissus*, in which the fibrillar system is polarized initially toward the two ends of the colpus. Again, in the majority of pollen grains, one site takes precedence as the fibrils converge principally to the potential germination aperture. Directly or indirectly, actin filaments must exert their influence on the migration of organelles to the vicinity of the germination pore. This could be accomplished by an actomyosin-based motility system; support for this view has come from the immunolocalization of myosin-coated organelles drifting toward the germination pore of hydrated pollen grains of *Nicotiana tabacum* (Tirlapur et al 1995).

A role for conformational changes of actin molecules in germination has been demonstrated in pollen grains of *P. communis* exposed to cytochalasin. An early study showed that actin, initially seen in pollen grains in the form of circular profiles, passes through an intermediate step as granules and is gradually replaced by fibrillar arrays that assume the form of an interapertural system of microfilaments traversing the three apertures of the pollen before they concentrate beneath the single aperture of tube emergence (Tiwari and Polito 1988a). Addition of cytochalasin was found to prevent the germination of pollen grains in which

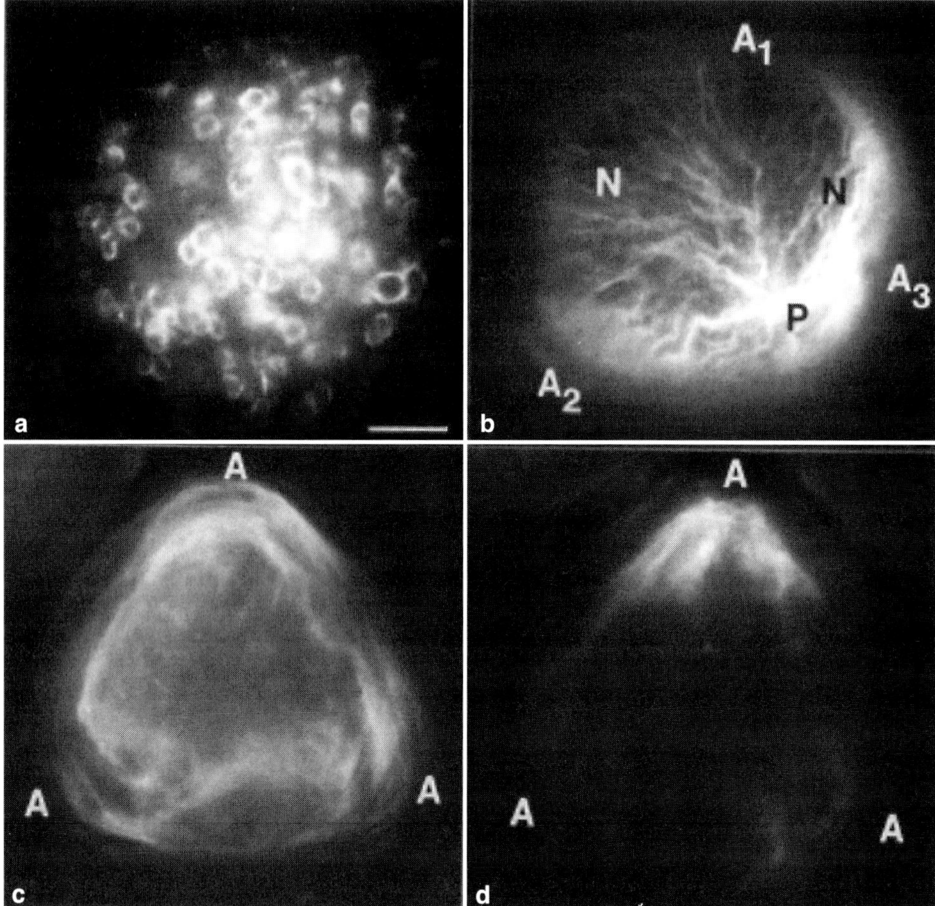

Figure 8.4. Immunofluorescence micrographs showing the organization of actin filaments during activation of pollen grains of *Pyrus communis*. **(a)** Circular profiles of actin in a pollen grain before activation. **(b)** A pollen grain after 10 minutes of incubation in the growth medium. The polar area (P) is tilted toward the right. The boundaries of the three apertures are indicated by A_1, A_2, and A_3; N, nonapertural surface. **(c)** A pollen grain after 15 minutes of incubation showing actin strands traversing the vegetative cell to the three apertures (A). **(d)** A pollen grain after 30 minutes of incubation showing actin concentrated beneath a single aperture (A). Scale bar = 6 μm. (From Tiwari and Polito 1990a. *Sex. Plant Reprod.* 3:121–129. © Springer-Verlag, Berlin. Photographs supplied by Dr. V. S. Polito.)

actin was present as dispersed fibrils, as filaments, or as filaments confined to the germination pore. These results show that a normal, continuous progression of actin changes is essential for pollen germination (Tiwari and Polito 1990a). In cultured pollen protoplasts of *Lilium longiflorum, Zea mays,* and *Gladiolus gandavensis,* which regenerate a partial or a complete wall, microfilaments undergo an ordered rearrangement with strict polarity preparatory to germination; treatment of the protoplast of *L. longiflorum* pollen with cytochalasin leads to fragmentation of the actin filaments and inhibition of germination (Zhou, Yang and Xu 1990; Tanaka and Wakabayashi 1992). Although the inhibitor experiments indicate the possibility that a reorganization of fibrils takes place in response to a specific signaling of the actin molecules, this suggestion, though a logical one, still has to be tested critically. The recent isolation of genes from maize pollen that encode ADF-like proteins offers great promise in broader studies of this problem (Lopez et al 1996).

Much less is known about microtubules and their role in pollen germination. In one study on this topic, it was noted that although microtubules are not present in the vegetative cell of dehydrated pollen of *P. communis*, they are a prominent feature of both cells of the pollen grain undergoing activation. During the period leading to germination, the microtubule complex of the vegetative cell displays an increasingly branching pattern throughout most of the pollen grain. However, unlike actin fibrils, the tubulin elements do not show any preferred orientation toward a potential germination site, but some short, branched microtubules converge around the base of the emerging pollen tube as a collar (Tiwari and Polito 1990b). In the absence of any demonstrable initiation sites, how the microtubules are assembled in the vegetative cell requires further study. The idea that microtubule nucleation may be triggered by clusters of unpolymerized tubulin floating in the vegetative cell is an attractive, but a highly speculative, possibility. In pollen protoplasts of *L. longiflorum*, microtubules are generally associated with the generative cell and, prior to germination, their polarized arrangement is similar to that of actin filaments. Indicative of a role for microtubules in pollen germination is the observation that their disruption by colchicine has little effect on microfilament organization, yet it inhibits pollen germination (Tanaka and Wakabayashi 1992).

Calcium Gradients during Activation. It is likely that free Ca^{2+} plays an important role in some as yet undetermined reactions during pollen activation. In a study involving pollen grains of 86 species including 79 genera representing 39 families, Brewbaker and Kwack (1963) demonstrated an almost universal requirement for Ca^{2+} in the medium to ensure pollen germination and pollen tube growth. Coming in the wake of the discovery of a Ca^{2+} gradient in the style as a chemotropic factor to encourage pollen tube growth toward the embryo sac (Mascarenhas and Machlis 1962a), this observation created a minor sensation among researchers in the area of pollen biology. Several attempts employing different approaches have been made to pinpoint the role of Ca^{2+} in pollen germination. In explorations done with a vibrating electrode, pollen grains of *Lilium longiflorum* were found to develop pollen tubes close to the direction of greatest inward current density, probably generated by Ca^{2+} (Weisenseel, Nuccitelli and Jaffe 1975). A requirement for Ca^{2+} for pollen germination has also been demonstrated in experiments using inhibitors like nifedipine (Reiss and Herth 1985; J. Heslop-Harrison and Heslop-Harrison 1992c), vanadate (Tirlapur and Cresti 1992), and verapamil (Bednarska 1989b; Tirlapur and Cresti 1992), which generally disrupt the expression of the ionic gradient by blocking Ca^{2+} channels, and in experiments using phenothiazine drugs, which block the calmodulin–calcium complex and inhibit pollen germination (Polito 1983; Estruch et al 1994). Another effort that has provided a striking result in correlating pollen germination with the distribution of membrane-bound Ca^{2+} is the monitoring of the ion cytochemically using the fluorescent probe chlorotetracycline. This work, from different laboratories, has shown that in activated pollen grains of *L. longiflorum* (Reiss and Herth 1978; Reiss, Herth and Nobiling 1985), *Pyrus communis* (Polito 1983), *Nicotiana tabacum* (Tirlapur and Cresti 1992), and *Gasteria verrucosa* (Tirlapur and Willemse 1992), there is a locally restricted influx of Ca^{2+} ions into the potential site of pollen tube emergence. Intense accumulation of calmodulin has been visualized at the germination site of activated pollen grains of *L. longiflorum* and *N. tabacum* by the use of fluorescent calmodulin inhibitors fluphenazine and chlorpromazine and by the use of anticalmodulin antibodies, respectively (Haußer, Herth and Reiss 1984; Tirlapur et al 1994).

Although the observations discussed support a role for Ca^{2+} in pollen germination, they do not identify the mode of action of this ion. In experiments with pollen grains of *Narcissus pseudonarcissus*, it was observed that the absence of Ca^{2+} in the medium does not interfere with activation, typically expressed as a circulatory movement of inclusions of the vegetative cell. However, in the absence of the ion, bursting of pollen grains was rampant and germination was severely inhibited in the survivors. Addition of nifedipine did not prevent activation-related movement in the vegetative cell, but it inhibited a subsequent polarized movement focused on the germination sites, indicating that a Ca^{2+} gradient in the pollen grains is clearly necessary for polarized germination. Similarities between the orientations of the actin fibril system and Ca^{2+} flux to the potential germination sites of pollen grains might indeed suggest a role for Ca^{2+} as a second messenger in controlling the spatial distribution of actin, but critical proofs have not yet been provided for this thesis (J. Heslop-Harrison and Heslop-Harrison 1992c).

Efflux of Metabolites. No account of the activation of pollen grains can be complete without reference to the efflux of proteins and other metabolites into the germination medium. It was mentioned in Chapter 7 that subsequent to pollination, recognition molecules are released from pollen grains onto

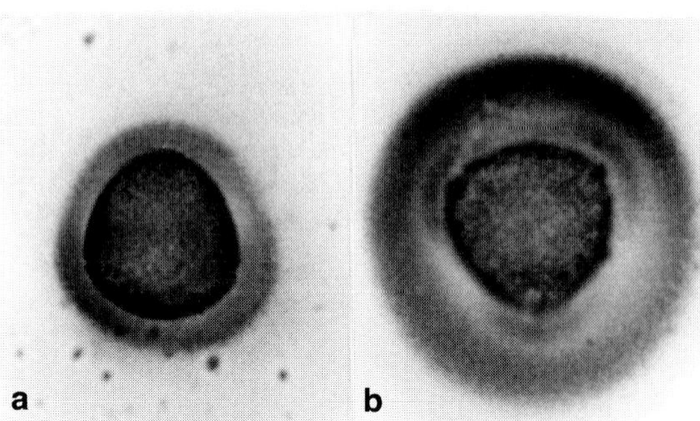

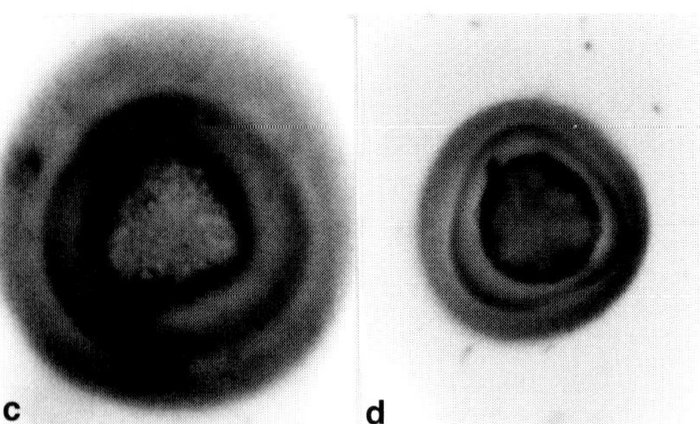

Figure 8.5. Emission of proteins from pollen grains of *Linum grandiflorum.* **(a)** Proteins diffusing from the entire surface of a thrum morph pollen grain about 60 seconds after immersion in the medium. **(b)** Same as (a), with three arcs marking the diffusion of proteins from the apertural intine. **(c)** Same as (a), after 10 minuutes in the medium, with an inner ring of proteins derived from the intine and an outer ring of exine proteins. **(d)** A pollen grain of the pin morph, after 7 minutes in the medium, showing proteins diffusing from the apertural intine sites and a surrounding zone of exine proteins. (From Dulberger 1990. *Sex. Plant Reprod.* 3:18–22. © Springer-Verlag, Berlin. Photographs supplied by Dr. R. Dulberger.)

the stigmatic surface. Although the identity of these molecules is known only in a few cases, there is overwhelming evidence indicating that the molecules participate in the adhesion of pollen grains to the papilla and in signal transduction mechanisms related to pollen acceptance or rejection. Pollen grains of various plants cultured in vitro release cell wall–held and cytoplasmic proteins and other chemicals into the medium. Some of the emissions have been characterized by precipitation reactions and by histochemical and biochemical analyses.

In a survey of incompatibility proteins of *Oenothera organensis,* Mäkinen and Lewis (1962) showed that serologically cross-reacting proteins diffuse from pollen grains into agar wells, where they form characteristic precipitation lines against antisera prepared from the pollen. As will be discussed at some length in Chapter 9, experiments of this type formed the basis for the thesis that incompatibility is due to an antigen–antibody type of reaction between a specific protein in the pollen grain and a homologous protein in the stigma or the style. Use of a "pollen print" technique, by which proteins emitted from single pollen grains are collected on agarose gels and made visible either by staining or by immunofluorescence (Figure 8.5), has made it possible to show that the exine-held proteins of sporophytic origin are first released within seconds after hydration of the pollen grain and are followed by the flow of intine proteins over a longer interval (J. Heslop-Harrison and Heslop-Harrison 1973; Howlett, Knox and Heslop-Harrison 1973; J. Heslop-Harrison, Knox and Heslop-Harrison 1974; J. Heslop-Harrison 1979a; Dulberger 1990).

Pollen grains of *Petunia hybrida* respond to hydration by emitting initially a flavonoid material, then following within five minutes with proteins, which constitute most of the secreted material. Protein secretion markedly increases up to 30 minutes, when the first pollen tubes appear. At this time, additional flavonoids and free nucleotides are present in the medium (Stanley and Linskens 1965; Kirby and Vasil 1979; Kamboj, Linskens and Jackson 1984). Other studies, using an improved pollen mass culture method, have shown that the efflux of amino acids, especially of proline, occurs at the same time or even before the flavonoids are released into the medium (Linskens and Schrauwen 1969; Zhang, Croes and Linskens 1982). Pollen cultures of *P. nyctaginiflora* release free sugars into a medium containing either sucrose or glucose as a substrate (Thomas and Dnyansagar 1975). Comparable experiments with lily pollen have shown that there is a leakage of soluble carbohydrates, but the problem is alleviated by including Ca^{2+} in the medium. Freeze-drying of pollen of lily results in increased leakage of carbohydrates, as well as of phosphate and ninhydrin-positive material, into the medium; this is not surprising since freeze-drying invariably damages the functional membranes of cells (Dickinson 1967; Davies and Dickinson 1971). Although many interacting factors may affect the efflux of solutes from plant cells, at least in the case of lily pollen this is related to the integrity of the membranes and does not appear to be developmentally modulated.

Electrophoretic analyses have further clarified the nature of pollen diffusion products (Figure 8.6). Diffusible proteins from pollen grains of maize are

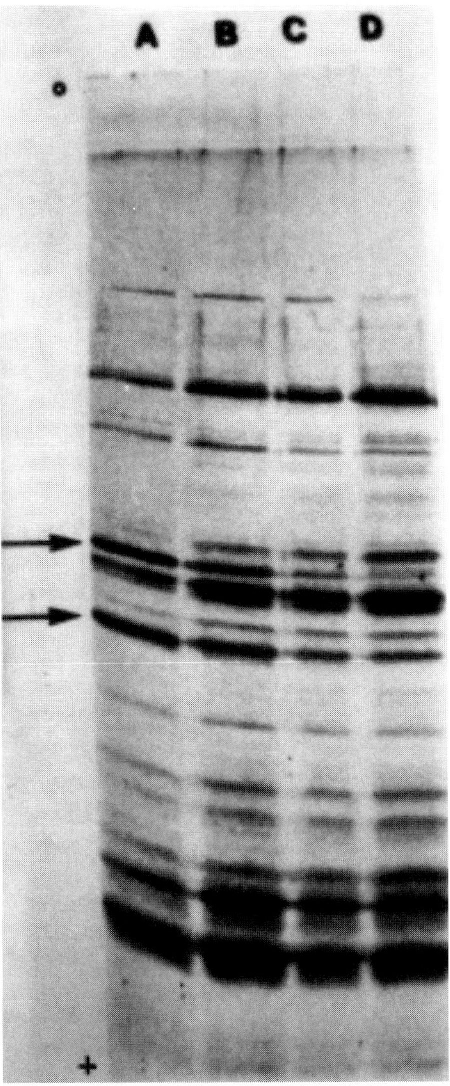

Figure 8.6. SDS-PAGE of proteins of *Zea mays* pollen diffusing into the medium at different times after sowing. A, 0–5 minutes; B, 5–10 minutes; C, 10–30 minutes; D, 30–90 minutes. Arrows indicate proteins that are eluted rapidly and decrease with time. (From Porter 1981.)

found to be heterogeneous on SDS-PAGE and differ qualitatively and quantitatively from their soluble proteins. A survey of diffusates from pollen grains of 40 races of maize with a very wide genetic background showed a surprising degree of conservation among the proteins as well as in the protein patterns. This observation was based on the relatively minor quantitative variations noted in the protein bands and the absence of new or missing bands in the electrophoretic profile (Porter 1981). Other investigations have drawn attention to the excretion of esterases, leucine aminopeptidases, catalases, amylases, and acid phosphatases by pollen of *Oenothera organensis* (Mäkinen and Brewbaker 1967); of cellulase and pectinase by pollen of *Hemerocallis fulva* (Konar and Stanley 1969); of cutinase by pollen of *Tropaeolum majus* (Shayk, Kolattukudy and Davis 1977); of hydrolases by pollen of *Pyrus communis* (Rosenfield and Matile 1979); of glyceraldehyde-3-phosphate dehydrogenase, alcohol dehydrogenase, malate dehydrogenase, and phosphoglucoisomerase by pollen of *Clarkia unguiculata* (Weeden and Gottlieb 1980); of nucleases and unspecified proteins by pollen of *Nicotiana tabacum* and other plants (Tupý, Hrabětová and Čapková-Balatková 1980; Čapková, Hrabětová and Tupý 1983; Matoušek and Tupý 1983, 1985); of invertase by pollen of *Lilium auratum* (Singh and Knox 1984); and of reducing substances that interfere with peroxidase activity by pollen of *N. tabacum* and *N. alata* (Žárský, Říhová and Tupý 1987). According to Stanley and Linskens (1974), approximately 80 enzymes have been detected in pollen extracts of diverse plants; whether they diffuse into the medium during a short germination time course is not known. Some of the pollen-held and diffusing proteins have been purified and their properties studied (Maiti, Kolattukudy and Shaykh 1979; Matoušek and Tupý 1984; Pressey and Reger 1989; Radłowski et al 1994).

Considering the universality of the phenomenon of efflux of metabolites from germinating pollen grains, remarkably little is known about the role of the released substances in the germination process. Although some of the diffusing ions might be involved in incompatibility reactions, this is not always the case. For example, in *Petunia hybrida*, compatible and incompatible pollen tube growth and seed set are not affected by the loss of substances diffusing from the pollen. This was shown by pollinating plants with pollen grains repeatedly washed and dried; the observation is remarkable considering the fact that the amount of proteins thus washed off may be as high as 18% of the total pollen weight (Gillissen and Brantjes 1978). In the context of pollination, it has been proposed that the enzymes released by pollen grains might favor the formation of special metabolites in the stylar tissues for pollen germination and pollen tube growth (Mäkinen and Brewbaker 1967). Substances such as flavonols may have a role in the incompatibility reactions, whereas the enzyme cutinase is too important to be ignored because of its obvious function. Although it would be anticipated that different materials released from the pollen would have different functions, some general principles governing the role of these materials in the physiology of pollen grains should be forthcoming from future research.

(iii) Pollen Tube Emergence

Once certain critical changes have occurred in the pollen grain during the hydration and activation phases, the new state becomes stable as it sets the stage for pollen tube appearance. Electron microscopic studies of pollen grains at this stage of germination show massive accumulation of vesicles near the germination pore, indicating that the Golgi apparatus is the most versatile component with a role in pollen tube emergence. The dictyosomes might be expected to participate to some extent in the secretion of polysaccharides necessary to form the new plasma membrane and cell wall. Another ultrastructural change observed in pollen grains in which the tube is poised to emerge is a polarized migration of organelles to the domain of the pore where the dictyosomes have already accumulated. Additionally, the newly regenerated vesicles may be characterized by differences in size sufficiently marked to allow the vesicles to be classified as large and small types (Larson 1965; Cresti et al 1977, 1985; Miki-Hirosige and Nakamura 1982; Malhó and Pais 1992).

The small size of the pollen grains and the presence of thick walls have limited the amount of information about cytoplasmic upheavals connected with pollen tube emergence that can be obtained from light microscopic studies of living materials. Nonetheless, indications are that within minutes after hydration, the cytoplasm of the pollen grain exhibits a rotational or streaming movement and possibly this provides the force necessary to channel the cytoplasm along the pathway leading to the germination pore (Iwanami 1956; Venema and Koopmans 1962; J. Heslop-Harrison and Heslop-Harrison 1992b).

The effect of cytoplasmic activities in the pollen grain is the opening of one of the germination apertures and the emergence of a hemispherical outgrowth. The innermost stratum of the intine, composed of cellulosic microfibrils embedded in an amorphous matrix of pectic materials, is recognized as a major player in the evolution of this outgrowth. In *Lychnis alba*, the intine is fortified by vesicular wall materials synthesized during a pregermination period in the pollen grain and it contributes in this form to the genesis of the pollen tube wall (Crang and Miles 1969). The transition of the hemispherical tip into a typical pollen tube has been described in exquisite detail in *Narcissus pseudonarcissus* (Y. Heslop-Harrison and Heslop-Harrison 1992). In brief, this involves change-over to cylindrical growth, migration of a few mitochondria and lipid globules from the pollen grain as the first organelle colonizers in the pollen tube, followed by rapid and polarized movement of the cytoplasm of the vegetative cell, and, finally, as growth continues, an "inverse fountain" pattern of cytoplasmic movement characteristic of an actively elongating tube. Formation of occluding plugs of callose, a characteristic feature of elongating pollen tubes, begins even before emergence of the tube or soon after it appears outside (Larson 1965; Cresti et al 1977, 1985; Y. Heslop-Harrison and Heslop-Harrison 1992).

Enzymatic processes are probably involved in the events associated with pollen tube emergence. It is uncertain where the enzymes are produced, but the conclusion that they are important and serve to weaken the site of pollen tube emergence is inescapable. One possibility is that lectin-binding sites on the pollen wall may be eliminated to give way to the pollen tube; this is based on the observation that the duration of the activation period of germination of lily pollen is considerably shortened by supplementing the medium with lectins like con-A and phytohaemagglutinin (PHA), which leads to an apparent promotion of germination (Southworth 1975; Pierson et al 1986).

2. BIOCHEMISTRY OF POLLEN GERMINATION

Historically, the biochemical study of pollen germination developed as an adjunct to the field of pollen physiology connected with the utilization of various carbohydrates, although in later years it has acquired an independent status. Biochemical parameters of pollen germination that have received attention may conveniently be classified into three small groups. First, there are the changes in respiratory activities; the second group of investigations comprises studies of nucleic acid and protein metabolism; and the third group represents changes in the activities of various enzymes.

(i) Respiratory Changes

As a quiescent cell, the pollen grain typically displays very low to no respiratory activity at all. Hydration triggers a dramatic increase in the respiratory rate of the pollen grain. The curve for respiration described in germinating pollen grains of *Lilium longiflorum* is representative of pollen of other species; here, a period of low respiration is sandwiched between two periods of high respiration, one before the outgrowth of the pollen tube and the other coinciding with the accelerated pollen tube growth (Dickinson 1965). The abundance of mitochondria in the dehydrated pollen and their structural transformation during hydration are certainly related to the speed with which respiratory metabolism picks up during germination. Additionally, the inhibition of respiration in lily pollen by oligomycin, and its reversal by 2,4-dinitrophenol, have indicated that the process is limited by the rate of oxidative phosphorylation in mitochondrial activity that results in variations in ATP synthesis (Dickinson 1966). It was mentioned earlier that mitochondria of tricellular pollen grains are more active than those of bicellular pollen. Another aspect of mitochondrial metabolism of pollen grains is highlighted by the observation that tricellular pollen grains maintain their rapid respiration over much shorter periods of time than bicellular pollen (Hoekstra and Bruinsma, 1975b, 1978). There are further data that argue for inherent differences in the respiratory metabolism between bicellular and tricellular pollen grains. For example, differences in respiratory rates are not related to water uptake, which is essentially the same in representative pollen grains of the two types over the initial period of hydration in an atmosphere of high humidity (Hoekstra and Bruinsma 1980). The slow attainment of respiratory competence by bicellular pollen grains is also unrelated to a requirement for mitochondrial protein synthesis because cycloheximide or chloramphenicol does not slow down the attainment of mitochondrial respiratory capacity in germinating pollen of *Typha latifolia* (Hoekstra 1979). Conceivably, both intrinsic biochemical and genetic effects at the cytoplasmic level modulate the activity of mitochondria in bicellular and tricellular pollen grains. The implications of this in the evolutionary strategy of the two types of pollen grains are briefly discussed by Hoekstra and Bruinsma (1978).

(ii) Evidence for Presynthesized mRNA

A major consideration in biochemical studies of nucleic acid and protein synthesis during pollen germination is the source of mRNA that codes for the first proteins of germination. It goes without saying that the transition of the pollen grain from quiescence to germination is accompanied by increased expression of many different genes, whose translation products provide for the vigorous metabolism associated with the emergence and growth of the pollen tube. A small number of proteins synthesized during the early germination period are probably involved in the subtle developmental aspects of germination, whereas the majority are associated with normal growth processes. What is the source of templates for these proteins? Are the first proteins of pollen germination synthesized on templates already present in the mature pollen grain, or are they translated on newly transcribed mRNAs?

It is now well established that mature pollen grains of many plants contain a complement of mRNAs that are used to code for proteins required for germination and early pollen tube growth. The starting point for investigations that led to this conclusion was the demonstration that when pollen grains of *Tradescantia paludosa* are germinated in a medium containing the transcription inhibitor actinomycin-D, germination, early pollen tube growth, and migration of the vegetative cell nucleus and the generative cell into the pollen tube occur normally, but the division of the generative cell is blocked (Mascarenhas 1966b; Lafleur and Mascarenhas 1978). Other studies using pollen grains of various plants have confirmed this general picture of actinomycin-D effects, although no additional details have been added (Dexheimer 1968, 1970, 1972; Lin, Chow and Lin 1971; Linskens, van der Donk and Schrauwen 1971; Sondheimer and Linskens 1974; Süss and Tupý 1976; Raghavan 1981b; Que and Tang 1988). Another generalization in support of the concept of stored mRNA is that ungerminated pollen grains contain variable amounts of polysomes that translate the information contained in the base sequence of mRNAs into proteins (Linskens 1967; Mascarenhas and Bell 1969; Tupý 1977). To the extent that a splicing mechanism is necessary for the processing of the primary transcripts, investigations using complementary probes have further revealed the presence of nuclear-based transcripts of small nuclear RNAs and their proteins in pollen grains (Concha et al 1995).

The ultimate test of the presence of stable mRNA in the quiescent pollen grain is the ability of the messages to prime the synthesis of proteins in a cell-free translation system. It has been clearly shown not only that poly(A)-RNA isolated from ungerminated pollen grains of *Tradescantia paludosa* (Frankis and Mascarenhas 1980) and *Zea mays*

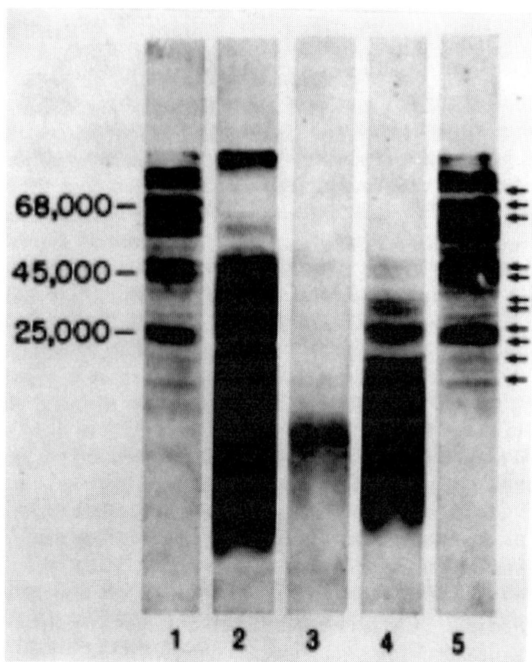

Figure 8.7. Fluorograms of SDS-PAGE gels of in vivo and in vitro synthesized proteins of pollen grains of *Tradescantia paludosa*. The lanes contain (1) molecular masses of marker proteins; (2) proteins synthesized in vivo by germinating pollen grains; (3) in vitro translated proteins using poly(A)-RNA from ungerminated pollen grains; (4) endogenous activity of the wheat germ translation system; (5) in vitro translated proteins using total RNA from ungerminated pollen grains. Common proteins synthesized in vivo and in vitro are indicated by arrows. (From R. Frankis and J. P. Mascarenhas 1980. Messenger RNA in the ungerminated pollen grain: a direct demonstration of its presence. *Ann. Bot.* 45:595–599, by permission of Academic Press Ltd., London.)

(Mascarenhas et al 1984) synthesizes proteins in vitro, but also that many of the proteins synthesized on templates of stored mRNA are similar to those synthesized by germinating pollen grains of the respective species (Figure 8.7). The control of synthesis of the first proteins of pollen germination by mRNA that is present as a holdover from an earlier stage in pollen ontogeny emerges as a novel idea in this experimental system, although comparable situations exist in seeds, fern spores, and fungal spores.

(iii) RNA Metabolism during Pollen Germination and Pollen Tube Growth

The attractiveness of the presynthesized mRNA concept has prompted the examination of other aspects of RNA metabolism during pollen germination and pollen tube growth. Since the classical view of the nucleus of the vegetative cell in the pollen tube as a degenerative organelle had never been seriously challenged, an early objective was to determine whether this nucleus is biochemically functional in the germinating pollen. Autoradiography of ^{3}H-uridine incorporation has been particularly useful in showing that in germinating pollen grains of *Tradescantia paludosa* (Mascarenhas 1966b), *Hyoscyamus niger* (Reynolds and Raghavan 1982), and *Secale cereale* (Haskell and Rogers 1985), nuclei of both vegetative and generative cells are active in synthesizing RNA during early periods of germination. Despite the physiological and structural differences between the development and germination of angiosperm and gymnosperm pollen grains, it is noteworthy that both vegetative cell (tube cell) and generative cell nuclei of *Pinus ponderosa* (Young and Stanley 1963) and *P. taeda* (Frankis 1990) pollen actively engage in RNA synthesis. A procedure to isolate large quantities of vegetative and generative cell nuclei from pollen tubes of *T. paludosa* has revealed significant differences between the base compositions of RNA synthesized by the two types of nuclei (LaFountain and Mascarenhas 1972).

As other biochemical studies have shown, knowledge of the types of RNA synthesized may hold the key to our understanding of the role of presynthesized and newly made mRNA in pollen germination and pollen tube growth. These investigations were, however, undertaken with homogenates of germinating pollen grains, in which the identity of the vegetative and generative nuclei had been lost. An interesting difference between the types of RNA made in the pollen grain and in the pollen tube of tobacco as shown by base ratio analysis suggested that RNA synthesized by pollen tubes might be of the messenger type; the key observation was that the base ratios of the newly synthesized RNA were close to those of tobacco DNA (Tano and Takahashi 1964). This view has found support in another work on RNA synthesis, this one in the pollen tubes of lily. Here, a combination of criteria, such as the difference between the base compositions of RNA associated with pollen tubes and typical rRNA, and the absence of nucleoli in the pollen tube nuclei, has led to the view that the RNA synthesized in the pollen tube is of the messenger type with little or no rRNA (Steffensen 1966).

Detailed studies on *T. paludosa* have shown that the failure to synthesize rRNA is clearly a fundamental characteristic of pollen tube growth. The

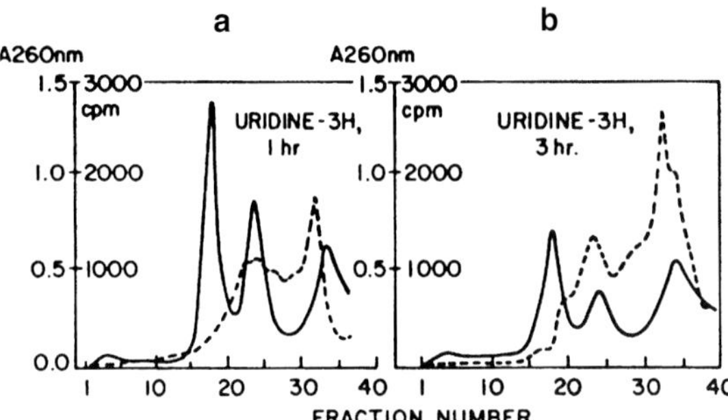

Figure 8.8. RNA synthesis during germination of pollen grains of *Tradescantia paludosa*. Pollen grains were sown in a medium containing ^{3}H-uridine for 1 hour **(a)** or 3 hours **(b)**. RNA was centrifuged in a sucrose gradient and fractions were collected to determine absorbance (solid line) and radioactivity (dotted line). The RNA peaks have sedimentation values (from left to right) of 25S, 16S, and 5S. (From Mascarenhas and Bell 1970.)

data of Mascarenhas and Bell (1970) illustrate several general horizons relating to this conclusion. In this work, RNA synthesized by pollen tubes at various times after germination is found to be polydisperse in size, sedimenting in the sucrose gradient between 2S and 30S, with a peak at about 8S (Figure 8.8). Besides its sedimentation properties, this RNA has a distinctive base ratio, characterized by low guanosine and cytosine and high adenylic acid, which rules it out as rRNA. A clear-cut answer to the question of whether the low-molecular-weight RNA synthesized by the pollen tubes includes any tRNA was sought by further analysis of fractions sedimenting between 2S and 9S. The unexpected finding that emerged was that the 2S–9S components consist of several different species of RNA but do not include any tRNA or 5S rRNA (Mascarenhas and Goralnick 1971). These results emphasize that label incorporation into RNA does not necessarily connote rRNA synthesis, as one would expect for a rapidly elongating cell. Participation of ribosomes and tRNA already present in the mature pollen grain appears, correspondingly, very likely in pollen tube growth.

Results of investigations on RNA synthesis during pollen germination and pollen tube growth in *Nicotiana tabacum*, however, conflict in important respects with those just described for *T. paludosa* and other species. Germinating tobacco pollen grains are biochemically unusual because of the large quantity of rRNA they synthesize, with the rate of 5S RNA synthesis being appreciably higher than that of 18S and 25S RNAs (Tupý 1977; Tupý, Hrabětová and Balatková 1977). Activation of tRNA genes also accounts for a small fraction of the newly synthesized RNA in germinating pollen of *N. alata* (Süss and Tupý 1976) and *N. tabacum* (Süss and Tupý1978). Following this initial surge in RNA synthesis, no further transcriptional activity is necessary for continued pollen tube growth in *N. tabacum*, which occurs even in the presence of transcription inhibitors (Tupý, Süss and Říhová 1986). Associated with the early stages of germination of pollen grains, there is an accelerated synthesis of poly(A)-RNA, followed by a decrease (Süss and Tupý 1979). Apparently, the newly synthesized poly(A)-RNA distributes into two functionally distinct compartments, namely, one for the assembly of ribosomes into polysomes, and the other for their translation into proteins. Another work has shown a broad correlation between the size distributions of poly(A)-RNA molecules of mature pollen grains and of germinating pollen; this evidence further substantiates the claim that a specific population of poly(A)-RNA synthesized during pollen development persists as stored mRNA in the mature pollen and codes for proteins at the onset of germination (Tupý 1982). Consistent with the results from tobacco pollen, the early phase of germination of pollen of *Malus domestica* (apple) is characterized by the activation of genes for rRNA and tRNA synthesis (Bagni, Adamo and Serafini-Fracassini 1981). The degree to which old and new mRNAs account for pollen germination will be better assessed when our knowledge of RNA metabolism is extended to pollen grains of other plants.

(iv) Protein Metabolism

Support for the concept that there is a requirement for protein synthesis for germination of pollen grains and for growth of pollen tubes can be drawn from several experiments, of which the use of inhibitors may be mentioned first. The most widely used inhibitor is cycloheximide, which has been shown to block in varying degrees the germination of pollen grains of *Lilium longiflorum*, *Clivia miniata* (Franke et al 1972; Li, Tsao and Linskens 1986), *Trigonella foenum-graecum* (Fabaceae) (Shivanna, Jaiswal and Mohan Ram 1974b), *Petunia hybrida* (Sondheimer and Linskens 1974), *Typha latifolia* (Hoekstra and Bruinsma 1979), *Nicotiana tabacum* (Čapková-Balatková, Hrabětová and Tupý 1980), and *Cucurbita moschata* (Que and Tang 1988). In several other plants, including *Tradescantia paludosa*, the effect of the drug is not so much on germination of pollen grains as it is on growth of the pollen tube (Lin, Chow and Lin 1971; Malik and Gupta 1976; Lafleur and Mascarenhas 1978;

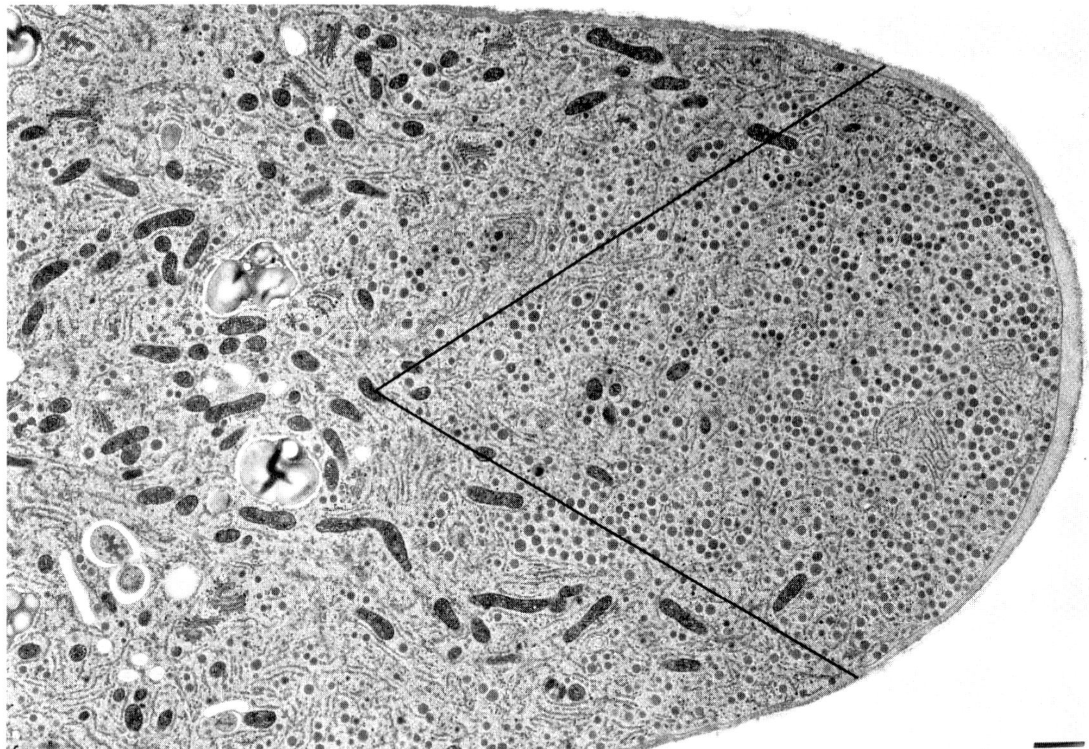

Figure 8.12. Electron micrograph of the tip region of the pollen tube of *Lilium longiflorum*. Organelles that extend into the apex along the flanks of the pollen tube are ER and mitochondria. The inverted cone-shaped area packed with vesicles, some random elements of ER, amyloplasts, and possibly lipids, is outlined. Scale bar = 1 μm. (From Lancelle and Hepler 1992; photograph supplied by Dr. P. K. Hepler.)

ever, revealed some variability in the origin of the polysaccharides of the matrix. In an investigation covering four genera, Larson (1965) noted that during the initial stages of pollen germination, the Golgi complex becomes secretory, pinching off numerous vesicles. One likely scenario is that the load of carbohydrates from the vesicles moves to the cell wall while the ghost membranes fuse with the plasma membrane. Studies on tobacco pollen have shown that besides a rapid synthesis and turnover of ER, dictyosomes and ER are both structurally continuous in the pollen tube, so that extensive membrane transformations or an elaborate dictyosome–ER membrane complex from which vesicles arise are involved (VanDerWoude and Morré 1968; Kappler, Kristen and Morré 1986; Noguchi and Morré 1991). In *Secale cereale* and *Pennisetum typhoideum*, the basis for polysaccharide contribution to the pollen tube wall has been sought through studies that compare polysaccharide-rich particles (P-particles) present in the ungerminated pollen grain with those present in the pollen tube. Suggestive of an apparent contribution of the P-particles to the extending pollen tube wall is the resemblance between the P-particles in the pollen grain and the polysaccharide-containing wall precursor bodies that aggregate in the tip of the pollen tubes (J. Heslop-Harrison and Heslop-Harrison 1982a). Another model for the origin of cell wall polysaccharide precursors, described in the pollen tube of *Lilium longiflorum*, has invoked a direct spatial relationship of the newly formed wall with starch and lipid bodies; these act in concert with the Golgi vesicles to package the polysaccharide skeleton of the pollen tube wall at the tip (Miki-Hirosige and Nakamura 1982). Although the vesicles are universally thought to contain polysaccharides, acid phosphatase activity has been localized in the membranes of the vesicles and ER at the tip of pollen tubes of *Prunus avium* (Lin, Uwate and Stallman 1977).

There is considerable evidence that the Golgi complex in the pollen tubes of *Tradescantia virginiana* is a dynamic system that produces a large num-

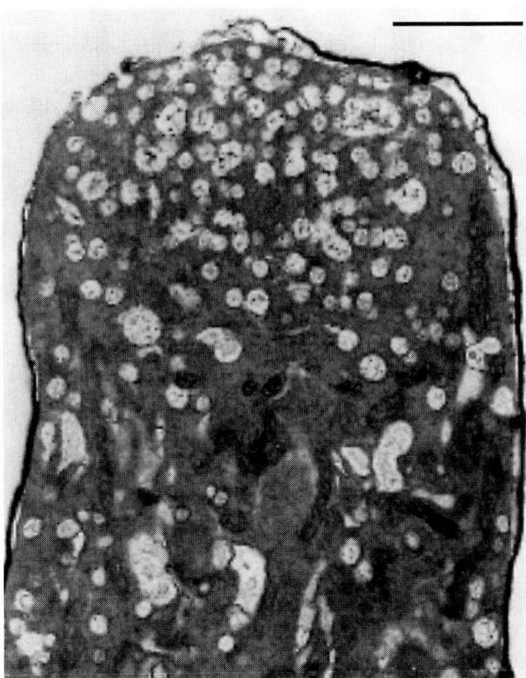

Figure 8.13. Electron micrograph of a longitudinal section through the tip of the pollen tube of *Tradescantia virginiana* showing the plasma membrane with fusion profiles and secretory vesicles at the apex. Scale bar = 1 μm. (From Steer and Steer 1989; photograph supplied by Dr. M. W. Steer.)

ber of secretory vesicles and coated vesicles. Experiments by Picton and Steer (1981, 1983c) showed that cytochalasin inhibition of pollen tube growth is accompanied by an accumulation of vesicles in the cytoplasm. This has been interpreted as possibly due to the inability of the cell to transport vesicles to the plasma membrane site while allowing full dictyosome activity (Shannon, Picton and Steer 1984). Using cytochalasin to provide estimates of vesicle production rates, it has been estimated that a growing pollen tube of *T. virginiana* produces a staggering 3,000–5,000 vesicles per minute. The validity of this method to determine the rate of vesicle production was confirmed by a method using antagonists of Ca^{2+} function, especially FITC. Since this compound inhibits fusion of vesicles at the plasma membrane without interfering with their production or transport, the number of vesicles can be determined as they pile up in the cytoplasm (Picton and Steer 1985). Treatment of pollen tubes with cycloheximide showed that the production of both types of vesicles is dependent upon continued protein synthesis, the coated vesicles being less sensitive to the inhibitor than the regular vesicles (Picton and Steer 1983b). A far-reaching aspect of dictyosome activity and vesicle production was revealed in pollen tubes grown in media containing two different concentrations of Ca^{2+}. Growth in length of the pollen tube was reduced by as much as 75% in the high-Ca^{2+} medium, yet there was little, if any, decrease in the rate of vesicle production, as determined by vesicle accumulation in the cytoplasm after cytochalasin treatment. This occurs in spite of the fact that vesicle requirements for growth are considerably different in pollen tubes growing slowly in high-Ca^{2+}-containing, and rapidly in low-Ca^{2+}-containing, media. The basis for continued vesicle production in the face of drastic inhibition of growth of pollen tubes in the Ca^{2+}-enriched medium has been attributed to a recycling of vesicles due to the lack of a mechanism for their efficient utilization (Picton and Steer 1983c). Another line of evidence for the functioning of secretory routes in the growth of pollen tubes has come from an analysis of the effect of monensin. This ionophore, which is known to cause the breakdown of Golgi-originating vesicular traffic with the plasma membrane, inhibits the germination of pollen grains and the growth of pollen tubes of *Malus domestica* and produces an increased ratio of synthesis of membrane proteins to soluble proteins. The increased synthesis of membrane proteins apparently reflects the effective action of monensin on the secretory route that delivers products from the dictyosomes to the tip of the pollen tube (Speranza and Calzoni 1992).

(ii) Pollen Tube Wall

The list of species in which the pollen tube wall has been studied is of course limited by the choice made by the investigators, but it is noteworthy that most of the observations have been made on members of Liliaceae, Solanaceae, and Poaceae. Although the structure of the pollen tube wall is a fascinating and complex subject in its own right, the intent here is to provide a basic foundation of the current understanding of the wall morphology and chemistry. The definitive structure of the mature wall can be observed in the proximal part of the pollen tube, and from there the layers can be followed to the growing tip. From extensive electron microscopic and cytochemical studies, there is strong evidence showing that the wall of the mature nongrowing part of the pollen tube is constituted of an outer pectic layer, a middle cellulosic layer, and an inner callosic sheath, with the growing tip itself being free of callose (Engels 1974b; Reynolds and Dashek 1976; Anderson et al 1987; J. Heslop-Harrison 1987). The pollen tube wall of *Nicotiana tabacum* is described as a two-layered structure; the wall formed at the tip is designated as the primary wall, and a later-formed, inner wall in the proximal region is designated as the secondary wall (Kroh and Knuiman 1982). By autoradiography of incorporation of ^{3}H-*myo*-inositol and ^{3}H-glucose into different parts of the wall of the pollen tube of *Pryus communis*, combined with enzyme digestion of pectic and cellulosic components, Roggen and Stanley (1971) have shown that the pollen tube tip is primarily composed of pectin, with cellulose predominating in the subtending wall of the tube. Additional evidence for the presence of cellulose and pectin in the pollen tube wall of *P. communis* has come from the observation that supplementation of the medium with enzymes such as cellulase and pectinase actually promotes the growth of pollen tubes. Much of the enzyme-mediated elongation is targeted to the tip of the pollen tube, whose plasticity is weakened by the enzymes (Roggen and Stanley 1969).

Immunocytochemistry has permitted localization of α-L-arabinofuranosyl residues in the outer layer, and of (1,3)-β-glucans in the inner layer, of the pollen tube wall of *N. alata* (Anderson et al 1987; Meikle et al 1991). The localization of arabinans and β-glucans is consistent with the presence of pectins and callose in the outer and inner layers, respectively. A monoclonal antibody, PCBC-3, recognizing arabinofuranosyl residues has been shown to bind to the surface of the pollen tubes of *N. alata* (Harris et al 1987). The outcome of immunocytochemical localization of polygalacturonic acid (unesterified pectin) and of methyl-esterified pectin in the walls of pollen tubes of 20 species of angiosperms using monoclonal antibodies JIM-5 (recognizes unesterified pectin epitope) and JIM-7 (recognizes esterified pectin epitope) might have an important bearing on the significance of pectin esterification during pollen tube growth. This work has identified two patterns of pectin deposition, one as periodic annular deposits found coating the pollen tube walls of species possessing solid styles, and the other as a more uniform sheath in pollen tubes of species with hollow styles (Figure 8.14). The differences in the labeling patterns between JIM-5 and JIM-7 indicate that the pectin layer in the tip of the pollen tube is richer in the esterified form of the carbohydrate than is its proximal part (Li et al 1994a; Geitmann, Li and Cresti 1995). These observations show that a complex developmental program is followed in the distribution of esterified and unesterified pectin in the pollen tube wall, with pectin esterification being related to tip wall loosening during pollen tube growth. In other investigations, MAC-207 and JIM-8, two monoclonal antibodies that are specific for arabinogalactan proteins on the plasma membranes of various plant cells, were found to label the inner callosic sheath of the wall of the pollen tube of *N. tabacum* in the vicinity of the plasma membrane. A remarkable periodic deposition of arabinogalactan epitopes in the callose lining of the pollen tube, similar to pectin deposition, was also demonstrated (Li et al 1992, 1995a). The callose sheath of maize pollen tube is the target of accumulation of a maize pollen-specific gene, *PEX-1*, with an extensin-like domain; it might be expected that proteins like extensin would ultimately be involved in supporting the growth of the pollen tube or in facilitating cell-to-cell signaling between the pollen tube and style (Rubinstein et al 1995b). A major role for pectins might be to set the stage for free passage of pollen tubes through the style; this function might also be accomplished by an adhesion event between the walls of the pollen tube and the transmitting tissue cells by a specific interaction involving pectins and secreted molecules such as arabinogalactans (Jauh and Lord 1996). These suggestions by no means exhaust the list of possibilities, but serve to emphasize that pollen tube growth through the style could stem from a variety of cellular interactions.

Alkali-resistant microfibrils are commonly observed in electron microscopic preparations of pollen tubes of diverse plants. Although it has been assumed that this rigid skeleton is cellulosic in nature, a detailed study of the pollen tube wall of *Lilium longiflorum* based on differential extraction procedures, microscopic techniques, and X-ray diffraction has revealed a different picture of the composition of the fibrils. It now appears that the microfibrils consist predominantly of a mixture of alkali-resistant crystalline (1,3)-β-polyglucan and common cellulosic (1,4)-β-glucan units. Models proposed to explain possible arrangements of the two components within the structural polysaccharide framework of the pollen tube wall have envisaged the cellulose fibrils embedded in the (1,3)-β-glucan fibrils, with the two fibril types alternating with one another, a central core of cellulose ensheathed by the (1,3)-β-glucan, and long (1,3)-β-glucan fibrils interspersed between small cellulosic units (Herth et al 1974).

It is something of a paradox that, despite the

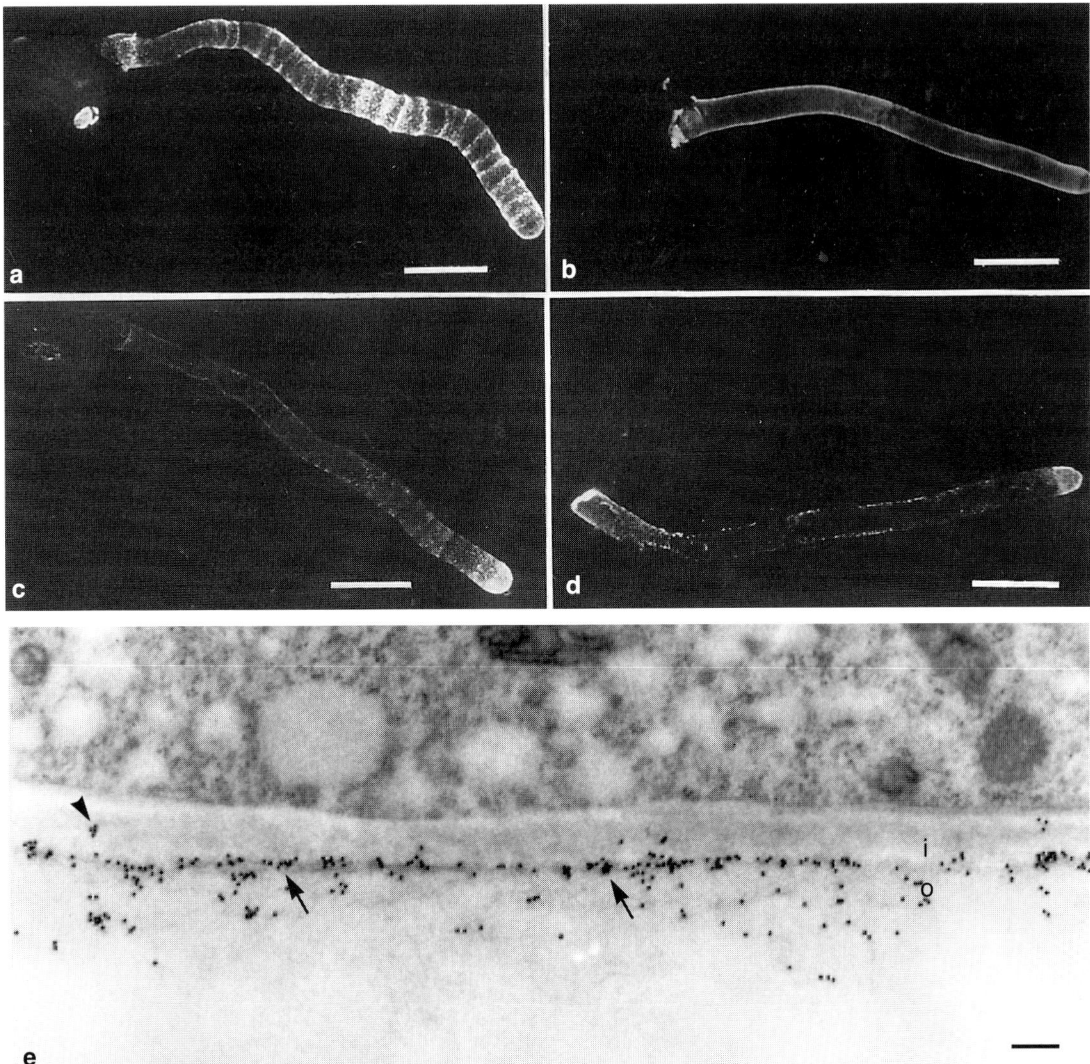

Figure 8.14. Immunolocalization of monoclonal antibodies JIM-5 and JIM-7 in pollen tubes of *Nicotiana tabacum* **(a, c)** and *Lilium longiflorum* **(b, d)**. (a, b) Pollen tubes labeled with JIM-5. (c, d) Pollen tubes labeled with JIM-7. All photographed by confocal laser scanning microscopy; scale bars = 25 μm (a, c), 50 μm (b, d). (From Li et al 1994a. *Sex. Plant Reprod.* 7:145–152. © Springer-Verlag, Berlin.) **(e)** Electron micrograph of the pollen tube wall of *N. tabacum* after treatment with JIM-5. The label (arrows) is seen in the inner part of the outer cell wall layer (o); the arrowhead indicates the inner layer (i). Scale bar = 0.2 μm. (From Geitmann, Li and Cresti 1995; all photographs supplied by Dr. M. Cresti and Dr. A. Geitmann.)

common end product synthesized, there is a diversity of views both on the proportion of the various carbohydrates that contribute to pollen tube wall formation and on the nature of their precursors. Cytochemical studies showed that pectin is a major constituent of the vesicles of the pollen tubes of *Lilium longiflorum* (Dashek and Rosen 1966; Rosen and Gawlik 1966). This was confirmed in experiments in which vesicles extracted from pollen tubes of lily were chemically analyzed and found to contain polysaccharides high in galacturonic acid (VanDerWoude, Morré and Bracker 1971). It has also been shown by chemical and cytochemical analyses that in grass pollen grains and growing pollen tubes, the principal polysaccharide of the P-particles is a pectin (J. Heslop-Harrison and Heslop-Harrison 1982a). In contrast, the presence of cellulose in Golgi vesicles isolated from pollen tubes of *Petunia hybrida* has provided a major argument for a significant role for this carbohydrate in pollen tube wall formation (Engels 1973, 1974a, c). A uridine phosphate–polysaccharide complex identified in the cytoplasm of pollen tubes of *Tradescantia paludosa* has been suggested as an intermediate in the biosynthesis of the pollen wall (Mascarenhas 1970). This envisages the transport of the precursor to the tip of the pollen tube, but whether the traditional vesicles or some other organelles are involved is not clear. A Golgi vesicle–enriched fraction isolated from pollen tubes of *P. hybrida* shows β-glucan synthetase activity and synthesizes cellulose subunits from uridine diphosphate (UDP)-glucose (Helsper, Veerkamp and Sassen 1977; Helsper 1979). These observations remain isolated, and much more work is needed before we can truly assess their generality and significance.

It was mentioned earlier that phytic acid is a major storage product of the pollen grains of certain angiosperms. A functional aspect of phytate metabolism in germinating pollen grains of *P. hybrida* is the mobilization of *myo*-inositol and inorganic phosphate, degradation products of phytic acid, for phospholipid and pectin biosynthesis. On a comparative level, the utilization of labeled *myo*-inositol by germinating pollen grains for phospholipid synthesis is about five times that for pectin synthesis (Helsper, Linskens and Jackson 1984). The results carry the implication that phosphatidylinositol and pectins that are required for pollen tube elongation are provided by the breakdown of phytic acid during germination.

Considerable uncertainty exists with regard to the macromolecular chemical composition of the pollen tube wall, especially because the predominance of some of the sugars is found to be different from that in somatic cells. Analyses of the monosaccharide composition of the walls of pollen tubes of *L. longiflorum* (VanDerWoude, Morré and Bracker 1971; Nakamura and Suzuki 1981; Li and Linskens 1983b), *P. hybrida* (Engels 1974a), *Camellia japonica, C. sasanqua, C. sinensis, Tulipa gesneriana* (Nakamura and Suzuki 1981), *N. alata* (Rae et al 1985), and *Malus domestica* (Calzoni et al 1993) have revealed an abundance of glucose and varying amounts of arabinose and galactose. In *N. alata*, (1,3)-β-glucan and an arabinan, together with small amounts of cellulose, constitute the major polysaccharides (Rae et al 1985). Glucose, arabinose, and galactose are found to be the most-labeled sugars when pollen tubes are incubated in radioactive precursors of carbohydrate biosynthesis (Mascarenhas 1970; Labarca and Loewus 1972). Proteins have been identified in the pollen tube wall of *L. longiflorum* (Li, Croes and Linskens 1983) and *N. alata* (Rae et al 1985), with total protein content ranging from 1.5% to 3% by weight. Although many details remain to be filled in, it is becoming increasingly clear that pollen tube walls are highly specific in the composition and proportions of the various sugars and proteins.

In terms of continued elongation of the pollen tube, synthesis of new wall materials occurs in tandem with considerable activity in the underlying plasma membrane. Tubular invaginations of the plasma membrane (plasmatubules) described in in vivo–grown pollen tubes of *N. sylvestris* are found at a reduced level in pollen tubes grown in vitro. These structures may function like wall ingrowths of transfer cells and may facilitate the uptake of nutrient substances for pollen tube growth (Kandasamy, Kappler and Kristen 1988). In pollen tubes of lily, the presence of hexagonal rosettes on the plasma membrane provides a setting that reveals a close analogy to cellulose microfibrils in plant cells. Accordingly, the view is advanced that the rosettes are involved in cellulose synthesis, although the low frequency of their occurrence suggests that the pollen tube can be expected to produce only a small amount of cellulose (Reiss, Herth and Schnepf 1985). The complexity of the plasma membrane has also become evident through biophysical techniques for the demonstration of proton pumps in pollen tubes (Weisenseel and Jaffe 1976; Tupý and Říhová 1984).

(iii) Pollen Tube Cytoskeleton

The cytoplasm confined to the tip of the pollen tube exhibits intense streaming. Because the pollen tube elongates at its tip, the general belief is that growth is related to cytoplasmic streaming. However, cytoplasmic streaming occurs in old, nongrowing parts of the pollen tube, as well as in tubes in which growth is inhibited by chemical treatments; thus, it is easy to believe that streaming is in no sense dependent on growth (J. Heslop-Harrison 1987). There is good evidence to show that microfilaments are present at the growing tip of the pollen tube. This is corroborated in some instances by findings based on cytochalasin inhibition of microfilament assembly that suggests that the contractile microfilaments are at the hub of the system directing cytoplasmic streaming and thus possibly play a role in pollen tube elongation. Indeed, the first convincing light and electron microscopic demonstrations of actin in an angiosperm were provided in the pollen tube.

Fibrils isolated from pollen protoplasts of *Amaryllis belladonna* by manipulation with a glass needle were found to bind muscle heavy meromyosin, indicating that the fibrils are actinlike in nature. The fact that all filaments originating from a disrupted fibril display the same polarity is considered to provide a molecular basis for the unidirectional movement of fibrils, which probably serve as guiding elements for cytoplasmic streaming, or for the transport of cell wall precursors to the pollen tube tip and for pollen tube growth (Condeelis 1974). Electron microscopic views of microfilaments in pollen tubes are based mainly on the work on *Lilium longiflorum* and *Clivia miniata* (Franke et al 1972; Miki-Hirosige and Nakamura 1982). The arresting feature of *L. longiflorum* is that the microfilaments are made up of aggregates of longitudinally apposed fibrillar structures found in close proximity to the Golgi-derived vesicles, ER, and mitochondria at the tip of the pollen tube. This observation has been confirmed and extended using freeze-substitution in place of chemical fixation to ensure preservation of the cytoskeleton. Some of the preparations using this method also consistently showed a network of microfilaments going all the way to the extreme tip of the pollen tube (Lancelle, Cresti and Hepler 1987; Tiwari and Polito 1988b; Lancelle and Hepler 1992).

Pioneering investigations using rhodamine-phalloidin staining showed that in the pollen tubes of *L. longiflorum, Petunia hybrida, Alstroemeria* sp. (Amaryllidaceae), and *Impatiens wallerana,* actin is present in the form of extended fibrils (Perdue and Parthasarathy 1985). This observation had such obvious impact that it spawned a whole series of similar studies in pollen tubes of *Nicotiana tabacum, Tradescantia virginiana* (Pierson, Derksen and Traas 1986; Pierson 1988; Åström, Virtanen and Raudaskoski 1991), *Pyrus communis* (Tiwari and Polito 1988b), *Iris pseudacorus* (J. Heslop-Harrison and Heslop-Harrison 1989b), *Narcissus pseudonarcissus* (J. Heslop-Harrison and Heslop-Harrison 1991a), and *Pinus densiflora* and *P. thunbergii* (Terasaka and Niitsu 1994). These studies have shown that major elements of the actin cytoskeleton system ramify the deeper parts of the pollen tube cytoplasm in the form of numerous, longitudinally oriented, dense arrays of fibrils extending up to the tip (Figure 8.15). Somewhat similar formations of actin are detected by immunogold labeling of *Nicotiana alata* and *N. tabacum* pollen tube microfilament bundles using monoclonal antibodies to actin (Tang, Lancelle and Hepler 1989; Grote et al 1995).

The reorganizational changes that occur in the actin filaments of protoplasts isolated from pollen tubes of *N. tabacum* reveal the dynamic character of this cytoskeleton. In freshly prepared protoplasts, the filaments are disorganized and fragmented, but during a recovery period lasting several hours, they become circumferentially aligned in parallel arrays, form bundles, converge into opposite ends as highly symmetrical files, and even form a cortical network. It would not be farfetched to equate these changes to the patterns of actin distribution seen in activated pollen grains (see pp. 214–217). That ordering of the actin filaments in pollen tube protoplasts is independent of the nucleus is a logical assumption, in view of the fact that it occurs in both karyoplasts (containing a vegetative cell nucleus and/or a generative cell) and cytoplasts (lacking nuclei but containing cytoplasmic organelles) (Rutten and Derksen 1990). A highly concentrated localization of release protein factor family guanosine triphosphatases (Rho family GTPases) in the pollen tubes of *Pisum sativum* has linked the actin-dependent pollen tube tip growth to a guanosine triphosphate (GTP)-binding protein signaling mechanism analogous to a molecular switch (Lin et al 1996).

Both electron microscopic and immunofluorescence studies of pollen tubes have revealed the presence of microtubules traversing the cortical cytoplasm below the plasma membrane. A sense of the complexity of their distribution is seen in the pollen tube of *L. longiflorum,* which was also the first to be investigated in detail. The main population of microtubules is present in longitudinal, regularly spaced groups in the cortical cytoplasm, except at the tip of the pollen tube; a few bundles and isolated microtubules are also found in the deeper layers of the cytoplasm. Cross-bridge connections link microtubules in each group, link microtubules with Golgi vesicles, and link membranes of each group of microtubules with the plasma membrane. Occasionally, the cortical microtubules are found to occur parallel to the microfilaments (Franke et al 1972). Association of cortical microtubules with microfilaments has also been shown by electron microscopy of chemically fixed or freeze-substituted pollen tubes of *Nicotiana alata* (Lancelle, Cresti and Hepler 1987; Tiezzi et al 1987; Lancelle and Hepler 1991, 1992), *N. tabacum,* and *L. longiflorum* (Pierson, Kengen and Derksen 1989), with or without fluorescent labeling of tubulin and actin. Other investigations of the disposition of microtubules in the pollen tubes of *Lycopersicon peruvianum, Malus domestica, Prunus avium* (Cresti,

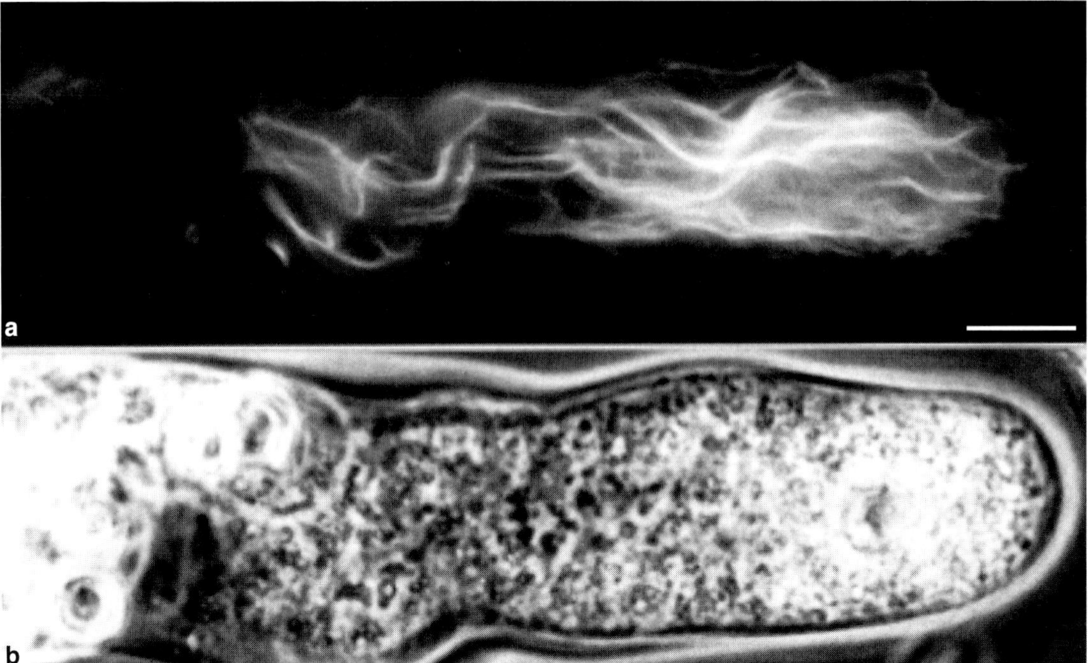

Figure 8.15. Actin distribution in the pollen tube of *Pinus densiflora.* **(a)** Immunofluorescence micrograph showing actin throughout the pollen tube. **(b)** A photograph of the same pollen tube using Nomarski optics. Scale bar = 10 µm. (From Terasaka and Niitsu 1994. *Sex. Plant Reprod.* 7:264–272. © Springer-Verlag, Berlin. Photographs supplied by Dr. O. Terasaka.)

Ciampolini and Kapil 1984), *N. tabacum* (Derksen, Pierson and Traas 1985; Raudaskoski et al 1987; Åström, Virtanen and Raudaskoski 1991), *N. sylvestris* (Joos, van Aken and Kristen 1994), *Lilium longiflorum* (Pierson, Derksen and Traas 1986), *L. auratum* (J. Heslop-Harrison and Heslop-Harrison 1988d), *Alopecurus pratensis* (J. Heslop-Harrison and Heslop-Harrison 1988c), *Pyrus communis* (Tiwari and Polito 1988b), *Endymion non-scriptus* (J. Heslop-Harrison et al 1988), *Zephyranthes grandiflora* (Zhou, Zee and Yang 1990), *Pinus densiflora*, and *P. thunbergii* (Terasaka and Niitsu 1994), detected electron microscopically and/or by immunofluorescence, support the existence of ordered arrays of cortical microtubules as a general feature of the cytoskeletal system.

Certainly, the spatial arrangement of microtubules in the examples cited is in keeping with the elongating nature of the pollen tubes. Using specific antibodies, Del Casino et al (1993) have observed that pollen tubes of *N. tabacum* generally lack the acetylated form of α-tubulin but contain the tyrosinated form concentrated at the tip. Although not unequivocal, this probably means that pollen tube growth is under the control of tyrosinated α-tubulin. As the growth of pollen tubes in culture gradually slows down, the longitudinal bands of microtubules are replaced by a mass of immunofluorescent granules and spicules mainly confined to the apical dome of the tube; this is nicely illustrated (Figure 8.16) in the pollen tube of *L. auratum* (J. Heslop-Harrison and Heslop-Harrison 1988d). Rutten and Derksen (1992) have reported that in karyoplasts and cytoplasts isolated from tobacco pollen tubes, the microtubule pattern does not resemble that found in the intact pollen tube; in protoplasts, microtubules do not colocalize with microfilaments.

The unambiguous demonstration of the presence of microfilaments and microtubules in the pollen tube raises an obvious question. Do these cytoskeletal elements participate in the streaming of the cytoplasm or in the elongation of the pollen tube? Based on inhibitor experiments, the prime movers in the cytoplasmic streaming of pollen tubes appear to be the actin filaments. When actively growing pollen tubes are treated with cytochalasin, the actin filaments disappear or change their morphology within minutes and cytoplasmic streaming stops, whereas colchicine affects

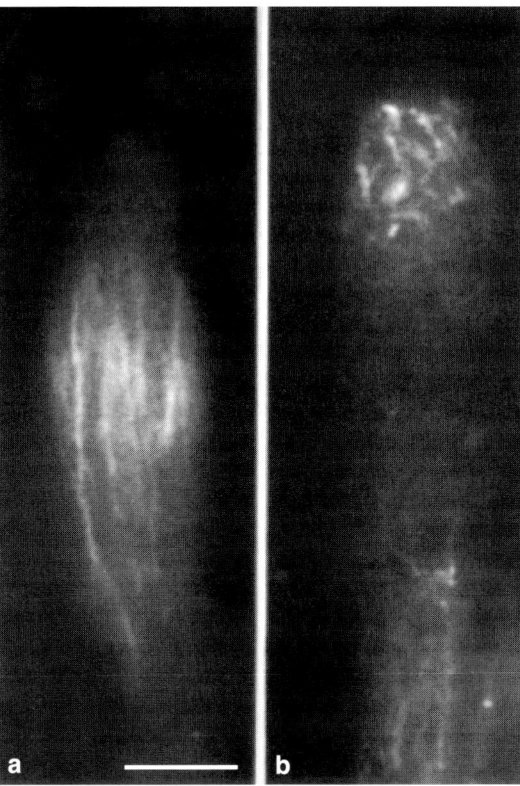

Figure 8.16. Immunofluorescence localization of tubulin in the pollen tubes of *Lilium longiflorum*. **(a)** A pollen tube during active growth, with focus at the cell surface, showing longitudinal bands of laterally apposed microtubules. **(b)** A pollen tube in which growth has ceased. The axially oriented bands of microtubules seen in (a) are no longer visible. Scale bar = 10 μm. (From J. Heslop-Harrison and Y. Heslop-Harrison 1988d. Sites of origin of the peripheral microtubule system of the vegetative cell of the angiosperm pollen tube. *Ann. Bot.* 62:455–461, by permission of Academic Press Ltd., London.)

neither the actin filaments nor the cytoplasmic streaming. Cytochalasin effects are reversible, and pollen tubes recover when the inhibitor is removed from the medium (Franke et al 1972; Mascarenhas and Lafountain 1972; Condeelis 1974; Perdue and Parthasarathy 1985; J. Heslop-Harrison and Heslop-Harrison 1989a; Tang, Lancelle and Hepler 1989; J. Heslop-Harrison et al 1991). Participation of actin microfilaments in the movement of organelles has also been shown by continuous observation of organelles and other cytoplasmic inclusions in the pollen tubes of *Iris pseudacorus* as they move along single, longitudinally oriented actin fibrils (J. Heslop-Harrison and Heslop-Harrison 1988a). Although these findings eliminate a role for microtubules in cytoplasmic streaming or in pollen tube growth, inhibition of pollen tube growth on the stigma of *Nicotiana sylvestris* treated with an antimicrotubule drug has suggested that microtubules are essential for in vivo pollen tube growth (Joos, van Aken and Kristen 1995). Similarly implicating both microfilaments and microtubules in pollen tube growth is an intriguing observation that cytochalasin and colchicine interfere with the pulsatory growth typically displayed by pollen tubes of *N. tabacum* and *Petunia hybrida* (Geitmann, Li and Cresti 1996). The merit of these views can only be judged by further experiments.

The role of actin in the movement of organelles in the pollen cytoplasm is based on the ability of this protein to combine with myosin molecules probably present on the surface of the migrating organelles. Apparently, an actomyosin sliding mechanism in the microfilaments, analogous to that found in smooth muscles, generates the force that directs the movement of the organelles. Analysis of actively growing pollen tubes of *Secale cereale* showed that the movement of individual amyloplasts and mitochondria is not coordinated, suggesting the absence of a general cyclotic flux, and consequently, the transport of these organelles along traffic lanes corresponding to the cytoskeletal elements (J. Heslop-Harrison and Heslop-Harrison 1987). The latter proposition has received strong support from immunofluorescence localization of myosin on organelles isolated from pollen tubes of various plants using an antibody to bovine muscle myosin (J. Heslop-Harrison and Heslop-Harrison 1989d). Biochemical evidence for the presence of myosin in pollen tubes of *Luffa cylindrica* (Cucurbitaceae) has also been obtained at the protein level by SDS-PAGE of total soluble proteins in the presence of rabbit myosin as the standard (Yan et al 1986). In other studies, following SDS-PAGE of proteins of pollen tubes of *N. alata*, *N. tabacum*, *Lilium longiflorum*, and *Tradescantia virginiana*, the gel was probed with antibodies made to the heavy chain of skeletal muscle or of rabbit myosin. A single band in the molecular mass range of heavy chain of myosin (about 170–175 kDa) in the pollen tube extract was found to bind to the antibodies. Suggestive of an association of myosin with organelles and membrane-bound vesicles, antimyosin immunofluorescence microscopy showed the presence of numerous fluorescent spots throughout the pollen tube, with an accumulation at the tip (Tang, Hepler and Scordilis 1989; Terasaka and Niitsu 1994; Tirlapur et al 1995; Yokota et al 1995).

It is clear that the heterologous antibodies used

problems that remain to be solved before any acceptable hypothesis on the mechanism of tip growth can be formulated. The machinery is needed because the amazingly rapid elongation of the pollen tube is controlled by reactions occurring at the tip. Based on the findings from metabolic and structural investigations, three main views to explain tip growth have held the ground for some time now. All three hypotheses possess parts that fit together in certain respects like parts of a jigsaw puzzle, but in other respects they do not have any common ground. Nonetheless, the existence of these hypotheses proves that we do not yet understand the mechanism of pollen tube tip growth.

The most commonly conceived notion about tip growth of the pollen tube is perhaps the one for which there is little experimental evidence. This envisions that cell wall synthesis in the growing tip of the pollen tube proceeds necessarily in exactly the same way as in typical somatic cells of plants. According to this view, many of the precursors that are used for construction of the wall at the tip are supplied from outside, and the actual interconversions and transformations of the precursors into the growing cell wall might occur outside the cell, inside the cell wall, or at the plasma membrane (J. Heslop-Harrison 1987). For efficient cell wall biosynthesis, appropriate enzymes must be present in the pollen tube; incorporation of exogenous precursors by particulate fractions of pollen tubes into (1,3)-β-glucan (Southworth and Dickinson 1975), cellulose (Helsper, Veerkamp and Sassen 1977), pectins (Kroh and Loewus 1968; Kroh et al 1970; Labarca and Loewus 1972, 1973; Rosenfield, Fann and Loewus 1978), and unidentified wall materials (Mascarenhas 1970) suggests that the synthesis of cell wall polysaccharides using common wall precursors may play a significant role in pollen tube growth. Although the wall may be synthesized from the outside, it is not known whether the rate of synthesis is sufficient to account for the massive elongation of the pollen tube. An unexpected result of kinematic analysis is that pollen tubes of species that display a steady state of growth over a period of time have a rather homogeneous distribution of cell wall components whereas others, which have a periodic, bandlike distribution of cell wall components, show pulsatory growth (Pierson et al 1995; Geitmann, Li and Cresti 1996). This observation has some theoretical interest because it underscores the importance of the deposition of cell wall materials in a particular configuration in effecting a specific type of growth.

Another postulate of pollen tube growth comes from observations of somatic cells in which a common force that drives their enlargement is turgor pressure. It has been reasoned that under turgor, as the lateral walls of the pollen tube resist the pressure, the relatively weak tip stretches forward. This hypothesis has benefited from the numerous reports that dictyosome activity at the tip of pollen tubes and the fusion of dictyosome vesicles help provide polysaccharide precursor materials for the synthesis of the plasma membrane and cell wall. These vesicles are transported to the tip by a microfilament-dependent exocytosis. In contrast, incorporation of plasma membrane–generated vesicles into the pollen tube tip (endocytosis) has not been conclusively demonstrated in pollen tubes (O'Driscoll et al 1993). A recent study has, however, revealed that clathrin-coated vesicles, which are active in intracellular membrane traffic in animal cells, are concentrated at the plasma membrane lining the tip of pollen tubes of *Lilium longiflorum* (Blackbourn and Jackson 1996). Thus, although it is clear that exocytosis cannot be the sole mechanism of vesicle accumulation in the pollen tube, there is no clear evidence for endocytosis. The role of Ca^{2+} in cross-linking pectin molecules argues for an interaction of the growing pollen tube with this cation in the medium, so that the tip of the pollen tube is strengthened to resist internal turgor effectively.

The third hypothesis, while conceding the development of turgor pressure in the growing pollen tube, invokes the Ca^{2+}-stabilized microfilaments of the cytoskeleton as providing the main mechanical resistance to pressure at the tip of the pollen tube. An explanation to account for the weakening of the cytoskeleton at the tip that is necessary for the pollen tube to extend forward has focused on the lowering of the local cytoplasmic Ca^{2+} concentration through uptake by mitochondria and/or ER immediately behind the tip. The presence of a tip-directed influx of extracellular Ca^{2+} might suggest that extension of the tip activates channels of ion influx and a loss of turgor abolishes them (Pierson et al 1994). As expounded by Picton and Steer (1982), this model is in some agreement with many experimental data on the physiology of growth of pollen tubes and the role of Ca^{2+} in the process.

5. GENERAL COMMENTS

Our picture of pollen germination on the stigma and pollen tube growth in the style has been strongly colored by the behavior of pollen grains under in vitro conditions in the laboratory. The advantages that initially drew investigators to pollen culture of a few model systems, namely, ease of manipulation of large populations, short germination time course, and the predictable events of gametogenesis, are now afforded by pollen grains of an ever-increasing number of species. Over the years, in vitro pollen germination and pollen tube growth have invited rigorous analytical investigations to provide insightful information on the nature of the milieu of the stigma and style that nurtures pollen grains and pollen tubes. Use of pollen culture combined with highly refined microscopical techniques has also improved our understanding of the mechanism of pollen tube growth. The fact that these studies have proved to be so illuminating underlines a curious irony; despite the importance of the pollen tube as the carrier of the sperm cells, the mechanism of its growth was slow to be recognized. Although it has long been known that there are important developmental differences between the growth of pollen tubes and other plant cells, it is now doubtful whether there are any similarities at all between them.

The concept of gene activation during pollen germination and pollen tube growth has evolved around the utilization of mRNAs stored in the mature pollen grain. One consequence of this is that no new informational type of RNA is synthesized by pollen grains to trigger germination and sustain most of pollen tube growth. Obviously, the pollen grain needs to make only minor adjustments in its genetic program to go from a quiescent state to germination and pollen tube elongation. In this scenario, provision for the availability of stylar nutrients for the continued growth of pollen tubes makes sense, as it probably ensures that sperm cells are delivered to the vicinity of the egg in prime condition. The fate of the pollen tube in the style will be taken up in relation to the process of double fertilization in Chapter 11.

Chapter 9

Developmental Biology of Incompatibility

The ability of a plant to achieve its full reproductive potential depends upon the completion of an uninterrupted cycle of sexual and asexual processes. However, there are certain instances in which physiological and genetic barriers converge to prevent completion of the component processes of sporogenesis, gametogenesis, fertilization, and embryogenesis, and thus thwart seed set. In Chapter 5 it was seen that in a wide range of plants, the arrest of normal pollen development results in male sterility. It is now well established that the molecular and cellular organization of the pollen grain and stigma provides effective recognition systems at the time of pollination for screening suitable gametes for fertilization; this theme permeated most of Chapter 7. This chapter considers the precise genetic control of cell recognition that operates in many plants and enables an individual flower to distinguish between self- and nonself-pollen grains once they land on the stigma and begin to germinate. It is now clear that the sporophytic tissues of the flower play leading roles in both the recognition and the rejection of male gametes that reinforce the outbreeding potential of the species. Although the practical importance of these phenomena has not been fully exploited, they are of great developmental and functional significance in the reproductive biology of angiosperms. Indeed, a case can be made that the evolution of angiosperms as the most advanced group of plants has been dependent upon the development of these built-in mechanisms of cell recognition, adhesion, and communication in the sporophytic tissues of the flower.

The need to maintain genetic variability within a particular species and to preserve the identity and stability of the species within a population defines in a general way the framework within which we can describe the control of pre- and post-pollination and postfertilization events in angiosperms. The genetic mechanisms involved are considered under the rubric of incompatibility, which, for purposes of discussion here, is defined as the inability of the functional gametes of a flower to affect sexual recombination and eventual seed set in a particular combination. Variability within a species is fostered by preventing unions that are too close, such as between gametes that originate from the same individual or, rarely, from other individuals of the same species; the preferred term used for this is self-incompatibility. On the other hand, the boundary of the biological species is maintained by precluding fusions that are too remote, such as between gametes that originate from different species or from different genera; this phenomenon is known as cross-incompatibility. There is considerable knowledge about the cytological and molecular aspects of self-incompatibility, and a number of laboratories are currently deep in these investigations. However, much less is known about cross-incompatibility. This imbalance is reflected in the treatment that follows.

The phenomenon of self-incompatibility, discussed here from the perspective of the pollen grain, operates prior to fertilization. It is considered as an intraspecific reproductive barrier common in flowering plants that results in the failure of fertilization by viable self-pollen. It does not include postzygotic events that lead to embryo lethality and failure of seed set. It may be pointed out with good reason that the requirement of angiosperms for outbreeding is an acute one, since the close proximity of the male and female parts on the same flower could facilitate self-pollination and inbreeding. In consequence, the genetic mechanisms of self-incompatibility, coupled with specializations in floral structure such as dioecism (segregation of sexes to different plants) and dichogamy (temporal separation of sexes or maturation of male and female sex organs on a flower at different times), must ultimately be interpreted in terms of promoting outcrossing. Research by East and Mangelsdorf (1925) provided the first fundamental understanding of the genetics of self-incompatibility in angiosperms. Their work on the self-incompatible *Nicotiana alata* and *N. forgetiana* led to the hypothesis that any pollen grain harboring what appears to be an incompatibility allele (represented by *S*) identical with the one present in the pistil will be rejected in an ensuing postpollination event. The genetic basis for incompatibility formulated in this dictum has gained wide acceptance through research conducted in the decades following its discovery; at the same time, new and more complex systems involving several gene loci have been discovered. Over the past 20 years, *S*-genes have been characterized at the molecular level; this has yielded information about the organization, structure, and nature of the DNA sequences regulating the expression of the genes and the protein products encoded by the genes.

Since the time of the great Darwin when it was recognized that different forms of flowers exist in individuals belonging to the same populations, two groups of self-incompatibility systems have been identified. They are conveniently known as heteromorphic and homomorphic types. In the heteromorphic type, in which cross-compatible plants have visible differences in floral morphology, more than one distinct type of flowers is produced on plants of the same species; in the homomorphic type, however, one type of flower is the rule. The recognition system in the heteromorphic type is often considered to be more complex than that of the homomorphic type.

The intent of this chapter is to focus on the developmental aspects of the two types of incompatibility in angiosperms with reference to the cellular interactions that occur during pollen rejection and inhibition of pollen germination and pollen tube growth; the next chapter will review the molecular studies. Background information on incompatibility is available from a number of sources; in particular, the book by de Nettancourt (1977) is a comprehensive account of the early work on incompatibility and a historical review of the subject is given by Arasu (1968).

1. HETEROMORPHIC SELF-INCOMPATIBILITY

Morphological analysis of heteromorphic systems has focused most of its attention on the relative length of the stamens and styles in the types of flowers produced. Depending upon the floral architecture, heteromorphic self-incompatibility has been attributed to distylous or dimorphic (producing two types of flowers) and tristylous or trimorphic (producing three types of flowers) conditions, with each individual plant producing only one type of flower. Following Huxley (1955), the term "morph" is used to denote the floral forms produced by heteromorphic systems. The complementary arrangement of floral parts is a highly evolved mechanism to promote insect pollination between anthers of one morph and styles of the same length of another morph and thus represents part of an overall strategy to improve the proficiency of cross-pollination.

There is a significant body of information on the breeding systems of heteromorphic plants, especially of the distylous species; yet, the specific mechanisms that plants utilize to activate the acceptance or rejection reactions between compatible and incompatible pollen are poorly understood. What follows below is an overview of our current knowledge of the developmental aspects of heterostyly in certain model systems. Reviews of research on various aspects of heterostyly are found in the book edited by Barrett (1992).

(i) Distyly

Distylous plants constitute a very interesting group to study, since the taxonomic value of the characters of the stamens and style has been more fully appreciated in this system than in the tristy-

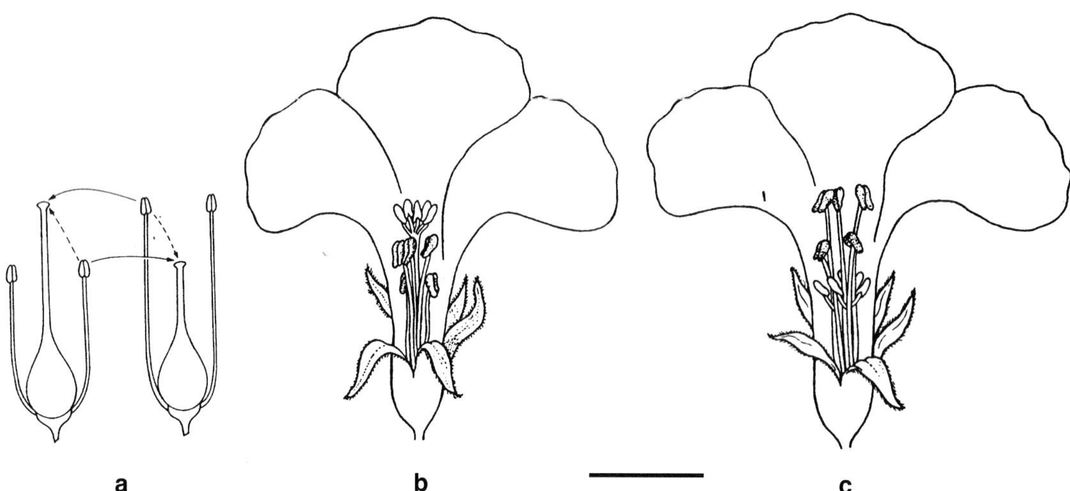

Figure 9.1. **(a)** Diagrammatic representation of distyly. ⟶, compatible; --->, incompatible. (From Shivanna 1982. In *Experimental Embryology of Vascular Plants*, ed. B. M. Johri, pp. 131–174. Berlin: Springer-Verlag. © Springer-Verlag, Berlin.) Pin flower **(b)** and thrum flower **(c)** of *Linum pubescens*. Scale bar = 10 mm. (From Dulberger 1973.)

lous species. In the typical distylous species, one morph is characterized by long style and short stamens and the other possesses short style and long stamens (Figure 9.1). The preferred terms used to designate the long-styled and short-styled morphs are pin and thrum, respectively. The salient feature of the breeding behavior of the two morphs is that successful pollination ensues between flowers of different morphs (intermorph), whereas incompatibility syndromes are unrestrained after pollination between flowers of the same morph (intramorph). Although most of the claims of distyly are based largely on stamen and style length, in many species the incompatibility package includes differences in pollen characters such as size, shape, and exine sculpturing, and in the morphology of the stigma, for example, the density and shape of the papillae.

Another situation in which two morphologically different flower types are borne on the same plant, goes by the name "cleistogamy." The hallmark of cleistogamy is the production of closed (cleistogamous) flowers, which self-pollinate in the bud and chasmogamous flowers, which open normally and outcross. Dimorphic flowers of plants like *Collomia grandiflora* (Polemoniaceae) show significant differences in pollen exine characteristics, structure of stigma papillae, and style lengths, very much as distylous flowers do. However, unlike in distylous plants, intermorph crosses between cleistogamous and chasmogamous flowers are incompatible (Lord and Eckard 1984, 1986). Since no genetic or other control mechanisms are known to prevent self-pollination in cleistogamous flowers, cleistogamy will not be taken up in this chapter again.

By far, one of the most extensively investigated distylous genera is *Linum* (Ockendon 1968; Dulberger 1973, 1974, 1981, 1987; Rogers 1979; Ghosh and Shivanna 1980a, b). In several species of *Linum*, the reciprocal arrangement of the stamens and style is associated with differences in the surface features of pollen grains and stigma. Generally, the pollen exine of thrum flowers is ornamented with excrescences of uniform size, whereas in pin morphs the outgrowths are of two or more types (Dulberger 1973, 1981; Rogers 1979). Similarly, the stigmatic papillae of the two morphs are of different sizes and exhibit differences bordering on the structure of the walls of the papillae. For example, in *L. grandiflorum* and *L. pubsecens*, the stigmatic papillae of pin morphs are nearly twice as long as those of thrum morphs; in the former, the cuticle covering the papillae is wavy, continuous, and thicker than in the thrum papillae, where it is highly irregular and nonuniform. A secondary effect of cuticular modification in the thrum morph is the accumulation of secretion products on the surface of the papillae, making the stigma a wet type (Dulberger 1974, 1987; Ghosh and Shivanna 1980a, b). Intermorph differences between exine ornamentation on pollen grains and the thickness of the papillar cuticle have also been described in several members of Plumbaginaceae (Baker 1966; Dulberger 1975). No differences are seen between the cellular organization of the stigmas of the pin and thrum morphs of *Primula vulgaris* (Y. Heslop-Harrison, Heslop-Harrison and Shivanna 1981). In *P. obconica*, however, intermorph differences are seen in the stigmatic papillae, which are small, smooth, and dry with a thin layer of exudate in the thrum morph, and large, uneven, and wet with copious exudate in the pin morph; aspects of dimorphism even extend to the nature of the enzymes localized in the exudates of the two types of stigmas (Schou 1984; Schou and Mattsson 1985). Besides obvious differences in the structure of the stigmatic papillae, morph differences in the stigma length have been reported in certain distylous genera of Rubiaceae (Murray 1990).

Physiological differences between pollen grains and stigmatic surfaces of floral morphs are important in controlling incompatibility responses of distylous plants. Since such differences are less obvious than the readily detected morphological and structural differences, they have not been subjected to critical investigations. Most of the efforts to identify physiological differences have taken place since the 1940s. Lewis (1943) called attention to the existence of a difference in osmolarity between the stigmatic cells and pollen grains of *L. grandiflorum* and suggested that this apparently accounts for the incompatibility response. It is generally observed that pollen grains of the pin morph do not absorb water from the papillae nor swell on the pin stigma due to low osmolarity of the pollen grain relative to that of the stigmatic cells. In contrast, thrum pollen grains imbibe water readily on the pin stigma due to their lower water potential compared to pin pollen. Pollen water economy appears to be a factor in intermorph differences in *P. vulgaris* also. This is indicated by a requirement for controlled hydration for normal germination and pollen tube growth of thrum pollen, whereas pin pollen is little affected by atmospheric conditions. The two types of pollen grains also display marked differences in their capacity to imbibe atmospheric moisture and swell, pin pollen being slower than thrum pollen in this respect. In a related way, the ability of thrum pollen to imbibe water is lost when surface materials of the exine are leached by organic solvents; modifying the surface of pollen grains of the pin morph does not interfere with the hydration process. Since absorption of water is the initial preparatory process for crossing the incompatibility barrier, membrane functions appear to be at the hub of the physiological processes controlling pollen acceptance or rejection reactions (Shivanna, Heslop-Harrison and Heslop-Harrison 1983). In *L. grandiflorum*, the stigma of the thrum morph is characterized by an activity of nonspecific esterases and acid phosphatases higher than that of pin morph. Both morphs also display differences in the stigmatic protein profile (Ghosh and Shivanna 1980a, b). However, differences in lipid components are nonexistent between the two types of flowers in *Forsythia intermedia* (Dumas 1977a). The most obvious physiological manifestation of dimorphism is in the external character of the stigma itself, but this has not lent itself to unequivocal conclusions. For example, in *P. obconica*, the thrum stigma is dry and the pin stigma is wet (Schou 1984), but the opposite is the condition in several species of *Linum* (Dulberger 1987).

(ii) Tristyly

Analysis of tristyly as an outbreeding mechanism has been pursued in a few members of Oxalidaceae, Pontederiaceae, and Lythraceae. The distinctive feature of tristyly is the presence of three floral morphs, namely, long-styled, mid-styled, and short-styled. The attachment of stamens of two different lengths in each morph, corresponding to the lengths of the styles in the other two morphs, adds to the complexity of the incompatibility system in tristylous plants. Thus, long-styled morphs have mid-length and short stamens, mid-styled have long and short stamens, and short-styled have long and mid-length stamens. In terms of compatibility relations, the general principle is that pollinations are successful between anthers and stigmas situated at an equivalent level in the different floral morphs (Figure 9.2). For example, in a long-styled morph, successful pollinations are feasible between the pollen of short stamens and the style of a short-styled morph, and between the pollen of mid-length stamens and the style of a mid-style morph (Devi 1964; Ornduff 1964, 1966; Dulberger 1970; Barrett 1977; Price and Barrett 1982; Glover and Barrett 1983; Richards and Barrett 1987; Scribailo and Barrett 1991a).

Tristyly in *Lythrum junceum* (Lythraceae)shows certain similarities to distyly, such as heteromorphism in the size of pollen grains and stigmatic papillae. A correlation is found between long-level anthers and large pollen grains, short-level anthers and small pollen grains, and mid-level anthers and pollen grains of intermediate size. Similarly, papillae of long-level stigmas are significantly longer than those of the mid-level and short-level stigmas (Dulberger 1970). Although long-level anthers of *Pontederia cordata*, *P. rotundifolia*, and *P. sagittata* alternate morph, in contrast, grow straight down in the style to their destination in the ovary (Lewis 1943). In an experiment wherein massive germination of self-pollen easily occurred on the thrum morph stigma of *Luculia gratissima* (Rubiaceae), it was possible to localize the incompatibility reaction zone in the stigma, because amputation of the stigma and upper part of the style eliminated the incompatibility reaction (Murray 1990). In contrast, in *Limonium vulgare*, incompatibility relies on the failure of pollen grains to germinate or of the pollen tubes to enter the stigma (Baker 1966). This rules out any possibility of the effect of stylar tissues in promoting incompatibility.

The perennial herb *Primula* has served as a superb material for study of the efficacy of pollen germination and the fate of pollen tubes during intra- and intermorph pollinations because of the transparency of the style and the small size, hardiness, and availability of the popular garden plant in many parts of the world. Although there are some conflicting reports on the fate of the pollen tube following intramorph pollinations, it is clear that the behavior of the pollen grain varies between the morphs. According to Lewis (1942), differential growth of pollen tubes characterizes intramorph pin and thrum pollinations in *P. obconica* and *P. sinensis*; the growth of the thrum pollen tube in the style of the same morph is poorer than in the pin × pin cross. In *P. obconica*, other reports have claimed the existence of rigorous incompatibility reactions characterized by the lack of pollen tube penetration of the stigma after pin and thrum self-pollinations, despite respectable levels of germination (Dowrick 1956; Stevens and Murray 1982). This result leaves no doubt that the expression of incompatibility occurs after germination. The breeding behavior of *P. veris* illustrates yet another aspect of the incompatibility phenomenon in this genus. Production of short, broad pollen tubes, coupled with their inability to make inroads into the style, accounts for incompatibility in pin × pin pollinations; in thrum × thrum pollinations, pollen germination is very low, and the pollen tubes are so unusually short and sinuous that it is doubtful that they could ever cross the stigmatic papillae and grow (Richards and Ibrahim 1982). From this it appears that, on the generic level, the progress of the incompatible pollen in *Primula* may be checked at several stages, such as during germination, during pollen tube penetration of the stigma, or during pollen tube growth in the style. Similar conclusions have been derived from a study of *P. vulgaris;* here a persuasive case has been made that incompatibility results from the additive effects of inhibition at different levels rather than at a single barrier (Shivanna, Heslop-Harrison and Heslop-Harrison 1981). In an investigation of within-morph crosses involving 52 species of *Primula*, pollen tube growth inhibition has been identified to occur in the stigma, style, or ovary, although most often failure of pollen germination at the stigmatic surface results in efficient incompatibility (Wedderburn and Richards 1990). Following pin × pin pollination in *Pulmonaria affinis* (Boraginaceae), a few pollen tubes make their way to the ovary wall but most end up in the style, as do pollen tubes from a thrum × thrum cross (Richards and Mitchell 1990). Inhibition of pollen tube growth in different regions of the stigma and style also occurs following incompatible pollinations of thrum and pin stigma with self-pollen in certain distylous species of Rubiaceae (Bawa and Beach 1983) and in *Averrhoa carambola* (Oxalidaceae) (Wong, Watanabe and Hinata 1994b).

Impairment of pollen tube growth in the stigma or in the upper part of the style probably accounts for incompatibility in the tristylous species, as seen from the analysis of intra- and intermorph pollinations in *Lythrum salicaria* and *L. junceum* (Esser 1953; Dulberger 1970). In *Pontederia sagittata* and *P. cordata*, pollen tubes make their way into a considerable length of the style, even into the ovary, before rejection occurs (Glover and Barrett 1983; Anderson and Barrett 1986; Scribailo and Barrett 1991b).

It is not possible to offer a general mechanism to explain the incompatibility behavior in heterostylous plants. A few ideas current in the literature are totally inadequate for more than a superficial insight into the factors involved in the rejection reactions. Unfortunately, there seems to have been no rigorous molecular studies on these intriguing systems.

Since water uptake is an early event in pollen germination, some perception of incompatibility may be gained from systems in which the rejection reaction is due to the failure of germination. A good example is *Linum grandiflorum*, in which there is a large-scale failure of pollen adhesion following pin × pin pollination, or failure of swelling of the few pollen grains that adhere. As indicated earlier, the failure of pollen grains to absorb water has been attributed to their low osmolarity compared to the stigmatic cells. Following thrum × thrum pollination in this species, the pollen tube bursts in the tissues of the style, presumably due to its higher osmotic potential relative to that of the cells of the style (Lewis 1943). Although determination of the relative osmotic potential of the pollen grain and stylar cell in *L. grandiflorum* supports a role for osmotic changes in pollen tube demise in the style, the work has not been extended to other species.

Another mechanism that has been proposed for incompatibility in heterostylous species is based on the differences in the distribution of the flavonols quercetin and rutin in the pollen grains of short-styled and long-styled plants of *Forsythia intermedia* (Moewus 1950). Quercetin is found almost exclusively in the pollen grains of pin morphs and rutin is found exclusively in the pollen grains of thrum morphs; paradoxically, the specific enzymes that degrade quercetin and rutin occur only in the styles of the opposite morphs. Because flavonols inhibit pollen germination, a model in which rutin of the thrum pollen is destroyed by the enzyme present in the pin style and quercetin of the pin pollen is destroyed by the enzyme found in the thrum style, would assuredly explain successful compatible intermorph pollinations. On the other hand, following incompatible pollinations, the flavonol inhibitors are not destroyed, due to lack of appropriate enzymes, and the result is the failure of pollen germination. Obviously, these effects of flavonols could serve as effective biochemical mechanisms to promote or impede pollen germination in vivo, but the hypothesis did not really take off, so to speak, and has not been useful for studying the mechanism of heteromorphic incompatibility for two reasons. One is the failure to identify the two specific enzymes that inactivate quercetin and rutin; second, and more importantly, is the reported failure to reproduce the results in populations of the species (Lewis 1954).

Because of their enzymatic nature, proteins have emerged as natural candidates to stop the growth of pollen tubes following intramorph pollinations. A recurrent view is that inhibition of growth of the incompatible pollen tube is somehow dependent upon continued protein synthesis, for when protein synthesis in the style is inhibited by cycloheximide, inhibition of pollen tube growth is partially reversed (Ghosh and Shivanna 1982a). Proteins present in the stylar extracts have also been shown to differentially inhibit in vitro growth of pollen tubes of pin and thrum morphs; pollen tubes of the same morph are more sensitive than those of the other morph to a stylar extract (Golynskaya, Bashkirova and Tomchuk 1976; Shivanna, Heslop-Harrison and Heslop-Harrison 1981). In the absence of refined biochemical studies, these observations fall short of a completely critical test for the involvement of protein synthesis in the incompatibility reactions. Suggesting a link between peroxidase activity and incompatibility response, an ultracytochemical study has revealed a marked accumulation of the enzyme in pin × pin pollinated styles of *Primula acaulis* and its absence in the cross-pollinated styles (Carraro, Lombardo and Gerola 1985). Differences reported in the soluble protein profiles of the styles of thrum and pin morphs of *Averrhoa carambola* are too insignificant to account for inhibition of pollen tube growth following pollination with the same-morph pollen (Wong, Watanabe and Hinata 1994a).

Based on a study of pollen–pistil interactions in *Pontederia sagittata*, it has been suggested that incompatibility in heterostylous systems is governed more by the interaction of heteromorphic characters than by an active rejection mechanism based on molecular specificities. Compelling evidence comes from the observation that incompatibility in certain pollen–pistil combinations involving pollen grains smaller than the legitimate size class is accompanied by an attrition of pollen tubes that occurs due to their inability to grow the required distance to the ovary. This, in turn, is attributed to the presence of storage reserves in the pollen grain insufficient for the tube to reach the ovule or due to impairment of an essential metabolic function (Scribailo and Barrett 1991b).

The problems relating to the physiology of incompatibility in heterostylous systems rest here at the present time. Admittedly, we do not have very much to say on the physiology of incompatibility in distylous species, and the basis of incompatibility in tristylous species stands out as a conspicuous unknown in this whole area of limited understanding. Not only is heterostyly an important genetic adaptation, but an understanding of it will no doubt inform us about aspects of the physiology of the pollen grains, stigma, and style that are crucial in the incompatibility response.

2. HOMOMORPHIC SELF-INCOMPATIBILITY

Homomorphic self-incompatibility is relatively more widespread than the heteromorphic type, and representative genera exhibiting homomorphic systems are found in more than half of the families of angiosperms, with prospects that many more will be identified (Brewbaker 1959). Whereas heteromorphic systems possess morphological mechanisms in the flower to discourage self-pollination, homomorphic plants are able to identify and reject self-pollen by genetic and physiological

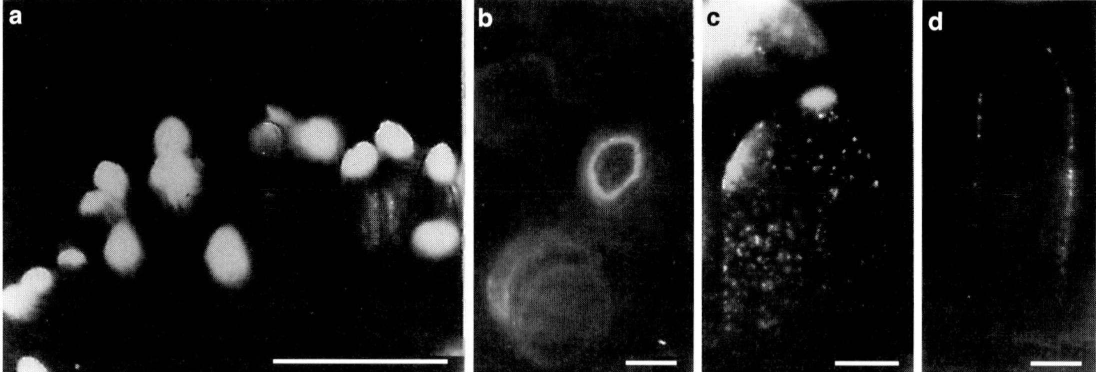

Figure 9.4. Fluorescence micrographs of aniline blue–stained stigmatic papillae of *Raphanus sativus* showing callose formation following an incompatible pollination. **(a)** Section of a portion of the stigma showing callose formed at the plasmalemma of each papilla. **(b)** Formation of annular callose in a papilla. **(c)** Diffuse callose reaction with dispersed granules in a papilla. **(d)** Same as in (c), focused to show callose in the plasmalemma. Scales bars = 100 μm (a) and 10 μm (b, c, d). (From Shivanna, Heslop-Harrison and Heslop-Harrison 1978; photographs supplied by Dr. K. R. Shivanna.)

as adhesion and hydration of pollen grains and erosion of pollen cell wall components, seem to be important in the rejection of self-pollen. Since self-pollen grains initially bind to the stigma less firmly than cross-pollen, there is reason to believe that components that bind the pollen to the stigma are differentially altered during self- and cross-pollinations (Stead, Roberts and Dickinson 1979). An early scanning electron microscopic study claimed that in *B. oleracea*, the incompatibility barrier is a waxy layer on the stigmatic surface on which pollen grains fail to adhere and germinate. Adhesion and germination are feats easily accomplished by the compatible pollen grains whose tubes penetrate the cuticle (Roggen 1972). However, once self-pollen grains stick to the stigmatic cells, they germinate, and pollen tubes that penetrate the base of the stigma have an equal chance as cross-pollen in effecting fertilization (Ockendon and Gates 1975).

Irrespective of the extent of pollen adhesion on the stigma, some rejection signal is transmitted from the stigmatic cells to the pollen grains. Symptoms of self-rejection are seen in the physical appearance of the pollen grains on the stigma and in the loss of their cellular integrity. Pollen grains of *Brassica oleracea* subject to a self-incompatibility episode at the stigmatic surface have higher levels of hydrolases (Dhaliwal and Malik 1982) and membrane-bound Ca^{2+} (Singh, Perdue and Paolillo 1989) than cross-pollen. Rejection signals evoke permissive conditions within the confining membrane of the pollen grain that sequester enzyme molecules and Ca^{2+} ions, although this can hardly be considered as the primary cause of pollen rejection.

Callose Deposition. One aspect of *S*-gene expression peculiar to sporophytic self-incompatibility systems is the reaction of the stigmatic papillae following self-pollination (Figure 9.4). When self-pollen grains make contact with the stigma, the cuticle of the papilla is eroded at the same time that the external layer of the cell is digested. This is a prelude to the deposition of callose at the tip of the papilla between the cell wall and the plasma membrane (Dickinson and Lewis 1973a; Knox 1973; J. Heslop-Harrison, Knox and Heslop-Harrison 1974; Vithanage and Knox 1977; Shivanna, Heslop-Harrison and Heslop-Harrison 1978). Callose deposition appears to be a Ca^{2+}-dependent process in self-pollinated *Brassica oleracea* because it is abolished by deprivation of the ion by Ca^{2+} channel antagonists and induced by Ca^{2+} ionophore. However, germination of self-pollen was prevented, and incompatibility prevailed when callose formation was abolished by treatment of stigmatic papillae with 2-deoxy-2-glucose. This observation seems to fit rather well with the view that callose formation in the stigma is not a requirement for the demise of self-pollen or for incompatibility functions (Singh and Paolillo 1990; Elleman and Dickinson 1996).

It has been claimed that the callose response is not pollen specific (Sood, Prabha and Gupta 1982); however, a number of observations made on *B. oleracea*, such as the induction of callose response on the stigmatic papillae by a diffusate of self-pollen but not by that of compatible pollen, activity of the diffusate at low concentrations, and the rapidity of the response, among others, seem to indicate that the callose plug formation on the stigmatic papillae is a dependable phenotypic expression of incompatible pollination (Kerhoas, Knox and Dumas 1983).

The importance of callose deposition in incompatibility mechanisms is dramatically demonstrated in the growth of pollen tubes following self-pollination in gametophytic systems (Figure 9.5). In many plants, callose deposition at regular intervals is a normal way of life for pollen tubes as they grow following a compatible mating. The suspected specific function of callose in the pollen tube is to isolate the cytoplasm harboring the sperm cells and the vegetative cell nucleus from the empty pollen grain. The distinctiveness of the incompatible pollen tube is the irregular nature of the callose deposits, often leading to the plugging of the entire tip of the pollen tube. Some pollen tubes accumulate so much callose that the tips swell until they burst to death within the style (Tupý 1959; Schlösser 1961; de Nettancourt et al 1974; J. Heslop-Harrison, Knox and Heslop-Harrison 1974; Sastri and Shivanna 1979; Vithanage, Gleeson and Clarke 1980); in others, callose is deposited even inside the pollen grain, beginning at what appears to be the site of pollen tube emergence (FranklinTong and Franklin 1992).

Ultrastructural Cytology. Our understanding of the structural basis of incompatibility reactions is linked almost exclusively to electron microscopic investigations of pollen tubes following self- and cross-pollinations. The foundation for fine structural study was laid by van der Pluijm and Linskens (1966), who noted that incompatible pollen tubes of *Petunia hybrida* have thickened walls and their organelles and cytoplasmic contents suffer a loss of identifiable form. A follow-up work showed that there is a scarcity of smooth tubular ER and polysomes in the incompatible pollen tubes; in addition, fibrous masses found in the nuclear zone of the compatible pollen tubes are absent in the corresponding region of the incompatible tubes. These observations led to the suggestion that incompatibility is probably due to the secretion and production of membrane components, synthesis of growth hormones, decreased protein synthesis, and callose accumulation (Cresti et al 1979). The anarchic ultrastructural changes seen in the incompatible pollen tubes are not surprising in view of the rapid rejection reactions that set in following self-pollination. Herrero and Dickinson (1981) confirmed many of these findings and showed that although both compatible and incompatible tubes of *P. hybrida* begin to grow rapidly upon entering the style, the growth of incompatible tubes slow downs. This might signify a metabolic imbalance in the incompatible pollen tube that leads to a decreased turgor pressure, rather than reflect a deposition of callose, as is generally assumed.

There are no major ultrastructural differences during the activation and germination of compatible and incompatible pollen grains of *Lycopersicon peruvianum* (Cresti, Ciampolini and Sarfatti 1980). However, modifications of the pollen tube tip remain an essential baseline for interpreting the demise of germinating pollen grains following incompatible pollination. Crucial changes in the tip of the incompatible pollen tube during its journey through the style are the appearance of numerous granules in the cytoplasm, configuration of the rough ER in the form of whorls of concentric parallel membranes, disappearance of the inner wall, thickening of the outer wall, and lysis of the pollen tube. Incompatibility reactions that have been proposed as causal for the growth arrest of the pollen tubes are cessation of protein synthesis and the binding of an incompatibility protein to some constituents of the inner wall to produce particles that accumulate in the cytoplasm (de Nettancourt et al 1973a, 1974). Two brief investigations on *Lilium longiflorum* seem to reflect the differences in metabolism and synthesis of nucleic acids and proteins by pollen tubes following self-compatible and self-incompatible pollinations and suggest that these changes may possibly influence pollen tube wall formation as well as pollen tube growth (Campbell and Ascher 1975; Li and Linskens 1983a).

The results clearly show that even though light microscopic work led to the belief that callose formation is a prelude to the commencement of incompatibility reaction, it is not the primary event that causes the arrest of pollen tube growth. Be that as it may, callose formation may well be necessary for the continuation of the rejection reaction. In grasses, it has been found that a thickening of the pollen tube tip by the apposition of microfibrillar pectic materials is the earliest morphological abnormality in the incompatible pollen tube and is initiated after the contact of the pollen grain with the stigma. This may be due to a dislocation of the normal pattern of wall growth, which involves the precise cross-linking of microfibrils in the emerging pollen tube. Callose is initially deposited behind the pollen tube and later extends even to its tip (Shivanna, Heslop-Harrison and Heslop-Harrison 1982). According to Geitmann et al (1995), during the incompatibility reaction in *Brugmansia suave-*

S-gene action in so far as it relates to sporophytic incompatibility.

A few comments about the organization of the S-locus are also relevant to understanding the functioning of the self-incompatibility gene. A model that can claim some validity is attributed to Lewis (1949, 1960), who proposed that the S-gene is composed of three closely linked parts. It has been reasoned that in order for the alleles of the same gene to be effective in both the pollen and the style for the incompatibility reaction, allelic specificity of both is determined by a common segment of the gene; accordingly, the first segment of the gene, referred to as the specificity part, is assumed to carry genetic information affecting pollen tube growth in the style following an incompatible cross. The second and third segments of the gene apparently determine the allelic specificity of the pollen and style, respectively. Although there are a number of lines of evidence to support this model, the most direct one is based on mutants that affect the pollen part without any effect on stylar behavior and that affect the style without interfering with the pollen activity part. Several questions about the molecular structure of the S-locus beg resolution, the most important one being the number of functional genes in the S-locus. Interpretations favoring the involvement of one, two, or even three genes to effect S-allele–specific recognition following incompatible crosses have been proposed, but hard evidence in support of them is scanty (Singh and Kao 1992).

(iii) Physiology and Biochemistry of Self-Incompatibility

Under this heading, several aspects need to be considered. First, what are the specific physiological changes triggered by self-compatible and incompatible pollinations? What is the nature of the S-allele products and where are they synthesized? What is the biochemical basis for self-incompatibility recognition? The events alluded to in these questions mark the early signs of change in the stigma and style, and represent a basic feature of the homomorphic self-incompatibility system.

Many investigators studying the physiology of cross- and self-pollination have looked at the post-pollination reactions of the flower, especially those of the pistil. One view of the physiological changes in the flower following pollination is based on the principle of translocation of organic substances from the site of their synthesis (source) to the site of their utilization (sink). In an unpollinated flower, stamens and carpels serve as the sink for attracting metabolites derived from the vegetative parts of the plant as well as from the calyx and corolla. According to Linskens (1975), in the unpollinated flower of *Petunia hybrida*, the anthers function as the primary sink for ^{14}C-protein hydrolyzate, with some label accumulating in the style also. Among pollination-induced changes is a strong influx of the label into the style and the ovary about 10 hours after self-pollination and a weak influx into the same organs of cross-pollinated flowers. After about 18 hours, the ovary continues to function as the sink in the cross-pollinated flower, whereas the self-pollinated flower suffers a major efflux of nutrients. Changes in phytic acid levels that occur in the flower can be considered in this context, since incompatible pollination leads to a more rapid decrease of phytic acid in the pistil than does compatible pollination (Jackson, Kamboj and Linskens 1983). Because the growth of the pollen tube in the self-pollinated flower stops about 12 hours after pollination, the efflux of nutrients after this time is not unexpected. Although significant metabolic changes are set in motion in the pistil following pollination, it has been difficult to identify specific differences between self- and cross-pollinated pistils in all instances. In general, the pistil of *P. hybrida* registers a dramatic increase in respiratory activity as the first of a cascade of changes following both compatible and incompatible matings, although the incompatibly pollinated pistil is unable to carry out normal respiration after about 12 hours (Linskens 1955). Some investigators have shown that an increase in CO_2 concentration in the air effectively eliminates the incompatibility reaction in *Brassica campestris* by overcoming the barrier to pollen germination (Nakanishi, Esashi and Hinata 1969; Nakanishi and Hinata 1973; Dhaliwal, Malik and Singh 1981; O'Neill, Singh and Knox 1988). Although no clues are at hand to help us understand the effect of CO_2 in this system in overcoming self-incompatibility, the possibility that the gas modifies the surface properties of the stigmatic papillae should not be discounted.

There are some reports of changes in the protein and nucleic acid contents of the style following self- and cross-pollinations. In *P. hybrida*, following cross-pollination, a single new glycoprotein fraction appears in the style. After selfing, however, two glycoprotein fractions with different mobilities are identified. Apparently, these glycoproteins are synthesized by contributions from both the pollen grains and the style (Linskens 1959). In this same species, there is a fall in the protein content of the style in the first 24 hours after both self- and cross-pollinations, but the fall is more marked in the selfed styles. The decrease in protein content does not result in a corresponding rise in the free amino acid pool, probably because the amino acids released from protein breakdown are used in the respiratory pathway (Linskens and Tupý 1966). Somewhat similar changes occur in the free amino acid pools of the pistils of *Nicotiana alata* following self- and cross-pollinations (Tupý 1961). Differences noted in the electrophoretically separated protein banding patterns between prefertilization ovaries of self- and cross-pollinated *Lotus corniculata* flowers suggest that biochemical self-recognition occurs in the ovary before fertilization (Dobrofsky and Grant 1980).

One of the enzymes implicated in incompatibility in a causal way is peroxidase. This assessment of the role of peroxidase was originally based on the work of Pandey (1967), who showed that a specific peroxidase isozyme is associated with the *S*-allele of *N. alata*. Since peroxidase is known to destroy IAA, it was logically thought that the enzyme might cause incompatibility by destroying IAA necessary for pollen tube growth. Although subsequent workers were unable to observe any apparent association between self-incompatibility and peroxidase activity, persistent differences in enzyme activity in some plants nonetheless occur between self- and cross-pollinated styles (Desborough and Peloquin 1968; Nasrallah, Barber and Wallace 1970; Nishio and Hinata 1977; Bredemeijer and Blaas 1980). For example, after one day, self- and cross-pollinated pistils of *N. alata* do not register a change in the total peroxidase activity compared to that of unpollinated pistils. In the next three days, there is a steady increase in the total enzyme activity, which is higher in cross-pollinated styles than in self-pollinated ones. After four days, enzyme activity remains more or less constant in self-pollinated styles but increases markedly in cross-pollinated ones (Bredemeijer 1974). A positive correlation also exists between peroxidase isozyme-10 and the intensity of the incompatible reaction, indicating that peroxidase-10 is in some way involved in pollen tube rejection. It has been proposed that after a compatible pollination, the pollen tube grows so fast that its tip marauds those parts of the style where peroxidase-10 is not present; on the other hand, after an incompatible pollination, because of the slow growth of the pollen tube, the tip is inextricably caught in the part of the style with a high peroxidase-10 activity and, as a result, growth is inhibited (Bredemeijer and Blaas 1975). Ultracytochemical analysis has shown that in *P. hybrida*, there is a substantial increase in the number of peroxidase-containing cells in the transmitting tissue of self-pollinated styles compared to that of unpollinated and cross-pollinated styles (Carraro, Lombardo and Gerola, 1986).

We have also some information on the relationship of isozymes of esterase to incompatibility. This is based on the observation that heat treatment of styles of self-incompatible genotypes of several species of *Nicotiana* and *Lilium longiflorum* leads to the disappearance of certain isozyme bands while others remain stable; the compatible genotypes show no changes in the isozyme pattern (Pandey 1973). Since heat treatment is known to eliminate incompatibility reactions in *L. longiflorum*, these observations support the concept of differential genetic control of the synthesis of a specific enzyme (Ascher and Peloquin 1966b; Dickinson, Moriarty and Lawson 1982). Studies on *P. hybrida* have shown a correlation between the appearance of isozymes of glutamate dehydrogenase and *S*-genotypes; activities of several glucan hydrolases register an increase that is greater in cross-pollinated styles than in selfed ones (Roggen 1967; Linskens et al 1969).

Other investigations on *P. hybrida* have revealed distinct differences in the synthesis of RNA and proteins and in the size of the free nucleotide pool in the styles after cross- and self-pollinations. Particularly striking is the observation that RNA synthesis is appreciably enhanced as early as 3 hours in cross-pollinated compared to self-pollinated styles; protein synthesis in the compatibly pollinated styles also peaks earlier than in their selfed counterparts (van der Donk 1974a, b; Deurenberg 1976). Electrophoretic analysis of proteins translated by polysomes isolated from self- and cross-pollinated pistils and injected into eggs of *Xenopus laevis* revealed some minor differences in the protein bands as early as 6 hours after pollination; the differences were accentuated by 24 hours (Deurenberg 1977). These synthetic activities are probably triggered by pollen recognition events and therefore reflect differential gene expression concerned with pollen acceptance or rejection. Observations such as these are of course inadequate as anything more than a first step in implicating nucleic acid and protein synthesis in the development of incompatibility reactions.

Another useful first step in ascertaining the role of nucleic acid and protein synthesis in self-incompatibility reactions is the use of metabolic inhibitors. One general approach employed con-

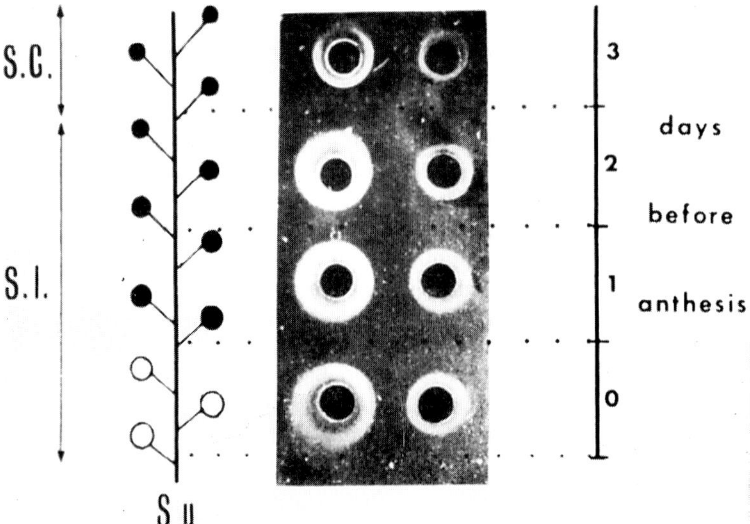

Figure 9.6. Immunodiffusion (central panel) showing precipitation patterns of stigmatic antigen along the inflorescences of incompatible (left) and compatible (right) homozygous lines of *Brassica oleracea*. The diagram on the left shows an incompatible inflorescence with self-compatible (S. C.) and self-incompatible (S. I.) zones; open circles, flowers; solid circles, buds 1, 2, and 3 days before anthesis. (From Nasrallah 1974.)

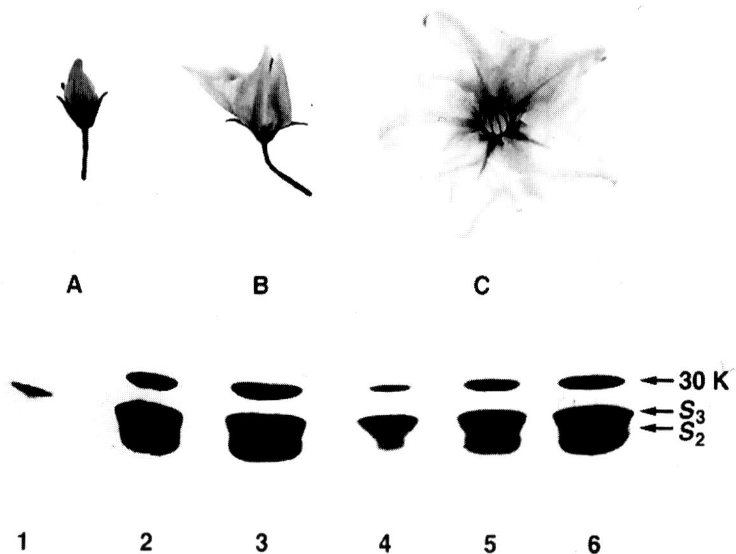

Figure 9.7. Development of *S*-allele–associated proteins in the style of *Lycopersicon peruvianum* (genotype S_2 S_3). The panel shows a segment of the SDS-PAGE of extracts of styles of flowers at the different stages of development indicated at the top. Lanes 1 and 4, green bud (A); lanes 2 and 5, yellow bud one day before opening (B); lanes 3 and 6, mature bud one day after opening (C). Lanes 1–3, extracts prepared from six styles; lanes 4–6, extracts with equal amounts of protein. (From Mau et al 1986. *Planta* 169:184–191. © Springer-Verlag, Berlin. Photographs supplied by Dr A. E. Clarke.)

The isolation and characterization of genes encoding components of the self-incompatibility response (Chapter 10) have led to the establishment of correlations between the accumulation of *S*-locus RNA and the acquisition of self-incompatibility. This was first shown in *B. oleracea* by Northern blot hybridization of RNA from stigmas of different stages of development with a cDNA clone containing sequences specifying an *S*-locus–specific glycoprotein. The level of the transcript was found to

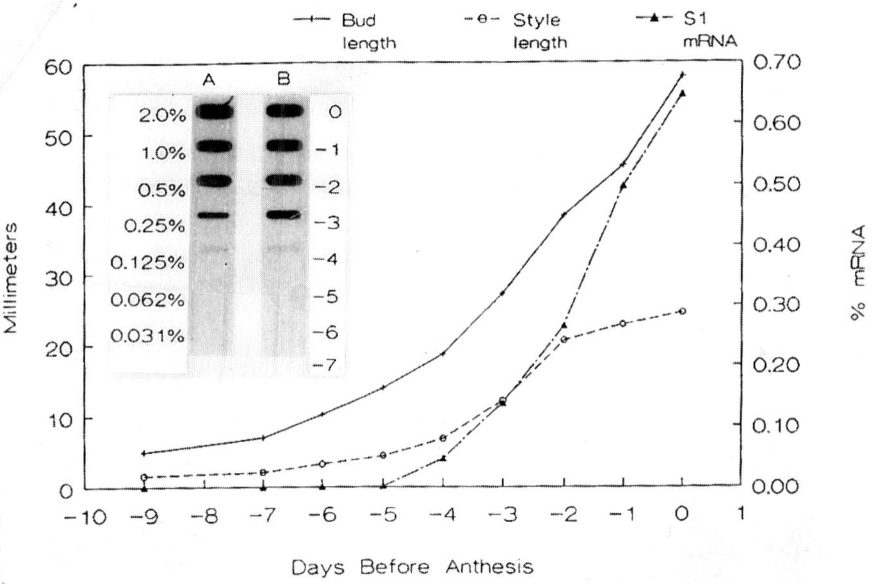

Figure 9.8. Developmental expression of *S*-locus mRNA in a self-incompatible homozygous line of *Petunia hybrida*. Inset: Total RNA isolated from floral buds of different stages or RNA transcribed in vitro from a cDNA clone (*PS-1B*) encoding the incompatibility proteins was hybridized with the same cDNA clone in a slot–blot apparatus. Lane A, hybridization of serially diluted RNA transcribed in vitro by *PS-1B*; lane B, total RNA isolated from styles of different ages as indicated by days prior to anthesis on the right. (Clark et al 1990. *Plant Cell* 2:815–826. © American Society of Plant Physiologists.)

increase as a function of maturity of the stigma and to reach maximum level with the onset of the self-incompatibility response (Nasrallah et al 1985). This was confirmed in a subsequent work, which involved in situ hybridization of stigma sections with a labeled probe (Nasrallah, Yu and Nasrallah 1988). Both Northern blot and in situ hybridization methods have shown that the expression of a cDNA clone encoding *S*-allele protein of *Nicotiana alata* correlates well with the development of incompatibility, transcripts for the gene being generally absent in the immature style, but appearing in the style of mature flowers (Anderson et al 1986; Cornish et al 1987). In *P. hybrida* (Figure 9.8), quantitative slot–blot assay revealed that the accumulation of *S*-locus mRNA parallels the acquisition of incompatibility, with a major accumulation of transcripts occurring during the transition from compatibility to incompatibility (Clark et al 1990). In *Solanum tuberosum*, the expression pattern of cloned mRNA transcripts fits with the functional requirement of providing the mature pistil with high concentrations of S-proteins (Kaufmann, Salamini and Thompson 1991).

It has not been determined whether the relevant mRNAs that code for recognition molecules in the style and stigma are produced at an early stage in floral ontogeny but remain masked until a late stage. It was mentioned earlier that in plants expressing self-incompatibility late in floral development, bud pollination results in the production of a full complement of seeds. Evidently, the components necessary for normal seed set are synthesized early in floral development and remain stable until anthesis, but the expression of incompatibility factors is confined to a narrow developmental window during floral ontogeny. Why do incompatibility factors remain unexpressed until late in the development of the flower? This is an intriguing question that cannot be answered at present.

(iv) Pollen Recognition Factors

Rejection of self-pollen in an incompatible reaction is initiated following contact of pollen grains with the stigmatic cells of a flower carrying the same *S*-allele. This interaction tacitly assumes a role for recognition factors of pollen grain origin in the rejection process. In *Brassica oleracea* and other plants exhibiting sporophytic self-incompatibility, substances responsible for the rejection reaction are presumably synthesized in the tapetum and transferred to the cavities of the sculptured layer of the pollen exine. Because these substances can be easily eluted from the exine, they have become prime candidates for the kind of rapid rejection reaction in the stigma following an incompatible pollination. It was mentioned earlier that in most cases of sporophytic incompatibility, self-pollinations lead to the failure of pollen germination and to the formation of callose at the point of pollen contact with the stigmatic papillae. Using callose formation as a bioassay, J. Heslop-Harrison, Knox and Heslop-Harrison (1974) induced a rejection reaction on the stigma of *Iberis sempervirens* by deposition of an agar film in which the exine proteins are collected. It was possible to duplicate the rejection reaction by apposing pieces of the intact tapetum isolated from incompatible anthers to the stigma. The presence of recognition substances in the pollen exine and the role of the substances in the incompatibility reaction have also been neatly demonstrated in another work. When anthers of *Raphanus sativus* were minced and centrifuged, a fraction enriched for the tryphine coating of the pollen was collected; tryphine thus extracted from incompatible pollen and adsorbed on a membrane was able to induce a callose reaction on the stigma in a manner analogous to that of an incompatible pollen tube. In contrast, tryphine from the cross-pollen failed to elicit the same reaction (Dickinson and Lewis 1973b).

Evidence for the role of exine proteins in the incompatibility reaction in *Cosmos bipinnatus* has come by another way. Here, self-incompatibility is partially overcome when flowers are pollinated following the application on the stigma of a diffusate from the compatible pollen. Apparently the diffusate from the compatible pollen acts as a "mentor" in stimulating germination and growth of self-pollen (Howlett et al 1975). As stated previously, after pollen grains are deposited on the stigma, responses of the stigmatic cells and style are also crucial in the recognition reaction. That the cells of the stigma of *B. oleracea* store incompatibility factors has been shown by the observation that a leachate of the stigma selectively inhibits germination of self-pollen without affecting cross-pollen; some inhibitors of pollen germination are not detected in unpollinated stigmas or in stigmas after a compatible pollination (Ferrari and Wallace 1975, 1976; Hodgkin and Lyon 1984). According to Doughty et al (1993), mixing crude stigmatic extract of *B. oleracea* with polypeptides extracted from pollen coats elicits the formation of an interaction product involving a 7-kDa pollen coat protein. Similar fusion proteins are also formed by interaction of pollen coat–borne peptides from *B. napus* with two stigmatic S-family glycoproteins from a compatible line of the same species (Hiscock et al 1995). These observations make it certain that a pollen–stigma dialogue involving proteins from both cell types is central to any hypothesis to explain self-incompatibility or self-compatibility in *Brassica* sp.

Gametophytic Systems. Considering the fact that the most common self-incompatibility system is of the gametophytic type, amazingly little is known about the location of pollen recognition factors in this group of plants. The problem of identifying factors of pollen grain origin in the gametophytic system is technically difficult because the rejection reaction results in the inhibition of growth of the pollen tube in the style. A suggestion that the outflow of proteins into the stigmatic papillae must be mainly from the intine sites has come from the observation that pollen grains of grasses are generally impoverished of substances of tapetal origin in the exine (J. Heslop-Harrison 1979b). Moreover, inhibition of pollen tube growth takes place at the stigmatic surface in grasses. In *Saccharum bengalense*, the timing of the rejection reaction is found to depend upon the way pollen grains establish contact with the stigmatic papillae. If pollen grains land with their germ pore in contact with the stigmatic papillae, germination is inhibited and the germ pore is occluded with callose. Pollen grains whose germ pore is away from the stigmatic papillae germinate and produce short tubes whose growth is inhibited as soon as they come in contact with the papillae. Assuming that the intine proteins are released when the germ pore is in contact with the papillae, these observations considerably bolster the role of intine proteins in the rejection reaction (Sastri and Shivanna 1979). In *Oenothera organensis* also, in which pollen tube growth is stymied in the surface cell zone of the stigma, an early release of the reactants lodged in the intine at the apertural site appears likely (J. Heslop-Harrison 1978).

Information on the location of incompatibility factors in gametophytic systems in which the inhibition occurs in the style is no more solid than in those cases in which stigmatic surface inhibition is the rule. The incompatibility response in *Lilium longiflorum* is eliminated by treating the whole, or parts of, the style at 50 °C in a water bath for 5–6 minutes and pollinating the stigma by self-pollen (Hopper, Ascher and Peloquin 1967; Fett, Paxton and Dickinson 1976). Heating the half of the style with the stigma has little effect, indicating that the recognition substances are confined to the lower half of the style and that the stigma does not have any profound effects on the incompatibility reactions (Fett, Paxton and Dickinson 1976). A critical test of the biological role of the recognition factors is that they should inhibit pollen tube growth with *S*-genotype specificity. Consistent with the postulated location of incompatibility factors in the style of *L. longiflorum* is the observation that an extract from the incompatible style inhibits in vitro growth of pollen tubes and that heat treatment shown to inactivate self-incompatibility eliminates two glycoprotein bands from the electrophoretic profile of the stigmatic extract (Dickinson, Moriarty and Lawson 1982).

Compared to pollen germination and pollen tube growth of compatible pollen grains of *Petunia hybrida* in an in vitro bioassay containing a homogenate of unpollinated pistils, germination and tube growth were markedly inhibited in a similar bioassay using incompatible pollen (Sharma and Shivanna 1982). In an extension of this work, it was found that incorporation of lectins such as con-A into the medium containing the pistil homogenate, or treatment of the stigma with lectin, effectively overcomes the inhibition of growth of self-pollen. A biochemical explanation of these results assumes that since sugar-containing molecules constitute the bulk of the recognition factors in the pistil, their binding with lectin makes them ineffective in recognizing self-pollen. Another experiment showed that when pollen grains are coated with either sugars or lectin and are cultured in a medium containing the pistil homogenate, sugar-coated pollen grains overcome the growth inhibition, but pollen treated with lectin do not. If lectinlike molecules of the pollen are involved in self-recognition, these results indicate that blocking lectin with sugars makes the molecules ineffective in an ensuing recognition reaction (Sharma and Shivanna 1983; Sharma, Bajaj and Shivanna 1985). Results of a later study showed that application of an extract from compatible pistils to the stigma before pollination promotes pollen germination and tube growth and overcomes incompatibility. It has been argued that treatment of the stigma with a compatible pistil extract apparently masks the recognition molecules of the pistil and thus overcomes incompatibility (Sharma and Shivanna 1986). The production and utilization of *S*-allele–specific proteins in *P. hybrida* have also been examined by monitoring the effects on pollen tube growth of proteins translated in the eggs of *Xenopus laevis* by RNA isolated from cross- and self-pollinated styles. The results, which showed an inhibition of pollen tube growth when the *S*-alleles of the style contributing the proteins matched *S*-alleles of the pollen, suggest that S-specific proteins are synthesized in the postpollination style (van der Donk 1975). In other experiments, specific inhibition of pollen tube growth has been demonstrated using crude pistil extracts in some incompatible genotypes of *Petunia inflata* (Brewbaker and Majumder 1961), *Papaver rhoeas* (Franklin-Tong, Lawrence and Franklin 1988), and *Malus domestica* (Speranza and Calzoni 1988). Purification of the stylar extract from *M. domestica* showed that inhibition is due to two glycoprotein fractions (Speranza and Calzoni 1990).

In the gametophytic self-incompatibility systems, the role of proteins bound to the pollen wall is not fully defined and it may be that such proteins are not responsible for triggering the incompatibility response. This was shown by experiments in which leaching out of substances loosely bound to the pollen walls of *Lilium longiflorum* (Fett, Paxton and Dickinson 1976) and *Petunia hybrida* (Gillissen and Brantjes 1978) was not found to alter the incompatibility reaction or seed set. Use of mentor pollen has shown that in some gametophytic self-incompatibility systems, it is possible to mask the effects of the self-pollen by mixing with nonviable cross-pollen (Pandey 1977; Sree Ramulu et al 1979; van Tuyl, Marcucci and Visser 1982). Whether the mentor pollen grains "fool" the incompatible pollen by providing recognition substances or stimulate germination by the release of nonspecific growth hormones is not known.

Characterization of *S*-Allele Products. Once the idea took hold that the signal involved in self-incompatibility is a chemical substance, identification of the substance and the unambiguous demonstration of its association with the expression of the *S*-allele loomed as important questions. The pioneering studies in this area began with immunolocalization of unique antigenic components in the pollen and pistil extracts of self-incompatible plants. Lewis (1952) found that pollen antiserum of incompatible genotypes of *Oenothera organensis* gave precipitation reactions with proteins from macerated pollen grains of the same genotype. In a later work, even single pollen grains, when placed on a thin layer of agar mixed with pollen antiserum, were shown to respond to the specific immunological reaction related to the incompatibility gene (Lewis, Burrage and Walls 1967). This points to a reaction produced by antigens diffusing from the pollen. The only other report of the presence of S-related molecules in the male gametophyte is from immunological studies of macerated pollen grains of *Petunia hybrida* that show a correlation between self-incompatibility reactions of the pollen and the presence of S-specific proteins (Linskens 1960).

Pistil proteins correlated with specific *S*-alleles were first identified in the sporophytic system of *Brassica oleracea*. In this work, antipistil antiserum from known genotypes was used to localize S-specific antibodies in diffusates of the intact pistil. That incompatibility is correlated with the appearance of particular antigens is also suggested from the fact that stigmas of immature buds that are genetically self-compatible lack the antigens (Nasrallah and Wallace 1967b). In other studies, these antigens were shown by electrophoresis and immunodiffusion to cosegregate with their respective alleles in genetic crosses. For example, an electrophoretic separation of stigmatic homogenates of several *S*-allele genotypes showed that each has a specific *S*-allele protein band. More to the point is the fact that in heterozygous plants, protein bands associated with both *S*-alleles could be resolved (Nasrallah, Barber and Wallace 1970). Analysis of stigma extracts of F1 and F2 populations of homozygous self-incompatible genotypes showed that certain antigens present in F1 and F2 progenies are precisely correlated with the segregation of *S*-alleles and with a functional incompatibility response (Nasrallah, Wallace and Savo 1972). Because of their agreement with immunodiffusion and electrophoretic data, these results suggest that S-antigens are specific protein products. In another work, Nasrallah (1974) showed that the degree of self-incompatibility in *B. oleracea* is dependent upon the relative concentrations of the S-protein, because self-incompatibility, partial self-incompatibility and self-compatibility displayed by parental genotypes and mutant phenotypes are accompanied by corresponding changes in the level of this protein. However, an apparent lack of cross-reaction between S-antigens also led to the view that they are not polypeptide variants of a protein molecule (Nasrallah 1979).

Results (Figure 9.9) that are completely in accord with the aforementioned have come from investigations of different genotypes of *B. oleracea* (Kučera and Polák 1975; Nishio and Hinata 1977). In a follow-up study involving isoelectric focusing of stigmatic proteins of 13 *S*-alleles in diallelic crosses of *B. campestris* and *B. oleracea*, Hinata and Nishio (1978) showed a provocative correlation among seven allelic families between *S*-genotypes determined by breeding analysis and specific proteins. Since the protein that corresponds to a particular *S*-allele cosegregates with that allele, the logical conclusion is that the protein is the product either of the *S*-allele or of a gene sequence closely

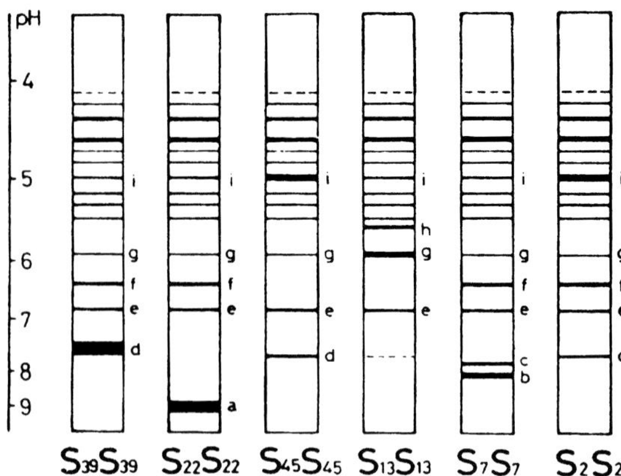

Figure 9.9. Diagram showing the separation of stigmatic protein bands in different homozygous lines of *Brassica oleracea*. The separation of bands labeled *a* to *i* differs in the different lanes. (From Nishio and Hinata 1977.)

linked to it. The evidence from PAS staining of the *S*-allele protein bands led to their identification as glycoproteins. The molecular mass of the proteins purified from stigma extracts ranged from 51 kDa to 65 kDa. The glycoproteins had basic PIs (pH 10.3–11.1) and showed immunological specificity like that of the crude stigma homogenates (Hinata, Nishio and Kimura 1982; Nishio and Hinata 1982). The glycoproteins were precipitated by con-A and were not detected in pollen grains or other plant tissues. In accordance with the established incompatibility reactions, low quantities of S-specific glycoproteins occur in the stigmas of self-compatible immature flowers and their concentration increases with flower maturity and attainment of self-incompatibility. As expected, stigmas of flowers homozygous for an *S*-allele have twice the amount of glycoproteins as heterozygotes (Sedgley 1974; Nishio and Hinata 1977). These studies have made *B. oleracea* a classic system for incompatibility studies, and the criteria used to establish the relationship between the *S*-allele and the protein product have gained general acceptance in molecular investigations of self-incompatibility in other plants.

If the glycoproteins are indeed *S*-gene products that interfere with the germination of incompatible pollen grains on the stigma, it should be possible to demonstrate the potency of the compounds in situ on the stigma; this has been done. Pretreatment of pollen grains of *B. oleracea* with a highly purified *S*-allele–specific glycoprotein isolated from stigmas induces an incompatible reaction leading to the inhibition of pollen germination on otherwise compatible stigmas bearing the same allele (Ferrari, Bruns and Wallace 1981). Since germination of similarly pretreated pollen grains of other *S*-allele genotypes is not prevented, the incompatibility reaction induced by the glycoprotein is considered to be specific to pollen containing the particular genotype.

Following these studies, using techniques similar to those described for *Brassica*, *S*-allele protein relationships have now been established in several species showing the gametophytic type of incompatibility (Figure 9.10). As will be seen in Chapter 10, a generous portion of the current work on the molecular biology of self-incompatibility has been done on *Nicotiana alata*. In this species, S-glycoproteins corresponding to three different *S*-alleles were first identified in both style and stigma (Bredemeijer and Blaas 1981). The proteins are generally present in the highest concentration in the zone of the pistil where inhibition of pollen tube growth occurs (Anderson et al 1986; Jahnen et al 1989; Kheyr-Pour et al 1990). Since in vitro growth of the pollen tube is inhibited by S-glycoproteins with *S*-genotype specificity, it is clear that they are gene products of *S*-alleles (Harris, Weinhandl and Clarke 1989; Jahnen, Lush and Clarke 1989).

Another genus in which considerable advance has been made in the analysis of S-proteins of self-incompatibility is *Petunia*. The most abundant protein of the pistil of self-incompatible *P. hybrida* is a glycoprotein with molecular mass in the range of 27–30 kDa and PI in the range of 8.3–9.3. A striking correlation of distribution of the glycoprotein with a given *S*-allele has been demonstrated by both one- and two-dimensional electrophoresis using homozygous and heterozygous lines segregating for four different *S*-alleles, their F1 progeny, and backcrosses. These proteins are generally localized in the stigma and in the upper part of the style, and their distribution pattern in flower buds of different ages is consistent with the development of self-incompatibility. They also display biological activity as indicated by the allele-specific inhibition of pollen tube growth (Kamboj and Jackson 1986; Broothaerts et al 1989, 1991). In *P. inflata* the proteins that cosegregate with their *S*-alleles have molecular masses of 24 and 25 kDa (Ai et al 1990). Stylar proteins identified from other solanaceous species, such as *Lycopersicon peruvianum* (Mau et al 1986), *Solanum tuberosum* (Kirch et al 1989), and *S. chacoense* (Xu et al 1990a), are also associated with *S*-alleles.

S-allele proteins have been identified with some success only in a few other plants. Of the two antigens detected in stylar components of *Prunus avium*, one (antigen-S) typically corresponds to the product of the *S*-allele and is thus a candidate for

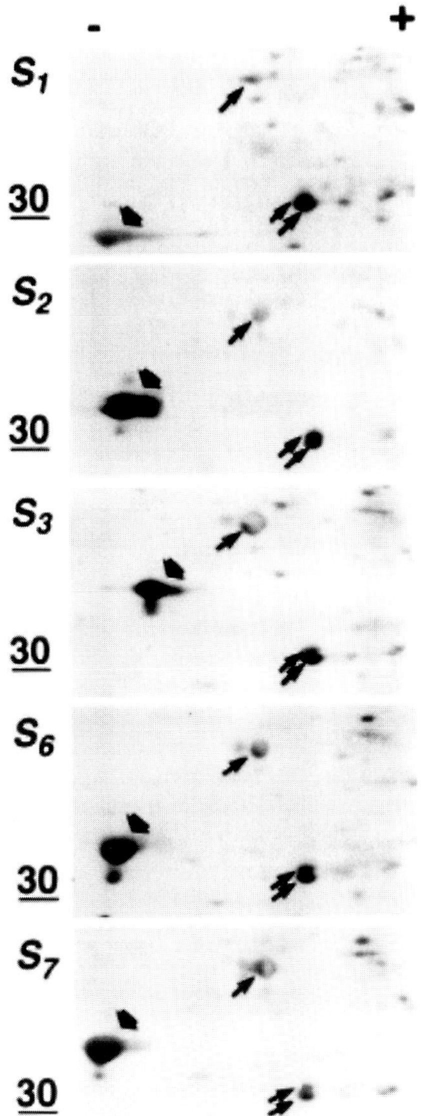

Figure 9.10. Two-dimensional SDS-PAGE of stylar proteins from mature flowers of different homozygous lines of *Nicotiana alata*. The part of each gel corresponding to the molecular mass range of about 25–40 kDa and to the neutral to basic pH range is shown. Arrows indicate proteins common to all genotypes; arrowheads indicate *S*-allele–specific proteins. (From Jahnen et al 1989. *Plant Cell* 1:493–499. © American Society of Plant Physiologists. Photograph supplied by Dr. A. E. Clarke.)

the *S*-gene product. It is restricted to the mature style, whereas the other (antigen-P) is present in styles of all stages of development in three other species of the genus: *P. cerasus*, *P. armeniaca*, and *P. persica* (Raff, Knox and Clarke 1981). The antigen is also secreted into the medium by callus cells derived from both leaf and stem of *P. avium*; rather than showing the ubiquitous nature of the antigen, this probably reflects on the totipotency of the callus cells and their ability to express the total genome of the plant (Raff and Clarke 1981). The S-antigen has been identified as a glycoprotein having a 16.3% carbohydrate content and consisting of two closely related components with an apparent molecular mass of 37–39 kDa. It is also a potent inhibitor of in vitro pollen tube growth of the same genotype of *P. avium* (Mau, Raff and Clarke 1982; Williams et al 1982). Two S-associated glycoprotein fractions of the style of a self-incompatible cultivar of *Malus domestica* selectively inhibit self-pollen germination and tube growth without any effect on cross-pollen. In some incompatible cultivars, varietal differences have been observed in the S-glycoproteins (Speranza and Calzoni 1990; Sassa et al 1994). In *Pyrus serotina*, sugar-containing basic proteins of molecular mass of about 30 kDa are associated with self-incompatibility. The glycoproteins are predominantly expressed in the style, have high RNase activity, and are considered homologous to those identified in Solanaceae (Sassa, Hirano and Ikehashi 1992, 1993). Demonstration of *S*-gene activity of a component isolated from the stigma of *Papaver rhoeas* stems from the work of Franklin-Tong et al (1989). When extracts from stigmatic tissues were fractionated on immobilized con-A, a fraction in the molecular mass range of 20–25 kDa bound to con-A was found to function as an *S*-allele-specific inhibitor of pollen tube growth. The uniqueness of this substance is that it appears as the component of a 22-kDa protein in SDS-PAGE of stigma extracts but is absent from similar preparations of buds, ovary, petal, leaf, and root. In the only published report of a link between incompatibility and stylar proteins in a member of Poaceae, Tan and Jackson (1988) showed that stigma extracts of various incompatible genotypes of *Phalaris coerulescens* exhibit differences in the pattern of proteins in the molecular mass range of 43–97 kDa. Since self-incompatibility in grasses is controlled by two unlinked loci, *S* and *Z*, it is also clear that their alleles control the expression of the stigma proteins; however, from the available data it is difficult to assign specific proteins to each allele.

(v) Models of *S*-Allele Action

The defining genetic feature of the self-incompatibility system enforcing outbreeding in angiosperms centers around the gene products of the stigma and style. Perhaps the most significant advance in our understanding of this system prior to the 1970s was in identifying and characterizing the *S*-allele recognition factors of the style. There are grounds for believing that in most self-incompatibility systems, the interaction between recognition factors contributes to physiological changes in the stigma or style that results in the demise of self-pollen. Models to explain the mechanism of self-incompatibility not only differ in the signaling substances thought to participate in inhibiting pollen germination and pollen tube growth, but also in the emphasis placed on the reactions themselves. However, without any data on the nature of the *S*-allele products of the pollen grain, it is premature to generalize the models proposed.

The tenet of the theory on pollen–pistil interaction is that through chemical signals, a basic compatibility between the two partners has evolved, and interactions controlled by *S*-genes are superimposed upon this compatibility. As a working hypothesis, Lewis (1965) proposed that following self-pollination, a specific interaction of gene products of each *S*-allele present in the pollen and in the style causes the synthesis of a dimer repressor. This compound, presumably a tetramer, functions as the incompatibility complex and serves to inhibit the growth of the pollen tube. In an interesting variant of this hypothesis, Ascher (1966, 1975) proposed that a regulatory system involving a repressor molecule generated from the interaction between identical *S*-locus products of the pollen grain and the style prevents the growth of the pollen tube in the style to an appreciable extent; the result is self-incompatibility. When there is a mismatch between *S*-allele products of the pollen grain and the style, a rapid-growth operon, which switches the pollen tube from a slow-rate metabolism to a fast-rate metabolism, is activated; this gives rise to a compatible reaction. A feature of the model, which is based on a long series of experiments on *Lilium longiflorum*, is the requirement for a special stylar substance to support compatible pollen tube growth; based on heat-treatment experiments, it is believed that this substance is synthesized beginning at anthesis and continuing until senescence of the flower.

Mulcahy and Mulcahy (1983) have proposed a heterosis model to explain gametophytic incompatibility as a passive mechanism. The basic premise of this model is that there are many incompatibility genes (supergenes), each containing one dominant and several deleterious recessive alleles. Pollen tube growth resulting in self-incompatibility or cross-compatibility is a manifestation of pollen–style interactions on a quantitative level. Accordingly, after cross-pollination, a large number of incompatibility loci would be heterozygous in the pollen and style; this raises the number of heterotic pollen–style interactions and leads to an increase in the rate of pollen tube growth. Following self-pollination, a high proportion of the supergenes would be homozygous and deleterious and would thus cause a reduction in the extent of pollen tube growth. The major thrust of this model is in asserting that self-incompatibility is not due to the action of *S*-allele–specific inhibitory molecules, but results from a slow growth of pollen tubes to effect fertilization before floral abscission. This model has been subject to some criticism (Lawrence et al 1985).

Extrapolating from the features of self-incompatibility system in grasses, J. Heslop-Harrison and Heslop-Harrison (1982b) proposed a model of pollen acceptance and rejection based on carbohydrate–protein interaction. Proteins with lectinlike properties seem to be involved on the pistil side, either binding to the complementary sugar sequences of the wall carbohydrates of the incompatible pollen tube or ignoring similar sequences on the compatible tube due to noncomplementarity. *S*-alleles in the pollen grain specify the sugar sequences that become ordered into distinct configurations in the polysaccharide matrix synthesized and deposited at the tip of the pollen tube. Binding of pistil S-factors to the specific sugars of the pectic wall components of the pollen tube tip may set off a train of determinative events that culminate in the disruption of pollen tube growth.

For sporophytic systems, a function for cutinase has been invoked to explain the self-incompatibility phenomenon. This might appear reasonable since the presence of a cuticularized dry stigma is one of the features that distinguish sporophytic systems from gametophytic systems. Although both direct and indirect lines of evidence exist for the presence of cutinase in germinating pollen grains, it is not understood how S-proteins affect the enzyme present in the responding self- and nonself-pollen grains (Linskens and Heinen 1962; Shayk, Kolattukudy and Davis 1977). S-proteins on an incompatible stigma might inactivate the cutinase present in the pollen or they might furnish permissive conditions on a compatible stigma for pollen cutinase or its precursors to be activated (Christ 1959).

The assumptions of the models described have been based on evidence primarily from genetic analysis and pollen behavior rather than from the chemical nature of the reactants or the reaction products. It is therefore not surprising that contradictory views have been presented against these models. The models are described here to stimulate the kind of research that will generate new information about the basic mechanism of self-incompatibility in the future. On a general level, it is safe to say that self-incompatibility covers a multiplicity of physiological processes in the stigma and style and that it would be unwise to focus exclusively on a single hypothesis.

3. CROSS-INCOMPATIBILITY

Gametic fusions between plants belonging to different species, and even between different genera, account for the introduction of beneficial alien genes despite natural reproductive barriers. Be that as it may, pressures to maintain the identity of species are so great that such fusions are impeded by cross-incompatibility mechanisms. This apparent contradiction between wide hybridization and cross-incompatibility is one of the reasons why there are so few reports of successful interspecific or intergeneric hybridizations in plants. Unlike self-incompatibility, which is a prefertilization barrier, cross-incompatibility can cause both pre- and postfertilization events to suffer. The magnitude of failure depends upon the stage at which the rejection reaction occurs; although prefertilization rejection generally results in the inhibition of pollen tube growth, postfertilization lesions cause more drastic effects, such as the arrest of embryo and endosperm development. This section will consider only examples of cross-incompatibility due to prefertilization blocks; an account of postfertilization blocks leading to cross-incompatibility is given in Chapter 14.

(i) Physiological Background

Many interspecific crosses are doomed to fail when self-incompatible pistillate parents are crossed with self-compatible staminate parents, although success is often assured in the reciprocal cross. This one-way isolation, known as unilateral incompatibility, is widespread in many cultivated plants and appears as an extension of self-incompatibility. In addition to occurring in interspecific crosses, unilateral incompatibility has been described in intergeneric matings between members of Solanaceae and even in crosses between members of two families (Lewis and Crowe 1958).

An analysis of unilateral incompatibility in *Solanum* illustrates some structural and developmental aspects of the phenomenon. Clones showing unilateral incompatibility exist among different species of *Solanum*, although some incompatible clones of a species could be successfully crossed

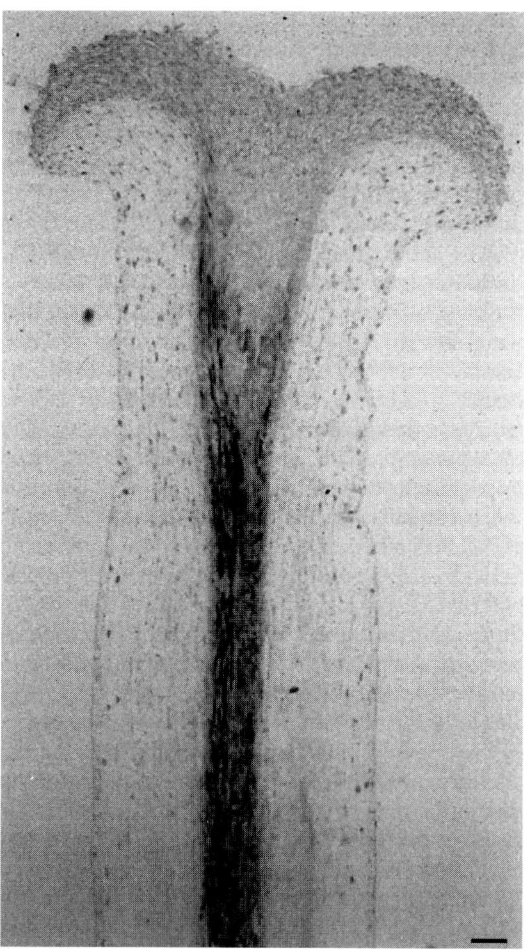

Figure 10.5. An immunostained, longitudinal section of the stigma and style of transgenic *Nicotiana tabacum* showing the distribution of an antigen encoded by *Brassica oleracea SLG-22* gene. Scale bar = 100 μm. (From Kandasamy et al 1990. *Plant Cell* 2:39–49. © American Society of Plant Physiologists. Photograph supplied by Dr. M. Kandasamy.)

genic plants and not in the stigmatic papillae (Figure 10.5). Similar results were obtained in tobacco transformed with the *SLG-22* gene in another version of this experiment, which also defined the spatial pattern of expression of the gene. The transgene-encoded protein product was localized in high concentrations in the style in the intercellular matrix of the transmitting tissue, with some label spilling over into the secretory matrix in the basal region of the stigma and into the placental epidermal cells adjoining the ovules (Kandasamy et al 1990). Localization of the *SLG-22*–encoded protein in transgenic tobacco coincides precisely with the expression of tobacco *S*-allele transcripts in the mature stigma of *N. alata* and traces the route taken by compatible pollen tubes during their growth to the ovary (Cornish et al 1987). Nishio et al (1992) have shown that self-compatible *B. napus* plants transformed with several *SLG* genes isolated from *B. oleracea* and *B. campestris* exhibit, at a reduced level, the same pattern of developmentally regulated and cell-specific gene expression as they do in the self-incompatible phenotype. Despite the clear evidence from these experiments for the similarity of the *cis*-acting elements of the donor and recipient plants, the expression of *SLG* genes, interestingly enough, failed to impart self-incompatibility in transgenic plants in the experiments just described. An instance in which introduction of an *SLG* gene perturbed self-incompatibility and induced full self-compatibility in the transgenic recipient was described by Toriyama et al (1991a) following transformation of an incompatible line of *B. oleracea* with the *S-8* allele of *B. campestris SLG* gene. Introduction of antisense RNA of the same gene led to the synthesis of SLG and the breakdown of self-incompatibility in transgenic *B. campestris* (Shiba et al 1995). In these experiments, the effect of the regular gene is less easy to explain than the effect of the antisense gene; the breakdown of the self-incompatibility system has been attributed to an unexpected interference by the antisense gene in the expression of the homologous host gene, which leads to a decrease in the levels of endogenous SLG. There is little doubt that other side effects of the antisense gene will be described in the future.

Although models of self-incompatibility predict the presence of *S*-locus components in the pollen, stigma, and style, it has not been established from studies reviewed here that pollen grains express *S*-locus transcripts or their protein products. On the surface, these results suggest that *S*-locus genes are very tightly regulated or are present in very low abundance in pollen grains. One strategy employed to compensate for the low abundance of putative *SLG* transcripts is to use the *SLG* promoter to activate very sensitive toxic genes or reporter genes in transgenic plants. In experiments with the heterologous host tobacco, activity of a promoter fragment isolated from the *SLG-13* gene was followed by fusing it to the *DT-A* gene; in parallel experiments, the promoter fused to the *GUS* gene was used to characterize the sites of gene expression by a standard histochemical assay. Since the *DT-A* gene is a potent inhibitor of protein synthesis and causes autonomous cell death, the effect of gene action was a genetical wipe-out of cells expressing the fusion gene. This catastrophic though specific effect of the *DT-A* gene caused seemingly obvious ablation of stylar tissues in most transformed plants and led to stunted growth of the style with a concomitant disruption of the transmitting tissue. The effect of the toxic gene was also targeted at the normally bilobed stigma, which became multilobed with abnormal papillae. Generally, the fusion gene was not fatal to the cells of young buds, but the abnormalities became severe in open flowers, so that the pistils were incapable of supporting pollination and pollen germination. Consistent with the gametophytic expression of the fusion gene, 50%–75% of pollen grains in the transgenic plant were empty and nonviable (Thorsness et al 1991). Altogether, it is not surprising that *DT-A* gene activity primed by *SLG* gene promoter affects those very cells and tissues that favor pollen capture, pollen germination, and pollen tube growth.

Similar phenotypic effects, confined to ablation of the stigmatic papillae and abnormal pollen development, are observed with impressive regularity in *Arabidopsis* and in self-fertile *B. napus* plants transformed with the same fusion gene. However, the effects of the fusion gene on the functional competence of the stigmatic papillae to support pollen germination and pollen tube growth are completely different in the two transformants. For example, a consequence of the toxic gene on *Arabidopsis* is the failure of self-pollen to germinate on the ablated papillar cells, whereas these same cells support germination and growth of pollen from untransformed plants; in contrast, ablation of the stigmatic papillae of *B. napus* almost completely blocks the ability of pollen grains of any genotype to germinate (Kandasamy et al 1993; Thorsness et al 1993). Apart from the obvious implication of these results that *B. napus* requires active papillar cell metabolism for successful pollination and *Arabidopsis* does not, the results underscore the complexity of the pollination system, even in self-fertile *B. napus*, and indicate an important function for the papillae in the initial recognition event.

GUS gene expression in tobacco plants transformed with *SLG-13/DT-A* constructs has resulted in the cytological detection of promoter activity at a high level in the stigmatic papillae and transmitting tissue of the style and at a reduced level in the placental epidermis of the ovary (Figure 10.6). This is just what one would expect in gametophytic self-incompatibility systems in which stylar inhibition of pollen tube growth is the norm. In the male gametophyte, *GUS* activity is expressed at the uninucleate microspore stage and it increases up to the binucleate stage. The percentage of pollen grains in the anther locule showing *GUS* activity is consistent with the gametophytic, rather than with the sporophytic, mode of self-incompatibility (Thorsness et al 1991). The spatial and temporal pattern of expression of *GUS* gene directed by a *B. oleracea SLR-1* gene promoter in transgenic tobacco is similar to that observed for *SLG–GUS* fusion gene. Unlike *SLG–GUS* fusion gene, *SLR-1–GUS* construct was not expressed in the placental epidermis of the ovary; despite this minor difference, the combined results suggest a close interaction between the different *S*-locus genes in pollination (Hackett, Lawrence and Franklin 1992).

When *SLG-13–GUS* construct was introduced into *Arabidopsis*, the pattern of expression of the *GUS* gene was somewhat like that of the *SLG-13* gene in untransformed *B. oleracea* (Figure 10.6). Activity was observed in the stigmatic papillae, but not in the underlying stylar tissue. Some transformants also showed activity in the tapetum during a narrow developmental window confined to the

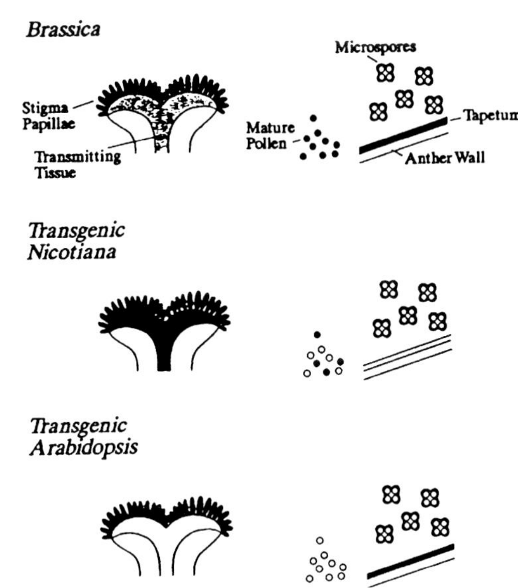

Figure 10.6. Diagrams showing the tissue specificity of *SLG-13* promoter activity in transgenic *Brassica napus*, *Nicotiana tabacum*, and *Arabidopsis*. Shaded areas indicate regions of expression. Patterns of expression seen are, in transgenic *B. napus*: stigmatic papillae, tapetum, pollen grains, and, to a lesser degree, the stylar transmitting tissue and placental epidermis; in transgenic *N. tabacum*: stigmatic papillae, transmitting tissue, placental epidermis, and 50% of pollen grains; and in transgenic *Arabidopsis*: stigmatic papillae and tapetum. (From Dzelzkalns, Nasrallah and Nasrallah 1992.)

period immediately after microsporocyte meiosis, but pollen grains did not express the reporter gene (Toriyama et al 1991b). This observation is in harmony with the view that in plants with the sporophytic type of self-incompatibility, *S*-locus expression occurs postmeiotically in the tapetum.

Although the main outlines of expression of the *SLG–GUS* fusion gene in transformed self-compatible *B. napus* and self-incompatible *B. oleracea* plants are similar to those described in transformed *Arabidopsis*, two unexpected findings have put the relationship between gametophytic and sporophytic types of self-incompatibility in a new perspective. One is that the *SLG–GUS* fusion gene is expressed not only in the stigmatic papillae, but also to a reduced extent in the stylar transmitting tissue and ovary of the transformed plants. The other is that the *SLG* promoter is active not only in the tapetum of the anther, but also in the haploid microspores (T. Sato et al 1991). Although the history of contention about the gametophytic and sporophytic types of self-incompatibility actually had to do with the timing of gene action, comments have also appeared periodically about the origin of the two systems. Results from transgenic *Brassica* plants showing features of a pollen recognition system based on gametophytic control in sporophytically self-incompatible plants might support the view that the gametophytic self-incompatibility system is the ancestral type and that the sporophytic system is derived from it.

The ability to break sequential parts of the promoter and to join them to a reporter gene, or to place promoter modules under the appropriate control so that they can be transcribed from a new promoter, allows one to determine the minimum sequence elements of the *SLG* promoter necessary for gene expression (Figure 10.7). Expression patterns of truncated versions of *SLG-13* promoter fused to *GUS* have shown that a 411-bp promoter fragment adjoining the translation initiation codon directs a normal level of *GUS* activity in the stigma and style and a reduced activity in the pollen grains of transgenic tobacco plants. A 196-bp region (–339 to –143 bp) is also sufficient to confer independent stigma- and style-specific expression at a reduced level. Although the 196-bp fragment does not direct expression in the pollen grains, two distinct but functionally redundant flanking domains (–415 to –291 bp, and –117 to +8 bp) allow expression of the gene in the pollen (Dzelzkalns et al 1993). It is clear that even when large fragments of the truncated promoter are expressed, selective mechanisms could operate to reduce the expression to unimpressively low levels in some tissues.

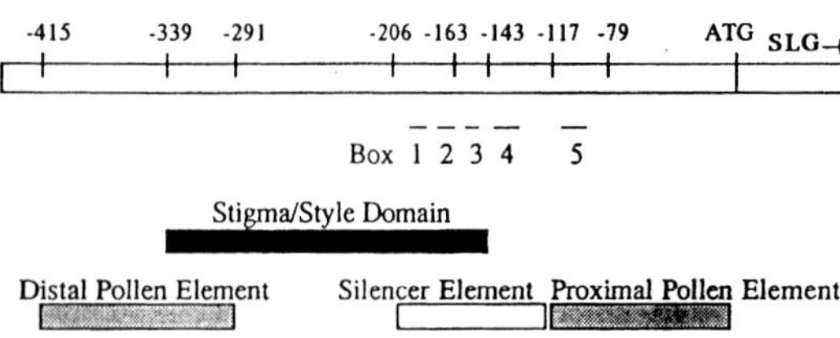

Figure 10.7. A diagram based on deletion analysis showing the arrangement of the functional elements of the *SLG-13* promoter. The deletion end points of the constructs are shown at the top. The positions of the stigma–style domain, the two pollen elements, and the pollen silencer element are shown relative to the positions of the sequence motifs found to be highly conserved in *SLG* and *SLR* genes (boxes 1 to 5). (From Dzelzkalns et al 1993. *Plant Cell* 5:855–863. © American Society of Plant Physiologists.)

2. THE *NICOTIANA* SYSTEM

The discovery of an abundant glycoprotein with a molecular mass of 32 kDa in the style of *Nicotiana alata* carrying *S-2* alleles was among the antecedents of attempts to isolate the *S-2* gene product and to clone the gene encoding the protein. The natural assumption that the protein segregates with *S-2* breeding behavior was confirmed by simple one-dimensional SDS-PAGE of stylar extracts of *S-2* genotypes. In addition, expression of this protein with the development of self-incompatibility in maturing styles and its concentration in the part of the style where inhibition of pollen tube growth occurs also seem to support the notion that the 32-kDa protein is a product of either the *S-2* allele or a gene tightly linked to it. In the approach followed to clone the gene encoding the self-incompatibility glycoprotein from the stigma–style complex of *N. alata*, the protein was isolated, purified, deglycosylated and its N-terminal sequence determined. From this sequence, a partial 30-bp DNA probe for the encoding gene was synthesized to screen a cDNA library prepared to poly(A)-RNA of mature styles of an *S-2/S-3* genotype of *N. alata*. Following differential screening with labeled cDNA made to the mRNA of mature styles, of immature styles and of ovaries, cDNA clones highly expressed in mature styles were further screened with the 30-bp synthetic oligonucleotide. A positive cDNA clone obtained by this method was used to screen another cDNA library to isolate a full-length cDNA clone of the glycoprotein. Sequencing of this cDNA clone revealed an open reading frame of 642 bp that encodes a 204-residue polypeptide with a classic signal sequence of 22 amino acids and a hydrophobic N-terminus of 15 amino acids. Other features of the predicted amino acid sequence of the protein are consistent with the view that it is a secreted glycoprotein whose transfer is directed by a signal sequence. In Northern blot hybridization, the cDNA clone was found to cross-hybridize intensely with RNA from mature styles bearing the *S-2* allele and weakly with RNA from nonhomologous alleles. No hybridization was detected with RNA of immature styles of any genotypes (Anderson et al 1986). Taken together, these data provide persuasive evidence that the cDNA characterized from *N. alata* encodes a protein product of the *S-2* allele.

Different approaches to screening cDNA libraries have been employed to isolate clones corresponding to the glycoproteins from other *S*-locus alleles of *N. alata*. Following differential screening of cDNA libraries with probes prepared from mRNA of *S-3* and *S-6* genotypes, potential *S-3* and *S-6* cDNA clones were identified by their ability to hybridize strongly to *S-3* and *S-6* probes and weakly to *S-6* and *S-3* probes, respectively (M. A. Anderson et al 1989). To isolate cDNAs from *Sz, Sf-11, S-1*, and *Sa* alleles, a cDNA library made to *Sz* pistil mRNA was differentially screened with pistil and leaf mRNA probes. A single clone representing the most abundant class of pistil-specific cDNA was used as a hybridization probe to screen cDNA libraries made to mRNA from *Sf-11, S-1*, and *Sa* genotypes (Kheyr-Pour et al 1990). A critical question about the deduced amino acid sequences of the alleles is the extent of variation and homology. Although the incompleteness of some of the cDNA clones has rendered estimates of amino acid sequences of limited accuracy, comparison between *S-2, S-3*, and *S-6* sequences has revealed about 56% overall homology at the amino acid level. The average homology was reduced to 54% when amino acid sequences of all seven *S*-locus alleles were compared. Variations are mainly due to amino acid substitutions, deletions, and insertions found throughout the sequence, but many of these are seemingly bunched together in four hydrophilic regions confined to the amino half of the sequence. The amino terminus, the cysteine residues, the multiple glycosylation sites, and the carboxy terminus are generally considered to be conserved regions and they account for the other half of the sequence. S-1, S-3, and S-6 proteins have 10 cysteine residues, whereas others, except Sa, have the traditional 9 residues (Figure 10.8). The Sa protein has only 7 of the 9 conserved cysteine clusters. When glycosylation sites are compared, 4 sites are found to be conserved among proteins of *S-2, S-3*, and *S-6* alleles, and the *S-3* allele has gained a new site (M. A. Anderson et al 1989; Woodward et al 1989; Kheyr-Pour et al 1990)). The occurrence of conserved cysteine residues and glycosylation sites in the alleles suggests specific structural roles for these molecules in the stability, conformation, and biological activity of the glycoproteins.

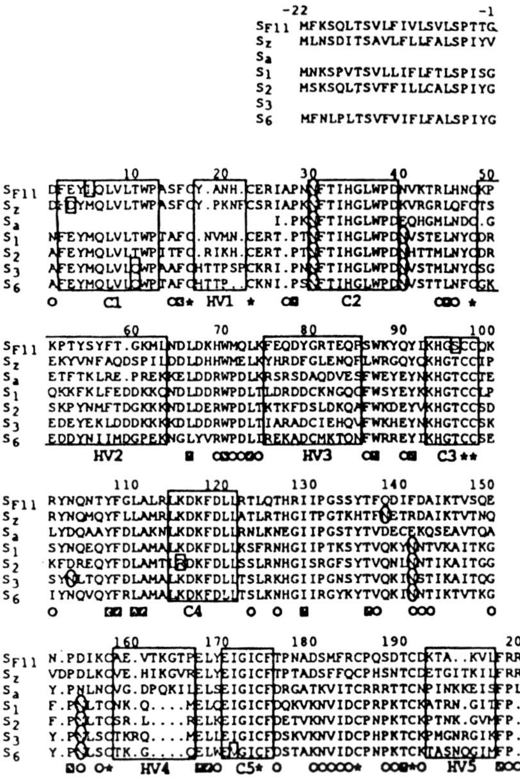

Figure 10.8. Alignment of the amino acid sequences of seven S-proteins of *Nicotiana alata*. The five conserved (C1 to C5) and five hypervariable (HV1 to HV5) regions are enclosed in solid lines. Small boxes within the conserved regions are residues that are different from the conserved residues at those sites. Open circles enclose residues that are different between the two homology classes but are conserved within one or two homology classes. The nine cysteine residues are indicated by asterisks. The circled asparagine residues are potential N-glycosylation sites. (From Kheyr-Pour et al 1990. *Sex. Plant Reprod.* 3:88–97. © Springer-Verlag, Berlin.)

Is the *SLG* gene of *N. alata* a member of a multigene family? When Southern blots of endonuclease-digested genomic DNA from *S-1, S-2, S-3, S-6*, and *S-7* allelic plants were probed with cDNAs corresponding to the alleles, the fragments hybridizing from different alleles were found to be dissimilar in size, and, in most cases, single diagnostic restriction fragments hybridized intensely with the probe. Each cDNA probe bound strongly to a DNA fragment of that allele and weakly or not at all to DNA fragments of other *S*-alleles. These observations are compatible with the view that the isolated cDNA clones are allelic products of the same S-locus and that this is a single-copy genetic system. Additional data confirming that the cDNA sequences are allelic products of the same S-locus have come from breeding experiments combined with DNA blot analysis, two-dimensional electrophoresis, and cation exchange liquid chromatography of style glycoproteins. It was found that the S-genotypes of individual plants determined from conventional breeding tests correlate precisely with their DNA blot analysis using S-allele–specific DNA probes and with the presence of S-allele–specific glycoprotein spots and protein elution profiles (Bernatzky, Anderson and Clarke 1988; M. A. Anderson et al 1989; Jahnen et al 1989). Of further interest is a report that a small fragment located at the 5'-end of the coding sequence of *S-2* DNA shows homology with mitochondrial DNA in Southern blots (Bernatzky, Mau and Clarke 1989). How the transfer of sequences between nuclear and mitochondrial DNA might have occurred, and what the role of this sequence is in the function and expression of self-incompatibility, remain to be explored.

(i) Ribonucleases as S-glycoproteins

A feature of the amino acid sequences of S-glycoproteins isolated from *N. alata* was their homology to extracellular fungal RNase. The homology itself was not particularly striking, as only 30 of the 122 amino acids conserved among *S-2, S-3,* and *S-6* allelic glycoproteins were aligned with identical molecules in the fungal enzyme. Another 22 amino acids in the glycoproteins aligned with closely related amino acids in the RNase, and 5 out of the 9 or 10 cysteine residues and 2 catalytic histidine residues were conserved between the two proteins. Nonetheless, this finding led to intensive research on the RNase activity of glycoproteins and eventually to the identification of glycoproteins as RNases. That RNase activity is an inherent component of the S-glycoproteins became clear from the fact that the enzyme is found to co-elute perfectly with the individual S-glycoproteins by ion-exchange chromatography and that it has a specific activity close to that of purified RNase T_2. In physiological terms, the specific activity of the enzyme from stylar extracts of the self-incompatible *N. alata* is found to be 100–1,000-fold higher than that of the self-compatible *N. tabacum*. This correlation implies a possible role for the stylar product of the self-incompatibility gene, designated as *S-RNase*, in self-pollen rejection reactions (McClure et al 1989). S-RNase also appears to have a bearing in pollen rejection associated with unilateral incompatibility in certain species combinations in *Nicotiana*. This was shown by up-regulating RNase expression in transgenic plants that acquire the capacity to reject normally compatible pollen grains (Murfett et al 1996).

A general mechanism by which RNase might function in the self-incompatibility response is by the degrading of pollen RNA, which results in pollen tube death in an incompatible pollination. Support for this view has come from the work of McClure et al (1990), who followed the integrity of RNA extracted from the style of *N. alata* of a defined genotype after compatible and incompatible pollinations with ^{32}P-labeled pollen grains. The results were straightforward in showing extensive RNA breakdown in incompatible styles but not in compatible ones (Figure 10.9). Furthermore, the results suggest that pollen tube rRNA could have suffered degradation in the incompatible style due to RNase cytotoxicity, because RNA appeared as a

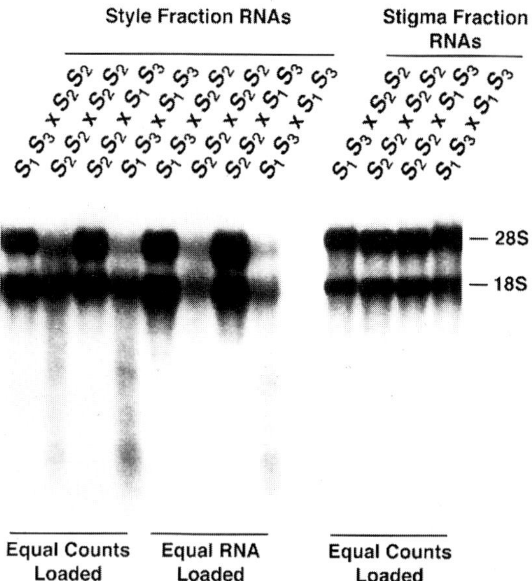

Figure 10.9. Agarose gel electrophoresis of ^{32}P-labeled RNA from compatible and incompatible crosses in *Nicotiana alata*. RNAs from style and stigma parts of the flower were treated separately. Compatible crosses represented are $S_1S_3 \times S_2S_2$ and $S_2S_2 \times S_1S_3$; incompatible crosses represented are $S_2S_2 \times S_2S_2$ and $S_1S_3 \times S_1S_3$. Mostly intact RNAs are seen in the styles of compatible crosses and in the stigmas of all crosses, whereas RNAs of the styles in the incompatible crosses are degraded, which results in the accumulation of low-molecular-mass species. (From McClure et al 1990. Reprinted with permission from *Nature* 347:757–760. © Macmillan Magazines Ltd. Photograph supplied by Dr. A. E. Clarke.)

smear of low-molecular-mass species in an agarose gel run under denaturing conditions. Results from other experiments have been marshalled to show that the inhibition of pollen tube growth in the style results from the entry of S-RNase and the consequent degradation of pollen tube RNA. For example, following autoradiography of ^{3}H-labeled S2-RNAse incorporation, the label is present in the pollen grain cytoplasm and, to a lesser extent, in the pollen tube. Immunogold electron microscopy using an antibody raised to a synthetic peptide corresponding to an allele-specific region of S2-RNase showed labeling in the walls of the germinated pollen grain and in the pollen tube cytoplasm. Treatment of pollen grains with S2-RNase inhibited pollen tube growth and, concomitantly, protein synthesis. S2-RNase also inhibited cell-free translation of proteins by self-pollen RNA to about 70% of the control rate (Gray et al 1991). In the absence of data on the comparative effects of S2-RNase on compatible pollen grains, the significance of these results leaves much to be desired. Despite this criticism, the results lead to the inference that RNA degradation is instituted soon after the incompatible pollen tubes reach the style, as a result of which their growth is stymied. Analyses of S-RNases recovered separately from the stigma and style have shown that the two enzymes are identical, although their respective transcripts differ in the length of the polyadenylate tail. This has raised the obvious question of why pollen tube growth is arrested in the style rather than in the stigma after self-pollination. Different mechanisms of enzyme uptake by the pollen tube from the stigma and stylar regions might underlie their differential sensitivity, but this remains to be further investigated (McClure et al 1993). In one working model proposed to explain the operation of the system, it is believed that in an incompatible pollination, pollen tubes carry *S*-locus–specific receptors that bind S-RNases. How these molecules find their way into the pollen tube is a mystery, but once inside, they decimate the endogenous RNA of the pollen tubes and inhibit their growth. Compatible pollen tubes survive the RNase onslaught since they have some way of preventing the uptake of the enzyme. Alternatively, it has been postulated that RNases are taken up by pollen tubes regardless of their specificity but, inside the pollen tube, specificity of action prevails due to an intracellular inactivation of the enzyme or a compartmentalization of the enzyme that makes the RNA pool inaccessible (Franklin-Tong and Franklin 1993; Dodds, Clarke and Newbigin 1996).

Attractive though these models are, the mechanism by which allelic specificity is controlled during the incompatibility episode remains unclear. Since the gametophytic self-incompatibility system is based on the expression of the *S*-gene in the pollen as well as in the style, the problem is compounded by the fact that *S*-gene products of the pollen have not been identified. The dilemma has been heightened with the discovery that the *S-RNase* gene is also expressed at a low level in the pollen grains of *N. alata*; it is difficult to visualize an incompatible reaction as a result of the interaction of two identical molecules (Dodds et al 1993). Based on the fine structural analysis of S-glycoproteins, N-linked carbohydrate chains are thought to be involved through posttranslational modification in the function of the *S*-alleles (Woodward et al 1992). This implies that the choice between the fates of *S*-alleles in the pollen grain and style is predetermined rather than stochastic. From observations showing that S-RNases are effective substrates for protein kinases of pollen tubes, a role for phosphorylation of S-RNase involving Ca^{2+}-dependent protein kinases in the signal transduction pathway of self-incompatibility appears as a distinct possibility (Kunz et al 1996).

(ii) Expression of *S*-Locus Glycoprotein Genes

An important aspect of the cloning of *S*-locus glycoprotein genes from *N. alata* concerns their temporal and spatial appearance. At what period are the transcripts of the gene expressed relative to the stage of floral development at which overt self-incompatibility syndromes appear? Does the accumulation of transcripts coincide with the site of pollen tube inhibition? Fortunately, these questions have generated complementary sets of data that indicate that the pattern of gene expression reflects the age of the flower when it becomes self-incompatible and the age of the tissues where the reactions occur. From Northern blot analysis of stylar RNA from flower buds of different ages probed with *S-2* cDNA, it would appear that gene expression is most intense in mature styles, weak in the ovary, and absent in immature styles. At the cellular level, the *S-2* gene is expressed in the stigmatic papillae and stylar transmitting tissue of mature flowers as a pathfinder for the pollen tube before it succumbs, but the gene is not prevalent in the immature stigma–style complex before it becomes self-incompatible. The weak expression of the gene detected in the ovary in Northern blot is seen by in situ hybridization as labeling in the epidermis of the placenta. The epidermal cells are probably engaged in arresting the growth of self-pollen tubes that escape inhibition in the style (Anderson et al 1986; Cornish et al 1987). Immunocytochemical studies have shown that S-glycoproteins accumulate in the same tissues in which the gene transcripts are found (M. A. Anderson et al 1989).

In an approach using transgenic plants to localize the sites of *S*-gene expression, self-compatible *N. tabacum* plants were transformed with constructs encoding the S2-glycoproteins, and gene expression was followed by protein gel blot and Northern blot analyses and by immunolocalization. The aspects of gene expression in transgenic tobacco plants that can be correlated with self-incompatibility in *N. alata* are the localization of S2-glycoproteins in the stylar transmitting tissues, an increase in the level of the glycoprotein with maturity of the style, and RNase activity of S2-glycoproteins. However, no change in the self-incompatibility phenotype of transformants was observed as they spontaneously self-pollinated (Murfett et al 1992). In another work, prompt rejection of self-pollen was seen when a high level of RNase expression was engineered in transgenic self-compatible hybrids between *N. alata* and *N. langsdorfii*. Conversely, hybrids of *N. plumbaginifolia* × *N. alata* transformed with antisense construct targeted to the *S2-RNase* gene of *N. alata* had low RNase activity and were compatible with *N. alata* pollen. These ingenious crosses were necessary to overcome the problems of the low expression of *S-RNase* genes in transgenic *N. alata* and the poor regeneration of this species in tissue culture (Murfett et al 1994, 1995; Murfett, Bourque and McClure 1995). Such problems will no doubt be examined critically as more *S-RNase* expressions are monitored in transgenic plants with the insight gained from the knowledge of the structure of RNase genes and their promoter sequences (Matton et al 1995). For the present, these experiments come close to providing a direct demonstration that the *S-RNase* gene is sufficient for self-pollen rejection.

3. *PETUNIA* AND OTHER SOLANACEAE

Petunia enjoys the distinction as one of the first genera to be used in self-incompatibility research, yet our knowledge of the molecular biology of self-incompatibility in this genus lags behind that of *Brassica* and *Nicotiana*. cDNA clones encoding *S*-allele–associated proteins were isolated and characterized almost simultaneously from *P. hybrida* and *P. inflata*. In studies of *P. hybrida*, cDNA libraries made to mRNA of mature styles of different *S*-alleles were screened with an oligonucleotide homologous to the conserved N-terminal region of *N. alata* glycoprotein to generate style-specific clones. These clones reacted specifically with RNA from the style, where the levels of mRNA homologous to the cloned cDNAs increased dramatically with the transition of the flower from self-compatibility to self-incompatibility. The clones were of low copy number and, consistent with their allelic nature, hybridized to polymorphic DNA restriction fragments (Clark et al 1990). A cDNA clone encoding *S*-locus glycoprotein in a *S-2* homozygote of *P. inflata* was selected from a cDNA library using *N. alata* cDNA as a probe; subsequently, cDNA clones corresponding to the glycoproteins of other *S*-locus alleles of *P. inflata* were obtained by probing with *S-2* cDNA clones (Ai et al 1990). Antisense technology has been successfully used to demonstrate a correlation between inhibition of S-protein synthesis in the style and enforced breakdown of self-incompatibility in transgenic *P. inflata* (Lee, Huang and Kao 1994).

Combining the predicted amino acid sequences of SLGs of *P. hybrida* and *P. inflata* allows the following conclusions to be drawn about their sequence conservation and variability. Like *N. alata* glycoproteins, *Petunia* proteins are about 220 amino acids in length and include a 20–amino acid signal peptide characteristic of secreted proteins. The sequences reveal a general pattern of several highly conserved domains, interspersed by regions of sequence variability. Among the individual residues that are highly conserved are eight cysteine clusters. Although SLGs of *N. alata* have several N-glycosylation sites, S-glycoproteins of *Petunia* have only one or, at best, two conserved residues that constitute potential N-glycosylation sites. The importance of RNase to the self-incompatibility mechanism in *Petunia* is underscored by the presence of conserved domains in the S-glycoproteins with strong homology to active site regions of fungal RNases (Broothaerts et al 1989, 1991; Ai et al 1990; Clark et al 1990; Ioerger et al 1991; Lee, Singh and Kao 1992). Additionally, SLGs from self-incompatible genotypes of *P. hybrida* and *P. inflata* have been purified and shown to be RNases, as are the SLGs of *N. alata* (Broothaerts et al 1991; Singh, Ai and Kao 1991). A further point of interest is that the carbohydrate moiety of the S-RNase is not required for the self-incompatibility reaction in *P. inflata*, suggesting that the *S*-allele specificity of the protein is determined by its amino acid sequence and not by the glycan chain (Karunanandaa, Huang and Kao 1994). Direct proof implicating RNase in self-incompatibility in *P. inflata* has come from an experiment in which a mutant *s*-gene that lacked RNase activity due to a site-directed mutagenesis of one of the catalytic histidines was introduced into transgenic *P. inflata* and failed to reject self-pollen bearing the same *S*-allele as the transgene (Huang et al 1994).

However, the extent to which stylar RNase activity in the self-incompatible lines of *Petunia* inhibits self-pollen tube growth is at present uncertain, and this gap in our knowledge has thwarted attempts to formulate a unitary concept of *S*-gene action in the gametophytic systems. The uncertainty stems from an examination of the levels of S-proteins and RNases in pseudo–self-compatible and self-compatible lines of *P. hybrida*. When RNase activity of these lines is compared, equal levels of activity are found in both (Clark et al 1990). Similarly, the S-proteins of self-compatible lines of *P. hybrida* display the same spatial and temporal pattern of expression as the SLGs of self-incompatible *P. inflata* lines. Proteins of both lines also have comparable levels of RNase activity (Ai, Kron and Kao 1991). This evidence begs the question of how S-RNase distinguishes between self- and nonself-pollen and how the recognition reaction leads to growth inhibition of the pollen tube.

Conflicting data exist about the expression of *S-RNase* transcripts in pollen grains of *Petunia*. Microprojectile bombardment showed that chimeric *S-RNase–GUS* gene accumulates to low levels in developing anthers of *P. hybrida* prior to anthesis (Clark and Sims 1994). However, the self-incompatibility trait of pollen grains of transgenic *P. inflata* is unaffected in transgenic plants whose phenotype is transformed from self-incompatibility to self-fertility (Lee, Huang and Kao 1994).

Some insight into the structure of the *S*-gene of *P. inflata* is provided by an analysis of genomic clones of two *S*-locus alleles. The structure associates both alleles with a single, short intron that is embedded in the region identified with high interallelic variability in SLGs and that shows small variations in size and position between alleles. The clones of both alleles share the same TATA-box structure and have short, untranslated leader sequences. In both, there is a lack of interallelic homology in the flanks of the coding regions, which themselves contain repeated sequences. Despite these and other general similarities showing that the clones are alleles of the same *S*-gene, the clones appear to be basically heterogeneous, except for conserved sites within the coding regions (Coleman and Kao 1992). The sequence divergence in the flanking regions might reflect an inevitable effect of an independent evolution of the two alleles.

Attempts to clone self-incompatibility genes from other solanaceous genera have used oligonucleotide probes or heterologous probes to screen cDNA libraries. Since conserved regions of SLGs thus far characterized display some homology, it might be reasonable to expect heterologous probes to identify sequences similar to the primary structural features of *S*-locus proteins from cDNA libraries of related genera. In one round of experiments, cDNA or genomic clones encoding *S*-locus proteins were isolated from *Solanum chacoense* (Xu et al 1990b; Després et al 1994; Saba-El-Leil et al 1994), *S. tuberosum* (Kaufmann, Salamini and Thompson 1991), and *Lycopersicon peruvianum* (Tsai et al 1992; Rivers et al 1993; Chung et al 1994). A rare, variant self-compatible line of *L. peruvianum* was used to illustrate the essentiality of RNases in preserving self-incompatibility in this species. Crosses between the variant line and a self-incompatible line showed that the mutation to self-compatibility is located at the *S*-locus and is associated with the absence of stylar RNase activity. The derived amino acid sequence for the glycoprotein from the compatible line was found to differ from that of the incompatible line by the loss of a histidine residue; as histidines are essential for the catalytic activity of the enzyme, this leads to the production of a protein that lacks RNase activity (Kowyama et al 1994; Royo et al 1994).

Comparison of the deduced amino acid sequences of *S*-locus proteins of *S. chacoense, S. tuberosum*, and *L. peruvianum* with SLGs of *Nicotiana alata, Petunia hybrida*, and *P. inflata* has shown a set of five conserved regions, including domains similar to those in the active site of fungal RNases. Equally significant is the presence of eight conserved cysteine residues in all *S*-locus proteins. Extreme sequence diversity between alleles of the same, as well as of different species, is also a fundamental property of the proteins characterized from the four genera of Solanaceae (Tsai et al 1992). Comparison of the amino acid sequences of *S*-locus proteins of 11 alleles from *N. alata, P. inflata*, and *S. chacoense* has revealed the interesting fact that sequence similarity is relatively low within species, whereas some interspecific homologies are higher than intraspecific homologies (Ioerger, Clark and Kao 1990). That extreme sequence diversity can be pervasive between populations of even a single species is apparent from a study in which the *S*-allele from *L. peruvianum* was used to probe DNA from several natural populations of the species (Rivers et al 1993). These and other observations have been put together in the form of a hypothesis about the phylogenetic relationships of the *S*-alleles that states that *S*-allele polymorphism has a common ancestry and that it arose prior to speciation in the four genera of Solanaceae (Ioerger, Clark and Kao 1990; Rivers et al 1993).

In common with the structure of the *S*-gene of *P. inflata*, genomic clones corresponding to two *S*-alleles of *S. tuberosum* have a single intron and a short, untranslated leader sequence (Kaufmann, Salamini and Thompson 1991). From the data available, no major differences are discernible between the gene structures of the alleles. Expression of an *S-RNase* gene from *S. tuberosum* under the control of the tomato pollen-specific *LAT-52* gene is developmentally regulated in the pollen grains of transgenic tobacco, but that alone does not appear sufficient to influence the compatibility phenotype of the transformants (Kirch et al 1995).

4. *PAPAVER RHOEAS* AND OTHER PLANTS

Molecular biology of self-incompatibility in *Papaver rhoeas* may provide a testing ground for concepts applied to gametophytic and sporophytic systems of self-incompatibility. Although the self-incompatibility system in this species is of the gametophytic type, inhibition of self-pollen occurs in the stigma, as in the sporophytic type. What makes this system interesting is that it is possible to reproduce the self-incompatibility reaction in vitro and that identification of the putative *S*-gene products has been made through their biological activity of the inhibition of pollen tube growth (Franklin-Tong, Lawrence and Franklin 1988; Franklin-Tong et al 1989). Transcript expression of the *S*-gene cloned from *P. rhoeas* correlates well spatially and temporally with the appearance of S-protein in the stigma and the onset of self-incompatibility. The functional significance of this gene is underscored by the fact that the recombinant protein isolated from *Escherichia coli* expressing the gene inhibits pollen tube growth in culture, as does the stigmatic extract (Foote et al 1994). Finally, some preliminary evidence has been obtained for the presence in mature *P. rhoeas* pollen grains of a plasma membrane–associated glycoprotein that binds to the stigma S-proteins in a nonspecific manner (Hearn, Franklin and Ride 1996). As is evident by now, these observations hold the key to the basic reactions in self-incompatibility in all plants.

In seeking a common denominator with other gametophytic self-incompatibility systems, it was found that the focus on RNase as a candidate for stigmatic *S*-locus protein in *P. rhoeas* was misplaced. This summation is based on several lines of evidence, such as the presence of negligible amounts of RNase in the stigma of *P. rhoeas*, the lack of correlation of enzyme activity with the development of incompatibility, the lack of RNase activity in purified *S*-locus protein, and the insensitivity of pollen grains to RNase in in vitro bioassays (Franklin-Tong et al 1991). On the other hand, the rapidity of the self-incompatibility reaction in *P. rhoeas* has led to the view that this is mediated by cytosolic Ca^{2+} acting as a second messenger. In support of this view, using laser scanning confocal microscopy and fluorescence ratio imaging of pollen tubes stained with the fluorescent dye Calcium Green-1 to localize Ca^{2+} fluxes, researchers have visualized transient increases in cytosolic Ca^{2+} levels in growing pollen tubes after the addition of partially purified, incompatible stigmatic protein extract or the recombinant S-protein. Evidently, changes in Ca^{2+} levels are far-reaching, for the growth of pollen tubes that generate a rise in Ca^{2+} when challenged with the stigmatic extract or S-protein is inhibited (Figure 10.10). There is no rise in the Ca^{2+} levels of pollen tubes after addition of either compatible or heat-denatured, incompatible stigmatic extract. Even when the level of endogenous Ca^{2+} is raised artificially by UV activation of caged Ca^{2+}, an inhibition of pollen tube growth is noted in the absence of incompatible stigmatic extract (Franklin-Tong et al 1993; Franklin-Tong, Ride and Franklin 1995). Conceivably, the transcriptional activation of genes necessary for the initiation of self-incompatibility reactions in *P. rhoeas* pollen might be the consequence of a rise in Ca^{2+} level that is induced by incompatible *S*-locus proteins and is accompanied by an increased phosphorylation of proteins, but we have a long way to go before the signal transduction pathway and the mode of action of gene products are defined (Franklin-Tong, Lawrence and Franklin 1990; Rudd et al 1996).

Brassicaceae, Solanaceae and Papaveraceae thus represent the three families in which *S*-genes and their products have been intensively studied at the molecular level; very little information is available on other plants. A beginning has been made toward the molecular analysis of self-incompatibility in members of Rosaceae by characterizing cDNA clones representing *S*-locus alleles of apple and *Pyrus serotina*. As is the case in the *S*-locus genes of other plants, specificity of expression of transcripts of apple cDNA clones in the stylar tissues is inher-

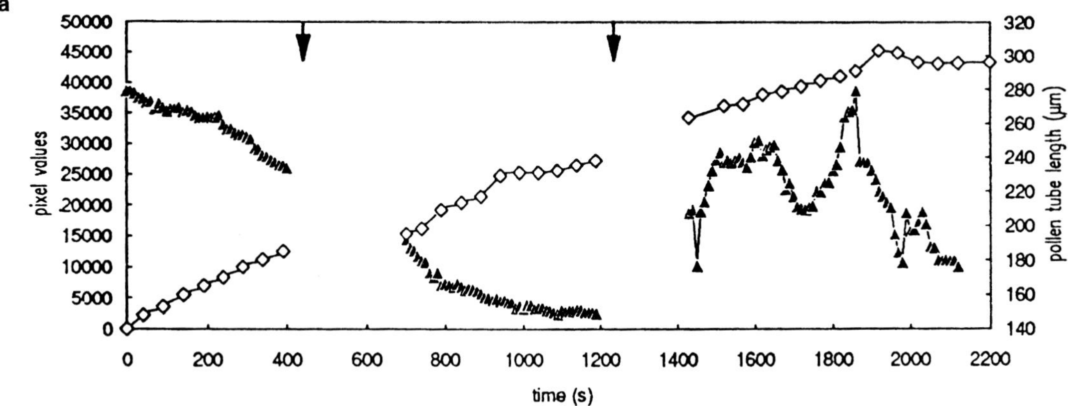

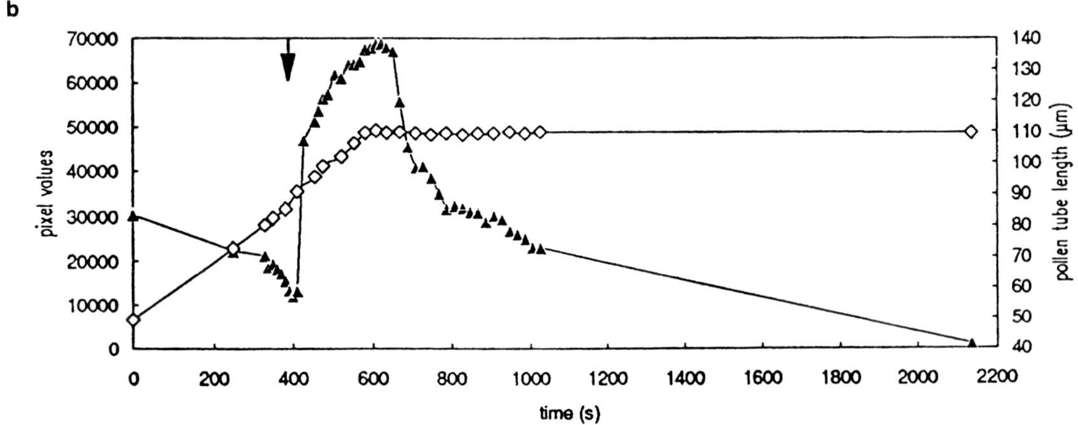

Figure 10.10. Changes in Ca^{2+} and growth of pollen tubes of *Papaver rhoeas* microinjected with Calcium Green-1. The filled triangles represent fluorescence (pixel values) from the first 100 µm of the pollen tube; the open squares indicate the relative length of the pollen tube at the corresponding time point. **(a)** Addition of compatible stigmatic S-glycoproteins (first arrow) followed by addition of incompatible stigmatic S-glycoproteins (second arrow). No increase in Ca^{2+} and no significant inhibition of pollen tube growth are observed after treatment with the compatible protein; after addition of the incompatible protein, there is a transient rise in Ca^{2+}, indicated by the two peaks, and a cessation of pollen tube growth around 2,000 seconds. **(b)** Photoactivation of caged Ca^{2+} (arrow) and cessation of pollen tube growth almost immediately. (From Franklin-Tong et al 1993.)

ent to their function, and, similar to S-proteins of members of Solanaceae, purified S-proteins from apple and *P. serotina* correspond to RNases (Broothaerts et al 1995; Sassa et al 1996). It appears that RNases are likely to mediate the self-incompatibility syndrome in certain lines of *Antirrhinum majus* also (Xue et al 1996). So, if a single regulatory molecule is required to activate the stylar self-pollen rejection mechanism, it would probably be the same molecule in members of Solanaceae, Rosaceae, and Scrophulariaceae. There is preliminary evidence showing that *SLG* and *SRK*-like genes modulate the sporophytic type of self-incompatibility in *Ipomoea trifida* (Kowyama et al 1995).

What appears to be the pollen component of the self-incompatibility allele has been cloned from the grass *Phalaris coerulescens* (X. Li et al 1994). Unlike the other gametophytic systems studied, self-incompatibility in *P. coerulescens* is under the control of two unlinked loci, *S* and *Z*, and self-pollen rejection is the rule when both *S* and *Z* alleles are present in the pollen as well as in the style. A cDNA clone identified by differential screening was found to be expressed only in mature pollen grains and was absent from all other tissues tested, including the stigma. The deduced amino acid sequence showed that the protein has a variable N-terminus that has no homology to known proteins and a conserved C-terminus with a thioredoxin domain. The absence of a signal sequence and the presence of thioredoxin sequences probably indicate that, unlike other S-locus proteins, this is a cytoplasmic protein. How this protein controls both recognition specificity and inhibition of pollen tube growth in the self-incompatibility reaction remains to be elucidated.

Considering the self-incompatibility systems for which good molecular data are available, it is clear the pollen self-rejection system evolved independently several times in the evolution of angiosperms. Millions of years have probably separated the different incompatibility systems. The molecular data from Brassicaceae, Solanaceae, and Papaveraceae show that there are some obvious, basic similarities, as well as differences, in the operation of signaling systems in the three families. A viewpoint that has gained momentum in recent times is that cellular responses unleashed during sporophytic self-incompatibility may share some components of host–pathogen responses in plants and that self-incompatibility is evolutionarily related to the system that conditions resistance to invading pathogens. Besides superficial similarities between the genetics, physiology, and biochemistry of the self-incompatibility and host–pathogen responses, the possible involvement of a receptor kinase in both strengthens the view that there is a connection between the two responses (Hodgkin, Lyon and Dickinson 1988; Lamb 1994).

5. GENERAL COMMENTS

Molecular analysis shows that specific characteristics of the self-incompatibility genes are varied in the four systems in which they have been studied intensively, and this makes it difficult to interpret the mechanism of action of the encoded proteins in the rejection of self-pollen. In a general sense, we know that in *Brassica* the structure of the proteins suggests a role in the signal transduction mechanism, whereas in *Nicotiana* and *Petunia* the protein that shows similarity to RNases functions by disabling incompatible pollen tubes. Particularly intriguing is that in *Papaver* the protein elicits a cellular response that results in a transient increase in Ca^{2+} concentration.

Dramatic progress is currently being made in understanding the way the different proteins work in the model incompatibility systems. Identification of the recognition sequences expressed by the pollen grains assumes an obvious significance in the expected flow of information. At this stage, we do not know whether the recognition sequences expressed by pollen grains are different from those expressed by the female part of the flower. In analyzing the conditions that lead to the incompatibility reactions, delineation of the parts of the incompatibility molecules important for recognition begs resolution. Equally significant is the role of other parts of the incompatibility protein, perhaps in the initiation of events that subsequently result in pollen tube collapse. It is encouraging to note that more than one model system of self-incompatibility has been investigated to a level insightful enough that realistic progress in our understanding of the molecular basis of the recognition reaction and its consequences can be expected in the foreseeable future.

Chapter 11

Fertilization: The Beginning of Sporophytic Growth

In the life cycle of plants, fertilization is invariably associated with sexual reproduction and therefore represents the genetic switch that initiates the diploid, or the sporophytic, phase of development. The history of research into fertilization in angiosperms may be said to have begun with the discovery of syngamy, or the actual fusion of the male and female gametes, by Strasburger (1884) and of double fertilization by Nawaschin (1898) and Guignard (1899). The latter two investigators called attention to the streamlining of the male gametophyte's economy by providing light microscopic evidence for the fusion of one sperm with the egg cell to form the diploid zygote and of the other sperm with the polar fusion nucleus of the central cell to generate the triploid endosperm. In spite of an abundance of studies on the development of the male and female gametophytes and on postfertilization events in angiosperms in the years following these discoveries, relatively few investigators have ventured into the area of fertilization research. This has led to the unenviable situation that, as a biological process, fertilization in angiosperms has remained as an unknown black box. The paucity of information on fertilization is undoubtedly related to the fact that events leading to the egg–sperm encounter take place in the privileged interior of the multicellular ovule, where, hidden from the eye, they are not prone to experimental assault and to the technical difficulties that have so far prevented the development of an *ex ovulo* in vitro fertilization system. However, during the past three decades, there has been considerable progress in our understanding of many facets of fertilization, particularly the ultrastructure of the male and female gametes and the sequence of events leading to gametic fusion. Extensive use of the electron microscope, experiments involving immunological and molecular approaches, and nondestructive isolation techniques have all contributed to this progress.

Chapter 6 described in some detail the structure of the egg at the time of fertilization, and so this material will not be considered here again except as it relates to the male gametophyte preparatory to gametic fusion. The primary focus of the present chapter will be on the structure of the sperm cells, their fate in the pollen tube, and their eventual fusion with the cells of the embryo sac. The final section of this chapter is devoted to a review of the work done on the manipulation of the fertilization process under controlled conditions. Until recently, the subject of fertilization has been reviewed almost always within the larger context of male and female gametogenesis (Linskens 1969; Jensen

1973; Kapil and Bhatnagar 1975; van Went and Willemse 1984), but more focused reviews (Knox and Singh 1987; Knox, Southworth and Singh 1988; Russell, Cresti and Dumas 1990; Russell 1992) are beginning to appear. These should serve as excellent sources of information on various aspects of the biology of fertilization.

1. CELLULAR NATURE OF THE SPERM

One point of special concern in the early studies on the structure of the sperm was whether the sperm is a true cell with all the trappings or just an orphaned nucleus. Although some light microscopic investigations on this question remained inconclusive, the cellular nature of the sperm was clearly established by the early 1940s. Subsequently, electron microscopy, particularly the pioneering investigations of Jensen and Fisher (1968b) on bicellular pollen grains of *Gossypium hirsutum*, and of Hoefert (1969b) and Cass (1973) on tricellular pollen grains of *Beta vulgaris* and *Hordeum vulgare*, respectively, provided the first direct evidence that the male gamete in angiosperms is not a naked nucleus but is a relatively simple cell, with its own cytoplasmic sheath containing a reduced complement of organelles. Exhibiting great simplicity, the sperm in *G. hirsutum* is reduced to little more than a bag for a prominent nucleus bathed in a thin layer of cytoplasm containing dictyosomes, ER, polysomes, vesicles, and organelles resembling mitochondria (Jensen and Fisher 1968b). Besides these cytoplasmic organelles, sperm cells of *B. vulgaris* and *H. vulgare* also harbor files of microtubules along their length, suggesting that the cytoskeletal elements might play a role in maintaining the shape of the sperm or in affecting sperm motility (Hoefert 1969b; Cass 1973).

Further information about the more intimate details of the structure of sperm cells has come from electron microscopy combined with three-dimensional reconstruction, freeze-fracture analysis, confocal laser scanning fluorescent microscopy, and immunochemistry. From these investigations emerges the following summary picture. Sperm cells generally extend in length to about 35 μm and are approximately 1–3 μm in width and 7–133 μm^3 in volume. The characteristic shape of sperm cells, as seen in mature pollen grains or in pollen tubes, is elongate or spindle shaped. The most conspicuous organelle of the sperm is the nucleus, which occupies an enlarged region at one end of the cell; one or more tapering cytoplasmic extensions are found at the other end. The nucleus of the mature sperm contains paternal genetic material, which appears as densely staining granular chromatin (Knox and Singh 1987; Knox, Southworth and Singh 1988). The nucleus is enclosed in a nuclear envelope in which the usual pores appear to be lacking or are few in number (Mogensen and Wagner 1987; Southworth, Platt-Aloia and Thomson 1988; Taylor et al 1989; Southworth 1990). Not much should be read into this last observation except that the nucleus of the mature sperm is transcriptionally inactive and negotiates a limited traffic with the surrounding cytoplasm.

It is within the framework of the subcellular structure that factors responsible for the differentiation of cells reside; however, in all cases examined, the cytoplasm of the male gamete appears truly undistinguished and does not give any hints about the specialized character of the cell (Figure 11.1). In classical descriptions, limited amounts of ER, ribosomes, polysomes, lipid and protein bodies, mitochondria, dictyosomes, microtubules, and vesicles appear to make up the bulk of the sperm cytoplasm, with the mitochondrion varying with respect to its morphology in certain species (Dumas, Knox and Gaude 1985; McConchie, Hough and Knox 1987; Charzyńska, Ciampolini and Cresti 1988; Hu and Yu 1988; Southworth and Knox 1989; Taylor et al 1989; Yu, Hu and Zhu 1989; Cresti, Murgia and Theunis 1990; Theunis 1990; P. E. Taylor et al 1991; Theunis, Cresti and Milanesi 1991; van der Maas et al 1994). Mitochondria are the most prominent organelles of the sperm of *Zea mays* and, as seen in three-dimensional reconstruction, they form a filamentous complex, each individual filament partially wrapping around itself (McConchie, Hough and Knox 1987; Mogensen, Wagner and Dumas 1990). At the other extreme are the conventional, small, ellipsoidal mitochondria found in the sperm cells of *Brassica campestris* and *B. oleracea* (McConchie et al 1987). An important question raised in these studies relates to the inheritance of plastids – whether plastids are paternally inherited or not – and hence to their presence in the cytoplasm of the sperm cells. The answer to this question appears equivocal, because sperm cells of some plants, like *Plumbago zeylanica* (Russell and Cass 1981; Russell 1983), *Rhododendron laetum*, and *R. macgregoriae* (Taylor et al 1989), come equipped with plastids in their cytoplasm, whereas those of *Spinacia oleracea* (Wilms, Leferink-ten Klooster and van Aelst 1986), *Z. mays* (McConchie, Hough and Knox 1987), *B. campestris, B. oleracea* (McConchie et al 1987), *B. napus* (Murgia et al 1991b), and *Nicotiana tabacum* (Yu, Hu and Zhu 1989) lack plastids.

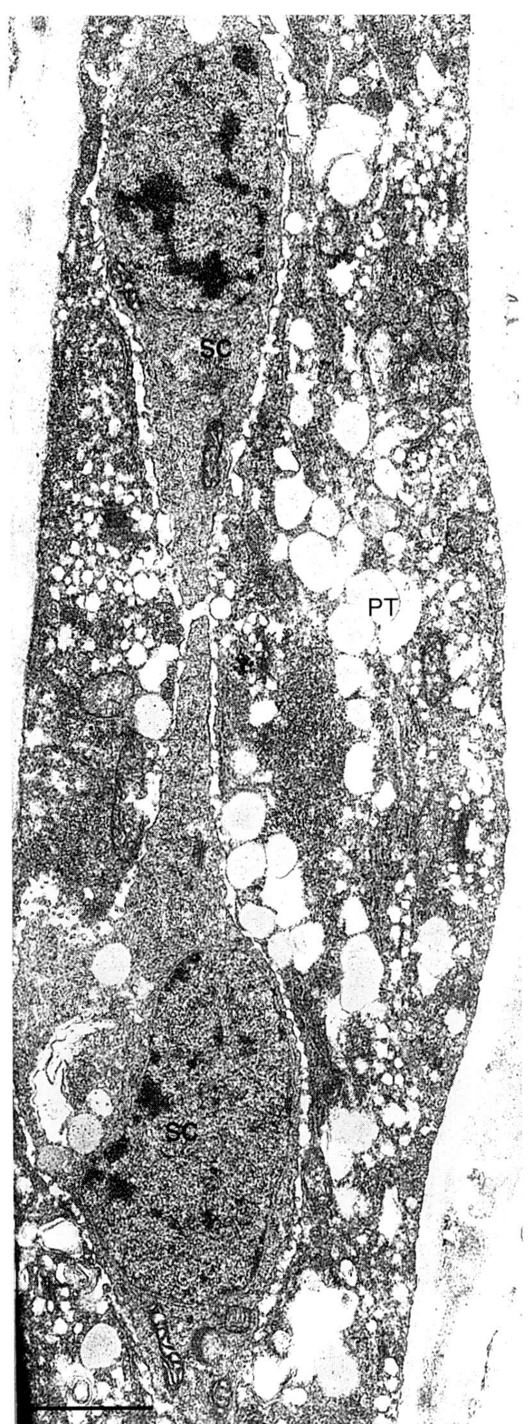

Figure 11.1. Structure of the sperm cell in the pollen tube of *Populus deltoides*. Two elongate sperm cells seen in the longitudinal section of the pollen tube are connected by narrow cytoplasmic projections. PT, pollen tube; SC, sperm cell. Scale bar = 1 µm. (From Rougier et al 1991; photograph supplied by Dr. C. Dumas.)

(i) The Sperm Cytoskeleton

Analysis of the cytoskeletal machinery of the sperm has followed a two-tier course, analogous to that employed in the study of the cytoskeleton of the vegetative and generative cells (Chapter 3). Early studies were conducted at the ultrastructural level, whereas later studies enlisted immunofluorescence methods. These investigations have provided a surprisingly uniform picture of the organization of sperm microtubules, which attain conformations compatible with the function of the sperm cells. Commonly, microtubules, singly or in bundles, are aligned longitudinally or helically in relation to the axis of the sperm (Hoefert 1969b; Cass 1973; Karas and Cass 1976; Zhu et al 1980; Dumas, Knox and Gaude 1985; Pierson, Derksen and Traas 1986; Charzyńska, Ciampolini and Cresti 1988; Taylor et al 1989; Cresti, Murgia and Theunis 1990; Del Casino et al 1992; Palevitz and Liu 1992). A striking feature of the sperm of *Tradescantia virginiana* is the presence of numerous interconnected helical to longitudinal branches of microtubules that appear to arise from the main microtubule bundle and invest the entire cell (Palevitz and Cresti 1988).

Whether the microtubule elements of the sperm are inherited from the interphase cytoskeletal network of the generative cell at the time of division or are assembled anew from stable precursors at the end of division is not clear. According to Palevitz and Cresti (1989), the entire array of microtubules of the sperm of *T. virginiana* is inherited from those seen in the late anaphase of the generative cell mitosis; these microtubule bundles are derived from the kinetochore fibers, which remain enmeshed with each other and with the surrounding microtubules during anaphase. Consistent with this view is the observation that the precise, stepwise aggregation of the kinetochore fibers at the poles of the generative cell is also reflected in the cytoskeleton of the young sperm (Palevitz 1990). However, circumstantial evidence such as the presence of fewer and less-prominent microtubule bundles in sperm cells compared to their progenitor cell, as seen in *Rhododendron laetum* and *R. macgregoriae* (Taylor et al 1989), argues against the inheritance of microtubules from the previous generation. On the whole, the situation presents an invitation for more experimentation about the origin of microtubules in the sperm cells of angiosperms.

In ultrastructural investigations of sperm cells, not much attention has been paid to the presence of actin-containing microfilaments, and, if microfilaments are present, they have clearly remained elusive. Experiments using rhodamine-phalloidin as a fluorescent detection probe have also failed to reveal the presence of microfilaments in sperm cells of *Lilium longiflorum, Nicotiana tabacum, Petunia hybrida,* and *Tradescantia virginiana* (Pierson, Derksen and Traas 1986; Palevitz and Cresti 1988, 1989). As indicated in Chapter 3, ultrastructural and immunofluorescence studies have failed to detect microfilaments in the generative cell; in this respect, sperm cells appear to share a unique property with their progenitor cell. The only reports of the presence of microfilaments are in sperm cells of wheat (Zhu et al 1980) and *Plumbago zeylanica* (Russell and Cass 1981), but these observations have not been corroborated by other researchers. One study using antiactin immunofluorescence as a specific probe has reported an impressive localization of actin in the generative cells and sperm cells of *Rhododendron laetum* and *R. macgregoriae* (Taylor et al 1989), but a subsequent work using the same and other materials and the same antiactin probe failed to substantiate this observation (Palevitz and Liu 1992). On the weight of the evidence, the latter investigators have concluded that microfilaments do not form part of the cytoskeleton of the sperm cells of angiosperms.

(ii) The Male Germ Unit

The electron microscopic image of the outer boundary of the male gametes in the pollen grain is that of protoplasm covered by a plasma membrane, with two sperm cells enclosed together in the inner plasma membrane of the vegetative cell. The membranes may be tightly appressed or may form a loose association with intramembrane pockets that enclose vesicles and osmiophilic droplets (Zhu et al 1980; Dumas, Knox and Gaude 1985; McConchie, Hough and Knox 1987; McConchie et al 1987; Southworth and Knox 1989). The paired nature of the outer membrane system of the sperm has been confirmed by visualizing cells by freeze-fracture, in which the frozen membranes are cleaved along their hydrophobic interior and the resulting surface is observed in the electron microscope; this approach did not reveal any connections or bridges between the membranes (Southworth, Platt-Aloia and Thomson 1988; Southworth 1990; van Aelst, Theunis and van Went 1990). This observation undoubtedly relegates the male gamete to the status of a natural protoplast; however, a rudimentary wall has been reported to be present in the sperm of wheat (Zhu et al 1980; Hu, Zhu and Xu 1981), *Plumbago zeylanica* (Russell and Cass 1981), and barley (Mogensen and Wagner 1987).

The contacts and physical associations of the two sperm cells born out of the division of a generative cell were occasionally commented upon in some of the early light and electron microscopic studies. Are the two sperm cells closely linked together as a pair, or are they aligned with the vegetative cell nucleus, or do they float aimlessly in the pollen grain or in the pollen tube cytoplasm? Although the occurrence of paired sperm cells and their closeness to the vegetative cell nucleus in the pollen tube of cotton was noted by Jensen and Fisher (1968b, 1970), it was the work of Russell and Cass (1981) that introduced a new level of understanding of the characteristics of the sperm cell association and the sperm–vegetative cell package. This work showed that in the tricellular pollen grain of *P. zeylanica*, the two sperm are directly linked to each other by a common lateral wall transgressed by plasmodesmata. It was also found that while both sperm are enclosed in the inner plasma membrane of the vegetative cell, one of them is interlocked with the nucleus of the vegetative cell in a complex way. The conformational relationship of the sperm cell to the vegetative cell nucleus is traced to a cytoplasmic extension that wraps around the periphery of the vegetative cell nucleus and is partially enclosed by convolutions of this nucleus (Figure 11.2). The sperm cell remains associated with the nucleus of the vegetative cell during the period of active growth of the pollen tube, but the connection is severed following the discharge of the pollen tube contents. Although it has been suggested that the sperm–vegetative cell nucleus assembly might facilitate gamete delivery, its importance is not fully understood because the association of the naked nucleus and the adhering cells cannot be easily manipulated or examined in situ.

An exciting concept demonstrated in a later work (Russell 1984) is that the two sperm cells in the generative cell, rather than being isomorphic, display varying degrees of dimorphism in size and organelle content. This conclusion was made possible by an examination of serial, ultrathin sections of sperm cells, by computer-aided three-dimensional reconstruction of their structure, and by the actual counting of organelles in sections. The dimorphism in shape is due to a membranous extension from the one sperm of the pair that superficially embraces the nucleus of the vegetative cell. This sperm appears to be the larger of the two; the smaller one, which lacks a cellular projection, is unassociated with the vegetative cell nucleus.

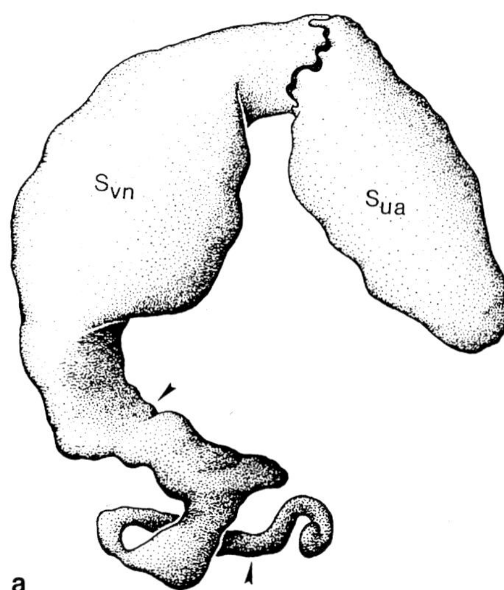

Figure 11.2. Reconstruction of the sperm cells and the male germ unit of *Plumbago zeylanica*. **(a)** The two sperm cells; the sperm cell associated with the vegetative cell nucleus is S_{vn}; the other sperm cell, not associated with the vegetative cell nucleus, is S_{ua}. **(b)** The physical association between S_{vn} and the nucleus of the vegetative cell (VN). The cellular protrusion of the sperm cell (arrowheads) is found within the clasping region of the lobed vegetative cell nucleus and extends through a shallow hole in the vegetative cell nucleus at one location (arrow). (From Russell 1984. *Planta* 162:385–391. © Springer-Verlag, Berlin.)

Perhaps the most dramatic difference between the two sperm lies in their relative content of heritable organelles. The large sperm is rich in mitochondria and is typically impoverished of plastids, whereas the small one is poor in mitochondria but has numerous plastids.

Since *P. zeylanica* is unusual in lacking synergids in the embryo sac, sperm–vegetative cell nucleus association and sperm dimorphism seen in this species raised objections that they may not be representative of the wide range of angiosperms characterized by the presence of two synergids flanking the egg cell. This objection was soon dispelled when similar features were found in sperm cells of *Brassica campestris* and *B. oleracea* (Dumas, Knox and Gaude 1985; McConchie, Jobson and Knox 1985; McConchie et al 1987), both of which possess synergids. In both, dimorphism in shape is due to the presence of an evagination on the sperm that is attached to the vegetative cell nucleus, which thus makes it the larger of the two sperm cells. The distinction between the two species lies in the fact that in *B. oleracea* the evagination on the sperm surrounds the surface of the vegetative cell nucleus superficially, whereas in *B. campestris* it penetrates the enclaves of the vegetative cell nucleus and emerges at the other end as a protuberance (Figure 11.3). In both species, the smaller sperm, which lacks a cytoplasmic outgrowth, is linked to the larger one by a common cell junction. The small sperm cells in both species of *Brassica* also appear unique in their paucity of mitochondria, with 4 in *B. campestris* and 6 in *B. oleracea*, compared to a maximum of 15 and 34 in the large sperm cells of *B. oleracea* and *B. campestris*, respectively. Neither

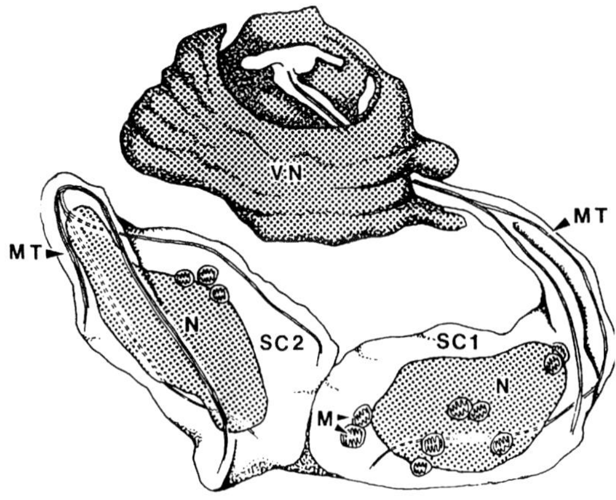

Fig. 11.3. A schematic diagram of the male germ unit of *Brassica campestris*. M, mitochondria; MT, microtubules; N, nucleus; VN, nucleus of the vegetative cell; SC1, SC2, sperm cells 1 and 2. (From McConchie, Jobson and Knox 1985.)

sperm cell contains any plastids. The published accounts of the structure of the sperm of *Spinacia oleracea* are much less detailed than those of *Brassica* sp., although differences in the sperms' size and association with the nucleus of the vegetative cell are obvious in the electron micrographs (Wilms, Leferink-ten Klooster and van Aelst 1986; Theunis, Cresti and Milanesi 1991).

Thanks to the skills of the electron microscopist and the availability of techniques for computer simulation of the data, these observations add a new dimension to our knowledge of sperm structure and organelle inheritance in angiosperms. Despite the minimal structural attributes displayed by the sperm, this cell appears to be highly differentiated in relation to its developmental tasks, especially to respond with preestablished specificity and to discriminate between the egg nucleus and the polar fusion nucleus during fertilization. The presence of significantly different numbers of heritable organelles in the two sperm cells of a pair suggests the possibility of sophisticated mechanisms of organelle transmission at fertilization. The differences in the sizes of the two sperm and in their organelle contents pose some important questions on differential gene activity during sperm cell differentiation.

The novel features of the structure of the sperm provided the impetus for further work, which brings the observations in line with the current thinking on the nature of the male gamete. Dumas et al (1984) designated the functional complex of the two sperm cells of common origin plus the nucleus of the vegetative cell in the pollen grain, or in the pollen tube, as the male germ unit; the rationale for this concept is the view that all the nuclear and cytoplasmic DNA of male heredity is linked together, forming a single functional unit for the recognition of target female cells, the transmission of hereditary materials, and fusion during double fertilization. Some investigators (Kaul et al 1987; Yu and Russell 1992) have referred to a complex association formed between the vegetative cell nucleus and the generative cell as the male germ unit; in other cases, the transient union noted between the two sperm after they are formed from the generative cell has been identified as the male germ unit (Levieil 1986; Charzyńska et al 1989; Charzyńska and Lewandowska 1990). It is a moot point whether such associations, however complex they are, should be included within the concept of the male germ unit, which was coined to convey a precise functional union involving the two sperm and the vegetative cell nucleus prior to double fertilization. This attention to detail seems justified because other electron microscopic studies, with or without three-dimensional reconstruction of serial sections, have demonstrated a physical relationship between the sperm and the vegetative cell nucleus; included in these studies are male germ units identified in pollen tubes of bicellular pollen grains of *Hippeastrum vittatum* (Mogensen 1986), *Petunia hybrida* (Wagner and Mogensen 1988), *Rhododendron laetum, R. macgregriae* (Taylor et al 1989), *Cyphomandra betacea* (Solanaceae) (Hu and Yu 1988), *Nicotiana tabacum* (Yu, Hu and Zhu 1989; Yu, Hu and Russell 1992; Yu and Russell 1994b), and *Populus deltoides* (Rougier et al 1991), and in tricellular pollen grains of *Spinacia oleracea* (Wilms, Leferink-ten Klooster and van Aelst 1986), *Catananche caerulea* (Asteraceae) (Barnes and Blackmore 1987), and *B. napus* (Murgia et al 1991b). In *C. betacea, N. tabacum,* and *P. hybrida*, the two sperm are linked together by a common transverse wall as in *Plumbago zeylanica*. In *N. tabacum*, the wall appears to be further specialized by the presence of fibrils, small tubules, and callose (Yu, Hu and Russell 1992).

The importance of three-dimensional reconstruction of the sperm–vegetative cell nucleus assembly lies in the demonstration that complex association between cells can be understood in simple graphic pictures. The impressive precision of these studies caused many to believe that it was only a matter of time before the concept of the male germ unit could be considered as a secure generalization applicable universally to angiosperms. This did not prove to be the case. From light and electron microscopic observations of pollen grains and pollen tubes of wheat (Chandra and Bhatnagar 1974), *Helleborus foetidus* (J. Heslop-Harrison, Heslop-Harrison and Heslop-Harrison 1986), *Euphorbia dulcis* (Murgia and Wilms 1988), and *Alopecurus pratensis* (J. Heslop-Harrison and Heslop-Harrison 1984), it appears that there is a spatial or temporal separation of the sperm pair from the nucleus of the vegetative cell. In live pollen tubes of *Secale cereale*, not only is there no clear-cut evidence of linkage between the sperm cells, but the cells have been observed to shoot past each other and move in opposite directions (J. Heslop-Harrison and Heslop-Harrison 1987). According to Weber (1988), who followed the division of the generative cell in tricellular pollen grains of *Galium mollugo, Trichodiadema setuliferum*, and *Avena sativa*, there is no linkage between sperm cells that separate from one another soon after they are formed. These observations, though drawing attention to the independent existence of sperm cells and the nucleus of the vegetative cell, do not take into account the fact that an association between them may be formed at a later stage, and so further work is called for before the issue can be considered settled. There is strong support for the view that in some plants the physical connection between sperm cells of a pair and between one sperm and the vegetative cell nucleus is ephemeral and would have been completely missed in routine electron microscopic observations. For example, by three-dimensional analysis of ultrathin sections, it was found that in the pollen grains of barley (Mogensen and Rusche 1985) and maize (McConchie, Hough and Knox 1987), the sperm cells are discrete and do not exist in physical association with the vegetative cell nucleus as the male germ unit. The major effort in a subsequent study on barley was to look for evidence of a physical association between sperm cells and the vegetative cell nucleus during hydration and germination of the pollen grain (Mogensen and Wagner 1987). This work showed that following controlled pollination, before sperm cells move into the emerging pollen tube, there is a close association between sperm cells and between them and the nucleus of the vegetative cell. Although the vegetative cell nucleus is not seen to migrate into the pollen tube, sperm cells remain attached in the pollen tube; as sperm cells approach the embryo sac, there is a further intimacy in their association and they appear to be connected at both ends. In maize pollen also, the vegetative cell nucleus is found to be separated from sperm cells, but it remains closely associated with sperm cells as a male germ unit in the course of postpollination journey (Rusche and Mogensen 1988; Mogensen 1992). In *Medicago sativa*, the association of sperm cells with the nucleus of the vegetative cell is tenuous in the pollen grain, but is well defined in the pollen tube throughout its growth (Zhu, Mogensen and Smith 1992). It has been argued from these observations that the male germ unit is not the developmental consequence of division of the generative cell; rather, it is induced as a result of pollen germination (Mogensen 1992). Since male germ unit formation is a morphogenetic process, it is also possible that, in some plants, the structure formed in the pollen grains is not stabilized before germination; germination is the trigger that stabilizes the male germ unit in the pollen tubes.

It was mentioned earlier that in certain plants that possess a male germ unit, the sperm cells show dimorphism in size and in the content of heritable organelles. But the presence of a male germ unit is not always associated with cytoplasmic asymmetry. Further work has shown that the high degree of cytoplasmic difference displayed by the sperm of *Plumbago zeylanica*, characterized by the presence of significantly different numbers of plastids and mitochondria, is somewhat unique and that far more widespread are cases in which organelle differences are restricted to mitochondrial number or nuclear size (McConchie, Hough and Knox 1987). Moreover, pollen grains of some plants with perfectly standard male germ units have revealed no appreciable differences in organelle contents by serial thin-sectioning, computer reconstruction, and quantification. Such isomorphic sperm cells have been found in barley (Mogensen and Rusche 1985), *Petunia hybrida* (Wagner and Mogensen 1988), *Rhododendron* sp. (Taylor et al 1989), *Medicago sativa* (Zhu, Mogensen and Smith 1992), and *Nicotiana tabacum* (Yu, Hu and Russell 1992). In the most detailed quantitative analysis of the sperm in *N. tabacum*, no statistically significant differences were found between the two sperm cells in several cellular parameters (Yu, Hu and Russell 1992; Yu and Russell 1994b).

In summary, it appears that in many angiosperms, part of the preprogramming of sperm cells involves the formation of the male germ unit with or without a concurrent asymmetry in the distribution of heritable organelles. How this specialized packaging of the sperm influences the fertilization process is not known at present. Dimorphism raises questions about the identity of the sperm cell that fuses with the egg and of the one that fuses with the polar fusion nucleus, with perhaps the hardest question to answer being, what is the nature of the factors that endow the specificity? We also need to know the mechanisms that cause the unequal partitioning of organelles and the trigger that induces it.

(iii) Isolation and Characterization of the Sperm

To gain insight into the molecular biology of double fertilization, it is necessary to know something about the dynamics of the participating gametes. Although it is relatively easy to monitor sperm cells microscopically, it is hard to investigate their metabolic activities experimentally. This objective is best achieved by isolating sperm from pollen grains or from pollen tubes. As the following survey shows, from a modest beginning using sperm cells extruded from pollen grains, the stage is set now for the large-scale isolation of viable sperm cells from pollen grains and pollen tubes of various plants. A brief review of the present state of the art is provided by Theunis, Pierson and Cresti (1991).

The first attempts to isolate sperm used tricellular pollen grains of barley (Cass 1973) and *Plumbago zeylanica* (Russell and Cass 1981). In recognition of the fact that the milieu of the pollen grain is in a state of high osmotic pressure, pollen contents were released into a medium of high osmolarity secured by the addition of 30% sucrose. Although the number of sperm cells isolated by this method was not large, some basic observations, such as the change in the shape of extruded sperm from spindle shaped to round or spherical, suggested a role for the cytoskeletal framework in the process. Similar methods, with variations in the concentrations of sucrose, were used subsequently for the small-scale isolation of sperm from tricellular pollen grains of *Brassica oleracea, Triticum aestivum* (Matthys-Rochon et al 1987), and *Zea mays* (Cass and Fabi 1988). The first successful isolation of sperm from bicellular types of pollen grains was accomplished by enzymatic hydrolysis of the pollen tube wall of *Rhododendron macgregoriae* and *Gladiolus gandavensis* (Shivanna et al 1988). For this purpose, pollen tubes allowed to grow through the style into a block of agar were treated with a mixture of macerozyme and cellulase, and sperm cells were separated by microfiltration.

Given the differences in density between the sperm and the rest of the pollen grain or pollen tube contents, one might expect to enrich a pollen fraction for sperm by differential centrifugation. This was achieved by subjecting osmotically shocked pollen grains of *P. zeylanica* to centrifugation in a sucrose gradient (Russell 1986). Moderately high yields of sperm essentially free of contamination by vegetative cell nuclei and starch and suitable for physiological manipulations were obtained by this method. Dupuis et al (1987) and Theunis and van Went (1989) used Percoll gradients to obtain high sperm yields from pollen grains of maize and spinach, respectively. Sperm cells at a concentration of 20,000/ml were obtained from pollen grains of *Gerbera jamesonii* by gentle homogenization followed by microfiltration (Southworth and Knox 1989). Both grinding and a two-step osmotic shock procedure have been claimed to give high yields of sperm from pollen grains of *B. campestris, B. napus, Z. mays,* and *Secale cereale* (Yang and Zhou 1989; Mo and Yang 1991). In the latter procedure, pollen grains hydrated in a high concentration of sucrose were shocked to burst and release the sperm by dilution of the medium. This is necessary when pollen cannot be burst osmotically; however, germinating the pollen grains for a short period of time to generate tubes was sufficient to allow sperm release by osmotic shock (P. E. Taylor et al 1991). All these methods for the large-scale isolation of sperm have used tricellular pollen grains; an osmotic shock combined with treatment with pectinase and cellulase has been devised to obtain high sperm yields from pollen tubes of several bicellular species (Mo and Yang 1992).

Viability of isolated sperm cells is routinely assessed by using Evans-blue as an exclusion dye (Russell 1986; Southworth and Knox 1989) or by fluorochrome reaction (Dupuis et al 1987; Yang and Zhou 1989; Mo and Yang 1991). Zhang et al (1992a, b) have advocated the use of flow cytometry as a highly sensitive method to evaluate the viability of isolated sperm cells. Not unexpectedly, viability was found to decrease with time of storage of isolated sperm cells (Russell 1986; Dupuis et al 1987; Theunis and van Went 1989; Yang and Zhou 1989; Mo and Yang 1991). Among the measures recommended to enhance the longevity and viability of isolated sperm of maize are the use of a buffered galactose solution or the use of a dilute mineral-salt solution especially low in calcium and magnesium (Zhang et al 1992b; Roeckel and Dumas 1993; G. Zhang et al 1995). A procedure for increasing the yield and viability of sperm cells isolated from *Lolium perenne* has used vitamin-C and fetal calf serum during isolation and storage (van der Maas et al 1993b). Maintaining isolated sperm in a prime condition is an issue to which attention needs be focused for sperm to fulfill their potential use in in vitro fertilization and as gene vectors in genetic manipulation. In other respects, such as in the absence of the common plasma membrane of the vegetative cell, isolated sperm cells resemble true protoplasts (Russell 1986; Dupuis et al 1987; Cass and Fabi 1988; Theunis 1990). Morphometric analysis and three-dimensional reconstruction of serial sections of isolated sperm of *Z. mays* have revealed that there is considerable heterogeneity in the volume and number of the complement of organelles in a given population of sperm (Wagner, Dumas and Mogensen 1989; Mogensen, Wagner and Dumas 1990). From a heuristic standpoint, this shows that strengthening the differences between two groups of sperm in a population is not an easy task.

Except in two instances (see Section 3), it has not been determined whether isolated sperm cells are functionally competent to fuse with an egg cell or with a central cell in an in vitro environment.

However, a few reported cases of fusion between isolated sperm cells of the same species and between sperm of different species and between isolated sperm and isolated microspore protoplasts have provided some hopeful hints of a wider application of the technique. This curious behavior has been described among sperm cells of *Gladiolus gandavensis, Zephyranthes candida, Hemerocallis minor, H. fulva,* and *Hippeastrum vittatum*; between sperm cells of *G. gandavensis* and *H. vittatum;* between sperm and microspore protoplasts of *H. minor*; and between sperm of *H. vittatum* and microspore protoplasts of *H. fulva* (Mo and Yang 1992).

The metabolic and biochemical activities of isolated sperm cells have hardly been investigated. Preliminary attempts have been made to characterize the surface antigens of sperm cells by employing monoclonal antibodies. Hough et al (1986) raised hybridoma cell lines against isolated male gametes of *Brassica campestris* and, by screening them for antibody specificity by a simplified immunofluorescence technique, identified a few sperm surface–specific antigens. In another investigation, Pennell et al (1987) screened hybridoma antibodies elicited by *Plumbago zeylanica* sperm against both sperm and pollen fractions. Whereas a number of antibodies were found to bind to contaminating fractions from the vegetative cell cytoplasm and soluble macromolecules, a high percentage of the reactants were also sperm specific. Although much remains to be done to characterize the antibodies in both species, this is a task that might eventually lead to insights into the basic problems of sperm recognition.

In isolated sperm cells of *P. zeylanica*, a beginning has also been made to study the protein profile of the sperm by SDS-PAGE with a view toward identifying recognition molecules during egg–sperm interaction (Geltz and Russell 1988). Obtaining evidence for the existence of sperm-specific proteins was difficult in this work because of contamination of the preparation with vegetative cell nuclei, organelles, and water-soluble proteins. Nonetheless, about 14% of the most conspicuous polypeptides were specific to sperm cells; whether these include cell surface determinants of male gametes remains to be investigated. In similar studies with isolated sperm cells of maize, the protein synthetic profile was found to change with time after isolation. The increased incorporation of radioactive precursors of RNA and protein synthesis into isolated sperm cells that was observed in these investigations evidently shows that changes in protein synthesis involve both transcription and translation (Zhang, Gifford and Cass 1993; Matthys-Rochon et al 1994). But the evidence is not really conclusive as to whether this is a reflection of the synthetic activities of sperm cells or is a reaction to isolation stress.

It is clear from this account that in recent years a major effort in the study of male gametes of angiosperms has been directed toward isolation procedures. Despite the fact that well-established procedures are already at hand for the biochemical characterization of sperm, such studies are limited. The special appeal of male gametes derives from the fact that each sperm of a pair in the pollen grain is programmed for different fusion events. Characterization of the recognition factors in the sperm cells is an essential prerequisite to understanding the physiological and molecular basis of double fertilization.

2. ODYSSEY OF THE SPERM AND DOUBLE FERTILIZATION

The fate of male gametes in the pollen tube growing in the style is certainly the most critical issue to be considered now, as it lies at the heart of much of our understanding of double fertilization. The pollen tube makes its way into the embryo sac, where it discharges its cargo for fertilization events. The tasks involved here are far too prodigious to be accomplished by sperm cells without some outside help. For convenience, the events leading up to pollen tube discharge are designated as the progamic phase; this is followed by the syngamic phase, in which the predominant event is a set of nuclear fusions.

(i) Growth of the Pollen Tube through the Style

In species with tricellular pollen grains, the sperm cells and the vegetative cell nucleus migrate into the emerging pollen tube, whereas in plants with bicellular pollen grains the generative cell divides in the pollen tube to give rise to sperm cells. The stigmatic tissue on which pollen grains land is composed of glandular cells that facilitate germination of pollen grains and growth of pollen tubes (see Chapter 7). These cells lead to the transmitting tissue of the style, which is also made up of cells with glandular properties. What is the mechanism by which the pollen tube is guided toward the ovary? It is now established that pollen tubes grow through the intercellular matrix of the style, and indications are that a variety of control mechanisms operate within the style to maintain the downward directionality of the pollen tube.

It has been argued quite convincingly that in some plants the guidance of the pollen tube is mechanical or electrical and that such cues interact with the intercellular growth machinery of the stylar matrix. Mechanical guidance begins to take hold when the tip of the pollen tube is properly oriented in the stylar matrix. To examine this proposition closely, Iwanami (1953) dusted the stigma of *Lilium auratum* with pollen and followed the course of the pollen tube. The finding in this work that the normal site of pollen tube extension was along the longitudinally oriented cells surrounding the stylar canal led to the view that the pollen tube is directed mechanically by these cells. That the pollen tube is not guided in this situation by some polarized gradient of a chemical substance was subsequently established in an extensive range of delicate surgical experiments on the styles of *L. auratum* and *L. longiflorum*. In one experiment, when cut ends of the style were grafted to its morphologically inverted pieces, the artificial style was found to serve admirably as a medium for the growth of pollen tubes (Iwanami 1959). This could not have happened if chemical gradients were at operation in directing pollen tube growth. Some investigations on pollen–stigma interaction in maize and other Poaceae are also consistent with the view that pollen tubes are oriented in the direction of the ovary by mechanical constraints (Y. Heslop-Harrison, Reger and Heslop-Harrison 1984b). Following germination of the pollen grain on the receptive stigmatic trichome in the species of Poaceae studied, the orientation of the pollen tube tip in the stylar tract seems to determine its unhindered growth toward the ovary. This occurs even in the absence of any visible structural adaptations directing the pollen tube toward the transmitting tracts. A long-distance influence, probably of a mechanical nature, on the direction of growth of the pollen tube through the style has also been indicated in *Nicotiana alata* (Mulcahy and Mulcahy 1987).

Reports of the responses of pollen tubes to applied electric fields have nurtured the view that endogenous currents may direct pollen tube growth. Since directional growth of pollen tubes in vitro toward both the anode (Wulff 1935; Wang, Rathore and Robinson 1989; Nozue and Wada 1993) and the cathode (Marsh and Beams 1945; Zeijlemaker 1956; Nakamura et al 1991) has been observed, it is difficult to evaluate the significance of the data. In *Narcissus pseudonarcissus*, the ovary is electrically negative by about 100 mV with respect to the stigma; the potential voltage difference may be the guiding force of the pollen tube to the ovary (Zeijlemaker 1956). Given the delicate nature of the pollen tube and the need for micromanipulative procedures, the mechanism directing electrical field–induced pollen tube growth is likely to remain mysterious for some time. The attraction of pollen tubes to a source of Ca^{2+}, to be described later, coupled with the evidence that calcium ions constitute one component of the endogenous current in the pollen tube (Weisenseel, Nuccitelli and Jaffe 1975; Weisenseel and Jaffe 1976) and that Ca^{2+} accumulates at the tip of the pollen tube (Jaffe, Weisenseel and Jaffe 1975; Reiss, Herth and Nobiling 1985), has engendered much speculation about the possible role of redistribution of these ions in influencing the direction of pollen tube growth through applied electrical fields.

The most commonly mentioned mechanism guiding the direction of growth of pollen tube through the style is the attraction toward unspecified chemical substances in the style itself. The glandular cells of the style have drawn attention as a possible source of chemotropic substances. Following upon the work of several earlier investigators going back to the end of the 19th century, Tsao (1949) demonstrated chemotropism of pollen tubes to pistil parts of 9 out of 36 species tested on artificial media; it was further shown that the active factor from the pistil of *L. superbum* was diffusible into agar blocks, which then attracted pollen tubes. A later study by Miki (1964) confirmed and extended these results using *L. longiflorum* and *L. japonicum* and showed that the chemotropic factor is a heat-labile, diffusible, small molecule. Linck and Blaydes (1960) demonstrated chemotropism in four additional species, of which *Clivia nobilis* was particularly noteworthy for the attraction of pollen tubes not only by slices of the carpel but also by those of sepal, petal, and leaf. Attempts to elucidate the mechanism regulating pollen tube pathway in other species of *Lilium* have also been made. For example, pollen tubes of several hybrids of *L. longiflorum* are found to respond positively to the upper part of the style, although the activity is weak or absent in the mid and lower segments; the ovary displays no activity at all (Rosen 1961). In *L. leucanthum* and *L. regale*, the glandular cells lining the stylar canal have been identified as the source of the putative chemotropic factor, which is apparently distributed in a basipetal gradient (Welk, Millington and Rosen 1965). An influence of the ovary on pollen tube growth in *Petunia hybrida* has been demonstrated using a "semi-vivo" technique in which the ovary is removed at different times after pollination. Whereas the growth of pollen tubes in the style detached from the ovary was inhibited irrespective of the time after pollination, their growth in the intact pistil occurred normally toward the ovary (Mulcahy and Mulcahy 1985). Despite the fact that most of these observations are based on in vitro experiments that may have little or no parallel to events in vivo, they nonetheless suggest a complex chemical interaction between the growing pollen tube, the stylar matrix, and an ovule-derived guidance mechanism controlling pollen tube growth (Hülskamp, Schneitz and Pruitt 1995). It is worth noting that use of the same methods described here has not led to the demonstration of pollen tube chemotropism in certain other plants (Cook and Walden 1967; Hepher and Boulter 1987; Kandasamy and Kristen 1987b).

Progress in our understanding of the interaction between the stylar matrix and the growing pollen tube has been bolstered by attempts to identify the chemotropic substance. A key piece of information in this context came from the work of Mascarenhas and Machlis (1962b), who showed that in *Antirrhinum majus* the positive chemotropic factor is a small, heat-stable, water-soluble molecule associated with larger molecules. This finding led to the search for inorganic ions and to the discovery that Ca^{2+} is chemotropically active for pollen tubes of *A. majus*, *Narcissus pseudonarcissus*, and *Clivia miniata* (Mascarenhas and Machlis 1962a, 1964). It was also found that the levels of Ca^{2+} present in chemotropically active parts of the pistil of *A. majus* were well within the physiological range of the ion active in bioassays of chemotropism. In the pistil of *Petunia hybrida*, membrane-bound Ca^{2+} is found in a decreasing gradient from the stigma to the base of the style before pollination, and in the reverse direction after pollination (Bednarska 1995). However, there is some skepticism as to whether Ca^{2+} is a universal chemoattractant for pollen tubes. In this context, in vitro bioassays showing that pollen tubes of *Oenothera longiflora* and other species grow in a random manner in the presence of calcium compounds have had a particularly powerful impact on the skeptics (Glenk, Schimmer and Wagner 1970). It is also relevant to point out that there is a lack of a gradient of ionic calcium in the style of *A. majus*, which seems to argue for the presence of an additional factor as a chemotropic agent (Mascarenhas 1966a). There has not been much research output on pollen tube chemotropism in the style in the period following the early studies and, until recently, other investigations on this topic have not been directed at chemotropic agents in the style. Pollen tubes of *Pennisetum glaucum* have been shown to respond positively to diffusates of the pistil tissue as well as to glucose; a totally different understanding of the in vivo metabolism of the style has emerged from the discovery that the pistil diffusate response is due to glucose produced by tissue-bound invertase action rather than to free glucose (Reger, Chaubal and Pressey 1992; Reger, Pressey and Chaubal 1992).

Instead of positing the growth of pollen tubes to be under a chemotropic influence, some investigations have led to a model of pollen tube growth in the style as a directional cell migration along a gradient of a factor in the secretory matrix of the style; this model is analogous to a biochemical recognition–adhesion system. The initial experimental support for this model came from the observation that latex beads approximating the size of pollen tube tips, when introduced into the stylar transmitting tracts of certain plants, traveled to the site of the ovary at rates similar to those of pollen tubes regardless of the stylar gravitational orientation (Sanders and Lord 1989). Since substrate adhesion molecules probably confer this ability on the beads to travel in the stylar matrix, attempts have been made to identify these substrate molecules and their receptors in the style. A candidate molecule identified in the substrate adhesion role is vitronectin. Immunolocalization of vitronectin-like proteins in cross sections of the pistil of *Vicia faba* showed strong reaction in the stylar transmitting tract, suggesting a possible role for these proteins in guiding pollen tube growth in a manner similar to cell migration in animals (Sanders et al 1991). Jauh and Lord (1995) have further shown that by excising different lengths of the profusely pollinated style of lily, it is possible to cause the tip of the pollen tube, in the absence of the remains of the pollen grain and the spent pollen tube, to convey the sperm cells to the ovules and effect fertilization. The tip of the pollen tube can be considered analogous to a population of migrating cells that adhere to one another and to the cells of the transmitting tissue of the style during its journey to the ovule. The role of adhesive substrates in pollen tube growth has unfolded further with the demonstration that glycoproteins isolated from the stylar transmitting tissue of tobacco adhere to the surface of pollen tubes and are incorporated into the wall of the pollen tubes growing in the style. It appears that within the style, pollen tube–bound enzymes deglycosylate the proteins and thus release sugar molecules that are presumably utilized by the growing pollen tubes (Wu, Wang and Cheung 1995). These observations suggest that stylar glycoproteins of tobacco serve an important role in matrix adhesion and in cell metabolism.

The general path of pollen tube entry into the ovule has been described convincingly in numerous classical studies. These works showed that after reaching the ovary, the pollen tube almost always grows downward over the outer integument, then between the inner ovary wall and the inner integument of the ovule, toward the placenta. When the placental region is reached, the pollen tube executes a sharp turn and continues its journey between the placenta and the outer integument to the micropyle. At the micropyle, the tube grows intercellularly through the nucellus to the embryo sac. Observations of this path reported from conventional histological studies were later confirmed in research using fluorescent microscopy to trace the pollen tube in the ovary (Y. Heslop-Harrison, Heslop-Harrison and Reger 1985; Reger, Chaubal and Pressey 1992). How does the pollen tube find its way to the micropyle from the base of the style? From indirect lines of evidence, the basic plan appears to be one guided by a chemotropic attractant produced at a site close to the micropyle. The presence of such substances has been demonstrated in the micropylar region of ovules of *Paspalum orbiculare* (Poaceae) (Chao 1971), *Ornithogalum caudatum* (Tilton 1980), maize (Y. Heslop-Harrison, Heslop-Harrison and Reger 1985), beet root (Bruun and Olesen 1989), *Gasteria verrucosa* (Franssen-Verheijen and Willemse 1993), and *Asclepias exaltata* (Asclepiadaceae) (Sage and Williams 1995), although the origin of the substances has not been clearly determined in all cases. The substance found in *P. orbiculare* is PAS positive and proteinaceous; it consists of secretion products of the nucellus and integuments, as well as the product of dissolution of cells of the outer and inner integuments close to the micropyle (Chao 1971, 1977). The substance in maize stains for proteins and pectic polysaccharides, although in the light microscope its origin is not easily resolved (Y. Heslop-Harrison, Heslop-Harrison and Reger 1985). In beet root, the middle lamella of the nucellar cells in the micropylar region comes apart during the maturation process and these cells probably secrete PAS-positive substances into the micropyle (Bruun and Olesen 1989). Proteins and RNA are the main components of the exudate secreted by the nucellar cap and inner integument of *O. caudatum* (Tilton 1980). The filiform apparatus is believed to produce most of the protein- and carbohydrate-rich exudate in *G. verrucosa*, with minor contributions from the micropylar cells of the nucellus (Franssen-Verheijen and Willemse 1993). The characteristics of the chemotropic systems guiding the pollen tube to the ovule that do not involve Ca^{2+} provide interesting comparisons with the growth of the pollen tube in the style, considered earlier, and its final push into the embryo sac to be considered next.

(ii) Passage of Sperm into the Embryo Sac

This part begins with a brief account of the ovary. As is well known, the ovary not only provides the physical protection for the enclosed ovules, but also harbors the tissues through which the pollen tube grows during the final phase of its existence. As described in Chapter 6, the embryo sac is formed within the ovule, which itself is born on the placenta. Before the pollen tube reaches the embryo sac, it encounters the placenta, funiculus, and integuments of the ovule. Light microscopic studies have revealed three general pathways through which the pollen tube enters the ovule and races to the embryo sac. In the most common route, known as porogamy, the pollen tube enters the ovule through the micropyle. In some plants, mucilaginous canals and ancillary structures such as obturator, embellum, ponticulus, and papillae that aid in the entry of the pollen tube into the micropyle have evolved (Kapil and Vasil 1963; Kapil and Bhatnagar 1975; Busri, Chapman and Greenham 1993; Martínez-Pallé and Herrero 1995). Two other preferred sites of entry of the pollen tube are at the chalazal end (chalazogamy) and on one side of the ovule (mesogamy), but they are used only sporadically (Maheshwari 1950). These observations suggest that pollen tubes may be directed to specific parts of the ovule by a gradient of chemical substances. Irrespective of the pathway taken by the pollen tube to the ovule, the tip of the pollen tube invariably makes contact with the embryo sac at its micropylar end.

This leads to the consideration of the entry of the pollen tube into the embryo sac and the discharge of sperm. It is an interesting commentary on the state of uncertainty in this field that up to the early 1960s, based on light microscopy, as many as six different modes of entry of the pollen tube into the embryo sac involving one or both synergids were described. As summarized by Kapil and Bhatnagar (1975), the pollen tube might (a) enter between the two intact synergids, (b) enter between the egg and the synergid, (c) enter between the embryo sac wall and one or both synergids, (d) penetrate one of the synergids, destroying it in the process, (e) destroy the synergid on contact, or (f) enter the cytoplasm of the synergid, which has degenerated before the arrival of the pollen tube. Route (d), by which the pollen tube enters the embryo sac and destroys one of the synergids in the process, has been described most frequently, whereas the other modes of entry have been documented in only a few cases. It should be pointed out that in the light microscope it is often difficult to distinguish between a healthy and a degenerating synergid in sections of the embryo sac clouded by pollen tube discharge.

The era of electron microscopy led to a new level of understanding of the passage of the pollen tube into the embryo sac. What appeared as a basic pattern of pollen tube entry at the electron microscope level was described by Jensen and Fisher (1968a) in *Gossypium hirsutum*, which to date remains as the one species in which events associated with fertilization are best known. In *G. hirsutum*, as in many other species, the two synergids are identical before pollination. However, one of the synergids begins to disintegrate prior to the entry of the pollen tube into the embryo sac. In fact, pollination triggers the degeneration process, which is most frequently initiated when the pollen tube is in the style. The first signs of synergid collapse are irregular thickening and darkening of the plasma membrane and of the outer membranes of adjacent plastids and mitochondria, accompanied by a general increase in the electron density of the cytoplasm. When the pollen tube reaches the embryo sac, the tip of the tube grows through the filiform apparatus into the degenerating synergid. The degeneration of the synergid initiated before pollen tube entry into the embryo sac is accelerated after entry of the pollen tube. Signs of accelerated degeneration of the synergid are transformation of the plastids and mitochondria into ghosts of their original structure, collapse of vacuoles, and breakdown and disappearance of the plasma membrane. As to the fate of the pollen tube, after passage through the cytoplasm of the degenerating synergid, pollen tube growth comes to a standstill and then the pollen tube discharges its contents of sperm and cytoplasm.

The observations on *G. hirsutum* imply that rather than being the result of mechanical intrusion of the pollen tube, synergid degeneration is caused by pollination and is actually required for normal pollen tube entry into the embryo sac. Fascinating though this thesis is, its possible significance in the fertilization process was not widely appreciated until there was a long list of plants in which it was established that synergid degeneration begins before the arrival of the pollen tube and that the pollen tube consistently favors the degenerating synergid. In subsequent years, other species, such as barley (Cass and Jensen 1970), *Zea mays* (Diboll 1968; van Lammeren 1986), *Epidendrum scutella* (Cocucci and Jensen 1969c), *Linum usitatissimum* (Vazart 1969; Russell and Mao 1990), *Quercus gambelii* (Mogensen 1972), *Proboscidea louisianica* (Mogensen 1978b), *Nicotiana tabacum* (Mogensen and Suthar 1979; Huang and Russell 1992), *Spinacia oleracea* (Wilms 1981b), *Crepis capillaris* (Kuroiwa 1989), *Glycine max* (Dute, Peterson and Rushing 1989), *Populus deltoides* (Russell, Rougier and Dumas 1990; Rougier et al 1991), and *Arabidopsis* (Murgia et al 1993), have gained general acceptance as examples in which synergid degeneration occurs in apparent response to pollination. The pattern of cytoplasmic reorganization observed in one of the synergids following pollination varies widely, from a series of disordered changes in the organelles, as in *G. hirsutum* (Jensen and Fisher 1968a), to only a difference in the density of the cytoplasm due to an increase in the number of ribosomes, as in *N. tabacum* (Mogensen and Suthar 1979). That the loss of synergid integrity is not an artifact of the chemical fixatives used in these investigations has been confirmed by observing similar changes in synergids of cryofixed ovules of *N. tabacum*, although the full extent of cellular degeneration is delayed until pollen tube penetration (Huang, Strout and Russell 1993).

Considerable variation in the condition of the synergid in relation to the timing of pollination and pollen tube growth has been observed in other plants. For example, in wheat, degeneration of both synergids is initiated in embryo sacs collected just prior to pollination; the degeneration cycle is completed in one synergid after entry of the pollen tube, whereas the other appears unaffected (You and Jensen 1985). In *Brassica campestris*, both synergids show degenerative changes in pollinated as well as in unpollinated flowers; this implies that degeneration of the synergid does not depend upon pollination (van Went and Cresti 1988a; Sumner 1992). In *Beta vulgaris* (Bruun 1987) and *Oryza sativa* (Dong and Yang 1989), one of the synergids readies to receive the pollen tube and begins to display signs of degeneration before anthesis of the flower. There are also reports of synergids remaining healthy even after pollination and pollen tube growth; apparently, when the pollen tube makes contact with the synergid, the degenerative program of the cell is set in motion. Examples in which this situation is found are *Capsella bursa-pastoris* (Schulz and Jensen 1968b), *Petunia hybrida* (van Went 1970c), *Ornithogalum caudatum* (van Rensburg and Robbertse 1988), and *Lilium longiflorum* (Janson and Willemse 1995). Although the paired synergids of *Helianthus annuus* do not show any structural differences before pollination, degenerative changes begin to surface in one of them gradually after pollination, whereas the other one remains perfectly healthy until after fertilization (Newcomb 1973a; Yan, Yang and Jensen 1991a). Despite the fact that ultrastructural analysis has not been carried out to the same depth in all of the aforementioned cases, sufficient details are at hand to establish the essential validity of the hypothesis that irrespective of the prior response of the synergid to pollination stimulus, the pollen tube grows into one of the two synergids and then, as the pollen tube discharge occurs, degenerative changes set in. It is not, however, easy to fit into this model genera that lack synergids, such as *Plumbago* and *Plumbagella*. According to Russell (1982), arrival of the pollen tube at the micropylar end of the embryo sac in *Plumbago zeylanica* is not heralded by the disarray of a receptor cell; rather, the pollen tube penetrates the wall ingrowths of the filiform apparatus at the base of the egg and grows in the intercellular space between the egg and the central cell until it reaches the vicinity of the nucleus near the chalazal end of the egg. A similar sequence of events occurs in *Plumbagella micrantha* (Russell and Cass 1988). It is clear from this that, whatever else is going on, cellular degeneration is not an indispensable part of successful pollen tube penetration.

By and large, the signal for pollen tube penetration into the embryo sac is provided by a synergid. The question arises, what is the nature of this signal? The most persistent hypothesis to account for the directional growth of the pollen tube is that it is attracted to a chemical substance. In the absence of direct evidence for the presence of a chemotactic substance in the synergids, a speculative view is that the release of Ca^{2+}, especially from the degenerating synergid, is associated with pollen tube penetration. Jensen (1965b) visualized inorganic ash in the vacuoles and cytoplasm of microincinerated sections of cotton ovules. The ash, which is largely absent from the egg, is assumed to contain Ca^{2+}. More recently, high-energy X-ray dispersion analysis and antimonate precipitation methods have been employed to localize gross Ca^{2+} and loosely sequestered exchangeable Ca^{2+}, respectively, in the embryo sac cells of cereal grains. The synergids were found by both methods to display high concentrations of Ca^{2+}, suggesting an active role for the ion as an erotropic substance for the attraction of pollen tubes (Chaubal and Reger 1990, 1992a). Since the pollen tube moves directionally to the synergid, the existence of a continuous Ca^{2+} gradient from the placenta to the synergid has been postulated (Chaubal and Reger 1990). In other cases, PAS-staining substances secreted by the synergid probably function as part of the pollen tube guidance system (Chao 1971; Bruun 1987; Willemse, Plyushch and Reinders 1995). However, as is true of most cells growing in isolation within a tissue of different origin, the movement of the pollen tube may be conditioned by the physical state of the neighboring cells. For example, compared to the intact cells of the female gametophyte, the degenerating synergid might offer a lower resistance to the growing pollen tube (Russell 1992). Moreover, when penetrating the synergid, the pollen tube conforms in a regimented way to the filiform apparatus. Since microtubules are associated with the filiform apparatus, the possibility that they may somehow be involved in giving directionality to the pollen tube also deserves attention (Cass and Karas 1974; B.-Q. Huang et al 1990).

The degeneration of the synergid before or after entry of the pollen tube is obviously an important step in the fertilization process. We do not yet understand how this happens, even in plants in which pollination triggers the degeneration program. One common view of the nature of the stimulus is that it is a hormonal substance. In support of this, it has been shown that in cotton ovules grown in culture, synergid degeneration occurs when a mixture of IAA and GA is added to the medium. As in ovules of pollinated flowers, only one synergid in cultured, unpollinated ovules responds to the hormonal signal (Jensen, Schulz and Ashton 1977). It has also been shown that supplementation of the medium with GA alone triggers the degeneration of the synergid in ovules of unpollinated flowers (Jensen, Ashton and Beasley 1983). These experiments by themselves do not prove that the hormone directly mediates the synergid response, but they alert us to such a possibility in any hypothesis to explain synergid degeneration. Another suggestion involving Ca^{2+} and related elements is that their loss after pollen tube discharge reduces the osmotic pressure of the synergid, leaving the way open for an efflux of water from the cell and for subsequent shrinkage of the cell (Chaubal and Reger 1992b). Since there are two synergids in most species investigated, the next question is whether degeneration is a random event or whether one of the synergids is more predisposed to degenerate than the other. Based on examination of a limited number of species, indeed it appears that synergid degeneration is a preprogrammed, nonrandom event. In *Helianthus annuus,* the synergids are isomorphic; however, one is found closer to the placenta than the other. Using this feature for orientation purposes, it was found that in 81% of the ovules examined, the degenerating synergid was the one aligned with the placenta (Yan, Yang and Jensen 1991a). In barley, Mogensen (1984) found that in one sampling there was a greater tendency for the synergid closer to the placental attachment of the ovule (proximal) to degenerate than for the distal one. In these instances, therefore, the force for degeneration is derived from the sporophytic tissues due to physiological differences between the two synergids, such as the transport of nutrients or erotropic substances through the placental vasculature to one of the synergids. On the other hand, in *Linum usitatissimum,* some kind of gametophytic control seems to be operating in determining which one of a pair of synergids degenerates (Russell and Mao 1990). Based on the orientation of synergids relative to the egg, there is a greater preference for the left than the right synergid to degenerate. Since the volume and the surface area of the filiform apparatus of the left synergid are significantly lower than those of the right, the model predicts a degree of control by intrinsic factors that affect synergid function.

A particularly fascinating and important aspect of the race of the pollen tube into the synergid is the fact that sperm cells keep pace with the distal end of the growing pollen tube. The mechanism by which sperm cells move down the pollen tube continues to attract the attention of investigators even to this day, but no satisfactory hypothesis has emerged. Because sperm cells share a number of properties with the generative cell, including shape and the cytoskeletal framework, comments made in Chapter 8 about the movement of the generative cell in the pollen tube are germane here. Although sperm cells have their own endowment of microtubules, it is doubtful whether they function as a motility system. Cass (1973) found that isolated sperm cells of barley show little or no directional movement, which should not be the case if their microtubules are called into service. In view of the absence of microfilaments in the sperm, it comes as no surprise that no one has implicated them in gamete movement. Because of the negative evidence, it has been speculated that, like the generative cell, the sperm cells are propelled along the longitudinally disposed actin cytoskeletal elements of the pollen tube cytoplasm (J. Heslop-Harrison and Heslop-Harrison 1987). It is hard to justify the presence of microtubules in the short-lived sperm cells if they do not have a role in locomotion; it might be that sperm movement in the pollen tube is due to the combined efforts of the sperm microtubules and the actin system of the pollen tube.

Once the pollen tube has made contact with the synergid, it grows some distance into this cell and readies itself to release sperm. Based on an examination of the cytoskeletal organization of the embryo sac of *Plumbago zeylanica* at various stages of pollen tube penetration, it has been proposed that positions of both microtubules and actin bundles change significantly to guide the pollen tube to the site of sperm delivery (B.-Q. Huang et al 1993). In other plants, the electron microscope has revealed several features of the final hours of the pollen tube and the life of the sperm as free cells. It is no accident that the first detailed description of the fate of the pollen tube contents in the synergid was provided in cotton, for this plant has proved to be remarkable for the highly regulated way in which fertilization-related events occur. In cotton, the arrest of pollen tube growth in the degenerating synergid is followed by the formation of a thick cap at the tip. The most dramatic event at this juncture is the appearance of a pore at the back of the tip through which the male gametes and the vegetative cell nucleus are released along with a generous amount of the pollen tube cytoplasm. Adding to the chaotic environment of the synergid is the discharge of polysaccharide spheres presumably constituting unused cell wall material (Jensen and Fisher 1968a). The force of the discharge, combined with changes in the physical properties of the discharged cytoplasm, apparently propels the vegetative cell nucleus and sperm cells to the chalazal end of the synergid, close to the egg nucleus. The end of the discharge is signaled by the formation of a callose plug over the pore, effectively preventing any cytoplasmic flow between the pollen tube and the synergid.

In other plants, such as *Capsella bursa-pastoris* (Schulz and Jensen 1968b), *Epidendrum scutella* (Cocucci and Jensen 1969c), *Petunia hybrida* (van Went 1970c), *Epilobium palustre* (Bednara 1977),

Quercus gambelii (Mogensen 1972), *Plumbago zeylanica* (Russell 1982), *Plumbagella micrantha* (Russell and Cass 1988), and *Brassica campestris* (Sumner 1992), the discharge of the pollen tube occurs through a terminal aperture. In *P. zeylanica*, growth of the pollen tube to the summit of the egg is often the prelude to pollen tube rupture. Sperm cells, vegetative cell nucleus, and a small amount of cytoplasm, which are released from the pollen tube, find a place between the egg and the central cell (Russell 1982). Various mechanisms have been proposed to account for the rupture of the pollen tube, but they seem incomplete in their explanations. One view is that as the pollen tube stops growing, osmotic pressure built within it causes rupture at the weakest point. In *P. zeylanica* the pollen tube wall is relatively thin at the tip near the point of rupture. Moreover, fibrillar elements near the aperture are less organized than in the proximal region, indicating both an apparent constraint in the synthesis of wall material and the onset of stress (Russell 1982). Another view is that changes in oxygen tension stop the growth of the pollen tube and cause its rupture (Stanley and Linskens 1967). This is based on the observation that low levels of oxygen promote bursting of pollen tubes in vitro and in the style; whether pollen tubes experience a similar stress in the embryo sac is not known. It has also been suggested that the cessation of pollen tube growth and the rupture of its tip may be caused by supraoptimal concentrations of Ca^{2+} in the synergid (Chaubal and Reger 1990).

(iii) Fusion of Nuclei

The observations terminating in sperm discharge lead to the consideration of the mechanism by which gametic fusion is accomplished. The range of the search for this mechanism is wide and involves such aspects of cell behavior as movement, recognition, and adhesion. The first issue concerns the actual transfer of sperm to the egg or the central cell. The two sperm cells are presumed to move separately, one to the egg and the other to the central cell. Electron microscopy has provided some pointers on this, although it is not possible to determine the mechanism that underlies the movement by inspection of electron micrographs alone. In *Torenia fournieri* (van der Pluijm 1964), the receptive synergid opens up a passage at a specific site at its chalazal end, providing direct access of the sperm to the egg and the central cell. In terms of the end result, the transfer of the sperm cells to the egg and the central cell in cotton falls into the same type as *T. fournieri*. Fusion is accomplished between the plasma membranes of the sperm and of the egg or the central cell, leading to the formation of a bridge through which the sperm slides. The chalazal part of the egg, where the cell wall is incomplete, serves as the window where the sperm fuses (Jensen and Fisher 1968a). In *Petunia hybrida*, contact between the sperm and the female reproductive cells is followed by cell fusion somewhat analogous to the fusion of protoplasts (van Went 1970c). Regarding the fate of the male nucleus, some observations suggest that the chromatin of the male nucleus becomes dispersed and indistinguishable from the chromatin of the egg nucleus (Hu and Zhu 1979). A specific role for the degenerating synergid in transporting sperm to the site of the egg and central cell has been described in *Triticum aestivum*. Transport is accomplished by a tubelike extension of the degenerating synergid containing the pollen tube discharge and the sperm nuclei, which penetrates into the space between the egg, central cell, and the persistent synergid (Gao et al 1992a).

Several events are played out in a compressed time frame during gametic fusion in the synergidless angiosperm *Plumbago zeylanica*. Immediately after their release from the pollen tube, sperm cells shed their investment of the inner plasma membrane of the vegetative cell and the incipient cell wall materials. At the same time, cell wall components of the egg are dispersed, thus exposing the egg cell membrane. The stage is now set for gametic fusion in which plasma membranes of the sperm cells become directly appressed to the plasma membranes of the egg and central cell. Fusion initiated at one location between the participating cells spreads in a wavelike fashion through the membrane and results in the transmission of both nucleus and cytoplasm. Aggregates of actin found at the sperm delivery site between the egg and the central cell, designated as a corona, have been assumed to cause the two sperm nuclei to align with their respective female nuclei (Russell 1983; B.-Q. Huang et al 1993). In the final act of fusion, the sperm cell is incorporated into the cytoplasm of the egg or the central cell, and the plasma membrane remaining from the fusion events dissipates as fragments in the cytoplasm of the female cells involved. The events described here occur within a time span of about 30 minutes. The actin corona is also present between the synergids, egg, and central cell, and near the chalazal end of the degenerating synergid, during fertilization in *N. tabacum* (Huang and Russell 1994). The corona has no counterpart in the embryo sac before one of the synergids begins to degenerate.

Nuclear fusions, or karyogamy, between the sperm and egg nuclei and between the sperm and central cell nuclei are at the heart of double fertilization and constitute a major reason for sustained interest in the process itself. Gerassimova-Navashina (1960) used the cyclic state of the nuclei as an elegantly simple criterion to identify two types of karyogamy involving the egg and sperm. In the most common type, known as the premitotic type, the sperm nucleus fuses immediately with the egg nucleus and, after a short hiatus, the zygote nucleus passes through S phase and continues through G2 and mitosis to complete the first zygotic division. Illustrative examples of this type of karyogamy are found in members of Asteraceae and Poaceae. Premitotic karyogamy has been nicely demonstrated recently (Figure 11.4) in isolated egg and zygote protoplasts of barley (Mogensen and Holm 1995). In the postmitotic type, frequently observed in members of Liliaceae, the male and female nuclei approach one another and remain in close proximity; then follows an increase in the size of the nucleus and nucleolus of the sperm. Although the nuclei enter mitosis independently, a common metaphase plate is formed, leading to the formation of a zygote nucleus. The underlying basis of postmitotic karyogamy in the gymnosperm *Ephedra trifurca* (Gnetales) entails only the completion of S phase of the cell cycle by the gametic nuclei during the period of courtship before fusion occurs (Friedman 1991). Intermediate situations also prevail, depending upon the duration of karyogamy. For example, in *Mirabilis jalapa, Impatiens glandulifera,* and *Tradescantia paludosa*, the sperm nucleus is already in the interphase stage at the time of fusion, and the first zygotic division follows without delay (Kapil and Bhatnagar 1975). In *Gnetum gnemon* (Gnetales), the gametic nuclei pass through the S phase and double their DNA content before they begin to court each other and fuse (Carmichael and Friedman 1995).

Electron microscopy has added a wealth of new details about karyogamy in a selected list of plants. In cotton, the sperm nucleus may enter the egg and travel the circumference of the egg before alignment with its nucleus preparatory to karyogamy. Karyogamy is initiated when the two nuclei come together and the bordering ER fuse at several points and join their outer nuclear membranes. This results in the formation of bridges between the nuclei. Next, the inner layers of the nuclear envelopes also fuse and the nucleoplasm becomes continuous at several points. When a fusion nucleus is formed, membranes of both nuclei contribute to the nuclear envelope. During nuclear fusion, pockets of cytoplasm containing fragments of organelles are trapped in the nuclear material, but they are probably forced out prior to the complete fusion of the nuclei (Jensen 1964; Jensen and Fisher 1967). Membrane fusion events similar to those observed during sexual karyogamy have been described during the fusion of the male nucleus with the polar fusion nucleus in *Capsella bursa-pastoris* (Schulz and Jensen 1973), *Spinacia oleracea* (Wilms 1981b), and *Triticale* (Hause and Schröder 1987). As shown in maize, karyogamy in the central cell occurs between the sperm nucleus and the diploid polar fusion nucleus or with one of the haploid polar nuclei (Mòl, Matthys-Rochon and Dumas 1994).

Tangible evidence for the completion of double fertilization in conventional embryo sacs is the presence of dark-staining oval structures, referred to as X-bodies, in the degenerate synergid. Although these structures were variously interpreted by early embryologists, present opinion favors the view that they are nuclei or fragments of nuclei, presumably of the vegetative cell and of the degenerate synergid (Fisher and Jensen 1969; Cass and Jensen 1970; van Went 1970c; Huang, Strout and Russell 1993).

A vexing question in angiosperm embryology is whether cell fusions during double fertilization are random or preferential events. In other words, are there traceable differences between sperm cells that fuse with the egg, on the one hand, and with the polar fusion nucleus, on the other? Based on light microscopic studies, the idea that both sperm cells are structurally identical and, hence, there is no preferential fusion, held center stage for several decades. The first dent in this model was administered by Roman (1948), who provided genetic evidence for the existence of nuclear dimorphism in the sperm cells of maize. It was found that when B-chromosomes do not separate during sperm cell formation, the sperm containing an extra set of B-chromosomes fuses in the majority of cases with the egg. Although not striking at first glance as a revolutionary insight, the significance of this observation in the context of double fertilization is that in maize, the presence of supernumerary B-chromosomes seems to confer a competitive edge on certain sperm cells as true male gametes. A fluorescence in situ hybridization method developed by Shi et al (1996) using a B-chromosome–specific DNA sequence to identify and track maize sperm cells containing B-chromosomes now makes it possible to follow up this question.

One of the most provocative pieces of evidence

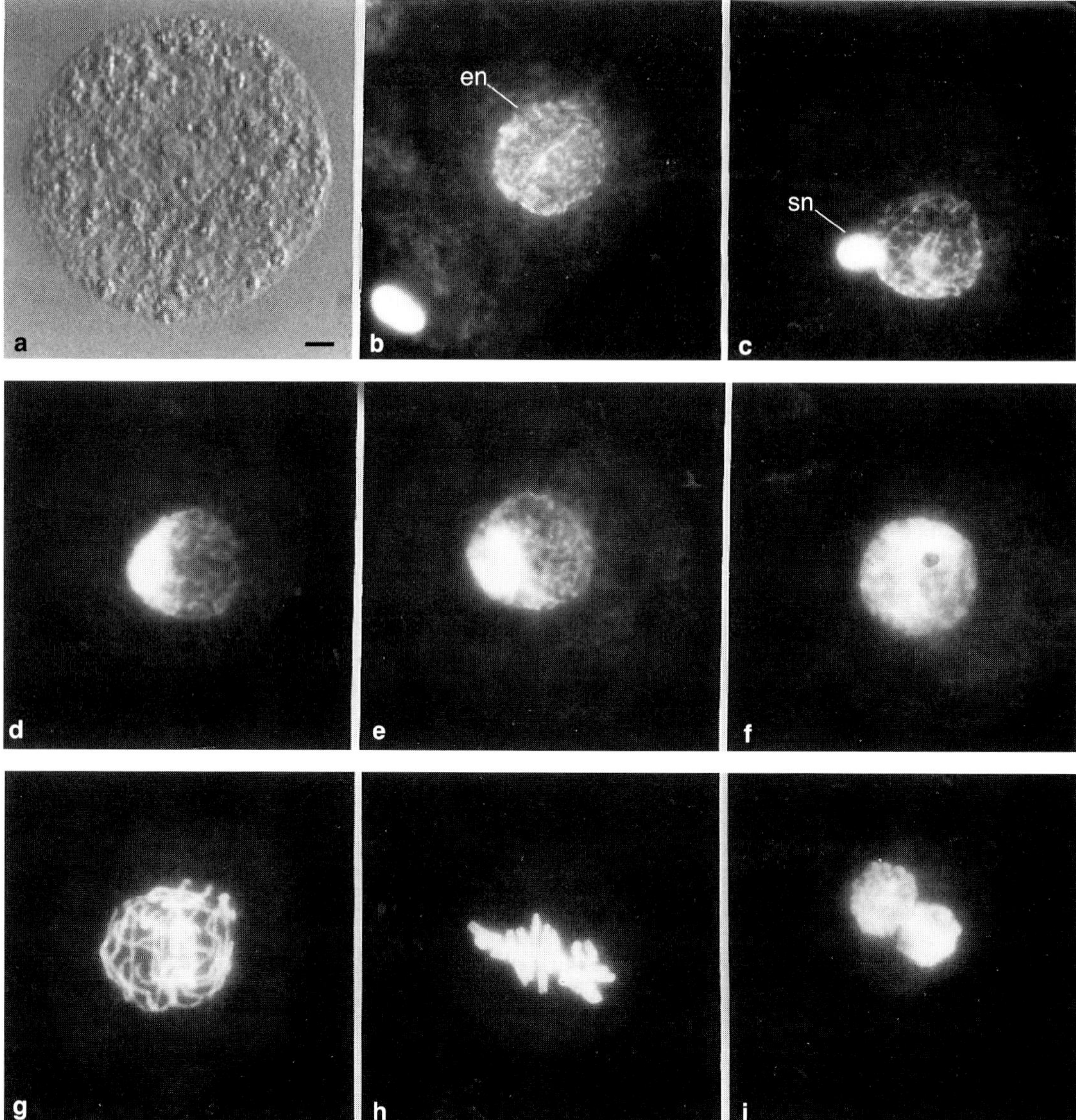

Figure 11.4. Premitotic karyogamy in isolated egg and zygote protoplasts of barley. **(a)** Nomarski optics photograph of an unfertilized egg. **(b)** Unfertilized egg, showing the fluorescence of DAPI-stained nucleus. **(c)** Egg showing the sperm nucleus beginning to fuse. **(d)** Fertilized egg showing the brightly fluorescing male chromatin. **(e)** Zygote showing the dispersion of the male and female chromatin. **(f)** Zygote before beginning of mitosis. **(g)** Zygote at the early prophase stage. **(h)** Zygote at the metaphase stage. **(i)** Late telophase stage of zygote mitosis. en, egg nucleus; sn, sperm nucleus. Scale bar = 5 μm. (From Mogensen and Holm 1995. *Plant Cell* 7:487–494. © American Society of Plant Physiologists. Photographs supplied by Dr. H. L. Mogensen.)

in support of preferential fusion was provided in *Plumbago zeylanica* (Russell 1985). It was pointed out earlier that the paired sperm cells of *P. zeylanica* display heteromorphism in size, shape, and organelle contents. Because plastids are virtually absent from the cytoplasm of the sperm that is in physical association with the nucleus of the vegetative cell, it is possible to use paternal plastids as markers to follow the fate of the two male gametes in the embryo sac. By painstaking identification and counting in electron microscopic profiles of the number of plastids of paternal origin in the egg after fertilization, it was established that the plastid-rich, mitochondrion-poor sperm fuses with the egg cell. Because fusion of the plastid-impoverished sperm with the central cell is as frequent as fusion of the plastid-rich sperm with the egg, this brings us tantalizingly close to considering the occurrence of a putative recognition event at the gametic level. In analogy to animal systems, this might involve the presence of different cell surface molecules on the two sperm cells that will recognize the egg and the polar fusion nucleus, respectively. Although this appears to be an attractive hypothesis, we should not lose sight of the fact that it is based on the study of one unusual species and that a great many problems have to be overcome before gamete surface recognition substances are identified. *P. zeylanica* has made an interesting contribution to our understanding of the structural basis of double fertilization, one that is far richer than has been apparent from studies of other angiosperms.

Little is known about the existence of any cryptic pattern of recognition system on the female side. Pennell et al (1991) have demonstrated the presence of a plasma membrane–bound arabinogalactan protein epitope in the paired sperm cells and in the egg cell, but not in the central cell, of *Brassica napus*; the differential accumulation of the marker protein in the egg and central cell may suggest a role for the arabinogalactan protein in female gametic cell recognition.

To round out this account, brief mention should be made of reports of fusion of a second sperm nucleus with egglike counterparts in the female gametophytes of *Ephedra* and *Gnetum*. These fusion events have been billed as double fertilization; in *E. trifurca*, in which it has been examined in detail, the product of second fertilization yields an embryo after a brief free-nuclear interlude (Friedman 1990, 1992; Carmichael and Friedman 1995). The direct analogy of the situation in the gymnosperms to that in angiosperms may be too facile a comparison, and the view of a likely origin of the endosperm in angiosperms from an embryolike structure in gymnosperms may be far-fetched. It does not take into account the fact that the product of the second fertilization event does not produce the nutritive tissue that is the hallmark of the angiosperms. In the absence of evidence to refute that the second fusion event is another manifestation of polyembryony, for which gymnosperms are notorious, the thesis that double fertilization, in the sense we understand the process in angiosperms, originated in gymnosperms cannot be supported.

(iv) Cytoplasmic Inheritance

The transfer of male cytoplasm along with the nucleus during fusion of the sperm with the egg raises an important question: Is the inheritance of plastids and mitochondria uniparental or biparental? Chapter 3 described several instances of the first haploid pollen mitosis, during which the generative cell is cut off with few or no plastids or mitochondria. This led to the view that in the majority of plants, cytoplasmic male DNA is not inherited, which results in uniparental maternal organelle inheritance. In a few instances in which sperm cells are endowed with plastids and mitochondria, mechanisms also operate to exclude them from the egg prior to fertilization. Rarely, plastids and mitochondria are transferred from the sperm to the egg during fertilization; however, as the fertilized egg begins to divide, it displays characteristic uniparental inheritance. How does this happen?

Mogensen (1996) has reviewed the elaborate ways by which cytoplasmic organelle transmission is restricted or enhanced during and following genetic recombination in angiosperms. The classical method of excluding sperm cytoplasm from the egg at fertilization is by allowing passage of only the nucleus into the egg cell, leaving the cytoplasm outside. This is probably the case in cotton, where no trace of the sperm cytoplasm in any form or shape is detected in the egg (Jensen and Fisher 1967, 1968a). Ultrastructural evidence for the exclusion of sperm cytoplasm and organelles during karyogamy has also been reported in barley (Mogensen 1982, 1988). In the barley embryo sac, an enucleate cytoplasmic body, approximating the size of a prefusion sperm cell and containing a generous number of mitochondria and a limited number of plastids, is frequently observed within the degenerate synergid adjacent to a recently fertilized egg cell. This body has been interpreted as the entire male cytoplasm left out after fusion of the egg and sperm. Since only one cytoplasmic body is observed in each postfertilization embryo sac examined, it is believed that uniparental inheritance is restricted to the zygote and that the endosperm receives the paternal cytoplasm. What causes the separation of the intact cytoplasm from the nucleus is difficult to comprehend, although cytoplasmic degeneration may certainly contribute to this effect. In *Populus deltoides*, the sperm cells develop large vacuoles during their descent into the pollen tube, and it is believed that this is the signal that leads to the separation of the cytoplasm from the nucleus at the time of fertilization

(Russell, Rougier and Dumas 1990). An unusual way of eliminating the male cytoplasm from the egg probably occurs in spinach in which plasma membranes of the two sperm cells initially fuse together to form a binucleate cell. It has been claimed that fusion of the male nuclei with the egg nucleus and with the polar fusion nucleus of the central cell is followed by sealing of the plasma membranes, thus excluding any remaining cytoplasm from the female cells (Wilms 1981b).

Just as the complete elimination of the sperm cytoplasm at the time of fertilization offers an explanation for maternal organelle inheritance, so also might a process of progressive cytoplasmic diminution during sperm cell maturation explain maternal inheritance. Mogensen and Rusche (1985) first reported that during maturation of barley sperm cells, lateral projections containing mitochondria are frequently pinched off, resulting in a conspicuous reduction in the surface area, volume, and mitochondrial number of the sperm. In *Nicotiana tabacum*, cytoplasmic diminution, which occurs in the sperm cells as they descend into pollen tubes, is accomplished by the formation of enucleate cytoplasmic bodies, the spontaneous vesiculation of the tip of cytoplasmic extensions, and the accumulation of vesiculate bodies in the periplasm of the sperm cells (Yu, Hu and Russell 1992). It is difficult to reconcile these observations with those of a subsequent work, which showed that during fertilization in *N. tabacum*, male cytoplasm including heritable organelles is transmitted into the female target cells (Yu, Huang and Russell 1994). Since sperm cells do not produce nuclei denuded completely of their enveloping cytoplasm, it is likely that processes such as those described in barley and tobacco, combined with cytoplasmic elimination at the time of fertilization, might offer an explanation for maternal organelle inheritance in the majority of plants in both species, with the caveat that occasional biparental inheritance also occurs.

Transmission of a part or all of the cytoplasm of the sperm cell in the act of fertilization certainly is the basis of biparental or paternal inheritance of organelles. In Chapter 3 it was reported that in some plants the generative cell is cut off with an appreciable number of plastids, which suggests that the plastids are transmitted to the sperm and subsequently into the egg cell during gametic fusion. This seems to be true where the fate of the sperm cell has been followed electron microscopically. Very clear-cut, indeed, is the evidence for the transmission of sperm cytoplasm into the egg and central cell in *Plumbago zeylanica*, in which, based on size and fluorescent-staining differences, it was possible to discriminate between plastids and mitochondria of sperm cell origin from those present in the egg at the time of fertilization (Russell 1980, 1987; Sodmergen et al 1995).

Electron microscopic evidence has augmented genetic data to help explain certain cases of uniparental, paternal inheritance of organelles. One such case is *Medicago sativa*, in which RFLP for plastid DNA has indicated a predominant paternal inheritance of this organelle in reciprocal crosses between two subspecies (Schumann and Hancock 1989). Paternal inheritance is also correlated with a high plastid content in the generative cell, although there are no significant differences in plastid abundance between genotypes known to differ in their male plastid transmission pattern (Zhu, Mogensen and Smith 1990, 1991). Likewise, there is no consistent correlation among genotypes between the mean number of plastid DNA aggregates (nucleoids) in the generative cell and the strength of male plastid transmission (Shi et al 1991). A quantitative, three-dimensional analysis of the plastid distribution patterns in egg cells of genotypes designated as weak plastid transmitters or strong plastid transmitters has provided an explanation of this paradox. Apparently, eggs of weak plastid transmitters are characterized by a largely micropylar plastid distribution, whereas plastids of the strong females are positioned in the chalazal part of the egg. This would ensure that following the first division of the zygote, the plastid population of the weak females is segregated to the basal micropylar cell, which becomes the transient suspensor, and the apical organogenetic part of the embryo will be blessed with the plastids delivered by the sperm. On the other hand, in strong plastid transmitters, the apical cell, which inherits the plastids at the time of the first division of the zygote, will perpetuate maternal or biparental plastid transmission (Zhu, Mogensen and Smith 1993; Rusche et al 1995). This viewpoint assumes that fertilization is a passive event and that organelles stay put in whichever locations they find themselves at the time of nuclear fusion. As described in Chapter 12, electron microscopic studies have revealed wholesale rearrangement in the distribution of organelles in the egg cells of several plants following fertilization.

A strong bias in favor of paternal inheritance of plastids also prevails in the interspecific cross *Daucus muricatus* × *D. carota*. The basis for this is the presence of an exact replica of restriction fragments of *D. carota* plastid DNA in self-pollinated F1 plants (Boblenz, Nothnagel and Metzlaff 1990). This is attributed to the presence of relatively few intact plastids in the egg of *D. muricatus* compared to the other species and to the presence of degenerate plastids (Hause 1991). However few in number the plastids may be, how these plastids are excluded from the zygote is an unsolved problem. Among the ideas suggested are the domination of paternal plastids and eventual elimination of the maternal plastid genome, differential plastid replication with the paternal plastids' holding an advantage, and the wholesale degeneration of maternal plastids. Investigations that take these ideas into account seem to be necessary to provide complete structural evidence for the genetic data for paternal plastid inheritance.

In recognition of the fact that some male plastids invariably find their way into the egg cytoplasm at the time of fertilization, efforts have been made to explain plastid behavior in the zygote that results in uniparental inheritance. According to Zhu, Mogensen and Smith (1991), who have collated this information, elimination of male plastids from the zygote occurs by (a) differential degradation of male plastid DNA within the zygote; (b) recombination between different plastomes in the zygote, with subsequent DNA repair using the strong plastome of the female parent as the template; (c) differential multiplication of plastids of the two parents within the zygote and proembryo, with the maternal plastids' holding an edge; and (d) segregation of maternal plastids to the apical cell and paternal plastids to the basal cell in embryo types in which the apical cell of the two-celled embryo becomes the organogenetic part and the basal cell becomes a short-lived suspensor. Perhaps symptomatic of the primitive state of our understanding of the fate of the paternal plastids following fertilization is the fact that there are no markers to follow them in the chaotic environment of the zygote.

3. IN VITRO APPROACHES TO THE STUDY OF FERTILIZATION

In the reproductive phase of angiosperms, fertilization and seed set are modulated by several external environmental and internal genetic factors. Among the former are female receptivity, the age and quality of pollen grains, and properties of the ambient atmosphere. In traditional plant breeding programs involving incompatible pollinations, prefertilization blocks such as the failure of pollen grains to germinate on the stigma, physiological disturbances in the directional growth of pollen tubes through the style toward the embryo sac, and failure of fertilization itself often thwart seed set. When transfer of beneficial alien genes across interspecific and intergeneric barriers is attempted, such as in wide hybridizations, it is seen that in spite of normal pollination and fertilization, deleterious genes carried forward into the zygote cause embryo lethality and seed collapse. Since success in agricultural practices would depend upon devising techniques to control reproductive processes in plants to our advantage, methods to overcome these natural bottlenecks to sexual recombination have attracted a great deal of attention. As described in Chapter 10, advances made in our understanding of the molecular biology of incompatibility offer potential for the development of methods for overcoming this barrier to crossability in plants. Methods to overcome embryo abortion in wide hybrids are briefly considered in Chapter 12. The purpose of this section is to provide an overview of the in vitro methods that have been developed to study fertilization in angiosperms under controlled conditions. Detailed protocols for most of the in vitro approaches currently in use are given by Yeung, Thorpe and Jensen (1981).

(i) Intraovarian Pollination

The primary objective of in vitro approaches to study fertilization is to secure the encounter of the egg and sperm in a relatively uncomplicated way that bypasses pollination and pollen tube growth in the style. An extensive discussion would be required to recount all of the early attempts in the experimental manipulation of the fertilization process. Only the highlights can be described here; refer to the brief review by Tilton and Russell (1984) for a historical perspective of this work. In one of the first published full investigations of this kind, Bosio (1940) was able to induce fertilization in ovules of three species of *Paeonia* by combining excision of the style with actual deposition of pollen grains on the upper part of the ovary. Because it effectively bypasses stigmatic pollination and stylar growth of pollen tubes in a distinctive way, the technique is christened as "intraovarian pollination." Unfortunately, neither the technique nor the observations sparked much interest until Kanta (1960) used a modified method to induce fertilization and seed set in *Papaver rhoeas*. The experimental design involved emasculating receptor flowers before anthesis and injecting into the ovary small quantities of a suspension of pollen grains prepared in boric acid. This resulted in the germination of pollen grains, in the growth of pollen tubes and their entry into the embryo sac, and eventually, in fertilization. Results acquired in terms of seed set demonstrated the success of intraovarian pollination in *P. rhoeas*, as well as in other members of Papaveraceae such as *P. somniferum*, *Argemone mexicana*, *A. ochroleuca*, and *Eschscholzia californica*. Although reciprocal crosses between the diploid *A. mexicana* and the tetraploid *A. ochroleuca* failed due to inhibition of pollen tube growth in the style, intraovarian pollination offered the potential for eliminating this roadblock to fertilization when either species was used as the female parent (Kanta and Maheshwari 1963a). An extension of this method is to inject sperm directly into the embryo sac; this approach holds promise in plants where the sperm cells do not reach the embryo sac after intraovarian pollination.

(ii) Test-Tube Fertilization

A step toward a more controlled encounter of the egg and sperm was achieved by culturing unpollinated receptive ovules of *P. somniferum* in a nutrient medium and dusting them with pollen grains. During the ensuing period of two to three weeks, pollen germination, pollen tube growth, fertilization, embryogenesis, and endosperm development occur, as attested by the transformation of cultured ovules into mature seeds that survive on the original medium. This was confirmed by histological preparations of ovules that established beyond doubt the occurrence of double fertilization in the test tube; hence, in analogy to animal systems, the technique is designated as "test-tube fertilization" (Kanta, Ranga Swamy and Maheshwari 1962). Using these general guidelines, the protocol was also successfully extended to other members of Papaveraceae, as well as to members of other families (Kanta and Maheshwari 1963b; Kameya, Hinata and Mizushima 1966; Zenkteler 1967; Zúbková and Sladký 1975). In a modified version of the protocol, whole pistils were cultured and pollinated in vitro with results similar to those obtained with cultured ovules; this is analogous to conducting stigmatic pollination in vitro. The technique was successful with, among others, *Petunia violacea* (Shivanna 1965), *Nicotiana tabacum* (Dulieu 1966), *N. rustica* (Rao and Rangaswamy 1972), *Antirrhinum majus* (Usha 1965), *Dicranostigma franchetianum* (Papaveraceae) (Rangaswamy and Shivanna 1969), and *Papaver nudicaule* (Olson and Cass 1981). In the last-mentioned example, the pollen tube enters the embryo sac by growing into the filiform apparatus of the synergid as it does during in vivo fertilization.

Test-tube fertilization also occurs by crossing interspecific and intergeneric barriers. One of the most convincing examples is fertilization of cultured ovules of *Melandrium album* with pollen from *M. rubrum*, *Silene schafta*, and *S. tatarica*, all of Caryophyllaceae (Zenkteler 1967). Indeed, this appears to be a choice method for raising interspecific hybrids because hybrid embryos complete their full-term development. Cultured ovules of *M. album* are also found to favor pollen grains of other genera of Caryophyllaceae such as *Cerastium arvense* and *Dianthus serotinus*, and fertilization ensues; however, the competence of zygotes to form full-term embryos is found to fade after a few divisions (Zenkteler, Misiura and Guzowska 1975).

Beginning with the works of Sladký and Havel (1976) and Gengenbach (1977a), a great effort has been made to identify the conditions for maximizing fertilization and kernel development under in vitro conditions in maize. These studies, which basically have involved culturing pieces of the cob and pollinating the exposed silk or dusting the nucellar tissue with pollen grains, have examined the influence of donor plant genotype (Gengenbach 1977b; Bajaj 1979; Higgins and Petolino 1988), the ratio of ovary to cob tissue (Sladký and Havel 1976; Higgins and Petolino 1988), pollen quantity (Raman, Walden and Greyson 1980), explant size (Higgins and Petolino 1988), silk length (Dupuis and Dumas 1989), tissue continuity of the silk (Booy, Krens and Bino 1992), stage of spikelet development (Dupuis and Dumas 1990a), and temperature stress (Mitchell and Petolino 1988; Dupuis and Dumas 1990b) on fertilization frequency and seed set. Using this system, it has been shown that compared to the mass pollination that occurs in nature, no more than a single pollen grain is required to affect fertilization of an ovule, although the frequency of successful fertilization events varies greatly in different trials (Raman, Walden and Greyson 1980; Hauptli and Williams 1988; Kranz and Lörz 1990). Since pollen tubes are able to travel a shorter distance within the style under in vitro conditions to affect fertilization than in vivo, it appears that sperm cells do not undergo an extended period of maturation in the pollen tube to become fusion competent (Dupuis and Dumas 1989). Havel and Novák (1981) used genetic color characters of embryos and kernels to demonstrate double fertilization in vitro. The development of kernels after in vitro pollination is not proof that the process of double fertilization has occurred or that subsequent developmental programs have been successful, because occasional absence of embryos and abnormalities in endosperm development are noted in apparently normal-looking kernels (Schel and Kieft 1986). In this case, the lesion is not in the fertilization processes but in the development of the embryo and endosperm.

One aspect of test-tube fertilization technique to be emphasized here concerns its potential use in eliminating self-incompatibility barriers. In the protocol used to overcome self-incompatibility in *Petunia axillaris*, a placenta with attached ovules, together with a short length of the pedicel, was cultured and the ovules were pollinated by dusting them with pollen. The result was prompt germination of pollen grains, rapid growth of pollen tubes, fertilization of the egg within about two days after pollination, and eventual seed set (Rangaswamy and Shivanna 1967). Additional manipulations have established that cultured ovules of self-incompatible plants do not show any preference to the pollen grains they encounter, as long as they belong to the same species. For example, when ovules on one placenta in the ovary are left unpollinated and those on another placenta in the same ovary are pollinated with self- or cross-pollen grains, the unpollinated ovules wilt, whereas fertilization and seed set occur in both selfed ovules and crossed ovules. The efficacy of self- and cross-pollinations for fertilization in vitro was demonstrated by the absence of differences in seed set when ovules on one placenta are selfed and those on another placenta in the same ovary are crossed (Rangaswamy and Shivanna 1971a). In yet another approach, when stigmatic pollination and placental pollination are performed on the same pistil in culture, the former is found to lead to seed set only with cross-pollination, whereas with the latter, both self- and cross-pollinations are found to be equally effective (Rangaswamy and Shivanna 1971b). These protocols are collectively designated as "placental pollination." Placental pollination has also been successfully used to overcome incompatibility barriers in *P. hybrida* (Niimi 1970; Wagner and Hess 1973), *Brassica campestris* (Zenkteler, Maheswaran and Williams 1987), and *Lilium longiflorum* (Janson 1993).

Zenkteler (1990) has compiled a list of plants in which fertilization and seed development have been achieved by the aforementioned methods. It should be noted that since fertilization events occur in eggs enclosed within ovules, these experiments fall short of a true in vitro fertilization system as understood in animal embryology.

(iii) In Vitro Fertilization

From the developmental point of view, in vitro fertilization with isolated, single gametes presents an ideal opportunity to analyze karyogamy at the level of the single cell, uninfluenced by the presence of other cells. From the long-range point of view, in vitro fertilization will provide a route for direct transformation of the egg using external sources of DNA and for transmission of alien cytoplasm to study organelle inheritance. Using the methods available for isolation of embryo sacs and sperm cells, a beginning has been made in the study of the conditions necessary for single gamete fusion and culture of artificial zygotes in maize. The initial strategy in this work is to use cleverly engineered microcapillary techniques to select sperm cells individually from ruptured pollen grains and to select eggs and other component cells from enzymatically treated embryo sacs. Selection is followed by the transfer of gametes to a microdrop of a fusion medium (0.55 M mannitol) placed on a coverslip; pairs of gametes are fused electrically. By this method, fusion has been accomplished not only between isolated, single egg and sperm cells, but also between single sperm cells and single synergids and central cells. The egg–sperm fusion products, when nurtured in a microculture surrounded by a suspension of feeder cells of maize embryo origin, develop into multicellular structures (Kranz, Bautor and Lörz 1991a, b). Electrofusion between isolated male and female gametes of wheat resulting in multicellular structures has also been reported (Kovács, Barnabás and Kranz 1995). In an alternative method, developed without the application of electrical pulses (Figure 11.5), the essential environment for the egg and sperm cell of maize to fuse was found to be a medium containing a high concentration of $CaCl_2$ at pH of 11.0 or a medium containing mannitol and $CaCl_2$ (Faure, Digonnet and Dumas 1994; Kranz and Lörz 1994). Later studies showed that the first few rounds of division of the artificially created zygote to form the multicellular mass and the transformation of the mass into an embryo bear obvious similarity to in vivo development of the zygote. Formation of plants occurred by germination of the bipolar embryos and, occasionally, by polyembryony and by organogenesis from the multicellular mass (Kranz and Lörz 1993; Kranz, von Wiegen and Lörz 1995).

What is the possibility that the plants regenerated are the products of spontaneous division of the egg in the absence of fusion with a sperm? This

SECTION III

Zygotic Embryogenesis

in reality concerned with the control of endosperm development.

Apart from the essential interest in it as a by-product of fertilization, the endosperm presents basic problems relevant to our understanding of progressive embryogenesis. The objective of this chapter is to consider the endosperm from such aspects that are germane to its function in the developing seed following double fertilization. First, attention will be paid to the period of division and differentiation of the primary endosperm nucleus. Second, there is a considerable amount of evidence showing that genetic characters play a major role in determining the final form of the endosperm in certain plants. The various genetic factors controlling endosperm development will be briefly considered. Third, information will be presented concerning gene expression pattern in the endosperm, with emphasis on the synthesis and accumulation of storage proteins. Much of the early work on the morphology, cytology, and genetics of the endosperm has been considered by Brink and Cooper (1947b) in an exhaustive article. Reviews of relatively recent vintage are by Bhatnagar and Sawhney (1981) and Vijayaraghavan and Prabhakar (1984) on endosperm developmental morphology; Olsen, Potter and Kalla (1992) on the cytology and molecular biology of cereal endosperms; Lopes and Larkins (1993) on gene regulation; and Shewry (1995) on biochemistry. These are immensely challenging topics addressed by a considerable accumulation of literature, and so some degree of subjective judgment has gone into the selection of material to be discussed here.

1. HISTOLOGY AND CYTOLOGY OF THE ENDOSPERM

The ontogeny of the endosperm has been followed in many different species to provide a descriptive basis for future biochemical and molecular investigations. These studies have reinforced the view that the endosperm is a relatively simple and amorphous tissue marked by the presence of only a few differentiated cell types. In spite of the fact that fusion events during double fertilization are almost simultaneous, it is generally acknowledged that the division of the primary endosperm nucleus takes precedence over that of the zygote. Little is known about the signal transduction system that potentiates the division of the primary endosperm nucleus ahead of the zygote or of the factors that induce the development of the primary endosperm nucleus into a compact tissue. Estimates of the development of the endosperm prior to division of the zygote range from the production of a few isolated nuclei to the formation of a fully developed tissue. A quantitative analysis of mitotic activity in the primary endosperm nucleus and the zygote in *Triticum aestivum* is particularly revealing of the shift in the balance of competing genomes. In plants reared at 20 °C, the primary endosperm nucleus begins to divide as early as 6 hours after pollination, whereas the division of the zygote is delayed until about 22 hours. The primary endosperm nucleus completes at least four rounds of nuclear divisions by the time the zygote divides, so that by 24 hours after pollination, at least 16 free endosperm nuclei are drifting randomly in the central cell around a two-celled proembryo. By the fifth day after pollination, the endosperm has already amassed more than 5,000 cells, in contrast to little more than 96 cells in the proembryo (Bennett et al 1973b). The same general trend in the rates of nuclear or cell doubling in the developing endosperm and embryo is maintained in other species of *Triticum*, in *Hordeum vulgare, H. bulbosum*, and *Secale cereale*, and in several genotypes of *Triticale*, with differences of similar magnitude in their cell numbers resulting (Bennett, Smith and Barclay 1975). To the extent that the endosperm possesses some nutrient value even at this early stage in its ontogeny, the interval between the division of the primary endosperm nucleus and the division of the zygote has significance in the nutrition of the zygote. Although the diploid polar fusion nucleus divides spontaneously in apomictic species, autonomous division of this nucleus may also occur in cultured, unpollinated ovules of certain amphimictic angiosperms (Mól, Betka and Wojciechowicz 1995). Ohad et al (1996) have isolated a female gametophyte mutation in *Arabidopsis* termed *fertilization-independent endosperm* (*fie*) that allows for replication of the diploid polar fusion nucleus and endosperm development without fertilization. Continued nuclear divisions in the central cell without the trigger of fertilization argue for a general molecular explanation based on cell cycle control of the fusion nucleus.

(i) Structure and Development of the Endosperm

The patterns of division of the primary endosperm nucleus have served as the basis for a simple classification of endosperms into nuclear, cellular, and helobial types. In the nuclear type, the

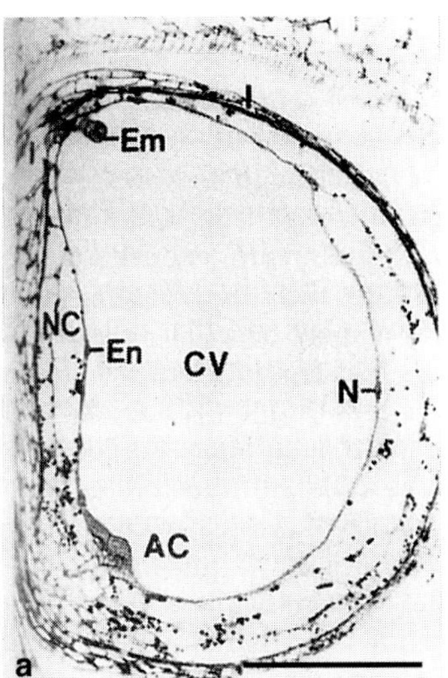

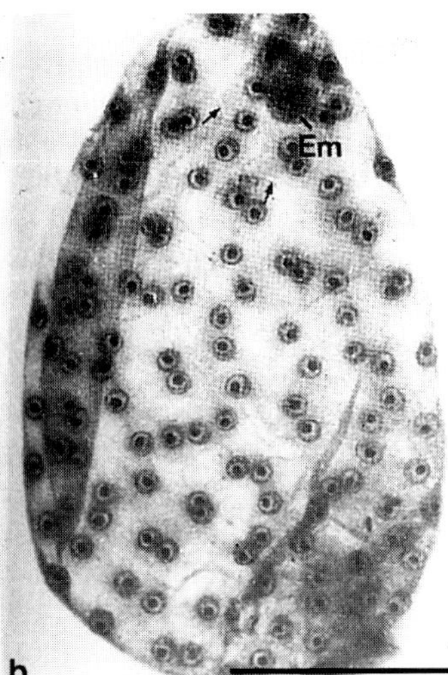

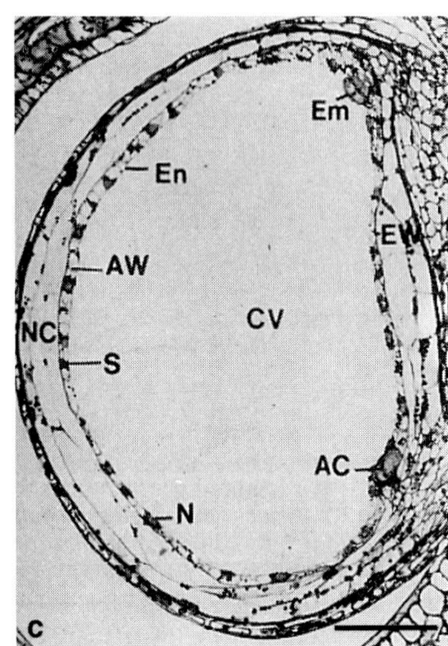

Figure 12.1. Endosperm development in *Ranunculus sceleratus.* **(a)** Longitudinal section of the ovule showing the embryo sac surrounded by two to three layers of nucellar cells and a single layer of cells of the integument. The vacuole is lined by a thin layer of endosperm cytoplasm containing free nuclei. **(b)** A whole mount of the embryo sac containing 128 endosperm nuclei. Arrows point to walls in the micropylar region. **(c)** In this longitudinal section of the ovule, the endosperm has been completely partitioned into uninucleate, open compartments. AC, antipodal complex; AW, anticlinal wall; CV, central vacuole; Em, embryo; En, endosperm; EW, embryo sac wall; I, integument; N, nucleus; NC, nucellus; S, starch. Scale bars = 100 μm (From Chitralekha and Bhandari 1993.)

first several rounds of division of the primary endosperm nucleus are unaccompanied by cytokinesis and thus result in a few thousand free nuclei in the central cell. During the early endosperm ontogeny, the nuclei are scattered throughout the central cell, or are pushed to its periphery by a burgeoning vacuole, or stream into incipient haustorial processes. Ultrastructural observations indicate that the cytoplasm of the central cell housing the free nuclei is undistinguished and is not enriched for any special organelles, although it is considered to be metabolically active (Marinos 1970a; Schulz and Jensen 1974, 1977; Ashley 1975a; Pacini, Simoncioli and Cresti 1975; Gori 1987; Dute and Peterson 1992; Chitralekha and Bhandari 1993). One might assume that as part of the distinction that characterizes the nuclear type of endosperm, the nuclei remain free throughout the ontogeny of the tissue, but this is not always the case. Indeed, close to the final stage of endosperm ontogeny in the majority of species examined, free nuclei and the incipient cytoplasmic domain around them are sequestered with walls and become cellular (Figure 12.1). Following the initial phase of cellularization, invaginations of the plasma membrane similar to transfer cells are seen on the wall of the endosperm cells surrounding the embryo in *Vigna sinensis* (Hu, Zhu and Zee 1983), *Zea mays* (Schel, Kieft and van Lammeren 1984; Davis, Smith and Cobb 1990; Charlton et al 1995), and *Vicia faba* (Johansson and Walles 1993a, 1994). To emphasize the structural specialization of the basal endosperm transfer cells of maize, it has been pointed out that transcripts and the protein product of a cDNA clone isolated from a 10-day-old endosperm are restricted to these transfer cells (Hueros et al 1995). Whereas the endosperm in the seeds of most legumes is absorbed during seed development, soybean and certain other legumes have a partial or fully developed cellular endosperm. In soybean, the outermost layer of the cellular endosperm is incorporated into the seed as a single layer of cells around the embryo (Yaklich and Herman 1995).

The flexibility of the nuclear type of endosperm is illustrated in cereals in which a coenocytic stage with free nuclei precedes a cellular configuration. Later, cells of the outermost layer of the endosperm differentiate into protein-rich aleurone cells, and the inner cells become filled with starch and storage proteins (Figure 12.2). This basic histology of endosperm development has been described repeatedly in wheat (Buttrose 1963a; Evers 1970; Mares, Norstog and Stone 1975; Morrison and O'Brien 1976; Mares et al 1977; Morrison, O'Brien and Kuo 1978; Briarty, Hughes and Evers 1979; Fineran, Wild and Ingerfeld 1982; van Lammeren 1988a), barley (Forster and Dale 1983; Bosnes et al 1987; Engell 1989; Bosnes, Weideman and Olsen

ing cellularization of the endosperm in *Phaseolus vulgaris* (XuHan and van Lammeren 1994). The consistent occurrence of microtubules delimiting the nucleo–cytoplasmic units has inspired the notion that rather than being a passive support system, the microtubular cytoskeleton undergoes a striking reorganization during the initial cellularization of the endosperm.

After the inception of the anticlinal wall from the embryo sac wall, the pattern of subsequent wall formation involves the building up of an incipient phragmoplast. Indeed, the formation of phragmoplasts and cell plates is easier to observe during this phase of endosperm development than during cellularization of the free nuclei. Some of the early studies using the wall-free liquid endosperm cells of *Haemanthus katherinae* have provided a dynamic picture of mitosis in the endosperm; a beautiful cinemicrographic record of the process, produced by Bajer (1965), continues to be used even to this day to illustrate mitosis in living plant cells. From the model of cell wall formation that incorporates data from polarized light, phase contrast, differential interference contrast, immunofluorescence, and electron microscopy, there seems little doubt that the phragmoplast is initially formed at late anaphase by the consolidation of small droplets rich in elements of the dictyosomes or ER, and that it coexists with a massive accumulation of microtubules. Not unlike the situation described in other plant cells, these droplets coalesce to form the cell plate (Hepler and Jackson 1968; Lambert and Bajer 1972; de Mey et al 1982). Schmit and Lambert (1987, 1990) have shown that endosperm cells of *H. katherinae* have a permanent system of microfilaments, and it is possible, and even likely, that actin may participate in the microtubule-guided transport of vesicles during phragmoplast formation.

(iii) DNA Amplification

Odd as it may seem, cellularization of the endosperm is not generally associated with a normal cycle of DNA synthesis and mitosis. Two conventional routes followed by the cells of the endosperm that escape a normal cell cycle are endoreduplication and polyploidy. D'Amato (1984) has reviewed the literature on the occurrence of these phenomena in the endosperms of diverse angiosperms; as tabulated therein, the range of endoreduplication and polyploidy displayed by endosperms may be considerable, depending upon the location of cells, the presence of haustoria, and the number of nuclear divisions completed after the formation of the primary endosperm nucleus. Nuclear activity is reflected in the extent to which endosperm haustoria with a limited number of nuclei, as compared to the multicellular tissue, undergo repeated endoreduplication. A striking example is the uninucleate chalazal endosperm haustorium of *Arum maculatum*, which exhibits a ploidy level of 24,576*N*, corresponding to 13 endoreduplication cycles, whereas cells of the endosperm proper remain triploid (Erbrich 1965).

Endosperms of various cereals offer considerable advantage for studying nuclear cytology due to the prolonged cellular phase when progressive changes in DNA content occur. The extent of DNA changes in the developing endosperm of maize is of special interest since a number of reports have characterized the nuclei extensively. The increase in nuclear volume that is often seen during polyploidization and endoreduplication has been cited as indicative of an increase in the DNA content of the endosperm ranging from 6C to 384C (Punnett 1953; Tschermak-Woess and Enzenberg-Kunz 1965; Cavallini et al 1995). Following upon an earlier observation that the DNA content of the endosperm of young maize grains detected by Feulgen microspectrophotometry increased up to 24C (Swift 1950), later studies showed that coincident with the decline in mitotic activity, there was a variable increase in the DNA content of the endosperm nucleus, followed by a decrease. In studies using flow cytometry of nuclear preparations of the entire endosperm, only a modest increase of about 12C in the average DNA content per nucleus was observed at the peak period (Kowles and Phillips 1985; Kowles, Srienc and Phillips 1990; Lur and Setter 1993). A tentative conclusion from these data is that preferential amplification or underreplication of part of the genome during endoreduplication is not a way of life for maize endosperm nuclei.

Studies of developing endosperms of hexaploid triticales, oat (Herz and Brunori 1985), wheat (Herz and Brunori 1985; Chojecki, Bayliss and Gale 1986; Brunori, Forino and Frediani 1989), rice (Ramachandran and Raghavan 1989), and barley (Giese 1992) by microspectrophotometry or autoradiography have also shown that some form of endoreduplication occurs in cell populations and causes a gradual rise in DNA content from 3C to 24C in triticales, oat, and wheat, and from 3C to more than 30C in rice (Figure 12.3). In the context of a general increase in DNA content of endosperm cells with time, there is a marked decrease in the

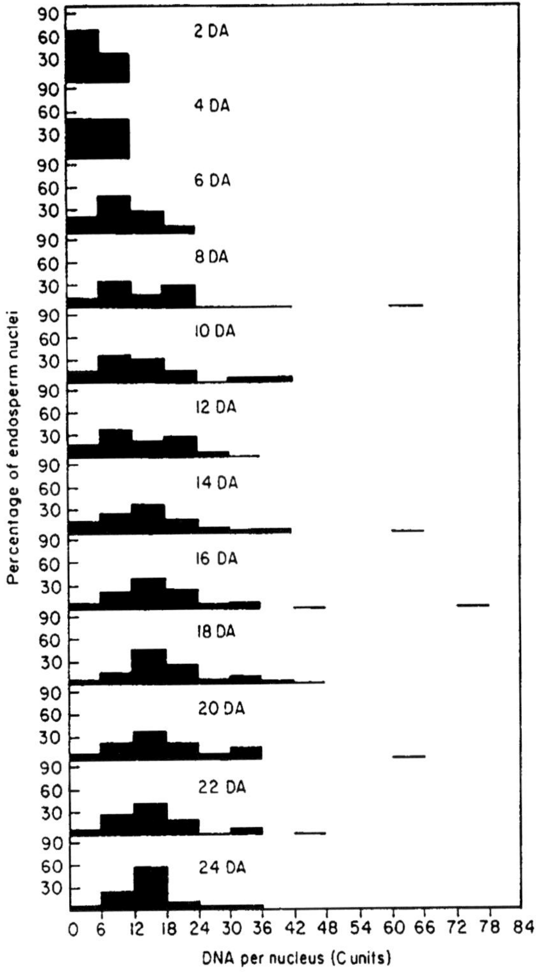

Figure 12.3. Distribution pattern of DNA content of nuclei in the developing endosperm of *Oryza sativa*. DA, days after anthesis. (From C. Ramachandran and V. Raghavan 1989. Changes in nuclear DNA content of endosperm cells during grain development in rice [*Oryza sativa*]. *Ann. Bot.* 64:459–468, by permission of Academic Press Ltd., London.)

number of nuclei with lower DNA values and a proportionate increase in nuclei with higher DNA values. A similar relationship exists during endosperm development in some noncereal monocotyledons (Marciniak 1993).

The basic changes in DNA values just described were elucidated mostly for the starchy cells or for both starchy and aleurone cells of cereal endosperms. Although assumed to remain unaltered during cytodifferentiation, the nuclear DNA content of developing aleurone cells reportedly varies when the cells are analyzed in the absence of contaminating starchy cells. It seems improbable, however, that aleurone cells belong to the same league as starchy cells, because only a simple doubling of DNA content, with some nongeometric fluctuations, has been observed in the examples studied (Maherchandani and Naylor 1971; Keown, Taiz and Jones 1977).

The mechanism of endoreduplication in the endosperm poses a fundamental problem, since it brings into focus the control systems for induction of the S phase and inactivation of the M phase of the cell cycle. Two key observations from a work on maize endosperm deserve to be mentioned in this connection. One is that DNA synthesis during the early stages of endosperm development is maintained by an increase in the amount and activity of S phase related protein kinases, which are implicated in the onset of the S phase in animal cells. Secondly, endoreduplicated cells contain an inhibitor that suppresses the activity of an M phase promoting factor (Grafi and Larkins 1995). The biological and functional significance of endoreduplication in the endosperm cells is still a matter of debate; according to a popular view, as in the cells of legume cotyledons, an increase in the DNA content of endosperm cells may be important for the synthesis and accumulation of storage products. Unfortunately, amplification of specific DNA sequences related to the function of the developing endosperm has not been demonstrated, because the evidence has turned out to be negative (Giese 1992).

2. NUTRITIONAL ROLE OF THE ENDOSPERM

Within the confines of the embryo sac, the free nuclei of the developing endosperm surround the embryo like a loose string of beads and eventually form a compact tissue around it. It is therefore obvious that some kind of interaction must take place between the endosperm and embryo. The best documented claims of interaction invoke a role for the endosperm in providing nurture and nutrients to the developing embryo. These claims, along with the structural modifications of developing embryos for nutrient uptake, will be discussed in Chapter 13.

This section considers the outgrowth of haustorial structures, the chemical composition of the endosperm, and the relationship between embryo abortion and endosperm activity in wide crosses – all of which constitute additional circumstantial evidence that relates the endosperm with embryo nutrition.

crosses, but afterward, the endosperm of the latter outstrips that of the former in its growth. The structural changes seen in the disintegrating endosperm cells are more dramatic than the decrease in cell number. In the hybrid ovule as early as 144 hours after pollination, cells of the endosperm at the chalazal end become vacuolate and less dense than those at the micropylar end. Although the endosperm survives for some time more, it undergoes no further divisions. Instead, as a result of the dissolution of cell walls and the fusion of the protoplasm and of the nuclei with those of contiguous cells, the endosperm is reduced to a few giant cells surrounding the embryo (Cooper and Brink 1945). In *Trifolium ambiguum* × *T. repens* cross, division of the primary endosperm nucleus ceases at different stages of development of the hybrid ovules, but in no case does the endosperm proceed beyond the 128-nucleate stage; in many hybrids, endosperm development is stymied at even earlier stages. In contrast, selfed *T. ambiguum* has a complete endosperm including a well-developed chalazal haustorium (Williams and White 1976). The ultrastructural changes in the hybrid endosperm cells are also impressive and seem to impinge on the synthetic activities of the cells and on nutrient transport into them. For example, in the endosperm cells of *Hibiscus costatus* × *H. aculeatus* and *H. costatus* × *H. furcellatus* hybrids, the rough ER appears as short irregular segments lacking ribosomes, whereas the nuclei contain multiple nucleoli (Ashley 1975a, b). Indubitable proof that endosperm abnormalities are related to the transport of nutrients is the demonstration that wall ingrowths characteristic of the normal endosperm are almost completely suppressed in the hybrid endosperm generated in crosses between normal diploid and autotetraploid maize (Charlton et al 1995). In the various examples studied, cells of the unhealthy endosperm have been found to show uninformative types of mitotic abnormalities such as lagging chromosomes, fused nuclei, and nuclei of high or low chromosome numbers. However, an understanding of the extent to which disturbances in the endosperm in inviable crosses may cause starvation of the embryo is certainly lacking at the present time. It must be concluded on logical grounds that the hybrid embryos are starved to death due to the inability of the physiologically disturbed endosperm to manufacture the exacting nutrients required for embryo growth or to absorb them from the surrounding maternal tissues and transmit them to the embryo. This means that the deleterious genes exert their primary effect on the endosperm rather than on the embryo.

An important factor governing the continued growth of the embryo and endosperm in a fertilized ovule is the nature of their interaction with the neighboring cells; therefore, some attention has been paid to the behavior of the nucellus and integuments in ovules of wide crosses. Several investigators, beginning with Renner (1914), have noted that the development of a weak endosperm and embryo abortion in inviable crosses are associated with the proliferative growth of a tissue from the nucellus. In a comparative histological examination of ovules from crosses between *Nicotiana* species, seed collapse was found to be associated with a retarded growth of the endosperm, which attained only a fraction of the growth attained in the selfed parent, and with a pronounced hyperplastic growth of the nucellus. The term "somatoplastic sterility" is used to describe the hyperplasia of the maternal tissue. It has been found that embryo failure is also associated with a deficiency in the development of the conducting elements within the seed, foreshadowing a suppression of nutrient transport to the endosperm; hyperplastic growth of the nucellus has been ascribed to the consequent abnormal distribution of nutrients (Brink and Cooper 1941).

A gene dosage effect leading to a change in the ratios of ploidy levels among the embryo, endosperm, and maternal tissues is seen in the ovules of several infertile crosses that show tumor formation from the inner epidermis of the integument (endothelium). In reciprocal crosses between 4*N* *L. pimpinellifolium* and 2*N* *L. peruvianum*, the endothelium assumes a densely cytoplasmic contour and surrounds the endosperm except for a small gap at the chalazal end. Tumor formation is initiated from the dorsal region of the endothelium near the chalazal end. As the tumor assumes form, it begins to grow without restraint and extends into the embryo sac as a burgeoning mass of cells. The cells of the endosperm surrounding the embryo deteriorate progressively to the point that the embryo is engulfed by the overgrown endothelial cells (Cooper and Brink 1945). It is no wonder that abortion of the already undernourished embryo is hastened by these changes.

An imbalance in the chromosome number between embryo and endosperm is postulated to cause abnormalities in the endosperm and consequent embryo abortion in crosses between diploid and tetraploid varieties of *Citrus* (Esen and Soost 1973). Considerations of the requirement for a balance between maternal and paternal genomes in the endosperm for seed set have led to a number of hypotheses to account for endosperm failure and embryo abortion in wide crosses (Kowles and Phillips 1988). A most useful observation based on data from a wide variety of intraspecific– and interspecific–intraploidy crosses is that the endosperm develops abnormally when the maternal–paternal ratio in the tissue deviates from 2:1. This has been incorporated into the notion of endosperm balance number (EBN) to explain endosperm development in crosses with different levels of ploidy. According to this hypothesis, the genome of each species is assigned an EBN on the basis of its crossing behavior with a standard species and, for crosses to be successful, the EBN has to conform to a 2:1 maternal–paternal ratio (Johnston et al 1980; Johnston and Hanneman 1982; Camadro and Masuelli 1995; Ehlenfeldt and Ortiz 1995). From the practical side, the attractiveness of this hypothesis is that the EBN can be manipulated to break the crossing barriers and to create new hybrids; from the theoretical side, it is unclear how the ratio of the maternal–paternal genomes is detected and how a deviant ratio causes endosperm malfunction. Using a mutant *indeterminate gametophyte* (*ig*) to create various ploidy level combinations in crosses in maize, Lin (1984) concluded that development of the endosperm is affected by the parental source of its chromosomes and that the 2:1 maternal–paternal chromosome number is critical in the development of a healthy endosperm. Although this hypothesis can readily explain why 2x × 4x or 4x × 2x crosses with 2:2 and 4:1 maternal–paternal chromosome ratios, respectively, in the endosperms will fail, it is not applicable to crosses in which endosperm development is not governed by a ratio of chromosome numbers.

The histology of the failure of endosperm development and its side effects on the maternal tissues of ovules of unsuccessful crosses are obviously in line with the view that gene action deprives hybrid embryos of a continuing supply of nutrients to complete development. The postulated nutritional relationship between the embryo and endosperm tends to suggest that the cryptic potentiality of the hybrid embryo to complete development may be realized if it is allowed to grow in an artificial medium supplied with nutrient substances that are normally identified with the endosperm. Early in the development of embryo culture as a research tool, Laibach (1925) demonstrated that progenies can be obtained from nonviable seeds of *Linum perenne* × *L. austriacum* hybrid by excision and culture of embryos before the embryos begin to disintegrate. Since this pioneering work, embryo culture methods have been widely used to obtain transplantable seedlings from aborted seeds of interspecific and intergeneric crosses that are traditionally condemned as being incapable of further growth. In certain cases, continued growth of the hybrid embryo is secured by implanting it on a normal endosperm, which is then cultured, thereby initiating a nurse culture. In a limited number of cases, efforts directed toward rescuing hybrid embryos by culturing ovules and ovaries have proved rewarding. Examples of hybrid embryo rescue by these different tissue culture protocols have been described or tabulated in published reviews (Raghavan 1977b, 1985, 1986b, 1994; Collins and Grosser 1984; Williams 1987) or books (Raghavan 1976b, 1986c), and the reader is referred to these sources for further information.

The appearance of haustorial structures on the endosperm, the presence in this tissue of chemical substances with potential growth promoting functions, and the rescue of hybrid embryos from ovules lacking a normal endosperm, present evidence of the potential role of the endosperm in the nutrition of the embryo. These observations, considered in conjunction with the various ultrastructural modifications of the embryo and suspensor cells for the absorption and translocation of metabolites from the surrounding cells (Chapter 13), make a compelling case for the nutrition of the embryo by endosperm metabolites.

3. GENE ACTION DURING ENDOSPERM ONTOGENY

It is natural to think of the development of the endosperm in terms of the synthesis of storage compounds and proteins necessary for the fundamental processes of cellular metabolism and housekeeping functions. Three major experimental approaches have been pursued to gain an understanding of how genes control the complex process of endosperm development in cereals. Historically, spontaneous kernel mutations in maize and barley have provided a tractable genetic system for monitoring the effects of marker genes on the development and differentiation of the endosperm. A large body of work has centered on the correlations of RNA and protein synthesis with endosperm development, and on the isolation of cDNA clones for specific proteins and the analysis of their expression in the developing endosperm. In recent years, extensive efforts have gone into the sequencing of genes regulating endosperm storage product synthesis and into the search for conserved nucleotide sequences of developmentally regulated genes and for the expression of promoter sequences of genes in transgenic plants. This section will deal with the genetic control of endosperm development, the regulation of RNA and protein synthesis, and the expression of genes for housekeeping proteins in the endosperms of cereals, in which they have been studied intensively. Due to the sheer volume of information, synthesis and accumulation of storage products in the endosperm will be considered in a separate section.

(i) Endosperm Mutants

The singular strength of maize as a genetic system among higher eukaryotes is the occurrence of morphologically identifiable mutants that affect virtually all stages of its life cycle. In considering the regulatory intervention of genes in kernel development in maize, most analyses have been performed to date with mutants that affect both the endosperm and the embryo and with those that are specific for endosperm development. Those in which growth of both embryo and endosperm is impaired are known as *defective kernel* (*dek*) mutants; included in the group that affects only endosperm development are *shrunken* (*sh*), *wx*, *brittle* (*bt*), *sugary* (*su*), *floury* (*fl*), and *opaque* (*op*) mutants. Since the latter group of mutations impinges on the nature of the storage reserves in the endosperm, it is appropriately considered in the next section.

The first *dek* mutants described turned out to be those in which the endosperm was normal but the grain did not have a viable embryo (Jones 1920). True *dek* mutants with defective endosperm were described by Mangelsdorf (1923), who found grains in which ovule development was arrested almost immediately after fertilization or after a small amount of endosperm had been formed. Additional *dek*-like mutants are designated as *reduced endosperm* (*re-1*, *re-2*) (Eyster 1931), *miniature seed* (*mn*) (Lowe and Nelson 1946), and *defective seed* (*de-17*) (Brink and Cooper 1947a). In general, mutant grains attain only a fraction of the weight of normal grains while maintaining a regular initial pattern of growth of the endosperm and embryo. Closer studies of *mn* mutant kernels have shown that subsequent disintegration of the vascular elements at the chalazal end of the ovary seemingly interrupts nutrient supply from the maternal tissues through the pedicel. Provision of the *MN* gene in the wild-type grain serves to encode an endosperm-specific cell wall invertase necessary for the mobilization of sucrose. Since the mutant kernels have an extremely low level of invertase, impairment of the movement of photosynthates between the pedicel and endosperm, and the consequent osmotic imbalance, offer a plausible explanation of the effects of the mutant gene on the endosperm (Lowe and Nelson 1946; Miller and Chourey 1992; Cheng, Taliercio and Chourey 1996). The basic histological manifestation of mutation in the *de-17* locus is the failure of endosperm cells facing the placenta to form a specialized absorbing tissue for transmitting nutrients from the maternal plant to the endosperm (Brink and Cooper 1947a).

A large number of *dek* mutants have been induced by pollinating maize ears with pollen grains mutagenized with ethyl methanesulfonate (EMS). The M1 plants raised from mature grains were self-pollinated and the harvested ears generated the stock of *dek* mutants for subsequent genetic analysis. The endosperms of the mutants varied considerably in their phenotypic appearance, with many having a collapsed or shrunken appearance; others were fully formed but were distinguished from the wild type by a floury or opaque appearance and a soft or fluid consistency. It was not unusual to find mutant kernels that were lighter in color than the normal due to blockage of carotenoid synthesis. Generally, the mutant phenotype is revealed within two weeks after pollination, at which time the growth of the defective endosperm slows down. Effects of the mutant gene on the embryo are seen even earlier than on the endosperm (Neuffer and Sheridan 1980; Sheridan and Neuffer 1980). Compared to the wild type, mutations impose serious restraints on the accumulation of dry matter, total proteins, and storage proteins in the endosperm. From these analyses, an informal recognition of three broad classes of mutants has emerged: (a) those showing a reduced accumulation of dry matter and proteins during the entire endosperm ontogeny; (b) those showing an initial normal rate of accumulation, followed by an early decline; and (c) those in which an initial lag in the rate of accumulation is coupled with an early termination of accumulation (Manzocchi, Daminati and Gentinetta 1980; Manzocchi et al 1980). In almost all cases examined, the average DNA content of the nucleus of mutant endosperm cells was found to be lower than that of the wild type; two idealized possibilities – a low proportion of nuclei in the higher DNA classes or the occurrence of fewer rounds of endoreduplication – can obviously tilt the balance in favor of a low DNA value (Kowles et al 1992).

An array of mutants equally diverse as those

just described has been obtained by the insertion of *mutator* (*mu*) transposons into the maize genome. This is accomplished by crossing standard lines with pollen from *mu* stock; the F1 plants grown from the resulting caryopses are then self-pollinated. Defective kernels are identified on the ears and are used for subsequent genetic dissection. Among the range of phenotypes observed, the most common are those containing a reduced endosperm and a small embryo, and others that have an empty pericarp and are almost always embryoless. Several mutants identified by transposon tagging have been found to be allelic to the *dek* mutants obtained by EMS mutagenesis (Scanlon et al 1994). These observations lead to the inference that following fusion of the sperm cell with the diploid polar fusion nucleus, many different genes must function in an orderly manner to effect the development of the endosperm into a functional nutritive tissue. The potential to gain deep insight into the genetic program governing endosperm development in maize by molecular cloning has been enhanced by the isolation of mutants from *mu* stocks.

Further evidence for the action of a multiplicity of genes during endosperm development has come from analyses of barley mutants with a defective endosperm. Deserving recognition for their importance are spontaneous endosperm mutants designated as *shrunken endosperm* (*se* and *seg*), in which shrunken kernels are caused by the expression of the mutant genes in the maternal tissues of the plant. Mutant *seg* endosperm types are distinguished from the wild type by a premature termination of grain filling caused by necrosis of the chalaza, or by upsets in the endosperm growth pattern that lead to characteristic abnormalities but permit normal development of the chalazal tissues (Jarvi and Eslick 1975; Felker, Peterson and Nelson 1984, 1985, 1987). Another group of mutants with a defective endosperm includes the chemically induced *shrunken endosperm expressing xenia* (*sex*) mutants, in which the mutant genes exert their effect on tissues of the entire grain. As in the case of maize, the ramifications of *sex* mutation on the endosperm of barley are extensive, ranging from the total lack of aleurone cells to the complete loss of starchy endosperm traits, and even to the presence of new cell types not found in the wild-type endosperm; the embryo may remain viable or nonviable. These effects are so profound as to suggest that the mutations cover genes participating in all phases of development of the endosperm (Bosnes et al 1987; Bosnes, Weideman and Olsen 1992).

(ii) Synthesis of Housekeeping Proteins

As a rapidly growing tissue, the endosperm must synthesize an enormous range of proteins necessary for cellular metabolism. However, in contrast to the extensive literature on storage protein synthesis, little information is available on the synthesis of metabolic proteins during endosperm development. The main reason for this is that synthetic activities of the widely investigated endosperm tissues are inextricably linked to the accumulation of storage products. Most importantly, the analysis of housekeeping proteins includes changes in the activities of enzymes for the metabolism of starch, proteins, and lipids, and in the levels of amino acids and nucleotides. Of necessity, much of the evidence from such investigations that relates to the synthesis of housekeeping proteins is correlative, and hence its significance is not fully understood. A few studies that have analyzed gene expression patterns in the endosperm dissociated from the synthesis of storage proteins will be described later.

Following fertilization, a readily available supply of substrates for the division of the endosperm nucleus must be present. This may be provided by transcription of the genome of the maternal tissues of the ovule or of the endosperm nucleus. Evidence based on autoradiography of ^{3}H-uridine incorporation shows that active transcription is initiated in the endosperm of barley as early as the free nuclear stage (Bosnes and Olsen 1992). The question of how soon after fertilization the genome of the endosperm nuclei begins transcription has been investigated in maize by crossing two strains with different electrophoretic isozymes for a particular locus and by screening the endosperm at different stages of development for expression of the variant enzymes. It was found that several enzymes of the maternal allele become active two or three days earlier than the gene products of the paternal allele. The earlier expression of enzymes from the maternal side has been attributed to their origin in the nucellus. However, other experiments showed that transcription of some isozyme genes in the endosperm nucleus begins relatively soon after fertilization but translation of the message is delayed until the endosperm becomes cellular (Chandlee and Scandalios 1983).

A study of catalase activity profiles has shown that continued development of the maize endosperm necessitates fundamental changes in the control of gene expression for the three unlinked genes of the enzyme (*CAT-1*, *CAT-2*, and *CAT-3*) in the different parts of the kernel. *CAT-1* is present throughout the 30-day developmental period analyzed, whereas *CAT-2* is expressed toward the period of kernel maturity and *CAT-3* appears only during the initial period of development. The appearance of isozymes correlates well with the expression of transcripts for the respective genes in RNA blots, indicating that the synthesis of the enzyme is controlled at the transcriptional level. The presence of *CAT-1* mRNA in the pericarp as well as in the starchy and aleurone cells of the endosperm, presence of *CAT-2* mRNA only in the aleurone layer, and presence of *CAT-3* mRNA only in the pericarp are evidence of strong tissue-specific regulation of the synthesis of these isozymes (Tsaftaris and Scandalios 1986; Wadsworth and Scandalios 1989). The reason for the unique, developmentally controlled pattern of expression of catalase genes is not at present understood, and investigations on the endosperms of other cereals will be valuable for comparative analysis.

Like catalase, pyruvate orthophosphate dikinase, a key enzyme in photosynthetic carbon dioxide fixation in C4 plants, illustrates a tissue-specific and developmentally regulated pattern of gene expression during endosperm development in wheat. The changes in the enzyme titer in the endosperm, with a peak at mid-maturation followed by a decrease, are correlated with changes in mRNA activity. At the tissue level, activity of the enzyme confined to the cells of the aleurone layer is paralleled by the appearance of mRNA in these cells (Aoyagi, Bassham and Greene 1984; Aoyagi and Chua 1988). Genes for glyceraldehyde-3-phosphate dehydrogenase and alcohol dehydrogenase (enzymes of the glucose–phosphate metabolism) show an increased expression correlated with progressive endosperm development in maize up to about 15 days after pollination, followed by a decrease (Russell and Sachs 1991). *Trans*-acting factors required for developmental expression of the maize *ADH-1* gene appear to be present in the endosperm of rice (Kyozuka et al 1994). As will become evident in the following section, the pattern of accumulation of storage proteins and their mRNAs in developing endosperms is not much different from that displayed by the photosynthetic and cytosolic enzymes.

4. SYNTHESIS AND ACCUMULATION OF ENDOSPERM STORAGE RESERVES

Because of the role of the endosperm as a major dietary source, the importance of the analysis of gene expression concerned with the synthesis of storage reserves in this tissue cannot be overemphasized. An understanding of the mechanism by which genes for the synthesis of the three principal storage products, namely, starch, proteins, and lipids, are regulated in the endosperm would be of great importance to the genetic engineering of cereal grains with improved nutritional qualities. A good deal of work, aided mostly by mutants in which the coordination and differential expression of endosperm-specific genes are disturbed, has been done to characterize the synthesis and deposition of storage products during endosperm development in our major cereal crops. This section will review the basic data on the nature of the endosperm storage products, mutants that have led to the identification of key steps in the biosynthetic pathway of storage products in the endosperm tissues of various plants, and the cell biology and biochemistry of the accumulation of proteins in the endosperm. A nutritional role for the storage proteins is supported by their presence in large quantities in the seed and by their rapid degradation during the early stages of germination as the embryo is transformed into a seedling. In seeds of some members of Fabaceae and Arecaceae, extensive deposition of polysaccharides like galactomannans and xyloglucans occurs on the inner cell walls of the endosperm and virtually occludes the cell contents. In addition, the cells contain small amounts of storage lipids and proteins (Meier and Grant Reid 1977; DeMason, Sexton and Grant Reid 1983; DeMason 1986; DeMason, Chandra Sekhar and Harris 1989; Edwards et al 1992). Although it is believed that these cell wall polymers serve as food materials, they are not further considered here.

(i) Mutants Defective in Starch Synthesis

This account will be prefaced by a brief reference to conventional biochemical studies that have identified the enzymes that link the synthesis of starch to its main substrate, sucrose. An important initial reaction in the sucrose–starch conversion is catalyzed by sucrose synthase, which splits sucrose to fructose and UDP-glucose. Another set of reactions involving hexokinase and hexose phosphate isomerase results in the transformation of fructose to glucose-6-phosphate. Formation of glucose-6-phosphate is also provoked from UDP-glucose by the action of pyrophosphorylase followed by a reaction mediated by a mutase. These changes, which occur in the cytoplasm, are followed by the conversion of glucose-6-phosphate to starch inside the amyloplast. This step is also catalyzed by a pyrophosphorylase and involves the formation of ADP-glucose, which donates its glucose to an increasing chain of glucose molecules. Starch exists as straight-chain amylose and as the branched-chain amylopectin, the latter being formed by the action of a branching enzyme. Although different biosynthetic pathways that lead to the same end products occur in maize and in some other plants, the basic steps described here could be rightfully considered as the most prevalent.

Several genes affect the type and quantity of carbohydrates in the mature endosperm of maize and, consequently, genetic approaches have been used extensively to determine the biochemical lesions caused by mutations. The *sh-1* mutation leads to a shrunken or collapsed endosperm and is attributed to the presence of a reduced amount of starch. At the protein level, the mutant endosperm is highly deficient in sucrose synthase; this in turn results in the accumulation of elevated levels of sucrose in the endosperm (Creech 1965; Chourey and Schwartz 1971; Chourey and Nelson 1976; Cobb and Hannah 1983). A special aspect of all *sh-1* mutants showing a substantial loss of sucrose synthase is that they retain a residual level of enzyme activity. The arguments about the normal enzyme and the residual enzyme have hinged around the question of whether the two enzymes are coded by the same or different genes; it is now resolved that the residual activity is due to a minor enzyme, similar in catalytic properties to the bulk enzyme, but encoded by a second gene, *SUCROSE SYNTHASE* (*SUS*) (Chourey 1981). A model of different in vivo functions for the two isozymes has gained currency with the demonstration that the isozymes are located in different parts of the endosperm. The enzyme encoded by *SH-1* generally accumulates in the bulk of starch-containing cells, whereas the *SUS*-encoded enzyme is restricted to the aleurone layer and the basal endosperm transfer cells (Chen and Chourey 1989; Heinlein and Starlinger 1989). Finally, transcripts of *SH-1* and *SUS* genes have been localized in the starchy endosperm and the aleurone layer, respectively (Springer et al 1986; Rowland and Chourey 1990). This finding indicates that the rate of transcription or message stability may be a controlling factor in the accumulation of the two isozymes. At the molecular level, both the *SH-1* and *SUS* genes have been cloned and characterized (Werr et al 1985; McCarty, Shaw and Hannah 1986; Gupta et al 1988).

Another mutant resulting from upsets in the carbohydrate metabolism of the endosperm is *wx*, which forms exclusively amylopectin, instead of a mixture of amylose and amylopectin as in the normal endosperm. The enzymatic basis for this difference has been attributed to the absence in the mutant endosperm of starch-granule–bound nucleoside diphosphate sugar transferase enzymes, which transfer glucose from ADP-glucose or UDP-glucose to starch (Nelson and Rines 1962; Nelson and Tsai 1964). As in the case of the *sh* mutant, the *wx* endosperm maintains some residual enzyme activity possibly coded by a different gene (Nelson, Chourey and Chang 1978). Cloning of the *WX* locus and characterization of its gene product have shown that some mutant phenotypes result from the production of a structurally altered protein; several other stable mutant phenotypes are caused by insertions or deletions within the *WX* gene transcription unit (Echt and Schwartz 1981; Shure, Wessler and Federoff 1983; Wessler and Varagona 1985).

Greatly reduced starch synthetic capability is associated with *sh-2* and *bt-2* mutants. The mutant block has been identified with a reduced activity of the enzyme ADP-glucose pyrophosphorylase in the endosperm; this enzyme catalyzes the formation of ADP-glucose from glucose-1-phosphate and ATP (Tsai and Nelson 1966). However, the presence of detectable levels of enzyme activity in several alleles of both mutants suggested that they may possess a pyrophosphorylase independent of *SH-2* and *BT-2* gene functions; this proved to be the case when it was found that 95% of the activity in the wild type exists in the heat-labile form and 5% in the heat-stable form. Both mutations reduce the heat-labile form appreciably and the heat-stable activity only marginally (Hannah and Nelson 1976; Hannah, Tuschall and Mans 1980). Although these observations point to the genes' preference for functioning in a nonstructural role in some post-translational steps, two later studies have strengthened the argument that *SH-2* is a structural gene. In one strategy, using the transposable element *dissociation* (*ds*), which has the ability to inhibit the function of *SH-2*, it was found that upon removal of the *ds* element, gene function is restored in the endosperm. Moreover, revertants that restore wild-type enzyme levels possess enzyme molecules with altered allosteric and kinetic parameters (Tuschall and Hannah 1982). Confirming that *SH-2* is indeed a structural gene for the maize endosperm ADP-glucose pyrophosphorylase, the deduced amino acid sequence of a *SH-2* cDNA clone has revealed much similarity to that of the bacterial enzyme (Bhave et al 1990). Using rice ADP-glucose pyrophosphorylase cDNA as a probe, a *BT-2* cDNA clone was isolated from maize endosperm and partially sequenced. When RNA isolated from various *bt-2* mutants was probed with the maize cDNA

synthesis and transport of storage proteins in the wheat endosperm revolve around the ER and Golgi vesicles, at least transiently, in a complex way.

It is unlikely that electron microscopy and immunolocalization can truly reflect the dynamics of the synthesis and transport of storage proteins in the endosperm, especially when the proteins are constituted of different subunits. Some new information about the role of the Golgi apparatus in protein transport has come from the ability of *Xenopus laevis* oocytes to sponsor the formation of -α and γ-subunits of gliadin when injected with the respective mRNAs. Labeling with radioactive precursor combined with electrophoresis was used to follow the accumulation of proteins in the injected oocytes and in the outside medium. The striking observation made in this experiment was that whereas α-gliadin accumulates inside the oocyte, γ-gliadin is secreted into the medium. Since oocytes of *X. laevis* do not have vacuoles, γ-gliadin is believed to be exported from the ER to the Golgi apparatus and secreted into the medium. This view was also supported by the fact that disruption of the Golgi apparatus by monensin halts the secretion of γ-gliadin from the oocyte (Simon et al 1990). Evidently, some activation event determined by the structural differences between α- and γ-gliadins must allow the gliadins to be distinguished for their assignment to markedly different subcellular transport routes. One final point in this context is that the expression of deletion mutants of γ-gliadin in the oocytes showed that the C-terminal region is responsible for secretion of the protein (Altschuler et al 1993).

Immunocytochemical evidence shows that during the transport of γ-secalins, which are synthesized in the ER, to the site of their accumulation, protein body formation in rye endosperm cells is analogous to that in wheat and involves the Golgi apparatus (Krishnan, White and Pueppke 1990). The biochemical similarity of endosperm prolamins of the wild grass *Haynaldia villosa* (Poaceae) to wheat gliadins has made it of interest to localize the protein using appropriate antibodies. Absence of label in the inclusions of the ER and the specific labeling of protein bodies found in the vacuoles and vesicles associated with the Golgi complexes provide strong support for a role for the Golgi apparatus in the transport of storage proteins, if not in their synthesis (Krishnan, White and Pueppke 1988).

The major mobilization of zein protein bodies in the maize endosperm commences when the cells are still undergoing division. The first change in the endosperm cell auguring its altered function is the appearance of small, membrane-enclosed, spherical deposits. The deposits begin to accumulate into dense protein bodies in vesicles produced by localized dilations of the ER that occur along the cisternae or at their enlarged ends. It is difficult to decide from the electron micrographs alone whether the protein is synthesized outside or inside the membranes (Khoo and Wolf 1970). Additional understanding of the role of ER in zein synthesis was provided by studies showing that membrane-bound polysomes and the mRNA population present in the rough ER of the endosperm direct the synthesis of virtually the same zein protein in vitro as they do in vivo (Larkins and Dalby 1975; Larkins, Bracker and Tsai 1976; Larkins, Jones and Tsai 1976; Larkins and Hurkman 1978; Melcher, 1979; Torrent et al 1986). Consistent with the view that fully formed protein bodies are surrounded by ER, a cell-free system that synthesizes zein using polysomes and mRNA derived from protein bodies as templates has also been described (Burr and Burr 1976; Burr et al 1978; Viotti et al 1978). A finding with implications for the role of ER in zein synthesis is that injection of zein mRNA into *X. laevis* oocytes potentiates the synthesis of proteins compartmentalized within ER vesicles (Larkins et al 1979; Hurkman et al 1981).

Any disadvantage in the use of endosperm from intact grains of maize for biochemical studies has been partially ameliorated by the successful culture of the endosperm in liquid suspension or on agar medium. The cultured cells maintain their endosperm phenotype and synthesize zeins that accumulate in protein bodies as in the endosperm cells of developing kernels (Shimamoto, Ackermann and Dierks-Ventling 1983; Felker 1987).

Although the classification and nomenclature of zeins have remained a contentious issue for a long time, a consensus seems to have emerged in recent years. Accordingly, these proteins are classified into five individual zeins of molecular masses of about 10, 15, 19, 22, and 28 kDa. These are also known as α-zeins (19 and 22 kDa), β-zein (15 kDa), γ-zein (28 kDa), and δ-zein (10 kDa). Immunolocalization has shown that the different classes of zeins are distributed heterogeneously in the protein bodies, with further variation dependent upon the location of the organelles within the endosperm cells. Generally, the protein bodies of the subaleurone cells are small and contain almost exclusively β- and γ-zeins, whereas those confined to the more central cells of the endosperm contain α-zeins also.

An interesting feature of the protein bodies containing, α-, β-, and γ-zeins is that the α-zein is confined to the central region of the organelles and the β- and γ-zeins become peripheral (Lending et al 1988; Lending and Larkins 1989).

Cells of the rice endosperm are unusual in harboring different types of protein bodies that differ in their morphology, the nature of accumulated proteins, and their subcellular origin. A model that has gained some acceptance recognizes two types of predominantly prolamin-containing spherical protein bodies and one type of glutelin-enriched, irregularly shaped crystalline body. Electron microscopic observations have supported the view that the three types of protein bodies share a common site of synthesis in the rough ER (Harris and Juliano 1977; Bechtel and Juliano 1980; Oparka and Harris 1982; Yamagata and Tanaka 1986). It has been claimed that different domains of the rough ER, designated as the protein body ER and cisternal type ER, are the site of the synthesis of prolamins and glutelins, respectively. The bases for this view are the clear identification of prolamin-derived protein bodies delimited by the rough ER and of glutelin-derived bodies delimited by single-lamellar membranes, and the differences in the content of prolamin and glutelin mRNAs in the two domains of the rough ER. Whereas the cisternal type of rough ER is enriched for glutelin mRNA, the protein body ER harbors mostly prolamin mRNA (Li, Franceschi and Okita 1993). Evidence has also appeared indicating that a special protein located in the ER lumen might bind to the nascent prolamins, promoting their folding and assembly into protein bodies and thus retaining them in the ER (X. Li et al 1993). Some studies have implicated the Golgi vesicles in a formative way in the final configuration of the prolamins and glutelins (Bechtel and Juliano 1980; Oparka and Harris 1982; Yamagata and Tanaka 1986). The definitive view of the role of the Golgi vesicles now appears to be that they mediate the transport of only the glutelin protein bodies synthesized in the cisternal type ER to the site of their deposition (Krishnan, Franceschi and Okita 1986; Li, Franceschi and Okita 1993). There is little evidence that can be marked out as suggestive of a mechanism that preferentially segregates glutelin mRNA and the protein product to the cisternal type of ER and the Golgi vesicles, respectively.

Much of the interest in protein body formation in endosperms of other plants hinges on the involvement of ER in the synthesis of the proteins and of Golgi bodies in their transport. In the developing endosperm of oat, globulins and avenin are deposited in spatially distinct subcellular domains. The main aggregation of globulin takes place within the vacuole, whereas most of the avenin appears to assemble directly within the rough ER, whence it migrates to the vacuole (Saigo, Peterson and Holy 1983; Lending et al 1989). Additional support for the synthesis of storage proteins within the rough ER comes from observations on endosperm development in barley (Cameron-Mills and von Wettstein 1980) and *Sorghum bicolor* (Taylor, Schüssler and Liebenberg 1985; Krishnan, White and Pueppke 1989). In barley, the proteins, following their synthesis in the ER, are transported to the vacuoles, where they form protein bodies; in sorghum, the proteins aggregate within the site of their synthesis. There is no evidence for the involvement of Golgi vesicles in the synthesis or transport of proteins in oat, barley, and sorghum endosperms. Other abundant proteins such as thionins, which are synthesized in the endosperm of wheat and barley, are detected around storage protein bodies as electron-dense spheroids; a role for ER in their synthesis has not been demonstrated (Carmona et al 1993). Although the endosperm of *Ricinus communis* (castor bean; Euphorbiaceae) stores abundant oil, there seems to be no constraints on the accumulation of proteins in the same cells. The protein body of the castor bean endosperm has a relatively complex structure, containing phytin globoids and protein crystalloids embedded in an amorphous proteinaceous matrix surrounded by a single membrane. The origin of protein bodies has been traced to small vacuoles that become filled with the protein (Tully and Beevers 1976; Gifford, Greenwood and Bewley 1982). Vacuolelike inclusions containing a particulate phase are also believed to be the site of oil body synthesis in the castor bean endosperm (Harwood et al 1971).

Protein Bodies of the Aleurone Cells. Unlike the cells of the starchy endosperm, the aleurone layer is made up of living cells. Although the aleurone layer is not a storage tissue like the starchy endosperm, the cells are filled with aleurone grains – spherosomes and protein granules of a chemical composition different from that of the storage proteins already considered. In wheat (Morrison, Kuo and O'Brien 1975), rice (Bechtel and Pomeranz 1977), barley (Jacobsen, Knox and Pyliotis 1971; Cameron-Mills and von Wettstein 1980; Olsen, Potter and Kalla 1992), and oat (Bechtel and Pomeranz 1981;

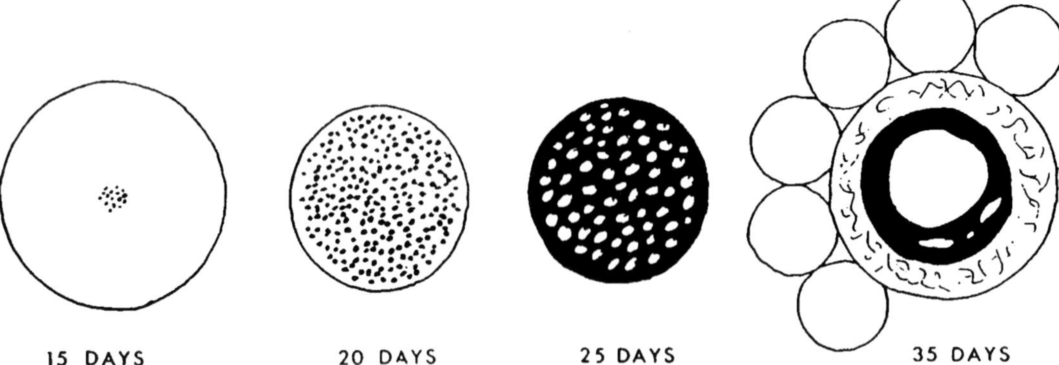

Figure 12.5. Diagrammatic representation of the formation of protein bodies in aleurone cells of the endosperm of *Zea mays*. Protein precipitates are first detected (15 days) in the vacuole, where they continue to accumulate (20 days). Later the proteins condense, forming a large, electron-transparent core (25 days). Spherosomes are apposed to the surface of the fully developed protein body (35 days). (From Kyle and Styles 1977. *Planta* 137:185–193. © Springer-Verlag, Berlin.)

Peterson, Saigo and Holy 1985), the aleurone grains are packaged as discrete particles of phytin, protein, and carbohydrate enclosed within a unit membrane. In electron micrographs, the phytin-containing area of the aleurone grain is recognized as one or more electron-dense globoids; another electron-dense area is constituted of a protein–carbohydrate complex that exists as a matrix or as inclusions. Lipid droplets or spherosomes surrounding the enclosing membrane like a ring complete the profile of the aleurone grain (Jones 1969; Jacobsen, Knox and Pyliotis 1971; Morrison, Kuo and O'Brien 1975). Buttrose (1963b) showed that aleurone grains of wheat originate in vacuoles as deposits of moderate electron density, with protein accumulation progressing with the maturity of the caryopsis. The assembly of aleurone grain in the oat endosperm differs significantly from that described in wheat. Although the vacuole remains as the site of its assembly, the aleurone is initiated by a sequential and progressive deposition of phytin and protein into the vacuole, with phytin deposition preceding that of protein. Compelling evidence in support of the view that the two types of deposits are targeted to different compartments is the early subdivision of the vacuole by internal membranes (Peterson, Saigo and Holy 1985). A similar vacuolar origin of the aleurone grain in the form of dense deposits (Figure 12.5) is the rule in endosperms of barley (Cameron-Mills and von Wettstein 1980) and maize (Kyle and Styles 1977).

The aleurone cells have provoked a great deal of research in connection with the mobilization of starch during germination of cereal grains. Much of the work done with barley has shown that α-amylase concerned with the digestion of starch in the endosperm is synthesized in the aleurone cells in response to GA coming from the embryo during the early hours of imbibition of the grain. Although the level of regulation of α-amylase by GA is attributable to changes in gene expression, it is not known what keeps the aleurone cells from expressing the genes in the endosperm of developing and mature grains.

Synthesis of carotenes and anthocyanins in the aleurone cells of maize endosperm has come under intense investigation from a developmental perspective, and several genes encoding enzymes for the production of these pigments have been cloned. The anthocyanin biosynthetic pathway in the aleurone cells has served as a model for studies in the genetic interactions between regulatory and structural genes. A discovery of great interest for investigations into the genetic and tissue-specific regulation of expression of genes involved in anthocyanin production is that such genes, introduced by microprojectile bombardment into aleurone cells of a mutant endosperm unable to synthesize the pigment, can faithfully program synthesis of the correct pigment exclusively in the target cells (Klein, Roth and Fromm 1989).

Metabolism of Storage Proteins. The changes in the pattern of accumulation of total storage proteins and their fractions during endosperm development have been followed in wheat, maize, barley, rice, rye, sorghum, and castor bean. In wheat, the production of most storage proteins occurs throughout the period of grain maturation; toward the latter phase of maturation, however, accumulation of low-molecular-mass proteins overtakes that of high-molecular-mass proteins (Flint, Ayers and Ries 1975; Pavlov et al 1975; Mecham, Fullington and Greene 1981; Singh and Shepherd 1987). Grain filling in maize is due to an increasing accumulation of globulins and zein from two weeks after pollination up to the mature stage, whereas albumins show a two-phase pattern, reaching a peak midway through endosperm development and declining further with time (Jones, Larkins and Tsai 1977; Dierks-Ventling 1981). In all essential respects, the pattern of accumulation of glutelin in the developing maize endosperm parallels that for zein (Ludevid et al 1984). In terms of their increasing molecular mass and other properties, the prolamins of barley endosperm are designated as B-, C-, and D-hordeins; B-hordeins are sulfur rich; C-hordeins are poor in sulfur; and high-molecular-mass proteins are D-hordeins. Based on genetic studies and protein characterization, three subfamilies, B1-, B3-, and γ-hordeins, have been recognized in the B-hordein group. During development, the relative amount of C-hordein is greatest in the young endosperm; it then decreases, remaining constant through grain maturity. In the B-hordein group, B1-hordein is found to increase throughout endosperm development whereas B3- and γ-hordeins behave like C-hordein. D-hordein appears to accumulate toward the final phase of endosperm development (Shewry et al 1979; Rahman, Shewry and Miflin 1982). Another protein of potential nutritional value found in the barley endosperm is thionin; accumulation of this protein follows a course almost parallel to that of D-hordein (Ponz et al 1983). The changes in the proportions of a 75-kDa γ-secalin and ω-secalin as the major and minor components, respectively, of the developing rye endosperm mirror to a large extent those for B1- and C-hordeins of barley (Shewry et al 1983). In rice, two groups of polypeptides (22–23 kDa and 37–39 kDa) composed of glutelin continue to increase coordinately during development, whereas the levels of the 57-kDa and 76-kDa polypeptides remain fairly constant throughout endosperm development (Figure 12.6). The relative proportion of glutelin to prolamins accumulated is high in 10-day-old grains but decreases in 25-day-old grains due to an increased accumulation of prolamins (Yamagata et al 1982; Li and Okita 1993). Total proteins of castor bean endosperm (Gifford, Greenwood and Bewley 1982) and prolamins of

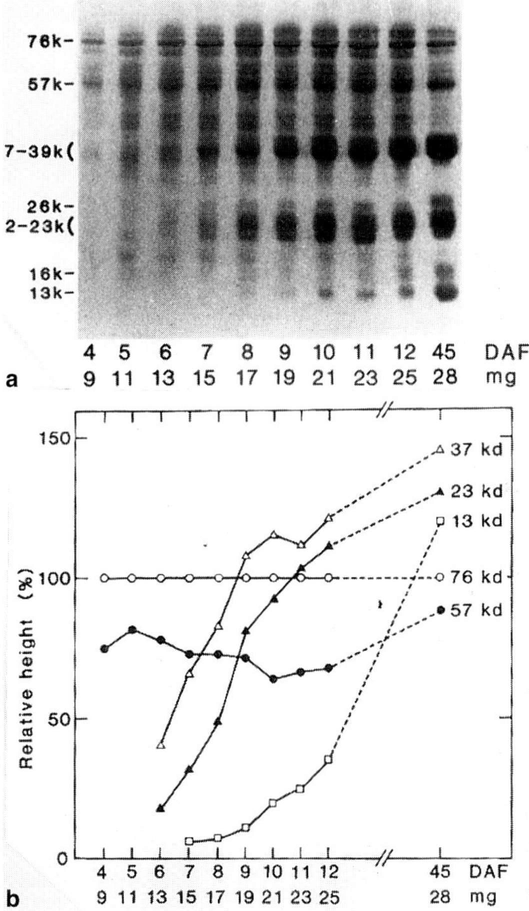

Figure 12.6. Protein composition of the endosperm of *Oryza sativa*. **(a)** SDS-PAGE of the extant proteins of the developing endosperm. Numbers at the bottom refer to the fresh weight of a grain (mg) at the indicated stage, days after flowering (DAF). The apparent molecular masses of the proteins are given in kDa on the left. **(b)** Changes in the relative amounts of the proteins given as peak heights of densitometric scanning of the gel. The peak height of the 76-kDa protein is designated as 100%. (From Yamagata et al 1982. *Plant Physiol.* 70:1094–1100. © American Society of Plant Physiologists.)

sorghum endosperm (Taylor, Schüssler and Liebenberg 1985) show the same patterns of accumulation characterized by an almost linear increase throughout development. Assay of the major storage albumins and oleosins of the endosperm of *Coriandrum sativum* (coriander; Apiaceae) using antibodies raised against similar proteins from carrot seeds showed that the albumins accumulate at a slightly earlier stage of development than the oleosins (Ross and Murphy 1992).

The dependence of continued accumulation of storage materials on protein synthesis has been demonstrated by the incorporation of labeled precursors in developing endosperms of rice (Luthe 1983; Zhu, Shen and Tang 1989; Li and Okita 1993), barley (Shewry et al 1979; Giese, Andersen and Doll 1983; Giese and Hejgaard 1984), wheat (Greene 1983), and oat (Luthe 1987). Because of the difficulty of harvesting endosperms at very early stages of development, these experiments have not generated a set of well-documented data on the pattern of synthesis of individual proteins during the developmental time span of the endosperm.

From the foregoing account it is not clear whether the inability of endosperm cells at early stages of development to accumulate storage reserves is due to a high turnover or due to lack of the components necessary for their synthesis. Along the same lines, it is also not determined whether differences exist in the turnover rates of the different storage proteins in the species studied.

5. ANALYSIS OF STORAGE PROTEIN GENE EXPRESSION

Studies on the expression of storage protein genes during endosperm ontogeny have considered in vitro protein synthesis by mRNA and polysomes, changes in mRNA accumulation and synthesis, and the expression of cloned genes. Since these topics are of vital importance in understanding the structure of endosperm storage protein genes and their expression in transgenic systems, they will be briefly reviewed here.

It was mentioned earlier that in vitro synthesis of zein is programmed by polysomes and mRNA isolated from the developing endosperm of maize; similar studies have reported the cell-free synthesis of globulins primed by polysomes of oat endosperm (Luthe and Peterson 1977; Fabijanski and Altosaar 1985; Fabijanski, Matlashewski and Altosaar 1985), hordeins and thionins primed by polysomes of barley endosperm (Brandt and Ingversen 1976; Matthews and Miflin 1980; Ponz et al 1983; Rahman et al 1984), prolamins primed by polysomes of sorghum endosperm (Johari, Mehta and Naik 1977), gliadin primed by polysomes and mRNA of wheat endosperm (Greene 1981; Okita and Greene 1982; Reeves, Krishnan and Okita 1986), secalin primed by mRNA of rye endosperm (Shewry et al 1983), and prolamins and glutelins primed by mRNA of rice endosperm (Yamagata and Tanaka 1986; Yamagata et al 1986). The storage proteins are generally detected in the translation products a few days before their detection in the endosperm cells in vivo (Figure. 12.7). Results of these studies in principle correlate well with those on protein metabolism in showing a coincidence between the periods of increased in vitro translational capacity of mRNA and increased protein synthesis or accumulation.

(i) Maize Storage Protein Gene Expression

As the major class of storage proteins of the maize endosperm, zein has been studied intensively with the focus on gene expression. Following the first detection of zein mRNAs 12 days after pollination, the accumulation of these mRNAs in the developing maize endosperm reaches a maximum between 18 and 22 days, and thereafter it declines somewhat until maturation. Using groups of cDNA clones corresponding to 10-, 15-, 19-, and 22-kDa zeins as probes, it was estimated that the efficiency of expression of members of different zein families varies somewhat; generally, transcription of the gene encoding 15-kDa zein is initiated earlier and is more active than that of 19- and 22-kDa zeins (Marks, Lindell and Larkins 1985a; Kirihara et al 1988). Patterns of accumulation of transcripts of individual zein genes also depend upon the growth conditions and genotypes of the plant (Langridge, Pintor-Toro and Feix 1982a, b; Kriz, Boston and Larkins 1987). At the cellular level, the noticeable characteristic of zein genes is their exclusive expression in the nucleus and cytoplasm of the endosperm cells, with the exception of the cells of the aleurone layer. As shown in Figure 12.8, a meaningful comparison of transcript localization allows one to recognize specific territories of expression for each zein mRNA within the developing endosperm (Dolfini et al 1992). Zein gene transcripts are also expressed in suspension cultures derived from the endosperm of maize, but expression occurs at a reduced level as compared to that of the endosperm of developing grains (Manzocchi, Bianchi and Viotti 1989; Ueda and Messing 1991). Transcripts of the much less investigated glutelin genes show a rate of accumulation in the developing endosperm similar to that of 19- and 22-kDa zeins (Pérez-Grau et al 1986).

The apparent lack of coordination in the expression of the individual zein genes raises the question of their control during endosperm development. Determination of the levels of the different zein mRNAs in run-off transcription assays from isolated endosperm nuclei of different ages has provided evidence that the levels of the two α-zein transcripts are generally higher than those of genes

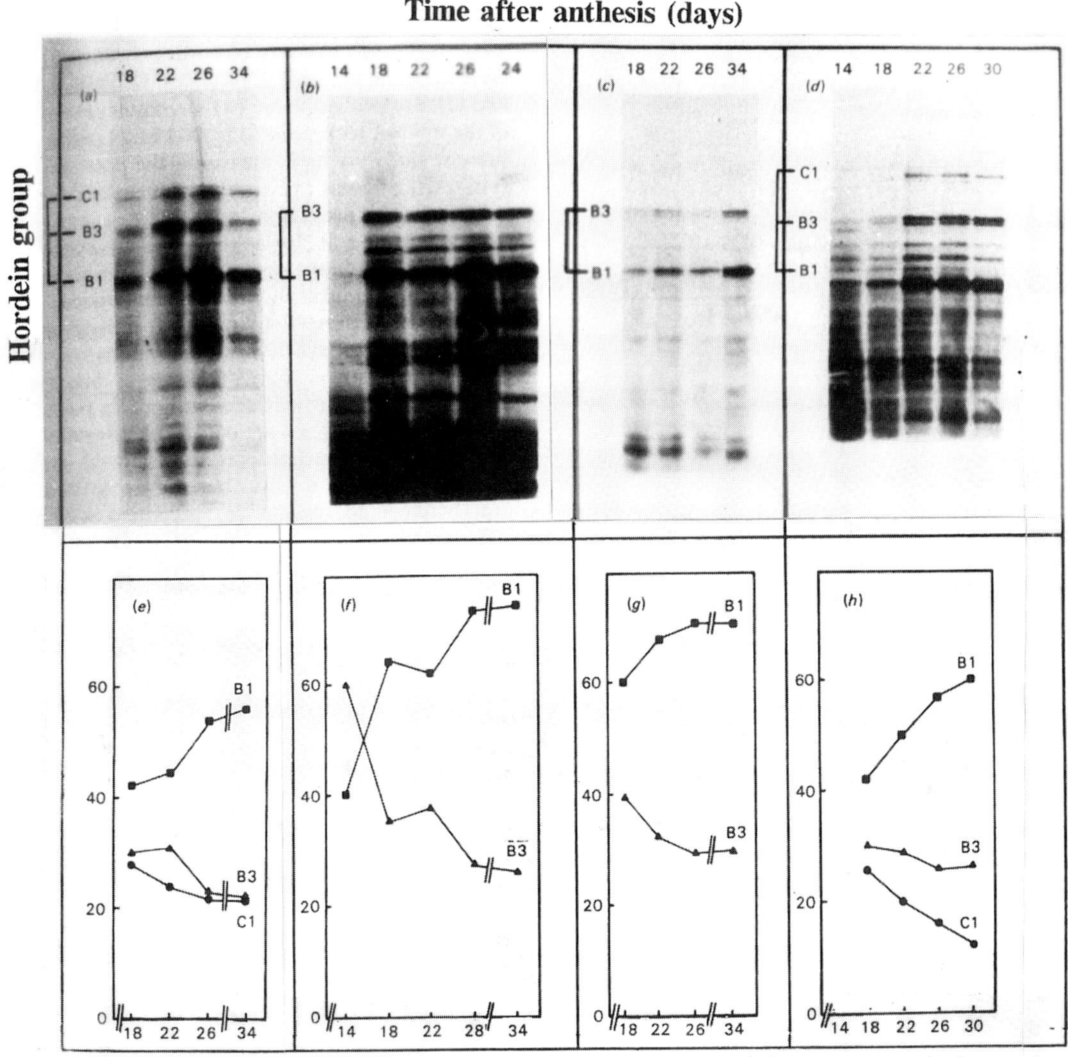

Figure 12.7. In vitro synthesis of hordeins by RNA isolated from the endosperm of *Hordeum vulgare*. Top panel: SDS-PAGE of proteins translated from membrane-bound polysomes **(a)**, polysomal poly(A)-RNA **(b)**, polysomal RNA **(c)**, and total RNA **(d)** at different days after anthesis. Bottom panel: Densitometric determination of the relative incorporation of the labeled amino acid into proteins; **(e–h)** correspond, respectively, to (a–d). (From Rahman et al 1984.)

encoding β-, γ-, and δ-zeins during all developmental periods studied. Among the two α-zeins, the one encoding the 22-kDa α-zein was transcribed more actively than that encoding the 19-kDa protein. Since the high transcriptional activity of the 22-kDa α-zein gene is not predicted by the low levels of its mRNA accumulation on polysomes, both transcriptional and posttranscriptional regulation of the zein genes appear to be operating during endosperm development (Kodrzycki, Boston and Larkins 1989).

Preferential accumulation of 10-kDa zein associated with the presence of a high level of its transcripts occurs in the endosperm of an inbred line of maize, BSSS-53, whereas an underexpression of this protein and its transcripts is seen in another inbred line, Mo-17 (Kirihara et al 1988; Schickler, Benner and Messing 1993). Evidence from analyses

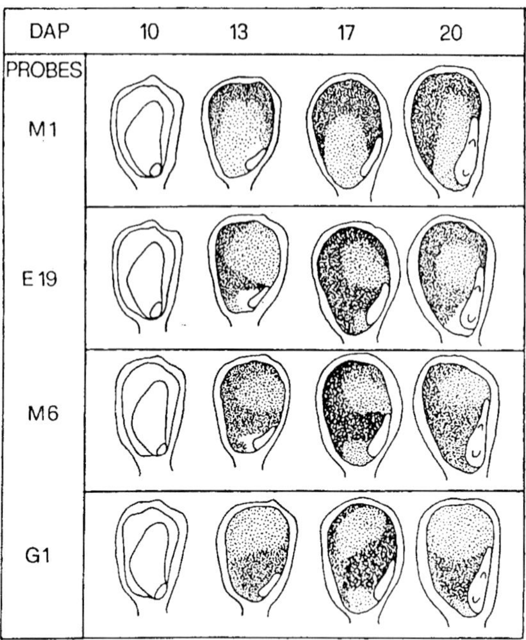

Figure 12.8. A diagrammatic representation of the pattern of accumulation of transcripts of zein and glutelin genes during the development of the endosperm in *Zea mays*. The diagram is based on in situ hybridization using probes for zein (M1, M6, and E19) and glutelin (G1) sequences. DAP, days after pollination. (From S. F. Dolfini et al 1992. Spatial regulation in the expression of structural and regulatory storage-protein genes in *Zea mays* endosperm. *Devel. Genet.* 13:264–276. © 1992. Reprinted by permission of Wiley-Liss Inc., a subsidiary of John Wiley & Sons Inc.)

of linkage relationships between the high mRNA levels and the BSSS-53 allele for the 10-kDa zein structural gene locus seems to indicate that a post-transcriptional regulation of the mRNA level by a *trans*-acting factor accounts for the increased accumulation of 10-kDa zein in the inbred line BSSS-53; a mechanism imposed by the lack of one or more *trans*-acting factors might account for the underexpression phenotype in Mo-17 (Cruz-Alvarez, Kirihara and Messing 1991; Schickler, Benner and Messing 1993; Chaudhuri and Messing 1994). Tissue-specific *trans*-acting factors also seem to determine the transcription of the 19-kDa and 15-kDa zein genes and the 28-kDa glutelin gene by binding to their 5'-flanking regions. Five sites in the promoter region of the 19-kDa zein gene capable of binding to a crude protein fraction of maize endosperm nuclei include a 22-bp region containing a 15-bp-long consensus sequence conserved in virtually all zein genes investigated (Maier et al 1987, 1988; So and Larkins 1991; Ponte et al 1994). The action of *trans*-acting factors carries the implication that they function to modulate transcript stability in some ways.

In considering the relationship between the transcriptional activity of zein genes and the accumulation of protein products in the endosperm, the occurrence of mutations credited with reducing zein synthesis needs to be taken into account. Two of these mutations characterized initially were *op-2* and *fl-2*, both of which cause a profound reduction in the synthesis of zein polypeptides while permitting an increase in the high-lysine-containing, nutritionally superior nonzein proteins (Mertz, Bates and Nelson 1964; Nelson, Mertz and Bates 1965). In the *op-2* mutant, depending upon the particular maize strain background, there is a partial to an almost complete inhibition of synthesis of 22-kDa zein at the same time that the levels of the lysine-rich protein and elongation factor-1α (EF-1α) are significantly increased (Lee et al 1976; Habben, Kirleis and Larkins 1993; Habben et al 1995). The endosperm of this mutant begins to synthesize zein several days later and ceases activity earlier than the wild-type endosperm (Murphy and Dalby 1971). The impact of EF-1α is that, as a lysine-rich protein that binds aminoacyl-tRNAs to ribosomes, it could be expected to contribute significantly to the increased lysine content. Kernels of the *defective endosperm-B30* (*de*-B30*) mutant also show a preferential reduction in the accumulation of 22-kDa zein, whereas another mutation, *op-7*, causes a greater reduction in the synthesis of 19-kDa than in the synthesis of 22-kDa zein. Other mutations, such as *op-6*, *fl-2*, and *mucronate* (*mc*), suppress the synthesis of all zeins to the same extent (Lee et al 1976; Jones 1978; di Fonzo et al 1979; Salamini et al 1983; Soave and Salamini 1984). A newly characterized mutant, *op-15*, shows a large reduction in the accumulation of 27-kDa zein, but contains high concentrations of 19- and 22-kDa zeins (Dannenhoffer et al 1995). Effects of the mutation extend to the level of the structure of the protein bodies and the distribution of the various classes of zeins in them. For example, the disorganized protein bodies with clumps of β- and γ-zeins interspersed within a mass of α-zein found in the endosperm cells of the *fl-2* mutant provide a striking visual contrast to the organized organelle structure in the cells of the wild-type endosperm (Lending and Larkins 1992).

The complete mechanism by which the mutations cause a reduction in zein synthesis is yet to be determined. Convincing evidence that the mutations affect the regulatory functions of genes rather than their organization was provided by restriction analysis of genomic DNA of *op-2*, *op-7*, and *fl-2* mutants with cloned cDNA probes for zein genes; this work showed that the coding sequences of the mutant genes had no deletions or structural alterations that might affect zein accumulation (Burr and Burr 1982). The patterns of accumulation of specific zein mRNAs in the mutant endosperms are relatively well described in several studies that show that the reduction of zein proteins is correlated with reduced amounts of the corresponding mRNAs (Pedersen et al 1980; Burr and Burr 1982; Langridge, Pintor-Toro and Feix 1982a; Marks, Lindell and Larkins 1985a; Kodrzycki, Boston and Larkins 1989; Lopes et al 1994; Dannenhoffer et al 1995; Neto et al 1995). In particular, during development of the *op-2* endosperm, none of the zein mRNAs can be detected at 12 days, as they would be in the normal endosperm. The mRNAs for 19-kDa zeins attain relatively normal levels by 28 days. Since the mRNAs arrive at that level slowly, it is clear that the mutation delays the expression of the 19-kDa zein gene (Figure 12.9). The appearance of 15-kDa zein mRNAs is also delayed and the transcript level is greatly reduced in the mature endosperm. Although the level of 22-kDa zein polypeptides is substantially reduced by the *op-2* mutation, deficiency in certain subgroups of 22-kDa zein transcripts, rather than in the whole range of transcripts that encode this protein, appears to cause the reduction (Burr and Burr 1982; Marks, Lindell and Larkins 1985a). This explanation of the effects of the *op-2* mutation on zein protein mRNA synthesis does not, however, preclude additional levels of regulation. In vitro culture studies have shown that the accumulation of 22-kDa zein and its mRNAs in the endosperm of the *op-2* mutant is favored when an excess of nitrogen is supplied in the medium (Balconi et al 1993). Thus, there appears to be a correlation between the level of nitrogen supply and the phenotypic effect of the mutation.

Two groups of investigators have cloned the *OP-2* gene by transposon tagging with the help of the mobile elements *suppressor mutator* (*spm*) and *activator* (*ac*). As expected, the gene is expressed in the wild-type endosperm but not in nonendospermic tissues or in the endosperm of the *op-2* mutant (Schmidt, Burr and Burr 1987; Motto et al 1988; Gallusci, Salamini and Thompson 1994). The theoretical translation of cDNA clones suggests that the gene encodes a transcription factor that contains a leucine zipper DNA-binding protein known as the bZIP domain (Hartings et al 1989; Schmidt et al 1990). The endosperm-specific expression of the *OP-2* gene correlates well with the expression of

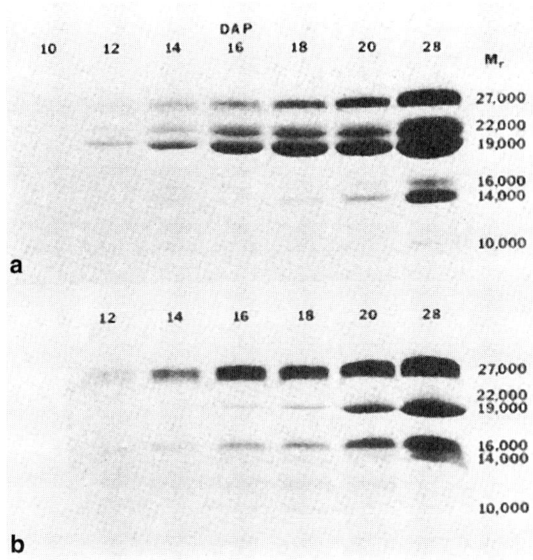

Figure 12.9. SDS-PAGE of zeins extracted from wild-type **(a)** and *op-2* mutant **(b)** of *Zea mays* to show the developmental regulation of zein gene expression. Developmental stages are indicated at the top as days after pollination (DAP). Molecular masses of the zein fractions are shown on the right. (From Kodrzycki, Boston and Larkins 1989. *Plant Cell* 1:105–114. © American Society of Plant Physiologists. Photograph supplied by Dr. B. A. Larkins.)

the OP-2 protein in the nuclei of maize endosperm cells (Varagona, Schmidt and Raikhel 1991).

A view advanced to explain the molecular basis for zein gene expression is that OP-2 protein activates zein gene transcription by binding to promoters. Some support for this idea has come from the demonstration that the protein product of the *OP-2* gene expressed in *Escherichia coli* can bind in vitro to upstream regions of a zein gene. Similarly, the OP-2 protein transiently expressed in the protoplasts of cultured maize endosperm cells or in the cells of the intact maize endosperm can transactivate the 22-kDa zein gene promoter (Schmidt et al 1990; Ueda et al 1992; Unger et al 1993). Conversely, the protein of an *op-2* mutant causing a disruption in zein gene expression is unable to bind to the same zein gene promoter (Aukerman et al 1991). To account for these complementary results, a 20-nucleotide target site has been identified in the promoter of the 22-kDa zein gene (Schmidt et al 1992). Because this sequence is not present in the promoters of other zein genes, this is prima facie evidence to explain the differential effect of the *op-2* mutation on the 22-kDa zein. Unraveling the ramifications of the mutation on zein gene expression is likely to assume complex proportions as other transcriptional regulatory proteins are identified (Yunes et al 1994; Holdsworth et al 1995). Another bZIP protein that might play a role in the regulation of the 22-kDa zein gene is the product of a maize endosperm cDNA clone (OHP-1) that binds to the *OP-2* target site in the promoter region of the 22-kDa zein gene (Pysh, Aukerman and Schmidt 1993). Considering the fact that the *OP-2* gene affects the expression of all subunits of zein, a finding with regulatory implications for the role of OP-2 protein is that this protein transactivates the promoters of β-zein of maize and a β-zein homologue from the related member of Poaceae, *Coix lacryma-jobi* (Neto et al 1995). Another relevant observation is that mutation in the *OP-2* gene down-regulates the accumulation of cytoplasmic pyruvate orthophosphate dikinase (PPDK) in the endosperm. Since PPDK is an enzyme involved in C4-photosynthesis, this finding might provide a meeting point for the roles of the *OP-2* gene in coordinating the expression of zein genes and genes for carbon partitioning (Maddaloni et al 1996).

The tide of reports of the isolation of transcription factors from the endosperm has continued to rise. Characterization of a salt-soluble 32-kDa protein (B-32) from the endosperm of wild-type maize and its absence in the *op-2* and *op-6* endosperms have led to the highly interesting idea that B-32 protein plays a role in the control of zein synthesis, perhaps as a transcription factor. The protein is coordinately expressed with zein during the development of the endosperm but is localized in the soluble cytoplasm, rather than in protein bodies. It has been speculated that B-32 protein is responsible for regulating zein gene expression in an indirect way. This view assumes that B-32 protein is encoded by the *OP-6* gene, which in turn regulates zein synthesis, and that the *OP-2* gene activates the *OP-6* gene (Soave et al 1981; di Fonzo et al 1986). It is, however, doubtful that B-32 protein is a product of the *OP-6* locus, and its abundance and location in the cytoplasm would argue against a direct interaction with zein or zein genes as a regulatory protein. The structure of the gene encoding B-32 protein and the deduced amino acid sequence have not given any indication of the presumed relationship of the protein to the *OP-6* gene or its translation product (di Fonzo et al 1988; Hartings et al 1990; Bass et al 1992). According to Lohmer et al (1991), the *B-32* gene is transactivated in the presence of the *OP-2* gene product, indicating that the interaction between the *OP-2* and *B-32* genes is at the transcriptional level and is not indirect through the activation of the *OP-6* gene. Admittedly, B-32 protein has received a good deal of attention as a regulator of zein gene expression, but its precise function remains unresolved.

Another contribution to our understanding of the action of *fl-2, de*-B30,* and *mc* mutations in maize showed that endosperms carrying these mutations overproduce a 70-kDa water-soluble protein (B-70) associated with both protein bodies and the rough ER (Galante et al 1983). Dramatic increases in the levels of both B-70 protein and its corresponding mRNA occur in the endosperms of these mutants between 14 and 20 days after pollination. This is followed by a decline in mRNA levels to the vanishing point, whereas B-70 protein levels remain high in the morphologically perturbed protein bodies (Boston et al 1991; Zhang and Boston, 1992). It appears that B-70 is a cytosolic chaperone-like protein with affinities to BiP and HSP-70 (Fontes et al 1991; Marocco et al 1991). Another candidate for a regulatory function in zein gene expression is an unusual 24-kDa protein associated with the zein fraction of the endosperm of the *fl-2* mutant (Lopes et al 1994). Results from the cloning of the gene for the 24-kDa protein and from its deduced amino acid sequence indicate that the gene is expressed as a 22-kDa α-zein with a defective signal peptide (Coleman et al 1995). Irrespective of the nature of the B-70 and 24-kDa proteins, there is circumstantial evidence to suggest they are involved in some ways in the disruption of zein processing and deposition in the endosperms of the mutants.

The phenotypic effect of the *op-2* mutation typically results in the formation of a soft, floury endosperm. A significant development affecting the phenotype of the *op-2* mutant is the modification of all or parts of the endosperm to a vitreous or glassy consistency. Genes that condition this phenotype are known as *OP-2 MODIFIERS*; the same genes are also active in changing the phenotype of the endosperm in other mutants such as *fl-2*. Wallace et al (1990) showed that the principal change in the endosperm storage protein of the modified *op-2* mutant is a two- to fourfold increase in the γ-zein content without any appreciable effect on the other classes of zeins. A central question in analyzing the effects of the various modifier genes is whether there are differences in the fundamental mechanisms of gene transcription or posttranslational events affecting protein synthesis. Experiments designed to answer this question have shown that opaque modifiers confer an increase in the level of γ-zein mRNAs; since this occurs in the absence of any increase in the number of gene copies, it appears that the increased gene expression is due to enhanced mRNA transcription or stability (Geetha et al 1991; Lopes and Larkins 1991). However, there is only a marginal difference between the transcription rates of the γ-zein gene in the *op-2* mutant and in the modified *op-2* genotypes. This has ruled out a direct role for the modifier genes in the transcription of the γ-zein gene and suggests a posttranscriptional role for them (Or, Boyer and Larkins 1993).

(ii) Barley Storage Protein Gene Expression

Some observations support the view that the amounts of different hordeins synthesized by the developing barley endosperm are determined by the abundance of corresponding mRNAs present, with little room for translational and posttranslational controls. Quantification of the relative amounts of B1-, B3-, and C-hordeins synthesized in vitro by mRNA preparations from developing endosperm showed that the amount of B1-hordein increased, and that of B3- and C-hordeins decreased, with age. The population of the corresponding mRNA species determined by dot–blot hybridization using cDNA clones also showed similar changes during endosperm development (Rahman et al 1984). It must be kept in mind that this kind of correlative evidence can never be considered complete because it is difficult to ascertain whether the amount of mRNA present at any given time is due to gene transcription or mRNA turnover. However, in vitro transcription rates of all hordein genes are more or less the same despite the fact that the steady-state levels of some hordein mRNAs are low; this indicates that the high levels of hordein mRNAs are due to posttranscriptional regulations (Sørensen, Cameron-Mills and Brandt 1989).

The use of cDNA clones isolated from aleurone cells of the barley endosperm has yielded vital information on the time course and pattern of accumulation of aleurone mRNAs. Transcripts of an aleurone-specific cDNA clone (*B11E*) and its genomic clone (*LTP-2*) show a nearly 200-fold increase in abundance exclusively in the aleurone cells beginning in the early-stage endosperm and extending to the stage of peak accumulation. During desiccation of the grain, transcript levels decrease; they disappear completely from the aleurone cells of the mature grain (Figure 12.10). In contrast, mRNAs complementary to other cDNA clones expressed in both the aleurone and the embryo are present more or less uniformly in the aleurone throughout grain development and decrease coincidentally with those of *B11E* (Jakobsen et al 1989; Olsen, Jakobsen and Schmelzer 1990; Liu et al 1992; Aalen et al 1994; Kalla et al 1994; Aalen 1995). Transcripts of another aleurone-specific recombinant (*LTP-1*) also accumulate characteristically in the aleurone cells during late grain development (Skriver et al 1992). Steady-state levels of mRNAs that encode a protein with some sequence homology to LEA proteins begin to accumulate in the aleurone and in the embryo midway through development and remain high in the aleurone cells of the mature endosperm (Hong, Barg and Ho 1992). Genes for chitinase isolated from barley endosperm display either aleurone-specific or both aleurone- and starchy endosperm–specific patterns of expression. This protein is generally reckoned as part of the grain's defense arsenal against invading pathogens (Swegle, Kramer and Muthukrishnan 1992; Leah et al 1994).

Mutants impaired in the synthesis and deposition of hordeins have been identified in cultivated varieties of barley. The mutant *Risø-56* suffers a 30% reduction in overall hordein accumulation and, in particular, the amount of B-hordein is reduced by about 75%. This reduction has been traced to the absence of B-hordein mRNA in the mutant endosperm due to a deletion covering most of the hordein-2 locus (Kreis et al 1983). In another mutant, *Risø-1508*, the amounts of B- and C-hordeins are reduced to about 20% and 7%, respectively, of those present in the parent line at the same time as the amount of D-hordein is increased fourfold. The changed amounts of B-, C-, and D-hordeins in this mutant are accompanied by reduced levels of B- and C-hordein mRNAs and a twofold increase in the amount of D-hordein mRNA, and by a dramatic reduction in the in vitro transcription activity of the genes encoding B- and C-hordeins (Thompson and Bartels 1983; Kreis et al 1984; Cameron-Mills and Brandt 1988; Sørensen, Cameron-Mills and Brandt 1989). Thus, the primary lesion caused by the mutations appears to be at the transcriptional level. The catastrophic effects of the mutation in *Risø-1508* have been further traced to transcriptional inactivation of the genes by methylation, whereas hypomethylation of the genes is correlated with their specific expression in the endosperm of the parent (Sørensen 1992).

(iii) Wheat Storage Protein Gene Expression

Storage protein mRNAs of the developing wheat endosperm are attached to both membrane-bound and free polysomes, with the former making up the major component of the storage protein

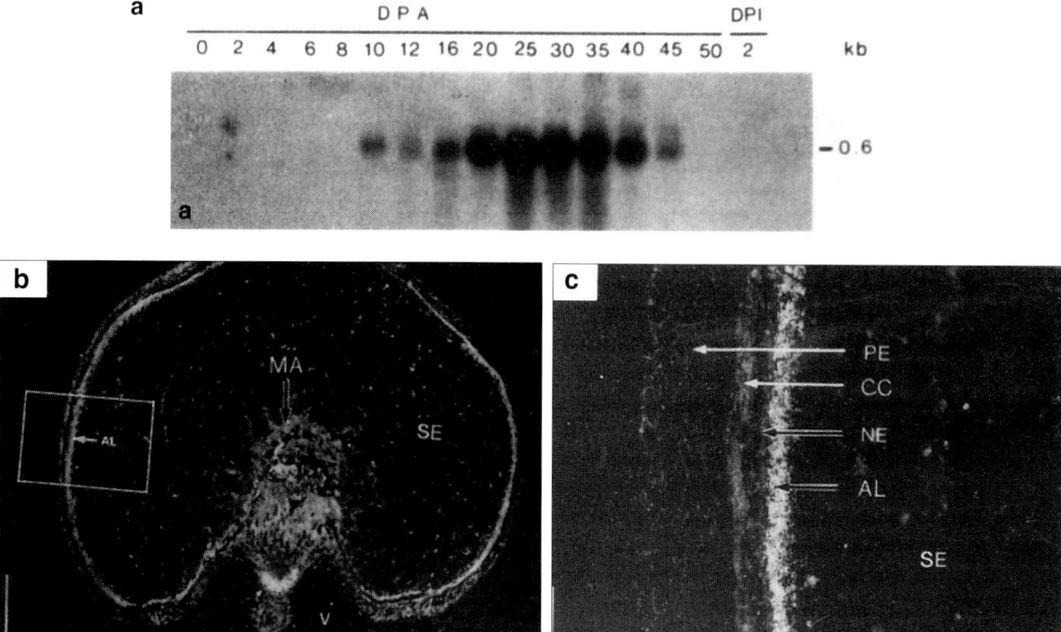

Figure 12.10. Temporal and spatial patterns of accumulation of transcripts of aleurone layer–specific genes during the development of the endosperm in *Hordeum vulgare*. **(a)** Northern blot hybridization showing steady-state levels of *B11E* mRNA in unfertilized ovules (0 days postanthesis, DPA), deembryonated grains (2–10 DPA), isolated aleurone–pericarp (12–45 DPA), mature aleurone (50 DPA), and mature aleurone imbibed for two days (2 days postimbibition, DPI). (From Olsen, Jakobsen and Schmelzer 1990. *Planta* 181:462–466. © Springer-Verlag, Berlin.) **(b, c)** In situ hybridization of a section of the grain with labeled *LTP-2* antisense strand. Dark-field micrographs showing hybridization of the probe to the peripheral aleurone cells, but not to the starchy cells of the endosperm. A portion boxed in (b) is magnified in (c). AL, aleurone layer; CC, cross-cell layer; MA, modified aleurone cells; NE, nucellar epidermis; PE, pericarp; SE, starchy endosperm. Scale bars = 100 μm (b) and 250 μm (c). (From Kalla et al 1994.)

messages. Based on hybridization kinetics, it has been estimated that both types of polysomes are constituted only of superabundant and intermediate abundant classes of mRNAs. Although there is no evidence for selective amplification of gliadin and glutenin genes in the developing wheat endosperm, it appears that more than half of the translated mRNAs direct the synthesis of the gliadins (Pernollet and Vaillant 1984). The results of changing patterns in RNA synthesis and in levels of the transcripts of different subunits of gliadins and glutenins have been taken to indicate that the subunit genes are expressed coordinately and that the amounts of mRNAs for the proteins increase sharply until the mid-maturation phase of endosperm development and then decline. The major RNA transcripts that accumulate during the peak period of endosperm development encode the α/β- and γ-gliadins. The pattern of accumulation of mRNAs of triticin, a globulinlike storage protein that constitutes only about 5% of the total wheat endosperm protein, is very similar to that of gliadins. As no protein is detected in the developing endosperm before its homologous mRNA is synthesized, a transcriptional control mechanism appears to coordinate storage protein synthesis in the wheat endosperm (Tercé-Laforgue and Pernollet 1982; Okita 1984; Bartels and Thompson 1986; Reeves, Krishnan and Okita 1986; Singh et al 1993). These data are in general accord with those described for maize and barley endosperms.

In attempts to characterize the protein factors that regulate transcription of gliadin genes, Vellanoweth and Okita (1993) have identified six sequences in the promoter of a α/β-gliadin gene that interact with different nuclear proteins from the developing endosperm (Figure 12.11). When nuclear extracts from endosperms of different stages of development were assayed, activities of four of the nuclear factors were found to appear at early stage to midstage, reach a maximum at midstage, and then decline later during endosperm ontogeny. The temporal changes in the activities of nuclear factors are coincident with the periods of

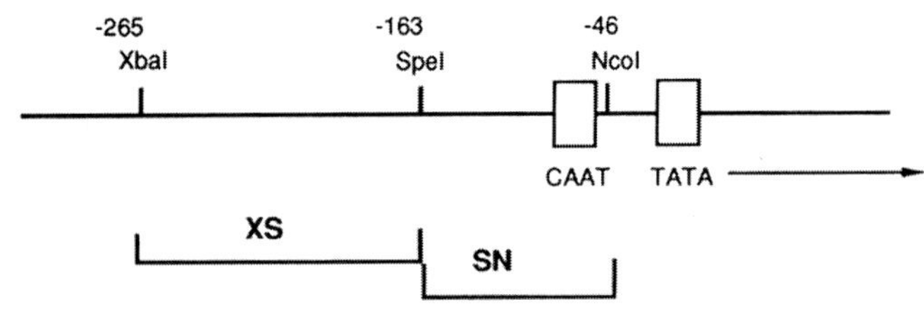

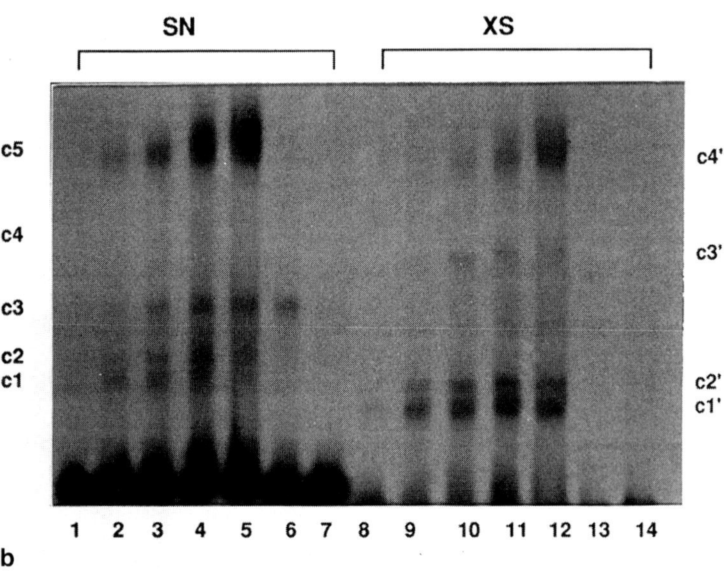

Figure 12.11. Interaction of nuclear proteins of *Triticum aestivum* with the α/β-gliadin promoter. **(a)** A schematic representation of the promoter region of the α/β-gliadin gene. Numbers refer to positions upstream of the transcriptional start site. TATA and CAAT elements are shown in boxes. The arrow indicates the direction of transcription. The probe and competitor fragments used in the experiment shown in (b) are indicated. **(b)** Electrophoretic mobility shift assay using restriction fragments SN (−163 to −46 bp) in lanes 1–7 and XS (−265 to −163 bp) in lanes 8–14. The amounts of crude nuclear proteins used were (μg): 0.5 (lanes 1, 8); 1 (lanes 2, 9); 2 (lanes 3, 10); 4 (lanes 4, 6, 7, 11, 13, 14); and 8 (lanes 5, 12). Lanes 6 and 13 and lanes 7 and 14 contained a 100-fold molar excess of unlabeled XS and SN DNA fragments, respectively. The five complexes formed with SN are indicated on the left as c1 to c5; those formed with XS are marked on the right as c1' to c4'. A sixth, low-mobility complex formed with SN is not seen in the figure. (From Vellanoweth and Okita 1993; photograph supplied by Dr. T. W. Okita.)

maximum accumulation and decrease of gliadin mRNA in the endosperm. Activity of a fifth nuclear factor was found at high levels only at mid-development; a sixth nuclear factor displayed a constant level of activity throughout endosperm development. Because the last-mentioned nuclear factor remains active even after transcription for gliadins has stopped, it has been surmised that some binding proteins may regulate gliadin expression by a negative *trans*-acting effect. An interaction between the putative *trans*-acting factors and sequences of the endosperm box (see p. 350) and GCN4-like (for General Control Nonderepressible) motifs that bind yeast regulatory proteins found in the low-molecular-mass glutenin has also been demonstrated by in vivo dimethylsulfate (DMS) footprinting. Not only is the binding highly specific to endosperm nuclear proteins, but the two types of proteins are found to bind sequentially, whereby the endosperm box motif becomes occupied during the early stages of development of the endosperm and the GCN4 sequence is coupled during the later stages (Hammond-Kosack, Holdsworth and Bevan 1993). From these results, a good case can be made that specific sequences in the promoter regions of the gliadin and glutenin genes combine with endosperm nuclear factors to regulate their temporal and spatial expression during development.

(iv) Storage Protein Gene Expression in Rice and Oat

Most of the evidence linking storage protein gene expression with endosperm development in rice comes from studies of glutelin and prolamin genes, the former representing a significantly higher proportion of the total proteins than prolamins. What is the nature of the control imposed on the relative expression of these proteins? In the absence of significant differences in the steady-state levels and transcriptional rates of glutelin and prolamin genes in the developing endosperm, it has been argued that differential recruitment of the respective mRNAs into polysomes might explain the differences in protein accumulation (Kim, Li and Okita 1993). In a similar way, there is a predominance of globulin over avenin in the endosperm of oat; yet, the amount of avenin mRNA is equal to or greater than the amount of globulin mRNA present during most periods of endosperm development (Chesnut et al 1989). Given the uncertainties about the involvement of nontranscriptional processes in the synthesis of proteins, the reasons for the translational efficiencies of these genes deserve further study.

The possibility that expression of glutelin and prolamin genes during rice endosperm development is noncoordinately regulated arises out of experiments on the transcriptional activities and transcript levels of several cDNA clones of these gene families. Comparative patterns of expression of 15.2-kDa prolamin messages belonging to two different homology classes showed that although both mRNAs increase during the 10 to 25 days following anthesis, there is a three- to fourfold increase in the abundance of one message during the early phase of endosperm development (Kim and Okita 1988b; Kim, Li and Okita 1993). Like the 15.2-kDa prolamin genes, the gene for a 13-kDa prolamin continues to be expressed at a high level up to the late stage of endosperm development (Masumura et al 1990). The expression of glutelin genes belonging to three subfamilies (*GT-1*, *GT-2*, and *GT-3*) shows that although transcripts of the homologous *GT-1*/*GT-2* genes steadily accumulate during endosperm development beginning 5 days after anthesis, those of *GT-3* accumulate maximally during the 5 to 10 days after anthesis, followed by a steady decline to imperceptible levels (Okita et al 1989; Takaiwa and Oono 1991; Takaiwa et al 1991; Kim, Li and Okita 1993). At different times during rice grain development, the glutelin gene is expressed not only in the endosperm but also in the embryo and pericarp (Ramachandran and Raghavan 1990). Izawa et al (1994) have characterized a bZIP transcription factor from the rice endosperm (RITA-1) that shows an affinity to ACGT motifs. The expression of the *RITA-1* gene in the aleurone and starchy endosperm cells of transgenic rice grain implies that this transcription factor might play a major role in the control of storage protein gene expression in the endosperm.

6. CLONING AND CHARACTERIZATION OF ENDOSPERM STORAGE PROTEIN GENES

Isolation and characterization of the relevant genes are necessary to gain insight into the molecular mechanisms of storage protein synthesis in the endosperm. The structure of the genes encoding many endosperm storage proteins is now known. Analyses have defined and delineated the differences among the different genes and have led to the recognition of a 7-nucleotide sequence (TGTAAAG), termed the –300 bp element, in the endosperm storage protein genes of several cereals. This sequence is frequently referred to as the "endosperm box" and is believed to program the endosperm-specific expression of the genes (Figure 12.12). Later progress in this area came with the dissection of promoter sequences of the genes, construction of deletion mutants, and introduction of chimeric genes into transgenic plants to follow the genes' spatial and temporal expression. This section presents an overview of recent findings pertaining to the various levels of characterization of endosperm storage protein genes.

(i) Zein Genes

Based on classical genetic approaches, nucleic acid hybridization analysis, and Southern blot hybridization, there have been several revisions of the estimated number of genes encoding zein. The consensus of the prevailing opinion is that α-zein is encoded by a large multigene family of 50–100 gene members, whereas β-, γ-, and δ-zeins are encoded by genes present in a few copies. Several cDNA and genomic clones encoding zeins have been characterized to provide a clear picture of the molecular biology of the gene. Genes characterized during the early period of zein gene cloning

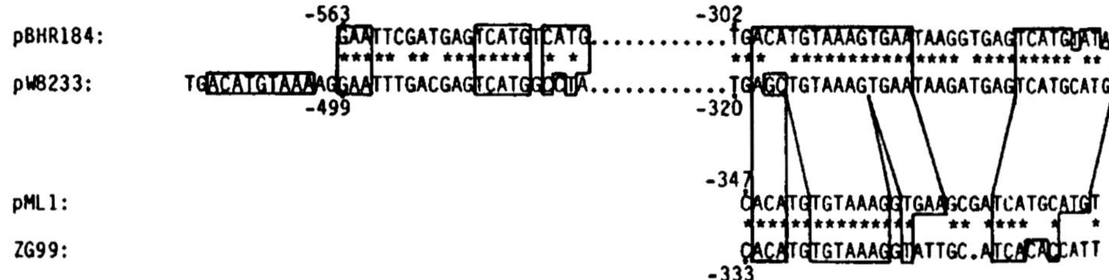

Figure 12.12. Conserved distal sequences of the –300 bp endosperm box in hordein (pBHR184), α-gliadin (pW8233), 21-kDa zein (pML1), and 19-kDa zein (ZG99) genes. The sequence of the repeat of the –300 bp element in the B1 hordein gene is incomplete because the repeat contains one of the two EcoR1 sites that define the boundaries of the sequenced fragment. Asterisks indicate identical residues in each pair of aligned sequences; those parts of the element that are common to the prolamin and zein genes are boxed. Numbering is relative to the ATG codon. (From Forde et al 1985a. *Nucl. Acids Res.* 13:7327–7339, by permission of Oxford University Press.)

belonged to the two major classes of α-zeins (19 kDa and 22 kDa). A list of cDNA or genomic clones of the two α-zeins includes those isolated by Weinand, Brüschke and Feix (1979), Geraghty et al (1981), Wienand, Langridge and Feix (1981), Burr et al (1982), Geraghty, Messing and Rubenstein (1982), Marks and Larkins (1982), Hu et al (1982), Spena, Viotti and Pirrotta (1982), Pedersen et al (1982), Heidecker and Messing (1983), Marks, Lindell and Larkins (1985b), Kriz, Boston and Larkins (1987), Wandelt and Feix (1989), and Liu and Rubenstein (1993). Clones encoding 15-kDa zein (Marks and Larkins 1982; Marks, Lindell and Larkins 1985b; Pedersen et al 1986), 27-kDa zein (Wang and Esen 1986; Das and Messing 1987), and 10-kDa zein (Kirihara, Petri and Messing 1988; Kirihara et al 1988), were isolated subsequently, completing the characterization of clones representing all zein subclasses. cDNA clones corresponding to some storage proteins of maize endosperm previously considered as zeins have been assigned to 28-kDa and 16-kDa glutelins because of their close homology in both nucleotide and amino acid sequences to gliadin and hordein (Prat et al 1985; Boronat et al 1986; Prat, Pérez-Grau and Puigdomènech 1987; Gallardo et al 1988; Reina et al 1990a, b).

Comments about the structure of the zein genes will be limited to general features that are important in their functioning. The zein genes belong to a rare class of eukaryotic genes that do not contain any introns, thus making their coding regions very long. All zein genes have repeated sequences, transcriptional regulatory sequences characterized by the presence of two or more promoters, as well as the –300 bp element and transcription initiation sites. The unusual promoter system may account for the high level of zein expression in the endosperm. Like typical eukaryotic genes, zene genes have consensus CAAT and TATA sequences in their 5'-noncoding region and one or more polyadenylation sites in the 3'-noncoding part. Differences between the different zein genes, which probably account for their operation, can be traced to the duplication of an original gene, the presence of repeated sequences, the insertion of sequences by duplication of small segments, and nucleotide changes. Based on the structure of individual genes, there is some interesting speculation that members of the zein multigene family have evolved from a common ancestral sequence. Protein structures of zeins deduced from cDNA clones show no homology to major storage proteins of other cereals. Among the different zeins, there is close similarity between the 19-kDa and 22-kDa zeins, as both contain a sequence of 20 amino acids repeated nine times. The amino acid repeats are not part of the sequence of the 15-kDa zein, which thus does not share homology with the 19-kDa and 22-kDa zeins. The mature 10-kDa has no homology to the other zeins characterized (Wilson and Larkins 1984; Heidecker and Messing 1986). Although the sequences and structure of the 28-kDa glutelins are different from those of zeins, the 5'- and 3'-flanking regions of the two groups of genes show considerable homology (Boronat et al 1986).

The transcriptional regulation of zein genes has been monitored by introducing individual genomic clones into heterologous tissues and examining their expression in the absence of other members of the multigene families. Transformed yeast cells used as a transcription system are able to recognize the promoter regions of 19- and 22-kDa zein genes and synthesize the appropriate transcripts (Langridge et al 1984). By integrating the gene into *Agrobacterium tumefaciens*, it was shown that the 5'-flanking regions of 15-kDa, 19-kDa, and 22-kDa zein genes contain sufficient information to accu-

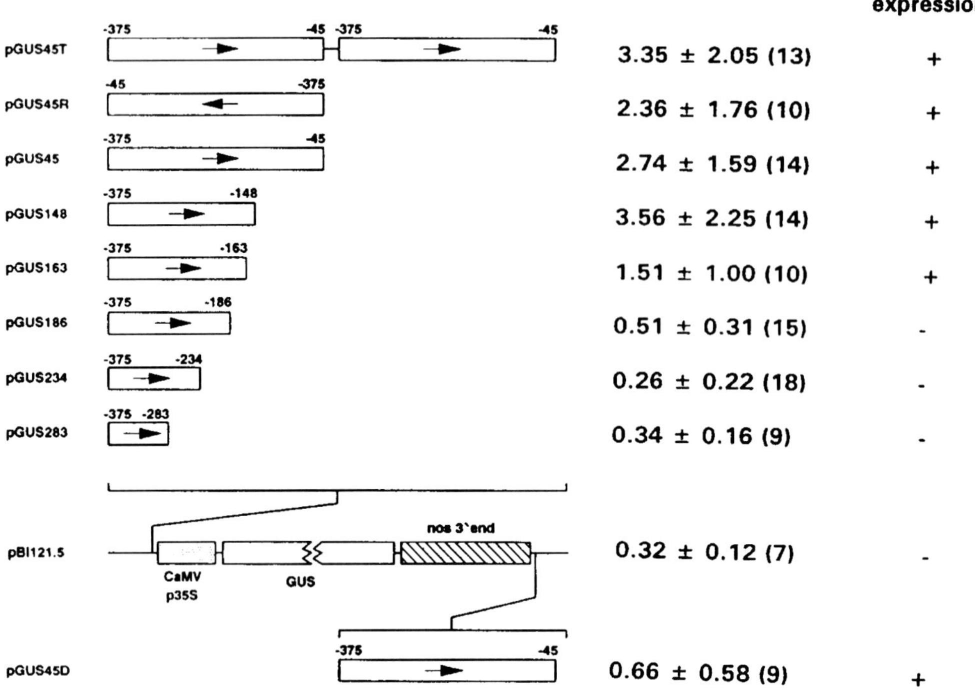

Figure 12.13. Deletion analysis of the promoter of a high-molecular-mass wheat glutenin gene in transgenic tobacco. The acceptor plasmid contained CaMV 35S promoter, the *GUS* gene, and the 3'-end of the *NOS* gene. The coordinates on the glutenin gene indicate the distance in bp of the end points from the initiation site for glutenin transcription. Polarity of the promoter fragments is shown by arrows. Data on the right indicate GUS activity (picomoles ± standard error) in the seeds of transgenic tobacco; numbers in parentheses are the number of plants used for GUS assay. Endosperm-specific expression is indicated by + or –. (From Thomas and Flavell 1990. *Plant Cell* 2:1171–1180. © American Society of Plant Physiologists.)

were found essential to potentiate endosperm-specific *GUS* gene expression in transgenic tobacco seeds (Figure 12.13) (Thomas and Flavell 1990). A functional analysis of the 5'-flanking region of a silent high-molecular-mass glutenin gene in transgenic tobacco has led to the proposal that single base changes in the –280 bp sequence immediately upstream of the transcription start site might result in the inactivation of the promoter. Sequences required for tissue and temporal specificity of the expressed high-molecular-mass glutenin gene in transgenic tobacco are also incorporated in the –280 bp region; since this includes only two core sequences of the conserved –300 bp element, there is some question about the essentiality of the latter for tissue-specific expression of the gene (Halford et al 1989).

In transient expression assays using protoplasts of *Nicotiana plumbaginifolia,* mammalian transcription factors of the leucine zipper class are found to stimulate transcriptional activity of promoters of the high-molecular-mass glutenin gene. This is apparently mediated by sequences found in the endosperm box of the glutenin promoter and suggests that such binding sites could act as *cis*-acting elements in regulating gene expression (Hilson et al 1990). The ability of *Escherichia coli* and yeast cells to express biologically active proteins has made possible an important advance in analyzing wheat storage protein gene expression (Bartels, Thompson and Rothstein 1985; Neill et al 1987; Galili 1989). Galili (1989) showed that the DNA sequence encoding the glutenin high-molecular-mass subunit cloned into a bacterial expression vector directs the synthesis of large amounts of the corresponding protein. Confirmation that the protein synthesized by the bacteria is identical to the native wheat protein subunit was provided by solubility characteristics. A practical consequence of this discovery is that it makes possible the analysis of the structure of the protein and the identification of its sequences that affect the quality of the dough.

(iv) Rice Glutelin and Other Genes

As indicated in the previous section, three subfamilies comprise the rice glutelin genes; each subfamily contains five to eight copies. Several cDNA clones and genomic clones of glutelin genes have been isolated and characterized (Takaiwa, Kikuchi and Oono 1986, 1987, 1989; Takaiwa et al 1987; Masumura et al 1989a; Okita et al 1989; Takaiwa and Oono 1991). Genes of all three classes contain three short introns precisely at the same positions in the coding regions. A comparison of genomic clones showed that the 5'-flanking regions of *GT-1* and *GT-2* are closely related, whereas *GT-3* has little or no homology upstream of –267 bp to either *GT-1* or *GT-2*. The derived sequence of the protein shows an amino terminus with features typical of a signal peptide. From molecular analyses of glutelin cDNA clones, sequence homology of the primary sequence of the gene to the 11S globulin gene of legume embryos has become apparent.

Analysis of rice prolamin cDNA clones encoding a 15.2-kDa protein revealed that they contain relatively long 5'-untranslated sequences of about 150–200 bp followed by coding regions of about 450 bp. The encoded proteins have a signal peptide but conspicuously lack repeated sequences (Kim and Okita 1988a, b). Prolamin cDNAs that encode 10-kDa and 13-kDa proteins show close similarity to the structure of cDNA clones that code for the 15.2-kDa protein (Masumura et al 1989b, 1990).

The oat globulin gene is a member of a small multigene family with a nucleotide sequence characterized by the presence of three introns and the lack of a –300 bp element. The protein product shares structural features with the 11S globulin of legume embryos (Walburg and Larkins 1986; Shotwell et al 1988, 1990; Schubert et al 1990). The nucleotide sequence of avenin, a member of a multigene family of about 25 genes, shows significant homology to the gliadins of wheat and the B-hordein of barley; the coding sequence lacks introns, and the promoter region has a –300 bp element (Chesnut et al 1989; Shotwell et al 1990). Genes for prolamins of rye (γ- and ω-secalin) (Kreis et al 1985; Hull et al 1991), sorghum (γ-kafirin) (DeRose et al 1989; de Freitas et al 1994), and *Coix lacryma-jobi* (coixin) (Ottoboni et al 1993) are also characterized by the lack of introns but they contain the –300 bp element. Both γ-kafirin and coixin show varying degrees of homology to different subunits of zein (de Barros et al 1991; Leite et al 1992; Ottoboni et al 1993).

Glutelin genes of rice contain functional promoters effective in directing transcription in isolated tobacco protoplasts and in transgenic tobacco. Three regions of the *GT-3* promoter, –945 to –726 bp, –346 to –263 bp, and –181 to +7 bp from the transcription start site, are required for the endosperm-specific expression and temporal regulation of chimeric genes in transgenic plants (Leisy et al 1989; Okita et al 1989; Zhao, Leisy and Okita 1994; Croissant-Sych and Okita 1996). A major difference between glutelin genes of the *GT-3* and *GT-1* classes is that in the 5'-flanking region of the latter, information required for both tissue specificity and temporal regulation is located in a single domain (between –441 and –237 bp) (Takaiwa, Oono and Kato 1991). A difference between the location of the promoter elements of the *GT-1* gene in heterologous and homologous hosts was also revealed when the gene was expressed in transgenic rice, suggesting limitations in the recognition of critical sequences by transcription factors in the heterologous host (Zheng et al 1993). The endosperm-specific expression of a rice prolamin gene in transgenic tobacco appears to be controlled by the upstream region from –680 to –18 bp (Zhou and Fan 1993). As detected by transient assay, endosperm-specific expression of the γ-kafirin gene is conserved in the grains of maize, sorghum, and *C. lacryma-jobi* and is caused by a –285 bp proximal region of the promoter (de Freitas et al 1994).

An aleurone-specific gene, *LTP-2*, from barley endosperm that encodes a nonspecific lipid transfer protein has been discussed earlier. Kalla et al (1994) have shown that expression of this gene is under strict temporal and spatial control in transgenic rice, directing transcription exclusively in the aleurone layer from shortly after the onset of aleurone cell differentiation to mid-maturity of the grain. At present, to account for the differential expression of this gene in the developing endosperm, it is assumed that the active promoter sequences contain *cis*-regulatory elements that bind to nuclear proteins. Although transcripts of the 12S oat globulin gene are found in both the embryo and the endosperm of transgenic tobacco, the protein product is localized exclusively in the endosperm. This suggests that posttranscriptional regulatory processes are involved in the expression of this gene in the heterologous host (Manteuffel and Panitz 1993; Schubert et al 1994).

As can be discerned from this account, there is a considerable industry for the isolation of various endosperm storage protein genes and their introduction into heterologous hosts. This work is fueled by the potential practical benefits that will the basal cell; during embryo formation, the first division of the terminal cell may be transverse, longitudinal, or, rarely, oblique. These facts have served as a basis for classifying different types of embryos in angiosperms. A widely used system of classification has identified two major groups, one in which the division of the terminal cell of the two-celled embryo is longitudinal, and the other in which the division wall is oriented in a transverse plane (Johansen 1950). Within these major groups, different embryo segmentation types are recognized and identified by the name of the family in which many examples of the type are found. However, it should be noted that in some families more than one type of embryo development, represented by one or two genera in each case, has also been observed. Accounts of the ontogeny of representatives of each segmentation type are given by Johansen (1950) and Maheshwari (1950), and so in this account ontogenetic information is passed over. In describing embryogenesis, the term "proembryo" is used to designate the early stages of development of the embryo; the initiation of cotyledons is considered as a good cut-off point for the end of the proembryo stage.

Longitudinal Division of the Terminal Cell. Depending upon the extent of the contribution of the basal cell of the two-celled proembryo to the formation of the organogenetic part of the mature embryo, Crucifer (or Onagrad) and Asterad types of embryo development have been recognized in the group exhibiting longitudinal division of the terminal cell. In the Crucifer type, exemplified by *Capsella bursa-pastoris* and *Arabidopsis*, the basal cell contributes very few derivatives to the formation of the organogenetic part of the embryo, which is thus generated almost entirely by the terminal cell. The cells derived from the basal cell form a short-lived, filamentous structure known as the suspensor; more will be said about the suspensor in a later section of this chapter. This type of embryogeny has also been described in plants belonging to Onagraceae, Bignoniaceae, Lamiaceae, Lythraceae, Papilionaceae, Ranunculaceae, Rutaceae, and Scrophulariaceae among the dicotyledons, and in Juncaceae and Liliaceae among the monocotyledons. In contrast, the embryo in the Asterad type is generated by the division of the terminal and basal cells of the two-celled proembryo. The description of embryogenesis in *Lactuca sativa* reveals a nice division of labor between the derivatives of the terminal cell and those of the basal cell in the formation of the embryo, the former giving rise to the cotyledons and plumule, and the latter to the root meristem, root cap, and hypocotyl; a small suspensor is also derived from the basal cell (Jones 1927). The Asterad type of embryo development or its variations have been described in members of Asteraceae, Geraniaceae, Lamiaceae, Oxalidaceae, Polygonaceae, Rosaceae, and Urticaceae (dicots), and Liliaceae and Poaceae (monocots) (Natesh and Rau 1984).

Transverse Division of the Terminal Cell. Three embryo developmental patterns (Solanad, Caryophyllad, and Chenopodiad) are recognized on the basis of the transverse division of the terminal cell and the participation of the division products of the terminal and basal cells of the two-celled proembryo in the formation of the mature embryo. The Solanad type, found in several species of *Nicotiana*, is characterized by the formation of the embryo from the terminal cell and by division of the basal cell to form a suspensor. Besides members of Solanaceae, members of Hydnoraceae, Linaceae, Papaveraceae, and Rubiaceae display the Solanad type of embryo formation. In the Caryophyllad type, the basal cell does not divide after it is cut off, and the organogenetic part of the embryo is derived exclusively from the division of the terminal cell. These characteristics are clearly evident during the embryogenesis of *Sagina procumbens* (Caryophyllaceae) (Souèges 1924). A suspensor is not a regular feature of the Caryophyllad type of embryos; if one is present, it is also derived from the terminal cell with contributions from the basal cell, as seen in *Saxifraga granulata* (Saxifragaceae) (Souèges 1936). The Caryophyllad type of embryo development has been described in plants included in Crassulaceae, Droseraceae, Fumariaceae, Holoragaceae, Portulacaceae, and Pyrolaceae (dicots), and Alismataceae, Araceae, Ruppiaceae, and Zannuchelliaceae (monocots). Another addition to the list is *Monotropa uniflora*, in which the basal cell collapses after it is cut off (Olson 1991). The Chenopodiad type has the distinction of having the division products of both the terminal and basal cells integrated into the organogenetic part of the embryo. Generally, the hypocotyl of the mature embryo has its origin in the basal cell, which also forms a short suspensor. In addition to being found in members of Chenopodiaceae, this type of embryo development is typical of plants included in Amaranthaceae and Polymoniaceae (Natesh and Rau 1984).

A type of embryo development in which the first division of the zygote itself is oblique or longitudinal (Piperad type) was slow to be recognized and was thought to be found only in a few families of dicotyledons, such as Piperaceae, Balanophoraceae, Dipsacaceae, and Loranthaceae. The genera and species assigned to this type are not many, and only occasionally has it been identified in monocots (Tohda 1974; Natesh and Rau 1984).

We see that descendants of the basal cell of the two-celled proembryo defy a neat packaging, because in some cases, part of this cell population becomes enmeshed in the organogenetic part of the embryo, the rest forming a suspensor, whereas in other cases, the basal cell population wholly forms the suspensor or simply wilts away. The general conclusion from these observations is that although the terminal cell is already determined as the progenitor of the embryo, the determinative state of the basal cell is at best described as being labile.

Embryo development in *Paeonia anomala, P. moutan,* and *P. wittmanniana,* first described by Yakovlev and Yoffe (1957), does not fit into any of the types just described, because the failure of wall formation during the first few rounds of division of the zygote results in embryogenesis that starts with numerous free nuclei suspended in a cellular bag. Later, the nuclei migrate toward the periphery of the cell, where wall formation is initiated to form a multicellular proembryo. A further remarkable change occurs as the marginal cells of the proembryo differentiate as embryo primordia. Although several primordia are born out of a proembryonal cluster, only one outlives the others and differentiates into a viable embryo. Independent investigations by other workers (Cave, Arnott and Cook 1961; Carniel 1967a) on additional species have essentially confirmed the unusual features of embryogenesis in this genus. A free nuclear phase is typical of embryogenesis in gymnosperms, and its occurrence in *Paeonia* appears to be the angiosperm equivalent of a gymnosperm feature.

(ii) Comparative Embryogenesis in Monocots and Dicots

The essence of embryogenesis is the establishment of tissue systems and primordial organs in the embryo. Obviously, a precise sequence of divisions of the zygote and proembryo, as well as a high degree of inductive interactions between the growing parts, bears heavily on the final form attained by the embryo. It is important to note that although mature embryos of dicots and monocots appear morphologically different, their early division sequences are basically similar. It is, therefore, expected that the modes of embryo development generally transcend the boundaries of dicots and monocots.

To understand how the zygote progresses from a single cell to a complex multicellular embryo in which fundamental tissues and organs are demarcated, embryogenesis of representatives of dicots and monocots will be considered. Any one of the many dicot and monocot embryos that have been studied over the years will illustrate the general principles of embryogenesis easily, but some plants have been favorites for this purpose. Embryogenesis in *Capsella bursa-pastoris* is almost a textbook example of the Crucifer type of embryogenesis in which the first division of the terminal cell of the two-celled proembryo is longitudinal (Figure 13.2). Each of the two cells thus formed again divides by longitudinal walls to yield a quadrant, followed by an octant stage by division of the four cells by transverse walls. Although the embryo is populated at this stage by a relatively small number of homogeneous cells, there are functional differences between the upper and lower tiers of cells. For example, the stem tip and cotyledons in the adult embryo are believed to be formed from the upper tier of cells, and the hypocotyl generated from the lower tier. In situ hybridization of embryos of *Brassica napus* with cloned genes for the storage proteins, cruciferin and napin, has shown that the determinative event established by the transverse division of the quadrant is reflected in the gene expression patterns in later stage embryos. In these embryos, cells derived from the lower tier accumulate high levels of transcripts whereas, except in the cotyledons, cells derived from the upper tier are devoid of transcripts (Fernandez, Turner and Crouch 1991). The point is important because the pattern of transcript accumulation does not appear to reflect tissue or cell types but follows the boundary set up by an earlier division. However, the developmental fate of a group of three or four layers of cells derived from the upper tier of the quadrant (designated as the epiphysis) and of the root apical meristem derived from the terminal cell of the suspensor (designated as the hypophysis) is expressed by their failure to accumulate storage protein transcripts. It appears that gene expression patterns in these groups of cells are set, not by the division boundaries, but by their biochemical and physiological characteristics.

The next round of division of cells of the octant embryo is periclinal and results in the formation of a globular embryo of 16 cells. That some covert changes may foreshadow the differential potencies of cells of the embryo at this early stage is shown by localization of transcripts of a Kunitz trypsin

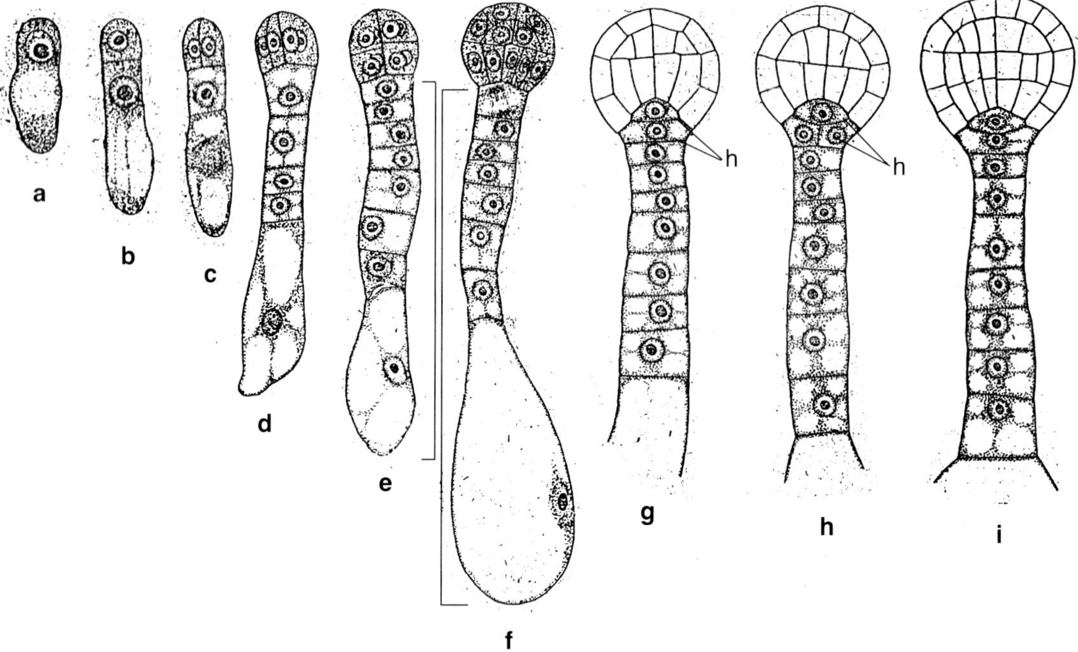

Figure 13.2. Stages in the division of the zygote and early development of the embryo and suspensor in *Capsella bursa-pastoris*. **(a)** Zygote. **(b)** Asymmetric division of the zygote. **(c)** Transverse division of the basal cell and longitudinal division of the terminal cell. **(d)** Quadrant stage embryo. **(e)** Octant stage embryo. **(f)** Early globular embryo. **(g–i)** Three stages of the globular embryo, showing the division of the hypophysis (h). Suspensor is outlined in the square bracket. The micropylar end is toward the bottom of the page. (From Souèges 1919.)

inhibitor gene (*KTI-3*) in the cells at the micropylar pole of globular embryos of soybean (Perez-Grau and Goldberg 1989). The expression of the *KTI-3* gene is one of the earliest indicators of polarity in the globular embryo. The globular embryo is phased into the heart-shaped stage when it attains bilateral symmetry due to a lateral expansion of the distal pole that gives rise to a pair of cotyledons. The elements of the future shoot apex, which remains essentially in a dormant state throughout the life of the embryo, are carved out from a few cells lying between the cotyledons. Based on the action of inhibitors of auxin transport on cultured embryos of *B. juncea* resulting in the formation of cotyledons as a fused collarlike structure around the apex, rather than as two discrete lateral outgrowths, a model proposed to account for the attainment of bilateral symmetry in embryos growing in vivo has invoked a requirement for polar auxin transport (Liu, Xu and Chua 1993b). Indeed, a small percentage of embryos of the *pin-formed* flower mutant (*pin-1.1*) of *Arabidopsis*, which has a defect in polar auxin transport, strikingly resemble treated embryos of *B. juncea* (Okada et al 1991). The patterns of growth displayed by isolated embryos of *Triticum aestivum* in the presence of auxin and inhibitors of auxin transport have supported a role for auxin in the establishment of bilateral symmetry in monocot embryos (Fischer and Neuhaus 1996).

The role of polar auxin transport in the maintenance of embryonic form was the focus of reports using the more easily accessible somatic embryos of carrot. Not only do polar auxin transport inhibitors block transitions in embryo development beginning as early as the globular stage, they also cause the formation of giant globular and oblong embryos (Schiavone and Cooke 1987). Microsurgical experiments on heart-shaped and torpedo-shaped somatic embryos have shed further light on the role of polar auxin supply and distribution by showing that the basal root half of the embryo grows more rapidly when cultured alone than when it is grafted back to the apical shoot half and cultured. However, when grafted somatic embryos are treated with a polar auxin transport inhibitor, the basal halves behave as if they have been cultured separately. Not surprisingly, treatment of isolated basal sections with IAA forced them to grow as if they were part of a graft union with the apical part (Schiavone 1988). These obser-

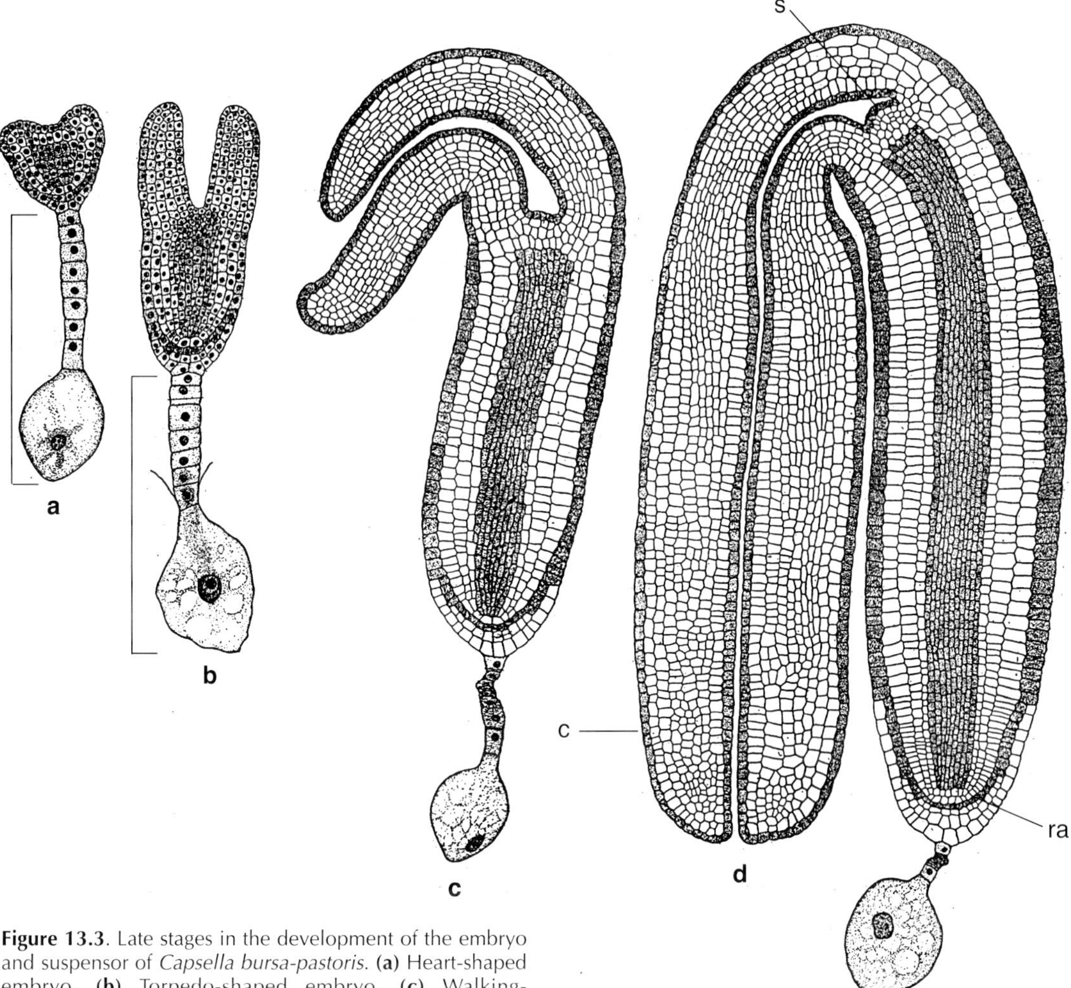

Figure 13.3. Late stages in the development of the embryo and suspensor of *Capsella bursa-pastoris*. **(a)** Heart-shaped embryo. **(b)** Torpedo-shaped embryo. **(c)** Walking-stick–shaped embryo. **(d)** Mature embryo. c, cotyledon; ra, root apex; s, shoot apex. Suspensor is outlined in the square bracket. The micropylar end is toward the bottom of the page. (From Schaffner 1906.)

vations also attest to a remarkable dependence of the embryonic root, rather than the shoot, on polar auxin transport.

After the cotyledons are initiated, division and differentiation of cells in the basal tier of the embryo give rise to the hypocotyl. There is considerable elongation of the cotyledons and hypocotyl at this stage, which gives rise to the torpedo-shaped embryo. A root apex delimited in the torpedo-shaped embryo is the progenitor of the embryonic radicle (Figure 13.3). During progressive embryogenesis, continued growth of the embryonic organs, compounded by spatial restrictions within the ovule, causes the embryo to curve at the tip to assume the walking-stick–shaped stage and finally to take on the shape of a horseshoe (Schaffner 1906; Souèges 1919). Although organs characteristic of the adult embryo differentiate as early as the heart-shaped or torpedo-shaped stage, subtle physiological and biochemical changes continue in the cells of the embryo and in the surrounding tissues before the maturation program sets in.

It was discussed earlier that a hallmark of the Crucifer type of embryo development is the ambiguity about the role of the descendants of the basal cell of the two-celled proembryo. That some of these cells are ultimately incorporated into the embryo of *C. bursa-pastoris* has been established by careful cell lineage studies (Schaffner 1906; Souèges 1919). Initially, the basal cell of the two-celled proembryo is partitioned transversely. Further divisions in this newly formed pair of cells are,

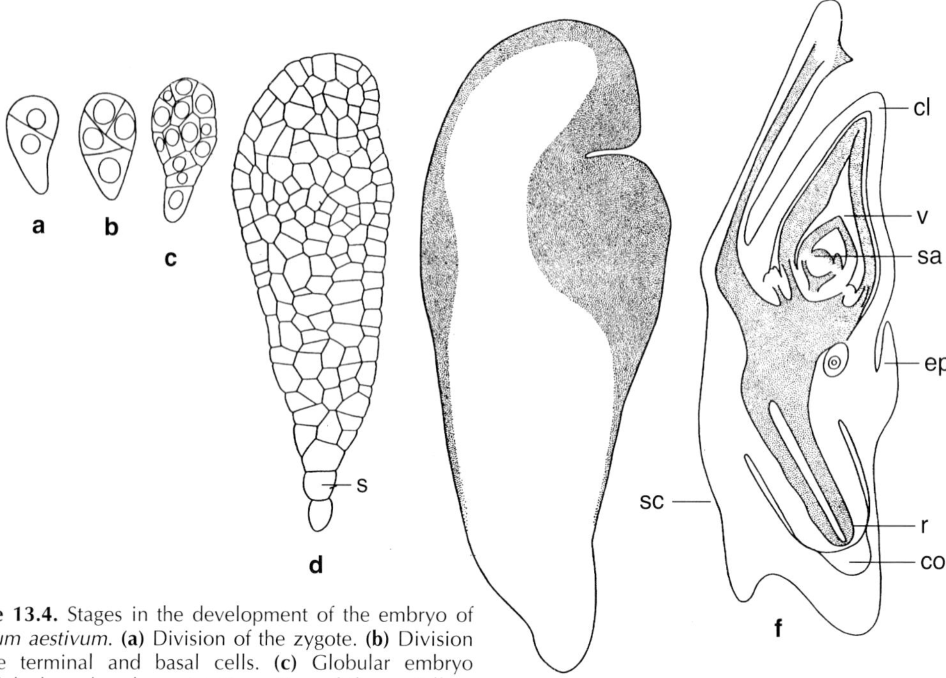

Figure 13.4. Stages in the development of the embryo of *Triticum aestivum*. **(a)** Division of the zygote. **(b)** Division of the terminal and basal cells. **(c)** Globular embryo **(d)** Club-shaped embryo. **(e)** Formation of the scutellum and coleoptile. **(f)** Diagrammatic section of the mature embryo. The micropylar end is toward the bottom of the page. cl, coleoptile; co, coleorhiza; ep, epiblast; r, root; s, suspensor; sa, shoot apex; sc, scutellum; v, first leaf of the plumule. (From Batygina 1969.)

however, restricted to the terminal cell, which produces, by transverse walls, 8–10 cells held together as a filament constituting the suspensor. The suspensor is terminated at the chalazal end by the organogenetic part of the embryo, and at the micropylar end by the basal cell, which has now become disproportionately large (Figures 13.2, 13.3). All the cells of the suspensor, except one, are eventually crushed and obliterated by the growing embryo; this fortunate cell (hypophysis) is in contact with the embryo and plays an important role in the formation of the root apex. The hypophysis divides transversely to form two cells, and the descendants of these cells contribute to the formation of the root cortex, the root cap, and the root epidermis of the mature embryo. It is as if a single cell from a group of about 10 cells born out of simple mitotic divisions of a mother cell is directed into a specialized channel of continued differentiation, while others perish along the way. Practically nothing is known about the mechanisms that cause changes in the developmental options of the cells of the suspensor.

In their morphological appearance, mature embryos of monocots are strikingly different from those of dicots because of the presence of only one cotyledon. Studies on the comparative ontogeny of monocot and dicot embryos were bedeviled by controversy for some time, but there is now general agreement that the development of the embryo up to the octant stage is almost identical in dicots and monocots. The main point of contention is with regard to the origin of the shoot apex and cotyledon in the monocots. For a period of time, the dominant view was that in monocots the shoot apex is lateral, and the cotyledon is terminal, in origin. Extensive ontogenetic studies (Swamy and Lakshmanan 1962a, b; Ba et al 1978; Swamy 1979; Guignard 1984), however, seem to support the view that both the shoot apex and the cotyledon share a common origin from the terminal cell of a three-celled proembryo. Accordingly, what is perceived as a lateral shoot apex is one that is displaced from its terminal position by the aggressive growth of the single cotyledon.

Sagittaria sagittaefolia (Alismaceae) exhibits a pattern of embryogenesis that is typical of many monocots. Here the differentiative potencies of the cells are established after the first transverse division of the zygote. The terminal cell gives rise to the cotyledon and shoot apex, whereas the hypocotyl–shoot axis and the entire suspensor are derived from the basal cell. The cells destined to form the shoot apex are initially sluggish, so that the rapidly growing cotyledon dwarfs the shoot apex. As the hypocotyl is organized in the lower part of the proembryo, there is a concomitant differentiation of the procambium and formation of root initials (Swamy 1980).

In their mature structure and ontogeny, embryos of Poaceae do not have much in common with other monocots. The main features of the grass embryo are the development of an absorptive organ known as the scutellum; the presence of caplike structures covering the radicle and plumule, known respectively as the coleorhiza and the coleoptile; and a flaplike outgrowth, called the epiblast, on one side of the coleorhiza.

Embryo development in *Triticum aestivum* will be considered as an example (Figure 13.4). Here, following a transverse-oblique division of the zygote, both cells divide by oblique walls to form a four-celled proembryo. Repeated divisions of these cells yield the mature embryo. Following a globular stage of 16–32 cells, the embryo becomes club shaped. It is at this stage that the scutellum appears as a vague elevation in the apical–lateral region of the embryo. At the same time, the opposite side of the embryo begins to enlarge, marking the differentiation of the shoot apex and leaf primordia. The final major event of embryogenesis is the differentiation of the radicle, which has an endogenous origin in the central zone of the embryo (Batygina 1969). In some other members of Poaceae, such as *Zea mays*, the pattern of cell divisions in the proembryo becomes unpredictable even as early as the second round (Randolph 1936). The use of the transcription factor–encoding homeobox gene *KNOTTED-1* (*KN-1*) has provided a promising experimental design to show that the first histologically recognizable sign of the shoot meristem in developing *Z. mays* embryos is foreshadowed by the expression of this marker gene. The scutellum, which is formed almost simultaneously with the shoot apical meristem, does not express the *KN-1* gene. Since this gene is known to be active throughout the postembryonic shoot ontogeny of maize, its transcription probably connotes a consistent feature of the shoot meristem from the time it is cut off in the embryo (Smith, Jackson and Hake 1995). A different result is obtained when the expression of a rice homeobox gene is followed in developing rice embryos. The gene is expressed in wild-type globular embryos well before organs are initiated, as well as in mutant embryos that lack detectable embryonic organs. On the face of it, this observation does not reveal that homeodomain proteins directly determine shoot development during embryogenesis (Sato et al 1996).

(iii) Reduced Embryo Types

It is clear that embryogenesis in angiosperms is a closely orchestrated process involving a great many cells whose activities are interlaced. As a result, the mature embryo attains the basic organization consisting of a bipolar axis with one or two cotyledons. The embryo is terminated at each pole by an apical meristem, and the point of attachment of the cotyledons separates the embryonic axis into a hypocotyl–root region and an epicotyl–plumule region. The former region has, at its lower end, the primordial root (radicle), whereas the primordial shoot (plumule) is attached to the stem part called the epicotyl. From the earliest investigations of embryogenesis, it has become apparent that embryo development does not proceed to completion in certain angiosperms. These plants thus shed seeds that harbor underdeveloped or reduced embryos (Figure 13.5). Members of Scrophulariaceae, Balanophoraceae, Ranunculaceae, Orobanchaceae, and Pyrolaceae among dicots, and Orchidaceae, Burmanniaceae, and Pandanaceae among monocots, are noteworthy in this respect, as are additional species belonging to at least 13 more families (Rangaswamy 1967; Natesh and Rau 1984). A paradigm of a seed with a reduced embryo is *Monotropa uniflora*. Here the embryo is generally embedded in the endosperm and consists of no more than two cells separated by a transverse wall (Olson 1980, 1991). Equally striking for its reduced embryo is the seed of *Burmannia pusilla* (Burmanniaceae), in which embryogenesis terminates at the quadrant stage of the apical cell (Arekal and Ramaswamy 1973). A higher level of embryo differentiation is found in the seeds of various orchids (Arditti 1979) and of root parasites like *Orobanche aegyptiaca* and *Cistanche tubulosa* (both of Orobanchaceae) (Rangaswamy 1967); at the time of shedding, seeds harbor globular embryos that do not show any differentiation into radicle, hypocotyl, cotyledons, epicotyl, and plumule. Mature seeds of *AlectΔ368*
ra vogelii and *Striga gesnerioides* (both of Scrophulariaceae) contain heart-shaped embryos with rudimentary cotyledons and a radicular pole (Okonkwo and Raghavan 1982). Despite the lack of morphological differentiation, it has become apparent that undifferentiated embryos of certain plants display subtle differences between shoot and root poles. For example, in both *O. aegyptiaca* and *C.*

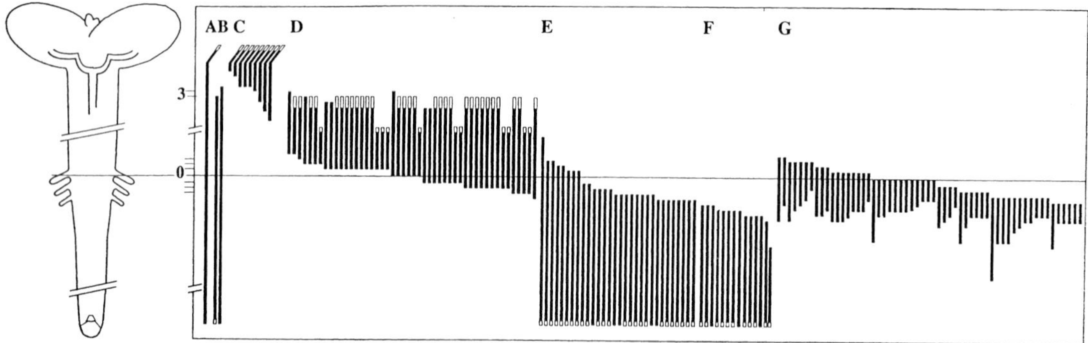

Figure 13.6. Mapping of the end points of colored sectors formed on the seedling of transgenic *Arabidopsis.* The end points of the large sectors demarcate the earliest divisions, and those of progressively smaller sectors mark later divisions. The sectors are indicated in relation to the seedling diagrammed at left. Black bars indicate the exact location of the sector terminus as determined; bars with white ends indicate estimated sector ends. (A) Sector demarcating the division of the upper and lower tiers of the globular embryo. (B) Root and hypocotyl sectors. (C) Cotyledon shoulder sectors. (D) Hypocotyl sectors. (E) Root sectors with apical ends overlapping basal ends of hypocotyl sectors. (F) Root sectors with apical ends disconnected from basal ends of hypocotyl sectors. (G) Intermediate zone sectors. On the vertical axis, the bars represent cell walls of adjacent epidermal cells. 0, lower wall of the first epidermal cell with root hair; 3, third epidermal cell below the junction of cotyledon vascular strands. (From Scheres et al 1994. *Development* 120:2475–2487. © Company of Biologists Ltd.)

roots in *Arabidopsis* (Benfey et al 1993). Findings from these investigations have to some extent altered previously held concepts about the relationship between meristems and contiguous regions; however, since these studies do not materially contribute to our understanding of the developmental fate of early division phase embryos, they will not be further considered.

Results from the culture of seed embryo segments have shown that the ability to produce working parts of the missing structures is generally restricted to segments containing the shoot apical meristem. Thus, whereas shoot segments develop into complete seedlings, root and hypocotyl segments retain their typical organization without regenerating shoots (Raghavan 1976b). As shown in an investigation using torpedo-shaped somatic embryos of carrot, region specification seems to operate even in the early stages of embryogenesis in a fashion similar to that seen in seed embryos. It appears that, following extirpation of the lower half of the embryo, virtually any part of the shoot consisting of portions of the hypocotyl or of the hypocotyl and root tissues was able to regenerate a new root, but remnants of the shoot with just the cotyledons and the apical notch did not possess root-forming potential. In contrast, restoration of a normal shoot from a root portion was a rare event and occurred only from root segments consisting of the extreme end of the root pole. During root regeneration from the shoot portion, re-formation of the missing regions of the embryo begins in the same order and polarity as in an uncut embryo, with the root meristem following the hypocotyl–root tissues (Schiavaone and Racusen 1990, 1991). From this it seems that genetic information required for directing the patterning and timing of events of later stages of embryogenesis is present in the cells of the shoot pole fragment.

The extent to which the undifferentiated cells of proembryos and the differentiated regions of slightly more advanced embryos are assigned to a particular fate probably depends upon a succession of events and upon changes in a variety of components. In the proembryo, the events are sustained until the cells are determined as the shoot or root regions. The position developed here with respect to the fate of the territories of early torpedo-shaped embryos is that the events continue until the meristems are fully established. As more cells are produced in the developing embryo, cellular interactions become increasingly complex due to unknown changes in the component processes and their effects on the differentiated cells. For example, based on the capacity of shoot and root segments to form the missing parts, one can conclude that different regulatory states exist in the fragments destined to regenerate. Since global statements such as these do not explain the mechanism of cell lineage and cell fate in developing embryos, one may hope that they will stimulate the formulation of cogent questions and experiments about the systems.

3. CELLULAR ASPECTS OF EMBRYOGENESIS

The foregoing account has been entirely concerned with embryo growth resulting in recognizable changes in embryo size and shape. However, size and shape changes, as indicators of the growth of a multicellular structure, leave much to be desired, since the tacit assumption remains that the embryo attains its form by spatial and temporal modulations in cell division, cell elongation, and cell differentiation. Histological diversity attained in the embryo might be presaged by detectable ultrastructural and cytochemical changes in critical cells and tissues. What patterns might these cellular activities follow in a growing embryo?

(i) Ultrastructural and Histochemical Characterization

It is hard to imagine that, following the distinct ultrastructural and histochemical differences that characterize the terminal and basal cells of the two-celled proembryo, further differentiation of the proembryo can occur without similar accompanying changes. There is no question that there are some subcellular differences in the distribution of organelles as the proembryo is repeatedly partitioned, but whether these changes are symptomatic of cell differentiation in the embryo is a debatable issue. In *Capsella bursa-pastoris,* as the basal cell of the two-celled proembryo divides transversely to form the suspensor, the terminal cell acquires more ribosomes and it stains more intensely for proteins and nucleic acids than do the suspensor cells. On the other hand, ER and dictyosomes are more abundant in the basal cell than in the terminal cell (Schulz and Jensen 1968c). As the terminal cell forms the globular embryo and the suspensor cell is partitioned transversely into a filament, ribosome density is distinctly greater in the cells of the embryo than in those of the suspensor. The presence of starch and lipid bodies in the terminal cell of the three-celled proembryo and their disappearance during the division of this cell have suggested that these storage bodies probably function as a source of energy for mitosis (Schulz and Jensen 1968a). When the hypophysis is cut off from the terminal cell of the suspensor in the globular embryo of *C. bursa-pastoris,* it acquires a basic subcellular profile different from that of cells of the embryo; this is evident in the highly vacuolate cytoplasm, the low ribosome density, and the generous amounts of ER, starch, and lipids in the cell. The developmental fate of the hypophysis is apparently determined after it is cut off, because ribosome density and concentrations of nucleic acids and proteins in the cells of the root cap and root epidermis cut off by the hypophysis appear intermediate between those of the embryo cells and suspensor cells. The earliest stage of embryogenesis in which cytological differences between cells are noted in anticipation of tissue and organ differentiation is the heart-shaped embryo. Here the cells of the procambium and ground meristem appear more highly vacuolate than those of the protoderm (Schulz and Jensen 1968a).

Ultrastructural studies of embryos of other angiosperms are less complete than those described in *C. bursa-pastoris* and contribute little to our understanding of cell specialization during embryogenesis. Early embryogenesis in *Hordeum vulgare* involves an increase in the population of ribosomes and mitochondria in all cells. As the embryo completes two or three rounds of divisions, there is a progressive decrease in the distribution of vacuoles in the cells of the suspensor (Norstog 1972a). Accumulation of vacuoles and multivesicular bodies in the cells of the apical part of the proembryo of *Secale cereale* is considered indicative of early stages of cell differentiation to form the scutellum (Stefaniak 1984). In *Helianthus annuus,* cells of the embryo have more ribosomes than those of the suspensor, giving these cells a more electron-dense appearance (Newcomb 1973b). According to Singh and Mogensen (1975), cells of the embryo and suspensor of *Quercus gambelii* seem to amplify the ultrastructural features of the terminal and basal cells, respectively, of the two-celled proembryo. As the embryo is phased out into the globular stage, suggestive of increased metabolic activity, the organelle population undergoes a complete reorganization. Increases in plastid number and complexity, mitochondrial number, dictyosomal activity, ribosomal aggregation, and the amount of ER are a conspicuous part of the processes that peak in the late globular stage embryo. By contrast, the number and distribution of these organelles in the suspensor cells remain unchanged. Besides the conventional subcellular structure, some unique changes that have been described are the disappearance of starch and lipid droplets and an increase in ribosome density and in the number of autophagic vacuoles in globular embryos of *Petunia hybrida* (Vallade 1980); the presence of extensive nuclear evaginations and vesicles in five- to eight-celled embryos of *Oryza sativa* (Jones and Rost 1989b); the appearance of large mitochondria in the cells of the globular embryo of *Arabidopsis* (Mansfield and Briarty 1991); and the presence of vacuoles with flocculent materials, membrane fragments, and weak electron-scattering bodies in the cells of the globular embryo of *Vicia faba* (Johansson and Walles 1993b).

Cells of the heart-shaped and older embryos have the normal complement of organelles typical of meristematic cells, and no generally characteristic ultrastructural changes augur the onset of tissue and organ differentiation in them. Clearly, in a situation in which newly formed cells are constituted into tissues in a developing organ, it is unrealistic to expect major upheavals in cell structure. Some exceptions are noteworthy, such as a 4- to 16-fold proliferation of the rough ER found in the cotyledons of *Phaseolus vulgaris* during the cell elongation phase, coincident with storage protein accumulation (Briarty 1980). At the histochemical level, tests on embryos of various plants have encountered higher concentrations of RNA in the cells of the presumptive cotyledons than in the rest of the embryo, though it is not surprising that transcriptional activity of the cells increases preparatory to differentiation (Rondet 1962; Pritchard 1964b; Schulz and Jensen 1968c; Vallade 1970; Norreel 1972; Syamasundar and Panchaksharappa 1976; Shah and Pandey 1978). Another measure of activity of developing embryos is the distribution of enzymes of respiratory metabolism. Determinations made in the embryos of *Gossypium hirsutum* show a striking correlation between tissue and organ differentiation and the activity of succinic dehydrogenase, a key enzyme in aerobic respiration. A high enzyme titer is found in all recognizable organs of the embryo, such as the suspensor, cotyledons, hypocotyl, shoot apex, subapical region, and the elongating radicle, during the period of their active growth (Forman and Jensen 1965). Alcohol dehydrogenase is involved in anaerobic metabolism, and multiple forms of the enzyme are generally present in cells. Differentiation of the axis and cotyledons of embryos of *Phaseolus vulgaris* is associated with the appearance of a fast-moving isozyme of alcohol dehydrogenase in these organs (Boyle and Yeung 1983). Since the embryo grows in an environment insulated from atmospheric oxygen, perhaps a case can be made for a functional role in normal cellular activities for organ-specific isozymes of the relevant enzymes of anaerobic metabolism. These observations not only give meaning to the process of embryogenesis, but also suggest that a cytochemical examination of embryogenesis with more refined techniques should prove instructive.

(ii) Cell Division and Cell Elongation

A generally accepted notion is that cell divisions predominate during the early period of embryogenesis and that cell enlargement prevails during the later period. A variation of this theme is observed in the embryos of *Capsella bursa-pastoris* and *Gossypium hirsutum*, in which the increase in cell number is interrupted by a lag period during cotyledon initiation; this results in a conventional S-shaped curve for cell number from zygote to the globular embryo, and another S-shaped curve from the heart-shaped to the mature embryo (Pollock and Jensen 1964). In the globular embryos of *Hordeum vulgare, Oryza sativa,* and *Triticum aestivum*, an exponential increase in cell number is maintained only up to the 20–40-celled stage, after which the rate of increase declines up to the 100–200-celled stage and again up to the 1,000-celled stage. This has been attributed to subtle changes in cell metabolism preparatory to changes in form in the embryos (Nagato 1978).

Changes in cell size are tightly linked to embryogenic divisions and, notwithstanding the lack of extensive quantitative studies, it is clear that a reduction in cell size accompanies the first several rounds of division of the zygote. In *G. hirsutum* and *C. bursa-pastoris*, the zygote remains the largest cell in the life of the embryo, and the average cell size is reduced to about one-fifth that of the zygote even as early as the first division. Cell size drops even further up to the 100-celled embryo stage in *G. hirsutum* and up to the young globular embryo stage in *C. bursapastoris* (Pollock and Jensen 1964). A distinctive reduction in cell size up to the midglobular stage of the embryo also occurs in *Arabidopsis*. The reduction in cell size is accompanied by progressive decreases in cell-to-nuclear volume ranging from 365 in the zygote to about 13 in the midglobular embryo (Mansfield and Briarty 1991). These changes probably help to establish a favorable balance between the nuclear size and cytoplasmic volume for rapid divisions.

The distribution of mitotic figures is uniform in very young embryos of *G. hirsutum* and *C. bursa-pastoris* and continually changes to reflect the onset of tissue and organ differentiation, assuming a U-shaped pattern in heart-shaped embryos and ending up in a Y-shaped pattern in torpedo-shaped embryos. Apparently, the onset of specific histogenic patterns during embryo development depends upon the attainment of a critical cell number before the shape change occurs, and is followed by asymmetric distribution of cell divisions. Thus, the transition of the globular embryo to the heart-shaped form seems to result from an increased mitotic activity in the distal part of the embryo (U-shaped pattern) that gives rise to the cotyledons. A different pattern of distribution of cell divisions (Y-shaped pattern), accompanied by cell elongation, might account for the transition of the heart-shaped embryo to the torpedo-shaped stage (Pollock and Jensen 1964). The key facts that may provide clues to a more explicit understanding of the factors that control differential cell divisions in a homogeneous tissue are, however, not available. One might suspect that some cytoplasmic gradient is being set up by cell division factors diffusing in the direction of the embryo where potential divisions occur, but as yet we have no evidence for this.

(iii) DNA Synthesis

The concept of constancy of DNA in a cell predicts that DNA synthesis must keep pace with the production of cells so that each new cell will have a fixed amount of DNA. One might hope that the precise division of the zygote to produce a multicellular embryo will depend upon an efficient DNA duplication mechanism, followed by cytokinesis. This picture is obscured in some studies because of complications introduced by endoreduplication. The DNA content of the zygote has been shown to range from the expected 2*C*–4*C* amounts in *Tradescantia paludosa* (Woodard 1956), *Hordeum vulgare*, and *H. vulgare* × *H. bulbosum* (Bennett and Smith 1976; Mogensen and Holm 1995), to attain a 3*C*–6*C* level in *Petunia hybrida* (Vallade et al 1978), or to become as high as 16*C* in *H. distichum* (Mericle and Mericle 1970). DNA amounts in the nuclei of two- to four-celled embryos of these species show a similar range, from a 2*C* amount in *T. paludosa* to 8*C* in *H. distichum*. In the globular stage embryo of *Vanda sanderiana*, there is a gradient of increasing DNA values from 2*C* to above 8*C* in cells from the distal meristematic tip to cells in the proximal parenchymatous region (Alvarez 1968). With further development, as in heart-shaped and torpedo-shaped embryos of *Glyine max* and *Vitis vinifera*, there is a stabilization of DNA content in the cells at the 2*C*–4*C* levels (Chamberlin, Horner and Palmer 1993b; Faure and Nougarède 1993). In *Gossypium hirsutum*, the amount of DNA per cell remains fairly constant from the early heart-shaped to the cotyledonary stage embryos, although there is nearly a 30-fold increase in cell number during this transition (Fisher and Jensen 1972). Analyses of embryos of *Lactuca sativa* (Brunori and D'Amato 1967), *Vicia faba* (Brunori 1967), and *Brassica napus* (Silcock et al 1990) in the middle or later periods of embryogenesis indicate that the DNA content of the cells does not deviate from the 2*C* level even as dehydration of embryos preparatory to quiescence of seeds sets in. It is possible that with progressive desiccation of the embryo, other factors, such as available water supply, might limit DNA synthesis.

A great deal of our present knowledge of the dynamics of DNA synthesis and accumulation in developing embryos has been acquired from analyses of cotyledons of legumes and other plants that accumulate storage proteins (Figure 13.7). After an initial period of growth by cell divisions, the subsequent growth of cotyledons is marked by cell expansion accompanied by the accumulation of enormous quantities of storage products, mostly proteins. Scharpé and van Parijs (1973) showed that cells of developing cotyledons of *Pisum sativum* continue to synthesize DNA after cell divisions have ceased; this leads to a progressive increase in the DNA amount in the cells of full-grown cotyledons to a level as high as 64*C*. The general pattern of changes in DNA levels of cotyledonary cells of *P. arvense* closely resembles that described in *P. sativum* (D. L. Smith 1973). These two studies have been augmented by reports of a continued increase in the nuclear DNA content of the parenchymatous cells of cotyledons of *Vicia faba* (Millerd and Whitfeld 1973), *Gossypium hirsutum* (Walbot and Dure 1976), *Glycine max* (Dhillon and Miksche 1983), *Phaseolus vulgaris* (Briarty 1980; Johnson and Sussex 1990), and *Sinapis alba* (Fischer et al 1988). In the embryos of *V. faba* (Millerd and Whitfeld 1973; Borisjuk et al 1995) and *G. hirsutum* (Walbot and Dure 1976), it has been established that the increase in DNA content of cells following completion of division is by endoreduplication; that endoreduplication occurs at the same time that cell elongation and storage protein accumulation proceed adds interest to the possible relationship among the three events.

The question as to whether the increase in DNA content of cotyledonary cells represents endoreduplication of the entire genome or selective amplification of certain sequences has engendered some discussion. According to Walbot and Dure (1976), the presence of identical reassociation kinetics of nonrepetitive DNA from cotton cotyledons before and after endoreduplication, the absence of satellites after analytical ultracentrifugation of the two DNA samples, and the constancy in the number of rRNA cistrons per unit of the two DNA samples argue against selective amplification. Results in

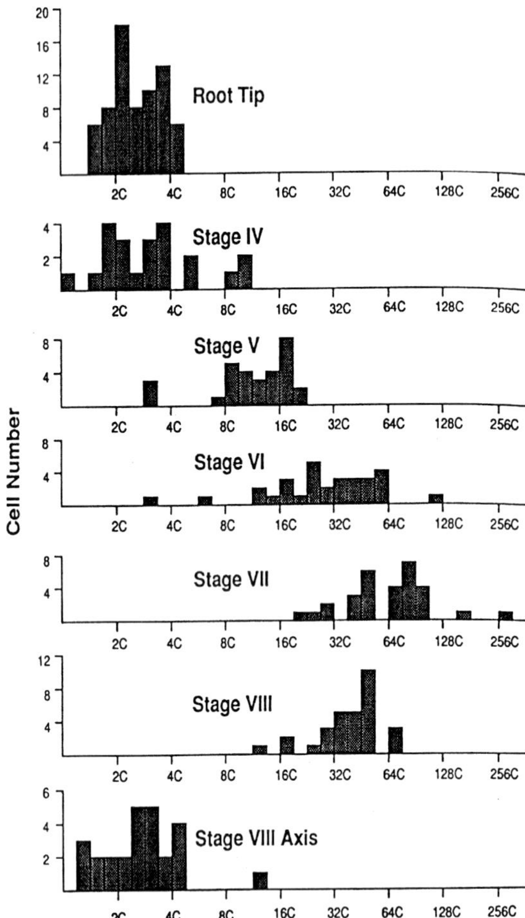

Figure 13.7. Distribution of nuclear DNA in the root tip cells, cells of developing cotyledons (stages IV to VIII), and stage VIII embryo axis of *Phaseolus vulgaris*. Stage IV is the youngest and stage VIII is the oldest embryo analyzed. Nuclear DNA contents are given in *C* values. (From Johnson and Sussex 1990. *Chromosoma* 99:223–230. © Springer-Verlag, Berlin.)

accord with this finding have come from a comparison of the hybridization kinetics of cloned soybean embryo storage protein gene with an excess of DNA isolated from embryo and leaf. The presence of identical smooth curves resulting from the annealing of the two DNA samples attests to the absence of selective gene amplification during continued DNA synthesis in the embryo (Goldberg et al 1981a). Additionally, Southern blot hybridization of DNA from the embryos and leaves of soybean with probes for glycinin, the major soybean storage protein, and for a 15-kDa protein showed similar patterns, confirming that selective gene amplification is not a factor to be reckoned with during endoreduplication (Fischer and Goldberg 1982). Results from the use of a wide selection of molecular probes in hybridization experiments have also discounted gene amplification and random synthesis of DNA as factors in DNA accumulation in the cells of cotyledons of *Phaseolus vulgaris* (Johnson and Sussex 1990). If storage protein genes or genes for other essential proteins are not selectively amplified in the cotyledons, it is a mystery why the cells contain a lot more genome than all the proteins it can code for. Conceivably, the extra DNA copies may be of restricted but strategic significance as a storage form of nucleotides during germination.

4. TISSUE AND MERISTEM DIFFERENTIATION DURING EMBRYOGENESIS

To give legitimacy to the embryo as the future sporophyte, the three primary meristems – namely, the protoderm, ground meristem, and procambium – and the shoot and root apical meristems are blocked out sequentially at different stages of embryogenesis. In the embryo of *Capsella bursa-pastoris*, the first meristem to differentiate is the protoderm, which is formed by a periclinal division of cells of the octant embryo. Because of the nature of this division, the packet of cells of the globular embryo can be neatly separated into eight external cells of the protoderm, sheltering an internal group of eight cells. After they are cut off, cells of the protoderm continue to divide anticlinally to keep pace with the increasing volume of the embryo and they gradually begin to acquire characters of the epidermis (Schulz and Jensen 1968a). That early during embryogenesis the peripheral layer of cells becomes superficially similar to epidermal cells is attested by the presence of cutin on globular and heart-shaped embryos (Rodkiewicz, Fyk and Szczuka 1994). The expression of a gene that encodes a candidate protein involved in cutin biosynthesis on the protoderm of early stage carrot embryos is consistent with this observation (Sterk et al 1991). At the heart of the timing of initiation of the protoderm lies the question whether all cells in the embryo are undetermined until the protoderm is carved out or whether genes for protoderm formation are turned on at an earlier stage. In *Citrus jambhiri*, the commitment to differentiate an epidermis is made by the zygote itself through the development of a cuticle around its outer surface; this trait is faithfully transmitted to the external layer of cells of the embryo. Failure of epidermal re-formation in embryos whose epidermis had been surgically removed has shown that both epidermal and subepidermal cell layers are already locked into their developmental fate along divergent pathways early during embryogenesis (Bruck and Walker 1985a, b).

The ground meristem and the procambium are derived from the inner core of cells of the globular embryo. The cells of the ground meristem may differentiate into a cortex or into both cortex and pith. In either case, the peripheral layer of cells of the central core of the proembryo is the progenitor of the cortex of the mature embryo. During the elongation of the embryo that begins around the torpedo-shaped stage, the cells of the peripheral layer divide periclinally to form three or four concentric layers of cells of the embryonic cortex. Formation of the pith from the ground meristem is foreshadowed by repeated divisions of the cells of the central region of the globular embryo, which continue as the pith of the hypocotyl part of the mature embryo. It was mentioned earlier that the presence of transcripts of the *KTI-3* gene is a molecular marker of the micropylar end of the soybean globular embryo. Because the transcripts persist in the cells of the ground meristem of heart-shaped and cotyledon stage embryos, it is conceivable that these cells have their origin in cells of the globular embryo harboring *KTI-3* transcripts (Perez-Grau and Goldberg 1989).

Procambium initiation is a critical feature of histogenesis of the embryo and occurs in the cells of the cortex or pith of globular embryos before the primordia of the cotyledons are formed. As shown in *Nerium oleander* (Apocynaceae), elements of the procambium are first seen in the innermost part of the cortex of the early heart-shaped embryo at the level of the presumptive cotyledons. More initials are laid down on either side of the original file and this leads to the formation of a complete ring of procambial cells. After the procambium is established in the hypocotyl of the developing embryo, it extends acropetally into the cotyledons and shoot apex, and basipetally into the root apex. Thus, by the time the embryo becomes torpedo shaped, it has a solid cylinder of procambium extending through its entire length and into the cotyledons (Mahlberg 1960). In *Rhizophora mangle* (Rhizophoraceae), files of procambium linking the shoot and root apices are apparently initiated independently, because they appear discontinuous for a brief period until they are connected to the hypocotyl procambium by cotyledonary strands (Juncosa 1982). Besides these selected examples, contributions have been made to the analysis of tissue differentiation in the embryos of several other plants, but limitations of space do not permit us to take up these examples. On the whole, the pattern of tissue formation is a crucial process of embryogenesis and one of enduring histogenetic interest, but one about whose mechanisms we know very little.

(i) Shoot and Root Apical Meristems

Differentiation of the shoot and root apical meristems in the embryo makes the preexisting, but cryptic, bipolarity of the embryo more obvious. Investigations on the histogenesis of embryos of *Nerium oleander* (Mahlberg 1960), *Downingia bacigalupi,* and *D. pulchella* (Campanulaceae) (Kaplan 1969) have shown that a precise pattern of cell divisions and differentiation distinguishes the formation of the shoot and root apical meristems. In *N. oleander*, the earliest indication of the shoot and root apices is the appearance of pockets of partially differentiated meristematic cells at both poles of the globular embryo. These cells are interpreted as direct descendants of the octant stage embryo rather than as newly generated cells with meristematic potential. In all three species, differentiation of the shoot apex is also augured by a change from the random distribution of cell divisions to their concentration in the apical half and by cytohistological differentiation into a peripheral and a central zone. It is to be noted that these changes occur before the appearance of the cotyledons, suggesting that the cotyledons are the first formative organs produced by the shoot apex. The visible differentiation of the shoot apex generally coincides with the change in the shape of the embryo from globular to heart shaped and is attributed to the continued meristematic activity of the protoderm and the underlying layer of cells in the apical part of the embryo. The formation of the central zone of the shoot apex is prefigured by the cytological differentiation of three or four layers of cells below the protoderm. Differentiation of the shoot apex is complete when it becomes dome shaped and lapses into a state of quiescence or dormancy in the mature embryo.

Depending upon the types of embryo development (described in a previous section), the root meristem and the root cap have their origin in the terminal cell or basal cell of the two-celled proembryo. For illustrative purposes, an example of the embryo in which the root apex is derived wholly from the terminal cell of a two-celled proembryo is that of *D. pulchella*. Limited transverse and longitudinal divisions of the terminal cell lead to the formation of a large proembryo consisting of four superimposed tiers of one to four cells in each tier. During the organization of the root apex, cells of the most basal tier divide periclinally to give rise to the root cap. Protodermal cells of the next tier, toward the apical part of the proembryo, also divide periclinally to contribute to the root cap. In addition, cells of this second tier generate derivatives that form the root cortex and meristem (Kaplan 1969). In *C. bursa-pastoris*, in which the embryonic root apex is derived from the hypophysis, the initial division of this cell is in the transverse plane. By further transverse and longitudinal divisions, the daughter cell close to the suspensor gives rise to the root cap and root epidermis, and the one close to the organogenetic part of the embryo generates the rest of the root apex, including the root meristem (Schaffner 1906; Souèges 1919). Whatever factors determine the extent to which the terminal and basal cells of the two-celled proembryo are involved in the formation of the root apex, the embryonic root apex – compared to the shoot apex – displays a remarkable degree of architectural plasticity that allows it to generate closely integrated final products.

(ii) Origin of the Quiescent Center

A feature of the angiosperm root meristem is the presence of a group of cells that divide rather infrequently or not at all. This packet of cells, known as the quiescent center, is characterized by low DNA, RNA, and protein synthetic activities; the combination of cellular properties has led to the suggestion that the quiescent center originates as a result of the decreasing metabolic activity of certain cells in the embryonic root meristem. Despite the merit of this suggestion, and the use of a variety of methods to monitor the formation of the quiescent center in developing embryos, the evidence in favor of it is at best ambiguous. Relevant data typically invoked to characterize the appearance and disappearance of the quiescent center in developing embryos include the tempo in mitotic activity, staining reactions, and the pattern of ^{3}H-thymidine incorporation (Sterling 1955; Rondet 1962; Vallade 1972; Jones 1977; Clowes 1978a, b). Localization of mRNA in developing embryos of *C. bursa-pastoris* using ^{3}H-poly(U) as a probe has shown that autoradiographic silver grains caused by annealing of the label are plentiful in the cells of the developing embryo, except in the hypophysis and its derivatives, which constitute the quiescent center in the root apex of the mature embryo. In seedling roots, the same cells that fail to anneal ^{3}H-poly(U) also fail to incorporate ^{3}H-thymidine, indicating their identity as the quiescent center. These correlations support the conclusion that the quiescent center in the root of *C. bursa-pastoris* has its origin in the hypophysis of the embryonic suspensor (Raghavan 1990a).

5. THE SUSPENSOR

In the study of angiosperm embryogenesis, interest at the morphological to molecular levels has been bestowed largely on the organogenetic part of the embryo, with little or no attention beyond anatomical descriptions given to the suspensor. This is mainly due to the well-grounded notion that the role of the suspensor is purely of a mechanical nature, namely, to push the growing embryo deep into the chalazal end of the embryo sac, close to the source of nutrients. During the past three decades, a number of ultrastructural and cytological studies, along with a few physiological and biochemical investigations, have led to a revival of interest in the structure and physiology of the suspensor and to a refreshing departure from the classical views about its function. The current view of the structure–function relations of the suspensor is that the suspensor is a dynamic part of the embryo complex, functioning in the absorption and short-distance translocation and exchange of metabolites necessary for the growth of the embryo.

Although a filament of 8–10 cells attached to the organogenetic part of the embryo, as seen in *Capsella bursa-pastoris* or *Arabidopsis*, is one of the most familiar images of the suspensor, comparative studies of embryogenesis in angiosperms have shown that suspensors come in various sizes and shapes. A few of these interesting variations in suspensor morphology have been collated, illustrated, and described (Natesh and Rau 1984; Raghavan 1986c). In many plants, the morphological complexity of the suspensor is attributed to the extension growth of its cells, which reach the extraembryonal tissues of the ovule and even the ovary. In embryology literature, these outgrowths are frequently referred to as haustoria, implying that they are specialized structural adaptations for the absorption of nutrient substances. In some families, such as Orchidaceae, Podostemaceae, and Trapaceae, in which complex suspensors with haustoria have been described, the ovules are characterized by the complete absence of endosperm or by the presence of a reduced amount of endosperm. This indicates that nutrients absorbed by the suspensor haustoria may provide an important source of energy for the growth of the embryo. However, direct evidence has been difficult to obtain, as most observations are based on the intimate association of the haustoria with the extraembryonal tissues and do not bear directly on the actual transfer of nutrients or on the ultrastructural modifications of the haustoria for nutrient uptake and transfer. As will be described, ultrastructural modifications suggest such a function for the regular, nonhaustorial suspensor cells of several plants.

(i) Ultrastructure of the Suspensor Cells

Despite the fact that the suspensor is an ephemeral organ that does not become a part of the mature embryo, structural adaptations for specific functions find their highest expression in the cells of the suspensor. With regard to the alignment of the organelles and the density of their distribution, suspensors of different species thus far investigated, as well as the different cells of a suspensor, are astonishingly diverse. Other aspects of the suspensor, such as the number of cells present and their life span, also vary in the different species.

The first detailed account of the ultrastructural cytology of the suspensor was provided in *Capsella bursa-pastoris*. For purposes of discussion, three stages have been identified in the life of the suspensor: (a) the octant embryo, whose suspensor consists of 6 cells; (b) the globular embryo, whose suspensor attains the maximum cell number and appears as a filament of 10 cells interposed between a large, vacuolate basal cell and the embryo; and (c) the heart-shaped embryo, whose suspensor attains the maximum length and lives out its genetically permissible life span. Throughout its development, the suspensor is functionally connected with both the embryo and the basal cell by numerous ER-containing plasmodesmata. Although cells of the suspensor harbor the usual array of organelles, ribosomes appear to be specially concerned with cellular interactions of this organ. In line with this view of the seemingly special role of ribosomes, it has been observed that at the octant stage of the embryo, the suspensor cells have a respectable number of ribosomes, which show an increasing gradient toward the chalazal end. In the globular embryo, ribosomes gradually dissipate in the cytoplasm; they completely disappear from the suspensor cells of the heart-shaped embryo. Symptomatic of the inevitable failure of the cellular machinery of the suspensor is the onset of cytoplasmic degeneration accompanied by the loss of nucleic acids and proteins in the heart-shaped embryo. Gradually, the middle lamella of cells begins to loosen, and eventually the suspensor is obliterated by the burgeoning embryo. The basal cell generally remains active for some time after the collapse of the other cells (Schulz and Jensen 1969).

The presence of a series of invaginations on the walls of cells of the suspensor of *C. bursa-pastoris* has been a focal point of ultrastructural study. The invaginations arise from the outer lateral walls of the mid-suspensor cells of the globular embryo and increase in number and complexity at the heart-shaped stage. The wall projections are a feature that the cells of the suspensor share with the basal cell. The inner wall of the micropylar and lateral parts of the basal cell are also thrown up into an elaborate network of invaginations projecting into the cytoplasm. In fact, the projections first appear in the wall at the micropylar end of the zygote and increase in number, peaking in the basal suspensor cell of the heart-shaped embryo (Schulz and Jensen 1968c, 1969).

As will be seen later, *Phaseolus* is a paradigm for defining and interpreting the nuclear cytology of suspensor cells in an angiosperm. In *P. coccineus*, the suspensor is polarized from the proembryo stage onwards with respect to mitotic activity. Although divisions mostly cease in the more basal cells of the suspensor by the proembryo stage, they continue in the cells toward the chalazal region. The suspensor attains its maximum number of about 200 cells in the heart-shaped embryo; at the cotyledon stage, it reaches its maximum size. The presence of a large number of cells in the suspensor, however, does not enhance its longevity (Yeung and Clutter 1978).

Cell specialization in the suspensor cells of *P. coccineus* begins with the appearance of wall infoldings as early as the proembryo stage. Soon after, the cells become committed to a developmental pathway characterized by a disproportionate increase in the number of ribosomes, plastids, mitochondria, dictyosomes, and smooth ER, which continues until the cotyledon stage. In considering the ultrastructure of the wall of developing suspensor cells, it is significant that an increase in the number of invaginations parallels the increase in organelle density and that a raft of organelles, particularly ER, dictyosomes, and mitochondria, lie in close proximity to the wall ingrowths. These observations further strengthen a role for the wall projections in nutrient exchange (Yeung and Clutter 1978).

Although the suspensor in *Stellaria media* is

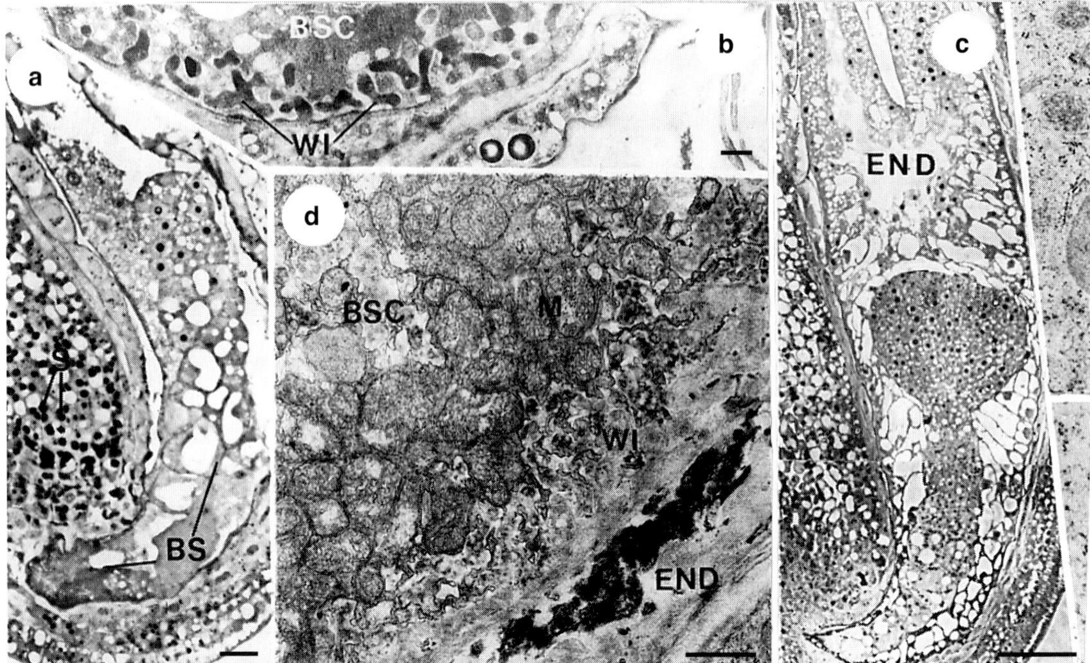

Figure 13.8. Morphology and ultrastructure of the suspensor of *Medicago scutellata.* **(a)** A globular embryo showing vacuolation in the chalazal and basal suspensor cells. **(b)** Electron micrograph of a portion of the basal suspensor cell of (a), showing wall ingrowths. **(c)** A heart-shaped embryo with suspensor. **(d)** Electron micrograph of a portion of the basal suspensor cell of (c), showing mitochondria associated with wall ingrowths. BS, basal cell; BSC, basal suspensor cell; END, endosperm; M, mitochondria; S, starch-containing integumentary cells; WI, wall ingrowth. Scale bar = 10 μm (a), 1μm (b, d), 100 μm (c). (From Sangduen, Kreitner and Sorensen 1983; photographs supplied by Dr. E. L. Sorensen.)

formed from the terminal cell of the two-celled proembryo (Caryophyllad type), the undivided basal cell presents a profile similar to, and no less complex than, that of the cells of the suspensor. One manifestation of the complexity portrayed in the basal cell is the appearance of an array of massive wall projections on the inner surface, along with a preponderance of mitochondria near them. Microbodies, plastids, and extensive whorls of tubular ER, which are generally absent from the cells of the embryo, become a conspicuous part of the organelle complement of the basal cell and the cells of the suspensor. The plastids of the suspensor cells have an unusual morphology with numerous tubules and electron-translucent inclusions. There is also a decreasing gradation in the complexity of plastids and in the number of microbodies from the basal cell to the chalazal suspensor cells (Newcomb and Fowke 1974). Plastids found in the suspensor cells of *Pisum sativum* (Marinos 1970b), *Phaseolus vulgaris, P. coccineus* (Schnepf and Nagl 1970; Yeung and Clutter 1979), *Ipomoea purpurea* (Ponzi and Pizzolongo 1973), *Tropaeolum majus* (Nagl and Kühner 1976), *Medicago sativa, M. scutellata* (Sangduen, Kreitner and Sorensen 1983a), *Alisma lanceolatum,* and *A. plantago-aquatica* (Alismaceae) (Bohdanowicz 1987) give every appearance of having undergone some degree of specialization unique to each species.

Wall labyrinths of the type just described appear to be a widespread phenomenon (Figure 13.8) and have been documented in suspensor cells of *P. vulgaris* (Schnepf and Nagl 1970), *Diplotaxis erucoides* (Simoncioli 1974), *T. majus* (Nagl 1976b), *Brassica napus* (Tykarska 1979), *Vigna sinensis* (Hu, Zhu and Zee 1983), *M. sativa, M. scutellata* (Sangduen, Kreitner and Sorensen 1983a), *Alyssum maritimum* (Brassicaceae) (Prabhakar and Vijayaraghavan 1983), *Alisma lanceolatum, A. plantago-aquatica* (Bohdanowicz 1987), *Glycine max* (Dute, Peterson and Rushing 1989; Chamberlin, Horner and Palmer 1993a), and *Vicia faba* (Johansson and Walles 1993b). There is now considerable and varied evidence based on the transfer cell morphology of the suspensor cells to support the solute absorption theory of suspensor function first articulated by Schulz and Jensen (1969). According to this view, metabolites from the endosperm or from the surrounding diploid cells of the ovule are delivered to the embryo through the suspensor cells, with the wall invaginations facilitating the transfer by increasing the surface area of absorption. The plasmodesmata perforating the embryo–suspensor boundary may serve to maintain an open channel of communication for the flow of solutes between the suspensor and the embryo. That the path of nutrients to the developing heart-shaped embryos of *P. vulgaris* and *P. coccineus* is through the suspensor has become evident from the work of Yeung (1980). The results showed that if ^{14}C-sucrose is administered through pods or to isolated embryos, much of the radioactivity appears in the suspensor and in the suspensor pole of the embryo. Somewhat similar results were obtained in the transport of a labeled polyamine, putrescine, in the ovules of *P. coccineus* (Nagl 1990). Although the new knowledge of the wall architecture of the cells of the suspensor of angiosperm embryos has pushed the frontiers of our understanding of the function of this organ to a higher level, until more is known about the gradient properties of the suspensor and its surrounding cells, the mechanism of nutrient incorporation into the embryo through the suspensor will remain unclear.

(ii) Nuclear Cytology of the Suspensor Cells

Investigations on the nuclear cytology of suspensor cells of various angiosperms have shown that endoreduplication coupled with the presence of polytene chromosomes distinguishes their development. A recurrent theme that has since emerged from these studies is that despite its short life span, the suspensor is made up of transcriptionally active cells that may play a regulatory role in the development of the embryo. Research on the cytology of suspensor cells has been reviewed before and will not be considered here again (Raghavan 1986c).

On the basis of nuclear volume or quantitative Feulgen microspectrophotometry, values for the DNA content of suspensor cells so far reported range from 16C in *Sophora flavescens* (Fabaceae) (Nagl 1978) to 8,192C in *Phaseolus coccineus* (Brady 1973). In *P. coccineus* (Brady 1973; Nagl 1974) and *Eruca sativa* (Corsi, Renzoni and Viegi 1973), the outcome of endoreduplication in cells in the different regions of the suspensor is somewhat different, but its consequences in terms of gene regulation are probably the same. In both these species, there is a gradient of DNA amounts, with low values in the cells close to the embryo, intermediate values in the neck region of the suspensor, and very high values in the cells of the basal region. Endoreduplication in the suspensors of *Alisma lanceolatum* and *A. plantago-aquatica* is nonconventional since it affects only the large basal cell. The nucleus of the cell rhythmically bloats in size and, based on the increase in nuclear volume, it has been estimated that in the latter species the nucleus attains a ploidy level as high as 1,024*N* (Bohdanowicz 1973, 1987). Although the chromosomal constitution of endoreduplicated suspensor cells of several species has revealed polyteny, the phenomenon has been most extensively studied in *P. coccineus* (Nagl 1967) and *P. vulgaris* (Nagl 1969a, b). Like polytene chromosomes in the salivary glands of dipteran larvae, those of the suspensor cells of the two species of *Phaseolus* also display puffs, which represent sites of DNA synthesis and transcription. However, some DNA-synthesizing puffs are found on chromosomes only at specific stages of embryogenesis, indicating the possibility of a stage-specific variation in DNA replication in the same polytene chromosome (Tagliasacchi et al 1984). A gene identified in the polytene chromosome of *P. coccineus* by in situ hybridization using a cloned probe encodes a protein that inhibits fungal endopolygalacturonase (Frediani et al 1993). Since this is a stress-related protein, it is too early to assess the significance of this observation.

As shown by the molecular hybridization of 25S, 18S, and 5S ^{3}H-labeled rRNA fractions to DNA of suspensor cells of *P. coccineus*, genes for rRNA are clustered at the nucleolus-organizing regions of polytene chromosomes (Durante et al 1977). However, only the ribosomal cistrons embedded in random nucleolus-organizing centers are functionally active and synthesize RNA in a given chromosome. In general, compared to the heterochromatic regions of the chromosomes, the bands engaged in RNA puffing are undermethylated; consequently, transcriptional activity of polytene chromosomes may rely on specific molecular components of DNA (Frediani et al 1986a, b). When the activities of the same chromosome puffs at different stages of embryogenesis are analyzed by autoradiography of ^{3}H-uridine incorporation, they are found to engage in RNA synthesis with different amplitudes (Forino et al 1992). Another intriguing observation is that differing labeling patterns appear in the polycistronic transcription units for 25S, 18S, and 5S rRNA genes when polytene chromosomes are annealed with fractions of DNA banded in the analytical ultracentrifuge, showing that rRNA cistrons of the same class are interspersed with different DNA sequences (Durante et al 1987). Given the not-too-low number of the chromosomes in the suspensor cells and given the complex, convoluted nature of the chromosomes, these cytological observations are illuminating and provide a good beginning toward the goal of achieving a complete understanding of the transcriptional apparatus of the polytene chromosomes.

The question as to whether polytene chromosomes undergo selective DNA amplification has been subject to intensive study. Autoradiography of ^{3}H-thymidine incorporation into suspensor cells of *P. coccineus* showed that in most cells the label is found exclusively in the heterochromatic region of the polytene chromosomes although in a small number of cells the label decorates the entire nucleus. These results have been interpreted to indicate that certain genes of the heterochromatic regions undergo DNA amplification, in addition to the scheduled DNA synthesis, due to endoreduplication (Avanzi, Cionini and D'Amato 1970). Consistent with the observation that cells in the different regions of the suspensor display varying degrees of endoreduplication, selective DNA amplification is found to prevail in the puffs of the highly polytene chromosomes of cells in the micropylar region of the suspensor, whereas cells produced during the early stages of suspensor development show uniform labeling over the nuclei due to endoreduplication (Cremonini and Cionini 1977). The occurrence of selective DNA amplification has been confirmed in the analytical ultracentrifuge from the density shifts in DNA isolated from suspensor cells of *P. coccineus* (Lima-de-Faria et al 1975). This work showed the presence in the suspensor cells of a class of DNA with buoyant density that has no parallel with DNA isolated from other parts of the plant. As shown in the same work by a lower saturation value in hybridization experiments with total suspensor DNA and rRNA, certain DNA sequences, particularly the genes for rRNA, are underreplicated in the genome of the suspensor cells. Thus, a seemingly fragile balance between amplification of certain genes and underreplication of other genes governs the nuclear activities of cells of the suspensor during a major part of its life. Although these observations stress a regulatory role for the polytene chromosomes of the suspensor cells, precise determination of the function of their gene products, given the small size and transient nature of the suspensor, requires a protracted study.

(iii) Physiology and Biochemistry of the Suspensor

Based on cytochemical, physiological, and biochemical information, the suspensor cells are considered to be metabolically active throughout their life span; the activities may be geared to the absorption of nutrients or for self-autolysis. In *Brassica campestris*, several enzymes of oxidative metabolism are found to be maximally active in the suspensor cells between the globular and torpedo-shaped embryos and to diminish thereafter (Malik, Singh and Thapar 1976; Malik, Vermani and Bhatia 1976; Bhalla, Singh and Malik 1980a, 1981b). The preponderance of enzymes in the actively growing suspensors presumably indicates the need for a high metabolism connected with nutrient uptake. However, the cessation of growth of the suspensor promptly initiates hydrolytic enzyme activities beginning in the cells at the micropylar end. The presence of specialized plastids in the suspensor cells of *Phaseolus coccineus* and *P. vulgaris* (Nagl 1977; Gärtner and Nagl 1980) has led to the view that the acid phosphatase of these organelles may possibly play a role in the autolysis of the suspensor. In *Tropaeolum majus*, whose suspensor produces haustoria, self-digestion of the suspensor is initiated by degenerative changes in the mitochondria coupled with the presence of high acid phosphatase activity in the haustorium (Nagl 1976c; Gärtner and Nagl 1980). Other enzymes, such as alkaline phosphatase and acetylesterase, are also active in the cells of the suspensor and in the haustorium, becoming maximally effective at the early cotyledon stage of the embryo when suspensor autolysis is well under way (Singh, Bhalla and Malik 1980).

Comparative analyses of RNA and protein metabolism of the suspensor and embryo of *P. coccineus* have provided evidence of higher transcriptional and translational activities in the suspensor cells than in the cells of the embryo. Expressed on a per cell basis, RNA content and the rate of RNA synthesis are low during the early development of the suspensor, increasing to a maximum at the late heart-shaped or early cotyledon embryo stage, and then declining. Protein content and the rate of protein synthesis also register low values in the suspensor cells of early stage embryos but increase substantially in the suspensors of late stage embryos. Characteristically, at most stages analyzed, cells of the suspensor contain more RNA and proteins and synthesize them more efficiently than do the cells of embryos of comparable age (Walbot et al 1972; Sussex et al 1973). Somewhat similar results have been obtained for the comparative biosynthetic activities of the embryo and suspensor cells of *T. majus* (Bhalla, Singh and Malik 1981a). Calculations based on the DNA content and rates of RNA synthesis in suspensor cells of *P. coccineus* at different stages of embryogeny have shown that the template activity of these cells in the late heart-shaped or early cotyledon stage embryos is higher than is expected of the diploid gene copies present in them (Clutter et al 1974). A clear expression of the protein synthetic activity of the suspensor geared to embryo nutrition is seen in the transient accumulation of the storage proteins, legumin and vicilin, in the suspensor cells of globular embryos of *Vicia faba* even before these proteins begin their massive, embryo-specific accumulation (Panitz et al 1995).

The focus on the synthesis of growth hormones by the suspensor and their relevance for the function of this organ represents a major and essentially novel addition to our knowledge of its physiology and biochemistry during the last 20 years. Auxins (Alpi, Tognoni and D'Amato 1975; Przybyllok and Nagl 1977), gibberellins (Alpi, Tognoni and D'Amato 1975; Alpi et al 1979; Picciarelli et al 1984; Picciarelli and Alpi 1987; Piaggesi et al 1989; Picciarelli, Piaggesi and Alpi 1991), cytokinins (Lorenzi et al 1978), and ABA (Perata, Picciarelli and Alpi 1990) have been identified in the suspensors of *P. coccineus* and other plants. The suspensor of *P. coccineus* has 10 times more gibberellin activity than the heart-shaped embryo and contains as many as six different gibberellins (Picciarelli and Alpi 1986; Piaggesi et al 1989). The high concentration of gibberellins is reflected in the ability of a cell-free extract prepared from homogenized suspensors to synthesize precursors of GA, as well as GA_1, GA_5, and GA_8, from radioactively labeled substrates (Ceccarelli, Lorenzi and Alpi 1979, 1981a, b). In the heart-shaped embryo of *P. coccineus*, cytokinins are also present in higher concentrations in the suspensor cells than in the cells of the embryo proper; a complete reversal of the status of both gibberellins and cytokinins occurs in the suspensor and embryo at the cotyledon stage (Alpi, Tognoni and D'Amato 1975; Lorenzi et al 1978). Changes in the auxin content of the suspensor have drawn little attention, although it appears that the suspensor contains higher amounts of auxin than does the embryo (Alpi, Tognoni and D'Amato 1975; Przybyllok and Nagl 1977). Very little ABA is present in the suspensor of *P. coccineus* during the initial stages of its development, but some increase, with two peaks, is noted subsequently (Perata, Picciarelli and Alpi 1990).

One prevalent notion about the role of hormones in the suspensor is that in the early stages of development, the embryo does not bear the burden of synthesizing the hormones necessary for its growth, but that this function is taken up by the suspensor (Cionini et al 1976; Bennici and Cionini 1979; Yeung and Sussex 1979). A role attributed to gibberellins is that, in the suspensor of *P. vulgaris*, the hormone increases the translational activity and maintains the protein level in the embryo. The role of the suspensor in enhancing translational activity was shown by a decrease in protein synthesis in embryos detached from the suspensor and cultured (Brady and Walthall 1985; Walthall and Brady 1986). A convincing demonstration of the effect of gibberellins in enhancing transcriptional activity of suspensor cells is the hormone-induced formation of RNA puffs and the increased RNA synthesis in certain regions of the polytene chromosomes of *P. coccineus* (Nagl 1974; Forino et al 1992).

The significance of the presence of other growth hormones in the suspensor remains largely unknown. In *Tropaeolum majus*, changes in the concentrations of peroxidase in the suspensor and the embryo are believed to ensure the regulation of endogenous auxin diffusing from the suspensor into the embryo. This correlation is somewhat typical of organs that synthesize or utilize IAA for growth (Singh, Bhalla and Malik 1979).

In conclusion, it is clear that the suspensor is an early embryonic organ that is programmed for terminal differentiation and, except in a few cases, it does not contribute any cells to the sporophytic generation of the plant. Yet, it cannot be naively assumed that the primary function of the suspensor is to orient the embryo in close proximity to the source of nutrition in the embryo sac or in the ovule. Important findings about the structure of the walls of suspensor cells implicate these cells in the absorption of metabolites from the surrounding tissues and in the transport of metabolites to the embryo. The cells of the suspensor, and especially their polytene chromosomes, are clearly involved in transcriptional activity, although the physiological role of the gene products synthesized is not understood. Finally, the accumulation of growth hormones by the suspensor has generated great interest in elucidating the possible function of hormones in promoting embryo growth. Despite these advances in our understanding of the suspensor, further studies are needed to gain new insights into the physiological and molecular bases for its terminally differentiated state and for its interaction with the embryo. The future directions in the study of the suspensor are briefly considered by Yeung and Meinke (1993) in a thoughtful essay.

6. NUTRITION OF THE EMBRYO

To a large extent, continued development of the embryo depends upon its interaction with the milieu of the embryo sac, the endosperm, and the cells of the ovule surrounding the embryo sac. Thus, in contrast to the static image conjured by pictures of embryos in isolation, it is reasonable to assume that developing embryos maintain a channel of communication with the external cellular environment. This traffic has in turn generated the reprise of a cherished objective of experimental plant embryologists, namely, to identify the substances that provide the nurture and nutriments to developing embryos, but this goal remains largely unattained.

(i) Nutrient Supply to the Zygote

It is noted that the primary endosperm nucleus generally begins to divide ahead of the zygote, thus ensuring that a mass of cells charged with storage products is in place when the zygote begins to divide. However, few would agree that the nutrition of a specialized cell like the zygote is, in the final analysis, grounded on the advance performance of another cell of a different genotype. This has led to a renewed interest in the ultrastructural analysis of the embryo sac, because an elucidation of the mechanism of zygote nutrition logically requires a consideration of the structure of the embryo sac before fertilization. It now appears that in a broad sense, the structural modifications of the embryo sac devoted to the nutrition of the egg persist after fertilization or find their counterpart in the embryo sac of the fertilized egg. It is likely that several other modifications of the embryo sac and its surrounding tissues for the nutrition of the female gametophyte described in Chapter 6 are carried forward unchanged after fertilization and have the same primary consequences for the zygote.

The number of investigations that encompass the structure of the embryo sac wall after fertilization is comparatively small, and invariably all refer to the presence of wall projections akin to transfer cells. The projections originate from the inner wall of the embryo sac at the micropylar end close to the zygote in *Euphorbia helioscopia* (Gori 1977), cotton (Schulz and Jensen 1977), soybean (Tilton, Wilcox and Palmer 1984; Chamberlin, Horner and Palmer 1993a), *Arabidopsis* (Mansfield and Briarty 1991), and *Butomus umbellatus* (Fernando and Cass 1996), or adjacent to the degenerating nucellus in wheat (Morrison, O'Brien and Kuo 1978). In *Vicia faba*, invaginations from the embryo sac wall are formed close to the zygote, as well as at the chalazal end opposite the surviving cells of the nucellus, referred to as the nucellar cap (Johansson and Walles 1993b). In an investigation to trace autoradiographically the fate of ^{14}C-labeled photosynthates in the ovules of soybean, it was found that at the zygotic stage the label was concentrated over the wall projections of the embryo sac and in the hypostase at the chalazal end. The interesting conclusion drawn from this study – that two pathways of solute transport, a primary one from the outer integument to the base of the zygote and to the central cell, and a secondary one through the hypostase, operate to supply nutrients to the zygote – obviously bolsters a role for the wall invaginations in the process of nutrient supply (Chamberlin, Horner and Palmer 1993a). Zygote nutrition is a problem with wide ramifications and of considerable interest; the stimuli and nutrients that this cell normally receives from the rest of the plant body deserve to be characterized.

(ii) Embryo–Endosperm Relations

From the time embryologists first began to understand the relevance of the endosperm in seed formation, the function of nurturing the embryo has traditionally been assigned to this tissue. Although evidence in favor of the role of the endosperm as a nurse tissue for the developing embryo is more circumstantial than experimental, the idea has gained great credibility over the years. The absence of the endosperm in mature seeds of some plants known to produce this tissue at early stages of ovule development seems to indicate that the endosperm is not a permanent source of nutrition but is apparently consumed by the growing embryo. In seeking structural adaptations in the proembryos that mediate the transfer of nutrients from the endosperm, as discussed in the previous section, much attention has centered on wall proliferations from the suspensor cells. The invaginations from the outer wall of the suspensor cells in contact with the endosperm, as seen in *Capsella bursa-pastoris* (Schulz and Jensen 1969, 1974), *Diplotaxis erucoides* (Simoncioli 1974), *Phaseolus coccineus* (Yeung and Clutter 1979), *Vigna sinensis* (Hu, Zhu and Zee 1983), and *Glycine max* (Chamberlin,

Horner and Palmer 1993a), and from the inner wall of the embryo sac at the micropylar or chalazal pole, as seen in pea (Marinos 1970a; Harris and Chaffey 1986), *Haemanthus katherinae* (Newcomb 1978), *Medicago sativa* (Sangduen, Kreitner and Sorensen 1983a), and *Vigna sinensis* (Hu, Zhu and Zee 1983), seem in particular to invite considerable speculation about the possible functions of the invaginations in nutrient transport into or from the endosperm. One investigation has shown that as the zygote of *G. max* is phased into a multicellular embryo, there is a sequential development of invaginations in at least four different sites in the embryo sac wall in addition to some wall ingrowths that appear in the cells of the inner integument and nucellus. This observation has drawn attention to the possibility that the wall of the embryo sac functions both as a sink and a source of apoplastic transport to nourish the growing embryo (Folsom and Cass 1986). In view of the postulated linkage between wall ingrowths and short-distance transport of solutes, it does not seem unreasonable to find evidence for the presence of ingrowths on the walls of proembryos (Dute, Peterson and Rushing 1989) or the walls of the embryonic cotyledons (Johansson and Walles 1993a). It is relevant to point out that a plasma membrane almost invariably follows the contour of the wall projections; this may not be coincidental, since the membrane greatly increases the surface area of the invaginations. Occasionally, concentrations of various organelles in the vicinity of the wall proliferations have been described with the implication that they serve to increase the active transport of solutes across the plasma membrane. However, the possible direction of the flow of solutes remains unclear because there is nothing to preclude the absorption of nutrients from the endosperm into the suspensor or the transport of metabolites from the degenerating suspensor into the endosperm. A sense of uncertainty prevails also about the direction of solute flow through the invaginations that arise from the inner wall of the suspensor cells.

The timing of the direct absorption of endosperm nutrients by the embryo may also be worth emphasizing at this point. Direct absorption generally happens only after the suspensor has degenerated to the point that individual cells are separate from one another. The direct utilization of the endosperm by the growing embryo is indicated by the appearance of a conspicuous zone of destroyed endosperm cells around the cotyledons, as in dicots (Erdelská 1980, 1985, 1986), or ahead of the tip of the scutellum and coleoptile, as in cereals (Smart and O'Brien 1983; Jones and Rost 1989b). In a similar way, the accumulation of legumin and vicilin in the endosperm cells surrounding the embryo of *Vicia faba* and their subsequent decay may be looked upon as processes involved in the supply of nutrients to the growing embryo (Panitz et al 1995). Significant increases in the cation-dependent phosphatase activity during tobacco seed development as detected biochemically and by ultracytochemical localization methods have been attributed to the active transport of assimilates to the endosperm and embryo (Mogensen and Pollak 1983; Mogensen 1985).

To the extent that the cytoplasm of lysed cells can serve as a source of metabolites, the expanding growth of the embryo sac after fertilization has profound effects on the nutrition of the embryo. Expansion of the embryo sac is evident mostly at its micropylar or chalazal end when cells of the nucellus or integument begin to disintegrate and release free cytoplasm and nuclei; some of the newly liberated organelles are even incorporated into the endosperm (Pacini, Cresti and Sarfatti 1972; Norstog 1974; Schulz and Jensen 1974; Pacini, Simoncioli and Cresti 1975; Hardham 1976). In *Capsella bursa-pastoris*, cells of the residual intact chalazal nucellus enlarge to form a tissue (chalazal proliferating tissue) that protrudes into the embryo sac. The cells of this tissue actively incorporate precursors of both DNA and RNA and accumulate poly(A)-RNA during the period of their active growth. However, by the time the embryo attains the globular stage, degeneration in the chalazal proliferating tissue is initiated in the cells close to the embryo sac and continues until all the cells have broken down (Pollock and Jensen 1967; Schulz and Jensen 1971; Raghavan 1990b). The general interpretation of these various observations is that proliferations on the embryo sac wall aid in the passage of cytoplasmic nutrients from the lysed ovular cells into the central cell.

(iii) Nutrient Supply from the Mother Plant

It is difficult to find structural evidence for the supply of nutrients from the vegetative parts of the plant to the growing embryo. From physiological experiments it is known that seeds that accumulate large quantities of storage reserves in the embryo or in the endosperm act as powerful sinks for metabolites from other parts of the plant. With legume seeds, the use of $^{14}CO_2$, which is incorporated into photosynthates, has given the most conteine, and tryptophan (all at 10.0 mg/l) is an effective substitute for coconut milk (Norstog and Smith 1963). The effectiveness of endosperm extracts of *Cucumis sativus*, *Cucurbita maxima*, and *C. moschata* in promoting the growth of their respective embryos, mentioned earlier, is similarly matched by manipulation of the hormonal and organic additives to the medium, because a mixture of IAA, 1,3-diphenylurea, and casein hydrolyzate supports growth of embryos to the same extent as does a medium containing the embryo factor (Nakajima 1962). The conclusion from these results is that growth induction in cultured proembryos by endosperm extracts is a measure of the effect of specific chemical components of the endosperm.

Effect of the Physical Parameters of Culture. Interest in the effect of the physical conditions of culture on the continued growth of proembryos was stimulated by the common observation that the amorphous liquid endosperm in which the proembryos are constantly bathed has a high osmolarity. Measurements have shown that the ovular sap surrounding proembryos has a very low osmotic potential value (negative), which substantially decreases (becomes more negative) with embryo maturity (Ryczkowski 1960; J. G. Smith 1973; Yeung and Brown 1982). This has led to the view that the osmotic pressure of the medium might play a role in promoting the growth of proembryos in vitro. The importance of this parameter of the culture medium in the control of the growth of proembryos has not been critically tested, although the balance of evidence is in favor of adjustment of the medium to be isotonic with the ovular sap for this purpose. As indicated earlier, the embryo factor from coconut milk fully met the requirements for continued growth of pre–heart-shaped embryos of *D. stramonium*. However, another work showed that whereas mature embryos of this species grow in the complete absence of sucrose in the medium, earlier-stage embryos require progressively higher concentrations of sucrose. For example, a medium containing 8%–12% sucrose is found optimal for continued growth and differentiation of globular and pre–heart-shaped embryos. Interestingly enough, these embryos grow to the same extent in a medium containing 2% sucrose plus enough mannitol to be isotonic with 8%–12% sucrose (Rietsema, Satina and Blakeslee 1953). From this it appears that the essentiality of sucrose in embryo culture medium is to be accounted for wholly or in part as an osmotic stabilizer rather than as a carbon energy source. The successful culture of proembryos of *Capsella bursa-pastoris* (Rijven 1952; Veen 1963), *D. tatula* (Matsubara 1964), *Linum usitatissimum* (Pretová 1974), and *Triticum aestivum* (Fischer and Neuhaus 1995) has also been linked to the provision of a high osmolarity in the medium. Although success has been sporadic, a new small-size–limit in proembryo culture in a medium of high osmolarity has been attained with 25–45-μm-diameter proembryos of *Arabidopsis* (Wu et al 1992).

The need to gradually reduce the osmolarity of the medium during the growth of embryos without sequential transfer from one medium to another has resulted in two technical modifications of the culture system. For the culture of embryos of *C. bursa-pastoris* as small as 50 μm in diameter, a continual change in the osmolarity of the medium was obtained by using two media of different compositions, solidified in juxtaposition in a Petri dish. During growth in the medium of high osmolarity, the embryo was subjected to continually changing osmotic conditions due to diffusion from the medium of low osmolarity (Monnier 1976a). A culture system comprised of two agar layers, with the top layer having a higher osmolarity than the bottom layer, was used to obtain continued growth and differentiation of 8–36-celled proembryos of *Brassica juncea*. Proembryos were embedded in the top layer, whose osmolarity decreased during culture (Liu, Xu and Chua 1993a). Although no illuminating ideas as to how osmolarity of the medium regulates the cells of the tiny embryos to engage their intrinsic ability for division and growth have been generated by these investigations, an effective control of the flow of metabolites and inorganic ions into the cells of the embryo by the osmoticum appears a distinct possibility.

Role of Hormones. Because of the widespread use of hormones as adjuvants in tissue culture media for growth induction in organ, tissue, and cell cultures of plants, it is not unexpected that proembryo culture has come under the spell of hormones. Embryos of *C. bursa-pastoris* have figured prominently in the investigations using hormones. Although some workers (Rijven 1952; Veen 1963) had successfully cultured heart-shaped embryos in an inorganic liquid medium of high osmolarity secured by the addition of 12%–18% sucrose, a later work showed that it is possible to culture these same embryos in an agar-solidified mineral-salt medium supplemented with 2% sucrose. Growth of still smaller embryos (up to about 55 μm long) was secured by fortifying this medium with IAA, kinetin, and adenine sulfate (Raghavan and Torrey 1963). The addition of the hormone mixture eliminated the need for a medium of high osmolarity, suggesting that the picture of osmotic regulation of the growth of proembryos is somewhat misleading. Nonetheless, the need for a properly balanced mixture of hormones for the growth of these embryos is in a large part met by an osmoticum or by an increase in the concentration of major salts by a factor of 10. The more the mineral salts of the medium became fully competent to support the growth of proembryos, the more they appeared likely to be the governing factor in regulating proembryo growth. From a comparative analysis of different inorganic salt media, Monnier (1976a) found that the high-salt medium of Murashige-Skoog is superior for the culture of proembryos of *C. bursa-pastoris*. A concurrent experiment also revealed the interesting point that excellent growth of proembryos ensued when they were cultured in a modification of this medium with increased levels of Ca^+ and K^+ and a decreased concentration of NH_4NO_3. In other approaches, enhanced growth of proembryos of *C. bursa-pastoris* was obtained by supporting them on a bed of polyacrylamide instead of agar (Monnier 1975) or by increasing the partial pressure of O_2 in the medium (Monnier 1976b). Although *C. bursa-pastoris* has dominated the field of proembryo culture involving manipulation of hormonal and other additives to the medium, a successful report of culture of proembryos of *Linum usitatissimum* by kinetin (Pretová, 1986) is noteworthy. As experiments in the culture of early division phase proembryos and of the single-celled zygotes of these plants are contemplated, simultaneous changes in the chemical and physical conditions of culture offer a means of channeling the work in new directions.

Role of the Suspensor. As discussed earlier, in the division of labor of the whole embryo, the suspensor with its wall projections for solute absorption and transport is presumed to channel exogenous stimuli for the growth of the organogenetic part of the embryo. This is especially true at the proembryo stage, when the suspensor is at the prime of its life before degradative changes set in. This has raised the question as to whether it is possible to demonstrate a causal relationship between the presence of the suspensor and the growth of proembryos in culture. The simple approach to interfering with the influence of the suspensor is to sever its connection with the embryo and follow the growth of the embryo in culture. This is no doubt an exacting task; it has been performed with only a few species, perhaps involving a limited number of embryos. From experiments using this approach, it was found that continued growth of proembryos of *Eruca sativa* (Corsi 1972), *Phaseolus coccineus* (Nagl 1974; Yeung and Sussex 1979), and *C. bursa-pastoris* (Monnier 1984) is more enhanced in the presence of an attached suspensor than in its absence. Growth of embryos of *P. coccineus* is promoted even by the presence of a detached suspensor kept in close proximity in the medium (Yeung and Sussex 1979).

A string of experiments with embryos of *P. coccineus* involving supplementation of the medium with growth hormones and determination of the growth hormone levels separately in the embryo and suspensor has provided indirect evidence showing that the alleged suspensor effect is due to the production of GA and cytokinins. Whereas intact pre–cotyledonary stage embryos grow in a mineral-salt medium, the chance of survival of embryos of the same age deprived of the suspensor is very remote unless GA is present in the medium. GA is, however, inimical to the growth of suspensorless post–cotyledonary stage embryos as compared to intact embryos of the same age grown in a hormone-free medium (Cionini et al 1976; Yeung and Sussex 1979). These responses of embryos to GA are correlated with a falling level of endogenous GA in the organogenetic part and a rising level in the suspensor of the pre–cotyledonary stage embryos. As the suspensor begins to undergo senescence in the post–cotyledonary stage embryos, there is a sharp decline in the concentration of GA in the suspensor and a corresponding increase in its titer in the embryo proper (Alpi, Tognoni and D'Amato 1975). Likewise, addition of zeatin or zeatin riboside fosters the growth of suspensorless pre–cotyledonary stage embryos of *P. coccineus*, but post–cotyledonary stage embryos scarcely respond to the cytokinins (Bennici and Cionini 1979). As prototypes of hormones that promote cell divisions, cytokinins like zeatin and its riboside might be expected to be present in the organogenetic part of the post–cotyledonary stage embryos in a quantity sufficient to promote embryo growth. Indeed, this has been found to be the case, in a manner that is comparable to the presence of the same cytokinins in the suspensor part of the pre–cotyledonary stage embryos (Lorenzi et al 1978). These results provide a strong argument for a role for hormonal gradients from the suspensor in promoting growth of proembryos. This concept might serve as baseline infor-

Chapter 14

Genetic and Molecular Analysis of Embryogenesis

Completion of embryogenesis in angiosperms results in the formation of a full-term embryo, which is an assemblage of organs whose cells and tissues are specialized for functions essential for the initiation and continuation of the sporophytic life of the plant. As seen in Chapter 13, our current understanding of embryogenesis owes much to extensive descriptive accounts gained from light and electron microscopic studies and experimental work relating to the culture of embryos. Based on these accounts, it is convenient to consider embryogenesis in terms of a series of developmental processes aimed at creating a recognizable morphological structure. Most important among these processes are the establishment of the precise spatial organization of cells derived from the first few rounds of division of the zygote, the differentiation of cells of common origin to create diversity in the resulting embryo, and biochemical preparations for embryo maturation, desiccation, and dormancy. As part of the mature seed, the embryo unleashes another developmental program to initiate germination; however, since germination-related events are not considered in this book, the topic will not be a part of the discussion here. The developmental episodes during embryogenesis have little in common, yet there is a common background because they are at the root of many multidimensional processes involving different levels of interactions. In arbitrary terms we can conceptualize embryo development by stating that the critical aspects of morphogenesis and differentiation are completed relatively early in the time course of embryogenesis, whereas changes connected with maturation are relegated to later periods.

The primary objective of this chapter is to cast the major events of embryogenesis within a genetic and molecular perspective and to explain them in terms of changes in gene expression; an overview of the research efforts that reflect on the role of genomic information during maturation, desiccation, and dormancy of the embryo is also the subject matter of this chapter. Some of the questions that will be considered here are, How is the basic body plan of the embryo generated and what are the mechanisms of cell specification that occur during the initial stages of embryo development? How do the different parts of a developing seed interact in the formation of a mature embryo? What are the genes and processes that regulate embryogenesis? What are the molecular changes that direct polarity and organ differentiation in the embryo? How does the embryo acquire desiccation tolerance? It must be admitted at the outset that complete answers to these questions and to others of equal significance are not presently at hand, but investigators have begun to tackle such questions. Moreover, considering the fact that on a conservative estimate at least 20,000 diverse genes are believed to be active during embryogenesis, the problem of sorting through them to identify genes active at landmark stages should prove to be a formidable task.

Although plant embryo development differs in many striking and fundamental respects from embryogenesis in animal systems, some comments about the approaches used to study the controlling mechanisms in the construction of the animal from the egg are appropriate here. The backbone of research in the early part of this century was provided by classical genetic studies using mutant genes as markers to study pattern formation, cellular determination, and the origin of diversely committed cell lineages. An alternative strategy was to isolate mutants that lead to developmental defects often reflected in anomalous patterns of morphogenesis and differentiation. The abnormalities were attributed to lesions in information flow, namely, the flow from genes to proteins. Blossoming of the field of developmental genetics led to the discovery of lethal mutants, which die after completing parts of the embryogenic program. Lethal mutants were of particular interest to embryologists because the availability of organisms whose development was arrested at relatively early stages of embryogenesis made it possible to identify genes that perform essential functions during this critical period. The advent of molecular biology has led to investigations on the expression of specific genes during embryogenesis by a multifaceted approach involving nucleic acid hybridization, construction of stage-specific cDNA libraries, analysis of the nucleotide sequence of developmentally regulated genes, identification of their protein products, and transformation of eggs and embryos with cloned genes and antisense RNAs. Following the lead of animal developmental biology, genetic analyses, aided mostly by mutational dissection and combined with molecular characterization at different levels of complexity, have been increasingly applied to study the fundamental genomic strategies utilized by developing angiosperm embryos. Indeed, few organs of the angiosperm plant body have recently benefited as much as embryos from the integrated genetic and molecular approach of investigators. However, identification and characterization of genes or gene products associated with embryogenesis have barely begun.

Useful complements to the contents of this chapter are the reviews by Dure (1985), Meinke (1986, 1991a, b, c), Sheridan and Clark (1987), Goldberg, Barker and Perez-Grau (1989), Reinbothe, Tewes and Reinbothe (1992a), Lindsey and Topping (1993), de Jong, Schmidt and de Vries (1993), Goldberg, de Paiva and Yadegari (1994), Jürgens, Torres Ruiz and Berleth (1994), and Jürgens (1995), from which this chapter has drawn heavily to achieve a broad coverage and to present the newer developments in their proper historical perspective.

1. GENETIC CONTROL OF EMBRYOGENESIS

A pervading role for genes in embryo development in angiosperms has been inferred in the past from analyses of embryos following wide hybridizations between plants, from analyses of the administration of ionizing radiations to developing ovules, and from investigations of spontaneous mutants with altered patterns of embryo development. Although studies described in minute detail the morphological and cytological abnormalities in developing embryos, they did not attempt to analyze gene action at the molecular level in arresting embryo growth. Moreover, many cases of cytological damage were not supported by breeding experiments designed to help understand the control of the aberrant morphology at the genetic level. Despite these limitations, results have provided important clues as to how the genotype of participating gametes determines the growth of embryos and how perturbation of the genome of certain cells affects the replicative activity and the final form of embryos.

(i) Embryo Abortion in Wide Hybrids

One of the key problems in present-day horticultural and breeding practices is that crosses between distantly related species and genera of plants (wide hybridizations) have not yielded many agriculturally beneficial hybrids. From a genetic point of view, this is primarily due to the fact that in attempts to transfer beneficial alien genes across interspecific and intergeneric barriers, many deleterious genes that are difficult to eliminate also find their way into the hybrid genome. The resulting disturbance in the equilibrium between growth of the maternal tissues, embryo, and endosperm leads to embryo lethality and seed collapse. In an extensive review of embryo–endosperm relations in plants, Brink and Cooper (1947b) listed six types of genetic disturbances that cause embryo abortion in hybrids. These are (a) enforced self-fertilization in a normally cross-fertilizing species; (b) crosses involving parents of the same ploidy level but belonging to different species and genera; (c) crosses between parents of the same species but of different ploidy levels; (d) crosses between parents of different species and ploidy levels; (e) unbalanced chromosome conditions, especially aneuploidy in the endosperm; and (f) maternal genotypes causing embryo arrest irrespective of the source of the male gamete. For effective hybridization, union of the egg and sperm from members of the same species with identical chromosome numbers appears a basic necessity, and even a slight asymmetry at the species level or chromosomal level can cause loss of embryos. Numerous examples of ovule development in wide hybrids have been described in previous publications (Raghavan 1976b, 1977b) and so only representative cases will be considered here.

What are the kinds of disturbances that occur during wide hybridization and how do they modify the normal development of the ovule and result in embryo abortion? Of interest is the fact that the deleterious genes affect not only the growth of the embryo but also that of the endosperm. Although accounts of the ontogeny of hybrid ovules differ in details, there are some close parallels in the patterns of embryogenesis, and virtually all investigators seem to agree in identifying an early stage when developmental anomalies in the embryo begin to surface. This was first shown in reciprocal crosses between *Oenothera biennis* × *O. muricata* and *O. biennis* × *O. lamarckiana*, in which the initial growth of embryos was not appreciably affected; however, the growth slowed down later, and eventually the embryo did not advance beyond a few cells. Development progressed further in some afflicted embryos, but survival was generally erratic and ovules enclosing underdeveloped embryos matured into the familiar, shrunken, aborted seeds (Renner 1914). Quantitative, time-course analysis of the growth of embryos and descriptive accounts of cellular changes associated with embryo abortion have further characterized the effects of genomic instability in hybrids. Seed development in a cross between *Trifolium ambiguum* and *T. repens* exemplifies the time-course changes associated with embryo growth in normal and hybrid plants. In selfed *T. ambiguum*, the zygote embarks on the embryogenic pathway soon after fertilization and proceeds through the globular, heart-shaped, torpedo-shaped, and mature stages of embryogenesis with precise regularity. Early growth of the hybrid embryo almost betrays its developmental fate, since it outgrows the selfed embryo during the first two days after pollination in the number of cells formed. Growth of the embryo subsequently slows down and, in the hybrid seeds, the embryo produces at best a tightly knit mass of about 1,000 cells and does not differentiate beyond the heart-shaped stage (Williams and White 1976). Hybrid embryos generated from *T. semipilosum* × *T. repens* cross maintain the same rate of growth as selfed *T. semipilosum* embryos up to four days after pollination, but their further growth bogs down by the globular stage (White and Williams 1976). Although embryos from *T. repens* × *T. medium* hybrid exhibit a regular pattern of growth up to four or five days after pollination, mitotic activity is slower in the hybrid embryos than in embryos of both parental species (Kazimierska 1978). The view that the action of the deleterious genes is initiated in the hybrid at the time of fertilization or shortly thereafter seems to be supported by this observation.

In crosses involving different species of *Phaseolus*, a gradation is found in the stages at which deleterious genes become active and block embryo growth. In *P. vulgaris* × *P. lunatus* and the reciprocal cross, embryos cease to grow at the pre–heart-shaped and four-celled stages, respectively (Mok, Mok and Rabakoarihanta 1978; Rabakoarihanta, Mok and Mok 1979). Embryos of hybrids from *P. vulgaris* × *P. acutifolius* and *P. coccineus* × *P. vulgaris* crosses begin development in the normal way and reach the adult form, but gene action prevents embryo maturation (Mok, Mok and Rabakoarihanta 1978; Shii et al 1982). A further extreme stage of hybrid embryo development is seen in *P. vulgaris* × *P. coccineus* cross, in which embryos escape the action of deleterious genes and attain maturity (Shii et al 1982). In wide hybridization studies using various lines of *Hordeum vulgare*, *Secale cereale*, and *Triticum aestivum* as female parents and different or closely related genera of Poaceae as pollen donors, embryos succumb to the action of alien genes mostly at the globular stage (Zenkteler and Nitzsche 1984). These observations suggest that embryo abortion in wide hybrids is clearly linked in some way to postfertilization events in the embryo sac.

Among the questions of general importance concerning gene action is, what are the cellular changes in the hybrid embryos that lead to their demise? In a comparative ultrastructural study of embryo development in selfed *Hibiscus costatus* and *H. aculeatus* and *H. costatus* × *H. furcellatus* hybrids, Ashley (1972) found that the embryogenic development of the selfed parent begins with a pronounced shrinkage of the zygote and a marked segregation of organelles toward the chalazal end. In contrast, the zygote of the hybrid hardly changes physically or cytologically. However, it is with regard to the subsequent fate of the organelles that the hybrid embryo presents in a microcosm a good example of the damage inflicted by deleterious genes. Whereas the cells of the selfed embryo are endowed with a generous supply of organelles, cells of hybrid embryos are characterized by a paucity of organelles and extensive vacuolation, so that cells of the two types of embryos do not bear even the most tenuous resemblance to each other. Early degenerative changes observed in embryos of *Medicago sativa* × *M. scutellata* hybrid are confined to the suspensor, which is much more reduced than suspensors of intraspecific hybrids in the number of cells formed and in the distribution of organelles in each cell. Gene action also affects the functional capacity of the suspensor, since the characteristic wall invaginations mediating in metabolic interchange at the cell surface are absent (Sangduen, Kreitner and Sorensen 1983b). These results provide a different perspective of the development of the hybrid embryo due to cellular and physiological malfunctions.

As described in Chapter 12, by and large, the most general problems relating to the genetics of wide hybrids are posed by the deterioration of the endosperm and by successful embryo rescue operations. In this sense, the alien genes do not cause lethality of hybrid embryos and their effects can be considered as transient. Unfortunately, genetic analysis of hybrids to identify the individual genes that account for embryo or endosperm abnormalities has not been possible because of the many genetic changes found in wide crosses.

(ii) Radiation-induced Mutations

Because ionizing radiations such as X-rays and γ-rays are powerful mutagenic agents, they have been used in attempts to induce chromosomal and gene mutations in embryos. This approach is based on the premise that a dose of radiation that destroys the integrity of cellular DNA will cause abnormalities in the orderly development of the embryo and that these can be related to gene action. However, this objective has hardly been realized, as investigations on the effects of ionizing radiations on embryos have yielded a bag of mixed results. A large body of research on the irradiation of seeds resulted in many chromosomal mutations and in the formation of seedlings with altered morphology of the vegetative parts. Irradiation of ovules with immature embryos led to the formation of full-term embryos with abnormal organs, whereas irradiation of ovules enclosing very young embryos resulted in embryo lethality associated with abnormal cellular changes. What was lacking in these studies was a focus on specific genetic changes in the embryos so that their relationship to gene function could be understood.

Impressive radiobiological contributions to embryogenesis have come from work with *Hordeum vulgare*. In these studies, spikes of an inbred line were exposed to acute X- or γ-irradiation and subsequent growth of embryos under normal conditions was monitored. Although the level of radiation used did not lead to lethality, the histological heterogeneity of embryos made them prime targets for radiation; indeed, the relatively homogeneous early stage proembryos were less radiosensitive than the midstage proembryos and differentiating embryos. Depending upon the developmental stage of embryos at the time of irradiation, X-rays induced profound cell and tissue abnormalities that led to aberrant growth of the scutellum, root, and shoot. These disturbances can be traced to differential cell behavior in specific organs, where certain cells are killed, others begin to divide at a low amplitude, and still others continue to divide at the same rate as before. Neither the radiosensitivity of specific embryonic organs nor the drastic nature of the cellular damage seems to be anticipated during the early period of recovery of embryos because many cell generations intervene between the actual exposure and the occurrence of microscopically detectable symptoms. A particular abnormality resulting in anomalous coleoptile formation, designated as "cleft coleoptile," was extremely stage specific and appeared only following irradiation of the spike at the midproembryo stage; irradiation before or after the critical period failed to produce the cleft. In the midproembryo stage, the scutellum and shoot are most radiosensitive, whereas in the differentiating embryos, these organs become insensitive and the ontogenetically younger root moves up on the list (Eunus 1955; Mericle and Mericle 1957, 1961). These observations are in line with the view that genes for the initiation of specific organs are activated at different times during embryo ontogeny, with those for coleoptile growth being very stage specific.

Results similar to those obtained with *H. vulgare* have been reported for embryo abortion and developmental abnormalities in other plants following irradiation. In *Arabidopsis*, by timing the stage at which flowers or ovules are X-irradiated, it is possible to control organ formation in embryos at any developmental period between fertilization of the egg and embryonic tissue differentiation. Embryo abortion occurs when flowers harboring ovules with egg, zygote, or two-celled proembryo are irradiated. The transition of the young embryo from radial to bilateral symmetry marks a sharp turning point in embryogenesis as this is the most radiosensitive phase. Among the abnormalities resulting from irradiation at this period are changes in the number and orientation of cotyledons, which lead to syncotyly, tricotyly, anisocotyly, tetracotyly, and the formation of multiple embryos. Surprisingly, full-grown embryos escape the debilitating effects of a dose of radiation that is effective in embryos at earlier stages (Reinholz 1959). Genetic alterations thus cause the loss of growth control by cells only at a vulnerable period, and the result is abnormal morphogenesis in the developing embryos.

Indirect evidence in support of the view that a gene product is required for continued growth of the organogenetic part of the embryo but not of the suspensor has been obtained from experiments involving exposure of immature ovules to ionizing radiations. In general, the suspensor cells of irradiated embryos show a tendency to divide under conditions that arrest the growth of the embryonal part itself. Induction of additional divisions in the suspensor cells is especially rapid when prefertilization ovules or ovules harboring proembryos are irradiated. Classic responses of the suspensor to irradiation are seen in *Nicotiana rustica* (Devreux and Scarascia Mugnozza 1962) and *Capsella bursa-pastoris* (Devreux 1963), in which the abnormal suspensors contain a file of 8–17 cells rather than the usual 4–7 cells; in *Arabidopsis*, with unusually long suspensors composed of several files of cells (Gerlach-Cruse 1969; Akhundova, Grinikh and Shevchenko 1978); and in *Eranthis hiemalis* (Ranunculaceae) (Haccius and Reichert 1964), in which cells of the suspensor burst into growth and establish new centers of mitosis as a prelude to forming adventive embryos. The requirement for a gene product for continued growth of embryos that was suggested by the irradiation experiments has been confirmed in recessive embryo-lethal mutants of *Arabidopsis* obtained by EMS mutagenesis and T-DNA insertional mutagenesis following transformation with *Agrobacterium tumefaciens*. In a few mutants, arrest of embryo growth is followed by abnormal divisions of the suspensor, resulting in a multi-tiered structure with as many as 150 cells in place of a normal suspensor of 6–8 cells (Marsden and Meinke 1985; Yadegari et al 1994). When the development of a mutant embryo is followed in concert with that of the suspensor, morphogenetic defects in the embryo appear to precede visible defects in the suspensor. This observation is consistent with the view that disruption of the growth of the embryo can cause anarchy in the suspensor, leading to uncontrolled growth. In these mutants, designated as *suspensor* (*sus*), defects are noted in the development of the embryo protoderm (*sus-2* and *sus-3*) and in the division of cells of the inner tier of the globular embryo (*sus-1*, *sus-2*, and *sus-3*) (Schwartz, Yeung and Meinke 1994). Another embryo-defective mutant is *twin* (*twn*), which produces a high percentage of twin seedlings (Figure 14.1); here, secondary embryos capable of surviving seed desiccation and germinating into seedlings originate by continued division of cells of the suspensor (Vernon and Meinke 1994). Two alternative predictions made by genetic models are that *SUS* genes are required for normal embryo development and suspensor identity and that these genes are involved in the production of an undefined signal for inducing the same morphogenetic changes. In any case, it is clear that continued growth of the suspensor is inhibited by the organogenetic part of the embryo and that when the inhibitory effect is removed by mutation or by treatments that arrest embryo growth, the suspensor is able to express its full developmental potential. In the case of the *twn* mutation, such a simplistic explanation is probably untenable because the mutation, besides promoting embryogenic potential of the suspensor, disrupts the normal development of the embryo in various ways.

In order to appreciate the mutagenic effects of ionizing radiations on embryos, it is worthwhile to remember that, in the experiments described, embryos were shielded during irradiation by the ovular tissues and endosperm. To what extent do the abnormalities result from the direct action of radiation on the embryo itself and how much is indirect? In experiments in which embryos excised

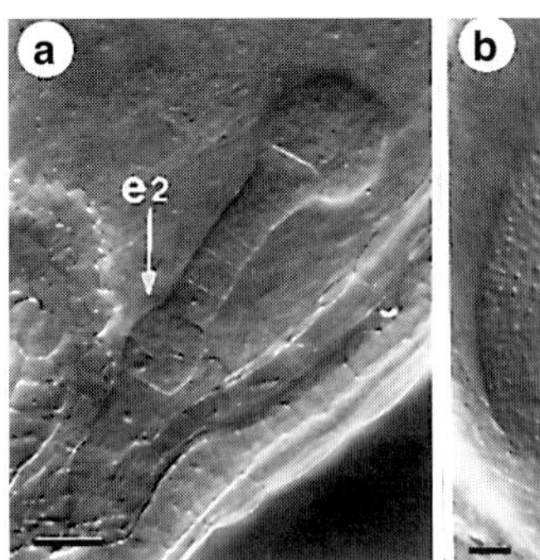

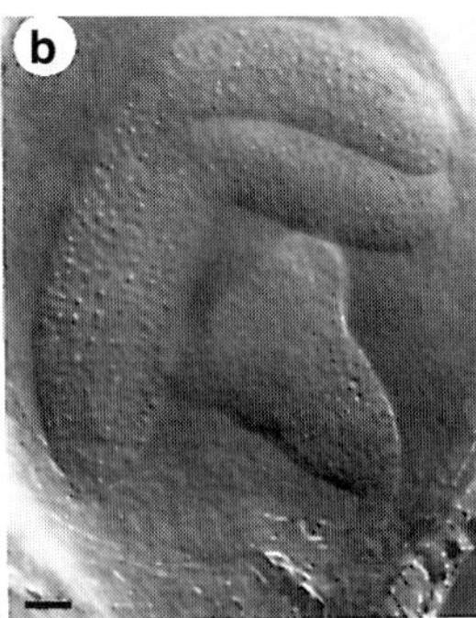

Figure 14.1. Suspensor development in mutants of *Arabidopsis*. **(a)** Initiation of a supernumerary embryo from the suspensor of a mutant seed. The cell adjacent to the basal cell of the suspensor has divided to form an octant stage embryo (e2). **(b)** Twin embryos in a mutant seed. The embryos are not linked with the suspensor. Scale bars = 20 μm. (From Vernon and Meinke 1994; photographs supplied by Dr. D. W. Meinke.)

from unirradiated cereal grains were transplanted into irradiated endosperm, one finds that the irradiated tissue is endowed with the same mutagenic power as direct irradiation to reduce the survival period of embryos and to cause chromosomal aberrations in their cells (Avanzi et al 1967; Floris, Meletti and D'Amato 1970). Along the same lines, other experiments have shown that the use of irradiated nutrient medium or the addition of irradiated sucrose to the nutrient medium causes cytological and morphological changes reminiscent of mutagenic effects in cultured seed embryos and somatic embryos (Swaminathan, Chopra and Bhaskaran 1962; Ammirato and Steward 1969).

Another approach in this research area has taken advantage of the finding that embryo growth is impaired following fertilization of a normal egg with sperm originating from irradiated pollen grains. Despite the fact that mitosis in the nucleus of the generative cell of the irradiated pollen grain is often abnormal, pollen grains germinate and produce pollen tubes, and sperm cells participate in double fertilization in the embryo sac. Cytological abnormalities are common in the cells of embryos developing from the union of one set of normal chromosomes of the egg and one set of irradiated chromosomes of the sperm (Brewbaker and Emery 1962). Irradiation of pollen grains of *Lilium formosanum* with a high dose of X-rays did not impair pollen germination, pollen tube growth, or production of functional sperm cells, but the treatment induced dominant lethality that results in the failure of normal zygote formation upon fertilization of an unirradiated egg. Although the endosperm nucleus undergoes a few rounds of division and thus succumbs later than the incipient zygote, the cells generated show more cytological abnormalities than the zygote (Brown and Cave 1953, 1954; Cave and Brown 1954). Comparative studies of the effects of irradiation of the male and female gametophytes on the development of the embryo in *Triticum durum* showed that irradiated sperm affected embryo development more than an irradiated egg did (Donini and Hussain 1968). Mutations did not result in the immediate death of the zygote, but degeneration of the proembryo due to necrosis of cells was frequently observed after fertilization of *T. durum* egg with irradiated sperm. The usefulness of these lethals to the study of the developmental effects of mutations is limited because they cannot be used for progeny tests.

(iii) Spontaneous Embryo Mutants

Spontaneous embryo mutants have been identified with precise regularity in some crop plants and are thus highly suitable for studying developmental problems of embryogenesis. Moreover, compared to the lethal mutants generated by radiation treatments, most defective embryo mutants are viable when maintained as heterozygotes. Many spontaneous embryo mutations have been described, ranging from those that impair basic physiological processes of embryos, such as their lapse into quiescence or dormancy, to those that affect biochemical characteristics such as pigment or storage protein accumulation. Mutations that interfere with the quiescence of embryos induce premature embryo germination. This happens during the maturation of the seed when the embryo begins to germinate and form a seedling although the seed is still attached to the mother plant. Such mutants are known as *viviparous* (*vp*), and the phenomenon is known as vivipary.

Vivipary. Due to its widespread occurrence, vivipary in maize is better understood than in other plants. Not only has the genetic control of vivipary been clearly demonstrated in maize, the system has also proved to be an attractive one for analyzing the various ramifications of gene action in regulating the maturation pathway of the embryo and endosperm in the grain. In maize, at least nine recessive genes (*vp-1, vp-2, vp-5, vp-7, vp-8, vp-9, w-3, y-9,* and *al*) that can induce vivipary are known and their respective genotypes have been described in detail. Genetic and physiological studies have shown that besides causing premature ger-

of two different genotypes, the *VP-1* gene seems to integrate the control of embryogenesis and germination in an unusual way.

It was mentioned earlier that the synthesis of globulins in viviparous embryos is related to the actions of the mutant gene and ABA. Northern blot analysis showed that whereas *GLB-1* and *GLB-2* transcripts are not detected in *vp-1* mutant embryos, they are present at a slightly reduced level in embryos of certain viviparous lines deficient in ABA (for example, *vp-5*). Since the expression of *GLB-1* and *GLB-2* transcripts is enhanced in the wild-type embryos by ABA, the function of the *VP-1* gene product could simply be to ensure the progress and completion of an ABA-regulated process in the maturation of the embryo (Kriz, Wallace and Paiva 1989; Paiva and Kriz 1994). Synthesis and accumulation of several LEA proteins in embryos of *vp* mutants are modulated differently than are those of globulins. The list includes a group of 23–25-kDa LEA proteins that appear in embryos of wild-type and certain ABA-deficient mutants (*vp-2, vp-5*, and *vp-7*) of maize. The proteins and their mRNAs are also precociously induced by ABA in young wild-type and mutant embryos (Pla et al 1989; Mao et al 1995). In the *vp-2* mutant, the pattern of transcript expression of the gene *RAB-28* (for responsive to ABA), which encodes a 27-kDa protein, is strikingly similar to that found in wild-type embryos. Although *RAB-28* mRNA is not detected in the embryos of *vp-1* mutant, it is inducible by low concentrations of ABA (Pla et al 1991). Another LEA protein, MLG-3, is detected in negligible amounts in *vp-5* mutant embryos, but the latter accumulate the protein when they are cultured in the presence of ABA. However, culturing *vp-1* mutant embryos in ABA does not result in the accumulation of MLG-3, whereas nongerminating *vp-1* embryos retained on the ear or cultured in a high osmoticum can accumulate this protein (Thomann et al 1992). The most direct interpretation of these results is that MLG-3 proteins, unlike globulins, have no specific requirement for the *VP-1* gene product, but other signals, such as ABA or osmotic changes, contribute to their expression in the mature embryo. On the other hand, the regulation of expression of the *EM* gene in *vp-1* mutant embryos is rigorously dependent upon the presence of ABA in the medium and upon its perception through a mechanism mediated by the *VP-1* gene, because no *EM* mRNA accumulates in mutant embryos of any age growing on the ear, or cultured in ABA, or subjected to an acute osmotic stress (Butler and Cuming 1993; Paiva and Kriz 1994).

Studies on vivipary in other plants have added few new dimensions to our present understanding of this phenomenon. In mangroves such as *Rhizophora mangle*, in which vivipary is a natural phenomenon, embryos have a high water content and are generally insensitive to inhibition by ABA in culture (Sussex 1975). This probably facilitates transition of the embryo from a developmental mode to a germination mode while still contained within the seed. The relationship between a low ABA level in the seed and vivipary was confirmed in an ABA-deficient mutant of *Arabidopsis* in which the seeds were induced to germinate in the fruit in an atmosphere of high humidity (Karssen et al 1983). In other mutants in which embryos maintain a high water content, seeds often showed vivipary without any further treatment (Koornneef et al 1989). This is what one would expect of embryos growing in ovules deficient in ABA. To the extent that water stress in plants is known to cause a rise in the ABA level of cells, it is of interest to note that vivipary is associated with a high water content in the embryos. A general conclusion is that ABA might regulate the metabolism of the embryo during the transitional period between the end of embryogenic development and onset of germination by its effects on the water potential of the cells. Screening a rice genomic library with maize *VP-1* gene as a probe has led to the isolation of a gene functionally homologous to the maize gene in transient expression experiments using rice protoplasts; transcripts of the gene are detected in rice embryos as early as 10 days after flowering and decrease toward maturity (Hattori, Terada and Hamasuna 1994). Isolation of genes for vivipary in rice has not hitherto been accomplished and offers great promise in broader studies.

Seed Mutants Deficient in Storage Products. Growers have known for a long time that inbred lines of some cultivated crops produce seeds that are genetically defective in the expression of specific storage products in the embryo. Indeed, it was the character of the pea seeds that affects the accumulation of storage products in the cotyledons (wrinkled and green seeds) that Mendel used in the famous experiments on the inheritance pattern involving the independent assortment of more than one gene. It is appropriate that this natural collection of mutant seeds has been studied to clarify the molecular basis for the derangement in storage product accumulation.

In pea, the production of wrinkled seeds is governed by two genetic loci, *r* (for rugosus) and *rb*, and it has often been argued that these genes separately affect starch metabolism of the embryo in a manner that results in a reduced starch content with a concomitant reduction in starch granule size and an increase in the proportion of the amylose content of the starch; alleles at the *rb* locus also profoundly affect the growth of the testa. Comparative analyses of the effects of the *r* locus on the growth and development of embryos of near-isogenic round (*RR*) and wrinkled (*rr*) lines of peas have revealed several physiological changes leading to a derangement in starch metabolism. One invariable change is the overall high fresh weight of embryos of the mutant line compared to embryos homozygous for the *R*-allele (Hedley et al 1986, 1994; Lloyd, Wang and Hedley 1996). Throughout most of the developmental period studied, cells of mutant embryos also showed a notable effect of the lesion in possessing a higher water content and in having a higher osmotic pressure than normal embryos (Wang et al 1987b). A genetic effect on cell size appeared somewhat late in embryogenesis and was reflected in the embryos of the wrinkled line having larger cells than those of the normal round line (Ambrose et al 1987). Other side effects of the mutant gene are seen in the lipid content and in the composition of the storage proteins of the embryos; generally, mature *rr* embryos contain more lipid and less of the storage protein legumin than *RR* embryos (Davies 1980; Coxon and Davies 1982). Some of the parameters of wrinkled seeds, such as starch content and composition and lipid content, could be reproduced in chemically induced mutants of peas (T. L. Wang et al 1990).

An important insight into the basis for the difference in the osmotic potential between *RR* and *rr* embryos was gained when it was found that wrinkled seeds have higher levels of free sucrose than normal seeds (Stickland and Wilson 1983). Since the increased sucrose content will lead to a high osmotic pressure in the cells, questions concerning this phenomenon center on the defect in the conversion of sucrose to starch. At the biochemical level, the *r* mutation is believed to lead to a reduction in the activity of the starch-branching enzyme (1,4–α–D-glucan, 1,4–α–D-glucan-6-glycosyltransferase), which results in the complete absence of one isoform of this enzyme (Matters and Boyer 1982; Smith 1988). This may cause a block in the pathway of starch synthesis in embryos of wrinkled peas and may lead to a consequent accumulation of sucrose.

Among other attempts made to relate metabolic consequences of gene action to starch synthesis in the embryos of *rr* seeds is the cloning of the *r* locus. Bhattacharyya et al (1990) used an antibody to the isoform of the starch-branching enzyme to clone a cDNA made to poly(A)-RNA of embryos of *RR* seeds and showed that transcripts encoding the enzyme are about 10-fold less abundant in the mutant embryos than in the normal ones. Furthermore, the absence of starch-branching enzyme in the mutant embryos has been attributed to an insertion in the gene encoding the enzyme and, because of the presence of inverted repeats at its termini, the inserted sequence is considered to be transposonlike.

Lectins are sugar-binding proteins and glycoproteins of wide occurrence in plants. Genetic defects that interfere with embryo lectin contents are common among cultivated varieties of soybeans. Molecular studies on a lectinless line of soybean have attributed the phenotypic condition to a virtual absence of lectin transcripts in the genome. The transcription lesion in the lectinless phenotype is apparently due to a silencing of the lectin gene caused by the transposonlike insertion of a 3.4-kb DNA segment (Vodkin 1981; Goldberg, Hoschek and Vodkin 1983; Vodkin, Rhodes and Goldberg 1983). It is of particular interest that seeds of transgenic tobacco plants harboring a chimeric lectin gene containing the 5'-half of the mutant gene and its promoter region and the 3'-half of the wild-type gene synthesize both lectin mRNA and protein. This shows that the control region of the mutant gene is transcriptionally competent but that insertion of the transposonlike element blocks gene transcription by silencing the promoter (Okamuro and Goldberg 1992). How the promoter functions are impaired by the transposon element remains unknown and should prove to be an exciting area of future research.

Mutation in a naturally occurring variety of soybean that lacks the α'-subunit of the 7S seed storage protein, β-conglycinin, has been found to be caused by a deletion that extends through most of the coding sequences of the gene (Ladin, Doyle and Beachy 1984). Another soybean seed protein mutation includes an inversion that separates the 5'- and 3'-regions of an aberrant glycinin gene, *GY-3* (Cho et al 1989). Because the synthesis of the respective transcripts predicted in the wild-type embryos is inhibited in the mutants by lesions in the corresponding genes, it follows that the mutations affect gene action at the transcriptional level.

A different picture emerges in considering the molecular basis of a null allele of a glycinin gene (*GY-4*) and a mutation in a *KTI* gene in soybean

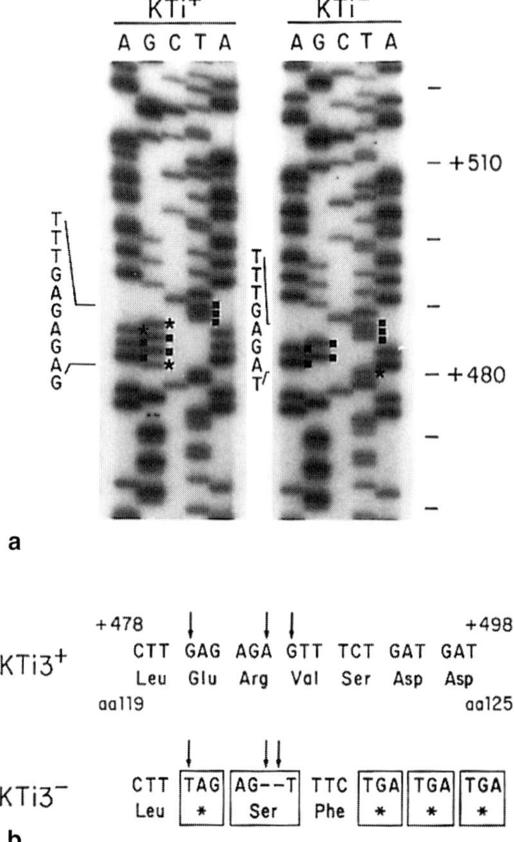

Figure 14.3. Nucleotide and amino acid sequence of Kunitz trypsin inhibitor genes in wild-type (KTi+) and mutant (KTi-) embryos of soybean. **(a)** DNA sequence ladders represent nucleotides +460 to +520 of the two genes. Nucleotides +481 to +490 of the wild type and the corresponding mutant sequences are shown to the left. Squares show the nucleotides that are conserved between the two genes; asterisks indicate nucleotides that are mutated in the KTi- gene. **(b)** Nucleotide and amino acid sequence comparisons of *KTI* genes of wild-type and mutant embryos. Only nucleotides +478 to +498 and translated amino acids 119 to 125 are shown. Arrows refer to the nucleotides mutated in the KTi- gene (shown as asterisks in [a]). The asterisks designate translational stop codons that result from the mutations. (From Jofuku, Schipper and Goldberg 1989. *Plant Cell* 1:427–435. © American Society of Plant Physiologists. Print supplied by Dr. R. B. Goldberg.)

embryos. The null phenotype is attributed to a point mutation that leads to the elimination of the translation start codon due to a single base change. Although this does not interfere with transcription, mRNA is not effectively retained on the polysomes like the functional mRNA (Scallon, Dickinson and Nielsen 1987). In the mutant defective in the gene encoding KTI, a frame-shift mutation due to alteration of three nucleotides causes drastically reduced accumulation of *KTI* mRNA and KTI in the embryos (Figure 14.3); it has also been found that the *KTI* gene is transcribed at the same relative rate in wild-type and mutant embryos (Jofuku, Schipper and Goldberg 1989). Evidently, some posttranscriptional modifications impose restraints on the stability of mRNAs and reduce the prevalence of mRNAs or of the protein products in the mutant embryos.

The lectin-nonproducing mutant seeds of soybean have a counterpart in certain genetic mutants of *Phaseolus vulgaris*. In cultivars like Pinto, mutations of the genes encoding the lectin, PHA, have the effect of producing embryos that are deficient in lectins. At the biochemical level, the effects of the mutation have been interpreted as the consequence of a reduced synthesis of PHA and an equally low level of PHA gene transcription (Staswick and Chrispeels 1984; Horowitz 1985; Chappel and Chrispeels 1986). A partial explanation for the molecular basis of the mutation was provided by the cloning and sequencing of the two PHA genes of Pinto called *PDLEC-1* and *PDLEC-2*. The latter follows very closely the sequence of the genes of the wild type and of *PDLEC-1* up to –240 bp relative to the transcription site. Compared to other genes, farther upstream, *PDLEC-2* suffers a 100-bp deletion that contains a 60-bp tandem repeat. Since the deleted segment might function as an enhancerlike element, this deletion could account for the low level of expression of *PDLEC-2* (Voelker, Staswick and Chrispeels 1986). In support of this hypothesis, it has been found that embryos of transgenic tobacco seeds harboring a *PDLEC-2* allele accumulate about 50 times less PHA than embryos of plants transformed with a gene from the normal cultivar (Voelker, Strum and Chrispeels 1987). But this clearly is not the whole explanation for the effect of the mutation, given the lack of knowledge about the causes for the low expression of *PDLEC-1*.

2. GENE EXPRESSION PATTERNS DURING EMBRYOGENESIS

Included among the genes expressed during embryogenesis are those that encode enzymes for housekeeping activities and mobilization of food reserves, structural proteins of the cytoskeletons and membranes, and storage proteins and embryo-specific genes that regulate specific aspects of embryo development. Although we are a long way from characterizing these genes and their products and analyzing the regulatory pathways of gene action, current experimental approaches have led to new insights into the molecular processes controlling gene expression pattern during embryogenesis.

(i) Diversity and Changes of mRNA Population during Embryogenesis

The modern global view of quantitative changes in the gene expression program during embryogenesis was developed by Goldberg et al (1981b), Galau and Dure (1981), and Morton et al (1983) from reassociation kinetics of nucleic acids of embryos of soybean, cotton, and pea, respectively. Because of the small size of the early stage embryos, which precludes their isolation in ample quantities for preparation of nucleic acids, these studies were undertaken with embryos in advanced stages of development; yet the power of this approach to the study of gene regulation is unmistakably demonstrated in these investigations.

As an intermediate between the gene and the protein it encodes, mRNA populations in embryos at different stages of development should provide an overview, down to the details of changing complexity, abundance distribution, and diversity of the genes active during embryogenesis. Hybridization experiments using an excess of mRNA from cotyledon stage (30 days after flowering) and midmaturation stage (75 days after flowering) embryos of soybean with cDNA prepared from corresponding mRNAs have shown stage-specific changes in mRNA complexity, ranging from 15,000 average-sized mRNAs in the cotyledon stage embryos to more than 30,000 in midmaturation stage embryos. Considering the simple morphology of the cotyledon stage embryos, the degree of specification at the level of mRNAs is apparently enormous. In terms of the number of molecules of mRNA present per cell per sequence, it is possible to delineate superabundant (>19,000 molecules), moderately prevalent (550–800 molecules), and rare (3–17 molecules) classes of mRNAs in the soybean embryo cells. Attesting to the diversity in the distribution of the different classes of mRNA is the fact that in the cotyledon stage embryos, mRNAs belong to the moderately prevalent and rare classes, whereas in the later-stage embryos they are predominantly of the superabundant class, with small amounts of the moderately prevalent and rare classes. Interestingly, the superabundant class, which accounts for nearly 50% of the mRNA mass of midmaturation stage embryos, contains only 6–10 diverse mRNAs. Despite the fact that morphological complexity of the embryo increases with development, a surprising finding is that cotyledon stage embryos and midmaturation stage embryo axes and cotyledons contain approximately the same total number (14,000–18,000) of diverse mRNAs. This signifies that in terms of the absolute number of structural genes expressed, there are no detectable changes in gene expression during morphological differentiation of the embryo and within morphologically distinct parts of the embryo (Goldberg et al 1981b).

Another approach to the analysis of mRNA changes during soybean embryo development was by the hybridization of kinetic fractions of midmaturation stage embryo-abundant and rare cDNA (selected by repeated cycles of hybridization with homologous mRNA, followed by hydroxyapatite fractionation) to an excess of cotyledon stage and midmaturation stage embryo mRNA. Comparison of the mRNA sequences present in embryos of the two developmental stages showed that almost the entire sequence comprising the mass and diversity of mRNAs of the older embryos is contained in the young embryos. Although this comparison is sensitive only to changes in at least hundreds of rare transcripts, the logical conclusion is that mRNAs of embryos of different ages tend to be qualitatively similar. The interesting possibility from these results is that in soybean the entire set of structural genes that encode proteins of embryogenesis is transcribed at a very early stage, regardless of whether the genes are translated or not. Finally, an intriguing feature of the fate of embryogenic mRNA sequences is that most of them persist throughout development and are stored in the cells of the mature seed embryo (Goldberg et al 1981b).

Quantitative changes in the gene expression program during embryogenesis in cotton have been followed by RNA-excess DNA–RNA hybridization studies that have identified mRNA sets from two-dimensional SDS-PAGE analyses of extant proteins, proteins synthesized in vivo and in vitro, and the changing mRNA populations in embryos of different ages. As in soybean, DNA–RNA hybridization with mRNA populations from young (50 mg fresh weight), maturation phase (110 mg fresh weight), mature (from seeds), and germinated embryos implies that 15,000–20,000 genes are expressed at the mRNA level during embryogenesis. However, only a few hundred genes encode mRNAs that are abundant enough to be detected in profiles of two-dimensional gels of in vitro translation products. The analysis of extant, in vivo and in vitro synthesized proteins allowed the changes in the abundant mRNAs and abundant proteins to be followed with

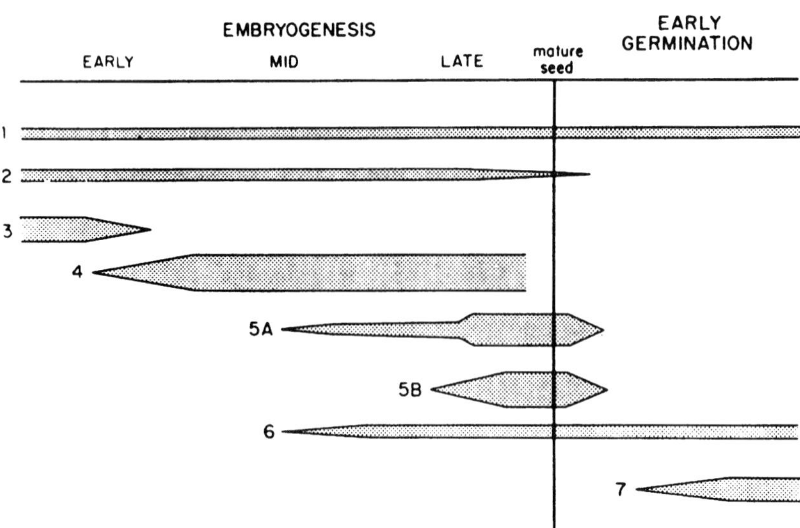

Figure 14.4. A diagrammatic representation of the groups of abundant mRNAs demonstrated in developing and germinating cotton embryos. (From Dure 1985. *Oxford Surv. Plant Mol. Cell Biol.* 2:179–197, by permission of Oxford University Press.)

a focus on two major questions. Does the population of extant proteins in the young embryo include some proteins whose synthesis is not required for continued differentiation of the embryo? Does a mature embryo synthesize proteins that are not needed for the survival of the young seedling? When the soluble protein fractions from cotyledons of embryos of different ages are analyzed, substantial modulations are found to occur in several sets of proteins whose synthesis is developmentally regulated; gene sets expressed in embryos of the same ages are deduced from in vitro translation analysis of their mRNAs. These data are diagrammatically presented in Figure 14.4 as groups of abundant mRNAs expressed during embryogenesis. mRNAs of groups 1 and 2 are present throughout embryogenesis; whereas group 1 mRNAs are expressed during germination as well, those of group 2 are turned off after embryogenesis is completed. Compared to group 2 mRNAs, which encode for only six proteins, group 1 mRNAs encode an impressively large set of proteins, including actin and tubulin. Another point of comparison is between the products of cell-free translation of mRNAs isolated from embryos of different ages and proteins synthesized in vivo by embryos of the same age; this comparison has shown that many translatable mRNAs of group 1 are detected only when their corresponding proteins are synthesized in vivo, indicating that gene activity with regard to the synthesis of these proteins is regulated at the transcriptional level. However, proteins encoded by group 2 mRNAs are synthesized in vitro by embryos of all ages.

Expression of group 3 mRNAs occurs coincident with the burst of mitotic activity during early embryogenesis but, as cell divisions cease, their genes are turned off in the maturation phase and in seed embryos. Proteins encoded by these transcripts accumulate extensively in embryos of all developmental stages, although they are synthesized only in young and maturation stage embryos. The most extensively changing mRNAs are those that code for storage proteins (group 4). These mRNAs are found in high levels in maturation stage embryos but vanish completely in mature embryos. Two other groups of mRNAs (groups 5 and 6), which become abundant during late embryogenesis and accumulate in seed embryos, differ in one important respect. Group 5 mRNAs, which consist of two subsets and their corresponding proteins, disappear from the abundant class during the early hours of germination but can be reinstated by incubating excised immature embryos in ABA. This is just what one would expect of a group of transcripts and their protein products whose synthesis in the embryos is regulated by endogenous ABA. The other group of mRNAs (group 6) and their proteins persist during normal germination of seed embryos, as well as during precocious germination of immature embryos. There is also one final group of mRNAs (group 7), which is not expressed during embryogenesis but appears during germination (Dure,

Greenway and Galau 1981; Galau and Dure 1981; Chlan and Dure 1983).

Comparisons of reassociation kinetics between cDNAs from cotton embryos of all three developmental stages and an excess of homologous mRNAs have identified the same groups of mRNAs whose abundant proteins are synthesized in vivo and in vitro, as well as several less abundant mRNAs. These mRNAs belong to abundance classes that show more than a twofold increase in their sequences in the seed embryos. Another type of analysis has involved hybridization of each cDNA clone to mRNAs isolated from other embryo stages. Such heterologous reassociations have shown that embryos of all ages contain most of the mRNA sequences, although sequences prevalent at one stage may be present at a reduced level in other stages. Studies of this type have also focused on some major differences in the concentrations of the more abundant mRNAs of embryos of different stages of development, the most dramatic being a stark reduction in the presumptive storage protein mRNAs in the mature embryo (Galau and Dure 1981).

The hybridization kinetics of mRNAs from cotyledons of developing pea embryos to cDNAs show much that is common with, and much that is different from, those of soybean and cotton. The total complexity of mRNA sequences of different abundance classes at the early stage of cotyledon development in pea (ca. 20,000) compares favorably with that of cotton (ca. 25,000) and soybean (ca. 32,000) embryos of slightly different ages. As noted in soybean, there is an increase in the very abundant class of mRNAs during embryo development when storage protein synthesis is initiated and sustained. However, compared to soybean and cotton, in pea the rare class of mRNAs completely disappears as embryogenesis proceeds, resulting in a 10-fold decrease in the number of mRNA sequences in embryos at the late stage of cotyledon development (Morton et al 1983). The obvious adaptive value of this strategy is that genes for the production of enzymes and structural proteins are no longer required and are switched off as the embryo begins to accumulate storage proteins. Since this reduces the number of active genes in the cell to about 200, some questions have been raised about the survival capacity of such cells (Dure 1985).

(ii) Gene Activity during Fertilization and Early Zygote Divisions

There have been very few investigations on the molecular biology of fertilization in gymnosperms and angiosperms, and the mechanism of gene activation during the early division phase of the zygote has completely eluded us. A major impediment to these studies is that events of fertilization and partitioning of the zygote take place within the confines of the archegonium in gymnosperms and the embryo sac in angiosperms. Reference was made in Chapter 13 to studies that seemed to indicate that fertilization of the egg and subsequent divisions of the zygote are accompanied by changes in the RNA content of the cells. Unfortunately, these investigations did not address issues such as the role of the maternal genome in fertilization and early embryo development. The concept of maternal message, which has its roots in animal embryology, assumes that the mature egg contains large stores of mRNAs that become available after fertilization to code for the first proteins of the zygote. There is no hard evidence for the existence of a legacy of template information in the gymnosperm or angiosperm egg, and the source of the genetic information parceled out for the division of the zygote is unknown. One study has shown that when rice and barley florets are cultured in a medium containing ^{3}H-uridine, no autoradiographically detectable incorporation of the label occurs in developing embryos of less than 100 cells. At the same time, ^{3}H-leucine is readily incorporated into embryos of all ages. Because the eggs are fertilized after the florets are placed in culture, assuming that the cells of the globular embryo do not possess a large uridine pool, it has been reasoned that protein synthesis in the absence of concurrent RNA synthesis probably involves the use of stored maternal mRNA (Nagato 1979). Based on in vitro translation of poly(A)-RNA isolated from ovules of *Pinus strobus* before and after fertilization, evidence was obtained for the transcription of a prominent mRNA at the time of fertilization and at the beginning of embryogenesis. A cDNA clone isolated from ovules containing globular embryos was found to encode a protein homologous to the 11S storage protein globulin. Since transcripts of this cDNA clone are found to accumulate in the megagametophytes of fertilized, but not of unfertilized, ovules, the trigger of the fertilization event in causing transcript abundance is obvious. Although transcripts are not detected in the developing embryos, this does not preclude a possible role for fertilization-induced signals from the embryo in transcript accumulation in the megagametophyte (Whitmore and Kriebel 1987; Baker et al 1996). Analysis of genes expressed in the embryo, with the exclusion of the female gametophyte and the sporophytic tissues that constitute the bulk of the ovule, may perhaps provide a clear picture of the embryo gene activation program during fertilization.

Based largely on the end result, namely, activation of the egg in the absence of fertilization (parthenogenesis), many claims have been made for an independent role for the maternal genome in directing early embryo development. These claims include the spontaneous division of the egg in the absence of a stimulus of fusion with the sperm, as in apomicts (Nogler 1984; Ramachandran and Raghavan 1992a); use of physical and chemical agents to inactivate the sperm without hindering the growth and penetration of the pollen tube and thus to stimulate the division of the egg (Lacadena 1974); and the purported origin of embryos from the eggs of cultured unfertilized ovules or ovaries (Yang and Zhou 1982; Yan, Yang and Jensen 1989; Ferrant and Bouharmont 1994). Haploid plants that arise by chromosome elimination in interspecific crosses between *Hordeum vulgare* and *H. bulbosum* also offer a basis for the assumption that maternal templates alone can support embryo development, albeit under somewhat abnormal conditions (Kasha and Kao 1970). Since the male chromosomes disappear even before they are integrated into the genome of the egg, the haploid embryo essentially develops by transcribing the stored maternal information. Although the examples considered here do not constitute a rigorous demonstration of the presence of stored mRNAs, there is probably some truth in the assertion that the angiosperm egg cell, whether it is fertilized or not, has the potential to initiate the developmental program of the embryo using maternal transcripts.

(iii) RNA Metabolism in Developing Embryos

Several early investigations have shown that progressive embryo development signals considerable changes in the temporal and spatial patterns of accumulation and synthesis of bulk RNA. In the wake of the isolation and characterization of embryo-specific genes, these studies can be considered to represent at best a gross approximation of gene activity in developing embryos. For this reason, only a small portion of this work will be discussed here.

One manifestation of RNA metabolism during embryogenesis is the continued accumulation of RNA, so that by the completion of embryogenesis, the amount of RNA in the mature embryo will have increased to many times the initial value of the early stage embryo. *Vicia faba* (Wheeler and Boulter 1967; Manteuffel et al 1976), *Phaseolus vulgaris* (Walbot 1971), *Hordeum distichum* (Duffus and Rosie 1975), *Oryza sativa* (Zhu, Shen and Tang 1980a), *Triticum vulgaris* (Zhu, Shen and Tang 1980b), and *Abelmoschus esculentus* (Malvaceae) (Bhalla, Singh and Malik 1980b), among others, are examples that illustrate a rising level of RNA through most periods of growth of the embryos. Investigations on *P. vulgaris* implicate the high water content of the embryo as the stimulus for increased RNA synthesis; as the water content of the embryo begins to decrease, the rate of RNA synthesis falls to a low value and remains at this level through the period of desiccation and dormancy (Walbot 1971, 1972). Other studies have shown that the RNA metabolism of embryos of different ages and that of specific parts of the same embryo are profoundly interrelated. When just the organogenetic part of embryos of different ages of *P. coccineus*, consisting of rapidly dividing cells, is analyzed, RNA content per cell and the rate of RNA synthesis per cell are highest in the early heart-shaped embryo and decline thereafter. In the numerically stabilized cell population of the suspensor, these parameters are low during early embryogenesis. The decrease in the RNA synthetic activity of the embryo cells probably reflects the inability of the nucleolus to synthesize new RNA to keep up with the rapid tempo of mitosis (Walbot et al 1972; Sussex et al 1973). Although the rate of RNA synthesis declines to near zero during desiccation of the embryo, protein accumulation is unabated; the possibility exists that the proteins are encoded by mRNAs synthesized at an earlier period in embryogenesis and stored for later use (Walbot 1971).

Investigations on the RNA metabolism of cotyledons of developing embryos have generally supported the results obtained with whole embryos. In the cotyledons of different species there is close agreement in the time of initiation of RNA synthesis, which generally follows DNA synthesis. RNA accumulates in cotyledons of near-normal water content in parallel with the period of rapid cell division activity; this is followed by a cessation or decrease of accumulation during desiccation. The major portion, if not all, of the RNA synthesized appears to be rRNA (Galitz and Howell 1965; Millerd and Whitfeld 1973; Poulson and Beevers 1973; Scharpé and van Parijs 1973; D. L. Smith 1973; Walbot 1973; Millerd and Spencer 1974). Considered together, the RNA metabolism of developing whole embryos and isolated cotyledons invariably indicates a changing biochemical pattern with time.

The rise and decline in RNA accumulation and

synthesis during embryogenesis reflect, to some extent, the availability of sufficient mRNA. Poulson and Beevers (1973) found that although the RNA content of pea cotyledons remains fairly constant during the seed maturation period, the capacity of the cotyledons for RNA synthesis declines considerably during this period. The abruptness of this change correlates well with the observation that ribosomal preparations made from cotyledons of mature embryos show a decreasing abundance of polysomes and are practically inefficient in supporting cell-free translation. Since the full cell-free translation potential of ribosomes is realized by the addition of an artificial messenger, experimental results are consistent with a declining mRNA component in the cotyledons with age (Beevers and Poulson 1972).

A genetic approach was used by Davies and Brewster (1975) to analyze RNA synthetic activity in developing pea cotyledons. The basis for this work was the observation that cotyledons of certain genotypes of pea show appreciable differences in their RNA contents. However, when plants with widely different rRNA contents are crossed, cotyledons of the F1 hybrid embryos are found to have an rRNA content identical to that of the maternal parent. Of the two models proposed to account for the essential silencing of a set of rRNA cistrons, one supposes that some aspect of gene activity, such as transcription or processing of RNA, is primarily under the control of the maternal genome. Alternatively, a selective activation of the maternal alleles might occur and result in an unusual situation in which one set of rRNA cistrons produces the same amount of RNA in the hybrid as two sets do in the parent. Clearly, both models are valid, but further work is necessary to choose between them.

Chapter 13 discussed the fact that in the cotyledons of legumes, the DNA content of cells increases during the first few days of cell expansion growth. Since the large increase in RNA content occurs almost in parallel with the increase in nuclear DNA amounts, a logical question is whether there is an apparent gene dosage effect on RNA synthesis in the cotyledons. One measure of the transcriptional activity of DNA is a change in the level of RNA polymerases. Experiments by Millerd and Spencer (1974) have proved illuminating by showing that cotyledons of a variety of pea do not synthesize additional RNA polymerase in proportion to the increasing amount of nuclear DNA template available. Since this occurs even when the template activity of chromatin in the cotyledons remains constant, a role for gene dosage, in the sense of loading the cytoplasm with transcription products in a quantitative way, does not appear to be valid. Investigations of Cullis (1976) with two other varieties of pea showed that an increase in the endogenous RNA polymerase activity of chromatin occurs even before the average DNA content increases in the majority of the cells and continues well into the period of DNA endoreduplication in most cells. This might suggest that some of the endoreduplicated DNA is used as a template for RNA synthesis. A new element of uncertainty was introduced into this issue when it was found in a later work that the extent of utilization of extra copies of DNA is associated with experimental conditions that affect the rate of growth of cotyledons. For example, in plants raised at 15 °C, which permits a tardy growth of cotyledons, RNA polymerase activity increases in proportion to the DNA content of cells over a wide developmental window. When cotyledons develop rapidly, as in plants grown at 30 °C, there is no increase in the level of chromatin-bound RNA polymerase activity in proportion to the DNA (Cullis 1978). The differences in the template activity of chromatin from two sets of cotyledons are interpretable in terms of differences in the number of transcriptionally active cistrons available or in the specific activity of the bound polymerase. These views are not inconsistent with the currently known mechanisms of regulation of DNA template activity.

In summary, it appears that there is no firm ground for the idea that, except under special conditions, transcriptional activity in the embryonic cotyledons is related to the availability of extra copies of DNA. Endoreduplication results in a multifold increase in the informational pool of the cells of the cotyledons, but the ultimate function and disposition of this information remain to be fully understood.

Gene expression pattern during embryogenesis has been deduced from changes in the most prevalent mRNA populations seen in cell-free translation profiles of embryos of different ages (Figure 14.5). The changes in the proteins synthesized and in the expression of their mRNAs during embryogenesis in cotton were described earlier. The existence of in vivo and in vitro synthesized polypeptide subsets that exhibit stage-specific expression appears to be a consistent feature of embryogenesis in maize (Sánchez-Martínez, Puigdomènech and Pagès 1986; Boothe and Walden 1990; Oishi and Bewley 1992), radish (Aspart et al 1984; Laroche-Raynal et al 1984), *Phaseolus vulgaris* (Misra and Bewley 1985b), rice (Mundy and Chua, 1988), and soybean

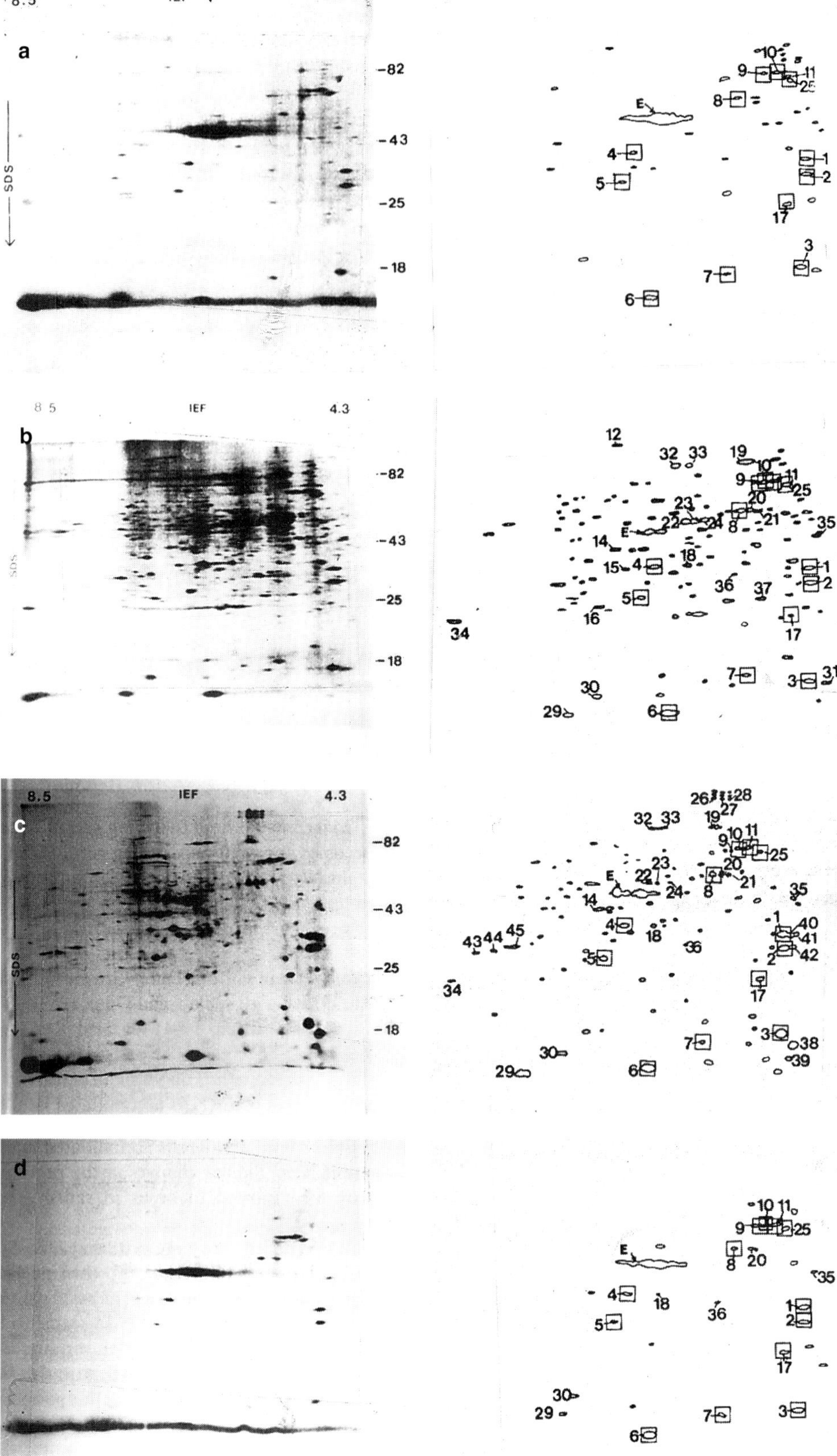

8.5
IEF
4.3
a
-82
-43
-25
-18
SDS
b
c
d

(Rosenberg and Rinne 1988). In both maize and radish, although the prevalent mRNAs show numerous changes during embryogenesis, two main subsets, one synthesized in young and immature embryos and the other accumulating in mature embryos, have been identified. Maturation of rice embryo is associated with the accumulation of mRNAs encoding about seven polypeptides ranging in molecular mass from 20 to 45 kDa. Most of the mRNA sequences of the mature embryos of these plants persist as stored mRNAs and, consequently, they are considered to be products of genes transcribed during embryogenesis. The concept of stored mRNAs in embryos of quiescent and dormant seeds of other angiosperms has been supported by a variety of experimental approaches. One important consideration about stored mRNAs is that they serve as templates for the synthesis of proteins of early germination, until the growing embryo begins to synthesize new mRNAs. In vitro translation products of mRNAs isolated from quiescent embryos of wheat (Helm and Abernethy 1990) and sorghum (Howarth 1990) have been shown to include transcripts of HSP genes. Additionally, cDNA libraries made to poly(A)-RNA of embryos excised from dry seeds of radish (Raynal et al 1989), *Vigna unguiculata* (Ishibashi and Minamikawa 1990; Ishibashi, Yamauchi and Minamikawa 1990), *V. radiata* (Manickam et al 1996), *Pinus thunbergii* (Masumori, Yamamoto and Sasaki 1992), and sunflower (Almoguera and Jordano 1992), and embryos excised from caryopses of rice (Ramachandran and Raghavan 1992b) and wheat (Kawashima et a 1992) have been differentially screened, and developmentally significant cDNA clones have been described. Of special interest is a gene that encodes a Zn-containing metallothionein-like protein isolated from mature wheat embryos. The expression of this protein is gene modulated in developing embryos by ABA and not by Zn^{2+}. In spite of much information about the structure and ubiquity of metallothioneins in animal cells, little is known about their function during plant embryogenesis.

Figure 14.5. Fluorographs showing the in vitro translation profiles of proteins synthesized by RNA from embryo axes of *Phaseolus vulgaris* of different stages of development. Schematic representations of the fluorographs are given on the right. E indicates the major translation products from the endogenous mRNA. Spots enclosed in boxes correspond to permanent polypeptides. **(a)** Embryo axes <12 days after pollination (DAP). **(b)** 22 DAP. **(c)** 32 DAP. **(d)** 40 DAP. (From Misra and Bewley 1985b. *J. Expt. Bot.* 36:1644–1652, by permission of Oxford University Press.)

(iv) Protein Metabolism in Developing Embryos

Investigations on the changes in protein accumulation and synthesis, polypeptide patterns, and enzyme titer have contributed in a limited way to our understanding of the utilization of genomic information during embryogenesis. Continued accumulation of proteins, principally attributable to large amounts of storage proteins, characterizes most periods of embryogenesis studied, except perhaps toward the end of seed maturity (Beevers and Poulson 1972; Scharpé and van Parijs 1973; Millerd and Spencer 1974; Manteuffel et al 1976; Zhu, Shen and Tang 1980a, b). In legumes such as *Phaseolus coccineus,* a translational control mechanism differentially regulates protein metabolism in the cells of the embryo and suspensor, the magnitude of the rate of protein accumulation in the cells of the embryo being appreciably lower than that in the suspensor cells (Walbot et al 1972; Sussex et al 1973). Toward the end of seed maturity, when protein accumulation in the cotyledons of *Pisum sativum* levels off, there is a concomitant decrease in the incorporation of labeled amino acids into proteins and in the amount of ribosomes present as polysomes (Beevers and Poulson 1972). The period leading up to embryo maturity in *Sinapis alba* (Fischer et al 1988) and *Lupinus luteus* (Gwóźdź and Deckert 1989) is marked by low levels of polysomes that progressively increase to the midcotyledon phase and then decrease. As seen earlier, the lack of sufficient mRNAs has, with some experimental justification, dominated thoughts about the limitations on protein synthesis in the cotyledons of mature embryos.

The results of experiments involving two-dimensional electrophoretic separation of proteins synthesized by embryos of *Zea mays* indicate that morphological and physiological events of progressive embryogenesis are correlated with the synthesis of different subsets of polypeptides (Sánchez-Martínez, Puigdomènech and Pagès 1986; Boothe and Walden 1990). Boothe and Walden have identified a subset of 2 polypeptides that are associated with the general maintenance of developmental functions and are synthesized throughout embryogenesis, another subset of 14 polypeptides synthesized during the period of morphological differentiation, and a third, smaller subset synthesized during the waning period of embryogenesis and concerned with the quiescence of the embryo (Figure 14.6). Although no function can be ascribed to the embryo polypeptides, studies reinforce the view that qualitative variations in gene products

appeared with large suspensors, distorted and fused cotyledons, or dwarfed hypocotyls, or they were devoid of hypocotyl and cotyledons. Seeds from some mutants completed embryogenesis, but germinated to form defective seedlings. The gametophytic expression of genes in several mutants at some point prior to fertilization was ascertained from the nonrandom distribution of aborted seeds in heterozygous fruits (siliques) or from the preferential location of aborted seeds toward the half of the silique close to the stigma (Meinke 1982, 1985). A library of nearly two hundred additional embryo mutants with varying degrees of abnormal development has been isolated following seed transformation of *Arabidopsis* with *Agrobacterium tumefaciens* (Errampalli et al 1991; Castle et al 1993; Yadegari et al 1994; Vernon and Meinke 1995). By genetic analysis of progeny of selfed F1 plants and by RFLP strategy, several genes essential for embryo development have been mapped with both visible and molecular markers (Patton et al 1991).

Further developmental consequences of these mutations were uncovered by cytological, biochemical, and tissue culture investigations. As reviewed in a previous section, some key mutants are characterized by abnormal growth of the suspensor at the same time that growth of the embryo proper is inhibited. In situ hybridization with several cell-specific mRNA markers, such as lipid transfer protein mRNA and storage protein mRNAs, showed that these transcripts are expressed in their normal spatial context in both mutant and wild-type embryos, as well as in the enlarged suspensor region (Yadegari et al 1994). The expression of marker genes carried under the control of promoters of a storage protein gene and a *LEA* gene in the embryos and suspensors of aborted seeds of some mutants also mirrors gene specificity seen in phenotypically normal seeds (Devic, Albert and Delseny 1996). These results are considered to provide molecular evidence for the view that defects in the embryo can cause the suspensor to take over the vital functions of the embryo and that the activity state of key genes is not necessarily dependent on organ formation in the embryo.

Because storage protein synthesis is largely confined to the maturation phase of embryogenesis, cellular and biochemical patterns of protein accumulation are reliable markers for the developmental arrest of mutant embryos. In their ultrastructural appearance, protein bodies of some mutant embryos are identical to those of the wild-type embryos, whereas in others they resemble immature protein bodies. This picture of protein body dimorphism is supported with the specific evidence that mutant embryos with morphologically altered protein bodies fail to accumulate the 12S and 2S storage proteins found in the wild-type embryos. These observations emphasize that mutant genes might disrupt either morphogenesis or both morphogenesis and cellular differentiation in embryos blocked at different stages of embryogenesis (Heath et al 1986; Patton and Meinke 1990). As storage protein accumulation represents only one of a large number of biochemical activities that lead to embryo maturation, examination of other parameters of the defective embryos is necessary to identify specific cellular and molecular defects of the mutation.

The culture of seeds enclosing mutant embryos in defined media has proved to be an excellent approach to recovering homozygous mutant plants for genetic and molecular studies, although the isolation of auxotrophs by this method has not been nearly as successful. In terms of general responses to identical culture conditions, it was found that mutant embryos arrested at the heart-shaped to cotyledonary stages of development can be reprogrammed in culture to produce respectable calluses that regenerate abnormal plants with defects in reproductive development, whereas embryos arrested at the globular or earlier stages fail to grow or they produce only slow-growing calluses incapable of complete plant regeneration. Embryos of one mutant (*emb-30*; also referred to as *112A-A*) appeared to be particularly recalcitrant to root-forming activity in culture although they produced abundant callus and even leaves (Baus, Franzmann and Meinke 1986; Franzmann, Patton and Meinke 1989). The presence of abnormal developmental syndromes in mutant plants rescued in culture opens up the possibility that many genes required for embryo development also have a role in the vegetative and reproductive growth of plants.

The response of seeds and embryos of one mutant (*bio-1*) that failed to respond to culture on a basal medium but produced normal plants in a medium enriched with biotin or desthiobiotin (a close precursor of biotin) is typical of that of a biotin auxotroph. Mutant embryos rescued in agar cultures were found to produce normal plants with viable seeds when they were transferred to soil and watered daily with either biotin or desthiobiotin (Schneider et al 1989). Further analysis showed that embryos of *bio-1* contain only trace amounts of biotin, whereas arrested embryos from other mutants with similar morphology have higher concentrations of biotin than the biotin auxotroph. A sharp definition of the biotin lesion was obtained by the demonstration that the *bio-1* mutant of *Arabidopsis* can be rendered prototrophic by transformation with a chimeric bacterial 7,8-diaminopelargonic acid aminotransferase gene that catalyzes the conversion of the biotin intermediate 7-keto-8-aminopelargonic acid to desthiobiotin (Shellhammer and Meinke 1990; Patton, Volrath and Ward 1996). These findings support the conclusion that *bio-1* embryos have a defective biotin synthetic pathway that disrupts the gene encoding 7,8-diaminopelargonic acid aminotransferase. As this is the first example of an embryo-lethal mutant with a defect in a defined biochemical pathway, further research on this system should assist in our understanding of the biosynthetic pathway of biotin in embryos and the role of biotin in embryogenesis.

Embryo-lethal Mutants in Maize. Just as *Arabidopsis* is an ideal plant among dicots for studying the genetic control of embryo development, maize has proved to be tailor-made for similar investigations among monocots. Of the many features of maize that contribute to this suitability, two are particularly noteworthy, namely, the availability of good background genetic and cytogenetic information and the presence of several well-characterized transposable elements. A number of single-gene recessive mutants in maize with lesions in their embryo and endosperm development were described between 1920 and 1950. Jones (1920) and Mangelsdorf (1923) analyzed several spontaneously occurring mutants characterized by a high degree of grain abortion due to impairment in the development of the embryo and endosperm. Although this analysis did not clearly identify the stage of embryo arrest, some of the mutant kernels were able to germinate into abnormal seedlings. This is probably suggestive of some defects during later stages of embryogenesis. True embryo-lethal mutants (*germless*) were described by Demerec (1923), who established that the germless condition is an inherited character. Histological studies of this line showed that aborted seeds generally contained a normal endosperm, but embryo development was arrested at various stages and was followed by degeneration prior to maturity (Sass and Sprague 1950). Clearly, these mutants did not show a loss of developmental potential in the endosperm at any time, or in the embryo up to the time that led to the arrested stage.

The genetic program of embryogenesis in maize was brought into sharp focus in later years by analyses of embryo-lethal mutants isolated by pollinating ears with EMS-mutagenized pollen grains or by tagging with *mu* transposable elements. These studies involved screening several thousand plants and examining an equally staggering number of kernels showing varying degrees of defects in both embryo and endosperm development; most kernels, however, were mutants with abnormal endosperm and nonviable embryo, or with normal endosperm and nonviable embryo. Taking advantage of maize translocation stocks in the crossing of mutants with plants of B–A translocation to produce kernels that contain combinations of a mutant hypoploid embryo and a nonmutant hyperploid endosperm, or a nonmutant hyperploid embryo and a mutant hypoploid endosperm, showed that a normal endosperm is unable to rescue a mutant embryo. It was also established from a reciprocal cross that even in the presence of a mutant endosperm, a normal embryo completes its development. These results provided the principal criteria to establish that lethality in a number of mutants studied is determined by the genotype of the embryo and not of the endosperm, and that a defective endosperm, created by previous interaction, does not prevent the completion of embryo development (Neuffer and Sheridan 1980).

A group of *dek* mutants isolated from maize are characterized by the presence of defective endosperm and of embryos arrested at different stages of development. Most of the mutant embryos develop at least to the point of differentiating one or more leaf primordia, whereas only a few are arrested at earlier stages of development and are unable to form leaf primordia. As shown in Figure 14.7, the impairment of growth in the latter group of embryos spans an array of stages, ranging from proembryo through club-shaped transition and coleoptile stages to the stage of initiation of leaf primordia (Sheridan and Neuffer 1980, 1982). Some of these mutant embryos have been characterized histologically, and the unique steps in the sequence of gene action essential for normal embryogenesis have been partly defined. Thus, a requirement for normal gene functions at the transition stage and at the stage of initiation of the coleoptile–shoot axis is implied by the developmental profiles of mutant embryos *dek-22* (E113A) and *dek-23* (E1428), respectively. Although the initial development of *dek-22* embryos is not different from that of wild-type embryos up to the transition stage, mutant embryos do not display any further morphogenetic changes necessary to pass to the coleoptile stage. Similarly, arrested embryos of *dek-23* mutants do

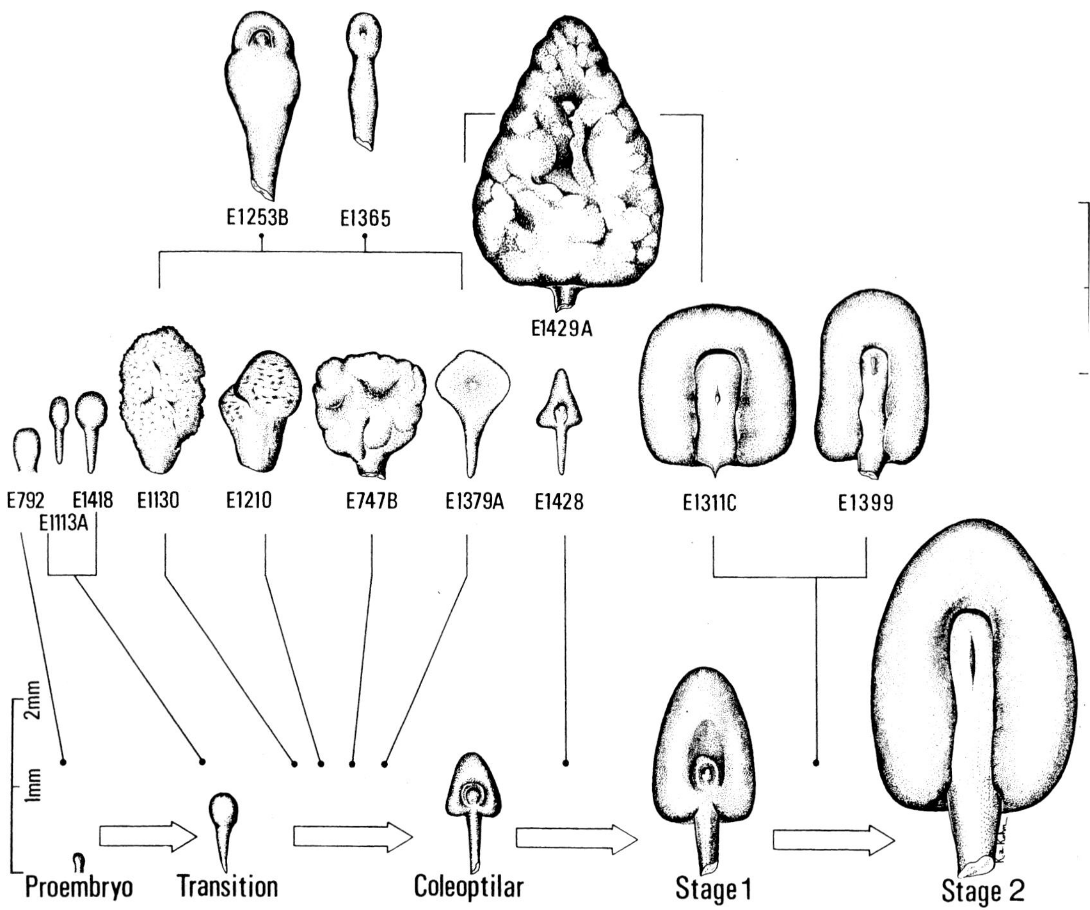

Figure 14.7. Stages of developmental arrest of embryos in *dek* mutants of maize. Stages of development of the normal embryo are shown in the lower half of the figure. The coleoptilar stage is characterized by the presence of a coleoptile ring and a shoot apical meristem, and stage 1 is characterized by the presence of the first leaf primordium. Besides exhibiting specific developmental defects, all mutants shown in the figure lacked leaf primordia. (From Sheridan and Neuffer 1982. *J. Hered.* 73:318–329. © 1982 by the American Genetic Association. Print supplied by Dr. W. F. Sheridan.)

not advance beyond an abnormal coleoptile stage characterized by the lack of a shoot apex and of a coleoptile ring found in wild-type embryos poised to form a coleoptile; eventually, the embryos begin to show signs of necrosis (Clark and Sheridan 1986; Dolfini and Sparvoli 1988). Similar defects are very common in mutants *bno* (E747B), *ptd* (E1130), and *cp* (E1418), which lag behind wild-type embryos in their development and are blocked at specific stages of embryogenesis (Sheridan and Thorstenson 1986). One distinguishing feature of embryogenesis in mutants *rgh* (E1210) and *fl* (E1253B) is the diversion of arrested embryos to abnormal pathways of morphogenesis that result first in the disruption of bilateral symmetry and apical meristem formation, and then in callus-like growth (*rgh*) and proliferation of the embryonic axis and surrounding regions, accompanied by hypertrophy of the scutellum (*fl*) (Clark and Sheridan 1988). It will be recalled from Chapter 12 that *dek* mutations affect the endosperm, indicating that they are pleiotropic in nature. How the same genes control developmental processes in two different tissues with differing gene dosages remains to be explored.

The genetically heterogeneous nature of these mutants was underscored by the growth of embryos in culture. Embryos of several mutants were rescued by the simple expedient of growing them on a mineral-salt medium with or without organic additives, whereas the developmental defect of some embryos could not be overcome even in the most enriched medium tested (Sheridan and Neuffer 1980; Clark and Sheridan 1986, 1988; Sheridan and Thorstenson 1986). Isolation of auxotrophs is a cherished hope of investigators in this field; similar to a spontaneously occurring mutant in maize, one of the induced mutations has been traced to a genetic block in the proline biosynthesis pathway (Racchi et al 1978; Sheridan and Neuffer 1980).

The morphogenetic effects of mutations in maize generated by transposon tagging appear to be specific to embryos and result in the same array of genetic lesions described in EMS-generated mutants. A general classification has placed these mutants, designated as *embryo-specific* (*emb*), into three groups: mutants in which (a) embryogenesis is blocked during the proembryo and transition stages; (b) the block occurs during the period of differentiation of the embryo axis and scutellum and initiation of the leaf primordium; and (c) the block persists during the period of elaboration of embryonic structures beginning at the late coleoptile stage. The majority of the 51 mutants characterized belong to the second group, leaving 10–12 in each of the other two groups (Figure 14.8). Correlated genetic and developmental studies have provided proof that many genes are essential for the ordered development of the maize embryo and that the *emb* phenotype reflects mutations in many loci of these genes. The conclusion that the *EMB* loci are essential to embryo development follows from the observations that the arrest of embryo growth occurs before kernel maturity and that mature kernels harboring mutant embryos do not germinate (Clark and Sheridan 1991; Sheridan and Clark 1993). However, this is only a part of the story because no understanding of the mechanism of action of *EMB* genes can be gained without molecular data.

Embryo-lethal Mutants in Other Plants. In broad terms, embryo-lethal mutants isolated from rice by chemical mutagenesis resemble transposon-tagged mutants of maize because the mutational effects are restricted primarily to the embryo and do not spread to the endosperm. In extreme cases, the mutant grain does not possess an embryo, which apparently disintegrates at the globular stage. Some mutants show defects in the differentiation of the shoot that result in a chaotic arrangement of leaf primordia at the apex and in abnormalities in their anatomy. Root initiation and development does not go awry as the shoot does, indicating that initiation of the two major embryonic organs is controlled by different loci (Nagato et al 1989; Tamura et al 1992; Hong et al 1995b). Another set of mutants has been separated into groups affecting three regulatory pathways of embryogenesis, namely, organ differentiation, positional regulation of organs, and size regulation of the embryo, that occur before embryo morphogenesis. Mutations in the first pathway produce club-shaped embryos that do not possess any visible organs, as well as embryos that do not produce any shoots but form normal roots; mutations in which both shoot and root are displaced more apically belong to the second group. Mutations affecting the size of embryos can lead either to reduction or to enlargement of the embryo (Kitano et al 1993; Hong et al 1996). This close relationship between mutations affecting early embryo development, without overlap, emphasizes the point that virtually every step in the embryogenesis program is gene controlled.

Both embryo and endosperm are affected in a negatively reciprocal way in a temperature-sensitive mutant of rice, *embryoless-1* (*eml-1*). Rearing plants at day/night temperatures of 30/25 °C following pollination causes the formation of grains with no embryos or of grains with malformed embryos but with a large quantity of endosperm (Figure 14.9). Grains with large embryos and poorly developed endosperm result when plants are reared at a constant temperature of 18 °C or 20 °C (Hong et al 1995a). The subtle effects of the mutant gene in the embryo and endosperm suggest a role for the *EML-1* gene product in the continued development of these parts of the grain, but the effects are mediated in some presently unknown way.

Pea is another plant that has been inducted into the ranks for analysis of the genetics of embryogenesis. Mutant embryos showing defects in cotyledon development and root initiation have been isolated, and genetic studies have shown that at least three loci participate in the development of the two cotyledons in their full multicellularity and structure (Johnson et al 1994). In one mutant, designated as *cytokinesis-defective* (*cyd*), cotyledon morphology is apparently altered due to a defect in the formation of the cell wall, resulting in partially formed cell plates and multinucleate cells (Liu, Johnson and Wang 1995). It will be interesting to see whether this deficiency could be corrected by culture of embryos in a medium with appropriate supplements.

(ii) Genetic Control of Embryonic Pattern Formation

The laying down of the basic body plan of the embryo, consisting of an upper shoot region,

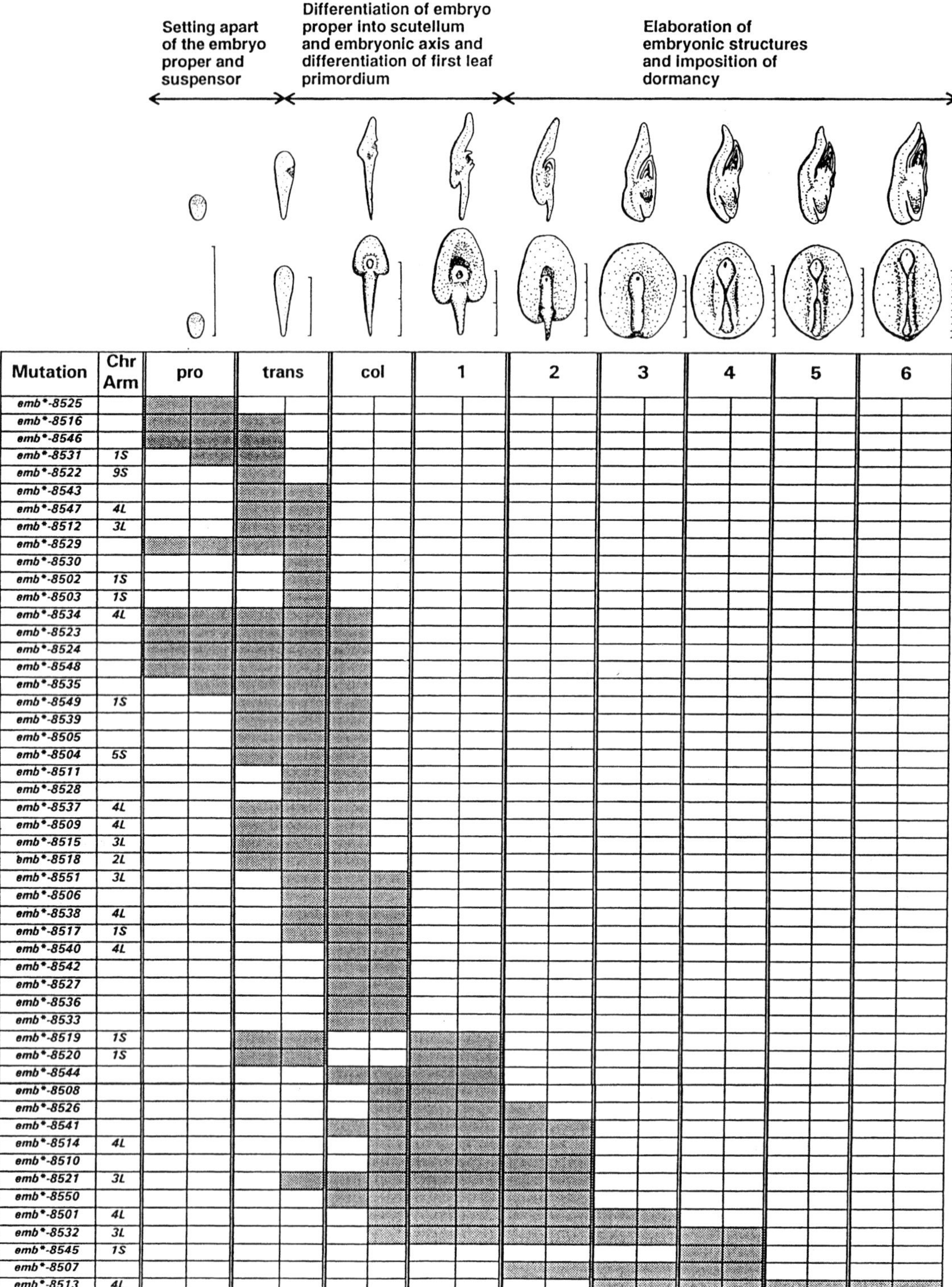

Mutation	Chr Arm	pro	trans	col	1	2	3	4	5	6
*emb**-8525										
*emb**-8516										
*emb**-8546										
*emb**-8531	1S									
*emb**-8522	9S									
*emb**-8543										
*emb**-8547	4L									
*emb**-8512	3L									
*emb**-8529										
*emb**-8530										
*emb**-8502	1S									
*emb**-8503	1S									
*emb**-8534	4L									
*emb**-8523										
*emb**-8524										
*emb**-8548										
*emb**-8535										
*emb**-8549	1S									
*emb**-8539										
*emb**-8505										
*emb**-8504	5S									
*emb**-8511										
*emb**-8528										
*emb**-8537	4L									
*emb**-8509	4L									
*emb**-8515	3L									
*emb**-8518	2L									
*emb**-8551	3L									
*emb**-8506										
*emb**-8538	4L									
*emb**-8517	1S									
*emb**-8540	4L									
*emb**-8542										
*emb**-8527										
*emb**-8536										
*emb**-8533										
*emb**-8519	1S									
*emb**-8520	1S									
*emb**-8544										
*emb**-8508										
*emb**-8526										
*emb**-8541										
*emb**-8514	4L									
*emb**-8510										
*emb**-8521	3L									
*emb**-8550										
*emb**-8501	4L									
*emb**-8532	3L									
*emb**-8545	1S									
*emb**-8507										
*emb**-8513	4L									

Figure 14.8. Summary of developmental blocks of the *emb* mutants in maize. Stages of wild-type embryos are shown at the top of the figure and specific developmental defects are shown immediately below. Stages (pro, proembryo; trans, transition; col, coleoptilar; and mature stages 1 to 6) at which each mutant is blocked are indicated by stippling. Chr arm, chromosome arm. Each division of the scale bar equals 0.5 mm. (From Sheridan and Clark 1993; print supplied by Dr. W. F. Sheridan.)

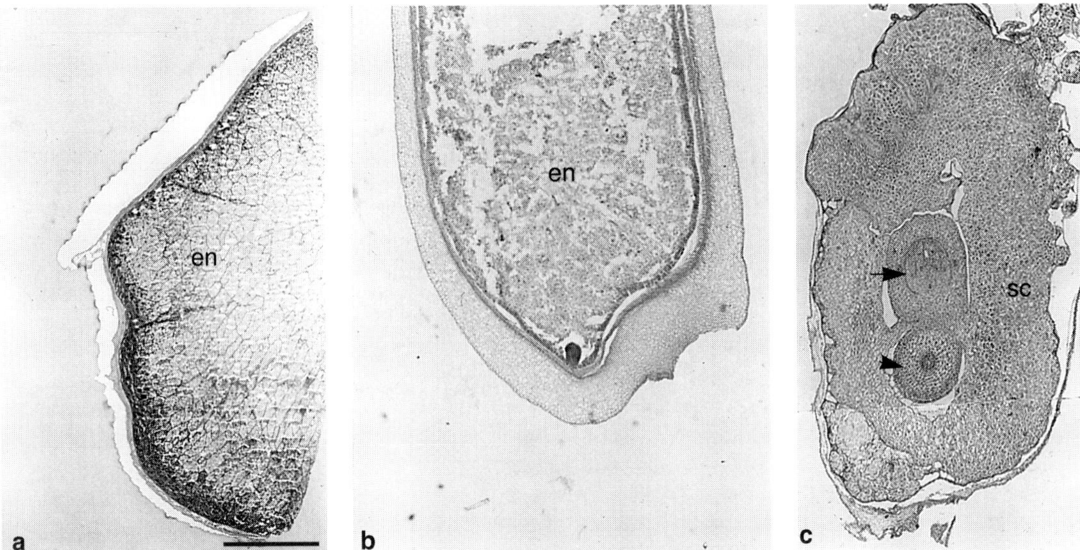

Figure 14.9. Sections of grains of temperature-sensitive mutants of rice. **(a)** An embryoless mutant. **(b)** A mutant containing an embryo lacking organ differentiation. **(c)** A mutant endospermless grain containing a giant embryo. Arrow indicates shoot and arrowhead indicates root. en, endosperm; sc, scutellum. Scale bar = 0.5 mm. (Reprinted from S. K. Hong et al 1995a. Temperature-sensitive mutation, *embryoless 1*, affects both embryo and endosperm development in rice. *Plant Sci.* 108:165–172, with kind permission of Elsevier Scientific-NL, Sara Burgerhartstraat 25, 1055 KV Amsterdam, The Netherlands. Photographs supplied by Dr. Y. Nagato.)

a middle hypocotyl, and a lower root segment, has been the subject of considerable study and some speculation. Descriptive embryology that addresses the organizational aspects of embryo development in plants has shown that these regions can be traced back to the terminal and basal cells of the two-celled proembryo. Speculation has centered mostly on the mechanism underlying pattern formation, which reflects deeply on the question as to whether the three regions of the embryo axis are specified autonomously and develop independent of each other.

A genetic approach to analyzing the mechanism of embryonic pattern formation has sought to identify mutant phenotypes in *Arabidopsis* that define different steps in the process and to characterize the genes involved. In saturation mutagenesis experiments using a chemical mutagen, embryonic pattern mutants were selected by large-scale screening of mutant seedlings for profound and specific defects in the apical–basal pattern, radial organization, and shape of the embryo. The rationale for this approach is that since mutants are isolated by their ability to recover from lethality during embryogenesis, any diagnostic defects detected in the seedlings are caused by patterning genes and not due to lesions in general cell metabolism. Complementation analysis has shown that nine loci selected are each represented on an average by eight mutant alleles. Included in the class of mutations with defects in the apical–basal organization of the seedling are (a) mutants that lack the apical region, including the cotyledons and the shoot meristem (*gurke* [*gu*]); (b) those that lack the hypocotyl and so have the cotyledons attached directly to the root (*fackel* [*fk*]); (c) those that lack both the hypocotyl and the root and so consist essentially of an apical piece of axis to subtend the cotyledons and shoot meristem (*monopteros* [*mp*]); and (d) those that lack both terminal pattern elements, including the cotyledons and the root, and so consist of a cellular mass with no obvious signs of apical–basal polarity (*gnom* [*gn*]); these are diagrammatically represented in Figure 14.10. Mutations in two genes, *KNOLLE* (*KN*) and *KEULE* (*KEU*), disrupt the radial pattern and result in seedlings with defects in the epidermis. Grossly abnormal seedlings are produced by mutations in three genes, *FASS* (*FS*), *KNOPF* (*KNF*), and *MICKEY* (*MIC*), that affect the shape of the embryo without eliminating any pattern elements (Jürgens et al 1991; Mayer et al 1991). Other genes implicated in the apical–basal organization of the seedling are *SHOOT MERISTEMLESS* (*STM*) and *WUSCHEL* (*WUS*), which are critical for the initiation of the shoot apical meristem, and *ROOTLESS* (*RTL*), which is critical for the initiation of the root apical meristem in the embryo (Barton and Poethig

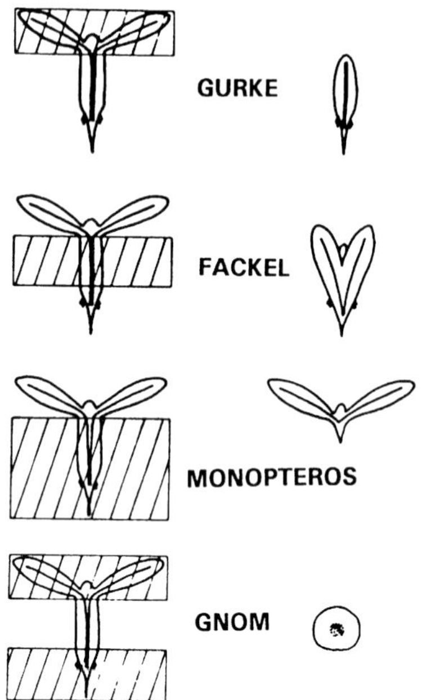

Figure 14.10. Schematic representations of the pattern deletions caused by mutations in *Arabidopsis* and the phenotypes formed. The wild-type seedling patterns with the deleted parts shaded are shown on the left, and the corresponding mutant patterns are on the right. (From Mayer et al 1991. Reprinted with permission from *Nature* 353:402–407. © Macmillan Magazines Ltd.)

1993; Laux et al 1996). Mature embryos also respond to the effects of the gene *CLAVATA3* (*CLV-3*), which causes abnormal enlargement of the vegetative and inflorescence meristems (Clark, Running and Meyerowitz 1995). Souer et al (1996) have described embryos of *Petunia* sp. carrying a *no apical meristem* (*nam*) mutation that fail to develop the shoot apical meristem.

Developmental profiles of embryos of *Arabidopsis* have shown that the mutant genes strike at the cellular level and disrupt the division planes that are circumscribed during early embryogenesis. The seedling phenotype of the *gn* mutant has been traced back to the first division of the zygote, which results in two nearly equal-sized cells instead of two unequal cells. Subsequent divisions of the terminal cell are anarchic in the mutant embryo, which also comes off without a hypophysis (Mayer, Büttner and Jürgens 1993). Formation of the *mp* mutant embryo is another case of abnormal early divisions, but abnormalities begin to appear only after the octant stage. They include random cell divisions in the lower tier of cells of the octant embryo during the globular to heart-shaped stage transition and a series of transverse divisions in the hypophysis (Figure 14.11). The defective divisions in the lower tier of cells do not interfere with the orderly pattern of divisions in the upper tier (Berleth and Jürgens 1993). Chaotic divisions beginning as early as the octant stage of the embryo and accompanied by incomplete cross wall formation play a major role in the evolution of the *kn* phenotype, but the genetic basis of the mutation also extends to the generation of multinucleate polyploid cells (Lukowitz, Mayer and Jürgens 1996). Effects of the *STM* gene can be identified only at the torpedo-shaped stage, long after other pattern elements are established; a mutation in this gene results in the inability of the presumptive cells of the shoot apical meristem to initiate divisions. Thus, in the hierarchy of gene actions responsible for specifying the embryo body plan, the *STM* gene acts only after the *GN* and *MP* genes (Barton and Poethig 1993). The first division of the zygote of the *fs* mutant is asymmetric, as in the wild-type embryo, indicating that the mutation does not interfere with polarity of the embryo; however, the *FS* gene plays a key role in the subsequent events of cell elongation and orientation of cell divisions in the embryo (Figure 14.11). This is reflected in the fact that by the time the wild-type embryo has turned heart shaped or torpedo shaped, the mutant embryo merely resembles an oversized globular embryo (Torres-Ruiz and Jürgens 1994). Implicating a derangement of auxin metabolism in the generation of *fs* mutants is the observation that wild-type *Arabidopsis* embryos cultured in a medium containing NAA are transformed into mutant phenocopies (Fisher et al 1996).

Possible molecular mechanisms to explain the action of pattern-forming genes have been sought by the cloning and characterization of *STM, NAM, GN*, and *KN* genes. The predicted STM protein is somewhat close to the protein encoded by the maize *KN-1* gene and, as in the latter, transcripts are expressed in the cells of the embryo slated to form the shoot apex (Long et al 1996). A novel protein that shares a conserved N-terminal domain with several other proteins of unknown function is apparently encoded by the *NAM* gene (Souer et al 1996). The *GN* gene has been found to be allelic to embryo-lethal gene *EMB-30*; mutation in the gene produces a fused cotyledon phenotype and does not promote root regeneration on embryos in culture (Meinke 1985; Baus, Franzmann and Meinke 1986; Mayer, Büttner and Jürgens 1993). An indication of the role of the *EMB-30/GN* gene in cellular functions at several stages of development of the

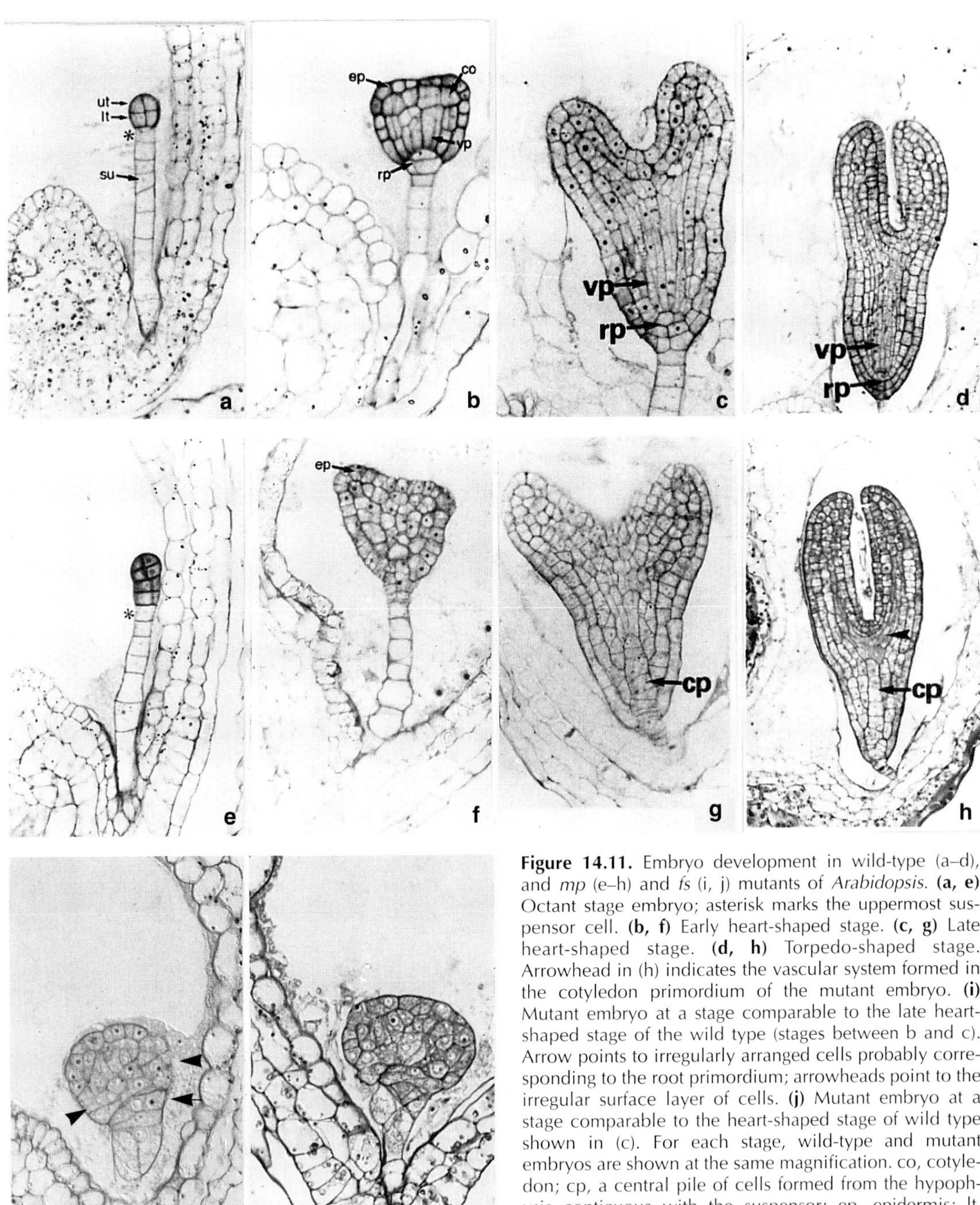

Figure 14.11. Embryo development in wild-type (a–d), and *mp* (e–h) and *fs* (i, j) mutants of *Arabidopsis.* **(a, e)** Octant stage embryo; asterisk marks the uppermost suspensor cell. **(b, f)** Early heart-shaped stage. **(c, g)** Late heart-shaped stage. **(d, h)** Torpedo-shaped stage. Arrowhead in (h) indicates the vascular system formed in the cotyledon primordium of the mutant embryo. **(i)** Mutant embryo at a stage comparable to the late heart-shaped stage of the wild type (stages between b and c). Arrow points to irregularly arranged cells probably corresponding to the root primordium; arrowheads point to the irregular surface layer of cells. **(j)** Mutant embryo at a stage comparable to the heart-shaped stage of wild type shown in (c). For each stage, wild-type and mutant embryos are shown at the same magnification. co, cotyledon; cp, a central pile of cells formed from the hypophysis contiguous with the suspensor; ep. epidermis; lt, lower tier of the octant; rp, root primordium with the cells of the hypophysis; su, suspensor; ut, upper tier of the octant; vp, procambium. ([a–h] from Berleth and Jürgens 1993. *Development* 118:575–587. © Company of Biologists Ltd. Photographs supplied by Dr. T. Berleth; [i, j] from Torres-Ruiz and Jürgens 1994. *Development* 120:2967–2978. © Company of Biologists Ltd. Photographs supplied by Dr. R. A. Torres-Ruiz.)

plant was provided, not surprisingly, by Northern blot analysis showing that gene transcripts are expressed in almost all organs of *Arabidopsis* tested. However, what was surprising was the discovery that the gene encodes a protein that has extensive similarity to the secretory (SEC-7) protein of yeast. SEC-7 is a cytosolic protein found in the Golgi complex and is probably involved in glycosylation, secretion, transport, and polysaccharide synthesis associated with the Golgi apparatus. In *Arabidopsis*, the functionally important component of the protein is restricted to a stretch of 157 amino acids, designated as the "Sec-7 domain" (Shevell et al 1994). Another proposal has assigned the GN protein a close homology to the translation product of a different yeast gene (*YEC-2*) with undefined functions, rather than to SEC-7 (Busch, Mayer and Jürgens 1996). The translation product of the *KN* gene has been found to be a member of the syntaxin family of proteins, whose main function is in directing vesicle traffic during cell division (Lukowitz, Mayer and Jürgens 1996). In any event, the principal conclusion from experimental observations is that *EMB-30/GN* and *KN* gene products may be concerned with cell division, cell expansion, and cell adhesion during embryogenesis; because of the ubiquitous nature of these proteins, they may affect related cellular events in developing embryos that are difficult or impossible to interpret in specific detail. As such, these results depict the *EMB-30/GN* and *KN* gene products as casting only a peripheral effect on pattern formation in embryos. The defective polarity in some of the mutant embryos suggests that a protein regulating cytoskeletal assembly might be sought as an implementing factor targeted by the mutation.

4. EXPRESSION OF GENES FOR HOUSEKEEPING AND STRUCTURAL PROTEINS

To understand the molecular processes that operate during embryogenesis, it is necessary to isolate a representative sample of the variety of genes active at different stages of embryo development and to study their regulation. This is easier said than done because of the sheer number of genes expressed at the mRNA level in embryos and the possibly bewildering complexity of their protein products. Presumably there are many genes that are unique to specific stages of development of embryos; a second group comprises genes for housekeeping proteins that are shared with other plant organs. Included in the first group are mainly storage protein genes and genes involved in pattern formation and morphogenesis of embryos, whereas genes encoding enzymes of general metabolism, respiration, and photosynthesis, and genes coding for proteins of cell wall biosynthesis, plant defense, desiccation, and cell division, for example, may be assigned to the second group. Following the discussion of genes involved in pattern formation and morphogenesis just completed, this section will consider genes for housekeeping proteins. Due to the large volume of information to be collated, gene action in the regulation of storage protein synthesis in the embryo will form the sole topic of Chapter 15.

(i) Genes for Glyoxylate Cycle and Related Enzymes

Investigations on the regulation of gene expression during seed germination and seedling growth of *Brassica napus* have led to the identification of cloned mRNA sequences that accumulate at different stages of embryogenesis. Most post-germination-abundant mRNAs selected from a cDNA library made to poly(A)-RNA of dark-grown 14-hour-old seedlings were found to accumulate in the cotyledons and axes of fully germinated seedlings. When mRNAs from staged embryos were hybridized in a Northern blot with cloned seedling probes, a surprising finding was the presence of cotyledon-abundant and axis-abundant gene sets at distinct stages of embryogenesis. Cotyledon-abundant mRNAs reach maximum levels in embryos poised for desiccation (35–45 days after flowering), whereas the axis-abundant mRNAs are most prevalent at 15–27 days after flowering and reach minimum levels by the time of embryo desiccation (Harada, Baden and Comai 1988). Further insight into regulation of the expression of cotyledon-abundant genes during embryogenesis has come from the work that showed that two cotyledon-abundant mRNAs encode the glyoxylate cycle enzymes ICL and MS, which participate in storage lipid mobilization (Comai, Baden and Harada 1989; Dietrich et al 1989). RNA blot hybridization showed that ICL and MS genes exhibit qualitatively similar expression patterns in the late stages of embryogenesis, but MS messages are more prevalent than ICL messages. Isolated nuclei provide a convenient system to study changes in the rates of transcription. Use of nuclei isolated from *B. napus* embryos of different ages to determine ICL and MS transcriptional run-off activity showed that genes for both enzymes and some other gene sets remain competent to be transcribed in the embryos, with the caveat that transcription to the level necessary for germination does not occur during embryogenesis (Comai et al 1989; Comai and Harada 1990). As predicted by these observations, the genes for MS and ICL are expressed during late embryogenesis in transgenic tomato and tobacco, respectively; furthermore, the transcripts encoded by the MS gene are distributed throughout the tissues of mature embryos, including the procambium, of both *B. napus* and transgenic tomato (Comai et al 1992; Zhang et al 1993). Deletion analysis using the ICL gene in transgenic *Arabidopsis* has also identified distinct DNA sequences that are primarily, but not exclusively, expressed during embryogenesis and during seedling growth (J. Z. Zhang et al 1996). These results reaffirm the well-established tenet that embryogenesis and germination represent different facets of the early sporophytic phase and call for the operation of distinct regulatory programs to activate and repress MS, ICL, and other similarly expressed genes. Transcripts of the ICL gene detected initially in maturing embryos of *Helianthus annuus* are maintained at a constant low level through seed desiccation and increase disproportionately during germination (Allen, Trelease and Thomas 1988).

To follow the regulation of an axis-abundant gene (*AX-92*) expressed at an early stage of embryogenesis, a chimeric gene construct consisting of 5'- and 3'-flanking regions of *AX-92* fused to *GUS* gene was expressed in transgenic *B. napus*. *GUS* gene expression was first detected predominantly in the root cortex of the torpedo-shaped embryo, the activity increasing up to the cotyledonary stage embryo and no further. Expression of deletion constructs in the embryos has identified DNA sequences located in the 3'-part of the protein coding region as critical for the activation of the chimeric gene in the embryos (Dietrich, Radke and Harada 1992). Since the same minimal gene construct induces *GUS* expression in the root cortex of seedlings, similar mechanisms appear to underlie the expression of the gene during embryogenesis and germination.

(ii) Heat-Shock Protein Genes

The synthesis of HSPs in developing embryos is the result of the accumulation of transcripts of HSP-related genes. Early stage embryos (16–28 days after flowering) of *Pisum sativum* transcribe an mRNA with appreciable homology to the highly expressed *HSP-70* gene, but homology decreases in the transcripts produced at later stages of embryogenesis (Domoney et al 1991). Consistent with the constitutive synthesis of HSPs, the expression of transcripts of *HSP-70* or small HSP genes is found to parallel protein accumulation in embryos of *P. sativum* (DeRocher and Vierling 1994, 1995), *Helianthus annuus* (Almoguera and Jordano 1992; Coca, Almoguera and Jordao 1994), *Nicotiana tabacum*, and *Arabidopsis* (Prändl, Kloske and Schöffl 1995).

In describing the heat-shock effects on developing embryos of *Zea mays*, Marrs et al (1993) have shown that whereas transcripts of some HSP genes (*HSP-81* and *HSP-90*) are present in embryos at normal temperatures, developmentally regulated expression of *HSP82* and *HSP-17.9* does not occur in uninduced embryos. However, the ability of *HSP-81* and *HSP-82* transcripts to respond to heat shock appears to be itself developmentally regulated, as neither transcript is thermally inducible until embryos have progressed to about 14 days after pollination. Thus, although the two genes are probably regulated by different mechanisms, both *HSP-81* and *HSP-82* produce the same end result, namely, enhanced heat-induced expression.

(iii) Expression of Other Genes

The expression of genes that play key roles in the developmental decisions of the embryo has remained surprisingly obscure. Two homeobox genes have provided the probes for an analysis of the expression of key genes in developing maize embryos. It may be significant that transcripts of these genes, which code for transcription factors, are activated during early embryogenesis and are confined to the meristematic cells, such as the shoot and root meristems, procambium, and leaf primordia (Klinge and Werr 1995). Consistent with their widely accepted role in controlling cell division, transcripts of the maize homologue of a cell cycle control protein kinase gene (*CDC-2*) are also found in actively dividing maize embryo cells (Colasanti, Tyers and Sundaresan 1991). Cell wall glycoproteins, histones, and chloroplast structural proteins must be active participants in the expression of genes for structural proteins during embryo development (Figure 14.12). A variety of methods, including Northern blot, Western blot, and in situ hybridization, have shown that both the protein and the mRNA of a gene for a hydroxyproline-rich glycoprotein component of *Z. mays* cell wall accumulate in the embryo axis but not in the scutellum

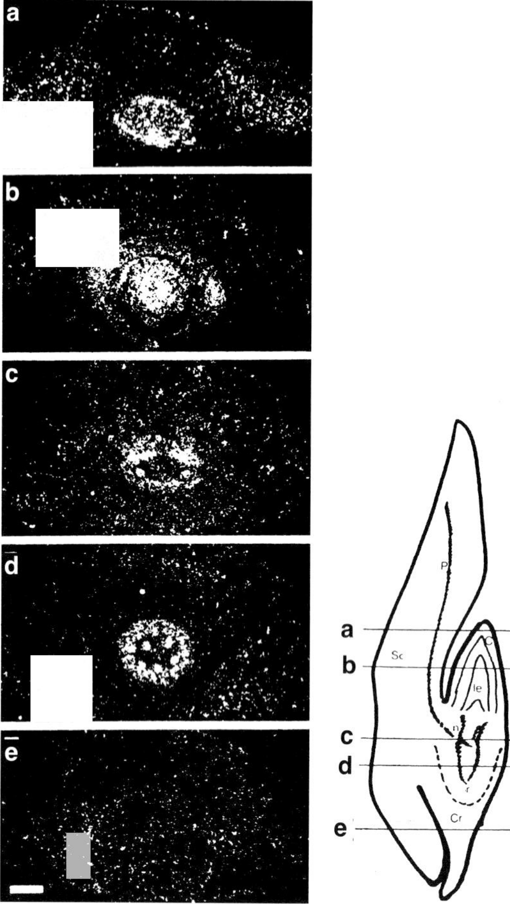

Figure 14.12. Autoradiographic localization of transcripts of a cell wall hydroxyproline-rich glycoprotein gene in the immature embryo of *Zea mays.* **(a–e),** Dark-field microscopic images of autoradiographs of sections cut at levels shown in the diagram of the embryo on the right. C, coleoptile; Cr, coleorhiza; le, leaf primordium; n, scutellar node; P, procambium; r, radicle; Sc, scutellum. Scale bar = 200 µm. (From Ruiz-Avila et al 1992.)

of immature embryos (Ruiz-Avila, Ludevid and Puigdomènech 1991; Ruiz-Avila et al 1992). However, transcripts of another gene from *Z. mays* that encode a proline-rich, hydrophobic protein accumulate both in the embryo axis and scutellum (José-Estanyol, Ruiz-Avila and Puigdomènech 1992). Although the functional significance of these observations is not clear, a general consensus would appear to be that cells of the embryo axis and scutellum differ in their wall protein composition.

Expression of certain histone genes during embryogenesis appears to be both division dependent and division independent. In immature embryos of *Z. mays*, transcripts of maize histone H4 gene are localized in the cells of the scutellum, leaf primordia, and vascular bundles of the coleoptile (Ruiz-Avila et al 1992). At different stages of development of embryos of *Oryza sativa*, pronounced localization of rice histone H3 gene transcripts is confined to the cells of the scutellum, coleoptile, first leaf, and root primordium, including the root cap, coleorhiza, and zone of elongation (Figure 14.13); however, transcripts are conspicuously absent from cells of the shoot apex, newly emerging leaf primordia, and the root quiescent center (Raghavan and Olmedilla 1989). As a speculative thought, the requirement for histone transcripts in DNA repair synthesis may explain their presence in the nonmeristematic cells of the embryo.

Chloroplast genes are expressed constitutively in developing embryos of *Gossypium hirsutum* (Borroto and Dure 1986), *Glycine max* (Saito et al 1989; Chang and Walling 1991, 1992), and *Arabidopsis* (Degenhardt, Fiebig and Link 1991). During maturation of soybean embryos, chloroplasts with highly differentiated grana are transformed into amyloplasts and proplastids; this transition is correlated with a decrease in *CAB* gene mRNAs. The transcription of different members of the *CAB* gene family during embryogenesis appears to be controlled by different factors, because *CAB-3* mRNAs are found to level off more slowly than mRNAs of other gene family members during embryo maturation (Saito et al 1989; Chang and Walling 1992). Transcripts of *PSB-A*, *RBC-L*, and *RBC-S* (nuclear-encoded small subunit of Rubisco) genes detected by in situ hybridization are found first in the proembryos of *Arabidopsis,* and the transcripts increase in abundance up to the mature stage. The increased concentration of the transcripts in the cotyledons of mature embryos would appear to correspond with the potential fate of cotyledons as the first photosynthetic organs of the seedling plant (Degenhardt, Fiebig and Link 1991). These data collectively suggest that the expression of chloroplast genes during embryogenesis in soybean and *Arabidopsis* is controlled developmentally, rather than by light signals. Another function of plastids in developing embryos is the synthesis of amylopectin involving the starch-branching enzyme. One isoform of this enzyme is highly expressed in the young embryos of pea, and another isoform attains maximum expression in old embryos. Differential expression of the enzyme isoforms is also correlated with the nature of amylopectin synthesized in developing embryos (Burton et al 1995).

One of the earliest molecular markers of

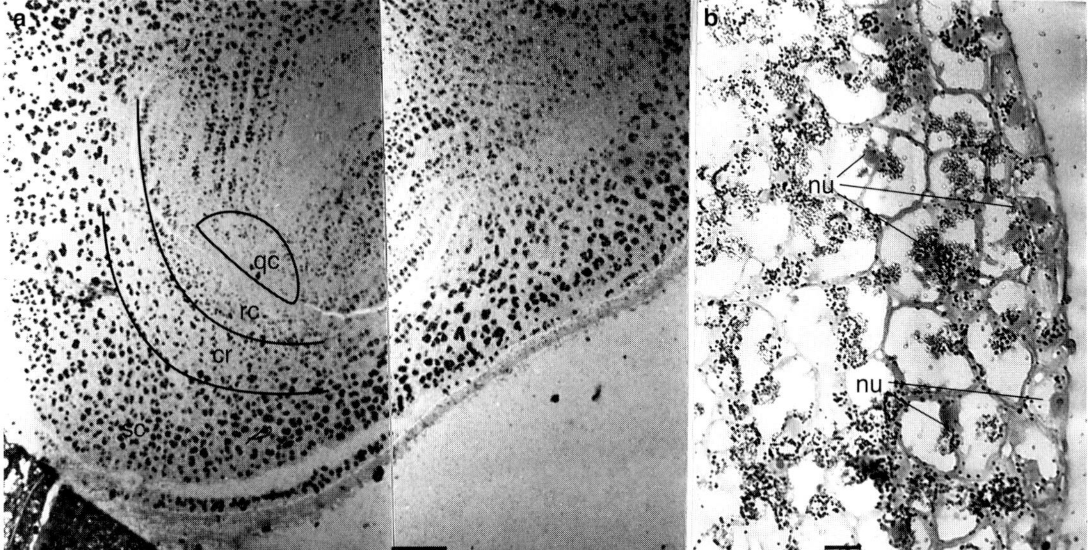

Figure 14.13. Autoradiographs of sections of rice embryos hybridized in situ with ³H-histone H3 RNA probe. **(a)** Section of the root apex of an 8-day-old embryo. Lines are drawn on the figure to indicate the boundaries of certain parts. **(b)** Section of the scutellum of a 6-day-old embryo showing the labeling pattern in individual cells. cr, coleorhiza; nu, nucleus; qc, quiescent center; rc, root cap; sc, scutellum. Scale bars = 50 μm (a) and 10 μm (b). (From V. Raghavan and A. Olmedilla 1989. Spatial patterns of histone mRNA expression during grain development and germination in rice. *Cell Diffn Devel.* 27:183–196, with kind permission of Elsevier Scientific-NL, Sara Burgerhartstraat 25, 1055 KV Amsterdam, The Netherlands.)

embryogenesis in *Daucus carota* is a gene for an extracellular protein (*EP-2*) that shows homology to plant lipid transfer proteins. Transcripts of this gene are first detected in the cells of the globular embryo, including the suspensor, but there is a progressive restriction of gene expression to the protoderm of heart-shaped and older embryos until transcripts are completely absent in mature embryos (Sterk et al 1991). The time course of accumulation of transcripts of maize and *Arabidopsis* lipid transfer protein genes in developing embryos of the respective species is similar to that seen in carrot embryos; at the cellular level, hybridization with the labeled maize probe is observed in the epidermal cells of the coleoptile and scutellum (Sossountzov et al 1991; Vroemen et al 1996). It is speculated that lipid transfer proteins might mediate in cutin biosynthesis in the embryo epidermis by transporting the precursors to the site of synthesis. Floral *MADS*-box genes code for proteins that serve as transcription factors; intense expression of transcripts of the *AGL-2* gene of *Arabidopsis* and *AGL-15* gene of *Arabidopsis* and *Brassica napus* in the globular and torpedo-shaped embryos and, to a lesser extent, in older embryos of the respective species might reflect the role of specific transcription factors in regulating gene expression during embryogenesis (Flanagan and Ma 1994; Heck et al 1995; Rounsley, Ditta and Yanofsky 1995).

cDNA clones hybridizing to mRNAs preferentially expressed in the embryo or in both the embryo and endosperm of barley have been useful in defining developmental events in embryogenesis. Transcripts corresponding to an embryo-specific clone (*pZE-40*) begin to appear in the club-shaped embryos of barley; with further differentiation, steady-state levels of transcripts are abundantly expressed in the scutellum, coleoptile, and coleorhiza, but the axial tissues and the scutellar epidermis are generally devoid of transcripts. This expression pattern does not change until desiccation of the grain, when the transcript levels precipitously decline (Smith et al 1992). Transcripts of some embryo- and endosperm-specific cDNA clones are also characteristically expressed in the scutellum of developing barley embryos, whereas others that encode oleosins are expressed in the scutellum or in the root–shoot axis (Olsen, Jakobsen and Schmelzer 1990; Aalen et al 1994). Identification of the protein products encoded by these genes is necessary to provide a consolidated picture of their role in the continued differentiation of the embryo.

Promoter sequences of some genes are active in

regulating embryo-specific expression of reporter genes in transgenic plants. These include promoters of chalcone flavanone isomerase gene (*CHI-A*) from *Petunia hybrida* in transgenic *P. hybrida* (van Tunen et al 1990); rice actin-1 gene in transgenic rice and maize (Zhang, McElroy and Wu 1991; Zhong et al 1996); maize *ADH* gene in transgenic rice (Kyozuka et al 1994); *CDC-2a*, mitotic cyclin (*CYC1-At*), and phytochrome genes from *Arabidopsis* in transgenic *Arabidopsis* (Hemerly et al 1993; Ferreira et al 1994; Somers and Quail 1995); and asparaginase gene from *Lupinus arboreus* in transgenic tobacco (Grant and Bevan 1994). Two conserved regions identified in the promoters of ABA-responsive genes of many plants are a palindromic sequence (*PA*, 5'-GCCACGTGGC-3') and motif I (*Iwt*, 5'-GTACGTGGCG-3'). When these sequences are inserted in appropriate constructs and expressed in transgenic tobacco, PA is found to confer expression in the embryonic root and Iwt is found to drive developmentally regulated expression in embryos of all ages (Salinas, Oeda and Chua 1992). A role for PA and Iwt sequences in differentially targeting *trans*-acting factors to regulate gene expression during embryogenesis is implied here.

The expression of two of the several oncogenes of T-DNA of root-inducing (Ri) plasmid of *Agrobacterium rhizogenes* has been followed during embryogenesis in transgenic tobacco by histochemical assay for *GUS* gene expression. Whereas stimulation of reporter gene expression by the promoter of the *ROL-B* gene occurs uniformly in all cells of the globular, heart-shaped, and torpedo-shaped stage embryos including the suspensor, gene activity in mature embryos is restricted to the vascular elements and the apical meristems. Since *ROL-B* gene promoter is controlled by auxin in plant tissues, these results invoke an association between *GUS* gene expression and the synthesis of endogenous auxin in the embryos. By contrast, *ROL-C* gene promoter activity is conspicuously absent in the early stage embryos but appears in the phloem tissue of the hypocotyl and cotyledons of late heart-shaped embryos and increases in intensity until the mature stage (Maurel et al 1990; Chichiricco, Constantino and Spanò 1992; Aspuria, Nagato and Uchimiya 1994). Phloem specificity of expression of the *GUS* gene under the control of *ROL-C* gene promoter widens the potential use of the promoter in monitoring vascular tissue differentiation during embryogenesis. Two main domains and several subdomains of the CaMV 35S promoter fused with a reporter gene are found to be expressed differentially in mature embryos of transgenic tobacco, but the stage of embryogenesis when stable promoter activation occurs is not determined (Benfey, Ren and Chua 1989, 1990).

5. EXPRESSION OF GENES RELATED TO DESICCATION TOLERANCE AND DORMANCY

During seed maturation, the embryo undergoes desiccation resulting in the loss of nearly 90% of the water from the cells and then lapses into a state of quiescence or dormancy. Despite its extremely low water content, the embryo remains viable during this period of enforced rest, only to spring forth and germinate later under appropriate environmental conditions. Although the mechanism that retains the embryo in a developmental mode during desiccation and dormancy is poorly understood, it is clear that as a physiological change, desiccation tolerance is a genetically programmed phenomenon. Besides the loss of water, an increase in the endogenous ABA level in the cells of the embryo marks this period of development. At the molecular level, there is a close relationship between the low embryo water potential and expression of specific genes encoding mRNAs that usually accumulate in the dehydrated embryo. These mRNAs, referred to as *LEA* mRNAs, are translated in vivo during embryo maturation and early germination and rapidly disappear thereafter. One sequence family amongst these genes encodes a set of proteins known as dehydrins.

Several aspects of the mechanism by which embryos sense the low water potential have been investigated, although we do not as yet have a complete picture of the cues that activate *LEA* genes. ABA has become an obvious candidate because it exercises a potent effect in inducing the precocious accumulation of *LEA* mRNAs in cultured immature embryos. Additionally, it has been demonstrated that the expression of a variety of *LEA* genes is activated in plant tissues subjected to osmotic or salt stress or to desiccation. Another pertinent fact is that in several plants the concentration of ABA peaks in embryos during the period of declining water potential, then a decrease in concentration follows during the final stages of maturation (Hsu 1979; Ackerson 1984a; Finkelstein et al 1985; Prevost and Le Page-Degivry 1985; Galau, Bijaisoradat and Hughes 1987; Wang et al 1987a; Robertson et al 1989; Chang and Walling 1991). These observations have placed embryo desiccation, and the correlated expression of *LEA* messages and proteins and increase in ABA content, as part of a general response of plants to a broad range of environmental stresses, of which water deficit is paramount.

The concept of *LEA* mRNAs arose from an analysis of the changing mRNA populations during different stages of embryogenesis in cotton based on comparisons of the extant, in vivo and in vitro synthesized proteins (Dure, Greenway and Galau 1981). This work showed that a subset of 14 polypeptides that became abundant in mature embryos was a good molecular marker of the onset of embryo maturation. The disappearance of these sequences during germination of mature embryos and their premature accumulation in young embryos cultured in a medium containing ABA were also established in this study. Galau, Hughes and Dure (1986) assigned the term "*LEA*" to denote these mRNA sequences that are significantly more abundant in maturation stage embryos than in younger embryos, with the caveat that the term does not imply their inducibility by ABA under any experimental conditions. Using the amino acid composition of six LEA proteins of cotton as a point of reference, it was shown that LEA proteins from different species share regions of homology with three of the cotton LEA proteins (Baker, Steele and Dure 1988; Dure et al 1989). Based on this analysis, a classification of LEA proteins has recognized three groups, namely, group 1 with cotton LEA protein D-19 and wheat EM protein, group 2 with cotton LEA protein D-11 and rice RAB-21 protein, and group 3 with cotton LEA protein D-7 and LEA proteins from barley, carrot, and rapeseed.

One of the most interesting developmental features of embryogenesis, and one that has received considerable attention in recent years, concerns the basis for the expression of *LEA* genes. It was pointed out earlier that the influence of *LEA* genes is felt as the embryo becomes dehydrated, and consequently, the gene products are presumed to facilitate desiccation survival of the embryo enclosed in the seed. Given the fact that desiccation stress affects the survival of every conceivable part of the plant, it is not surprising that expression of *LEA* genes or their protein products is commonly observed in leaves and other organs subjected to this stress. Viewed from another angle, it is significant that the promoter region of a desiccation-related gene cloned from leaves of an extremely desiccation-tolerant plant (*Craterostigma plantagineum*) is activated in the embryos of transgenic tobacco as the seeds become dehydrated (Michel et al 1993). It is also of interest that the deduced amino acid sequence of a cDNA clone isolated from dormant spores of the fern *Onoclea sensibilis* reveals some homology to LEA proteins. Typical of *LEA* gene expression, transcripts of this cDNA clone begin to accumulate during the maturation and desiccation phases of spores before they lapse into dormancy (Raghavan and Kamalay 1993). Structural features of LEA proteins suggested by their amino acid sequences (in particular by the presence of a tandemly repeated sequence) and their hydrophilic nature in some cases (Baker, Steele and Dure 1988; Galau, Wang and Hughes 1992b; Dure 1993), by their homology to water-stress–related proteins in other cases (Galau, Wang and Hughes, 1993), and by their uniform distribution in the cytosol of all cells of the embryo in high molar concentrations (J. K. Roberts et al 1993), are considered to be ideal to protect the embryo from desiccation damage or to repair such damage. In short, most of the speculative attention given to the function of LEA proteins has focused on desiccation survival, with ABA thrown in as a signal transducer, but further work is necessary to place these ideas in a tangible context.

The intent of this section is to deal with the isolation, characterization, and expression of *LEA* and other genes involved in the desiccation and dormancy of embryos of different species of plants and to discuss how their expression is regulated by ABA. An unusual breadth of information into many general features of *LEA* gene expression arose from studies with model systems such as embryos of cotton and wheat. For this reason, these systems are treated in greater detail than others.

(i) *LEA* Genes from Cotton and Other Dicots

cDNAs for a total of 18 different *LEA* genes have been cloned and characterized from cotyledons of cotton. Based on sequence data, 6 of the clones have been assigned to small multigene families; each of the 6 genes contains the standard features, including TATA-box, poly(A) addition sites, and one or more introns (Baker, Steele and Dure 1988; Galau and Close 1992). As expected, mRNAs of all *LEA* sequences are abundant in the mature embryos and most, but not all, are modulated to varying degrees by ABA in excised and cultured embryos. However, the accumulation kinetics of *LEA* mRNAs during embryogenesis have placed them in two groups of coordinately regulated mRNAs (Figure 14.14). These two groups are (a) *LEA* genes, with 12 members whose mRNAs exhibit their maximum level of accumulation in

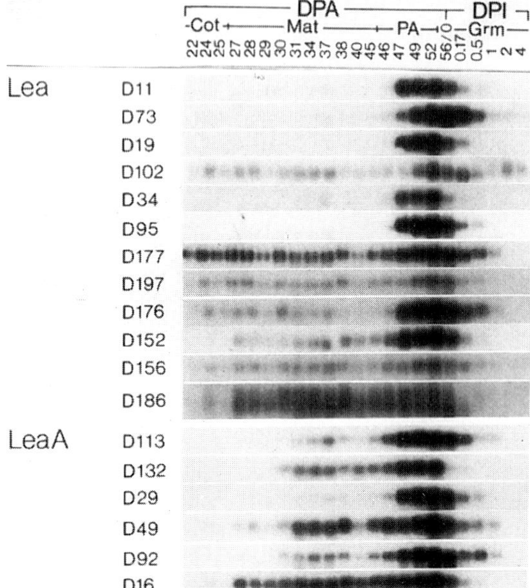

Figure 14.14. Gel–blot hybridization of total RNA from cotyledons of cotton embryos with two groups of cDNA clones encoding LEA proteins. Days indicated at the top are DPA, days postanthesis; DPI, days postimbibition. Stages of embryo development indicated are Cot, cotyledon; Mat, maturation; PA, postabscission; Grm, germination. (From Hughes and Galau 1989; photograph supplied by Dr. G. A. Galau.)

embryos after ovules have severed their connection to the maternal plant at the funiculus (postabscission stage, an 8-day period between maturation and desiccation, beginning 45–46 days postanthesis); and (b) *LEA-A* genes, with 6 members whose mRNAs, in addition to postabscission stage expression, are also modulated by ABA (Galau, Hughes and Dure 1986; Galau, Bijaisordat and Hughes 1987; Hughes and Galau 1989). Since *Gossypium hirsutum*, used in these investigations, is a natural allotetraploid, it may be possible that the complex kinetics of *LEA* and *LEA-A* mRNA expression are the sum of two alloalleles encoding them. However, the fact that transcripts of a few randomly selected *LEA* and *LEA-A* alloallele pairs display similar expressions at the mRNA or protein levels excludes this possibility (Hughes and Galau 1987; Galau and Hughes 1987). Several cotton *LEA* and *LEA-A* mRNAs also accumulate in the postabscission stage embryos of other dicots (Jakobsen, Hughes and Galau 1994).

Harada et al (1989) have identified a cDNA clone whose mRNA accumulates at low to undetectable levels in immature embryos of *Brassica napus* but is inducible by ABA; however, it is highly prevalent in mature embryos and disappears during the early hours of germination. In the context of stress tolerance, a further novel feature of this recombinant is that mild, nonlethal freezing of plants triggers the premature synthesis of its transcripts in embryos during subsequent growth at room temperature (Johnson-Flanagan et al 1992). Typical of other LEA proteins, the deduced structure of the protein encoded by the gene shows extreme hydrophilicity and the presence of tandemly organized direct repeats. A somewhat similar expression pattern has been demonstrated for transcripts of a gene isolated from a cDNA library made to poly(A)-RNA of mature embryos of *Raphanus sativus*; the protein deduced from the nucleotide sequence shares good homology with a cotton LEA protein and wheat EM protein (Raynal et al 1989).

Investigations on precocious germination of cultured immature embryos of *Pisum sativum* showed that addition of ABA to the medium increases the production of two low-molecular-mass proteins, ABR-17 (molecular mass 17.2 kDa) and ABR-18 (molecular mass 26 kDa). Although both of these proteins and their mRNAs are detected in mature embryos, in other respects the proteins differ in their spatial and temporal expression patterns. Especially interesting are the accumulation of ABR-18 protein in the testa during early seed development and the continued synthesis of ABR-17 protein during germination. Amino acid analysis showed that both proteins have more homology with pea disease-resistant response proteins than to any of the LEA proteins. However, other data relating to the accumulation of these proteins in aging tissues and in response to various stimuli unrelated to pathogenicity suggest that the expression of genes for these proteins in the pea embryos appropriately fits the bill for desiccation response (Barratt, Domoney and Wang 1989; Barratt and Clark 1991, 1993). A strong candidate for the desiccation-related protein of pea embryos is encoded by a cDNA clone isolated from a pea embryo expression library using maize dehydrin antiserum. Not only is this protein nearly homologous to dehydrins from other plants, but, typical of desiccation-related genes, transcripts of the gene encoding it are expressed at a high level during embryo maturity (Roberton and Chandler 1992). Similar evidence for the presence of LEA proteins has come from analyses of sunflower embryos (Almoguera and Jordano 1992) and embryos of *Cicer arietinum* (chick pea;

Fabaceae) (Colorado, Nicolás and Rodríguez 1995; Colorado et al 1995).

A quite different category of LEA proteins is represented in sunflower embryos by ubiquitin. The characteristic abundance and up-regulation of expression patterns of transcripts of a cDNA clone encoding tetraubiquitin during embryogenesis in sunflower are very similar to those of *LEA* genes. The suggestion that degradation of denatured proteins produced during desiccation of embryos might involve a ubiquitin-mediated step is consistent with the known functions of this protein (Almoguera, Coca and Jordano 1995).

In addition to the accumulation of a set of proteins, a feature that endows desiccation tolerance in embryos of soybean is an increase in the concentration of sugars, especially stachyose (Blackman et al 1991; Blackman, Obendorf and Leopold 1992). There is thus circumstantial evidence for the view that proteins may function in concert with oligosaccharides in the development of desiccation tolerance. Several cDNA clones, corresponding to different maturation proteins ranging in molecular mass from 18 kDa to 70 kDa, have been isolated from soybean embryos, and the expression of transcripts of these clones has been shown to increase not only in mature embryos but also in precociously matured embryos (Hsing et al 1990; Hsing and Wu 1992). The derived amino acid sequences of two of the clones analyzed show many homologies to LEA proteins (Chen et al 1992; Hsing, Chen and Chow 1992).

Beyond the examples described in which *LEA* mRNAs and proteins have been isolated and delineated from mature embryos of dicots, some investigators have followed in developing embryos of *Arabidopsis* the expression of *LEA* transcripts isolated from other plants. These studies include localization of *EM*-like genes and cotton *D-19* gene in mature embryos by in situ hybridization and Northern blot, respectively (Gaubier et al 1993; Finkelstein 1993) and expression of maize *RAB-17* gene in the embryos of transgenic *Arabidopsis* (Vilardell et al 1994). Transcripts of several clones homologous to *LEA* genes and some novel genes isolated from developing siliques of *Arabidopsis* are found to be prevalent in seeds undergoing desiccation, indicating that they are *LEA* transcripts (Parcy et al 1994). In the case of one of the genes, a functional analysis of the promoter in a transgenic setting has indicated that putative regulatory elements similar to those of *EM* and storage protein β-phaseolin genes govern the embryo specificity of its expression (Hull et al 1996).

(ii) *LEA* Genes from Wheat and Other Monocots

Among the recombinants identified from a cDNA library made to poly(A)-RNA of wheat embryos was one representing about 200 bases of a message that declined rapidly during germination (Cuming 1984). The identity of this sequence, designated as *EM*, as a member of *LEA* group 1 was established in related studies that led to the isolation of a near full-length clone, the determination of its nucleotide and amino acid sequences, and demonstration of its presence in low levels early in embryogenesis, its abundance in late stage embryos, and its inducibility by ABA (Williamson, Quatrano and Cuming 1985; Litts et al 1987, 1991; Williamson and Quatrano 1988; Morris et al 1990). Another wheat *LEA* gene characterized by high expression in mature embryos and inducibility by ABA is *PMA-2005*; it encodes a protein product of molecular mass 23 kDa and is assigned to *LEA* group 3 (Curry, Morris and Walker-Simmons 1991). In attempts to determine the mechanism of ABA regulation of *EM* gene expression in developing embryos, it was found that a construct containing a 646-bp 5'-fragment of the *EM* gene fused with *GUS* gene causes a sharp increase in gene expression in near mature seeds of transgenic tobacco relative to seeds of other ages. Expression of the construct was inducible in embryos of immature seeds by incubation in ABA (Marcotte, Russell and Quatrano 1989). A protein that belongs to the leucine zipper class of transcription factors and interacts with an 8-bp core (T/CACGTGGC) consensus sequence of the ABA-response element of the *EM* gene has been identified (Figure 14.15); this development and the delineation of a transcript with homology to protein kinases that increases with embryo maturity and ABA treatment are considered important steps in determining signal transduction in the ABA-induced expression of the *EM* gene (Guiltinan, Marcotte and Quatrano 1990; Anderberg and Walker-Simmons 1992). Lam and Chua (1991) have shown that a mutated sequence from the wheat histone H3 promoter confers a high level of expression in embryos of transgenic tobacco. Proof for the probable function of this sequence as an ABA-responsive element has come from the observation that the expression of the promoter is diminished upon germination and is reinstated by ABA. An unidentified tobacco nuclear factor has been proposed as a candidate element that interacts with the mutated sequence.

Attempts to identify *LEA* mRNA and proteins in

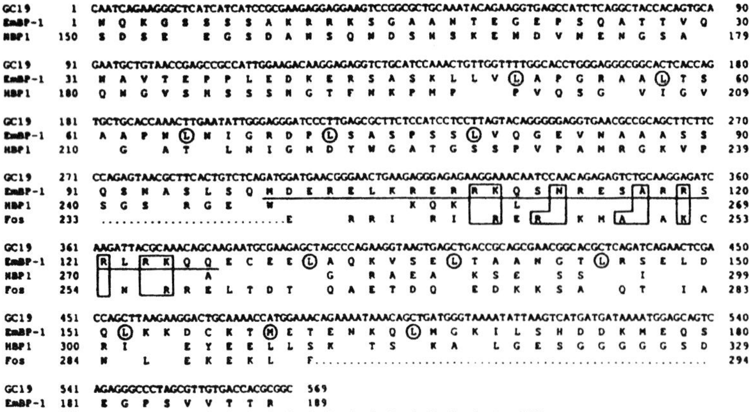

Figure 14.15. Nucleotide sequence of the wheat *EM* gene (GC19) and its deduced amino acid sequence. Numbering is from the first nucleotide and encoded amino acid, respectively. A comparison is made with a portion of the amino acid sequence of wheat histone DNA-binding protein HBP-1, and the residues in HBP-1 that differ are indicated. The basic domain is underlined and the periodic leucines and the single methionine are circled. Boxes indicate the residues conserved among most leucine zipper proteins. (Reprinted with permission from M. J. Guiltinan, W. R. Marcotte, Jr. and R. S. Quatrano 1990. A plant leucine zipper protein that recognizes an abscisic acid responsive element. *Science* 250:267–271. © 1990. American Association for the Advancement of Science.)

barley embryos go back to experiments on the acquisition of desiccation tolerance. Transition from desiccation susceptibility in 12-day-old embryos to desiccation tolerance at 18 days is associated with the synthesis of a set of 25–30 proteins (Bartels, Singh and Salamini 1988). Isolation and characterization of a gene temporally linked with desiccation tolerance in embryos showed that it is homologous to mammalian genes encoding an NADPH-dependent aldose reductase (Bartels et al 1991). This enzyme is considered to play an important role in the synthesis of the common cell osmolyte, sorbitol, in animal cells; osmolytes generally help to balance the changes in the osmolarity of the cytoplasm caused by adverse environmental conditions. It is interesting to note that the same gene holds the key to the mobilization of defense against common traumatic effects in both plants and animals. Another gene with a potential function in desiccation tolerance that becomes transcriptionally active in developing barley embryos has a protein sequence homologous to bacterial glucose dehydrogenase and ribitol dehydrogenase (Alexander et al 1994).

Three additional *LEA* mRNAs (*B-19.1, B-19.3,* and *B-19.4*) were identified by using a cotton *LEA* cDNA probe to screen a barley cDNA library made to poly(A)-RNA from ABA-treated immature embryos, and a fourth one was identified by screening the library with a *B-19.3* cDNA probe. The temporal accumulation pattern of transcripts is essentially the same for all genes and, beginning about 25 days after anthesis, it shows accumulation to high levels throughout embryogenesis. However, based on the distinct patterns of accumulation of transcripts of three members of the *B-19* gene family in embryos of different ages in response to mannitol, ABA, and NaCl, it seems that different signal transduction pathways operate in their expression (Espelund et al 1992, 1995). Several properties of the putative B-19 proteins, especially their strong hydrophilic nature and the internal 20–amino acid sequence repeated one to four times, support their strong similarity to the EM protein from wheat and to LEA proteins from several dicots (Hollung, Espelund and Jakobsen 1994).

Studies on the regulation of gene expression during embryogenesis in maize have shown that a set of 23–25-kDa polypeptides appears during embryo maturation and disappears during germination (Sánchez-Martínez, Puigdomènech and Pagès 1986). Further characterization of these proteins involved the isolation of cDNA clones from mature maize embryos that hybridize with specific mRNAs, the demonstration of the absence of transcripts of cDNA clones in young embryos and their abundance in mature embryos, and the ABA inducibility of the transcripts in immature embryos (Goday et al 1988; Gómez et al 1988). Transcripts of one of the clones (*PMAH-9*) are restricted in distribution to the two outer layers of cells of the scutellum and to the epidermal cells of the embryonic leaves. The deduced protein of this clone has an unusually high glycine content, like other proteins with putative structural and protective functions, and based on its affinity to uridine- and guanosine-rich RNAs, it is considered to be an RNA-binding protein (Gómez et al 1988; Ludevid et al 1992). Another clone (*MA-12*) also encodes a glycine-rich protein (molecular mass 17 kDa) with conserved tracts of amino acid sequences as in other LEA proteins. Because of the extensive homology of this protein to RAB-21 from rice – a homology evidenced especially by the presence of glycine-rich domains and conserved sequences and by the C-terminal regions – it is referred to as RAB-17. The state of phosphorylation of RAB-17 protein changes during embryogenesis, because it becomes highly phosphorylated in the desiccated embryo (Vilardell et al 1990). Structural and functional analyses have demonstrated that 5'-upstream sequences of the *RAB-17* gene confer developmental and hormonal expression of *CAT* reporter gene in embryos of transgenic tobacco. Developmentally, expression of *RAB-17* gene and of *RAB-28*, another maize stress-protein gene, is restricted to the mature embryos and, hormonally, expression is induced in immature embryos by ABA (Vilardell et al 1991; Niogret et al 1996). Moreover, the proximal promoter region of *RAB-28* contains a motif homologous to the conserved ABA-responsive element (T/CACGTGGC) of wheat embryos; DNA fragments containing this motif have been shown to bind to a *trans*-acting factor from the embryo nuclei (Pla et al 1993). Changes in steady-state levels of mRNA of a cDNA clone, *EMB-564*, during embryogenesis in maize are similar to those observed with *PMAH-9* and *RAB-17*. *EMB-564* encodes a protein that shows extensive similarity to protein products of wheat *EM* gene and cotton and radish *LEA* genes (Williams and Tsang 1991).

Transcriptional and translational expression of the *RAB-17* gene occurs in all cells of the maize embryo, including those of the scutellum, embryo axis, and provascular tissues. With the progress of embryo maturation, gene transcripts are increasingly confined to the outermost cells of the embryo. Both the cytoplasm and nucleus have been identified as the subcellular targets of RAB-17 protein accumulation. Furthermore, a specific binding of RAB-17 protein to a nuclear localization signal binding protein has been demonstrated (Asghar et al 1994; Goday et al 1994). In the larger view, this implies that nuclear protein transport relies heavily on RAB-17 protein during stress, but this idea is in its infancy and requires further investigation.

A pivotal feature of the maturation program of rice embryos is the progressive accumulation of transcripts of the *RAB-21* gene, which is also expressed in roots, leaves, and suspension cells under stress. The gene encodes a low-molecular-mass (16 kDa), basic, glycine-rich protein that accumulates only in the cytosol. The protein contains two domains with an A and a B repeat in each; these features characterize the gene and its products as a member of the LEA family (Mundy and Chua 1988). *RAB-21* is a member of a small multigene family, and four members of this family that have been characterized are referred to as *RAB-16A* to *RAB-16D*. Northern blot analysis with gene-specific probes showed that mRNAs of only *A*, *B*, and *C* accumulate in mature embryos (Yamaguchi-Shinozaki, Mundy and Chua 1989). An indispensable region of the *B* genome, which confers a developmentally regulated expression in embryos of transgenic tobacco, is restricted to 482 bp of 5'-sequences (Yamaguchi-Shinozaki et al 1990). Two genes isolated from rice embryos encoding LEA proteins and a basic leucine zipper binding protein are found to display similar characteristics, such as control by ABA and activation by the hormone in transient expression assays (Hattori, Terada and Hamasuna 1995; Nakagawa, Ohmiya and Hattori 1996).

(iii) Is Embryo Desiccation Developmentally Regulated by ABA?

One of the most intriguing problems in the physiology of the growth of embryos concerns the role of ABA as a developmental regulator of embryogenesis and a mediator of the response of embryos to stress. Earlier, it was mentioned that there is an increase in the concentration of ABA in embryos during the period of their declining water content and that the addition of ABA promotes the accumulation of *LEA* mRNAs in cultured immature embryos. In principle, two types of models can be invoked to explain the role of ABA: (a) the hormone promotes *LEA* mRNA accumulation that obviously results in a cascade effect leading to desiccation, and (b) the increase in the hormone level is a secondary effect of desiccation-related events in the embryo that are potentiated by other factors. Since much of the argument in favor a specific function for ABA in embryogenesis has followed from its effects on the accumulation of *LEA* mRNAs or their protein products in cultured immature embryos, this work will be reviewed here.

Many more *LEA* mRNA markers have been followed in cotton embryos than in embryos of any other angiosperm to study ABA effects, and yet there is no convincing picture of how the hormone is required during maturation and subsequent events of embryogenesis. The work of Dure, Greenway and Galau (1981) initially demonstrated that a subset of proteins that are synthesized late in embryogenesis but disappear during germination can be induced prematurely in excised young embryos cultured in ABA. This study was followed by another work, which showed that out of 18 cloned *LEA* mRNA sequences, as many as 13 are clearly regulated by ABA in cultured immature embryos (Galau, Hughes and Dure 1986). Subsequent investigations have shown that addition of ABA to the medium regulates the accumulation of a selection of *LEA* mRNAs and/or proteins in cultured immature embryos of wheat (Williamson, Quatrano and Cuming 1985; Morris et al 1990), rapeseed (Harada et al 1989; Jakobsen, Hughes and Galau 1994), barley (Bartels, Singh and Salamini 1988; Espelund et al 1992; Hollung, Espelund and Jakobsen 1994), maize (Sánchez-Martínez, Puigdomènech and Pagès 1986; Goday et al 1988; Gómez et al 1988; Williams and Tsang 1991; Thomann et al 1992; Williamson and Scandalios 1992), pea (Barratt, Domoney and Wang 1989; Barratt and Clark 1991), tobacco, soybean (Jakobsen, Hughes and Galau 1994), and rice (Hattori, Terada and Hamasuna 1995). In some cases, various types of osmotic stress and putative plant hormones such as jasmonic acid mimic the effects of ABA on the induction of *LEA* mRNAs and proteins (Espelund et al 1992; S. Reinbothe et al 1992; Thomann et al 1992; Barratt and Clark 1993; Hollung, Espelund and Jakobsen 1994).

As described in a previous section, some suggestive evidence for the role of ABA as a regulator of embryogenesis is provided by cases of vivipary. Induction of the germination mode seen in embryos of viviparous mutants and the analysis of ABA-insensitive mutants have played a role in articulating the idea that vivipary is analogous to precocious germination in vivo. However, the few cases described neither provide absolute confirmation that vivipary is related to the failure of ABA synthesis in the embryos nor assert that the hormone regulates the accumulation of *LEA* mRNAs or proteins.

A perspective that ABA and water stress function as environmental inducers, rather than as developmental regulators, of embryogenesis has been provided by Galau, Jakobsen and Hughes (1991), who have also pointed out many interpretational pitfalls in embryo culture experiments. The significant observations supporting their model are that culture of immature embryos of cotton in a mineral-salt medium under standard conditions executes a maturation program in the molecular sense and leads to an increase in *LEA* mRNA amounts at rates very similar to those seen in embryos in vivo up to the postabscission stage; this mRNA induction is insensitive to the addition of fluridone. Thus, evidence is tenuous that ABA is involved in the accumulation of *LEA* mRNAs, and the picture is complicated by the observation that the hormone or a high osmoticum does indeed cause increased accumulation of *LEA* messages in cultured immature embryos (Hughes and Galau 1991; Jakobsen, Hughes and Galau 1994). Finkelstein (1993) has shown that a cotton *LEA* gene is expressed in the seeds of an ABA-deficient mutant of *Arabidopsis* at levels found in the wild-type seeds; this observation does not point to the necessity of endogenous ABA as an essential control signal for *LEA* gene expression. A hypothesis proposed to explain the postabscission program of embryogenesis includes ovule abscission that prepares the embryo for desiccation and an environmentally responsive mechanism resulting in water stress. Obviously, this may not seem to be an attractive hypothesis to those with a bent toward hormonal explanations, as this discounts a role for endogenous ABA in embryo desiccation.

The discovery of *LEA* genes and the focus on the meaning of their proteins in the desiccation tolerance of embryos represent a major breakthrough in the molecular embryology of angiosperms in the last 10 years. The next areas of impact will be the determination of how desiccation leads to gene expression and an understanding of the pathways involved in desiccation tolerance of the embryos.

(iv) Genes Involved in Dormancy Regulation and Seedling Growth

In a great many plants, desiccation of the embryo is a prelude to the seed's lapse into dormancy. Several critical metabolic events are elaborated in the embryo during the final stages of embryogenesis, which lead to the dormant state. Some of these events include changes in water content, acquisition of desiccation tolerance, changes in pigmentation, synthesis of ABA, and accumulation of LEA and storage proteins. Mutations in genes that regulate these events produce embryos with defects in the following areas: storage product accumulation, pigmentation, maturation program, and seed morphology; the embryos also have multiple lesions in the regulation of the final stages of development. Analysis of these mutations has made feasible a genetic approach to the identification of the genes involved in the dormancy of embryos.

Since ABA affects many aspects of late embryo development, including dormancy, water relations, germination, and stress tolerance, mutational approaches to the identification of genes regulating late embryogenesis have logically related the defects to disturbance in the ABA metabolism of embryos. Using *ABA-deficient* (*aba*) mutants of *Arabidopsis*, it was initially found that ABA plays an important part in the induction of seed dormancy, because the mutant seeds germinate even without undergoing a period of dormancy; in contrast, seeds of the wild type that contain ABA remain dormant (Karssen et al 1983). In a later work, which screened a large number of mutants for seed germination in the presence of inhibitory concentrations of ABA, three *ABA-insensitive* (*abi*) loci, designated as *abi-1*, *abi-2*, and *abi-3*, were identified. These mutants differed in their vegetative growth and seed development; *abi-1* and *abi-2* were characterized by wilting due to disturbed water relations but had wild-type levels of storage proteins and corresponding mRNAs, whereas the *abi-3* mutants had reduced amounts of these macromolecules. In severe mutant alleles of *abi-3*, the embryos are found to lapse directly into the germination mode while they are still enclosed within the seed coats (Koornneef, Reuling and Karssen 1984; Finkelstein and Somerville 1990; Nambara et al 1995). When ABA deficiency and hormone insensitivity are combined in double mutants, *aba/abi-3* recombinants are found to generate seeds that do not desiccate normally, that remain green, and that germinate without lapsing into dormancy; moreover, seeds of the double mutant are deficient in some of the major storage proteins and other low-molecular-mass polypeptides (Koornneef et al 1989; Meurs et al 1992).

The foregoing observations suggest that seed development in *Arabidopsis* is much more sensitive to ABA action than is induction of dormancy. To further explore the role of the *ABI-3* gene in regulating the late embryogenesis program, uniconazol, an inhibitor of GA biosynthesis, was used to isolate mutants defective in the normal germination program. One of the lines isolated was insensitive to the inhibitor due to a mutation in the *ABI-3* gene and produced seeds that were green, desiccation intolerant, and nondormant. The mutation reduced the amount of, but did not eliminate, the principal seed storage proteins (Nambara, Naito and McCourt 1992). Most of these features were also noted in the seed phenotypes of two additional mutants that are characterized as severely ABA insensitive (Ooms et al 1993). Hence, the data suggest that some products of the *ABI-3* gene regulate an ABA-dependent stage of embryogenesis that prevents a germinative type of development and induces dormancy. In support of this view is the observation that the *ABI-3* gene encodes a protein that shares regions of sequence homology with *VP-1* gene of maize (Giraudat et al 1992). As noted earlier (p. 401), VP-1 protein acts as a transcription factor and, by implication, the *ABI-3* gene product can be considered to regulate embryo dormancy at the transcriptional level.

An understanding of the precise roles of *ABI-1*, *ABI-2*, and *ABI-3* genes in inducing seed dormancy is complicated by the fact that seeds of all three mutants skip the usual period of dormancy and germinate, yet embryos of *abi-1* and *abi-2* mutants continue to synthesize storage proteins. Do these genes regulate the same pathways leading to dormancy or do the genes constitute independent regulatory networks with and without the involvement of ABA? An answer to this question may lie in the complete molecular characterization of the *ABI* locus and the proteins it encodes. The *ABI-1* gene has been cloned independently by two groups of investigators and shown to code for a Ca^{2+}-modulated phosphatase having a carboxy-terminal domain with high affinity to serine-threonine protein phosphatases. The involvement of serine or threonine phosphatases in signaling processes makes the *ABI-1* gene product a prime candidate to integrate ABA- and Ca^{2+}-dependent stimuli with the phosphorylation status of the target membranes of embryos (Leung et al 1994; Meyer, Leube and Grill 1994). Since seeds of *abi-1* mutants germinate and produce long roots in the presence of inhibitory concentrations of ABA, it has been suggested that ABI-1 protein might neutralize the various phosphorylation events necessary for cell division in the root meristem, ensuring that the embryo remains dormant (Leung et al 1994). Thus, it seems that *ABI-1* and *ABI-3* gene products, which provide the link with the genome, prepare the embryo for dormancy by regulating two strategically different pathways.

The genetic approach to the study of the embryo maturation program has been augmented by the isolation and characterization of mutant phenotypes defective in maturation-associated embryo morphology. The *leafy cotyledon* (*lec-1*) phenotype isolated by screening seeds of transgenic *Arabidopsis* plants produced by T-DNA insertion is a paradigm of such a mutant with striking defects in the maturation-specific events of embryogenesis. Besides being desiccation intolerant and nondormant, the mutant embryos are defective in storage product accumulation and lack the protein and lipid bodies characteristic of wild-type embryos. However, the distinguishing feature of this mutant is the reversion of cotyledons to a leaflike state by the development of trichomes and stomata on their adaxial surface, by a complex pattern of vascularization associated with the formation of differentiated tissues, and by the resemblance of cotyledonary parenchymatous cells to mesophyll (Meinke 1992; Meinke et al 1994). Three other mutants with related phenotypes that produce trichomes on cotyledons but that generally differ from each other in certain other aspects of embryo maturation have allowed delineation of a group of regulatory genes that function during late stages of embryogenesis. Mutant phenotypes of *lec-1*, *lec-1.2* (an allele of *lec-1*), *lec-2*, and *fusca* (*fus-3*) are the most closely related because these mutations affect similar features of the late embryogenesis program. Differences between *lec-1* and *fus-3* embryos are restricted mostly to the frequency of stomata and the position of anthocyanin accumulation. In contrast, *lec-2* embryos are desiccation tolerant, do not germinate precociously, and show a pattern of storage product accumulation closer to the wild-type embryos than to *lec-1* or *fus-3* embryos (Keith et al 1994; Meinke et al 1994). These characteristics indicate that a functional *LEC* gene is necessary for the development of cotyledons from the primordia initiated early in embryogenesis and that genes active in desiccation and dormancy are independent of those that specify cotyledon characters.

In a unifying model, West et al (1994) have suggested a specific role for the *LEC-1* gene not only in the regulation of late embryogenic development but also in the transition of embryos to a germination mode. The basis for this model is the observation that although *lec-1* mutation affects desiccation tolerance and storage protein accumulation associated with late embryogenesis, some postgerminative traits, such as the expression of lipid transfer protein gene at a level characteristic of seedlings, are initiated prematurely in the mutant embryos. Interestingly, some aspects of the maturation and desiccation phases, such as the temporal and spatial expression of cruciferin, oleosin, and *LEA* genes, are not affected by the mutation. The implication is that a battery of genes controlling independent regulatory pathways may operate concurrently to control the embryo maturation program and germination and that a large majority of these genes remain unidentified.

Examination of other mutants has confirmed that activation of a set of genes during late embryogenesis initiates strategies for dealing with the growth of seedlings after germination. Several newly isolated *fus* mutants fall into the group showing defects in the embryo and the seedling; embryos of these mutants go through a normal program of development but have the peculiar virtue of overproduction of anthocyanins in their cotyledons (Bäumlein et al 1994; Castle and Meinke 1994; Miséra et al 1994). In addition, seedling lethality resulting in limited seedling development is an important pervasive factor throughout the early stages of germination of seeds of *fus* mutants. Since several *fus* seedlings respond to sugars and hormones differently than do the wild-type seedlings and since they do not become etiolated in the dark, it has been suggested that gene products needed for normal responses to carbohydrates, hormones, and light are synthesized during embryogenesis. Characterization of one of the *FUS* gene products showed that it belongs to a new class of proteins that is hydrophilic, α-helical, with several consensus protein kinase-C phosphorylation sites. Since some *fus* mutants are alleles of other *Arabidopsis* mutants defective in light-regulated responses, FUS proteins have been implicated in regulating signal transduction pathways during seedling growth (Castle and Meinke 1994; Miséra et al 1994). Bäumlein et al (1994) have shown that when storage protein and nonstorage protein promoters are tested in transgenic *Arabidopsis* plants in a *fus-3* genetic background, transcription of only the storage protein promoters is inhibited; this has led to the view that the *FUS-3* gene product might function as a transcription regulator of storage protein genes.

6. REGULATION OF EXPRESSION OF DEFENSE-RELATED GENES

The role of specialized proteins synthesized by embryos in preventing bacteria and fungi from attacking seeds was not recognized until relatively recently. Because the production of these proteins exploits the genetic information possessed and expressed by cells of the embryo, some attention has been devoted to cloning the genes involved and identifying their protein products.

(i) Lectin Gene Expression in Embryos of Cereals and Grasses

One of the most extensively studied defense-related proteins in embryos is lectin. Although a role for lectins in the defense against bacterial and fungal infection of plants has been mired in controversy, the bulk of the evidence is in favor of such a role (Stinissen, Peumans and Carlier 1983; Raikhel and Lerner 1991). The importance of other proteins, such as catalase and superoxide dismutase, in the defense of embryos against stress tolerance has barely been commented upon and, consequently, the regulation of expression of their genes deserves to be investigated (Acevedo and Scandalios 1990; M. Suzuki et al 1995).

Lectin of wheat embryos, known as wheat germ agglutinin (WGA), is an embryo-specific protein whose synthesis commences about midway during grain development (Triplett and Quatrano 1982; Morris et al 1985). At this point in development, WGA is first detected immunocytochemically in the epidermal cells of the radicle, root cap, and coleorhiza. In the mature embryo, WGA is also found in the epiblast and in the surface layers of the coleoptile and scutellum. At the subcellular level, it is confined to the matrix of the protein bodies, the periphery of electron-translucent regions of the cytoplasm, and the cell wall–protoplast interface (Mishkind et al 1982; Raikhel and Quatrano 1986). Thus, lectin is localized in those cells and tissues of the embryo that establish direct contact with the soil during germination and its presence in the cell wall conforms to the expectation of it as a component of the defense arsenal of the embryo.

Molecular studies of expression of a cDNA clone (*WGA-B*) encoding WGA isolectin B in many respects supports this view of a rather restricted temporal and spatial presence of lectin. Northern blot hybridization revealed that *WGA-B* transcripts increase significantly in the embryos between 10 and 40 days after anthesis and decrease markedly in mature embryos. The major sites of persistent accumulation of *WGA-B* transcripts are the epidermal layers of the radicle and coleorhiza of developing embryos (Figure 14.16). The expression patterns of *WGA-B* mRNA and WGA have much in common because both are present in the same cell types of the embryos (Raikhel and Wilkins 1987; Raikhel, Bednarek and Wilkins 1988). The expression pattern of barley lectin mRNA transcripts in the outermost layer of cells of the embryonic root tip and coleorhiza is similar to that for lectin accumulation (Mishkind et al 1983; Lerner and Raikhel 1989).

Posttranslational modifications play a key role in the processing of rice lectin from a molecular mass of 36 kDa into small polypeptides of molecular masses 18, 10, and 8 kDa. The large molecule is synthesized on the rough ER whence it is transported to the site of accumulation. It is believed that the processing to the mature form occurs outside the site of synthesis (Stinissen, Peumans and Chrispeels 1984). A surprising observation made with regard to rice lectin gene is that its expression is governed by two mRNA transcripts (0.9 kb and 1.1 kb). Both genes begin to function at high levels

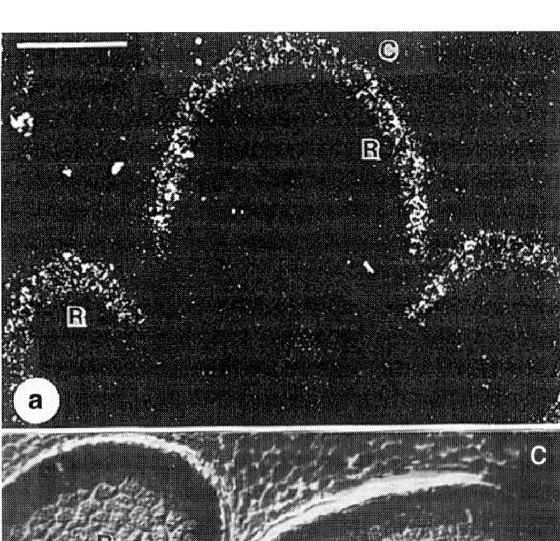

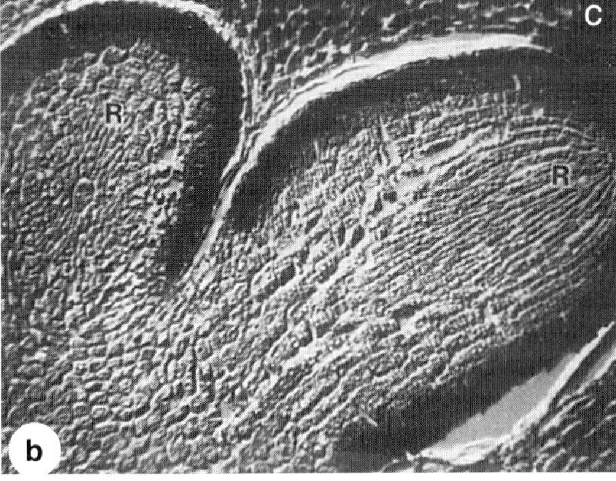

Figure 14.16. Autoradiographic localization of transcripts of *WGA-B* in the embryo of dry grain of *Triticum aestivum* and immunolocalization of WGA in the embryo of a 40-day-old caryopsis. **(a)** Localization of mRNA in the epidermal layer of the radicle; dark-field photograph. **(b)** Localization of WGA in the epidermal cells of the radicle and coleorhiza; Nomarski optics photograph. C, coleorhiza; R, radicle. Scale bar = 100 μm. (From Raikhel, Bednarek and Wilkins 1988. *Planta* 176:406–414. © Springer-Verlag, Berlin.)

early during embryogenesis in a tissue-specific manner and decline slightly at embryo maturation. A localized pattern of expression of the transcripts confined to the cells of the root cap, to the outer cell layers of the coleorhiza, radicle, and scutellum, and to all of the coleoptile is first observed in embryos as early as five days after anthesis. What fraction of the lectin expression can be attributed to each mRNA during embryogenesis? Data from gel–blot analysis of RNA isolated from embryos of different stages of development show that whereas both lectin mRNAs are present in high levels in developing embryos, the 1.1-kb species is the more prevalent one in mature embryos (Wilkins and Raikhel 1989).

The selective accumulation of lectins in maturing embryos in a manner analogous to LEA proteins supports the notion that lectin gene expression is controlled by ABA in a developmental context. Additional support for this has come from the demonstration that exogenous ABA enhances the synthesis and tissue-specific accumulation of WGA and its mRNA in immature wheat embryos (Triplett and Quatrano 1982; Raikhel and Quatrano 1986; Mansfield and Raikhel 1990). As revealed by nuclear run-off transcription assay, the distinguishing feature of the action of ABA is the transcriptional induction of lectin steady-state mRNA levels (Mansfield and Raikhel 1990).

(ii) Expression of Defense-related Genes in Embryos of Other Plants

The initial diversification of organs in developing embryos of legumes is accompanied by the synthesis of storage proteins and lectin in the protein bodies of cells of the cotyledons. In *Phaseolus vulgaris*, the storage protein phaseolin accounts for about 50% of the total embryo protein, whereas the lectin PHA accounts for 5%–10% of the total proteins. PHAs are tetrameric proteins composed of two different subunits known as PHA-E (erythrocyte-agglutinating) and PHA-L (lymphocyte agglutinating), which are encoded respectively by two tandemly arrayed genes, *DLEC-1* and *DLEC-2*. These genes are intronless and their coding regions exhibit 90% homology (Hoffman an Donaldson 1985). The presence of regulatory factors in transgenic tobacco that recognize the promoter of *DLEC-2* for the proper spatial and temporal expression of the gene has been established by using the reporter gene luciferase to assay for the activity of the promoter and by deletion analysis with respect to expression in the seeds of transgenic tobacco (Riggs, Voelker and Chrispeels 1989; Riggs et al 1989). The latter approach has identified positive regulatory elements between –550 bp and –125 bp in the upstream region of the *DLEC-2* gene that lead to enhanced levels of the gene product. Since gel retardation assays have indicated that this region contains nuclear protein binding sites, it is possible that the cellular machinery of the tobacco embryo recognizes the *DLEC-2* gene and modulates its expression by protein–DNA interactions (Riggs, Voelker and Chrispeels 1989).

Two other products of the lectin gene are an α-amylase inhibitor purified from cotyledons of *P. vulgaris* (Moreno and Chrispeels 1989) and arcelin, found in certain wild relatives of *P. vulgaris* (Osborn et al 1988; Anthony, Vonder Haar and Hall 1991). Additional insight into PHA formation has been derived from the observation that in *P. vulgaris* and *P. coccineus*, PHA genes are expressed abundantly at the mRNA level in the cotyledons during seed maturation and at very low levels in the embryo axis (Hoffman, Ma and Barker 1982; Chrispeels, Vitale and Staswick 1984; Voss, Schumann and Nagl 1992).

Biochemical and electron microscopic characterization has provided a nearly complete, but complex, picture of the synthesis, processing, and accumulation of PHA. The protein is synthesized exclusively on polysomes attached to the ER; biosynthesis is complete with the cotranslational glycosylation of the protein within the lumen of the ER. The Golgi apparatus mediates in the transport of PHA from the ER to the vacuole for storage; this transport involves dense vesicles with an electron-dense matrix (Bollini and Chrispeels 1978; Chrispeels 1983). Immunocytochemical localization has shown that PHA is associated with the protein bodies of the cotyledonary cells, although in the cells of the embryo axis the same protein is found floating in the cytoplasm. This is surprising since the biosynthetic steps in the formation of the mature protein body are the same in the cotyledons and the embryo axis (Manen and Pusztai 1982; Chrispeels, Vitale and Staswick 1984).

The fact that constructs of the *DLEC-2* gene result in the developmentally regulated expression of the protein in embryos of transgenic tobacco seeds evidently shows that the gene product is correctly glycosylated, assembled into tetramers, and targeted to the vacuole-based protein bodies; this has been confirmed by isolation and biochemical analysis of protein bodies from transgenic tobacco seeds (Sturm et al 1988). To test the possible role of glycan side chains in the targeting of proteins, one or both of the glycosylation sites were disrupted by site-directed mutagenesis. When tobacco plants were transformed with the mutated gene, the protein was correctly targeted and localized in the protein body of the embryo as with the unmodified gene (Voelker, Herman and Chrispeels 1989). Thus, the immediate function of glycans does not appear to be one of targeting the signal of PHA to the vacuole, although other functions might be unearthed later.

Like PHA, soybean lectin is also localized in protein bodies of the cotyledonary cells (Vodkin and Raikhel 1986). The nucleotide sequence of the gene encoding soybean seed lectin (*LE-1*) shows that the gene is devoid of introns and directs the synthesis of a 1.1-kb mRNA. Gene transcripts are most prevalent at the midmaturation stage of the embryo, representing about 0.75% of its total mRNA mass, and are present only in insignificant amounts in late maturation stage embryo. This is similar to the regulation of other seed protein genes of soybean that are transcriptionally inactive or weakly transcribed in the maturation stage embryos (Goldberg, Hoschek, and Vodkin 1983; Vodkin, Rhodes and Goldberg 1983; Walling, Drews and Goldberg 1986). When the soybean lectin gene is introduced into the tobacco genome, lectin transcripts are found to show the same type of quantitative fluctuation in transformed seeds as in developing soybean embryos (Okamuro, Jofuku and Goldberg 1986; Lindstrom et al 1990). In any event, it should not seem peculiar that these genes no longer transcribe efficiently in the embryos of normal or transgenic plants after the cells are filled with all the gene products they can hold.

Additional insight into the transcriptional apparatus that regulates lectin gene expression during soybean embryogenesis has been derived from experiments on lectin DNA-binding activity of proteins extracted from nuclei isolated from embryos of different stages of development (Figure 14.17). From the outset, a high degree of correlation was observed between lectin gene transcription and the presence of lectin DNA-binding activity. Thus, embryo stages showing a high degree of lectin gene transcription have high DNA-binding protein activity; in embryos with low levels of lectin gene expression, the protein activity is reduced accordingly. In fact, the lectin gene binds to a 60-kDa nuclear protein that exhibits a developmentally regulated program paralleling the lectin gene transcription rate. This protein is able to bind to a fragment of lectin gene sequences −77 bp to −217 bp relative to the transcription start site (Jofuku, Okamuro and Goldberg 1987). It follows that the binding of the lectin gene to the nuclear protein may be among the parameters that determine its highly regulated expression during embryogenesis; a new basic understanding of the regulation of lectin gene expression during embryogenesis could come from the identification of this protein.

The lectin gene in pea embryos is encoded by a small multigene family but, due to the presence of many stop codons, only one gene is considered functional (Gatehouse et al 1987; Kaminski, Buffard and Strosberg 1987). The lectin is localized in the protein bodies of the cotyledons and embryo axis (van Driessche et al 1981). De Pater et al (1993) showed that nuclear proteins from leaves and embryos of pea bind to a small fragment of the promoter of the functional lectin gene containing three overlapping TGAC-like motifs. Introduction of constructs containing this fragment into tobacco confers a high level of expression in transgenic embryos and, to a lesser degree, in the endosperm. These results show that embryo-specific expression of the lectin gene in pea can be identified with a group of specific nucleotides.

Proteinase inhibitors are widely distributed in angiosperms, including legumes, in which they are localized in the embryos. These compounds defy attempts to define an accurate function during embryogenesis, but are apparently effective in roles as widely disparate as storage proteins, regulators of endogenous proteinases, and factors that confer disease resistance. KTI is a common proteinase inhibitor found in the protein bodies of cotyledonary cells of soybean embryos in association with storage proteins and lectins. It is encoded by a multigene family consisting of at least 10 members. Three *KTI* genes characterized (*KTI-1*, *KTI-2*, and *KTI-3*) and sequenced are intronless and encode distinct proteins, of which the one encoded by *KTI-3* is identical to the major KTI found in embryos. General features of the expression programs of these three genes have been extensively investigated and the current picture can be summarized as follows. All three genes follow the same temporal profiles of expression in soybean embryos and in seeds of transformed tobacco plants, but their mRNAs accumulate to different levels during embryogenesis and seed formation. *KTI-3* mRNA is the most prevalent and is followed by *KTI-1* and *KTI-2* in abundance. Other differences delineated by in situ localization, such as the greater abundance of *KTI-3* mRNA compared to *KTI-1* and *KTI-2* transcripts in the embryo axis, the appearance of *KTI-3* transcripts in the globular and heart-shaped embryos

Chapter 15

Storage Protein Synthesis in Developing Embryos

As discussed in Chapter 14, the embryo responds to changing genetic and physiological pressures by initiating a series of developmental and molecular changes. These are reflected in the abrupt switch from a period of cell divisions to one of cell expansion, desiccation of cells, changes in growth hormone levels, and synthesis and accumulation of defense-related proteins and storage compounds. Storage materials are constituted primarily of an acervate complex of proteins known as storage proteins, in addition to starch and lipids. Like the endosperm storage products discussed in Chapter 12, embryo storage compounds play an important role in seedling survival by providing the source of carbon and nitrogen skeletons to the embryo during seed germination.

Many aspects of the synthesis of storage proteins during embryogenesis have been investigated using embryos of various agronomically important plants, such as bean, pea, soybean, and rapeseed. What these embryos have in common is that they represent cases in which cellular metabolism of the cotyledons is deflected from a programmed synthesis of housekeeping proteins to one concerned with the production of storage proteins. In recent years, considerable progress has been made toward understanding the regulation of the expression of storage protein genes in plant embryos. Progress was made possible not only by the desire to decipher the often complex biosynthetic mechanisms of the accumulation of embryo storage proteins that constitute a significant part of our diet, but also by the hope that the isolation and characterization of storage protein genes would provide clues to genetically engineer their synthesis in other plants. This chapter will explore the present state of knowledge on the cytology of storage protein accumulation and the biochemistry and molecular biology of storage protein synthesis in developing embryos.

General accounts of the molecular control of storage protein synthesis during seed development are found in the reviews by Larkins (1981), Brown, Ersland and Hall (1982), Higgins (1984), Müntz (1987), and Shotwell and Larkins (1989), and in the volumes by Shannon and Chrispeels (1986) and Bewley and Black (1994); reviews by Casey, Domoney and Ellis (1986) and Gatehouse et al (1986) are addressed to the problems of gene expression during storage protein synthesis in seeds of legumes. A useful review covering the state of embryo storage protein research before the recent flurry of gene cloning work is by Millerd (1975).

1. CYTOLOGY OF RESERVE DEPOSITION

The major embryo storage proteins fall into the globulin and albumin groups. The two components of globulins, which differ in molecular mass and sedimentation coefficients, are the 7S and 11S proteins. Both are holoproteins and, like endosperm storage proteins, are sometimes given names derived from the family, genus, or species from which they were first described. Thus, names such as vicilin, convicilin, and legumin (from *Vicia faba, Pisum sativum*); conglycinin and glycinin (from *Glycine max*); phaseolin (from *Phaseolus vulgaris*); cruciferin (from *Brassica napus*); alfin and medicagin (from *Medicago sativa*); canavalin and the lectin con-A (from *Canavalia* sp., Fabaceae); and helianthinin (from *Helianthus annuus*) are in common use. A comparison of the coding nucleotide and amino acid sequences of several globulin storage proteins of angiosperm embryos has revealed that they have vestigial sequence homology to either the 7S or 11S legume storage proteins; this has led to the view that as a group, globulin storage proteins have evolved from two genes that existed at the beginning of angiosperm evolution (Borroto and Dure 1987).

Although albumins were long considered as metabolic proteins, they have now been upgraded to storage proteins because of their wide distribution, especially in embryos of oil seeds, and their amino acid composition with a high amide content. Albumins generally migrate with a 2S sedimentation coefficient and are different from globulins in having a high cysteine content (Youle and Huang 1981). As will become evident later in this chapter, napin, an albumin from embryos of *B. napus*, has been studied extensively at the molecular level. Ferns are another group of vascular plants that contain albuminlike storage proteins. A 2S storage protein designated as matteuccin has been characterized from haploid spores of the fern *Matteuccia struthiopteris* and shown to be closely related to napin and a 2S albumin from sunflower embryos (Templeman, DeMaggio and Stetler 1987; Templeman, Stein and DeMaggio 1988; Rödin and Rask 1990).

(i) Origin of Protein Bodies

Modern investigations into the cytology of the deposition of storage reserves in the cells of developing embryos began with the advent of electron microscopy. Initial results showed that in the embryos of legumes and other plants, abundant supplies of reserve proteins are stored in the parenchyma cells of the cotyledons in the form of large protein bodies. The protein body consists of an amorphous protein matrix bounded by two close, nearly parallel electron-dense phospholipid layers of a limiting unit membrane. Lipid or oil bodies found in the cells of embryos storing fat contain a matrix of triglycerides surrounded by a unit or by a half-unit membrane of one phospholipid layer. Carbohydrates are generally stored as free starch grains in the cytoplasm or in the plastids. Considerable anatomical and ultrastructural information has been published that supports these general statements about the final form of the reserve bodies in the cells of embryos of various angiosperms.

However, there is some controversy as to how the protein and lipid bodies are formed and how they are released from the site of their deposition. Common to most plants examined is the initial deposition of the storage protein body in the central vacuole of the cells of the cotyledons. The protein body may increase in size to fill most of the vacuole. Alternatively, numerous protein bodies appear in the cells later in embryogenesis, probably as the large vacuole becomes chopped up into smaller ones by cytoplasmic strands (Harris and Boulter 1976; Wenzel et al 1993). The enormity of this change has been quantified in the developing cotyledons of *Pisum sativum*. About 8 days after flowering, coincident with the appearance of proteinaceous materials, cells of the cotyledon contain one or two vacuoles of an average diameter of 39 μm. During the next 12 days, the average vacuole diameter falls to 1 μm, with a concomitant increase to 175,000 per cell in the number of vacuoles enclosing protein bodies. The reality of a massive deposition of protein bodies is also underscored by the observation that during embryogenesis the surface area of vacuoles increases from an initial value of 5,500 μm^2 to a final value of 550,000 μm^2 (Craig, Goodchild and Hardham 1979). The flow of events connected with protein body formation described in *P. sativum* can be considered as representative of other plants.

The first indication of storage protein synthesis is the appearance of small clumps of dense staining material on the inner surface of the vacuolar membrane. Later, this material becomes diffuse and almost fills the entire vacuole; eventually it assumes a dense form as the protein body. An increase in protein body number might also occur by budding from the vacuole (Craig, Goodchild and Hardham 1979; Craig, Millerd and Goodchild

1980; Hinz et al 1995). A dual origin of protein bodies has been described in developing cotyledons of *Glycine max*. During the early stages of cotyledon development, protein bodies directly aggregate in the vacuoles from precursor proteins, whereas in maturing cotyledons the lumen of rough ER is filled with proteinaceous material. Subsequently, this is transformed into a large number of small protein bodies. It appears that irrespective of whether they are fabricated in the vacuole or in the lumen of the ER, no further transport of protein bodies is involved (Zheng et al 1990, 1992). Although in some plants both ER and Golgi bodies have been implicated in the synthesis of storage proteins, the prevalent view is that they are synthesized on polysomes bound to the ER and transferred to the vacuole by the Golgi system. Supporting a role for ER in reserve protein synthesis, close correlations have been observed between the proliferation of rough ER and the onset of storage protein accumulation in the cotyledons of *Crambe abyssinica* (Brassicaceae) (Smith 1974), *Vicia faba* (Neumann and Weber 1978; Harris 1979; Adler and Müntz 1983), *Lupinus albus* (Davey and van Staden 1978), *Phaseolus vulgaris* (Briarty 1980), *Sinapis alba* (Rest and Vaughan 1972; Bergfeld, Kühnl and Schopfer 1980), *Arabidopsis* (Mansfield and Briarty 1992), and *Picea glauca* (Krasowski and Owens 1993). In legumes such as *P. vulgaris* (Baumgartner, Tokuyasu and Chrispeels 1980; Greenwood and Chrispeels 1985b), *V. faba* (Bailey, Cobb and Boulter 1970; zur Nieden et al 1982), and *G. max* (Nishizawa et al 1994), the structural analysis has been combined with immunocytochemical and autoradiographic techniques to support the origin of storage proteins in or close to the ER. More direct evidence linking rough ER with storage protein synthesis has come from studies showing that the main components of storage proteins of *P. vulgaris* (Bollini and Chrispeels 1979), *Pisum sativum* (Chrispeels et al 1982), and *Cucurbita* sp. (Hara-Nishimura, Nishimura and Akazawa 1985) are made exclusively on membrane-bound polysomes that can be recovered from rough ER by sucrose-gradient centrifugation. There is evidence to show that in developing cotyledons of *P. vulgaris*, the newly synthesized phaseolin is transiently stored in the lumen of the ER before it commences its journey to the protein bodies (Bollini, van der Wilden and Chrispeels 1982).

Storage proteins thus far characterized are associated with a signal sequence that is necessary for targeting proteins into the lumen of the ER as the first step in intracellular transport. During protein transport, or once the protein is in the ER lumen, the signal sequence is cleaved off. Cotranslational glycosylation by the addition of mannose-rich sugar residues is another intrinsic modification that changes the structure of the protein molecules before their transport to the site of storage. The occurrence of glycosylation has been supported by experiments in which incubation of cotyledons with the glycosylation inhibitor tunicamycin was found to cause a decrease in glycosylated polypeptides (Badenoch-Jones et al 1981; Bollini, Vitale and Chrispeels 1983). The presence of unglycosylated translation products in *Xenopus laevis* eggs injected with phaseolin mRNA and tunicamycin also argues for the involvement of a glycosylation step in the final assembly of the storage protein (Matthews, Brown and Hall 1981). Another line of evidence that leads to the view of glycosylation as a general feature of embryo storage protein synthesis is the presence of glycosylation enzymes such as glycosyltransferases in particulate fractions of the embryos of pea (Nagahashi and Beevers 1978), soybean (Bailey et al 1980), and bean (Davies and Delmer 1981; Chrispeels 1985), among others. Glycosylation may be an important clue to the function of the protein or to its stabilization during desiccation of the embryo in the seed, although at present we do not know the full meaning of this step.

The synthesis of storage proteins in the ER or other organelles naturally raises the question as to how the proteins are transported to the vacuoles for storage. As in the case of lectins, discussed in Chapter 14, one idea that has good experimental support is that the relocation of storage proteins from the ER to the protein bodies is mediated by the Golgi apparatus. The general experimental strategies upon which a role for the Golgi apparatus is based are electron microscopy to demonstrate the presence of electron-dense vesicles in close proximity to the Golgi complex and protein bodies, as in developing embryos of *Capsella bursa-pastoris*, *Arachis hypogaea* (peanut; Fabaceae) (Dieckert and Dieckert 1976), and *Fagus sylvatica* (Fagaceae) (Collada et al 1993); high-resolution electron microscopic immunochemistry to demonstrate the presence of storage proteins in the Golgi apparatus and associated transport vesicles, as in *Phaseolus vulgaris* (Greenwood and Chrispeels 1985b), *Pisum sativum* (Craig and Goodchild 1984a), *Vicia faba* (zur Nieden et al 1984), and *Glycine max* (Nishizawa et al 1994); and biochemical techniques to show the presence in the Golgi complex of specific enzymes for the glycosylation of storage proteins, as in *P.*

vulgaris (Chrispeels 1985). Craig and Goodchild (1984b) have shown that the ionophores monensin and nigericin interfere with the transport of newly synthesized vicilin in the cotyledons of *P. sativum*, causing its buildup in the Golgi apparatus and its diversion from the vacuole tonoplast to the plasmalemma. From these observations and by analogy to protein secretion in animal systems, a similar mechanism, involving the Golgi bodies, might be envisaged to account for protein transport in developing plant embryos. Although the transport of lectins and storage proteins using the Golgi system suggests common pathways, only further work will reveal whether the behaviors of the two kinds of molecules have distinctive features.

The complete sequence of events by which the storage protein becomes visible in the cells of the mature embryo involves several changes that are not detected cytologically. Nonetheless, they have been analyzed at the molecular level, so that a model of the assembly of the protein has emerged. Soon after glycosylation, both 11S and 7S precursors are forged into trimers within the ER and transported to the protein bodies. A further change in the molecule occurs at this stage by the addition or deletion of sugar residues from the oligosaccharide side chains. In the vacuole, the trimer is proteolytically cleaved, which leads to the separation of the acidic and basic subunits. In general, the 11S trimers are fashioned into hexamers whereas the 7S protein is assembled as a trimer, and both appear as spherical structures enveloped by a single membrane (Sengupta et al 1981; Dickinson, Hussein and Nielsen 1989).

(ii) Origin of Lipids

Based on the close association of nascent lipid or oil bodies in developing embryos with ER (Frey-Wyssling, Grieshaber and Mühlethaler 1963), plastids (Bergfeld et al 1978), or both (Mansfield and Briarty 1992), current views on the ontogeny of lipid bodies are that precursors for their synthesis are provided either by one or by both of these organelles. However, it is doubtful whether these organelles are the site of synthesis of lipids; it appears more likely that during storage lipid formation, oil bodies are secreted as small, electron-dense, particulate masses into the cytoplasm, where they fuse to form larger oil bodies (Smith 1974; Herman 1987; Murphy, Cummins and Kang 1989). Because of the presence of the proteinaceous membrane (or oleosin), oil bodies that store triglycerides are different from the lipid or oil bodies referred to previously (Herman 1987; Murphy and Cummins 1989a; Tzen et al 1990). Interestingly enough, like storage proteins, oleosins appear to be synthesized in the ER, and a conformational change presumably accounts for their integration as a membrane around the oil body (Loer and Herman 1993). The fairly constant association between oleosins and phospholipids of the unit membrane surrounding the triglyceride-containing oil bodies implies a role for oleosins in stabilizing the oil bodies as discrete organelles and in safeguarding them from coalescence (Tzen and Huang 1992).

2. REGULATION OF STORAGE PROTEIN SYNTHESIS

The regulation of storage protein synthesis can be viewed as part of an overall process controlled by gene action. At a gross level, the effects of gene action are seen in the timing and kinetics of storage protein accumulation in different genotypes of bean (Mutschler, Bliss and Hall 1980), and in the variations in protein characteristics, including the electrophoretic patterns of legumin and vicilin in different varieties of pea (Thomson and Schroeder 1978; Davies 1980; Schroeder 1982). Molecular approaches necessary to develop a conceptual understanding of the varietal and genotypic effects on storage protein synthesis and accumulation have not, however, been developed.

This section of the chapter will discuss the temporal and spatial patterns of protein deposition and synthesis in developing embryos and the expression of their genes. The multisubunit combinations of storage proteins have offered a rich source of materials for such studies.

(i) Developmental Changes in Storage Protein Synthesis and Accumulation

Deposition of storage proteins in developing embryos has been followed quantitatively by the incorporation of radioactive precursors, and qualitatively by SDS-PAGE analysis of accumulated polypeptides and by immunochemical methods. From various data it appears that developing cotyledons of *Pisum sativum* accumulate 11S legumin and 7S vicilin to a considerable extent, with the synthesis of vicilin generally preceding that of legumin. In terms of the timing, legumin accumulates throughout embryogenesis, whereas several periods of intensive accumulation of vicilin are spread over the whole time course. It is also found that, in contrast to vicilin subunits, which are pre-

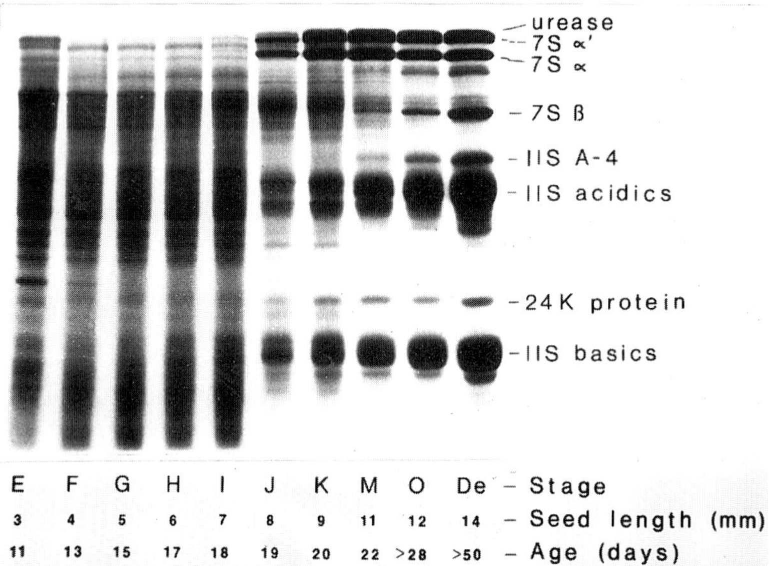

Figure 15.1. Changes in the extant storage protein profiles of soybean embryos of different stages of development. Samples E–I and J–De were prepared from whole seeds and from isolated cotyledons, respectively. Stages of seed or embryo development are indicated in the lower part of the figure. (From Meinke, Chen and Beachy 1981. *Planta* 153:130–139. © Springer-Verlag, Berlin. Photograph supplied by Dr. D. W. Meinke.)

sent in appreciable amounts at the early stages of embryogenesis, legumin and a 7S-type convicilin are prominent later in embryogenesis. Another apparent change in protein accumulation pattern includes a depression in the synthesis of all major storage polypeptides in the cotyledons with maturity of embryos (Millerd and Spencer 1974; Guldager 1978; Millerd, Thomson and Schroeder 1978; Gatehouse et al 1982; Chandler et al 1984). Several 2S albumins are also major components of the cotyledons and embryo axes of pea. Like globulins, they are broken down during germination, suggesting that they have a storage function (Higgins et al 1986).

Vicilin is formed prior to legumin in developing embryos of *Vicia faba*; however, starting in a modest way, legumin synthesis overtakes vicilin production, thereby accounting for the ability of cotyledons to accumulate nearly four times more legumin than vicilin. Most legumin accumulation occurs during the period of cell elongation between embryo lengths of 10 and 25 mm, and embryos less than 4 mm do not contain any legumin at all. Fractionation of legumin and vicilin into component polypeptides showed that they consist of three and four subunits, respectively. The subunit composition of vicilin is found to change during embryogenesis, whereas that of legumin remains constant. This suggests that the synthesis of individual subunits is under separate genetic control (Graham and Gunning 1970; Millerd, Simon and Stern 1971; Wright and Boulter 1972; Bassüner et al 1983).

Both temporal and rate differences are apparent in the accumulation of three major fractions of seed proteins and their subunits during soybean embryogenesis. The bulk of the storage proteins, accounted for by 7S conglycinin and 11S glycinin, are synthesized late during embryo development; a minor 2S component predominates at very early stages, decreasing proportionately throughout embryogenesis. Two α-subunits of conglycinin and most of the acidic and basic subunits of glycinin begin to appear shortly after the termination of cell divisions in the developing cotyledons, whereas β-subunits of conglycinin and a cluster of acidic subunits of glycinin begin to accumulate still later, coincident with the maturation of the cotyledons (Figure 15.1). An interesting inference about conglycinin accumulation is that the turnover of this protein is very rapid during the phase of active cell divisions in the cotyledons and that the protein becomes stable later in development. The embryo axis, unlike the cotyledons, is not committed to the accumulation of massive amounts of storage protein subunits and, due to rapid degradation, the β-subunit of conglycinin is virtually undetectable in the embryo axis. The development of the embryo axis and cotyledons entails other changes, which result in the former's containing different isomers of conglycinin from those found in the cotyledons. In addition, the spatial pattern of accumulation of storage proteins is somewhat selective, with higher levels of glycinin and conglycinin being present in the abaxial, as compared to the adaxial, part of the cotyledons (Hill and Breidenbach 1974; Gayler and Sykes 1981; Meinke, Chen and Beachy 1981; Sengupta et al 1981; Shuttuck-Eidens and Beachy 1985; Ladin et al 1987; Sugimoto et al 1987; Tang et al 1993). Synthesis of storage proteins slightly increases during heat-shock treatment of developing soybean embryos; no explanation is known for this phenomenon, other than that HSPs and storage proteins share common control mechanisms (Altschuler and Mascarenhas 1982).

Qualitative and quantitative developmental changes in the synthesis and accumulation of storage proteins also occur during embryogenesis in *Zea mays* (Khavkin et al 1977, 1978; Cross and Adams 1983; Sánchez-Martínez et al 1987; Kriz and Schwartz 1986; Kriz 1989), *Vigna unguiculata* (Carasco et al 1978; Khan, Gatehouse and Boulter 1980), *Phaseolus vulgaris* (Hall, McLeester and Bliss 1972; Sun et al 1978; Murray and Kennard 1984), *Gossypium hirsutum* (Dure and Chlan 1981; Dure and Galau 1981), *Brassica napus* (Crouch and Sussex 1981; Murphy and Cummins 1989b; Murphy, Cummins and Ryan 1989; Höglund et al 1992), *Helianthus annuus* (Allen, Nessler and Thomas 1985; This et al 1988), *Triticum durum* (Brinegar, Stevens and Fox 1985), *Canavalia gladiata* (Yamauchi and Minamikawa 1986, 1987), *Sinapis alba* (Fischer et al 1988), *Medicago sativa* (Xu et al 1991; Krochko, Pramanik and Bewley 1992), *Amaranthus hypochondriacus* (Amaranthaceae) (Raina and Datta 1992), and *Cicer arietinum* (Mandaokar et al 1993), and in the gymnosperms *Pinus monticola* (Gifford 1988), *Picea abies* (Hakman et al 1990), *P. glauca* (Misra and Green 1991), and *P. glauca*/*P. engelmannii* (Flinn et al 1991). Despite the inevitable variations encountered with species and with each storage protein, it can be stated in a general way that the synthesis and the accumulation of storage proteins go hand in hand beginning with the early development of embryos and tend to level off as embryos enter the maturation phase.

Only limited information is available on the timing and pattern of accumulation of storage lipids in developing embryos. Typically, in embryos of *Brassica napus*, the most rapid phase of storage lipid formation occurs before the onset of storage protein deposition; subsequently, it levels off during the desiccation phase of embryos. The synthesis of a 19-kDa oleosin associated with triglycerides is initiated late in embryogenesis, after the synthesis of cruciferin and napin is under way (Figure 15.2). The deposition of oleosin overlaps with the period of maximum storage protein accumulation, but is independent of the synthesis of triglycerides (Murphy and Cummins 1989a; Murphy, Cummins and Kang 1989). The time course of storage lipid synthesis in maize embryos closely parallels increases in embryo fresh weight, although 90%–95% of the lipids and lipid bodies are localized in the scutellum, the remainder being confined to the embryo axis. It has been claimed that embryos of maize, rapeseed, and certain other plants begin to accumulate triglycerides and oleosins concomitantly during the initial stages of development; this observation is in agreement with a model that implicates ER as the site of their syn-

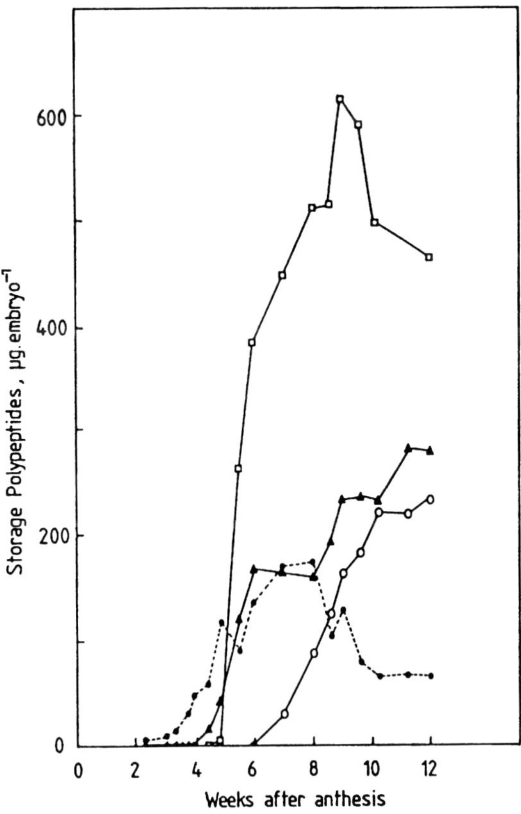

Figure 15.2. Synthesis of storage proteins and oleosins in developing embryos of *Brassica napus*. □, cruciferin; ▲, napin; ○, oleosin; ●, other proteins. (From Murphy and Cummins 1989b.)

thesis (Qu et al 1986; Tzen et al 1993). In conifer embryos, triglycerides are found to accumulate continuously during the course of development and to account for nearly 50% of the fresh weight of the mature embryos (Feirer, Conkey and Verhagen 1989). Clearly, before a general picture of the gene expression program for storage lipids can emerge, more work is needed to document the temporal and spatial patterns of lipid accumulation in developing embryos of additional species.

Lipoxygenase is an enzyme that catalyzes the oxidation of polyunsaturated fatty acids and other compounds and is abundant in seeds of leguminous plants. In pea (Domoney et al 1990) and soybean (Hildebrand, Versluys and Collins 1991), the enzyme is expressed in the embryo only during the cell expansion phase, coincident with the accumulation of storage proteins. Suggestive of gene duplication is the observation that three isozymes of lipoxygenase are coordinately expressed in the cotyledons and embryo axes of soybean.

The ability of cultured cotyledons or cotyledons of cultured embryos to synthesize storage proteins is consistent with the notion that the switching on of genetic information for this activity is independent of any signal provided by the parent plant. Cotyledons excised from young embryos of pea, containing no trace of vicilin, begin to synthesize this protein in culture. However, legumin does not accumulate in excised and cultured cotyledons unless its synthesis is already under way before culture. Although a trigger provided by the pod was invoked to account for legumin synthesis in isolated cotyledons, a later work showed that cotyledons of cultured young embryos synthesize and accumulate legumin in higher quantities than cotyledons of embryos of similar fresh weight growing in vivo (Millerd et al 1975; Domoney, Davies and Casey 1980). Immature embryos of *Theobroma cacao* (A. J. Kononowicz and Janick 1984), *Simmondsia chinensis* (Buxaceae) (Wang and Janick 1986), and *Brassica napus* (Avjioglu and Knox 1989) cultured in vitro synthesize storage lipids in response to high concentrations of sucrose in the medium. Thus, for both storage proteins and lipids, the use of an embryo culture system is more amenable to control than embryos growing in vivo.

Analysis of the DNA content of cells and storage protein accumulation in the cotyledons of cultured pea embryos has provided additional evidence showing that there is no selective amplification of the genes for storage proteins (Chapter 13). In the cotyledonary cells of cultured embryos, the maximum cellular DNA level increases, yet the minimum DNA level at which vicilin is detected remains at the 5C level, as in embryos growing in vivo. Evidence is also at hand showing that when embryos are treated with aphidicolin, an inhibitor of DNA replication, even cells of the cotyledons at the 4C DNA level synthesize vicilin (Corke, Hedley and Wang 1990a, b). From these observations, it seems likely that the cessation of mitosis, along with accompanying cell expansion, rather than elevated cellular DNA levels, accounts for the onset of vicilin synthesis in the cells of cotyledons.

(ii) Regulation of Storage Protein mRNAs

Given the background of the pattern of accumulation of storage proteins in developing embryos, it is now possible to examine the relationship of the accumulation patterns to the levels of mRNA species and assess the extent of operation of transcriptional and translational controls. Changes in storage protein mRNAs of developing embryos have been assayed by cell-free translation of poly(A)-RNA, by competitive nucleic acid hybridization with labeled cDNA clones, by analysis of run-off transcripts from isolated nuclei, and by Northern blot, dot–blot, and in situ hybridization techniques.

Polysome fractions or poly(A)-RNA isolated from developing embryos of *Glycine max* (Beachy, Thompson and Madison 1978; Beachy 1980; Beachy, Jarvis and Barton 1981; Sengupta et al 1981; Tumer, Thanh and Nielsen 1981), *Phaseolus vulgaris* (Hall et al 1978; Sun et al 1978; Chappell and Chrispeels 1986), *Pisum sativum* (Evans et al 1979; Croy et al 1980a, b; Gatehouse et al 1981; Higgins and Spencer 1981; Higgins et al 1986; Domoney and Casey 1987; Yang et al 1990), *Gossypium hirsutum* (Dure and Galau 1981; Dure et al 1983), *Vicia faba* (Bassüner et al 1983), *Cucurbita* sp. (Hara-Nishimura, Nishimura and Akazawa 1985), *Lupinus angustifolius* (Johnson, Knight and Gayler 1985), *Triticum durum* (Brinegar, Stevens and Fox 1985), *Raphanus sativus* (Laroche-Raynal and Delseny 1986), *Canavalia gladiata* (Yamauchi and Minamikawa 1987), *Brassica napus* (Rödin et al 1990), and *Picea glauca* (Leal and Misra 1993a) have been shown to synthesize the appropriate storage proteins or their precursors when translated in a cell-free system (Figure 15.3). In some of these investigations, immunoprecipitation with antibodies combined with polypeptide mapping has led to the characterization of mRNA fractions coding for specific polypeptide subunits of storage proteins. A general conclusion from these studies is that mRNAs coding for storage proteins increase

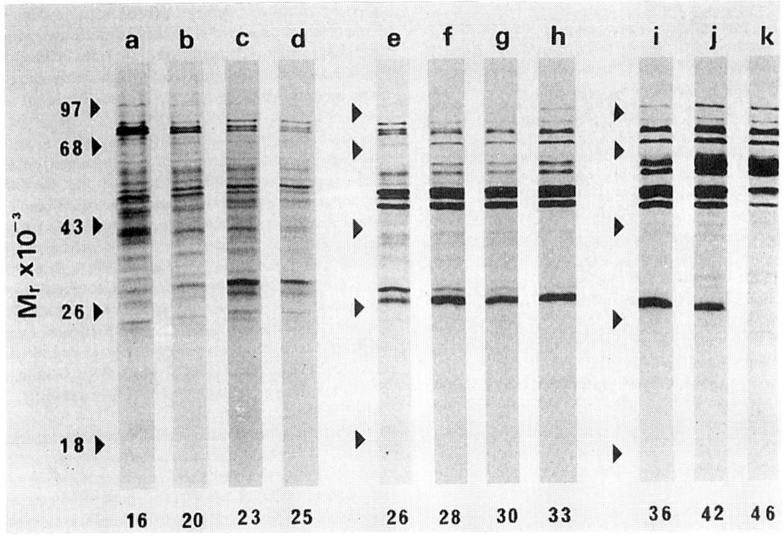

Figure 15.3. In vitro translation profiles of proteins of pea embryos of different stages of development. Arrowheads indicate the position of proteins of the different molecular masses given on the left. Age of embryos is given at the bottom as days after flowering. (From Yang et al 1990. *J. Expt. Bot.* 41:283–288, by permission of Oxford University Press.)

from very low levels to become the dominant species in the cells of the cotyledons during the period of peak protein accumulation, and then they decline to low, near undetectable levels at later stages of maturity.

Measurements of cDNA–RNA reassociation kinetics have shown that the decrease in storage protein synthesis during late embryogenesis in cotton cotyledons occurs simultaneously with the loss of a superabundant class of mRNAs. In this experiment, a cDNA probe made to storage protein mRNA was reassociated with mRNAs isolated from embryos of different stages of development. Results showed that mRNA sequences for the abundant storage protein decrease and are almost totally obliterated as the embryo progresses from the cell division phase to phases of cell expansion and maturation, indicating that the storage protein mRNA is active only during a transient phase of embryogenesis (Dure and Galau 1981; Dure et al 1983). In a similar way, using cloned cDNA probes made to storage protein mRNAs of soybean embryos, Goldberg et al (1981a, b) found that the appearance of complementary mRNA sequences in high frequency closely corresponds to the period of maximum storage protein accumulation and that the messages decay late in embryogenesis as dehydration sets in and embryos become mature. These superabundant messages, which number about 7–10, are probably transcribed exclusively in the embryos, because they are undetectable in the leaf polysomes.

The pattern of changing levels of storage protein mRNAs shows that the amount of protein synthesized is determined by the amount of its mRNA and suggests that storage protein synthesis in developing embryos is controlled at the transcriptional level. This idea has been pursued by extensive hybridization analyses using cloned probes made to different storage protein genes isolated from embryos of plants belonging to Fabaceae and Brassicaceae.

Fabaceae. Changes in the abundance of transcripts for storage proteins during embryogenesis in pea have been well characterized by Northern and dot–blot hybridizations using cDNA clones for vicilin (Croy et al 1982), convicilin (Domoney and Casey 1983), and legumin (Chandler et al 1983; A. J. Thompson et al 1991). The different mRNAs that encode storage proteins are present at very low concentrations during early stages of embryo development. They reach their maximum at slightly different times during embryogenesis and subsequently decrease in a manner that correlates with the accumulation of their respective proteins; vicilin mRNAs, in general, accumulate more rapidly and decrease earlier than either legumin or convicilin mRNAs (Gatehouse et al 1982, 1988; Chandler et al 1984; Boulter et al 1987; Domoney and Casey 1987; Bown, Ellis and Gatehouse 1988; Yang et al 1990; A. J. Thompson et al 1991). From assays for legumin and vicilin in runoff transcripts of nuclei isolated from cotyledons at different stages of embryogenesis, it is seen that the synthesis of vicilin transcripts precedes that of legumin during early embryogenesis, whereas legumin transcripts, in the virtual absence of vicilin, predominate during later stages. These results essentially confirm the changes in specific mRNA levels during embryogenesis in pea noted previously and suggest that storage protein synthesis in this system is dominated by transcription-driven controls (Evans et al 1984). This does not, however, rule out some posttranscriptional control of the level of expression of these proteins. The basis for this assertion is another work, which revealed no clear correlation between the steady-state levels of mRNAs for the different classes of legumin genes and the transcription rates of these genes in embryos of different ages. This suggested that the accumulation of vicilin and legumin mRNA in pea embryos is a two-step process in which temporal increases in transcription rates are followed by some posttranscriptional effects (Thompson et al 1989). At the cellular level, legumin and vicilin mRNAs are expressed in the storage parenchyma cells of the cotyledons and, depending upon the growing conditions of the plant, they may be further restricted to the nondividing cells. An association of legumin mRNA with the ER has been noted at the electron microscope level using a biotinylated probe complexed with avidin-ferritin (Harris and Croy 1986; Harris et al 1989; Hauxwell et al 1990).

In soybean, genes for glycinin and conglycinin are transcriptionally activated to peak levels during the midmaturation stage of embryogenesis and are repressed in the mature embryos. Although mRNAs for both genes are detected early in the development of the embryo, in the timing of mRNA appearance, conglycinin transcription precedes that of glycinin by a few days (Goldberg et al 1981a; Meinke, Chen and Beachy 1981; Scallon et al 1985; Walling, Drews and Goldberg 1986). We do not have a comparative picture of the cellular localization of glycinin and conglycinin in the developing embryos, but it has been shown that conglycinin mRNA first accumulates not only in the cotyledons but also in the embryo axis (Perez-Grau and Goldberg 1989).

Expression patterns of members of the glycinin and conglycinin gene families have provided some surprising findings. Five different glycinin genes, designated *GY-1, GY-2, GY-3, GY-4,* and *GY-5,* accumulate and decay coordinately during soybean embryogenesis within the temporal framework already noted (Nielsen et al 1989). In contrast, mRNAs encoding the α'/α- and β-subunits of the conglycinin gene family accumulate and decay at different times during embryogenesis; compared to the α'/α-subunit, the β-subunit typically exhibits delayed accumulation and decay. When the transcription rates of the α'/α- and β-subunits of conglycinin in embryos of different ages were determined, genes encoding the two subunits were found to be transcribed at roughly the same relative rates, indicating that despite changes in the mRNA levels, the cells appear to use the templates at the same rate (Figure 15.4). Thus, it has been argued that embryos rely on posttranscriptional mechanisms in the accumulation of mRNAs for the two conglycinin subunits (Naito, Dubé and Beachy 1988; Harada, Barker and Goldberg 1989).

In developing cotyledons of *Phaseolus vulgaris,* both transcriptional and posttranscriptional events such as mRNA stability appear to be important in controlling gene expression for the 7S protein phaseolin. This is based on the observation that the transcription rate of phaseolin is high early in embryo development and then declines, although

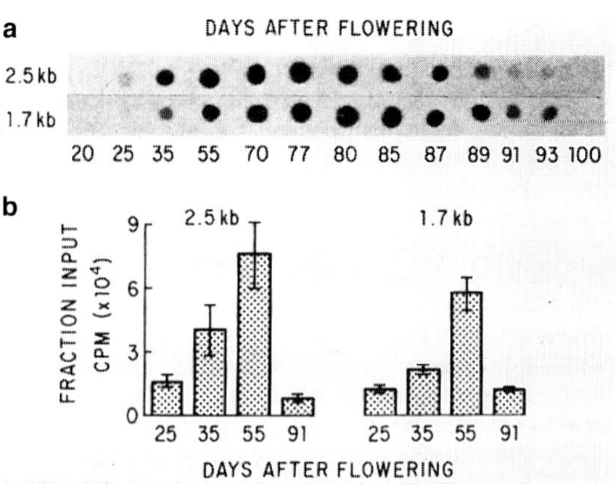

Figure 15.4. Regulation of conglycinin gene expression in developing soybean embryos. **(a)** A dot–blot hybridization showing differential accumulation of α'/α- (2.5 kb) and β-subunits (1.7 kb) of β-conglycinin mRNAs. RNA blotted on nitrocellulose filter was hybridized with labeled sequences complementary to the two conglycinin subunits. Days after flowering are indicated below the dot–blots. **(b)** Transcription of α'/α- and β-conglycinin genes from nuclei isolated from soybean embryos of different ages. The relative transcription rates were quantified by hybridizing labeled nuclear RNAs with filter-bound linearized sequences complementary to the two glycinin subunits. Ranges of values are indicated by the vertical lines. (From Harada, Barker and Goldberg 1989. *Plant Cell* 1:415–425. © American Society of Plant Physiologists.)

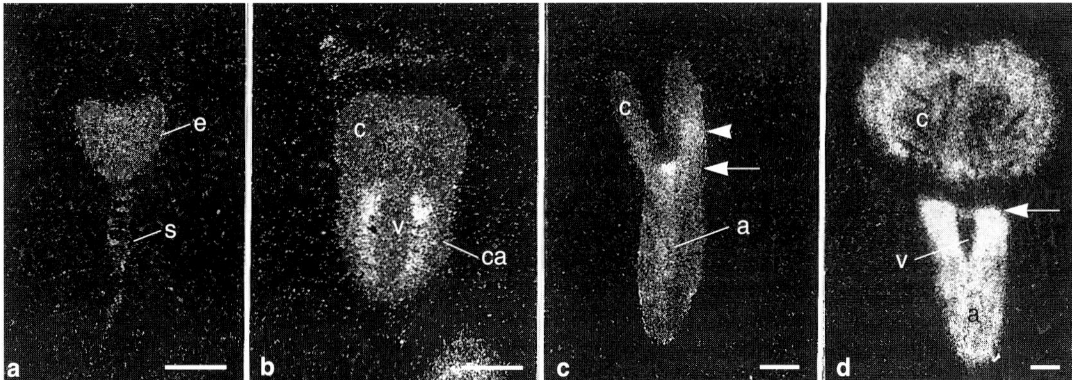

Figure 15.5. Dark-field images of autoradiographs of sections of *Brassica napus* embryos hybridized with ^{35}SUTP-labeled napin antisense RNA probes. **(a)** Early heart-shaped embryo. **(b)** Late heart-shaped stage embryo. Label is seen in the cortex of the axis (ca). **(c)** Torpedo-shaped embryo. Label is in the axis, immediately below the incipient shoot apex (arrow) and in the outer face (arrowhead) of the cotyledons. **(d)** Another torpedo-shaped embryo showing the absence of label in the provascular tissue and in a set of cells in the upper axis (above the arrow). a, axis; c, cotyledon; e, embryo; s, suspensor; v, provascular tissue. Scale bars = 100 μm. (From Fernandez, Turner and Crouch 1991. *Development* 111:299–313. © Company of Biologists Ltd. Photographs supplied by Dr. D. E. Fernandez.)

high mRNA levels are maintained (Chappell and Chrispeels 1986). During embryo development in *Canavalia gladiata*, mRNA levels for canavalin and con-A are differently regulated, with canavalin transcript expression preceding that of con-A. The accumulation of transcripts of canavalin also ceases early in embryogenesis even as the expression of con-A messages continues (Yamauchi et al 1988). The abundance of mRNAs for the 11S protein medicagin, 7S protein alfin, and a 2S protein in developing embryos of *Medicago sativa*, as determined by Northern hybridization using probes for pea vicilin and legumin and a synthetic oligomer for the 2S protein, appears to coincide with the periods of maximum synthesis of the respective proteins (Krochko, Pramanik and Bewley 1992). According to Bassüner et al (1988), transcripts of a gene that encode an unknown seed protein are present during early development of embryos of *Vicia faba* with a time profile similar to that of vicilin mRNA. However, the gene for narbonin, a 2S protein from *V. narbonensis*, is expressed throughout most of the period of embryogenesis (Nong et al 1995).

Brassicaceae. Through the use of cDNA clones, mRNA for 1.7S napin is detected by dot–blot hybridization in embryos of *Brassica napus* several days before 12S cruciferin mRNA. Transcripts for both genes share similarities with glycinin and conglycinin expression in soybean embryos in accumulating approximately halfway through embryogenesis and then declining to barely detectable levels in the mature embryos (Finkelstein et al 1985; DeLisle and Crouch 1989). In situ localization profiles are in agreement with blot hybridization data in showing temporal differences in the first appearance of cruciferin and napin transcripts in developing embryos and in showing their simultaneous disappearance in mature embryos (Figure 15.5). Beginning with the detection of transcripts in the cortex of heart-shaped (napin) and torpedo-shaped embryos (cruciferin), transcript accumulation spreads to the outer and inner faces of the cotyledons until most cells and tissues of the embryo are enriched for both transcripts (Fernandez, Turner and Crouch 1991; Sjödahl et al 1993). Comparison of the abundance of mRNAs for napin and cruciferin in developing embryos of *Raphanus sativus* has shown that napin mRNA is synthesized a few days ahead of cruciferin message, but the latter is slow to decay in the mature embryos (Laroche-Raynal and Delseny 1986).

Investigations of the transcription rates of napin and cruciferin genes in *B. napus* embryos have highlighted the fact that cruciferin mRNA, which accumulates to a higher level, is more stable than napin mRNA during the peak periods of their accumulation. Later in embryogenesis, stability of both mRNAs is eroded because comparable transcription rates result in reduced levels of both mRNAs (DeLisle and Crouch 1989). Since the relative stability of mRNA sequences determines their abundance in the cell, it appears that cruciferin and napin genes are controlled differently at different stages of embryogenesis.

There are three major subunits of cruciferin (*CRU-1, CRU-2/3,* and *CRU-4*), each encoded by a small gene family of about 5 members, whereas napin is encoded by a multigene family of at least 16 members. Regulatory trends in the study of individual members of the cruciferin family and members of the three different subunits have revealed similar developmental patterns of expression during embryogenesis of *B. napus* (Breen and Crouch 1992; Sjödahl et al 1993); on the other hand, the expression of members of one class of the napin gene peaks and declines earlier than expression of the other members of the family (Blundy, Blundy and Crouch 1991). The reasons for the different expression patterns for members of the napin gene family in embryos of *B. napus* are not presently known. It is possible that expression patterns of subunits of the same gene are genus specific, because two members of a napin gene family are coordinately expressed during embryogenesis in *R. sativus* (Raynal et al 1991).

The time courses of expression of transcripts of three 12S globulin storage protein genes and four 2S albumin genes in embryos of *Arabidopsis* are basically invariant from one another, showing a peak midway through embryogenesis and declining during embryo maturation. Whereas transcripts of 12S proteins genes appear equally abundant at different stages of embryogenesis, those of albumin genes *AT2S-2* and *AT2S-3* are expressed at significantly higher levels than are those of *AT2S-1* and *AT2S-4*. The spatial pattern of expression of *AT2S-1* is unique and, unlike the other members of this gene family, which are expressed both in the cotyledons and in the embryo axis, *AT2S-1* is expressed mostly in the embryo axis (Pang, Pruitt and Meyerowitz 1988; Guerche et al 1990b). How the four closely related genes in a family encoding the same protein are differentially regulated is puzzling, although the use of different combinations of *cis*-acting elements to achieve this has been suggested.

Other Plants. The principal storage proteins of cotton embryos have apparent molecular masses of 48 kDa and 52 kDa and are encoded by three subfamilies of genes, one subfamily coding for 48-kDa proteins and two subfamilies coding for 52-kDa proteins. Dot–blot hybridization with cDNA clones representing the three mRNA subsets showed that mRNAs for the storage proteins increase rapidly in young embryo cotyledons, maintain a high steady-state level during the period of rapid cotyledon growth, and fall precipitously thereafter. Since no significant changes are detected in the pattern of accumulation of the three transcripts, the transcription of the genes appears to be regulated coordinately (Dure et al 1983; Galau, Chlan and Dure 1983; Galau, Bijaisoradat and Hughes 1987).

The maximum level of accumulation of mRNAs of a 12S protein helianthinin in the embryos of sunflower is observed several days before the protein level peaks (Allen, Nessler and Thomas 1985; This et al 1988). A gene that encodes a 35-kDa protein rich in essential amino acids isolated from seeds of *Amaranthus hypochondriacus* follows a course of expression characterized by transcript synthesis during early embryogenesis, reaching a maximum at the midmaturation stage and decreasing during maturation (Raina and Datta 1992). During the prolonged period of development (extending to nearly six months) of embryos of *Prunus amygdalus,* expression of transcripts corresponding to the globulin, prunin, occurs in the cotyledons around 110 days after flowering (Garcia-Mas et al 1995, 1996). Unlike storage proteins in the seeds of legumes and other plants considered here, the storage proteins of *Bertholletia excelsa* (Brazil nut; Lecythidaceae), constituted of 2S albumins, accumulate in the embryonic hypocotyl, and expression of the genes for these proteins is restricted to embryos that approach maturity (Gander et al 1991). Another pattern of gene expression confined to most of the embryo developmental period is displayed by globulin genes *GLB-1* and *GLB-2* of maize embryos (Belanger and Kriz 1989; Kriz and Wallace 1991). Transcripts of vicilin, albumin, and a crystalloid storage protein are present at maximum levels in the developing embryos of *Picea glauca* and *P. glauca/P. engelmannii* complex preceding the period of rapid accumulation of the proteins (Flinn et al 1993; Leal and Misra 1993a). In *Pseudotsuga menziesii* (Pinaceae), mRNAs corresponding to a leguminlike storage protein begin to accumulate in the early to midstage embryos several days after the peak accumulation is observed in the megagametophytic tissues and then decline during embryo maturation (Leal and Misra 1993b).

Storage Lipid Transcripts. What is the relationship between the rate of transcription and the expression of genes of the enzymes involved in the synthesis of oil body fatty acids and of the oleosins that enclose the oil bodies? In *Brassica napus*, a pattern of a relatively early peak in transcript expression and the gradual loss of transcripts as embryos become desiccated has been observed for genes for ACP (de Silva et al 1990), enoyl-ACP-reductase (Kater et al 1991), stearoyl-ACP-desaturase (Slocombe et al 1992), and acyl-ACP thioesterase (Loader et al 1993). Transcripts for these genes are detected before the onset of, or coincident with, the peak of storage lipid synthesis in the embryos, and their disappearance corresponds with the attainment of steady-state levels of lipids. Late transcription, apparently independent of the transcription of genes for oil body enzymes, is seen for the oleosin gene. Early stage embryos lack oleosin transcripts, which are found at only a low level until the embryo reaches its maximum fresh weight and begins to desiccate (Keddie et al 1992b). At the organ level, transcripts of stearoyl-ACP-desaturase and oleosin genes are restricted to the embryonic cotyledons and radicle. Consistent with their temporal mode of expression, transcripts of stearoyl-ACP-desaturase come into the limelight in the embryonic organs earlier than those of oleosins and they disappear when the oleosin message level is still high (Slocombe et al 1992; Cummins et al 1993). Changes in mRNA levels of the 16-kDa and 18-kDa oleosin isoforms of maize embryo fall into a clearly recognizable category of developmental regulation as they increase during embryo development and, following a peak at midpoint of embryogenesis, decrease to low values. The close parallel noted between the rate of transcription of the 16-kDa oleosin isoform gene and the steady-state level of the corresponding mRNA in developing embryos (Figure 15.6) suggests that the transcript level of this gene is controlled at least in part by the rate of transcription (Vance and Huang 1988; Qu, Vance and Huang 1990). A similar temporal pattern of expression is seen for oleosin mRNAs from cotton embryos (Hughes and Galau 1989). The pattern of accumulation of oleosin mRNAs in developing embryos of *Prunus amygdalus* follows that for storage proteins, although oleosin transcripts are still detectable at late embryo developmental stages (Garcia-Mas et al 1995).

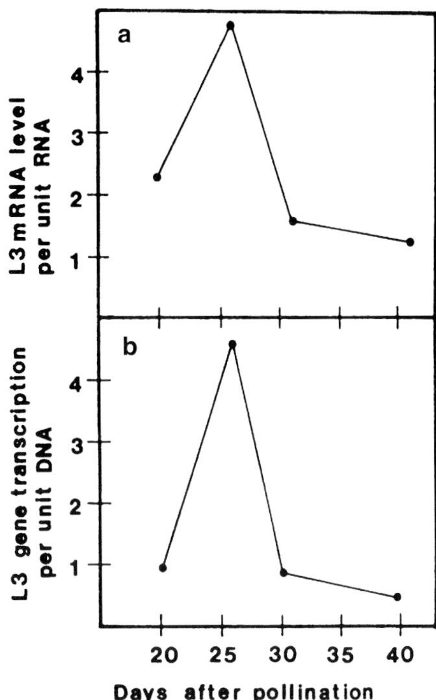

Figure 15.6. Time course of transcript accumulation of the 16-kDa oleosin gene (L-3 gene) and the rate of gene transcription in developing maize embryos. (**a**) Densitometric scanning of a Northern blot using labeled probe for the 16-kDa oleosin gene. (**b**) Relative levels of gene transcription determined by run-off transcription assays using nuclei isolated from embryos at the stages shown. (From Vance and Huang 1988.)

(iii) Modulation of Storage Protein Synthesis by Abscisic Acid

There are several indications that the deposition of storage reserves in maturing embryos is linked to changes in their ABA content. Although the bulk of evidence for this has come from studies on cultured immature embryos, this discussion will begin by referring to a work done in vivo (Schroeder 1984). The possible intervention of exogenous hormones in storage protein synthesis in the embryos appeared to be an attractive idea that could be tested by injecting the hormones into developing pea pods. This experiment showed that the total protein content and, in particular, the vicilin content of embryos increase in response to ABA injections whereas other hormones, such as NAA and benzylaminopurine, induce increases in the legumin content. The most extensive series of experiments to establish a connection between ABA and storage protein synthesis has been done with cultured embryos of *Brassica napus*. The results are basically straightforward in showing that when cultured in media containing ABA or a high osmoticum secured by the addition of sucrose or mannitol, embryos at the early cotyledon stage proceed through a normal developmental program and continue to synthesize cruciferin and napin; in contrast, in the absence of ABA or high osmotic environment in the medium, cultured embryos germinate precociously and synthesize storage proteins at barely detectable levels. The effects of these same treatments on the developmental regulation of storage lipid synthesis were less clear-cut (Crouch and Sussex 1981; Finkelstein and Crouch 1984; Finkelstein and Somerville 1989). Consistent with a role for ABA in stimulating cruciferin and napin synthesis, the mRNA levels for these proteins and the message transcription rates are appreciably higher in embryos cultured in ABA than in those grown in the basal medium (DeLisle and Crouch 1989). Seeds of transgenic tobacco harboring the promoter of a napin gene linked to the *GUS* gene also respond to ABA by showing enhanced reporter gene activity (Jiang, Abrams and Kermode 1996). However, in assigning a role to ABA in mediating gene expression for cruciferin and napin synthesis, we are confronted with two other observations. One is that the hormone could not effectively sustain the accumulation of cruciferin or its mRNA in cultured older embryos that are on the verge of desiccation. Secondly, osmotic effects on cruciferin synthesis are not associated with elevated levels of ABA in the embryos. Results like this have led to the conclusion that ABA plays an indirect role in storage protein accumulation, perhaps by inhibiting water uptake (Finkelstein et al 1985; Finkelstein and Crouch 1986). How this is translated into an effect on gene expression is difficult to comprehend at present.

Some of the best evidence for the role of ABA in the synthesis of storage proteins comes from the work on cultured embryos of soybean. As in the case of *B. napus*, the effects of ABA on the synthesis of storage proteins and their mRNAs in cultured soybean embryos are resoundingly correlated with their stage of development. The presence of ABA in the medium potentiates the expression of genes for 7S and 11S storage proteins in midmaturation stage embryos, but the hormone has the opposite effect in cultured early cotyledon stage embryos. The decline in 11S protein mRNA level seen in cultured late maturation stage embryos is not, however, completely reversed by ABA (Eisenberg and Mascarenhas 1985). Although the ABA effect must involve much more than mere differences in the age of embryos, it would appear from these results that ABA is necessary to maintain the high levels of storage protein mRNAs. Another work showed that synthesis of the different subunits of 7S protein is not coordinately regulated by ABA in cultured cotyledon explants, because the hormone specifically promotes accumulation of the β-subunit and its mRNA without any effect on the α- and α'-subunits. The involvement of ABA in the synthesis of the β-subunit is readily confirmed by the addition of fluridone, which reverses the effect of the hormone and causes a decrease in the β-subunit level (Bray and Beachy 1985).

The promotive effects of ABA on the synthesis of storage proteins and their mRNAs have been carefully documented in embryos of germinating grains of *Zea mays* (Vance and Huang 1988; Qu, Vance and Huang 1990); cultured embryos of *Sinapis alba* (Fischer, Bergfeld and Schopfer 1987), *Triticum aestivum* (Williamson and Quatrano 1988), and *Z. mays* (Paiva and Kriz 1994); cultured cotyledons of *Vicia faba* (Barratt 1986a, b), *Helianthus annuus* (Goffner, This and Delseny 1990), and *Medicago sativa* (Xu and Bewley 1995a, b, c); and cultured embryo axes of *Phaseolus vulgaris* (Long, Dale and Sussex 1981). The results with *V. faba* cotyledons are especially interesting in showing that not only does a high osmotic medium mimic the effect of ABA in promoting the accumulation of legumin and vicilin, but the osmotic effect is reversed by fluridone (Barratt 1986a).

The effect of ABA on the fate of oleosins of germinating embryos has been the focus of some investigations. That the expression of a maize 18-kDa oleosin gene is positively regulated by ABA is seen from the observation that a decrease in the transcript level of this gene noted in germinating embryos is reversed by the addition of ABA (Vance and Huang 1988; Qu, Vance and Huang 1990). According to van Rooijen et al (1992), the expression of oleosin genes in embryos of *B. napus* is sensitive to both ABA and osmotic stress, although the level of mRNA accumulation under osmotic stress is higher than with ABA. Dramatic increases in oil body proteins and oleosin mRNA, comparable to those obtained with ABA, are also observed in embryos of *B. napus* grown in the presence of jasmonic acid (Wilen et al 1991; van Rooijen et al 1992). Culture of embryos of transgenic *B. napus* plants harboring the oleosin gene promoter from *Arabidopsis* in media containing ABA, jasmonic acid, and osmoticum has given some interesting results. Basically, this work showed that transgenic embryos respond to both ABA and osmoticum by increased *GUS* gene expression, with an additive effect when both are applied together, but no appreciable response is seen with jasmonic acid (Plant et al 1994). Although these results affirm that ABA may not be the sole mediator of osmotically induced gene expression, they also show that any effect of jasmonic acid may be indirect or posttranscriptional. One of the few studies on the role of ABA on lipid synthesis in embryos has shown that in a medium containing sucrose, the addition of

ABA or osmoticum restores the ability of cultured wheat embryos to conserve and accumulate triacylglycerol (Rodriguez-Sotres and Black 1994).

The hypothesis favoring a role for ABA in modulating gene activity for storage protein synthesis did not remain uncontested and, on the balance, the available evidence is conflicting as to whether or not endogenous ABA is involved in storage protein synthesis in the embryos. In retrospect, it seems that in some cases the ability of ABA to prevent precocious germination and induce a maturation program in cultured immature embryos misled investigators to link the hormonal effect to storage protein synthesis. A few observations in favor of the negative view of ABA's effect on storage protein synthesis will be considered here. Reference was made earlier to a study that showed that a high osmoticum reinstates storage protein accumulation in cultured embryos of *B. napus* without increasing the level of endogenous ABA. Somewhat similar results were obtained with cotyledon explants of *H. annuus* incubated in a high osmoticum (Goffner, This and Delseny 1990). These observations are inconsistent with any possible consequence of ABA on storage protein synthesis and gene expression. Contrary to the effects of ABA on storage protein synthesis in cultured soybean embryos, in pea embryos grown in a high osmoticum sufficient to forestall precocious germination, addition of ABA does not enhance storage protein accumulation (Davies and Bedford 1982). Similarly, although ABA prevents precocious germination of cotton embryos, the hormone does not reinstate the levels of 48-kDa and 52-kDa storage proteins. This is not surprising, since synthesis of these proteins is seen to begin, and to reach a maximum, when ABA levels are low in the embryo and to decline during the period of ABA increase (Dure and Galau 1981; Galau, Bijaisoradat and Hughes 1987). Finally, analysis of several *Arabidopsis* mutants deficient in ABA or insensitive to ABA has shown that these mutations are not sufficient in themselves to alter storage protein gene expression, which appears comparable to that in the wild type (Pang, Pruitt and Meyerowitz 1988).

It will be recalled that Chapter 14 discussed at some length the problems in assigning a role to ABA as a developmental regulator of those aspects of embryogenesis related to the accumulation of *LEA* mRNAs and proteins. Clearly, based on the information presented here, we are not in a position to generalize on the relationship between ABA and storage protein synthesis and gene expression in the embryos. The reasons for this uncertainty seem to reside in the multiple effects of ABA, its relation to water stress, and the variety of proteins whose synthesis is affected by the hormone.

3. ISOLATION AND CHARACTERIZATION OF STORAGE PROTEIN GENES

A number of embryo storage protein genes have been cloned and sequenced, and important information about the regulatory constraints for their expression has been gathered. The cloning of storage protein genes, like the cloning of other developmentally regulated genes that has been considered in this book, has traditionally begun with the isolation of their corresponding cDNAs. For characterization of the genes, cDNA clones are used for the isolation of genomic clones and for determination of the amino acid sequence of proteins and the number of gene copies encoding the proteins. As discussed in the previous section, the cDNA clones have been used to monitor the spatial and temporal expression of their transcripts in the developing embryos.

To date, storage protein genes from embryos of approximately 10 species have been cloned, and their genetic organization has been defined to varying degrees. The following account represents, on a case-by-case basis, a summary of the more important characteristics of these genes, their products, and their regulatory mechanisms.

(i) Phaseolin Genes

Phaseolin is one of the first storage proteins well characterized at the DNA and protein levels. An initial study reported the isolation and partial nucleotide sequence of a phaseolin genomic clone and a cloned cDNA that contained about 40% of the mRNA transcript (Sun, Slightom and Hall 1981). The complete nucleotide sequence of the genomic clone subsequently showed that it includes 1,990 bp of DNA, consisting of 80 bp of 5'-untranslated sequence, 1,263 bp of protein-encoding region interrupted by five introns, 135 bp of 3'-untranslated sequence, a promoter sequence, and two AT-rich regions. The protein encoded by the gene has a very hydrophobic amino-terminal segment typical of signal peptides (Slightom, Sun and Hall 1983). On the basis of solution hybridization kinetics and Southern blot analysis, it was estimated that phaseolin is encoded as a small multigene family of about seven members and, considering the enormous amount of phaseolin synthesized during growth of the cotyledons, these genes appear to be overworked (Talbot et al 1984). However, organization of the gene varies among members of the phaseolin multigene family. Complete and partial nucleotide sequences of several phaseolin cDNA clones have shown that the protein is encoded by two distinct subfamilies that differ in their coding regions by the presence or absence of direct repeats of two different sizes (Slightom et al 1985). This probably provides a reason for the heterogenous nature of the proteins encoded by the phaseolin gene.

The phaseolin gene has been widely used to study the temporal and spatial regulation of expression of a cloned storage protein gene in transgenic plants. A native phaseolin genomic fragment consisting of the 782-bp upstream, 1,990-bp coding, and 1,100-bp downstream regions fused to the T-DNA gene of *Agrobacterium tumefaciens* was shown to induce expression in tumor tissues regenerated on infected sunflower stems (Murai et al 1983). When the same gene was integrated into the tobacco genome, levels of phaseolin in the embryonic tissues of seeds of transformed plants are found to be enormously higher than those in the endosperm and seed coat, thereby attesting to the tissue-specific and developmentally regulated expression of the gene (Sengupta-Gopalan et al 1985). Immunocytochemical techniques revealed that, as in the case of bean embryos, the protein is targeted primarily to the protein bodies of the embryos of transgenic tobacco seeds (Greenwood and Chrispeels 1985a). Other experiments have demonstrated expression of the normal phaseolin gene in the cells of an insect, *Spodoptera frugiperda* (Bustos et al 1988), and expression of modified phaseolin genes in callus tissues and seeds of transgenic tobacco (Cramer, Lea and Slightom 1985; Chee, Klassy and Slightom 1986; Hoffman, Donaldson and Herman 1988; Chee, Jones and Slightom 1991; van der Geest et al 1995) and alfalfa (Bagga et al 1992), in caryopses of rice (Zheng et al 1995), and in seeds of tobacco plants regenerated from callus cultures of transgenic plants (Frisch et al 1995). In all cases, the native and modified genes were fully integrated into the genomes of the respective heterologous hosts and they invariably specified the phaseolin transcripts or their protein products.

Considerable work has been done to analyze the regulatory elements that control embryo-specific expression of phaseolin in transgenic systems. An initial investigation showed that an 802-bp fragment of phaseolin genomic clone 5'-flanking sequences that extend from –782 bp to +20 bp confers the correct spatial and temporal regulation of *GUS* gene activity in developing embryos of transgenic tobacco, similar to the pattern characteristic of phaseolin accumulation in bean embryos. Subsequently it was found by 5'-deletion analysis that the upstream region of the phaseolin gene has at least three positive and two negative regulatory elements that interact to confer embryo specificity in transient assays using bean protoplasts and in stable assays of seeds of transformed tobacco. This work also showed that sequences between –295 bp to +20 bp from the upstream DNA region of the gene suffice to provide the genomic regulatory elements required for specific developmental activation and detectably high expression in both assays (Bustos et al 1989, 1991).

How the developmental control of the phaseolin gene follows from the interplay of positive and negative elements has turned out to be a complex story in itself. Analysis of the level of expression of phaseolin mRNA and protein in the seeds of tobacco plants transformed with additional 5'-deletion mutants has identified a major positive regulatory element (–295 bp to –228 bp), a minimal promoter (–64 bp to –14 bp), a minor negative element (–422 bp to –296 bp), and a minor positive element (–782 bp to –423 bp) in the upstream sequence. A negative role has also been assigned to a stretch of nucleotides between –295 bp and –107 bp in repressing the expression of phaseolin mRNA and protein in the stems and roots of transgenic tobacco. These results have raised the possibility that a boost in the accumulation of phaseolin mRNA and protein in the embryo is the net result of a combination of transcriptional activation and repression, such as promotion by a strong positive element (–295 bp to –228 bp) and a minimal promoter (–64 bp to –14 bp) in the targeted embryo cells, and inhibition of expression in the nontargeted vegetative parts by a negative element (–295 bp to –107 bp) (Burow et al 1992).

As a diagnostic characteristic of *cis*-acting DNA regulatory elements, four distinct nuclear proteins that recognize in vitro the –295 bp to +45 bp sequences of the phaseolin promoter have been identified (Figure 15.7). One DNA-binding protein, designated as CAN, binds motifs present in three locations, CACGTG (–248 bp to –243 bp), CACCTG (–163 bp to –158 bp), and CATATG (–100 bp to –95 bp), in the 5'-proximal region. A second protein, AG-1, binds to the two AT-rich sequences at –376 bp to –367 bp and –356 bp to –347 bp and to a nearly complementary sequence at –191 bp to –182 bp. Of these, the CACGTG motif or the G-box core

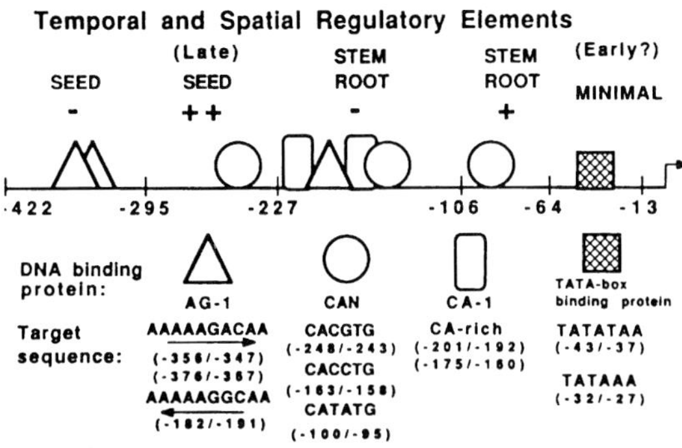

Figure 15.7. Regulatory elements in the phaseolin gene. A major positive element (++) (–295/–228) and a minimal promoter (–64/–14) are shown; others not shown are a minor positive element (–783/–423) and a negative element (–422/–296). The sequence upstream of –107 contains a negative element repressing phaseolin expression in the stem and root. A positive element (+) for stem and root expression lies within the –106/–65 region. Also indicated is a temporal element for late expression and a possible early expression element (–64/–14). DNA-binding proteins, AG-1, CAN, and CA-1, and TATA-box–binding protein and their target sequences are shown at the bottom. (From Kawagoe, Campbell and Murai 1994.)

sequence has been proposed as a major positive *cis*-acting element in the phaseolin gene (Kawagoe and Murai 1992). Introduction of plasmid constructs with substitution mutations in the G-box into protoplasts derived from bean cotyledons substantially reduces the G-box activity in *GUS* gene assay. Substitution mutations in the AG-1 binding sequences act both as positive and as negative regulators, depending upon the position to which they bind. Other results have revealed that the G-box and CACCTG motif affect each other's expression synergistically. The implication of the substitution mutation experiments is that both CAN and AG-1 are transcription factors, although the mechanism of their interaction is at present unknown (Kawagoe, Campbell and Murai 1994). A cDNA clone for one of the G-box–binding proteins has been shown to encode a basic region/helix-loop-helix protein (bHLH), a member of a family of ubiquitous transcription factors found in many eukaryotes (Kawagoe and Murai 1996). The identification of this new transcription factor is consistent with current models of gene regulation in eukaryotes involving the binding of a multiplicity of proteins to different regulatory sequences in the gene.

The molecular evidence relating to the role of the upstream regions of the phaseolin gene will now be considered. That the functions mediated by this segment of the gene make an important contribution toward gene expression has come from the observation that a 55-bp, AT-rich sequence motif in the 5'-region of the gene (–682 bp to –628 bp) interacts with nuclear proteins from immature bean cotyledons and enhances the expression of a chimeric *GUS* gene in transgenic tobacco embryos (Bustos et al 1989). A later work showed that this enhancer is located in a 1,047-bp fragment in the 5'-region of the phaseolin transcription initiation site and that sequences 3'- of the polyadenylation site of the gene also contain motifs that bind to matrix preparations from tobacco nuclei. The presence of the matrix attachment regions easily accounts for the high expression in transformed tobacco seeds of phaseolin promoter–*GUS* gene constructs containing these elements (van der Geest et al 1994).

Very little attention has been paid to the endogenous factors in developing embryos that enhance or repress storage protein gene expression. A gene (*PVALF*) (for *Phaseolus vulgaris* ABAI-3-like factor) whose transcripts are expressed in abundance in bean embryos during the few days preceding the accumulation of phaseolin and later during seed desiccation has been shown to up-regulate phaseolin promoter in transient assays of embryos. The explanation of this effect doubtless lies in the transcriptional activation domain of the protein product of the gene in promoting phaseolin expression during embryogenesis. The functional homology of the protein to ABI-3 and VP-1 proteins seems to assign it a role as a regulatory factor during embryo maturation processes, including storage protein synthesis (Bobb, Eiben and Bustos 1995). A regulator that is isolated from embryonic nuclei and that in part constitutes a basic/leucine-zipper protein by binding to a hybrid G-box/C-box (GCCACGTCAG or CACACGTCAA) on the phaseolin gene antagonizes the transactivation of the phaseolin promoter by *PVALF* (Chern, Eiben and Bustos, 1996). This observation helps explain the silencing of storage protein gene transcription during late embryogenesis.

The preceding account of the regulation of genes that encode phaseolin may provide a preview of the complexity of the transcriptional apparatus of other storage protein genes. Both G-box and matrix attachment regions are common to the genes isolated from many plant and animal systems and, although their functions are not completely understood, available evidence indicates that they make the entire genetic information easily accessible for replication and transcription. The presence of these motifs in the phaseolin gene suggests that when the mechanism of action of the gene is fully known, it may not be very different from that which regulates other genes.

(ii) Glycinin and Conglycinin Genes

Five glycinin genes (*GY-1, GY-2, GY-3, GY-4*, and *GY-5*) from soybean embryos, including one (*GY-2*) from embryos of wild soybean (*Glycine soja*), have been cloned, sequenced, and compared in attempts to understand their structure and the mechanisms that regulate their expression. Each member of this family is encoded by two to four genes constituting a small multigene family. The genes are separated into two subfamilies, designated as Group I (*GY-1, GY-2, GY-3*) and Group II (*GY-4, GY-5*) based on percent sequence homology. All five genes share a common structure characterized by the presence of four exons interrupted by three introns. Potential developmental control sequences identified in the 5'-regions of the genes include a TATA-box and CAAT-like boxes located 25–30 bp and about 100 bp, respectively, upstream from the transcription start sites and multiple polyadenylation sequences (Fischer and Goldberg 1982; Scallon et al 1985; Nielsen et al 1989; Xue et al 1992; Weng et al 1995). Before it is cotranslationally processed, each glycinin gene encodes a precursor polypeptide that contains an amino-terminal leader sequence followed by an acidic polypeptide, a short peptide linker, and a basic polypeptide (Ereken-Tumer, Richter and Nielsen 1982; Marco et al 1984; Staswick, Hermodson and Nielson 1984). The difference in the amino acid sequences of glycinin encoded by the different glycinin genes that is most impressive is that the amino-terminal sequences show only about 50% homology (Moreira et al 1979, 1981).

Genes encoding α-, α'-, and β-subunits of soybean conglycinin have been cloned and analyzed (Schuler, Schmitt and Beachy 1982; Schuler et al 1982; Tierney et al 1987). Soybean conglycinin belongs to a multigene family that contains about 15–20 genes encoding the different subunits (Ladin, Doyle and Beachy 1984; Harada, Barker and Goldberg 1989). The overall organization of the subunits has revealed close homologies between members of the gene families throughout the 3'-terminal half of the coding sequences and the 3'-noncoding sequences. From these homologies it has been suggested that the α- and α'-subunits of conglycinin diverged from a common ancestral gene. The major difference in size between the α- and α'-subunits has been attributed to a recombinational event, such as a large insertion in the first exon of the α'-gene (Schuler, Schmitt and Beachy 1982; Doyle et al 1986; Schuler et al 1982). Amino acid composition also illustrates considerable similarity between the α- and α'-subunits in their being rich in aspartate, glutamate, leucine, and arginine, and low in methionine and cysteine. In comparison to the α- and α'-subunits, the β-subunit is devoid of methionine and cysteine, and is more hydrophobic in nature (Thanh and Shibasaki 1977).

The degree to which soybean protein genes are expressed in transgenic systems in terms of tissue specificity and developmental regulation has been studied extensively following the report that transcripts of the α'-subunit of conglycinin accumulate temporally in developing embryos of *Petunia hybrida* (Beachy et al 1985). Other investigations that closely followed this work demonstrated expression of the β-subunit of conglycinin in embryos of tobacco and *P. hybrida* (Bray et al 1987; Barker, Harada and Goldberg 1988), and expression of glycinin genes in embryos of tobacco (Lelievre, Oliveira and Nielsen 1992; Itoh et al 1993) and soybean (Iida, Nagasawa and Oeda 1995), and in transgenic potato tubers (Utsumi et al 1994). When the α'- and β-subunit genes of conglycinin were introduced into *P. hybrida* in a single construct, time-course studies showed that the order of expression of their transcripts in developing embryos of the transgenic plant mimics the expression pattern during soybean embryogenesis (Naito, Dubé and Beachy 1988). The predictable outcome of experimental operations using the α'-subunit promoter fused to the *GUS* gene in the antisense orientation was the suppression of reporter gene activity in the seeds of transgenic tobacco during the period when expression of the α'-promoter was high (Fujiwara, Lessard and Beachy 1992). According to Fujiwara et al (1992), accumulation of the β-subunit of conglycinin in embryos of transgenic *P. hybrida* is repressed by application of exogenous methionine to developing pods and enhanced by subjecting plants to sulfur deficiency in a manner similar to that in which embryos of soybean respond to the same nutritional stimuli. Exogenously applied methionine down-regulates expression of the β-subunit gene in the seeds of wild-type transgenic *Arabidopsis* and in seeds of mutants of transgenic plants that overaccumulate methionine; accumulation of both β-subunit mRNAs and their protein products is enhanced, however, in the seeds of transgenic plants grown under sulfur deficiency (Hirai et al 1994; Naito et al

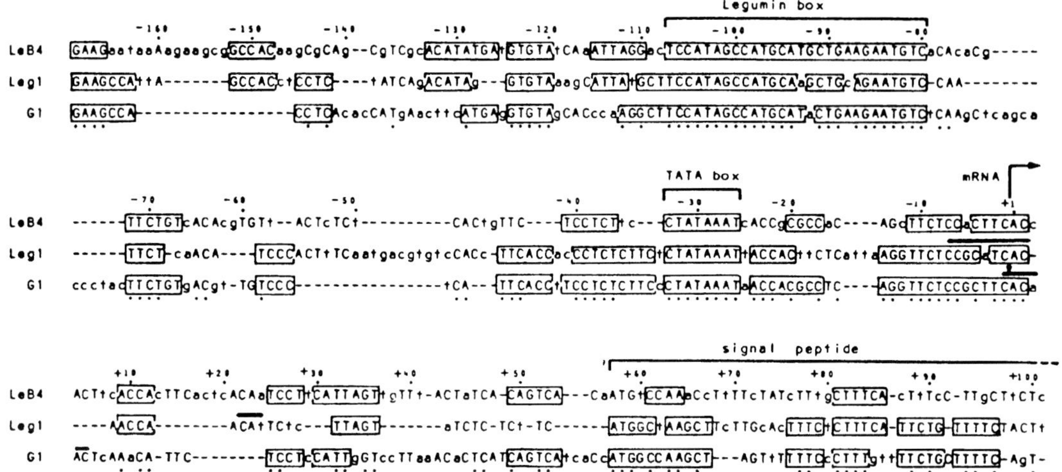

Figure 15.8. Major conserved elements, including the legumin-box, in the 5'-flanking regions of three legumin genes. Matches of four or more bases between any two genes are boxed; all matches are capitalized. Substitutions are shown in lower case; deletions are shown as dashes. Nucleotides represented in all three genes are indicated by dots below the *G1* (glycinin 1) sequence. Brackets identify the signal peptide coding region, the TATA element, and the legumin-box, as indicated. Heavy overlining indicates the experimentally determined 5'-ends of the *LEG1* and *G1* genes. The proposed mRNA start site of *LEB4* (nucleotide +1) corresponds to the major 5'-end of the *G1* gene. (From Bäumlein et al 1986. *Nucl. Acids Res.* 14:2707–2720, by permission of Oxford University Press.)

1994a, b). These observations suggest that the inductive or repressive specificity of exogenous chemicals in the regulation of storage protein synthesis can function even in transgenic systems whose storage proteins are not regulated by these chemicals.

A deletion series of conglycinin α'-subunit promoter analyzed in transgenic *P. hybrida* has revealed that the first 257-bp segment contiguous to the transcription start site contains sufficient regulatory information to program gene expression to a level almost equal to that of the complete promoter (Chen, Schuler and Beachy 1986). In addition, a fragment at –78 bp to –257 bp in the upstream region of the α'-promoter was identified as a *cis*-acting element enhancing embryo-specific activity in transgenic tobacco when inserted into the constitutive CaMV 35S promoter (Chen, Pan and Beachy 1988; Chen et al 1989). In vitro studies using crude nuclear extracts from immature soybean embryos have shown the presence of several proteins that interact with upstream sequences of the α'- and β-subunits of conglycinin. Two of these proteins, designated as soybean embryo factor-3 (SEF-3) and SEF-4, bind to the promoter fragment of the α'-subunit that has been shown to confer full gene expression in transgenic tobacco. During soybean embryo development, activities of SEF-3 and SEF-4 generally parallel the accumulation of α'-subunit mRNA (Allen et al 1989; Lessard et al 1991). Specific mutations introduced into the sequences that bind SEF-3 and SEF-4 generally abolish the in vitro binding of these embryo factors (Fujiwara and Beachy 1994). Seen in the light of these results, it appears that certain SEFs play a specific role in the expression of the conglycinin gene.

A novel aspect of the regulation of transcription of globulin genes in embryos of leguminous plants is the possible enhancer role of a 28-bp DNA element known as the legumin-box (Figure 15.8) about –45 bp to –65 bp upstream of the TATA-box identified in the promoter region of all globulin genes so far sequenced (Lycett et al 1985; Bäumlein et al 1986). The key sequences in the legumin-box responsible for regulating the expression of glycinin genes in transformed tobacco have been found to be 5'-CATGCAT-3', 5'-CATGCAC-3', and 5'-CATGCATG-3' motifs known as RY-repeats (Dickinson, Evans and Nielsen 1988; Lelievre, Oliveira and Nielsen 1992; Itoh et al 1993). For conglycinin, the presence of the legumin-box and a core of the 3'-part of the promoter of the α'-subunit of the gene appears to be hardly required to direct a low level of embryo-specific gene expression in transgenic tobacco. Disruption of the RY-repeat elements alone also reduces expression of the gene (Chamberland, Daigle and Bernier 1992; Fujiwara and Beachy 1994).

(iii) Legumin, Vicilin, and Convicilin Genes

Genes for legumin and vicilin have been isolated mostly from embryos of pea and broad bean, in which they occur in families of 8–12 per haploid genome. Four members of the legumin A-type gene family sequenced thus far from pea (*LEG-A, LEG-A1-LEG-A2, LEG-B,* and *LEG-C*) have nearly identical promoter regions characterized by TATA- and CAAT-boxes and sequences homologous to an adenovirus enhancer. The sequences of the coding regions of the genes that are interrupted by three introns, however, show slight differences due to insertions or deletions (Croy et al 1982; Lycett et al 1984, 1985; Rerie, Whitecross and Higgins 1991). A fifth legumin A-type gene (*LEG-D*) is relegated to the status of a pseudogene because of the presence of two stop codons that could impede the synthesis of any functional polypeptides (Bown et al 1985). In addition, genes for two minor B-type subunits of legumin, *LEG-J* and *LEG-K,* have also been fairly well characterized (Gatehouse et al 1988; A. J. Thompson et al 1991). The predicted amino acid sequences of *LEG-A, LEG-B,* and *LEG-C* genes generally matched the sequences obtained by direct protein sequencing, and no clear evidence was obtained for the presence of signal peptides in the proteins (Croy et al 1982; Lycett et al 1984, 1985). The structure of two B-type legumin genes (*LEB-4* and *LELB-3*) cloned from embryos of *Vicia faba* differs from that of A-type legumin genes cloned from the same species and from pea in having only two introns in the coding region; moreover, the predicted polypeptide of *LEB-4,* as well as that of the gene that codes for an unknown seed protein in *V. faba* embryos, comes equipped with signal sequences (Bäumlein et al 1986; Wobus et al 1986; Bassüner et al 1988; Heim, Bäumlein and Wobus 1994). The amino acid sequences of leguminlike storage proteins deduced from cDNA clones isolated from embryos of *Magnolia salicifolia* (Fischer et al 1995) and seeds of *Pseudotsuga menziesii* (Leal and Misra 1993b), *Ginkgo biloba* (Ginkgoales) (Arahira and Fukazawa 1994; Häger et al 1995), and *Calocedrus decurrens* (Cupressaceae) (Häger and Dank 1996) show reasonable homology with pea legumin and with 12S globulins from other plants.

Vicilin subunits of pea embryos are encoded by a small multigene family. Sequence comparison of three vicilin genes has revealed more than 80% homology among them, with most of the differences being confined to the sequences coding for the post-translational processing sites. Similarity in the basic structure of the genes also extends to the presence of five introns. Unlike legumin, vicilin has a predicted amino acid sequence that allows for a signal peptide (Gatehouse et al 1983; Lycett et al 1983; Domoney and Casey 1985; Higgins et al 1988). Comparable to the legumin-box, a highly conserved C-rich region at approximately 120 bp from the transcription start site in the vicilin genes of *Pisum sativum* and *Phaseolus vulgaris,* designated as the vicilin-box, is another example of a conserved upstream sequence in the storage protein gene determining tissue specificity of expression (Gatehouse et al 1986; Bown, Ellis and Gatehouse 1988).

The coding sequence of a convicilin gene (*CVC-A*) cloned from pea is similar to that of vicilin and is interrupted by five introns. The major difference between the two sequences is the insertion of a very hydrophobic 121–amino acid sequence near the N-terminus of convicilin. The predicted amino acid sequence of convicilin also confirms the presence of a leader sequence (Bown, Ellis and Gatehouse 1988; Newbigin et al 1990).

In attempts to determine the controlling sequences of the legumin gene, the promoter of *LEG-A* gene from pea was found to function only marginally in an in vitro transcription system (Evans et al 1985). However, *LEG-A,* vicilin, and convicilin genes from pea, and *LEB-4* gene and the gene encoding an unknown seed protein in broad bean that contains most of the 5'-flanking sequences are able to direct the synthesis of seed-specific proteins in transgenic *Nicotiana tabacum* and *N. plumbaginifolia* (Bäumlein et al 1987, 1988, 1991a; Croy et al 1988; Ellis et al 1988; Higgins et al 1988; Rerie, Whitecross and Higgins 1991; Manteuffel and Panitz 1993). Perhaps an exciting result with potential application in increasing the level of sulfur-containing amino acids of embryo storage proteins is the demonstration of stable expression in transgenic tobacco seeds of a vicilin gene from *Vicia faba* modified with eight additional methionine residues (Saalbach et al 1995). In transient assay, vicilin gene driven by barley hordein B-1 gene promoter is also expressed in the endosperm of barley (Heim et al 1995). A full-length *LEG-A* cDNA from pea engineered to encode legumin in yeast cells has been shown to direct the protein to Golgi-associated vesicles of the heterologous host (Croy et al 1988).

Whereas the 5'-flanking sequence of pea *LEG-A* gene containing the complete legumin-box sequence failed to initiate expression in transgenic systems, correct spatial and temporal expression was detected in a promoter construct that contained 549 bp of 5'-flanking sequence. The evidence for the presence of controlling elements in the 5'-flanking sequence demonstrated by transgenic plant expression has been strengthened by the identification of sequence-specific interactions between pea embryo nuclear proteins and the specific promoter fragment (Shirsat et al 1989; Shirsat, Meakin and Gatehouse 1990; Meakin and Gatehouse 1991). This supports the idea that the legumin-box is not the sole promoter determinant in *LEG-A* gene expression. An extensive analysis of the expression of functionally important DNA sequences in the 5'-flanking region of legumin gene *LEB-4* from *V. faba* in transgenic *N. tabacum* has yielded results similar to those obtained with *LEG-A* gene, implying that the legumin-box core element CATGCATG functions properly in cooperation with additional upstream promoter elements (Bäumlein et al 1991b, 1992). Finally, is promotion of growth of the embryo linked with the expression of storage protein genes? The extreme case of an experimental approach to answer this question was the introduction of a pea vicilin gene promoter fused to *DT-A* gene into transgenic *Arabidopsis* and tobacco; the results showed that the final form of the embryos was only marginally affected by the toxic gene, which drastically affected storage protein accumulation (Czakó et al 1992).

(iv) Cruciferin and Napin Genes

Although enormous quantities of cruciferin are synthesized during embryogenesis in members of Brassicaceae, there are no more than three gene families active in the process. As expected, each major group of cruciferin subunits is encoded by a distinct gene family (*CRU-1, CRU-2/3*, and *CRU-4*). Our knowledge of the molecular structure of the cruciferin gene is based on nucleotide sequence analyses of several cDNA clones of the three subunits and two genomic clones isolated from *Brassica napus*. Comparative data from cDNA clones have shown a high degree of similarity in the 5'-flanking and 3'-untranslated regions between gene members of a family and between members of different families (Simon et al 1985; Rödin et al 1990; Sjödahl, Rödin and Rask 1991; Breen and Crouch 1992). Features typical of a eukaryotic genome, such as the TATA-box, transcription start site, and polyadenylation signals, are present in the genomic clones of *CRU-1*, which, however, lack the legumin-box (Ryan et al 1989; Rödin et al 1992). Transformation of tobacco with deletion constructs made from the *CRU-1* gene has implicated different elements of the promoter in regulating transcriptional activity in transgenic embryos during early stages and late stages of development (Sjödahl et al 1995). The three major groups of cruciferin genes are also represented in radish seeds and, based on sequence comparison of the members of the two gene families with one another and with rapeseed genes, it appears that there is more homology between cruciferin genes from the two different species than between the two genes from radish (Depigny-This et al 1992). Analysis of partial or complete nucleotide sequences of four genomic clones isolated from *Arabidopsis* supports the view that the clones are members of three cruciferin subfamilies; the presence of putative introns in the *Arabidopsis* sequence in the same positions as in the pea legumin sequence, and their absence in the homologous *B. napus* sequence, seem to indicate the close relationship of *Arabidopsis* cruciferins to the 12S globulins of other angiosperms (Pang, Pruitt and Meyerowitz 1988). Expression of a cruciferin–*GUS* chimeric gene in embryos of transgenic tobacco appears to follow a pattern similar to that of the native gene in *B. napus* embryos. Analysis of the expression of the truncated gene promoter showed that positive regulatory elements containing sequences sufficient for seed-specific activity and a region for enhanced activity in the root meristem of transgenic embryos are dispersed over the 5'-flanking sequence (Bilodeau, Lafontaine and Bellemare 1994).

Napin is a well-characterized albumin-type of seed storage protein and cDNA and/or genomic clones have been isolated and characterized from seeds of *Brassica napus* (Crouch et al 1983; Ericson et al 1986, 1991; Josefsson et al 1987; Scofield and Crouch 1987; Baszczynski and Fallis 1990), *Raphanus sativus* (Raynal et al 1991), and *Arabidopsis* (Krebbers et al 1988; van der Klei et al 1993). It is a member of a small to large gene family, with estimates of the number of genes in the family ranging from 5 in *Arabidopsis* to more than 10 in *B. napus* and *R. sativus*. An important conclusion from nucleotide sequence data is that, as in other 2S storage proteins, there are no introns. Repeated CATGCA sequences, sometimes reiterated as overlapping tandem repeats, and stretches of TACACAT consensus motifs have been identified in many napin genes. The presence of these sequences may have gross molecular consequences; in line with this possibility, the TACACAT and other sequences of a napin gene were found to interact with a nuclear protein from developing seeds of *B. napus* (Ericson et al 1991; Gustavsson et al 1991). Although napin sequence data available for the three genera have afforded an opportunity to trace the evolution of this gene family in Brassicaceae, analysis of the coding and 3'-noncoding sequences has led to contradictory models (Raynal et al 1991).

Tagged napin genes are expressed in the embryos of transformed *B. napus* but are silent in the leaves; the amount of napin is reduced in embryos of transgenic plants transformed with an antisense napin gene. These results are in agreement with the essential role of the promoter in the differential regulation of gene activity (Radke et al 1988; Kohno-Murase 1994). Based on 5'-deletion analysis, two portions of the 5'-upstream region of the promoter of a napin gene have been identified as important for the transcriptional regulation of the gene. These include an AT-rich region between –309 bp and –211 bp that confers strong expression of *GUS* gene in transgenic tobacco and conserved sequences in the region between –148 bp and –120 bp that are necessary for temporal and spatial, as well as quantitative, regulation of the gene. Whereas globular and heart-shaped embryos of transgenic plants exhibited no *GUS* gene activity, it was expressed in the cotyledons and axes of late stage embryos. Through the use of gene constructs containing a 196-bp fragment between –152 bp and +44 bp, napin-gene-promoter-directed *GUS* gene was also detected in the cells of the endosperm of transgenic seeds at all stages of development (Stålberg et al 1993). It appears that the same motifs that enhance endosperm-specific expression of the zein gene in maize endosperm are responsible for napin gene expression in the endosperm of transgenic tobacco seeds.

(v) Storage Protein Genes of Other Plants

Consistent with the specialized role of storage proteins in the nutrition of the embryo during its development and germination, the structure of storage protein genes isolated from embryos of other plants is not radically different from those described already. Based on analyses of cDNA and genomic clones and deduced protein sequences, three sets of genes representing two gene families are believed to encode storage protein genes of cotton embryos. These families, known as the α-globulin family and the β-globulin family, are further subdivided into two subfamilies each. Mapping of genomic DNA has revealed that members of the two subfamilies of the α-globulin family are arranged in tandem, with genes of each subfamily showing considerable divergence (Chlan et al 1986, 1987; Galau, Wang and Hughes 1991, 1992a). The existence of genes from both subfamilies adjacent to each other in the genome is traced to the amphidiploid nature of cotton. Although there is considerable homology between nucleotide sequences of cotton α-globulins and vicilins from legumes, cotton genes contain but four introns, compared to the presence of five introns in vicilin genes.

The primary difference between the structure of genes encoding canavalin and that of genes encoding con-A in embryos of *Canavalia ensiformis* and *C. gladiata* lies in the presence of five introns in canavalin and the absence of introns in con-A (Takei, Yamauchi and Minamikawa 1989; Yamauchi and Minamikawa 1990; Ng, Ko and McPherson 1993). The DNA sequences of canavalin and con-A genes contain a number of A/T-rich regions and the core motif of the legumin-box (Figure 15.9); assay of transient *GUS* gene expression by deletion analysis of 5'-upstream regions of the genes has suggested that the A/T-rich regions and legumin-box contain positive regulatory elements for transcriptional activation (Yamamoto et al 1995).

Genes encoding helianthinin (Vonder Haar et al 1988; Bogue et al 1990) and a 2S albumin (Allen et al 1987; Kortt et al 1991) from embryos of *Helianthus annuus*, 7S globulins *GLB-1* and *GLB-2* from maize embryos (Belanger and Kriz 1989; Wallace and Kriz 1991), a 7S protein from embryos of *Theobroma cacao* (Spencer and Hodge 1992), and narbonin from *Vicia narbonensis* (Nong et al 1995) belong to small families; indeed, there is no evidence for the presence of more than one 7S protein gene in *T. cacao*. Cloning and sequencing of ubiquitin from alfalfa embryos has shown that, despite its well-known role as a carrier protein, ubiquitin has a high degree of homology to 2S albumin from Brazil nut embryos (Pramanik and Bewley 1993). Characterization of two allelic variants (*GLB-1L, GLB-1S*) and a null allele (*GLB-10*) of *GLB-1* gene from maize embryos has provided new insight into the molecular basis for allelic polymorphism of this gene; to the extent that the nature of this polymorphism is reflected in identifiable defects in the gene, *GLB-1L* and *GLB-10* differ from the functional *GLB-1S* allele by the presence of small nucleotide insertions. These changes result in the production of a deranged protein in *GLB-1L* and in a translational frameshift that introduces a premature termination codon in *GLB-10*. However, the low steady-state level of *GLB-10* transcripts in embryos homozygous for the *GLB-10* allele does not reconcile with the observation that

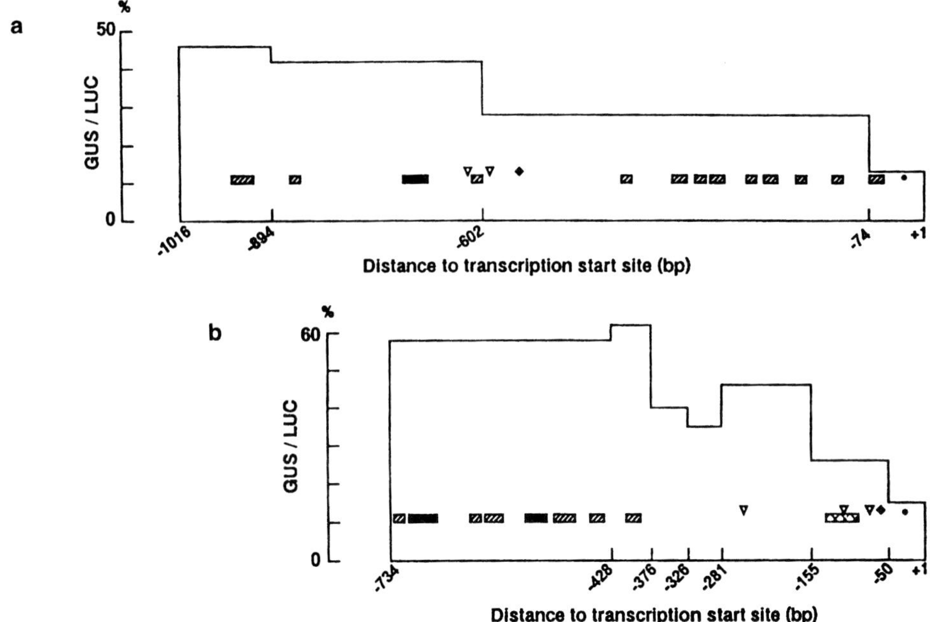

Figure 15.9. A diagrammatic representation of the correlation between 5'-deletion of the con-A gene **(a)** and canavalin gene **(b)** and transient *GUS* gene expression levels in embryos of *Canavalia gladiata*. Stippled boxes and black boxes represent regions containing more than 15 bp and more than 30 bp of AT-sequences, respectively. The sequence motifs are CATGCAT/A (black diamond); CANNTG (triangle); vicilin-box (crossed box); TATA-box (black dot). (From Yamamoto et al 1995.)

both *GLB-10* and *GLB-1L* genes maintain comparable levels of transcription as determined by nuclear run-off transcription assays. It might appear that mRNA instability impairs the proper functioning of the *GLB-10* transcripts (Belanger and Kriz 1991). A 404-bp DNA fragment located 322 bp upstream of a helianthinin gene (*HAG-3D*) has been shown to contain nuclear protein binding sites and to enhance *GUS* gene expression in embryos of transgenic tobacco (Jordano, Almoguera and Thomas 1989). Bogue et al (1990) found that transfer of 5'-flanking sequences of another helianthinin gene (*HAG-3A*) is sufficient to confer developmentally regulated helianthinin expression in embryos of transgenic tobacco and to correctly target the protein to protein bodies of the transformed embryo. In both *HAG-3A* and *HAG-3D* genes, a region designated as the proximal promoter region contains identical sequences that confer significant reporter gene expression in the embryos of transgenic tobacco; identification of sequence motifs in the proximal region that bind to seed nuclear proteins has taken us a step closer to determining the basis of expression of sunflower storage protein genes (Nunberg et al 1994). The observations described here emphasize again that the ease of manipulation of chimeric genes in transgenic plants offers a fruitful approach to studying the control of gene expression during embryogenesis.

Molecular approaches combined with transgene technology have been employed to identify *cis*-regulatory domains involved in differential expression of *Arabidopsis* 2S storage protein genes and to offer a shortcut to alter the amino acid composition of transformed embryos. De Clercq et al (1990a) showed that the promoter of the *AT2S-1* gene, which encodes a 2S albumin isoform of *Arabidopsis*, drives expression of a reporter gene in both transformed *Brassica napus* and *Nicotiana tabacum*. When the entire *AT2S-1* gene was transferred into tobacco, not only was gene expression found to be tissue specific and developmentally regulated, but the proteins were correctly processed and targeted to the protein bodies. In a previous section, it was pointed out that in contrast to *AT2S-2*, which is expressed throughout the embryo, *AT2S-1* is expressed only in the embryo axis of *Arabidopsis*. That a specific sequence in the genome is responsible for this differential expression pattern was indicated by the demonstration that replacement of a 67-bp region within a conserved 300 bp 5' of the initiation codon of *AT2S-1* with a corresponding fragment from *AT2S-2* was sufficient for expression of the chimeric *AT2S-1* gene throughout the embryo. A significant feature of the 67-bp cotyledon regulatory region of *AT2S-2* is the CATGCA motif found in the legumin-box of seed protein genes. Abolition of cotyledon expression by the *AT2S-1* gene has been traced to a mutation in the CATGCA motif to CAAGTA (Conceição and Krebbers 1994; Conceição, van Vliet and Krebbers 1994).

Cloning of genes encoding a protein rich in methionine from embryos of Brazil nut has afforded exceptional opportunities to manipulate the amino acid composition of embryo storage proteins of other plants (Altenbach et al 1987; de Castro et al 1987; Gander et al 1991). A beginning in the development of a system for the genetic manipulation of the 2S albumin gene of *Arabidopsis* has been made with the demonstration that the level of methionine-containing proteins of transgenic embryos can be raised by chimeric genes consisting of parts of *AT2S-1* fused with Brazil nut 2S albumin gene (de Clercq et al 1990b; Conceição, van Vliet and Krebbers 1994). Chimeric genes driven by regulatory regions of phaseolin gene, soybean lectin gene, or CaMV 35S promoter attached to the cDNA clone encoding Brazil nut protein have been similarly used to change the quality of the proteins expressed in transformed embryos of legumes and rapeseed (Altenbach et al 1989, 1992; Guerche et al 1990a; Saalbach et al 1994). The potential benefits of reshuffling genes in combinations not possible in nature to limit or to increase the levels of specific amino acids of storage proteins in embryos is enormous.

Reference was made earlier to changes in transcript abundance of a gene that encodes an unknown protein in embryos of *Vicia faba*. Although it is doubtful whether the protein has storage functions, characterization of the gene has shown that it has two introns and belongs to a family of 10–20 members. The gene is expressed in the embryos of tobacco and *Arabidopsis* transformed with a fragment containing 637 bp of the 5'-flanking region together with the complete 51 bp of the 5'-untranslated sequence (Bäumlein et al 1991b). Based on the expression of smaller deletion constructs made from the promoter region in seeds of transgenic tobacco, it appears that the promoter function is mediated in a combinatorial way by the interaction of several positive and negative elements. Surprisingly, a CATGCATG motif of the legumin-box found in the promoter acts as a negative regulatory element controlling gene expression (Fiedler et al 1993).

(vi) Genes for Enzymes of Lipid Body Biosynthesis and for Oleosins

Understanding the genetic basis of the regulation of storage lipid synthesis in developing embryos has been possible by cloning the genes of enzymes involved in the fatty acid biosynthesis machinery and by manipulating the genes in transgenic systems. The structure of the genes characterized and the ease of their manipulation support the general conclusions drawn previously from analyses of storage protein genes. cDNA clones for ACP genes have been isolated from embryos of *Brassica campestris* (Rose et al 1987; Scherer et al 1992) and *B. napus* (Safford et al 1988). ACP in embryos of *B. napus* is encoded by a heterogeneous, multigene family of some 35 genes per haploid genome. Immunocytochemical studies have shown that the enzyme is localized exclusively in the plastids of the embryo cells of *B. napus*; to accommodate this observation, it is believed that the protein is probably encoded by the nuclear DNA and directed into plastids (Safford et al 1988; de Silva et al 1990). Nucleotide sequences of three genomic clones encoding ACP in *B. napus* and *B. campestris* appear to show a high degree of homology in the coding and 5'-noncoding regions and in the location of three introns in identical positions. The promoter region of the *B. campestris* gene is characterized by the presence of three different types of repeats. The coding sequences of ACP genes of *B. napus* and *B. campestris* have much in common, implying that the genes may have evolved from a common ancestral gene (de Silva et al 1990; Scherer et al 1992). Four other enzymes of the fatty acid synthetase family from embryos for which some nucleotide data are available are lipoxygenases from *Pisum sativum* (Casey, Domoney and Nielsen 1985) and soybean (Shibata et al 1988; Yenofsky, Fine and Liu 1988); enoyl-ACP reductase and acyl-ACP thioesterase from *B. napus* (Kater et al 1991; Loader et al 1993); and stearoyl-ACP desaturase from *B. napus* (Slocombe et al 1992) and *Carthamus tinctorius* (safflower; Asteraceae) (G. A. Thompson et al 1991). The most striking feature of the genes isolated from embryos is their considerable identity to sequences of the same genes characterized from nonembryonic organs of other plants. When soybean embryo lipoxygenase gene was transferred into tobacco, not only was the gene expressed in the seeds, but it also affected fatty acid oxidative metabolism in the leaves of the transformed plants (Deng et al 1992). The time course of expression of the promoter of stearoyl-ACP desaturase gene in embryos of trans-

```
-949                                                CTTTCGGGATAAAGCAATCACCTGGCGATTCAACGTGGTCGGATCATGA
-900 CGCTTCCAGAAAACATCGAGCAAGCTCTCGAAGCAACCAAAGCTGACCTCTTTCGGATCGTACAGAACCCGAACAATCTCGTTATGTCCC
-810 GTCGTCTCCGAACAGACATCCTCGTAGCTCGGATTATCGACGAATCCATGGCTATACCCAACCTCCGTCTTCGTCACGCCTGGAACCCTC
-720 TGGTACGCCAATTCCGCTCCCCAGAAGCAACCGGCGCCGAATTGCGCGAATTGCTGACCTGGAGACGGAACATCGTCGTCGGGTCCTTGC
-630 GCGATTGCGGCGGAAGCCGGGTCGGGTTGGGGACGAGACCCGAATCCGAGCCTGGTGAAGAGGTTGTTCATCGGAGATTTATAGACGGAG
-540 ATGGATCGAGCGGTTTTGGGGAAAGGGGAAGTGGGTTTGGCTCTTTTGGATAGAGAGAGTGCAGCTTTGGAGAGAGACTGGAGAGGTTTA
-450 GAGAGAGACGCGGCGGAGATTACCGGAGAGAGGCGACGAGAGATAGCATTATCGAAGGGAAGGGAGAAAGAGTGACGTGGAGAAATAGAA
-360 AACCGTTAAGAGTCGGATATTTATCATATTAAAAGCCCAATGGGCCTGAACCCATTTAAACAAGACAGATAAATGGGCCGTGTGTTAAGT
-270 TAACAGAGTGTTAACGTTCGGTTTCAAATGCCAACGCCATAGGAACAAAACAAACGTGTCCTCAAGTAAACCCCTGCCGTTTACACCTCA
-180 ATGCGTGCATGTGAAGCCATTAACACGTGGCGTAGGATGCATGACGACGCCATTGACACCTGACTCTCTTCCCTT TCTTCATATATCTC
 -90 TAATCAATTCAACTACTCATTGTCATAGCTATTCGGAAAATACATACACATCCTTTTCTCTTCGATCTCTCTCAATTCACAAGAAGCAAA
   1 ATGACGGATACAGCTAGAACCCATCACGATATCACAAGTCGAGATCAGTATCCCCGAGACCGAGACCAGTATTCTATGATCGGTCGAGAC
     M  T  D  T  A  R  T  H  H  D  I  T  S  R  D  Q  Y  P  R  D  R  D  Q  Y  S  M  I  G  R  D
  91 CGAGACAAGTATTCCATGATTGGCCGAGGCCGAGACCAGTACAACATGTATGGTCGAGACTACTCCAAGTCTAGACAGATTGCTAAGGCT
     R  D  K  Y  S  M  I  G  R  D  R  D  Q  Y  N  M  Y  G  R  D  Y  S  K  S  R  Q  I  A  K  A
 181 GTTACCGCAGTCACGGCCGGTGGGTCCCTTCTTGTCCTCTCCAGTCTCACCCTTGTCGGAACTGTCATTGCTCTGACTGTTGCGACTCCT
     V  T  A  V  T  A  G  G  S  L  L  V  L  S  S  L  T  L  V  G  T  V  I  A  L  T  V  A  T  P
 271 CTGCTTGTTATCTTTAGTCCAATCCTTGTCCCTGCTCTCATCACCGTTGCATTGCTCATCACCGGCTTTCTCTCCTCTGGTGGCTTTGGC
     L  L  V  I  F  S  P  I  L  V  P  A  L  I  T  V  A  L  L  I  T  G  F  L  S  S  G  G  F  G
 361 ATTGCAGCTATAACCGTCTTCTCTTGGATCTACAAgtaagtggacatttaaacatatatttcaagttgtacaatatgttttaagaagcgg
     I  A  A  I  T  V  F  S  W  I  Y  K
 451 taatttatttttttttttttttgaattttaagaattcagggtttccccaaaggctttctaggcccaaaggactggtcccccttcctggcg
 541 ctgacgagctccatgtaataatgcccoagtggcccgagagaattgtttgcagcgtgaggcttcgaacccgggcgtattgggaagcggtaa
 631 tttattaaataaaacataatggttgaatatagcgacatgccttgtgaggggaaaaaaagtacaaaccataaaattatacataaccgacaa
 721 gtggattttagatattacattaaaatgccgttttacatcatcattttggctagctatacacaagacttgacttagctagcttgatacgta
 811 cgtgtagtatatgtagcatgcacgtgtgtgtgaattgtgatgaatagGTATGCAACGGGAGAGCACCCACAAGGGTCAGATAAACTGGAC
                                                      Y  A  T  G  E  H  P  Q  G  S  D  K  L  D
 901 AGTGCAAGGATGAAGCTGGGAGGCAAAGTTCAGGATATGAAGGACAGAGCTCAGTACTATGGACAACAGCAAACAGGTGGGAAACGAC
     S  A  R  M  K  L  G  G  K  V  Q  D  M  K  D  R  A  Q  Y  Y  G  Q  Q  Q  T  G  G  E  H  D
 991 CGTGACCGTACCCGTGGAACCCAGCACACTACCTAAATTACGCCATGACTATTTTCATAGTCCAATAAGGCTGATGTCGGGAGTCCAGTT
     R  D  R  T  R  G  T  Q  H  T  T  *
1081 TATGAGCAATAAGGTGTTTAGAATTTGATCAATGTTTATAATAAAAGGGGGAAGATGATATCACAGTCTTTTTTTCTTTTTGGCTTTTGT
1171 TAAATTTGTGTGTTTCTATTTGTAAACCTTCTGTATATGTTGTACTTCTTTCCCTTTTTAAGTGGTATCGTCTATATGGTAAAACGTTAT
1261 GATTGGTCTTTCCTTTTCTCTGTTTAGGATAAAAAGACTGCATGTTTTATCTTTAGTTATATTATGTTGAGTAAATGAACTTTCATAGAT
1351 CTGGTTCCGTAGAGTAGACTAGCAGCCGAGTTGAGCTGAACTGAACTGCTGGCAATGTGAACACTGGATGCAAGATCAGATGTGAAGATC
1441 TCTAATATGGTGGTGGGATTGAACATATCGTGTCTATATTTTTGTTGGCATTAAGCTCTTAACATAGATATAACTGATGCAGTCATTGGT
1531 TCATACACATATATAGTAAGGAATTACAATGGCAACCCAAACTTCAAAAACAGTAGGCCACCTGAATTGCCTTATCGAATAAGAGTTTGT
1621 TTCCCCCCACTTCATGGGATGTAATACATGGGATTTGGGAGTTTGAATGAACGTTGAGACATGGCAGAACC
```

Figure 15.10. A 2,640-bp oleosin gene from *Brassica napus*. The transcription initiation codon, ATG, is numbered as the first three nucleotides. The putative regulatory sequences such as TATA-box, CAAT-boxes, RY-repeats, GC-boxes, octamer, putative ABA-binding sites (ACGTGGCGT and ACGTGTCCT), and polyadenylation signal are underlined. The intron sequence is given in lower case. The predicted amino acid sequence, including the sequence determined by peptide microsequencing, is underlined. (From Lee and Huang 1991. *Plant Physiol.* 96:1395–1397. © American Society of Plant Physiologists.)

genic tobacco shows that activity of the gene increases during the early period of development in a way similar to the accumulation of transcripts in embryos of *B. napus* (Slocombe et al 1994).

Oleosin cDNA or genomic clones have been isolated and characterized from embryos of maize (Vance and Huang 1987; Qu and Huang 1990; Qu, Vance and Huang 1990; Lee and Huang 1994), *Glycine max* (Kalinski et al 1991), *B. napus* (Lee and Huang 1991; Keddie et al 1992a, b, 1994), *Helianthus annuus* (Cummins and Murphy 1992; Thoyts et al 1995), *Arabidopsis* (van Rooijen, Terning and Moloney 1992), *Gossypium hirsutum* (Hughes, Wang and Galau 1993), and *Hordeum vulgare* (Aalen 1995). The conclusion drawn from a comparison of the sequences of maize, *B. napus*, and *Arabidopsis* is that versions of the gene both with and without introns are known; whereas the maize genes are intronless, there is just one intron in the genes of the other two genera. A striking property of oleosin genes from maize and *Arabidopsis* is the presence of RY-repeat sequences, usually a characteristic displayed by legume storage protein genes (Qu and Huang 1990; van Rooijen, Terning and Moloney 1992; Lee and Huang 1994). It is also worth noting that promoters of maize (Qu, Vance and Huang 1990), *B. napus* (Lee and Huang 1991; Keddie et al 1992a), and *Arabidopsis* (van Rooijen, Terning and Moloney 1992) oleosin genes have sequences that match the ABA-responsive element from the 5'-regulatory region of wheat *EM* gene (Chapter 14); the interspecific ubiquity of this element suggests that it might play a fundamental role in the expression of ABA-regulated genes (Figure 15.10).

Chimeric constructs of 5'-flanking sequences of the *Arabidopsis* oleosin gene introduced into *B. napus* plants after transcriptional fusions to *GUS* gene have exposed both positive and negative regulatory elements involved in modulating quantitative levels of expression. Assay of embryos of transgenic plants for reporter gene activity has identified sequences between –1,100 bp to –600 bp and –400 bp to –200 bp as positive regulatory components and those between –600 bp to –400 bp as down-regulators of gene expression. A clear confirmation that the oleosin gene is expressed during early embryogenesis is provided by the demonstration that its promoter directs intense gene expression in both cotyledons and the embryo axis beginning at the heart-shaped stage (Plant et al 1994). A similar analysis with the promoter of a *B. campestris* ACP gene showed that the direct repeats are not essential for its embryo-specific expression in transgenic plants (Scherer et al 1992). Illustrative of the developmentally modulated expression of the oleosin gene of *B. napus* is the observation that its promoter fragment fused with *GUS* gene is expressed specifically in the embryo and endosperm of transgenic tobacco throughout seed ontogeny (Keddie et al 1994). Other experiments with chimeric oleosin genes and genes involved in fatty acid biosynthesis have been directed toward engineering the fatty acid composition of embryos by antisense expression (Knutzon et al 1992), targeting of heterologous oleosin proteins (Lee et al 1991; Batchelder, Ross and Murphy 1994), production of recombinant protein bodies (van Rooijen and Moloney 1994), and overexpression of bacterial fatty acid synthetase genes (Verwoert et al 1995) in transgenic plants.

4. GENERAL COMMENTS

Of the variety of metabolic changes that occur in developing angiosperm embryos, storage protein synthesis is very much a central one. The voluminous data reviewed in this chapter show that embryos vary enormously in the type of proteins they stockpile in the cells; yet, all storage proteins are probably synthesized in the rough ER and stored in protein bodies. Although there is some evidence justifying a role for Golgi vesicles in the transport of proteins from the site of their synthesis to the site of storage, the precise way this occurs in the large majority of embryos remains to be elucidated. Once the synthesis of storage macromolecules is set in motion, measurements of mRNA synthesis, polysome formation, and a variety of other parameters indicate that cells of the embryo committed to the accumulation of storage proteins are metabolically different from the rest of the cells of the embryo.

Embryos of most of our economically important pulses, cereals, and oilseeds have provided the experimental material for a series of interactive investigations into storage gene expression patterns. Foremost among these has been the isolation and characterization of genes encoding storage proteins. The transcriptional activation of these genes, whether mediated by hormones such as ABA or some other mechanisms, essentially accounts for the massive accumulation of storage proteins in the embryos. In this sense the genes can be considered to function as the ultimate controlling agent in the synthesis of proteins in the conventional way, by the interaction of *cis*-acting sequences with DNA-binding factors. However, transcriptional control alone does not fully explain the mechanism controlling storage protein synthesis. Although not emphasized in this chapter, a cascade of posttranscriptional and translational control mechanisms also regulate the expression of storage protein genes studied.

Expression of storage protein genes in transgenic plants has been a direct result of research on the cloning and sequencing of these genes. The formation of protein bodies in the embryo cells of heterologous transformants exhibits all the basic phenomena associated with their development in embryos of homologous mother plants: the synthesis of proteins in the rough ER, their targeting to vacuoles, and the embryo specificity of their expression. Because storage proteins constitute the major nitrogen reserve of the embryos of seeds that form an integral part of the diet of humans and animals, in one important way the expression of storage protein genes in transgenic plants promises immense potential of a practical nature. This relates to the means to produce genetically engineered seeds with desirable storage protein contents.

SECTION IV

Adventive Embryogenesis

Chapter 16

Somatic Embryogenesis

This chapter will examine the cell and molecular biology of the embryogenic development of somatic cells of angiosperms and gymnosperms, or the phenomenon of somatic embryogenesis. Embryogenic development of somatic cells can be contrasted to the rigorously programmed development of the embryo from the zygote (zygotic embryogenesis) insofar as virtually any somatic cell of the plant body can, under certain experimental conditions, behave like a zygote and faithfully replay a developmental program leading to the production of embryolike structures while remaining innocent of sex. Thus, somatic embryogenesis provides the most clear-cut demonstration of the dictum that all plant cells except those that have undergone irreversible differentiation are totipotent and retain the developmental potential to proliferate into an adult plant. Compared to the limited number of embryos arising from gametic fusion and the difficulty of extracting them from the confines of the ovule, the enormous number of somatic cells potentially capable of embryogenic development ensures the availability of an equally staggering number of embryolike structures by simple experimental manipulations. Despite the fact that zygotic embryos and embryolike structures formed from somatic cells are identical in appearance and possess the same morphogenetic potential, to emphasize the divergent pathways through which they have evolved, the term "embryoid" is generally used to refer to the latter.

Although embryogenic development of somatic cells appears simple, unhindered by the sorting of genes that occurs in the zygote and limited by the genetic information parceled out to the cells from generation to generation, there are a number of cellular and physiological factors that influence unfolding of the developmental program. Some of these factors will be discussed in the early part of the consideration of somatic embryogenesis in this chapter. Later, the biochemical and molecular changes of somatic cells that enable them to organize into stable tissues and organs of embryos will be examined. The reader is referred to the reviews by Merkle, Parrott and Flinn (1995) and Nomura and Komamine (1995) for a comprehensive discussion of the physiological and tissue-culture aspects of this topic, and to those by Dudits, Bögre and Györgyey (1991), de Jong, Schmidt and de Vries (1993), and Dudits et al (1995) for a discussion of the cellular and molecular investigations. However, first the discussion will navigate swiftly through the story of the breakthrough that led to the discovery of somatic embryogenesis.

1. HISTORICAL BACKGROUND

The discovery of somatic embryogenesis in plants is ineluctably tied to the demonstration of totipotency of plant cells. Totipotency has been viewed as the quintessential problem in plant cell biology ever since Haberlandt prophesied at the turn of this century that it would be possible to grow embryos from the vegetative cells of plants. The pioneering experiments that led to a demonstration of the totipotency of plant cells were initiated by Steward and co-workers (Steward, Mapes and Smith 1958; Steward, Mapes and Mears 1958) using the secondary phloem of domestic carrot. These investigators found that culturing slabs of the carrot tissue in a solidified tissue culture medium that supplied the requirements of essential salts, vitamins, and organic nutrients supplemented with coconut milk typically produces a proliferating mass of callus constituted of simple parenchymatous cells. Whereas previous investigators using the same system were preoccupied with the formulation of culture conditions for the indefinite growth of the callus, Steward's group found that it was possible to produce a suspension culture of individual cells and small cell clusters by growing the callus with gentle agitation in a liquid medium of the same composition as the solid medium. Although the origin of the cell clusters appeared to be a vexing problem at first, careful examination of a range of cells and cell aggregates established that individual cells dissociating from the callus divide and the daughter cells formed remain attached to one another as cell clusters. Subsequent growth of this totally disorganized population of cells and cell clumps in the liquid medium without subculture leads to lignification of the inner cells of the clump, the formation of cambiumlike derivatives, and, eventually, the appearance of lateral root primordia. A normal carrot plant is assembled in the culture flask when the rooted aggregate is subsequently grown on a solid medium lacking coconut milk. In these studies, the arduous task of monitoring a single cell in isolation during its transformation into a plantlet was not attempted; it remained for Vasil and Hildebrandt (1965) to show that a single cell of a hybrid tobacco nurtured in isolation from other cells in a defined medium forms a completely organized plant. This cell, grown in a drop of the liquid medium in a microculture chamber, divides repeatedly to form a callus that is subsequently induced to form roots and shoots on a solid medium.

The sweeping transformation of a single cell into a whole new plant is a remarkable developmental feat. In contrast to the preceding, Reinert (1959) found that a callus originating from a strain of carrot root, following a long period of culture in a medium containing coconut milk and IAA, differentiates bipolar embryolike structures upon transfer to a synthetic medium enriched with an elaborate mixture of amino acids, amides, vitamins, hypoxanthine, and IAA. Although it was hard to specify from this work what caused some callus cells to behave differently from the rest of the proliferated mass and whether the embryoids originated from single somatic cells, the observation nonetheless dispelled the notion that the formation of embryolike structures is the monopoly of the zygote. It also took us a step closer to fulfilling Haberlandt's prophesy.

A natural consequence of these observations was an intensification of the efforts to induce an embryogenic type of development in identifiable, single somatic cells by manipulations of culture techniques and medium composition. The breakthrough occurred when, almost simultaneously, Steward (1963) and Wetherell and Halperin (1963) reported that single cells and cell clusters in a cell suspension culture of carrot regenerate an enormous number of embryoids, which are faithful replicas of zygotic embryos. In Steward's original method, embryos isolated from seeds of domestic or wild carrot are allowed to germinate in a medium containing coconut milk. The free cells that slough off from the seedling hypocotyl are a lavish source of totipotent cells that exhibit a typical embryogenic type of development. The effect is even more dramatic when a cell suspension originating from an immature embryo of wild carrot is plated on a nutrient agar plate; here, virtually every cell of the suspension yields an embryoid (Steward et al 1964). Wetherell and Halperin (1963) extended these observations to calluses originating from other parts of wild carrot, such as the root, petiole, and peduncle. A follow-up work also explicitly showed that coconut milk is not a factor in inducing embryogenic development of the somatic cells of carrot and that cells of the callus reared in a medium containing a moderate dose of 2,4-D form embryoids upon transfer to the same medium containing a reduced concentration of auxin. This new idea introduced interpretations that led for a period of time to lively disagreements in the literature on somatic embryogenesis. At about the same time that these discoveries came to light, in a note that has often been ignored, Kato and Takeuchi (1963) described a balanced embryogenic sequence of development in single cells of carrot root callus sloughed off into the growth medium. Finally, the transformation of a single cell of the callus, nurtured in isolation from other cells, into an embryoid was also demonstrated, reinforcing the conclusion that embryoids formed in cell suspension cultures indeed have their origin in single cells (Backs-Hüsemann and Reinert 1970).

One of the important questions considered in the wake of the discovery of somatic embryogenesis was the criteria to define an embryoid. In the work on carrot just described, structures appearing in the cell suspension were designated as embryoids on the basis of their bipolarity and their close resemblance to the organogenetic part of typical zygotic embryos of dicotyledons. Since superficial resemblance alone may be deceptive, it was suggested that besides acquiring early bipolarity, an embryoid should have had its origin in a single cell and should not be connected to a preexisting vascular strand in the mother tissue (Street and Withers 1974; Haccius 1978). However, these criteria have generally been ignored in the majority of contemporary publications that describe somatic embryogenesis in cultured plant cells, tissues, and organs; even to the present day, it is the acquisition of bipolarity that is used by investigators to describe as embryoids structures that originate in tissue culture.

The division sequences during the transformation of somatic cells into embryoids appear to be different from the pattern of divisions of the zygote to form an embryo. As seen in carrot, these differences place limitations on the reliance on external morphology alone in comparative studies of zygotic and somatic embryogenesis in a given species. Development of the embryo of carrot follows a pattern typified by the Solanad type and begins with an unequal transverse division of the zygote. Both resulting cells divide by transverse walls to form a file of four cells. A third round of transverse divisions in each of the four cells gives rise to eight cells held together in a filament. The first longitudinal divisions occur in the three cells farthest from the micropyle, whose descendants eventually generate the organogenetic part of the embryo except the root tip. A limited number of divisions in the other five cells of the filament produce the root tip and a massive suspensor (Borthwick 1931; Lackie and Yeung 1996). Unraveling the division sequences of somatic cells of carrot during their transformation into embryoids appeared to be difficult because it had not been possible to identify by any unique cytochemical or ultrastructural features those cells in a suspension culture that were destined to divide in the embryogenic pathway. Moreover, the occurrence of potential embryogenic cells as part of a cell cluster and as single cells seemed to demand the presence of more than one division sequence to effect the transformation into embryoids. Halperin (1966) found that in carrot cell suspension culture, the presumptive embryoid appears as a globular structure from a disorganized cellular mass without displaying any definite sequence of cell lineage associated with early segmentation. Although a suspensor comparable in stature to that subtending the zygotic embryo is absent, cells of the original aggregate often remain attached to the globular embryoid as a suspensorlike appendage. In the only instance in which a single somatic cell of carrot was followed in isolation during embryogenesis, it was found that in contrast to the zygote, which gives rise to a polarized bicellular structure by an asymmetric division, the somatic cell yields a conglomerate of cells by random divisions; the embryoid emerges from this cellular mass by an additional series of random divisions (Backs-Hüsemann and Reinert 1970). The main problem with these investigations was that it was not possible to establish the normal fates of the division products of the cell as the cell was transformed into an embryoid. This problem was overcome in a detailed study of the early development of somatic embryos of carrot by McWilliam, Smith and Street (1974). This work showed that although the first division of the presumed embryogenic initial in a cell aggregate is in the transverse plane, the daughter cells that are formed diverge along different pathways. The terminal cell divides longitudinally to form the embryo proper, and the cell closest to the cellular aggregate divides transversely to form an incipient suspensorlike structure. This is enough discrimination for our purposes to say that early division sequences of the somatic embryo of carrot appear to have more in common with the Crucifer type than with the Solanad type of embryo formation. Although these results point to a fundamental difference in the segmentation patterns of zygotic and somatic embryos of carrot, it is important to recognize that both sequences create developmentally programmed cell lineages that result in identical structures. The fact that embryolike structures are induced in free cells of carrot by simple manip-

proembryos by continuous subculture in a buffered medium (pH 4.0) containing 1–5 mM NH_4^+ as the sole source of nitrogen. The determining factor that allows continued embryogenic development of the cell clusters appears to be an increase in the pH of the medium to 4.5 or above (Smith and Krikorian 1988, 1989, 1990). Other options available to induce somatic embryogenesis in carrot in a hormone-free medium include culture of seedling meristems in media of high osmolarity or in those containing heavy metal ions, although under these conditions, embryoids are formed without a visible callus phase (Kamada et al 1989; Kiyosue et al 1990b). A possible mechanism based on the release of cells from positional or chemical restraints, which allows the cells to express their innate embryogenic potential, is suggested to account for these results. Later in this chapter it will be seen that the role of auxin in the establishment and maintenance of embryogenic cell cultures of carrot has occasioned intense debate. The embryogenic cell culture developed in the hormone-free medium offers a new level of opportunity for a renewed investigation of the seemingly intractable problem of auxin action in somatic embryogenesis.

The Medicago System. The development of protocols for regeneration of somatic embryos in *Medicago sativa* illustrates the need for protracted studies using various genotypes, explants, and medium formulations to obtain suspension cultures that yield embryoids reproducibly and in large numbers. The first report of somatic embryogenesis in *M. sativa* showed that a callus originating from cultured embryos regenerates embryoids upon transfer to a medium containing an optimum level of 12.5 mM NH_4^+ (Walker and Sato 1981). Coming in the wake of a previous report of plant regeneration by organogenesis from calluses initiated from immature ovaries and anthers cultured in a medium containing kinetin and/or 2,4-D and NAA and transferred to a medium lacking auxin and cytokinin, this suggested that shoots observed in the earlier study were actually embryoids arrested in development (Saunders and Bingham 1972). Later investigations with different genotypes of *M. sativa* showed that the genetic background of the donor plants clearly betrays the regeneration response of the explants (Mitten, Sato and Skokut 1984; Brown and Atanassov 1985). This principle was recognized in breeding experiments in which it was possible to transfer the trait for somatic embryogenesis within populations of plants (Brown 1988; Wan, Sorensen and Liang 1988; Hernández-Fernández and Christie 1989). Apparently, when a favorable genotype is selected, the nature of the explant is of secondary importance in determining the potential for somatic embryogenesis, and a wide range of explants – including cotyledons, hypocotyl, shoot tip, petiole, leaf, root, and stem – are responsive in culture (Brown, Finstad and Watson 1995).

There have been several experimental studies on the choice of the culture medium and the specific components of the medium necessary to induce embryogenic development in cultured explants of *M. sativa*. There is evidence to show that the growth hormone component and nitrogen supply in the medium are the critical factors that affect somatic embryogenesis. The explants become most susceptible to the inducing effects of an auxin such as 2,4-D, 2,4,5-trichlorophenoxyacetic acid, or (2,4-dichlorophenoxy) propanoic acid, and a cytokinin like kinetin, which in combination can increase appreciably the number of cells becoming embryogenic (Meijer and Brown 1987b; Brown, Finstad and Watson 1995). Turning to nitrogen supply, both induction of embryoids and their subsequent development appear to be controlled by the quality and quantity of nitrogen supplied in the medium. The nitrogen requirement can be met by optimizing the concentration with an amino acid in combination with NH_4^+ or by maintaining an appropriate ratio of NH_4^+:NO_3^- in the medium (Stuart and Strickland 1984a, b; Meijer and Brown 1987a).

Based on the data, a widely used protocol to obtain embryogenic cell suspensions and embryoids of *M. sativa* for biochemical and molecular investigations (Figure 16.2) involves the induction of a callus in a medium containing a weak auxin (such as NAA) and kinetin, maintenance of the tissue as a suspension of microcalluses in the same medium, initiation of embryogenesis by a short pulse treatment with 2,4-D, and the subsequent growth of embryoids in a hormone-free medium (Dudits, Bögre and Györgyey 1991). In another protocol, callus initiated on a solid medium containing 2,4-D and kinetin is transferred to a suspension culture in which kinetin is replaced by NAA. Embryogenesis is initiated in cell suspensions collected on a 200-μm mesh by transfer to a hormone-free medium (Pramanik, Krochko and Bewley 1992).

Cereals and Grasses. Because of their economic importance, members of Poaceae have been special targets of attempts to induce somatic embryogenesis. In the early period of this work, success or its

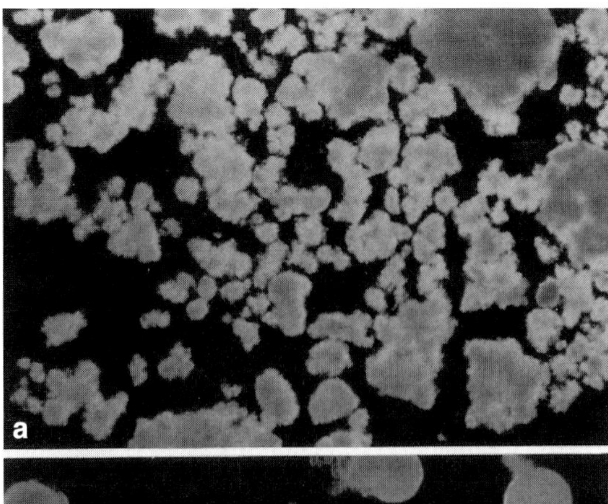

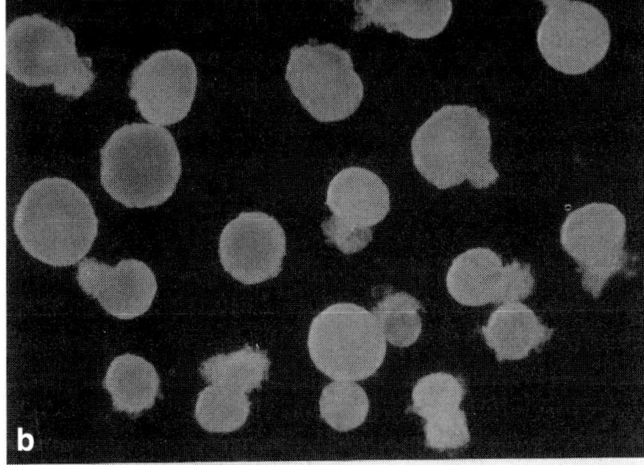

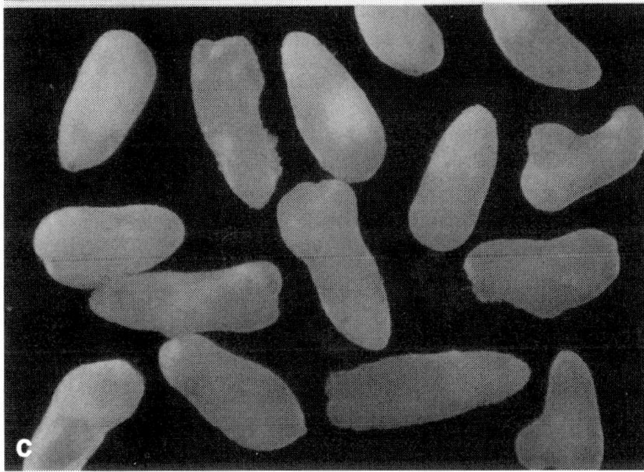

Figure 16.2. Induction of somatic embryos in alfalfa. **(a)** A microcallus suspension grown in the presence of NAA and kinetin. **(b)** A purified fraction of early globular stage embryoids formed in a hormone-free medium following pulse treatment with 2,4-D. **(c)** A purified fraction of late torpedo-shaped embryoids. (Photographs supplied by Dr. D. Dudits.)

lack thereof with cereals and grasses was critically noted, but less so today since it is now a routine practice to obtain continuously growing embryogenic cell suspensions from a wide range of cereals and grasses and to induce somatic embryogenesis in them. In no case are the specific requirements of explant types, medium composition, and hormonal additives that promote callus growth and subsequent embryogenesis completely understood; indeed, a feature that gripped the imagination of workers in the field was that the first notable success was achieved by judicious choice of explants, use of selected hormonal additives, and unorthodox manipulation of the medium at critical stages of culture. The example that vividly illustrates this is *Pennisetum americanum*. Culture of immature embryos excised from 10–15-day-old caryopses or culture of segments of unopened inflorescence on a solid medium supplemented with 2,4-D is found to regenerate a callus. Transfer of the callus to a liquid medium supplemented additionally with coconut milk produces a suspension culture consisting predominantly of richly cytoplasmic embryogenic cells that spontaneously develop into globular or heart-shaped structures. Even at this stage, the multicellular units are not fully committed to the embryogenic pathway; embryoids with the characteristic morphology of cereal embryos are obtained when the suspension culture is plated on an agar medium containing ABA but lacking 2,4-D (Vasil and Vasil 1981, 1982b). In an alternative pathway of somatic embryogenesis displayed by cultured immature embryos, some of the superficial cells of the scutellum divide repeatedly to form embryoids or they regenerate an embryogenic callus from which embryoids subsequently arise (Vasil and Vasil 1982a). With some minor adjustments in protocol, surgically separated parts of mature embryos of *P. americanum* consisting of the shoot apex and subjacent leaf primordia are also found to yield an embryogenic callus and embryoids in culture (Botti and Vasil 1983).

There are indications in other members of Poaceae that the maturation of embryoids is a critical stage in embryogenesis that is controlled by a set of nutrients different from those that induce embryogenesis. This may be illustrated by somatic embryogenesis in wheat. Embryoids are formed with relative ease on calluses raised from young inflorescence axes and immature embryos grown in a medium containing 2,4-D, but precocious germination resulting in a leafy scutellum and multiple shoots is rampant in these embryoids. However, an embryogenic callus regenerated from explants cultured in a double-strength mineral-salt medium supplemented with *myo*-inositol, glutamine, casein hydrolyzate, and 2,4-D readily forms normal, full-term embryoids (Ozias-Akins and Vasil 1982, 1983). During the years of, or closely following, these investigations, embryogenesis was induced in calluses derived from a variety of explants of a host of other members of Poaceae by subtle manipulations of the medium composition. These reports have become far too numerous to catalogue here, but they have been tabulated elsewhere recently (Krishnaraj and Vasil 1995). The successful reports of somatic embryogenesis in Poaceae have led to the view that cultured explants produce different types of calluses, which differ in their regenerative potential, and that selection of the appropriate callus (embryogenic callus) is an important step in inducing embryogenesis. Two aspects of somatic embryogenesis in this group of plants that we do not understand at all and that have limited the use of the system in biochemical and molecular investigations are the slow growth of the callus and the rapid loss of its embryogenic competence.

(ii) Direct Somatic Embryogenesis

The induction of a callus on the cultured explant and the transfer of the loosely dissociated cells to a medium of a different composition to induce embryogenesis present a complicated situation, particularly in light of the bewildering hormonal requirements of the different species thus far investigated and the need in some cases to select embryogenic calluses. The justification for the continued use of carrot and a few other plants is partly that they provide elegant embryoid induction systems but also that they offer favorable material for molecular studies. However, the feasibility of somatic embryogenesis directly on the cultured explant has simplified the system and offers the prospect of gaining a better understanding of the cell-level controls that operate during embryogenic induction. This has been demonstrated in a most dramatic way by the culture of young flower buds of *Ranunculus sceleratus* in a medium supplemented with coconut milk and IAA; the dense callus that forms on the cut end of the explant subsequently differentiates numerous embryoids. Some of the embryoids germinate precociously into seedlings and bear a second generation of embryoids freely exposed along the entire length of the hypocotyl and stem (Konar and Nataraja 1965). Much of what we know about medium requirements for direct somatic embryogenesis has been gleaned from other studies, which have used explants of leaves, ovules, nucellar tissues, embryos, somatic embryos, and haploid pollen embryoids of various plants. The summary of evidence as it stands now suggests that the culture of explants in a medium containing coconut milk or in a medium supplemented with 2,4-D alone or in combination with a cytokinin, elicits the formation of embryoids directly from the superficial differentiated cells of the explant or from an incipient callus regenerated on the cut surface of the explant. In leaf explants of *Dactylis glomerata*, anatomical studies have unequivocally demonstrated that the embryoid arises by dedifferentiation of a differentiated cell of the explant (Conger et al 1983). In most other cases, direct evidence of this kind is lacking.

There is a further aspect to the analysis of *D. glomerata* that provides an underlying basis for direct somatic embryogenesis and embryogenesis through the intervention of a callus. This aspect is in evidence when the medium is confronted with explants taken from different parts of the same leaf or with segments from the same region of different leaves. Segments taken from the more basal parts of the leaf are so dominated by a callus-forming habit that they never display direct somatic embryogenesis; by contrast, those excised from the distal portions form embryoids directly. Comparison between the responses of leaves of two different ages shows that the potential for direct embryogenesis shifts slightly toward the base of the older of the two leaves. These observations carry important implications for the role of the meristematic tissues of the leaf in promoting callus growth versus embryogenesis. A juvenile grass leaf has an increasing basipetal gradient of meristematic activity, with the meristematic area in a mature leaf very much restricted to its base. Apparently, direct somatic embryogenesis is perpetuated by a recrudescence of growth in the fully differentiated cells in the more apical part of the leaf, whereas cells that are actively dividing tend to lapse into a callus-forming mode (Conger et al 1983). It is a prediction from this work that the cytological state of the leaf cells at the time of culture is more important than any special additives to the medium in choosing the pathways toward morphogenesis. This said, the process of direct somatic embryogenesis still remains obscure because we do not know the genetic, epigenetic, and physiological changes in the cultured leaf segments that affect the outcome.

(iii) Recurrent Embryogenesis

Current views concerning the formation of repeated cycles of embryoids from a single explant are derived from the observations on *Ranunculus sceleratus* referred to previously. Although this work did not explicitly demonstrate the formation of embryoids for more than one cycle from the cultured explant, subsequent work has shown that in some cases, the primary somatic embryos fail to mature but give rise to successive cycles of embryoids directly from the superficial cells. This phenomenon of recurrent or secondary embryogenesis is well studied in *Medicago sativa*, in which the initiation of recurrent embryogenesis requires transferring a callus formed in a hormone-enriched medium to a hormone-free medium. When the regenerated somatic embryos are maintained in the same medium or transferred to fresh hormone-free medium, senescence of the cotyledons and formation of a second generation of embryoids from the hypocotyl are observed. The newly formed embryoids can be used to start the cyclic regeneration of fresh crops of embryoids almost indefinitely (Lupotto 1983, 1986). There are aspects of this process that do not provide easy explanations, notably the relationship between the senescence of the cotyledons and the activation of cells of the hypocotyl. This might involve the loss of intercellular coordination due to cotyledonary senescence or a block in the maturation pathway of the embryoid that leads to embryogenic divisions in the hypocotyl cells.

(iv) Embryogenesis from Isolated Protoplasts

Embryoid development from protoplasts isolated directly from somatic cells of explants and from embryogenic cells maintained in suspension cultures has been studied under standard tissue culture conditions. In the first reports of embryogenesis from isolated protoplasts, pieces of carrot root and aliquots of suspension cultures from root and petiole of carrot seedling were used as starting materials (Grambow et al 1972; Kameya and Uchimiya 1972). The protoplasts were cleared by centrifugation and suspended in an isotonic culture medium. They were subsequently cultured in the same medium, the osmolarity of which was reduced in a step-wise fashion with the progress of culture. The naked protoplasts formed embryoids only after they re-formed a cell wall. Generally, the period up to the formation of a new cell wall was marked by changes in the size and shape of the protoplasts and, irrespective of the protoplasts' origin, the first one or two divisions in the isolated protoplasts were observed in about six days after culture. Various fates can befall a re-formed cell, one of which is repeated divisions to form a cell cluster and the transformation of daughter cells into embryolike structures. Characteristically, the requirements for embryogenic development of cells originating from either cell type by way of isolated protoplasts reflect those of the respective parent cells. Cell clusters originating from protoplasts isolated from suspension cultures regenerate embryoids upon transfer to a medium lacking 2,4-D, whereas those formed from protoplasts of root cells require a medium enriched with coconut milk. Overall, these results show that despite the completely different pressures to which isolated protoplasts are exposed, once they re-form a wall their subsequent morphogenesis is as highly precise and predictable as that of a totipotent cell.

The amazing progress in protoplast technology has led to the development of protocols to generate calluses and plantlets from protoplasts isolated from a number of plants. In many cases, protoplasts isolated from leaf mesophyll cells have been found to undergo embryogenic development with relative ease. However, it has not been clearly established that, in the majority of species thus far studied, isolated protoplasts regenerate plantlets by embryogenesis.

3. CELL BIOLOGY AND PHYSIOLOGY OF SOMATIC EMBRYOGENESIS

Two essential facts about somatic embryogenesis that have withstood the test of time well are that (a) dedifferentiation and division of the cell induced by a strong auxin, such as 2,4-D, is the invariant first step in the process, and (b) embryogenic development is initiated in the absence of auxin or in the presence of a reduced concentration of auxin. Although auxin carries the overtone of a producer of proembryogenic masses in the sense of assembling the newly formed cells into a cluster, it has become clear that the role of auxin is somewhat limited. A simple illustration of this is provided by the observation that if carrot explants are exposed to 2,4-D for a short period of time, potentially embryogenic single cells can be subsequently induced to differentiate into proembryogenic masses in a medium lacking auxin; treatment of the explant for a prolonged period with auxin also results in the formation of proembryogenic masses. Thus, a knowledge of the source of the embryo mother cell, particularly with regard to its encounter with auxin, becomes an important consideration in the study of the cell biology of somatic embryogenesis. Most of the work in this area has been done with cell suspensions propagated by and calmodulin, the distribution patterns of these markers do not reveal a critical role in embryogenic induction (Timmers, de Vries and Schel 1989; Emons, Mulder and Kieft 1993; Timmers, Kieft and Schel 1995).

Analysis of the nuclear cytology of certain carrot cell lines has indicated their preference for alternatives to simple mitotic divisions in culture. These abnormal divisions, referred to as reductional grouping or prophase chromosome reduction, mimic meiosis and generate a limited number of cells with a reduced number of chromosomes, mostly at the haploid level (Nuti Ronchi et al 1992a, b). If these haploid cells are in fact the potential embryogenic cells, as segregation at the DNA level indicates, we can assume that reduction in chromosome number must be an essential first step in the process whereby somatic cells assume an embryogenic role (Giorgetti et al 1995). Other studies have noted correlations between the tetraploid state attained by carrot cells during prolonged serial subculture and their inability to produce embryoids (Smith and Street 1974; Coutos-Thevenot et al 1990). Although the possible effect of the ploidy level of cells cannot be dismissed lightly, it would appear that the cells in a callus or suspension culture are very dissimilar from one another and therefore need not follow identical nuclear changes during embryogenesis.

(ii) Attainment of Polarity

It was seen in Chapter 13 that during zygotic embryogenesis, polarity is already established in the zygote by the concentration of cellular components toward one end of the cell. The first division of the cell separates an organelle-enriched terminal cell that forms the embryo proper from a basal cell that forms the suspensor. It is certain that polarity is already inherent in embryogenic cells attached to a proembryogenic mass or in superficial cells of an explant that directly differentiate into embryoids; here, daughter cells formed away from the point of attachment of the mother cells invariably differentiate into embryoids. Since no signs of polarity are evident in the free-floating embryogenic cells, polarity is believed to arise during subsequent differentiation of the cell into an embryoid.

Attempts to demonstrate polarity in embryogenic single cells have not been entirely successful because no consistent changes in the generation of electric current or in the intracellular redistribution of specific ions that are important in the fixation of polar axis have been documented. The symmetric, globular embryoid is the earliest recognizable stage of somatic embryogenesis in carrot in which electrical polarity is first detected. Currents are found to enter the apical pole of the embryoid and to leave the region near the presumptive radicle pole; this pattern remains constant up to the torpedo-shaped stage. The traffic of electric current through the cells apparently results in ionic gradients, the inward current at the apical pole of the embryoid being carried by K^+ influx and the outward current at the radicle end resulting from active H^+ efflux (Brawley, Wetherell and Robinson 1984; Rathore, Hodges and Robinson 1988). If one considers that ionic currents are important in the accumulation of specific metabolites or in the facilitation of the polar transport of auxin, their role in the fixation of polarity in the embryoids becomes clear. In related studies, the application of a low-voltage electrical field and an increase in Ca^{2+} in the medium have been shown to greatly increase embryoid formation from alfalfa mesophyll protoplasts and carrot cell suspension cultures, respectively (Dijak et al 1986; Jansen et al 1990). The simplest interpretation of these observations is that the electrical pulse might impose polarity in the cells, which then leads to embryogenic development, and the redistribution of a specific ion that accompanies the fixation of polarity is attained by perturbing the Ca^{2+} concentration in the cells. The attractiveness of the role of Ca^{2+} in fixing polarity prompted a further study, which showed that globular embryoids arising from proembryogenic masses of carrot cells had increased levels of this ion (Timmers et al 1996). Electrical polarity similar to that detected in embryogenic cell clusters and embryoids also prevails in cell clusters of carrot proliferating in a medium containing 2,4-D. In molecular terms, this may mean that the cell clusters are embryogenically determined but are restrained from further development by the presence of 2,4-D in the medium (Gorst, Overall and Wernicke 1987). The distribution of activated calmodulin in cell clusters growing in a medium containing 2,4-D and the localized presence of calmodulin in a polar fashion in embryoids also suggest that embryogenic determination is already attained in the cell clusters (Timmers, de Vries and Schel 1989).

The distribution of plasma membrane–associated arabinogalactan protein is a reliable indicator of polarity in embryogenic cells and embryoids of carrot, although the function of this protein is not well understood. The protein epitope defined by a monoclonal antibody is first expressed in the surface cells of proembryogenic masses and thus identifies cells in terms of their position. The incipient polarity in the globular stage embryoid is reflected in the presence of the protein in several surface layers at the potential shoot end (Stacey, Roberts and Knox 1990). These and the other data considered earlier may constitute the beginnings of a mechanistic explanation of how certain cells of the proembryogenic mass begin to function as embryoid mother cells.

Examination of some other parameters that characterize somatic embryogenesis gives few helpful insights into the basis of polarity. Serial observations have shown that in single embryogenic cells of carrot, the first division is unequal, giving rise to a small cytoplasmic cell and a large vacuolate cell. Further divisions are confined to the cytoplasmic cell and lead to the formation of a proembryogenic mass. Polarity of DNA synthesis and mRNA accumulation is observed in the proembryogenic masses when they are transferred to the embryoid-inducing medium (Nomura and Komamine 1986a, b). In the globular embryoid of carrot, polarity is first evident at the shoot pole by the appearance of esterase activity (Caligo, Nuti Ronchi and Nozzolini 1985), whereas in transformed carrot cells harboring *GUS* gene under the control of *Agrobacterium tumefaciens* T-DNA gene-5 promoter, polarity is first noted in the root pole of the globular embryoid by the expression of GUS activity (Mattsson, Borkird and Engström 1992). Enzyme systems sensitive to regulatory functions concerned with establishment of the polarity of the shoot and of the root poles thus appear to be different.

(iii) Nutritional Factors

The requirements for the most rapid growth of an embryogenic callus may not necessarily be the same as those that promote somatic embryogenesis in the tissue. Nevertheless, some components of the medium are important in causing enhanced growth of the callus and, secondarily, in increasing the number of embryoids formed. Nearly all the work on somatic embryogenesis has involved continued subculture of the callus in a medium that has mineral salts and sucrose as its base. The importance of the type of nitrogen present in the mineral-salt mix used in the medium in inducing somatic embryogenesis has generated much discussion. Even from the early days of plant tissue culture, the secondary phloem of carrot was famous for its ability to proliferate uncontrollably in a medium containing 2,4-D and nitrate as the sole source of nitrogen. Halperin and Wetherell (1965) showed that this callus displays little inclination to form embryoids when challenged in an induction medium unless it was originally grown in a medium containing a small amount of NH_4^+ as well. Using petiole-derived callus of carrot, it was found that cells from the callus grown in a medium containing NO_3^- fail to form embryoids in the induction medium even if the medium contains NH_4^+, whereas callus cells grown in an ammonium-containing medium form embryoids in large numbers in the induction medium with or without NH_4^+. The conclusion from these results was that NH_4^+ is essential for the expression of embryogenic competence in carrot cells, representing a contrast with NO_3^-, which probably facilitates the subsequent development of embryoids. But the role of NH_4^+ in somatic embryogenesis came under suspicion when it was claimed that the actual concentration of nitrogen in the medium, and not its form, is important in the somatic embryogenesis of carrot (Reinert, Tazawa and Semenoff 1967). A subsequent work clearly demonstrated the importance of reduced nitrogen by showing the contrast between very low embryogenesis in the presence of 5–95 mM KNO_3 and high embryogenesis in a medium containing 12–40 mM KNO_3 and 10 mM NH_4Cl. Ammonium ion alone is also effective when the pH of the medium is kept constant (Wetherell and Dougall 1976; Dougall and Verma 1978). Additional evidence for a specific requirement for NH_4^+ in the medium for somatic embryogenesis has come from work on *Medicago sativa* and other legumes (Walker and Sato 1981; Meijer and Brown 1987a; Greinwald and Czygan 1991).

Several exogenously supplied amino acids regulate somatic embryogenesis in carrot to varying degrees, the most effective one being α-alanine. Glutamine, asparagine, glutamic acid, arginine, proline, and serine also promote initiation and development of embryoids to some extent, whereas histidine, leucine, valine, and methionine are considered inhibitory (Kamada and Harada 1979b; Nuti Ronchi et al 1984). Based on the lack of accumulation of α-alanine and the presence of an active alanine amino transferase system in carrot cells supplied with α-alanine, the effectiveness of this amino acid as a nitrogen source for somatic embryogenesis is attributed to its rapid incorporation and its utilization for protein synthesis during the period of cell divisions (Kamada and Harada 1984a, b). As compared to glycine and glutamic acid, α-alanine and proline induced a 5–10-fold increase in the number of embryoids in embryogenic cell cultures of *Medicago sativa*. NMR determination of the utilization of nitrogen by the cells has led to the inter-

GA in the embryoids than in the nonembryogenic cells proliferating in the auxin-containing medium. In contrast, the nonembryogenic cells had high levels of polar GA (Noma et al 1982). On a per embryoid basis, an increase in the concentration of free and water-soluble GA-like substances is found during embryogenesis in a hybrid grape cell suspension culture (Takeno et al 1983). From these observations one is left with the feeling that changes induced during somatic embryogenesis of carrot and grape may be targeted at some aspect of GA metabolism, converting the hormone to a form that permits embryogenesis.

We are also confronted with a set of conflicting data on the effects of ethylene on somatic embryogenesis because both promotory and inhibitory effects of the hormone have been noted in the same system. The main point of contention is that in carrot cell suspension, some ethylene-releasing agents stimulate embryogenesis whereas others inhibit it; similarly, some presumed inhibitors of ethylene biosynthesis promote embryogenesis whereas others are inhibitory (Roustan, Latché and Fallot 1989a, b, 1990; Nissen 1994). Nissen (1994) has pointed out the pitfalls of using inhibitors of ethylene biosynthesis to monitor somatic embryogenesis because their effects may not be related to the inhibition of ethylene production. If ethylene stimulates somatic embryogenesis as claimed, it is clear that inhibition of embryogenesis by 2,4-D and other auxins cannot be due to auxin-induced ethylene production. Ethylene-induced inhibition and promotion of somatic embryogenesis have been reported in cell cultures of *Hevea brasiliensis* (Auboiron, Carron and Michaux-Ferrière 1990) and *Medicago sativa* (Kępczyński, McKersie and Brown 1992), respectively.

The deepening understanding of the role of ABA during maturation and dormancy of seed embryos has underlined the importance of this hormone in somatic embryogenesis. As anticipated from the work on seed embryos, exogenous addition of ABA inhibits somatic embryogenesis in carrot cell clusters (Fujimura and Komamine 1975; Kamada and Harada 1979a, 1981) and prevents precocious germination of *Brassica napus* embryoids (Maquoi, Hanke and Deltour 1993). Suggestive of the involvement of protein kinase or protein phosphatase in the ABA signal transduction pathway, the levels of several acidic phosphoproteins are found to decline when embryogenic carrot cell suspension cultures are treated with ABA (Koontz and Choi 1993). The majority of embryoids formed in a cell suspension of *Carum carvi* (caraway; Apiaceae) have an abnormal morphology, although a few are normal or near normal. Abnormalities have probably arisen by imbalances in the hormonal constituents of the medium or in the culture environment. A considerable restoration of the form of the embryoids to that of normal seed embryos is achieved if cells are subcultured in the dark in a medium containing ABA; in the presence of ABA in the light, restoration is less complete (Ammirato 1974). This interesting result led to studies that demonstrated an interaction between ABA, zeatin, and GA during the normal maturation of somatic embryos from caraway cells (Ammirato 1977). Supplementation of the medium with ABA also causes an increase in the number of morphologically normal embryoids formed in somatic cell cultures of carrot (Kamada and Harada 1981), *Picea abies* (von Arnold and Hakman 1988), wheat (Carman 1988), and *Digitalis lanata* (Scrophulariaceae) (Reinbothe, Diettrich and Luckner 1990). There have been attempts to relate globally the changes in ABA content of embryogenic cells and embryoids to different phases of somatic embryogenesis, but the data do not support a connection (Kamada and Harada 1981; Rajasekaran, Vine and Mullins 1982; Kiyosue et al 1992a; Etienne et al 1993a, b).

(v) Role of Secreted Molecules

The centerpiece of somatic embryogenesis is the utilization by cultured cells of metabolites supplied in the medium to evoke totipotent development. Yet, it is ironic that as single cells and cell clusters metabolize nutrient substances, they release into the medium components that favor cell proliferation and somatic embryogenesis. This was first borne out by the observation that there is a quantitative reduction in embryogenesis when carrot cell suspension is diluted to low cell density and that a partial alleviation of this is achieved by the addition of conditioned cell suspension medium (Hari 1980). Later years have witnessed intense activity to characterize the secreted molecules, and a role for extracellular proteins in somatic embryogenesis in carrot and other plants is now firmly entrenched.

Glycoproteins such as peroxidase and phosphatase have been identified among the compounds released by embryogenic carrot cells into the medium, although it is doubtful whether they have any regulatory function in embryogenesis (Ciarrocchi, Cella and Nielsen 1981). An interesting relationship exists between the release of a 65-kDa glycoprotein and somatic embryogenesis in a carrot cell suspension, whereas nonembryogenic cells

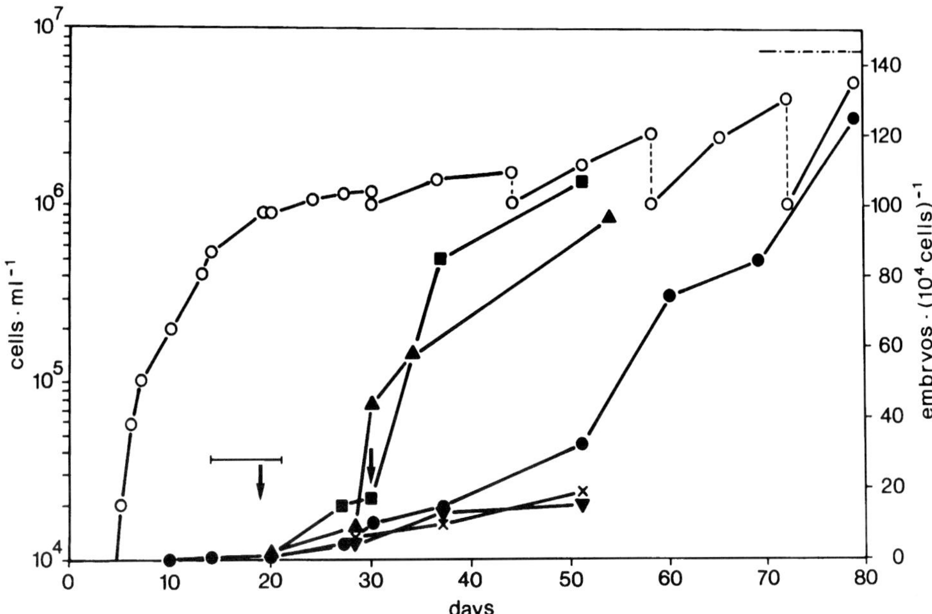

Figure 16.3. Rescue of embryogenic potential of a carrot cell suspension by conditioned medium. Open symbols represent the number of suspension cells per ml of culture medium containing 2,4-D. Closed symbols represent the embryogenic potential expressed as the number of embryoids of all stages produced by 10^4 suspension cells with additions as indicated: ●, no addition; ■, addition at day 19 of 50% (v/v) medium conditioned by a 7-day-old embryogenic suspension culture; ▲, addition at day 19 of a 50% (v/v) equivalent high-molecular-mass fraction (over 5 kDa) obtained from the conditioned medium of 7-day-old embryogenic suspension culture; ▼, addition at day 19 of a 50% (v/v) equivalent boiled high-molecular-mass fraction obtained from the conditioned medium of 7-day-old embryogenic suspension culture; ✖, addition at day 19 of 50% (v/v) conditioned medium devoid of material over 5 kDa, obtained from a 7-day-old embryogenic suspension culture. The arrow at day 19 indicates removal of the hypocotyl explant in all cultures; at day 30 (arrow), all cultures were transferred to fresh medium with 2,4-D. (From de Vries et al 1988b. *Planta* 176:196–204. © Springer-Verlag, Berlin. Print supplied by Dr. S. C. de Vries.)

excrete a 57-kDa protein into the medium (Satoh et al 1986). The most promising attempt to implicate extracellular proteins in somatic embryogenesis in carrot cell suspension was the demonstration that addition of a crude, high-molecular-mass, heat-labile protein from a conditioned medium accelerates the process (Figure 16.3). Cell lines impaired in somatic embryogenesis fail to excrete one or more of these proteins, identified as glycoproteins, but their inability to regenerate embryoids is not inevitable, because it is partially reversed by the addition of a protein preparation from embryogenic cell lines (de Vries et al 1988a, b).

The use of the glycosylation inhibitor tunicamycin illustrates a fruitful approach to identifying the excreted protein. This fungal antibiotic inhibits somatic embryogenesis, but not unorganized proliferation of carrot cells cultured in an induction medium (Lo Schiavo, Quesada-Allue and Sung 1986). Embryogenic competence of the cells is restored when the inhibitor is supplied along with an adequate dose of the concentrated conditioned medium. If a protein present in the conditioned medium were to alleviate the effect of the inhibitor, the protein may have the same composition as the wild-type metabolite. The component in the conditioned medium that restored embryogenic competence in tunicamycin-inhibited cells was identified as a 38-kDa cationic peroxidase whose catalytic properties, rather than the carbohydrate moiety, were found to be decisive in the complementation assay. An explanation sought at the cellular level for the tunicamycin effect and its reversal by peroxidase is along the lines of the promotion of cell expansion by the inhibitor and its prevention by the secreted peroxidase (Cordewener et al 1991).

Additional evidence linking extracellular proteins to somatic embryogenesis is provided by the analysis of a temperature-sensitive carrot variant cell line (*ts-11c*) isolated by chemical mutagenesis. Cells of the variant line usually grow and regenerate embryoids when propagated at 24 °C but are unable to do so at 32 °C, in which environment they do not progress beyond the globular stage. Moreover, embryoids formed at the nonpermissive temperature display pronounced abnormalities in their morphology. Defects in the variant cells are overcome simply by growing them at the nonpermissive temperature in a medium conditioned by wild-type somatic embryos (Lo Schiavo et al 1990). Purification of the conditioned medium has led to the demonstration that rescue of embryoids in the *ts-11c* cell lines is accomplished by a 32-kDa glycosylated acidic endochitinase, which is apparently secreted at a reduced level at the nonpermissive temperature (de Jong et al 1992, 1995).

The specific role of endochitinase in overcoming the effect of the mutation has not yet been defined. Experiments in which compounds that contain oligomers of *N*-acetyl glucosamine were tested for their ability to promote embryoid formation in *ts-11c* cells at the nonpermissive temperature showed that the metabolic products of *Rhizobium leguminosarum* nodulation (*NOD*) genes, identified as lipooligosaccharides, are effective in mimicking the effect of endochitinase (de Jong et al 1993). But it is far from clear how chitinase and the nodulation factors lead to organized growth in somatic cells, a problem that is difficult to investigate because of the possible involvement of unidentified signal molecules. A set of arabinogalactan proteins secreted into the medium by carrot embryoids have been shown to restore embryogenic potential in a nonembryogenic cell line and to accelerate the production of embryoids in embryogenic lines. The activity of this crude mixture depends upon specific fractions that inhibit or promote embryogenesis (Kreuger and van Holst 1993, 1995). A role for secreted xyloglucan endotransglycosylase in somatic embryogenesis has been inferred from its ability to contribute to cell elongation through wall loosening. A corollary to the idea that a carrot cell suspension enriched for nonexpanding cells, such as those undergoing embryogenesis, secretes less of this enzyme into the medium is that expanding nonembryogenic cells should secrete more of the enzyme, as shown by Hetherington and Fry (1993). This observation is consistent with a model that substantiates the requirement for rapid cell divisions during the initial phase of embryogenesis. In summary, somatic embryogenesis in carrot depends on the activity of more than one secreted glycoprotein that probably helps to maintain the proper balance between cell division and cell enlargement in single cells and cell clusters in a suspension culture.

Mention should also be made of secreted proteins that are not considered to be directly involved in promoting somatic embryogenesis in carrot cell cultures because their addition does not alleviate the inhibitory effects of tunicamycin. One such protein, designated as EP-1, belongs to a family of heterogeneous proteins and is found localized specifically in the pectin-containing wall material of the vacuolate, nonembryogenic cells. At the nucleotide level, *EP-1* shows limited homology to the *SLG* and *SLR* genes of the self-incompatibility series and to genes encoding *S*-locus putative receptor kinases (van Engelen et al 1991, 1993). Another extracellular protein (EP-2) shows homology to plant lipid transfer proteins. The involvement of EP-2 protein in embryogenesis can be seen by the expression of its transcripts in seed embryos and in the proembryogenic mass of cells when they decide whether to become embryoids or to continue to proliferate. Localized expression of transcripts of the gene encoding EP-2 protein in the peripheral protoderm cells of globular and heart-shaped embryoids has suggested a role for lipid transfer proteins in the transport of cutin monomers. The requirement of embryoids to maintain a proper osmotic balance in the liquid medium in which they are bathed explains the presence of cutinized walls (Sterk et al 1991). Nonembryogenic cells of carrot cell suspension culture also secrete a 47-kDa glycoprotein (EP-4). This protein is localized specifically in the walls of the clustered cells, with the highest level in the cross walls. The gene that encodes this protein has a sequence similarity with the early nodulin gene from alfalfa (van Engelen et al 1995). Although one can argue that these proteins do not have a direct role in embryogenesis, an alternative is to suppose that they specify the type of cells produced in a suspension culture (Figure 16.4).

Extracellular proteins have been identified during somatic embryogenesis in barley (Nielsen and Hansen 1992), grape (Coutos-Thevenot et al 1992), *Citrus aurantium* (Gavish, Vardi and Fluhr 1991, 1992), *Digitalis lanata* (C. Reinbothe et al 1992), *Pinus caribaea* (Domon et al 1994), *Picea glauca* (Dong and Dunstan 1994), and *P. abies* (Egertsdotter and von Arnold 1995). Growth of cell suspension cultures of grapevine is associated with an ordered process of protein secretion that depends more upon the presence of auxin in the medium than upon its absence (Figure 16.5). In formal terms, the grapevine cell suspension may be described as one bearing the closest resemblance to the carrot cell culture system; as in the carrot system, the two extracellular proteins secreted by embryogenic

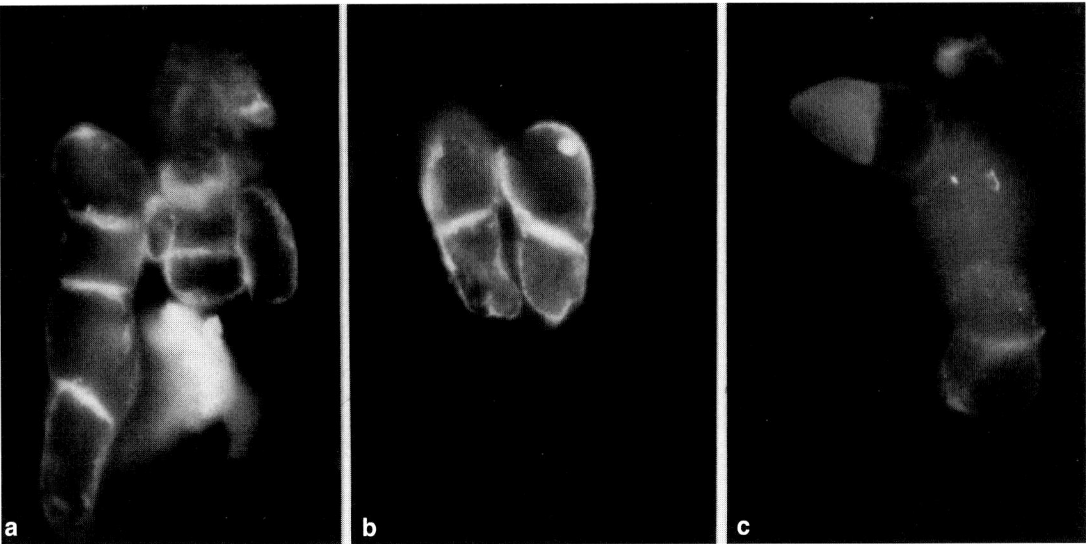

Figure 16.4. Immunofluorescence localization of EP-4 protein in the embryogenic cells of carrot. **(a, b)** Serum produced against EP-4 protein. **(c)** Preimmune serum. (From van Engelen et al 1995; photographs supplied by Dr S. C. de Vries.)

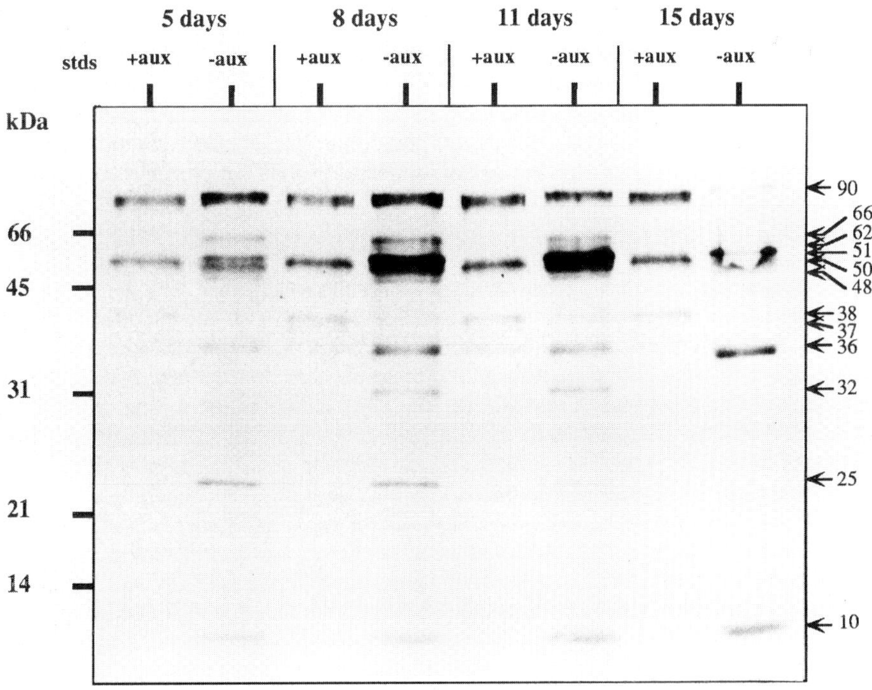

Figure 16.5. A Western blot showing the accumulation of proteins secreted into the medium by embryogenic (-aux) and nonembryogenic (+aux) cells of grape. Proteins from the culture media were separated by SDS-PAGE and were revealed by a peroxidase-labeled con-A staining protocol. Arrows on the right indicate positions of prominent bands. stds, molecular mass standards. (Reprinted from P. Coutos-Thevenot et al 1992. Extracellular protein patterns of grapevine cell suspensions in embryogenic and non-embryogenic situations. *Plant Sci.* 86:137–145, with kind permission of Elsevier Scientific-NL, Sara Burgerhartstraat 25, 1055 KV Amsterdam, The Netherlands. Photograph supplied by Dr. P. Coutos-Thevenot.)

cells nurtured in an auxin-free medium are 32-kDa cationic peroxidase and 10-kDa lipid transfer protein (Coutos-Thevenot et al 1992, 1993). By secreting endochitinases into the medium, embryogenic cells of barley also bear some similarities to carrot cells (Kragh et al 1991). Compared to nonembryogenic cells, embryogenic cells of *P. caribaea* have a highly specific protein secretion profile in which the synthesis of at least five proteins identified in two-dimensional gels is modulated at different stages of early embryogenesis. One of these proteins is analogous to germin, a glycosylated protein synthesized during germination of wheat embryos (Domon et al 1994, 1995). Germinlike proteins are probably the first markers of somatic embryogenesis in *P. caribaea*, and their synthesis during the period of the most active growth of somatic embryos is significant in the general context of cell division and cell elongation. By taking advantage of an observation that in *P. abies*, concentrated extracellular proteins from embryogenic cell lines influence the morphology of less-developed somatic embryos, researchers found an arabinogalactan protein and a protein showing antifungal activity to be the active components stimulating the development of embryoids of different morphological stages (Egertsdotter, Mo and von Arnold 1993; Egertsdotter and von Arnold 1995; Mo, Egertsdotter and von Arnold 1996). Embryogenic cells of maize and carrot secrete mucilaginous polysaccharides into the medium, but it is not established whether they have a role in embryogenic induction (Everett, Wach and Ashworth 1985; Kikuchi et al 1995).

Although the terms in which somatic embryogenesis is viewed here attribute a promotory role for secreted proteins, not all secreted proteins promote embryogenesis. Nucellar cell cultures of *C. aurantium* accumulate several glycoproteins in the medium in a stage-dependent manner; of these, certain 53–57-kDa proteins secreted by proembryogenic masses inhibit embryogenesis in newly initiated proembryogenic masses and prevent the subsequent transition of embryoid initials to globular embryoids (Gavish, Vardi and Fluhr 1991, 1992). This is unlikely to cause practical difficulties in the continued growth of somatic embryos that is generally accomplished by transfer of the proembryogenic masses to another medium, in which secretion of inhibitory proteins ceases.

Summarizing, the potential of the extracellular proteins is that it is possible to purify the crude extracts and test the proteins for their ability to promote or inhibit embryogenesis in vitro. Undoubtedly, these proteins are the products of gene activity of the cells, but only in a few cases have the genes been isolated and characterized.

4. BIOCHEMISTRY OF SOMATIC EMBRYOGENESIS

A suspension of embryogenic cells is an excellent model of a differentiating system in which all biochemical activities associated with somatic embryogenesis can be studied in a clonal population, but the goal of understanding the biochemistry of somatic embryogenesis has hardly been achieved. Much emphasis in the biochemical analysis of somatic embryogenesis has been placed on comparing the synthesis of nucleic acids and proteins during embryogenic and during nonembryogenic development of the cells. The results have made it possible to delve into the causal mechanisms of the process and to provide a baseline level of information for investigations of somatic embryogenesis at the molecular level.

(i) Changes in Macromolecule Synthesis

Some early experiments give us a look at the metabolism of DNA, RNA, and proteins during the growth and embryogenic development of carrot cells in suspension culture. Because the most fundamental event in the cell suspension culture is an increase in cell number, DNA synthesis is essential for the process. This conclusion is based on data on changes in the DNA content of cells and the incorporation of labeled precursors of DNA synthesis. In cells growing in a medium lacking auxin, a rapid replication of chromatin and other nuclear materials is a biochemical transition point in anticipation of embryogenic induction (Wochok 1973b; Verma and Dougall 1978; Fujimura, Komamine and Matsumoto 1980; Kononowicz and Janick 1988).

One interesting question is the extent to which changes in cytoplasmic DNA are associated with the acquisition of embryogenic potency. Consistent with a body of work showing that novel mitochondrial genome reorganization is seen in long-term plant tissue cultures, there is some evidence suggesting that similar changes may be correlated with embryogenic development of somatic cells. Changes are generally reflected in the abundance of certain forms of mitochondrial genome, a decreased abundance of certain hybridizing fragments and an amplification of others, or the appearance or disappearance of some hybridizing bands (Rode et al 1988; DeVerno, Charest and Bonen 1994).

The specialization of carrot cells in the embryogenic pathway involves not only increased DNA synthesis but also increased RNA and protein synthesis. The augmentation of RNA and protein synthesis is noted beginning 2–4 hours after the transfer of cells to an auxin-free medium, although the amounts synthesized are too small to have an impact on their accumulation in the cells. Because no new cells are formed in the suspension culture during the first 12 hours of growth in the auxin-free medium, it is clear that the increases noted are due to cellular RNA and protein synthesis. Another small peak in the RNA synthetic activity of embryogenic cells is seen at 96 hours, coincident with the formation of the first embryoids (Fujimura, Komamine and Matsumoto 1980; Sengupta and Raghavan 1980a). Accompanying active RNA synthesis is also an accelerated synthesis of enzymes of the pyrimidine nucleotide pathway (Ashihara, Fujimura and Komamine 1981). On the whole, these data are merely suggestive and do not prove that RNA and protein synthesis is essential for the transition of carrot cells to the embryogenic mode.

A determination of the RNA species synthesized by carrot cells during the early hours of embryogenic induction is important in establishing the role of mRNA in the process. It was found that as early as 6 hours after the transfer of cells to a hormone-free medium, there is a decrease in the synthesis of rRNA and a concomitant increase in the synthesis of minor species of RNA in the 12S and 18S regions. Prompted by the discontinuation of hormone treatment, embryogenic cells synthesize more mRNA than do nonembryogenic cells, and the periods of observed increase correlate well with embryogenic induction, which is when RNA synthesis appears to pick up (Sengupta and Raghavan 1980b). That no significant replication of rRNA genes occurs during somatic embryogenesis in carrot is supported by the results of experiments in which the amounts of rDNA in embryogenic and nonembryogenic cells were determined by DNA–rRNA hybridization (Masuda, Kikuta and Okazawa 1984).

The experimental evidence favors the view that removal of auxin from the medium controls the biochemical events of embryogenesis by eliciting the synthesis of mRNA. Yet, it has not become unequivocally clear that gene activity for embryogenesis in carrot cells is initiated upon transfer of the cells to a medium lacking auxin. The issue is whether proteins synthesized in cells nurtured in the auxin-free medium are coded on newly formed mRNA or on mRNA transcribed when the cells are in contact with auxin. Based on the striking similarities noted between the in vitro translation products of mRNA populations of proembryogenic masses and torpedo-shaped embryoids of carrot, Wilde et al (1988) have suggested that the gene expression program for embryogenesis occurs when cells are growing in the auxin-containing medium. As will be discussed later, many proteins necessary for embryogenic induction are already synthesized in the proembryogenic masses, probably using mRNA transcribed under the influence of auxin (Figure 16.6).

(ii) Role of Embryonic Proteins

Analysis of protein metabolism during somatic embryogenesis has concentrated on isozyme changes and on the characteristic appearance and disappearance of a number of proteins and enzymes during embryogenesis. Results of these investigations have an important bearing on understanding the molecular events that trigger embryogenic development in somatic cells and in the formation of embryoids. An early electrophoretic analysis showed different patterns of expression of isozymes of peroxidase, esterase, and glutamic dehydrogenase in embryogenic and nonembryogenic cells of carrot. Especially significant was the presence of only slow-migrating isozymes of glutamic dehydrogenase in the embryogenic cells as compared to the presence of additional, fast-moving ones in the nonembryogenic counterparts (Lee and Dougall 1973). In this context there is now evidence showing that activities of protease (Carlberg et al 1984) and nitrate reductase (Rambaud and Rambour 1989) and isozymes of esterase (Chibbar et al 1988) and isoperoxidase (Joersbo et al 1989) can be used as potential markers of somatic embryogenesis in carrot cells. The phenomenon of isozyme changes is perhaps best characterized in an embryogenic line of *Citrus sinensis* that shows an upsurge of peroxidase activity concomitant with the appearance of embryoids. At the time when the rise in enzyme activity is under way, a new band typifying embryogenic activity appears in the isozyme profile (Kochba, Lavee and Spiegel-Roy 1977). Selective expression or repression of proteins or of isoforms of various enzymes has been noted during somatic embryogenesis in the following: explants of *Brassica alboglabra* (Chinese celery) petiole (Zee, Wu and

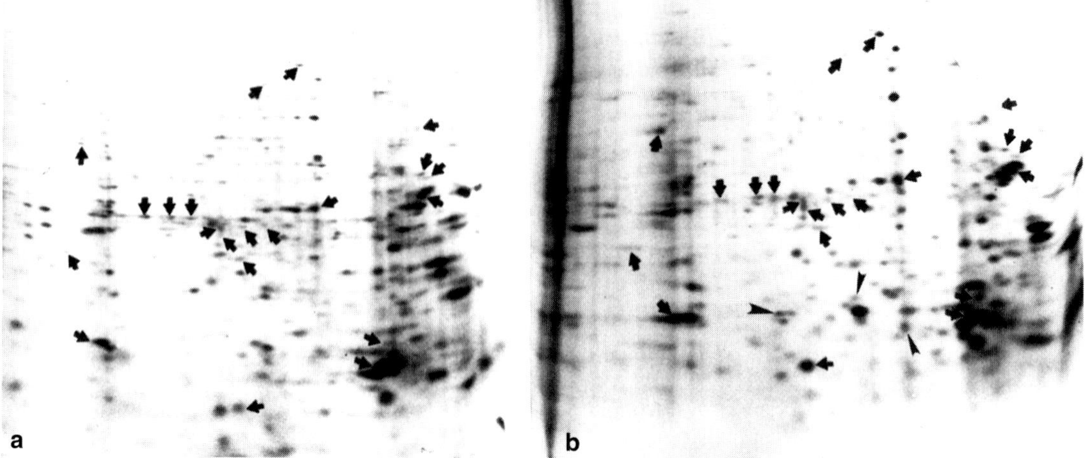

Figure 16.6. Two-dimensional SDS-PAGE of in vitro translated proteins from **(a)** proembryogenic masses and **(b)** torpedo-shaped embryoids of carrot. Closed arrows show translation products present in both; arrowheads show proteins specific to torpedo-shaped embryoids. (From Wilde et al 1988. *Planta* 176:205–211. © Springer-Verlag, Berlin. Photographs supplied by Dr. T. L. Thomas.)

Yue 1979); embryos of maize (Everett, Wach and Ashworth 1985; Fransz, de Ruijter and Schel 1989), rice (Abe and Futsuhara 1989), barley (Coppens and Dewitte 1990), soybean (W. Liu et al 1991; Stejskal and Griga 1995), and wheat (Bapat, Rawal and Mascarenhas 1992); immature maize glumes (Rao, Suprasanna and Reddy 1990); lettuce cotyledons (Zhou et al 1992); ovules of *Cercis canadensis* (Fabaceae) (Buckley and Trigiano 1994); and apical meristem of date palm (Baaziz et al 1994). The relevance of biotin enzymes in somatic embryogenesis in carrot has been highlighted by the observation that the activity of acetyl-coA carboxylase increases substantially during embryogenesis. This could well play an important role in the accumulation of biotin-containing polypeptides associated with the transformation of embryogenic cell clusters into embryoids (Wurtele and Nikolau 1992).

The concept of embryonic proteins as early developmental markers of somatic embryogenesis arose from a comparative two-dimensional SDS-PAGE analysis of newly synthesized proteins of carrot cells growing in the presence or absence of 2,4-D in the medium. Comparison of the spectrum of proteins synthesized in cells growing for 12 days showed no pronounced differences in the nearly two hundred or so polypeptides spotted on the gels, except that the embryogenic cells had two additional proteins, E-1 and E-2, designated as embryonic proteins. The surprising finding was that regardless of the presence or absence of auxin in the medium, these two proteins were synthesized in the cells as early as 4 hours of growth in fresh media but, in the presence of 2,4-D, they gradually diminished and finally disappeared (Sung and Okimoto 1981). From this it appears that the synthesis of embryonic proteins is an early event of somatic embryogenesis triggered by auxin; however, by its very presence in the medium, auxin also bars the continued synthesis of these proteins. Among other biochemical functions that change their expression during callus and embryogenic modes of growth in carrot cells is the synthesis of two callus-specific proteins (C-1 and C-2) by cells grown in high-density cultures in a medium containing 2,4-D. Particular insight into the functions of embryonic proteins and callus-specific proteins has been provided by the protein synthetic profile of a mutant cell line of carrot that is resistant to cycloheximide in both callus and embryogenic modes of growth. Despite the inhibition of protein synthesis, the mutant cells growing as callus in the presence of 2,4-D in the medium synthesize the embryonic proteins, but not the callus-specific proteins (Sung and Okimoto 1981, 1983; Borkird, Choi and Sung 1986). The corollary to these findings is that callus growth can proceed even in the absence of synthesis of callus-specific proteins, whereas the synthesis of embryonic proteins does not necessarily induce embryogenesis. The evidence based on these studies supports the concept that the callus trait and the embryogenic trait are coordinately expressed by a common mechanism, the activation of one function being accompanied by the elimination of the other. If this interpretation is valid, it might appear that in somatic cells of carrot, the action of a single gene provides the signal either for continued proliferative growth or for embryogenesis.

Specific proteins have been identified as molecular markers during somatic embryogenesis in pea (Stirn and Jacobsen 1987), rice (Chen and Luthe 1987), *Dactylis glomerata* (Hahne, Mayer and Lörz 1988), *Trifolium rubens* and *T. pratense* (McGee et al 1989), soybean (Dahmer, Hildebrandt and Collins 1992), *Nicotiana plumbaginifolia* (Reinbothe, Tewes and Reinbothe 1992b), *Digitalis lanata* (C. Reinbothe et al 1992), *Cichorium intybus* (Hilbert, Dubois and Vasseur 1992; Boyer, Hilbert and Vasseur 1993), and *Coffea arabica* (coffee; Rubiaceae) (Yuffá, de García and Nieto 1994). In callus lines of rice, several polypeptides in the 40–44-kDa range are detected more abundantly in the embryogenic callus than in the nonembryogenic callus; these polypeptides also have counterparts in the extant proteins of grain embryos of rice (Chen and Luthe 1987). In leaves of chicory, in which embryoids are formed directly by the transformation of mesophyll cells, synthesis of the first embryonic proteins occurs within two days of culture of the explant, before any overt morphological or cytological signs of change are detected (Figure 16.7). During a five-day induction period when the mesophyll cells acquire embryogenic competence, at least 15 new proteins that appear transiently or accumulate steadily have been identified (Boyer, Hilbert and Vasseur 1993). A rather complex pattern of protein changes is seen during somatic embryogenesis in callus cultures of *Dactylis glomerata*. On the basis of an analysis of extant, in vivo synthesized and in vitro translated proteins, undifferentiated single cells and embryoids are found to possess their own set of characteristic proteins, which do not change during proliferative and embryogenic modes of growth (Hahne, Mayer and Lörz 1988). Thus, if embryonic marker proteins are synthesized, they are made very early, at a stage that could not be identified precisely. A major reason for this is that, unlike in carrot, somatic embryos develop spontaneously in *D. glomerata* and cannot be reliably induced or suppressed by changes in the medium composition.

The production and accumulation of histone and nonhistone chromosomal proteins (Gregor, Reinert and Matsumoto 1974; Fujimura, Komamine and Matsumoto 1981; Sanchez-Pina, Kieft and Schel 1989), cytoskeletal proteins (Cyr et al 1987), embryoid-specific proteins (Dong, Satoh and Fujii 1988; Racusen and Schiavone 1988; Reinbothe, Tewes and Reinbothe 1992b; C. Reinbothe et al 1992), protein related to the onset of cell divisions (Smith et al 1988), membrane proteins (Slay, Grimes and Hodges 1989), LEA proteins (Reinbothe, Tewes and Reinbothe 1992b), oleosins (Ross and Murphy 1993), lipids (Turnham and Northcote 1982; Feirer, Conkey and Verhagen 1989; Dutta et al 1991), and storage proteins (Crouch 1982; Shoemaker, Christofferson and Galbraith 1987; Hakman et al 1990; Roberts et al 1990; Dahmer, Hildebrand and Collins 1992; Krochko, Pramanik and Bewley 1992; Lai, Senaratna and McKersie 1992; Misra et al 1993; Burns and Wetzstein 1994; Krochkov et al 1994) are major chores of progressive somatic embryogenesis in cell and callus cultures of different species. The transient appearance and disappearance of subsets of proteins in globular and late stage embryoids formed in cell suspensions of *Nicotiana plumbaginifolia* and *Digitalis lanata* provide an unusually clear picture of the transcriptional and posttranscriptional controls of protein synthesis during somatic embryogenesis. Changes in only less than 10% out of the approximately 250 polypeptides identified in the in vivo protein synthetic profile of both species were associated with an altered in vitro translation profile (Figure 16.8). Although these proteins are under developmental regulation at the transcriptional level, a posttranscriptional control operates for the majority of proteins of embryogenesis whose synthesis does not correlate with corresponding changes in in vitro translated mRNAs (Reinbothe, Tewes and Reinbothe 1992b; C. Reinbothe et al 1992). When the apical and basal halves of torpedo-shaped embryoids of carrot are analyzed separately, the stage-specific proteins are found to be asymmetrically distributed in the two segments. This observation might be reconciled with the existence of biochemical differences between the two cell types or spatially polarized gene expression patterns in the developing embryoid (Racusen and Schiavone 1988).

Compared to seed embryos, somatic embryos of different species show varying degrees of defects in storage protein accumulation that generally result in lower amounts of protein in the embryoids (Crouch 1982; Shoemaker, Christofferson and Galbraith 1987; Roberts et al 1990; Flinn, Roberts and Taylor 1991; Dahmer, Hildebrand and Collins 1992; Misra et al 1993; Burns and Wetzstein 1994; Krochko et al 1994). Treatment of somatic embryos with glutamine (Lai, Senaratna and McKersie 1992), maltose and sucrose (Burns and Wetzstein 1994), and more reproducibly with ABA (Dutta

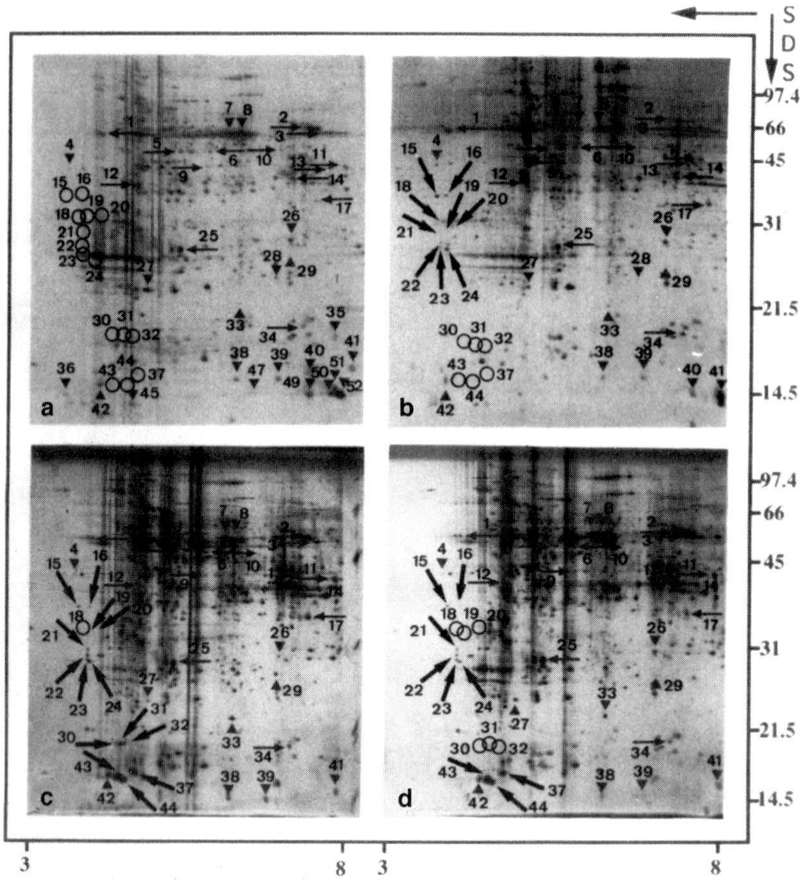

Figure 16.7. SDS-PAGE showing the changes in extant proteins associated with somatic embryogenesis in leaf segments of *Cichorium intybus.* **(a)** Noninduced leaves. **(b)** Leaves induced for 3 days. **(c)** Leaves induced for 5 days. **(d)** 5-day induced leaves after transfer to an expression medium (same as induction medium, but without glycerol) for 3 days. ▲, polypeptides showing increased accumulation; ▼, polypeptides showing decreased accumulation or not detected; →, polypeptides that remained constant during embryogenesis. Embryogenesis-related polypeptides are indicated by large arrows and their absence is noted by circles. Polypeptides of general interest are numbered 1 to 51. (Reprinted from C. Boyer, J.-L. Hilbert and J. Vasseur 1993. Embryogenesis-related protein synthesis and accumulation during early acquisition of somatic embryogenesis competence in *Cichorium. Plant Sci.* 93:41–53, with kind permission of Elsevier Scientific-NL, Sara Burgerhartstraat 25, 1055 KV Amsterdam, The Netherlands.)

and Appelqvist 1989; Roberts et al 1990; Kim and Janick 1991; Maquoi, Hanke and Deltour 1993; Misra et al 1993) leads to accumulation of storage products at levels closely approaching those of mature seed embryos.

(iii) Polyamine Metabolism during Embryogenesis

Polyamines, of which putrescine, spermidine, and spermine are the best known, are characteristically basic in nature and are distinguished from amino acids by the absence of a carboxyl group. Their ability to bind with acidic groups, especially with those in nucleic acids, have made them prime candidates for gene regulation in a variety of physiological processes in plants. The importance of polyamine metabolism in somatic embryogenesis was highlighted by the observation that induction of embryogenesis in carrot cells is accompanied by increased levels of the three polyamines within one to two days of their transfer to the hormone-free medium. On the other hand, a mutant cell line of carrot that fails to form embryoids but exhibits a growth rate similar to that of the wild-type cells does not show increased polyamine levels (Montague, Koppenbrink and Jaworski 1978; Fienberg et al 1984; Mengoli et al 1989). In line with

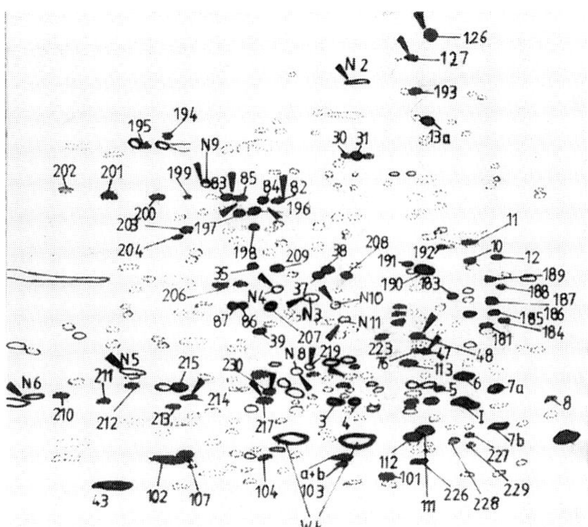

Figure 16.8. Composite diagram of polypeptides translated in a cell-free system by RNA isolated from somatic embryos of *Digitalis lanata*. Filled spots indicate polypeptides for which an exact counterpart was detected in the extant protein profile of embryoids; numbering is arbitrary. All other polypeptides synthesized in vitro had no counterpart in the extant protein profile. In vitro translated polypeptides whose mRNA levels changed during somatic embryogenesis are indicated by N and a number. Polypeptides indicated by arrowheads represent mRNAs that were developmentally regulated and reached maximum level in the embryoids. WK indicates polypeptides translated in the absence of exogenous RNA. (From C. Reinbothe et al 1992; print supplied by Dr. S. Reinbothe.)

these observations, activities of arginine decarboxylase (an enzyme involved in the synthesis of putrescine from arginine) and of *S*-adenosylmethionine decarboxylase (which catalyzes the decarboxylation of *S*-adenosylmethionine in the putrescine-to-spermidine and the spermidine-to-spermine pathways) are found to be higher in the embryogenic cells than in their nonembryogenic counterparts (Montague, Armstrong and Jaworski 1979; Fienberg et al 1984). A role for arginine decarboxylase and *S*-adenosylmethionine decarboxylase in somatic embryogenesis, suggested by these results, has also been substantiated by experiments that showed a significant reduction in embryogenic potential, with a concomitant decrease in the levels of polyamines, in carrot cells treated with α-difluoromethylarginine (an inhibitor of arginine decarboxylase) or methylglyoxal bis(guanylhydrazone) (an inhibitor of *S*-adenosylmethionine decarboxylase). The inhibition of embryogenesis cannot be considered an indirect effect of inhibitors because it occurs without any decrease in the growth rate of cells and is reversible by the simultaneous addition of putrescine with α-difluoromethylarginine and of spermidine with methylglyoxal bis(guanylhydrazone) (Feirer, Mignon and Litvay 1984; Minocha et al 1991). Another investigation showed that difluoromethylornithine, an inhibitor of ornithine decarboxylase, has the opposite effect on somatic embryogenesis in carrot because it accelerates the development of embryoids in a medium containing inhibitory levels of 2,4-D. This is accompanied by an increased accumulation of polyamines, especially putrescine, and by high arginine decarboxylase activity in the cells (Robie and Minocha 1989). An approach involving the expression of a mammalian ornithine decarboxylase gene in transgenic carrot cells has shown that the increased biosynthesis of putrescine leads to the improved ability of cells to regenerate somatic embryos (Figure 16.9). The presumed mechanism here is the provision of an effective alternative, provided by the mammalian enzyme, to the usual arginine decarboxylase pathway for the biosynthesis of putrescine (Bastola and Minocha 1995). Other examples of embryogenic systems in which polyamine levels and/or levels of corresponding enzymes correlate highly with somatic embryogenesis are *Medicago sativa* (Meijer and Simmonds 1988), *Hevea brasiliensis* (El Hadrami et al 1989), *Vitis vinifera* (Faure et al 1991), *Picea abies* (Santanen and Simola 1992), *Cichorium intybus* (Helleboid et al 1995), and *P. glauca* × *P. engelmannii* (Amarasinghe, Dhami and Carlson 1996).

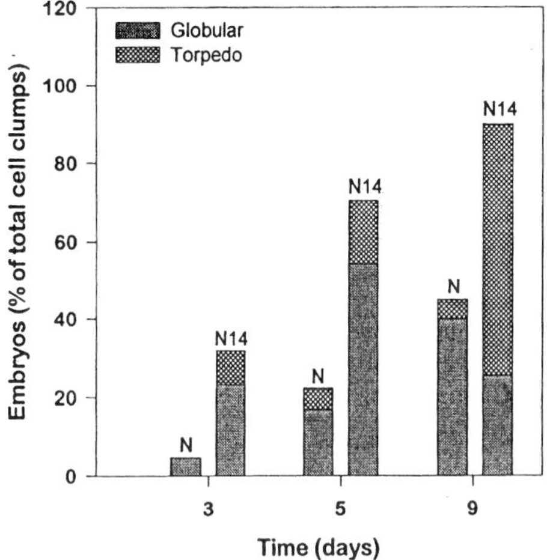

Figure 16.9. Number of globular and torpedo-shaped embryoids in a control (N) and in a transgenic line of carrot (N14) transformed with a mouse ornithine decarboxylase cDNA. The cells were grown in a liquid medium without 2,4-D and embryoids were counted at different times after transfer. (From Bastola and Minocha 1995. *Plant Physiol.* 109:63–71. © American Society of Plant Physiologists.)

Since the effects of polyamines on somatic embryogenesis in carrot cells appear to be antagonistic to those of ethylene, a close physiological link between these two compounds and somatic embryogenesis has been suspected. In a carrot cell line sensitive to ethylene, ethylene-induced inhibition of somatic embryogenesis is accompanied by a reduction in spermidine and spermine synthesis and in the activities of polyamine biosynthetic enzymes (Roustan, Latché and Fallot 1992). Consequently, ethylene might be considered to inhibit somatic embryogenesis in carrot indirectly by interfering with polyamine metabolism.

5. REGULATION OF GENE EXPRESSION

The logistical problems of somatic embryogenesis in the widely investigated carrot system suggest that one of the traits immediately affected during embryogenic induction of cells is the differential utilization of genetic information, which leads to the synthesis of new mRNAs and proteins. The complex morphology attained by embryoids also demands the operation of a program of continued gene expression with long-term implications. The carrot somatic embryo system, used extensively to help analyze questions concerning gene regulation, has proved to be unusual in one fundamental respect: Given the extensive overlap between in vivo synthesized proteins of nonembryogenic and embryogenic cells, there is the possibility that only minor changes in gene expression accompany embryogenic induction. Another option is to suppose that genes involved in embryogenesis are already expressed in cells during their proliferative growth. Both these possibilities are borne out by the identification and characterization of genes from embryogenic cells and somatic embryos of carrot.

Despite the uncertainty about the relevance to the induction process of genes isolated from embryogenic cells, much of the thinking about gene activation during somatic embryogenesis has been focused on these genes. A group of three cDNA clones that are homologous to mRNAs preferentially expressed in embryoids of carrot were first isolated by screening an expression library with embryoid-enriched antibodies. The transcript levels of these clones are severalfold higher in the embryoids than in nonembryogenic cell clusters. A further embryo-related aspect of these recombinants concerns the similarity in the expression patterns of their transcripts in somatic and seed embryos of carrot (Choi et al 1987). The increasing accumulation of transcripts of two of the clones (*DC-8* and *DC-59*) in the heart-shaped and older embryoids has emphasized the developmental regulation of these genes. Although the genes are first expressed in the globular embryoids, they are not required for the formation of globular proembryogenic masses in the auxin-containing medium (Borkird et al 1988).

Subsequent molecular and genetic analyses have shown that *DC-8* and *DC-59* genes both encode proteins showing characteristics of LEA proteins. Thus, the *DC-8* gene encodes a hydrophilic protein that does not possess an obvious signal sequence at the N-terminal end (Franz et al 1989). As with other *LEA* genes, the embryoid-specific expression of both *DC-8* and *DC-59* genes is enhanced by the presence of ABA in the medium (Hatzopoulos, Fong and Sung 1990; Hatzopoulos et al 1990; Goupil et al 1992). Additional *LEA* genes isolated from embryogenic cells of carrot include *DC-3* (Wilde et al 1988; Vivekananda, Drew and Thomas 1992), *ECP-31*, *ECP-40* (Kiyosue et al 1992c, 1993), and *EMB-1* (Wurtele et al 1993). The isolation and characterization of other genes – represented by clones *DC-5* and *DC-13* (Wilde et al 1988), *DC-2.15*, *DC-7.1*, and *DC-9.1* (Aleith and Richter 1990), *AX-110* (Nagata et al 1993), *CEM-1* (Kawahara et al 1994), and *CEM-6* (Sato et al 1995) – storage protein genes (Newton, Flinn and Sutton 1992), and genes for protein kinase (Lindzen and Choi 1995) and for GTP-binding proteins (Kiyosue and Shinozaki 1995) allow comparisons to be made between the expression patterns of different mRNA populations and their protein products during somatic embryogenesis.

(i) Genes of Somatic Embryogenesis and Their Proteins

As indicated, the early investigations on gene expression that seemed pertinent to somatic embryogenesis concerned the two *LEA* genes, *DC-8* and *DC-59*. The latter encodes a basic polypeptide homologous to the lipid body membrane protein L-3 (16-kDa oleosin isomer) from maize embryos. Molecular analysis has shown that DNA-binding factors from embryoids interact with sequences within the 5'-flanking regions of the *L-3* gene. These segments show similarity to the ABA-responsive sequences of the *DC-8* gene; their relationship with *LEA* genes is probably similar, if not identical, to that seen in seed embryos (Hatzopoulos, Fong and Sung 1990; Hatzopoulos et al 1990). Clear evidence that initiation and maintenance of the embryoid gene expression program occurs under conditions that allow cell proliferation in the auxin-containing medium was provided by Northern blot hybridization of RNA with probes made to cDNA clones *DC-3*, *DC-5*, and *DC-13*. The pattern of expression of these recombinants is similar in the proembryogenic masses and embryoids of different ages. It is a prediction of the embryoid-specific nature of the transcripts that they would be expressed only in the embryogenic cells and embryoids and not in other parts of the plant; transcripts of the *DC-3* gene meet this requirement (de Vries et al 1988b; Wilde et al 1988). *DC-3* is a member of group 3 *LEA* genes belonging to a small gene family. Suggestive of the physiological role of DC-3 proteins in conferring desiccation tolerance in a way analogous to LEA proteins, expression of *GUS* gene linked to the promoter of *DC-3* gene is found to increase in transgenic tobacco seedlings in response to water deficit and ABA application (Seffens et al 1990; Vivekananda, Drew and Thomas 1992).

The *LEA* genes *ECP-31* and *ECP-40*, which belong to small gene families, have much in common between them. Nucleotide and amino acid sequences showed that *ECP-31*–encoded protein is homologous to a cotton LEA protein, whereas the *ECP-40* gene has regions of homology in the ABA-inducible *RAB-16* gene from rice. Transcripts of both genes are found at high levels in the embryogenic cells and at low levels in developing somatic embryos; treatment of somatic embryos with ABA leads to an increase in the accumulation of transcripts of *ECP-31* and *ECP-40*. The peripheral cells of carrot proembryogenic masses are the sites of accumulation of transcripts or protein products of these genes (Kiyosue et al 1990a, 1991, 1992b, c, 1993). Another *LEA* gene that is highly expressed in the somatic embryos of carrot, but is virtually undetectable in the proembryogenic masses, is *EMB-1*. Transcripts of this gene are absent in non-embryogenic cells invariably carried along with embryogenic cells into the induction medium, indicating that they are not induced by the transfer of cells to an auxin-free medium but are directly related to embryogenesis (Figure 16.10). The in situ pattern of accumulation of the transcripts of this gene is very similar in both seed embryos and embryoids. The protein encoded by the *EMB-1* gene displays homologies to both EM protein of wheat and a LEA protein of cotton (Ulrich, Wurtele and Nikolau 1990; Wurtele et al 1993). Corre et al (1996) have shown that the acquisition of potency for somatic embryogenesis by wheat embryo cal-

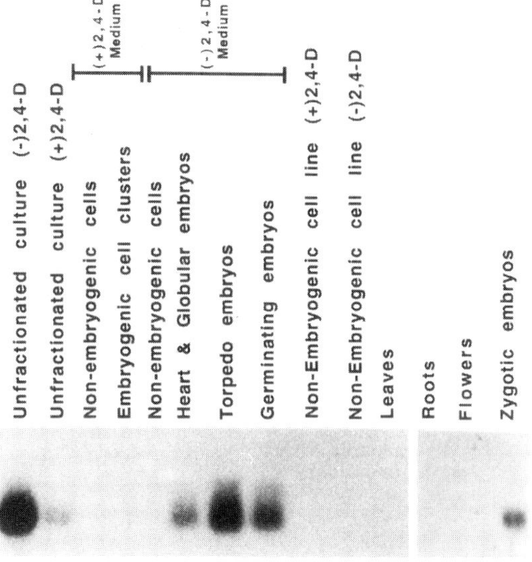

Figure 16.10. Northern blot showing the accumulation of *EMB-1* mRNA during somatic embryogenesis in carrot cell suspension. Poly(A)-RNA was obtained from cell cultures and parts of the carrot plant as indicated. Cells lacking embryogenic capacity were grown to the log phase in media with or without 2,4-D. A single *EMB-1* mRNA is present in somatic and zygotic embryos. (From Wurtele et al 1993. *Plant Physiol.* 102:303–312. © American Society of Plant Physiologists. Photograph supplied by Dr. E. S. Wurtele.)

lus coincides with the accumulation of transcripts homologous to the wheat *EM* gene.

By hybridizing RNA gel blots with oligonucleotide probes, relatively few differences are found in the temporal expression pattern of *LEA* transcripts during somatic embryogenesis in *Nicotiana plumbaginifolia* and *Digitalis lanata*. In both species, transcript levels peak in globular embryoids and decline during subsequent embryoid development; this implies the possible existence of common regulatory pathways in the expression of *LEA* genes even in taxonomically unrelated species (Reinbothe, Tewes and Reinbothe 1992b). A gene isolated from embryogenic cells of *Betula pendula* was found to be homologous to carrot *DC-8* gene at the protein level; despite this, no positive signals of the gene were seen in *B. pendula* cells representing different stages of somatic embryogenesis (Puupponen-Pimiä et al 1993). According to Reinbothe et al (1994), *LEA* transcripts are induced upon treatment of leaf segments of *N. plumbaginifolia* with jasmonic acid; this has led to the suggestion that this compound alone, or along with ABA, may have a specific role in regulating the expression of *LEA* genes during somatic embryogenesis.

The transfer of proembryogenic cells of carrot to

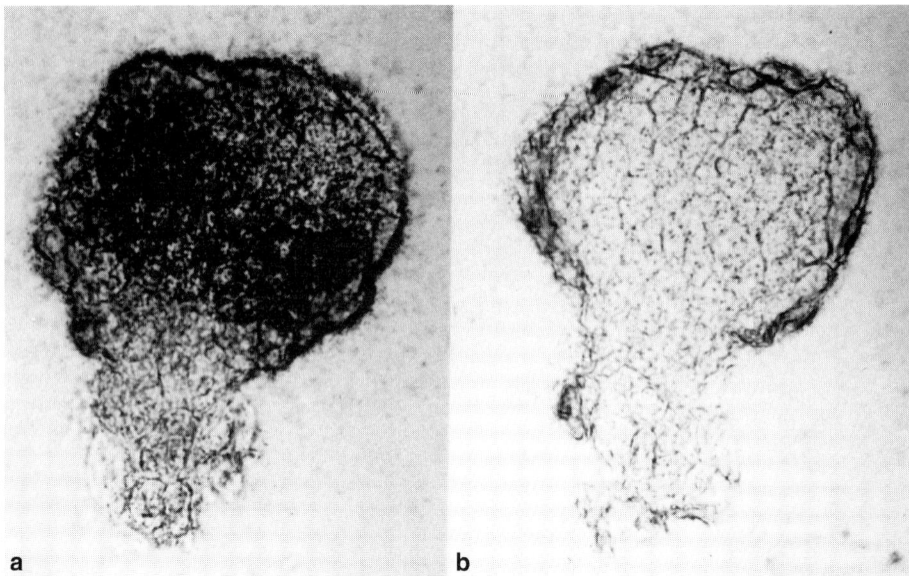

Figure 16.11. Autoradiographs of globular embryoids of carrot showing localization of ^{35}S-labeled antisense **(a)** and sense **(b)** *CEM-1* RNA by in situ hybridization. (From Kawahara et al 1992; photographs supplied by Dr. R. Kawahara.)

a hormone-free medium also elicits transient expression of genes that probably have a role in triggering embryogenesis. The number of these genes is quite small and they even bear similarity to those expressed in the proembryogenic mass. Transcripts of several clones isolated from a cDNA library made to poly(A)-RNA of carrot cells grown for 8 days in the hormone-free medium show a transient accumulation from 3 days up to 16 days after transfer, coinciding with the globular or, rarely, with the heart-shaped embryoids. Thereafter, the prevalence of transcripts reaches such low values as to render them nonfunctional. The deduced proteins of three clones from this group (DC-2.15, DC-7.1, and DC-9.1) show unusual amino acid sequences, such as the presence of the core of a proline-rich domain in DC-2.15 and a large number of glycine residues in DC.7.1 and DC.9.1. Although these proteins bear some resemblance to cell wall proteins, the lack of clearly demonstrable function raises the question of their relevance to the onset of embryogenesis (Aleith and Richter 1990).

Screening a cDNA library of 3-day-old proembryogenic masses of carrot has also identified genes encoding a glycine-rich protein (*CEM-6*) as well as a protein homologous to the eukaryotic translation elongation factor 1α (*CEM-1*) (Kawahara et al 1992; Sato et al 1995). Although the latter protein functions in the interaction of the amino acyl tRNA with ribosomes during the synthesis of proteins for housekeeping chores in the cell, transcription of the gene is developmentally regulated and is confined to the cells of embryoids that are in an active state of division (Kawahara et al 1992, 1994). This makes a good case that large amounts of the elongation factor are necessary to support cell divisions during embryogenesis although they may not be involved in triggering the signals for embryogenesis (Figure 16.11). Genes encoding a variety of proteins, including LEA, EM, and HSPs, have been isolated from somatic embryos of *Picea glauca*; preliminary evidence shows that these proteins function in the same way in somatic embryos as they do in zygotic embryos (Dong and Dunstan 1996).

(ii) Storage Protein and Other Genes

The expression of storage protein genes in developing seed embryos of various angiosperms and gymnosperms was discussed in Chapter 15. The pattern of expression of storage protein genes in developing somatic embryos differs in some respects from that described in the corresponding seed embryos. Analysis of mRNA accumulation and synthesis of storage proteins in somatic embryos of *Medicago sativa* showed that although the periods of accumulation of 7S and 11S storage proteins generally coincide with the availability of their mRNAs, the accumulation of transcripts of 2S storage protein precedes the first appearance of the protein by at least seven days (Krochko, Pramanik and Bewley 1992). The perfectly normal translation of 2S storage protein mRNA in a cell-free system confirmed that there is no internal damage in the messengers; from this, the repression of 2S storage protein synthesis during the early stages of somatic embryogenesis in *M. sativa* has been visualized as a translational control effect (Pramanik, Krochko and Bewley 1992).

In contrast to the earliest expression of napin gene transcripts in the torpedo-shaped seed embryos of *Brassica napus*, transcripts of this gene tend to be present in somatic embryos of all stages. However, the up-regulation of napin gene expression in somatic embryos by ABA is similar to that seen in seed embryos (Fleming and Hanke 1993). The effects of ABA in stimulating and maintaining steady-state levels of various storage protein mRNAs are retained in somatic embryos of certain species of the gymnosperms (Flinn et al 1993; Leal et al 1995). These results were anticipated to some extent from reports of physiological studies of somatic embryos. The work reviewed has some theoretical interest since it shows that somatic embryos simulate zygotic embryos even at the molecular level.

Although carrot embryoids of certain stages respond to a heat shock by synthesizing specific proteins, an observation that heat shock to globular embryoids can permanently arrest their development whereas embryoids of other developmental stages subjected to the same stress recover fully, led to a detailed analysis of the regulation of heat-shock genes during somatic embryogenesis in this system. Using a cloned heat-shock gene encoding a small HSP, it was found that globular embryoids synthesize and accumulate significantly less heat-shock mRNAs than do either embryoids of other developmental stages or undifferentiated callus cells. Despite this, globular embryoids synthesize the same spectrum of HSPs as those produced by cell suspensions or embryoids of other stages of development. The failure of globular embryoids to synthesize heat-shock mRNA has been attributed to the absence of transcriptional induction of the heat-shock gene. Globular embryoids appear to respond to heat shock by translating the existing heat-shock mRNAs at an elevated rate; in all probability, the ability of these embryoids to regulate mRNA sequestration or availability compensates for the low level of available mRNAs (Pitto et al 1983; Zimmerman et al 1989; Apuya and Zimmerman 1992). In the general context of a sensitive developmental window, it is the dominance of translational control without the production of essential stage-specific transcripts that seems most significant in protecting globular embryoids from heat shock while curtailing their development. Transcripts of two HSP cDNAs isolated from alfalfa callus share the property of being induced in developing embryoids at noninductive temperatures although their accumulation is also temperature dependent (Györgyey et al 1991).

Transcripts of both actin and tubulin genes are present in nonembryogenic cells, and the requirement for the cytoskeletal proteins associated with embryogenic induction is generally accompanied by an increased transcript accumulation (Cyr et al 1987; Reinbothe et al 1994). Studies on histone gene expression during somatic embryogenesis in alfalfa have revealed significant differences in the regulation of the transcript levels of two clones of histone *H3* gene. In callus suspension cultures, histone *H3-1* gene transcripts showed a sharp increase one day after a short pulse in a high concentration of 2,4-D and then decreased to very low levels, although differentiated embryoids contained significant levels of transcripts. Despite the presence of consistently high levels of *H3-11* gene transcripts in the noninduced cells, 2,4-D treatment activated higher levels of expression of this gene in induced cells and in embryoids (Wu et al 1988; Kapros et al 1992). An underlying element of similarity in the expression of the two variant genes is that it is broadly confined to the cell cycle, but distinct mechanisms of regulation of expression probably prevail. A key regulator of the eukaryotic cell cycle is a 34-kDa phosphoprotein; during somatic embryogenesis in alfalfa, the time course of expression of a cDNA encoding an alfalfa protein kinase homologous to the yeast cell division cycle phosphoprotein ($p34^{cdc2}$) coincides with the expression of *H3-1* gene (Hirt et al 1991).

The protein product of a homeotic gene has been implicated in soybean somatic embryogenesis. Homeotic genes of the type described in floral organ initiation (Chapter 2) contain the consensus homeobox domain and are also important in early embryo development in insects, amphibians, and mammals. A cDNA was isolated by screening a soybean somatic embryo cDNA library with a maize *KN-1* cDNA clone. Although the deduced protein of the isolated clone shares a high amino acid identity with the KN-1 protein, it has also some unique features of its own, such as the presence of a long leader sequence prior to the translation start site of the protein (Figure 16.12). The

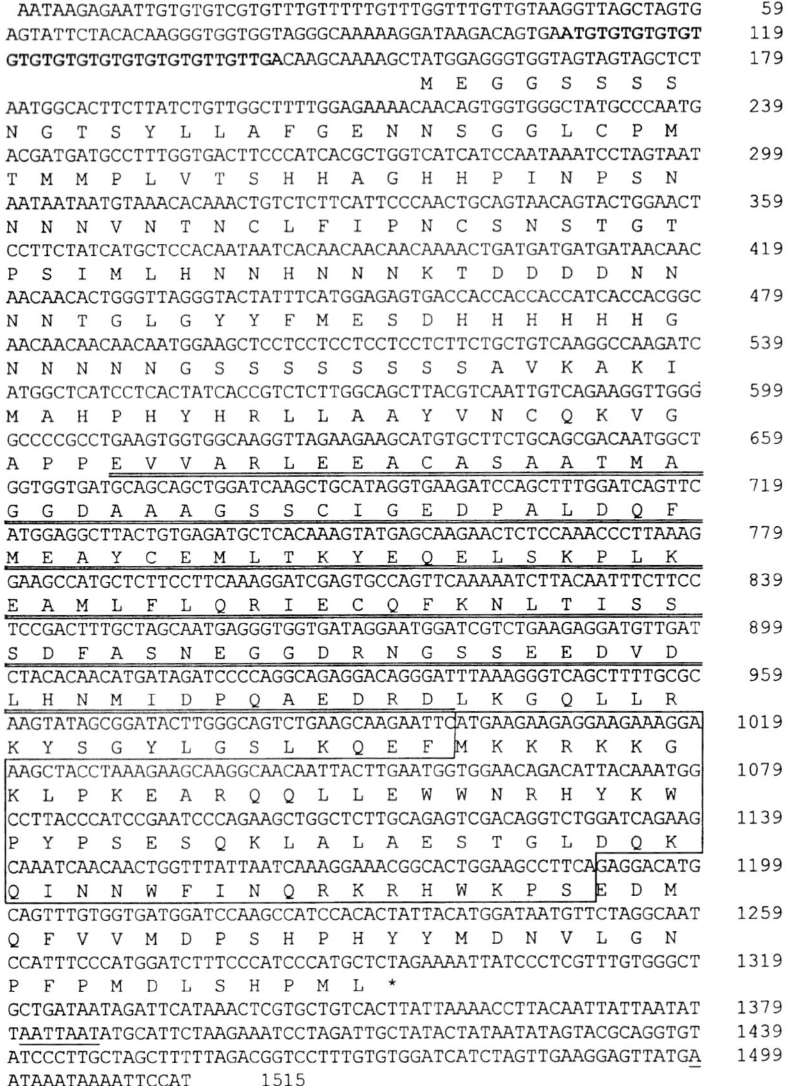

Figure 16.12. Nucleotide sequence and the deduced amino acid sequence of a homeobox-containing gene (*SBH-1*) from somatic embryos of soybean. The homeobox nucleic acid sequence and the homeodomain protein are boxed. Putative polyadenylation signals are underlined. The suggested acidic region is noted by double underline. The nucleic acid sequence that possibly encodes the short leader peptide is in bold type. (From Ma, McMullen and Finer 1994; print supplied by Dr. J. J. Finer.)

expression of transcripts of this gene is relatively specific for embryoids, in which it is developmentally regulated. It has been suggested that the protein encoded by the gene might play an important part as a transcription factor during embryoid development, especially at the critical globular to heart-shaped transition stage (Ma, McMullen and Finer 1994).

A potentially interesting aspect of cell suspension cultures is that they can be used to follow foreign gene expression during the proliferative growth of cells and during embryogenic development. With carrot cell suspension cultures, an obvious hypothesis that can be tested is that certain genes are expressed in transformed embryoids but not in the nonembryogenic cells. This was found to be true in a transgenic carrot cell suspension harboring the chimeric gene construct containing the *ROL-C* promoter of the Ri plasmid of *Agrobacterium rhizogenes* and the *GUS* gene. Transformed embryoids showing high levels of reporter gene expression were produced when cells were transferred to a hormone-free medium, whereas proliferative growth of cells in a medium containing 2,4-D led to decrease in gene activity with increasing time in culture (Fujii and Uchimiya 1991). Studies on *GUS* gene expression using different promoter constructs demonstrated that a combination of several regulatory regions in the *ROL-C* promoter is required for the activation of *GUS* gene during somatic embryogenesis (Fujii, Yokoyama and Uchimiya 1994). These results show that activation of the *ROL-C* promoter occurs as a consequence of the developmental switch that leads to embryogenic divisions of the somatic cells. A highly active expression pattern is also directed by the promoter of *A. tumefaciens* T-DNA gene-5 during somatic embryogenesis in transformed carrot cells (Mattsson, Borkird and Engström 1992).

(iii) Genetic Analysis of Somatic Embryogenesis

The foundations of our present concepts on the utilization of genetic information by developing systems rest not only on the isolation and characterization of genes and the protein products involved in the specification of organs, but also on the analyses of sophisticated mutant phenotypes. As seen in previous chapters, some of the best insights into the gene expression programs during floral organogenesis, pattern formation in embryos, progressive embryogenesis, and endosperm development have been gleaned from the study of mutants. Despite the obvious advantages of a mutational approach to studying gene action during somatic embryogenesis, no true genetic mutants have been isolated from embryogenic cell suspension cultures. In a broader sense, this is due to the spectrum of genetic and epigenetic variations that one encounters in the currently used somatic embryogenesis systems that obscure interesting mutant phenotypes (Zimmerman 1993).

In an attempt to understand the genetic basis of somatic embryogenesis, some attention has been paid to variant cell lines and conditional mutants generated in the carrot system. In a previous section, reference was made to 5-methyltryptophan-resistant cells of carrot, which present a striking contrast to the wild-type cells in their inability to regenerate embryoids in the induction medium. How can this effect on embryogenesis be explained by biochemical alterations modulated by a genetic mutation? This happens simply because in the variant cell line, the enzyme anthranilate synthetase, which catalyzes the synthesis of tryptophan, is resistant to feedback inhibition by the end product, tryptophan. Consequently, there is an accumulation of tryptophan, in turn leading to an increase in the endogenous level of IAA in the cells to the point at which embryogenesis is impaired (Sung 1979). It seems quite likely that similar mechanisms may exist in other auxin-resistant, variant cell lines that are impaired in embryogenesis; however, direct evidence is lacking (Filippini et al 1992).

The extent of alterations in cellular properties during somatic embryogenesis as a function of gene regulation is best revealed by the responses to cycloheximide of wild-type cells and cycloheximide-resistant cells of carrot. In the former cell line, callus proliferation but not somatic embryogenesis is sensitive to the drug, whereas in the variant cell line both functions are insensitive to the drug. A central question is whether there are fundamental differences in the cycloheximide insensitivity of the two cell lines or whether a common mechanism operates differently in the two lines. It has been shown that cycloheximide resistance in the embryoids of both cell lines and in the callus of the variant line is caused by internal cellular mechanisms that inactivate the drug (Sung, Lazar and Dudits 1981). This observation falls in line with the notion that the general form of organization of the gene encoding for cycloheximide resistance is the same, but its expression is regulated differently in the two cell types. In α-amanitin-resistant carrot cell lines that inactivate the drug during both the proliferative and the embryogenic phases of growth, the lesion appears to be due to changes in the properties of RNA polymerase II (Vergara et al 1982; Pitto, Lo Schiavo and Terzi 1985).

Another mutational approach for the genetic dissection of somatic embryogenesis is the isolation of conditional mutants that are unable to produce embryoids at high temperature but that remain wild type at low temperature (Breton and Sung 1982; Giuliano, Lo Schiavo and Terzi 1984; Schnall, Cooke and Cress 1988). The growth of carrot cell suspension culture at 32 °C has allowed the retrieval of three primary classes of cells, which are impaired at different stages of embryogenesis (Figure 16.13). They are distinguished when nurtured in the embryogenic medium at the restrictive temperature by their phenotypes, such as continued growth as callus without producing embryoids and growth blocked at the globular stage of embryogenesis (Breton and Sung 1982). A systematic analysis of the variant cell line blocked at the

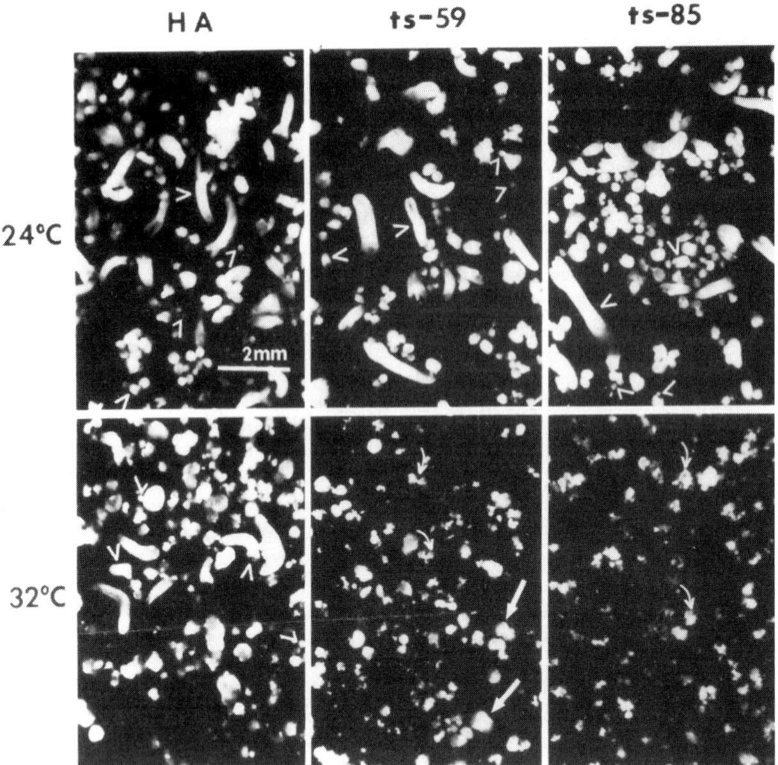

Figure 16.13. Temperature-sensitive mutants of carrot impaired at different stages of embryogenesis during growth in media lacking 2,4-D. Parental line HA produces embryoids at both temperatures when transferred to a medium lacking 2,4-D, although later stages of embryoids are slightly affected at the higher temperature. Phenotypes of *ts-59* and *ts-85* are similar to the wild type at 24 °C in media without 2,4-D. At 34 °C, *ts-59* embryoids are arrested at the globular stage whereas *ts-85* cells continue to grow as callus in media without 2,4-D. >, embryoids at globular and heart-shaped stages; >, embryoids at torpedo-shaped stage; →, undifferentiated callus; →, enlarged globular embryoids. (From Breton and Sung 1982.)

globular stage showed that the developmental block can be imposed by exposure to the high temperature either at the globular stage or at the globular to heart-shaped transition stage. The impact of this mutation is seen in the HSP profiles of the wild-type and mutant cells because the latter apparently fail to phosphorylate some HSPs in the 36–45-kDa range. Whether the differential modification of HSPs is timed to coincide with the arrest in development of the embryoids is not known (Lo Schiavo, Giuliano and Sung 1988). It is, however, doubtful whether genetic changes in the temperature-sensitive mutants can be discerned by protein analysis. Schnall et al (1991) showed that differences in the protein profiles of two conditional mutants generally reflect cell line differences, age differences, or random variations, rather than variations due to genetic lesions.

6. GENERAL COMMENTS

An important concept that emerged in the wake of the discovery of somatic embryogenesis is the totipotency of plant cells. Totipotency invokes the ingrained genetic information in the cell as an essential element in the cell's ability to recapitulate the stages of embryogenesis in a different environment. If one compares the origin and position of a seed embryo and an embryoid, there is no commonality between them. Moreover, as the progenitor of the zygote and embryo, the egg contains specialized proteins that determine its functional competence after fertilization, whereas the somatic cell is almost at the end of its life as a differentiated cell. Yet, the final forms of embryos and embryoids are identical, suggesting that attainment of form in terms of the hierarchy of decisions is similar in both systems. Thus, the implications of the genomic information in the nucleus and its mediation through the environment of the cell in somatic embryogenesis are immense, but have not been properly appreciated.

Carrot cell suspension culture continues to retain preeminence as the workhorse in much of the research into the physiology and molecular biology of somatic embryogenesis. That gene expression for embryogenesis is initiated even when the cells are undergoing proliferative growth instead of embryogenic growth is a particularly valuable insight into the mode of function of the genes in the carrot system. The expression of additional genes, which specify the function of the embryogenic cells, appears to proceed when the cells are induced in the embryogenic pathway. Also expressed are genes for the array of secreted proteins that have become recognized as a common feature of several somatic embryogenesis systems.

On the applied front, the virtually unlimited production of embryoids in some of our economically important crop species has stimulated interest in plant propagation using automated systems to generate somatic embryos by cloning. Considerable progress has also been made in the production of synthetic seeds and desiccation-tolerant somatic embryos that can undergo dehydration without loss of viability. In recent years the technology of somatic embryogenesis from protoplasts is finding an increased application for the production of transgenic plants by the introduction of foreign DNA. Although numerous refinements await to be made in the manipulation of cell and protoplast culture systems to suit particular crops and in the encapsulation systems to produce mature, quiescent somatic embryos that functionally mimic seeds, new information is also needed to understand the basis of somatic embryogenesis in a large number of plants.

Chapter 17

Embryogenic Development of Pollen Grains

In the reproductive biology of angiosperms, the microspore and pollen grain represent paradigms for studies on cell differentiation. As described in Chapter 4, an important reason for this is that by a program of gene expression regulated in space and time, the unicellular microspore matures into the pollen grain and embarks on a pathway leading to terminal differentiation by the production of two cells with divergent developmental potential. A second reason is that the induction of embryogenic divisions in microspores or pollen grains of certain plants by tissue culture methods makes it possible to gain insight into mechanisms involved in the deflection of a typical gametophytic program into an atypical sporophytic pathway. This chapter will focus on this alternative developmental pathway of microspores and pollen grains, which gives rise to embryoids and plantlets with the haploid or gametic number of chromosomes. This phenomenon is known as androgenesis, haploid embryogenesis, or pollen embryogenesis, but the last-mentioned term is preferred for use in this book. Irrespective of how the morphological and cytological changes converge to cause gametophytic or sporophytic types of growth in pollen grains, it seems certain that the same genetic blueprint is utilized to meet the informational demands of these transformation episodes. However, as will be shown in the following account, the biochemical and molecular basis for the embryogenic growth of pollen grains is much less understood than that for gametophytic development.

The topic as developed here is based on the work done on representative species, and no attempt is made to cover the numerous reports of the induction of sporophytic divisions in pollen grains of diverse species or the manipulations of tissue culture systems required to induce such divisions or improve the efficiency of pollen embryogenesis. This chapter will evaluate the subcellular, biochemical, and molecular changes that are triggered as pollen grains enter the sporophytic program and also the intracellular and extracellular control mechanisms that modulate embryogenic induction. For comprehensive reviews of the tissue culture aspects of pollen embryogenesis, the reader is referred to Bajaj (1983, 1990); although these reviews give access to the names of genera and species in which pollen embryogenesis has been demonstrated, it should be noted that the list has grown substantially since these reviews were written. Condensed treatments of some segments of the topic covered here can be found in several other surveys (Raghavan 1986a; Sangwan and Sangwan-Norreel 1987b; Vicente, Benito Moreno and

Heberle-Bors 1991; Ferrie, Palmer and Keller 1995).

Since this chapter follows immediately the one on somatic embryogenesis, it is reasonable to hope that embryogenic development of somatic cells may help to illuminate the development of pollen grains along a similar pathway. Compared to somatic embryogenesis, pollen grains employ a different strategy to become embryogenic, but the final products generated by dedifferentiated somatic cells and pollen grains are tantalizingly similar.

1. HISTORICAL BACKGROUND

The history of research on the transformation of pollen grains into embryoids began with a seminal discovery by Guha and Maheshwari (1964). These investigators found that when excised anthers of *Datura innoxia* at the pollen grain stage were cultured in a mineral-salt medium supplemented either with casein hydrolyzate, IAA, and kinetin or with coconut milk, grape juice, or plum juice, embryolike structures appeared from the sides of the anther in about six or seven weeks. Although it was initially suspected that the embryoids might have had their origin in the somatic tissues of the anther, persuasive ontogenetic evidence of their origin from the pollen grains, and hence of their haploid nature, was soon forthcoming (Guha and Maheshwari 1966). It was found that in cultured anthers, a variable but substantial number of pollen grains acquired a selective advantage over the rest of the population and began to divide repeatedly to form multicellular masses. Later, the exine gave way, releasing the cellular masses that organized into typical bipolar embryoids. Moreover, had the embryoids originated from the somatic tissues of the anther, they would be diploid, rather than haploid, as was the case.

A variation of the mode of pollen sporophytic development described in *D. innoxia* was reported in cultured anthers of *Oryza sativa*. Niizeki and Oono (1968) found that when anthers of rice were cultured at the mature pollen grain stage in a medium supplemented with IAA, 2,4-D, and kinetin, multicellular bodies appeared from within the anther in about four to eight weeks. These subsequently yielded dense calluses. Cytological examination of the callus confirmed its haploid nature and, hence, its origin from the pollen grain. Transfer of the callus to a medium containing IAA and kinetin elicited the regeneration of shoot and root systems and the formation of plantlets by organogenesis.

One problem with anther culture as a means of inducing the embryogenic development of pollen grains is that they are encased within the anther locule during the crucial period when the initial divisions in the embryogenic pathway occur. This makes it difficult to dissociate any influence of the anther wall and tapetum on embryogenic divisions from the more obvious effects of the components of the medium. For an analysis of the sporophytic development of pollen grains in the absence of any influence from the somatic tissues of the anther, the introduction of a method for inducing embryogenesis in isolated pollen grains proved to be a significant improvement of the anther culture system (Nitsch and Norreel 1973). In this method, pollen grains from the anthers of flower buds of *D. innoxia* pretreated at 3 °C for 48 hours are released into an isotonic medium and cleaned by filtration and centrifugation. Aliquots of the pollen suspension are cultured in a liquid medium conditioned by the addition of an extract of cultured anthers of *D. innoxia*. Under these conditions, embryogenic pollen grains in the aliquot divide and form structures resembling zygotic embryos.

Viewed from a developmental angle, these observations offer tangible evidence showing that differentiation of the angiosperm pollen grain is not accompanied by an irreversible change in the genome; rather, the same genome is reprogrammed to evoke totipotent development. The acquisition of stable dedifferentiative changes by animal nuclei irreversibly blocked from inducing morphogenesis was demonstrated by the nuclear transplantation experiments of King and Briggs (1953). In these studies, it was shown that nuclei removed from a lethal frog hybrid (*Rana pipiens* × *R. catesbeiana*) and transplanted into enucleate eggs of *R. pipiens* replicate in the alien cytoplasm and promote hybrid development. Like the lethally blocked frog nuclei, the block to morphogenesis in the terminally differentiating pollen grain is removed by the simple expedient of culturing anthers in a mineral-salt medium. The ability to produce haploid plants from pollen grains has made it possible to obtain stable inbred lines in large numbers in a single step for breeding purposes. Homozygous diploid lines are generated by treating the haploid plants with colchicine.

(i) A Survey of Pollen Sporophytic Development

During the last three decades, research in pollen embryogenesis has proceeded along the three major lines just described and has generated a long enough to single out individual cultivars. A trend toward species specificity is observed with regard to the percentage of embryogenic pollen grains in anthers of different species of the same genus. This observation indicates that although genotypic differences may not account for all known cases of failure to obtain pollen dedifferentiation in cultured anthers, this is a factor to be reckoned with in highly heterogeneous populations of plants.

(ii) Stage of Pollen Development

It is now accepted that the cytological stage of the pollen grain in the anther at the time of culture is critical for the induction of sporophytic divisions. Although this stage varies between species, the developmental window lies between the early unicellular and the bicellular stage of the pollen grain. In *Nicotiana tabacum*, anthers are generally responsive when they are cultured at periods beginning with the liberation of microspores from the tetrad and ending with bicellular pollen grains, but embryoids are readily formed in high numbers when anthers are cultured at the unicellular microspore stage or as the microspore begins to divide (Sunderland and Wicks 1971). However, with shed pollen cultures and isolated pollen cultures of tobacco, bicellular pollen grains yield embryoids with high frequency (Sunderland and Roberts 1977; Heberle-Bors and Reinert 1979). The cytologically responsive stage in *Beta vulgaris* extends from close to the end of meiosis to the early unicellular microspore (van Geyt, D'Halluin and Jacobs 1985), whereas the mid-unicellular stage is optimal for barley (Wheatley, Marsolais and Kasha 1986). When embryoid formation was quantified as a function of pollen developmental stage, the best results are obtained at the unicellular stage of pollen development in cultured anthers of *Hyoscyamus niger* (Raghavan 1978), *Peltophorum pterocarpum* (Fabaceae) (Rao and De 1987), *Brassica napus* (Kott, Polsoni and Beversdorf 1988), *Simocalamus latiflora* (bamboo; Poaceae) (Tsay, Yeh and Hsu 1990), and *Arachis hypogaea* (Willcox et al 1991), and rarely have anthers containing pollen grains at earlier or later stages of development been as bountiful. This raises the question of whether the embryogenic fate of pollen grains is determined by the stage of the mitotic cycle of the first haploid microspore mitosis. It has been claimed that in *B. napus*, embryoids are induced in isolated pollen grains cultured from G1 through G2 phases, but the development of the dividing cells was not followed far enough to relate this to embryogenesis (Kott, Polsoni and Beversdorf 1988; Pechan and Keller 1988; Binarova et al 1993). Ultrastructural studies combined with various cytochemical, immunochemical, and in situ hybridization approaches have shown that at the most favorable stage for embryogenic induction, the microspore nucleus in *Capsicum annuum* has features associated with active transcription (González-Melendi et al 1995). A challenging problem is to establish these relationships in a satisfactory way for pollen embryogenesis in a wide range of plants.

(iii) Medium Composition

According to our current understanding, anthers of a few species respond to embryogenic induction when they are cultured in a simple medium containing only macronutrient salts, micronutrients, vitamins, *myo*-inositol, and sucrose. Despite its widespread use as a carbon energy source in supporting plant regeneration in tissue cultures, sucrose has not been favored in anther cultures of barley, potato, and wheat, in which it has been replaced by maltose. The differences in the metabolism of sucrose and of maltose in microspore cultures of barley have been used to weigh the reasons for the superiority of maltose over sucrose. It appears that compared to the slow metabolism of maltose, sucrose is metabolized rapidly, leading to the accumulation of toxic amounts of ethanol within the microspores (Scott, Lyne and ap Rees 1995).

It is now known that anthers of the vast majority of plants studied require either an auxin or a cytokinin or a combination of both in the medium. There is strong evidence indicating the existence of wide variations in the kind and concentrations of both these groups of hormones that are necessary for inducing and sustaining the pollen sporophytic type of growth in cultured anthers of various species. It is of interest to note that when pollen grains of cultured anthers form calluses instead of embryoids, they do so in a medium enriched with auxins and cytokinins; subsequent regeneration of plants by organogenesis occurs when the callus is transferred to the basal medium, to a medium containing cytokinin with a low level of auxin, or to one containing a weak auxin. The hormones thus perform a morphogenetic function, since their absence from the medium does not provoke pollen dedifferentiation.

A new level of understanding of the role of auxins and cytokinins in inducing the embryogenic development of pollen grains in cultured anthers has been achieved in some cases in which these hormones have been found to modify the process. For example, although auxins such as IAA, NAA, and indolebutyric acid (IBA), and cytokinins such as zeatin, kinetin, and benzylaminopurine, are effective in enhancing the production of pollen embryoids in cultured anthers of *Datura innoxia*, the best response is obtained with zeatin and kinetin. This has led to the suggestion that embryogenic divisions are related to the maintenance of a certain level of endogenous hormones in the pollen grain (Sopory and Maheshwari 1976). Pollen grains in cultured anthers of *Hyoscyamus niger* respond to moderately high concentrations of 2,4-D in the medium by regenerating calluses rather than embryoids. In cultured anthers, compared with the basal medium, auxin does not affect the number of pollen grains dividing in the sporophytic pathway or the patterns of early division sequences involving the vegetative and generative cells. However, after release from the exine, undifferentiated growth prevails in the cellular mass and results in a small nodule of callus (Raghavan 1978). These results augment the general impression that 2,4-D does not interfere with the embryogenic induction of the pollen grain but affects the subsequent growth of embryoids. In contrast, the role of cytokinins such as kinetin, benzylaminopurine, zeatin, zeatin riboside, and 2iP is one of reducing pollen efficiency in cultured anthers of *H. niger*. During prolonged growth of anthers in media containing cytokinins, there is a tendency for the transformation of all or parts of the surviving embryoids into calluses (Raghavan and Nagmani 1989). The deleterious effects of cytokinins on pollen efficiency in cultured anthers of *H. niger* raise questions about the wisdom of their widespread and arbitrary use in anther culture media.

Several workers have suggested a role for ethylene in pollen embryogenesis in cultured anthers (Dunwell 1979; Babbar and Gupta 1986; Reynolds 1987; Biddington, Sutherland and Robinson 1988; Cho and Kasha 1989). In *Datura metel*, the ethylene precursor, methionine, and the ethylene-releasing compound, ethrel, were found to stimulate embryoid formation, whereas ethylene antagonists like Co^{++} and Ag^{+} decreased anther productivity in culture (Babbar and Gupta 1986). According to Reynolds (1984a, 1986), the presence of a moderate concentration of IAA in the medium leads to embryogenic development in pollen grains of cultured anthers of *Solanum carolinense*, whereas the addition of 2,4-D causes pollen callus formation. The effect of IAA was attributed to auxin-stimulated ethylene production in cultured anthers (Reynolds 1987). Unlike IAA, 2,4-D does not stimulate ethylene biosynthesis by cultured anthers, nor does ethylene cause callus growth in the absence of 2,4-D in the medium (Reynolds 1989). Since artificially increasing the concentration of intracellular free Ca^{2+} in cultured anthers by the ionophore A23187 enhances IAA action in inducing embryogenic divisions in pollen grains and in partially substituting for 2,4-D in promoting pollen callus growth, the involvement of Ca^{2+} in signal transduction during pollen sporophytic growth has been suggested (Reynolds 1990).

Among other additives to the culture medium, activated charcoal has beneficial effects on pollen embryogenesis that seem to be well established. The most convincing explanation of these effects stems from the assumption that charcoal might adsorb inhibitors that are released from the anther, or present in the agar, or produced by the degradation of other metabolites present in the medium. Phenolic substances appear to be prime candidates for the role of anther-derived inhibitors. In the anther culture of *Anemone canadensis* (Ranunculaceae), the concentration of phenolic substances is found to be high in the culture medium lacking activated charcoal. The addition of activated charcoal, which reduces the concentration of phenolics to less than 20% of the original value, also enhances embryogenesis in cultured anthers (Johansson 1983). It is likely that an increase in the frequency of pollen embryogenesis reported to be produced by polyvinylpolypyrrolidone in *Datura innoxia* anthers is attributable to its well-known action in adsorbing phenolic compounds (Tyagi, Rashid and Maheshwari 1981). Whether this compound, like activated carbon, can be used on a wider level to enhance anther productivity remains to be seen.

(iv) Thermal Stress

In several plants, temperature shocks are effective in inducing a parthenogenetic activation of the egg that leads to the production of haploid plants. Similarly, a cold treatment given to plants has been known to alter the plane of division of the pollen grains and to result in symmetrical cells. In the wake of the discovery of pollen embryogenesis, the administration of temperature shock has become a routine procedure to enhance the embryogenic response of pollen grains in cultured anthers. The first reports of favorable effects of low temperature on pollen embryogenesis were concerned with the induction of embryogenic divisions in the isolated plants play a key role in the evolution of embryogenic competence in pollen grains. Generally, plants raised under short days (8-hour photoperiod) and low temperature (18 °C) produce small anthers that have a high percentage of embryogenically competent pollen grains, which are referred to as premitotic pollen (P-grains); high yields of embryoids are obtained when these anthers, or pollen grains isolated from them, are cultured (Heberle-Bors and Reinert 1980, 1981). It turns out that P-grains in tobacco are functionally sterile pollen grains and that the frequency of their formation is in some way linked to the development of sexuality in flowers. It has been proposed that P-grains are produced under conditions that induce high pollen sterility and that when the sex balance of the flower is tipped toward femaleness there is also a tendency for large-scale pollen sterility (Heberle-Bors 1982a, b, 1983). Although not widely investigated, genetic factors are also perhaps responsible for P-grain formation in anthers. The primary reason for the presence of high frequencies of P-grains in anthers of different genotypes of wheat has been traced to certain cms lines with different cytoplasmic and nuclear backgrounds (Heberle-Bors and Odenbach 1985).

All this is impressive in support of the concept of embryogenic predetermination of pollen grains in anthers. In the light of this hypothesis, reports of the spontaneous formation of multicellular pollen grains in anthers of hybrids of *Solanum* (Ramanna 1974; Ramanna and Hermsen 1974), *Narcissus biflorus* (Koul and Karihaloo 1977), and *Paeonia* (Li 1982) and the induction of multiple pollen grains by treating plants with chemicals, such as ethrel in wheat (Bennett and Hughes 1972), can be considered in a new light. These represent cases in which embryogenically competent pollen grains formed in the anther have apparently escaped from the intracellular control system and express incipient sporophytic features. On the whole, evidence for pollen predetermination is scanty in the large majority of plants that yield embryoids or calluses from cultured anthers with relative ease, and so it cannot be said that the concept of embryogenic predetermination of pollen grains has been proved in a rigorous sense.

3. CYTOLOGY OF POLLEN EMBRYOGENESIS

In most species studied, the vulnerable stage in pollen development – the one at which the gametophytic program is deflected toward the sporophytic pathway – is the unicellular stage. The first division of the pollen grain in culture is cytologically and morphologically very similar to the division that occurs in vivo. However, from this point onwards, there are striking differences in the manner in which the daughter cells participate in embryoid or callus formation. As one might expect, several ultrastructural changes suggestive of the replacement of the gametophytic program with a sporophytic imprint should also be occurring in the microspores during this period. Several key studies that are largely responsible for our current understanding of the cellular and subcellular cytology of pollen embryogenesis are reviewed in this section.

(i) Pollen Division Pathways

Based on the assumption that a vegetative cell and a generative cell are born out of the first haploid mitosis, current theory considers three general routes to the multicellularity of bicellular pollen grains to be important: (a) repeated divisions of the vegetative cell; (b) repeated divisions of the generative cell; and (c) divisions of both vegetative and generative cells.

Division of the Vegetative Cell. The essential cytological details of the origin of embryoids from the vegetative cell were first traced in a cultivar of *Nicotiana tabacum*. By about the sixth day after culture of the anther, the vegetative cell loses its morphogenetic individuality and is partitioned by a series of internal walls until a mass of cells typical of somatic cell size is produced. Unable to contain the burgeoning growth within, the exine breaks open, liberating the cellular mass into the anther locule where it passes through the typical stages of globular, heart-shaped, torpedo-shaped, cotyledon initiation, and mature embryoids. The culmination of this developmental feat is the germination of the embryoid and its appearance outside the anther wall as a plantlet. A restrictive feature of this pathway is that the generative cell either remains undivided or undergoes but a few divisions and, without contributing to the formation of an embryoid, these cells simply wilt away (Sunderland and Wicks 1971). With minor differences in detail, embryogenesis by the division of the vegetative cell has been described in other cultivars of tobacco (Bernard 1971; Mii 1973), *Datura metel* (Iyer and Raina 1972), *D. innoxia* (Sunderland, Collins and Dunwell 1974), *Capsicum annuum* (George and Narayanaswamy 1973), *Solanum surattense, Luffa cylindrica* (Sinha, Jha and Roy 1978), *Brassica napus*

(Fan, Armstrong and Keller 1988), and wheat (Reynolds 1993).

The reality of the vegetative cell giving rise to a callus was first described in anther cultures of a cultivar of barley, in which the initial dispositions of the vegetative and generative cells are the same as in tobacco. The cellular mass released by the breakage of the exine continues unorganized growth as a callus rather than differentiates into an embryoid (Clapham 1971). Investigations that followed this work furnished additional examples of pollen callus formation from the vegetative cell in other cultivars of barley (Sun 1978; Wilson, Mix and Foroughi-Wehr 1978; Zhou and Yang 1980), as well as in cultured anthers of wheat (Wang et al 1973; Pan, Gao and Ban 1983; Szakács and Barnabás 1988), rice (Chen 1977; Yang and Zhou 1979), maize (Miao et al 1978), and rye (Sun 1978).

Extremes of behavior of the vegetative cell during its transformation into a callus have been noted in some cases. One is the appearance of micronuclei in the division products of this cell in pollen grains of wheat (Zhu, Sun and Wang 1978). In a cultivar of barley, nuclear divisions in the vegetative cell are not accompanied by cytokinesis and lead to the formation of multinucleate pollen grains (Zhou and Yang 1980). In another cultivar, fusion between two or three nuclei occurs during the intermediate free nuclear stage of the vegetative cell, or sometimes the nucleus of the generative cell is incorporated into a derivative of the vegetative cell by fusion (Sunderland and Evans 1980). These examples draw attention to the extent of cytological abnormalities in the division of the vegetative cell that may contribute to the formation of a mixoploid callus from a single pollen grain.

Pollen sporophytic growth in cultured anthers by the activity of a functional vegetative cell is designated as the A pathway. From a detailed analysis of pollen embryogenesis in cultured anthers of tobacco, a further subdivision of this pathway into A-1, A-2, and A-3 has been proposed. The primary considerations for this are the different fates of the generative cell in the final products formed and the occurrence of an intermediate free nuclear or cellular phase during the division of the vegetative cell (Misoo, Yoshida and Mastubayashi 1979). Unfortunately, these variations have been little investigated in other plants and thus we do not know for sure whether they are widely applicable.

Divisions of the Generative Cell or of Both Generative and Vegetative Cells. The role of the generative cell in pollen sporophytic growth was slow to be recognized, but it appears that this cell is capable of independent divisions as an embryoid or a callus mother cell. The first hint about the developmental significance of the generative cell in pollen embryogenesis came from observations on the division patterns of pollen grains in cultured anthers of *Nicotiana tabacum* (Bernard 1971; Sunderland and Wicks 1971; Nitsch 1972; Rashid and Street 1974), *N. sylvestris* (Rashid and Street 1974), and *D. metel* (Iyer and Raina 1972) showing that as the embryoid is formed from the vegetative cell, the generative cell divides at a slow pace to produce a small population of free nuclei or cells. Conclusive evidence for the formation of embryoids by the division of the generative cell came from the work on cultured anthers and anther segments of *Hyoscyamus niger*. Periodic cytological examination revealed the occurrence of embryogenic divisions delimiting the generative and vegetative cells as early as 12 hours after culture. Further divisions are confined to the generative cell, which initially forms a group of cells within the exine. Following its release from the exine, the multicellular mass faithfully recapitulates the stages of zygotic embryogenesis to form embryoids and plantlets. Various fates can befall the vegetative cell, one of which is to form a two- to many-celled suspensor-like structure subtending the organogenetic part of the embryoid formed by the division of the generative cell (Raghavan 1976a, 1978). The concept of an active role for the generative cell in the presence of a passive or less active vegetative cell in pollen sporophytic growth has received support from anther culture investigations in maize (Guo et al 1978; Miao et al 1978), barley (Sunderland et al 1979), wheat (Pan, Gao and Ban 1983), and rice (Qu and Chen 1984). In cultured anthers of *H. niger* (Raghavan 1978), *N. tabacum* (Anand, Arekal and Swamy 1980), cereals (Bouharmont 1977; Miao et al 1978; Sun 1978; Sunderland et al 1979), and *Cocos nucifera* (Monfort 1985), it is also not unusual to find embryogenesis by the repeated divisions of both generative and vegetative cells. The resulting structures, being amalgams of unlike parts, are chimeras at the cellular level. Although nuclear divisions are invariably followed by cytokinesis, the vegetative cell or generative cell may occasionally produce several free nuclei, which are incorporated into the embryoid or callus (Sun 1978; Sunderland et al 1979). Pollen sporophytic growth by the exclusive division of the generative cell or by the division of both generative and vegetative cells is included in the E pathway.

A unique type of pollen embryogenesis involving both the generative and vegetative cells or their nuclei occurs in cultured anthers of *Datura innoxia*. What we find here are not independent divisions of these cells in a strict sense, but their fusion to form an abnormal cell; both nuclei then divide simultaneously on a common spindle. The fusion might occur between one or two haploid vegetative cell nuclei and a haploid or endopolyploid generative cell nucleus. Fusions of this kind allow the relationship between parental genomes to change, so that nuclei with nonhaploid chromosome numbers are formed (Sunderland, Collins and Dunwell 1974). This pathway of embryogenesis, known as the C pathway, may have widespread significance in accounting for the frequent occurrence of embryoids with nonhaploid chromosome numbers.

Symmetrical Division of Pollen Grains. In some culture systems, a symmetrical division of the pollen grain to produce two identical cells or nuclei is the cue that sets the division products off into the differentiative pathway, known as the B pathway. As described in cultured anthers of *Atropa belladonna* (Solanaceae) (Narayanaswamy and George 1972; Rashid and Street 1973), wheat (Ouyang et al 1973; Pan, Gao and Ban 1983; Reynolds 1993), rice (Guha-Mukherjee 1973), *Datura innoxia* (Sunderland, Collins and Dunwell 1974), *Nicotiana sylvestris, N. tabacum* (Rashid and Street 1974), *Hordeum distichum, H. vulgare* (Matsubayashi and Kuranuki 1975; Zhou and Yang 1980), *Hyoscyamus niger* (Raghavan 1978), *Solanum surattense, Luffa cylindrica,* and *L. echinata* (Sinha, Jha and Roy 1978), the two cells divide repeatedly to form the embryoid; the contributions of neither cell can be considered more important than those of the other. It should be added here that it has not been determined whether there are any cytochemical or histochemical differences between the two similar-looking cells or nuclei in the examples cited. Just because two cells look alike does not mean that they are not different at another level of organization. In cultured anthers of *Triticale* (Sun, Wang and Chu 1974) and *H. vulgare* (Wilson, Mix and Foroughi-Wehr 1978), despite the close similarity of the two nuclei formed, cytochemical and physiological observations seem to indicate that multicellular pollen grain is generated by the division of the nucleus, which appears to be similar to the nucleus of the vegetative cell of an asymmetrically partitioned pollen grain. This makes pollen grains with symmetrical cells prime suspects for housing disguised generative and vegetative cells. In cultured anthers of wheat, the symmetrical division of microspores may underlie their potential to develop into embryoids, whereas asymmetrically dividing microspores almost always regenerate calluses from the vegetative cell (Szakács and Barnabás 1988, 1995). Since colchicine treatment has been shown to lead to an increase in the number of symmetrically dividing embryogenic pollen grains in anther and pollen cultures of *Brassica napus*, it has been suggested that cytokinetic symmetry is an important factor in deflecting the gametophytic program of the pollen grain toward the embryogenic pathway (Zaki and Dickinson 1991, 1995).

Another scenario, which is not yet fully documented, appears to be a variation of the B pathway and has been described in cultured anthers of wheat. Here, the two identical nuclei divide repeatedly to generate a cluster of free nuclei, but it is not certain whether cells are forged out of these nuclei and whether the multicellular mass so produced proceeds to form an embryoid or a callus (Zhu, Sun and Wang 1978; Pan, Gao and Ban 1983). This is known as the D pathway.

The disposition of cells and nuclei in the different pathways of pollen embryogenesis described here is illustrated in Figure 17.3. In cultured anthers, pollen sporophytic units originate by multiple pathways. The relative abundance of the units formed by a specific division sequence varies not only in anthers of different flower buds of the same plant but also in anthers of the same flower bud. What prompts pollen grains to choose a particular division sequence is not understood.

(ii) Ultrastructural Cytology

The ultrastructure of pollen embryogenesis has been studied in detail in only four species (*Nicotiana tabacum, Datura innoxia, Hyoscyamus niger,* and *Brassica napus*); in consequence, most of what can be said is based on these investigations. Despite this limitation, the starting premise is that there are no cellular changes more fundamental to pollen dedifferentiation than those that can be discerned in the electron microscope. An important line of enquiry is the fate of the microspore cytoplasm programmed for gametophytic differentiation. The distinctive ultrastructural features of the vegetative and generative cells as they are incorporated into an embryoid or a callus are also of interest.

Nicotiana tabacum. Few plants have rivaled *N. tabacum* as a target for ultrastructural investigations of pollen embryogenesis. However, for all of the virtues of tobacco as a good experimental system,

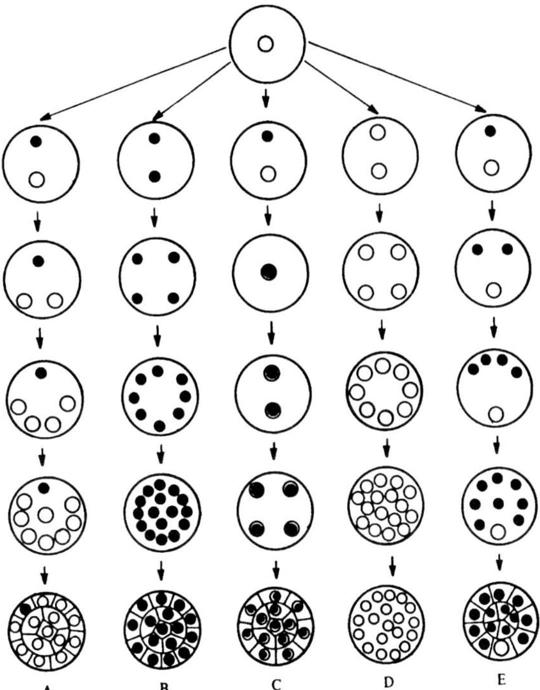

Figure 17.3. Diagrammatic representation of the different pathways of pollen embryogenesis. In A, C, and E, solid circles represent nuclei of the generative cell or its division products; hollow circles represent nuclei of the vegetative cell or its division products. In C, solid circles enclosed in hollow circles indicate fusions between the two nuclei. In B and D, circles represent symmetrical nuclei born out of the first haploid mitosis and their division products. (From Raghavan 1986c.)

contradictory observations exist, depending upon whether ultrastructural changes are followed in pollen grains of cultured anthers or in *ab initio* pollen cultures. In the sequence of changes described, no obvious features associated with embryogenesis are seen in pollen grains during the first five days of culture of anthers; rather, during this period, the gametophytic imprint appears more and more prominent in the vegetative and generative cells. This period in culture may well be critical in establishing that gametophytic differentiation proceeds to the maximum permissible extent before the sporophytic program takes over. A major structural upheaval associated with embryogenesis is seen seven to eight days after culture as a regression of the cytoplasmic organization of the vegetative cell resulting in the elimination of much of the gametophytic program. During this period, the organelles appear in complete disarray and in various states of disintegration; this is coupled with the disappearance of ribosomes and the appearance of zones of multivesiculate bodies analogous to lysozomes (Vazart 1971; Dunwell and Sunderland 1974a, b). Cytoplasmic regression stimulates a series of alterations in the structure, function, and developmental program of the vegetative cell. The earliest change seen in this cell is its division in the embryogenic pathway. As the newly formed cells accumulate their own cytoplasm, it becomes populated by a fresh array of organelles, including ribosomes arranged in polysomal profiles, lipid centers, and starch-containing plastids (Dunwell and Sunderland 1975). More than any other single event, the sequential disappearance and appearance of the cytoplasmic organelles has led to the view that embryogenic pollen is incarnated from the gametophytic pollen by a programmed destruction of the cytoplasm of the vegetative cell, which thereby depletes this cell of a specific phenotype, followed by the repopulation of the cell by a new set of organelles. Because the generative cell eventually degenerates, its ultrastructure has not been followed in much detail.

Other studies have shown that subcellular changes during the embryogenic development of tobacco pollen grains must be viewed in the context of the cold stress accorded to flower buds to increase pollen efficiency. In *ab initio* pollen cultures originating from anthers of cold-stressed flower buds, embryogenic divisions begin with the laying down of a convoluted fibrillar wall between the plasma membrane and the inner layer of the intine, thus insulating the entire pollen cytoplasm. Embryogenic pollen grains do not display ultrastructural features that could be construed as organelle regression; rather, they appear to possess the hallmarks of metabolically repressed cells, such as an attenuated cytoplasm, condensed mitochondria, and sparsity of ribosomes. As the vegetative cell divides in the embryogenic pathway, the newly formed cells regain the ultrastructural profile of metabolically active cells; this is evidenced by an increase in the ribosome population in the cytoplasm sufficient to crowd out other organelles and by a change in the mitochondrial morphology (Rashid, Siddiqui and Reinert 1981, 1982). The presence of an increased number of ribosomes and complex mitochondria is also characteristically seen during pollen sporophytic development in *Aesculus hippocastanum* (Radojević, Zylberberg and Kovoor 1980) and *Hordeum vulgare* (Idzikowska 1981). These observations support the notion that the metabolic requirements of the pollen grains during the initiation of embryogenic divisions are very demanding.

One plausible explanation for the conflicting reports on the ultrastructural changes during pollen embryogenesis in tobacco is that the reports do not take into account the effect of low-temperature stress on the donor plants in preventing organelle regression or in causing other ultrastructural anomalies in the cytoplasm of cultured pollen grains. A previous section referred to the changes in the vacuole in pollen grains of cold-stressed anthers of *Datura metel* (Camefort and Sangwan 1979; Sangwan and Camefort 1984). Similar changes that might obscure the regression of organelles due to embryogenic induction cannot be ruled out in tobacco pollen grains from cold-stressed plants. In this context, attention may be drawn to the fact that signs of regression, such as chromatin condensation, a decrease in nucleolar size, and the loss of nuclear pores, are frequent in the vegetative cell of isolated pollen grains of tobacco subjected to a starvation diet as a prelude to embryogenesis (Garrido et al 1995). A comparative analysis of the ultrastructural changes associated with embryogenesis in *ab initio* cultures of tobacco pollen grains collected from plants grown under standard greenhouse conditions and from plants subjected to temperature stress will go a long way toward resolving this controversy.

Datura innoxia. In contrast to the results from tobacco, there is no evidence for a breakdown of the gametophytic cytoplasm during the ultrastructural displacement that inevitably accompanies pollen embryogenesis in cultured anthers of *D. innoxia*. Prior to, or at the time of, culture of anthers, the enclosed pollen grains display an undistinguished ultrastructural profile characterized by the presence of structurally simple mitochondria and proplastids distributed in the sparse cytoplasm surrounding a large central vacuole. Although this profile remains essentially unchanged in the unicellular embryogenic pollen grains of cultured anthers, both vegetative and generative cells are enriched with ribosomes, plastids, and mitochondria following the first haploid mitosis. Changes in proplastids, such as their transformation into amyloplasts, and the appearance of numerous cristae in the mitochondria have been identified as being specially indicative of embryogenic induction of pollen grains. It is this basic subcellular organization of the vegetative cell that is partitioned repeatedly as embryogenic divisions are initiated (Dunwell and Sunderland 1976; Sangwan-Norreel 1978; Sangwan and Sangwan-Norreel 1987a, b). In *D. metel*, curious organelles resembling groups of polysomes and rough ER surface as early as the 2–4-celled stage of the embryoid, but they disappear at the globular or heart-shaped stage. Although its relevance is not determined, the tonoplast of embryogenic pollen grains of *D. metel* possesses a tanninlike deposit that is absent in normal pollen grains and persists up to the stage of the globular embryoid (Sangwan and Camefort 1982, 1983). This change in the embryogenic pollen grain is of interest as a possible marker for the induction process.

Hyoscyamus niger. By combining ultrastructural observations with stereological analysis, Reynolds (1984b) has shown that the early features that distinguish the embryogenic pollen grain from the normal pollen are an increased volume of the granular zone in the nucleolus and an increased amount of dispersed chromatin in the nucleus. These changes, visible in the unicellular pollen grains as early as six hours after culture of the anther, are indicative of enhanced nuclear and nucleolar activities. Following the first haploid mitosis, the nuclear and nucleolar changes are perpetuated in the generative cell, which functions as the embryo mother cell. Some differences are also noted between the volume–area fractions of the cytoplasm occupied by mitochondria, plastids, rough ER, and Golgi cisternae in the generative cells of embryogenic pollen grains and those in the nonembryogenic pollen grains. Since these changes occur in the absence of any noticeable regression of the cytoplasm in the generative cell, it has been emphasized that metabolic quiescence is not associated with embryogenic induction; rather, it appears that ultrastructurally there is a gradual assimilation of the gametophytic program into one that ensures a sporophytic mode of development. Another study by Reynolds (1985) has identified some pollen grains that divide symmetrically to yield two apparently identical nuclei, both of which contribute to the formation of the embryoid; in these pollen grains there is an abrupt cessation of gametophytic features before embryogenic division is initiated. Comparing the subcellular cytology of pollen grains that divide symmetrically and asymmetrically, a case can be made that the specific pathways of pollen embryogenesis in cultured anthers of *H. niger* are correlated with the timing and degree of repression of gene activity for gametophytic development.

Other Plants. As one considers the ultrastructure of pollen embryogenesis in other plants, no common thread of events or sequence of changes appears to emerge. In cultured anthers of *Brassica napus*, the major changes that augur the division of the embryogenic pollen grain to form two identical cells are the movement of the nucleus from the peripheral to the central position, starch assembly by plastids, the formation of a thick fibrillar wall beneath the intine, an increase in ribosome number, and the appearance of large cytoplasmic aggregates of an unidentified globular material. Speculatively, the cytoplasmic granules have been considered as heat-shock granules that probably shield mRNAs from damage (Zaki and Dickinson 1990; Telmer, Newcomb and Simmonds 1993, 1995; Testillano et al 1995). The cytoskeletal elements of the pollen grain do not seem to have a primary function in embryogenic induction, because the microtubules and microfilaments show the expected pattern of distribution that coincides with the position of the nuclei during embryogenic divisions (Hause et al 1993). In a continuation of the trends established in the daughter cells born out of the first haploid mitosis, organelles are distributed uniformly in all cells of the globular embryoid enclosed within the exine. Following the rupture of the exine, there is an asymmetric distribution of starch granules in the cells of the embryoid that is the first sign of polarity. Granules generally disappear from cells at the site of the exine break but they persist at the opposite end, which eventually becomes the root apex (Hause et al 1994).

The main finding of light and electron microscopic studies of cultured anthers of barley is that embryogenic pollen grains display a proliferation of cytoplasmic organelles, followed by division to form the vegetative and generative cells or two symmetrical cells (Chen, Kasha and Marsolais 1984; Chen et al 1984). In barley anthers subjected to a cold stress, prior to the first haploid pollen mitosis, the cytoplasm of embryogenic pollen grains is depleted of storage materials and organelles, which for the most part display little differentiation (Huang 1986). Comparison of the subcellular cytology of pollen grains from normal and cold-stressed anthers shows that the pollen cytoplasm is not stable enough to withstand the stress. A highly differentiated cytoplasm in the vegetative cell sets the stage for embryogenic divisions in pollen grains of cultured anthers of *Zea mays* (Barnabás, Fransz and Schel 1987).

The formation of new cells in a rapid burst of mitotic activity, and the subsequent burgeoning growth, begin to stretch the exine and, eventually, only remnants of sporopollenin are seen on the pollen wall. The frequent occurrence of binucleate and multinucleate pollen grains in cultured anthers suggests that nuclear divisions during pollen embryogenesis are independent of cytokinesis. In such pollen grains, cell walls are formed subsequently by centripetal ingrowth from the intine or from the plasma membrane (Reynolds 1985; Huang 1986; Barnabás, Fransz and Schel 1987). Phragmoplasts with attached vesicles and attached spindle microtubules and with plasmodesmata-rich walls have been noted in electron microscopic examination of pollen embryoids, indicating that the newly formed cells are partitioned by walls formed by conventional cell plates (Huang 1986; Barnabás, Fransz and Schel 1987; Zaki and Dickinson 1990).

It is clear from the account presented in this section that ultrastructural analysis has not identified a subcellular marker of the sporophytic transformation of the pollen grain. Perhaps it is time now to take a close look at the first haploid mitosis that sets the stage for embryogenic division of the pollen grain. For example, are there changes in the positioning of the cytoskeletal elements of the pollen grain during normal gametophytic development and during induced embryogenesis? Is there a differential distribution of cytoskeletal proteins in the pollen grain during its developmental transformations? These questions are worth exploring if only to highlight what we need to know about the early cytological signals of embryogenic induction in the pollen grains.

4. MOLECULAR BIOLOGY OF POLLEN DEDIFFERENTIATION

The fundamental premise of the molecular biology of pollen embryogenesis is that pollen grains programmed for terminal differentiation into gametes are diverted to a pathway of continued divisions and increasingly complex morphology of an embryolike structure. From what we know about the basis for cell differentiation, it is an ingrained idea that these changes depend on the input of newly synthesized mRNAs and proteins. Unfortunately, little information is available on the nucleic acid and protein metabolism of the vegetative and generative cells of embryogenic pollen grains, and investigations that take into account the biochemical basis for the altered developmental expression seem to be a major lacuna in this topic. This is attributable in part to the small number of pollen grains in a given anther that become embryogenic and in part to the different pathways of pollen embryogenesis observed in one and the same anther in culture. It was therefore natural in the early stages to rely on cytochemical and autoradiographic data to evaluate the macromolecular changes associated with pollen sporophytic development. In recent years, a beginning has been made in the use of *ab initio* pollen cultures to follow the pattern of gene expression during the embryogenic transformation of microspores. For now, attention will be directed to the pattern of nucleic acid and protein synthesis in the pollen grains that heralds embryogenic development.

(i) Dynamics of DNA, RNA, and Protein Metabolism

At the heart of the division of the vegetative cell or the generative cell in the embryogenic pathway is the renewed DNA replication in these cells. Microspectrophotometric measurements of the DNA content of embryogenic pollen grains in cultured anthers of tobacco have shown that DNA doubling occurs in the vegetative cell before any overt signs of dedifferentiation; in contrast, the vegetative cell of the normal gametophytic pollen grain retains the 1C DNA amount. The culture of isolated tobacco pollen grains in a starvation medium before transfer to an enriched medium for embryogenesis affects neither renewed DNA synthesis in the vegetative cell nor ongoing DNA synthesis in the generative cell (Chu, Liu and Du 1982; Žárský et al 1992). In similar studies with cultured anthers of *Datura innoxia*, variations are noted in the DNA content of the nuclei of embryogenic pollen grains. Generally, unicellular embryogenic pollen grains have 1C and 2C DNA amounts, whereas each of the two identical nuclei in a bicellular pollen grain has 1C DNA. A modal 1C DNA is also measured in the nuclei of vegetative and generative cells. The generation from cultured anthers of *D. innoxia* of plants with differing ploidy levels has led to the suggestion that fusion between nuclei might account for this phenomenon (Sangwan-Norreel 1979, 1981, 1983). Somewhat similar conclusions were reached from observations of the changes in the DNA contents of embryogenic pollen grains of wheat and *Petunia* sp. (Raquin et al 1982).

The cytological view of embryogenesis by the division of the generative cell in cultured anthers of *Hyoscyamus niger* has been confirmed by autoradiography of ^{3}H-thymidine incorporation. Failure of DNA synthesis or cytokinesis in the vegetative cell and its division products is typical of embryoids formed by the exclusive division of the generative cell. When the embryoid is derived from both generative and vegetative cells, silver grain density due to label incorporation is relatively low in the derivatives of the vegetative cell in comparison to that found in the generative cell derivatives (Raghavan 1977a). Overall, the patterns of DNA replication in the vegetative and generative cells provide useful models to establish the respective roles of these cells as embryoid mother cells.

RNA Metabolism. An early step in the embryogenic development of pollen grains is the transcription of DNA to form RNA. This might be linked to the initiation of the sporophytic program or might be directed to the elimination of the existing gametophytic program. In cultured anthers of tobacco, depending upon the temperature of incubation, there is a lag period of 6–12 days before the vegetative cell divides in the embryogenic pathway. The stainable RNA content of the pollen grains stabilizes during this period to serve as a good marker for distinguishing between gametophytic and embryogenic pollen grains. Generally, the former display a four- to sixfold increase in RNA content, whereas the latter have a low level of RNA. It appears that suppression of the gametophytic program, as reflected in the low RNA level, is necessary to ensure that the genes for embryogenic induction are fully expressed without being masked by the simultaneous expression of genes for pollen maturation and germination (Bhojwani, Dunwell and Sunderland 1973). Cytochemical study of pollen embryogenesis in *Datura innoxia* has shown that compared to nonembryogenic pollen grains, embryogenic pollen grains exhibit an increased stainability for cytoplasmic RNA prior to the first haploid mitosis (Sangwan-Norreel 1978).

Results from autoradiography of the incorporation of ^{3}H-uridine show that embryogenically determined pollen grains of *Hyoscyamus niger* begin to synthesize RNA as early as the first hour of culture of anthers. Although the generative cell of normal bicellular pollen grains does not incorporate any appreciable amount of the label into its RNA, there is an accelerated RNA synthesis in the generative cell of the embryogenic bicellular pollen grains. During progressive embryogenesis, RNA synthetic activity is finely balanced between the fates of the generative and vegetative cells (Figure 17.4). This ensures that pollen grains in which RNA synthesis occurs either in the generative cell or in both the generative and vegetative cells divide fur-

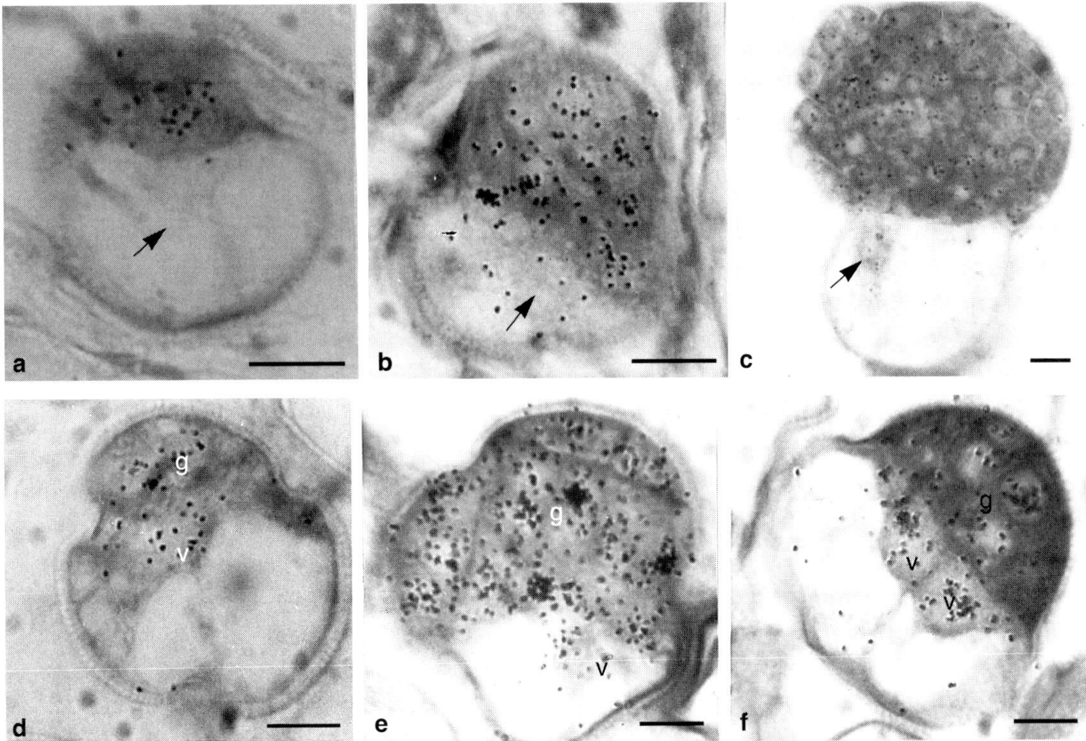

Figure 17.4. Autoradiographic incorporation of ^{3}H-uridine into pollen grains of cultured anthers of *Hyoscyamus niger*. **(a)** A multicellular pollen grain with label in the division products of the generative cell nucleus, 72 hours after culture. **(b)** Another multicellular pollen grain, 96 hours after culture. **(c)** A globular pollen embryoid, 144 hours after culture. In all cases, the vegetative cell nucleus is indicated by the arrow. **(d)** A 3-celled pollen grain with label in the division products of the generative cell and in the nucleus of the vegetative cell, 96 hours after culture. **(e)** A multicellular embryoid, 120 hours after culture. **(f)** Another multicellular embryoid with two nuclei formed by the division of the vegetative cell, 144 hours after culture. In d–f, division products of both vegetative (v) and generative (g) cells are labeled. Scale bars = 10 μm. (From Raghavan 1979b.)

ther along the embryogenic pathway, whereas those in which RNA synthesis occurs exclusively in the vegetative cell become starch filled and nonembryogenic (Raghavan 1979a, b; Reynolds and Raghavan 1982). Thus, transcription in the generative cell is a more important prerequisite for embryogenic divisions of pollen grains of *H. niger* than transcription in the vegetative nucleus.

In spite of their small representation in the genome, mRNAs accumulate in a characteristic way in the embryogenic pollen grains of *H. niger*. By in situ hybridization using ^{3}H-poly(U) as a probe, it was found that although unicellular pollen grains do not bind the probe during their normal ontogeny, the embryogenically determined pollen grains are labeled within a few hours of culture of anthers. This labeling is sensitive to actinomycin-D, which supports the idea that mRNA is newly synthesized by the pollen grains as the anther establishes contact with the nutrient medium. Compared to the generative and vegetative cells of normal pollen grains, which are only transiently active in binding ^{3}H-poly-(U), embryogenic bicellular pollen grains exhibit the striking feature of continued transcriptional activity in the generative cell. The vegetative cell of embryogenic bicellular pollen grains is generally starved for mRNAs and pollen grains that accumulate mRNAs exclusively in this cell pass their time in culture as starch-filled, nonembryogenic pollen (Raghavan 1981b). It would appear from these results that in cultured anthers, a small number of the enclosed pollen grains synthesize new mRNAs that apparently code for the proteins necessary to induce the first embryogenic mitosis. The trigger that sets in motion the synthesis of new mRNAs is possibly the trauma of the excision and culture of anthers.

In vitro translation profiles of mRNAs isolated

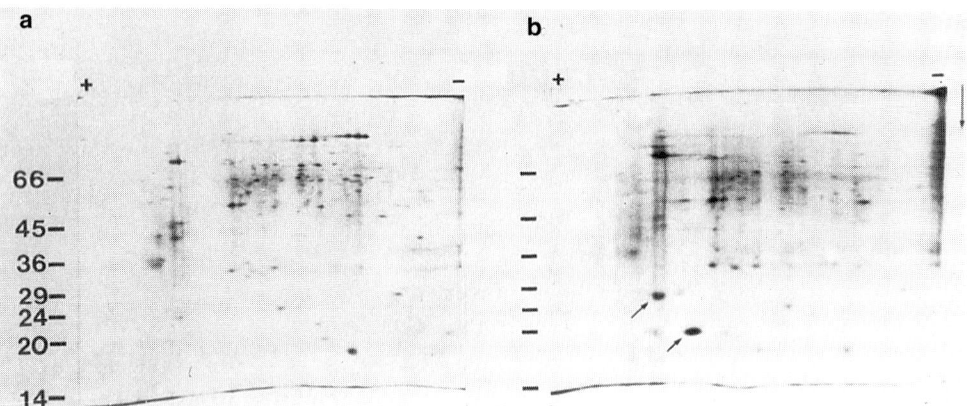

Figure 17.5. In vitro translation profiles of proteins from **(a)** bicellular and **(b)** embryogenic pollen grains of tobacco. Autoradiograms of translation products separated by two-dimensional SDS-PAGE are shown. Arrows in (B) point to two major proteins not seen in (A). Molecular mass markers are shown on the left. (From Garrido et al 1993. *Sex. Plant Reprod.* 6:40–45. © Springer-Verlag, Berlin. Slide supplied by Dr. O. Vicente.)

from embryogenic pollen grains and embryoids generally support the results from cytological studies. Late unicellular to early bicellular pollen grains of *Brassica napus* undergo embryogenic divisions after a thermal shock at 32 °C for 8–12 hours. Pechan et al (1991) have shown that during the first 8 hours of high-temperature stress, a number of mRNAs that are absent, or present at low levels, in freshly isolated microspores appear in the embryogenic counterparts. Since some of these messages probably code for proteins homologous to HSPs of the 70-kDa class, their relevance to the embryogenic induction of pollen grains remains to be clarified (Cordewener et al 1995). That the synthesis of new mRNA is associated with embryogenic induction is suggested by the appearance of two abundant mRNAs when isolated pollen grains of tobacco are induced to form embryoids in an enriched medium after a period of starvation (Figure 17.5) (Garrido et al 1993). Some differences confined to the relative abundance of certain messages have also been noted between pollen embryoids of two different stages of development isolated from cultured anthers of barley (Higgins and Bowles 1990).

Protein Metabolism. Information available on the protein metabolism of embryogenic pollen grains lags behind that on RNA metabolism. The previously noted decrease in RNA accumulation in the embryogenic pollen grains of tobacco, suggestive of the elimination of the gametophytic program, is associated with an apparent lack of protein accumulation, whereas pollen grains that mature and complete the gametophytic program in culture accumulate a large amount of protein (Bhojwani, Dunwell and Sunderland 1973). In the embryogenic bicellular pollen grains of *D. innoxia*, the nucleus of the generative cell is more enriched for histones than that of the vegetative cell; a possible role for histones in preventing the division of the generative cell in the embryogenic pathway is implied by this observation (Sangwan-Norreel 1978). A change in the spatial accumulation of 70-kDa HSPs may be achieved in the microspores of *B. napus* during the thermal shock preparatory to embryogenic induction (Figure 17.6). Compared to normal unicellular and bicellular pollen grains, the embryogenic counterparts display intense and differential labeling of nuclei with anti–HSP-70 antibodies, showing that embryogenic induction depends critically upon stage-specific changes in the intracellular distribution of HSP-70 (Cordewener et al 1995). The role demanded of the embryogenic proteins suggests that they should be present at the critical stage of induction. However, the hope of identifying specific proteins that are synthesized sufficiently early to have a role in the embryogenic determination and induction of microspore embryogenesis appears unrealistic. A two-dimensional electrophoretic analysis did not reveal any qualitative differences between the in vivo synthesized proteins of embryogenic and nonembryogenic pollen grains of *B. napus*; by a careful quantitative study, it was, however, found that some polypeptides are synthesized at a slightly higher rate in the embryogenic than in the nonembryogenic pollen grains (Cordewener et al 1994; Custers et al 1994). The most optimistic outlook is that the new proteins are synthesized transiently or in such small quantities

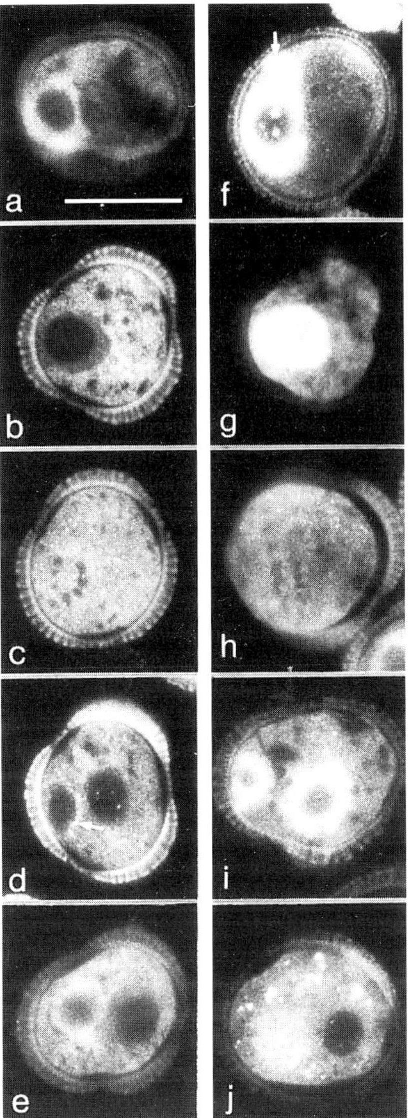

Figure 17.6. Immunofluorescence localization of 70-kDa HSP in pollen grains of *Brassica napus* immediately after isolation (a–e) and after culture for 8 hours at 32 °C under embryogenic conditions (f–j). **(a)** Vacuolate, late unicellular pollen grain with label in the nucleoplasm and weak label in the cytoplasm. **(b)** Late unicellular pollen grain with label in the cytoplasm, less in the nucleoplasm, and none in the nucleolus. **(c)** Mitotic microspore with label throughout the cytoplasm and negative staining on the chromosomes. **(d)** Early bicellular pollen grain with the cytoplasm of both the vegetative and generative cells and the wall between both cells (arrow) labeled. **(e)** Late bicellular pollen grain with slightly more intense label in the cytoplasm of the generative cell than in the vegetative cell. **(f)** Stage corresponding to (a), with intense label in the nucleoplasm and weak label in the cytoplasm. Arrow points to an intensely labeled area. **(g)** Stage corresponding to (b), with intense label in the nucleoplasm and weak label in the cytoplasm. **(h)** Stage corresponding to (c), with label in the cytoplasm. **(i)** Stage corresponding to (d), with strong label in both the generative cell and vegetative cell nuclei. **(j)** Stage corresponding to (e), with diffuse label in the cytoplasm and nucleoplasm of both the generative and vegetative cells. Intensely labeled spots are also seen in the cytoplasm of the latter. Scale bar = 10 μm. (From Cordewener et al 1995. *Planta* 196:747–755. © Springer-Verlag, Berlin. Photographs supplied by Dr. M. M. van Lookeren Campagne.)

that they cannot be detected by the techniques used.

In *H. niger*, the embryogenic development of pollen grains is linked to the synthesis of certain general and basic proteins. Whereas unicellular pollen grains of intact anthers do not incorporate any ^{3}H-tryptophan or ^{3}H-leucine into proteins, culture of anthers is associated with the incorporation of these amino acids, as well as with the accelerated incorporation of ^{3}H-arginine and ^{3}H-lysine, into unicellular, potentially embryogenic pollen grains. Following the first haploid mitosis, these pollen grains incorporate ^{3}H-leucine, ^{3}H-arginine, and ^{3}H-lysine into the generative cell or into both the generative and vegetative cells. The course of development pursued by the bicellular pollen grains is markedly affected by the protein synthetic activity in the constituent cells. Pollen grains that incorporate labeled arginine, leucine, and lysine into the vegetative cell fail to become embryogenic but continue the gametophytic program. When the generative cell or when both the generative and vegetative cells incorporate the labeled precursors, pollen grains continue embryogenic divisions. In the bicellular pollen grains, the incorporation of ^{3}H-tryptophan is sparse and is found only in the vegetative cell (Raghavan 1984). The pattern of protein synthesis in the unicellular and bicellular pollen grains of *H. niger* is remarkably similar to the pattern of RNA synthesis.

It is to be expected that increased protein synthesis detected autoradiographically in the embryogenic pollen grains will be matched at the biochemical level. In embryogenic pollen grains of *Nicotiana tabacum*, new sets of membrane-bound phosphoproteins and soluble and insoluble glycoproteins are synthesized, whereas different sets of these classes of proteins appear in the nonembryogenic pollen (Kyo and Harada 1990; Kyo and Ohkawa 1991; Říhová, Čapková and Tupý 1996). In a restricted analysis, attempts to identify changes in mitosis-specific phosphoproteins during embryogenic induction of isolated microspores of *B. napus* were not, however, successful (Hause et al 1995). Increased levels of arginine and polyamines characteristically appear during embryogenic development of pollen grains of *N. tabacum* and *D. innoxia* (Villanueva, Mathivet and Sangwan 1985; Garrido, Chibi and Matilla 1995). The transient nature of the phosphoproteins and the large increases in the arginine and polyamine levels in the embryogenic pollen grains give these biochemical parameters some legitimacy as part of the induction process in an as yet undetermined way.

Clark et al (1991) found that a 17-kDa embryo-specific polypeptide reliably predicts the transition of pollen embryoids in cultured anthers of barley to the stage of embryonic organ production. The appearance of this protein in itself is not concerned with the regulation of pollen embryogenesis, because near-identical proteins are seen in seed embryos of barley and other cereals. The temporal distribution of storage proteins (Crouch 1982; Taylor et al 1990; Holbrook et al 1991), storage lipids (Pomeroy et al 1991; D. C. Taylor et al 1991; Wiberg et al 1991), and activities of enzymes involved in the glycolytic pathway (Sangwan et al 1992) and in fatty acid biosynthesis (Weselake et al 1991, 1993; Taylor et al 1992, 1993) in pollen embryoids of *Brassica napus* and *B. campestris* generally resemble their patterns seen in zygotic embryos.

Certain stages of the dedifferentiation of pollen grains in cultured anthers of barley are associated with the appearance of specific isozymes of acid phosphatase, esterase, NADH dehydrogenase, and peroxidase. The most distinct isozyme changes are observed concomitant with the transition of the disorganized cells into heart-shaped embryoids and later with the formation of incipient embryonic organs (Pedersen and Andersen 1993). Potentially, isozymes could act as markers of the embryogenic transformation of microspores, but this possibility has not been considered as widely in pollen embryogenesis as in somatic embryogenesis.

(ii) Genes Expressed during Pollen Embryogenesis

A global view of gene expression during pollen embryogenesis has been derived from studies of the patterns of the appearance and disappearance of transcripts of genes for storage proteins in developing embryoids of *Brassica napus*. As with seed embryos, there are indications that the development of pollen embryoids is associated with the accumulation of storage protein gene transcripts and that it is strongly influenced by ABA. Embryoids accumulate little or no napin and cruciferin mRNAs until they reach the globular or heart-shaped stages. From this period onward, transcript levels also begin to increase through the inclusion of ABA in the medium. The accumulation of cruciferin transcripts is particularly tardy, and even in torpedo-shaped embryoids the transcripts attain detectable levels only in the presence of ABA (Wilen et al 1990). Jasmonic acid is another hormonal substance that functions like ABA in inducing napin and cruciferin mRNA accumulation. Moreover, the natural occurrence of jasmonic acid in seed embryos and pollen embryoids of *B. napus* suggests that subtleties in the action of this hormone are important in the regulation of storage protein synthesis during pollen embryogenesis (Wilen et al 1991). Oleosin mRNAs respond in a similar way to the treatment of embryoids with ABA, sorbitol, or jasmonic acid (van Rooijen et al 1992; Zou et al 1995). Embryogenically induced microspores of *B. napus* also express a gene for acetohydroxyacid synthase, which codes for an essential enzyme in branched-chain amino acid biosynthesis (Ouellet, Rutledge and Miki 1992). On the other hand, transcripts of cloned nitrate reductase genes are activated in the embryoids no earlier than the heart- or torpedo-shaped stages. Since gene expression is independent of the presence of nitrate in the medium, some nitrate-sensitive mechanism appears to be turned on in the embryoids in a stage-specific manner (Fukuoka et al 1996).

There is good evidence linking the expression of a cloned napin gene with the embryogenic induction of isolated microspores of *B. napus*. This evidence is based on the observation that transcripts of the gene that are present in small amounts in microspores at the time of culture are expressed at elevated levels after an inductive heat shock at 32 °C for four days. Particularly interesting is the fact that napin transcripts decrease in abundance as the microspores divide in the embryogenic pathway and they do not reappear until the embryoids are phased into the cotyledonary stage of development. High expression of a *GUS*–napin gene construct in embryogenically induced microspores of transgenic plants is considered a rigorous proof of the role of the napin gene as an early marker of pollen dedifferentiation (Boutilier et al 1994). In a similar way, following an inductive in vitro starvation treatment, there is a dramatic increase in the level of a low-molecular-mass heat-shock protein gene transcript in embryogenic tobacco pollen grains (Žárský et al 1995). Whether or not the napin and heat-shock protein genes have a functional role

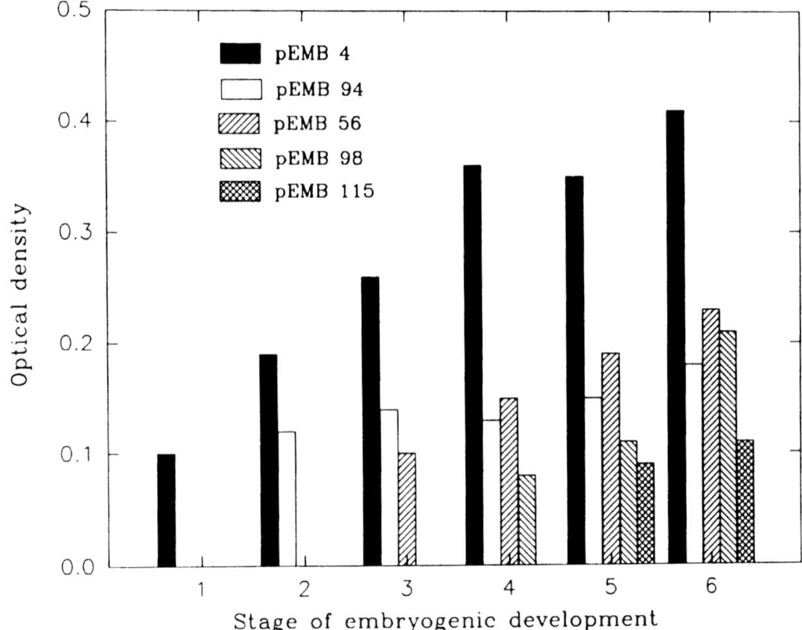

Figure 17.7. Expression of embryoid-abundant mRNAs during pollen embryogenesis in cultured anthers of *Triticum aestivum*. Data are based on densitometry of autoradiograms of Northern blot of RNAs from embryoids of different stages hybridized with cDNA clones. Stages represented: (1), unicellular pollen grain at the time of culture; (2), multicellular pollen grain, 2 days after culture; (3), proembryoid, 7 days after culture; (4), globular embryoid, 10 days after culture; (5), bipolar embryoid, 14 days after culture; (6), young embryoid, 21 days after culture. Clones *pEMB-4* and *pEMB-94* are considered embryoid specific. (From Reynolds and Kitto 1992. *Plant Physiol.* 100:1744–1750. © American Society of Plant Physiologists. Print supplied by Dr. T. L. Reynolds.)

in embryogenesis in the respective systems is not clear at present.

Analysis of cDNA clones from a library made to poly(A)-RNA from young pollen embryoids of *Triticum aestivum* promises to provide new insight into the genes that are activated during the early stages of pollen embryogenesis (Figure 17.7). Two embryoid-specific recombinants isolated by differential screening are expressed early during embryogenesis and are not detected in the nonembryogenic pollen grains or in the vegetative tissues of the plant (Reynolds and Kitto 1992). The same library also yielded an embryoid-abundant gene that encodes a metallothionein-like protein. Expression of the gene is restricted to embryogenic microspores, embryoids, and zygotic embryos and is modulated by ABA (Reynolds and Crawford 1996). This is attractive preliminary evidence indicating that some genes, which are not part of the gametophytic program, are associated with the dedifferentiation phase of the pollen grains. It is possible that in this species, in which pollen grains are not encumbered by thermal shock or starvation stress for embryogenic induction, an overlap between the expression of low-abundance embryoid-specific genes and that of stress-related genes can be avoided.

5. GENERAL COMMENTS

The concept that the vegetative and generative cells of the angiosperm pollen grain possess the innate potential to develop into whole new plants by a process of simulated embryogenesis has arisen largely from anther and pollen culture investigations reviewed in this chapter. To initiate embryogenic divisions, pollen grains enclosed within the anther, or isolated from it, must be cultured in a medium and sufficient time must elapse under specified conditions to allow the gametophytic program gradually to wane. We cannot identify with certainty any hormonal or chemical substances in the medium that stimulate embryogenic divisions of pollen grains. In the two most widely exploited systems, subjecting pollen grains to temperature or nutritional stress appears to be associated with the acquisition of totipotency by the vegetative cell. There is preliminary evidence that connects stress-imposed embryogenic divisions to the activation of genes for storage protein and heat-shock protein synthesis but their relevance to the pollen embryogenic induction process remains to be established. Moreover, the changeover from the gametophytic program to the sporophytic program in pollen grains is so profound that it is not unreasonable to look for genes that might qualify as embryonic genes, as in somatic embryogenesis.

A highly interesting finding is that in several species investigated, the stage of development of the pollen grains at the time of culture is critical in determining whether the pollen grains will continue as gametophytes or evolve into sporophytic units. Hence, the ultimate control of embryogenic induction might lie in the genes that are expressed during a narrow developmental window centering around the time of the first haploid mitosis. Determination of the parameters of this stage in pollen cytology as it relates to the cell cycle is necessary to achieve synchrony in pollen development which can lead to increased yields of embryoids and plantlets.

Before concluding this chapter, it is important to emphasize that the ease of production of isogenic diploids by anther culture has had a significant impact on genetic and breeding experiments in many plants. Although this area of work has not been stressed in the preceding pages, it should be kept in mind that haploid plants and isogenic diploids have been generated in most of our crops by anther culture technology and improved strains of plants using doubled haploids in breeding programs are already under cultivation in many parts of the world.

SECTION V

Applications

Chapter 18

Genetic Transformation of Embryos

One of the most promising approaches currently available to improve the quality, diversity, and yield of our crop plants is genetic engineering. In a broad sense, the genetic engineering of plants connotes a manipulation of plant cells and organs at the molecular level leading to the introduction, integration, and expression of specific and useful segments of foreign genetic material in a host plant. The application of genetic manipulative techniques has resulted in the production of plants with altered metabolic pathways and useful agronomic traits such as insect, viral, or herbicide resistance; it has also generated custom-made male-sterile plants useful in hybrid seed production. Other genetic engineering approaches, such as improving the post-harvest qualities of fruits and vegetables and creating crops that synthesize useful vaccines, pharmaceuticals, or chemicals, are already at an advanced stage on the drawing board or are well into the developmental phase.

Historically, engineering of the first transgenic plants relied on the ability of the natural vector *Agrobacterium tumefaciens* to introduce recombinant DNA molecules into plant cells; in later years, this has become the most widely used procedure for the genetic transformation of various dicotyledonous plants. Cocultivation of isolated protoplasts with bacteria was the method of choice in some of the early attempts at *Agrobacterium*-mediated transformation. Currently, the less cumbersome and more reliable method of leaf disc transformation is routinely used to obtain transgenic dicots under standard tissue culture conditions. The limited susceptibility of monocotyledons to *A. tumefaciens* infection has severely restricted the use of this vector to transform monocots, particularly the cereals. Successful gene transfer methods developed for monocots involve direct DNA uptake by isolated protoplasts through osmotic shock, DNA transfer by electroporation of protoplasts and cells, and microprojectile-mediated DNA transfer into cells, tissues, and organs capable of regeneration by organogenesis or somatic embryogenesis.

In the context of the discussion on various aspects of male and female gametogenesis and embryogenesis in flowering plants given in the previous chapters, the objective here is to review the work done on plant transformation using male and female sexual cells, zygotes, developing embryos, embryogenic cells, and embryoids instead of leaf discs as the starting material. It is now a relatively simple exercise in genetic engineering to produce transgenic tobacco plants by transplanting the appropriate cloned gene by *Agrobacterium*-mediated leaf disc transformation. The sexual cells, embryogenic cells, and embryos thus far investigated are not comparable to the tobacco leaf in its simplicity and efficiency in the uptake and integration of foreign genes. However, the original protocols referred to previously and their modifications to suit particular situations, used alone or in combination, have resulted in the introduction of alien genes into many plants through the sexual route.

As will be seen later, not all genes that have been successfully introduced are agronomically important; moreover, in many cases, only the expression of reporter genes has been followed by transient assays of regenerants. A further caveat regarding these investigations is that field trials to determine the stability of gene integration and expression through several sexual crosses have been conducted in only a few cases. The elegance of the use of the sexual route for plant transformation resides in the freedom to choose a single-celled sperm, egg, zygote, or microspore for the delivery and integration of foreign DNA. Embryogenic somatic cells and microspores offer potential targets for transformation comparable to the zygote in their ability to recapitulate embryogenic development with the introduced genes. Reviews by Christou (1995), Jähne, Becker and Lörz (1995), Ritala et al (1995), Sági et al (1995b), and Sautter et al (1995) published in a symposium volume are considered especially relevant to the topics discussed here.

1. TRANSFORMATION THROUGH POLLEN GRAINS AND OVULES

There is considerable latitude for the investigator using pollen grains as vectors for transformation purposes because DNA can be transferred to recipient plants in a variety of ways. These include mixing foreign DNA with pollen grains, mixing *Agrobacterium* containing the desired gene with pollen grains, the microinjection of DNA into pollen cells, the electroporation of germinating pollen grains in the presence of DNA, the introduction of DNA into pollen grains by particle bombardment, and allowing the uptake of DNA by pollen tubes. The pollen grains and pollen tubes are subsequently used for pollination and fertilization in the normal reproductive cycle. The possibility of introducing foreign genes into plants through pollen grains was suggested by experiments on the transfer of anthocyanin genes within and between species of *Petunia* (Hess 1980). Later investigators obtained some putative transformants using the male gametophyte as a carrier of exogenous DNA into the genome of the egg or zygote of maize using the normal fertilization cycle (de Wet et al 1985; Ohta 1986; Roeckel et al 1988). Either the total genomic DNA of plants possessing dominant alleles of a few marker genes or plasmid DNA harboring a gene expressing kanamycin resistance was used in the transformation experiments. Since no molecular controls of transformation were included in the protocol, or controls turned out to be negative, the case for the unambiguous incorporation of foreign DNA remains unproven in these experiments. Some investigators have reported a total lack of success in attempted pollen-mediated transformation in maize (Sanford, Skubik and Reisch 1985; Booy, Krens and Huizing 1989).

Two protocols using electrotransformed pollen to produce transgenic plants have been described (Matthews, Abdul-Baki and Saunders 1990; Smith et al 1994). Plasmid constructs carrying the *GUS* gene were incorporated by electroporation into germinating pollen grains of *Nicotiana gossei*. Positive evidence of the production of transformed plants by pollination with the electroporated pollen has come mainly from Southern blot of genomic DNA of the putative transformants.

Another method tested involved targeting male sporogenous cells, instead of pollen grains, as the recipients for gene transfer. This was done by the particle bombardment of tassels of maize with *GUS* gene or anthocyanin marker genes at the time of the initiation of stamen primordia and the culturing of the tassels to produce functional pollen grains in vitro. The positive GUS activity that developed in the tassels of the mature anthers showed that DNA integration could be facilitated during anther ontogeny (Dupuis and Pace 1993). Assuming that in vitro matured and transformed pollen grains will be as effective as normal pollen grains in the fertilization process, this technique offers some advantages when the collection of pollen grains from small flowers presents a drawback.

An indirect pollen delivery system used with some success followed from the demonstration of the transfer of the tumor-inducing gene from *A. tumefaciens* into progenies of *Petunia hybrida* (Hess and Dressler 1989) and the *A. tumefaciens*–mediated transfer of the *GUS* gene into pollen grains (Süssmuth, Dressler and Hess 1991). In the protocol based on the *Agrobacterium* infection of pollen grains to transfer the kanamycin-resistance gene to wheat, the pollen–bacterial suspension was pipetted into the spikelets, which were allowed to complete the fertilization cycle and form mature caryopses. The kanamycin-resistant seedlings obtained were fully fertile and showed evidence of the integration of parts of the donor DNA in the genome of the F2 generation (Hess, Dressler and Nimmrichter 1990). However, problems concerning the stability of the foreign DNA were not completely eliminated in this work and have continued to plague the approach. Another group of investigators have now claimed that the apparent transformation of maize and other cereals by T-DNA from *A. tumefaciens* through the male sexual route is an artifact of the procedure and does not rigorously prove transformation of the plant nuclear genome (Langridge et al 1992).

The possibility of directly introducing genes into the egg, the zygote, or the endosperm nucleus through the pollen tubes is inferred from studies in which exogenous DNA was applied to the styles of pollinated flowers of cotton, rice, and wheat. It is believed that the injected DNA is taken up by the pollen tubes and transferred to the embryo sac before or after fertilization (Gong et al 1988; G. Y. Zhou et al 1988; F.-Y. Wang et al 1995). Changes observed in the protein profiles of endosperms of lines of wheat transformed with sorghum DNA suggest that the donor genes are probably expressed (F.-Y. Wang et al 1995). That DNA penetrates the germ cells of rye was shown by injecting the inflorescence, before male and female meiosis, with plasmids harboring a gene coding for resistance to kanamycin. Although molecular signals established the transgenic nature of the small number of kanamycin-resistant seedlings recovered, Mendelian inheritance of the gene was not demonstrated (de la Peña, Lörz and Schell 1987).

In another approach, followed by Luo and Wu (1988), plasmid DNA coding for neomycin phosphotransferase (NPTII) under the control of the CaMV 35S promoter was applied to decapitated stylar ends of rice florets after normal pollination, and both phenotypic and molecular proofs to support claims of transformation were observed. It has been proposed that the exogenous DNA is taken up by the pollen tubes and integrated into the egg during fertilization. A critique of this work, however, has pointed out that instead of integrating into the rice genome as a unique, high-molecular mass, the foreign DNA is found in Southern blots at different locations in different transformants (Potrykus 1990). So, although we have some glimpse of the fate of the donor DNA, it remains to be established whether it becomes part of the host genome without undergoing denaturation or degradation.

2. TRANSFORMATION OF DEVELOPING EMBRYOS

There are indications that seed embryos of certain plants can be rapidly transformed by coculturing with *A. tumefaciens*. In the protocol developed for the transfer of two reporter genes into *Arabidopsis*, the critical step for efficient transformation was a three-day preculture of immature or mature embryos in a medium supplemented with an auxin and a cytokinin, followed by incubation with the bacterial vector for two days. The kanamycin- or hygromycin-resistant calluses that appeared on a regeneration medium were isolated and allowed to undergo organogenesis in a suitable medium, and the plantlets formed were reared to maturity. The expression of reporter genes in the transformants, presence of integrated copies of the inserted DNA in their genome, ability of explants of transformants to form calluses and roots on the selective medium, and transmission of the resistance gene according to Mendelian principles are considered to provide evidence of transformation (Sangwan, Bourgeois and Sangwan-Norreel 1991). Through variations of this protocol, transgenic plants have been regenerated from seed embryos of *Vigna unguiculata* (Penza, Lurquin and Filippone 1991), *Cicer arietinum* (Fontana et al 1993), *Pisum sativum* (Schroeder et al 1993), *Datura innoxia* (Ducrocq, Sangwan and Sangwan-Norreel 1994), and *Panax ginseng* (ginseng; Araliaceae) (Lee et al 1995). Investigations on *D. innoxia* and ginseng have taken advantage of the ability of the cells of the infected cultured embryos to regenerate somatic embryos directly on the explant or through a callus phase and thus produce a large crop of transformed plants.

With the demonstration of the *Agrobacterium*-mediated transfer of viral sequence to the stem of maize seedlings (agroinfection), the competence of maize embryos for agroinfection has been established. Although the susceptibility of embryos varies with the genotype, the stage of embryogenesis endowed with a functional shoot apical meristem is thought to be the major factor conferring readiness for successful T-DNA transfer (Schläppi and Hohn 1992). Transgenic rice tissues were recovered from mature rice embryos inoculated with *Agrobacterium*, but the regeneration of plants was not reported from these cultures (Raineri et al 1990). A breakthrough in the *Agrobacterium*-mediated transformation of rice was made in two independent investigations using embryo explants to transfer marker genes. Chan et al (1993) obtained a few transgenic plants by infecting immature embryos with bacteria carrying chimeric *GUS* and *NPTII* genes under the control of promoters of the rice α-amylase gene and the *Agrobacterium* nopaline synthase gene, respectively. Calluses regenerated from embryos gave rise to plants by organogenesis. A higher rate of recovery of transgenic plants was obtained when calluses initiated from excised scutella of mature embryos were cocultured with the bacteria (Hiei et al 1994). Both works clearly demonstrated the stable integration of foreign DNA into the rice genome, as well as the sexual inheritance of the genes in a Mendelian fashion. This technology has now been extended to maize embryos to produce high frequencies of transformants hitherto not possible with biolistic methods (Ishida et al 1996). Although the reasons for the success that had eluded previous investigators are not clear, the long controversy about the ability of *Agrobacterium* to transform monocotyledons may now be put to rest, at least for two major crops.

Transformation using pollen embryoids has rarely been attempted, but its potential is great. The transformation of developing embryoids in cultured anthers will eliminate the need for the cumbersome isolation procedures usually necessary for seed embryos. Since the embryoids grow directly into plantlets, it is possible to bypass the problems of the regeneration of plantlets from transgenic calluses. The first, and the most direct, attempt to transform pollen embryoids was by Neuhaus et al (1987), who delivered DNA by microinjection into globular pollen embryoids of *Brassica napus*. The very nature of the technique, targeting only one cell for injection, resulted in numerous chimeric events in the regenerated embryoids and plantlets. When secondary embryoids, presumably formed from transformed single cells, were analyzed, it was possible to demonstrate the predicted segregation of the marker gene. Microinjection of chimeric DNA into unicellular microspores of *B. napus* has shown that the relatively small amount of DNA transferred is insufficient to attain stable transformation (Jones-Villeneuve et al 1995). Direct DNA uptake mediated either by polyethylene glycol or by electroporation was found to cause transient expression of reporter genes in microspores of barley (Kuhlmann et al 1991) and maize (Fennell and Hauptmann 1992), but the regeneration of transgenic plants was not achieved. Present indications are that *Agrobacterium*-mediated transformation is probably a more versatile way to transform pollen embryoids than either microinjection or direct DNA uptake. This became evident from the successful infection of proembryoids of *B. napus* by bacteria containing selectable genes for antibiotic resistance, although no transgenic plants showing integration of T-DNA into the nuclear genome were recovered (Pechan 1989). The production of transgenic haploid plants of *Datura innoxia* and *Nicotiana tabacum* seems to provide good examples of the ease of transforming pollen embryoids by the cocultivation of anthers or embryogenic pollen grains with the bacterial vector. The data from this work imply that with both species, competence for transformation is correlated with the onset of the growth of cotyledons on the embryoids, because no transformants are recovered from embryogenic pollen grains or proembryoids cocultured with the bacteria (Sangwan, Ducrocq and Sangwan-Norreel 1993).

(i) DNA Delivery by Microprojectiles

By direct microprojectile delivery of DNA into embryogenic pollen or embryoids, transgenic plants of barley, wheat, and tobacco have been produced. When freshly isolated pollen grains of barley were used as target cells, the transferred *GUS* genes and *BAR* genes (encoding phosphinothricin acetyltransferase [PAT], an enzyme that inactivates the herbicidal compound L-phosphinothricin [L-PPT]) were inherited in all progeny plants of pollen grain origin, indicating the homozygous nature of the primary regenerants (Jähne et al 1994). In experiments targeting pollen embryoids of barley, the possibility of producing herbicide-resistant plants was clearly in view, although only resistant calluses were obtained in the limited trials conducted (Wan and Lemaux 1994). In work with tobacco, pollen grains induced to form embryoids by starvation were bombarded immediately before or after the first division with plasmid DNA bearing a *GUS*-CaMV 35S promoter construct. The two transgenic plants obtained were analyzed by Southern blot to establish stable integration of the foreign DNA, and selfing and backcrosses of one diploidized progeny clearly demonstrated that the transgene was transmitted to the offspring in accordance with Mendelian predictions (Stöger et al 1995). Compared to CaMV 35S promoter, *GUS* gene under the control of maize ubiquitin promoter, when delivered to early stage pollen embryoids of wheat, produced a high level of expression in regenerated plants (Loeb and Reynolds 1994).

The delivery of foreign DNA into developing seed embryos by microprojectile bombardment has been attempted mainly in cereals. In some dicots that are easily infected by *Agrobacterium tumefaciens*, regeneration by leaf discs is notoriously difficult and slow. Here also, as in the cereals, developing embryos have been used as the starting material for direct introduction of foreign genes.

In some of the early works, using immature embryos of barley, pearl millet, and wheat, success was limited to demonstrating the transient expression of reporter genes under the control of appropriate promoters (Kartha et al 1989; Chibbar et al 1991; Taylor and Vasil 1991; Taylor, Vasil and Vasil 1993). Based on these results, transformation systems for cereals of major agricultural importance have been developed that involve the microprojectile bombardment of embryos, the subsequent culture of bombarded embryos to promote the formation of callus, the selection and maintenance of the embryogenic callus, the regeneration of plants, and segregation analysis of the progeny containing the introduced gene. Included in the list of plants used for transformation by this route are sorghum (Casas et al 1993), several varieties of wheat (Lonsdale, Önde and Cuming 1990; Vasil et al 1992, 1993; Weeks, Anderson and Blechl 1993; Becker, Brettschneider and Lörz 1994; Iglesias et al 1994; Nehra et al 1994; Blechl and Anderson 1996), maize (Klein et al 1988; Koziel et al 1993), rice (Christou, Ford and Kofron 1991; L. Li et al 1993), and barley (Wan and Lemaux 1994; Ritala et al 1994). Most of the manipulations were carried out with embryos excised from grains midway through development. Embryos were bombarded for a brief period immediately after excision, or after culture, but before the initiation of callus. Microprojectiles were delivered from the scutellar side or, rarely, from the side of the embryo axis; in some cases, excised pieces of the scutellum were bombarded and cultured. Some of the results obtained bear out the prediction that it is possible to engineer insecticide or herbicide tolerance in plants by transferring into immature embryos the appropriate genes. For example, European corn borer, which attacks maize, is susceptible to insecticidal proteins produced by *Bacillus thuringiensis*. Koziel et al (1993) introduced into maize embryos a chimeric *CRY-1A* gene encoding a truncated version of the protein derived from *B. thuringiensis*. Transformed plants containing the chimeric gene were found to express high levels of the insecticide protein and to offer resistance to repeated heavy infestations by the European corn borer. In wheat transformation experiments, the marker gene *BAR* was introduced; stably transformed plants selected using L-PPT, an ingredient of the commercial herbicide basta, showed resistance to topical applications of the herbicide (Vasil et al 1993; Nehra et al 1994). Rice (Christou, Ford and Kofron 1991), wheat (Weeks, Anderson and Blechl 1993), sorghum (Casas et al 1993), and barley (Wan and Lemaux 1994) plants similarly transformed with the *BAR* gene were selected in the presence of another L-PPT–containing herbicide, bialaphos, and were found to be resistant to further applications of this compound. The feasibility of manipulating the levels and types of storage proteins in wheat endosperm appears close to reality by the introduction of a high-molecular-mass glutenin gene into embryos and by the subsequent regeneration of transgenic lines (Blechl and Anderson 1996). These findings have considerably broadened our understanding of the usefulness of organized tissues such as the embryo to act as transformation targets by particle acceleration. The protocol may also prove highly efficient in introducing other useful traits into additional varieties of these crops. Since regeneration of many crops in tissue culture has a strong genotype-dependent component, the development of a genotype-independent transformation system for each crop would be a significant achievement.

Among dicots, transgenic soybeans were the first to be regenerated by the particle bombardment of excised embryos. Plasmid DNA containing an *NPTII* coding region under the control of CaMV 35S promoter was introduced into the shoot apical meristems of immature embryo axes by particle acceleration. The explants were then cultured on a medium containing a high level of benzylaminopurine to induce shoot regeneration. Although the plantlets obtained were chimeric, the foreign gene was expressed in the recovered first generation offspring (McCabe et al 1988). In another work, using cotyledons of immature embryos of six varieties of soybeans, it was found that transient expression of the reporter gene driven by CaMV 35S promoter was dependent upon the stage of cotyledon development, with the younger stages being less responsive than the older ones, but the genotype had an overriding effect over the age of the embryo (Moore, Moore and Collins 1994). The development of a transformation system for cotton embryonic axes involved a protocol similar to that used for soybean and resulted in the production of transformants in which *GUS* gene was integrated and was transmitted to the progeny in a Mendelian fashion (McCabe and Martinell 1993). Stable transformation of embryos of *Carica papaya* (papaya) (Fitch et al 1990; Cabrera-Ponce, Vegas-Garcia and Herrera-Estrella 1995), bean (Aragao et al 1993), sunflower (Hunold et al 1995), *Eucalyptus globulus* (Rochange et al 1995; Serrano et al 1996), and *Brassica juncea* (Kost et al 1996) has been achieved following DNA delivery by particle bombardment, although regenerants were recovered only in papaya. In one exciting development in this line of investigation, the regeneration of transgenic papaya plants resistant to the papaya ring spot virus was possible by the bombardment of 2,4-D-treated immature seed embryos with the gene for the ring spot virus coat protein (Fitch et al 1992). In another work, after particle bombardment of embryos, bean plants transgenic for *GUS, BAR*, and bean golden mosaic virus coat genes were characterized for over five generations of selfing without any loss of introduced genes (Russell et al 1993).

(ii) Other Methods of Embryo Transformation

In a continuing search for methods to transform embryos of specific plants, other successful reports may bear a closer examination. Soyfer (1980) found that injection of milky stage endosperm DNA of a nonwaxy mutant of barley into milky stage grains of a *wx* mutant caused normal starch production in the offspring. It may be that integration of the donor DNA into the proembryo occurred in this experiment; however, because of the reversion of the transformed state, the conditions for maintaining the stability of the transformants are in need of definition. Electroporation is another technique that has been used to introduce DNA into whole embryos. Fertile transgenic plants harboring the *NPTII* gene were regenerated after electroporation both of immature maize embryos wounded either enzymatically or mechanically (D'Halluin et al 1992) and of mature half-embryos of rice (Xu and Li 1994). Gene transfer has been achieved into isolated rice embryos and into scutella of isolated wheat embryos electroporated without any pretreatment, although no transgenic plants have been regenerated by this method (Klöti et al 1993; Rao 1995). In its simplest form, the transient expression of chimeric genes is possible when dry embryos excised from cereals and legumes are imbibed in a solution containing DNA (Töpfer et al 1989; Yoo and Jung 1995); the uptake of DNA is thought to occur due to changes in the physico-chemical properties of plant cell membranes when dehydrated embryos are hydrated. This method, however, does not seem to have caught on.

In short, considering the results obtained with dicots and monocots, it can be said that the particle acceleration method of introducing new genes into embryos has proved to be an efficient method of producing transgenic plants. No doubt, the presence of meristems makes isolated embryos an interesting target system for microprojectile bombardment. The ability of embryos to grow into new plants in culture directly, or through the intervention of a callus, or by somatic embryogenesis, is also a unique property of isolated embryos that makes them suitable starting material for genetic engineering.

3. TRANSFORMATION OF EMBRYOGENIC CELLS AND SOMATIC EMBRYOS

Embryogenic cells that regenerate somatic embryos possess features characteristic of meristematic cells and hence have served as favorite targets for DNA delivery. In a few instances, somatic embryos have been used for the transformation of regenerable plants. Microprojectile bombardment is the most exploited of the techniques for gene transfer into embryogenic cells and embryoids. This technique has scored over *Agrobacterium*-mediated transfer and electroporation because of the relatively high transformation frequencies obtained.

(i) Monocots and Dicots

Experiments on a few species such as maize, in which embryogenic suspension cells are bombarded with the *BAR* gene, illustrate the ease of obtaining fertile transgenic plants by this route. Like transgenic wheat plants described in the previous section, transgenic maize plants regenerated from embryogenic cells by organogenesis or somatic embryogenesis are protected against basta and other L-PPT–containing herbicides (Fromm et al 1990; Gordon-Kamm et al 1990; Weymann et al 1993). A similar protocol was used to obtain transformed lines by selecting for resistance to hygromycin and chlorsulfuron (Fromm et al 1990; Walters et al 1992). There is now an impressive list of plants in which transformants for resistance to herbicides or antibiotics have been selected after microprojectile bombardment of embryogenic cells. Among monocots, this includes oat (Somers et al 1992), rice (Cao et al 1992), and barley (Wan and Lemaux 1994) selected for herbicide resistance; rice (L. Li et al 1993; S. Zhang et al 1996), *Festuca arundinacea, F. rubra* (Poaceae) (Spangenberg et al 1995a), *Lolium perenne* (Spangenberg et al 1995b), and *Musa* sp. (Musaceae) (Sági et al 1995a) selected for hygromycin resistance; and sugarcane (Bower and Birch 1992) and *Triticum* × *Hordeum* amphidiploid hybrid (Barcelo et al 1994) selected for geneticin resistance. In other monocots, such as *Agrostis palustris* (Poaceae) (Zhong et al 1993) and *Cenchrus ciliaris* (Poaceae) (Ross, Manners and Birch 1995), transgenic plants have been recovered without any selection pressures after bombardment of embryogenic calluses. It appears likely that the efficiency of regeneration in each case may be related to the state of the callus at the time of the bombardment effecting DNA transfer process. Alternative methods to obtain transgenic plants include the incorporation of plasmid DNA by electroporation into untreated embryogenic calluses of cereals or of calluses wounded either enzymatically or mechanically and the incorporation by direct uptake following laser microbeam puncture of cells (D'Halluin et al 1992; Guo, Liang and Berns 1995; Pescitelli and Sukhapinda 1995).

Attempts to introduce DNA by particle bombardment into embryogenic calluses or embryoids of dicotyledons are limited to a few species, such as *Gossypium hirsutum* (Finer and McMullen 1990), *Glycine max* (Sato et al 1993; Hadi, McMullen and Finer 1996), *Vitis* sp. (Hébert et al 1993; Scorza et al 1995; Kikkert et al 1996), *Coffea* sp. (van Boxtel et al 1995), *Lathyrus sativus* (Barna and Mehta 1995), *Carica papaya* (Cabrera-Ponce, Vegas-Garcia and Herrera-Estrella 1995), *Citrus reticulata* × *C. paradisi* (Yao et al 1996), and *Manihot esculenta* (Schöpke et al 1996). In *V. vinifera*, recovery of transgenic plants resistant to kanamycin was facilitated when somatic embryos initially bombarded with uncoated microprojectile particles were cocultivated with *Agrobacterium tumefaciens* or *A. rhizogenes* housing *GUS* or *NPT11* genes (Scorza et al 1995). Direct cocultivation with *A. tumefaciens* containing genes with suitable expression signals has proved to be reliable and efficient with other embryogenic calluses or somatic embryos for regeneration of transgenic plants. Examples of transgenic plants obtained by this means whose transformation status has been established are *Juglans regia* (McGranahan et al 1988, 1990), *Carica papaya* (Fitch et al 1993; Cabrera-Ponce, Vegas-Garcia and Herrera-Estrella 1996), *Asparagus officinalis* (Delbreil, Guerche and Jullien 1993), *V. rupestris* (Mullins, Tang and Facciotti 1990), and *V. berlandieri* × *V. rupestris* (Le Gall et al 1994). However, transgenic plants of *V. vinifera* recovered after cocultivation of embryogenic callus with an engineered strain of *A. rhizogenes* harboring both *GUS* and *NPTII* genes exhibited various phenotypes, including normal plants and plants with wrinkled leaves and abundant roots (Nakano, Hoshino and Mii 1994). In the approach taken with *Dendrathema grandiflora* (Asteraceae), leaf discs transformed with *A. tumefaciens* were induced to regenerate transgenic somatic embryos and plants in a suitable medium (Pavingerová et al 1994). Although transformation rates obtained by cocultivation with bacteria were somewhat low, this was compensated for by the high regenerative ability of the embryogenic cells and by the large number of transformed somatic embryos.

(ii) Gymnosperms

The development of protocols for somatic embryogenesis in gymnosperm genera has led to the exploitation of transformation techniques to study the regulation of gene expression in stably transformed cells. Biolistic approaches have been used to monitor the expression of *GUS* or *NPTII* genes in the embryogenic cells and somatic embryos of *Picea abies* (Robertson et al 1992; Yibrah et al 1994; Clapham et al 1995); in the embryogenic callus, somatic embryos, and seedlings of *P. glauca* and *P. marina* (Bommineni et al 1993; Charest et al 1993; Ellis et al 1993; Charest, Devantier and Lachance 1996); and in the embryogenic cells of *Pinus radiata* (Walter et al 1994). As other methods of gene delivery are tested with embryogenic cells, it should be possible to tailor one of them to generate a high frequency of regeneration in gymnosperms.

The formulation of a routine method to obtain stably transformed gymnosperms on a level comparable to that achieved in angiosperms is not on the horizon. Two major reasons can be cited for this pessimistic view. Although somatic embryos of gymnosperms possess the capacity to regenerate whole plants, the process is relatively slow. Attempts to establish the inheritance of foreign genes have to be seen in terms of the number of years required for plants to reach the cone-bearing age. The limitation of performing controlled pollination experiments is also a strong reason for believing that many years of hard work lie ahead on the way to reaching the goal of demonstrating sexual inheritance of the introduced gene.

4. GENERAL COMMENTS

Critical insights into the fundamental processes of sexual reproduction in angiosperms, combined with the use of recombinant DNA techniques, have helped to pioneer the production of transgenic plants by the introduction of external genetic material into the genotype of the embryo. The transformation systems described in this chapter have the advantage that they will allow efficient gene transfer into germ cells, embryos, and embryoids, which can regenerate directly into transgenic plants. It has not been convincingly established that naked DNA can be taken up by egg cells or integrated into them through pollen tubes and that the altered phenotypes of the transformed offspring are heritable. To some extent this is compensated for by the rigorous approaches employed to transform pollen embryoids, zygotic embryos, embryogenic cells, and somatic embryos so that they can regenerate genetically engineered phenotypes of some of our agricultural crops. When techniques for the transformation of the egg or zygote are perfected, they may provide a strategic testing ground for the genetic engineering of the plant phenotype beginning with the first cell of the sporophyte. Therefore, systems that may be uniquely suited to this approach deserve to be explored widely.

References

Note: For a given author, all references with one or two coauthors are first listed alphabetically. All *et al* references are listed alphabetically after the last three-author reference for a given author.

Aalen, R.B. 1995. The transcripts encoding two oleosin isoforms are both present in the aleurone and in the embryo of barley (*Hordeum vulgare* L.) seeds. *Plant Mol. Biol.* 28:583–588.

Aalen, R.B., Opsahl-Ferstad, H.-G., Linnestad, C., and Olsen, O.-A. 1994. Transcripts encoding an oleosin and a dormancy-related protein are present in both the aleurone layer and the embryo of developing barley (*Hordeum vulgare* L.) seeds. *Plant J.* 5:385–396.

Aarts, M.G.M., Dirkse, W.G., Stiekema, W.J., and Pereira, A. 1993. Transposon tagging of male sterility gene in *Arabidopsis*. *Nature* 363:715–717.

Aarts, M.G.M., Keijzer, C.J., Stiekema, W.J., and Pereira, A. 1995. Molecular characterization of the *CER1* gene of *Arabidopsis* involved in epicuticular wax biosynthesis and pollen fertility. *Plant Cell* 7:2115–2127.

Abad, A.R., Mehrtens, B.J., and Mackenzie, S.A. 1995. Specific expression in reproductive tissues and fate of a mitochondrial sterility-associated protein in cytoplasmic male-sterile bean. *Plant Cell* 7:271–285.

Abbott, A.G., Ainsworth, C.C., and Flavell, R.B. 1984. Characterization of anther differentiation in cytoplasmic male sterile maize using a specific isozyme system (esterase). *Theor. Appl. Genet.* 67:469–473.

Abdalla, M.M.F., and Hermsen, J.G.T. 1972. Unilateral incompatibility: hypothesis, debate and its implications for plant breeding. *Euphytica* 21:32–47.

Abe, T., and Futsuhara, Y. 1989. Selection of higher regenerative callus and change in isozyme pattern in rice (*Oryza sativa* L.). *Theor. Appl. Genet.* 78:648–652.

Acevedo, A., and Scandalios, J.G. 1990. Expression of the catalase and superoxide dismutase genes in mature pollen in maize. *Theor. Appl. Genet.* 80:705–711.

Ackerson, R.C. 1984a. Regulation of soybean embryogenesis by abscisic acid. *J. Expt. Bot.* 35:403–413.

Ackerson, R.C. 1984b. Abscisic acid and precocious germination in soybeans. *J. Expt. Bot.* 35:414–421.

Adler, K., and Müntz, K. 1983. Origin and development of protein bodies in cotyledons of *Vicia faba*. Proposal for a uniform mechanism. *Planta* 157:401–410.

Agarwal, S. 1962. Embryology of *Quinchamalium chilense* Lam. In *Plant Embryology – A Symposium*, pp. 162–169. New Delhi: Council of Scientific & Industrial Research.

Aguirre, P.J., and Smith, A.G. 1993. Molecular characterization of a gene encoding a cysteine-rich protein preferentially expressed in anthers of *Lycopersicon esculentum*. *Plant Mol. Biol.* 23:477–487.

Ahokas, H. 1978. Cytoplasmic male sterility in barley. II. Physiology and anther cytology of *msm 1*. *Hereditas* 89:7–21.

Ahokas, H. 1980. Cytoplasmic male sterility in barley. V. Physiological characterization of the msm1-Rfm1a system. *Physiol. Plant.* 48:231–238.

Ahokas, H. 1982a. Cytoplasmic male sterility in barley. VIII. Lipoxygenase activity and anther amino nitrogen in the *msm1-Rfm1a* system. *Plant Physiol.* 69:268–272.

Ahokas, H. 1982b. Cytoplasmic male sterility in barley: evidence for the involvement of cytokinins in fertility restoration. *Proc. Natl Acad. Sci. USA* 79:7605–7608.

Ai, Y., Kron, E., and Kao, T.-H. 1991. S-alleles are retained and expressed in a self-compatible cultivar of *Petunia hybrida*. *Mol. Gen. Genet.* 230:353–358.

Ai, Y., Singh, A., Coleman, C.E., Ioerger, T.R., Kheyr-Pour, A., and Kao, T.-H. 1990. Self-incompatibility in *Petunia inflata*: isolation and characterization of cDNAs encoding three S-allele-associated proteins. *Sex. Plant Reprod.* 3:130–138.

Ainsworth, C., Crossley, S., Buchanan-Wollaston, V., Thangavelu, M., and Parker, J. 1995. Male and female flowers of the dioecious plant sorrel show different patterns of MADS box gene expression. *Plant Cell* 7:1583–1598.

Akatsuka, T., and Nelson, O.E. 1966. Starch granule-bound adenosine diphosphate glucose-starch glucosyltransferase of maize seeds. *J. Biol. Chem.* 241:2280–2285.

Akhundova, G.G., Grinikh, L.I., and Shevchenko, V.V. 1978. The embryogenesis in *Arabidopsis thaliana* following the γ-irradiation of the plants in the generative phase. *Ontogenez* 9:514–519.

Alam, S., and Sandal, P.C. 1969. Electrophoretic analyses of anther proteins from male-fertile and male-sterile Sudangrass, *Sorghum vulgare* var. *sudanense* (Piper). *Crop Sci.* 9:157–159.

Albani, D., Altosaar, I., Arnison, P.G., and Fabijanski, S.F. 1991. A gene showing sequence similarity to pectin esterase is specifically expressed in developing pollen of *Brassica napus*. Sequences in its 5' flanking region are conserved in other pollen-specific promoters. *Plant Mol. Biol.* 16:501–513.

Albani, D., Robert, L.S., Donaldson, P.A., Altosaar, I., Arnison, P.G., and Fabijanski, S.F. 1990. Characterization of a pollen-specific gene family from *Brassica napus* which is activated during early microspore development. *Plant Mol. Biol.* 15:605–622.

Albani, D., Sardana, R., Robert, L.S., Altosaar, I., Arnison, P.G., and Fabijanski, S.F. 1992. A *Brassica napus* gene family which shows sequence similarity to ascorbate oxidase is expressed in developing pollen. Molecular characterization and analysis of promoter activity in transgenic tobacco plants. *Plant J.* 2:331–342.

Albertini, L. 1965. Étude autoradiographique des synthèses d'acide ribonucléique (RNA) au cours de la microsporogenèse chez le *Rhoeo discolor* (Hance). *Compt. Rend. Acad. Sci. Paris* 260:651–653.

Albertini, L. 1967. Étude autoradiographique des synthèse de protéines au cours de la microsporogenèse chez le *Rhoeo discolor* (Hance). *Compt. Rend. Acad. Sci. Paris* 264D:2773–2776.

Albertini, L. 1971. Les acides nucléiques et les protéines au cours de la microsporogénèse chez le *Rhoeo discolor* Hance. Etude autoradiographique et cytophotométrique. *Rev. Cytol. Biol. Végét.* 34:49–92.

Albertini, L. 1975. Étude autoradiographique de l'incorporation du tryptophane-^{3}H dans les microsporocytes et dans le tapis du *Rhoeo discolor* Hance. *Caryologia* 28:445–458.

Albertini, L., Grenet-Auberger, H., and Souvré, A. 1981. Étude autoradiographique de l'incorporation de précurseurs des polysaccharides et des lipides dans les microsporocytes et le tapis du *Rhoeo discolor* Hance. *Caryologia* 34:53–68.

Albertini, L., Souvré, A., and Audran, J.C. 1987. Le tapis de l'anthère et ses relations avec les microsporocytes et les grains de pollen. *Rev. Cytol. Biol. Végét.-Bot.* 10:211–242.

Albertsen, M.C., and Palmer, R.G. 1979. A comparative light- and electron-microscopic study of microsporogenesis in male sterile (MS_1) and male fertile soybeans (*Glycine max* (L.) Merr.). *Am. J. Bot.* 66:253–265.

Albertsen, M.C., and Phillips, R.L. 1981. Developmental cytology of 13 genetic male sterile loci in maize. *Can. J. Genet. Cytol.* 23:195–208.

Alché, J.D., Fernández, M.C., and Rodríguez-García, M.I. 1994. Cytochemical features common to nucleoli and cytoplasmic nucleoloids of *Olea europaea* meiocytes: detection of rRNA by *in situ* hybridization. *J. Cell Sci.* 107:621–629.

Aldrich, H.C., and Vasil, I.K. 1970. Ultrastructure of the postmeiotic nuclear envelope in microspores of *Podocarpus macrophyllus*. *J. Ultrastr. Res.* 32:307–315.

Aleith, F., and Richter, G. 1990. Gene expression during induction of somatic embryogenesis in carrot cell suspensions. *Planta* 183:17–24.

Alexander, R., Alamillo, J.M., Salamini, F., and Bartels, D. 1994. A novel embryo-specific barley cDNA clone encodes a protein with homologies to bacterial glucose and ribitol dehydrogenase. *Planta* 192:519–525.

Allen, R.D., Bernier, F., Lessard, P.A., and Beachy, R.N. 1989. Nuclear factors interact with a soybean β-conglycinin enhancer. *Plant Cell* 1:623–631.

Allen, R.D., Nessler, C.L., and Thomas, T.L. 1985. Developmental expression of sunflower 11S storage protein genes. *Plant Mol. Biol.* 5:165–173.

Allen, R.D., Trelease, R.N., and Thomas, T.L. 1988. Regulation of isocitrate lyase gene expression in sunflower. *Plant Physiol.* 86:527–532.

Allen, R.D., Cohen, E.A., Vonder Haar, R.A., Adams, C.A., Ma, D.P., Nessler, C.L., and Thomas, T.L. 1987. Sequence and expression of a gene encoding an albumin storage protein in sunflower. *Mol. Gen. Genet.* 210:211–218.

Allen, R.L., and Lonsdale, D.M. 1992. Sequence analysis of three members of the maize polygalacturonase gene family expressed during pollen development. *Plant Mol. Biol.* 20:343–345.

Allen, R.L., and Lonsdale, D.M. 1993. Molecular characterization of one of the maize polygalacturonase gene family members which are expressed during late pollen development. *Plant J.* 3:261–271.

Allison, D.C., and Fisher, W.D. 1964. A dominant gene for male-sterility in upland cotton. *Crop Sci.* 4:548–549.

Almoguera, C., Coca, M.C., and Jordano, J. 1995. Differential accumulation of sunflower tetraubiquitin mRNAs during zygotic embryogenesis and developmental regulation of their heat-shock response. *Plant Physiol.* 107:765–773.

Almoguera, C., and Jordano, J. 1992. Developmental and environmental concurrent expression of sunflower dry-seed-stored low-molecular-weight heat-shock protein and Lea mRNAs. *Plant Mol. Biol.* 19:781–792.

Alpi, A., Tognoni, F., and D'Amato, F. 1975. Growth regulator levels in embryo and suspensor of *Phaseolus coccineus* at two stages of development. *Planta* 127:153–162.

Alpi, A., Lorenzi, R., Cionini, P.G., Bennici, A., and D'Amato, F. 1979. Identification of gibberellin A_1 in the embryo suspensor of *Phaseolus coccineus*. *Planta* 147:225–228.

Altenbach, S.B., Kuo, C.-C., Staraci, L.C., Pearson, K.W., Wainwright, C., Georgescu, A., and Townsend, J. 1992. Accumulation of a Brazil nut albumin in seeds of transgenic canola results in enhanced levels of seed protein methionine. *Plant Mol. Biol.* 18:235–245.

Altenbach, S.B., Pearson, K.W., Leung, F.W., and Sun, S.S.M. 1987. Cloning and sequence analysis of a cDNA encoding a Brazil nut protein exceptionally rich in methionine. *Plant Mol. Biol.* 8:239–250.

Altenbach, S.B., Pearson, K.W., Meeker, G., Staraci, L.C., and Sun, S.S.M. 1989. Enhancement of the methionine content of seed proteins by the expression of a chimeric gene encoding a methionine-rich protein in transgenic plants. *Plant Mol. Biol.* 13:513–522.

Altschuler, M., and Mascarenhas, J.P. 1982. Heat shock proteins and effects of heat shock in plants. *Plant Mol. Biol.* 1:103–115.

Altschuler, Y., Rosenberg, N., Harel, R., and Galili, G. 1993. The N- and C-terminal regions regulate the transport of wheat γ–gliadin through the endoplasmic reticulum in *Xenopus* oocytes. *Plant Cell* 5:443–450.

Alvarez, M.R. 1968. Quantitative changes in nuclear DNA accompanying postgermination embryonic development in *Vanda* (Orchidaceae). *Am. J. Bot.* 55:1036–1041.

Alvarez, M.R., and Sagawa, Y. 1965a. A histochemical study of embryo sac development in *Vanda* (Orchidaceae). *Caryologia* 18:241–249.

Alvarez, M.R., and Sagawa, Y. 1965b. A histochemical study of embryo development in *Vanda* (Orchidaceae). *Caryologia* 18:251–261.

Amarasinghe, V., Dhami, R., and Carlson, J.E. 1996. Polyamine biosynthesis during somatic embryogenesis in interior spruce (*Picea glauca* × *P. engelmannii* complex). *Plant Cell Rep.* 15:495–499.

Ambrose, M.J., Wang, T.L., Cook, S.K., and Hedley, C.L. 1987. An analysis of seed development in *Pisum sativum*. IV. Cotyledon cell populations *in vivo* and *in vitro*. *J. Expt. Bot.* 38:1909–1920.

Ammirato, P.V. 1974. The effects of abscisic acid on the development of somatic embryos from cells of caraway (*Carum carvi* L.). *Bot. Gaz.* 135:328–337.

Ammirato, P.V. 1977. Hormonal control of somatic embryo development from cultured cells of caraway. Interactions of abscisic acid, zeatin, and gibberellic acid. *Plant Physiol.* 59:579–586.

Ammirato, P.V., and Steward, F.C. 1969. Indirect effects of irradiation: morphogenetic effects of irradiated sucrose. *Devel Biol.* 19:87–106.

An, Y.-Q., Huang, S., McDowell, J.M., McKinney, E.C., and Meagher, R.B. 1996. Conserved expression of the *Arabidopsis ACT1* and *ACT3* actin subclass in organ primordia and mature pollen. *Plant Cell* 8:15–30.

Anand, V.V., Arekal, G.D., and Swamy, B.G.L. 1980. Chimeral embryoids of pollen origin in tobacco. *Curr. Sci.* 49:603–604.

Anderberg, R.J., and Walker-Simmons, M.K. 1992. Isolation of a wheat cDNA clone for an abscisic acid-inducible transcript with homology to protein kinases. *Proc. Natl Acad. Sci. USA* 89:10183–10187.

Anderson, J.M., and Barrett, S.C.H. 1986. Pollen tube growth in tristylous *Pontederia cordata* (Pontederiaceae). *Can. J. Bot.* 64:2602–2607.

Anderson, M.A., Sandrin, M.S., and Clarke, A.E. 1984. A high proportion of hybridomas raised to a plant extract secrete antibody to arabinose or galactose. *Plant Physiol.* 75:1013–1016.

Anderson, M.A., Cornish, E.C., Mau, S.-L., Williams, E.G., Hoggart, R., Atkinson, A., Bonig, I., Grego, B., Simpson, R., Roche, P.J., Haley, J.D., Penschow, J.D., Niall, H.D., Tregear, G.W., Coghlan, J.P., Crawford, R.J., and Clarke, A.E. 1986. Cloning of cDNA for a stylar glycoprotein associated with expression of self-incompatibility in *Nicotiana alata*. *Nature* 321:38–44.

Anderson, M.A., Harris, P.J., Bonig, I., and Clarke, A.E. 1987. Immuno-gold localization of α–L-arabinofuranosyl residues in pollen tubes of *Nicotiana alata* Link et Otto. *Planta* 171:438–442.

Anderson, M.A., McFadden, G.I., Bernatzky, R., Atkinson, A., Orpin, T., Dedman, H., Tregear, G., Fernley, R., and Clarke, A.E. 1989. Sequence variability of three alleles of the self-incompatibility gene of *Nicotiana alata*. *Plant Cell* 1:483–491.

Anderson, O.D., Greene, F.C., Yip, R.E., Halford, N.G., Shewry, P.R., and Malpica-Romero, J.-M. 1989. Nucleotide sequences of the two high-molecular-weight glutenin genes from the D-genome of a hexaploid bread wheat, *Triticum aestivum* L. cv. Cheyenne. *Nucl. Acids Res.* 17:461–462.

Anderson, O.D., Litts, J.C., Gautier, M.-F., and Greene, F.C. 1984. Nucleic acid sequence and chromosome assignment of a wheat storage protein gene. *Nucl. Acids Res.* 12:8129–8144.

Angenent, G.C., Busscher, M., Franken, J., Dons, H.J.M., and van Tunen, A.J. 1995a. Functional interaction between the homeotic genes *fbp1* and *pMADS1* during *Petunia* floral organogenesis. *Plant Cell* 7:507–516.

Angenent, G.C., Busscher, M., Franken, J., Mol, J.N.M., and van Tunen, A.J. 1992. Differential expression of two MADS box genes in wild-type and mutant *Petunia* flowers. *Plant Cell* 4:983–993.

Angenent, G.C., Franken, J., Busscher, M., Colombo, L., and van Tunen, A.J. 1993. Petal and stamen formation in *Petunia* is regulated by the homeotic gene *fbp1*. *Plant J.* 4:101–112.

Angenent, G.C., Franken, J., Busscher, M., van Dijken, A., van Went, J.L., Dons, H.J.M., and van Tunen, A.J. 1995b. A novel class of MADS box genes is involved in ovule development in *Petunia*. *Plant Cell* 7:1569–1582.

Angold, R.E. 1967. The ontogeny and fine structure of the pollen grain of *Endymion non-scriptus*. *Rev. Palaeobot. Palynol.* 3:205–212.

Angold, R.E. 1968. The formation of the generative cell in the pollen grain of *Endymion non-scriptus* (L.). *J. Cell Sci.* 3:573–578.

Anthony, J.L., Vonder Haar, R.A., and Hall, T.C. 1991. Nucleotide sequence of a genomic clone encoding arcelin, a lectin-like seed protein from *Phaseolus vulgaris*. *Plant Physiol.* 97:839–840.

Aoyagi, K., Bassham, J.A., and Greene, F.C. 1984. Pyruvate orthophosphate dikinase gene expression in developing wheat seeds. *Plant Physiol.* 75:393–396.

Aoyagi, K., and Chua, N.-H. 1988. Cell-specific expression of pyruvate, Pi dikinase. *In situ* mRNA hybridization and immunolocalization labeling of protein in wheat seed. *Plant Physiol.* 86:364–368.

Appels, R., Bouchard, R.A., and Stern, H. 1982. cDNA clones from meiotic-specific poly(A)$^+$RNA in *Lilium*: homology with sequences in wheat, rye, and maize. *Chromosoma* 85:591–602.

Apuya, N.R., and Zimmerman, J.L. 1992. Heat shock gene expression is controlled primarily at the translational level in carrot cells and somatic embryos. *Plant Cell* 4:657–665.

Aragao, F.J.L., Grossi de Sa, M.-F., Davey, M.R., Brasileiro, A.C.M., Faria, J.C., and Rech, E.L. 1993. Factors influencing transient gene expression in bean (*Phaseolus vulgaris* L.) using an electrical particle acceleration device. *Plant Cell Rep.* 12:483–490.

Arahira, M., and Fukazawa, C. 1994. *Ginkgo* 11S seed storage protein family mRNA: unusual Asn-Asn linkage as post-transitional cleavage site. *Plant Mol. Biol.* 25:597–605.

Arasu, N.T. 1968. Self-incompatibility in angiosperms: a review. *Genetica* 39:1–24.

Arditti, J. 1979. Aspects of the physiology of orchids. *Adv. Bot. Res.* 7:421–655.

Arekal, G.D., and Ramaswamy, S.N. 1973. Embryology of *Burmannia pusilla* (Wall. ex Miers) THW. and its taxonomic status. *Beitr. Biol. Pflanz.* 49:35–45.

Arnison, P.G., Donaldson, P., Ho, L.C.C., and Keller, W.A. 1990. The influence of various physical parameters on anther culture of broccoli (*Brassica oleracea* var. *italica*). *Plant Cell Tissue Organ Cult.* 20:147–155.

Arrigoni, O., de Gara, L., Tommasi, F., and Liso, R. 1992. Changes in the ascorbate system during seed development of *Vicia faba* L. *Plant Physiol.* 99:235–238.

Artschwager, E. 1947. Pollen degeneration in male-sterile sugar beets, with special reference to the tapetal plasmodium. *J. Agr. Res.* 75:191–197.

Aryan, A.P., An, G., and Okita, T.W. 1991. Structural and functional analysis of promoter from gliadin, an endosperm-specific storage protein gene of *Triticum aestivum* L. *Mol. Gen. Genet.* 225:65–71.

Asahi, T., Kumashiro, T., and Kubo, T. 1988. Constitution of mitochondrial and chloroplast genomes in male sterile tobacco obtained by protoplast fusion of *Nicotiana tabacum* and *N. debneyi*. *Plant Cell Physiol.* 29:43–49.

Ascher, P.D. 1966. A gene action model to explain gametophytic self-incompatibility. *Euphytica* 15:179–183.

Ascher, P.D. 1971. The influence of RNA-synthesis inhibitors on *in vivo* pollen tube growth and the self-incompatibility reaction in *Lilium longiflorum* Thunb. *Theor. Appl. Genet.* 41:75–78.

Ascher, P.D. 1975. Special stylar property required for compatible pollen-tube growth in *Lilium longiflorum* Thunb. *Bot. Gaz.* 136:317–321.

Ascher, P.D. 1977. Localization of the self- and the interspecific-incompatibility reactions in style sections of *Lilium longiflorum*. *Plant Sci. Lett.* 10:199–203.

Ascher, P.D., and Drewlow, L.W. 1975. The effect of prepollination injection with stigmatic exudate on interspecific pollen tube growth in *Lilium longiflorum* Thunb. styles. *Plant Sci. Lett.* 4:401–405.

Ascher, P.D., and Peloquin, S.J. 1966a. Effect of floral aging on the growth of compatible and incompatible pollen tubes in *Lilium longiflorum*. *Am. J. Bot.* 53:99–102.

Ascher, P.D., and Peloquin, S.J. 1966b. Influence of temperature on incompatible and compatible pollen tube growth in *Lilium longiflorum*. *Can. J. Genet. Cytol.* 8:661–664.

Ascher, P.D., and Peloquin, S.J. 1968. Pollen tube growth and incompatibility following intra- and inter-specific pollinations in *Lilium longiflorum*. *Am. J. Bot.* 55:1230–1234.

Asghar, R., Fenton, R.D., DeMason, D.A., and Close, T.J. 1994. Nuclear and cytoplasmic localization of maize embryo and aleurone dehydrin. *Protoplasma* 177:87–94.

Ashihara, H., Fujimura, T., and Komamine, A. 1981. Pyrimidine nucleotide biosynthesis during somatic embryogenesis in a carrot cell suspension culture. *Z. Pflanzenphysiol.* 104:129–137.

Ashley, T. 1972. Zygote shrinkage and subsequent development in some *Hibiscus* hybrids. *Planta* 108:303–317.

Ashley, T. 1975a. Fine structure of early endosperm development in *Hibiscus*. *Caryologia* 28:63–71.

Ashley, T. 1975b. Alterations in the fine structure of developing endosperm of *Hibiscus* hybrids. *Caryologia* 28:73–80.

Asker, S.E., and Jerling, L. 1992. *Apomixis in Plants*. Boca Raton: CRC Press.

Aspart, L., Meyer, Y., Laroche, M., and Penon, P. 1984. Developmental regulation of the synthesis of proteins encoded by stored mRNA in radish embryos. *Plant Physiol.* 76:664–673.

Aspinall, G.O., and Rosell, K.-G. 1978. Polysaccharide component in the stigmatic exudate from *Lilium longiflorum*. *Phytochemistry* 17:919–922.

Aspuria, E.T., Nagato, Y., and Uchimiya, H. 1994. Histochemical localization of a chimeric gene (*rolC-GUS*) expression in zygotic embryos of transgenic tobacco plants. *Ann. Bot.* 73:465–469.

Åström, H., Sorri, O., and Raudaskoski, M. 1995. Role of microtubules in the movement of the vegetative nucleus and generative cell in tobacco pollen tubes. *Sex. Plant Reprod.* 8:61–69.

Åström, H., Virtanen, I., and Raudaskoski, M. 1991. Cold-stability in the pollen tube cytoskeleton. *Protoplasma* 160:99–107.

Atkinson, A.H., Heath, R.L., Simpson, R.J., Clarke, A.E., and Anderson, M.A. 1993. Proteinase inhibitors in *Nicotiana alata* stigmas are derived from a precursor protein which is processed into five homologous inhibitors. *Plant Cell* 5:203–213.

Atkinson, B.G., Raizada, M., Bouchard, R.A., Frappier, J.R.H., and Walden, D.B. 1993. The independent stage-specific expression of the 18-kDa heat shock protein genes during microsporogenesis in *Zea mays* L. *Devel Genet.* 14:15–26.

Attia, M.S. 1950. The nature of incompatibility in cabbage. *Proc. Am. Soc. Hort. Sci.* 56:369–371.

Auboiron, E., Carron, M.-P., and Michaux-Ferrière, N. 1990. Influence of atmospheric gases, particularly ethylene, on somatic embryogenesis of *Hevea brasiliensis*. *Plant Cell Tissue Organ Cult.* 21:31–37.

Audran, J.C., and Dicko-Zafimahova, L.D. 1992. Aspects ultrastructuraux et cytochimiques du tapis staminal chez *Calotropis procera* (Asclepiadaceae). *Grana* 31:253–272.

Aukerman, M.J., Schmidt, R.J., Burr, B., and Burr, F.A. 1991. An arginine to lysine substitution in the bZIP domain of an *opaque-2* mutant in maize abolishes specific DNA binding. *Genes Devel.* 5:310–320.

Avanzi, S., Cionini, P.G., and D'Amato, F. 1970. Cytochemical and autoradiographic analyses on the embryo suspensor cells of *Phaseolus coccineus*. *Caryologia* 23:605–638.

Avanzi, S., Corsi, G., D'Amato, F., Floris, C., and Meletti, P. 1967. The chromosome breaking effect of the irradiated endosperm in water-soaked seeds of *durum* wheat. *Mutation Res.* 4:704–707.

Avjioglu, A., and Knox, R.B. 1989. Storage lipid accumulation by zygotic and somatic embryos in culture. *Ann. Bot.* 63:409–420.

Ba, L.T., Cavé, G., Henry, M., and Guignard, J.-L. 1978. Embryogénie des Potamogétonacées. Étude en microscopie électronique à balayage de l'origine du cotylédon chez *Potamogeton lucens* L. *Compt. Rend. Acad. Sci. Paris* 286D:1351–1353.

Baaziz, M., Aissam, F., Brakez, Z., Bendiab, K., El Hadrami, I., and Cheikh, R. 1994. Electrophoretic patterns of acid soluble proteins and active isoforms of peroxidase and polyphenoloxidase typifying calli and somatic embryos of two reputed date palm cultivars in Morocco. *Euphytica* 76:159–168.

Babbar, S.B., and Gupta, S.C. 1980. Chilling induced androgenesis in anthers of *Petunia hybrida* without any culture medium. *Z. Pflanzenphysiol.* 100:279–283.

Babbar, S.B., and Gupta, S.C. 1986. Putative role of ethylene in *Datura metel* microspore embryogenesis. *Physiol. Plant.* 68:141–144.

Bacic, A., Gell, A.C., and Clarke, A.E. 1988. Arabinogalactan proteins from stigmas of *Nicotiana alata*. *Phytochemistry* 27:679–684.

Backs-Hüsemann, D., and Reinert, J. 1970. Embryobildung durch isolierte Einzelzellen aus Gewebekulturen von *Daucus carota*. *Protoplasma* 70:49–60.

Badenoch-Jones, J., Spencer, D., Higgins, T.J.V., and Millerd, A. 1981. The role of glycosylation in storage-protein synthesis in developing pea seeds. *Planta* 153:201–209.

Bae, J.M., Giroux, M., and Hannah, L. 1990. Cloning and characterization of the *brittle-2* gene of maize. *Maydica* 35:317–322.

Baerson, S.R., and Lamppa, G.K. 1993. Developmental regulation of an acyl carrier protein gene promoter in vegetative and reproductive tissues. *Plant Mol. Biol.* 22:255–267.

Baerson, S.R., Vander Heiden, M.G., and Lamppa, G.K. 1994. Identification of domains in an *Arabidopsis* acyl carrier protein gene promoter required for maximal organ-specific expression. *Plant Mol. Biol.* 26:1947–1959.

Bagga, S., Adams, H., Kemp, J.D., and Sengupta-Gopalan, C. 1995. Accumulation of 15-kilodalton zein in novel protein bodies in transgenic tobacco. *Plant Physiol.* 107:13–23.

Bagga, S., Sutton, D., Kemp, J.D., and Sengupta-Gopalan, C. 1992. Constitutive expression of the β-phaseolin gene in different tissues of transgenic alfalfa does not ensure phaseolin accumulation in non-seed tissue. *Plant Mol. Biol.* 19:951–958.

Bagni, N., Adamo, P., and Serafini-Fracassini, D. 1981. RNA, proteins and polyamines during tube growth in germinating apple pollen. *Plant Physiol.* 68:727–730.

Bai, S., and Sung, Z.R. 1995. The role of *EMF1* in regulating the vegetative and reproductive transition in *Arabidopsis thaliana* (Brassicaceae). *Am. J. Bot.* 82:1095–1103.

Bailey, C.J., Cobb, A., and Boulter, D. 1970. A cotyledon slice system for the electron autoradiographic study of the synthesis and intracellular transport of the seed storage protein of *Vicia faba. Planta* 95:103–118.

Bailey, D.S., DeLuca, V., Dürr, M., Verma, D.P.S., and Maclachlan, G.A. 1980. Involvement of lipid-linked oligosaccharides in synthesis of storage glycoproteins in soybean seeds. *Plant Physiol.* 66:1113–1118.

Bailey-Serres, J., Dixon, L.K., Liddell, A.D., and Leaver, C.J. 1986a. Nuclear-mitochondrial interactions in cytoplasmic male-sterile *Sorghum. Theor. Appl. Genet.* 73:252–260.

Bailey-Serres, J., Hanson, D.K., Fox, T.D., and Leaver, C.J. 1986b. Mitochondrial genome rearrangement leads to extension and relocation of the cytochrome *c* oxidase subunit I gene in *Sorghum. Cell* 47:567–576.

Bajaj, Y.P.S. 1974a. Isolation and culture studies on pollen tetrad and pollen mother-cell protoplasts. *Plant Sci. Lett.* 3:93–99.

Bajaj, Y.P.S. 1974b. Induction of repeated cell division in isolated pollen mother cells of *Atropa belladonna. Plant Sci. Lett.* 3:309–312.

Bajaj, Y.P.S. 1979. Test-tube fertilization and development of maize (*Zea mays* L.) plants. *Indian J. Expt. Biol.* 17:475–478.

Bajaj, Y.P.S. 1983. *In vitro* production of haploids. In *Handbook of Plant Cell Culture,* Vol. 1, ed. D.A. Evans, W.R. Sharp, P.V. Ammirato, and Y. Yamada, pp. 228–287. New York: Macmillan Publishing Co.

Bajaj, Y.P.S. 1990. *In vitro* production of haploids and their use in cell genetics and plant breeding. In *Biotechnology in Agriculture and Forestry 12. Haploids in Crop Improvement* I, ed. Y.P.S. Bajaj, pp. 3–44. Berlin: Springer-Verlag.

Bajer, A. 1965. Cine micrographic analysis of cell plate formation in endosperm. *Expt. Cell Res.* 37:376–398.

Baker, H.G. 1966. The evolution, functioning and breakdown of heteromorphic incompatibility systems. I. The Plumbaginaceae. *Evolution* 20:349–368.

Baker, J., Steele, C., and Dure, L., III. 1988. Sequence and characterization of 6 *Lea* proteins and their genes from cotton. *Plant Mol. Biol.* 11:277–291.

Baker, S.S., Rugh, C.L., Whitmore, F.W., and Kamalay, J.C. 1996. Genes encoding 11S-globulin-like proteins are expressed in the megagametophyte soon after fertilization in eastern white pine (*Pinus strobus* L.). *Int. J. Plant Sci.* 157:453–461.

Bal, A.K., and De, D.N. 1961. Developmental changes in the submicroscopic morphology of the cytoplasmic components during microsporogenesis in *Tradescantia. Devel Biol.* 3:241–254.

Balconi, C., Rizzi, E., Motto, M., Salamini, F., and Thompson, R. 1993. The accumulation of zein polypeptides and zein mRNA in cultured endosperms of maize is modulated by nitrogen supply. *Plant J.* 3:325–334.

Baldan, B., Frattini, C., Guzzo, F., Branca, C., Terzi, M., Mariani, P., and Lo Schiavo, F. 1995. A stage-specific block is produced in carrot somatic embryos by 1,2-benzisoxazole-3-acetic acid. *Plant Sci.* 108:85–92.

Baldi, B.G., Franceschi, V.R., and Loewus, F.A. 1987. Preparation and properties of pollen sporoplasts. *Protoplasma* 141:47–55.

Baldwin, T.C., Coen, E.S., and Dickinson, H.G. 1992. The *ptl1* gene expressed in the transmitting tissue of *Antirrhinum* encodes an extensin-like protein. *Plant J.* 2:733–739.

Balsamo, R.A., Wang, J.-L., Eckard, K.J., Wang, C.-S., and Lord, E.M. 1995. Immunogold localization of a developmentally regulated, tapetal-specific, 15 kDa lily anther protein. *Protoplasma* 189:17–25.

Baltz, R., Domon, C., Pillay, D.T.N., and Steinmetz, A. 1992a. Characterization of a pollen-specific cDNA from sunflower encoding a zinc finger protein. *Plant J.* 2:713–721.

Baltz, R., Evrard, J.-L., Domon, C., and Steinmetz, A. 1992b. A LIM motif is present in a pollen-specific protein. *Plant Cell* 4:1465–1466.

Banaś, M., Tirlapur, U.K., Charzyńska, M., and Cresti, M. 1996. Some events of mitosis and cytokinesis in the generative cell of *Ornithogalum virens* L. *Planta* 190:202–208.

Banerjee, U.C., Rowley, J.R., and Alessio, M. 1965. Exine plasticity during pollen grain maturation. *J. Palynol.* 1:70–89.

Banga, S.S., Labana, K.S., and Banga, S.K. 1984. Male sterility in Indian mustard (*Brassica juncea* (L.) Coss.) – a biochemical characterization. *Theor. Appl. Genet.* 67:515–519.

Bapat, S.A., Rawal, S.K., and Mascarenhas, A.F. 1992. Isozyme profiles during ontogeny of somatic embryos in wheat (*Triticum aestivum* L.). *Plant Sci.* 82:235–242.

Barcelo, P., Hagel, C., Becker, D., Martin, A., and Lörz, H. 1994. Transgenic cereal (Tritordeum) plants obtained at high efficiency by microprojectile bombardment of inflorescence tissue. *Plant J.* 5:583–592.

Barker, S.J., Harada, J.J., and Goldberg, R.B. 1988. Cellular localization of soybean storage protein mRNA in transformed tobacco seeds. *Proc. Natl Acad. Sci. USA* 85:458–462.

Barna, K.S., and Mehta, S.L. 1995. Genetic transformation and somatic embryogenesis in *Lathyrus sativus. J. Plant Biochem. Biotech.* 4:67–71.

Barnabás, B., Fransz, P.F., and Schel, J.H.N. 1987. Ultrastructural studies on pollen embryogenesis in maize (*Zea mays* L.). *Plant Cell Rep.* 6:212–215.

Barnabás, B., and Fridvalszky, L. 1984. Adhesion and germination of differentially treated maize pollen grains on the stigma. *Acta Bot. Hung.* 30:329–332.

Barnabás, B., and Kovács, G. 1992. *In vitro* pollen maturation and successful seed production in detached spikelet cultures in wheat (*Triticum aestivum* L). *Sex. Plant Reprod.* 5:286–291.

Barnes, S.H., and Blackmore, S. 1987. Preliminary observations on the formation of the male germ unit in

Catananche caerulea L. (Compositae: Lactuceae). Brief report. *Protoplasma* 138:187–189.

Barratt, D.H.P. 1986a. Modulation by abscisic acid of storage protein accumulation in *Vicia faba* L. cotyledons cultured *in vitro*. *Plant Sci.* 46:159–167.

Barratt, D.H.P. 1986b. Regulation of storage protein accumulation by abscisic acid in *Vicia faba* L. cotyledons cultured *in vitro*. *Ann. Bot.* 57:245–256.

Barratt, D.H.P., and Clark, J.A. 1991. Proteins arising during the late stages of embryogenesis in *Pisum sativum* L. *Planta* 184:14–23.

Barratt, D.H.P., and Clark, J.A. 1993. A stress-induced, developmentally regulated, highly polymorphic protein family in *Pisum sativum* L. *Planta* 191:7–17.

Barratt, D.H.P., Domoney, C., and Wang, T.L. 1989. Purification and partial characterisation of two abscisic-acid-responsive proteins induced in cultured embryos of *Pisum sativum* L. *Planta* 180:16–23.

Barratt, D.H.P., and Flavell, R.B. 1975. Alterations in mitochondria associated with cytoplasmic and nuclear genes concerned with male sterility in maize. *Theor. Appl. Genet.* 45:315–321.

Barratt, D.H.P., Whitford, P.N., Cook, S.K., Butcher, G., and Wang, T.L. 1989. An analysis of seed development in *Pisum sativum*. VIII. Does abscisic acid prevent precocious germination and control storage protein synthesis? *J. Expt. Bot.* 40:1009–1014.

Barrett, S.C.H. 1977. The breeding system of *Pontederia rotundifolia* L., a tristylous species. *New Phytol.* 78:209–220.

Barrett, S.C.H., ed. 1992. *Evolution and Function of Heterostyly*. Berlin: Springer-Verlag.

Bartalesi, A., Del Casino, C., Moscatelli, A., Cai, G., and Tiezzi, A. 1991. Confocal laser scanning microscopy of the microtubular system of dividing generative cell in *Nicotiana tabacum* pollen tube. *Giorn. Bot. Ital.* 125:21–28.

Bartels, D., Singh, M., and Salamini, F. 1988. Onset of desiccation tolerance during development of the barley embryo. *Planta* 175:485–492.

Bartels, D., and Thompson, R.D. 1983. The characterization of cDNA clones coding for wheat storage proteins. *Nucl. Acids Res.* 11:2961–2977.

Bartels, D., and Thompson, R.D. 1986. Synthesis of mRNAs coding for abundant endosperm proteins during wheat grain development. *Plant Sci.* 46:117–125.

Bartels, D., Thompson, R.D., and Rothstein, S. 1985. Synthesis of a wheat storage protein subunit in *Escherichia coli* using novel expression vectors. *Gene* 35:159–167.

Bartels, D., Altosaar, I., Harberd, N.P., Barker, R.F., and Thompson, R.D. 1986. Molecular analysis of γ–gliadin gene families at the complex *Gli-1* locus of bread wheat (*T. aestivum* L.). *Theor. Appl. Genet.* 72:845–853.

Bartels, D., Engelhardt, K., Roncarati, R., Schneider, K., Rotter, M., and Salamini, F. 1991. An ABA and GA modulated gene expressed in the barley embryo encodes an aldose reductase related protein. *EMBO J.* 10:1037–1043.

Barton, M.K., and Poethig, R.S. 1993. Formation of the shoot apical meristem in *Arabidopsis thaliana*: an analysis of development in the wild type and in the *shoot meristemless* mutant. *Development* 119:823–831.

Bashe, D., and Mascarenhas, J.P. 1984. Changes in potassium ion concentrations during pollen dehydration and germination in relation to protein synthesis. *Plant Sci. Lett.* 35:55–60.

Bass, H.W., Webster, C., OBrian, G.R., Roberts, J.K.M., and Boston, R.S. 1992. A maize ribosome-inactivating protein is controlled by the transcriptional activator *opaque-2*. *Plant Cell* 4:225–234.

Bassett, C.L., Mothershed, C.P., and Galau, G.A. 1988. Floral-specific polypeptides of the Japanese morning glory. *Planta* 175:221–228.

Bassüner, R., Bäumlein, H., Huth, A., Jung, R., Wobus, U., Rapoport, T.A., Saalbach, G., and Müntz, K. 1988. Abundant embryonic mRNA in field bean (*Vicia faba* L.) codes for a new class of seed proteins: cDNA cloning and characterization of the primary translation product. *Plant Mol. Biol.* 11:321–334.

Bassüner, R., Manteuffel, R., Müntz, K., Püchel, M., Schmidt, P., and Weber, E. 1983. Analysis of *in vivo* and *in vitro* globulin formation during cotyledon development of field beans (*Vicia faba* L. var. *minor*). *Biochem. Physiol. Pflanz.* 178:665–684.

Bastola, D.R., and Minocha, S.C. 1995. Increased putrescine biosynthesis through transfer of mouse ornithine decarboxylase cDNA in carrot promotes somatic embryogenesis. *Plant Physiol.* 109:63–71.

Baszczynski, C.L., and Fallis, L. 1990. Isolation and nucleotide sequence of a genomic clone encoding a new *Brassica napus* napin gene. *Plant Mol. Biol.* 14:633–635.

Batchelder, C., Ross, J.H.E., and Murphy, D.J. 1994. Synthesis and targeting of *Brassica napus* oleosin in transgenic tobacco. *Plant Sci.* 104:39–47.

Bateman, A.J. 1952. Self-incompatibility systems in angiosperms. I. Theory. *Heredity* 6:285–310.

Bateman, A.J. 1954. Self-incompatibility systems in angiosperms. II. *Iberis amara*. *Heredity* 8:305–332.

Bateman, A.J. 1955. Self-incompatibility systems in angiosperms. III. Cruciferae. *Heredity* 9:53–68.

Batygina, T.B. 1969. On the possibility of separation of a new type of embryogenesis in angiosperms. *Rev. Cytol. Biol. Végét.* 32:335–341.

Baumgartner, B., Tokuyasu, K.T., and Chrispeels, M.J. 1980. Immunocytochemical localization of reserve protein in the endoplasmic reticulum of developing bean (*Phaseolus vulgaris*) cotyledons. *Planta* 150:419–425.

Bäumlein, H., Boerjan, W., Nagy, I., Bassüner, R., van Montagu, M., Inzé, D., and Wobus, U. 1991a. A novel seed protein gene from *Vicia faba* is developmentally regulated in transgenic tobacco and *Arabidopsis* plants. *Mol. Gen. Genet.* 225:459–467.

Bäumlein, H., Boerjan, W., Nagy, I., Panitz, R., Inzé, D., and Wobus, U. 1991b. Upstream sequences regulating legumin gene expression in heterologous transgenic plants. *Mol. Gen. Genet.* 225:121–128.

Bäumlein, H., Miséra, S., Luerßen, H., Kölle, K., Horstmann, C., Wobus, U., and Müller, A.J. 1994. The *FUS3* gene of *Arabidopsis thaliana* is a regulator of gene expression during late embryogenesis. *Plant J.* 6:379–387.

Bäumlein, H., Müller, A.J., Schiemann, J., Helbing, D., Manteuffel, R., and Wobus, U. 1987. A legumin B gene of *Vicia faba* is expressed in developing seeds of transgenic tobacco. *Biol. Zentral.* 106:569–575.

Bäumlein, H., Müller, A.J., Schiemann, J., Helbing, D., Manteuffel, R., and Wobus, U. 1988. Expression of a *Vicia faba* legumin B gene in transgenic tobacco plants: gene dosage-dependent protein accumulation. *Biochem. Physiol. Pflanz.* 183:205–210.

Bäumlein, H., Nagy, I., Villarroel, R., Inzé, D., and Wobus, U. 1992. *Cis*-analysis of a seed protein gene promoter: the conservative RY repeat CATGCATG within the legumin box is essential for tissue-specific expression of a legumin gene. *Plant J.* 2:233–239.

Bäumlein, H., Wobus, U., Pustell, J., and Kafatos, F.C. 1986. The legumin gene family: structure of a B type gene of *Vicia faba* and a possible legumin gene specific regulatory element. *Nucl. Acids Res.* 14:2707–2720.

Baus, A.D., Franzmann, L., and Meinke, D.W. 1986. Growth *in vitro* of arrested embryos from lethal mutants of *Arabidopsis thaliana. Theor. Appl. Genet.* 72:577–586.

Bawa, K.S., and Beach, J.H. 1983. Self-incompatibility systems in the Rubiaceae of a tropical lowland wet forest. *Am. J. Bot.* 70:1281–1288.

Bayliss, M.W., and Dunn, S.D.M. 1979. Factors affecting callus formation from embryos of barley (*Hordeum vulgare*). *Plant Sci. Lett.* 14:311–316.

Beachy, R.N. 1980. *In vitro* synthesis of the α and α' subunits of the 7S storage proteins (conglycinins) of soybean seeds. *Plant Physiol.* 65:990–994.

Beachy, R.N., Jarvis, N.P., and Barton, K.A. 1981. Biosynthesis of subunits of the soybean 7S storage protein. *J. Mol. Appl. Genet.* 1:19–27.

Beachy, R.N., Thompson. J.F., and Madison, J.T. 1978. Isolation of polyribosomes and messenger RNA active in *in vitro* synthesis of soybean seed proteins. *Plant Physiol.* 61:139–144.

Beachy, R.N., Chen, Z.-L., Horsch, R.B., Rogers, S.G., Hoffmann, N.J., and Fraley, R.T. 1985. Accumulation and assembly of soybean β-conglycinin in seeds of transformed *Petunia* plants. *EMBO J.* 4:3047–3053.

Beadle, G.W. 1932. Genes in maize for pollen sterility. *Genetics* 17:413–431.

Bechtel, D.B., and Barnett, B.D. 1986. A freeze-fracture study of storage protein accumulation in unfixed wheat starchy endosperm. *Cereal Chem.* 63:232–240.

Bechtel, D.B., Gaines, R.L., and Pomeranz, Y. 1982. Early stages in wheat endosperm formation and protein body initiation. *Ann. Bot.* 50:507–518.

Bechtel, D.B., and Juliano, B.O. 1980. Formation of protein bodies in the starchy endosperm of rice (*Oryza sativa* L.): a re-investigation. *Ann. Bot.* 45:503–509.

Bechtel, D.B., and Pomeranz, Y. 1977. Ultrastructure of the mature ungerminated rice (*Oryza sativa*) caryopsis. The caryopsis coat and the aleurone cells. *Am. J. Bot.* 64:966–973.

Bechtel, D.B., and Pomeranz, Y. 1981. Ultrastructure and cytochemistry of mature oat (*Avena sativa* L.) endosperm. The aleurone layer and starchy endosperm. *Cereal Chem.* 58:61–69.

Becker, D., Brettschneider, R., and Lörz, H. 1994. Fertile transgenic wheat from microprojectile bombardment of scutellar tissue. *Plant J.* 5:299–307.

Bedinger, P.A., and Edgerton, M.D. 1990. Developmental staging of maize microspores reveals a transition in developing microspore proteins. *Plant Physiol.* 92:474–479.

Bednara, J. 1977. Female gametophyte and pollen tube of *Epilobium palustre* L. *Acta Soc. Bot. Polon.* 46:603–615.

Bednara, J., van Lammeren, A.A.M., and Willemse, M.T.M. 1988. Microtubular configurations during meiosis and megasporogenesis in *Gasteria verrucosa* and *Chamaenerion angustifolium. Sex. Plant Reprod.* 1:164–172.

Bednara, J., Willemse, M.T.M., and van Lammeren, A.A.M. 1990. Organization of the actin cytoskeleton during megasporogenesis in *Gasteria verrucosa* visualized with fluorescent-labelled phalloidin. *Acta Bot. Neerl.* 39:43–48.

Bednarska, E. 1981. Autoradiographic studies of DNA and histone synthesis in successive differentiation stages of pollen grain in *Hyacinthus orientalis* L. *Acta Soc. Bot. Polon.* 50:367–380.

Bednarska, E. 1984. Ultrastructural and metabolic transformations of differentiating *Hyacinthus orientalis* L. pollen grain cells. I. RNA and protein synthesis. *Acta Soc. Bot. Polon.* 53:145–158.

Bednarska, E. 1989a. Localization of calcium on the stigma surface of *Ruscus aculeatus* L. Studies using chlorotetracycline and X-ray microanalysis. *Planta* 179:11–16.

Bednarska, E. 1989b. The effect of exogenous Ca^{2+} ions on pollen grain germination and pollen tube growth. Investigations with $^{45}Ca^{2+}$ together with verapamil, La^{3+}, and ruthenium red. *Sex. Plant Reprod.* 2:53-58.

Bednarska, E. 1991. Calcium uptake from the stigma by germinating pollen in *Primula officinalis* L. and *Ruscus aculeatus* L. *Sex. Plant Reprod.* 4:36–38.

Bednarska, E. 1993. Localization of calcium-dependent ATPase in germinating pollen grain and pollen tube in *Vicia faba. Folia Histochem. Cytobiol.* 31:147–151.

Bednarska, E. 1995. Localization of membrane-associated calcium in unpollinated and pollinated pistil of *Petunia hybrida* Hort. *Acta Soc. Bot. Polon.* 64:19–24.

Bednarska, E., and Butowt, R. 1994. Calcium in pollen-pistil interaction in *Petunia hybrida* Hort. I. Localization of Ca^{2+} ions in mature pollen grain using pyroantimonate and autoradiographic methods. *Folia Histochem. Cytobiol.* 32:265–269.

Bednarska, E., and Butowt, R. 1995a. Calcium in pollen–pistil interaction in *Petunia hybrida* Hort. II. Localization of Ca^{2+} ions and Ca^{2+}-ATPase in unpollinated pistil. *Folia Histochem. Cytobiol.* 33:43–52.

Bednarska, E. and Butowt, R. 1995b. Calcium in pollen–pistil interaction in *Petunia hybrida* Hort. III. Localization of Ca^{2+} ions and Ca^{2+}-ATPase in pollinated pistil. *Folia Histochem. Cytobiol.* 33:125–132.

Bednarska, E., and Karbowska, J. 1990. Localization of Ca^{2+} and Ca-ATPase in papillae of maturing stigma in *Pharbitis nil* L. *Phytomorphology* 40:323–329.

Bednarska, E., and Tretyn, A. 1989. Ultrastructural localization of acetylcholinesterase activity in the stigma of *Pharbitis nil. Cell Biol. Int. Rep.* 13:275–281.

Beerhues, L., Forkmann, G., Schöpker, H., Stotz, G., and Wiermann, R. 1989. Flavanone 3-hydroxylase and dihydroflavonol oxygenase activities in anthers of *Tulipa*. The significance of the tapetum fraction in flavonoid metabolism. *J. Plant Physiol.* 133:743–746.

Beevers, L., and Poulson, R. 1972. Protein synthesis in cotyledons of *Pisum sativum* L. I. Changes in cell-free amino acid incorporation capacity during seed development and maturation. *Plant Physiol.* 49:476–481.

Belanger, F.C., and Kriz, A.L. 1989. Molecular characterization of the major maize embryo globulin encoded by the *Glb1* gene. *Plant Physiol.* 91:636–643.

Belanger, F.C., and Kriz, A.L. 1991. Molecular basis for allelic polymorphism of the maize *globulin-1* gene. *Genetics* 129:863–872.

Bell, J., and Hicks, G. 1976. Transmitting tissue of the pistil of tobacco: light and electron microscopic observations. *Planta* 131:187–200.

Belliard, G., Vedel, F., and Pelletier, G. 1979.

siently associated with the endoplasmic reticulum of developing *Phaseolus vulgaris* cotyledons. *Physiol. Plant.* 55:82–92.

Bollini, R., Vitale, A., and Chrispeels, M.J. 1983. *In vivo* and *in vitro* processing of seed reserve protein in the endoplasmic reticulum: evidence for two glycosylation steps. *J. Cell Biol.* 96:999–1007.

Bolz, G. 1961. Genetisch-züchterische Untersuchungen bei *Tagetes*. II. Herstellung, Genetik und Verwendung röntgeninduzierter Mutanten in der Züchtung. *Z. Pflanzenzüchtg* 45:121–142.

Bommineni, V.R., and Greyson, R.I. 1987. *In vitro* culture of ear shoots of *Zea mays* and the effect of kinetin on sex expression. *Am. J. Bot.* 74:883–890.

Bommineni, V.R., and Greyson, R.I. 1990. Regulation of flower development in cultured ears of maize (*Zea mays* L.). *Sex. Plant Reprod.* 3:109–115.

Bommineni, V.R., Atkinson, B.G., Greyson, R.I., and Walden, D.B. 1990. Polypeptides synthesized during the maturation of flower organs from tassel and ear inflorescences of *Zea mays* L. *Maydica* 35:195–201.

Bommineni, V.R., Chibbar, R.N., Datla, R.S.S., and Tsang, E.W.T. 1993. Transformation of white spruce (*Picea glauca*) somatic embryos by microprojectile bombardment. *Plant Cell Rep.* 13:17–23.

Bonhomme, S., Budar, F., Férault, M., and Pelletier, G. 1991. A 2.5 kb *NcoI* fragment of Ogura radish mitochondrial DNA is correlated with cytoplasmic male-sterility in *Brassica* cybrids. *Curr. Genet.* 19:121–127.

Bonhomme, S., Budar, F., Lancelin, D., Small, I., Defrance, M.-C., and Pelletier, G. 1992. Sequence and transcript analysis of the *Nco2.5* Ogura-specific fragment correlated with cytoplasmic male sterility in *Brassica* cybrids. *Mol. Gen. Genet.* 235:340–348.

Boothe, J.G., and Walden, D.B. 1990. Gene expression in embryos and seedlings of maize. *Maydica* 35:187–194.

Booy, G., Krens, F.A., and Bino, R.J. 1992. Analysis of pollen-tube growth in cultured maize silks. *Sex. Plant Reprod.* 5:227–231.

Booy, G., Krens, F.A., and Huizing, H.J. 1989. Attempted pollen-mediated transformation of maize. *J. Plant Physiol.* 135:319–324.

Bopp-Hassenkamp, G. 1960. Elektronenmikroskopische Untersuchungen an Pollenschläuchen zweier Liliaceen. *Z. Naturfors.* 15b:91–94.

Borisjuk, L., Weber, H., Panitz, R., Manteuffel, R., and Wobus, U. 1995. Embryogenesis of *Vicia faba* L.: histodifferentiation in relation to starch and storage protein synthesis. *J. Plant Physiol.* 147:203–218.

Borkird, C., Choi, J.H., and Sung, Z.R. 1986. Effect of 2,4-dichlorophenoxyacetic acid on the expression of embryogenic program in carrot. *Plant Physiol.* 81:1143–1146.

Borkird, C., Choi, J.H., Jin, Z.-H., Franz, G., Hatzopoulos, P., Chorneau, R., Bonas, U., Pelegri, F., and Sung, Z.R. 1988. Developmental regulation of embryonic genes in plants. *Proc. Natl Acad. Sci. USA* 85:6399–6403.

Boronat, A., Martínez, M.C., Reina, M., Puigdomènech, P., and Palau, J. 1986. Isolation and sequencing of a 28 kD glutelin-2 gene from maize. Common elements in the 5' flanking regions among zein and glutelin genes. *Plant Sci.* 47:95–102.

Borroto, K., and Dure, L., III. 1987. The globulin seed storage proteins of flowering plants are derived from two ancestral genes. *Plant Mol. Biol.* 8:113–131.

Borroto, K.E., and Dure, L., III. 1986. The expression of chloroplast genes during cotton embryogenesis. *Plant Mol. Biol.* 7:105–113.

Borthwick, H.A. 1931. Development of the macrogametophyte and embryo of *Daucus carota*. *Bot. Gaz.* 92:23–44.

Bosio, M.G. 1940. Ricerche sulla fecondazione intraovarica in *Helleborus* e *Poeonia*. *Nuovo Giorn. Bot. Ital.* 47:591–598.

Bosnes, M., and Olsen, O.-A. 1992. The rate of nuclear gene transcription in barley endosperm syncytia increases sixfold before cell-wall formation. *Planta* 186:376–383.

Bosnes, M., Weideman, F., and Olsen, O.-A. 1992. Endosperm differentiation in barley wild-type and *sex* mutants. *Plant J.* 2:661–674.

Bosnes, M., Harris, E., Aigeltinger, L., and Olsen, O.-A. 1987. Morphology and ultrastructure of 11 barley shrunken endosperm mutants. *Theor. Appl. Genet.* 74:177–187.

Boston, R.S., Becwar, M.R., Ryan, R.D., Goldsbrough, P.B., Larkins, B.A., and Hodges, T.K. 1987. Expression from heterologous promoters in electroporated carrot protoplasts. *Plant Physiol.* 83:742–746.

Boston, R.S., Fontes, E.B.P., Shank, B.B., and Wrobel, R.L. 1991. Increased expression of the maize immunoglobulin binding protein homolog b-70 in three zein regulatory mutants. *Plant Cell* 3:497–505.

Botti, C., and Vasil, I.K. 1983. Plant regeneration by somatic embryogenesis from parts of cultured mature embryos of *Pennisetum americanum* (L.) K. Schum. *Z. Pflanzenphysiol.* 111:319–325.

Bouchard, R.A. 1990. Characterization of expressed meiotic prophase repeat transcript clones of *Lilium*: meiosis-specific expression, relatedness, and affinities to small heat shock protein genes. *Genome* 33:68–79.

Bouchard, R.A., and Stern, H. 1980. DNA synthesized at pachytene in *Lilium*: a non-divergent subclass of moderately repetitive sequences. *Chromosoma* 81:349–363.

Bouharmont, J. 1977. Cytology of microspores and calli after anther culture in *Hordeum vulgare*. *Caryologia* 30:351–360.

Boulter, D., Evans, I.M., Ellis, J.R., Shirsat, A., Gatehouse, J.A., and Croy, R.R.D. 1987. Differential gene expression in the development of *Pisum sativum*. *Plant Physiol. Biochem.* 25:283–289.

Bouthyette, P.-Y., Spitsberg, V., and Gregory, P. 1985. Mitochondrial interaction with *Helminthosporium maydis* race T toxin: blocking by dicyclohexylcarbodiimide. *J. Expt. Bot.* 36:511–528.

Boutilier, K.A., Ginés, M.-J., DeMoor, J.M., Huang, B., Baszczynski, C.L., Iyer, V.N., and Miki, B.L. 1994. Expression of the BnmNAP subfamily of napin genes coincides with the induction of *Brassica* microspore embryogenesis. *Plant Mol. Biol.* 26:1711–1723.

Boutry, M., and Briquet, M. 1982. Mitochondrial modifications associated with the cytoplasmic male sterility in faba beans. *Eur. J. Biochem.* 127:129–135.

Bower, R., and Birch, R.G. 1992. Transgenic sugarcane plants via microprojectile bombardment. *Plant J.* 2:409–416.

Bowman, J.L., Drews, G.N., and Meyerowitz, E.M. 1991. Expression of the *Arabidopsis* floral homeotic gene *AGAMOUS* is restricted to specific cell types late in flower development. *Plant Cell* 3:749–758.

Bowman, J.L., Smyth, D.R., and Meyerowitz, E.M. 1989.

Genes directing flower development in *Arabidopsis*. *Plant Cell* 1:37–52.

Bowman, J.L., Smyth, D.R., and Meyerowitz, E.M. 1991. Genetic interactions among floral homeotic genes of *Arabidopsis*. *Development* 112:1–20.

Bowman, J.L., Alvarez, J., Weigel, D., Meyerowitz, E.M., and Smyth, D.R. 1993. Control of flower development in *Arabidopsis thaliana* by *APETALA1* and interacting genes. *Development* 119:721–743.

Bowman, J.L., Sakai, H., Jack, T., Weigel, D., Mayer, U., and Meyerowitz, E.M. 1992. *SUPERMAN*, a regulator of floral homeotic genes in *Arabidopsis*. *Development* 114:599–615.

Bown, D., Ellis, T.H.N., and Gatehouse, J.A. 1988. The sequence of a gene encoding convicilin from pea (*Pisum sativum* L.) shows that convicilin differs from vicilin by an insertion near the *N*-terminus. *Biochem. J.* 251:717–726.

Bown, D., Levasseur, M., Croy, R.R.D., Boulter, D., and Gatehouse, J.A. 1985. Sequence of a pseudogene in the legumin gene family of pea (*Pisum sativum* L.). *Nucl. Acids Res.* 13:4527–4538.

Boyer, C., Hilbert, J.-L., and Vasseur, J. 1993. Embryogenesis-related protein synthesis and accumulation during early acquisition of somatic embryogenesis competence in *Cichorium*. *Plant Sci.* 93:41–53.

Boyes, D.C., and Nasrallah, J.B. 1993. Physical linkage of the *SLG* and *SRK* genes at the self-incompatibility locus of *Brassica oleracea*. *Mol. Gen. Genet.* 236:369–373.

Boyes, D.C., and Nasrallah, J.B. 1995. An anther-specific gene encoded by an *S* locus haplotype of *Brassica* produces complementary and differentially regulated transcripts. *Plant Cell* 7:1283–1294.

Boyes, D.C., Chen, C.-H., Tantikanjana, T., Esch, J.J., and Nasrallah, J.B. 1991. Isolation of a second *S*-locus-related cDNA from *Brassica oleracea*: genetic relationships between the *S* locus and two related loci. *Genetics* 127:221–228.

Boyle, S.A., and Yeung, E.C. 1983. Embryogeny of *Phaseolus*: developmental pattern of lactate and alcohol dehydrogenases. *Phytochemistry* 22:2413–2416.

Brace, J., Ockendon, D.J., and King, G.J. 1993. Development of a method for the identification of S alleles in *Brassica oleracea* based on digestion of PCR-amplified DNA with restriction endonucleases. *Sex. Plant Reprod.* 6:133–138.

Bradley, D., Carpenter, R., Sommer, H., Hartley, N., and Coen, E. 1993. Complementary floral homeotic phenotypes result from opposite orientations of a transposon at the *plena* locus of *Antirrhinum*. *Cell* 72:85–95.

Brady, T. 1973. Feulgen cytophotometric determination of the DNA content of the embryo proper and suspensor cells of *Phaseolus coccineus*. *Cell Diffn* 2:65–75.

Brady, T., and Walthall, E.D. 1985. The effect of the suspensor and gibberellic acid on *Phaseolus vulgaris* embryo protein content. *Devel Biol.* 107:531–536.

Brander, K.A., and Kuhlemeier, C. 1995. A pollen-specific DEAD-box protein related to translation initiation factor eIF-4A from tobacco. *Plant Mol. Biol.* 27:637–649.

Brandt, A., and Ingversen, J. 1976. *In vitro* synthesis of barley endosperm proteins on wild type and mutant templates. *Carlsberg Res. Commun.* 41:311–320.

Brandt, A., Montembault, A., Cameron-Mills, V., and Rasmussen, S.K. 1985. Primary structure of a B1 hordein gene from barley. *Carlsberg Res. Commun.* 50:333–345.

Braun, C.J., Siedow, J.N., and Levings, C.S., III. 1990. Fungal toxins bind to the URF13 protein in maize mitochondria and *Escherichia coli*. *Plant Cell* 2:153–161.

Braun, C.J., Siedow, J.N., Williams, M.E., and Levings, C.S., III. 1989. Mutations in the maize mitochondrial T-*urf13* gene eliminate sensitivity to a fungal pathotoxin. *Proc. Natl Acad. Sci. USA* 86:4435–4439.

Brawley, S.H., Wetherell, D.F., and Robinson, K.R. 1984. Electrical polarity in embryos of wild carrot precedes cotyledon differentiation. *Proc. Natl Acad. Sci. USA* 81:6064–6067.

Bray, E.A., and Beachy, R.N. 1985. Regulation by ABA of β-conglycinin expression in cultured developing soybean cotyledons. *Plant Physiol.* 79:746–750.

Bray, E.A., Naito, S., Pan, N.-S., Anderson, E., Dubé, P., and Beachy, R.N. 1987. Expression of the β-subunit of β-conglycinin in seeds of transgenic plants. *Planta* 172:364–370.

Bredemeijer, G.M.M. 1974. Peroxidase activity and peroxidase isoenzyme composition in self-pollinated, cross-pollinated and unpollinated styles of *Nicotiana alata*. *Acta Bot. Neerl.* 23:149–157.

Bredemeijer, G.M.M., and Blaas, J. 1975. A possible role of a stylar peroxidase gradient in the rejection of incompatible growing pollen tubes. *Acta Bot. Neerl.* 24:37–48.

Bredemeijer, G.M.M., and Blaas, J. 1980. Do S allele-specific peroxidase isoenzymes exist in self-incompatible *Nicotiana alata*? *Theor. Appl. Genet.* 57:119–123.

Bredemeijer, G.M.M., and Blaas, J. 1981. S-specific proteins in styles of self-incompatible *Nicotiana alata*. *Theor. Appl. Genet.* 59:185–190.

Breen, J.P., and Crouch, M.L. 1992. Molecular analysis of a cruciferin storage protein gene family of *Brassica napus*. *Plant Mol. Biol.* 19:1049–1055.

Breiteneder, H., Pettenburger, K., Bito, A., Valenta, R., Kraft, D., Rumpold, H., Scheiner, O., and Breitenbach, M. 1989. The gene coding for the major birch pollen allergen *BetvI*, is highly homologous to a pea disease resistance response gene. *EMBO J.* 8:1935–1938.

Breslavetz, L.P. 1935. Abnormal development of pollen in different races and grafts of hemp. *Genetica* 17:154–169.

Breton, A.M., and Sung, Z.R. 1982. Temperature-sensitive carrot variants impaired in somatic embryogenesis. *Devel Biol.* 90:58–66.

Brettell, R.I.S., Thomas, E., and Ingram, D.S. 1980. Reversion of Texas male-sterile cytoplasm maize in culture to give fertile, T-toxin resistant plants. *Theor. Appl. Genet.* 58:55–58.

Brewbaker, J.L. 1957. Pollen cytology and self-incompatibility systems in plants. *J. Hered.* 48:271–277.

Brewbaker, J.L. 1959. Biology of the angiosperm pollen grain. *Indian J. Genet. Plant Breed.* 19:121–133.

Brewbaker, J.L. 1967. The distribution and phylogenetic significance of binucleate and trinucleate pollen grains in the angiosperms. *Am. J. Bot.* 54:1069–1083.

Brewbaker, J.L., and Emery, G.C. 1962. Pollen radiobotany. *Radiation Bot.* 1:101–154.

Brewbaker, J.L., and Gorrez, D.D. 1967. Genetics of self-incompatibility in the monocot genera, *Ananas* (pineapple) and *Gasteria*. *Am. J. Bot.* 54:611–616.

Buttrose, M.S. 1963a. Ultrastructure of the developing wheat endosperm. *Aust. J. Biol. Sci.* 16:305–317.

Buttrose, M.S. 1963b. Ultrastructure of the developing aleurone cells of wheat grain. *Aust. J. Biol. Sci.* 16:768–774.

Bystedt, P.-A. 1990. The transmitting tract in *Trimezia fosteriana* (Iridaceae). I. Ultrastructure in the stigma, style and ovary. *Nord. J. Bot.* 9:507–518.

Cabrera-Ponce, J.L., Vegas-Garcia, A., and Herrera-Estrella, L. 1995. Herbicide resistant transgenic papaya plants produced by an efficient particle bombardment transformation method. *Plant Cell Rep.* 15:1–7.

Cabrera-Ponce, J.L., Vegas-Garcia, A., and Herrera-Estrella, L. 1996. Regeneration of transgenic papaya plants via somatic embryogenesis induced by *Agrobacterium rhizogenes*. *In Vitro Cell Devel Biol.* 32:86–90.

Cadic, A., and Sangwan-Norreel, B.S. 1983. Modifications ultrastructurales provoquées par des traitements promoteurs de l'androgenèse chez le *Datura innoxia* Mill. *Ann. Sci. Nat. Bot.* Ser. XIII 5:97–114.

Cai, G., Bartalesi, A., Del Casino, C., Moscatelli, A., Tiezzi, A., and Cresti, M. 1993. The kinesin-immunoreactive homologue from *Nicotiana tabacum* pollen tubes: biochemical properties and subcellular localization. *Planta* 191:496–506.

Cai, G., Moscatelli, A., Del Casino, C., and Cresti, M. 1996. Cytoplasmic motors and pollen tube growth. *Sex. Plant Reprod.* 9:59–64.

Cai, X., Dong, Y., and Sodmergen. 1995. Actin organization and distribution during hydration, activation and germination of pollen in *Amaryllis vittata* Ait. using TRITC-phalloidin as a probe. *Cytologia* 60:133–139.

Caiola, M.G., Banas, M., and Canini, A. 1993. Ultrastructure and germination percentage of *Crocus biflorus* Miller subsp. *biflorus* (Iridaceae) pollen. *Bot. Acta* 106:488–495.

Caligo, M.A., Nuti Ronchi, V., and Nozzolini, M. 1985. Proline and serine affect polarity and development of carrot somatic embryos. *Cell Diffn* 17:193–198.

Callis, J., and Bedinger, P. 1994. Developmentally regulated loss of ubiquitin and ubiquitinated proteins during pollen maturation in maize. *Proc. Natl Acad. Sci. USA* 91:6074–6077.

Callis, J., Raasch, J.A., and Vierstra, R.D. 1990. Ubiquitin extension proteins of *Arabidopsis thaliana*. Structure, localization, and expression of their promoters in transgenic tobacco. *J. Biol. Chem.* 265:12486–12493.

Calzoni, G.L., Speranza, A., Li, Y.Q., Ciampolini, F., and Cresti, M. 1993. Wall biosynthesis and wall polysaccharide composition in pollen tubes of *Malus domestica* growing at low rate after monensin treatment. *Acta Bot. Neerl.* 42:473–480.

Camadro, E.L., and Masuelli, R.W. 1995. A genetic model for the endosperm balance number (EBN) in the wild potato *Solanum acaule* Bitt. and two related diploid species. *Sex. Plant Reprod.* 8:283–288.

Camefort, H., and Sangwan, R.S. 1979. Effets d'un choc thermique sur certaines ultrastructures des grains de pollen embryogènes du *Datura metel* L. *Compt. Rend. Acad. Sci. Paris* 288D:1383–1386.

Cameron-Mills, V., and Brandt, A. 1988. A γ-hordein gene. *Plant Mol. Biol.* 11:449–461.

Cameron-Mills, V., and von Wettstein, D. 1980. Protein body formation in the developing barley endosperm. *Carlsberg Res. Commun.* 45:577–594.

Campbell, R.J., and Ascher, P.D. 1975. Incorporation of radioactive label into nucleic acids of compatible and incompatible pollen tubes of *Lilium longiflorum* Thunb. *Theor. Appl. Genet.* 46:143–148.

Campbell, W.P., Lee, J.W., O'Brien, T.P., and Smart, M.G. 1981. Endosperm morphology and protein body formation in developing wheat grain. *Aust. J. Plant Physiol.* 8:5–19.

Campenot, M.K., Zhang, G., Cutler, A.J., and Cass, D.D. 1992. *Zea mays* embryo sacs in culture. I. Plant regeneration from 1 day after pollination embryos. *Am. J. Bot.* 79:1368–1373.

Cañas, L., and Malmberg, R. 1992. Isolation and characterization of three monoclonal antibodies specific for pistil and stamen tissues of tobacco (*Nicotiana tabacum* L.). *Plant Sci.* 83:195–203.

Cañas, L.A., Busscher, M., Angenent, G.C., Beltrán, J.-P., and van Tunen, A.J. 1994. Nuclear localization of the *Petunia* MADS box protein FBP1. *Plant J.* 6:597–604.

Cao, J., Duan, X., McElroy, D., and Wu, R. 1992. Regeneration of herbicide resistant transgenic rice plants following microprojectile-mediated transformation of suspension culture cells. *Plant Cell Rep.* 11:586–591.

Čapková, V., Hrabětová, E., and Tupý, J. 1983. Reduction of leucine efflux and protein release from tobacco pollen tubes in culture by actinomycin D in the absence of calcium. *Biochem. Physiol. Pflanz.* 178:521–527.

Čapková, V., Hrabětová, E., and Tupý, J. 1987. Protein changes in tobacco pollen culture; a newly synthesized protein related to pollen tube growth. *J. Plant Physiol.* 130:307–314.

Čapková, V., Hrabětová, E., and Tupý, J. 1988. Protein synthesis in pollen tubes: preferential formation of new species independent of transcription. *Sex. Plant Reprod.* 1:150–155.

Čapková, V., Zbrožek, J., and Tupý, J. 1994. Protein synthesis in tobacco pollen tubes: preferential synthesis of cell-wall 69-kDa and 66-kDa glycoproteins. *Sex. Plant Reprod.* 7:57–66.

Čapková, V., Hrabětová, E., Tupý, J., and Říhová, L. 1983. Amino acid uptake and protein synthesis in cultured tobacco pollen. *Biochem. Physiol. Pflanz.* 178:511–520.

Čapková-Balatková, V., Hrabětová, E., and Tupý, J. 1980. Effects of cycloheximide on pollen of *Nicotiana tabacum* in culture. *Biochem. Physiol. Pflanz.* 175:412–420.

Carasco, J.F., Croy, R., Derbyshire, E., and Boulter, D. 1978. The isolation and characterization of the major polypeptides of the seed globulin of cowpea (*Vigna unguiculata* L. Walp) and their sequential synthesis in developing seeds. *J. Expt. Bot.* 29:309–323.

Carlberg, I., Söderhäll, K., Glimelius, K., and Eriksson, T. 1984. Protease activities in non-embryogenic and embryogenic carrot cell strains during callus growth and embryo formation. *Physiol. Plant.* 62:458–464.

Carman, J.G. 1988. Improved somatic embryogenesis in wheat by partial simulation of the *in-ovulo* oxygen, growth-regulator and desiccation environments. *Planta* 175:417–424.

Carman, J.G., Crane, C.F., and Riera-Lizarazu, O. 1991. Comparative histology of cell walls during meiotic and apomeiotic megasporogenesis in two hexaploid Australasian *Elymus* species. *Crop Sci.* 31:1527–1532.

Carmichael, J.S., and Friedman, W.E. 1995. Double fertilization in *Gnetum gnemon*: the relationship between the cell cycle and sexual reproduction. *Plant Cell* 7:1975–1988.

Carmona, M.J., Hernández-Lucas, C., San Martín, C., González, P., and García-Olmedo, F. 1993. Subcellular localization of type I thionins in the endosperms of wheat and barley. *Protoplasma* 173:1–7.

Carniel, K. 1952. Das Verhalten der Kerne im Tapetum der Angiospermen mit besonderer Berücksichtigung von Endomitosen und sogenannten Endomitosen. *Österr. Bot. Z.* 99:318–362.

Carniel, K. 1961. Das Antherentapetum von *Zea mays*. *Österr. Bot. Z.* 108:89-96.

Carniel, K. 1967a. Über die Embryobildung in der Gattung *Paeonia*. *Österr. Bot. Z.* 114:4–19.

Carniel, K. 1967b. Licht- und elektronenmikroskopische Untersuchung der Ubischkörperentwicklung in der Gattung *Oxalis*. *Österr. Bot. Z.* 114:490–501.

Carpenter, J.L., Ploense, S.E., Snustad, D.P., and Silflow, C.D. 1992. Preferential expression of an α-tubulin gene of *Arabidopsis* in pollen. *Plant Cell* 4:557–571.

Carpenter, R., and Coen, E.S. 1990. Floral homeotic mutations produced by transposon-mutagenesis in *Antirrhinum majus*. *Genes Devel.* 4:1483–1493.

Carraro, L., and Lombardo, G. 1976. Tapetal ultrastructural changes during pollen development. II. Studies on *Pelargonium zonale* and *Kalanchoë obtusa*. *Caryologia* 29:339–344.

Carraro, L., Lombardo, G., and Gerola, F.M. 1985. Electron-cytochemical localization of peroxidase in self- and cross-pollinated styles of *Primula acaulis*. *Caryologia* 38:83–94.

Carraro, L., Lombardo, G., and Gerola, F.M. 1986. Stylar peroxidase and incompatibility reactions in *Petunia hybrida*. *J. Cell Sci.* 82:1–10.

Casas, A.M., Kononowicz, A.K., Zehr, U.B., Tomes, D.T., Axtell, J.D., Butler, L.G., Bressan, R.A., and Hasegawa, P.M. 1993. Transgenic sorghum plants via microprojectile bombardment. *Proc. Natl Acad. Sci. USA* 90:11212–11216.

Casey, R., Domoney, C., and Ellis, N. 1986. Legume storage proteins and their genes. *Oxford Surv. Plant Mol. Cell Biol.* 3:1–95.

Casey, R., Domoney, C., and Nielsen, N.C. 1985. Isolation of a cDNA clone for pea (*Pisum sativum*) seed lipoxygenase. *Biochem. J.* 232:79–85.

Cass, D.D. 1972. Occurrence and development of a filiform apparatus in the egg of *Plumbago capensis*. *Am. J. Bot.* 59:279–283.

Cass, D.D. 1973. An ultrastructural and Nomarski-interference study of the sperms of barley. *Can. J. Bot.* 51:601–605.

Cass, D.D., and Fabi, G.C. 1988. Structure and properties of sperm cells isolated from the pollen of *Zea mays*. *Can. J. Bot.* 66:819–825.

Cass, D.D., and Jensen, W.A. 1970. Fertilization in barley. *Am. J. Bot.* 57:62–70.

Cass, D.D., and Karas, I. 1974. Ultrastructural organization of the egg of *Plumbago zeylanica*. *Protoplasma* 81:49–62.

Cass, D.D., and Karas, I. 1975. Development of sperm cells in barley. *Can. J. Bot.* 53:1051–1062.

Cass, D.D., and Peteya, D.J. 1979. Growth of barley pollen tubes *in vivo*. I. Ultrastructural aspects of early tube growth in the stigmatic hair. *Can. J. Bot.* 57:386–396.

Cass, D.D., Peteya, D.J., and Robertson, B.L. 1985. Megagametophyte development in *Hordeum vulgare*. 1. Early megagametogenesis and the nature of cell wall formation. *Can. J. Bot.* 63:2164–2171.

Cass, D.D., Peteya, D.J., and Robertson, B.L. 1986. Megagametophyte development in *Hordeum vulgare*. 2. Later stages of wall development and morphological aspects of megagametophyte cell differentiation. *Can. J. Bot.* 64:2327–2336.

Castle, L.A., and Meinke, D.W. 1994. A *FUSCA* gene of *Arabidopsis* encodes a novel protein essential for plant development. *Plant Cell* 6:25–41.

Castle, L.A., Errampalli, D., Atherton, T.L., Franzmann, L.H., Yoon, E.S., and Meinke, D.W. 1993. Genetic and molecular characterization of embryonic mutants identified following seed transformation in *Arabidopsis*. *Mol. Gen. Genet.* 241:504–514.

Cavallini, A., Natali, L., Balconi, C., Rizzi, E., Motto, M., Cionini, G., and D'Amato, F. 1995. Chromosome endoreduplication in endosperm cells of two maize genotypes and their progenies. *Protoplasma* 189:156–162.

Cave, M.S., Arnott, H.J., and Cook, S.A. 1961. Embryogeny in the California peonies with reference to their taxonomic position. *Am. J. Bot.* 48:397–404.

Cave, M.S., and Brown, S.W. 1954. The detection and nature of dominant lethals in *Lilium*. II. Cytological abnormalities in ovules after pollen irradiation. *Am. J. Bot.* 41:469–483.

Cebrat, J., and Zadęcka, A. 1978. Development of anthers in three male-sterile lines of rye (*Secale cereale* L.). *Genet. Polon.* 19:25–31.

Ceccarelli, N., Lorenzi, R., and Alpi, A. 1979. Kaurene and kaurenol biosynthesis in cell-free system of *Phaseolus coccineus* suspensor. *Phytochemistry* 18:1657–1658.

Ceccarelli, N., Lorenzi, R., and Alpi, A. 1981a. Gibberellin biosynthesis in *Phaseolus coccineus* suspensor. *Z. Pflanzenphysiol.* 102:37–44.

Ceccarelli, N., Lorenzi, R., and Alpi, A. 1981b. Kaurene metabolism in cell-free extracts of *Phaseolus coccineus* suspensors. *Plant Sci. Lett.* 21:325–332.

Chamberland, S., Daigle, N., and Bernier, F. 1992. The legumin boxes and the 3' part of a soybean β-conglycinin promoter are involved in seed gene expression in transgenic tobacco plants. *Plant Mol. Biol.* 19:937–949.

Chamberlin, M.A., Horner, H.T., and Palmer, R.G. 1993a. Nutrition of ovule, embryo sac, and young embryo in soybean: an anatomical and autoradiographic study. *Can. J. Bot.* 71:1153–1168.

Chamberlin, M.A., Horner, H.T., and Palmer, R.G. 1993b. Nuclear size and DNA content of the embryo and endosperm during their initial stages of development in *Glycine max* (Fabaceae). *Am. J. Bot.* 80:1209–1215.

Chan, M.-T., Chang, H.-H., Ho, S.-L., Tong, W.-F., and Yu, S.-M. 1993. *Agrobacterium*-mediated production of transgenic rice plants expressing a chimeric α-amylase promoter/β-glucuronidase gene. *Plant Mol. Biol.* 22:491–506.

Chandlee, J.M., and Scandalios, J.G. 1983. Gene expression during early kernel development in *Zea mays*. *Devel Genet.* 4:99–115.

Chandlee, J.M., and Scandalios, J.G. 1984. Regulation of *Cat1* gene expression in the scutellum of maize during early sporophytic development. *Proc. Natl Acad. Sci. USA* 81:4903–4907.

Chandler, P.M., Higgins, T.J.V., Randall, P.J., and Spencer,

D. 1983. Regulation of legumin levels in developing pea seeds under conditions of sulfur deficiency. Rates of legumin synthesis and levels of legumin mRNA. *Plant Physiol.* 71:47–54.

Chandler, P.M., Spencer, D., Randall, P.J., and Higgins, T.J.V. 1984. Influence of sulfur nutrition on developmental patterns of some major pea seed proteins and their mRNAs. *Plant Physiol.* 75:651–657.

Chandra, S., and Bhatnagar, S.P. 1974. Reproductive biology of *Triticum*. II. Pollen germination, pollen tube growth, and its entry into the ovule. *Phytomorphology* 24:11–217.

Chandra Sekhar, K.N., and Heij, E.G. 1995. Changes in proteins and peroxidases induced by compatible pollination in the ovary of *Nicotiana tabacum* L. ahead of the advancing pollen tubes. *Sex. Plant Reprod.* 8:369–374.

Chandra Sekhar, K.N., and Williams, E.G. 1992. Nonradioactive *in situ* localization of poly(A)+RNA during pollen development in anthers of tobacco (*Nicotiana tabacum* L.). *Protoplasma* 169:9–17.

Chang, M.T., and Neuffer, M.G. 1989. Maize microsporogenesis. *Genome* 32:232–244.

Chang, Y.C., and Walling, L.L. 1991. Abscisic acid negatively regulates expression of chlorophyll *a/b* binding protein genes during soybean embryogeny. *Plant Physiol.* 97:1260–1264.

Chang, Y.C., and Walling, L.L. 1992. Chlorophyll *a/b*-binding protein genes are differentially expressed during soybean development. *Plant Mol. Biol.* 19:217–230.

Chao, C.-Y. 1971. A periodic acid-Schiff's substance related to the directional growth of pollen tube into the embryo sac in *Paspalum* ovules. *Am. J. Bot.* 58:649–654.

Chao, C.-Y. 1977. Further cytological studies of a periodic acid-Schiff's substance in the ovules of *Paspalum orbiculare* and *P. longifolium*. *Am. J. Bot.* 64:921–930.

Chapman, G.P. 1987. The tapetum. *Int. Rev. Cytol.* 107:111–125.

Chapman, G.P., and Busri, N. 1994. Apomixis in *Pennisetum*: an ultrastructural study. *Int. J. Plant Sci.* 155:492–497.

Chappell, J., and Chrispeels, M.J. 1986. Transcriptional and posttranscriptional control of phaseolin and phytohemagglutinin gene expression in developing cotyledons of *Phaseolus vulgaris*. *Plant Physiol.* 81:50–54.

Chardard, R. 1958. L'ultrastructure des grains de pollen d'Orchidées. *Rev. Cytol. Biol. Végét.* 19:223–235.

Chardard, R. 1962. Recherches sur les cellules-mères des microspores des Orchidées. Étude au microscope électronique. *Rev. Cytol. Biol. Végét.* 24:1–148.

Charest, P.J., Devantier, Y., and Lachance, D. 1996. Stable genetic transformation of *Picea marina* (Black spruce) via particle bombardment. *In Vitro Cell Devel Biol.* 32:91–99.

Charest, P.J., Caléro, N., Lachance, D., Datla, R.S.S., Duchêsne, L.C., and Tsang, E.W.T. 1993. Microprojectile-DNA delivery in conifer species: factors affecting assessment of transient gene expression using the β-glucoronidase reporter gene. *Plant Cell Rep.* 12:189–193.

Charlton, W.L., Keen, C.L., Merriman, C., Lynch, P., Greenland, A.J., and Dickinson, H.G. 1995. Endosperm development in *Zea mays*; implication of gametic imprinting and paternal excess in regulation of transfer layer development. *Development* 121:3089–3097.

Charne, D.G., and Beversdorf, W.D. 1988. Improving microspore culture as a rapeseed breeding tool: the use of auxins and cytokinins in an induction medium. *Can. J. Bot.* 66:1671–1675.

Charzyńska, M., Ciampolini, F., and Cresti, M. 1988. Generative cell division and sperm cell formation in barley. *Sex. Plant Reprod.* 1:240–247.

Charzyńska, M., and Cresti, M. 1993. Early events in division of the generative cell of *Ornithogalum virens*. *Protoplasma* 172:77–83.

Charzyńska, M., and Lewandowska, E. 1990. Generative cell division and sperm cell association in the pollen grain of *Sambucus nigra*. *Ann. Bot.* 65:685–689.

Charzyńska, M., and Maleszka, J. 1978. ^{3}H-thymidine incorporation into the microspores and pollen grains nuclei in excised *Tradescantia* stamens. *Acta Soc. Bot. Polon.* 47:163–171.

Charzyńska, M., Murgia, M., and Cresti, M. 1989. Ultrastructure of the vegetative cell of *Brassica napus* pollen with particular reference to microbodies. *Protoplasma* 152:22–28.

Charzyńska, M., Murgia, M., and Cresti, M. 1990. Microspore of *Secale cereale* as a transfer cell type. *Protoplasma* 158:26–32.

Charzyńska, M., and Pannenko, I. 1976. Inhibition of cytokinesis in the microspores of *Tradescantia bracteata* Small. by caffeine. *Acta Soc. Bot. Polon.* 45:469–476.

Charzyńska, M., Murgia, M., Milanesi, M., and Cresti, M. 1989. Origin of sperm cell association in the "male germ unit" of *Brassica* pollen. *Protoplasma* 149:1–4.

Chase, C.D. 1994. Expression of CMS-unique and flanking mitochondrial DNA sequences in *Phaseolus vulgaris* L. *Curr. Genet.* 25:245–251.

Chase, C.D., and Ortega, V.M. 1992. Organization of ATPA coding and 3' flanking sequences associated with cytoplasmic male sterility in *Phaseolus vulgaris* L. *Curr. Genet.* 22:147–153.

Chaubal, R., and Reger, B.J. 1990. Relatively high calcium is localized in synergid cells of wheat ovaries. *Sex. Plant Reprod.* 3:98–102.

Chaubal, R., and Reger, B.J. 1992a. Calcium in the synergids and other regions of pearl millet ovaries. *Sex. Plant Reprod.* 5:34–46.

Chaubal, R., and Reger, B.J. 1992b. The dynamics of calcium distribution in the synergid cells of wheat after pollination. *Sex. Plant Reprod.* 5:206–213.

Chaubal, R., and Reger, B.J. 1993. Prepollination degeneration in mature synergids of pearl millet: an examination using antimonate fixation to localize calcium. *Sex. Plant Reprod.* 6:225–238.

Chaubal, R., and Reger, B.J. 1994. Dynamics of antimonate-precipitated calcium and degeneration in unpollinated pearl millet synergids after maturity. *Sex. Plant Reprod.* 7:122–134.

Chaudhuri, S., and Messing, J. 1994. Allele-specific parental imprinting of *drz1*, a posttranscriptional regulator of zein accumulation. *Proc. Natl Acad. Sci. USA* 91:4867–4871.

Chaudhury, A.M., Craig, S., Bloemer, K.C., Farrell, L., and Dennis, E.S. 1992. Genetic control of male fertility in higher plants. *Aust. J. Plant Physiol.* 19:419–426.

Chaudhury, A.M., Lavithis, M., Taylor, P.E., Craig, S., Singh, M.B., Signer, E.R., Knox, R.B., and Dennis, E.S. 1994. Genetic control of male fertility in *Arabidopsis thaliana*: structural analysis of premeiotic developmental mutants. *Sex. Plant Reprod.* 7:17–28.

Chauhan, S.V.S., and Kinoshita, T. 1979. Histochemical localization of histones, DNA and proteins in anthers of male-fertile and male-sterile plants. *Jap. J. Breed.* 29:287–293.

Chauhan, S.V.S., and Singh, S.P. 1966. Pollen abortion in male-sterile hexaploid wheat (`Norin') having *Aegilops ovata* L. cytoplasm. *Crop Sci.* 6:532–535.

Chaumont, F., Bernier, B., Buxant, R., Williams, M.E., Levings, C.E., III, and Boutry, M. 1995. Targeting the maize T-*urf13* product into tobacco mitochondria confers methomyl sensitivity to mitochondrial respiration. *Proc. Natl Acad. Sci. USA* 92: 1167–1171.

Chay, C.H., Buehler, E.G., Thorn, J.M., Whelan, T.M., and Bedinger, P.A. 1992. Purification of maize pollen exines and analysis of associated proteins. *Plant Physiol.* 100:756–761.

Chee, P.P., Jones, J.M., and Slightom, J.L. 1991. Expression of bean storage protein minigene in tobacco seeds. Introns are not required for seed specific expression. *J. Plant Physiol.* 137:402–408.

Chee, P.P., Klassy, R.C., and Slightom, J.L. 1986. Expression of a bean storage protein `phaseolin minigene' in foreign plant tissues. *Gene* 41:47–57.

Chen, C. 1977. *In vitro* development of plants from microspores of rice. *In Vitro* 13:484–489.

Chen, C.C., Kasha, K.J., and Marsolais, A. 1984. Segmentation patterns and mechanisms of genome multiplication in cultured microspores of barley. *Can. J. Genet. Cytol.* 26:475–483.

Chen, C.C., Howarth, M.J., Peterson, R.L., and Kasha, K.J. 1984. Ultrastructure of androgenic microspores of barley during the early stages of anther culture. *Can. J. Genet. Cytol.* 26:484–491.

Chen, C.-G., Cornish, E.C., and Clarke, A.E. 1992. Specific expression of an extensin-like gene in the style of *Nicotiana alata*. *Plant Cell* 4:1053–1062.

Chen, C.-G., Mau, S.-L., and Clarke, A.E. 1993. Nucleotide sequence and style-specific expression of a novel proline-rich protein gene from *Nicotiana alata*. *Plant Mol. Biol.* 21:391–395.

Chen, C.-H., and Nasrallah, J.B. 1990. A new class of S sequences defined by a pollen recessive self-incompatibility allele of *Brassica oleracea*. *Mol. Gen. Genet.* 222:241–248.

Chen, K., Johal, S., and Wildman, S.G. 1977. Phenotypic markers for chloroplast DNA genes in higher plants and their use in biochemical genetics. In *Nucleic Acids and Protein Synthesis in Plants*, ed. L. Bogorad and J. H. Weil, pp. 183–194. New York: Plenum Press.

Chen, K., and Meyer, V.G. 1979. Mutation in chloroplast DNA coding for the large subunit of fraction 1 protein correlated with male sterility in cotton. *J. Hered.* 70:431–433.

Chen, L.-J., and Luthe, D.S. 1987. Analysis of proteins from embryogenic and non- embryogenic rice (*Oryza sativa* L.) calli. *Plant Sci.* 48:181–188.

Chen, R., Aguirre, P.J., and Smith, A.G. 1994. Characterization of an anther- and tapetum-specific gene encoding a glycine-rich protein from tomato. *J. Plant Physiol.* 143:651–658.

Chen, R., and Smith, A.G. 1993. Nucleotide sequence of a stamen- and tapetum-specific gene from *Lycopersicon esculentum*. *Plant Physiol.* 101:1413.

Chen, Y.-C., and Chourey, P.S. 1989. Spatial and temporal expression of the two sucrose synthase genes in maize: immunohistological evidence. *Theor. Appl. Genet.* 78:553–559.

Chen, Z., Schertz, K.F., Mullet, J.E., DuBell, A., and Hart, G.E. 1995. Characterization and expression of *rpoC2* in CMS and fertile lines of sorghum. *Plant Mol. Biol.* 28:799–809.

Chen, Z.-K., Wang, F.-H., and Zhou, F. 1988. On the origin, development and ultrastructure of the orbicules and pollenkitt in the tapetum of *Anemarrhena asphodeloides* (Liliaceae). *Grana* 27:273–282.

Chen, Z.-L., Pan, N.-S., and Beachy, R.N. 1988. A DNA sequence element that confers seed-specific enhancement to a constitutive promoter. *EMBO J.* 7:297–302.

Chen, Z.-L., Schuler, M.A., and Beachy, R.N. 1986. Functional analysis of regulatory elements in a plant embryo-specific gene. *Proc. Natl Acad. Sci. USA* 83:8560–8564.

Chen, Z.-L., Naito, S., Nakamura, I., and Beachy, R.N. 1989. Regulated expression of genes encoding soybean β-conglycinins in transgenic plants. *Devel Genet.* 10:112–122.

Chen, Z.-Y., Hsing, Y.-I.C., Lee, P.-F., and Chow, T.-Y. 1992. Nucleotide sequences of a soybean cDNA encoding an 18 kilodalton late embryogenesis abundant protein. *Plant Physiol.* 99:773–774.

Cheng, P.C., Greyson, R.I., and Walden, D.B. 1979. Comparison of anther development in genic male-sterile (*ms*10) and in male-fertile corn (*Zea mays*) from light microscopy and scanning electron microscopy. *Can. J. Bot.* 57:578–596.

Cheng, W.-H., Taliercio, E.W., and Chourey, P.S. 1996. The *miniature1* seed locus of maize encodes a cell wall invertase required for normal development of endosperm and maternal cells in the pedicel. *Plant Cell* 8:971–983.

Chern, M.-S., Eben, H.G., and Bustos, M.M. 1996. The developmentally regulated bZIP factor ROM1 modulates transcription from lectin and storage protein genes in bean embryos. *Plant J.* 10:135–148.

Chernyshev, A.I., Davletova, S.K., Bashkirov, V.I., Shakhmanov, N.B., Mekhedov, S.L., and Anan'ev, E.V. 1989. Nucleotide sequence of the B1 hordein gene of barley (*Hordeum vulgare* L.). *Soviet Genet.* 25:865–869.

Chesnut, R.S., Shotwell, M.A., Boyer, S.K., and Larkins, B.A. 1989. Analysis of avenin proteins and the expression of their mRNAs in developing oat seeds. *Plant Cell* 1:913–924.

Cheung, A.Y., Wang, H., and Wu, H.-M. 1995. A floral transmitting tissue-specific glycoprotein attracts pollen tubes and stimulates their growth. *Cell* 82:383–393.

Cheung, A.Y., May, B., Kawata, E.E., Gu, Q., and Wu, H.-M. 1993. Characterization of cDNAs for stylar transmitting tissue-specific proline-rich proteins in tobacco. *Plant J.* 3:151–160.

Cheung, A.Y., Zhan, X.-Y., Wang, H., and Wu, H.-M. 1996. Organ-specific and agamous-regulated expression and glycosylation of a pollen tube growth-promoting protein. *Proc. Natl Acad. Sci. USA* 93:3853–3858.

Chibbar, R.N., Kartha, K.K., Leung, N., Qureshi, J., and Caswell, K. 1991. Transient expression of marker genes in immature zygotic embryos of spring wheat (*Triticum aestivum*) through microprojectile bombardment. *Genome* 34:453–460.

Chibbar, R.N., Shyluk, J., Georges, F., Mallard, C.S., and Constabel, F. 1988. Esterase isozymes as markers of somatic embryogenesis in cultured carrot cells. *J. Plant Physiol.* 133:367–370.

Chibi, F., Angosto, T., Garrido, D., and Matilla, A. 1993. Requirement of polyamines for *in-vitro* maturation of the mid-binucleate pollen of *Nicotiana tabacum*. *J. Plant Physiol.* 142:452–456.

Chichiricco, G., Constantino, P., and Spanò, L. 1992. Expression of the *rolB* oncogene from *Agrobacterium rhizogenes* during zygotic embryogenesis in tobacco. *Plant Cell Physiol.* 33:827–832.

Childers, W.R. 1952. Male sterility in *Medicago sativa* L. *Sci. Agr.* 32:351–364.

Chitralekha, P., and Bhandari, N.N. 1991. Post-fertilization development of antipodal cells in *Ranunculus sceleratus* Linn. *Phytomorphology* 41:200–212.

Chitralekha, P., and Bhandari, N.N. 1992. Cellularization of endosperm in *Asphodelus tenuifolius* Cav. *Phytomorphology* 42:185–193.

Chitralekha, P., and Bhandari, N.N. 1993. Cellularization of free-nuclear endosperm in *Ranunculus sceleratus* Linn. *Phytomorphology* 43:165–183.

Chlan, C.A., and Dure, L., III. 1983. Plant seed embryogenesis as a tool for molecular biology. *Mol. Cell. Biochem.* 55:5–15.

Chlan, C.A., Borroto, K., Kamalay, J.A., and Dure, L., III. 1987. Developmental biochemistry of cottonseed embryogenesis and germination. XIX. Sequences and genomic organization of the α–globulin (vicilin) genes of cottonseed. *Plant Mol. Biol.* 9:533–546.

Chlan, C.A., Pyle, J.B., Legocki, A.B., and Dure, L., III. 1986. Developmental biochemistry of cottonseed embryogenesis and germination. XVIII. cDNA and amino acid sequences of members of the storage protein families. *Plant Mol. Biol.* 7:475–489.

Cho, T.-J., Davies, C.S., Fischer, R.L., Turner, N.E., Goldberg, R.B., and Nielsen, N.C. 1989. Molecular characterization of an aberrant allele for the Gy_3 glycinin gene: a chromosomal rearrangement. *Plant Cell* 1:339–350.

Cho, U.-H., and Kasha, K.J. 1989. Ethylene production and embryogenesis from anther cultures of barley (*Hordeum vulgare*). *Plant Cell Rep.* 8:415–417.

Choi, J.H., Liu, L.-S., Borkird, C., and Sung, Z.R. 1987. Cloning of genes developmentally regulated during plant embryogenesis. *Proc. Natl Acad. Sci. USA* 84:1906–1910.

Choinski, J.S., Jr., and Trelease, R.N. 1978. Control of enzyme activities in cotton cotyledons during maturation and germination. II. Glyoxysomal enzyme development in embryos. *Plant Physiol.* 62:141–145.

Choinski, J.S., Jr., Trelease, R.N., and Doman, D.C. 1981. Control of enzyme activities in cotton cotyledons during maturation and germination. III. *In vitro* embryo development in the presence of abscisic acid. *Planta* 152:428–435.

Chojecki, A.J.S., Bayliss, M.W., and Gale, M.D. 1986. Cell production and DNA accumulation in the wheat endosperm, and their association with grain weight. *Ann. Bot.* 58:809–817.

Chourey, P.S. 1981. Genetic control of sucrose synthetase in maize endosperm. *Mol. Gen. Genet.* 184:372–376.

Chourey, P.S., and Nelson, O.E. 1976. The enzymatic deficiency conditioned by the *shrunken-1* mutations in maize. *Biochem. Genet.* 14:1041–1055.

Chourey, P.S., and Schwartz, D. 1971. Ethyl methanesulfonate-induced mutations of the Sh_1 protein in maize. *Mutation Res.* 12:151–157.

Chowdhury, J.B., and Das, K. 1968. Cyto-morphological studies on male sterility in *Brassica campestris* L. *Cytologia* 33:195–199.

Chrispeels, M.J. 1983. The Golgi apparatus mediates the transport of phytohemagglutinin in the protein bodies in bean cotyledons. *Planta* 158:140–151.

Chrispeels, M.J. 1985. UDP-GlcNAc:glycoprotein GlcNAc-transferase is located in the Golgi apparatus of developing bean cotyledons. *Plant Physiol.* 78:835–838.

Chrispeels, M.J., Vitale, A., and Staswick, P. 1984. Gene expression and synthesis of phytohemagglutinin in the embryonic axes of developing *Phaseolus vulgaris* seeds. *Plant Physiol.* 76:791–796.

Chrispeels, M.J., Higgins, T.J.V., Craig, S., and Spencer, D. 1982. Role of the endoplasmic reticulum in the synthesis of reserve proteins and the kinetics of their transport to protein bodies in developing pea cotyledons. *J. Cell Biol.* 93:5–14.

Christ, B. 1959. Entwicklungsgeschichtliche und physiologische Untersuchungen über die Selbststerilität von *Cardamine pratensis* L. *Z. Bot.* 47:88–112.

Christensen, J.E., and Horner, H.T., Jr. 1974. Pollen pore development and its spatial orientation during microsporogenesis in the grass *Sorghum bicolor*. *Am. J. Bot.* 61:604–623.

Christensen, J.E., Horner, H.T., Jr., and Lersten, N.R. 1972. Pollen wall and tapetal orbicular wall development in *Sorghum bicolor* (Gramineae). *Am. J. Bot.* 59:43–58.

Christianson, M.L. 1986. Fate map of the organizing shoot apex in *Gossypium*. *Am. J. Bot.* 73:947–958.

Christou, P. 1995. Strategies for variety-independent genetic transformation of important cereals, legumes and woody species utilizing particle bombardment. *Euphytica* 85:13–27.

Christou, P., Ford, T.L., and Kofron, M. 1991. Production of transgenic rice (*Oryza sativa* L.) plants from agronomically important indica and japonica varieties via electric discharge particle acceleration of exogenous DNA into immature zygotic embryos. *Biotechnology* 9:957–962.

Chu, C., Liu, H.-T., and Du, R.-H. 1982. Microphotometric determination of DNA contents of early developmental pollen grains in tobacco anther culture. *Acta Bot. Sinica* 24:1–9.

Chung, I.-K., Ito, T., Tanaka, H., Ohta, A., Nan, H.G., and Takagi, M. 1994. Molecular diversity of three *S*-allele cDNAs associated with gametophytic self-incompatibility in *Lycopersicon peruvianum*. *Plant Mol. Biol.* 26:757–762.

Chung, Y.-Y., Kim, S.-R., Kang, H.-G., Noh, Y.-S., Park, M.C., Finkel, D., and An, G. 1995. Characterization of two rice MADS box genes homologous to *GLOBOSA*. *Plant Sci.* 109:45–56.

Ciampolini, F., Cresti, M., and Kapil, R.N. 1983. Fine structural and cytochemical characteristics of style and stigma in olive. *Caryologia* 36:211–230.

Ciampolini, F., Moscatelli, A., and Cresti, M. 1988. Ultrastructural features of *Aloe ciliaris* pollen. I. Mature grain and its activation *in vitro*. *Sex. Plant Reprod.* 1:88–96.

Ciampolini, F., Shivanna, K.R., and Cresti, M. 1991. High humidity and heat stress causes dissociation of endoplasmic reticulum in tobacco pollen. *Bot. Acta* 104:110–116.

Ciampolini, F., Cresti, M., Sarfatti, G., and Tiezzi, A. 1981. Ultrastructure of the stylar canal cells of *Citrus limon* (Rutaceae). *Plant Syst. Evol.* 138:263–274.

Ciarrocchi, G., Cella, R., and Nielsen, E. 1981. Release of nucleotide-cleaving acid phosphatase from carrot cells grown in suspension culture. *Physiol. Plant.* 53:375–377.

Cionini, P.G., Bennici, A., Alpi, A., and D'Amato, F. 1976. Suspensor, gibberellin and *in vitro* development of *Phaseolus coccineus* embryos. *Planta* 131:115–117.

Clapham, D. 1971. *In vitro* development of callus from the pollen of *Lolium* and *Hordeum*. *Z. Pflanzenzüchtg* 65:285–292.

Clapham, D., Manders, G., Yibrah, H.S., and von Arnold, S. 1995. Enhancement of short- and medium-term expression of transgenes in embryogenic suspensions of *Picea abies* (L.) Karst. *J. Expt. Bot.* 46:655–662.

Clapham, D.H., and Östergren, G. 1984. Immunocytochemistry of tubulin at meiosis in *Tradescantia* by a protein-A gold method. *Hereditas* 101:137–142.

Clark, A.J., Higgins, P., Martin, H., and Bowles, D.J. 1991. An embryo-specific protein of barley (*Hordeum vulgare*). *Eur. J. Biochem.* 199:115–121.

Clark, E., Gafni, Y., and Izhar, S. 1988. Loss of CMS-specific mitochondrial DNA arrangement in fertile segregants of *Petunia* hybrids. *Plant Mol. Biol.* 11:249–253.

Clark, E.M., Izhar, S., and Hanson, M.R. 1985. Independent segregation of the plastid genome and cytoplasmic male sterility in *Petunia* somatic hybrids. *Mol. Gen. Genet.* 199:440–445.

Clark, F.J. 1940. Cytogenetic studies of divergent meiotic spindle formation in *Zea mays*. *Am. J. Bot.* 27:547–559.

Clark, J.K., and Sheridan, W.F. 1986. Developmental profiles of the maize embryo-lethal mutants *dek22* and *dek23*. *J. Hered.* 77:83–92.

Clark, J.K., and Sheridan, W.F. 1988. Characterization of the two maize embryo-lethal defective kernel mutants *rgh*-1210* and *fl*-1253B*: effects on embryo and gametophyte development. *Genetics* 120:279–290.

Clark, J.K., and Sheridan, W.F. 1991. Isolation and characterization of 51 *embryo-specific* mutations of maize. *Plant Cell* 3:935–951.

Clark, K.R., and Sims, T.L. 1994. The S-ribonuclease gene of *Petunia hybrida* is expressed in nonstylar tissue, including immature anthers. *Plant Physiol.* 106:25–36.

Clark, K.R., Okuley, J.J., Collins, P.D., and Sims, T.L. 1990. Sequence variability and developmental expression of S-alleles in self-incompatible and pseudo-selfcompatible *Petunia*. *Plant Cell* 2:815–826.

Clark, S.E., Running, M.P., and Meyerowitz, E.M. 1995. *CLAVATA3* is a specific regulator of shoot and floral meristem development affecting the same processes as *CLAVATA1*. *Development* 121:2057–2067.

Clarke, A., Gleeson, P., Harrison, S., and Knox, R.B. 1979. Pollen-stigma interactions: identification and characterization of surface components with recognition potential. *Proc. Natl Acad. Sci. USA* 76:3358–3362.

Clarke, A.E., Abbot, A., Mandel, T.E., and Pettitt, J.M. 1980. Organization of the wall layers of the stigmatic papillae of *Gladiolus gandavensis*: a freeze-fracture study. *J. Ultrastr. Res.* 73:269–281.

Clarke, A.E., Considine, J.A., Ward, R., and Knox, R.B. 1977a. Mechanism of pollination in *Gladiolus*: roles of the stigma and pollen-tube guide. *Ann. Bot.* 41:15–20.

Clarke, A.E., Harrison, S., Knox, R.B., Rapp, J., Smith, P., and Marchalonis, J.P. 1977b. Common antigens and male-female recognition in plants. *Nature* 265:161–163.

Clarke, E., and Steer, M.W. 1983. Cytoplasmic structure of germinated and ungerminated pollen grains of *Tradescantia virginiana*. *Caryologia* 36:299–305.

Clauhs, R.P., and Grun, P. 1977. Changes in plastid and mitochondrion content during maturation of generative cells of *Solanum* (Solanaceae). *Am. J. Bot.* 64:377–383.

Clément, C., and Audran, J.C. 1995. Anther wall layers control pollen sugar nutrition in *Lilium*. *Protoplasma* 187:172–181.

Clowes, F.A.L. 1978a. Origin of the quiescent centre in *Zea mays*. *New Phytol.* 80:409–419.

Clowes, F.A.L. 1978b. Origin of quiescence at the root pole of pea embryos. *Ann. Bot.* 42:1237–1239.

Clutter, M., Brady, T., Walbot, V., and Sussex, I. 1974. Macromolecular synthesis during plant embryogeny. Cellular rates of RNA synthesis in diploid and polytene cells in bean embryos. *J. Cell Biol.* 63:1097–1102.

Cobb, B.G, and Hannah, L.C. 1983. Development of wild type, *shrunken-1* and *shrunken-2* maize kernels grown *in vitro*. *Theor. Appl. Genet.* 65:47–51.

Coca, M.A., Almoguera, C., and Jordano, J. 1994. Expression of sunflower low-molecular-weight heat-shock proteins during embryogenesis and persistence after germination: localization and possible functional implications. *Plant Mol. Biol.* 25:479–492.

Cochrane, M.P. 1994. Observations on the germ aleurone of barley. Morphology and histochemistry. *Ann. Bot.* 73:113–119.

Cochrane, M.P., and Duffus, C.M. 1980. The nucellar projection and modified aleurone in the crease region of developing caryopses of barley (*Hordeum vulgare* L. var. *distichum*). *Protoplasma* 103:361–375.

Cochrane, M.P., and Duffus, C.M. 1981. Endosperm cell number in barley. *Nature* 289:399–401.

Cocucci, A., and Jensen, W.A. 1969a. Orchid embryology: the mature megagametophyte of *Epidendrum scutella*. *Kurtziana* 5:23–38.

Cocucci, A., and Jensen, W.A. 1969b. Orchid embryology: pollen tetrads of *Epidendrum scutella* in the anther and on the stigma. *Planta* 84:215–229.

Cocucci, A., and Jensen, W.A. 1969c. Orchid embryology: megagametophyte of *Epidendrum scutella* following fertilization. *Am. J. Bot.* 56:629–640.

Coen, E.S., and Meyerowitz, E.M. 1991. The war of the whorls: genetic interactions controlling flower development. *Nature* 353:31–37.

Colasanti, J., Tyers, M., and Sundaresan, V. 1991. Isolation and characterization of cDNA clones encoding a functional $p34^{cdc2}$ homologue from *Zea mays*. *Proc. Natl Acad. Sci. USA* 88:3377–3381.

Cole, K. 1959. Inheritance of male sterility in green sprouting broccoli. *Can. J. Genet. Cytol.* 1:203–207.

Coleman, C.E., and Kao, T.-H. 1992. The flanking regions of two *Petunia inflata S* alleles are heterogeneous and contain repetitive sequences. *Plant Mol. Biol.* 18:725–737.

Coleman, C.E., Lopes, M.A., Gillikin, J.W., Boston, R.S., and Larkins, B.A. 1995. A defective signal peptide in the maize high-lysine mutant floury 2. *Proc. Natl Acad. Sci. USA* 92:6828–6831.

Colhoun, C.W., and Steer, M.W. 1981. Microsporogenesis and the mechanism of cytoplasmic male sterility in maize. *Ann. Bot.* 48:417–424.

Colhoun, C.W., and Steer, M.W. 1983. The cytological effects of the gametocides ethrel and RH-531 on microsporogenesis in barley (*Hordeum vulgare* L.). *Plant Cell Environ.* 6:21–29.

Collada, C., Allona, I., Aragoncillo, P., and Aragoncillo, C. 1993. Development of protein bodies in cotyledons of *Fagus sylvatica*. *Physiol. Plant.* 89:354–359.

germination and early tube formation. *Am. J. Bot.* 72:719–727.

Cresti, M., Ciampolini, F., Pacini, E., and Sarfatti, G. 1978a. Phytoferritin in plastids of the style of *Olea europaea* L. *Acta Bot. Neerl.* 27:417–423.

Cresti, M., Ciampolini, F., Pacini, E., Sarfatti, G., van Went, J.L., and Willemse, M.T.M. 1979. Ultrastructural differences between compatible and incompatible pollen tubes in the stylar transmitting tissue of *Petunia hybrida. J. Submicros. Cytol.* 11:209–219.

Cresti, M., Ciampolini, F., Pacini, E., Sree Ramulu, K., and Devreux, M. 1978b. Gamma irradiation of *Prunus avium* L. flower buds: effects on stylar development – an ultrastructural study. *Acta Bot. Neerl.* 27:97–106.

Cresti, M., Ciampolini, F., van Went, J.L., and Wilms, H.J. 1982. Ultrastructure and histochemistry of *Citrus limon* (L.) stigma. *Planta* 156:1–9.

Cresti, M., Hepler, P.K., Tiezzi, A., and Ciampolini, F. 1986a. Fibrillar structures in *Nicotiana* pollen: changes in ultrastructure during pollen activation and tube emission. In *Biotechnology and Ecology of Pollen*, ed. D.L. Mulcahy, G.B. Mulcahy, and E. Ottaviano, pp. 283–288. New York: Springer-Verlag.

Cresti, M., Keijzer, C.J., Tiezzi, A., Ciampolini, F., and Focardi, F. 1986b. Stigma of *Nicotiana*: ultrastructural and biochemical studies. *Am. J. Bot.* 73:1713–1722.

Cresti, M., Milanesi, C., Salvatici, P., and van Aelst, A.C. 1990. Ultrastructural observations of *Papaver rhoeas* mature pollen grains. *Bot. Acta* 103:349–354.

Cresti, M., Milanesi, C., Tiezzi, A., Ciampolini, F., and Moscatelli, A. 1988. Ultrastructure of *Linaria vulgaris* pollen grains. *Acta Bot. Neerl.* 37:379–386.

Cresti, M., Pacini, E., Ciampolini, F., and Sarfatti, G. 1977. Germination and early tube development *in vitro* of *Lycopersicum peruvianum* pollen: ultrastructural features. *Planta* 136:239–247.

Cresti, M., van Went, J.L., Pacini, E., and Willemse, M.T.M. 1976. Ultrastructure of transmitting tissue of *Lycopersicon peruvianum* style: development and histochemistry. *Planta* 132:305–312.

Crété, P. 1963. Embryo. In *Recent Advances in the Embryology of Angiosperms*, ed. P. Maheshwari, pp. 171–220. Delhi: International Society of Plant Morphologists.

Croissant-Sych, Y., and Okita, T.W. 1996. Identification of positive and negative regulatory *cis*-elements of the rice glutelin *Gt3* promoter. *Plant Sci.* 116:27–35.

Cross, J.W., and Adams, W.R., Jr. 1983. Embryo-specific globulins from *Zea mays* L. and their subunit composition. *J. Agr. Food Chem.* 31:534–538.

Crossley, S.J., Greenland, A.J., and Dickinson, H.G. 1995. The characterisation of tapetum-specific cDNAs isolated from a *Lilium henryi* L. meiocyte subtractive cDNA library. *Planta* 196:523–529.

Crouch, M.L. 1982. Non-zygotic embryos of *Brassica napus* L. contain embryo-specific storage proteins. *Planta* 156:520–524.

Crouch, M.L., and Sussex, I.M. 1981. Development and storage-protein synthesis in *Brassica napus* L. embryos *in vivo* and *in vitro*. *Planta* 153:64–74.

Crouch, M.L., Tenbarge, K.M., Simon, A.E., and Ferl, R. 1983. cDNA clones for *Brassica napus* seed storage proteins: evidence from nucleotide sequence analysis that both subunits of napin are cleaved from a precursor polypeptide. *J. Mol. Appl. Genet.* 2:273–283.

Croy, R.R.D., Evans, I.M., Yarwood, J.N., Harris, N., Gatehouse, J.A., Shirsat, A.H., Kang, A., Ellis, J.R., Thompson, A., and Boulter, D. 1988. Expression of pea legumin sequences in pea, *Nicotiana* and yeast. *Biochem. Physiol. Pflanz.* 183:183–197.

Croy, R.R.D., Gatehouse, J.A., Evans, I.M., and Boulter, D. 1980a. Characterisation of the storage protein subunits synthesised *in vitro* by polyribosomes and RNA from developing pea (*Pisum sativum* L.). I. Legumin. *Planta* 148:49–56.

Croy, R.R.D., Gatehouse, J.A., Evans, I.M., and Boulter, D. 1980b. Characterisation of the storage protein subunits synthesised *in vitro* by polyribosomes and RNA from developing pea (*Pisum sativum* L.). II. Vicilin. *Planta* 148:57–63.

Croy, R.R.D., Lycett, G.W., Gatehouse, J.A., Yarwood, J.N., and Boulter, D. 1982. Cloning and analysis of cDNAs encoding plant storage protein precursors. *Nature* 295:76–79.

Cruz-Alvarez, M., Kirihara, J.A., and Messing, J. 1991. Post-transcriptional regulation of methionine content in maize kernels. *Mol. Gen. Genet.* 225:331–339.

Cui, X., Wise, R.P., and Schnable, P.S. 1996. The *rf2* nuclear restorer gene of male-sterile T-cytoplasm maize. *Science* 272:1334–1336.

Cullis, C.A. 1976. Chromatin-bound DNA-dependent RNA polymerase in developing pea cotyledons. *Planta* 131:293–298.

Cullis, C.A. 1978. Chromatin-bound DNA-dependent RNA polymerase in developing pea cotyledons. II. Polymerase activity and template availability under different growth conditions. *Planta* 144:57–62.

Cuming, A.C. 1984. Developmental regulation of gene expression in wheat embryos. Molecular cloning of a DNA sequence encoding the early-methionine-labelled (E_m) polypeptide. *Eur. J. Biochem.* 145:351–357.

Cummings, D.P., Green, C.E., and Stuthman, D.D. 1976. Callus induction and plant regeneration in oats. *Crop Sci.* 16:465–470.

Cummins, I., and Murphy, D.J. 1992. cDNA sequence of a sunflower oleosin and transcript tissue specificity. *Plant Mol. Biol.* 19:873–876.

Cummins, I., Hills, M.J., Ross, J.H.E., Hobbs, D.H., Watson, M.D., and Murphy, D.J. 1993. Differential, temporal and spatial expression of genes involved in storage oil and oleosin accumulation in developing rapeseed embryos: implication for the role of oleosins and the mechanisms of oil-body formation. *Plant Mol. Biol.* 23:1015–1027.

Curry, J., Morris, C.F., and Walker-Simmons, M.K. 1991. Sequence analysis of a cDNA encoding a group 3 LEA mRNA inducible by ABA or dehydration stress in wheat. *Plant Mol. Biol.* 16:1073–1076.

Custers, J.B.M., and Bergervoet, J.H.W. 1990. *In vitro* culture of embryos of *Cucumis* spp.: heart-stage embryos have a higher ability of direct plant formation than advanced-stage embryos. *Sex. Plant Reprod.* 3:152–159.

Custers, J.B.M., Cordewener, J.H.G., Nöllen, Y., Dons, H.J.M., and van Lookeren Campagne, M.M. 1994. Temperature controls both gametophytic and sporophytic development in microspore cultures of *Brassica napus*. *Plant Cell Rep.* 13:267–271.

Cyr, R.J., Bustos, M.M., Guiltinan, M.J., and Fosket, D.E. 1987. Developmental modulation of tubulin protein and mRNA levels during somatic embryogenesis in cultured carrot cells. *Planta* 171:365–376.

Czakó, M., Jang, J.-C., Herr, J.M., Jr., and Márton, L. 1992. Differential manifestation of seed mortality induced by seed-specific expression of the gene for diphtheria toxin A chain in *Arabidopsis* and tobacco. *Mol. Gen. Genet.* 235:33–40.

D'Alascio-Deschamps, R. 1972. Le sac embryonnaire du lin après la fécondation. *Botaniste* 50:273–288.

D'Alascio-Deschamps, R. 1973. Organisation du sac embryonnaire du *Linum catharticum* L., espèce récoltée en station naturelle; étude ultrastructurale. *Bull. Soc. Bot. Fr.* 120:189–200.

D'Alascio-Deschamps, R. 1981. Embryologie du *Linum catharticum* L. Le zygote: étude ultrastructurale. *Bull. Soc. Bot. Fr. Lett. Bot.* 128:269–278.

D'Amato, F. 1984. Role of polyploidy in reproductive organs and tissues. In *Embryology of Angiosperms*, ed. B.M. Johri, pp. 519–566. Berlin: Springer-Verlag.

D'Amato, F., Devreux, M., and Scarascia Mugnozza, G.T. 1965. The DNA content of the nuclei of the pollen grains in tobacco and barley. *Caryologia* 18:377–382.

Dahlgren, K.V.O. 1934. Die Embryologie von *Impatiens roylei*. *Svensk. Bot. Tidskr.* 28:103–125.

Dahmer, M.L., Hildebrand, D.F., and Collins, G.B. 1992. Comparative protein accumulation patterns in soybean somatic and zygotic embryos. *In Vitro Cell Devel Biol.* 28P:106–114.

Dale, P.J. 1975. Pollen dimorphism and anther culture in barley. *Planta* 127:213–220.

Dale, P.J., and Deambrogio, E. 1979. A comparison of callus induction and plant regeneration from different explants of *Hordeum vulgare*. *Z. Pflanzenphysiol.* 94:65–77.

Dannenhoffer, J.M., Bostwick, D.E., Or, E., and Larkins, B.A. 1995. Opaque-15, a maize mutation with properties of a defective opaque-2 modifier. *Proc. Natl Acad. Sci. USA* 92:1931–1935.

Das, N.K. 1965. Inactivation of the nucleolar apparatus during meiotic prophase in corn anthers. *Expt. Cell Res.* 40:360–364.

Das, O.P., and Messing, J.W. 1987. Allelic variation and differential expression at the 27-kilodalton zein locus in maize. *Mol. Cell. Biol.* 7:4490–4497.

Dasgupta, J., and Bewley, J.D. 1982. Desiccation of axes of *Phaseolus vulgaris* during development of a switch from a development pattern of protein synthesis to a germination pattern. *Plant Physiol.* 70:1224–1227.

Dashek, W.V., Erickson, S.S., Hayward, D.M., Lindbeck, G., and Mills, R.R. 1979. Peroxidase in cytoplasm and cell wall of germinating lily pollen. *Bot. Gaz.* 140:261–265.

Dashek, W.V., and Harwood, H.I. 1974. Proline, hydroxyproline, and lily pollen tube elongation. *Ann. Bot.* 38:947–959.

Dashek, W.V., and Rosen, W.G. 1966. Electron microscopical localization of chemical components in the growth zone of lily pollen tubes. *Protoplasma* 61:192–204.

Dashek, W.V., Thomas, H.R., and Rosen, W.G. 1971. Secretory cells of lily pistils. II. Electron microscope cytochemistry of canal cells. *Am. J. Bot.* 58:909–920.

Datta, S.K. 1987. Plant regeneration by pollen embryogenesis from cultured whole spikes of barley (*Hordeum vulgare*). *Theor. Appl. Genet.* 74:121–124.

Dauphin-Guerin, B., Teller, G., and Durand, D. 1980. Different endogenous cytokinins between male and female *Mercurialis annua* L. *Planta* 148:124–129.

Davey, J.E., and van Staden, J. 1978. Ultrastructural aspects of reserve protein deposition during cotyledonary cell development in *Lupinus albus*. *Z. Pflanzenphysiol.* 89:259–271.

Davies, D.R. 1980. The r_a locus and legumin synthesis in *Pisum sativum*. *Biochem. Genet.* 18:1207–1219.

Davies, D.R., and Bedford, I.D. 1982. Abscisic acid and storage protein accumulation in *Pisum sativum* embryos grown *in vitro*. *Plant Sci. Lett.* 27:337–343.

Davies, D.R., and Brewster, V. 1975. Studies of seed development in *Pisum sativum*. II. Ribosomal RNA contents in reciprocal crosses. *Planta* 124:303–309.

Davies, H.M., and Delmer, D.P. 1981. Two kinds of protein glycosylation in a cell-free preparation from developing cotyledons of *Phaseolus vulgaris*. *Plant Physiol.* 68:284–291.

Davies, M.D., and Dickinson, D.B. 1971. Effects of freeze-drying on permeability and respiration of germinating lily pollen. *Physiol. Plant.* 24:5–9.

Davies, M.D., and Dickinson, D.B. 1972. Properties of uridine diphosphoglucose dehydrogenase from pollen of *Lilium longiflorum*. *Arch. Biochem. Biophys.* 152:53–61.

Davis, B.D. 1983. Growth of excised pea embryonic axes on different sugars. *Am. J. Bot.* 70:816–820.

Davis, R.W., Smith, J.D., and Cobb, B.G. 1990. A light and electron microscope investigation of the transfer cell region of maize caryopses. *Can. J. Bot.* 68:471–479.

Dawe, R.K., and Freeling, M. 1990. Clonal analysis of the cell lineages in the male flower of maize. *Devel Biol.* 142:233–245.

Dawe, R.K., and Freeling, M. 1992. The role of initial cells in maize anther morphogenesis. *Development* 116:1077–1085.

Dawe, R.K., Lachmansingh, A.R., and Freeling, M. 1993. Transposon-mediated mutations in the untranslated leader of maize *Adh1* that increase and decrease pollen-specific gene expression. *Plant Cell* 5:311–319.

Dawe, R.K., Sedat, J.W., Agard, D.A., and Cande, W.Z. 1994. Meiotic chromosome pairing in maize is associated with a novel chromatin organization. *Cell* 76:901–912.

Dawson, J., Wilson, Z.A., Aarts, M.G.M., Braithwaite, A.F., Briarty, L.G., and Mulligan, B.J. 1993. Microspore and pollen development in six male-sterile mutants of *Arabidopsis thaliana*. *Can. J. Bot.* 71:629–638.

De, D.N. 1961. Autoradiographic studies of nucleoprotein metabolism during the division cycle. *Nucleus* 4:1–24.

de Barros, E.G., Takasaki, K., Kirleis, A.W., and Larkins, B.A. 1991. Nucleotide sequence of a cDNA clone encoding γ-kafirin protein from *Sorghum bicolor*. *Plant Physiol.* 97:1606–1607.

de Block, M., and Debrouwer, D. 1993. Engineered fertility control in transgenic *Brassica napus* L.: histochemical analysis of anther development. *Planta* 189:218–225.

de Boer-de Jeu, M.J. 1978. Megasporogenesis. A comparative study of the ultrastructural aspects of megasporogenesis in *Lilium, Allium* and *Impatiens*. *Meded. Landbouwhoges. Wageningen* 78-16:1–127.

de Castro, L.A.B., Lacerda, Z., Aramayo, R.A., Sampaio, M.J.A.M., and Gander, E.S. 1987. Evidence for a precursor molecule of Brazil nut 2S seed proteins from biosynthesis and cDNA analysis. *Mol. Gen. Genet.* 206:338–343.

de Clercq, A., Vandewiele, M., de Rycke, R., van Damme, J., van Montagu, M., Krebbers, E., and Vandekerckhove, J. 1990a. Expression and processing of an *Arabidopsis* 2S albumin in transgenic tobacco. *Plant Physiol.* 92:899–907.

de Clercq, A., Vandewiele, M., van Damme, J., Guerche, P., van Montagu, M., Vandekerckhove, J., and Krebbers, E. 1990b. Stable accumulation of modified 2S albumin seed storage proteins with higher methionine contents in transgenic plants. *Plant Physiol.* 94:970–979.

de Freitas, F.A., Yunes, J.A., da Silva, M.J., Arruda, P., and Leite, A. 1994. Structural characterization and promoter activity analysis of the γ-kafirin gene from sorghum. *Mol. Gen. Genet.* 245:177–186.

de Jong, A.J., Schmidt, E.D.L., and de Vries, S.C. 1993. Early events in higher-plant embryogenesis. *Plant Mol. Biol.* 22:367–377.

de Jong, A.J., Cordewener, J., Lo Schiavo, F., Terzi, M., Vandekerckhove, J., van Kammen, A., and de Vries, S.C. 1992. A carrot somatic embryo mutant is rescued by chitinase. *Plant Cell* 4:425–433.

de Jong, A.J., Heidstra, R., Spaink, H.P., Hartog, M.V., Meijer, E.A., Hendriks, T., Lo Schiavo, F., Terzi, M., Bisseling, T., van Kammen, A., and de Vries, S.C. 1993. *Rhizobium* lipooligosaccharides rescue a carrot somatic embryo mutant. *Plant Cell* 5:615–620.

de Jong, A.J., Hendriks, T., Meijer, E.J., Penning, M., Lo Schiavo, F., Terzi, M., van Kammen, A., and de Vries, S.C. 1995. Transient reduction in secreted 32 kD chitinase prevents somatic embryogenesis in the carrot (*Daucus carota* L.) variant *ts11*. *Devel Genet.* 16:332–343.

de la Peña, A. 1986. '*In vitro*' culture of isolated meiocytes of rye, *Secale cereale* L. *Environ. Expt. Bot.* 26:17–23.

de la Peña, A., Lörz, H., and Schell, J. 1987. Transgenic rye plants obtained by injecting DNA into young floral tillers. *Nature* 325:274–276.

de Mey, J., Lambert, A.M., Bajer, A.S., Moeremans, M., and de Brabander, M. 1982. Visualization of microtubules in interphase and mitotic plant cells of *Haemanthus* endosperm with immuno-gold staining method. *Proc. Natl Acad. Sci. USA* 79:1898–1902.

de Nettancourt, D. 1977. *Incompatibility in Angiosperms.* Berlin: Springer-Verlag.

de Nettancourt, D., Devreux, M., Bozzini, A., Cresti, M., Pacini, E., and Sarfatti, G. 1973a. Ultrastructural aspects of the self-incompatibility mechanism in *Lycopersicum peruvianum* Mill. *J. Cell Sci.* 12:403–419.

de Nettancourt, D., Devreux, M., Laneri, U., Cresti, M., Pacini, E., and Sarfatti, G. 1974. Genetical and ultrastructural aspects of self and cross incompatibility in interspecific hybrids between self-compatible *Lycopersicum esculentum* and self-incompatible *L. peruvianum. Theor. Appl. Genet.* 44:278–288.

de Nettancourt, D., Devreux, M., Laneri, U., Pacini, E., Cresti, M., and Sarfatti, G. 1973b. Ultrastructural aspects of unilateral interspecific incompatibility between *Lycopersicum peruvianum* and *L. esculentum. Caryologia* 25: Suppl. 207–217.

de Paepe, R., Forchioni, A., Chétrit, P., and Vedel, F. 1993. Specific mitochondrial proteins in pollen: presence of an additional ATP synthase β–subunit. *Proc. Natl Acad. Sci. USA* 90:5934–5938.

de Pater, S., Pham, K., Chua, N.-H., Memelink, J., and Kijne, J. 1993. A 22-bp fragment of the pea lectin promoter containing essential TGAC-like motifs confers seed-specific gene expression. *Plant Cell* 5:877–886.

de Silva, J., Loader, N.M., Jarman, C., Windust, J.H.C., Hughes, S.G., and Safford, R. 1990. The isolation and sequence analysis of two seed-expressed acyl carrier protein genes from *Brassica napus. Plant Mol. Biol.* 14:537–548.

de Vries, A.P., and Ie, T.S. 1970. Electron-microscopy on anther tissue and pollen of male sterile and fertile wheat (*Triticum aestivum* L.). *Euphytica* 19:103–120.

de Vries, S.C., Booij, H., Janssens, R., Vogels, R., Saris, L., Lo Schiavo, F., Terzi, M., and van Kammen, A. 1988a. Carrot somatic embryogenesis depends on the phytohormone-controlled presence of correctly glycosylated extracellular proteins. *Genes Devel.* 2:462–476.

de Vries, S.C., Booij, H., Meyerink, P., Huisman, G., Wilde, H.D., Thomas, T.L., and van Kammen, A. 1988b. Acquisition of embryogenic potential in carrot cell–suspension cultures. *Planta* 176:196–204.

de Wet, J.M.J., Bergquist, R.R., Harlan, J.R., Brink, D.E., Cohen, C.E., Newell, C.A., and de Wet, A.E. 1985. Exogenous gene transfer in maize (*Zea mays*) using DNA-treated pollen. In *The Experimental Manipulation of Ovule Tissues*, ed. G.P. Chapman, S.H. Mantell, and R.W. Daniels, pp. 197–209. New York: Longman.

Degenhardt, J., Fiebig, C., and Link, G. 1991. Chloroplast and nuclear transcripts for plastid proteins in *Arabidopsis thaliana*: tissue distribution in mature plants and during seedling development and embryogenesis. *Bot. Acta* 104:455–463.

Del Casino, C., Li, Y.-Q., Moscatelli, A., Scali, M., Tiezzi, A., and Cresti, M. 1993. Distribution of microtubules during the growth of tobacco pollen tubes. *Biol. Cell.* 79:125–132.

Del Casino, C., Tiezzi, A., Wagner, V.T., and Cresti, M. 1992. The organization of the cytoskeleton in the generative cell and sperms of *Hyacinthus orientalis. Protoplasma* 168:41–50.

Del Rosario, A.G., and de Guzman, E.V. 1976. The growth of coconut "Makapuno" embryos *in vitro* as affected by mineral composition and sugar level of the medium during the liquid and solid cultures. *Philipp. J. Sci.* 105:215–222.

Delannay, X. 1979. Evolution of male sterility mechanisms in gynodioecious and dioecious species of *Cirsium* (Cynareae, Compositae). *Plant Syst. Evol.* 132:327–332.

Delannay, X., and Palmer, R.G. 1982. Genetics and cytology of the ms_4 male-sterile soybean. *J. Hered.* 73:219–223.

Delbreil, B., Guerche, P., and Jullien, M. 1993. *Agrobacterium*-mediated transformation of *Asparagus officinalis* L. long-term embryogenic callus and regeneration of transgenic plants. *Plant Cell Rep.* 12:129–132.

DeLisle, A.J., and Crouch, M.L. 1989. Seed storage protein transcription and mRNA levels in *Brassicfa napus* during development and in response to exogenous abscisic acid. *Plant Physiol.* 91:617–623.

DeLong, A., Calderon-Urrea, A., and Dellaporta, S.L. 1993. Sex determination gene *TASSELSEED2* of maize encodes a short-chain alcohol dehydrogenase required for stage-specific floral organ abortion. *Cell* 74:757–768.

Delorme, V., Giranton, J.-L., Hatzfeld, Y., Friry, A., Heizmann, P., Ariza, M.J., Dumas, C., Gaude, T., and Cock, J.M. 1995. Characterization of the *S* locus genes, *SLG* and *SRK*, of the *Brassica* S_3 haplotype:

identification of a membrane-localized protein encoded by the *S* locus receptor kinase gene. *Plant J.* 7:429–440.

Delvallée, I., and Dumas, C. 1988. Anther development in *Zea mays*: changes in protein, peroxidase, and esterase patterns. *J. Plant Physiol.* 132:210–217.

DeMason, D.A. 1986. Endosperm structure and storage reserve histochemistry in the palm, *Washingtonia filifera. Am. J. Bot.* 73:1332–1340.

DeMason, D.A., and Chandra Sekhar, K.N. 1988. The breeding system in the date palm (*Phoenix dactylifera* L.) and its recognition by early cultivators. *Adv. Econ. Bot.* 6:20–35.

DeMason, D.A., Chandra Sekhar, K.N., and Harris, M. 1989. Endosperm development in the date palm (*Phoenix dactylifera*) (Arecaceae). *Am. J. Bot.* 76:1255–1265.

DeMason, D.A., Sexton, R., and Grant Reid, J.S. 1983. Structure, composition and physiological state of the endosperm of *Phoenix dactylifera* L. *Ann. Bot.* 52:71–80.

Demerec, M. 1923. Heritable characters of maize. XV. Germless seeds. *J. Hered.* 14:297–300.

Demerec, M. 1924. A case of pollen dimorphism in maize. *Am. J. Bot.* 11:461–464.

Deng, W., Grayburn, W.S., Hamilton-Kemp, T.R., Collins, G.B., and Hildebrand, D.F. 1992. Expression of soybean-embryo lipoxygenase 2 in transgenic tobacco tissue. *Planta* 187:203–208.

Denis, M., Delourme, R., Gourret, J.-P., Mariani, C., and Renard, M. 1993. Expression of engineered nuclear male sterility in *Brassica napus. Plant Physiol.* 101:1295–1304.

Depigny-This, D., Raynal, M., Aspart, L., Delseny, M., and Grellet, F. 1992. The cruciferin gene family in radish. *Plant Mol. Biol.* 20:467–479.

Derksen, J., Pierson, E.S., and Traas, J.A. 1985. Microtubules in vegetative and generative cells of pollen tubes. *Eur. J. Cell Biol.* 38:142–148.

Derksen, J., Rutten, T., van Amstel, T., de Win, A., Doris, F., and Steer, M. 1995. Regulation of pollen tube growth. *Acta Bot. Neerl.* 44:93–119.

DeRocher, A., and Vierling, E. 1995. Cytoplasmic HSP70 homologues of pea: differential expression in vegetative and embryonic organs. *Plant Mol. Biol.* 27:441–456.

DeRocher, A.E., and Vierling, E. 1994. Developmental control of small heat shock protein expression during pea seed maturation. *Plant J.* 5:93–102.

DeRose, R.T., Ma, D.-P., Kwon, I.-S., Hasnain, S.E., Klassy, R.C., and Hall, T.C. 1989. Characterization of the kafirin gene family from sorghum reveals extensive homology with zein from maize. *Plant Mol. Biol.* 12:245–256.

Desborough, S., and Peloquin, S.J. 1968. Disc-electrophoresis of proteins and enzymes from styles, pollen and pollen tubes of self-incompatible cultivars of *Lilium longiflorum. Theor. Appl. Genet.* 38:327–331.

Deschamps, R. 1969. Premiers stades du développement de l'embryon et de l'albumen du lin: etude au microscope électronique. *Rev. Cytol. Biol. Végét.* 32:379–390.

Deshusses, J., Gumber, S.C., and Loewus, F.A. 1981. Sugar uptake in lily pollen. A proton symport. *Plant Physiol.* 67:793–796.

Després, C., Saba-El-Leil, M., Rivard, S.R., Morse, D., and Cappadocia, M. 1994. Molecular cloning of two *Solanum chacoense* S-alleles and a hypothesis concerning their evolution. *Sex. Plant Reprod.* 7:169–176.

Desprez, B., Chupeau, M.-C., Vermeulen, A., Delbreil, B., Chupeau, Y., and Bourgin, J.-P. 1995. Regeneration and characterization of plants produced from mature tobacco pollen protoplasts via gametosomatic hybridization. *Plant Cell Rep.* 14:204–209.

Detchepare, S., Heizmann, P., and Dumas, C. 1989. Changes in protein patterns and protein synthesis during anther development in *Brassica oleracea. J. Plant Physiol.* 135:129–137.

Deurenberg, J.J.M. 1976. *In vitro* protein synthesis with polysomes from unpollinated, cross- and self-pollinated *Petunia* ovaries. *Planta* 128:29–33.

Deurenberg, J.J.M. 1977. Differentiated protein synthesis with polysomes from *Petunia* ovaries before fertilization. *Planta* 133:201–206.

DeVerno, L.L., Charest, P.J., and Bonen, L. 1994. Mitochondrial DNA variation in somatic embryogenic cultures of *Larix. Theor. Appl. Genet.* 88:727–732.

Devi, P.M. 1964. Heterostyly in *Biophytum sensitivum* DC. *J. Genet.* 59:41–48.

Devic, M., Albert, S., and Delseny, M. 1996. Induction and expression of seed-specific promoters in *Arabidopsis* embryo-defective mutants. *Plant J.* 9:205–215.

Devreux, M. 1963. Effets de l'irradiation gamma chronique sur l'embryogenèse de *Capsella bursa-pastoris* Moench. In *L'Energia Nucleare in Agricoltura, VI Congr. Nucl. (Roma)*, pp. 199–217. Vallecchi: Comitato Nazionale Energia Nucleare.

Devreux, M., and Scarascia Mugnozza, G.T. 1962. Action des rayons gamma sur les premiers stades de developpement de l'embryon de *Nicotiana rustica* L. *Caryologia* 15:279–291.

Dewey, R.E., Levings, C.S., III, and Timothy, D.H. 1986. Novel recombinations in the maize mitochondrial genome produce a unique transcriptional unit in the Texas male-sterile cytoplasm. *Cell* 44:439–449.

Dewey, R.E., Timothy, D.H., and Levings, C.S., III. 1987. A mitochondrial protein associated with cytoplasmic male sterility in the T cytoplasm of maize. *Proc. Natl Acad. Sci. USA* 84:5374–5378.

Dewey, R.E., Siedow, J.N., Timothy, D.H., and Levings, C.S., III. 1988. A 13-kilodalton maize mitochondrial protein in *E. coli* confers sensitivity to *Bipolaris maydis* toxin. *Science* 239:293–295.

DeWitt, N.D., Harper, J.F., and Sussman, M.R. 1991. Evidence for a plasma membrane proton pump in phloem cells of higher plants. *Plant J.* 1:121–128.

Dexheimer, J. 1965. Sur les structures cytoplasmiques dans les grains de pollen de *Lobelia erinus. Compt. Rend. Acad. Sci. Paris* 260:6963–6965.

Dexheimer, J. 1968. Sur la synthèse d'acide ribonucléique par les tubes polliniques en croissance. *Compt. Rend. Acad. Sci. Paris* 267D:2126–2128.

Dexheimer, J. 1970. Recherches cytophysiologiques sur les grains de pollen. *Rev. Cytol. Biol. Végét.* 33:169–233.

Dexheimer, J. 1972. Etude expérimentale des grains de pollen des angiospermes. *Rev. Cytol. Biol. Végét.* 35:17–39.

Dhaliwal, A.S., and Malik, C.P. 1982. Localization of some hydrolases in the pollen and stigma of *Brassica campestris* following self- and cross-pollination. *Phytomorphology* 32:37–41.

Dhaliwal, A.S., Malik, C.P., and Singh, M.B. 1981. Overcoming incompatibility in *Brassica campestris* L.

by carbon dioxide, and dark fixation of the gas by self- and cross-pollinated pistils. *Ann. Bot.* 48:227–233.

D'Halluin, K., Bonne, E., Bossut, M., de Beuckeleer, M., and Leemans, J. 1992. Transgenic maize plants by tissue electroporation. *Plant Cell* 4:1495–1505.

Dhillon, S.S., and Miksche, J.P. 1983. DNA, RNA, protein and heterochromatin changes during embryo development and germination of soybean (*Glycine max* L.). *Histochem. J.* 15:21–37.

di Fonzo, N., Gentinetta, E., Salamini, F., and Soave, C. 1979. Action of the *opaque-7* mutation on the accumulation of storage products in maize endosperm. *Plant Sci. Lett.* 14:345–354.

di Fonzo, N., Hartings, H., Brembilla, M., Motto, M., Soave, C., Navarro, E., Palau, J., Rhode, W., and Salamini, F. 1988. The b-32 protein from maize endosperm, an albumin regulated by the *O2* locus: nucleic acid (cDNA) and amino acid sequences. *Mol. Gen. Genet.* 212:481–487.

di Fonzo, N., Manzocchi, L., Salamini, F., and Soave, C. 1986. Purification and properties of an endospermic protein of maize associated with the *opaque-2* and *opaque-6* genes. *Planta* 167:587–594.

Diboll, A.G. 1968. Fine structural development of the megagametophyte of *Zea mays* following fertilization. *Am. J. Bot.* 55:797–806.

Diboll, A.G., and Larson, D.A. 1966. An electron microscopic study of the mature megagametophyte in *Zea mays. Am. J. Bot.* 53:391–402.

Dickinson, C.D., Evans, R.P., and Nielsen, N.C. 1988. RY repeats are conserved in the 5'-flanking regions of legume seed-protein genes. *Nucl. Acids Res.* 16:371.

Dickinson, C.D., Hussein, E.H.A., and Nielsen, N.C. 1989. Role of posttranslational cleavage in glycinin assembly. *Plant Cell* 1:459–469.

Dickinson, D.B. 1965. Germination of lily pollen: respiration and tube growth. *Science* 150:1818–1819.

Dickinson, D.B. 1966. Inhibition of pollen respiration by oligomycin. *Nature* 210:1362–1363.

Dickinson, D.B. 1967. Permeability and respiratory properties of germinating pollen. *Physiol. Plant.* 20:118–127.

Dickinson, D.B., and Davies, M.D. 1971. Nucleoside diphosphate kinase from lily pollen. *Plant Cell Physiol.* 12:157–160.

Dickinson, H.G. 1970a. Ultrastructural aspects of primexine formation in the microspore tetrad of *Lilium longiflorum. Cytobiologie* 1:437–449.

Dickinson, H.G. 1970b. Membrane investing the microsporangium of *Pinus banksiana. New Phytol.* 69:1065–1068.

Dickinson, H.G. 1971a. The role played by sporopollenin in the development of pollen in *Pinus banksiana*. In *Sporopollenin*, ed. J. Brooks, P.R. Grant, M. Muir, P. van Gijzel, and G. Shaw, pp. 31–67. London: Academic Press.

Dickinson, H.G. 1971b. Nucleo–cytoplasmic interaction following meiosis in the young microspores of *Lilium longiflorum*; events at the nuclear envelope. *Grana* 11:117–127.

Dickinson, H.G. 1973. The role of plastids in the formation of pollen grain coatings. *Cytobios* 8:25–40.

Dickinson, H.G. 1976. Common factors in exine deposition. In *The Evolutionary Significance of the Exine*, ed. I.K. Ferguson and J. Muller, pp. 67–89. London: Academic Press.

Dickinson, H.G. 1981. The structure and chemistry of plastid differentiation during male meiosis in *Lilium henryi. J. Cell Sci.* 52:223–241.

Dickinson, H.G. 1990. Self-incompatibility in flowering plants. *Bioessays* 12:155–161.

Dickinson, H.G., and Andrews, L. 1977. The rôle of membrane-bound cytoplasmic inclusions during gametogenesis in *Lilium longiflorum* Thumb. *Planta* 134:229–240.

Dickinson, H.G., and Bell, P.R. 1970. Nucleocytoplasmic interaction at the nuclear envelope in post meiotic microspores of *Pinus banksiana. J. Ultrastr. Res.* 33:356–359.

Dickinson, H.G., and Bell, P.R. 1972a. The rôle of the tapetum in the formation of sporopollenin-containing structures during microsporogenesis in *Pinus banksiana. Planta* 107:205–215.

Dickinson, H.G., and Bell, P.R. 1972b. Structures resembling nuclear pores at the orifice of nuclear invaginations in developing microspores of *Pinus banksiana. Devel Biol.* 27:425–429.

Dickinson, H.G., and Bell, P.R. 1976. The changes in the tapetum of *Pinus banksiana* accompanying formation and maturation of the pollen. *Ann. Bot.* 40:1101–1109.

Dickinson, H.G., and Elleman. C.J. 1985. Structural changes in the pollen grain of *Brassica oleracea* during dehydration in the anther and development on the stigma as revealed by anhydrous fixation techniques. *Micron Micros. Acta* 16:255–270.

Dickinson, H.G., and Heslop-Harrison, J. 1968. Common mode of deposition for the sporopollenin of sexine and nexine. *Nature* 220:926–927.

Dickinson, H.G., and Heslop-Harrison, J. 1970a. The behaviour of plastids during meiosis in the microsporocyte of *Lilium longiflorum* Thunb. *Cytobios* 2:103–118.

Dickinson, H.G., and Heslop-Harrison, J. 1970b. The ribosome cycle, nucleoli, and cytoplasmic nucleoloids in the meiocytes of *Lilium. Protoplasma* 69:187–200.

Dickinson, H.G., and Heslop-Harrison, J. 1971. The mode of growth of the inner layer of the pollen-grain exine in *Lilium. Cytobios* 4:233–243.

Dickinson, H.G., and Heslop-Harrison, J. 1977. Ribosomes, membranes and organelles during meiosis in angiosperms. *Phil. Trans. Roy. Soc. Lond.* 277B:327–342.

Dickinson, H.G., and Lawson, J. 1975. Pollen tube growth in the stigma of *Oenothera organensis* following compatible and incompatible intraspecific pollinations. *Proc. Roy. Soc. Lond.* 188B:327–344.

Dickinson, H.G., and Lewis, D. 1973a. Cytochemical and ultrastructural differences between intraspecific compatible and incompatible pollinations in *Raphanus. Proc. Roy. Soc. Lond.* 183B:21–38.

Dickinson, H.G., and Lewis, D. 1973b. The formation of the tryphine coating the pollen grains of *Raphanus*, and its properties relating to the self-incompatibility system. *Proc. Roy. Soc. Lond.* 184B:149–165.

Dickinson, H.G., and Lewis, D. 1974. Changes in the pollen grain wall of *Linum grandiflorum* following compatible and incompatible intraspecific pollinations. *Ann. Bot.* 38:23–29.

Dickinson, H.G., Moriarty, J., and Lawson, J. 1982. Pollen–pistil interaction in *Lilium longiflorum*: the role of the pistil in controlling pollen tube growth following cross- and self-pollinations. *Proc. Roy. Soc. Lond.* 215B:45–62.

Dickinson, H.G., and Potter, U. 1975. Post meiotic nucleo–cytoplasmic interaction in *Pinus banksiana*: the secretion of RNA by the nucleus. *Planta* 122:99–104.

Dickinson, H.G., and Potter, U. 1976. The development of patterning in the alveolar sexine of *Cosmos bipinnatus*. *New Phytol*. 76:543–550.

Dickinson, H.G., and Potter, U. 1978. Cytoplasmic changes accompanying the female meiosis in *Lilium longiflorum* Thunb. *J. Cell Sci*. 29:147–169.

Dickinson, H.G., and Potter, U. 1979. Post-meiotic nucleo–cytoplasmic interaction in *Cosmos bipinnatus*. Early events at the nuclear envelope. *Planta* 145:449–457.

Dickinson, H.G., and Sheldon, J.M. 1984. A radial system of microtubules extending between the nuclear envelope and the plasma membrane during early male haplophase in flowering plants. *Planta* 161:86–90.

Dickinson, H.G., and Willson, C. 1983. Two stages in the redifferentiation of amyloplasts in the microspores of *Lilium*. *Ann. Bot*. 52:803–810.

Dickinson, H.G., and Willson, C. 1985. Behaviour of nucleoli and cytoplasmic nucleoloids during meiotic divisions in *Lilium henryi*. *Cytobios* 43:349–365.

Dieckert, J.W., and Dieckert, M.C. 1976. The chemistry and cell biology of the vacuolar proteins of seeds. *J. Food Sci*. 41:475–482.

Dierks-Ventling, C. 1981. Storage proteins in *Zea mays* (L.): interrelationship of albumins, globulins and zeins in the opaque-2 mutation. *Eur. J. Biochem*. 120:177–182.

Diers, L. 1963. Elektronenmikroskopische Beobachtungen an der generativen Zelle von *Oenothera hookeri* Torr. et Gray. *Z. Naturfors*. 18b:562–566.

Dieterich, K. 1924. Über Kultur von Embryonen außerhalb des Samens. *Flora* 117:379–417.

Dietrich, P.S., Bouchard, R.A., Casey, E.S., and Sinibaldi, R.M. 1991. Isolation and characterization of a small heat shock protein gene from maize. *Plant Physiol*. 96:1268–1276.

Dietrich, R.A., Radke, S.E., and Harada, J.J. 1992. Downstream DNA sequences are required to activate a gene expressed in the root cortex of embryos and seedlings. *Plant Cell* 4:1371–1382.

Dietrich, R.A., Maslyar, D.J., Heupel, R.C., and Harada, J.J. 1989. Spatial patterns of gene expression in *Brassica napus* seedlings: identification of a cortex-specific gene and localization of mRNAs encoding isocitrate lyase and a polypeptide homologous to proteinases. *Plant Cell* 1:73–80.

Dieu, P., and Dunwell, J.M. 1988. Anther culture with different genotypes of opium poppy (*Papaver somniferum* L.): effect of cold treatment. *Plant Cell Tissue Organ Cult*. 12:263–271.

Dijak, M., and Simmonds, D.H. 1988. Microtubule organization during early direct embryogenesis from mesophyll protoplasts of *Medicago sativa* L. *Plant Sci*. 58:183–191.

Dijak, M., Smith, D.L., Wilson, T.J., and Brown, D.C.W. 1986. Stimulation of direct embryogenesis from mesophyll protoplasts of *Medicago sativa*. *Plant Cell Rep*. 5:468–470.

Dinis, A.M., and Mesquita, J.F. 1993. The F-actin distribution during microsporogenesis in *Magnolia soulangeana* Soul. (Magnoliaceae). *Sex. Plant Reprod*. 6:57–63.

Dixon, L.K., and Leaver, C.J. 1982. Mitochondrial gene expression and cytoplasmic male sterility in *Sorghum*. *Plant Mol. Biol*. 2:89–102.

Dixon, L.K., Leaver, C.J., Brettell, R.I.S., and Gengenbach, B.G. 1982. Mitochondrial sensitivity to *Drechslera maydis* T-toxin and the synthesis of a variant mitochondrial polypeptide in plants derived from maize tissue cultures with Texas male-sterile cytoplasm. *Theor. Appl. Genet*. 63:75–80.

Dobrofsky, S., and Grant, W.F. 1980. Electrophoretic evidence supporting self-incompatibility in *Lotus corniculatus*. *Can. J. Bot*. 58:712–716.

Dodds, K.S., and Simmonds, N.W. 1946. A cytological basis of sterility in *Tripsacum laxum*. *Ann. Bot*. 10:109–116.

Dodds, P.N., Clarke, A.E., and Newbigin, E. 1996. A molecular perspective on pollination in flowering plants. *Cell* 85:141–144.

Dodds, P.N., Bönig, I., Du, H., Rödin, J., Anderson, M.A., Newbigin, E., and Clarke, A.E. 1993. S-RNase gene of *Nicotiana alata* is expressed in developing pollen. *Plant Cell* 5:1771–1782.

Dolfini, S.F., and Sparvoli, F. 1988. Cytological characterization of the embryo-lethal mutant *dek-1* of maize. *Protoplasma* 144:142–148.

Dolfini, S.F., Landoni, M., Tonelli, C., Bernard, L., and Viotti, A. 1992. Spatial regulation in the expression of structural and regulatory storage-protein genes in *Zea mays* endosperm. *Devel Genet*. 13:264–276.

Domon, C., and Steinmetz, A. 1994. Exon shuffling in anther-specific genes from sunflower. *Mol. Gen. Genet*. 244:312–317.

Domon, C., Evrard, J.-L., Herdenberger, F., Pillay, D.T.N., and Steinmetz, A. 1990. Nucleotide sequence of two anther-specific cDNAs from sunflower (*Helianthus annuus* L.). *Plant Mol. Biol*. 15:643–646.

Domon, C., Evrard, J.-L., Pillay, D.T.N., and Steinmetz, A. 1991. A 2.6 kb intron separates the signal peptide coding sequence of an anther-specific protein from the rest of the gene in sunflower. *Mol. Gen. Genet*. 229:238–244.

Domon, J.-M., Dumas, B., Lainé, E., Meyer, Y., David, A., and David, H. 1995. Three glycosylated polypeptides secreted by several embryogenic cell cultures of pine show highly specific serological affinity to antibodies directed against the wheat germin apoprotein monomer. *Plant Physiol*. 108:141–148.

Domon, J.-M., Meyer, Y., Faye, L., David, A., and David, H. 1994. Extracellular (glyco)proteins in embryogenic and non-embryogenic cell lines of Carribbean pine. Comparison between phenotypes of stage one somatic embryos. *Plant Physiol. Biochem*. 32:137–147.

Domoney, C., and Casey, R. 1983. Cloning and characterization of complementary DNA for convicilin, a major seed storage protein in *Pisum sativum* L. *Planta* 159:446–453.

Domoney, C., and Casey, R. 1985. Measurement of gene number for seed storage proteins in *Pisum*. *Nucl. Acids Res*. 13:687–699.

Domoney, C., and Casey, R. 1987. Changes in legumin messenger RNAs throughout seed development in *Pisum sativum* L. *Planta* 170:562–566.

Domoney, C., Davies, D.R., and Casey, R. 1980. The initiation of legumin synthesis in immature embryos of *Pisum sativum* L. grown *in vivo* and *in vitro*. *Planta* 149:454–460.

Domoney, C., Ellis, N., Turner, L., and Casey, R. 1991. A developmentally regulated early-embryogenesis protein in pea (*Pisum sativum* L.) is related to the heat-shock protein (HSP70) gene family. *Planta* 184:350–355.

Domoney, C., Firmin, J.L., Sidebottom, C., Ealing, P.M., Slabas, A., and Casey, R. 1990. Lipoxygenase heterogeneity in *Pisum sativum*. *Planta* 181:35–43.

Dong, J., and Yang, H.-Y. 1989. An ultrastructural study of embryo sac in *Oryza sativa* L. *Acta Bot. Sinica* 31:81–88.

Dong, J.G., Satoh, S., and Fujii, T. 1988. Variation in endoplasmic-reticulum-associated glycoproteins of carrot cells cultured *in vitro*. *Planta* 173:419–423.

Dong, J.-Z., and Dunstan, D.I. 1994. Growth parameters, protein and DNA synthesis of an embryogenic suspension culture of white spruce (*Picea glauca*). *J. Plant Physiol.* 144:201–208.

Dong, J.-Z., and Dunstan, D.I. 1996. Expression of abundant mRNAs during somatic embryogenesis of white spruce [*Picea glauca* (Moench) Voss]. *Planta* 199:459–466.

Donini, B., and Hussain, S. 1968. Development of the embryo in *Triticum durum* following irradiation of male or female gamete. *Radiation Bot.* 8:289–295.

Dooner, H.K. 1985. *Viviparous-1* mutation in maize conditions pleiotropic enzyme deficiencies in the aleurone. *Plant Physiol.* 77:486–488.

Dooner, H.K., and Nelson, O.E. 1979. Interaction among *C*, *R* and V_p in the control of the *Bz* glucosyltransferase during endosperm development in maize. *Genetics* 91:309–315.

Dos Santos, A.V.P., Cutter, E.G., and Davey, M.R. 1983. Origin and development of somatic embryos in *Medicago sativa* L. (alfalfa). *Protoplasma* 117:107–115.

Dougall, D.K., and Verma, D.C. 1978. Growth and embryo formation in wild-carrot suspension cultures with ammonium ion as a sole nitrogen source. *In Vitro* 14:180–182.

Doughty, J., Hedderson, F., McCubbin, A., and Dickinson, H. 1993. Interaction between a coating-borne peptide of the *Brassica* pollen grain and stigmatic *S* (self-incompatibility)-locus-specific glycoproteins. *Proc. Natl Acad. Sci. USA* 90:467–471.

Dover, G.A. 1972. The organization and polarity of pollen mother cells of *Triticum aestivum*. *J. Cell Sci.* 11:699–711.

Dow, D.A., and Mascarenhas, J.P. 1991a. Optimization of conditions for *in situ* hybridization and determination of the relative number of ribosomes in the cells of the mature embryo sac of maize. *Sex. Plant Reprod.* 4:244–249.

Dow, D.A., and Mascarenhas, J.P. 1991b. Synthesis and accumulation of ribosomes in individual cells of the female gametophyte of maize during its development. *Sex. Plant Reprod.* 4:250–253.

Dowrick, V.P.J. 1956. Heterostyly and homostyly in *Primula obconica*. *Heredity* 10:219–236.

Doyle, J.J., Schuler, M.A., Godette, W.D., Zenger, V., and Beachy, R.N. 1986. The glycosylated seed storage proteins of *Glycine max* and *Phaseolus vulgaris*. Structural homologies of genes and proteins. *J. Biol. Chem.* 261:9228–9238.

Dresselhaus, T., Lörz, H., and Kranz, E. 1994. Representative cDNA libraries from few plant cells. *Plant J.* 5:605–610.

Drews, G.N., Bowman, J.L., and Meyerowitz, E.M. 1991. Negative regulation of the *Arabidopsis* homeotic gene *AGAMOUS* by the *APETALA2* product. *Cell* 65:991–1002.

Drews, G.N., Beals, T.P., Bui, A.Q., and Goldberg, R.B. 1992. Regional and cell-specific gene expression patterns during petal development. *Plant Cell* 4:1383–1404.

Driscoll, C.J. 1986. Nuclear male sterility systems in seed production of hybrid varieties. *Crit. Rev. Plant Sci.* 3:227–256.

Dryanovska, O.A. 1981. Cytochemical and autoradiographic studies of meiosis and microsporanogenesis in *Tradescantia paludosa*. *Compt. Rend. Acad. Sci. Bulg.* 34:1169–1172.

Du, H., Simpson, R.J., Clarke, A.E., and Bacic, A. 1996. Molecular characterization of a stigma-specific gene encoding an arabinogalactan-protein (AGP) from *Nicotiana alata*. *Plant J.* 9:313–323.

Du, H., Simpson, R.J., Moritz, R.L., Clarke, A.E., and Bacic, A. 1994. Isolation of the protein backbone of an arabinogalactan-protein from the styles of *Nicotiana alata* and characterization of a corresponding cDNA. *Plant Cell* 6:1643–1653.

Dubald, M., Barakate, A., Mandaron, P., and Mache, R. 1993. The ubiquitous presence of exopolygalacturonase in maize suggests a fundamental cellular function for this enzyme. *Plant J.* 4:781–791.

Dubey, D.K., and Singh, S.P. 1965. Mechanism of pollen abortion in three male sterile lines of flax (*Linum usitatissimum* L.). *Crop Sci.* 5:121–124.

Dubois, T., Guedira, M., Dubois, J., and Vasseur, J. 1990. Direct somatic embryogenesis in roots of *Cichorium*: is callose an early marker? *Ann. Bot.* 65:539–545.

Dubois, T., Guedira, M., Dubois, J., and Vasseur, J. 1991. Direct somatic embryogenesis in leaves of *Cichorium*. A histological and SEM study of early stages. *Protoplasma* 162:120–127.

Duchenne, M., Lejeune, B., Fouillard, P., and Quetier, F. 1989. Comparison of the organization and expression of mtDNA of fertile and male-sterile sugar beet varieties (*Beta vulgaris* L.). *Theor. Appl. Genet.* 78:633–640.

Duck, N., McCormick, S., and Winter, J. 1989. Heat shock protein hsp70 cognate gene expression in vegetative and reproductive organs of *Lycopersicon esculentum*. *Proc. Natl Acad. Sci. USA* 86:3674–3678.

Duck, N.B., and Folk, W.R. 1994. Hsp70 heat shock protein cognate is expressed and stored in developing tomato pollen. *Plant Mol. Biol.* 26:1031–1039.

Ducker, S.C., and Knox, R.B. 1976. Submarine pollination in seagrasses. *Nature* 263:705–706.

Ducrocq, C., Sangwan, R.S., and Sangwan-Norreel, B.S. 1994. Production of *Agrobacterium*-mediated transgenic fertile plants by direct somatic embryogenesis from immature zygotic embryos of *Datura innoxia*. *Plant Mol. Biol.* 25:995–1009.

Dudareva, N., Evrard, J.-L., Pillay, D.T.N., and Steinmetz, A. 1994. Nucleotide sequence of a pollen-specific cDNA from *Helianthus annuus* L. encoding a highly basic protein. *Plant Physiol.* 106:403–404.

Dudits, D., Bögre, L. and Györgyey, J. 1991. Molecular and cellular approaches to the analysis of plant embryos development from somatic cells *in vitro*. *J. Cell Sci.* 99:473–482.

Dudits, D., Györgyey, J., Bögre, L., and Bako, L. 1995. Molecular biology of somatic embryogenesis. In In Vitro *Embryogenesis in Plants*, ed. T.A. Thorpe, pp. 267–308. Dordrecht: Kluwer Academic Publishers.

Duffus, C.M., and Rosie, R. 1975. Biochemical changes during embryogeny in *Hordeum distichum*. *Phytochemistry* 14:319–323.

Dulberger, R. 1970. Tristyly in *Lythrum junceum*. *New Phytol*. 69:751–759.

Dulberger, R. 1973. Distyly in *Linum pubescens* and *L. mucronatum*. *Bot. J. Linn. Soc*. 66:117–126.

Dulberger, R. 1974. Structural dimorphism of stigmatic papillae in distylous *Linum* species. *Am. J. Bot*. 61:238–243.

Dulberger, R. 1975. Intermorph structural differences between stigmatic papillae and pollen grains in relation to incompatibility in Plumbaginaceae. *Proc. Roy. Soc. Lond*. 188B:257–274.

Dulberger, R. 1981. Dimorphic exine sculpturing in three distylous species of *Linum* (Linaceae). *Plant Syst. Evol*. 139:113–119.

Dulberger, R. 1987. Fine structure and cytochemistry of the stigma surface and incompatibility in some distylous *Linum* species. *Ann. Bot*. 59:203–217.

Dulberger, R. 1989. The apertural wall in pollen of *Linum grandiflorum*. *Ann. Bot*. 63:421–431.

Dulberger, R. 1990. Release of proteins from the pollen wall of *Linum grandiflorum*. *Sex. Plant Reprod*. 3:18–22.

Dulieu, H.L. 1966. Pollination of excised ovaries and culture of ovules of *Nicotiana tabacum* L. *Phytomorphology* 16:69–75.

Dumas, C. 1973. Contribution à l'étude cyto-physiologique du stigmate. III. Evolution et rôle du réticulum endoplasmique au cours de la sécrétion chez *Forsythia intermedia* Z.; étude cytochimique. *Z. Pflanzenphysiol*. 70:119–130.

Dumas, C. 1974a. Contribution à l'étude cyto-physiologique du stigmate. VIII. Les associations réticulum endoplasmique-plaste et la sécrétion stigmatique. *Botaniste* 56:81–102.

Dumas, C. 1974b. Some aspects of stigmatic secretion in *Forsythia*. In *Fertilization in Higher Plants*, ed. H.F. Linskens, pp. 119–126. Amsterdam: North-Holland Publishing Co.

Dumas, C. 1977a. Lipochemistry of the progamic stage of a self-incompatible species: neutral lipids and fatty acids of the secretory stigma during its glandular activity, and of the solid style, the ovary and the anther in *Forsythia intermedia* Zab. (heterostylic species). *Planta* 137:177–184.

Dumas, C. 1977b. Établissement d'un modèle de la cinétique de la sécrétion lipo-polyphénolique du stigmate de *Forsythia intermedia* Zabel. *Compt. Rend. Acad. Sci. Paris* 284D:1777–1779.

Dumas, C. 1978. Stigmates sécréteurs et lipides neutres sécrétés. *Bull. Soc. Bot. Fr. Actual. Bot*. 125:61–68.

Dumas, C. and Gaude, T. 1981. Stigma–pollen recognition and pollen hydration. *Phytomorphology* 31:191–201.

Dumas, C., Knox, R.B., and Gaude, T. 1985. The spatial association of the sperm cells and vegetative nucleus in the pollen grain of *Brassica*. *Protoplasma* 124:168–174.

Dumas, C., Knox, R.B., McConchie, C.A., and Russell, S.D. 1984. Emerging physiological concepts in fertilization. *What's New in Plant Physiol*. 15:17–20.

Dumas, C., Rougier, M., Zandonella, P., Ciampolini, F., Cresti, M., and Pacini, E. 1978. The secretory stigma in *Lycopersicum peruvianum* Mill.: ontogenesis and glandular activity. *Protoplasma* 96:173–187.

Dunbar, A. 1973. Pollen ontogeny in some species of Campanulaceae. A study by electron microscopy. *Bot. Not*. 126:277–315.

Dundas, I.S., Saxena, K.B., and Byth, D.E. 1981. Microsporogenesis and anther wall development in male-sterile and fertile lines of pigeon pea (*Cajanus cajan* (L.) Millsp.). *Euphytica* 30:431–435.

Dundas, I.S., Saxena, K.B., and Byth, D.E. 1982. Pollen mother cell and anther wall development in a photoperiod-insensitive male-sterile mutant of pigeon pea (*Cajanus cajan* (L.) Millsp.). *Euphytica* 31:371–375.

Dunstan, D.I., Tautorus, T.E., and Thorpe, T.A. 1995. Somatic embryogenesis in woody plants. In In Vitro *Embryogenesis in Plants*, ed. T.A. Thorpe, pp. 471–538. Dordrecht: Kluwer Academic Publishers.

Dunwell, J.M. 1979. Anther culture of *Nicotiana tabacum*: the role of the culture vessel atmosphere in pollen embryo induction and growth. *J. Expt. Bot*. 30:419–428.

Dunwell, J.M. 1981. Influence of genotype and environment on growth of barley embryos *in vitro*. *Ann. Bot*. 48:535–542.

Dunwell, J.M., Cornish, M., and de Courcel, A.G.L. 1985. Influence of genotype, plant growth temperature and anther incubation temperature on microspore embryo production in *Brassica napus* spp. *oleifera*. *J. Expt. Bot*. 36:679–689.

Dunwell, J.M., and Sunderland, N. 1974a. Pollen ultrastructure in anther cultures of *Nicotiana tabacum*. I. Early stages of culture. *J. Expt. Bot*. 25:352–361.

Dunwell, J.M., and Sunderland, N. 1974b. Pollen ultrastructure in anther cultures of *Nicotiana tabacum*. II. Changes associated with embryogenesis. *J. Expt. Bot*. 25:363–373.

Dunwell, J.M., and Sunderland, N. 1975. Pollen ultrastructure in anther cultures of *Nicotiana tabacum*. III. The first sporophytic division. *J. Expt. Bot*. 26:240–252.

Dunwell, J.M., and Sunderland, N. 1976. Pollen ultrastructure in anther cultures of *Datura innoxia*. I. Division of the presumptive vegetative cell. *J. Cell Sci*. 22:469–480.

Dupuis, F. 1972. Formation de la paroi de séparation entre la cellule génératrice et la cellule végétative dans le pollen d'*Impatiens balsamina* L. *Bull. Soc. Bot. Fr*. 119:41–50.

Dupuis, I., and Dumas, C. 1989. *In vitro* pollination as a model for studying fertilization in maize (*Zea mays* L.). *Sex. Plant Reprod*. 2:265–269.

Dupuis, I., and Dumas, C. 1990a. Biochemical markers of female receptivity in maize (*Zea mays* L.) assessed using *in vitro* fertilization. *Plant Sci*. 70:11–19.

Dupuis, I., and Dumas, C. 1990b. Influence of temperature stress on *in vitro* fertilization and heat shock protein synthesis in maize (*Zea mays* L.) reproductive tissues. *Plant Physiol*. 94:665–670.

Dupuis, I., and Pace, G.M. 1993. Gene transfer to maize male reproductive structure by particle bombardment of tassel primordia. *Plant Cell Rep*. 12:607–611.

Dupuis, I., Roeckel, P., Matthys-Rochon, E., and Dumas, C. 1987. Procedure to isolate viable sperm cells from corn (*Zea mays* L.) pollen grains. *Plant Physiol*. 85:876–878.

Durand, B. 1966. Action d'une kinétine sur les caractères sexuels de *Mercurialis annua* L. ($2n$ = 16). *Compt. Rend. Acad. Sci. Paris* 263D:1309–1311.

Durante, M., Cionini, P.G., Avanzi, S., Cremonini, R., and D'Amato, F. 1977. Cytological localization of the genes for the four classes of ribosomal RNA (25S, 18S, 5.8S and 5S) in polytene chromosomes of *Phaseolus coccineus*. *Chromosoma* 60:269–282.

Durante, M., Cremonini, R., Tagliasacchi, A.M., Forino,

L.M.C., and Cionini, P.G. 1987. Characterization and chromosomal localization of fast renaturing and satellite DNA sequences in *Phaseolus coccineus*. *Protoplasma* 137:100–108.

Dure, L., III. 1985. Embryogenesis and gene expression during seed formation. *Oxford Surv. Plant Mol. Cell Biol.* 2:179–197.

Dure, L., III. 1993. A repeating 11-mer amino acid motif and plant desiccation. *Plant J.* 3:363–369.

Dure, L., III., and Chlan, C. 1981. Developmental biochemistry of cottonseed embryogenesis and germination. XII. Purification and properties of principal storage proteins. *Plant Physiol.* 68:180–186.

Dure, L., III., and Galau, G.A. 1981. Developmental biochemistry of cotton seed embryogenesis and germination. XIII. Regulation of biosynthesis of principal storage proteins. *Plant Physiol.* 68:187–194.

Dure, L., III., Greenway, S.C., and Galau, G.A. 1981. Developmental biochemistry of cottonseed embryogenesis and germination: changing messenger ribonucleic acid populations as shown by *in vitro* and *in vivo* protein synthesis. *Biochemistry* 20:4162–4168.

Dure, L., III, Crouch, M., Harada, J., Ho, T.-H.D., Mundy, J., Quatrano, R., Thomas, T., and Sung, Z.R. 1989. Common amino acid sequence domains among the LEA proteins of higher plants. *Plant Mol. Biol.* 12:475–486.

Dure, L., III, Pyle, J.B., Chlan, C.A., Baker, J.C., and Galau, G.A., 1983. Developmental biochemistry of cottonseed embryogenesis and germination. XVII. Developmental expression of genes for the principal storage proteins. *Plant Mol. Biol.* 2:199–206.

Dure, L.S., and Jensen, W.A. 1957. The influence of gibberellic acid and indoleacetic acid on cotton embryos cultured *in vitro*. *Bot. Gaz.* 118:254–261.

Dute, R.R., and Peterson, C.M. 1992. Early endosperm development in ovules of soybean, *Glycine max* (L.) Merr. (Fabaceae). *Ann. Bot.* 69:263–271.

Dute, R.R., Peterson, C.M., and Rushing, A.E. 1989. Ultrastructural changes of the egg apparatus associated with fertilization and proembryo development of soybean, *Glycine max* (Fabaceae). *Ann. Bot.* 64:123–135.

Dutta, P.C., and Appelqvist, L.-A. 1989. The effects of different cultural conditions on the accumulation of depot lipids notably petroselinic acid during somatic embryogenesis in *Daucus carota* L. *Plant Sci.* 64:167–177.

Dutta, P.C., Appelqvist, L.-A., Gunnarsson, S., and von Hofsten, A. 1991. Lipid bodies in tissue culture, somatic and zygotic embryo of *Daucus carota* L.: a qualitative and quantitative study. *Plant Sci.* 78:259–267.

Dwyer, K.G., Balent, M.A., Nasrallah, J.B., and Nasrallah, M.E. 1991. DNA sequences of self-incompatibility genes from *Brassica campestris* and *B. oleracea*: polymorphism predating speciation. *Plant Mol. Biol.* 16:481–486.

Dwyer, K.G., Chao, A., Cheng, B., Chen, C.H., and Nasrallah, J.B. 1989. The *Brassica* self-incompatibility multigene family. *Genome* 31:969–972.

Dwyer, K.G., Kandasamy, M.K., Mahosky, D.I., Acciai, J., Kudish, B.I., Miller, J.E., Nasrallah, M.E., and Nasrallah, J.B. 1994. A superfamily of *S* locus-related sequences in *Arabidopsis*: diverse structures and expression patterns. *Plant Cell* 6:1829–1843.

Dwyer, K.G., Lalonde, B.A., Nasrallah, J.B., and Nasrallah, M.E. 1992. Structure and expression of *AtS1*, an *Arabidopsis thaliana* gene homologous to the *S*-locus related genes of *Brassica*. *Mol. Gen. Genet.* 231:442–448

Dzelskalns, V.A., Nasrallah, J.B., and Nasrallah, M.E. 1992. Cell–cell communication in plants: self incompatibility in flower development. *Devel Biol.* 153:70–82.

Dzelskalns, V.A., Thorsness, M.K., Dwyer, K.G., Baxter, J.S., Balent, M.A., Nasrallah, M.E. and Nasrallah, J.B. 1993. Distinct *cis*-acting elements direct pistil-specific and pollen-specific activity of the *Brassica S* locus glycoprotein gene promoter. *Plant Cell* 5:855–863.

Eady, C., Lindsey, K., and Twell, D. 1994. Differential activation and conserved vegetative cell-specific activity of a late pollen promoter in species with bicellular and tricellular pollen. *Plant J.* 5:543–550.

Eady, C., Lindsey, K., and Twell, D. 1995. The significance of microspore division and division symmetry for vegetative cell-specific transcription and generative cell differentiation. *Plant Cell* 7:65–74.

East, E.M. 1934. Norms of pollen-tube growth in incompatible matings of self-sterile plants. *Proc. Natl Acad. Sci. USA* 20:225–230.

East, E.M., and Mangelsdorf, A.J. 1925. A new interpretation of the hereditary behavior of self-sterile plants. *Proc. Natl Acad. Sci. USA* 11:166–171.

Ebert, P.R., Anderson, M.A., Bernatzky, R., Altschuler, M., and Clarke, A.E. 1989. Genetic polymorphism of self-incompatibility in flowering plants. *Cell* 56:255–262.

Echlin, P. 1972. The ultrastructure and ontogeny of pollen in *Helleborus foetidus* L. IV. Pollen grain maturation. *J. Cell Sci.* 11:111–129.

Echlin, P., and Godwin, H. 1968a. The ultrastructure and ontogeny of pollen in *Helleborus foetidus* L. I. The development of the tapetum and Ubisch bodies. *J. Cell Sci.* 3:161–174.

Echlin, P., and Godwin, H. 1968b. The ultrastructure and ontogeny of pollen in *Helleborus foetidus* L. II. Pollen grain development through the callose special wall stage. *J. Cell Sci.* 3:175–186.

Echt, C.S., and Schwartz, D. 1981. Evidence for the inclusion of controlling elements within the structural gene at the waxy locus in maize. *Genetics* 99:275–284.

Edwards, M., Scott, C., Gidley, M.J., and Grant Reid, J.S. 1992. Control of mannose/galactose ratio during galactomannan formation in developing legume seeds. *Planta* 187:67–74.

Edwardson, J.R. 1956. Cytoplasmic male-sterility. *Bot. Rev.* 22:696–738.

Edwardson, J.R. 1970. Cytoplasmic male sterility. *Bot. Rev.* 36:341–420.

Edwardson, J.R., Bond, D.A., and Christie, R.G. 1976. Cytoplasmic sterility factors in *Vicia faba* L. *Genetics* 82:443–449.

Edwardson, J.R., and Warmke, H.E. 1967. Fertility restoration in cytoplasmic male-sterile *Petunia*. *J. Hered.* 58:195–196.

Egertsdotter, U., Mo, L.H., and von Arnold, S. 1993. Extracellular proteins in embryogenic suspension cultures of Norway spruce (*Picea abies*). *Physiol. Plant.* 88:315–321.

Egertsdotter, U., and von Arnold, S. 1995. Importance of arabinogalactan proteins for the development of somatic embryos of Norway spruce (*Picea abies*). *Physiol. Plant.* 93:334–345.

Ehlenfeldt, M.K., and Ortiz, R. 1995. Evidence on the nature and origins of endosperm dosage requirements in *Solanum* and other angiosperm genera. *Sex. Plant Reprod.* 8:189–196.

Eisenberg, A.J., and Mascarenhas, J.P. 1985. Abscisic acid and the regulation of synthesis of specific seed proteins and their messenger RNAs during culture of soybean embryos. *Planta* 166:505–514.

El-Ghazaly, G., and Jensen, W. 1986a. Studies of the development of wheat (*Triticum aestivum*) pollen: formation of the pollen aperture. *Can. J. Bot.* 64:3141–3154.

El-Ghazaly, G., and Jensen, W.A. 1986b. Studies of the development of wheat (*Triticum aestivum*) pollen. I. Formation of the pollen wall and Ubisch bodies. *Grana* 25:1–29.

El-Ghazaly, G.A., and Jensen, W.A. 1990. Development of wheat (*Triticum aestivum*) pollen wall before and after effect of a gametocide. *Can. J. Bot.* 68:2509–2516.

El-Ghazaly, G.A., and Nilsson, S. 1991. Development of tapetum and orbicules of *Catharanthus roseus* (Apocynaceae). In *Pollen and Spores*, ed. S. Blackmore and S.H. Barnes, pp. 317–329. Oxford: Clarendon Press.

El Hadrami, I., Michaux-Ferrière, N., Carron, M.-P., and d'Auzac, J. 1989. Les polyamines, facteur limitant possible de l'embryogenèse somatique chez l'*Hevea brasiliensis. Compt. Rend. Acad. Sci. Paris* Ser. III 308:205–211.

Elleman, C.J., and Dickinson, H.G. 1986. Pollen–stigma interactions in *Brassica*. IV. Structural reorganization in the pollen grains during hydration. *J. Cell Sci.* 80:141–157.

Elleman, C.J., and Dickinson, H.G. 1990. The role of the exine coating in pollen–stigma interactions in *Brassica oleracea* L. *New Phytol.* 114:511–518.

Elleman, C.J., and Dickinson, H.G. 1996. Identification of pollen components regulating pollination-specific responses in the stigmatic papillae of *Brassica oleracea. New Phytol.* 133:197–205.

Elleman, C.J., Franklin-Tong, V., and Dickinson, H.G. 1992. Pollination in species with dry stigmas: the nature of the early stigmatic response and the pathway taken by pollen tubes. *New Phytol.* 121:413–424.

Elleman, C.J., Willson, C.E., and Dickinson, H.G. 1987. Fixation of *Brassica oleracea* pollen during hydration; a comparative study. *Pollen Spores* 29:273–290.

Elleman, C.J., Willson, C.E., Sarker, R.H., and Dickinson, H.G. 1988. Interaction between the pollen tube and stigmatic cell wall following pollination in *Brassica oleracea. New Phytol.* 109:111–117.

Elliott, R.C., Betzner, A.S., Huttner, E., Oakes, M.P., Tucker, W.Q.J., Gerentes, D., Perez, P., and Smyth, D.R. 1996. *AINTEGUMENTA*, an *APETALA2*-like gene of *Arabidopsis* with pleiotropic roles in ovule development and floral organ growth. *Plant Cell* 8:155–168.

Ellis, D.D., McCabe, D.E., McInnis, S., Ramachandran, R., Russell, D.R., Wallace, K.M., Martinell, B.J., Roberts, D.R., Raffa, K.L., and McCown, B.H. 1993. Stable transformation of *Picea glauca* by particle acceleration. *Biotechnology* 11:84–89.

Ellis, J.R., Gates, P.J., and Boulter, D. 1987. Storage-protein deposition in the developing rice caryopsis in relation to the transport tissues. *Ann. Bot.* 60:663–670.

Ellis, J.R., Shirsat, A.H., Hepher, A., Yarwood, J.N., Gatehouse, J.A., Croy, R.R.D., and Boulter, D. 1988. Tissue-specific expression of a pea legumin gene in seeds of *Nicotiana plumbaginifolia. Plant Mol. Biol.* 10:203–214.

Emons, A.M.C. 1994. Somatic embryogenesis: cell biological aspects. *Acta Bot. Neerl.* 43:1–14.

Emons, A.M.C., and Kieft, H. 1990. Comparison of embryogenic and non-embryogenic suspension cells of maize by means of freeze-fracturing. *Micron Micros. Acta* 21:255–256.

Emons, A.M.C., Mulder, M.M., and Kieft, H. 1993. Pyrolysis mass spectrometry of developmental stages of maize embryos. *Acta Bot. Neerl.* 42:319–339.

Emons, A.M.C., Vos, J.W., and Kieft, H. 1992. A freeze fracture analysis of the surface of embryogenic and non-embryogenic suspension cells of *Daucus carota. Plant Sci.* 87:85–97.

Engell, K. 1989. Embryology of barley: time course and analysis of controlled fertilization and early embryo formation based on serial sections. *Nord. J. Bot.* 9:265–280.

Engell, K. 1994. Embryology of barley. IV. Ultrastructure of the antipodal cells of *Hordeum vulgare* L. cv. Bomi before and after fertilization of the egg cell. *Sex. Plant Reprod.* 7:333–346.

Engels, F.M. 1973. Function of Golgi vesicles in relation to cell wall synthesis in germinating *Petunia* pollen. I. Isolation of Golgi vesicles. *Acta Bot. Neerl.* 22:6–13.

Engels, F.M. 1974a. Function of Golgi vesicles in relation to cell wall synthesis in germinating *Petunia* pollen. II. Chemical composition of Golgi vesicles and pollen tube wall. *Acta Bot. Neerl.* 23:81–89.

Engels, F.M. 1974b. Function of Golgi vesicles in relation to cell wall synthesis in germinating *Petunia* pollen. III. The ultrastructure of the tube wall. *Acta Bot. Neerl.* 23:201–207.

Engels, F.M. 1974c. Function of Golgi vesicles in relation to cell wall synthesis in germinating *Petunia* pollen. IV. Identification of cellulose in pollen tube walls and Golgi vesicles by X-ray diffraction. *Acta Bot. Neerl.* 23:209–215.

Engvild, K.C., Linde-Laursen, I., and Lundqvist, A. 1972. Anther cultures of *Datura innoxia*: flower bud stage and embryoid level of ploidy. *Hereditas* 72:331–332.

Enjuto, M., Lumbreras, V., Marin, C., and Boronat, A. 1995. Expression of the *Arabidopsis HMG2* gene, encoding 3-hydroxy-3-methylglutaryl coenzyme A reductase, is restricted to meristematic and floral tissues. *Plant Cell* 7:517–527.

Entwistle, J. 1988. Primary structure of a C-hordein gene from barley. *Carlsberg Res. Commun.* 53:247–258.

Erbrich, P. 1965. Über Endopolyploidie und Kernstrukturen in Endospermhaustorien. *Österr. Bot. Z.* 112:197–262.

Erdelská, O. 1966. Einfluss der Befruchtung auf die Entwicklung der Antipodenkerne der Gerste. *Biológia* 21:857–864.

Erdelská, O. 1980. Some structural aspects of flax embryo nutrition. *Biológia* 35:243–249.

Erdelská, O. 1985. Dynamics of the development of embryo and endosperm. II. (*Linum usitatissimum, Capsella bursa-pastoris*). *Biológia* 40:849–857.

Erdelská, O. 1986. Cytolysis of the endosperm in different types of correlation in endosperm and endosperm development. *Acta Bot. Neerl.* 35:437–441.

Erdelská, O., and Klasova, A. 1978. La région micropylaire du sac embryonnaire de *Jasione montana* L. avant et après la fécondation. *Bull. Soc. Bot. Fr. Actual. Bot.* 125:249–252.

Ereken-Tumer, N., Richter, J.D., and Nielsen, N.C. 1982. Structural characterization of the glycinin precursors. *J. Biol. Chem.* 257:4016–4018.

Erickson, L., Grant, I., and Beversdorf, W. 1986a. Cytoplasmic male sterility in rapeseed (*Brassica napus* L.). 1. Restriction patterns of chloroplast and mitochondrial DNA. *Theor. Appl. Genet.* 72:145–150.

Erickson, L., Grant, I., and Beversdorf, W. 1986b. Cytoplasmic male sterility in rapeseed (*Brassica napus* L.). 2. The role of a mitochondrial plasmid. *Theor. Appl. Genet.* 72:151–157.

Erickson, R.O. 1948. Cytological and growth correlations in the flower bud and anther of *Lilium longiflorum*. *Am. J. Bot.* 35:729–739.

Ericson, M.L., Murén, E., Gustavsson, H.-O., Josefsson, L.-G., and Rask, L. 1991. Analysis of the promoter region of napin genes from *Brassica napus* demonstrates binding of nuclear protein *in vitro* to a conserved sequence motif. *Eur. J. Biochem.* 197:741–746.

Ericson, M.L., Rödin, J., Lenman, M., Glimelius, K., Josefsson, L.-G., and Rask, L. 1986. Structure of the rapeseed 1.7S storage protein, napin, and its precursor. *J. Biol. Chem.* 261:14576–14581.

Errampalli, D., Patton, D., Castle, L., Mickelson, L., Hansen, K., Schnall, J., Feldmann, K., and Meinke, D. 1991. Embryonic lethals and T-DNA insertional mutagenesis in *Arabidopsis*. *Plant Cell* 3:149–157.

Eschrich, W. 1961. Untersuchungen über den Ab- und Aufbau der Callose. (III. Mitteilungüber Callose). *Z. Bot.* 49:153–218.

Esen, A., and Soost, R.K. 1973. Seed development in *Citrus* with special reference to 2X × 4X crosses. *Am. J. Bot.* 60:448–462.

Espelie, K.E., Loewus, F.A., Pugmire, R.J., Woolfenden, W.R., Baldi, B.G., and Given, P.H. 1989. Structural analysis of *Lilium longiflorum* sporopollenin by ^{13}C NMR spectroscopy. *Phytochemistry* 28:751–753.

Espelund, M., de Bedout, J.A., Outlaw, W.H., Jr., and Jakobsen, K.S. 1995. Environmental and hormonal regulation of barley late-embryogenesis-abundant (*Lea*) mRNAs is via different signal transduction pathways. *Plant Cell Environ.* 18:943–949.

Espelund, M., Sæbøe-Larssen, S., Hughes, D.W., Galau, G.A., Larsen, F., and Jakobsen, K.S. 1992. Late embryogenesis-abundant genes encoding proteins with different numbers of hydrophilic repeats are regulated differentially by abscisic acid and osmotic stress. *Plant J.* 2:241–252.

Esser, K. 1953. Genomverdopplung und Pollenschlauchwachstum bei Heterostylen. *Z. Indukt. Abstamm. Vererbungs.* 85:28–50.

Estelle, M.A., and Somerville, C. 1987. Auxin-resistant mutants of *Arabidopsis thaliana* with an altered morphology. *Mol. Gen. Genet.* 206:200–206.

Estruch, J.J., Kadwell, S., Merlin, E., and Crossland, L. 1994. Cloning and characterization of a maize pollen-specific calcium-dependent calmodulin-independent protein kinase. *Proc. Natl Acad. Sci. USA* 91:8837–8841.

Etienne, H., Sotta, B., Montoro, P., Miginiac, E., and Carron, M.-P. 1993a. Relations between exogenous growth regulators and endogenous indole-3-acetic acid and abscisic acid in the expression of somatic embryogenesis in *Hevea brasiliensis* (Müll. Arg.). *Plant Sci.* 88:91–96.

Etienne, H., Sotta, B., Montoro, P., Miginiac, E., and Carron, M.-P. 1993b. Comparison of endogenous ABA and IAA contents in somatic and zygotic embryos of *Hevea brasiliensis* (Müll. Arg.) during ontogenesis. *Plant Sci.* 92:111–119.

Ettinger, W.F., and Harada, J.J. 1990. Translational or post-translational processes affect differentially the accumulation of isocitrate lyase and malate synthase proteins and enzyme activities in embryos and seedlings of *Brassica napus*. *Arch. Biochem. Biophys.* 281:139–143.

Eunus, A.M. 1955. The effects of X-rays on the embryonal growth and development of *Hordeum vulgare* L. *J. Expt. Bot.* 6:409–421.

Evans, D.E., Rothnie, N.E., Palmer, M.V., Burke, D.G., Sang, J.P., Knox, R.B., Williams, E.G., Hilliard, E.P., and Salisbury, P.A. 1987. Comparative analysis of fatty acids in pollen and seed of rapeseed. *Phytochemistry* 26:1895–1897.

Evans, D.E., Sang, J.P., Cominos, X., Rothnie, N.E., and Knox, R.B. 1990. A study of phospholipids and galactolipids in pollen of two lines of *Brassica napus* L. (rapeseed) with different ratios of linoleic to linolenic acid. *Plant Physiol.* 92:418–424.

Evans, D.E., Taylor, P.E., Singh, M.B., and Knox, R.B. 1991. Quantitative analysis of lipids and protein from the pollen of *Brassica napus* L. *Plant Sci.* 73:117–126.

Evans, D.E., Taylor, P.E., Singh, M.B., and Knox, R.B. 1992. The interrelationship between the accumulation of lipids, protein and the level of acyl carrier protein during the development of *Brassica napus* L. pollen. *Planta* 186:343–354.

Evans, I.M., Bown, D., Lycett, G.W., Croy, R.R.D., Boulter, D., and Gatehouse, J.A. 1985. Transcription of a legumin gene from pea (*Pisum sativum* L.) *in vitro*. *Planta* 165:554–560.

Evans, I.M., Croy, R.R.D., Hutchinson, P., Boulter, D., Payne, P.I., and Gordon, M.E. 1979. Cell free synthesis of some storage protein subunits by polyribosomes and RNA isolated from developing seeds of pea (*Pisum sativum* L.). *Planta* 144:455–462.

Evans, I.M., Gatehouse, J.A., Croy, R.R.D., and Boulter, D. 1984. Regulation of the transcription of storage-protein mRNA in nuclei isolated from developing pea (*Pisum sativum* L.) cotyledons. *Planta* 160:559–568.

Evans, P.T., Holaway, B.L., and Malmberg, R.L. 1988. Biochemical differentiation in the tobacco flower probed with monoclonal antibodies. *Planta* 175:259–269.

Evans, P.T., and Malmberg, R.L. 1989. Alternative pathways of tobacco placental development: time of commitment and analysis of a mutant. *Devel Biol.* 136:273–283.

Everett, N.P., Wach, M.J., and Ashworth, D.J. 1985. Biochemical markers of embryogenesis in tissue cultures of the maize inbred B73. *Plant Sci.* 41:133–140.

Evers, A.D. 1970. Development of the endosperm of wheat. *Ann. Bot.* 34:547–555.

Evrard, J.-L., Jako, C., Saint-Guily, A., Weil, J.-H., and Kuntz, M. 1991. Anther-specific, developmentally regulated expression of genes encoding a new class of proline-rich proteins in sunflower. *Plant Mol. Biol.* 16:271–281.

Eyal, Y., Curie, C., and McCormick, S. 1995. Pollen specificity elements reside in 30 bp of the proximal promoters of two pollen-expressed genes. *Plant Cell* 7:373–384.

Eyster, W.H. 1931. Heritable characters of maize. XLII. Reduced endosperm. *J. Hered.* 22:250–252.

Fabergé, A.C. 1937. The cytology of the male sterile *Lathyrus odoratus*. *Genetica* 19:423–430.

Fabijanski, S., and Altosaar, I. 1985. Evidence for translational control of storage protein biosynthesis during embryogenesis of *Avena sativa* L. (oat endosperm). *Plant Mol. Biol.* 4:211–218.

Fabijanski, S., Matlashewski, G.J., and Altosaar, I. 1985. Characterization of developing oat seed mRNA: evidence for many globulin mRNAs. *Plant Mol. Biol.* 4:205–210.

Falk, A., Ek, B., and Rask, L. 1995. Characterization of a new myrosinase from *Brassica napus*. *Plant Mol. Biol.* 27:863–874.

Fan, Z., Armstrong, K.C., and Keller, W.A. 1988. Development of microspores *in vivo* and *in vitro* in *Brassica napus* L. *Protoplasma* 147:191–199.

Faure, J.-E., Digonnet, C., and Dumas, C. 1994. An *in vitro* system for adhesion and fusion of maize gametes. *Science* 263:1598–1600.

Faure, J.-E., Mogensen, H.L., Dumas, C., Lörz, H., and Kranz, E. 1993. Karyogamy after electrofusion of single egg and sperm cell protoplasts from maize: cytological evidence and time course. *Plant Cell* 5:747–755.

Faure, J.-E., Mogensen, H.L., Kranz, E., Digonnet, C., and Dumas, C. 1992. Ultrastructural characterization and three-dimensional reconstruction of isolated maize (*Zea mays* L.) egg cell protoplasts. *Protoplasma* 171:97–103.

Faure, O., and Nougarède, A. 1993. Nuclear DNA content of somatic and zygotic embryos of *Vitis vinifera* cv. Grenache noir at the torpedo stage. Flow cytometry and *in situ* DNA microspectrophotometry. *Protoplasma* 176:145–150.

Faure, O., Mengoli, M., Nougarede, A., and Bagni, N. 1991. Polyamine pattern and biosynthesis in zygotic and somatic embryo stages of *Vitis vinifera*. *J. Plant Physiol.* 138:545–549.

Fauron, C.M.-R., Havlik, M., and Brettell, R.I.S. 1990. The mitochondrial genome organization of a maize fertile cmsT revertant line is generated through recombination between two sets of repeats. *Genetics* 124:423–428.

Fauron, C.M.-R., Casper, M., Gesteland, R., and Albertsen, M. 1992. A multi-recombination model for the mtDNA rearrangements seen in maize cmsT regenerated plants. *Plant J.* 2:949–958.

Favre-Duchartre, M. 1978. Oogenèses chez les angiospermes et autres plantes ovulées. *Rev. Cytol. Biol. Végét.-Bot.* 1:79–95.

Feirer, R.P., Conkey, J.H., and Verhagen, S.A. 1989. Triglycerides in embryogenic conifer calli: a comparison with zygotic embryos. *Plant Cell Rep.* 8:207–209.

Feirer, R.P., Mignon, G., and Litvay, J.D. 1984. Arginine decarboxylase and polyamines required for embryogenesis in the wild carrot. *Science* 223:1433–1435.

Felker, F.C. 1987. Ultrastructure of maize endosperm suspension cultures. *Am. J. Bot.* 74:1912–1920.

Felker, F.C., Peterson, D.M., and Nelson, O.E. 1984. Development of tannin vacuoles in chalaza and seed coat of barley in relation to early chalazal necrosis in the *seg*1 mutant. *Planta* 161:540–549.

Felker, F.C., Peterson, D.M., and Nelson, O.E. 1985. Anatomy of immature grains of eight maternal effect shrunken endosperm barley mutants. *Am. J. Bot.* 72:248–256.

Felker, F.C., Peterson, D.M., and Nelson, O.E. 1987. Early grain development of the *seg2* maternal-effect shrunken-endosperm mutant of barley. *Can. J. Bot.* 65:943–948.

Fennell, A., and Hauptmann, R. 1992. Electroporation and PEG delivery of DNA into maize microspores. *Plant Cell Rep.* 11:567–570.

Fernandez, D.E., Turner, F.R., and Crouch, M.L. 1991. *In situ* localization of storage protein mRNAs in developing meristems of *Brassica napus* embryos. *Development* 111:299–313.

Fernando, D.D., and Cass, D.D. 1994. Plasmodial tapetum and pollen wall development in *Butomus umbellatus* (Butomaceae). *Am. J. Bot.* 81:1592–1600.

Fernando, D.D., and Cass, D.D. 1996. Development and structure of ovule, embryo sac, embryo, and endosperm in *Butomus umbellatus* (Butomaceae). *Int. J. Plant Sci.* 157:269–279.

Ferrant, V., and Bouharmont J. 1994. Origin of gynogenetic embryos of *Beta vulgaris* L. *Sex. Plant Reprod.* 7:12–16.

Ferrari, T.E., Bruns, D., and Wallace, D.H. 1981. Isolation of a plant glycoprotein involved with control of intercellular recognition. *Plant Physiol.* 67:270–277.

Ferrari, T.E., and Wallace, D.H. 1975. Germination of *Brassica* pollen and expression of incompatibility *in vitro*. *Euphytica* 24:757–765.

Ferrari, T.E., and Wallace, D.H. 1976. Pollen protein synthesis and control of incompatibility in *Brassica*. *Theor. Appl. Genet.* 48:243–249.

Ferreira, P.C.G., Hemerly, A.S., de Almeida Engler, J., van Montagu, M., Engler, G., and Inzé, D. 1994. Developmental expression of the *Arabidopsis* cyclin gene *cyc1At*. *Plant Cell* 6:1763-1774.

Ferrie, A.M.R., Palmer, C.E., and Keller, W.A. 1995. Haploid embryogenesis. In *In Vitro Embryogenesis in Plants*, ed. T.A. Thorpe, pp. 309–344. Dordrecht: Kluwer Academic Publishers.

Fett, W.F., Paxton, J.D., and Dickinson, D.B. 1976. Studies on the self-incompatibility response of *Lilium longiflorum*. *Am. J. Bot.* 63:1104–1108.

Feys, B.J.F., Benedetti, C.E., Penfold, C.N., and Turner, J.G. 1994. *Arabidopsis* mutants selected for resistance to the phytotoxin coronatine are male sterile, insensitive to methyl jasmonate, and resistant to a bacterial pathogen. *Plant Cell* 6:751–759.

Fiedler, U., Filistein, R., Wobus, U., and Bäumlein, H. 1993. A complex ensemble of *cis*-regulatory elements controls the expression of a *Vicia faba* non-storage seed protein gene. *Plant Mol. Biol.* 22:669–679.

Fienberg, A.A., Choi, J.H., Lubich, W.P., and Sung, Z.R. 1984. Developmental regulation of polyamine metabolism in growth and differentiation of carrot culture. *Planta* 162:532–539.

Filion, W.G., and Christie, B.R. 1966. The mechanism of male sterility in a clone of orchardgrass (*Dactylis glomerata* L.). *Crop Sci.* 6:345–347.

Filippini, F., Terzi, M., Cozzani, F., Vallone, D., and Lo Schiavo, F. 1992. Modulation of auxin-binding proteins in cell suspensions. II. Isolation and initial characterization of carrot cell variants impaired in somatic embryogenesis. *Theor. Appl. Genet.* 84:430–434.

Finer, J.J., and McMullen, M.D. 1990. Transformation of cotton (*Gossypium hirsutum* L.) via particle bombardment. *Plant Cell Rep.* 8:586–589.

Fineran, B.A., Wild, D.J.C., and Ingerfeld, M. 1982. Initial wall formation in the endosperm of wheat, *Triticum aestivum*: a reevaluation. *Can. J. Bot.* 60:1776–1795.

Finkelstein, R., and Somerville, C. 1989. Abscisic acid or

high osmoticum promote accumulation of long-chain fatty acids in developing embryos of *Brassica napus*. *Plant Sci.* 61:213–217.

Finkelstein, R.R. 1993. Abscisic acid-insensitive mutations provide evidence for stage-specific signal pathways regulating expression of an *Arabidopsis* late embryogenesis-abundant (*lea*) gene. *Mol. Gen. Genet.* 238:401–408.

Finkelstein, R.R., and Crouch, M.L. 1984. Precociously germinating rapeseed embryos retain characteristics of embryogeny. *Planta* 162:125–131.

Finkelstein, R.R., and Crouch, M.L. 1986. Rapeseed embryo development in culture on high osmoticum is similar to that in seeds. *Plant Physiol.* 81:907–912.

Finkelstein, R.R., and Somerville, C.R. 1990. Three classes of abscisic acid (ABA)-insensitive mutations of *Arabidopsis* define genes that control overlapping subsets of ABA responses. *Plant Physiol.* 94:1172–1179.

Finkelstein, R.R., Tenbarge, K.M., Shumway, J.E., and Crouch, M.L. 1985. Role of ABA in maturation of rapeseed embryos. *Plant Physiol.* 78:630–636.

Fischer, C., and Neuhaus, G. 1995. *In vitro* development of globular zygotic wheat embryos. *Plant Cell Rep.* 15:186–191.

Fischer, C., and Neuhaus, G. 1996. Influence of auxin on the establishment of bilateral symmetry in monocots. *Plant J.* 9:659–669.

Fischer, H., Haake, V., Horstmann, C., and Jensen, U. 1995. Characterization and evolutionary relationships of *Magnolia* legumin-encoding cDNAs representing two divergent gene subfamilies. *Eur. J. Biochem.* 229:645–650.

Fischer, R.L., and Goldberg, R.B. 1982. Structure and flanking regions of soybean seed protein genes. *Cell* 29:651–660.

Fischer, W., Bergfeld, R., and Schopfer, P. 1987. Induction of storage protein synthesis in embryos of mature plant seeds. *Naturwissenschaften* 74:86–88.

Fischer, W., Bergfeld, R., Plachy, C., Schäfer, R., and Schopfer, P. 1988. Accumulation of storage materials, precocious germination and development of desiccation tolerance during seed maturation in mustard (*Sinapis alba* L.). *Bot. Acta* 101:344–354.

Fisher, D.B., and Jensen, W.A. 1969. Cotton embryogenesis: the identification, as nuclei, of the X-bodies in the degenerated synergid. *Planta* 84:122–133.

Fisher, D.B., and Jensen, W.A. 1972. Nuclear and cytoplasmic DNA synthesis in cotton embryos: a correlated light and electron microscope autoradiographic study. *Histochemie* 32:1–22.

Fisher, D.B., Jensen, W.A., and Ashton, M.E. 1968. Histochemical studies of pollen: storage pockets in the endoplasmic reticulum (ER). *Histochemie* 13:169–182.

Fisher, R.H., Barton, M.K., Cohen, J.D., and Cooke, T.J. 1996. Hormonal studies of *fass*, an *Arabidopsis* mutant that is altered in organ elongation. *Plant Physiol.* 110:1109–1121.

Fitch, M.M.M., Manshardt, R.M., Gonsalves, D., and Slightom, J.L. 1993. Transgenic papaya plants from *Agrobacterium*-mediated transformation of somatic embryos. *Plant Cell Rep.* 12:245–249.

Fitch, M.M.M., Manshardt, R.M., Gonsalves, D., Slightom, J.L., and Sanford, J.C. 1990. Stable transformation of papaya via microprojectile bombardment. *Plant Cell Rep.* 9:189–194.

Fitch, M.M.M., Manshardt, R.M., Gonsalves, D., Slightom, J.L., and Sanford, J.C. 1992. Virus resistant papaya plants derived from tissues bombarded with the coat protein gene of papaya ringspot virus. *Biotechnology* 10:1466–1472.

Fitzgerald, M.A., and Knox, R.B. 1995. Initiation of primexine in freeze-substituted microspores of *Brassica campestris*. *Sex. Plant Reprod.* 8:99–104.

Fitzgerald, M.A., Barnes, S.H., Blackmore, S., Calder, D.M., and Knox, R.B. 1994. Exine formation in the pollinium of *Dendrobium*. *Protoplasma* 179:121–130.

Flamand, M.-C., Goblet, J.-P., Duc, G., Briquet, M., and Boutry, M. 1992. Sequence and transcription analysis of mitochondrial plasmids isolated from cytoplasmic male-sterile lines of *Vicia faba*. *Plant Mol. Biol.* 19:913–923.

Flanagan, C.A., and Ma, H. 1994. Spatially and temporally regulated expression of the MADS-box gene *AGL2* in wild-type and mutant *Arabidopsis* flowers. *Plant Mol. Biol.* 26:581–595.

Flavell, R. 1974. A model for the mechanism of cytoplasmic male sterility in plants, with special reference to maize. *Plant Sci. Lett.* 3:259–263.

Fleming, A.J., and Hanke, D.E. 1993. The regulation of napin gene expression in secondary embryos of *Brassica napus*. *Physiol. Plant.* 87:396–402.

Flinn, B.S., Roberts, D.R., and Taylor, I.E.P. 1991. Evaluation of somatic embryos of interior spruce. Characterization and developmental regulation of storage proteins. *Physiol. Plant.* 82:624–632.

Flinn, B.S., Roberts, D.R., Newton, C.H., Cyr, D.R., Webster, F.B., and Taylor, I.E.P. 1993. Storage protein gene expression in zygotic and somatic embryos of interior spruce. *Physiol. Plant.* 89:719–730.

Flinn, B.S., Roberts, D.R., Webb, D.T., and Sutton, B.C.S. 1991. Storage protein changes during zygotic embryogenesis in interior spruce. *Tree Physiol.* 8:71–81.

Flint, D., Ayers, G.S., and Ries, S.K. 1975. Synthesis of endosperm proteins in wheat seed during maturation. *Plant Physiol.* 56:381–384.

Floris, C., Meletti, P., and D'Amato, F. 1970. Further observations on embryo–endosperm relations in irradiated water soaked seeds of *durum* wheat. *Mutation Res.* 10:253–255.

Folsom, M.W., and Cass, D.D. 1986. Changes in transfer cell distribution in the ovule of soybean after fertilization. *Can. J. Bot.* 64:965–972.

Folsom, M.W., and Cass, D.D. 1988. The characteristics and fate of the soybean inner nucellus. *Acta Bot. Neerl.* 37:387–393.

Folsom, M.W., and Cass, D.D. 1989. Embryo sac development in soybean: ultrastructure of megasporogenesis and early megagametogenesis. *Can. J. Bot.* 67:2841–2849.

Folsom, M.W., and Cass, D.D. 1990. Embryo sac development in soybean: cellularization and egg apparatus expansion. *Can. J. Bot.* 68:2135–2147.

Folsom, M.W., and Cass, D.D. 1992. Embryo sac development in soybean: the central cell and aspects of fertilization. *Am. J. Bot.* 79:1407–1417.

Folsom, M.W., and Peterson, C.M. 1984. Ultrastructural aspects of the mature embryo sac of soybean, *Glycine max* (L.) Merr. *Bot. Gaz.* 145:1–10.

Fong, F., Smith, J.D., and Koehler, D.E. 1983. Early events in maize seed development. 1-Methyl-3-phenyl-5-(3-[trifluoromethyl]phenyl)-4-(1*H*)-pyridinone induction of vivipary. *Plant Physiol.* 73:899–901.

Fontana, G.S., Santini, L., Caretto, S., Frugis, G., and Mariotti, D. 1993. Genetic transformation in the grain legume *Cicer arietinum* L. (chickpea). *Plant Cell Rep.* 12:194–198.

Fontes, E.B.P., Shank, B.B., Wrobel, R.L., Moose, S.P., OBrian, G.R., Wurtzel, E.T., and Boston, R.S. 1991. Characterization of an immunoglobulin binding protein homolog in the maize *floury-2* endosperm mutant. *Plant Cell* 3:483–496.

Foote, H.C.C., Ride, J.P., Franklin-Tong, V.E., Walker, E.A., Lawrence, M.J., and Franklin, F.C.H. 1994. Cloning and expression of a distinctive class of self-incompatibility (*S*) gene from *Papaver rhoeas* L. *Proc. Natl Acad. Sci. USA* 91:2265–2269.

Forde, B.G., and Leaver, C.J. 1980. Nuclear and cytoplasmic genes controlling synthesis of variant mitochondrial polypeptides in male-sterile maize. *Proc. Natl Acad. Sci. USA* 77:418–422.

Forde, B.G., Oliver, R.J.C., and Leaver, C.J. 1978. Variation in mitochondrial translation products associated with male-sterile cytoplasms in maize. *Proc. Natl Acad. Sci. USA* 75:3841–3845.

Forde, B.G., Heyworth, A., Pywell, J., and Kreis, M. 1985a. Nucleotide sequence of a B1 hordein gene and the identification of possible upstream regulatory elements in endosperm storage protein genes from barley, wheat and maize. *Nucl. Acids Res.* 13:7327–7339.

Forde, B.G., Kreis, M., Williamson, M.S., Fry, R.P., Pywell, J., Shewry, P.R., Bunce, N., and Miflin, B.J. 1985b. Short tandem repeats shared by B- and C-hordein cDNAs suggest a common evolutionary origin for two groups of cereal storage protein genes. *EMBO J.* 4:9–15.

Forde, B.G., Oliver, R.J.C., Leaver, C.J., Gunn, R.E., and Kemble, R.J. 1980. Classification of normal and male-sterile cytoplasms in maize. I. Electrophoretic analysis of variation in mitochondrially synthesized proteins. *Genetics* 95:443–450.

Forde, J., Forde, B.G., Fry, R.P., Kreis, M., Shewry, P.R., and Miflin, B.J. 1983. Identification of barley and wheat cDNA clones related to the high-M_r polypeptides of wheat gluten. *FEBS Lett.* 162:360–366.

Forde, J., Malpica, J.-M., Halford, N.G., Shewry, P.R., Anderson, O.D., Greene, F.C., and Miflin, B.J. 1985. The nucleotide sequence of a HMW glutenin subunit gene located on chromosome 1A of wheat (*Triticum aestivum* L.). *Nucl. Acids Res.* 13:6817–6832.

Forino, L.M.C., Tagliasacchi, A.M., Cavallini, A., Cionini, G., Giraldi, E., and Cionini, P.G. 1992. RNA synthesis in the embryo suspensor of *Phaseolus coccineus* at two stages of embryogenesis, and the effect of supplied gibberellic acid. *Protoplasma* 167:152–158.

Forman, M., and Jensen, W.A. 1965. Respiration and embryogenesis in cotton. *Plant Physiol.* 40:765–769.

Forster, B.P., and Dale, J.E. 1983. A comparative study of early seed development in genotypes of barley and rye. *Ann. Bot.* 52:603–612.

Fosket, D.E., and Morejohn, L.C. 1992. Structural and functional organization of tubulin. *Annu. Rev. Plant Physiol. Plant Mol. Biol.* 43:201–240.

Foster, G.D., Robinson, S.W., Blundell, R.P., Roberts, M.R., Hodge, R., Draper, J., and Scott, R.J. 1992. A *Brassica napus* mRNA encoding a protein homologous to phospholipid transfer proteins, is expressed specifically in the tapetum and developing microspores. *Plant Sci.* 84:187–192.

Foster, T.S., and Stern, H. 1959. The accumulation of soluble deoxyribosidic compounds in relation to nuclear division in anthers of *Lilium longiflorum*. *J. Biophys. Biochem. Cytol.* 5:187–192.

Fougère-Rifot, M. 1975. L'édification de l'appareil filiforme et l'évolution cytoplasmique des synergides du sac embryonnaire d'*Aquilegia vulgaris*. *Compt. Rend. Acad. Sci. Paris* 280D:2445–2447.

Fowke, L.C., Attree, S.M., Wang, H., and Dunstan, D.I. 1990. Microtubule organization and cell division in embryogenic protoplast cultures of white spruce (*Picea glauca*). *Protoplasma* 158:86–94.

Franchi, G.G., and Pacini, E. 1980. Wall projections in the vegetative cell of *Parietariaofficinalis* L. pollen. *Protoplasma* 104:67–74.

Francis, R.R., and Bemis, W.P. 1970. A cytomorphological study of male sterility in a mutant of *Cucurbita maxima* Dutch. *Econ. Bot.* 24:325–332.

Franke, W.W., Herth, W., VanDerWoude, W.J., and Morré, D.J. 1972. Tubular and filamentous structures in pollen tubes: possible involvement as guide elements in protoplasmic streaming and vectorial migration of secretory vesicles. *Planta* 105:317–341.

Frankel, O.H. 1940. Studies in *Hebe*. II. The significance of male sterility in the genetic system. *J. Genet.* 40:171–184.

Frankel, R., and Galun, E. 1977. *Pollination Mechanisms, Reproduction and Plant Breeding*. Berlin: Springer-Verlag.

Frankel, R., Izhar, S., and Nitsan, J. 1969. Timing of callase activity and cytoplasmic male sterility in *Petunia*. *Biochem. Genet.* 3:451–455.

Frankel, R., Scowcroft, W.R., and Whitfeld, P.R. 1979. Chloroplast DNA variation in isonuclear male-sterile lines of *Nicotiana*. *Mol. Gen. Genet.* 169:129–135.

Frankis, R., and Mascarenhas, J.P. 1980. Messenger RNA in the ungerminated pollen grain: a direct demonstration of its presence. *Ann. Bot.* 45:595–599.

Frankis, R.C., Jr. 1990. RNA and protein synthesis in germinating pine pollen. *J. Expt. Bot.* 41:1469–1473.

Frankis, R.C., Jr., and Grayson, G.K. 1990. Heat-shock response in germinating pine pollen. *Sex. Plant Reprod.* 3:195–199.

Franklin, F.C.H., Lawrence, M.J., and Franklin-Tong, V.E. 1995. Cell and molecular biology of self-incompatibility in flowering plants. *Int. Rev. Cytol.* 158:1–64.

Franklin-Tong, V.E., and Franklin, F.C.H. 1992. Gametophytic self-incompatibility in *Papaver rhoeas* L. *Sex. Plant Reprod.* 5:1–7.

Franklin-Tong, V.E., and Franklin, F.C.H. 1993. Gametophytic self-incompatibility: contrasting mechanisms for *Nicotiana* and *Papaver*. *Trends Cell Biol.* 3:340–345.

Franklin-Tong, V.E., Lawrence, M.J., and Franklin, F.C.H. 1988. An *in vitro* bioassay for the stigmatic product of the self-incompatibility gene in *Papaver rhoeas* L. *New Phytol.* 110:109–118.

Franklin-Tong, V.E., Lawrence, M.J., and Franklin, F.C.H. 1990. Self-incompatibility in *Papaver rhoeas* L.: inhibition of incompatible pollen tube growth is dependent on pollen gene expression. *New Phytol.* 116:319–324.

Franklin-Tong, V.E., Ride, J.P., and Franklin, F.C.H. 1995. Recombinant stigmatic self-incompatibility (S-) protein elicits a Ca^{2+} transient in pollen of *Papaver rhoeas*. *Plant J.* 8:299–307.

Franklin-Tong, V.E., Atwal, K.K., Howell, E.C., Lawrence, M.J., and Franklin, F.C.H. 1991. Self-incompatibility in *Papaver rhoeas*: there is no evidence for the involvement of stigmatic ribonuclease activity. *Plant Cell Environ.* 14:423–429.

Franklin-Tong, V.E., Ride, J.P., Read, N.D., Trewavas, A.J., and Franklin, F.C.H. 1993. The self-incompatibility

Fraley, R.T. 1989. Isolation of tissue-specific cDNAs from tomato pistils. *Plant Cell* 1:15–24.

Gatehouse, J.A., Bown, D., Evans, I.M., Gatehouse, L.N., Jobes, D., Preston, P., and Croy, R.R.D. 1987. Sequence of the seed lectin gene from pea (*Pisum sativum* L.). *Nucl. Acids Res.* 15:7642.

Gatehouse, J.A., Bown, D., Gilroy, J., Levasseur, M., Castleton, J., and Ellis, T.H.N. 1988. Two genes encoding `minor' legumin polypeptides in pea (*Pisum sativum* L.). Characterization and complete sequence of the *legJ* gene. *Biochem. J.* 250:15–24.

Gatehouse, J.A., Croy, R.R.D., Morton, H., Tyler, M., and Boulter, D. 1981. Characterisation and subunit structures of the vicilin storage proteins of pea (*Pisum sativum* L.). *Eur. J. Biochem.* 118:627–633.

Gatehouse, J.A., Evans, I.M., Bown, D., Croy, R.R.D., and Boulter, D. 1982. Control of storage-protein synthesis during seed development in pea (*Pisum sativum* L.). *Biochem. J.* 208:119–127.

Gatehouse, J.A., Evans, I.M., Croy, R.R.D., and Boulter, D. 1986. Differential expression of genes during legume seed development. *Phil. Trans. Roy. Soc. Lond.* 314B:367–384.

Gatehouse, J.A., Lycett, G.W., Delauney, A.J., Croy, R.R.D., and Boulter, D. 1983. Sequence specificity of the post-translational proteolytic cleavage of vicilin, a seed storage protein of pea (*Pisum sativum* L.). *Biochem. J.* 212:427–432.

Gaubier, P., Raynal, M., Hull, G., Huestis, G.M., Grellet, F., Arenas, C. Pagès, M., and Delseny, M. 1993. Two different *Em*-like genes are expressed in *Arabidopsis thaliana* seeds during maturation. *Mol. Gen. Genet.* 238:409–418.

Gaude, T., and Dumas, C. 1984. A membrane-like structure on the pollen wall surface in *Brassica*. *Ann. Bot.* 54:821–825.

Gaude, T., and Dumas, C. 1986. Organization of stigma surface components in *Brassica*: a cytochemical study. *J. Cell Sci.* 82:203–216.

Gaude, T., and Dumas, C. 1987. Molecular and cellular events of self-incompatibility. *Int. Rev. Cytol.* 107:333–366.

Gaude, T., Friry, A., Heizmann, P., Mariac, C., Rougier, M., Fobis, I., and Dumas, C. 1993. Expression of a self-incompatibility gene in a self-compatible line of *Brassica oleracea*. *Plant Cell* 5:75–86.

Gaude, T., Rougier, M., Heizmann, P., Ockendon, D.J., and Dumas, C. 1995. Expression level of the *SLG* gene is not correlated with the self-incompatibility phenotype in the class II *S* haplotypes of *Brassica oleracea*. *Plant Mol. Biol.* 27:1003–1014.

Gavish, H., Vardi, A., and Fluhr, R. 1991. Extracellular proteins and early embryo development in *Citrus* nucellar cell cultures. *Physiol. Plant.* 82:606–616.

Gavish, H., Vardi, A., and Fluhr, R. 1992. Suppression of somatic embryogenesis in *Citrus* cell cultures by extracellular proteins. *Planta* 186:511–517.

Gayler, K.R., and Sykes, G.E. 1981. β-Conglycinins in developing soybean seeds. *Plant Physiol.* 67:958–961.

Geetha, K.B., Lending, C.R., Lopes, M.A., Wallace, J.C., and Larkins, B.A. 1991. *opaque-2 modifiers* increase γ-zein synthesis and alter its spatial distribution in maize endosperm. *Plant Cell* 3:1207–1219.

Geitmann, A., Li, Y.-Q., and Cresti, M. 1995. Ultrastructural immunolocalization of periodic pectin depositions in the cell wall of *Nicotiana tabacum* pollen tubes. *Protoplasma* 187:168–171.

Geitmann, A., Li, Y.Q., and Cresti, M. 1996. The role of the cytoskeleton and dictyosome activity in the pulsatory growth of *Nicotiana tabacum* and *Petunia hybrida* pollen tubes. *Bot. Acta.* 109:102–109.

Geitmann, A., Hudák, J., Vennigerholz, F., and Walles, B. 1995. Immunogold localization of pectin and callose in pollen grains and pollen tubes of *Brugmansia suaveolens* - implications for the self-incompatibility reaction. *J. Plant Physiol.* 147:225–235.

Geli, M.I., Torrent, M., and Ludevid, D. 1994. Two structural domains mediate two sequential events in γ-zein targeting: protein endoplasmic reticulum retention and protein body formation. *Plant Cell* 6:1911–1922.

Gell, A.C., Bacic, A., and Clarke, A.E. 1986. Arabinogalactan-proteins of the female sexual tissue of *Nicotiana alata*. I. Changes during flower development and pollination. *Plant Physiol.* 82:885–889.

Geltz, N.R., and Russell, S.D. 1988. Two-dimensional electrophoretic studies of the proteins and polypeptides in mature pollen grains and the male germ unit of *Plumbago zeylanica*. *Plant Physiol.* 88:764–769.

Genevès, L. 1966. Évolution des infrastructures du cytoplasme dans le tissu sporifère des anthères de *Ribes rubrum* L. (Grossulariacées). *Compt. Rend. Acad. Sci. Paris* 262D:72–75.

Gengenbach, B.G. 1977a. Development of maize caryopses resulting from *in-vitro* pollination. *Planta* 134:91–93.

Gengenbach, B.G. 1977b. Genotypic influences on *in vitro* fertilization and kernel development of maize. *Crop Sci.* 17:489–492.

Gengenbach, B.G., Green, C.E., and Donovan, C.M. 1977. Inheritance of selected pathotoxin resistance in maize plants regenerated from cell cultures. *Proc. Natl Acad. Sci. USA* 74:5113–5117.

Gengenbach, B.G., Connelly, J.A., Pring, D.R., and Conde, M.F. 1981. Mitochondrial DNA variation in maize plants regenerated during tissue culture selection. *Theor. Appl. Genet.* 59:161–167.

Gengenbach, B.G., Miller, R.J., Koeppe, D.E., and Arntzen, C.J. 1973. The effect of toxin from *Helminthosporium maydis* (race T) on isolated corn mitochondria: swelling. *Can. J. Bot.* 51:2119–2125.

Genovesi, A.D., and Collins, G.B. 1982. *In vitro* production of haploid plants of corn via anther culture. *Crop Sci.* 22:1137–1144.

George, L., and Narayanaswamy, S. 1973. Haploid *Capsicum* through experimental androgenesis. *Protoplasma* 78:467–470.

Geraghty, D., Peifer, M.A., Rubenstein, I., and Messing, J. 1981. The primary structure of a plant storage protein: zein. *Nucl. Acids Res.* 9:5163–5174.

Geraghty, D.E., Messing, J., and Rubenstein, I. 1982. Sequence analysis and comparison of cDNAs of the zein multigene family. *EMBO J.* 1:1329–1335.

Gerassimova-Navashina, G. 1960. A contribution to the cytology of fertilization in flowering plants. *Nucleus* 3:111–120.

Gerlach-Cruse, D. 1969. Embryo- und Endospermentwicklung nach einer Röntgenbestrahlung der Fruchtknoten von *Arabidopsis thaliana* (L.) Heynh. *Radiation Bot.* 9:433–442.

Gerster, J., Allard, S., and Robert, L.S. 1996. Molecular characterization of two *Brassica napus* pollen-expressed genes encoding putative arabinogalactan proteins. *Plant Physiol.* 110:1231–1237.

Ghosh, B., Perry, M.P., and Marsh, D.G. 1991. Cloning the

cDNA encoding the AmbtV allergen from giant ragweed (*Ambrosia trifida*) pollen. *Gene* 101:231–238.

Ghosh, S., and Shivanna, K.R. 1980a. Pollen–pistil interaction in *Linum grandiflorum*. Scanning electron microscopic observations and proteins of the stigma surface. *Planta* 149:257–261.

Ghosh, S., and Shivanna, K.R. 1980b. Pollen–pistil interaction in *Linum grandiflorum*: stigma-surface proteins and stigma receptivity. *Proc. Indian Natl Sci. Acad.* 46B:177–183.

Ghosh, S., and Shivanna, K.R. 1982a. Studies on pollen–pistil interaction in *Linum grandiflorum*. *Phytomorphology* 32:385–395.

Ghosh, S., and Shivanna, K.R. 1982b. Anatomical and cytochemical studies on the stigma and style in some legumes. *Bot. Gaz.* 143:311–318.

Gibbs, P.E., and Bianchi, M. 1993. Post-pollination events in species of *Chorisia* (Bombacaceae) and *Tabebuia* (Bignoniaceae) with late-acting self-incompatibility. *Bot. Acta* 106:64–71.

Giese, H. 1992. Replication of DNA during barley endosperm development. *Can. J. Bot.* 70:313–318.

Giese, H., Andersen, B., and Doll, H. 1983. Synthesis of the major storage protein, hordein, in barley. Pulse-labeling study of grain filling in liquid-cultured detached spikes. *Planta* 159:60–65.

Giese, H., and Hejgaard, J. 1984. Synthesis of salt-soluble proteins in barley. Pulse-labeling study of grain filling in liquid-cultured detached spikes. *Planta* 161:172–177.

Gifford, D.J. 1988. An electrophoretic analysis of the seed proteins from *Pinus monticola* and eight other species of pine. *Can. J. Bot.* 66:1808–1812.

Gifford, D.J., Greenwood, J.S., and Bewley, J.D. 1982. Deposition of matrix and crystalloid storage proteins during protein body development in the endosperm of *Ricinus communis* L. cv. Hale seeds. *Plant Physiol.* 69:1471–1478.

Giles, K.L., Bassett, H.C.M., and Eastin, J.D. 1975. The structure and ontogeny of the hilum region in *Sorghum bicolor*. *Aust. J. Bot.* 23:795–802.

Gilissen, L.J.W. 1977. The influence of relative humidity on the swelling of pollen grains *in vitro*. *Planta* 137:299–301.

Gillissen, L.J.W., and Brantjes, N.B.M. 1978. Function of the pollen coat in different stages of the fertilization process. *Acta Bot. Neerl.* 27:205–212.

Giorgetti, L., Vergara, M.R., Evangelista, M., Lo Schiavo, F., Terzi, M., and Nuti Ronchi, V. 1995. On the occurrence of somatic meiosis in embryogenic carrot cell cultures. *Mol. Gen. Genet.* 246:657–662.

Giovinazzo, G., Manzocchi, L.A., Bianchi, M.W., Coraggio, I., and Viotti, A. 1992. Functional analysis of the regulatory region of a zein gene in transiently transformed protoplasts. *Plant Mol. Biol.* 19:257–263.

Giranton, J.-L., Ariza, M.J., Dumas, C., Cock, J.M., and Gaude, T. 1995. The *S* locus receptor kinase gene encodes a soluble glycoprotein corresponding to the SRK extracellular domain in *Brassica oleracea*. *Plant J.* 8:827–834.

Giraudat, J., Hauge, B.M., Valon, C., Smalle, J., Parcy, F., and Goodman, H.M. 1992. Isolation of the *Arabidopsis ABI3* gene by positional cloning. *Plant Cell* 4:1251–1261.

Giuliano, G., Lo Schiavo, F., and Terzi, M. 1984. Isolation and developmental characterization of temperature-sensitive carrot cell variants. *Theor. Appl. Genet.* 67:179–183.

Giuliano, G., Rosellini, D., and Terzi, M. 1983. A new method for the purification of the different stages of carrot embryoids. *Plant Cell Rep.* 2:216–218.

Glab, N., Wise, R.P., Pring, D.R., Jacq, C., and Slonimski, P. 1990. Expression in *Saccharomyces cerevisiae* of a gene associated with cytoplasmic male sterility from maize: respiratory dysfunction and uncoupling of yeast mitochondria. *Mol. Gen. Genet.* 223:24–32.

Glavin, T.L., Goring, D.R., Schafer, U., and Rothstein, S.J. 1994. Features of the extracellular domain of the *S*-locus receptor kinase from *Brassica*. *Mol. Gen. Genet.* 244:630–637.

Gleeson, P.A., and Clarke, A.E. 1979. Structural studies on the major component of *Gladiolus* style mucilage, arabinogalactan-protein. *Biochem. J.* 181:607–621.

Gleeson, P.A., and Clarke, A.E. 1980a. Comparison of the structures of the major components of the stigma and style secretions of *Gladiolus*: the arabino-3,6-galactans. *Carbohydr. Res.* 83:187–192.

Gleeson, P.A., and Clarke, A.E. 1980b. Antigenic determinants of a plant proteoglycan, the *Gladiolus* style arabinogalactan-protein. *Biochem. J.* 191:437–447.

Gleeson, P.A., and Clarke, A.E. 1980c. Arabinogalactans of sexual and somatic tissues of *Gladiolus* and *Lilium*. *Phytochemistry* 19:1777–1782.

Glenk, H.-O., Schimmer, O., and Wagner, W. 1970. Die Calcium-Verteilung in *Oenothera*-Pflanzen und ihr möglicher Einfluß auf den Chemotropismus der Pollenschläuche und auf die Befruchtung. *Phyton Austria* 14:97–111.

Glover, D.E., and Barrett, S.C.H. 1983. Trimorphic incompatibility in Mexican populations of *Pontederia sagittata* Presl. (Pontederiaceae). *New Phytol.* 95:439–455.

Goblet, J.-P., Flamand, M.-C., and Briquet, M. 1985. A mitochondrial plasmid specifically associated with male sterility and its relation with other mitochondrial plasmids in *Vicia faba* L. *Curr. Genet.* 9:423–426.

Goblet, J.-P., Boutry, M., Duc, G., and Briquet, M. 1983. Mitochondrial plasmid-like molecules in fertile and male sterile *Vicia faba* L. *Plant Mol. Biol.* 2:305–309.

Goday, A., Jensen, A.B., Culiáñez-Macià, F.A., Mar Albà, M., Figueras, M., Serratosa, J., Torrent, M., and Pagès, M. 1994. The maize abscisic acid-responsive protein Rab17 is located in the nucleus and interacts with nuclear localization signals. *Plant Cell* 6:351–360.

Goday, A., Sánchez-Martínez, D., Gómez, J., Puigdomènech, P., and Pagès, M. 1988. Gene expression in developing *Zea mays* embryos: regulation by abscisic acid of a highly phosphorylated 23- to 25-kD group of proteins. *Plant Physiol.* 88:564–569.

Godfrey, C.A., and Linskens, H.F. 1968. Nucleic acid estimations in pollinated styles of *Petunia hybrida*. *Planta* 80:185–190.

Godineau, J.-C. 1966. Ultrastructure du sac embryonnaire du *Crepis tectorum* L.: les cellules du pôle micropylaire. *Compt. Rend. Acad. Sci. Paris* 263D:852–855.

Godineau, J.-C. 1969. Ultra-structure des synergides chez quelques composées. *Rev. Cytol. Biol. Végét.* 32:209–226.

Godwin, H. 1968. The origin of the exine. *New Phytol.* 67:667–676.

Godwin, H., Echlin, P., and Chapman, B. 1967. The development of the pollen grain wall in *Ipomoea purpurea* (L.) Roth. *Rev. Palaeobot. Palynol.* 3:181–195.

Goffner, D., This, P., and Delseny, M. 1990. Effects of abscisic acid and osmotica on helianthinin gene expression in sunflower cotyledons *in vitro*. *Plant Sci.* 66:211–219.

Guiltinan, M.J., Marcotte, W.R., Jr., and Quatrano, R.S. 1990. A plant leucine zipper protein that recognizes an abscisic acid responsive element. *Science* 250:267–271.

Guldager, P. 1978. Immunoelectrophoretic analysis of seed proteins from *Pisum sativum* L. *Theor. Appl. Genet.* 53:241–250.

Guo, F.L., and Hu, S.Y. 1995. Cytological evidence of biparental inheritance of plastids and mitochondria in *Pelargonium*. *Protoplasma* 186:201–207.

Guo, Y., Liang, H., and Berns, M.W. 1995. Laser-mediated gene transfer in rice. *Physiol. Plant.* 93:19–24.

Guo, Z.-S., Sun, A.-C., Wang, Y.-Y., Gui, Y.-L., Gu, S.-R., and Miao, S.-H. 1978. Studies on induction of pollen plants and androgenesis in maize. *Acta Bot. Sinica* 20:204–209.

Gupta, M., Chourey, P.S., Burr, B., and Still, P.E. 1988. cDNAs of two non-allelic sucrose synthase genes in maize: cloning, expression, characterization and molecular mapping of the *sucrose synthase*-2 gene. *Plant Mol. Biol.* 10:215–224.

Gupta, S.C., and Babbar, S.B. 1980. Enhancement of plantlet formation in anther cultures of *Datura metel* L. by pre-chilling of buds. *Z. Pflanzenphysiol.* 96:465–470.

Gustafson-Brown, C., Savidge, B., and Yanofsky, M.F. 1994. Regulation of the *Arabidopsis* floral homeotic gene *APETALA1*. *Cell* 76:131–143.

Gustavsson, H.-O., Ellerström, M., Stålberg, K., Ezcurra, I., Koman, A., Höglund, A.-S., Rask, L., and Josefsson, L.-G. 1991. Distinct sequence elements in a napin promoter interact *in vitro* with DNA-binding proteins from *Brassica napus*. *Physiol. Plant.* 82:205–212.

Guzzo, F., Baldan, B., Levi, M., Sparvoli, E., Lo Schiavo, F., Terzi, M., and Mariani, P. 1995. Early cellular events during induction of carrot explants with 2,4-D. *Protoplasma* 185:28–36.

Guzzo, F., Baldan, B., Mariani, P., Lo Schiavo, F., and Terzi, M. 1994. Studies on the origin of totipotent cells in explants of *Daucus carota* L. *J. Expt. Bot.* 45:1427–1432.

Gwóźdź, E.A., and Deckert, J.E. 1989. The formation and translational activity of polysomes from developing lupin seeds. *Physiol. Plant.* 75:208–214.

Györgyey, J., Gartner, A., Németh, K., Magyar, Z., Hirt, H., Heberle-Bors, E., and Dudits, D. 1991. Alfalfa heat shock genes are differentially expressed during somatic embryogenesis. *Plant Mol. Biol.* 16:999–1007.

Habben, J.E., Kirleis, A.W., and Larkins, B.A. 1993. The origin of lysine-containing proteins in *opaque-2* maize endosperm. *Plant Mol. Biol.* 23:825–838.

Habben, J.E., Moro, G.L., Hunter, B.G., Hamaker, B.R., and Larkins, B.A. 1995. Elongation factor 1 concentration is highly correlated with the lysine content of maize endosperm. *Proc. Natl Acad. Sci. USA* 92:8640–8644.

Haccius, B. 1978. Question of unicellular origin of non-zygotic embryos in callus cultures. *Phytomorphology* 28:74–81.

Haccius, B., and Reichert, H. 1964. Restitutionserscheinungen an pflanzlichen Meristemen nach Röntgenbestrahlung. II. Advetiv-Embryonie nach Samenbestrahlung von *Eranthis hiemalis*. *Planta* 62:355–372.

Hack, E., Lin, C., Yang, H., and Horner, H.T. 1991. T-URF13 protein from mitochondria of Texas male-sterile maize (*Zea mays* L.). Its purification and submitochondrial localization, and immunogold labeling in anther tapetum during microsporogenesis. *Plant Physiol.* 95:861–870.

Hackett, R.M., Lawrence, M.J., and Franklin, F.C.H. 1992. A *Brassica S*-locus related gene promoter directs expression in both pollen and pistil of tobacco. *Plant J.* 2:613–617.

Hadi, M.Z., McMullen, M.D., and Finer, J.J. 1996. Transformation of 12 different plasmids into soybean via particle bombardment. *Plant Cell Rep.* 15:500–505.

Hagemann, R., and Schröder, M.-B. 1989. The cytological basis of the plastid inheritance in angiosperms. *Protoplasma* 152:57–64.

Häger, K.-P., and Dank, N. 1996. Seed storage proteins of Cupressaceae are homologous to legumins from angiosperms: molecular characterization of cDNAs from incense cedar (*Calocedrus decurrens* [Torr.] Florin). *Plant Sci.* 116:85–96.

Häger, K.-P., Braun, H., Czihal, A., Müller, B., and Bäumlein, H. 1995. Evolution of seed storage protein genes: legumin genes of *Ginkgo biloba*. *J. Mol. Evol.* 41:457–466.

Hahne, G., Mayer, J.E., and Lörz, H. 1988. Embryogenic and callus-specific proteins in somatic embryogenesis of the grass, *Dactylis glomerata* L. *Plant Sci.* 55:267–279.

Haig, D. 1990. New perspectives on the angiosperm female gametophyte. *Bot. Rev.* 56:236–274.

Håkansson, A. 1957. Notes on the giant chromosomes of *Allium nutans*. *Bot. Not.* 110:196–200.

Håkansson, G., Glimelius, K., and Bonnett, H.T. 1990. Respiration in cells and mitochondria of male-fertile and male-sterile *Nicotiana* spp. *Plant Physiol.* 93:367–373.

Hakman, I., Fowke, L.C., von Arnold, S., and Eriksson, T. 1985. The development of somatic embryos in tissue cultures initiated from immature embryos of *Picea abies* (Norway spruce). *Plant Sci.* 38:53–59.

Hakman, I., Stabel, P., Engström, P., and Eriksson, T. 1990. Storage protein accumulation during zygotic and somatic embryo development in *Picea abies* (Norway spruce). *Physiol. Plant.* 80:441–445.

Halford, N.G., Forde, J., Anderson, O.D., Greene, F.C., and Shewry, P.R. 1987. The nucleotide and deduced amino acid sequences of an HMW glutenin subunit gene from chromosome 1B of bread wheat (*Triticum aestivum* L.) and comparison with those genes from chromosomes 1A and 1D. *Theor. Appl. Genet.* 75:117–126.

Halford, N.G., Forde, J., Shewry, P.R., and Kreis, M. 1989. Functional analysis of the upstream regions of a silent and an expressed member of a family of wheat seed protein genes in transgenic tobacco. *Plant Sci.* 62:207–216.

Hall, T.C., McLeester, R.C., and Bliss, F.A. 1972. Electrophoretic analysis of protein changes during the development of the French bean fruit. *Phytochemistry* 11:647–649.

Hall, T.C., Ma, Y., Buchbinder, B.U., Pyne, J.W., Sun, S.M., and Bliss, F.A. 1978. Messenger RNA for G1 protein of French bean seeds: cell-free translation and product characterization. *Proc. Natl Acad. Sci. USA* 75:3196–3200.

Halperin, W. 1966. Alternative morphogenetic events in cell suspensions. *Am. J. Bot.* 53:443–453.

Halperin, W., and Jensen, W.A. 1967. Ultrastructural changes during growth and embryogenesis in carrot cell cultures. *J. Ultrastr. Res.* 18:428–443.

Halperin, W., and Wetherell, D.F. 1965. Ammonium requirement for embryogenesis *in vitro*. *Nature* 205:519–520.

Hamilton, D.A., Bashe, D.M., Stinson, J.R., and Mascarenhas, J.P. 1989. Characterization of a pollen-specific genomic clone from maize. *Sex. Plant Reprod.* 2:208–212.

Hamilton, D.A., Roy, M., Rueda, J., Sindhu, R.K., Sanford, J., and Mascarenhas, J.P. 1992. Dissection of a pollen-specific promoter from maize by transient transformation assays. *Plant Mol. Biol.* 18:211–218.

Hammond-Kosack, M.C.U., Holdsworth, M.J., and Bevan, M.W. 1993. *In vivo* footprinting of a low molecular weight glutenin gene (LMWG-1D1) in wheat endosperm. *EMBO J.* 12:545–554.

Hannah, L.C., and Nelson, O.E., Jr. 1976. Characterization of ADP-glucose pyrophosphorylase from *shrunken-2* and *brittle-2* mutants of maize. *Biochem. Genet.* 14:547–560.

Hannah, L.C., Tuschall, D.M., and Mans, R.J. 1980. Multiple forms of maize endosperm ADP-glucose pyrophosphorylase and their control by shrunken-2 and brittle-2. *Genetics* 95:961–970.

Hannig, E. 1904. Zur Physiologie pflanzlicher Embryonen. Ueber die Cultur von Cruciferen-Embryonen ausserhalb der Embryosacks. *Bot. Ztg* 62:45–80.

Hansen, D.J., Bellman, S.K., and Sacher, R.M. 1976. Gibberellic acid-controlled sex expression of corn tassels. *Crop. Sci.* 16:371–374.

Hansen, G., Estruch, J.J., Sommer, H., and Spena, A. 1993. *NTGLO*: a tobacco homologue of the *GLOBOSA* floral homeotic gene of *Antirrhinum majus*: cDNA sequence and expression pattern. *Mol. Gen. Genet.* 239:310–312.

Hanson, D.D., Hamilton, D.A., Travis, J.L., Bashe, D.M., and Mascarenhas, J.P. 1989. Characterization of a pollen-specific cDNA clone from *Zea mays* and its expression. *Plant Cell* 1:173–179.

Hanson, M.R., and Conde, M.F. 1985. Functioning and variation of cytoplasmic genomes: lessons from cytoplasmic-nuclear interactions affecting male fertility in plants. *Int. Rev. Cytol.* 94:213–267.

Hara-Nishimura, I., Nishimura, M., and Akazawa, T. 1985. Biosynthesis and intracellular transport of 11S globulin in developing pumpkin cotyledons. *Plant Physiol.* 77:747–752.

Harada, J.J., Baden, C.S., and Comai, L. 1988. Spatially regulated genes expressed during seed germination and postgerminative development are activated during embryogeny. *Mol. Gen. Genet.* 212:466–473.

Harada, J.J., Barker, S.J., and Goldberg, R.B. 1989. Soybean β-conglycinin genes are clustered in several DNA regions and are regulated by transcriptional and posttranscriptional processes. *Plant Cell* 1:415–425.

Harada, J.J., DeLisle, A.J., Baden, C.S., and Crouch, M.L. 1989. Unusual sequence of an abscisic acid-inducible mRNA which accumulates late in *Brassica napus* seed development. *Plant Mol. Biol.* 12:395–401.

Hardenack, S., Ye, D., Saedler, H., and Grant, S. 1994. Comparison of MADS box gene expression in developing male and female flowers of the dioecious plant white campion. *Plant Cell* 6:1775–1787.

Hardham, A.R. 1976. Structural aspects of the pathways of nutrient flow to the developing embryo and cotyledons of *Pisum sativum* L. *Aust. J. Bot.* 24:711–721.

Hari, V. 1980. Effect of cell density changes and conditioned media on carrot cell embryogenesis. *Z. Pflanzenphysiol.* 96:227–231.

Harikrishna, K., Jampates-Beale, R., Milligan, S.B., and Gasser, C.S. 1996. An endochitinase gene expressed at high levels in the stylar transmitting tissue of tomatoes. *Plant Mol. Biol.* 30:899–911.

Haring, V., Gray, J.E., McClure, B.A., Anderson, M.A., and Clarke, A.E. 1990. Self-incompatibility: a self-recognition system in plants. *Science* 250:937–941.

Harris, N. 1979. Endoplasmic reticulum in developing seeds of *Vicia faba*. A high voltage electron microscope study. *Planta* 146:63–69.

Harris, N., and Boulter, D. 1976. Protein body formation in cotyledons of developing cowpea (*Vigna unguiculata*) seeds. *Ann. Bot.* 40:739–744.

Harris, N., and Chaffey, N.J. 1986. Plasmatubules - real modifications of the plasmalemma. *Nord. J. Bot.* 6:599–607.

Harris, N., and Croy, R.R.D. 1986. Localization of mRNA for pea legumin: *in situ* hybridization using a biotinylated cDNA probe. *Protoplasma* 130:57–67.

Harris, N., and Juliano, B.O. 1977. Ultrastructure of endosperm protein bodies in developing rice grains differing in protein content. *Ann. Bot.* 41:1–5.

Harris, N., Grindley, H. Mulchrone, J., and Croy, R.R.D. 1989. Correlated *in situ* hybridisation and immunochemical studies of legumin storage protein deposition in pea (*Pisum sativum* L.). *Cell Biol. Int. Rep.* 13:23–35.

Harris, P.J., Weinhandl, J.A., and Clarke, A.E. 1989. Effect on *in vitro* pollen growth of an isolated style glycoprotein associated with self-incompatibility in *Nicotiana alata*. *Plant Physiol.* 89:360–367.

Harris, P.J., Freed, K., Anderson, M.A., Weinhandl, J.A., and Clarke, A.E. 1987. An enzyme-linked immunosorbent assay (ELISA) for *in vitro* pollen growth based on binding of a monoclonal antibody to pollen tube surface. *Plant Physiol.* 84:851–855.

Hartings, H., Lazzaroni, N., Marsan, P.A., Aragay, A., Thompson, R., Salamini, F., di Fonzo, N., Palau, J., and Motto, M. 1990. The b-32 protein from maize endosperm: characterization of genomic sequences encoding two alternative central domains. *Plant Mol. Biol.* 14:1031–1040.

Hartings, H., Maddaloni, M., Lazzaroni, N., di Fonzo, N., Motto, M., Salamini, F., and Thompson, R. 1989. The *O2* gene which regulates zein deposition in maize endosperm encodes a protein with structural homologies to transcriptional activators. *EMBO J.* 8:2795–2801.

Harwood, J.L., Sodja, A., Stumpf, P.K., and Spurr, A.R. 1971. On the origin of oil droplets in maturing castor bean seeds, *Ricinus communis*. *Lipids* 6:851–854.

Hasenkampf, C., Qureshi, M., Horsch, A., and Riggs, C.D. 1992. Temporal and spatial distribution of meiotin-1 in anthers of *Lilium longiflorum*. *Devel Genet.* 13:425–434.

Hasitschka, G. 1956. Bildung von Chromosomenbündeln nach Art der Speicheldrüsenchromosomen, spiralisierte Ruhekernchromosomen und andere Struktureigentümlichkeiten in den endopolyploiden Riesenkernen der Antipoden von *Papaver rhoeas*. *Chromosoma* 8:87–113.

Hasitschka-Jenschke, G. 1957. Die Entwicklung der Samenanlage von *Allium ursinum* mit besonderer Berücksichtigung der endopolyploiden Kerne in Synergiden und Antipoden. *Österr. Bot. Z.* 104:1–24.

Hasitschka-Jenschke, G. 1958. Zur Karyologie der Samenanlage dreier *Allium*-Arten. *Österr. Bot. Z.* 105:71–82.

Hasitschka-Jenschke, G. 1959. Vergleichende karyologische Untersuchungen an Antipoden. *Chromosoma* 10:229–267.

Hasitschka-Jenschke, G. 1962. Notizen über endopolyploide Kerne im Bereich der Samenanlage von Angiospermen. *Österr. Bot. Z.* 109:125–137.

Haskell, D.W., and Rogers, O.M. 1985. RNA synthesis by vegetative and sperm nuclei of trinucleate pollen. *Cytologia* 50:805–809.

Hatton, D., Sablowski, R., Yung, M.-H., Smith, C., Schuch, W., and Bevan, M. 1995. Two classes of *cis* sequences contribute to tissue-specific expression of a *PAL2* promoter in transgenic tobacco. *Plant J.* 7:859–876.

Hattori, T., Terada, T., and Hamasuna, S.T. 1994. Sequence and functional analyses of the rice gene homologous to the maize *Vp1*. *Plant Mol. Biol.* 24:805–810.

Hattori, T., Terada, T., and Hamasuna, S. 1995. Regulation of *Osem* gene by abscisic acid and the transcriptional activator VP1: analysis of *cis*-acting promoter elements required for regulation by abscisic acid and VP1. *Plant J.* 7:913–925.

Hattori, T., Vasil, V., Rosenkrans, L., Hannah, L.C., McCarty, D.R., and Vasil, I.K. 1992. The *viviparous-1* gene and abscisic acid activate the *C1* regulatory gene for anthocyanin biosynthesis during seed maturation in maize. *Genes Devel.* 6:609–618.

Hatzopoulos, P., Fong, F., and Sung, Z.R. 1990. Abscisic acid regulation of DC8, a carrot embryonic gene. *Plant Physiol.* 94:690–695.

Hatzopoulos, P., Franz, G., Choy, L., and Sung, R.Z. 1990. Interaction of nuclear factors with upstream sequences of a lipid body membrane protein gene from carrot. *Plant Cell* 2:457–467.

Hauffe, K.D., Paszkowski, U., Schulze-Lefert, P., Hahlbrock, K., Dangl, J.L., and Douglas, C.J. 1991. A parsley 4CL-1 promoter fragment specifies complex expression patterns in transgenic tobacco. *Plant Cell* 3:435–443.

Haughn, G.W., Schultz, E.A., and Martinez-Zapater, J.M. 1995. The regulation of flowering in *Arabidopsis thaliana*: meristems, morphogenesis and mutants. *Can. J. Bot.* 73:959–981.

Haughn, G.W., and Somerville, C.R. 1988. Genetic control of morphogenesis in *Arabidopsis*. *Devel Genet.* 9:73–89.

Hauptli, H., and Williams, S. 1988. Maize *in vitro* pollination with single pollen grains. *Plant Sci.* 58:231–237.

Hause, B., Hause, G., Pechan, P., and van Lammeren, A.A.M. 1993. Cytoskeletal changes and induction of embryogenesis in microspore and pollen cultures of *Brassica napus* L. *Cell Biol. Int.* 17:153–168.

Hause, B., van Veenendaal, W.L.H., Hause, G., and van Lammeren, A.A.M. 1994. Expression of polarity during early development of microspore-derived and zygotic embryos of *Brassica napus* L. cv. Topas. *Bot. Acta* 107:407–415.

Hause, G. 1986. Organelle distribution during pollen development of *Pisum sativum* L. *Biol. Zentral.* 105:283–288.

Hause, G. 1991. Ultrastructural investigations of mature embryo sacs of *Daucus carota, D. aureus* and *D. muricatus* - possible cytological explanations of paternal plastid inheritance. *Sex. Plant Reprod.* 4:288–292.

Hause, G., Hause, B., and van Lammeren, A.A.M. 1992. Microtubular and actin filament configurations during microspore and pollen development in *Brassica napus* cv. Topas. *Can. J. Bot.* 70:1369–1376.

Hause, G., and Schröder, M.-B. 1986. Reproduction in *Triticale*. I. The structure of the mature embryo sac. *Biol. Zentral.* 105:511–517.

Hause, G., and Schröder, M.-B. 1987. Reproduction in *Triticale*. 2. Karyogamy. *Protoplasma* 139:100–104.

Hause, G., Cordewener, J.H.G., Ehrmanova, M., Hause, B., Binarova, P., van Lookeren Campagne, M.M., and van Lammeren, A.A.M. 1995. Cell cycle dependent distribution of phosphorylated proteins in microspores and pollen of *Brassica napus* L., detected by the monoclonal antibody MPM-2. *Protoplasma* 187:117–126.

Haußer, I., Herth, W., and Reiss, H.-D. 1984. Calmodulin in tip-growing plant cells, visualized by fluorescing calmodulin-binding phenothiazines. *Planta* 162:33–39.

Hauxwell, A.J., Corke, F.M.K., Hedley, C.L., and Wang, T.L. 1990. Storage protein gene expression is localised to regions lacking mitotic activity in developing pea embryos. An analysis of seed development in *Pisum sativum* XIV. *Development* 110:283–289.

Havel, L., and Novák, F.J. 1981. *In vitro* pollination of maize (*Zea mays* L.) - proof of double fertilization. *Plant Cell Rep.* 1:26–28.

Hayman, D.L. 1956. The genetical control of incompatibility in *Phalaris coerulescens* Desf. *Aust. J. Biol. Sci.* 9:321–331.

He, C., Tirlapur, U., Cresti, M., Peja, M., Crone, D.E., and Mascarenhas, J.P. 1996. An *Arabidopsis* mutant showing aberrations in male meiosis. *Sex. Plant Reprod.* 9:54–57.

He, C.-P., and Yang, H.-Y. 1992. Ultracytochemical localization of calcium in the embryo sac of sunflower. *Chinese J. Bot.* 4:99–106.

He, S., Lyznik, A., and Mackenzie, S. 1995. Pollen fertility restoration by nuclear gene *Fr* in CMS bean: nuclear-directed alteration of a mitochondrial population. *Genetics* 139:955–962.

Hearn, M.J., Franklin, F.C.H., and Ride, J.P. 1996. Identification of a membrane glycoprotein in pollen of *Papaver rhoeas* which binds stigmatic self-incompatibility (S-) proteins. *Plant J.* 9:467–475.

Heath, J.D., Weldon, R., Monnot, C., and Meinke, D.W. 1986. Analysis of storage proteins in normal and aborted seeds from embryo-lethal mutants of *Arabidopsis thaliana*. *Planta* 169:304–312.

Heberle-Bors, E. 1982a. *In vitro* pollen embryogenesis in *Nicotiana tabacum* L. and its relation to pollen sterility, sex balance, and floral induction of the pollen donor plants. *Planta* 156:396–401.

Heberle-Bors, E. 1982b. On the time of embryogenic pollen grain induction during sexual development of *Nicotiana tabacum* L. plants. *Planta* 156:402–406.

Heberle-Bors, E. 1983. Induction of embryogenic pollen grains *in situ* and subsequent *in vitro* pollen embryogenesis in *Nicotiana tabacum* by treatments of the pollen donor plants with feminizing agents. *Physiol. Plant.* 59:67–72.

Heberle-Bors, E. 1989. Isolated pollen culture in tobacco: plant reproductive development in a nutshell. *Sex. Plant Reprod.* 2:1–10.

Heberle-Bors, E., and Odenbach, W. 1985. *In vitro* pollen embryogenesis and cytoplasmic male sterility in *Triticum aestivum*. *Z. Pflanzenzüchtg* 95:14–22.

Heberle-Bors, E., and Reinert, J. 1979. Androgenesis in

isolated pollen cultures of *Nicotiana tabacum*: dependence upon pollen development. *Protoplasma* 99:237–245.

Heberle-Bors, E., and Reinert, J. 1980. Isolated pollen cultures and pollen dimorphism. *Naturwissenschaften* 67:311.

Heberle-Bors, E., and Reinert, J. 1981. Environmental control and evidence for predetermination of pollen embryogenesis in *Nicotiana tabacum* pollen. *Protoplasma* 109:249–255.

Hébert, D., Kikkert, J.R., Smith, F.D., and Reisch, B.I. 1993. Optimization of biolistic transformation of embryogenic grape cell suspensions. *Plant Cell Rep.* 12:585–589.

Hecht, N.B., and Stern, H. 1971. A late replicating DNA protein complex from cells in meiotic prophase. *Expt. Cell Res.* 69:1–10.

Heck, G.R., Perry, S.E., Nichols, K.W., and Fernandez, D.E. 1995. AGL15, a MADS domain protein expressed in developing embryos. *Plant Cell* 7:1271–1282.

Hedley, C.L., Lloyd, J.R., Ambrose, M.J., and Wang, T.L. 1994. An analysis of seed development in *Pisum sativum*. XVII. The effect of the *rb* locus alone and in combination with *r* on the growth and development of the seed. *Ann. Bot.* 74:365–371.

Hedley, C.L., Smith, C.M., Ambrose, M.J., Cook, S., and Wang, T.L. 1986. An analysis of seed development in *Pisum sativum*. II. The effect of the *r*-locus on the growth and development of the seed. *Ann. Bot.* 58:371–379.

Heidecker, G., and Messing, J. 1983. Sequence analysis of zein cDNAs obtained by an efficient mRNA cloning method. *Nucl. Acids Res.* 11:4891–4906.

Heidecker, G., and Messing, J. 1986. Structural analysis of plant genes. *Annu. Rev. Plant Physiol.* 37:439–466.

Heim, U., Bäumlein, H., and Wobus, U. 1994. The legumin gene family: A reconstructed *Vicia faba* legumin gene encoding a high-molecular-weight subunit is related to type B genes. *Plant Mol. Biol.* 25:131–135.

Heim, U., Manteuffel, R., Bäumlein, H., Steinbiss, H.-H., and Wobus, U. 1995. Transient expression of a lysine-rich vicilin gene of *Vicia faba* in barley endosperm detected by immunological tissue printing after particle bombardment. *Plant Cell Rep.* 15:125–138.

Heinlein, M., and Starlinger, P. 1989. Tissue- and cell-specific expression of the two sucrose synthase isoenzymes in developing maize kernels. *Mol. Gen. Genet.* 215:441–446.

Helleboid, S., Couillerot, J.-P., Hilbert, J.-L., and Vasseur, J. 1995. Inhibition of direct somatic embryogenesis by α-difluoromethylarginine in a *Cichorium* hybrid: effects on polyamine content and protein patterns. *Planta* 196:571–576.

Helm, K.W., and Abernethy, R.H. 1990. Heat shock proteins and their mRNAs in dry and early imbibing embryos of wheat. *Plant Physiol.* 93:1626–1633.

Helsper, J.P.F.G. 1979. The possible role of lipid intermediates in the synthesis of β-glucans by a membrane fraction from pollen tubes of *Petunia hybrida*. *Planta* 144:443–450.

Helsper, J.P.F.G., Linskens, H.F., and Jackson, J.F. 1984. Phytate metabolism in *Petunia* pollen. *Phytochemistry* 23:1841–1845.

Helsper, J.P.F.G., Veerkamp, J.H., and Sassen, M.M.A. 1977. β-Glucan synthetase activity in Golgi vesicles of *Petunia hybrida*. *Planta* 133:303–308.

Hemerly, A.S., Ferreira, P., de Almeida Engler, J., van Montagu, M., Engler, G., and Inzé, D. 1993. *cdc2a* expression in *Arabidopsis* is linked with competence for cell division. *Plant Cell* 5:1711–1723.

Hensgens, L.A.M., de Bakker, E.P.H.M., van Os-Ruygrok, E.P., Rueb, S., van de Mark, F., van der Maas, H., van der Veen, S., Kooman-Gersmann, M., 't Hart, L., and Schilperoort, R.A. 1993. Transient and stable expression of *gusA* fusions with rice genes in rice, barley and perennial ryegrass. *Plant Mol. Biol.* 23:643–669.

Hepher, A., and Boulter, M.E. 1987. Pollen tube growth and fertilization efficiency in *Salpiglossis sinuata*: implications for the involvement of chemotropic factors. *Ann. Bot.* 60:595–601.

Hepler, P.K., and Jackson, W.T. 1968. Microtubules and early stages of cell-plate formation in the endosperm of *Haemanthus katherinae* Baker. *J. Cell Biol.* 38:437–446.

Herd, Y.R., and Beadle, D.J. 1980. The site of the self-incompatibility mechanism in *Tradescantia pallida*. *Ann. Bot.* 45:251–256.

Herd, Y.R., and Steer, M.W. 1984. Microsporogenesis in genic male-sterile lines of barley (*Hordeum vulgare*). *Can. J. Bot.* 62:1127–1135.

Herdt, E., Sütfeld, R., and Wiermann, R. 1978. The occurrence of enzymes involved in phenylpropanoid metabolism in the tapetum fraction of anthers. *Eur. J. Cell Biol.* 17:433–441.

Herman, E.M. 1987. Immunogold-localization and synthesis of an oil-body membrane protein in developing soybean seeds. *Planta* 172:336–345.

Hernández-Fernández, M.M., and Christie, B.R. 1989. Inheritance of somatic embryogenesis in alfalfa (*Medicago sativa* L.). *Genome* 32:318–321.

Hernould, M., Suharsono, S., Litvak, S., Araya, A., and Mouras, A. 1993. Male-sterility induction in transgenic tobacco plants with an unedited *atp9* mitochondrial gene from wheat. *Proc. Natl Acad. Sci. USA* 90:2370–2374.

Hérouart, D., van Montagu, M., and Inzé, D. 1994. Developmental and environmental regulation of the *Nicotiana plumbaginifolia* cytosolic Cu/Zn-superoxide dismutase promoter in transgenic tobacco. *Plant Physiol.* 104:873–880.

Herrero, M., and Dickinson, H.G. 1979. Pollen–pistil incompatibility in *Petunia hybrida*: changes in the pistil following compatible and incompatible intraspecific crosses. *J. Cell Sci.* 36:1–18.

Herrero, M., and Dickinson, H.G. 1980. Ultrastructural and physiological differences between buds and mature flowers of *Petunia hybrida* prior to and following pollination. *Planta* 148:138–145.

Herrero, M., and Dickinson, H.G. 1981. Pollen tube development in *Petunia hybrida* following compatible and incompatible intraspecific matings. *J. Cell Sci.* 47:365–383.

Herth, W. 1978. Ionophore A 23187 stops tip growth, but not cytoplasmic streaming, in pollen tubes of *Lilium longiflorum*. *Protoplasma* 96:275–282.

Herth, W., Franke, W.W., Bittiger, H., Kuppel, A., and Keilich, G. 1974. Alkali-resistant fibrils of β-1,3- and β-1,4-glucans: structural polysaccharides in the pollen tube wall of *Lilium longiflorum*. *Cytobiologie* 9:344–367.

Herz, M., and Brunori, A. 1985. Nuclear DNA content in the endosperm of developing grain of hexaploid Triticales and parental species. *Z. Pflanzenzüchtg* 95:336–341.

Heslop-Harrison, J. 1957. The experimental modification of sex expression in flowering plants. *Biol. Rev.* 32:38–90.

Heslop-Harrison, J. 1959. Photoperiod and fertility in *Rottboellia exaltata* L.f. *Ann. Bot.* 23:345–349.

Heslop-Harrison, J. 1962. Origin of exine. *Nature* 195:1069–1071.

Heslop-Harrison, J. 1963a. An ultrastructural study of pollen wall ontogeny in *Silene pendula. Grana Palynol.* 4:1–24.

Heslop-Harrison, J. 1963b. Ultrastructural aspects of differentiation in sporogenous tissue. *Symp. Soc. Expt. Biol.* 17:315–340.

Heslop-Harrison, J. 1964. Cell walls, cell membranes and protoplasmic connections during meiosis and pollen development. In *Pollen Physiology and Fertilization*, ed. H.F. Linskens, pp. 39–47. Amsterdam: North-Holland Publishing Co.

Heslop-Harrison, J. 1966a. Cytoplasmic continuities during spore formation in flowering plants. *Endeavour* 25:65–72.

Heslop-Harrison, J. 1966b. Cytoplasmic connexions between angiosperm meiocytes. *Ann. Bot.* 30:221–230.

Heslop-Harrison, J. 1968a. Ribosome sites and *S* gene action. *Nature* 218:90–91.

Heslop-Harrison, J. 1968b. Pollen wall development. *Science* 161:230–237.

Heslop-Harrison, J. 1968c. Some fine structural features of intine growth in the young microspore of *Lilium henryi. Portug. Acta Biol.* 10A:235–246.

Heslop-Harrison, J. 1968d. The formation of the generative cell in massulate orchids. *J. Cell Sci.* 3:457–466.

Heslop-Harrison, J. 1968e. Anther carotenoids and the synthesis of sporopollenin. *Nature* 220:605.

Heslop-Harrison, J. 1968f. Tapetal origin of pollen-coat substances in *Lilium. New Phytol.* 67:779–786.

Heslop-Harrison, J. 1968g. Wall development within the microspore tetrad of *Lilium longiflorum. Can. J. Bot.* 46:1185–1192.

Heslop-Harrison, J. 1969a. The origin of surface features of the pollen wall of *Tagetes patula* as observed by scanning electron microscopy. *Cytobios* 2:177–186.

Heslop-Harrison, J. 1969b. An acetolysis-resistant membrane investing tapetum and sporogenous tissue in the anthers of certain Compositae. *Can. J. Bot.* 47:541–542.

Heslop-Harrison, J. 1971a. Sporopollenin in the biological context. In *Sporopollenin*, ed. J. Brooks, P.R. Grant, M. Muir, P. van Gijzel and G. Shaw, pp. 1–30. London: Academic Press.

Heslop-Harrison, J. 1971b. Wall pattern formation in angiosperm microsporogenesis. *Symp. Soc. Expt. Biol.* 25:277–300.

Heslop-Harrison, J. 1972. Sexuality of angiosperms. In *Plant Physiology. A Treatise*, Vol. VIC, ed. F.C. Steward, pp. 133–289. New York: Academic Press.

Heslop-Harrison, J. 1975a. The physiology of the pollen grain surface. *Proc. Roy. Soc. Lond.* 190B:275–299.

Heslop-Harrison, J. 1975b. Incompatibility and the pollen–stigma interaction. *Annu. Rev. Plant Physiol.* 26:403–425.

Heslop-Harrison, J. 1976. The adaptive significance of the exine. In *The Evolutionary Significance of the Exine*, ed. I.K. Ferguson and J. Muller, pp. 27–37. London: Academic Press.

Heslop-Harrison, J. 1978. Genetics and physiology of angiosperm incompatibility systems. *Proc. Roy. Soc. Lond.* 202B:73–92.

Heslop-Harrison, J. 1979a. Aspects of the structure, cytochemistry and germination of the pollen of rye (*Secale cereale* L.). *Ann. Bot.* 44: Suppl. 1, 1–47.

Heslop-Harrison, J. 1979b. Pollen-stigma interaction in grasses: a brief review. *New Zealand J. Bot.* 17:537–546.

Heslop-Harrison, J. 1979c. An interpretation of the hydrodynamics of pollen. *Am. J. Bot.* 66:737–743.

Heslop-Harrison, J. 1982. Pollen–stigma interaction and cross-incompatibility in the grasses. *Science* 215:1358–1364.

Heslop-Harrison, J. 1987. Pollen germination and pollen-tube growth. *Int. Rev. Cytol.* 107:1–78.

Heslop-Harrison, J., and Dickinson, H.G. 1967. A cycle of spherosome aggregation and disaggregation correlated with the meiotic divisions in *Lilium. Phytomorphology* 17:195–199.

Heslop-Harrison, J., and Dickinson, H.G. 1969. Time relationships of sporopollenin synthesis associated with tapetum and microspores in *Lilium. Planta* 84:199–214.

Heslop-Harrison, J., Heslop-Harrison, J.S., and Heslop-Harrison, Y. 1986. The comportment of the vegetative nucleus and generative cell in the pollen and pollen tubes of *Helleborus foetidus* L. *Ann. Bot.* 58:1–12.

Heslop-Harrison, J., and Heslop-Harrison, Y. 1973. Pollen-wall proteins: `gametophytic' and `sporophytic' fractions in the pollen walls of the Malvaceae. *Ann. Bot.* 37:403–412.

Heslop-Harrison, J., and Heslop-Harrison, Y. 1975. Enzymic removal of the proteinaceous pellicle of the stigma papilla prevents pollen tube entry in the Caryophyllaceae. *Ann. Bot.* 39:163–165.

Heslop-Harrison, J., and Heslop-Harrison, Y. 1980a. Cytochemistry and function of the Zwischenkorper in grass pollens. *Pollen Spores* 22:5–10.

Heslop-Harrison, J., and Heslop-Harrison, Y. 1980b. The pollen–stigma interaction in the grasses. I. Fine-structure and cytochemistry of the stigmas of *Hordeum* and *Secale. Acta Bot. Neerl.* 29:261–276.

Heslop-Harrison, J., and Heslop-Harrison, Y. 1981. The pollen–stigma interaction in the grasses. 2. Pollen-tube penetration and the stigma response in *Secale. Acta Bot. Neerl.* 30:289–307.

Heslop-Harrison, J., and Heslop-Harrison, Y. 1982a The growth of the grass pollen tube: 1. Characteristics of the polysaccharide particles ("P-particles") associated with apical growth. *Protoplasma* 112:71–80.

Heslop-Harrison, J., and Heslop-Harrison, Y. 1982b. The pollen–stigma interaction in the grasses. 4. An interpretation of the self-incompatibility response. *Acta Bot. Neerl.* 31:429–439.

Heslop-Harrison, J., and Heslop-Harrison, Y. 1982c. Pollen–stigma interaction in the Leguminosae: constituents of the stylar fluid and stigma secretion of *Trifolium pratense* L. *Ann. Bot.* 49:729–735.

Heslop-Harrison, J., and Heslop-Harrison, Y. 1984. The disposition of gamete and vegetative-cell nuclei in the extending pollen tubes of a grass species, *Alopecurus pratensis* L. *Acta Bot. Neerl.* 33:131–134.

Heslop-Harrison, J., and Heslop-Harrison, Y. 1985. Germination of stress-tolerant *Eucalyptus* pollen. *J. Cell Sci.* 73:135–157.

Heslop-Harrison, J., and Heslop-Harrison, Y. 1987. An analysis of gamete and organelle movement in the pollen tube of *Secale cereale* L. *Plant Sci.* 51:203–213.

Heslop-Harrison, J., and Heslop-Harrison, Y. 1988a.

Organelle movement and fibrillar elements of the cytoskeleton in the angiosperm pollen tube. *Sex. Plant Reprod.* 1:16–24.

Heslop-Harrison, J., and Heslop-Harrison, Y. 1988b. Some permeability properties of angiosperm pollen grains, pollen tubes and generative cells. *Sex. Plant Reprod.* 1:65–73.

Heslop-Harrison, J., and Heslop-Harrison, Y. 1988c. Tubulin and male-gamete interconnections in the pollen tubes of the grass *Alopecurus pratensis*. *Ann. Bot.* 61:249–254.

Heslop-Harrison, J., and Heslop-Harrison, Y. 1988d. Sites of origin of the peripheral microtubule system of the vegetative cell of the angiosperm pollen tube. *Ann. Bot.* 62:455–461.

Heslop-Harrison, J., and Heslop-Harrison, Y. 1989a. Cytochalasin effects on structure and movement in the pollen tube of *Iris*. *Sex. Plant Reprod.* 2:27–37.

Heslop-Harrison, J., and Heslop-Harrison, Y. 1989b. Actomyosin and movement in the angiosperm pollen tube: an interpretation of some recent results. *Sex. Plant Reprod.* 2:199–207.

Heslop-Harrison, J., and Heslop-Harrison, Y. 1989c. Conformation and movement of the vegetative nucleus of the angiosperm pollen tube: association with the actin cytoskeleton. *J. Cell Sci.* 93:299–308.

Heslop-Harrison, J., and Heslop-Harrison, Y. 1989d. Myosin associated with the surfaces of organelles, vegetative nuclei and generative cells in angiosperm pollen grains and tubes. *J. Cell Sci.* 94:319–325.

Heslop-Harrison, J., and Heslop-Harrison, Y. 1990. Dynamic aspects of apical zonation in the angiosperm pollen tube. *Sex. Plant Reprod.* 3:187–194.

Heslop-Harrison, J., and Heslop-Harrison, Y. 1991a. The actin cytoskeleton in unfixed pollen tubes following microwave-accelerated DMSO-permeabilisation and TRITC-phalloidin staining. *Sex. Plant Reprod.* 4:6–11.

Heslop-Harrison, J., and Heslop-Harrison, Y. 1991b. Structural and functional variation in pollen intines. In *Pollen and Spores*, ed. S. Blackmore and S.H. Barnes, pp. 331–343. Oxford: Clarendon Press.

Heslop-Harrison, J., and Heslop-Harrison, Y. 1992a. Cyclical transformations of the actin cytoskeleton of hyacinth pollen subjected to recurrent vapor-phase hydration and dehydration. *Biol. Cell.* 75:245–252.

Heslop-Harrison, J., and Heslop-Harrison, Y. 1992b. Intracellular motility, the actin cytoskeleton and germinability in the pollen of wheat (*Triticum aestivum* L.). *Sex. Plant Reprod.* 5:247–255.

Heslop-Harrison, J., and Heslop-Harrison, Y. 1992c. Germination of monocolpate angiosperm pollen: effects of inhibitory factors and the Ca^{2+}-channel blocker, nifedipine. *Ann. Bot.* 69:395–403.

Heslop-Harrison, J., Heslop-Harrison, Y., and Barber, J. 1975. The stigma surface in incompatibility responses. *Proc. Roy. Soc. Lond.* 188B:287–297.

Heslop-Harrison, J., Knox, R. B., and Heslop-Harrison, Y. 1974. Pollen-wall proteins: exine-held fractions associated with the incompatibility response in Cruciferae. *Theor. Appl. Genet.* 44:133–137.

Heslop-Harrison, J., and Mackenzie, A. 1967. Autoradiography of soluble [2-^{14}C]thymidine derivatives during meiosis and microsporogenesis in *Lilium* anthers. *J. Cell Sci.* 2:387–400.

Heslop-Harrison, J., Heslop-Harrison, Y., Cresti, M., and Ciampolini, F. 1991. Ultrastructural features of pollen tubes of *Endymion non-scriptus* modified by cytochalasin D. *Sex. Plant Reprod.* 4:73–80.

Heslop-Harrison, J., Heslop-Harrison, Y., Cresti, M., Tiezzi, A., and Ciampolini, F. 1986. Actin during pollen germination. *J. Cell Sci.* 86:1–8.

Heslop-Harrison, J., Heslop-Harrison, Y., Cresti, M., Tiezzi, A., and Moscatelli, A. 1988. Cytoskeletal elements, cell shaping and movement in the angiosperm pollen tube. *J. Cell Sci.* 91:49–60.

Heslop-Harrison, J., Heslop-Harrison, Y., Knox, R.B., and Howlett, B. 1973. Pollen-wall proteins: 'gametophytic' and 'sporophytic' fractions in the pollen walls of the Malvaceae. *Ann. Bot.* 37:403–412.

Heslop-Harrison, J., f, R.B., Heslop-Harrison, Y., and Mattsson, O. 1975. Pollen-wall proteins: emission and role in incompatibility responses. In *The Biology of the Male Gamete*, ed. J.G. Duckett and P.A. Racey, pp. 189–202. London: Academic Press.

Heslop-Harrison, J.S., Heslop-Harrison, J., Heslop-Harrison, Y., and Reger, B.J. 1985. The distribution of calcium in the grass pollen tube. *Proc. Roy. Soc. Lond.* 225B:315–327.

Heslop-Harrison, Y. 1976. Localisation of concanavalin A binding sites on the stigma surface of a grass species. *Micron* 7:33–36.

Heslop-Harrison, Y. 1977. The pollen–stigma interaction: pollen-tube penetration in *Crocus*. *Ann. Bot.* 41:913–922.

Heslop-Harrison, Y. 1981. Stigma characteristics and angiosperm taxonomy. *Nord. J. Bot.* 1:401–420.

Heslop-Harrison, Y., and Heslop-Harrison, J. 1982a. Pollen–stigma interactions in the Leguminosae: the secretory system of the style in *Trifolium pratense* L. *Ann. Bot.* 50:635–645.

Heslop-Harrison, Y., and Heslop-Harrison, J. 1982b. The micro-fibrillar component of the pollen intine: some structural features. *Ann. Bot.* 50:831–842.

Heslop-Harrison, Y., and Heslop-Harrison, J. 1992. Germination of monocolpate angiosperm pollen: evolution of the actin cytoskeleton and wall during hydration, activation and tube emergence. *Ann. Bot.* 69:385–394.

Heslop-Harrison, Y., Heslop-Harrison, J., and Reger, B.J. 1985. The pollen–stigma interaction in the grasses. 7. Pollen-tube guidance and the regulation of tube number in *Zea mays* L. *Acta Bot. Neerl.* 34:193–211.

Heslop-Harrison, Y., Heslop-Harrison, J., and Shivanna, K.R. 1981. Heterostyly in *Primula*. 1. Fine-structural and cytochemical features of the stigma and style in *Primula vulgaris* Huds. *Protoplasma* 107:171–187.

Heslop-Harrison, Y., Heslop-Harrison, J.S., and Heslop-Harrison, J. 1986. Germination of *Corylus avellana* L. (hazel) pollen: hydration and the function of the oncus. *Acta Bot. Neerl.* 35:265–284.

Heslop-Harrison, Y., and Reger, B.J. 1988. Tissue organisation, pollen receptivity and pollen tube guidance in normal and mutant stigmas of the grass *Pennisetum typhoides* (Burm.) Stapf et Hubb. *Sex. Plant Reprod.* 1:182–193.

Heslop-Harrison, Y., Reger, B.J., and Heslop-Harrison, J. 1984a. The pollen–stigma interaction in the grasses. 5. Tissue organisation and cytochemistry of the stigma ("silk") of *Zea mays* L. *Acta Bot. Neerl.* 33:81–99.

Heslop-Harrison, Y., Reger, B.J., and Heslop-Harrison, J. 1984b. The pollen–stigma interaction in the grasses. 6. The stigma ('silk') of *Zea mays* L. as host to the pollens of *Sorghum bicolor* (L.) Moench and *Pennisetum americanum* (L.) Leeke. *Acta Bot. Neerl.* 33:205–227.

Heslop-Harrison, Y., and Shivanna, K.R. 1977. The receptive surface of the angiosperm stigma. *Ann. Bot.* 41:1233–1258.

Hess, D. 1980. Investigations on the intra- and interspecific transfer of anthocyanin genes using pollen as vectors. *Z. Pflanzenphysiol.* 98:321–337.

Hess, D., and Dressler, K. 1989. Tumor transformation of *Petunia hybrida* via pollen co-cultured with *Agrobacterium tumefaciens. Bot. Acta* 102:202–207.

Hess, D., Dressler, K., and Nimmrichter, R. 1990. Transformation experiments by pipetting *Agrobacterium* into the spikelets of wheat (*Triticum aestivum* L.). *Plant Sci.* 72:233–244.

Hess, J.R., and Carman, J.G. 1993. Normalizing development of cultured *Triticum aestivum* L. embryos. I. Low oxygen tensions and exogenous ABA. *J. Expt. Bot.* 44:1067–1073.

Hess, M.W. 1993. Cell-wall development in freeze-fixed pollen: intine formation of *Ledebouria socialis* (Hyacinthaceae). *Planta* 189:139–149.

Hess, M.W., and Frosch, A. 1994. Subunits of forming pollen exine and Ubisch bodies as seen in freeze substituted *Ledebouria socialis* Roth (Hyacinthaceae). *Protoplasma* 182:10–14.

Hess, M.W., and Hesse, M. 1994. Ultrastructural observations on anther development of freeze-fixed *Ledebouria socialis* Roth (Hyacinthaceae). *Planta* 192:421–430.

Hesse, M., Pacini, E., and Willemse, M., ed. 1993. *The Tapetum. Cytology, Function, Biochemistry and Evolution.* Wien: Springer-Verlag.

Hetherington, P.R., and Fry, S.C. 1993. Xyloglucan endotransglycosylase activity in carrot cell suspensions during cell elongation and somatic embryogenesis. *Plant Physiol.* 103:987–992.

Hicks, G.S. 1975. Carpelloids on tobacco stamen primordia *in vitro. Can. J. Bot.* 53:77–81.

Hicks, G.S., Browne, R., and Sand, S.A. 1981. Organogenesis from cultured floral meristems of a male sterile tobacco hybrid. *Can. J. Bot.* 59:1665–1670.

Hicks, G.S., and Sand, S.A. 1977. *In vitro* culture of the stamen primordia from a male sterile tobacco. *Plant Sci. Lett.* 10:257–263.

Hicks, G.S., and Sussex, I.M. 1970. Development *in vitro* of excised flower primordia of *Nicotiana tabacum. Can. J. Bot.* 48:133–139.

Hiei, Y., Ohta, S., Komari, T., and Kumashiro, T. 1994. Efficient transformation of rice (*Oryza sativa* L.) mediated by *Agrobacterium* and sequence analysis of the boundaries of the T-DNA. *Plant J.* 6:271–282.

Higgins, P., and Bowles, D.J. 1990. Comparative analysis of translatable mRNA populations in zygotic and pollen-derived embryos of barley (*Hordeum vulgare* L.). *Plant Sci.* 69:239–247.

Higgins, R.K., and Petolino, J.F. 1988. *In vitro* pollination fertilization of maize: influence of explant factors on kernel development. *Plant Cell Tissue Organ Cult.* 12:21–30.

Higgins, T.J.V. 1984. Synthesis and regulation of major proteins in seeds. *Annu. Rev. Plant Physiol.* 35:191–221.

Higgins, T.J.V., and Spencer, D. 1981. Precursor forms of pea vicilin subunits. Modification by microsomal membranes during cell-free translation. *Plant Physiol.* 67:205–211.

Higgins, T.J.V., Chandler, P.M., Randall, P.J., Spencer, D., Beach, L.R., Blagrove, R.J., Kortt, A.A., and Inglis, A.S. 1986. Gene structure, protein structure, and regulation of the synthesis of a sulfur-rich protein in pea seeds. *J. Biol. Chem.* 261:11124–11130.

Higgins, T.J.V., Newbigin, E.J., Spencer, D., Llewellyn, D.J., and Craig, S. 1988. The sequence of a pea vicilin gene and its expression in transgenic tobacco plants. *Plant Mol. Biol.* 11:683–695.

Hilbert, J.-L., Dubois, T., and Vasseur, J. 1992. Detection of embryogenesis-related proteins during somatic embryo formation in *Cichorium. Plant Physiol. Biochem.* 30:733–741.

Hildebrand, D.F., Versluys, R.T., and Collins, G.B. 1991. Changes in lipoxygenase isozyme levels during soybean embryo development. *Plant Sci.* 75:1–8.

Hill, J.E., and Breidenbach, R.W. 1974. Proteins of soybeans seeds. II. Accumulation of the major protein components during seed development and maturation. *Plant Physiol.* 53:747–751.

Hill, J.P., and Lord, E.M. 1987. Dynamics of pollen tube growth in the wild radish *Raphanus raphanistrum* (Brassicaceae). II. Morphology, cytochemistry and ultrastructure of transmitting tissues, and path of pollen tube growth. *Am. J. Bot.* 74:988–997.

Hill, J.P., and Lord, E.M. 1989. Floral development in *Arabidopsis thaliana*: a comparison of the wild type and the homeotic pistillata mutant. *Can. J. Bot.* 67:2922–2936.

Hilson, P., de Froidmont, D., Lejour, C., Hirai, S.-I., Jacquemin, J.-M., and Yaniv, M. 1990. Fos and Jun oncogenes transactivate chimeric or native promoters containing AP1/GCN4 binding sites in plant cells. *Plant Cell* 2:651–658.

Hinata, K., and Nishio, T. 1978. S-allele specificity of stigma proteins in *Brassica oleracea* and *B. campestris. Heredity* 41:93–100.

Hinata, K., Nishio, T., and Kimura, J. 1982. Comparative studies on S-glycoproteins purified from different S-genotypes in self-incompatible *Brassica* species. II. Immunological specificities. *Genetics* 100:649–657.

Hinata, K., Watanabe, M., Toriyama, K., and Isogai, A. 1993. A review of recent studies on homomorphic self-incompatibility. *Int. Rev. Cytol.* 143:257–296.

Hinz, G., Hoh, B., Hohl, I., and Robinson, D.G. 1995. Stratification of storage proteins in the protein storage vacuole of developing cotyledons of *Pisum sativum* L. *J. Plant Physiol.* 145:437–442.

Hirai, M.Y., Fujiwara, T., Goto, K., Komeda, Y., Chino, M., and Naito, S. 1994. Differential regulation of soybean seed storage protein gene promoter-GUS fusions by exogenously applied methionine in transgenic *Arabidopsis thaliana. Plant Cell Physiol.* 35:927–934.

Hirano, H.-Y., and Sano, Y. 1991. Molecular characterization of the *waxy* locus of rice (*Oryza sativa*). *Plant Cell Physiol.* 32:989–997.

Hirano, H.-Y., Tabayashi, N., Matsumura, T., Tanida, M., Komeda, Y., and Sano, Y. 1995. Tissue-dependent expression of the rice wx^{+} gene promoter in transgenic rice and *Petunia. Plant Cell Physiol.* 36:37–44.

Hird, D.L., Worrall, D., Hodge, R., Smartt, S., Paul, W., and Scott, R. 1993. The anther-specific protein encoded by the *Brassica napus* and *Arabidopsis thaliana* A6 gene displays similarity to β-1,3-glucanases. *Plant J.* 4:1023–1033.

Hirt, H., Páy, A., Györgyey, J., Bakó, L., Németh, K., Bögre, L., Schweyen, R.J., Heberle-Bors, E., and Dudits, D. 1991. Complementation of a yeast cell cycle mutant by an alfalfa cDNA encoding a protein kinase homologous to p34[cdc2]. *Proc. Natl Acad. Sci. USA* 88:1636–1640.

Hiscock, S.J., Doughty, J., and Dickinson, H.G. 1995. Synthesis and phosphorylation of pollen proteins during the pollen–stigma interaction in self-compatible *Brassica napus* L. and self-incompatible *Brassica oleracea* L. *Sex. Plant Reprod.* 8:345–353.

Hiscock, S.J., Dewey, F.M., Coleman, J.O.D., and Dickinson, H.G. 1994. Identification and localization of an active cutinase in the pollen of *Brassica napus* L. *Planta* 193:377–384.

Hiscock, S.J., Doughty, J., Willis, A.C., and Dickinson, H.G. 1995. A 7-kDa pollen coating-borne peptide from *Brassica napus* interacts with S-locus glycoprotein and S-locus-related glycoprotein. *Planta* 196:367–374.

Hodgkin, T., and Lyon, G.D. 1984. Pollen germination inhibitors in extracts of *Brassica oleracea* stigmas. *New Phytol.* 96:293–298.

Hodgkin, T., Lyon, G.D., and Dickinson, H.G. 1988. Recognition in flowering plants: a comparison of the *Brassica* self-incompatibility system and plant pathogen interactions. *New Phytol.* 110:557–569.

Hoecker, U., Vasil, I.K., and McCarty, D.R. 1995. Integrated control of seed maturation and germination programs by activator and repressor functions of viviparous-1 of maize. *Genes Devel.* 9:2459–2469.

Hoefert, L.L. 1969a. Ultrastructure of *Beta* pollen. I. Cytoplasmic constituents. *Am. J. Bot.* 56:363–368.

Hoefert, L.L. 1969b. Fine structure of sperm cells in pollen grains of *Beta*. Brief report. *Protoplasma* 68:237–240.

Hoekstra, F.A. 1979. Mitochondrial development and activity of binucleate and trinucleate pollen during germination *in vitro*. *Planta* 145:25–36.

Hoekstra, F.A., and Bruinsma, J. 1975a. Viability of Compositae pollen: germination *in vitro* and influences of climatic conditions during dehiscence. *Z. Pflanzenphysiol.* 76:36–43.

Hoekstra, F.A., and Bruinsma, J. 1975b. Respiration and vitality of binucleate and trinucleate pollen. *Physiol. Plant.* 34:221–225.

Hoekstra, F.A., and Bruinsma, J. 1978. Reduced independence of the male gametophyte in angiosperm evolution. *Ann. Bot.* 42:759–762.

Hoekstra, F.A., and Bruinsma, J. 1979. Protein synthesis of binucleate and trinucleate pollen and its relationship to tube emergence and growth. *Planta* 146:559–566.

Hoekstra, F.A., and Bruinsma, J. 1980. Control of respiration of binucleate and trinucleate pollen under humid conditions. *Physiol. Plant.* 48:71–77.

Hoffman, L.M., and Donaldson, D.D. 1985. Characterization of two *Phaseolus vulgaris* phytohemagglutinin genes closely linked on the chromosome. *EMBO J.* 4:883–889.

Hoffman, L.M., Donaldson, D.D., and Herman, E.M. 1988. A modified storage protein is synthesized, processed, and degraded in the seeds of transgenic plants. *Plant Mol. Biol.* 11:717–729.

Hoffman, L.M., Ma, Y., and Barker, R.F. 1982. Molecular cloning of *Phaseolus vulgaris* lectin mRNA and use of cDNA as a probe to estimate lectin transcript levels in various tissues. *Nucl. Acids Res.* 10:7819–7828.

Hoffman, L.M., Donaldson, D.D., Bookland, R., Rashka, K., and Herman, E.M. 1987. Synthesis and protein body deposition of maize 15-kd zein in transgenic tobacco seeds. *EMBO J.* 6:3213–3221.

Hogan, C.J. 1987. Microtubule patterns during meiosis in two higher plant species. *Protoplasma* 138:126–136.

Hogenboom, N.G. 1973. A model for incongruity in intimate partner relationships. *Euphytica* 22:219–233.

Hoggart, R.M., and Clarke, A.E. 1984. Arabinogalactans are common components of angiosperm styles. *Phytochemistry* 23:1571–1573.

Höglund, A.-S., Lenman, M., Falk, A., and Rask, L. 1991. Distribution of myrosinase in rapeseed tissues. *Plant Physiol.* 95:213–221.

Höglund, A.-S., Rödin, J., Larsson, E., and Rask, L. 1992. Distribution of napin and cruciferin in developing rape seed embryos. *Plant Physiol.* 98:509–515.

Höhler, B., and Börner, T. 1980. Vergleichende Untersuchungen von Isoenzymen des Antherengewebes fertiler und cytoplasmatisch männlich steriler Weizenpflanzen. *Biochem. Physiol. Pflanz.* 175:562–569.

Holbrook, L.A., van Rooijen, G.J.H., Wilen, R.W., and Moloney, M.M. 1991. Oilbody proteins in microspore-derived embryos of *Brassica napus*. Hormonal, osmotic, and developmental regulation of synthesis. *Plant Physiol.* 97:1051–1058.

Holden, M.J., and Sze, H. 1984. *Helminthosporium maydis* T toxin increased membrane permeability to Ca^{2+} in susceptible corn mitochondria. *Plant Physiol.* 75:235–237.

Holden, M.J., and Sze, H. 1987. Dissipation of the membrane potential in susceptible corn mitochondria by the toxin of *Helminthosporium maydis*, Race T, and toxin analogs. *Plant Physiol.* 84:670–676.

Holdsworth, M.J., Muñoz-Blanco, J., Hammond-Kosack, M., Colot, V., Schuch, W., and Bevan, M.W. 1995. The maize transcription factor opaque-2 activates a wheat germ glutenin promoter in plant and yeast cells. *Plant Mol. Biol.* 29:711–720.

Hollung, K., Espelund, M., and Jakobsen, K.S. 1994. Another *Lea* B19 gene (group 1 *Lea*) from barley containing a single 20 amino acid hydrophilic motif. *Plant Mol. Biol.* 25:559–564.

Holm, P.B. 1977. The premeiotic DNA replication of euchromatin and heterochromatin in *Lilium longiflorum* (Thunb.). *Carlsberg Res. Commun.* 42:249–281.

Holm, P.B, Knudsen, S., Mouritzen, P., Negri, D., Olsen, F.L., and Roué, C. 1994. Regeneration of fertile barley plants from mechanically isolated protoplasts of the fertilized egg cell. *Plant Cell* 6:531–543.

Holm, P.B., Knudsen, S., Mouritzen, P., Negri, D., Olsen, F.L., and Roué, C. 1995. Regeneration of the barley zygote in ovule culture. *Sex. Plant Reprod.* 8:49–59.

Hong, B., Barg, R., and Ho, T.-H.D. 1992. Developmental and organ-specific expression of an ABA- and stress-induced protein in barley. *Plant Mol. Biol.* 18:663–674.

Hong, S.K., Aoki, T., Kitano, H., Satoh, H., and Nagato, Y. 1995a. Temperature-sensitive mutation, *embryoless 1*, affects both embryo and endosperm development in rice. *Plant Sci.* 108:165–172.

Hong, S.K., Aoki, T., Kitano, H., Satoh, H., and Nagato, Y. 1995b. Phenotypic diversity of 188 rice embryo mutants. *Devel Genet.* 16:298–310.

Hong, S.K., Kitano, H., Satoh, H., and Nagato, Y. 1996. How is embryo size genetically regulated in rice? *Development* 122:2051–2058.

Hong-Qi, Z., Croes, A.F., and Linskens, H.F. 1984. Qualitative changes in protein synthesis in germinating pollen of *Lilium longiflorum* after a heat shock. *Plant Cell Environ.* 7:689–691.

Hopf, N., Plesofsky-Vig, N., and Brambl, R. 1992. The heat

shock response of pollen and other tissues of maize. *Plant Mol. Biol.* 19:623–630.

Hopper, J.E., Ascher, P.D., and Peloquin, S.J. 1967. Inactivation of self-incompatibility following temperature pretreatments of styles of *Lilium longiflorum*. *Euphytica* 16:215–220.

Hopping, M.E., and Jerram, E.M. 1979. Pollination of kiwifruit (*Actinidia chinensis* Planch.): stigma-style structure and pollen tube growth. *New Zealand J. Bot.* 17:233–240.

Horn, R., Köhler, R.H., and Zetsche, K. 1991. A mitochondrial 16 kDa protein is associated with cytoplasmic male sterility in sunflower. *Plant Mol. Biol.* 17:29–36.

Horn, R., Hustedt, J.E.G., Horstmeyer, A., Hahnen, J., Zetsche, K., and Friedt, W. 1996. The CMS-associated 16 kDa protein encoded by *orfH522* in the PET1 cytoplasm is also present in other male-sterile cytoplasms of sunflower. *Plant Mol. Biol.* 30: 523–538.

Horner, H.T., Jr. 1977. A comparative light- and electron-microscopic study of microsporogenesis in male-fertile and cytoplasmic male-sterile sunflower (*Helianthus annuus*). *Am. J. Bot.* 64:745–759.

Horner, H.T., Jr., and Lersten, N.R. 1971. Microsporogenesis in *Citrus limon* (Rutaceae). *Am. J. Bot.* 58:72–79.

Horner, H.T., Jr., and Pearson, C.B. 1978. Pollen wall and aperture development in *Helianthus annuus* (Compositae: Heliantheae). *Am. J. Bot.* 65:293–309.

Horner, H.T., Jr., and Rogers, M.A. 1974. A comparative light and electron microscopic study of microsporogenesis in male-fertile and cytoplasmic male-sterile pepper (*Capsicum annuum*). *Can. J. Bot.* 52:435–441.

Horner, M., and Mott, R.L. 1979. The frequency of embryogenic pollen grains is not increased by *in vitro* anther culture in *Nicotiana tabacum* L. *Planta* 147:156–158.

Horner, M., and Pratt, M.L. 1979. Amino acid analysis of *in vivo* and androgenic anthers of *Nicotiana tabacum*. *Protoplasma* 98:279–282.

Horner, M., and Street, H.E. 1978. Pollen dimorphism - origin and significance in pollen plant formation by anther culture. *Ann. Bot.* 42:763–771.

Horovitz, A., Galil, J., and Portnoy, L. 1972. Soluble sugars in the stigmatic exudate of *Yucca aloifolia* L. *Phyton Argentina* 29:43–46.

Horowitz, J. 1985. A *Phaseolus* mutation results in a reduced level of lectin mRNA. *Mol. Gen. Genet.* 198:482-485.

Hoshikawa, K. 1976. Histology of endosperm development and reserve substance accumulation in cereal grains. In *Gamma Field Symposium*, No. 13, pp. 1–15. Japan: Institute of Radiation Breeding.

Hotta, Y., and Hecht, N. 1971. Methylation of *Lilium* DNA during the meiotic cycle. *Biochim. Biophys. Acta* 238:50–59.

Hotta, Y., Ito, M., and Stern, H. 1966. Synthesis of DNA during meiosis. *Proc. Natl Acad. Sci. USA* 56:1184–1191.

Hotta, Y., Parchman, L.G., and Stern, H. 1968. Protein synthesis during meiosis. *Proc. Natl Acad. Sci. USA* 60:575–582.

Hotta, Y., and Shepard. J. 1973. Biochemical aspects of colchicine action on meiotic cells. *Mol. Gen. Genet.* 122:243–260.

Hotta, Y., and Stern, H. 1961a. Deamination of deoxycytidine and 5-methyldeoxycytidine in developing anthers of *Lilium longiflorum* (var. Croft). *J. Biophys. Biochem. Cytol.* 9:279–284.

Hotta, Y., and Stern, H. 1961b. Transient phosphorylation of deoxyribosides and regulation of deoxyribonucleic acid synthesis. *J. Biophys. Biochem. Cytol.* 11:311–319.

Hotta, Y., and Stern, H. 1963a. Synthesis of messenger-like ribonucleic acid and protein during meiosis in isolated cells of *Trillium erectum*. *J. Cell Biol.* 19:45–58.

Hotta, Y., and Stern, H. 1963b. Inhibition of protein synthesis during meiosis and its bearing on intracellular regulation. *J. Cell Biol.* 16:259–279.

Hotta, Y., and Stern, H. 1963c. Molecular facets of mitotic regulation, I. Synthesis of thymidine kinase. *Proc. Natl Acad. Sci. USA* 49:648–654.

Hotta, Y., and Stern, H. 1963d. Molecular facets of mitotic regulation, II. Factors underlying the removal of thymidine kinase. *Proc. Natl Acad. Sci. USA* 49:861–865.

Hotta, Y., and Stern, H. 1965a. Inducibility of thymidine kinase by thymidine as a function of interphase stage. *J. Cell Biol.* 25:99–108.

Hotta, Y., and Stern, H. 1965b. Polymerase and kinase activities in relation to RNA synthesis during meiosis. *Protoplasma* 60:218–232.

Hotta, Y., and Stern, H. 1971a. A DNA-binding protein in meiotic cells of *Lilium*. *Devel Biol.* 26:87–99.

Hotta, Y., and Stern, H. 1971b. Analysis of DNA synthesis during meiotic prophase in *Lilium*. *J. Mol. Biol.* 55:337–355.

Hotta, Y., and Stern, H. 1974. DNA scission and repair during pachytene in *Lilium*. *Chromosoma* 46:279–296.

Hotta, Y., and Stern, H. 1976. Persistent discontinuities in late replicating DNA during meiosis in *Lilium*. *Chromosoma* 55:171–182.

Hotta, Y., and Stern, H. 1978. DNA unwinding protein from meiotic cells of *Lilium*. *Biochemistry* 17:1872–1880.

Hotta, Y., and Stern, H. 1979. The effect of dephosphorylation on the properties of a helix-destabilizing protein from meiotic cells and its partial reversal by a protein kinase. *Eur. J. Biochem.* 95:31–38.

Hotta, Y., and Stern, H. 1981. Small nuclear RNA molecules that regulate nuclease accessibility in specific chromatin regions of meiotic cells. *Cell* 27:309–319.

Hotta, Y., and Stern, H. 1984. The organization of DNA segments undergoing repair synthesis during pachytene. *Chromosoma* 89:127–137.

Hotta, Y., Tabata, S., and Stern, H. 1984. Replication and nicking of zygotene DNA sequences. Control by a meiosis-specific protein. *Chromosoma* 90:243–253.

Hotta, Y., Bennett, M.D., Toledo, L.A., and Stern, H. 1979. Regulation of R-protein and endonuclease activities in meiocytes by homologous chromosome pairing. *Chromosoma* 72:191–201.

Hotta, Y., Tabata, S., Bouchard, R.A., Piñon, R., and Stern, H. 1985a. General recombination mechanisms in extracts of meiotic cells. *Chromosoma* 93:140–151.

Hotta, Y., Tabata, S., Stubbs, L., and Stern, H. 1985b. Meiosis-specific transcripts of a DNA component replicated during chromosome pairing: homology across the phylogenetic spectrum. *Cell* 40:785–793.

Hough, T., Singh, M.B., Smart, I.J., and Knox, R.B. 1986. Immunofluorescent screening of monoclonal antibodies to surface antigens of animal and plant cells bound to polycarbonate membranes. *J. Immuno. Meth.* 92:103–107.

Howarth, C. 1990. Heat shock proteins in *Sorghum bicolor* and *Pennisetum americanum*. II. Stored mRNA in sorghum seed and its relationship to heat shock protein synthesis during germination. *Plant Cell Environ.* 13:57–64.

Howell, S.H., and Hecht, N.B. 1971. The appearance of polynucleotide ligase and DNA polymerase during the synchronous mitotic cycle in *Lilium* microspores. *Biochim. Biophys. Acta* 240:343–352.

Howell, S.H., and Stern, H. 1971. The appearance of DNA breakage and repair activities in the synchronous meiotic cycle of *Lilium*. *J. Mol. Biol.* 55:357–378.

Howlett, B.J., Knox, R.B., and Heslop-Harrison, J. 1973. Pollen-wall proteins: release of the allergen antigen E from intine and exine sites in pollen grains of ragweed and *Cosmos*. *J. Cell Sci.* 13:603–619.

Howlett, B.J., Knox, R.B., Paxton, J.D., and Heslop-Harrison, J. 1975. Pollen-wall proteins: physicochemical characterization and role in self-incompatibility in *Cosmos bipinnatus*. *Proc. Roy. Soc. Lond.* 188B:167–182.

Hsing, Y.-I.C., Chen, Z.-Y., and Chow, T.-Y. 1992. Nucleotide sequences of a soybean complementary DNA encoding a 50-kilodalton late embryogenesis abundant protein. *Plant Physiol.* 99:354–355.

Hsing, Y.-I.C., and Wu, S.-J. 1992. Cloning and characterization of cDNA clones encoding soybean seed maturation polypeptides. *Bot. Bull. Acad. Sinica* 33:191–199.

Hsing, Y.-I.C., Rinne, R.W., Hepburn, A.G., and Zielinski, R.E. 1990. Expression of maturation-specific genes in soybean seeds. *Crop Sci.* 30:1343–1350.

Hsu, F.C. 1979. Abscisic acid accumulation in developing seeds of *Phaseolus vulgaris* L. *Plant Physiol.* 63:552–556.

Hu, N.-T., Peifer, M.A., Heidecker, G., Messing, J., and Rubenstein, I. 1982. Primary structure of a genomic zein sequence of maize. *EMBO J.* 1:1337–1342.

Hu, S.Y. 1964. Morphological and cytological observations on the process of fertilization of the wheat. *Sci. Sinica* 13:925–936.

Hu, S.Y., and Chu, Z.C. 1964. A histochemical study of DNA in sexual cells of some angiosperms. *Acta Bot. Sinica* 12:333–345.

Hu, S.-Y., Li, L.-G., and Zhou, C. 1985. Isolation of viable embryo sacs and their protoplasts of *Nicotiana tabacum*. *Acta Bot. Sinica* 27:337–344.

Hu, S.-Y., Wang, Y., and Yuan, Z.-F. 1993. The regional aggregation of plastids and mitochondria in microsporocyte and microspore of *Gossypium hirsutum*. *Chinese J. Bot.* 5:11–17.

Hu, S.-Y., and Yu, H.-S. 1988. Preliminary observations on the formation of the male germ unit in pollen tubes of *Cyphomandra betacea* Sendt. *Protoplasma* 147:55–63.

Hu, S.-Y., and Zhu, C. 1979. The fusion of male and female nuclei in fertilization of higher plants. *Acta Bot. Sinica* 21:1–10.

Hu, S.-Y., Zhu, C., and Xu, L.-Y. 1981. Ultrastructure of male gametophyte in wheat. II. Formation and development of sperm cell. *Acta Bot. Sinica* 23:85–91.

Hu, S.-Y., Zhu, C., and Zee, S.Y. 1983. Transfer cells in suspensor and endosperm during early embryogeny of *Vigna sinensis*. *Acta Bot. Sinica* 25:1–7.

Hu, S.-Y., Zhu, C., Xu, L.-Y., Li, X.-R., and Shen, J.-H. 1979. Ultrastructure of male gametophyte in wheat. I. The formation of generative and vegetative cells. *Acta Bot. Sinica* 21:208–214.

Huala, E., and Sussex, I.M. 1992. *LEAFY* interacts with floral homeotic genes to regulate *Arabidopsis* floral development. *Plant Cell* 4:901–913.

Huang, B. 1986. Ultrastructural aspects of pollen embryogenesis in *Hordeum*, *Triticum* and *Paeonia*. In *Haploids in Higher Plants*, ed. H. Hu and H. Yang, pp. 91–117. Beijing: China Academic Publishers.

Huang, B. 1992. Genetic manipulation of microspores and microspore-derived embryos. *In Vitro Cell Devel Biol.* 28P:53–58.

Huang, B., and Sunderland, N. 1982. Temperature-stress pretreatment in barley anther culture. *Ann. Bot.* 49:77–88.

Huang, B.-Q., and Russell, S.D. 1990. Isolation of fixed and viable eggs, central cells, and embryo sacs from ovules of *Plumbago zeylanica*. *Plant Physiol.* 90:9–12.

Huang, B.Q., and Russell, S.D. 1992. Synergid degeneration in *Nicotiana*: a quantitative, fluorochromatic and chlorotetracycline study. *Sex. Plant Reprod.* 5:151–155.

Huang, B.Q., and Russell, S.D. 1993. Polarity of nuclear and organellar DNA during megasporogenesis and megagametogenesis in *Plumbago zeylanica*. *Sex. Plant Reprod.* 6:205–211.

Huang, B.-Q., and Russell, S.D. 1994. Fertilization in *Nicotiana tabacum*: cytoskeletal modifications in the embryo sac during synergid degeneration. A hypothesis for short-distance transport of sperm cells prior to gamete fusion. *Planta* 194:200–214.

Huang, B.-Q., and Sheridan, W.F. 1994. Female gametophyte development in maize: microtubular organization and embryo sac polarity. *Plant Cell* 6:845–861.

Huang, B.-Q., Strout, G.W., and Russell, S.D. 1993. Fertilization in *Nicotiana tabacum*: ultrastructural organization of propane-jet-frozen embryo sacs *in vivo*. *Planta* 191:256–264.

Huang, B.-Q., Pierson, E.S., Russell, S.D., Tiezzi, A., and Cresti, M. 1992. Video microscopic observations of living, isolated embryo sacs of *Nicotiana* and their component cells. *Sex. Plant Reprod.* 5:156–162.

Huang, B.-Q., Pierson, E.S., Russell, S.D., Tiezzi, A., and Cresti, M. 1993. Cytoskeletal organisation and modification during pollen tube arrival, gamete delivery and fertilisation in *Plumbago zeylanica*. *Zygote* 1:143–154.

Huang, B.-Q., Russell, S.D., Strout, G.W., and Mao, L.-J. 1990. Organization of isolated embryo sacs and eggs of *Plumbago zeylanica* (Plumbaginaceae) before and after fertilization. *Am. J. Bot.* 77:1401–1410.

Huang, H., Mizukami, Y., Hu, Y., and Ma, H. 1993. Isolation and characterization of the binding sequences for the product of the *Arabidopsis* floral homeotic gene *AGAMOUS*. *Nucl. Acids Res.* 21:4769–4776.

Huang, H., Tudor, M., Weiss, C.A., and Ma, H. 1995. The *Arabidopsis* MADS-box gene *AGL3* is widely expressed and encodes a sequence-specific DNA-binding protein. *Plant Mol. Biol.* 28:549–567.

Huang, J., Lee, S.-H., Lin, C., Medici, R., Hack, E., and Myers, A.M. 1990. Expression in yeast of the T-URF13 protein from Texas male-sterile maize mitochondria confers sensitivity to methomyl and to Texas-cytoplasm-specific fungal toxins. *EMBO J.* 9:339–347.

Huang, S., Lee, H.-S., Karunanandaa, B., and Kao, T.-H. 1994. Ribonuclease activity of *Petunia inflata* S proteins is essential for rejection of self-pollen. *Plant Cell* 6:1021–1028.

Hudák, J., Walles, B., and Vennigerholz, F. 1993. The transmitting tissue in *Brugmansia suaveolens* L.: ultrastructure of the stylar transmitting tissue. *Ann. Bot.* 71:177–186.

Hueros, G., Varotto, S., Salamini, F., and Thompson, R.D. 1995. Molecular characterization of *BET1*, a gene expressed in the endosperm transfer cells of maize. *Plant Cell* 7:747–757.

Hughes, D.W., and Galau, G.A. 1987. Translation efficiency of *Lea* mRNAs in cotton embryos: minor changes during embryogenesis and germination. *Plant Mol. Biol.* 9:301–313.

Hughes, D.W., and Galau, G.A. 1989. Temporally modular gene expression during cotyledon development. *Genes Devel.* 3:358–369.

Hughes, D.W., and Galau, G.A. 1991. Developmental and environmental induction of *Lea* and *LeaA* mRNAs and the postabscission program during embryo culture. *Plant Cell* 3:605–618.

Hughes, D.W., Wang, H.Y.-C., and Galau, G.A. 1993. Cotton (*Gossypium hirsutum*) *MatP6* and *MtP7* oleosin genes. *Plant Physiol.* 101:697–698.

Hull, G.A., Bies, N., Twell, D., and Delseny, M. 1996. Analysis of the promoter of an abscisic acid responsive late embryogenesis abundant gene of *Arabidopsis thaliana. Plant Sci.* 114:181–192.

Hull, G.A., Halford, N.G., Kreis, M., and Shewry, P.R. 1991. Isolation and characterisation of genes encoding rye prolamins containing a highly repetitive sequence motif. *Plant Mol. Biol.* 17:1111–1115.

Hülskamp, M., Schneitz, K., and Pruitt, R.E. 1995. Genetic evidence for a long-range activity that directs pollen tube guidance in *Arabidopsis. Plant Cell* 7:57–64.

Hülskamp, M., Kopczak, S.D., Horejsi, T.F., Kihl, B.K., and Pruitt, R.E. 1995. Identification of genes required for pollen–stigma recognition in *Arabidopsis thaliana. Plant J.* 8:703–714.

Hunold, R., Burrus, M., Bronner, R., Duret, J.-P., and Hahne, G. 1995. Transient gene expression in sunflower (*Helianthus annuus* L.) following microprojectile bombardment. *Plant Sci.* 105:95–109.

Hurkman, W.J., Smith, L.D., Richter, J., and Larkins, B.A. 1981. Subcellular compartmentalization of maize storage proteins in *Xenopus* oocytes injected with zein messenger RNAs. *J. Cell Biol.* 89:292–299.

Hussey, P.J., Lloyd, C.W., and Gull, K. 1988. Differential and developmental expression of β-tubulins in a higher plant. *J. Biol. Chem.* 263:5474–5479.

Huxley, J. 1955. Morphism and evolution. *Heredity* 9:1–52.

Idzikowska, K. 1981. Mitochondria during androgenesis in *Hordeum vulgare. Acta Soc. Bot. Polon.* 50:359–366.

Idzikowska, K., Ponitka, A., and Młodzianowski, F. 1982. Pollen dimorphism and androgenesis in *Hordeum vulgare. Acta Soc. Bot. Polon.* 51:153–156.

Iglesias, V.A., Gisel, A., Bilang, R., Ledue, N., Potrykus, I., and Sautter, C. 1994. Transient expression of visible marker genes in meristem cells of wheat embryos after ballistic micro-targeting. *Planta* 192:84–91.

Ihle, J.N., and Dure, L., III. 1970. Hormonal regulation of translation inhibition requiring RNA synthesis. *Biochem. Biophys. Res. Commun.* 38:995–1001.

Ihle, J.N., and Dure, L.S., III. 1972. The developmental biochemistry of cottonseed embryogenesis and germination. III. Regulation of the biosynthesis of enzymes utilized in germination. *J. Biol. Chem.* 247:5048–5055.

Iida, A., Nagasawa, A., and Oeda, K. 1995. Positive and negative *cis*-regulatory regions in the soybean glycinin promoter identified by quantitative transient gene expression. *Plant Cell Rep.* 14:539–544.

Ioerger, T.R., Clark, A.G., and Kao, T.-H. 1990. Polymorphism at the self-incompatibility locus in Solanaceae predates speciation. *Proc. Natl Acad. Sci. USA* 87:9732–9735.

Ioerger, T.R., Gohlke, J.R., Xu, B., and Kao, T.-H. 1991. Primary structural features of the self-incompatibility protein in Solanaceae. *Sex. Plant Reprod.* 4:81–87.

Irish, E.E., Langdale, J.A., and Nelson, T.M. 1994. Interactions between *Tassel seed* genes and other sex determining genes in maize. *Devel Genet.* 15:155–171.

Irish, E.E., and Nelson, T.M. 1993. Development of tassel seed 2 inflorescences in maize. *Am. J. Bot.* 80:292–299.

Irish, V.F., and Sussex, I.M. 1990. Function of the *apetala-1* gene during *Arabidopsis* floral development. *Plant Cell* 2:741–753.

Isaac, P.G., Jones, V.P., and Leaver, C.J. 1985. The maize cytochrome *c* oxidase subunit I gene: sequence, expression and rearrangement in cytoplasmic male sterile plants. *EMBO J.* 4:1617–1623.

Ishibashi, N., and Minamikawa, T. 1990. Molecular cloning and characterization of stored mRNA in cotyledons of *Vigna unguiculata. Plant Cell Physiol.* 31:39–44.

Ishibashi, N., Yamauchi, D., and Minamikawa, T. 1990. Stored mRNA in cotyledons of *Vigna unguiculata* seeds: nucleotide sequence of cloned cDNA for a stored mRNA and induction of its synthesis by precocious germination. *Plant Mol. Biol.* 15:59–64.

Ishida, Y., Saito, H., Ohta, S., Hiei, Y., Komari, T., and Kumashiro, T. 1996. High efficiency transformation of maize (*Zea mays* L.) mediated by *Agrobacterium tumefaciens. Nature Biotech.* 14:745–750.

Isogai, A., Takayama, S., Shiozawa, H., Tsukamoto, C., Kanbara, T., Hinata, K., Okazaki, K., and Suzuki, A. 1988. Existence of a common glycoprotein homologous to *S*-glycoproteins in two self-incompatible homozygotes of *Brassica campestris. Plant Cell Physiol.* 29:1331–1336.

Isogai, A., Takayama, S., Tsukamoto, C., Ueda, Y., Shiozawa, H., Hinata, K., Okazaki, K., and Suzuki, A. 1987. *S*-locus-specific glycoproteins associated with self-incompatibility in *Brassica campestris. Plant Cell Physiol.* 28:1279–1291.

Isogai, A., Yamakawa, S., Shiozawa, H., Takayama, S., Tanaka, H., Kono, T., Watanabe, M., Hinata, K., and Suzuki, A. 1991. The cDNA sequence of NS_1 glycoprotein of *Brassica campestris* and its homology to S-locus-related glycoproteins of *B. oleracea. Plant Mol. Biol.* 17:269–271.

Israel, H.W., and Sagawa, Y. 1964. Post-pollination ovule development in *Dendrobium* orchids. II. Fine structure of the nucellar and archesporial phases. *Caryologia* 17:301–316.

Ito, M. 1973. Studies on the behavior of meiotic protoplasts. I. Isolation from microsporocytes of liliaceous plants. *Bot. Mag. Tokyo* 86:133–141.

Ito, M., and Hotta, Y. 1973. Radioautography of incorporated ^{3}H-thymidine and its metabolism during meiotic prophase in microsporocytes of *Lilium. Chromosoma* 43:391–398.

Ito, M., Hotta, Y., and Stern, H. 1967. Studies of meiosis *in vitro*. II. Effect of inhibiting DNA synthesis during meiotic prophase on chromosome structure and behavior. *Devel Biol.* 16:54–77.

Ito, M., and Maeda, M. 1973. Fusion of meiotic protoplasts in liliaceous plants. *Expt. Cell Res.* 80:453–456.

Ito, M., and Stern, H. 1967. Studies of meiosis *in vitro*. I. *In vitro* culture of meiotic cells. *Devel Biol.* 16:36–53.

Ito, M., and Takegami, M.H. 1982. Commitment of mitotic cells to meiosis during the G_2 phase of premeiosis. *Plant Cell Physiol.* 23:943–952.

Itoh, Y., Kitamura, Y., Arahira, M., and Fukazawa, C. 1993. *Cis*-acting regulatory regions of the soybean seed storage 11S globulin gene and their interactions with seed embryo factors. *Plant Mol. Biol.* 21:973–984.

Iwabuchi, M., Kyozuka, J., and Shimamoto, K. 1993. Processing followed by complete editing of an altered mitochondrial *atp6* RNA restores fertility of cytoplasmic male sterile rice. *EMBO J.* 12:1437–1446.

Iwanami, Y. 1953. Physiological researches of pollen. (V). On the inductive tissue and the growth of the pollen tube in the style. *Bot. Mag. Tokyo* 66:189–196.

Iwanami, Y. 1956. Protoplasmic movement in pollen grains and tubes. *Phytomorphology* 6:288–295.

Iwanami, Y. 1959. Physiological studies of pollen. *J. Yokohama Municip. Univ.* (C34 Biol. 13) 116:1–137.

Iyer, R.D., and Raina, S.K. 1972. The early ontogeny of embryoids and callus from pollen and subsequent organogenesis in anther cultures of *Datura metel* and rice. *Planta* 104:146–156.

Izawa, T., Foster, R., Nakajima, M., Shimamoto, K., and Chua, N.-H. 1994. The rice bZIP transcriptional activator RITA-1 is highly expressed during seed development. *Plant Cell* 6:1277–1287.

Izhar, S. 1975. The timing of temperature effect on microsporogenesis in cytoplasmic male-sterile *Petunia*. *J. Hered.* 66:313–314.

Izhar, S. 1978. Cytoplasmic male sterility in *Petunia*. III. Genetic control of microsporogenesis and male fertility restoration. *J. Hered.* 69:22–26.

Izhar, S., and Frankel, R. 1971. Mechanism of male sterility in *Petunia*: the relationship between pH, callase activity in the anthers, and the breakdown of the microsporogenesis. *Theor. Appl. Genet.* 41:104–108.

Izhar, S., and Frankel, R. 1973a. Mechanism of male sterility in *Petunia*. II. Free amino acids in male fertile and male sterile anthers during microsporogenesis. *Theor. Appl. Genet.* 43:13–17.

Izhar, S., and Frankel, R. 1973b. Duration of meiosis in *Petunia* anthers *in vivo* and in floral bud culture. *Acta Bot. Neerl.* 22:14–22.

Izhar, S., Schlicter, M., and Swartzberg, D. 1983. Sorting out of cytoplasmic elements in somatic hybrids of *Petunia* and the prevalence of the heteroplasmon through several meiotic cycles. *Mol. Gen. Genet.* 190:468–474.

Jack, T., Brockman, L.L., and Meyerowitz, E.M. 1992. The homeotic gene *APETALA3* of *Arabidopsis thaliana* encodes a MADS box and is expressed in petals and stamens. *Cell* 68:683–697.

Jack, T., Fox, G.L., and Meyerowitz, E.M. 1994. *Arabidopsis* homeotic gene *APETALA3* ectopic expression: transcriptional and posttranscriptional regulation determine floral organ identity. *Cell* 76:703–716.

Jackson, J.F., Jones, G., and Linskens, H.F. 1982. Phytic acid in pollen. *Phytochemistry* 21:1255-1258.

Jackson, J.F., Kamboj, R.K., and Linskens, H.F. 1983. Localization of phytic acid in the floral structure of *Petunia hybrida* and relation to the incompatibility genes. *Theor. Appl. Genet.* 64:259–262.

Jackson, J.F., and Linskens, H.F. 1982. Phytic acid in *Petunia hybrida* pollen is hydrolysed during germination by a phytase. *Acta Bot. Neerl.* 31:441–447.

Jacobsen, J.V., Knox, R.B., and Pyliotis, N.A. 1971. The structure and composition of aleurone grains in the barley aleurone layer. *Planta* 101:189–209.

Jacobsen, S.E., and Olszewski, N.E. 1991. Characterization of the arrest in anther development associated with gibberellin deficiency of the *gib-1* mutant of tomato. *Plant Physiol.* 97:409–414.

Jacobsen, S.E., Shi, L., Xin, Z., and Olszewski, N.E. 1994. Gibberellin-induced changes in the translatable mRNA populations of stamens and shoots of gibberellin-deficient tomato. *Planta* 192:372–378.

Jaffe, L.A., Weisenseel, M.H., and Jaffe, L.F. 1975. Calcium accumulations within the growing tips of pollen tubes. *J. Cell Biol.* 67:488–492.

Jähne, A., Becker, D., and Lörz, H. 1995. Genetic engineering of cereal crop plants: a review. *Euphytica* 85:35–44.

Jähne, A., Becker, D., Brettschneider, R., and Lörz, H. 1994. Regeneration of transgenic, microspore-derived fertile, barley. *Theor. Appl. Genet.* 89:525–533.

Jahnen, W., Lush, W.M., and Clarke, A.E. 1989. Inhibition of *in vitro* pollen tube growth by isolated *S*-glycoproteins of *Nicotiana alata*. *Plant Cell* 1:501–510.

Jahnen, W., Batterham, M.P., Clarke, A.E., Moritz, R.L., and Simpson, R.J. 1989. Identification, isolation, and N-terminal sequencing of style glycoproteins associated with self-incompatibility in *Nicotiana alata*. *Plant Cell* 1:493–499.

Jain, S.M., Gupta, P.K., and Newton, R.J., ed. 1995. *Somatic Embryogenesis in Woody Plants, Vol. 3. Gymnosperms.* Dordrecht: Kluwer Academic Publishers.

Jakobsen, K., Klemsdal, S.S., Aalen, R.B., Bosnes, M., Alexander, D., and Olsen, O.-A. 1989. Barley aleurone cell development: molecular cloning of aleurone-specific cDNAs from immature grains. *Plant Mol. Biol.* 12:285–293.

Jakobsen, K.S., Hughes, D.W., and Galau, G.A. 1994. Simultaneous induction of postabscission and germination mRNAs in cultured dicotyledonous embryos. *Planta* 192:384–394.

Jalouzot, M.-F. 1971. Aspects ultrastructuraux de la megasporogenese d'*Oenothera lamarckiana* en rapport avec les depots callosiques observes. *Ann. Univ. A.R.E.R.S.* 9:36–45.

Jalouzot, M.-F. 1975. Aspects ultrastructuraux du sac embryonnaire d'*Oenothera lamarckiana*. *Compt. Rend. Acad. Sci. Paris* 281D:1305–1308.

Jalouzot, R. 1969a. Differenciation nucleaire et cytoplasmique du grain de pollen de *Lilium candidum*. *Expt. Cell Res.* 55:1–8.

Jalouzot, R. 1969b. Aspects cytochimiques des deux cellules du grain de pollen de *Crocus longiflorus*. *Rev. Cytol. Biol. Végét.* 32:115–120.

James, M.G., Robertson, D.S., and Myers, A.M. 1995. Characterization of the maize gene *sugary1*, a determinant of starch composition in kernels. *Plant Cell* 7:417–429.

Janaki-Ammal, E.K. 1941. The breakdown of meiosis in a male-sterile *Saccharum*. *Ann. Bot.* 5:83–87.

Jansen, M.A.K., Booij, H., Schel, J.H.N., and de Vries, S.C. 1990. Calcium increases the yield of somatic embryos in carrot embryogenic suspension cultures. *Plant Cell Rep.* 9:221–223.

Janska, H., and Mackenzie, S.A. 1993. Unusual mitochondrial genome organization in cytoplasmic male sterile common bean and the nature of cytoplasmic reversion to fertility. *Genetics* 135:869–879.

Janson, J. 1993. Placental pollination in *Lilium longiflorum* Thunb. *Plant Sci.* 90:105–115.

Janson, J., and Willemse, M.T.M. 1995. Pollen tube penetration and fertilization in *Lilium longiflorum* (Liliaceae). *Am. J. Bot.* 82:186–196.

Jardinaud, M.-F., Souvré, A., and Alibert, G. 1993. Transient GUS gene expression in *Brassica napus* electroporated microspores. *Plant Sci.* 93:177–184.

Jardinaud, M.F., Souvré, A., Alibert, G., and Beckert, M. 1995. *uidA* gene transfer and expression in maize microspores using the biolistic method. *Protoplasma* 187:138–143.

Jarvi, A.J., and Eslick, R.F. 1975. Shrunken endosperm mutants in barley. *Crop Sci.* 15:363–366.

Jauh, G.Y., and Lord, E.M. 1995. Movement of the tube cell in the lily style in the absence of the pollen grain and the spent pollen tube. *Sex. Plant Reprod.* 8:168–172.

Jauh, G.Y., and Lord, E.M. 1996. Localization of pectins and arabinogalactan-proteins in lily (*Lilium longiflorum* L.) pollen tube and style, and their possible roles in pollination. *Planta* 199:251–261.

Jayakaran, M. 1972. Suppression of stamens in *Capsicum annuum* by a morphactin (EMD-IT 7839). *Curr. Sci.* 41:849–850.

Jegla, D.E., and Sussex, I.M. 1989. Cell lineage patterns in the shoot meristem of the sunflower embryo in the dry seed. *Devel Biol.* 131:215–225.

Jensen, W.A. 1963. Cell development during plant embryogenesis. In *Meristems and Differentiation. Brookhaven Symp. Biol.* 16:179–202.

Jensen, W.A. 1964. Observations on the fusion of nuclei in plants. *J. Cell Biol.* 23:669–672.

Jensen, W.A. 1965a. The composition and ultrastructure of the nucellus in cotton. *J. Ultrastr. Res.* 13:112–128.

Jensen, W.A. 1965b. The ultrastructure and histochemistry of the synergids of cotton. *Am. J. Bot.* 52:238–256.

Jensen, W.A. 1965c. The ultrastructure and composition of the egg and central cell of cotton. *Am. J. Bot.* 52:781–797.

Jensen, W.A. 1968. Cotton embryogenesis: the zygote. *Planta* 79:346–366.

Jensen, W.A. 1973. Fertilization in flowering plants. *BioScience* 23:21–27.

Jensen, W.A., Ashton, M., and Heckard, L.R. 1974. Ultrastructural studies of the pollen of subtribe Castilleiinae, family Scrophulariaceae. *Bot. Gaz.* 135:210–218.

Jensen, W.A., Ashton, M.E., and Beasley, C.A. 1983. Pollen tube-embryo sac interactions in cotton. In *Pollen: Biology and Implications for Plant Breeding*, ed. D.L. Mulcahy and E. Ottaviano, pp. 67–72. New York: Elsevier.

Jensen, W.A., and Fisher, D.B. 1967. Cotton embryogenesis: double fertilization. *Phytomorphology* 17:261–269.

Jensen, W.A., and Fisher, D.B. 1968a. Cotton embryogenesis: the entrance and discharge of the pollen tube in the embryo sac. *Planta* 78:158–183.

Jensen, W.A., and Fisher, D.B. 1968b. Cotton embryogenesis: the sperm. *Protoplasma* 65:277–286.

Jensen, W.A., and Fisher, D.B. 1969. Cotton embryogenesis: the tissues of the stigma and style and their relation to the pollen tube. *Planta* 84:97–121.

Jensen, W.A., and Fisher, D.B. 1970. Cotton embryogenesis: the pollen tube in the stigma and style. *Protoplasma* 69:215–235.

Jensen, W.A., Fisher, D.B., and Ashton, M.E. 1968. Cotton embryogenesis: the pollen cytoplasm. *Planta* 81:206–228.

Jensen, W.A., Schulz, P., and Ashton, M.E. 1977. An ultrastructural study of early endosperm development and synergid changes in unfertilized cotton ovules. *Planta* 133:179–189.

Ji, L.-H., and Langridge, P. 1994. An early meiosis cDNA clone from wheat. *Mol. Gen. Genet.* 243:17–23.

Jiang, L., Abrams, S.R., and Kermode, A.R. 1996. Vicilin and napin storage-protein gene promoters are responsive to abscisic acid in developing transgenic tobacco seed but lose sensitivity following premature desiccation. *Plant Physiol.* 110:1135–1144.

Jobson, S., Knox, R.B., Kenrick, J., and Dumas, C. 1983. Plastid development and ferritin content of stigmas of the legumes *Acacia, Lotus* and *Trifolium. Protoplasma* 116:213–218.

Joersbo, M., Andersen, J.M., Okkels, F.T., and Rajagopal, R. 1989. Isoperoxidases as markers of somatic embryogenesis in carrot cell suspension cultures. *Physiol. Plant.* 76:10–16.

Jofuku, K.D., and Goldberg, R.B. 1989. Kunitz trypsin inhibitor genes are differentially expressed during the soybean life cycle and in transformed tobacco plants. *Plant Cell* 1:1079–1093.

Jofuku, K.D., Okamuro, J.K., and Goldberg, R.B. 1987. Interaction of an embryo DNA binding protein with a soybean lectin gene upstream region. *Nature* 328:734–737.

Jofuku, K.D., Schipper, R.D., and Goldberg, R.B. 1989. A frameshift mutation prevents Kunitz trypsin inhibitor mRNA accumulation in soybean embryos. *Plant Cell* 1:427–435.

Jofuku, K.D., den Boer, B.G.W., van Montagu, M., and Okamuro, J.K. 1994. Control of *Arabidopsis* flower and seed development by the homeotic gene *APETALA2*. *Plant Cell* 6:1211–1225.

Johansen, D.A. 1950. *Plant Embryology. Embryogeny of the Spermatophyta.* Waltham: Chronica Botanica Co.

Johansson, L. 1983. Effects of activated charcoal in anther cultures. *Physiol. Plant.* 59:397–403.

Johansson, M., and Walles, B. 1993a. Functional anatomy of the ovule in broad bean (*Vicia faba* L.). I. Histogenesis prior to and after pollination. *Int. J. Plant Sci.* 154:80–89.

Johansson, M., and Walles, B. 1993b. Functional anatomy of the ovule in broad bean, *Vicia faba* L. II. Ultrastructural development up to early embryogenesis. *Int. J. Plant Sci.* 154:535–549.

Johansson, M., and Walles, B. 1994. Functional anatomy of the ovule in broad bean (*Vicia faba* L.): ultrastructural seed development and nutrient pathways. *Ann. Bot.* 74:233–244.

Johari, R.P., Mehta, S.L., and Naik, M.S. 1977. Protein synthesis and changes in nucleic acids during grain development of *Sorghum. Phytochemistry* 16:19–24.

John, M.E., and Petersen, M.W. 1994. Cotton (*Gossypium hirsutum* L.) pollen-specific polygalacturonase mRNA: tissue and temporal specificity of its promoter in transgenic tobacco. *Plant Mol. Biol.* 26:1989–1993.

Johns, C., Lu, M., Lyznik, A., and Mackenzie, S. 1992. A mitochondrial DNA sequence is associated with abnormal pollen development in cytoplasmic male sterile bean plants. *Plant Cell* 4:435–449.

Johns, C., Nickels, R., McIntosh, L., and Mackenzie, S. 1993. The expression of alternative oxidase and alternative respiratory capacity in cytoplasmic male sterile common bean. *Sex. Plant Reprod.* 6:257–265.

Johnson, C.M., Mulcahy, D.L., and Galinat, W.C. 1976. Male gametophyte in maize: influences of the gametophytic genotype. *Theor. Appl. Genet.* 48:299–303.

Johnson, E.D., Knight, J., and Gayler, K.R. 1985. Biosynthesis and processing of legumin-like storage proteins in *Lupinus angustifolius* (lupin). *Biochem. J.* 232:673–679.

Johnson, K.A., and Sussex, I.M. 1990. Genomic amplification in the cotyledon parenchyma of common bean. *Chromosoma* 99:223–230.

Johnson, L.E.B., Wilcoxson, R.D., and Frosheiser, F.I. 1975. Transfer cells in tissues of the reproductive system of alfalfa. *Can. J. Bot.* 53:952–956.

Johnson, S., Liu, C.-M., Hedley, C.L., and Wang, T.L. 1994. An analysis of seed development in *Pisum sativum*. XVIII. The isolation of mutants defective in embryo development. *J. Expt. Bot.* 45:1503–1511.

Johnson-Flanagan, A.M., Huiwen, Z., Geng, X.-M., Brown, D.C.W., Nykiforuk, C.L., and Singh, J. 1992. Frost, abscisic acid, and desiccation hasten embryo development in *Brassica napus*. *Plant Physiol.* 99:700–706.

Johnsson, H. 1944. Meiotic aberrations and sterility in *Alopecurus myosuroides* Huds. *Hereditas* 30:469–565.

Johnston, S.A., and Hanneman, R.E., Jr. 1982. Manipulations of endosperm balance number overcome crossing barriers between diploid *Solanum* species. *Science* 217:446–448.

Johnston, S.A., den Nijs, T.P.M., Peloquin, S.J., and Hanneman, R.E., Jr. 1980. The significance of genic balance to endosperm development in interspecific crosses. *Theor. Appl. Genet.* 57:5–9.

Johri, B.M., Ambegaokar, K.B., and Srivastava, P.S. 1992. *Comparative Embryology of Angiosperms*, Vols. 1 & 2. Berlin: Springer-Verlag.

Johri, B.M., and Vasil, I.K. 1961. Physiology of pollen. *Bot. Rev.* 27:325–381.

Johri, M.M., and Coe, E.H., Jr. 1983. Clonal analysis of corn plant development. I. The development of the tassel and the ear shoot. *Devel Biol.* 97:154–172.

Jones, D.F. 1920. Heritable characters of maize. IV. A lethal factor-defective seeds. *J. Hered.* 11:160–167.

Jones, H.A. 1927. Pollination and life history studies of lettuce (*Lactuca sativa* L.). *Hilgardia* 2:425–479.

Jones, P.A. 1977. Development of the quiescent center in maturing embryonic radicles of pea (*Pisum sativum* L. cv. Alaska). *Planta* 135:233–240.

Jones, R.A. 1978. Effects of *floury-2* locus on zein accumulation and RNA metabolism during maize endosperm development. *Biochem. Genet.* 16:27–38.

Jones, R.A., Larkins, B.A., and Tsai, C.Y. 1977. Storage protein synthesis in maize. III. Developmental changes in membrane-bound polyribosome composition and *in vitro* protein synthesis of normal and *opaque*-2 maize. *Plant Physiol.* 59:733–737.

Jones, R.L. 1969. The fine structure of barely aleurone cells. *Planta* 85:359–375.

Jones, T.J., and Rost, T.L. 1989a. The developmental anatomy and ultrastructure of somatic embryos from rice (*Oryza sativa* L.) scutellum epithelial cells. *Bot. Gaz.* 150:41–49.

Jones, T.J., and Rost, T.L. 1989b. Histochemistry and ultrastructure of rice (*Oryza sativa*) zygotic embryogenesis. *Am. J. Bot.* 76:504–520.

Jones-Villeneuve, E., Huang, B., Prudhomme, I., Bird, S., Kemble, R., Hattori, J., and Miki, B. 1995. Assessment of microinjection for introducing DNA into uninuclear microspores of rapeseed. *Plant Cell Tissue Organ Cult.* 40:97–100.

Joos, U., van Aken, J., and Kristen, U. 1994. Microtubules are involved in maintaining the cellular polarity in pollen tubes of *Nicotiana sylvestris*. *Protoplasma* 179:5–15.

Joos, U., van Aken, J., and Kristen, U. 1995. The anti-microtubule drug carbetamide stops *Nicotiana sylvestris* pollen tube growth in the style. *Protoplasma* 187:182–191.

Joppa, L.R., McNeal, F.H., and Welsh, J.R. 1966. Pollen and anther development in cytoplasmic male sterile wheat (*Triticum aestivum* L.). *Crop Sci.* 6:296–297.

Jordaan, A., and Kruger, H. 1993. Pollen wall ontogeny of *Felicia muricata* (Asteraceae: Astereae). *Ann. Bot.* 71:97–105.

Jordano, J., Almoguera, C., and Thomas, T.L. 1989. A sunflower helianthinin gene upstream sequence ensemble contains an enhancer and sites of nuclear protein interaction. *Plant Cell* 1:855–866.

Josè-Estanyol, M., Ruiz-Avila, L., and Puigdomènech, P. 1992. A maize embryo-specific gene encodes a proline-rich and hydrophobic protein. *Plant Cell* 4:413–423.

Josefsson, L.-G., Lenman, M., Ericson, M.L., and Rask, L. 1987. Structure of a gene encoding the 1.7 S storage protein, napin, from *Brassica napus*. *J. Biol. Chem.* 262:12196–12201.

Juliano, J.B., and Aldama, M.J. 1937. Morphology of *Oryza sativa* Linnaeus. *Philipp. Agr.* 26:1–76.

Juncosa, A.M. 1982. Developmental morphology of the embryo and seedling of *Rhizophora mangle* L. (Rhizophoraceae). *Am. J. Bot.* 69:1599–1611.

Jürgens, G. 1995. Axis formation in plant embryogenesis: cues and clues. *Cell* 81:467–470.

Jürgens, G., Mayer, U., Torres Ruiz, R.A., Berleth, T., and Miséra, S. 1991. Genetic analysis of pattern formation in the *Arabidopsis* embryo. *Development* Suppl. 1:27–38.

Jürgens, G., Torres Ruiz, R.A., and Berleth, T. 1994. Embryonic pattern formation in flowering plants. *Annu. Rev. Genet.* 28:351–371.

Kadej, A.J., Wilms, H.J., and Willemse, M.T.M. 1985. Stigma and stigmatoid tissue of *Lycopersicon esculentum* Mil. *Acta Bot. Neerl.* 34:95–103.

Kadowaki, K.-I., Suzuki, T., and Kazama, S. 1990. A chimeric gene containing 5′ portion of *atp6* is associated with cytoplasmic male-sterility of rice. *Mol. Gen. Genet.* 224:10–16.

Kahn, T.L., and DeMason, D.A. 1986. A quantitative and structural comparison of *Citrus* pollen tube development in cross-compatible and self-incompatible gynoecia. *Can. J. Bot.* 64:2548–2555.

Kalinski, A., Loer, D.S., Weisemann, J.M., Matthews, B.F., and Herman, E.M. 1991. Isoforms of soybean seed oil body membrane protein 24 kDa oleosin are encoded by closely related cDNAs. *Plant Mol. Biol.* 17:1095–1098.

Kalla, R., Shimamoto, K., Potter, R., Nielsen, P.S., Linnestad, C., and Olsen, O.-A. 1994. The promoter of the barley aleurone-specific gene encoding a putative 7 kDa lipid transfer protein confers aleurone cell-specific expression in transgenic rice. *Plant J.* 6:849–860.

Kaltsikes, P.J. 1973. Early seed development in hexaploid triticale. *Can. J. Bot.* 51:2291–2300.

Kamada, H., and Harada, H. 1979a. Studies on the organogenesis in carrot tissue cultures. I. Effects of growth regulators on somatic embryogenesis and root formation. *Z. Pflanzenphysiol.* 91:255–266.

Kamada, H., and Harada, H. 1979b. Studies on the organogenesis in carrot tissue cultures. II. Effects of amino acids and inorganic nitrogenous compounds on somatic embryogenesis. *Z. Pflanzenphysiol.* 91:453–463.

Kamada, H., and Harada, H. 1981. Changes in the

endogenous level and effects of abscisic acid during somatic embryogenesis of *Daucus carota* L. *Plant Cell Physiol.* 22:1423–1429.

Kamada, H., and Harada, H. 1984a. Studies on nitrogen metabolism during somatic embryogenesis in carrot. I. Utilization of α-alanine as a nitrogen source. *Plant Sci. Lett.* 33:7–13.

Kamada, H., and Harada, H. 1984b. Changes in endogenous amino acid compositions during somatic embryogenesis in *Daucus carota* L. *Plant Cell Physiol.* 25:27–38.

Kamada, H., Kobayashi, K., Kiyosue, T., and Harada, H. 1989. Stress induced somatic embryogenesis in carrot and its application to synthetic seed production. *In Vitro Cell Devel Biol.* 25:1163–1166.

Kamalay, J.C., and Goldberg, R.B. 1980. Regulation of structural gene expression in tobacco. *Cell* 19:935–946.

Kamalay, J.C., and Goldberg, R.B. 1984. Organ-specific nuclear RNAs in tobacco. *Proc. Natl Acad. Sci. USA* 81:2801–2805.

Kamboj, R.K., and Jackson, J.F. 1986. Self-incompatibility alleles control a low molecular weight, basic protein in pistils of *Petunia hybrida. Theor. Appl. Genet.* 71:815–819.

Kamboj, R.K., Linskens, H.F., and Jackson, J.F. 1984. Energy-driven protein release from germinating pollen of *Petunia hybrida. Ann. Bot.* 54:647–652.

Kameya, T., Hinata, K., and Mizushima, U. 1966. Fertilization *in vitro* of excised ovules treated with calcium chloride in *Brassica oleracea* L. *Proc. Jap. Acad.* 42:165–167.

Kameya, T., and Uchimiya, H. 1972. Embryoids derived from isolated protoplasts of carrot. *Planta* 103:356–360.

Kaminski, P.A., Buffard, D., and Strosberg, A.D. 1987. The pea lectin gene family contains only one functional gene. *Plant Mol. Biol.* 9:497–507.

Kandasamy, M.K., Kappler, R., and Kristen, U. 1988. Plasmatubules in the pollen tubes of *Nicotiana sylvestris. Planta* 173:35–41.

Kandasamy, M.K., and Kristen, U. 1987a. Developmental aspects of ultrastructure, histochemistry and receptivity of the stigma of *Nicotiana sylvestris. Ann. Bot.* 60:427–437.

Kandasamy, M.K., and Kristen, U. 1987b. Pollen tube growth in the style of *Nicotiana sylvestris* is neither influenced by the ovary nor directed by a gradient. *J. Plant Physiol.* 131:495–500.

Kandasamy, M.K., and Kristen, U. 1989. Ultrastructural responses of tobacco pollen tubes to heat shock. *Protoplasma* 153:104–110.

Kandasamy, M.K., and Kristen, U. 1990. Developmental aspects of ultrastructure and histochemistry of the stylar transmitting tissue of *Nicotiana sylvestris. Bot. Acta* 103:384–391.

Kandasamy, M.K., Nasrallah, J.B., and Nasrallah, M.E. 1994. Pollen–pistil interactions and developmental regulation of pollen tube growth in *Arabidopsis. Development* 120:3405–3418.

Kandasamy, M.K., Parthasarathy, M.V., and Nasrallah, M.E. 1991. High pressure freezing and freeze substitution improve immunolabeling of S-locus specific glycoproteins in the stigma papillae of *Brassica. Protoplasma* 162:187–191.

Kandasamy, M.K., and Vivekanandan, M. 1983. Biochemical composition of stigmatic exudate of *Eichhornia crassipes* (Mart.) Solms. *Aquatic Bot.* 16:41–47.

Kandasamy, M.K., Dwyer, K.G., Paolillo, D.J., Doney, R.C., Nasrallah, J.B., and Nasrallah, M.E. 1990. *Brassica* S-proteins accumulate in the intercellular matrix along the path of pollen tubes in transgenic tobacco pistils. *Plant Cell* 2:39–49.

Kandasamy, M.K., Paolillo, D.J., Faraday, C.D., Nasrallah, J.B., and Nasrallah, M.E. 1989. The S-locus specific glycoproteins of *Brassica* accumulate in the cell wall of developing stigma papillae. *Devel Biol.* 134:462–472.

Kandasamy, M.K., Thorsness, M.K., Rundle, S.J., Goldberg, M.L., Nasrallah, J.B., and Nasrallah, M.E. 1993. Ablation of papillar cell function in *Brassica* flowers results in the loss of stigma receptivity to pollination. *Plant Cell* 5:263–275.

Kang, H.-G., Noh, Y.-S., Chung, Y.-Y., Costa, M.A., An, K., and An, G. 1995. Phenotypic alterations of petal and sepal by ectopic expression of a rice MADS box gene in tobacco. *Plant Mol. Biol.* 29:1–10.

Kanno, T., and Hinata, K. 1969. An electron microscopic study of the barrier against pollen-tube growth in self-incompatible Cruciferae. *Plant Cell Physiol.* 10:213–216.

Kanta, K. 1960. Intra-ovarian pollination in *Papaver rhoeas* L. *Nature* 188:683–684.

Kanta, K., and Maheshwari, P. 1963a. Intraovarian pollination in some Papaveraceae.*Phytomorphology* 13:215–230.

Kanta, K., and Maheshwari, P. 1963b. Test-tube fertilization in some angiosperms. *Phytomorphology* 13:230–237.

Kanta, K., Ranga Swamy, N.S., and Maheshwari, P. 1962. Test-tube fertilization in a flowering plant. *Nature* 194:1214–1217.

Kapil, R.N., and Bhatnagar, A.K. 1975. A fresh look at the process of double fertilization in angiosperms. *Phytomorphology* 25:334–368.

Kapil, R.N., and Bhatnagar, A.K. 1981. Ultrastructure and biology of female gametophyte in flowering plants. *Int. Rev. Cytol.* 70:291–341.

Kapil, R.N., and Tiwari, S.C. 1978a. Plant embryological investigations and fluorescence microscopy: an assessment of integration. *Int. Rev. Cytol.* 53:291–331.

Kapil, R.N., and Tiwari, S.C. 1978b. The integumentary tapetum. *Bot. Rev.* 44:457–490.

Kapil, R.N., and Vasil, I.K. 1963. Ovule. In *Recent Advances in the Embryology of Angiosperms*, ed. P. Maheshwari, pp. 41–67. Delhi: International Society of Plant Morphologists.

Kaplan, D.R. 1969. Seed development in *Downingia. Phytomorphology* 19:253–278.

Kapoor, M. 1959. Influence of growth substances on the ovules of *Zephyranthes. Phytomorphology* 9:313–315.

Kappler, R., Kristen, U., and Morré, D.J. 1986. Membrane flow in plants: fractionation of growing pollen tubes of tobacco by preparative free-flow electrophoresis and kinetics of labeling of endoplasmic reticulum and Golgi apparatus with [^{3}H]leucine. *Protoplasma* 132:38–50.

Kapros, T., Bögre, L., Németh, K., Bakó, L., Györgyey, J., Wu, S.C., and Dudits, D. 1992. Differential expression of histone H3 gene variants during cell cycle and somatic embryogenesis in alfalfa. *Plant Physiol.* 98:621–625.

Karas, I., and Cass, D.D. 1976. Ultrastructural aspects of sperm cell formation in rye: evidence for cell plate involvement in generative cell division. *Phytomorphology* 26:36–45.

Karim, M.A., Mehta, S.L., and Singh, M.P. 1984. Studies on esterase isoenzyme pattern in anthers and seeds of male sterile wheats. *Z. Pflanzenzüchtg* 93:309–319.

Karper, R.E., and Stephens, J.C. 1936. Floral abnormalities in sorghum. *J. Hered.* 27:183–194.

Karssen, C.M., Brinkhorst-van der Swan, D.L.C., Breekland, A.E., and Koornneef, M. 1983. Induction of dormancy during seed development by endogenous abscisic acid: studies on abscisic acid deficient genotypes of *Arabidopsis thaliana* (L.) Heynh. *Planta* 157:158–165.

Kartha, K.K., Chibbar, R.N., Georges, F., Leung, N., Caswell, K., Kendall, E., and Qureshi, J. 1989. Transient expression of chloramphenicol acetyltransferase (CAT) gene in barley cell cultures and immature embryos through microprojectile bombardment. *Plant Cell Rep.* 8:429–432.

Karunanandaa, B., Huang, S., and Kao, T.-H. 1994. Carbohydrate moiety of the *Petuniainflata* S_3 protein is not required for self-incompatibility interactions between pollen and pistil. *Plant Cell* 6:1933–1940.

Karunanandaa, B., Singh, A., and Kao, T.-H. 1994. Characterization of a predominantly pistil-expressed gene encoding a γ-thionin-like protein of *Petunia inflata*. *Plant Mol. Biol.* 26:459–464.

Kasarda, D.D., Okita, T.W., Bernardin, J.E., Baecker, P.A., Nimmo, C.C., Lew, E.J.-L., Dietler, M.D., and Greene, F.C. 1984. Nucleic acid (cDNA) and amino acid sequences of α-type gliadins from wheat (*Triticum aestivum*). *Proc. Natl Acad. Sci. USA* 81:4712–4716.

Kasembe, J.N.R. 1967. Phenotypic restoration of fertility in a male-sterile mutant by treatment with gibberellic acid. *Nature* 215:668.

Kasha, K.J., and Kao, K.N. 1970. High frequency haploid production in barley (*Hordeum vulgare* L.). *Nature* 225:874–876.

Kater, M.M., Koningstein, G.M., Nijkamp, H.J.J., and Stuitje, A.R. 1991. cDNA cloning and expression of *Brassica napus* enoyl-acyl carrier protein reductase in *Escherichia coli*. *Plant Mol. Biol.* 17:895–909.

Kato, H. 1968. The serial observations of the adventive embryogenesis in the microculture of carrot tissue. *Sci. Papers Coll. Gen. Edn Univ. Tokyo* 18:191–197.

Kato, H., and Takeuchi, M. 1963. Morphogenesis *in vitro* starting from single cells of carrot root. *Plant Cell Physiol.* 4:243–245.

Katti, R.Y., Giddanavar, H.S., Naik, S., Agadi, S.N., and Hegde, R.R. 1994. Persistence of callose and tapetum in the microsporogenesis of genic male sterile *Cajanus cajan* (L.) Millsp. with well formed endothecium. *Cytologia* 59:65–72.

Kaufmann, H., Salamini, F., and Thompson, R.D. 1991. Sequence variability and gene structure at the self-incompatibility locus of *Solanum tuberosum*. *Mol. Gen. Genet.* 226:457–466.

Kaul, C.L., and Singh, S.P. 1966. Studies in male-sterile barley. II. Pollen abortion. *Crop Sci.* 6:539–541.

Kaul, M.L.H. 1988. *Male Sterility in Higher Plants*. Berlin: Springer-Verlag.

Kaul, M.L.H., and Nirmala, C. 1993. Male sterility in pea. II. Male sex specific dyssynapsis. *Cytologia* 58:67–76.

Kaul, V., Theunis, C.H., Palser, B.F., Knox, R.B., and Williams, E.G. 1987. Association of the generative cell and vegetative nucleus in pollen tubes of *Rhododendron*. *Ann. Bot.* 59:227–235.

Kawagoe, Y., Campbell, B.R., and Murai, N. 1994. Synergism between CACGTG (G-box) and CACCTG *cis*-elements is required for activation of the bean seed storage protein *β-phaseolin* gene. *Plant J.* 5:885–890.

Kawagoe, Y., and Murai, N. 1992. Four distinct nuclear proteins recognize *in vitro* the proximal promoter of the bean seed storage protein *β-phaseolin* gene conferring spatial and temporal control. *Plant J.* 2:927–936.

Kawagoe, Y., and Murai, N. 1996. A novel basic region/helix-loop-helix protein binds to a G-box motif CACGTG of the bean seed storage protein β-phaseolin gene. *Plant Sci.* 116:47–57.

Kawahara, R., Sunabori, S., Fukuda, H., and Komamine, A. 1992. A gene expressed preferentially in the globular stage of somatic embryogenesis encodes elongation-factor 1α in carrot. *Eur. J. Biochem.* 209:157–162.

Kawahara, R., Sunabori, S., Fukuda, H., and Komamine, A. 1994. Analysis by *in situ* hybridization of the expression of elongation factor 1α in the carrot cells during somatic embryogenesis. *J. Plant Res.* 107:361–364.

Kawashima, I., Kennedy, T.D., Chino, M., and Lane, B.G. 1992. Wheat E_c metallothionein genes. Like mammalian Zn^{2+} metallothionein genes, wheat Zn^{2+} metallothionein genes are conspicuously expressed during embryogenesis. *Eur. J. Biochem.* 209:971–976.

Kazimierska, E.M. 1978. Embryological studies of cross compatibility in the genus *Trifolium* L. II. Fertilization, development of embryo and endosperm in crossing *T. repens* L. with *T. medium* L. *Genet. Polon.* 19:15–24.

Keddie, J.S., Edwards, E.-W., Gibbons, T., Shaw, C.H., and Murphy, D.J. 1992a. Sequence of an oleosin cDNA from *Brassica napus*. *Plant Mol. Biol.* 19:1079–1083.

Keddie, J.S., Hübner, G., Slocombe, S.P., Jarvis, R.P., Cummins, I., Edwards, E.-W., Shaw, C.H., and Murphy, D.J. 1992b. Cloning and characterisation of an oleosin gene from *Brassica napus*. *Plant Mol. Biol.* 19:443–453.

Keddie, J.S., Tsiantis, M., Piffanelli, P., Cella, R., Hatzopoulos, P., and Murphy, D.J. 1994. A seed-specific *Brassica napus* oleosin promoter interacts with a G-box-specific protein and may be bi-directional. *Plant Mol. Biol.* 24:327–340.

Keeler, S.J., Sanders, P., Smith, J.K., and Mazur, B.J. 1993. Regulation of tobacco acetolactate synthase gene expression. *Plant Physiol.* 102:1009–1018.

Kehrel, B., and Wiermann, R. 1985. Immunochemical localization of phenylalanineammonia-lyase and chalcone synthase in anthers. *Planta* 163:183–190.

Keijzer, C.J. 1987. The processes of anther dehiscence and pollen dispersal. II. The formation and the transfer mechanism of pollenkitt, cell-wall development of the loculus tissues and a function of orbicules in pollen dispersal. *New Phytol.* 105:499–507.

Keijzer, C.J., and Cresti, M. 1987. A comparison of anther tissue development in male sterile *Aloe vera* and male fertile *Aloe ciliaris*. *Ann. Bot.* 59:533–542.

Keijzer, C.J., and Willemse, M.T.M. 1988. Tissue interactions in the developing locule of *Gasteria verrucosa* during microgametogenesis. *Acta Bot. Neerl.* 37:475–492.

Keith, K., Kraml, M., Dengler, N.G., and McCourt, P. 1994. *fusca3*: a heterochronic mutation affecting late embryo development in *Arabidopsis. Plant Cell* 6:589–600.

Keller, W.A., and Armstrong, K.C. 1978. High frequency production of microspore-derived plants from *Brassica napus* anther cultures. *Z. Pflanzenzüchtg* 80:100–108.

Keller, W.A., and Armstrong, K.C. 1979. Stimulation of embryogenesis and haploid production in *Brassica campestris* anther cultures by elevated temperature treatments. *Theor. Appl. Genet.* 55:65–67.

Kemble, R.J., and Bedbrook, J.R. 1980. Low molecular weight circular and linear DNA in mitochondria from normal and male-sterile *Zea mays* cytoplasm. *Nature* 284:565–566.

Kemble, R.J., Flavell, R.B., and Brettell, R.I.S. 1982. Mitochondrial DNA analyses of fertile and sterile maize plants derived from tissue culture with the Texas male sterile cytoplasm. *Theor. Appl. Genet.* 62:213–217.

Kemble, R.J., Gunn, R.E., and Flavell, R.B. 1980. Classification of normal and male-sterile cytoplasms in maize. II. Electrophoretic analysis of DNA species in mitochondria. *Genetics* 95:451–458.

Kemble, R.J., and Mans, R.J. 1983. Examination of the mitochondrial genome of revertant progeny from S *cms* maize with cloned S-1 and S-2 hybridization probes. *J. Mol. Appl. Genet.* 2:161–171.

Kemp, C.L. 1964. The effects of inhibitors of RNA and protein synthesis on cytological development during meiosis. *Chromosoma* 15:652–665.

Kemp, C.L. 1966. Electron microscope autoradiographic studies of RNA metabolism in *Trillium erectum* microspores. *Chromosoma* 19:137–148.

Kempin, S.A., Mandel, M.A., and Yanofsky, M.F. 1993. Conversion of perianth into reproductive organs by ectopic expression of the tobacco floral homeotic gene *NAG1. Plant Physiol.* 103:1041–1046.

Kennell, J.C., and Horner, H.T. 1985. Megasporogenesis and megagametogenesis in soybean, *Glycine max. Am. J. Bot.* 72:1553–1564.

Kennell, J.C., and Pring, D.R. 1989. Initiation and processing of *atp6*, T-*urf13* and *ORF221* transcripts from mitochondria of T cytoplasm maize. *Mol. Gen. Genet.* 216:16–24.

Kennell, J.C., Wise, R.P., and Pring, D.R. 1987. Influence of nuclear background on transcription of a maize mitochondrial region associated with Texas male sterile cytoplasm. *Mol. Gen. Genet.* 210:399–406.

Kenrick, J., Kaul, V., and Williams, E.G. 1986. Self-incompatibility in *Acacia retinodes*: site of pollen-tube arrest is the nucellus. *Planta* 169:245–250.

Kenrick, J., and Knox, R.B. 1981a. Post-pollination exudate from stigmas of *Acacia* (Mimosaceae). *Ann. Bot.* 48:103–106.

Kenrick, J., and Knox, R.B. 1981b. Structure and histochemistry of the stigma and style of some Australian species of *Acacia. Aust. J. Bot.* 29:733–745.

Kenrick, J., and Knox, R.B. 1985. Self-incompatibility in the nitrogen-fixing tree, *Acacia retinodes*: quantitative cytology of pollen tube growth. *Theor. Appl. Genet.* 69:481–488.

Kent, N., and Brink, R.A. 1947. Growth *in vitro* of immature *Hordeum* embryos. *Science* 106:547–548.

Keown, A.C., Taiz, L., and Jones, R.L. 1977. The nuclear DNA content of developing barley aleurone cells. *Am. J. Bot.* 64:1248–1253.

Kępczyński, J., McKersie, B.D., and Brown, D.C.W. 1992. Requirement of ethylene for growth of callus and somatic embryogenesis in *Medicago sativa* L. *J. Expt. Bot.* 43:1199–1202.

Kerhoas, C., Gay, G., and Dumas, C. 1987. A multidisciplinary approach to the study of the plasma membrane of *Zea mays* pollen during controlled dehydration. *Planta* 171:1–10.

Kerhoas, C., Knox, R.B., and Dumas, C. 1983. Specificity of the callose response in stigmas of *Brassica. Ann. Bot.* 52:597–602.

Kermode, A.R., and Bewley, J.D. 1985. The role of maturation drying in the transition from seed development to germination. I. Acquisition of desiccation-tolerance and germinability during development of *Ricinus communis* L. seeds. *J. Expt. Bot.* 36:1906–1915.

Kermode, A.R., and Bewley, J.D. 1988. The role of maturation drying in the transition from seed development to germination. V. Responses of the immature castor bean embryo to isolation from the whole seed: a comparison with premature desiccation. *J. Expt. Bot.* 39:487–497.

Khan, M.R.I., Gatehouse, J.A., and Boulter, D. 1980. The seed proteins of cowpea (*Vigna unguiculata* L. Walp.). *J. Expt. Bot.* 31:1599–1611.

Khavkin, E.E., Misharin, S.I., Ivanov, V.N., and Danovich, K.N. 1977. Embryonal antigens in maize caryopses: the temporal order of antigen accumulation during embryogenesis. *Planta* 135:225–231.

Khavkin, E.E., Misharin, S.I., Markov, Y.Y., and Peshkova, A.A. 1978. Identification of embryonal antigens of maize: globulins as primary reserve proteins of the embryo. *Planta* 143:11–20.

Kheyr-Pour, A., Bintrim, S.B., Ioerger, T.R., Remy, R., Hammond, S.A., and Kao, T.-H. 1990. Sequence diversity of pistil S-proteins associated with gametophytic self-incompatibility in *Nicotiana alata. Sex. Plant Reprod.* 3:88–97.

Khoo, U., and Stinson, H.T., Jr. 1957. Free amino acid differences between cytoplasmic male sterile and normal fertile anthers. *Proc. Natl Acad. Sci. USA* 43:603–607.

Khoo, U., and Wolf, M.J. 1970. Origin and development of protein granules in maize endosperm. *Am. J. Bot.* 57:1042–1050.

Kiang, A.-S., and Kavanagh, T.A. 1996. Cytoplasmic male sterility (CMS) in *Lolium perenne* L. 2. The mitochondrial genome of a CMS line is rearranged and contains a chimaeric *atp* 9 gene. *Theor. Appl. Genet.* 92:308–315.

Kiesselbach, T.A. 1949. The structure and reproduction of corn. *Univ. Nebraska Agr. Expt. Station Bull.* 161:3–96.

Kikkert, J.R., Hébert-Soulé, D., Wallace, P.G., Striem, M., and Reisch, B.I. 1996. Transgenic plantlets of `Chancellor' grapevine (*Vitis* sp.) from biolistic transformation of embryogenic cell suspensions. *Plant Cell Rep.* 15:311–316.

Kikuchi, A., Satoh, S., Nakamura, N., and Fujii, T. 1995. Differences in pectic polysaccharides between carrot embryogenic and non-embryogenic calli. *Plant Cell Rep.* 14:279–284.

Kim, M.Z., and Raghavan, V. 1988. Induction of pollen plantlets in rice by spikelet culture. *Plant Cell Rep.* 7:560–563.

Kim, S.-R., Kim, Y., and An, G. 1993. Molecular cloning and characterization of anther-preferential cDNA encoding a putative actin-depolymerizing factor. *Plant Mol. Biol.* 21:39–45.

Kim, S.-R., Finkel, D., Chung, Y.-Y., and An, G. 1994. Abundance patterns of lily pollen cDNAs: characterization of three pollen-preferential cDNA clones. *Sex. Plant Reprod.* 7:76–86.

Kim, W.T., Li, X., and Okita, T.W. 1993. Expression of storage protein multigene families in developing rice endosperm. *Plant Cell Physiol.* 34:595–603.

Kim, W.T., and Okita, T.W. 1988a. Nucleotide and primary sequence of a major rice prolamine. *FEBS Lett.* 231:308–310.

Kim, W.T., and Okita, T.W. 1988b. Structure, expression, and heterogeneity of the rice seed prolamines. *Plant Physiol.* 88:649–655.

Kim, W.T., Franceschi, V.R., Krishnan, H.B., and Okita, T.W. 1988. Formation of wheat protein bodies: involvement of the Golgi apparatus in gliadin transport. *Planta* 176:173–182.

Kim, Y., and An, G. 1992. Pollen-specific expression of the *Arabidopsis thaliana* α_1-tubulin promoter assayed by β-glucuronidase, chloramphenicol acetyltransferase and diphtheria toxin reporter genes. *Transgen. Res.* 1:188–194.

Kim, Y.-H., and Janick, J. 1991. Abscisic acid and proline improve desiccation tolerance and increase fatty acid content of celery somatic embryos. *Plant Cell Tissue Organ Cult.* 24:83–89.

Kindiger, B., Beckett, J.B., and Coe, E.H., Jr. 1991. Differential effects of specific chromosomal deficiencies on the development of the maize pollen grain. *Genome* 34:579–594.

King, R.W. 1976. Abscisic acid in developing wheat grains and its relationship to grain growth and maturation. *Planta* 132:43–51.

King, T.J., and Briggs, R. 1953. The transplantability of nuclei of arrested hybrid blastulae (*R. pipiens* ♀ × *R. catesbeiana* ♂). *J. Expt. Zool.* 123:61–78.

Kini, A.V., Seetharam, A., and Joshi, S.S. 1994. Mechanism of pollen abortion in cytoplasmic male sterile line of sunflower. *Cytologia* 59:121–124.

Kirby, E.G., and Vasil, I.K. 1979. Effect of pollen protein diffusates on germination of eluted pollen samples of *Petunia hybrida in vitro*. *Ann. Bot.* 44:361–367.

Kirch, H.-H., Li, Y.-Q., Seul, U., and Thompson, R.D. 1995. The expression of a potato (*Solanum tuberosum*) S-RNase gene in *Nicotiana tabacum* pollen. *Sex. Plant Reprod.* 8:77–84.

Kirch, H.H., Uhrig, H., Lottspeich, F., Salamini, F., and Thompson, R.D. 1989. Characterization of proteins associated with self-incompatibility in *Solanum tuberosum*. *Theor. Appl. Genet.* 78:581–588.

Kirihara, J.A., Petri, J.B., and Messing, J. 1988. Isolation and sequence of a gene encoding a methionine-rich 10-kDa zein protein from maize. *Gene* 71:359–370.

Kirihara, J.A., Hunsperger, J.P., Mahoney, W.C., and Messing, J.W. 1988. Differential expression of a gene for a methionine-rich storage protein in maize. *Mol. Gen. Genet.* 211:477–484.

Kirk, J.T.O., and Tilney-Bassett, R.A.E. 1978. *The Plastids. Their Chemistry, Structure, Growth and Inheritance.* 2nd ed. Amsterdam: Elsevier/North-Holland Biomedical Press.

Kishitani, S., Yomoda, A., Konno, N., and Tanaka, Y. 1993. Involvement of phenylalanine ammonia-lyase in the development of pollen in broccoli (*Brassica oleracea* L.). *Sex. Plant Reprod.* 6:244–248.

Kitada, K., Kurata, N., Satoh, H., and Omura, T. 1983. Genetic control of meiosis in rice, *Oryza sativa* L. I. Classification of meiotic mutants induced by MNU and their cytogenetical characteristics. *Jap. J. Genet.* 58:231–240.

Kitano, H., Tamura, Y., Satoh, H., and Nagato, Y. 1993. Hierarchical regulation of organ differentiation during embryogenesis in rice. *Plant J.* 3:607–610.

Kiyosue, T., and Shinozaki, K. 1995. Cloning of a carrot cDNA for a member of the family of ADP-ribosylation factors (ARFs) and characterization of the binding of nucleotides by its product after expression in *E. coli*. *Plant Cell Physiol.* 36:849–856.

Kiyosue, T., Dong, J.G., Satoh, S., Kamada, H., and Harada, H. 1990a. Detection of an embryogenic cell antigen in carrot. *Plant Cell Physiol.* 31:947–950.

Kiyosue, T., Nakajima, M., Yamaguchi, I., Satoh, S., Kamada, H., and Harada, H. 1992a. Endogenous levels of abscisic acid in embryogenic cells, non-embryogenic cells and somatic embryos of carrot (*Daucus carota* L.). *Biochem. Physiol. Pflanz.* 188:343–347.

Kiyosue, T., Nakayama, J., Satoh, S., Isogai, A., Suzuki, A., Kamada, H., and Harada, H. 1992b. Partial amino-acid sequence of ECP31, a carrot embryogenic-cell protein, and enhancement of its accumulation by abscisic acid in somatic embryos. *Planta* 186:337–342.

Kiyosue, T., Satoh, S., Kamada, H., and Harada, H. 1991. Purification and immunohistochemical detection of an embryogenic cell protein in carrot. *Plant Physiol.* 95:1077–1083.

Kiyosue, T., Takano, K., Kamada, H., and Harada, H. 1990b. Induction of somatic embryogenesis in carrot by heavy metal ions. *Can. J. Bot.* 68:2301–2303.

Kiyosue, T., Yamaguchi-Shinozaki, K., Shinozaki, K., Higashi, K., Satoh, S., Kamada, H., and Harada, H. 1992c. Isolation and characterization of a cDNA that encodes ECP31, an embryogenic-cell protein from carrot. *Plant Mol. Biol.* 19:239–249.

Kiyosue, T., Yamaguchi-Shinozaki, K., Shinozaki, K., Kamada, H., and Harada, H. 1993. cDNA cloning of ECP40, an embryogenic-cell protein in carrot, and its expression during somatic and zygotic embryogenesis. *Plant Mol. Biol.* 21:1053–1068.

Klein, H.D. 1969. Male sterility in *Pisum*. *Nucleus* 12:167–172.

Klein, H.D., and Milutinović, M. 1971. Genbedingte Störungen der Infloreszenz- und Blütenbildung. *Theor. Appl. Genet.* 41:255–258.

Klein, T.M., Roth, B.A., and Fromm, M.E. 1989. Regulation of anthocyanin biosynthetic genes introduced into intact maize tissues by microprojectiles. *Proc. Natl Acad. Sci. USA* 86:6681–6685.

Klein, T.M., Gradziel, T., Fromm, M.E., and Sanford, J.C. 1988. Factors influencing gene delivery into *Zea mays* cells by high-velocity microprojectiles. *Biotechnology* 6:559–563.

Kleman-Mariac, C., Rougier, M., Cock, J.M., Gaude, T., and Dumas, C. 1995. S-locus glycoproteins are expressed along the path of the pollen tubes in *Brassica* pistils. *Planta* 196:614–621.

Klemsdal, S.S., Hughes, W., Lönneborg, A., Aalen, R.B., and Olsen, O.-A. 1991. Primary structure of a novel barley gene differentially expressed in immature aleurone layers. *Mol. Gen. Genet.* 228:9–16.

Klinge, B., and Werr, W. 1995. Transcription of the *Zea mays* homeobox (*ZmHox*) genes is activated early

1995. SLG/SRK-like genes are expressed in the reproductive tissues of *Ipomoea trifida*. *Sex. Plant Reprod.* 8:333–338.

Kowyama, Y., Kunz, C., Lewis, I., Newbigin, E., Clarke, A.E., and Anderson, M.A. 1994. Self-compatibility in a *Lycopersicon peruvianum* variant (LA2157) is associated with a lack of style S-RNase activity. *Theor. Appl. Genet.* 88:859–864.

Kozar, F. 1974. Ultrastructure of pollen of *Opuntia polyacantha*. *Can. J. Bot.* 52:313–315.

Koziel, M.G., Beland, G.L., Bowman, C., Carozzi, N.B., Crenshaw, R., Crossland, L., Dawson, J., Desai, N., Hill, M., Kadwell, S., Launis, K., Lewis, K., Maddox, D., McPherson, K., Meghji, M.R., Merlin, E., Rhodes, R., Warren, G.W., Wright, M., and Evola, S.V. 1993. Field performance of elite transgenic maize plants expressing an insecticidal protein derived from *Bacillus thuringiensis*. *Biotechnology* 11:194–200.

Kragh, K.M., Jacobsen, S., Mikkelsen, J.D., and Nielsen, K.A. 1991. Purification and characterization of three chitinases and one β-1,3-glucanase accumulating in the medium of cell suspension cultures of barley (*Hordeum vulgare* L.). *Plant Sci.* 76:65–77.

Kranz, E., Bautor, J., and Lörz, H. 1991a. *In vitro* fertilization of single, isolated gametes of maize mediated by electrofusion. *Sex. Plant Reprod.* 4:12–16.

Kranz, E., Bautor, J., and Lörz, H. 1991b. Electrofusion-mediated transmission of cytoplasmic organelles through the *in vitro* fertilization process, fusion of sperm cells with synergids and central cells, and cell reconstitution in maize. *Sex. Plant Reprod.* 4:17–21.

Kranz, E., and Lörz, H. 1990. Micromanipulation and *in vitro* fertilization with single pollen grains of maize. *Sex. Plant Reprod.* 3:160–169.

Kranz, E., and Lörz, H. 1993. *In vitro* fertilization with isolated, single gametes results in zygotic embryogenesis and fertile maize plants. *Plant Cell* 5:739–746.

Kranz, E., and Lörz, H. 1994. *In vitro* fertilisation of maize by single egg and sperm cell protoplast fusion mediated by high calcium and high *p*H. *Zygote* 2:125–128.

Kranz, E., von Wiegen, P., and Lörz, H. 1995. Early cytological events after induction of cell division in egg cells and zygote development following *in vitro* fertilization with angiosperm gametes. *Plant J.* 8:9–23.

Krasowski, M.J., and Owens, J.N. 1993. Ultrastructural and histochemical postfertilization megagametophyte and zygotic embryo development of white spruce (*Picea glauca*) emphasizing the deposition of seed storage products. *Can. J. Bot.* 71:98–112.

Krebbers, E., Herdies, L., de Clercq, A., Seurinck, J., Leemans, J., van Damme, J., Segura, M., Gheysen, G., van Montagu, M., and Vandekerckhove, J. 1988. Determination of the processing sites of an *Arabidopsis* 2S albumin and characterization of the complete gene family. *Plant Physiol.* 87:859–866.

Kreis, M., Forde, B.G., Rahman, S., Miflin, B.J., and Shewry, P.R. 1985. Molecular evolution of the seed storage proteins of barley, rye and wheat. *J. Mol. Biol.* 183:499–502.

Kreis, M., Shewry, P.R., Forde, B.G., Rahman, S., Bahramian, M.B., and Miflin, B.J. 1984. Molecular analysis of the effects of the *lys 3a* gene on the expression of *Hor* loci in developing endosperms of barley (*Hordeum vulgare* L.). *Biochem. Genet.* 22:231–255.

Kreis, M., Shewry, P.R., Forde, B.G., Rahman, S., and Miflin, B.J. 1983. Molecular analysis of a mutation conferring the high-lysine phenotype on the grain of barley (*Hordeum vulgare*). *Cell* 34:161–167.

Kress, W.J., and Stone, D.E. 1982. Nature of the sporoderm in monocotyledons, with special reference to the pollen grains of *Canna* and *Heliconia*. *Grana* 21:129–148.

Kress, W.J., Stone, D.E., and Sellers, S.C. 1978. Ultrastructure of exine-less pollen: *Heliconia* (Heliconiaceae). *Am. J. Bot.* 65:1064–1076.

Kreuger, M., and van Holst, G.-J. 1993. Arabinogalactan proteins are essential in somatic embryogenesis of *Daucus carota* L. *Planta* 189:243–248.

Kreuger, M., and van Holst, G.-J. 1995. Arabinogalactan-protein epitopes in somatic embryogenesis of *Daucus carota* L. *Planta* 197:135–141.

Kriete, G., Niehaus, K., Perlick, A.M., Pühler, A., and Broer, I. 1996. Male sterility in transgenic tobacco plants induced by tapetum-specific deacetylation of the externally applied non-toxic compound *N*-acetyl-L-phosphinothricin. *Plant J.* 9:809–818.

Krishnan, H.B., Franceschi, V.R., and Okita, T.W. 1986. Immunochemical studies on the role of the Golgi complex in protein-body formation in rice seeds. *Planta* 169:471–480.

Krishnan, H.B., and White, J.A. 1995. Morphometric analysis of rice seed protein bodies. Implication for a significant contribution of prolamine to the total protein content of rice endosperm. *Plant Physiol.* 109:1491–1495.

Krishnan, H.B., White, J.A., and Pueppke, S.G. 1988. Immunogold localization of prolamines in developing *Haynaldia villosa* endosperm. *Protoplasma* 144:25–33.

Krishnan, H.B., White, J.A., and Pueppke, S.G. 1989. Immunocytochemical analysis of protein body formation in seeds of *Sorghum bicolor*. *Can. J. Bot.* 67:2850–2856.

Krishnan, H.B., White, J.A., and Pueppke, S.G. 1990. Immunocytochemical evidence for the involvement of the Golgi apparatus in the transport of the vacuolar protein, γ-secalin, in rye (*Secale cereale*) endosperm. *Cereal Chem.* 67:360–366.

Krishnaraj, S., and Vasil, I.K. 1995. Somatic embryogenesis in herbaceous monocots. In In Vitro *Embryogenesis in Plants*, ed. T. A. Thorpe, pp. 417–470. Dordrecht: Kluwer Academic Publishers.

Krishnasamy, S., Grant R.A., and Makaroff, C.A. 1994. Subunit 6 of the F_0-ATP synthase complex from cytoplasmic male-sterile radish: RNA editing and NH_2-terminal protein sequencing. *Plant Mol. Biol.* 24:129–141.

Krishnasamy, S., and Makaroff, C.A. 1993. Characterization of the radish mitochondrial *orfB* locus: possible relationship with male sterility in Ogura radish. *Curr. Genet.* 24:156–163.

Krishnasamy, S., and Makaroff, C.A. 1994. Organ-specific reduction in the abundance of a mitochondrial protein accompanies fertility restoration in cytoplasmic male-sterile radish. *Plant Mol. Biol.* 26:935–946.

Kristen, U. 1977. Granulocrine Ausscheidung von Narbensekret durch Vesikel des Endoplasmatischen Retikulums bei *Aptenia cordifolia*. *Protoplasma* 92:243–251.

Kristen, U., Biedermann, M., Liebezeit, G., Dawson, R., and Böhm, L. 1979. The composition of stigmatic exudate and the ultrastructure of the stigma papillae in *Aptenia cordifolia*. *Eur. J. Cell Biol*. 19:281–287.

Kriz, A.L. 1989. Characterization of embryo globulins encoded by the maize *Glb* genes. *Biochem. Genet*. 27:239–251.

Kriz, A.L., Boston, R.S., and Larkins, B.A. 1987. Structural and transcriptional analysis of DNA sequences flanking genes that encode 19 kilodalton zeins. *Mol. Gen. Genet*. 207:90–98.

Kriz, A.L., and Schwartz, D. 1986. Synthesis of globulins in maize embryos. *Plant Physiol*. 82:1069–1075.

Kriz, A.L., and Wallace, N.H. 1991. Characterization of the maize *Globulin-2* gene and analysis of two null alleles. *Biochem. Genet*. 29:241–254.

Kriz, A.R., Wallace, M.S., and Paiva, R. 1989. Globulin gene expression in embryos of maize *viviparous* mutants. Evidence for regulation of the *Glb1* gene by ABA. *Plant Physiol*. 92:538–542.

Krizek, B.A., and Meyerowitz, E.M. 1996a. The *Arabidopsis* homeotic genes *APETALA3* and *PISTILLATA* are sufficient to provide the B class organ identity function. *Development* 122:11–22.

Krizek, B.A., and Meyerowitz, E.M. 1996b. Mapping the protein regions responsible for the functional specificities of the *Arabidopsis* MADS domain organ-identity proteins. *Proc. Natl Acad. Sci. USA* 93:4063–4070.

Krochko, J.E., Pramanik, S.K., and Bewley, J.D. 1992. Contrasting storage protein synthesis and messenger RNA accumulation during development of zygotic and somatic embryos of alfalfa (*Medicago sativa* L.). *Plant Physiol*. 99:46–53.

Krochko, J.E., Bantroch, D.J., Greenwood, J.S., and Bewley, J.D. 1994. Seed storage proteins in developing somatic embryos of alfalfa: defects in accumulation compared to zygotic embryos. *J. Expt. Bot*. 45:699–708.

Kroh, M. 1966. Reaction of pollen after transfer from one stigma to another (Contribution to the character of the incompatibility mechanism in Cruciferae). *Züchter* 36:185–189.

Kroh, M. 1967a. Fine structure of *Petunia* pollen germinated *in vivo*. *Rev. Palaeobot. Palynol*. 3:197–203.

Kroh, M. 1967b. Bildung und Transport des Narbensekrets von *Petunia hybrida*. *Planta* 77:250–260.

Kroh, M., Gorissen, M.H., and Pfahler, P.L. 1979. Ultrastructural studies on styles and pollen tubes of *Zea mays* L. General survey on pollen tube growth *in vivo*. *Acta Bot. Neerl*. 28:513–518.

Kroh, M., and Knuiman, B. 1982. Ultrastructure of cell wall and plugs of tobacco pollen tubes after chemical extraction of polysaccharides. *Planta* 154:241–250.

Kroh, M., and Loewus, F. 1968. Biosynthesis of pectic substance in germinating pollen: labeling with myoinositol-2-^{14}C. *Science* 160:1352–1354.

Kroh, M., and Munting, A.J. 1967. Pollen-germination and pollen tube growth in *Diplotaxis tenuifolia* after cross-pollination. *Acta Bot. Neerl*. 16:182–187.

Kroh, M., and van Bakel, C.H.J. 1973. Incorporation of label into the intercellular substance of stylar transmitting tissue from *Petunia* pistils labeled with tritiated myo-inositol. An electron microscopic autoradiographic study. *Acta Bot. Neerl*. 22:106–111.

Kroh, M., Miki-Hirosige, H., Rosen, W., and Loewus, F. 1970. Incorporation of label into pollen tube walls from myoinositol-labeled *Lilium longiflorum* pistils. *Plant Physiol*. 45:92–94.

Krug, C.A., and Bacchi, O. 1943. Triploid varieties of *Citrus*. *J.Hered*. 34:277–283.

Kubień, E. 1968. Cytological processes during the development of antipodals in *Ammophila arenaria* Link. *Acta Biol. Cracow* Ser. Bot. 11:21–29.

Kučera, V., and Polák, J. 1975. The serological specificity of S alleles of homozygous incompatible lines of the marrow-stem kale (*Brassica oleracea* var. *acephala* DC). *Biol. Plant*. 17:50–54.

Kuehnle, A.R., and Earle, E.D. 1989. *In vitro* selection for methomyl resistance in CMS-T maize. *Theor. Appl. Genet*. 78:672–682.

Kuhlmann, U., Foroughi-Wehr, B., Graner, A., and Wenzel, G. 1991. Improved culture system for microspores of barley to become a target for DNA uptake. *Plant Breed*. 107:165–168.

Kühtreiber, W.M., and Jaffe, L.F. 1990. Detection of extracellular calcium gradients with a calcium-specific vibrating electrode. *J. Cell Biol*. 110:1565–1573.

Kulikauskas, R., Hou, A., Muschietti, J., and McCormick, S. 1995. Comparisons of diverse plant species reveal that only grasses show drastically reduced levels of ubiquitin monomer in mature pollen. *Sex. Plant Reprod*. 8:326–332.

Kumamaru, T., Satoh, H., Iwata, N., Omura, T., Ogawa, M., and Tanaka, K. 1988. Mutants for rice storage proteins. 1. Screening of mutants for rice storage proteins of protein bodies in the starchy endosperm. *Theor. Appl. Genet*. 76:11–16.

Kumar, V., and Trick, M. 1993. Sequence complexity of the *S* receptor kinase gene family in *Brassica*. *Mol. Gen. Genet*. 241:440–446.

Kumar, V., and Trick, M. 1994. Expression of the *S*-locus receptor kinase multigene family in *Brassica oleracea*. *Plant J*. 6:807–813.

Kunce, C.M., and Trelease, R.N. 1986. Heterogeneity of catalase in maturing and germinated cotton seeds. *Plant Physiol*. 81:1134–1139.

Kunst, L., Klenz, J.E., Martinez-Zapater, J., and Haughn, G.W. 1989. *AP2* gene determines the identity of perianth organs in flowers of *Arabidopsis thaliana*. *Plant Cell* 1:1195–1208.

Kunz, C., Chang, A., Faure, J.-D., Clarke, A.E., Polya, G.M., and Anderson, M.A. 1996. Phosphorylation of style *S*-RNases by Ca^{2+}-dependent protein kinases from pollen tubes. *Sex. Plant Reprod*. 9:25–34.

Kuran, H. 1972. Callose localization in the walls of megasporocytes and megaspores in the course of development of monospore embryo sacs. *Acta Soc. Bot. Polon*. 41:519–534.

Kurata, N., and Ito, M. 1978. Electron microscope autoradiography of ^{3}H-thymidine incorporation during the zygotene stage in microsporocytes of lily. *Cell Struct. Funct*. 3:349–356.

Kuroiwa, H. 1989. Ultrastructural examination of embryogenesis in *Crepis capillaris* (L.) Wallr.: 1. The synergid before and after pollination. *Bot. Mag. Tokyo* 102:9–24.

Kuroiwa, H., and Kuroiwa, T. 1992. Giant mitochondria in the mature egg cell of *Pelargonium zonale*. *Protoplasma* 168:184–188.

Kush, A., Brunelle, A., Shevell, D., and Chua, N.-H. 1993. The cDNA sequence of two MADS box proteins in *Petunia*. *Plant Physiol*. 102:1051–1052.

Kvaale, A., and Olsen, O.A. 1986. Rates of cell division in

developing barley endosperms. *Ann. Bot.* 57:829–833.

Kyle, D.J., and Styles, E.D. 1977. Development of aleurone and sub-aleurone layers in maize. *Planta* 137:185–193.

Kyo, M., and Harada, H. 1985. Studies on conditions for cell division and embryogenesis in isolated pollen culture of *Nicotiana rustica*. *Plant Physiol.* 79:90–94.

Kyo, M., and Harada, H. 1986. Control of the developmental pathway of tobacco pollen *in vitro*. *Planta* 168:427–432.

Kyo, M., and Harada, H. 1990. Specific phosphoproteins in the initial period of tobacco pollen embryogenesis. *Planta* 182:58–63.

Kyo, M., and Ohkawa, T. 1991. Investigation of subcellular localization of several phosphoproteins in embryogenic pollen grains of tobacco. *J. Plant Physiol.* 137:525–529.

Kyozuka, J., Olive, M., Peacock, W.J., Dennis, E.S., and Shimamoto, K. 1994. Promoter elements required for developmental expression of the maize *Adh1* gene in transgenic rice. *Plant Cell* 6:799–810.

Kysely, W., and Jacobsen, H.-J. 1990. Somatic embryogenesis from pea embryos and shoot apices. *Plant Cell Tissue Organ Cult.* 20:7–14.

La Cour, L.F. 1949. Nuclear differentiation in the pollen grain. *Heredity* 3:319–337.

Labarca, C., Kroh, M., and Loewus, F. 1970. The composition of stigmatic exudate from *Lilium longiflorum*. Labeling studies with *myo*-inositol, D-glucose, and L-proline. *Plant Physiol.* 46:150–156.

Labarca, C., and Loewus, F. 1972. The nutritional role of pistil exudate in pollen tube wall formation in *Lilium longiflorum*. I. Utilization of injected stigmatic exudate. *Plant Physiol.* 50:7–14.

Labarca, C., and Loewus, F. 1973. The nutritional role of pistil exudate in pollen tube wall formation in *Lilium longiflorum*. II. Production and utilization of exudate from stigma and stylar canal. *Plant Physiol.* 52:87–92.

Lacadena, J.-R. 1974. Spontaneous and induced parthenogenesis and androgenesis. In *Haploids in Higher Plants. Advances and Potential*, ed. K.J. Kasha, pp. 13–32. Guelph: University of Guelph.

Lackie, S., and Yeung, E.C. 1996. Zygotic embryo development in *Daucus carota*. *Can. J. Bot.* 74:990–998.

Ladin, B.F., Doyle, J.J., and Beachy, R.N. 1984. Molecular characterization of a deletion mutation affecting the α′-subunit of β-conglycinin of soybean. *J. Mol. Appl. Genet.* 2:372–380.

Ladin, B.F., Tierney, M.L., Meinke, D.W., Hosángadi, P., Veith, M., and Beachy, R.N. 1987. Developmental regulation of β-conglycinin in soybean axes and cotyledons. *Plant Physiol.* 84:35–41.

Lafleur, G.J., and Mascarenhas, J.P. 1978. The dependence of generative cell division in *Tradescantia* pollen tubes on protein and RNA synthesis. *Plant Sci. Lett.* 12:251–255.

LaFountain, J.R., and LaFountain, K.L. 1973. Comparison of density of nuclear pores on vegetative and generative nuclei in pollen of *Tradescantia*. *Expt. Cell Res.* 78:472–476.

LaFountain, K.L., and Mascarenhas, J.P. 1972. Isolation of vegetative and generative nuclei from pollen tubes. *Expt. Cell Res.* 73:233–236.

Lagriffol, J., and Monnier, M. 1985. Effects of endosperm and placenta on development of *Capsella* embryos in ovules cultivated *in vitro*. *J. Plant Physiol.* 118:127–137.

Lai, F.-M., Senaratna, T., and McKersie, B.D. 1992. Glutamine enhances storage protein synthesis in *Medicago sativa* L. somatic embryos. *Plant Sci.* 87:69–77.

Laibach, F. 1925. Das Taubwerden von Bastardsamen und künstliche Aufzucht früh absterbender Bastardembryonen. *Z. Bot.* 17:417–459.

Lalonde, B.A., Nasrallah, M.E., Dwyer, K.G., Chen, C.-H., Barlow, B., and Nasrallah, J.B. 1989. A highly conserved *Brassica* gene with homology to the S-locus-specific glycoprotein structural gene. *Plant Cell* 1:249–258.

Lam, E., and Chua, N.-H. 1991. Tetramer of a 21-base pair synthetic element confers seed expression and transcriptional enhancement in response to water stress and abscisic acid. *J. Biol. Chem.* 266:17131–17135.

Lamb, C.J. 1994. Plant disease resistance genes in signal perception and transduction. *Cell* 76:419–422.

Lambert, A.-M., and Bajer, A.S. 1972. Dynamics of spindle fibers and microtubules during anaphase and phragmoplast formation. *Chromosoma* 39:101–144.

Lancelle, S.A., Cresti, M., and Hepler, P.K. 1987. Ultrastructure of the cytoskeleton in freeze-substituted pollen tubes of *Nicotiana alata*. *Protoplasma* 140:141–150.

Lancelle, S.A., and Hepler, P.K. 1991. Association of actin with cortical microtubules revealed by immunogold localization in *Nicotiana* pollen tubes. *Protoplasma* 165:167–172.

Lancelle, S.A., and Hepler, P.K. 1992. Ultrastructure of freeze-substituted pollen tubes of *Lilium longiflorum*. *Protoplasma* 167:215–230.

Lang, J.D., Ray, S., and Ray, A. 1994. *sin1*, a mutation affecting female fertility in *Arabidopsis*, interacts with *mod1*, its recessive modifier. *Genetics* 137:1101–1110.

Langdale, J.A., Irish, E.E., and Nelson, T.M. 1994. Action of the *tunicate* locus on maize floral development. *Devel Genet.* 15:176–187.

Langridge, P., and Feix, G. 1983. A zein gene of maize is transcribed from two widely separated promoter regions. *Cell* 34:1015–1022.

Langridge, P., Pintor-Toro, J.A., and Feix, G. 1982a. Transcriptional effects of the opaque-2 mutation of *Zea mays* L. *Planta* 156:166–170.

Langridge, P., Pintor-Toro, J.A., and Feix, G. 1982b. Zein precursor mRNAs from maize endosperm. *Mol. Gen. Genet.* 187:432–438.

Langridge, P., Brettschneider, R., Lazzeri, P., and Lörz, H. 1992. Transformation of cereals via *Agrobacterium* and the pollen pathway: a critical assessment. *Plant J.* 2:631–638.

Langridge, P., Brown, J.W.S., Pintor-Toro, J.A., Feix, G., Neuhaus, G., Neuhaus-Url, G., and Schweiger, H.-G. 1985. Expression of zein genes in *Acetabularia mediterranea*. *Eur. J. Cell Biol.* 39:257–264.

Langridge, P., Eibel, H., Brown, J.W.S., and Feix, G. 1984. Transcription from maize storage protein gene promoters in yeast. *EMBO J.* 3:2467–2471.

Larkins, B.A. 1981. Seed storage proteins: characterization and biosynthesis. In *The Biochemistry of Plants. A Comprehensive Treatise*, Vol. 6, ed. A. Marcus, pp. 449–489. New York: Academic Press.

Larkins, B.A., Bracker, C.E., and Tsai, C.Y. 1976. Storage protein synthesis in maize. Isolation of zein-synthesizing polyribosomes. *Plant Physiol.* 57:740–745.

Larkins, B.A., and Dalby, A. 1975. *In vitro* synthesis of zein-like protein by maize polyribosomes. *Biochem. Biophys. Res. Commun.* 66:1048–1054.

Larkins, B.A., and Hurkman, W.J. 1978. Synthesis and deposition of zein in protein bodies of maize endosperm. *Plant Physiol.* 62:256–263.

Larkins, B.A., Jones, R.A., and Tsai, C.Y. 1976. Isolation and *in vitro* translation of zein messenger ribonucleic acid. *Biochemistry* 15:5506–5511.

Larkins, B.A., Pedersen, K., Handa, A.K., Hurkman, W.J., and Smith, L.D. 1979. Synthesis and processing of maize storage proteins in *Xenopus laevis* oocytes. *Proc. Natl Acad. Sci. USA* 76:6448–6452.

Laroche-Raynal, M., and Delseny, M. 1986. Identification and characterization of the mRNA for major storage proteins from radish. *Eur. J. Biochem.* 157:321–327.

Laroche-Raynal, M., Aspart, L., Delseny, M., and Penon, P. 1984. Characterization of radish mRNA at three developmental stages. *Plant Sci. Lett.* 35:139–146.

Larsen, K. 1977. Self-incompatibility in *Beta vulgaris* L. I. Four gametophytic, complementary S-loci in sugar beet. *Hereditas* 85:227–248.

Larsen, P.B., Ashworth, E.N., Jones, M.L., and Woodson, W.R. 1995. Pollination-induced ethylene in carnation. Role of pollen tube growth and sexual compatibility. *Plant Physiol.* 108:1405–1412.

Larson, D.A. 1963. Cytoplasmic dimorphism within pollen grains. *Nature* 200:911–912.

Larson, D.A. 1965. Fine-structural changes in the cytoplasm of germinating pollen. *Am. J. Bot.* 52:139–154.

Larson, D.A., and Lewis, C.W., Jr. 1963. Pollen wall development in *Parkinsonia aculeata. Grana Palynol.* 3 (3):3–19.

Larson, D.A., Skvarla, J.J., and Lewis, C.W., Jr. 1962. An electron microscope study of exine stratification and fine structure. *Pollen Spores* 4:232–246.

Laser, K.D., and Lersten, N.R. 1972. Anatomy and cytology of microsporogenesis of cytoplasmic male sterile angiosperms. *Bot. Rev.* 38:425–454.

Laurain, D., Trémouillaux-Guiller, J., and Chénieux, J.-C. 1993. Embryogenesis from microspores of *Ginkgo biloba* L., a medicinal woody species. *Plant Cell Rep.* 12:501–505.

Laux, T., Mayer, K.F.X., Berger, J., and Jürgens, G. 1996. The WUSCHEL gene is required for shoot and floral meristem integrity in *Arabidopsis. Development* 122:87–96.

Laver, H.K., Reynolds, S.J., Monéger, F., and Leaver, C.J. 1991. Mitochondrial genome organization and expression associated with cytoplasmic male sterility in sunflower (*Helianthus annuus*). *Plant J.* 1:185–193.

Lawrence, M.J. 1975. The genetics of self-incompatibility in *Papaver rhoeas*. *Proc. Roy. Soc. Lond.* 188B:275–285.

Lawrence, M.J., Marshall, D.F., Curtis, V.E., and Fearon, C.H. 1985. Gametophytic self-incompatibility re-examined: a reply. *Heredity* 54:131–138.

Le Coq, C. 1972a. La mégasporogénèse chez l'*Iris pseudacorus* L. I. Etude cytologique qualitative. *Rev. Cytol. Biol. Végét.* 35:41–163.

Le Coq, C. 1972b. La mégasporogénèse chez l'*Iris pseudacorus* L. II. Etude cytologique quantitative. *Rev. Cytol. Biol. Végét.* 35:303–329.

Le Gall, O., Torregrosa, L., Danglot, Y., Candresse, T., and Bouquet, A. 1994. *Agrobacterium*-mediated genetic transformation of grapevine somatic embryos and regeneration of transgenic plants expressing the coat protein of grapevine chrome mosaic nepovirus (GCMV). *Plant Sci.* 102:161–170.

Leah, R., Skriver, K., Knudsen, S., Ruud-Hansen, J., Raikhel, N.V., and Mundy, J. 1994. Identification of an enhancer/silencer sequence directing the aleurone-specific expression of a barley chitinase gene. *Plant J.* 6:579–589.

Leal, I., and Misra, S. 1993a. Developmental gene expression in conifer embryogenesis and germination. III. Analysis of crystalloid protein mRNAs and desiccation protein mRNAs in the developing embryo and megagametophyte of white spruce (*Picea glauca* (Moench) Voss). *Plant Sci.* 88:25–37.

Leal, I., and Misra, S. 1993b. Molecular cloning and characterization of a legumin-like storage protein cDNA of Douglas fir seeds. *Plant Mol. Biol.* 21:709–715.

Leal, I., Misra, S., Attree, S.M., and Fowke, L.C. 1995. Effect of abscisic acid, osmoticum and desiccation on 11S storage protein gene expression in somatic embryos of white spruce. *Plant Sci.* 106:121–128.

Leduc, N., Matthys-Rochon, E., and Dumas, C. 1995. Deleterious effect of minimal enzymatic treatments on the development of isolated maize embryo sacs in culture. *Sex. Plant Reprod.* 8:313–317.

Leduc, N., Matthys-Rochon, E., Rougier, M., Mogensen, L., Holm, P., Magnard, J.-L., and Dumas, C. 1996. Isolated maize zygotes mimic *in vivo* embryonic development and express microinjected genes when cultured *in vitro*. *Devel Biol.* 177:190–203.

Lee, C.H., and Power, J.B. 1988. Intraspecific gametosomatic hybridisation in *Petunia hybrida. Plant Cell Rep.* 7:17–18.

Lee, D.W., and Dougall, D.K. 1973. Electrophoretic variation in glutamate dehydrogenase and other isozymes in wild carrot cells cultured in the presence and absence of 2,4-dichlorophenoxyacetic acid. *In Vitro* 8:347–352.

Lee, H.-S., Huang, S., and Kao, T.-H. 1994. S proteins control rejection of incompatible pollen in *Petunia inflata. Nature* 367:560–563.

Lee, H.-S., Singh, A., and Kao, T.-H. 1992. RNase X2, a pistil-specific ribonuclease from *Petunia inflata*, shares sequence similarity with solanaceous S proteins. *Plant Mol. Biol.* 20:1131–1141.

Lee, H.S., Kim, S.W., Lee, K.-W., Eriksson, T., and Liu, J.R. 1995. *Agrobacterium*-mediated transformation of ginseng (*Panax ginseng*) and mitotic stability of the inserted β-glucuronidase gene in regenerants from isolated protoplasts. *Plant Cell Rep.* 14:545–549.

Lee, K., and Huang, A.H.C. 1991. Genomic nucleotide sequence of a *Brassica napus* 20-kilodalton oleosin gene. *Plant Physiol.* 96:1395–1397.

Lee, K., and Huang, A.H.C. 1994. Genes encoding oleosins in maize kernel of inbreds Mo17 and B73. *Plant Mol. Biol.* 26:1981–1987.

Lee, K.H., Jones, R.A., Dalby, A., and Tsai, C.Y. 1976. Genetic regulation of storage protein content in maize endosperm. *Biochem. Genet.* 14:641–650.

Lee, S.-H., Karunanandaa, B., McCubbin, A., Gilroy, S., and Kao, T.-H. 1996. PRK1, a receptor-like kinase of *Petunia inflata*, is essential for postmeiotic development of pollen. *Plant J.* 9:613–624.

Lee, S.-L.J., Earle, E.D., and Gracen, V.E. 1980. The cytology of pollen abortion in S cytoplasmic male-sterile corn anthers. *Am. J. Bot.* 67:237–245.

Lee, S.-L.J., Gracen, V.E., and Earle, E.D. 1979. The cytology of pollen abortion in C-cytoplasmic male-sterile corn anthers. *Am. J. Bot.* 66:656–667.

Lee, S.-L.J., and Warmke, H.E. 1979. Organelle size and number in fertile and T-cytoplasmic male-sterile corn. *Am. J. Bot.* 66:141–148.

Lee, W.S., Tzen, J.T.C., Kridl, J.C., Radke, S.E., and Huang, A.H.C. 1991. Maize oleosin is correctly targeted to seed oil bodies in *Brassica napus* transformed with the maize oleosin gene. *Proc. Natl Acad. Sci. USA* 88:6181–6185.

Leisy, D.J., Hnilo, J., Zhao, Y., and Okita, T.W. 1989. Expression of a rice glutelin promoter in transgenic tobacco. *Plant Mol. Biol.* 14:41–50.

Leite, A., Yunes, J.A., Turcinelli, S.R., and Arruda, P. 1992. Cloning and characterization of a cDNA encoding a sulfur-rich coixin. *Plant Mol. Biol.* 18:171–174.

Lelievre, J.-M., Oliveira, L.O., and Nielsen, N.C. 1992. 5′-CATGCAT-3′ elements modulate the expression of glycinin genes. *Plant Physiol.* 98:387–391.

Lending, C.R., and Larkins, B.A. 1989. Changes in the zein composition of protein bodies during maize endosperm development. *Plant Cell* 1:1011–1023.

Lending, C.R., and Larkins, B.A. 1992. Effect of the *floury-2* locus on protein body formation during maize endosperm development. *Protoplasma* 171:123–133.

Lending, C.R., Chesnut, R.S., Shaw, K.L., and Larkins, B.A. 1989. Immunolocalization of avenin and globulin storage proteins in developing endosperm of *Avena sativa* L. *Planta* 178:315–324.

Lending, C.R., Kriz, A.L., Larkins, B.A., and Bracker, C.E. 1988. Structure of maize protein bodies and immunocytochemical localization of zeins. *Protoplasma* 143:51–62.

Léon-Kloosterziel, K.M., Keijzer, C.J., and Koornneef, M. 1994. A seed shape mutant of *Arabidopsis* that is affected in integument development. *Plant Cell* 6:385–392.

Lerner, D.R., and Raikhel, N.V. 1989. Cloning and characterization of root-specific barley lectin. *Plant Physiol.* 91:124–129.

Lessard, P.A., Allen, R.D., Bernier, F., Crispino, J.D., Fujiwara, T., and Beachy, R.N. 1991. Multiple nuclear factors interact with upstream sequences of differentially regulated β-conglycinin genes. *Plant Mol. Biol.* 16:397–413.

Leung, D.W.M. 1992. Involvement of plant chitinase in sexual reproduction of higher plants. *Phytochemistry* 31:1899–1900.

Leung, J., Bouvier-Durand, M., Morris, P.-C., Guerrier, D., Chefdor, F., and Giraudat, J. 1994. *Arabidopsis* ABA response gene *ABI1*: features of a calcium-modulated protein phosphatase. *Science* 264:1448–1452.

Levanony, H., Rubin, R., Altschuler, Y., and Galili, G. 1992. Evidence for a novel route of wheat storage proteins to vacuoles. *J. Cell Biol.* 119:1117–1128.

Levieil, C. 1986. Évolution de l'association des cellules mâles dans le tube pollinique et dans le sac embryonnaire chez *Cichorium intybus* L. *Compt. Rend. Acad. Sci. Paris* Ser. III 303:769–774.

Levin, J.Z., and Meyerowitz, E.M. 1995. *UFO*: an *Arabidopsis* gene involved in both floral meristem and floral organ development. *Plant Cell* 7:529–548.

Levings, C.S., III. 1983. Cytoplasmic male sterility. In *Genetic Engineering of Plants. An Agricultural Perspective*, ed. T. Kosuge, C.P. Meredith and A. Hollaender, pp. 81–92. New York: Plenum Press.

Levings, C.S., III. 1993. Thoughts on cytoplasmic male sterility in *cms-T* maize. *Plant Cell* 5:1285–1290.

Levings, C.S., III., Kim, B.D., Pring, D.R., Conde, M.F., Mans, R.J., Laughnan, J.R., and Gabay-Laughnan, S.J. 1980. Cytoplasmic reversion of *cms-S* in maize: association with a transpositional event. *Science* 209:1021–1023.

Levings, C.S., III., and Pring, D.R. 1976. Restriction endonuclease analysis of mitochondrial DNA from normal and Texas cytoplasmic male-sterile maize. *Science* 193:158–160.

Lewandowska, E., and Charzyńska, M. 1977. *Tradescantia bracteata* pollen *in vitro*; pollen tube development and mitosis. *Acta Soc. Bot. Polon.* 46:587–598.

Lewis, D. 1942. The physiology of incompatibility in plants. I. The effect of temperature. *Proc. Roy. Soc. Lond.* 131B:13–26.

Lewis, D. 1943. The physiology of incompatibility in plants. II. *Linum grandiflorum*. *Ann. Bot.* 7:115–122.

Lewis, D. 1949. Structure of the incompatibility gene. II. Induced mutation rate. *Heredity* 3:339–355.

Lewis, D. 1952. Serological reactions of pollen incompatibility substances. *Proc. Roy. Soc. Lond.* 140B:127–135.

Lewis, D. 1954. Comparative incompatibility in angiosperms and fungi. *Adv. Genet.* 6:235–285.

Lewis, D. 1960. Genetic control of specificity and activity of the *S* antigen in plants. *Proc. Roy. Soc. Lond.* 151B:468–477.

Lewis, D. 1965. A protein dimer hypothesis on incompatibility. In *Genetics Today*, Vol. 3, ed. S.J. Geerts, pp. 657–663. Oxford: Pergamon Press.

Lewis, D., Burrage, S., and Walls, D. 1967. Immunological reactions of single pollen grains, electrophoresis and enzymology of pollen protein exudates. *J. Expt. Bot.* 18:371–378.

Lewis, D., and Crowe, L.K. 1958. Unilateral interspecific incompatibility in flowering plants. *Heredity* 12:233–256.

Lewis, D., and Jones, D.A. 1992. The genetics of heterostyly. In *Evolution and Function of Heterostyly*, ed. S.C.H. Barrett, pp. 129–150. Berlin: Springer-Verlag.

Li, H.-J., and Sodmergen. 1995. Maternal cytoplasmic inheritance and pollen nucleolytic activities in some Poaceae species. *Cytologia* 60:173–181.

Li, J., and Liu, Y.-N. 1983. Chloroplast DNA and cytoplasmic male-sterility. *Theor. Appl. Genet.* 64:231–238.

Li, L., Qu, R., de Kochko, A., Fauquet, C., and Beachy, R.N. 1993. An improved rice transformation system using the biolistic method. *Plant Cell Rep.* 12:250–255.

Li, M.-X. 1982. Pollen dimorphism and androgenesis of *Paeonia in vivo*. *Acta Bot. Sinica* 24:17–20.

Li, W.-D., Xu, H., Cheng, X.-F., and Ma, F.-S. 1991. The behaviour of pollen tubes on the stigma in the intersectional crosses in *Populus* and its relation to seed-setting. *Chinese J. Bot.* 3:102–109.

Li, Y.-Q., Croes, A.F., and Linskens, H.F. 1983. Cell-wall proteins in pollen and roots of *Lilium longiflorum*: extraction and partial characterization. *Planta* 158:422–427.

Li, Y.-Q., and Linskens, H.F. 1983a. Wall-bound proteins of pollen tubes after self- and cross-pollination in *Lilium longiflorum*. *Theor. Appl. Genet.* 67:11–16.

Li, Y.-Q., and Linskens, H.F. 1983b. Neutral sugar composition of pollen tube walls of *Lilium longiflorum*. *Acta Bot. Neerl.* 32:437–445.

Li, Y.-Q., and Tsao, T.H. 1985. Covalently bound wall proteins of pollen grains and pollen tubes grown *in vitro* and in styles after self- and cross-pollination in *Lilium longiflorum*. *Theor. Appl. Genet*. 71:263–267.

Li, Y.-Q., Tsao, T.H., and Linskens, H.F. 1986. Dependence of *Lilium* pollen germination and tube growth on protein synthesis and glycosylation after inhibitor treatments. *Proc. Koninkl. Nederl. Akad. Wetensch.* 89C:61–73.

Li, Y.-Q., Bruun, L., Pierson, E.S., and Cresti, M. 1992. Periodic deposition of arabinogalactan epitopes in the cell wall of pollen tubes of *Nicotiana tabacum*. *Planta* 188:532–538.

Li, Y.-Q., Chen, F., Linskens, H.F., and Cresti, M. 1994a. Distribution of unesterified and esterified pectins in cell walls of pollen tubes of flowering plants. *Sex. Plant Reprod*. 7:145–152.

Li, Y.-Q., Faleri, C., Geitmann, A., Zhang, H.-Q., and Cresti, M. 1995a. Immunogold localization of arabinogalactan proteins, unesterified and esterified pectins in pollen grains and pollen tubes of *Nicotiana tabacum* L. *Protoplasma* 189:26–36.

Li, Y.Q., Faleri, C., Thompson, R.D., Tiezzi, A., Eijlander, R., and Cresti, M. 1994b. Cytochemical immunolocalization of the abundant pistil protein S_{k2} in potato (*Solanum tuberosum*). *Sex. Plant Reprod*. 7:164–168.

Li, Y.Q., Southworth, D., Linskens, H.F., Mulcahy, D.L., and Cresti, M. 1995b. Localization of ubiquitin in anthers and pistils of *Nicotiana*. *Sex. Plant Reprod*. 8:123–128.

Li, X., Franceschi, V.R., and Okita, T.W. 1993. Segregation of storage protein mRNAs on the rough endoplasmic reticulum membranes of rice endosperm cells. *Cell* 72:869–879.

Li, X., Nield, J., Hayman, D., and Langridge, P. 1994. Cloning a putative self-incompatibility gene from the pollen of the grass *Phalaris coerulescens*. *Plant Cell* 6:1923–1932.

Li, X., Wu, Y., Zhang, D.-Z., Gillikin, J.W., Boston, R.S., Franceschi, V.R., and Okita, T.W. 1993. Rice prolamine protein body biogenesis: a BiP-mediated process. *Science* 262:1054–1056.

Li, X., and Okita, T.W. 1993. Accumulation of prolamines and glutelins during rice seed development: a quantitative evaluation. *Plant Cell Physiol*. 34:385–390.

Liang, X., Dron, M., Schmid, J., Dixon, R.A., and Lamb, C.J. 1989. Developmental and environmental regulation of a phenylalanine ammonia-lyase-β-glucuronidase gene fusion in transgenic tobacco plants. *Proc. Natl Acad. Sci. USA* 86:9284–9288.

Lima-de-Faria, A. 1965. Labeling of the cytoplasm and the meiotic chromosomes of *Agapanthus* with H^3-thymidine. *Hereditas* 53:1–11.

Lima-de-Faria, A., Pero, R., Avanzi, S., Durante, M., Ståhle, U., D'Amato, F., and Granström, H. 1975. Relation between ribosomal RNA genes and the DNA satellites of *Phaseolus coccineus*. *Hereditas* 79:5–19.

Lin, B.-Y. 1984. Ploidy barrier to endosperm development in maize. *Genetics* 107:103–115.

Lin, J., Uwate, W.J., and Stallman, V. 1977. Ultrastructural localization of acid phosphatase in the pollen tube of *Prunus avium* L. (sweet cherry). *Planta* 135:183–190.

Lin, J.-J., and Dickinson, D.B. 1984. Ability of pollen to germinate prior to anthesis and effect of desiccation on germination. *Plant Physiol*. 74:746–748.

Lin, J.-J., Dickinson, D.B., and Ho, T.-H.D. 1987. Phytic acid metabolism in lily (*Lilium longiflorum* Thunb.) pollen. *Plant Physiol*. 83:408–413.

Lin, Y., Wang, Y., Zhu, J.-K., and Yang, Z. 1996. Localization of a Rho GTPase implies a role in tip growth and movement of the generative cell in pollen tubes. *Plant Cell* 8:293–303.

Lin, Y.-T., Chow, T.-Y., and Lin, C.-Y. 1971. Nitrogen metabolism associated with pollen grain germination. *Taiwania* 16:67–84.

Linck, A.J., and Blaydes, G.W. 1960. Demonstration of the chemotropism of pollen tubes *in vitro* in four plant species. *Ohio J. Sci*. 60:274–278.

Lind, J.L., Bacic, A., Clarke, A.E., and Anderson, M.A. 1994. A style-specific hydroxyproline-rich glycoprotein with properties of both extensins and arabinogalactan proteins. *Plant J*. 6:491–502.

Lind, J.L., Bönig, I., Clarke, A.E., and Anderson, M.A. 1996. A style-specific 120-kDa glycoprotein enters pollen tubes of *Nicotiana alata in vivo*. *Sex. Plant Reprod*. 9:75–86.

Lindsey, K., and Topping, J.F. 1993. Embryogenesis: a question of pattern. *J. Expt. Bot*. 44:359–374.

Lindstrom, J.T., Vodkin, L.O., Harding, R.W., and Goeken, R.M. 1990. Expression of soybean lectin gene deletions in tobacco. *Devel Genet*. 11:160–167.

Lindzen, E., and Choi, J.H. 1995. A carrot cDNA encoding an atypical protein kinase homologous to plant calcium-dependent protein kinases. *Plant Mol. Biol*. 28:785–797.

Linnestad, C., Lönneborg, A., Kalla, R., and Olsen, O.-A. 1991. Promoter of a lipid-transfer protein gene expressed in barley aleurone cells contains similar *myb* and *myc* recognition sites as the maize *Bz-McC* allele. *Plant Physiol*. 97:841–843.

Linskens, H. 1955. Physiologische Untersuchungen der Pollenschlauch-Hemmung selbststeriler Petunien. *Z. Bot*. 43:1–44.

Linskens, H.F. 1958. Physiologische Untersuchungen zur Reifeteilung. II. Mitteilung. Über die Änderung des Nukleinsäurengehaltes während der Pollenmeiose und Pollenentwicklung von *Lilium henryi*. *Acta Bot. Neerl*. 7:61–68.

Linskens, H.F. 1959. Zur Frage der Entstehung der Abwehrkörper bei der Inkompatibilitätsreaktion von *Petunia*. *Ber. Deut. Bot. Ges*. 72:84–92.

Linskens, H.F. 1960. Zur Frage der Entstehung der Abwehr-Körper bei der Inkompatibilitätsreaktion von *Petunia*. III. Mitteilung: serologische Teste mit Leitgewebs- und Pollen-Extrakten. *Z. Bot*. 48:126–135.

Linskens, H.F. 1964. Pollen physiology. *Annu. Rev. Plant Physiol*. 15:255–270.

Linskens, H.F. 1966. Die Änderung des Protein- und Enzym-Musters während der Pollenmeiose und Pollenentwicklung. Physiologische Untersuchungen zur Reifeteilung. *Planta* 69:79–91.

Linskens, H.F. 1967. Isolation of ribosomes from pollen. *Planta* 73:194–200.

Linskens, H.F. 1969. Fertilization mechanisms in higher plants. In *Fertilization*, Vol. II, ed. C.B. Metz and A. Monroy, pp. 189–253. New York: Academic Press.

Linskens, H.F. 1974. Study of growth of *Petunia* styles. *Soviet Plant Physiol*. 21:878–882.

Linskens, H.F. 1975. Incompatibility in *Petunia*. *Proc. Roy. Soc. Lond*. 188B:299–311.

Linskens, H.F., and Heinen, W. 1962. Cutinase-Nachweis in Pollen. *Z. Bot.* 50:338–347.

Linskens, H.F., Kochuyt, A.S.L., and So, A. 1968. Regulation der Nucleinsäuren-Synthese durch Polyamine in keimendem Pollen von *Petunia. Planta* 82:111–122.

Linskens, H.F., and Schrauwen, J. 1968a. Änderung des Ribosomen-Musters während der Meiose. *Naturwissenschaften* 55:91.

Linskens, H.F., and Schrauwen, J. 1968b. Quantitative nucleic acid determinations in the microspore and tapetum fractions of lily anthers. *Proc. Koninkl. Nederl. Akad. Wetensch.* 71C:267–279.

Linskens, H.F., and Schrauwen, J. 1969. The release of free amino acids from germinating pollen. *Acta Bot. Neerl.* 18:605–614.

Linskens, H.F., Schrauwen, J.A.M., and Konings, R.N.H. 1970. Cell-free protein synthesis with polysomes from germinating *Petunia* pollen grains. *Planta* 90:153–162.

Linskens, H.F., and Tupý, J. 1966. The amino acids pool in the style of self-incompatible strains of *Petunia* after self- and cross-pollination. *Züchter* 36:151–158.

Linskens, H.F., van der Donk, J.A.W.M., and Schrauwen, J. 1971. RNA synthesis during pollen germination. *Planta* 97:290–298.

Linskens, H.F., Havez, R., Linder, R., Salden, M., Randoux, A., Laniez, D., and Coustaut, D. 1969. Etude des glycanne-hydrolases au cours de la croissance du pollen chez *Petunia hybrida* auto-incompatible. *Compt. Rend. Acad. Sci. Paris* 269D:1855–1857.

Lintilhac, P.M. 1974. Differentiation, organogenesis, and the tectonics of cell wall orientation. II. Separation of stresses in a two-dimensional model. *Am. J. Bot.* 61:135–140.

Lippi, M.M., Cimoli, F., Maugini, E., and Tani, G. 1994. A comparative study of the tapetal behaviour in male fertile and male sterile *Iris pallida* Lam. during microsporogenesis. *Caryologia* 47:109–120.

Lippmann, B., and Lippmann, G. 1984. Induction of somatic embryos in cotyledonary tissue of soybean, *Glycine max* L. Merr. *Plant Cell Rep.* 3:215–218.

Litts, J.C., Colwell, G.W., Chakerian, R.L., and Quatrano, R.S. 1987. The nucleotide sequence of a cDNA clone encoding the wheat E_m protein. *Nucl. Acids Res.* 15:3607–3618.

Litts, J.C., Colwell, G.W., Chakerian, R.L., and Quatrano, R.S. 1991. Sequence analysis of a functional member of the Em gene family from wheat. *DNA Sequence* 1:263–274.

Liu, B., and Palevitz, B.A. 1991. Kinetochore fiber formation in dividing generative cells of *Tradescantia.* Kinetochore reorientation associated with the transition between lateral microtubule interactions and end-on kinetochore fibers. *J. Cell Sci.* 98:475–482.

Liu, B., and Palevitz, B.A. 1992. Anaphase chromosome separation in dividing generative cells of *Tradescantia.* Changes in microtubule organization and kinetochore distribution visualized by antitubulin and CREST immunocytochemistry. *Protoplasma* 166:122–133.

Liu, C.-M., Johnson, S., and Wang, T.L. 1995. *cyd*, a mutant of pea that alters embryo morphology is defective in cytokinesis. *Devel Genet.* 16:321–331.

Liu, C.-M., Xu, Z.-H., and Chua, N.-H. 1993a. Proembryo culture: *in vitro* development of early globular-stage zygotic embryos from *Brassica juncea. Plant J.* 3:291–300.

Liu, C.-M., Xu, Z.-H., and Chua, N.-H. 1993b. Auxin polar transport is essential for the establishment of bilateral symmetry during early plant embryogenesis. *Plant Cell* 5:621–630.

Liu, C.-N., and Rubenstein, I. 1993. Transcriptional characterization of an -zein gene cluster in maize. *Plant Mol. Biol.* 22:323–336.

Liu, G.-Q., Cai, G., Del Casino, C., Tiezzi, A., and Cresti, M. 1994. Kinesin-related polypeptide is associated with vesicles from *Corylus avellana* pollen. *Cell Mot. Cytoskel.* 29:155–166.

Liu, Q., Golubovskaya, I., and Cande, W.Z. 1993. Abnormal cytoskeletal and chromosome distribution in *po, ms4* and *ms6;* mutant alleles of *polymitotic* that disrupt the cell cycle progression from meiosis to mitosis in maize. *J. Cell Sci.* 106:1169–1178.

Liu, R., Olsen, O.-A., Kreis, M., and Halford, N.G. 1992. Molecular cloning of a novel barley seed protein gene that is repressed by abscisic acid. *Plant Mol. Biol.* 18:1195–1198.

Liu, W., Hildebrand, D.F., Grayburn, W.S., Phillips, G.C., and Collins, G.B. 1991. Effects of exogenous auxins on expression of lipoxygenases in cultured soybean embryos. *Plant Physiol.* 97:969–976.

Liu, X., and Meyerowitz, E.M. 1995. *LEUNIG* regulates *AGAMOUS* expression in *Arabidopsis* flowers. *Development* 121:975–991.

Liu, X.C., and Dickinson, H.G. 1989. Cellular energy levels and their effect on male cell abortion in cytoplasmically male sterile lines of *Petunia hybrida. Sex. Plant Reprod.* 2:167–172.

Liu, X.C., Jones, K., and Dickinson, H.G. 1987. DNA synthesis and cytoplasmic differentiation in tapetal cells of normal and cytoplasmically male sterile lines of *Petunia hybrida. Theor. Appl. Genet.* 74:846–851.

Liu, X.C., Jones, K., and Dickinson, H.G. 1988. Cytoplasmic male sterility in *Petunia hybrida*: factors affecting mitochondrial ATP export in normal and cytoplasmically male sterile plants. *Theor. Appl. Genet.* 76:305–310.

Liu, X.-Y., Rocha-Sosa, M., Hummel, S., Willmitzer, L., and Frommer, W. B. 1991. A detailed study of the regulation and evolution of the two classes of patatin genes in *Solanum tuberosum* L. *Plant Mol. Biol.* 17:1139–1154.

Lloyd, J.R., Wang, T.L., and Hedley, C.L. 1996. Analysis of seed development in *Pisum sativum.* XIX. Effect of mutant alleles at the *r* and *rb* loci on starch grain size and on the content and composition of starch in developing pea seeds. *J. Expt. Bot.* 47:171–180.

Lo Schiavo, F., Giuliano, G., and Sung, Z.R. 1988. Characterization of a temperature-sensitive carrot cell mutant impaired in somatic embryogenesis. *Plant Sci.* 54:157–164.

Lo Schiavo, F., Quesada-Allue, L.A., and Sung, Z.R. 1986. Tunicamycin affects somatic embryogenesis but not cell proliferation of carrot. *Plant Sci.* 44:65–71.

Lo Schiavo, F., Filippini, F., Cozzani, F., Vallone, D., and Terzi, M. 1991. Modulation of auxin-binding proteins in cell suspensions. I. Differential responses of carrot embryo cultures. *Plant Physiol.* 97:60–64.

Lo Schiavo, F., Giuliano, G., de Vries, S.C., Genga, A., Bollini, R., Pitto, L., Cozzani, F., Nuti-Ronchi, V., and Terzi, M. 1990. A carrot cell variant temperature sensitive for somatic embryogenesis reveals a defect in the glycosylation of extracellular proteins. *Mol. Gen. Genet.* 223:385–393.

Lo Schiavo, F., Pitto, L., Giuliano, G., Torti, G., Nuti-Ronchi, V., Marazziti, D., Vergara, R., Orselli, S., and Terzi, M. 1989. DNA methylation of embryogenic carrot cell cultures and its variations as caused by mutation, differentiation, hormones and hypomethylating drugs. *Theor. Appl. Genet.* 77:325–331.

Loader, N.M., Woolner, E.M., Hellyer, A., Slabas, A.R., and Safford, R. 1993. Isolation and characterization of two *Brassica napus* embryo acyl-ACP thioesterase cDNA clones. *Plant Mol. Biol.* 23:769–778.

Loeb, T.A., and Reynolds, T.L. 1994. Transient expression of the *uidA* gene in pollen embryoids of wheat following microprojectile bombardment. *Plant Sci.* 104:81–91.

Loer, D.S., and Herman, E.M. 1993. Cotranslational integration of soybean (*Glycine max*) oil body membrane protein oleosin into microsomal membranes. *Plant Physiol.* 101:993–998.

Lohmer, S., Maddaloni, M., Motto, M., di Fonzo, N., Hartings, H., Salamini, F., and Thompson, R.D. 1991. The maize regulatory locus *opaque-2* encodes a DNA-binding protein which activates the transcription of the *b-32* gene. *EMBO J.* 10:617–624.

Lolle, S.J., and Cheung, A.Y. 1993. Promiscuous germination and growth of wildtype pollen from *Arabiopsis* and related species on the shoot of the *Arabidopsis* mutant, *fiddlehead. Devel Biol.* 155:250–258.

Lombardo, G., and Carraro, L. 1976a. Tapetal ultrastructural changes during pollen development. I. Studies on *Antirrhinum majus. Caryologia* 29:113–125.

Lombardo, G., and Carraro, L. 1976b. Tapetal ultrastructural changes during pollen development. III. Studies on *Gentiana acaulis. Caryologia* 29:345–349.

Lombardo, G., and Gerola, F.M. 1968a. Cytoplasmic inheritance and ultrastructure of the male generative cell of higher plants. *Planta* 82:105–110.

Lombardo, G., and Gerola, F.M. 1968b. Ultrastructure of the pollen grain and taxonomy. *Giorn. Bot. Ital.* 102:353–380.

Long, J.A., Moan, E.I., Medford, J.I., and Barton, M.K. 1996. A member of the KNOTTED class of homeodomain proteins encoded by the *STM* gene of *Arabidopsis. Nature* 379:66–69.

Long, S.R., Dale, R.M.K., and Sussex, I.M. 1981. Maturation and germination of *Phaseolus vulgaris* embryonic axes in culture. *Planta* 153:405–415.

Lonsdale, D., Önde, S., and Cuming, A. 1990. Transient expression of exogenous DNA in intact, viable wheat embryos following particle bombardment. *J. Expt. Bot.* 41:1161–1165.

Lopes, M.A., Coleman, C.E., Kodrzycki, R., Lending, C.R., and Larkins, B.A. 1994. Synthesis of an unusual α-zein protein is correlated with the phenotypic effects of the *floury2* mutation in maize. *Mol. Gen. Genet.* 245:537–547.

Lopes, M.A., and Larkins, B.A. 1991. Gamma-zein content is related to endosperm modification in quality protein maize. *Crop Sci.* 31:1655–1662.

Lopes, M.A., and Larkins, B.A. 1993. Endosperm origin, development and function. *Plant Cell* 5:1383–1399.

Lopez, I., Anthony, R.G., Maciver, S.K., Jiang, C.-J., Khan, S., Weeds, A.G., and Hussey, P.J. 1996. Pollen specific expression of maize genes encoding actin depolymerizing factor-like proteins. *Proc. Natl Acad. Sci. USA* 93:7415–7420.

Lord, E.M., and Eckard, K.J. 1984. Incompatibility between the dimorphic flowers of *Collomia grandiflora*, a cleistogamous species. *Science* 223:695–696.

Lord, E.M., and Eckard, K.J. 1986. Ultrastructure of the dimorphic pollen and stigmas of the cleistogamous species, *Collomia grandiflora* (Polemoniaceae). *Protoplasma* 132:12–22.

Lord, E.M., and Heslop-Harrison, Y. 1984. Pollen–stigma interaction in the Leguminosae: stigma organization and the breeding system in *Vicia faba* L. *Ann. Bot.* 54:827–836.

Lord, E.M., and Kohorn, L.U. 1986. Gynoecial development, pollination, and the path of pollen tube growth in the tepary bean, *Phaseolus acutifolius. Am. J. Bot.* 73:70–78.

Lord, E.M., and Webster, B.D. 1979. The stigmatic exudate of *Phaseolus vulgaris* L. *Bot. Gaz.* 140:266–271.

Lorenzi, R., Bennici, A., Cionini, P.G., Alpi, A., and D'Amato, F. 1978. Embryo-suspensor relations in *Phaseolus coccineus*: cytokinins during seed development. *Planta* 143:59–62.

Lotan, T., Ori, N., and Fluhr, R. 1989. Pathogenesis-related proteins are developmentally regulated in tobacco flowers. *Plant Cell* 1:881–887.

Louis, J.-P., Augur, C., and Teller, G. 1990. Cytokinins and differentiation processes in *Mercurialis annua.* Genetic regulation, relations with auxins, indoleacetic acid oxidases, and sexual expression patterns. *Plant Physiol.* 94:1535–1541.

Löve, A. 1943. A *Y*-linked inheritance of asynapsis in *Rumex acetosa. Nature* 152:358–359.

Lowe, J., and Nelson, O.E., Jr. 1946. Miniature seed – a study in the development of a defective caryopsis in maize. *Genetics* 31:525–533.

Lu, Y., and Rutger, J.N. 1984. Cytological observations on induced genetic male sterile mutants in rice (*Oryza sativa* L.). *Sci. Sinica* 27B:322–331.

Lu, Z.-X., Wu, M., Loh, C.-S., Yeong, C.-Y., and Goh, C.-J. 1993. Nucleotide sequence of a flower-specific MADS box cDNA clone from orchid. *Plant Mol. Biol.* 23:901–904.

Luck, B.T., and Jordan, E.G. 1980. The mitochondria and plastids during microsporogenesis in *Hyacinthoides non-scripta* (L.) Chouard. *Ann. Bot.* 45:511–514.

Ludevid, M.D., Freire, M.A., Gómez, J., Burd, C.G., Albericio, F., Giralt, E., Dreyfuss, G., and Pagès, M. 1992. RNA binding characteristics of a 16 kDa glycine-rich protein from maize. *Plant J.* 2:999–1003.

Ludevid, M.D., Torrent, M., Martinez-Izquierdo, J.A., Puigdomènech, P., and Palau, J. 1984. Subcellular localization of glutelin-2 in maize (*Zea mays* L.) endosperm. *Plant Mol. Biol.* 3:227–234.

Luegmayr, E. 1993. The generative cell and its close association with the endoplasmic reticulum of the vegetative cell in pollen of *Cyrtandra pendula* (Gesneriaceae). *Protoplasma* 177:73–81.

Lund, H.A. 1956. Growth hormones in the styles and ovaries of tobacco responsible for fruit development. *Am. J. Bot.* 43:562–568.

Lundqvist, A. 1956. Self-incompatibility in rye. I. Genetic control in the diploid. *Hereditas* 42:293–348.

Lundqvist, A. 1975. Complex self-incompatibility systems in angiosperms. *Proc. Roy. Soc. Lond.* 188B:235–245.

Lundqvist, A. 1993. The self-incompatibility system in *Lotus tenuis* (Fabaceae). *Hereditas* 119:59–66.

Lukowitz, W., Mayer, U., and Jürgens, G. 1996. Cytokinesis in the *Arabidopsis* embryo involves the syntaxin-related KNOLLE gene product. *Cell* 84:61–71.

Luo, Z.-X., and Wu, R. 1988. A simple method for the transformation of rice via the pollen-tube pathway. *Plant Mol. Biol. Rep.* 6:165–174.

Lupotto, E. 1983. Propagation of an embryogenic culture of *Medicago sativa* L. *Z. Pflanzenphysiol.* 111:95–104.

Lupotto, E. 1986. The use of single somatic embryo culture in propagating and regenerating lucerne (*Medicago sativa* L.). *Ann. Bot.* 57:19–24.

Lur, H.-S., and Setter, T.L. 1993. Role of auxin in maize endosperm development. Timing of nuclear DNA endoreduplication, zein expression, and cytokinin. *Plant Physiol.* 103:273–280.

Luthe, D.S. 1983. Storage protein accumulation in developing rice (*Oryza sativa* L.) seeds. *Plant Sci.* 32:147–158.

Luthe, D.S. 1987. Storage protein synthesis during oat (*Avena sativa* L.) seed development. *Plant Physiol.* 84:337–340.

Luthe, D.S., and Peterson, D.M. 1977. Cell-free synthesis of globulin by developing oat (*Avena sativa* L.) seeds. *Plant Physiol.* 59:836–841.

Lutz, R.W., and Sjolund, R.D. 1973. Development of the generative cell wall in *Monotropa uniflora* L. pollen. *Plant Physiol.* 52:498–500.

Lycett, G.W., Croy, R.R.D., Shirsat, A.H., and Boulter, D. 1984. The complete nucleotide sequence of a legumin gene from pea (*Pisum sativum* L.). *Nucl. Acids Res.* 12:4493–4506.

Lycett, G.W., Croy, R.R.D., Shirsat, A.H., Richards, D.M., and Boulter, D. 1985. The 5'-flanking regions of three pea legumin genes: comparison of the DNA sequences. *Nucl. Acids Res.* 13:6733–6743.

Lycett, G.W., Delauney, A.J., Gatehouse, J.A., Gilroy, J., Croy, R.R.D., and Boulter, D. 1983. The vicilin gene family of pea (*Pisum sativum* L.): a complete cDNA coding sequence for preprovicilin. *Nucl. Acids Res.* 11:2367–2380.

Ma, H. 1994. The unfolding drama of flower development: recent results from genetic and molecular analyses. *Genes Devel.* 8:745–756.

Ma, H., McMullen, M.D., and Finer, J.J. 1994. Identification of a homeobox-containing gene with enhanced expression during soybean (*Glycine max* L.) somatic embryo development. *Plant Mol. Biol.* 24:465–473.

Ma, H., Yanofsky, M.F., and Meyerowitz, E.M. 1991. *AGL1-AGL6*, an *Arabidopsis* gene family with similarity to floral homeotic and transcription factor genes. *Genes Devel.* 5:484–495.

Mackenzie, A., Heslop-Harrison, J., and Dickinson, H.G. 1967. Elimination of ribosomes during meiotic prophase. *Nature* 215:997–999.

MacKenzie, C.J., Yoo, B.Y., and Seabrook, J.E.A. 1f990. Stigma of *Solanum tuberosum* cv Shepody: morphology, ultrastructure, and secretion. *Am. J. Bot.* 77:1111–1124.

Mackenzie, S.A. 1991. Identification of a sterility-inducing cytoplasm in a fertile accession line of *Phaseolus vulgaris* L. *Genetics* 127:411–416.

Mackenzie, S.A., and Chase, C.D. 1990. Fertility restoration is associated with a loss of a portion of the mitochondrial genome in cytoplasmic male-sterile common bean. *Plant Cell* 2:905–912.

Mackenzie, S.A., Pring, D.R., Bassett, M.J., and Chase, C.D. 1988. Mitochondrial DNA rearrangement associated with fertility restoration and cytoplasmic reversion to fertility in cytoplasmic male sterile *Phaseolus vulgaris* L. *Proc. Natl Acad. Sci. USA* 85:2714–2717.

Maddaloni,, M., Donini, G., Balconi, C., Rizzi, E., Gallusci, P., Forlani, F., Lohmer, S., Thompson, R., Salamini, F., and Motto, M. 1996. The transcriptional activator *opaque-2* controls the expression of a cytosolic form of pyruvate orthophosphate dikinase-1 in maize endosperm. *Mol. Gen. Genet.* 250:647–654.

Maddock, S.E., Lancaster, V.A., Risiott, R., and Franklin, J. 1983. Plant regeneration from cultured immature embryos and inflorescences of 25 cultivars of wheat (*Triticum aestivum*). *J. Expt. Bot.* 34:915–926.

Maeda, E., and Maeda, K. 1990. Ultrastructure of egg apparatus of rice (*Oryza sativa*) after anthesis. *Jap. J. Crop Sci.* 59:179–197.

Maeda, M., Yoshioka, M., and Ito, M. 1979. Studies on the behavior of meiotic protoplasts. IV. Protoplasts isolated from microsporocytes of liliaceous plants. *Bot. Mag. Tokyo* 92:111–121.

Magoon, M.L., Jos, J.S., and Vasudevan, K.N. 1968. Male sterile cassava. *Nucleus* 11:1–6.

Maherchandani, N.J., and Naylor, J.M. 1971. Variability in DNA content and nuclear morphology of the aleurone cells of *Avena fatua* (wild oats). *Can. J. Genet. Cytol.* 13:578–584.

Maheshwari, N. 1958. *In vitro* culture of excised ovules of *Papaver somniferum. Science* 127:342.

Maheshwari, N., and Lal, M. 1961. *In vitro* culture of excised ovules of *Papaver somniferum* L. *Phytomorphology* 11:307–314.

Maheshwari, P. 1950. *An Introduction to the Embryology of Angiosperms*. New York: McGraw-Hill Book Co.

Maheshwari, S.C., and Prakash, R. 1965. Physiology of anther development in *Agave americana. Physiol. Plant.* 18:841–852.

Maheswaran, G., and Williams, E.G. 1984. Direct somatic embryoid formation on immature embryos of *Trifolium repens, T. pratense* and *Medicago sativa*, and rapid clonal propagation of *T. repens. Ann. Bot.* 54:201–211.

Maheswaran, G., and Williams, E.G. 1985. Origin and development of somatic embryoids formed directly on immature embryos of *Trifolium repens in vitro. Ann. Bot.* 56:619–630.

Mahlberg, P.G. 1960. Embryogeny and histogenesis in *Nerium oleander* L. I. Organization of primary meristematic tissues. *Phytomorphology* 10:118–131.

Maier, U.-G., Brown, J.W.S., Schmitz, L.M., Schwall, M., Dietrich, G., and Feix, G. 1988. Mapping of tissue-dependent and independent protein binding sites to the 5' upstream region of a zein gene. *Mol. Gen. Genet.* 212:241–245.

Maier, U.-G., Brown, J.W.S., Toloczyki, C., and Feix, G. 1987. Binding of a nuclear factor to a consensus sequence in the 5' flanking region of zein genes from maize. *EMBO J.* 6:17–22.

Maiti, I.B., Kolattukudy, P.E., and Shaykh, M. 1979. Purification and characterization of a novel cutinase from nasturtium (*Tropaeolum majus*) pollen. *Arch. Biochem. Biophys* 196:412–423.

Majewska, A., and Rodríguez-García, M.I. 1996. rRNA distribution during microspore development in anthers of *Beta vulgaris* L.: quantitative *in situ* hybridization analysis. *J. Cell Sci.* 109:859–866.

Majewska-Sawka, A., Rodríguez-García, M.I., Nakashima, H., and Jassen, B. 1993. Ultrastructural expression of cytoplasmic male sterility in sugar beet (*Beta vulgaris* L.). *Sex. Plant Reprod.* 6:22–32.

Makaroff, C.A., Apel, I.J., and Palmer, J.D. 1989. The *atp6* coding region has been disrupted and a novel reading frame generated in the mitochondrial genome of cytoplasmic male-sterile radish. *J. Biol. Chem.* 264:11706–11713.

Makaroff, C.A., and Palmer, J.D. 1988. Mitochondrial DNA rearrangements and transcriptional alterations in the male-sterile cytoplasm of Ogura radish. *Mol. Cell. Biol.* 8:1474–1480.

Mäkinen, Y., and Brewbaker, J.L. 1967. Isoenzyme polymorphism in flowering plants. I. Diffusion of enzymes out of intact pollen grains. *Physiol. Plant.* 20:477–482.

Mäkinen, Y.L.A., and Lewis, D. 1962. Immunological analysis of incompatibility (S) proteins and of cross-reacting material in a self-compatible mutant of *Oenothera organensis*. *Genet. Res.* 3:352–363.

Malhó, R., and Pais, M.S.S. 1992. Kinetics and hydrodynamics of *Agapanthus umbellatus* pollen-tube growth: a structural and stereological study. *Sex. Plant Reprod.* 5:163–168.

Malhó, R., Read, N.D., Pais, M.S., and Trewavas, A.J. 1994. Role of cytosolic free calcium in the reorientation of pollen tube growth. *Plant J.* 5:331–341.

Malhó, R., Read, N.D., Trewavas, A.J., and Pais, M.S. 1995. Calcium channel activity during pollen tube growth and reorientation. *Plant Cell* 7:1173–1184.

Malhotra, K., and Maheshwari, S.C. 1977. Enhancement by cold treatment of pollen embryoid development in *Petunia hybrida*. *Z. Pflanzenphysiol.* 85:177–180.

Malik, C.P., and Gupta, S.C. 1976. Electrophoretic analysis of protein from pollen and pollen tubes of *Calotropis procera* R. Br. *Indian J. Expt. Biol.* 14:688–690.

Malik, C.P., Singh, M., and Thapar, N. 1976. Physiology of sexual reproduction. IV. Histochemical characteristics of suspensor of *Brassica campestris*. *Phytomorphology* 26:384–389.

Malik, C.P., and Singh, M.B. 1977. Dehydrogenase and isocitrate lyase activity during pollen germination in *Calotropis procera*. *Proc. Indian Acad. Sci.* 86B:371–374.

Malik, C.P., Tewari, H.B., and Sood, P.P. 1969. On the functional significance of certain phosphatases in the germinating pollen grains of *Portulaca grandiflora*. *Portug. Acta Biol.* 11A:245–252.

Malik, C.P., and Vermani, S. 1975. Physiology of sexual reproduction. I. A histochemical study of the embryo sac development in *Zephyranthes rosea* and *Lagenaria vulgaris*. *Acta Histochem.* 53:244–280.

Malik, C.P., Vermani, S., and Bhatia, D.S. 1976. Physiology of sexual reproduction. III. Histochemical characteristics of suspensor during embryo development in *Brassica campestris* Linn. var., Sarson. *Acta Histochem.* 57:178–182.

Malti, and Shivanna, K.R. 1984. Structure and cytochemistry of the pistil of *Crotalaria retusa* L. *Proc. Indian Natl Sci. Acad.* 50B:92–102.

Mandaokar, A.D., Koundal, K.R., Kansal, R., and Bansal, H.C. 1993. Characterization of vicilin seed storage protein of chickpea (*Cicer arietinum* L.). *J. Plant Biochem. Biotech.* 2:35–38.

Mandaron, P., Niogret, M.F., Mache, R., and Monéger, F. 1990. *In vitro* protein synthesis in isolated microspores of *Zea mays* at several stages of development. *Theor. Appl. Genet.* 80:134–158.

Mandel, M.A., and Yanofsky, M.F. 1995. The *Arabidopsis AGL8* MADS box gene is expressed in inflorescence meristems and is negatively regulated by *APETALA1*. *Plant Cell* 7:1763–1771.

Mandel, M.A., Bowman, J.L., Kempin, S.A., Ma, H., Meyerowitz, E.M., and Yanofsky, M.F. 1992a. Manipulation of flower structure in transgenic tobacco. *Cell* 71:133–143.

Mandel, M.A., Gustafson-Brown, C., Savidge, B., and Yanofsky, M.F. 1992b. Molecular characterization of the *Arabidopsis* floral homeotic gene *APETALA1*. *Nature* 360:273–277.

Manen, J.F., and Pusztai, A. 1982. Immunocytochemical localisation of lectins in cells of *Phaseolus vulgaris* L. seeds. *Planta* 155:328–334.

Mangelsdorf, P.C. 1923. The inheritance of defective seeds in maize. *J. Hered.* 14:119–126.

Manickam, A., van Damme, E.J.M., Kaliselvi, K., Verhaert, P., and Peumans, W.J. 1996. Isolation and cDNA cloning of an Em-like protein from mung bean (*Vigna radiata*) axes. *Physiol. Plant.* 97:524–530.

Mann, V., McIntosh, L., Theurer, C., and Hirschberg, J. 1989. A new cytoplasmic male sterile genotype in the sugar beet *Beta vulgaris* L.: a molecular analysis. *Theor. Appl. Genet.* 78:293–297.

Mansfield, M.A., and Raikhel, N.V. 1990. Abscisic acid enhances the transcription of wheat-germ agglutinin mRNA without altering its tissue-specific expression. *Planta* 180:548–554.

Mansfield, S.G., and Briarty, L.G. 1991. Early embryogenesis in *Arabidopsis thaliana*. II. The developing embryo. *Can. J. Bot.* 69:461–476.

Mansfield, S.G., and Briarty, L.G. 1992. Cotyledon cell development in *Arabidopsis thaliana* during reserve deposition. *Can. J. Bot.* 70:151–164.

Mansfield, S.G., Briarty, L.G., and Erni, S. 1991. Early embryogenesis in *Arabidopsis thaliana*. I. The mature embryo sac. *Can. J. Bot.* 69:447–460.

Manteuffel, R., and Panitz, R. 1993. *In situ* localization of faba bean and oat legumin-type proteins in transgenic tobacco seeds by a highly sensitive immunological tissue print technique. *Plant Mol. Biol.* 22:1129–1134.

Manteuffel, R., Müntz, K., Püchel, M., and Scholz, G. 1976. Phase-dependent changes of DNA, RNA and protein accumulation during the ontogenesis of broad bean seeds (*Vicia faba* L., var. *minor*). *Biochem. Physiol. Pflanz.* 169:595–605.

Manzocchi, L.A., Bianchi, M.W., and Viotti, A. 1989. Expression of zein in long term cultures of wildtype and *opaque-2* maize endosperms. *Plant Cell Rep.* 7:639–643.

Manzocchi, L.A., Daminati, M.G., and Gentinetta, E. 1980. Viable defective endosperm mutants in maize. II. Kernel weight, nitrogen and zein accumulation during endosperm development. *Maydica* 25:199–210.

Manzocchi, L.A., Daminati, M.G., Gentinetta, E., and Salamini, F. 1980. Viable defective endosperm mutants in maize. I. Kernel weight, protein fractions and zein subunits in mature endosperms. *Maydica* 25:105–116.

Mao, Z., Paiva, R., Kriz, A.L., and Juvik, J.A. 1995. Dehydrin gene expression in normal and *viviparous* embryos of *Zea mays* during seed development and germination. *Plant Physiol. Biochem.* 33:649–653.

Maquoi, E., Hanke, D.E., and Deltour, R. 1993. The effects of abscisic acid on the maturation of *Brassica napus* somatic embryos. An ultrastructural study. *Protoplasma* 174:147–157.

Marciniak, K. 1993. DNA endoreplication level in

A., and Clarke, A.E. 1986. Style proteins of a wild tomato (*Lycopersicon peruvianum*) associated with expression of self-incompatibility. *Planta* 169:184–191.

Maurel, C., Brevet, J., Barbier-Brygoo, H., Guern, J., and Tempé, J. 1990. Auxin regulates the promoter of the root-inducing *rolB* gene of *Agrobacterium rhizogenes* in transgenic tobacco. *Mol. Gen. Genet.* 223:58–64.

May, K.W., and Kasha, K.J. 1980. The cytological expression and inheritance of desynapsis in a clone of diploid timothy (*Phleum nodosum* L.). *Euphytica* 29:233–240.

Mayer, U., Büttner, G., and Jürgens, G. 1993. Apical-basal pattern formation in the *Arabidopsis* embryo: studies on the role of the *gnom* gene. *Development* 117:149–162.

Mayer, U., Torres Ruiz, R.A., Berleth, T., Miséra, S., and Jürgens, G. 1991. Mutations affecting body organization in the *Arabidopsis* embryo. *Nature* 353:402–407.

Maze, J., and Bohm, L.R. 1974. Embryology of *Agrostis interrupta* (Gramineae). *Can. J. Bot.* 52:365–379.

Maze, J., and Bohm, L.R. 1977. Embryology of *Festuca microstachys* (Gramineae). *Can. J. Bot.* 55:1768–1782.

Maze, J., and Lin, S.-C. 1975. A study of the mature megagametophyte of *Stipa elmeri*. *Can. J. Bot.* 53:2958–2977.

McCabe, D.E., and Martinell, B.J. 1993. Transformation of elite cotton cultivars via particle bombardment of meristems. *Biotechnology* 11:596–598.

McCabe, D.E., Swain, W.F., Martinell, B.F., and Christou, P. 1988. Stable transformation of soybean (*Glycine max*) by particle acceleration. *Biotechnology* 6:923–926.

McCarty, D.R., and Carson, C.B. 1991. The molecular genetics of seed maturation in maize. *Physiol. Plant.* 81:267–272.

McCarty, D.R., Shaw, J.R., and Hannah, L.C. 1986. The cloning, genetic mapping, and expression of the constitutive sucrose synthase locus of maize. *Proc. Natl Acad. Sci. USA* 83:9099–9103.

McCarty, D.R., Carson, C.B., Lazar, M., and Simonds, S.C. 1989a. Transposable element-induced mutations of the *viviparous-1* gene in maize. *Devel Genet.* 10:473–481.

McCarty, D.R., Carson, C.B., Stinard, P.S., and Robertson, D.S. 1989b. Molecular analysis of *viviparous-1*: an abscisic acid-insensitive mutant of maize. *Plant Cell* 1:523–532.

McCarty, D.R., Hattori, T., Carson, C.B., Vasil, V., Lazar, M., and Vasil, I.K. 1991. The *viviparous-1* developmental gene of maize encodes a novel transcriptional activator. *Cell* 66:895–905.

McClure, B.A., Du, H., Liu, Y.-H., and Clarke, A.E. 1993. *S*-locus products in *Nicotiana alata* pistils are subject to organ-specific post-transcriptional processing but not post-translational processing. *Plant Mol. Biol.* 22:177–181.

McClure, B.A., Gray, J.E., Anderson, M.A., and Clarke, A.E. 1990. Self-incompatibility in *Nicotiana alata* involves degradation of pollen rRNA. *Nature* 347:757–760.

McClure, B.A., Haring, V., Ebert, P.E., Anderson, M.A., Simpson, R.J., Sukiyama, F., and Clarke, A.E. 1989. Style self-incompatibility gene products of *Nicotiana alata* are ribonucleases. *Nature* 342:955–957.

McConchie, C.A., Hough, T., and Knox, R.B. 1987. Ultrastructural analysis of the sperm cells of mature pollen of maize, *Zea mays*. *Protoplasma* 139:9–19.

McConchie, C.A., Jobson, S., and Knox, R.B. 1985. Computer-assisted reconstruction of the male germ unit in pollen of *Brassica campestris*. *Protoplasma* 127:57–63.

McConchie, C.A., Knox, R.B., and Ducker, S.C. 1982. Pollen wall structure and cytochemistry in the seagrass *Amphibolis griffithii* (Cymodoceaceae). *Ann. Bot.* 50:729–732.

McConchie, C.A., Russell, S.D., Dumas, C., Tuohy, M., and Knox, R.B. 1987. Quantitative cytology of the sperm cells of *Brassica campestris* and *B. oleracea*. *Planta* 170:446–452.

McConn, M., and Browse, J. 1996. The critical requirement for linolenic acid is pollen development, not photosynthesis, in an *Arabidopsis* mutant. *Plant Cell* 8:403–416.

McCormick, S. 1991. Molecular analysis of male gametogenesis in plants. *Trends Genet.* 7:298–303.

McCoy, K., and Knox, R.B. 1988. The plasma membrane and generative cell organization in pollen of the mimosoid legume, *Acacia retinodes*. *Protoplasma* 143:85–92.

McDaniel, C.N., and Poethig, R.S. 1988. Cell-lineage patterns in the shoot apical meristem of the germinating maize embryo. *Planta* 175:13–22.

McDonnell, R.E., and Conger, B.V. 1984. Callus induction and plantlet formation from mature embryo explants of Kentucky bluegrass. *Crop Sci.* 24:573–578.

McGee, J.D., Williams, E.G., Collins, G.B., and Hildebrand, D.F. 1989. Somatic embryogenesis in *Trifolium*: protein profiles associated with high- and low-frequency regeneration. *J. Plant Physiol.* 135:306–312.

McGranahan, G.H., Leslie, C.A., Uratsu, S.L., and Dandekar, A.M. 1990. Improved efficiency of the walnut somatic embryo gene transfer system. *Plant Cell Rep.* 8:512–516.

McGranahan, G.H., Leslie, C.A., Uratsu, S.L., Martin, L.A., and Dandekar, A.M. 1988. *Agrobacterium*-mediated transformation of walnut somatic embryos and regeneration of transgenic plants. *Biotechnology* 6:800–804.

McGuire, D.C., and Rick, C.M. 1954. Self-incompatibility in species of *Lycopersicon* sect. Eriopersicon and hybrids with *L. esculentum*. *Hilgardia* 23:101–124.

McHughen, A. 1980. The regulation of tobacco floral organ initiation. *Bot. Gaz.* 141:389–395.

McWilliam, A.A., Smith, S.M., and Street, H.E. 1974. The origin and development of embryoids in suspension cultures of carrot (*Daucus carota*). *Ann. Bot.* 38:243–250.

Meakin, P.J., and Gatehouse, J.A. 1991. Interaction of seed nuclear proteins with transcriptionally-enhancing regions of the pea (*Pisum sativum* L.) *legA* gene promoter. *Planta* 183:471–477.

Mecham, D.K., Fullington, J.G., and Greene, F.C. 1981. Gliadin proteins in the developing wheat seed. *J. Sci. Food Agr.* 32:773–780.

Medford, J.I., and Sussex, I.M. 1989. Regulation of chlorophyll and Rubisco levels in embryonic cotyledons of *Phaseolus vulgaris*. *Planta* 179:309–315.

Medina, F.J., and Risueño, M.C. 1981. Nucleolar structure and dynamics during meiotic prophase in pea ovules. *Biol. Cell.* 42:79–86.

Medina, F.J., Risueño, M.C., and Rodriguez-García, M.I. 1981. Evolution of the cytoplasmic organelles during female meiosis in *Pisum sativum* L. *Planta* 151:215–225.

Medina, F.J., Risueño, M.C., Rodriguez-García, M.I., and Sanchez-Pina, M.A. 1983. The nucleolar organizer (NOR) and fibrillar centers during plant gametogenesis. *J. Ultrastr. Res.* 85:300–310.

Meier, H., and Grant Reid, J.S. 1977. Morphological aspects of the galactomannan formation in the endosperm of *Trigonella foenum-graecum* L. (Leguminosae). *Planta* 133:243–248.

Meijer, E.G.M., and Brown, D.C.W. 1987a. Role of exogenous reduced nitrogen and sucrose in rapid high frequency somatic embryogenesis in *Medicago sativa. Plant Cell Tissue Organ Cult.* 10:11–19.

Meijer, E.G.M., and Brown, D.C.W. 1987b. A novel system for rapid high frequency somatic embryogenesis in *Medicago sativa. Physiol. Plant.* 69:591–596.

Meijer, E.G.M., and Simmonds, J. 1988. Polyamine levels in relation to growth and somatic embryogenesis in tissue cultures of *Medicago sativa* L. *J. Expt. Bot.* 39:787–794.

Meikle, P. J., Bonig, I., Hoogenraad, N.J., Clarke, A.E., and Stone, B.A. 1991. The location of (1→3)-β-glucans in the walls of pollen tubes of *Nicotiana alata* using a (1→3)-β-glucan-specific monoclonal antibody. *Planta* 185:1–8.

Meinke, D.W. 1982. Embryo-lethal mutants of *Arabidopsis thaliana*: evidence for gametophytic expression of the mutant genes. *Theor. Appl. Genet.* 63:381–386.

Meinke, D.W. 1985. Embryo-lethal mutants of *Arabidopsis thaliana*: analysis of mutants with a wide range of lethal phases. *Theor. Appl. Genet.* 69:543–552.

Meinke, D.W. 1986. Embryo-lethal mutants and the study of plant embryo development. *Oxford Surv. Plant Mol. Cell Biol.* 3:122–165.

Meinke, D.W. 1991a. Embryonic mutants of *Arabidopsis thaliana. Devel Genet.* 12:382–392.

Meinke, D.W. 1991b. Genetic analysis of plant development. In *Plant Physiology. A Treatise*, Vol. X, ed. F.C. Steward and R.G.S. Bidwell, pp. 437–490. San Diego: Academic Press.

Meinke, D.W. 1991c. Perspectives on genetic analysis of plant embryogenesis. *Plant Cell* 3:857–866.

Meinke, D.W. 1992. A homeotic mutant of *Arabidopsis thaliana* with leafy cotyledons. *Science* 258:1647–1650.

Meinke, D.W., Chen, J., and Beachy, R.N. 1981. Expression of storage-protein genes during soybean seed development. *Planta* 153:130–139.

Meinke, D.W., and Sussex, I.M. 1979a. Embryo-lethal mutants of *Arabidopsis thaliana*. A model system for genetic analysis of plant embryo development. *Devel Biol.* 72:50–61.

Meinke, D.W., and Sussex, I.M. 1979b. Isolation and characterization of six embryo-lethal mutants of *Arabidopsis thaliana. Devel Biol.* 72:62–72.

Meinke, D.W., Franzmann, L.H., Nickle, T.C., and Yeung, E.C. 1994. *Leafy cotyledon* mutants of *Arabidopsis. Plant Cell* 6:1049–1064.

Melcher, U. 1979. *In vitro* synthesis of a precursor to the methionine-rich polypeptide of the zein fraction of corn. *Plant Physiol.* 63:354–358.

Melchers, G., Mohri, Y., Watanabe, K., Wakabayashi, S., and Harada, K. 1992. One-step generation of cytoplasmic male sterility by fusion of mitochondrial-inactivated tomato protoplasts with nuclear-inactivated *Solanum* protoplasts. *Proc. Natl Acad. Sci. USA* 89:6832-6836.

Mena, M., Mandel, M.A., Lerner, D.R., Yanofsky, M.F., and Schmidt, R.J. 1995. A characterization of the MADS-box gene family in maize. *Plant J.* 8:845–854.

Mengoli, M., Bagni, N., Luccarini, G., Nuti Ronchi, V., and Serafini-Fracassini, D. 1989. *Daucus carota* cell cultures: polyamines and effect of polyamine biosynthesis inhibitors in the preembryogenic phase and different embryo stages. *J. Plant Physiol.* 134:389–394.

Menzel, G., Apel, K., and Melzer, S. 1995. Isolation and analysis of SaMADS C, the apetala 1 cDNA homolog from mustard. *Plant Physiol.* 108:853–854.

Mepham, R.H., and Lane, G.R. 1969. Formation and development of the tapetalperiplasmodium in *Tradescantia bracteata. Protoplasma* 68:175–192.

Mepham, R.H., and Lane, G.R. 1970. Observations on the fine structure of developing microspores of *Tradescantia bracteata. Protoplasma* 70:1–20.

Mericle, L.W., and Mericle, R.P. 1957. Irradiation of developing plant embryos. I. Effects of external irradiation (X rays) on barley embryogeny, germination and subsequent seedling development. *Am. J. Bot.* 44:747–756.

Mericle, L.W., and Mericle, R.P. 1961. Radiosensitivity of the developing plant embryo. In *Fundamental Aspects of Radiosensitivity. Brookhaven Symp. Biol.* 14:262–286.

Mericle, L.W., and Mericle, R.P. 1970. Nuclear DNA complement in young proembryos of barley. *Mutation Res.* 10:515–518.

Merkle, S.A., Parrott, W.A., and Flinn, B.S. 1995. Morphogenetic aspects of somatic embryogenesis. In In Vitro *Embryogenesis in Plants*, ed. T.A. Thorpe, pp. 155–203. Dordrecht: Kluwer Academic Publishers.

Mertz, E.T., Bates, L.S., and Nelson, O.E. 1964. Mutant gene that changes protein composition and increases lysine content of maize endosperm. *Science* 145:279–280.

Meurs, C., Basra, A.S., Karssen, C.M., and van Loon, L.C. 1992. Role of abscisic acid in the induction of desiccation tolerance in developing seeds of *Arabidopsis thaliana. Plant Physiol.* 98:1484–1493.

Meyer, K., Leube, M.P., and Grill, E. 1994. A protein phosphatase 2C involved in ABA signal transduction in *Arabidopsis thaliana. Science* 264:1452–1455.

Meyer, V.G. 1969. Some effects of genes, cytoplasm, and environment on male sterility of cotton (*Gossypium*). *Crop Sci.* 9:237–242.

Meyer, V.G., and Buffet, M. 1962. Cytoplasmic effects on external-ovule production in cotton. *J. Hered.* 53:251–253.

Meyerowitz, E.M. 1994. The genetics of flower development. *Sci. Am.* 271(5):40–47.

Meyerowitz, E.M., Bowman, J.L., Brockman, L.L., Drews, G.N., Jack, T., Sieburth, L., and Weigel, D. 1991. A genetic and molecular model for flower development in *Arabidopsis thaliana. Development* Suppl. 1:157–167.

Mian, H.R., Kuspira, J., Walker, G.W.R., and Muntjewerff, N. 1974. Histological and cytochemical studies on five genetic male-sterile lines of barley (*Hordeum vulgare*). *Can. J. Genet. Cytol.* 16:355–379.

Miao, S.-H., Kuo, C.-S., Kwei, Y.-L., Sun, A.-T., Ku, S.-Y., Lu, W.-L., Wang, Y.-Y., Chen, M.-L., Wu, M.-K., and Hang, L. 1978. Induction of pollen plants of maize and observations on their progeny. In *Proceedings of Symposium on Plant Tissue Culture*, pp. 23–33. Peking: Science Press.

Michalczuk, L., Cooke, T.J., and Cohen, J.D. 1992. Auxin levels at different stages of carrot somatic embryogenesis. *Phytochemistry* 31:1097–1103.

Mogensen, H.L., and Rusche, M.L. 1985. Quantitative ultrastructural analysis of barley sperm. I. Occurrence and mechanism of cytoplasm and organelle reduction and the question of sperm dimorphism. *Protoplasma* 128:1–13.

Mogensen, H.L., and Suthar, H.K. 1979. Ultrastructure of the egg apparatus of *Nicotiana tabacum* (Solanaceae) before and after fertilization. *Bot. Gaz.* 140:168–179.

Mogensen, H.L., and Wagner, V.T. 1987. Associations among components of the male germ unit following *in vivo* pollination in barley. *Protoplasma* 138:161–172.

Mogensen, H.L., Wagner, V.T., and Dumas, C. 1990. Quantitative, three-dimensional ultrastructure of isolated corn (*Zea mays*) sperm cells. *Protoplasma* 153:136–140.

Mogensen, H.L., Leduc, N., Matthys-Rochon, E., and Dumas, C. 1995. Nuclear DNA amounts in the egg and zygote of maize (*Zea mays* L.). *Planta* 197:641–645.

Mohan Ram, H.Y., and Jaiswal, V.S. 1971. Feminization of male flowers of *Cannabis sativa* L. by a morphactin. *Naturwissenschaften* 58:149–150.

Mohapatra, S.S., and Knox, R.B., ed. 1996. *Pollen Biotechnology. Gene Expression and Allergen Characterization*. New York: Chapman & Hall.

Mohapatra, S.S., Hill, R., Astwood, J., Ekramoddoullah, A.K.M., Olsen, E., Silvanovitch, A., Hatton, T., Kisil, F., and Sehon, A.H. 1990. Isolation and characterization of a cDNA clone encoding an IgE-binding protein from Kentucky bluegrass (*Poa pratensis*) pollen. *Int. Arch. Allergy Appl. Immunol.* 91:362–368.

Mohr, S., Schulte-Kappert, E., Oldenbach, W., Oettler, G., and Kück, U. 1993. Mitochondrial DNA of cytoplasmic male-sterile *Triticum timopheevi*: rearrangement of upstream sequences of the *atp6* and *orf25* genes. *Theor. Appl. Genet.* 86:259–268.

Mok, D.W.S., Mok, M.C., and Rabakoarihanta, A. 1978. Interspecific hybridization of *Phaseolus vulgaris* with *P. lunatus* and *P. acutifolius*. *Theor. Appl. Genet.* 52:209–215.

Mól, R. 1986. Isolation of protoplasts from female gametophytes of *Torenia fournieri*. *Plant Cell Rep.* 5:202–206.

Mól, R., Betka, A., and Wojciechowicz, M. 1995. Induction of autonomous endosperm in *Lupinus luteus*, *Helleborus niger* and *Melandrium album* by *in vitro* culture of unpollinated ovaries. *Sex. Plant Reprod.* 8:273–277.

Mòl, R., Matthys-Rochon, E., and Dumas, C. 1993. *In vitro* culture of fertilized embryo sacs of maize: zygotes and two-celled proembryos can develop into plants. *Planta* 189:213–217.

Mòl, R., Matthys-Rochon, E., and Dumas, C. 1994. The kinetics of cytological events during double fertilization in *Zea mays* L. *Plant J.* 5:197–206.

Mòl, R., Matthys-Rochon, E., and Dumas, C. 1995. Embryogenesis and plant regeneration from maize zygotes by *in vitro* culture of fertilized embryo sacs. *Plant Cell Rep.* 14:743–747.

Monéger, F., Smart, C.J., and Leaver, C.J. 1994. Nuclear restoration of cytoplasmic male sterility in sunflower is associated with the tissue-specific regulation of a novel mitochondrial gene. *EMBO J.* 13:8–17.

Monéger, F., Mandaron, P., Niogret, M.-F., Freyssinet, G., and Mache, R. 1992. Expression of chloroplast and mitochondrial genes during microsporogenesis in maize. *Plant Physiol.* 99:396–400.

Monfort, S. 1985. Androgenesis of coconut: embryos from anther culture. *Z. Pflanzenzüchtg* 94:251–254.

Monnier, M. 1975. Action d'un gel de polyacrylamide employé comme support pour la culture de l'embryon immature de *Capsella bursa-pastoris. Compt. Rend. Acad. Sci. Paris* 280D:705–708.

Monnier, M. 1976a. Culture *in vitro* de l'embryon immature de *Capsella bursa-pastoris* Moench. *Rev. Cytol. Biol. Végét.* 39:1–120.

Monnier, M. 1976b. Action de la pression partielle d'oxygène sur le développement de l'embryon de *Capsella bursa-pastoris* cultivé *in vitro*. *Compt. Rend. Acad. Sci. Paris* 282D:1009–1012.

Monnier, M. 1984. Survival of young immature *Capsella* embryos cultured *in vitro*. *J. Plant Physiol.* 115:105–113.

Montague, M.J., Armstrong, T.A., and Jaworski, E.G. 1979. Polyamine metabolism in embryogenic cells of *Daucus carota*. II. Changes in arginine decarboxylase activity. *Plant Physiol.* 63:341–345.

Montague, M.J., Koppenbrink, J.W., and Jaworski, E.G. 1978. Polyamine metabolism in embryogenic cells of *Daucus carota*. I. Changes in intracellular content and rates of synthesis. *Plant Physiol.* 62:430–433.

Moore, C.W., and Creech, R.G. 1972. Genetic fine structure analysis of the *amylose-extender* locus in *Zea mays* L. *Genetics* 70:611–619.

Moore, H.M., and Nasrallah, J.B. 1990. A *Brassica* self-incompatibility gene is expressed in the stylar transmitting tissue of transgenic tobacco. *Plant Cell* 2:29–38.

Moore, P.J., Moore, A.J., and Collins, G.B. 1994. Genotypic and developmental regulation of transient expression of a reporter gene in soybean zygotic cotyledons. *Plant Cell Rep.* 13:556–560.

Moreira, M.A., Hermodson, M.A., Larkins, B.A., and Nielsen, N.C. 1979. Partial characterization of the acidic and basic polypeptides of glycinin. *J. Biol. Chem.* 254:9921–9926.

Moreira, M.A., Hermodson, M.A., Larkins, B.A., and Nielsen, N.C. 1981. Comparison of the primary structure of the acidic polypeptides of glycinin. *Arch. Biochem. Biophys* 210:633–642.

Moreno, J., and Chrispeels, M.J. 1989. A lectin gene encodes the α-amylase inhibitor of the common bean. *Proc. Natl Acad. Sci. USA* 86:7885–7889.

Morgan, J.M. 1980. Possible role of abscisic acid in reducing seed set in water-stressed wheat plants. *Nature* 285:655–657.

Morris, D.A. 1978. Germination inhibitors in developing seeds of *Phaseolus vulgaris* L. *Z. Pflanzenphysiol.* 86:433–441.

Morris, P.C., Kumar, A., Bowles, D.J., and Cuming, A.C. 1990. Osmotic stress and abscisic acid induce expression of the wheat Em genes. *Eur. J. Biochem.* 190:625–630.

Morris, P.C., Maddock, S.E., Jones, M.G.K., and Bowles, D.J. 1985. Changes in the levels of wheat- and barley-germ agglutinin during embryogenesis *in vivo, in vitro* and during germination. *Planta* 166:407–413.

Morris, P.C., Weiler, E.W., Maddock, S.E., Jones, M.G.K., Lenton, J.R., and Bowles, D.J. 1988. Determination of endogenous abscisic acid levels in immature cereal embryos during *in vitro* culture. *Planta* 173:110–116.

Morrison, I.N., Kuo, J., and O'Brien, T.P. 1975. Histochemistry and fine structure of developing wheat aleurone cells. *Planta* 123:105–116.

Morrison, I.N., and O'Brien, T.P. 1976. Cytokinesis in the developing wheat grain; division with and without a phragmoplast. *Planta* 130:57–67.

Morrison, I.N., O'Brien, T.P., and Kuo, J. 1978. Initial cellularization and differentiation of the aleurone cells in the ventral region of the developing wheat grain. *Planta* 140:19–30.

Morrison, R.A., and Evans, D.A. 1987. Gametoclonal variation. *Plant Breed. Rev.* 5:359–391.

Morrison, R.A., and Evans, D.A. 1988. Haploid plants from tissue culture: new plant varieties in a shortened time frame. *Biotechnology* 6:684–690.

Morton, C.M., Lawson, D.L., and Bedinger, P. 1989. Morphological study of the maize male sterile mutant *ms7*. *Maydica* 34:239–245.

Morton, H., Evans, I.M., Gatehouse, J.A., and Boulter, D. 1983. Sequence complexity of messenger RNA in cotyledons of developing pea (*Pisum sativum*) seed. *Phytochemistry* 22:807–812.

Morton, R.K., Palk, B.A., and Raison, J.K. 1964. Intracellular components associated with protein synthesis in developing wheat endosperm. *Biochem. J.* 91:522–528.

Moscatelli, A., Del Casino, C., Lozzi, L., Cai, G., Scali, M., Tiezzi, A., and Cresti, M. 1995. High molecular weight polypeptides related to dynein heavy chains in *Nicotiana tabacum* pollen tubes. *J. Cell Sci.* 108:1117–1125.

Moses, M.J., and Taylor, J.H. 1955. Desoxypentose nucleic acid synthesis during microsporogenesis in *Tradescantia*. *Expt. Cell Res.* 9:474–488.

Moss, G.I., and Heslop-Harrison, J. 1967. A cytochemical study of DNA, RNA, and protein in the developing maize anther. II. Observations. *Ann. Bot.* 31:555–572.

Moss, G.I., and Heslop-Harrison, J. 1968. Photoperiod and pollen sterility in maize. *Ann. Bot.* 32:833–846.

Motto, M., Maddaloni, M., Ponziani, G., Brembilla, M., Marotta, R., di Fonzo, N., Soave, C., Thompson, R., and Salamini, F. 1988. Molecular cloning of the *o2-m5* allele of *Zea mays* using transposon marking. *Mol. Gen. Genet.* 212:488–494.

Mouritzen, P., and Holm, P.B. 1995. Isolation and culture of barley megasporocyte protoplasts. *Sex. Plant Reprod.* 8:321–325.

Mu, J.-H., Lee, H.-S., and Kao, T.-H. 1994. Characterization of a pollen-expressed receptor-like kinase gene of *Petunia inflata* and the activity of its encoded kinase. *Plant Cell* 6:709–721.

Mu, J.-H., Stains, J.P., and Kao, T.-H. 1994. Characterization of a pollen-expressed gene encoding a putative pectin esterase of *Petunia inflata*. *Plant Mol. Biol.* 25:539–544.

Mulcahy, D.L., and Mulcahy, G.B. 1983. Gametophytic self-incompatibility reexamined. *Science* 220:1247–1251.

Mulcahy, D.L., Mulcahy, G.B., and Robinson, R.W. 1979. Evidence for postmeiotic genetic activity in pollen of *Cucurbita* species. *J. Hered.* 70:365–368.

Mulcahy, D.L., Robinson, R.W., Ihara, M., and Kesseli, R. 1981. Gametophytic transcription for acid phosphatases in pollen of *Cucurbita* species hybrids. *J. Hered.* 72:353–354.

Mulcahy, G.B., and Mulcahy, D.L. 1985. Ovarian influence on pollen tube growth, as indicated by the semivivo technique. *Am. J. Bot.* 72:1078–1080.

Mulcahy, G.B., and Mulcahy, D.L. 1987. Induced pollen tube directionality. *Am. J. Bot.* 74:1458–1459.

Müller, A.J. 1963. Embryonentest zum Nachweis rezessiver Letalfaktoren bei *Arabidopsis thaliana*. *Biol. Zentral.* 82:133–163.

Müller-Röber, B., la Cognata, U., Sonnewald, U., and Willmitzer, L. 1994. A truncated version of an ADP-glucose pyrophosphorylase promoter from potato specifies guard cell-selective expression in transgenic plants. *Plant Cell* 6:601–612.

Mullins, M.G., Tang, F.C.A., and Facciotti, D. 1990. *Agrobacterium*-mediated genetic transformation of grapevines: transgenic plants of *Vitis rupestris* Scheele and buds of *Vitis vinifera* L. *Biotechnology* 8:1041–1045.

Mundy, J., and Chua, N.-H. 1988. Abscisic acid and water-stress induce the expression of a novel rice gene. *EMBO J.* 7:2279–2286.

Munjal, S.V., and Narayan, R.K.J. 1995. Restriction analysis of the mitochondrial DNA of male-sterile and maintainer lines of pearl millet, *Pennisetum americanum* L. *Plant Breed.* 114:256–258.

Munksgaard, D., Mattsson, O., and Okkels, F.T. 1995. Somatic embryo development in carrot is associated with an increase in levels of S-adenosylmethionine, S-adenosylhomocysteine and DNA methylation. *Physiol. Plant.* 93:5–10.

Muñoz, C.A.P., Webster, B.D., and Jernstedt, J.A. 1995. Spatial congruence between exine pattern, microtubules and endomembranes in *Vigna* pollen. *Sex. Plant Reprod.* 8:147–151.

Müntz, K. 1987. Developmental control of storage protein formation and its modulation by some internal and external factors during embryogenesis in plant seeds. *Biochem. Physiol. Pflanz.* 182:93–116.

Murai, N., Sutton, D.W., Murray, M.G., Slightom, J.L., Merlo, D.J., Reichert, N.A., Sengupta-Gopalan, C., Stock, C.A., Barker, R.F., Kemp, J.D., and Hall, T.C. 1983. Phaseolin gene from bean is expressed after transfer to sunflower via tumor-inducing plasmid vectors. *Science* 222:476–482.

Murfett, J., Bourque, J.E., and McClure, B.A. 1995. Antisense suppression of *S*-RNase expression in *Nicotiana* using RNA polymerase II- and III-transcribed gene constructs. *Plant Mol. Biol.* 29:201–212.

Murfett, J., Atherton, T.L., Mou, B., Gasser, C.S., and McClure, B.A. 1994. S-RNase expressed in transgenic *Nicotiana* causes S-allele-specific pollen rejection. *Nature* 367:563–566.

Murfett, J., Cornish, E.C., Ebert, P.R., Bönig, I., McClure, B.A., and Clarke, A.E. 1992. Expression of a self-incompatibility glycoprotein (S_2-ribonuclease) from *Nicotiana alata* in transgenic *Nicotiana tabacum*. *Plant Cell* 4:1063–1074.

Murfett, J., Ebert, P.R., Haring, V., and Clarke, A.E. 1995. An *S*-RNase promoter from *Nicotiana alata* functions in transgenic *N. alata* plants but not in *Nicotiana tabacum*. *Plant Mol. Biol.* 28:957–963.

Murfett, J., Strabala, T.J., Zurek, D.M., Mou, B., Beecher, B., and McClure, B.A. 1996. S RNase and interspecific pollen rejection in the genus *Nicotiana*: multiple pollen-rejection pathways contribute to unilateral incompatibility between self-incompatible and self-compatible species. *Plant Cell* 8:943–958.

Murgia, M., and Wilms, H.J. 1988. Three-dimensional image and mitochondrial distribution in sperm cells of *Euphorbia dulcis*. In *Plant Sperms Cells as Tools for Biotechnology*, ed. H.J. Wilms and C.J. Keijzer, pp. 75–79. Wageningen: Pudoc.

Murgia, M., Charzyńska, M., Rougier, M., and Cresti, M.

Panicum maximum. Sex. Plant Reprod. 8:107–204.

Nave, E.B., and Sawhney, V.K. 1986. Enzymatic changes in post-meiotic anther development in *Petunia hybrida*. I. Anther ontogeny and isozyme analyses. *J. Plant Physiol.* 125:451–465.

Nawaschin, S.G. 1898. Resultate einer Revision der Befruchtungsvorgänge bei *Lilium martagon* und *Fritillaria tenella*. *Bull. Acad. Imp. Sci. St. Petersburg* 9:377–382.

Negruk, V.I., Cherny, D.I., Nififorova, I.D., Aleksandrov, A.A., and Butenko, R.G. 1982. Isolation and characterization of minicircular DNAs found in mitochondrial fraction of *Vicia faba*. *FEBS Lett.* 142:115–117.

Nehra, N.S., Chibbar, R.N., Leung, N., Caswell, K., Mallard, C., Steinhauer, L., Baga, M., and Kartha, K.K. 1994. Self-fertile transgenic wheat plants regenerated from isolated scutellar tissues following microprojectile bombardment with two distinct gene constructs. *Plant J.* 5:285–297.

Neill, J.D., Litts, J.C., Anderson, O.D., Greene, F.C., and Stiles, J.I. 1987. Expression of a wheat α-gliadin gene in *Saccharomyces cerevisiae*. *Gene* 55:303–317.

Neill, S.J., Horgan, R., and Parry, A.D. 1986. The carotenoid and abscisic acid content of viviparous kernels and seedlings of *Zea mays* L. *Planta* 169:87–96.

Neill, S.J., Horgan, R., and Rees, A.F. 1987. Seed development and vivipary in *Zea mays* L. *Planta* 171:358–364.

Nelson, O.E., Chourey, P.S., and Chang, M.T. 1978. Nucleoside diphosphate sugar-starch glucosyl transferase activity of *wx* starch granules. *Plant Physiol.* 62:383–386.

Nelson, O.E., Mertz, E.T., and Bates, L.S. 1965. Second mutant gene affecting the amino acid pattern of maize endosperm proteins. *Science* 150:1469–1470.

Nelson, O.E., and Rines, H.W. 1962. The enzymatic deficiency in the waxy mutant of maize. *Biochem. Biophys. Res. Commun.* 9:297–300.

Nelson, O.E., and Tsai, C.Y. 1964. Glucose transfer from adenosine diphosphate-glucose to starch in preparations of waxy seeds. *Science* 145:1194–1195.

Nelson, P.M., and Rossman, E.C. 1958. Chemical induction of male sterility in inbred maize by use of gibberellins. *Science* 127:1500–1501.

Nepi, M., Ciampolini, F., and Pacini, E. 1995. Development of *Cucurbita pepo* pollen: ultrastructure and histochemistry of the sporoderm. *Can. J. Bot.* 73:1046–1057.

Neto, G.C., Yunes, J.A., da Silva, M.J., Vettore, A.L., Arruda, P., and Leite, A. 1995. The involvement of opaque 2 on β-prolamin gene regulation in maize and *Coix* suggests a more general role for this transcriptional activator. *Plant Mol. Biol.* 27:1015–1029.

Neuffer, M.G., and Sheridan, W.F. 1980. Defective kernel mutants of maize. I. Genetic and lethality studies. *Genetics* 95:929–944.

Neuhaus, G., Spangenberg, G., Scheid, O.M., and Schweiger, H.-G. 1987. Transgenic rapeseed plants obtained by the microinjection of DNA into microspore-derived embryoids. *Theor. Appl. Genet.* 75:30–36.

Neumann, D., and Weber, E. 1978. Formation of protein bodies in ripening seeds of *Vicia faba* L. *Biochem. Physiol. Pflanz.* 173:167–180.

Newbigin, E., Anderson, M.A., and Clarke, A.E. 1993. Gametophytic self-incompatibility systems. *Plant Cell* 5:1315–1324.

Newbigin, E.J., deLumen, B.O., Chandler, P.M., Gould, A., Blagrove, R.J., March, J.F., Kortt, A.A., and Higgins, T.J.V. 1990. Pea convicilin: structure and primary sequence of the protein and expression of a gene in the seeds of transgenic tobacco. *Planta* 180:461–470.

Newcomb, W. 1973a. The development of the embryo sac of sunflower *Helianthus annuus* before fertilization. *Can. J. Bot.* 51:863–878.

Newcomb, W. 1973b. The development of the embryo sac of sunflower *Helianthus annuus* after fertilization. *Can. J. Bot.* 51:879–890.

Newcomb, W. 1978. The development of cells in the coenocytic endosperm of the African blood lily *Haemanthus katherinae*. *Can. J. Bot.* 56:483–501.

Newcomb, W., and Fowke, L.C. 1973. The fine structure of the change from the free-nuclear to cellular condition in the endosperm of chickweed *Stellaria media*. *Bot. Gaz.* 134:236–241.

Newcomb, W., and Fowke, L.C. 1974. *Stellaria media* embryogenesis: the development and ultrastructure of the suspensor. *Can. J. Bot.* 52:607–614.

Newcomb, W., and Steeves, T.A. 1971. *Helianthus annuus* embryogenesis: embryo sac wall projections before and after fertilization. *Bot. Gaz.* 132:367–371.

Newton, C.H., Flinn, B.S., and Sutton, B.C.S. 1992. Vicilin-like seed storage proteins in the gymnosperm interior spruce (*Picea glauca/engelmanii*). *Plant Mol. Biol.* 20:315–322.

Ng, J.D., Ko, T.-Z., and McPherson, A. 1993. Cloning, expression, and crystallization of jack bean (*Canavalia ensiformis*) canavalin. *Plant Physiol.* 101:713–728.

Nielsen, K.A., and Hansen, I.B. 1992. Appearance of extracellular proteins associated with somatic embryogenesis in suspension cultures of barley (*Hordeum vulgare* L.). *J. Plant Physiol.* 139:489–497.

Nielsen, N.C., Dickinson, C.D., Cho, T.-J., Thanh, V.H., Scallon, B.J., Fischer, R.L., Sims, T.L., Drews, G.N., and Goldberg, R.B. 1989. Characterization of the glycinin gene family in soybean. *Plant Cell* 1:313–328.

Niimi, Y. 1970. *In vitro* fertilization in the self-incompatible plant, *Petunia hybrida*. *J. Jap. Soc. Hort. Sci.* 39:346–352.

Niitsu, T., Hanaoka, A., and Uchiyama, K. 1972. Reinvestigations of the spindle body and related problems. II. Effects of cycloheximide upon the development of kinetochore fibers studied in the pollen mitosis of *Ornithogalum virens in vivo*. *Cytologia* 37:143–154.

Niizeki, H., and Oono, K. 1968. Induction of haploid rice plant from anther culture. *Proc. Jap. Acad.* 44:554–557.

Ninnemann, H., and Epel, B. 1973. Inhibition of cell division by blue light. *Expt. Cell Res.* 79:318–326.

Niogret, M.F., Culiáñez-Macià, Goday, A., Albà, M.M., and Pagès, M. 1996. Expression and cellular localization of *rab*28 mRNA and RAB28 protein during maize embryogenesis. *Plant J.* 9:549–557.

Niogret, M.-F., Dubald, M., Mandaron, P., and Mache, R. 1991. Characterization of pollen polygalacturonase encoded by several cDNA clones in maize. *Plant Mol. Biol.* 17:1155–1164.

Nirmala, C., and Kaul, M.L.H. 1994. Male sterility in pea. V. Gene action during heterotypic and homotypic divisions. *Cytologia* 59:43–50.

Nishihara, M., Ito, M., Tanaka, I., Kyo, M., Ono, K., Irifune, K., and Morikawa, H. 1993. Expression of the β-glucuronidase gene in pollen of lily (*Lilium longiflorum*), tobacco (*Nicotiana tabacum*), *Nicotiana rustica*, and peony (*Paeonia lactiflora*) by particle bombardment. *Plant Physiol.* 102:357–361.

Nishio, T., and Hinata, K. 1977. Analysis of S-specific proteins in stigma of *Brassica oleracea* L. by isoelectric focusing. *Heredity* 38:391–396.

Nishio, T., and Hinata, K. 1982. Comparative studies on S-glycoproteins purified from different S-genotypes in self-incompatible *Brassica* species. I. Purification and chemical properties. *Genetics* 100:641–647.

Nishio, T., Toriyama, K., Sato, T., Kandasamy, M.K., Paolillo, D.J., Nasrallah, J.B., and Nasrallah, M.K. 1992. Expression of S-locus glycoprotein genes from *Brassica oleracea* and *B. campestris* in transgenic plants of self-compatible *B. napus* cv Westar. *Sex. Plant Reprod.* 5:101–109.

Nishizawa, N.K., Mori, S., Watanabe, Y., and Hirano, H. 1994. Ultrastructural localization of the basic 7S globulin in soybean (*Glycine max*) cotyledons. *Plant Cell Physiol.* 35:1079–1085.

Nissen, P. 1994. Stimulation of somatic embryogenesis in carrot by ethylene: effects of modulators of ethylene biosynthesis and action. *Physiol. Plant.* 92:397–403.

Nitsch, C. 1974a. Pollen culture - a new technique for mass production of haploid and homozygous plants. In *Haploids in Higher Plants. Advances and Potential*, ed. K.J. Kasha, pp. 123–135. Guelph: The University of Guelph.

Nitsch, C. 1974b. La culture de pollen isolé sur milieu synthétique. *Compt. Rend. Acad. Sci. Paris* 278D:1031–1034.

Nitsch, C., and Norreel, B. 1973. Effet d'un choc thermique sur le pouvoir embryogène du pollen de *Datura innoxia* cultivé dans l'anthère ou isolé de l'anthère. *Compt. Rend. Acad. Sci. Paris* 276D:303–306.

Nitsch, J.P. 1972. Haploid plants from pollen. *Z. Pflanzenzüchtg* 67:3–18.

Nitsch, J.P., and Nitsch, C. 1969. Haploid plants from pollen grains. *Science* 163:85–87.

Nivison, H.T., and Hanson, M.R. 1989. Identification of a mitochondrial protein associated with cytoplasmic male sterility in *Petunia. Plant Cell* 1:1121–1130.

Nivison, H.T., Sutton, C.A., Wilson, R.K., and Hanson, M.R. 1994. Sequencing, processing, and localization of the petunia cms-associated mitochondrial protein. *Plant J.* 5:613–623.

Nobiling, R., and Reiss, H.-D. 1987. Quantitative analysis of calcium gradients and activity in growing pollen tubes of *Lilium longiflorum. Protoplasma* 139:20–24.

Nogler, G.A. 1984. Gametophytic apomixis. In *Embryology of Angiosperms*, ed. B.M. Johri, pp. 475–518. Berlin: Springer-Verlag.

Noguchi, T. 1990. Consumption of lipid granules and formation of vacuoles in the pollen tube of *Tradescantia reflexa. Protoplasma* 156:19–28.

Noguchi, T., and Morré, D.J. 1991. Membrane flow in plants: preparation and kinetics of labelling of plasma membranes from growing pollen tubes of tobacco. *Protoplasma* 163:34–42.

Noguchi, T., and Ueda, K. 1990. Structure of pollen grains of *Tradescantia reflexa* with special reference to the generative cell and the ER around it. *Cell Struct. Funct.* 15:379–384.

Noher de Halac, I. 1980a. Fine structure of nucellar cells during development of the embryo sac in *Oenothera biennis* L. *Ann. Bot.* 45:515–521.

Noher de Halac, I. 1980b. Callose deposition during megagametogenesis in two species of *Oenothera. Ann. Bot.* 46:473–477.

Noher de Halac, I., Cismondi, I.A., and Harte, C. 1990. Pollen ontogenesis in *Oenothera*: a comparison of genotypically normal anthers with the male-sterile mutant *sterilis*. *Sex. Plant Reprod.* 3:41–53.

Noher de Halac, I., and Harte, C. 1977. Different patterns of callose wall formation during megasporogenesis in two species of *Oenothera* (Onagraceae). *Plant Syst. Evol.* 127:23–38.

Noma, M., Huber, J., Ernst, D., and Pharis, R.P. 1982. Quantitation of gibberellins and the metabolism of $[^3H]$gibberellin A_1 during somatic embryogenesis in carrot and anise cell cultures. *Planta* 155:369–376.

Nomura, K., and Komamine, A. 1985. Identification and isolation of single cells that produce somatic embryos at a high frequency in a carrot suspension culture. *Plant Physiol.* 79:988–991.

Nomura, K., and Komamine, A. 1986a. Polarized DNA synthesis and cell division in cell clusters during somatic embryogenesis from single carrot cells. *New Phytol.* 104:25–32.

Nomura, K., and Komamine, A. 1986b. *In situ* hybridization on tissue sections. A method for detection of poly(A)$^+$RNA. *Plant Tissue Cult. Lett. Japan* 3:92–94.

Nomura, K., and Komamine, A. 1995. Physiological and biochemical aspects of somatic embryogenesis. In In Vitro *Embryogenesis in Plants*, ed. T.A. Thorpe, pp. 249–265. Dordrecht: Kluwer Academic Publishers.

Nong, V.H., Schlesier, B., Bassüner, R., Repik, A., Horstmann, C., and Müntz, K. 1995. Narbonin, a novel 2S protein from *Vicia narbonensis* L. seeds: cDNA, gene structure and developmentally regulated formation. *Plant Mol. Biol.* 28:61–72.

Nørgaard, J.V., and Krogstrup, P. 1991. Cytokinin induced somatic embryogenesis from immature embryos of *Abies nordmanniana* Lk. *Plant Cell Rep.* 9:509–513.

Norreel, B. 1972. Etude comparative de la répartition des acides ribonucléiques au cours de l'embryogenèse zygotique et de l'embryogenèse androgénétique chez le *Nicotiana tabacum* L. *Compt. Rend. Acad. Sci. Paris* 275D:1219–1222.

Norstog, K. 1961. The growth and differentiation of cultured barley embryos. *Am. J. Bot.* 48:876–884.

Norstog, K. 1972a. Early development of the barley embryo: fine structure. *Am. J. Bot.* 59:123–132.

Norstog, K. 1972b. Factors relating to precocious germination in cultured barley embryos. *Phytomorphology* 22:134–139.

Norstog, K. 1974. Nucellus during early embryogeny in barley: fine structure. *Bot. Gaz.* 135:97–103.

Norstog, K., and Klein, R.M. 1972. Development of cultured barley embryos. II. Precocious germination and dormancy. *Can. J. Bot.* 50:1887–1894.

Norstog, K., and Smith, J.E. 1963. Culture of small barley embryos on defined media. *Science* 142:1655–1656.

Nozue, K., and Wada, M. 1993. Electrotropism of *Nicotiana* pollen tubes. *Plant Cell Physiol.* 34:1291–1296.

Nunberg, A.N., Li, Z., Bogue, M.A., Vivekananda, J., Reddy, A.S., and Thomas, T.L. 1994. Developmental and hormonal regulation of sunflower helianthinin genes: proximal promoter sequences confer regionalized seed expression. *Plant Cell* 6:473–486.

Nuti Ronchi, V., Caligo, M.A., Nozzolini, M., and Luccarini, G. 1984. Stimulation of carrot somatic

Pacini, E., and Juniper, B.E. 1979b. The ultrastructure of pollen grain development in the olive (*Olea europaea*). 2. Secretion by the tapetal cells. *New Phytol.* 83:165–174.

Pacini, E., and Juniper, B.E. 1983. The ultrastructure of the formation and development of the amoeboid tapetum in *Arum italicum* Miller. *Protoplasma* 117:116–129.

Pacini, E., and Keijzer, C.J. 1989. Ontogeny of intruding non-periplasmodial tapetum in the wild chicory, *Chichorium intybus* (Compositae). *Plant Syst. Evol.* 167:149–164.

Pacini, E., Simoncioli, C., and Cresti, M. 1975. Ultrastructure of nucellus and endosperm of *Diplotaxis erucoides* during embryogenesis. *Caryologia* 28:525–538.

Pacini, E., Taylor, P.E., Singh, M.B., and Knox, R.B. 1992. Development of plastids in pollen and tapetum of rye-grass, *Lolium perenne* L. *Ann. Bot.* 70:179–188.

Pais, M.S., and Feijo, J.A. 1987. Microbody proliferation during the microsporogenesis of *Ophrys lutea* Cav. (Orchidaceae). *Protoplasma* 138:149–155.

Paiva, R., and Kriz, A.L. 1994. Effect of abscisic acid on embryo-specific gene expression during normal and precocious germination in normal and *viviparous* maize (*Zea mays)* embryos. *Planta* 192:332–339.

Pakendorf, K.W. 1970. Male sterility in *Lupinus mutabilis* Sweet. *Z. Pflanzenzüchtg* 63:227–236.

Palevitz, B.A. 1990. Kinetochore behavior during generative cell division in *Tradescantia virginiana*. *Protoplasma* 157:120–127.

Palevitz, B.A. 1993. Organization of the mitotic apparatus during generative cell division in *Nicotiana tabacum*. *Protoplasma* 174:25–35.

Palevitz, B.A., and Cresti, M. 1988. Microtubule organization in the sperm of *Tradescantia virginiana*. *Protoplasma* 146:28–34.

Palevitz, B.A., and Cresti, M. 1989. Cytoskeletal changes during generative cell division and sperm formation in *Tradescantia virginiana*. *Protoplasma* 150:54–71.

Palevitz, B.A., and Liu, B. 1992. Microfilaments (F-actin) in generative cells and sperm: an evaluation. *Sex. Plant Reprod* 5:89–100.

Palevitz, B.A., and Tiezzi, A. 1992. Organization, composition, and function of the generative cell and sperm cytoskeleton. *Int. Rev. Cytol.* 140:149–185.

Palmer, J.D., Shields, C.R., Cohen, D.B., and Orton, T.J. 1983. An unusual mitochondrial DNA plasmid in the genus *Brassica*. *Nature* 301:725–728.

Palmer, R.G. 1971. Cytological studies of ameiotic and normal maize with reference to premeiotic pairing. *Chromosoma* 35:233–246.

Palmer, R.G., Johns, C.W., and Muir, P.S. 1980. Genetics and cytology of the ms_3 male-sterile soybean. *J. Hered.* 71:343–348.

Palser, B.F., Rouse, J.L., and Williams, E.G. 1992. A scanning electron microscope study of the pollen tube pathway in pistils of *Rhododendron*. *Can. J. Bot.* 70:1039–1060.

Pan, D., and Nelson, O.E. 1984. A debranching enzyme deficiency in endosperms of the *sugary-1* mutants of maize. *Plant Physiol.* 74:324–238.

Pan, J.-L., Gao, G.-H., and Ban, H. 1983. Initial patterns of androgenesis in wheat anther culture. *Acta Bot. Sinica* 25:34–39.

Panchaksharappa, M.G., and Syamasundar, J. 1975. A cytochemical study of ovule development in *Dipcadi montanum* Dalz. *Cytologia* 40:141–149.

Pandey, K.K. 1958. Time of the S allele action. *Nature* 181:1220–1221.

Pandey, K.K. 1960. Incompatibility in *Abutilon `hybridum.'* *Am. J. Bot.* 47:877–883.

Pandey, K.K. 1964. Elements of the S-gene complex. *Genet. Res.* 5:397–409.

Pandey, K.K. 1967. Origin of genetic variability: combinations of peroxidase isozymes determine multiple allelism of the S gene. *Nature* 213:669–672.

Pandey, K.K. 1970. Time and site of the *S*-gene action, breeding systems and relationships in incompatibility. *Euphytica* 19:364–372.

Pandey, K.K. 1973. Heat sensitivity of esterase isozymes in the styles of *Lilium* and *Nicotiana*. *New Phytol.* 72:839–850.

Pandey, K.K. 1977. Mentor pollen: possible role of wall-held pollen growth promoting substances in overcoming intra- and interspecific incompatibility. *Genetica* 47:219–229.

Pang, P.P., Pruitt, R.E., and Meyerowitz, E.M. 1988. Molecular cloning, genomic organization, expression and evolution of 12S seed storage protein genes of *Arabidopsis thaliana*. *Plant Mol. Biol.* 11:805–820.

Panitz, R., Borisjuk, L., Manteuffel, R., and Wobus, U. 1995. Transient expression of storage-protein genes during early embryogenesis of *Vicia faba*: synthesis and metabolization of vicilin and legumin in the embryo, suspensor and endosperm. *Planta* 196:765–774.

Pankow, H. 1957. Über den Pollenkitt bei *Galanthus nivalis* L. *Flora* 146:240–253.

Parchman, L.G., and Lin, K.-C. 1972. Nucleolar RNA synthesis during meiosis of lily microsporocytes. *Nature New Biol.* 239:235–237.

Parchman, L.G., and Roth, T.F. 1971. Pachytene synaptonemal complexes and meiotic achiasmatic chromosomes. *Chromosoma* 33:129–145.

Parchman, L.G., and Stern, H. 1969. The inhibition of protein synthesis in meiotic cells and its effect on chromosome behavior. *Chromosoma* 26:298–311.

Parcy, F., Valon, C., Raynal, M., Gaubier-Comella, P., Delseny, M., and Giraudat, J. 1994. Regulation of gene expression programs during *Arabidopsis* seed development: roles of the *ABI3* locus and of endogenous abscisic acid. *Plant Cell* 6:1567–1582.

Pareddy, D.R., and Greyson, R.I. 1985. *In vitro* culture of immature tassels of an inbred field variety of *Zea mays*, cv. Oh43. *Plant Cell Tissue Organ Cult.* 5:119–128.

Pareddy, D.R., and Petolino, J.F. 1992. Maturation of maize pollen *in vitro*. *Plant Cell Rep.* 11:535–539.

Paris, D., Rietsema, J., Satina, S., and Blakeslee, A.F. 1953. Effect of amino acids, especially aspartic and glutamic acid and their amides, on the growth of *Datura stramonium* embryos *in vitro*. *Proc. Natl Acad. Sci. USA* 39:1205–1212.

Parker, M.L. 1982. Protein accumulation in developing endosperm of a high-protein line of *Triticum dicoccoides*. *Plant Cell Environ.* 5:37–43.

Parker, M.L., and Hawes, C.R. 1982. The Golgi apparatus in developing endosperm of wheat (*Triticum aestivum* L.). *Planta* 154:277–283.

Patton, D.A., Franzmann, L.A., and Meinke, D.W. 1991. Mapping genes essential for embryo development in *Arabidopsis thaliana*. *Mol. Gen. Genet.* 227:337–347.

Patton, D.A., and Meinke, D.W. 1990. Ultrastructure of arrested embryos from lethal mutants of *Arabidopsis thaliana*. *Am. J. Bot.* 77:653–661.

Patton, D.A., Volrath, S., and Ward, E.R. 1996. Complementation of an *Arabidopsis thaliana* biotin auxotroph with an *Escherichia coli* biotin biosynthetic gene. *Mol. Gen. Genet.* 251:261–266.

Paul, A.-L., and Ferl, R.J. 1994. *In vivo* footprinting identifies an activating element of the maize *Adh2* promoter specific for root and vascular tissues. *Plant J.* 5:523–533.

Paul, W., Hodge, R., Smartt, S., Draper, J., and Scott, R. 1992. The isolation and characterisation of the tapetum-specific *Arabidopsis thaliana* A9 gene. *Plant Mol. Biol.* 19:611–622.

Pavingerová, D., Dostál, J., Bisková, R., and Benetka, V. 1994. Somatic embryogenesis and *Agrobacterium*-mediated transformation of chrysanthemum. *Plant Sci.* 97:95–101.

Pavlov, A.N., Konarev, V.G., Kolesnik, T.I., and Shayakhmetov, I.F. 1975. Gliadins of the caryopsis of wheat during the process of its development. *Soviet Plant Physiol.* 22:63–67.

Payne, G., Kono, Y., and Daly, J.M. 1980. A comparison of purified host specific toxin from *Helminthosporium maydis*, race T, and its acetate derivative on oxidation by mitochondria from susceptible and resistant plants. *Plant Physiol.* 65:785–791.

Pearson, O.H. 1981. Nature and mechanisms of cytoplasmic male sterility in plants. *HortScience* 16:482–487.

Pechan, P.M. 1989. Successful cocultivation of *Brassica napus* microspores and proembryos with *Agrobacterium*. *Plant Cell Rep.* 8:387–390.

Pechan, P.M., and Keller, W.A. 1988. Identification of potentially embryogenic microspores in *Brassica napus*. *Physiol. Plant.* 74:377–384.

Pechan, P.M., Bartels, D., Brown, D.C.W., and Schell, J. 1991. Messenger-RNA and protein changes associated with induction of *Brassica* microspore embryogenesis. *Planta* 184:161–165.

Peddada, L.B., and Mascarenhas, J.P. 1975. The synthesis of 5S ribosomal RNA during pollen development. *Devel. Growth Diffn* 17:1–8.

Pedersen, K., Argos, P., Naravana, S.V.L., and Larkins, B.A. 1986. Sequence analysis and characterization of a maize gene encoding a high-sulfur zein protein of M_r 15,000. *J. Biol. Chem.* 261:6279–6284.

Pedersen, K., Bloom, K.S., Anderson, J.N., Glover, D.V., and Larkins, B.A. 1980. Analysis of the complexity and frequency of zein genes in the maize genome. *Biochemistry* 19:1644–1650.

Pedersen, K., Devereux, J., Wilson, D.R., Sheldon, E., and Larkins, B.A. 1982. Cloning and sequence analysis reveal structural variation among related zein genes in maize. *Cell* 29:1015–1026.

Pedersen, S., and Andersen, S.B. 1993. Developmental expression of isozymes during embryogenesis in barley anther culture. *Plant Sci.* 91:75–86.

Pedersen, S., Simonsen, V., and Loeschcke, V. 1987. Overlap of gametophytic and sporophytic gene expression in barley. *Theor. Appl. Genet.* 75:200–206.

Peirson, B.N., Owen, H.A., Feldmann, K.A., and Makaroff, C.A. 1996. Characterization of three male-sterile mutants of *Arabidopsis thaliana* exhibiting alterations in meiosis. *Sex. Plant Reprod.* 9:1–10.

Pelletier, G., and Ilami, M. 1972. Les facteurs de l'androgénèse *in vitro* chez *Nicotiana tabacum*. *Z. Pflanzenphysiol.* 68:97–114.

Pennell, R.I., and Roberts, K. 1990. Sexual development in the pea is presaged by altered expression of arabinogalactan protein. *Nature* 344:547–549.

Pennell, R.I., Geltz, N.R., Koren, E., and Russell, S.D. 1987. Production and partial characterization of hybridoma antibodies elicited to the sperm of *Plumbago zeylanica*. *Bot. Gaz.* 148:401–406.

Pennell, R.I., Janniche, L., Kjellbom, P., Scofield, G.N., Peart, J.M., and Roberts, K. 1991. Developmental regulation of a plasma membrane arabinogalactan protein epitope in oilseed rape flowers. *Plant Cell* 3:1317–1326.

Pennell, R.I., Janniche, L., Scofield, G.N., Booij, H., de Vries, S.C., and Roberts, K. 1992. Identification of a transitional cell state in the developmental pathway to carrot somatic embryogenesis. *J. Cell Biol.* 119:1371–1380.

Pennell, R.I., Knox, J.P., Scofield, G.N., Selvendran, R.R., and Roberts, K. 1989. A family of abundant plasma membrane-associated glycoproteins related to the arabinogalactan proteins is unique to flowering plants. *J. Cell Biol.* 108:1967–1977.

Pental, D., Mukhopadhyay, A., Grover, A., and Pradhan, A.K. 1988. A selection method for the synthesis of triploid hybrids by fusion of microspore protoplasts (n) with somatic cell protoplasts (2n). *Theor. Appl. Genet.* 76:237–243.

Penza, R., Lurquin, P.F., and Filippone, E. 1991. Gene transfer by cocultivation of mature embryos with *Agrobacterium tumefaciens*: application to cowpea (*Vigna unguiculata* Walp). *J. Plant Physiol.* 138:39–43.

Perata, P., Picciarelli, P., and Alpi, A. 1990. Pattern of variations in abscisic acid content in suspensors, embryos, and integuments of developing *Phaseolus coccineus* seeds. *Plant Physiol.* 94:1776–1780.

Perdue, T.D., Loukides, C.A., and Bedinger, P.A. 1992. The formation of cytoplasmic channels between tapetal cells in *Zea mays*. *Protoplasma* 171:75–79.

Perdue, T.D., and Parthasarathy, M.V. 1985. *In situ* localization of F-actin in pollen tubes. *Eur. J. Cell Biol.* 39:13–20.

Pereira, A.S.R., and Linskens, H.F. 1963. The influence of glutathione and glutathione antagonists on meiosis in excised anthers of *Lilium henryi*. *Acta Bot. Neerl.* 12:302–314.

Perez, M., Ishioka, G.Y., Walker, L.E., and Chesnut, R.W. 1990. cDNA cloning and immunological characterization of the rye grass allergen *Lol p* I. *J. Biol. Chem.* 265:16210–16215.

Perez-Grau, L., and Goldberg, R.B. 1989. Soybean seed protein genes are regulated spatially during embryogenesis. *Plant Cell* 1:1095–1109.

Pérez-Grau, L., Cortadas, J., Puigdomènech, P., and Palau, J. 1986. Accumulation and subcellular localization of glutelin-2 transcripts during maturation of maize endosperm. *FEBS Lett.* 202:145–148.

Pérez-Muñoz, C.A., Jernstedt, J.A., and Webster, B.D. 1993a. Pollen wall development in *Vigna vexillata*. I. Characterization of wall layers. *Am. J. Bot.* 80:1183–1192.

Pérez-Muñoz, C.A., Jernstedt, J.A., and Webster, B.D. 1993b. Pollen wall development in *Vigna vexillata*. II. Ultrastructural studies. *Am. J. Bot.* 80:1193–1202.

Periasamy, K., and Amalathas, J. 1991. Absence of callose and tetrad in the microsporogenesis of *Pandanus odoratissimus* with well-formed pollen exine. *Ann. Bot.* 67:29–33.

Pernollet, J.-C., and Vaillant, V. 1984. Characterization and complexity of wheat developing endosperm mRNAs. *Plant Physiol.* 76:187–190.

Pescitelli, S.M., and Sukhapinda, K. 1995. Stable transformation via electroporation into maize type II callus and regeneration of fertile transgenic plants. *Plant Cell Rep.* 14:712–716.

Peterson, D.M., Saigo, R.H., and Holy, J. 1985. Development of oat aleurone cells and their protein bodies. *Cereal Chem.* 62:366–371.

Peterson, P.A., Flavell, R.B., and Barratt, D.H.P. 1975. Altered mitochondrial membrane activities associated with cytoplasmically-inherited disease sensitivity in maize. *Theor. Appl. Genet.* 45:309–314.

Petolino, J.F., and Jones, A.M. 1986. Anther culture of elite genotypes of maize. *Crop Sci.* 26:1072–1074.

Petolino, J.F., and Thompson, S.A. 1987. Genetic analysis of anther culture response in maize. *Theor. Appl. Genet.* 74:284–286.

Pettitt, J.M. 1976. Pollen wall and stigma surface in the marine angiosperms *Thalassia* and *Thalassodendron*. *Micron* 7:21–32.

Pettitt, J.M. 1980. Reproduction in seagrasses: nature of the pollen and receptive surface of the stigma in the Hydrocharitaceae. *Ann. Bot.* 45:257–271.

Pettitt, J.M. 1981. Reproduction in seagrasses: pollen development in *Thalassia hemprichii, Halophila stipulacea* and *Thalassodendron ciliatum*. *Ann. Bot.* 48:609–622.

Phatak, S.C., Wittwer, S.H., Honma, S., and Bukovac, M.J. 1966. Gibberellin-induced anther and pollen development in a stamen-less tomato mutant. *Nature* 209:635–636.

Piaggesi, A., Picciarelli, P., Lorenzi, R., and Alpi, A. 1989. Gibberellins in embryo-suspensor of *Phaseolus coccineus* seeds at the heart stage of embryo development. *Plant Physiol.* 91:362–366.

Picciarelli, P., and Alpi, A. 1986. Gibberellins in suspensors of *Phaseolus coccineus* L. seeds. *Plant Physiol.* 82:298–300.

Picciarelli, P., and Alpi, A. 1987. Embryo-suspensor of *Tropaeolum majus*: identification of gibberellin A_{63}. *Phytochemistry* 26:329–330.

Picciarelli, P., Piaggesi, A., and Alpi, A. 1991. Gibberellins in suspensor, embryo and endosperm of developing seeds of *Cytisus laburnum*. *Phytochemistry* 30:1789–1792.

Picciarelli, P., Alpi, A., Pistelli, L., and Scalet, M. 1984. Gibberellin-like activity in suspensors of *Tropaeolum majus* L. and *Cytisus laburnum* L. *Planta* 162:566–568.

Picton, J.M., and Steer, M.W. 1981. Determination of secretory vesicle production rates by dictyosomes in pollen tubes of *Tradescantia* using cytochalasin D. *J. Cell Sci.* 49:261–272.

Picton, J.M., and Steer, M.W. 1982. A model for the mechanism of tip extension in pollen tubes. *J. Theor. Biol.* 98:15–20.

Picton, J.M., and Steer, M.W. 1983a. Evidence for the role of Ca^{2+} ions in tip extension in pollen tubes. *Protoplasma* 115:11–17.

Picton, J.M., and Steer, M.W. 1983b. The effect of cycloheximide on dictyosome activity in *Tradescantia* pollen tubes determined using cytochalasin D. *Eur. J. Cell Biol.* 29:133–138.

Picton, J.M., and Steer, M.W. 1983c. Membrane recycling and the control of secretory activity in pollen tubes. *J. Cell Sci.* 63:303–310.

Picton, J.M., and Steer, M.W. 1985. The effects of ruthenium red, lanthanum, fluorescein isothiocyanate and trifluoperazine on vesicle transport, vesicle fusion and tip extension in pollen tubes. *Planta* 163:20–26.

Pierson, E.S. 1988. Rhodamine-phalloidin staining of F-actin in pollen after dimethylsulphoxide permeabilization. *Sex. Plant Reprod.* 1:83–87.

Pierson, E.S., Derksen, J., and Traas, J.A. 1986. Organization of microfilaments and microtubules in pollen tubes grown *in vitro* or *in vivo* in various angiosperms. *Eur. J. Cell Biol.* 41:14–18.

Pierson, E.S., Kengen, H.M.P., and Derksen, J. 1989. Microtubules and actin filaments co-localize in pollen tubes of *Nicotiana tabacum* L. and *Lilium longiflorum* Thunb. *Protoplasma* 150:75–77.

Pierson, E.S., Lichtscheidl, I.K., and Derksen, J. 1990. Structure and behaviour of organelles in living pollen tubes of *Lilium longiflorum*. *J. Expt. Bot.* 41:1461–1468.

Pierson, E.S., Li, Y.Q., Zhang, H.Q., Willemse, M.T.M., Linskens, H.F., and Cresti, M. 1995. Pulsatory growth of pollen tubes: investigation of a possible relationship with the periodic distribution of cell wall components. *Acta Bot. Neerl.* 44:121–128.

Pierson, E.S., Miller, D.D., Callaham, D.A., Shipley, A.M., Rivers, B.A., Cresti, M., and Hepler, P.K. 1994. Pollen tube growth is coupled to the extracellular calcium ion flux and the intracellular calcium gradient: effect of BAPTA-type buffers and hypertonic media. *Plant Cell* 6:1815–1828.

Pierson, E.S., Willekens, P.G.M., Maessen, M., and Helsper, J.P.F.G. 1986. The effect of lectins on germinating pollen of *Lilium longiflorum*. I. Effect on pollen germination, pollen tube growth and organization of microfilaments. *Acta Bot. Neerl.* 35:249–256.

Pipkin, J.L., Jr., and Larson, D.A. 1972. Characterization of the `very' lysine-rich histones of active and quiescent anther tissues of *Hippeastrum belladonna*. *Expt. Cell Res.* 71:249–260.

Pipkin, J.L., Jr., and Larson, D.A. 1973. Changing patterns of nucleic acids, basic and acidic proteins in generative and vegetative nuclei during pollen germination and pollen tube growth in *Hippeastrum belladonna*. *Expt. Cell Res.* 79:28–42.

Pirrie, A., and Power, J.B. 1986. The production of fertile, triploid somatic hybrid plants (*Nicotiana glutinosa* (n) + *N. tabacum* (2n)) via gametic:somatic protoplast fusion. *Theor. Appl. Genet.* 72:48–52.

Pitto, L., Lo Schiavo, F., and Terzi, M. 1985. α-amanitin resistance is developmentally regulated in carrot. *Proc. Natl Acad. Sci. USA* 82:2799–2803.

Pitto, L., Lo Schiavo, F., Giuliano, G., and Terzi, M. 1983. Analysis of the heat-shock protein pattern during somatic embryogenesis of carrot. *Plant Mol. Biol.* 2:231–237.

Pla, M., Goday, A., Vilardell, J., Gómez, J., and Pagès, M. 1989. Differential regulation of ABA-induced 23-25 kDa proteins in embryo and vegetative tissues of the *viviparous* mutants of maize. *Plant Mol. Biol.* 13:385–394.

Pla, M., Gómez, J., Goday, A., and Pagès, M. 1991. Regulation of the abscisic acid-responsive gene *rab28* in maize *viviparous* mutants. *Mol. Gen. Genet.* 230:394–400.

Pla, M., Vilardell, J., Guiltinan, M.J., Marcotte, W.R., Niogret, M.F., Quatrano, R.S., and Pagès, M. 1993. The *cis*-regulatory element CCACGTGG is involved in ABA and water-stress responses of the maize gene *rab*28. *Plant Mol. Biol.* 21:259–266.

Plant, A.L., van Rooijen, G.J.H., Anderson, C.P., and Moloney, M.M. 1994. Regulation of an *Arabidopsis* oleosin gene promoter in transgenic *Brassica napus*. *Plant Mol. Biol*. 25:193–205.

Platt-Aloia, K.A., Lord, E.M., DeMason, D.A., and Thomson, W.W. 1986. Freeze-fracture observations on membranes of dry and hydrated pollen from *Collomia, Phoenix* and *Zea*. *Planta* 168:291–298.

Plaut, W.S. 1953. DNA synthesis in the microsporocytes of *Lilium henryi*. *Hereditas* 39:438–444.

Pnueli, L., Abu-Abeid, M., Zamir, D., Nacken, W., Schwarz-Sommer, Z., and Lifschitz, E. 1991. The MADS box gene family in tomato: temporal expression during floral development, conserved secondary structures and homology with homeotic genes from *Antirrhinum* and *Arabidopsis*. *Plant J*. 1:255–266.

Pnueli, L., Hareven, D., Broday, L., Hurwitz, C., and Lifschitz, E. 1994a. The TM5 MADS box gene mediates organ differentiation in the three inner whorls of tomato flowers. *Plant Cell* 6:175–186.

Pnueli, L., Hareven, D., Rounsley, S.D., Yanofsky, M.F., and Lifschitz, E. 1994b. Isolation of the tomato *AGAMOUS* gene *TAG1* and analysis of its homeotic role in transgenic plants. *Plant Cell* 6:163–173.

Poddubnaya-Arnoldi, V.A. 1967. Comparative embryology of the Orchidaceae. *Phytomorphology* 17:312–320.

Poethig, R.S., Coe, E.H., Jr., and Johri, M.M. 1986. Cell lineage patterns in maize embryogenesis: a clonal analysis. *Devel Biol*. 117:392–404.

Polito, V.S. 1983. Membrane-associated calcium during pollen grain germination: a microfluorometric analysis. *Protoplasma* 117:226–232.

Pollak, P.E. 1992. Cytological differences between a cytoplasmic male sterile tobacco cybrid and its fertile counterpart during early anther development. *Am. J. Bot*. 79:937–945.

Pollak, P.E., Hansen, K., Astwood, J.D., and Taylor, L.P. 1995. Conditional male fertility in maize. *Sex. Plant Reprod*. 8:231–241.

Pollak, P.E., Vogt, T., Mo, Y., and Taylor, L.P. 1993. Chalcone synthase and flavonol accumulation in stigmas and anthers of *Petunia hybrida*. *Plant Physiol*. 102:925–932.

Pollock, E.G., and Jensen, W.A. 1964. Cell development during early embryogenesis in *Capsella* and *Gossypium*. *Am. J. Bot*. 51:915–921.

Pollock, E.G., and Jensen, W.A. 1967. Ontogeny and cytochemistry of the chalazal proliferating cells of *Capsella bursa-pastoris* (L.) Medic. *New Phytol*. 66:413–417.

Polowick, P.L., and Greyson, R.I. 1982. Anther development, meiosis and pollen formation in *Zea* tassels cultured in defined liquid medium. *Plant Sci. Lett*. 26:139–145.

Polowick, P.L., and Greyson, R.I. 1984. The relative efficiency of cytokinins in the development of normal spikelets on cultured tassels of *Zea mays*. *Can. J. Bot*. 62:830–834.

Polowick, P.L., and Sawhney, V.K. 1985. Temperature effects on male fertility and flower and fruit development in *Capsicum annuum* L. *Sci. Hort*. 25:117–127.

Polowick, P.L., and Sawhney, V.K. 1987. A scanning electron microscopic study on the influence of temperature on the expression of cytoplasmic male sterility in *Brassica napus*. *Can. J. Bot*. 65:807–814.

Polowick, P.L., and Sawhney, V.K. 1988. High temperature induced male and female sterility in canola (*Brassica napus* L.). *Ann. Bot*. 62:83–86.

Polowick, P.L., and Sawhney, V.K. 1990. Microsporogenesis in a normal line and in the *ogu* cytoplasmic male-sterile line of *Brassica napus*. I. The influence of high temperatures. *Sex. Plant Reprod*. 3:263–276.

Polowick, P.L., and Sawhney, V.K. 1991. Microsporogenesis in a normal line and in the *ogu* cytoplasmic male-sterile line of *Brassica napus*. II. The influence of intermediate and low temperatures. *Sex. Plant Reprod*. 4:22–27.

Polowick, P.L., and Sawhney, V.K. 1992. Ultrastructural changes in the cell wall, nucleus and cytoplasm of pollen mother cells during meiotic prophase I in *Lycopersicon esculentum* (Mill.). *Protoplasma* 169:139–147.

Polowick, P.L., and Sawhney, V.K. 1993a. Differentiation of the tapetum during microsporogenesis in tomato (*Lycopersicon esculentum* Mill.), with special reference to the tapetal cell wall. *Ann. Bot*. 72:595–605.

Polowick, P.L., and Sawhney, V.K. 1993b. An ultrastructural study of pollen development in tomato (*Lycopersicon esculentum*). I. Tetrad to early binucleate microspore stage. *Can. J. Bot*. 71:1039–1047.

Polowick, P.L., and Sawhney, V.K. 1993c. An ultrastructural study of pollen development in tomato (*Lycopersicon esculentum*). II. Pollen maturation. *Can. J. Bot*. 71:1048–1055.

Polowick, P.L., and Sawhney, V.K. 1995. Ultrastructure of the tapetal cell wall in the stamenless-2 mutant of tomato (*Lycopersicon esculentum*): correlation between structure and male sterility. *Protoplasma* 189:249–255.

Polsoni, L., Kott, L.S., and Beversdorf, W.D. 1988. Large-scale microspore culture technique for mutation-selection studies in *Brassica napus*. *Can. J. Bot*. 66:1681–1685.

Polya, G.M., Micucci, V., Rae, A.L., Harris, P.J., and Clarke, A.E. 1986. Ca^{2+}-dependent protein phosphorylation in germinated pollen of *Nicotiana alata*, an ornamental tobacco. *Physiol. Plant*. 67:151–157.

Pomeroy, M.K., Kramer, J.K.G., Hunt, D.J., and Keller, W.A. 1991. Fatty acid changes during development of zygotic and microspore-derived embryos of *Brassica napus*. *Physiol. Plant*. 81:447–454.

Ponte, I., Guillén, P., Debón, R.M., Reina, M., Aragay, A., Espel, E., di Fonzo, N., and Palau, J. 1994. Narrow A/T-rich zones present at the distal 5′-flanking sequences of the zein genes *Zc1* and *Zc2* bind a unique 30 kDa HMG-like protein. *Plant Mol. Biol*. 26:1893–1906.

Ponz, F., Paz-Ares, J., Hernández-Lucas, C., Carbonero, P., and García-Olmedo, F. 1983. Synthesis and processing of thionin precursors in developing endosperm from barley (*Hordeum vulgare* L.). *EMBO J*. 2:1035–1040.

Ponzi, R., and Pizzolongo, P. 1973. Ultrastructure of plastids in the suspensor cells of *Ipomoea purpurea* Roth. *J. Submicros. Cytol*. 5:257–263.

Ponzi, R., and Pizzolongo, P. 1976. *Cytinus hypocistis* L. embryogenesis: ultrastructural aspects of megasporogenesis and megagametogenesis. *J. Submicros. Cytol*. 8:327–336.

Ponzi, R., and Pizzolongo, P. 1984. Ultrastructural study of change from free-nuclear to cellular endosperm in *Ipomoea purpurea* Roth. and *Cytinus hypocistis* L. *Giorn. Bot. Ital*. 118:147–154.

Porath, D., and Galun, E. 1967. *In vitro* culture of hermaphrodite floral buds of *Cucumis melo* L.: Microsporogenesis and ovary formation. *Ann. Bot.* 31:283–290.

Porter, E.K. 1981. Origins and genetic nonvariability of the proteins which diffuse from maize pollen. *Environ. Health Pers.* 37:53–59.

Porter, E.K., Bird, J.M., and Dickinson, H.G. 1982. Nucleic acid synthesis in microsporocytes of *Lilium* cv. Cinnabar: events in the nucleus. *J. Cell Sci.* 57:229–246.

Porter, E.K., Parry, D., and Dickinson, H.G. 1983. Changes in poly(A)$^+$RNA during male meiosis in *Lilium. J. Cell Sci.* 62:177–186.

Porter, E.K., Parry, D., Bird, J., and Dickinson, H.G. 1984. Nucleic acid metabolism in the nucleus and cytoplasm of angiosperm meiocytes. *Symp. Soc. Expt. Biol.* 38:363–379.

Portnoi, L., and Horovitz, A. 1977. Sugars in natural and artificial pollen germination substrates. *Ann. Bot.* 41:21–27.

Potrykus, I. 1990. Gene transfer to plants: assessment and perspectives. *Physiol. Plant.* 79:125–134.

Potz, H., and Tatlioglu, T. 1993. Molecular analysis of cytoplasmic male sterility in chives (*Allium schoenoprasum* L.). *Theor. Appl. Genet.* 87:439–445.

Poulson, R., and Beevers, L. 1973. RNA metabolism during the development of cotyledons of *Pisum sativum* L. *Biochim. Biophys. Acta* 308:381–389.

Powling, A. 1981. Species of small DNA molecules found in mitochondria from sugarbeet with normal and male sterile cytoplasms. *Mol. Gen. Genet.* 183:82–84.

Powling, A. 1982. Restriction endonuclease analysis of mitochondrial DNA from sugarbeet with normal and male-sterile cytoplasms. *Heredity* 49:117–120.

Powling, A., and Ellis, T.H.N. 1983. Studies on the organelle genomes of sugarbeet with male-fertile and male-sterile cytoplasms. *Theor. Appl. Genet.* 65:323–328.

Prabhakar, K., and Vijayaraghavan, M.R. 1983. Histochemistry and ultrastructure of suspensor cells in *Alyssum maritimum. Cytologia* 48:389–402.

Prahl, A.-K., Springstubbe, H., Grumbach, K., and Wiermann, R. 1985. Studies on sporopollenin biosynthesis: the effect of inhibitors of carotenoid biosynthesis on sporopollenin accumulation. *Z. Naturfors.* 40c:621–626.

Pramanik, S.K., and Bewley, J.D. 1993. A ubiquitin carrier protein cDNA from developing alfalfa embryos. *Plant Physiol.* 102:1049–1050.

Pramanik, S.K., Krochko, J.E., and Bewley, J.D. 1992. Distribution of cytosolic mRNAs between polysomal and ribonucleoprotein complex fractions in alfalfa embryos. *Plant Physiol.* 99:1590–1596.

Prändl, R., Kloske, E., and Schöffl, F. 1995. Developmental regulation and tissue-specific differences of heat shock gene expression in transgenic tobacco and *Arabidopsis* plants. *Plant Mol. Biol.* 28:73–82.

Prasad, K. 1977. Histochemistry of anther and ovule in *Farsetia hamiltonii* (Royle) and *Eruca sativa* Mill (Cruciferae). *J. Indian Bot. Soc.* 56:90–99.

Prat, S., Pérez-Grau, L., and Puigdomènech, P. 1987. Multiple variability in the sequence of a family of maize endosperm proteins. *Gene* 52:41–49.

Prat, S., Cortadas, J., Puigdomènech, P., and Palau, J. 1985. Nucleic acid (cDNA) and amino acid sequences of the maize endosperm protein glutelin-2. *Nucl. Acids Res.* 13:1493–1504.

Preiss, J., Lammel, C., and Sabraw, A. 1971. A unique adenosine diphosphoglucose pyrophosphorylase associated with maize embryo tissue. *Plant Physiol.* 47:104–108.

Pressey, R. 1991. Polygalacturonase in tree pollens. *Phytochemistry* 30:1753–1755.

Pressey, R., and Reger, B.J. 1989. Polygalacturonase in pollen from corn and other grasses. *Plant Sci.* 59:57–62.

Pretová, A. 1974. The influence of the osmotic potential of the cultivation medium on the development of excised flax embryos. *Biol. Plant.* 16:14–20.

Pretová, A. 1986. Growth of zygotic flax embryos *in vitro* and influence of kinetin. *Plant Cell Rep.* 3:210–211.

Preuss, D., Rhee, S.Y., and Davis, R.W. 1994. Tetrad analysis possible in *Arabidopsis* with mutation of the *QUARTET* (*QRT*) genes. *Science* 264:1458–1460.

Preuss, D., Lemieux, B., Yen, G., and Davis, R.W. 1993. A conditional sterile mutation eliminates surface components from *Arabidopsis* pollen and disrupts cell signaling during fertilization. *Genes Devel.* 7:974–985.

Prevost, I., and Le Page-Degivry, M.T. 1985. Changes in abscisic acid content in axis and cotyledons of developing *Phaseolus vulgaris* embryos and their physiological consequences. *J. Expt. Bot.* 36:1900–1905.

Price, S.D., and Barrett, S.C.H. 1982. Tristyly in *Pontederia cordata* (Pontederiaceae). *Can. J. Bot.* 60:897–905.

Priestley, D.A., and de Kruijff, B. 1982. Phospholipid motional characteristics in a dry biological system. A ^{31}P-nuclear magnetic resonance study of hydrating *Typha latifolia* pollen. *Plant Physiol.* 70:1075–1078.

Pring, D.R., Conde, M.F., and Levings, C.S., III. 1980. DNA heterogeneity within the C group of maize male-sterile cytoplasms. *Crop Sci.* 20:159–162.

Pring, D.R., Conde, M.F., and Schertz, K.F. 1982. Organelle genome diversity in *Sorghum*: male-sterile cytoplasms. *Crop Sci.* 22:414–421.

Pring, D.R., and Levings, C.S., III. 1978. Heterogeneity of maize cytoplasmic genomes among male-sterile cytoplasms. *Genetics* 89:121–136.

Pring, D.R., Conde, M.F., Schertz, K.F., and Levings, C.S., III. 1982. Plasmid-like DNAs associated with mitochondria of cytoplasmic male-sterile *Sorghum. Mol. Gen. Genet.* 186:180–184.

Pring, D.R., Levings, C.S., III., Hu, W.W.L., and Timothy, D.H. 1977. Unique DNA associated with mitochondria in the "S"-type cytoplasm of male-sterile maize. *Proc. Natl Acad. Sci. USA* 74:2904–2908.

Pritchard, A.J., and Hutton, E.M. 1972. Anther and pollen development in male-sterile *Phaseolus atropurpureus. J. Hered.* 63:280–282.

Pritchard, H.N. 1964a. A cytochemical study of embryo sac development in *Stellaria media. Am. J. Bot.* 51:371–378.

Pritchard, H.N. 1964b. A cytochemical study of embryo development in *Stellaria media. Am. J. Bot.* 51:472–479.

Pruitt, K.D., and Hanson, M.R. 1989. Cytochrome oxidase subunit II sequences in *Petunia* mitochondria: two intron-containing genes and an intron-less pseudogene associated with cytoplasmic male sterility. *Curr. Genet.* 16:281–291.

Pruitt, K.D., and Hanson, M.R. 1991. Transcription of the *Petunia* mitochondrial CMS-associated *Pcf* locus in male sterile and fertility-restored lines. *Mol. Gen. Genet.* 227:348–355.

Przybyllok, T., and Nagl, W. 1977. Auxin concentration in the embryo and suspensors of *Tropaeolum majus*, as determined by mass fragmentation (single ion detection). *Z. Pflanzenphysiol.* 84:463–465.

Punnett, H.H. 1953. Cytological evidence of hexaploid cells in maize endosperm. *J. Hered.* 44:257–259.

Puupponen-Pimiä, R., Saloheimo, M., Vasara, T., Ra, R., Gaugecz, J., Kurtén, U., Knowles, J.K.C., Keränen, S., and Kauppinen, V. 1993. Characterization of a birch (*Betula pendula* Roth.) embryogenic gene, *BP8*. *Plant Mol. Biol.* 23:423–428.

Pwee, K.-H., and Gray, J.G. 1993. The pea plastocyanin promoter directs cell-specific but not full light-regulated expression in transgenic tobacco plants. *Plant J.* 3:437–449.

Pysh, L.D., Aukerman, M.J., and Schmidt, R.J. 1993. OHP1: a maize basic domain/ leucine zipper protein that interacts with opaque2. *Plant Cell* 5:227–236.

Qiu, X., and Erickson, L. 1996. A pollen-specific polygalacturonase-like cDNA from alfalfa. *Sex. Plant Reprod.* 9:123–124.

Qu, R., and Huang, A.H.C. 1990. Oleosin KD 18 on the surface of oil bodies in maize. Genomic and cDNA sequences and the deduced protein structure. *J. Biol. Chem.* 265:2238–2243.

Qu, R., Vance, V.B., and Huang, A.H.C. 1990. Expression of genes encoding oleosin isoforms in the embryos of maturing maize kernels. *Plant Sci.* 72:223–232.

Qu, R., Wang, S.-M., Lin, Y.-H., Vance, V.B., and Huang, A.H.C. 1986. Characteristics and biosynthesis of membrane proteins of lipid bodies in the scutella of maize (*Zea mays* L.). *Biochem. J.* 235:57–65.

Qu, R.-D., and Chen, Y. 1984. Pathways of androgenesis and observations on cultured pollen grains in rice (*Oryza sativa* subsp. Keng). *Acta Bot. Sinica* 26:580–587.

Quattrocchio, F., Tolk, M.A., Coraggio, I., Mol. J.N.M., Viotti, A., and Koes, R.E. 1990. The maize zein gene zE19 contains two distinct promoters which are independently activated in endosperm and anthers of transgenic *Petunia* plants. *Plant Mol. Biol.* 15:81–93.

Quayle, T.J.A., Hetz, W., and Feix, G. 1991. Characterization of a maize endosperm culture expressing zein genes and its use in transient transformation assays. *Plant Cell Rep.* 9:544–548.

Que, Q.-D., and Tang, X.-H. 1988. The changes of protein content and synthetic activity in squash pollen during germination. *Acta Bot. Sinica* 30:501–507.

Rabakoarihanta, A., Mok, D.W.S., and Mok, M.C. 1979. Fertilization and early embryo development in reciprocal interspecific crosses of *Phaseolus*. *Theor. Appl. Genet.* 54:55–59.

Racchi, M.L., Gavazzi, G., Monti, D., and Manitto, P. 1978. An analysis of the nutritional requirements of the *pro* mutant in *Zea mays*. *Plant Sci. Lett.* 13:357–364.

Racusen, R.H., and Schiavone, F.M. 1988. Detection of spatially- and stage-specific proteins in extracts from single embryos of the domesticated carrot. *Development* 103:665–674.

Radding, C.M. 1982. Homologous pairing and strand exchange in genetic recombination. *Annu. Rev. Genet.* 16:405–437.

Radke, S.E., Andrews, B.M., Moloney, M.M., Crouch, M.L., Kridl, J.C., and Knauf, V.C. 1988. Transformation of *Brassica napus* L. using *Agrobacterium tumefaciens*: developmentally regulated expression of a reintroduced napin gene. *Theor. Appl. Genet.* 75:685–694.

Radłowski, M., Kalinowski, A., Kròlikowski, Z., and Bartkowiak, S. 1994. Protease activity from maize pollen. *Phytochemistry* 35:853–856.

Radojević, L., Zylberberg, L., and Kovoor, J. 1980. Etude ultrastructurale des embryons androgenetiques d'*Aesculus hippocastanum* L. *Z. Pflanzenphysiol.* 98:255–261.

Rae, A.L., Harris, P.J., Bacic, A., and Clarke, A.E. 1985. Composition of the cell walls of *Nicotiana alata* Link et Otto pollen tubes. *Planta* 166:128–133.

Raeber, J.G., and Bolton, A. 1955. A new form of male sterility in *Nicotiana tabacum* L. *Nature* 176:314–315.

Rafalski, J.A. 1986. Structure of wheat gamma-gliadin genes. *Gene* 43:221–229.

Rafalski, J.A., Scheets, K., Metzler, M., Peterson, D.M., Hedgcoth, C., and Söll, D.G. 1984. Developmentally regulated plant genes: the nucleotide sequence of a wheat gliadin genomic clone. *EMBO J.* 3:1409–1415.

Raff, J.W., and Clarke, A.E. 1981. Tissue-specific antigens secreted by suspension-cultured callus cells of *Prunus avium* L. *Planta* 153:115–124.

Raff, J.W., Knox, R.B., and Clarke, A.E. 1981. Style antigens of *Prunus avium* L. *Planta* 153:125–129.

Raff, J.W., Pettitt, J.M., and Knox, R.B. 1981. Cytochemistry of pollen tube growth in stigma and style of *Prunus avium*. *Phytomorphology* 31:214–231.

Rafnar, T., Griffith, I.J., Kuo, M.-C., Bond, J.F., Rogers, B.L., and Klapper, D.G. 1991. Cloning of *Amb a I* (antigen E), the major allergen family of short ragweed pollen. *J. Biol. Chem.* 266:1229–1236.

Raghavan, P., and Philip, V.J. 1982. Morphological and histochemical changes in the egg and zygote of *Lagerstroemia speciosa*. I. Cell size, vacuole and insoluble polysaccharides. *Proc. Indian Acad. Sci. (Plant Sci.)* 91:465–472.

Raghavan, V. 1976a. Role of the generative cell in androgenesis in henbane. *Science* 191:388–389.

Raghavan, V. 1976b. *Experimental Embryogenesis in Vascular Plants*. London: Academic Press.

Raghavan, V. 1977a. Patterns of DNA synthesis during pollen embryogenesis in henbane. *J. Cell Biol.* 73:521–526.

Raghavan, V. 1977b. Applied aspects of embryo culture. In *Applied and Fundamental Aspects of Plant Cell, Tissue, and Organ Culture*, ed. J. Reinert and Y.P.S. Bajaj, pp. 375–397. Berlin: Springer-Verlag.

Raghavan, V. 1978. Origin and development of pollen embryoids and pollen calluses in cultured anther segments of *Hyoscyamus niger* (henbane). *Am. J. Bot.* 65:984–1002.

Raghavan, V. 1979a. Embryogenic determination and ribonucleic acid synthesis in pollen grains of *Hyoscyamus niger* (henbane). *Am. J. Bot.* 66:36–39.

Raghavan, V. 1979b. An autoradiographic study of RNA synthesis during pollen embryogenesis in *Hyoscyamus niger* (henbane). *Am. J. Bot.* 66:784–795.

Raghavan, V. 1981a. A transient accumulation of poly(A)-containing RNA in the tapetum of *Hyoscyamus niger* during microsporogenesis. *Devel Biol.* 81:342–348.

Raghavan, V. 1981b. Distribution of poly(A)-containing RNA during normal pollen development and during induced pollen embryogenesis in *Hyoscyamus niger*. *J. Cell Biol.* 89:593–606.

Raghavan, V. 1984. Protein synthetic activity during normal pollen development and during induced pollen embryogenesis in *Hyoscyamus niger*. *Can. J. Bot.* 62:2493–2513.

Raghavan, V. 1985. The applications of embryo rescue in agriculture. In *Biotechnology in International Agricultural Research*, pp. 189–197. Manila: International Rice Research Institute.

Raghavan, V. 1986a. Pollen developmental biology in cultured anthers. In *Cell Culture and Somatic Cell Genetics of Plants*, Vol. 3, ed. I.K. Vasil, pp. 275–304. Orlando: Academic Press.

Raghavan, V. 1986b. Variability through wide crosses and embryo rescue. In *Cell Culture and Somatic Cell Genetics of Plants*, Vol. 3, ed. I.K. Vasil, pp. 612–633. Orlando: Academic Press.

Raghavan, V. 1986c. *Embryogenesis in Angiosperms. A Developmental and Experimental Study*. Cambridge: Cambridge University Press.

Raghavan, V. 1988. Anther and pollen development in rice (*Oryza sativa*). *Am. J. Bot.* 75:183–196.

Raghavan, V. 1989. mRNAs and a cloned histone gene are differentially expressed during anther and pollen development in rice (*Oryza sativa* L.). *J. Cell Sci.* 92:217–229.

Raghavan, V. 1990a. Origin of the quiescent center in the root of *Capsella bursa-pastoris* (L.) Medik. *Planta* 181:62–70.

Raghavan, V. 1990b. Spatial distribution of mRNA during pre-fertilization ovule development in *Capsella bursa-pastoris*. *Sex. Plant Reprod.* 3:170–178.

Raghavan, V. 1990c. *Hyoscyamus* spp.: anther culture studies. In *Biotechnology in Agriculture and Forestry 12. Haploids in Crop Improvement* I, ed. Y.P.S. Bajaj, pp. 290–305. Berlin: Springer-Verlag.

Raghavan, V. 1991. Some perspectives on the development of embryos in angiosperms. A synthesis of old ideas and new concepts. *Rev. Cytol. Biol. Végét.-Bot.* 14:163–174.

Raghavan, V. 1993. Embryo culture: methods and applications. In *Plant Biotechnology. Commercial Prospects and Problems*, ed. J. Prakash and R.L.M. Pierik, pp. 143–168. New Delhi: Oxford & IBH Publishing Co.

Raghavan, V. 1994. *In vitro* methods for the control of fertilization and embryo development. In *Plant Cell and Tissue Culture*, ed. I.K. Vasil and T.A. Thorpe, pp. 173–194. Dordrecht: Kluwer Academic Publishers.

Raghavan, V. 1995. Manipulation of pollen grains for gametophytic and sporophytic types of growth. *Meth. Cell Biol.* 49:367–375.

Raghavan, V., and Goh, C.J. 1994. DNA synthesis and mRNA accumulation during germination of embryos of the orchid *Spathoglottis plicata*. *Protoplasma* 183:137–147.

Raghavan, V., Jiang, C., and Bimal, R. 1992. Cell- and tissue-specific expression of rice histone gene transcripts during anther and pollen development in henbane (*Hyoscyamus niger*). *Am. J. Bot.* 77:778–783.

Raghavan, V., and Kamalay, J.C. 1993. Expression of two cloned mRNA sequences during development and germination of spores of the sensitive fern, *Onoclea sensibilis* L. *Planta* 189:1–9.

Raghavan, V., and Nagmani, R. 1989. Cytokinin effects on pollen embryogenesis in cultured anthers of *Hyoscyamus niger*. *Can. J. Bot.* 67:247–257.

Raghavan, V., and Olmedilla, A. 1989. Spatial patterns of histone mRNA expression during grain development and germination in rice. *Cell Diffn Devel.* 27:183–196.

Raghavan, V., and Sharma, K.K. 1995. Zygotic embryogenesis in gymnosperms and angiosperms. In In Vitro *Embryogenesis in Plants*, ed. T.A. Thorpe, pp. 73–115. Dordrecht: Kluwer Academic Press.

Raghavan, V., and Torrey, J.G. 1963. Growth and morphogenesis of globular and older embryos of *Capsella* in culture. *Am. J. Bot.* 50:540–551.

Raghavan, V., and Torrey, J.G. 1964. Effects of certain growth substances on the growth and morphogenesis of immature embryos of *Capsella* in culture. *Plant Physiol.* 39:691–699.

Rahman, S., Shewry, P.R., and Miflin, B.J. 1982. Differential protein accumulation during barley grain development. *J. Expt. Bot.* 33:717–728.

Rahman, S., Kreis, M., Forde, B.G., Shewry, P.R., and Miflin, B.J. 1984. Hordein-gene expression during development of the barley (*Hordeum vulgare*) endosperm. *Biochem. J.* 223:315–322.

Raikhel, N.V., Bednarek, S.Y., and Wilkins, T.A. 1988. Cell-type-specific expression of a wheat-germ agglutinin gene in embryos and young seedlings of *Triticum aestivum*. *Planta* 176:406–414.

Raikhel, N.V., and Lerner, D.R. 1991. Expression and regulation of lectin genes in cereals and rice. *Devel Genet.* 12:255–260.

Raikhel, N.V., and Quatrano, R.S. 1986. Localization of wheat-germ agglutinin in developing wheat embryos and those cultured in abscisic acid. *Planta* 168:433–440.

Raikhel, N.V., and Wilkins, T.A. 1987. Isolation and characterization of a cDNA clone encoding wheat germ agglutinin. *Proc. Natl Acad. Sci. USA* 84:6745–6749.

Raina, A., and Datta, A. 1992. Molecular cloning of a gene encoding a seed-specific protein with nutritionally balanced amino acid composition from *Amaranthus*. *Proc. Natl Acad. Sci. USA* 89:11774–11778.

Raineri, D.M., Bottino, P., Gordon, M.P., and Nester, E.W. 1990. *Agrobacterium*-mediated transformation of rice (*Oryza sativa* L.). *Biotechnology* 8:33–38.

Rajam, M.V. 1989. Restriction of pollen germination and tube growth in lily pollen by inhibitors of polyamine metabolism. *Plant Sci.* 59:53–56.

Rajasekaran, K., Vine, J., and Mullins, M.G. 1982. Dormancy in somatic embryos and seeds of *Vitis*: changes in endogenous abscisic acid during embryogeny and germination. *Planta* 154:139–144.

Rajasekaran, K., Hein, M.B., Davis, G.C., Carnes, M.G., and Vasil, I.K. 1987. Endogenous growth regulators in leaves and tissue cultures of *Pennisetum purpureum* Schum. *J. Plant Physiol.* 130:13–25.

Rajasekhar, E.W. 1973. Nuclear divisions in protoplasts isolated from pollen tetrads of *Datura metel*. *Nature* 246:223–224.

Rajora, O.P., and Zsuffa, L. 1986. Sporophytic and gametophytic gene expression in *Populus deltoides* Marsh., *P. nigra* L., and *P. maximowiczii* Henry. *Can. J. Genet. Cytol.* 28:476–482.

Ramachandran, C., and Raghavan, V. 1989. Changes in nuclear DNA content of endosperm cells during grain development in rice (*Oryza sativa*). *Ann. Bot.* 64:459–468.

Ramachandran, C., and Raghavan, V. 1990. Intracellular localization of glutelin mRNA during grain development in rice. *J. Expt. Bot.* 41:393–399.

Ramachandran, C., and Raghavan, V. 1992a. Apomixis in distant hybridization. In *Distant Hybridization in*

Crop Plants, ed. G. Kalloo and J.B. Chowdhury, pp. 106–121. Berlin: Springer-Verlag.

Ramachandran, C., and Raghavan, V. 1992b. Regulation of gene expression during rice grain development and germination. *Trans. Malaysian Soc. Plant Physiol.* 3:217–227.

Ramaer, H. 1935. Cytology of *Hevea*. *Genetica* 17:193–236.

Raman, K., Walden, D.B., and Greyson, R.I. 1980. Fertilization in *Zea mays* by cultured gametophytes. *J. Hered.* 71:311–314.

Ramanna, M.S. 1974. The origin and *in vivo* development of embryoids in the anthers of *Solanum* hybrids. *Euphytica* 23:623–632.

Ramanna, M.S., and Hermsen, J.G.T. 1974. Embryoid formation in the anthers of some interspecific hybrids in *Solanum*. *Euphytica* 23:423–427.

Ramaswamy, S.N., and Arekal, G.D. 1982. Embryology of *Eriocaulon xeranthemum* Mart. (Eriocaulaceae). *Acta Bot. Neerl.* 31:41–54.

Ramaswamy, S.N., Swamy, B.G.L., and Govindappa, D.A. 1981. From zygote to seedling in *Eriocaulon robusto-brownianum* Ruhl. (Eriocaulaceae). *Beitr. Biol. Pflanz.* 55:179–188.

Rambaud, C., and Rambour, S. 1989. Partial characterization of nitrate reductase in carrot cells: changes in enzymatic activity during somatic embryogenesis. *Plant Physiol. Biochem.* 27:235–243.

Randolph, L.F. 1936. Developmental morphology of the caryopsis in maize. *J. Agr. Res.* 53:881–916.

Rangaswamy, N.S. 1967. Morphogenesis of seed germination in angiosperms. *Phytomorphology* 17:477–487.

Rangaswamy, N.S., and Shivanna, K.R. 1967. Induction of gamete compatibility and seed formation in axenic cultures of a diploid self-incompatible species of *Petunia*. *Nature* 216:937–939.

Rangaswamy, N.S., and Shivanna, K.R. 1969. Test-tube fertilization in *Dicranostigma franchetianum* (Prain) Fedde. *Curr. Sci.* 38:257–259.

Rangaswamy, N.S., and Shivanna, K.R. 1971a. Overcoming self-incompatibility in *Petunia axillaris*. II. Placental pollination *in vitro*. *J. Indian Bot. Soc.* 50A:286–296.

Rangaswamy, N.S., and Shivanna, K.R. 1971b. Overcoming self-incompatibility in *Petunia axillaris*. III. Two-site pollinations *in vitro*. *Phytomorphology* 21:284–289.

Rao, K.V. 1995. Transient gene expression in electroporated immature embryos of rice (*Oryza sativa* L.). *J. Plant Physiol.* 147:71–74.

Rao, K.V., Suprasanna, P., and Reddy, G.M. 1990. Biochemical changes in embryogenic and non-embryogenic calli of *Zea mays* L. *Plant Sci.* 66:127–130.

Rao, M.K., and Devi, K.U. 1983. Variation in expression of genic male sterility in pearl millet. *J. Hered.* 74:34–38.

Rao, M.K., and Koduru, P.R.K. 1978. Cytogenetics of a factor for syncyte formation and male sterility in *Pennisetum americanum*. *Theor. Appl. Genet.* 53:1–7.

Rao, P.S., and Rangaswamy, N.S. 1972. *In vitro* development of the pollinated pistils of *Nicotiana rustica* L. *Bot. Gaz.* 133:350–355.

Rao, P.V.L., and De, D.N. 1987. Haploid plants from *in vitro* anther culture of the leguminous tree, *Peltophorum pterocarpum* (DC) K. Hayne (copper pod). *Plant Cell Tissue Organ Cult.* 11:167–177.

Raquin, C., Amssa, M., Henry, Y., de Buyser, J., and Essad, S. 1982. Origine des plantes polyploïdes obtenues par culture d'anthères. Analyse cytophotométrique *in situ* et *in vitro* des microspores de *Petunia* et blé tendre. *Z. Pflanzenzüchtg* 89:265–277.

Rasch, E., and Woodard, J.W. 1959. Basic proteins of plant nuclei during normal and pathological cell growth. *J. Biophys. Biochem. Cytol.* 6:263–276.

Rashid, A., and Reinert, J. 1981. *In vitro* differentiation of embryogenic pollen, control by cold treatment and embryo formation in *ab initio* pollen cultures of *Nicotiana tabacum* var. Badischer Burley. *Protoplasma* 109:285–294.

Rashid, A., and Reinert, J. 1983. Factors affecting high-frequency embryo formation in *ab initio* pollen cultures of *Nicotiana*. *Protoplasma* 116:155–160.

Rashid, A., Siddiqui, A.W., and Reinert, J. 1981. Ultrastructure of embryogenic pollen of *Nicotiana tabacum* var. Badischer Burley. *Protoplasma* 107:375–385.

Rashid, A., Siddiqui, A.W., and Reinert, J. 1982. Subcellular aspects of origin and structure of pollen embryos of *Nicotiana*. *Protoplasma* 113:202–208.

Rashid, A., and Street, H.E. 1973. The development of haploid embryoids from anther cultures of *Atropa belladonna* L. *Planta* 113:263–270.

Rashid, A., and Street, H.E. 1974. Segmentations in microspores of *Nicotiana sylvestris* and *Nicotiana tabacum* which lead to embryoid formation in anther cultures. *Protoplasma* 80:323–334.

Rasmussen, J., and Hanson, M.R. 1989. A NADH dehydrogenase subunit gene is co-transcribed with the abnormal *Petunia* mitochondrial gene associated with cytoplasmic male sterility. *Mol. Gen. Genet.* 215:332–336.

Rasmussen, N., and Green, P.B. 1993. Organogenesis in flowers of the homeotic green pistillate mutant of tomato (*Lycopersicon esculentum*). *Am. J. Bot.* 80:805–813.

Rasmussen, S.K., and Brandt, A. 1986. Nucleotide sequences of cDNA clones for C-hordein polypeptides. *Carlsberg Res. Commun.* 51:371–379.

Rastogi, R., and Sawhney, V.K. 1988a. Flower culture of a male sterile stamenless-2 mutant of tomato (*Lycopersicon esculentum*). *Am. J. Bot.* 75:513–518.

Rastogi, R., and Sawhney, V.K. 1988b. Suppression of stamen development by CCC and ABA in tomato floral buds cultured *in vitro*. *J. Plant Physiol.* 133:620–624.

Rastogi, R., and Sawhney, V.K. 1990a. Polyamines and flower development in the male sterile stamenless-2 mutant of tomato (*Lycopersicon esculentum* Mill.). I. Level of polyamines and their biosynthesis in normal and mutant flowers. *Plant Physiol.* 93:439–445.

Rastogi, R., and Sawhney, V.K. 1990b. Polyamines and flower development in the male sterile stamenless-2 mutant of tomato (*Lycopersicon esculentum* Mill.). II. Effects of polyamines and their biosynthetic inhibitors on the development of normal and mutant floral buds cultured *in vitro*. *Plant Physiol.* 93:446–452.

Rathburn, H., Song, J., and Hedgcoth, C. 1993. Cytoplasmic male sterility and fertility restoration in wheat are not associated with rearrangements of mitochondrial DNA in the gene regions for *cob, coxII*, or *coxI*. *Plant Mol. Biol.* 21:195–201.

Rathore, K.S., Cork, R.J., and Robinson, K.R. 1991. A cytoplasmic gradient of Ca^{2+} is correlated with the growth of lily pollen tubes. *Devel Biol.* 148:612–619.

Rathore, K.S., Hodges, T.K., and Robinson, K.R. 1988.

Ionic basis of currents in somatic embryos of *Daucus carota*. *Planta* 175:280–289.

Raudaskoski, M., Aström, H., Pertillä, K., Virtanen, I., and Louhelainen, J. 1987. Role of the microtubule cytoskeleton in pollen tubes: an immunocytochemical and ultrastructural approach. *Biol. Cell.* 61:177–188.

Ray, A., Robinson-Beers, K., Ray, S., Baker, S.C., Lang, J.D., Preuss, D., Milligan, S.B., and Gasser, C.S. 1994. *Arabidopsis* floral homeotic gene *BELL* (*BEL1*) controls ovule development through negative regulation of AGAMOUS gene (*AG*). *Proc. Natl Acad. Sci. USA* 91:5761–5765.

Raynal, M., Depigny, D., Cooke, R., and Delseny, M. 1989. Characterization of a radish nuclear gene expressed during late seed maturation. *Plant Physiol.* 91:829–836.

Raynal, M., Depigny, D., Grellet, F., and Delseny, M. 1991. Characterization and evolution of napin-encoding genes in radish and related crucifers. *Gene* 99:77–86.

Reddy, B.V.S., Green, J.M., and Bisen, S.S. 1978. Genetic male sterility in pigeon pea. *Crop Sci.* 18:362–364.

Reddy, J.T., Dudareva, N., Evrard, J.-L., Kräuter, R., Steinmetz, A., and Pillay, D.T.N. 1995. A pollen-specific gene from sunflower encodes a member of the leucine-rich-repeat protein superfamily. *Plant Sci.* 111:81–93.

Reddy, V.M., and Daynard, T.B. 1983. Endosperm characteristics associated with rate of grain filling and kernel size in corn. *Maydica* 28:339–355.

Reeves, C.D., Krishnan, H.B., and Okita, T.W. 1986. Gene expression in developing wheat endosperm. Accumulation of gliadin and ADPglucose pyrophosphorylase messenger RNAs. *Plant Physiol.* 82:34–40.

Reeves, C.D., and Okita, T.W. 1987. Analyses of α/β-type gliadin genes from diploid and hexaploid wheats. *Gene* 52:257–266.

Regan, S.M., and Moffatt, B.A. 1990. Cytochemical analysis of pollen development in wild-type *Arabidopsis* and a male-sterile mutant. *Plant Cell* 2:877–889.

Reger, B.J., Chaubal, R., and Pressey, R. 1992. Chemotropic responses by pearl millet pollen tubes. *Sex. Plant Reprod.* 5:47–56.

Reger, B.J., Pressey, R., and Chaubal, R. 1992. *In vitro* chemotropism of pearl millet pollen tubes to stigma tissue: a response to glucose produced in the medium by tissue-bound invertase. *Sex. Plant Reprod.* 5:201–205.

Rehm, S. 1952. Male sterile plants by chemical treatment. *Nature* 170:38–39.

Reina, M., Guillén, P., Ponte, I., Boronat, A., and Palau, J. 1990a. DNA sequence of the gene encoding the Zc1 protein from *Zea mays* W64 A. *Nucl. Acids Res.* 18:6425.

Reina, M., Ponte, I., Guillén, P., Boronat, A., and Palau, J. 1990b. Sequence analysis of a genomic clone encoding a Zc2 protein from *Zea mays* W64 A. *Nucl. Acids Res.* 18:6426.

Reinbothe, C., Diettrich, B., and Luckner, M. 1990. Regeneration of plants from somatic embryos of *Digitalis lanata*. *J. Plant Physiol.* 137:224–228.

Reinbothe, C., Tewes, A., and Reinbothe, S. 1992a. Comparative molecular analysis of gene expression during plant embryogenesis: do evolutionarily conserved mechanisms control early plant development? *AgBiotech News Inform.* 4:381N–397N.

Reinbothe, C., Tewes, A., and Reinbothe, S. 1992b. Altered gene expression during somatic embryogenesis in *Nicotiana plumbaginifolia* and *Digitalis lanata*. *Plant Sci.* 82:47–58.

Reinbothe, C., Tewes, A., Lehmann, J., Parthier, B., and Reinbothe, S. 1994. Induction by methyl jasmonate of embryogenesis-related proteins and mRNAs in *Nicotiana plumbaginifolia*. *Plant Sci.* 104:59–70.

Reinbothe, C., Tewes, A., Luckner, M., and Reinbothe, S. 1992. Differential gene expression during somatic embryogenesis in *Digitalis lanata* analyzed by *in vivo* and *in vitro* protein synthesis. *Plant J.* 2:917–926.

Reinbothe, S., Machmudova, A., Wasternack, C., Reinbothe, C., and Parthier, B. 1992. Jasmonate-induced proteins in cotton: immunological relationship to the respective barley proteins and homology of transcripts to late embryogenesis abundant (Lea) mRNAs. *J. Plant Growth Regul.* 11:7–14.

Reinert, J. 1959. Uber die Kontrolle der Morphogenese und die Induktion von Adventivembryonen an Gewebekulturen aus Karotten. *Planta* 53:318–333.

Reinert, J., Tazawa, M., and Semenoff, S. 1967. Nitrogen compounds as factors of embryogenesis *in vitro*. *Nature* 216:1215–1216.

Reinholz, E. 1959. Beeinflussung der Morphogenese embryonaler Organe durch ionisierende Strahlungen. I. Keimlingsanomalien durch Röntgenbestrahlung von *Arabidopsis thaliana*-Embryonen in verschiedenen Entwicklungsstadien. *Strahlentherapie* 109:537–553.

Reinold, S., Hauffe, K.D., and Douglas, C.J. 1993. Tobacco and parsley 4-coumarate:coenzyme A ligase genes are temporally and spatially regulated in a cell type-specific manner during tobacco flower development. *Plant Physiol.* 101:373–383.

Reiser, L., Modrusan, Z., Margossian, L., Samach, A., Ohad, N., Haughn, G.W., and Fischer, R.L. 1995. The *BELL1* gene encodes a homeodomain protein involved in pattern formation in the *Arabidopsis* ovule primordium. *Cell* 83:735–742.

Reiss, H.-D., and Herth, W. 1978. Visualization of the Ca^{2+}-gradient in growing pollen tubes of *Lilium longiflorum* with chlorotetracycline fluorescence. *Protoplasma* 97:373–377.

Reiss, H.-D., and Herth, W. 1979a. Calcium ionophore A 23187 affects localized wall secretion in the tip region of pollen tubes of *Lilium longiflorum*. *Planta* 145:225–232.

Reiss, H.-D., and Herth, W. 1979b. Calcium gradients in tip growing plant cells visualized by chlorotetracycline fluorescence. *Planta* 146:615–621.

Reiss, H.-D., and Herth, W. 1980. Broad-range effects of ionophore X-537A on pollen tubes of *Lilium longiflorum*. *Planta* 147:295–301.

Reiss, H.-D., and Herth, W. 1982. Disoriented growth of pollen tubes of *Lilium longiflorum* Thunb. induced by prolonged treatment with the calcium-chelating antibiotic, chlorotetracycline. *Planta* 156:218–225.

Reiss, H.-D., and Herth, W. 1985. Nifedipine-sensitive calcium channels are involved in polar growth of lily pollen tubes. *J. Cell Sci.* 76:247–254.

Reiss, H.-D., Herth, W., and Nobiling, R. 1985. Development of membrane- and calcium-gradients during pollen germination of *Lilium longiflorum*. *Planta* 163:84–90.

Reiss, H.-D., Herth, W., and Schnepf, E. 1985. Plasmamembrane "rosettes" are present in the lily pollen tube. *Naturwissenschaften* 72:276.

Reiss, H.-D., and McConchie, C.A. 1988. Studies of *Najas*

pollen tubes. Fine structure and the dependence of chlorotetracycline fluorescence on external free ions. *Protoplasma* 142:25–35.

Reiss, H.-D., and Nobiling, R. 1986. Quin-2 fluorescence in lily pollen tubes. Distribution of free cytoplasmic calcium. *Protoplasma* 131:244–246.

Reiss, H.-D., Herth, W., Schnepf, E., and Nobilin, R. 1983. The tip-to-base calcium gradient in pollen tubes of *Lilium longiflorum* measured by proton-induced X-ray emission (PIXE). *Protoplasma* 115:153–159.

Rembert, D.H., Jr. 1971. Phylogenetic significance of megaspore tetrad patterns in Leguminales. *Phytomorphology* 21:1–9.

Rembur, J., Nougarède, A., Rondet, P., and Francis, D. 1992. Floral-specific polypeptides in *Silene coeli-rosa*. *Can. J. Bot.* 70:2326–2633.

Remy, R., and Ambard-Bretteville, F. 1983. Two dimensional analysis of chloroplast proteins from normal and cytoplasmic male sterile *Brassica napus*. *Theor. Appl. Genet.* 64:249–253.

Renner, O. 1914. Befruchtung und Embryobildung bei *Oenothera lamarckiana* und einigen verwandten Arten. *Flora* 107:115–150.

Rerie, W.G., Whitecross, M., and Higgins, T.J.V. 1991. Developmental and environmental regulation of pea legumin genes in transgenic tobacco. *Mol. Gen. Genet.* 225:148–157.

Rest, J.A., and Vaughan, J.G. 1972. The development of protein and oil bodies in the seed of *Sinapis alba* L. *Planta* 105:245–262.

Reynolds, J.D., and Dashek, W.V. 1976. Cytochemical analysis of callose localization in *Lilium longiflorum* pollen tubes. *Ann. Bot.* 40:409–416.

Reynolds, T.L. 1984a. Callus formation and organogenesis in anther cultures of *Solanum carolinense* L. *J. Plant Physiol.* 117:157–161.

Reynolds, T.L. 1984b. An ultrastructural and stereological analysis of pollen grains of *Hyoscyamus niger* during normal ontogeny and induced embryogenic development. *Am. J. Bot.* 71:490–504.

Reynolds, T.L. 1985. Ultrastructure of anomalous pollen development in embryogenic anther cultures of *Hyoscyamus niger*. *Am. J. Bot.* 72:44–51.

Reynolds, T.L. 1986. Pollen embryogenesis in anther cultures of *Solanum carolinense* L. *Plant Cell Rep.* 5:273–275.

Reynolds, T.L. 1987. A possible role for ethylene during IAA-induced pollen embryogenesis in anther cultures of *S. carolinense* L. *Am. J. Bot.* 74:967–969.

Reynolds, T.L. 1989. Ethylene effects on pollen callus formation and organogenesis in anther cultures of *Solanum carolinense* L. *Plant Sci.* 61:131–136.

Reynolds, T.L. 1990. Interactions between calcium and auxin during pollen androgenesis in anther cultures of *Solanum carolinense* L. *Plant Sci.* 72:109–114.

Reynolds, T.L. 1993. A cytological analysis of microspores of *Triticum aestivum* (Poaceae) during normal ontogeny and induced embryogenic development. *Am. J. Bot.* 80:569–576.

Reynolds, T.L., and Crawford, R.L. 1996. Changes in abundance of an abscisic acid-responsive, early cysteine-labeled metallothionein transcript during pollen embryogenesis in bread wheat (*Triticum aestivum*). *Plant Mol. Biol.* 32:823–829.

Reynolds, T.L., and Kitto, S.L. 1992. Identification of embryoid-abundant genes that are temporally expressed during pollen embryogenesis in wheat anther cultures. *Plant Physiol.* 100:1744–1750.

Reynolds, T.L., and Raghavan, V. 1982. An autoradiographic study of RNA synthesis during maturation and germination of pollen grains of *Hyoscyamus niger*. *Protoplasma* 111:177–188.

Reznickova, S.A., and Bogdanov, Y.F. 1972. Meiosis in excised anthers of *Lilium candidum*. *Biol. Zentral.* 91:409–428.

Reznickova, S.A., and Dickinson, H.G. 1982. Ultrastructural aspects of storage lipid mobilization in the tapetum of *Lilium hybrida* var. Enchantment. *Planta* 155:400–408.

Reznickova, S.A., and Willemse, M.T.M. 1980. Formation of pollen in the anther of *Lilium*. II. The function of the surrounding tissues in the formation of pollen and pollen wall. *Acta Bot. Neerl.* 29:141–156.

Richards, A.J., and Ibrahim, H.B. 1982. The breeding system in *Primula veris* L. II. Pollen tube growth and seed-set. *New Phytol.* 90:305–314.

Richards, A.J., and Mitchell, J. 1990. The control of incompatibility in distylous *Pulmonaria affinis* Jordan (Boraginaceae). *Bot. J. Linn. Soc.* 104:369–380.

Richards, J.H., and Barrett, S.C.H. 1984. The developmental basis of tristyly in *Eichhornia paniculata* (Pontederiaceae). *Am. J. Bot.* 71:1347–1363.

Richards, J.H., and Barrett, S.C.H. 1987. Development of tristyly in *Pontederia cordata* (Pontederiaceae). I. Mature floral structure and patterns of relative growth of reproductive organs. *Am. J. Bot.* 74:1831–1841.

Rick, C.M. 1945. A survey of cytogenetic causes of unfruitfulness in the tomato. *Genetics* 30:347–362.

Rick, C.M. 1948. Genetics and development of nine male-sterile tomato mutants. *Hilgardia* 18:599–633.

Rietsema, J., Satina, S., and Blakeslee, A.F. 1953. The effect of sucrose on the growth of *Datura stramonium* embryos *in vitro*. *Am. J. Bot.* 40:538–545.

Rifot, M. 1973. Evolution structurale du pôle chalazien du sac embryonnaire d'*Aquilegia vulgaris* en liaison avec son activité trophique. *Compt. Rend. Acad. Sci. Paris* 277D:1313–1316.

Rigau, J., Capellades, M., Montoliu, L., Torres, M.A., Romera, C., Martinez-Izquierdo, J.A., Tagu, D., and Puigdomènech, P. 1993. Analysis of a maize α-tubulin gene promoter by transient expression and in transgenic tobacco plants. *Plant J.* 4:1043–1050.

Riggs, C.D., and Hasenkampf, C.A. 1991. Antibodies directed against a meiosis-specific, chromatin-associated protein identify conserved meiotic epitopes. *Chromosoma* 101:92–98.

Riggs, C.D., Voelker, T.A., and Chrispeels, M.J. 1989. Cotyledon nuclear proteins bind to DNA fragments harboring regulatory elements of phytohemagglutinin genes. *Plant Cell* 1:609–621.

Riggs, C.D., Hunt, D.C., Lin, J., and Chrispeels, M.J. 1989. Utilization of luciferase fusion genes to monitor differential regulation of phytohemagglutinin and phaseolin promoters in transgenic tobacco. *Plant Sci.* 63:47–57.

Říhová, L., Čapková, V., and Tupý, J. 1996. Changes in glycoprotein patterns associated with male gametophyte development and with induction of pollen embryogenesis in *Nicotiana tabacum* L. *J. Plant Physiol.* 147:573–581.

Rijven, A.H.G.C. 1952. *In vitro* studies on the embryo of *Capsella bursa-pastoris*. *Acta Bot. Neerl.* 1:157–200.

Rijven, A.H.G.C. 1956. Glutamine and asparagine as nitrogen sources for the growth of plant embryos *in vitro*: a comparative study of 12 species. *Aust. J. Biol. Sci.* 9:511–527.

Riley, R., and Bennett, M.D. 1971. Meiotic DNA synthesis. *Nature* 230:182–185.

Risueño, M.C., Giménez-Martín, G., López-Sáez, J.F., and Rodríguez-García, M.I. 1969. Origin and development of sporopollenin bodies. *Protoplasma* 67:361–374.

Ritala, A., Aikasalo, R., Aspegren, K., Salmenkallio-Marttila, M., Akerman, S., Mannonen, L., Kurtén, U., Puupponen-Pimiä, R., Teeri, T.H., and Kauppinen, V. 1995. Transgenic barley by particle bombardment. Inheritance of the transferred gene and characteristics of transgenic barley plants. *Euphytica* 85:81–88.

Ritala, A., Aspegren, K., Kurtén, U., and Salmenkallio-Marttila, M. 1994. Fertile transgenic barley by particle bombardment of immature embryos. *Plant Mol. Biol.* 24:317–325.

Rittscher, M., and Wiermann, R. 1983. Occurrence of phenylalanine ammonia-lyase (PAL) in isolated tapetum cells of *Tulipa* anthers. *Protoplasma* 118:219–224.

Rivers, B.A., and Bernatzky, R. 1994. Protein expression of a self-compatible allele from *Lycopersicon peruvianum*: introgression and behavior in a self-incompatible background. *Sex. Plant Reprod.* 7:357–362.

Rivers, B.A., Bernatzky, R., Robinson, S.J., and Jahnen-Dechent, W. 1993. Molecular diversity at the self-incompatibility locus is a salient feature in natural populations of wild tomato (*Lycopersicon peruvianum*). *Mol. Gen. Genet.* 238:419–427.

Rivin, C.J., and Grudt, T. 1991. Abscisic acid and the developmental regulation of embryo storage proteins in maize. *Plant Physiol.* 95:358–365.

Roath, W.W., and Hockett, E.A. 1971. Genetic male sterility in barley. III. Pollen and anther characteristics. *Crop Sci.* 11:200–203.

Robert, L.S., Thompson, R.D., and Flavell, R.B. 1989. Tissue-specific expression of a wheat high molecular weight glutenin gene in transgenic tobacco. *Plant Cell* 1:569–578.

Robert, L.S., Allard, S., Franklin, T.M., and Trick, M. 1994a. Sequence and expression of endogenous S-locus glycoprotein genes in self-compatible *Brassica napus*. *Mol. Gen. Genet.* 242:209–216.

Robert, L.S., Allard, S., Gerster, J.L., Cass, L., and Simmonds, J. 1993. Isolation and characterization of a polygalacturonase gene highly expressed in *Brassica napus* pollen. *Plant Mol. Biol.* 23:1273–1278.

Robert, L.S., Allard, S., Gerster, J.L., Cass, L., and Simmonds, J. 1994b. Molecular analysis of two *Brassica napus* genes expressed in the stigma. *Plant Mol. Biol.* 26:1217–1222.

Robert, L.S., Gerster, J., Allard, S., Cass, L., and Simmonds, J. 1994c. Molecular characterization of two *Brassica napus* genes related to oleosins which are highly expressed in the tapetum. *Plant J.* 6:927–933.

Roberton, M., and Chandler, P.M. 1992. Pea dehydrins: identification, characterisation and expression. *Plant Mol. Biol.* 19:1031–1044.

Roberts, D.R., Flinn, B.S., Webb, D.T., Webster, F.B., and Sutton, B.C.S. 1990. Abscisic acid and indole-3-butyric acid regulation of maturation and accumulation of storage proteins in somatic embryos of interior spruce. *Physiol. Plant.* 78:355–360.

Roberts, I.N., Harrod, G., and Dickinson, H.G. 1984a. Pollen–stigma interactions in *Brassica oleracea* I. Ultrastructure and physiology of the stigmatic papillar cells. *J. Cell Sci.* 66:241–253.

Roberts, I.N., Harrod, G., and Dickinson, H.G. 1984b. Pollen-stigma interactions in *Brassica oleracea*. II. The fate of stigma surface proteins following pollination and their role in the self-incompatibility response. *J. Cell Sci.* 66:255–264.

Roberts, I.N., Stead, A.D., Ockendon, D.J., and Dickinson, H.G. 1979. A glycoprotein associated with the acquisition of the self-incompatibility system by maturing stigmas of *Brassica oleracea*. *Planta* 146:179–183.

Roberts, J.K., DeSimone, N.A., Lingle, W.L., and Dure, L., III. 1993. Cellular concentrations and uniformity of cell-type accumulation of two Lea proteins in cotton embryos. *Plant Cell* 5:769–780.

Roberts, M.R., Boyes, E., and Scott, R.J. 1995. An investigation of the role of the anther tapetum during microspore development using genetic cell ablation. *Sex. Plant Reprod.* 8:299–307.

Roberts, M.R., Hodge, R., and Scott, R. 1995. *Brassica napus* pollen oleosins possess a characteristic C-terminal domain. *Planta* 195:469–470.

Roberts, M.R., Foster, G.D., Blundell, R.P., Robinson, S.W., Kumar, A., Draper, J., and Scott, R. 1993a. Gametophytic and sporophytic expression of an anther-specific *Arabidopsis thaliana* gene. *Plant J.* 3:111–120.

Roberts, M.R., Hodge, R., Ross, J.H.E., Sorensen, A., Murphy, D.J., Draper, J., and Scott, R. 1993b. Characterization of a new class of oleosins suggests a male gametophyte-specific lipid storage pathway. *Plant J.* 3:629–636.

Roberts, M.R., Robson, F., Foster, G.D., Draper, J., and Scott, R.J. 1991. A *Brassica napus* mRNA expressed specifically in developing microspores. *Plant Mol. Biol.* 17:295–299.

Robertson, D., Weissinger, A.K., Ackley, R., Glover, S., and Sederoff, R.R. 1992. Genetic transformation of Norway spruce (*Picea abies* (L.) Karst) using somatic embryo explants by microprojectile bombardment. *Plant Mol. Biol.* 19:925–935.

Robertson, D.S. 1952. The genotype of the endosperm and embryo as it influences vivipary in maize. *Proc. Natl Acad. Sci. USA* 38:580–583.

Robertson, D.S. 1955. The genetics of vivipary in maize. *Genetics* 40:745–760.

Robertson, M., Walker-Simmons, M., Munro, D., and Hill, R.D. 1989. Induction of α-amylase inhibitor synthesis in barley embryos and young seedlings by abscisic acid and dehydration stress. *Plant Physiol.* 91:415–420.

Robichaud, C., and Sussex, I.M. 1986. The response of viviparous-1 and wild type embryos of *Zea mays* to culture in the presence of abscisic acid. *J. Plant Physiol.* 126:235–242.

Robichaud, C., and Sussex, I.M. 1987. The uptake and metabolism of [2-^{14}C]-ABA by excised wild type and viviparous-1 embryos of *Zea mays* L. *J. Plant Physiol.* 130:181–188.

Robichaud, C.S., Wong, J., and Sussex, I.M. 1980. Control of *in vitro* growth of viviparous embryo mutants of maize by abscisic acid. *Devel Genet.* 1:325–330.

Robie, C.A., and Minocha, S.C. 1989. Polyamines and somatic embryogenesis in carrot. I. The effects of difluoromethylornithine and difluoromethylarginine. *Plant Sci.* 65:45–54.

Robinson-Beers, K., Pruitt, R.E., and Gasser, C.S. 1992.

Ovule development in wild-type *Arabidopsis* and two female-sterile mutants. *Plant Cell* 4:1237–1249.

Rochange, F., Serrano, L., Marque, C., Teulières, C., and Boudet, A.-M. 1995. DNA delivery into *Eucalyptus globulus* zygotic embryos through biolistics: optimization of the biological and physical parameters of bombardment for two different particle guns. *Plant Cell Rep*. 14:674–678.

Rode, A., Hartmann, C., de Buyser, J., and Henry, Y. 1988. Evidence for a direct relationship between mitochondrial genome organization and regeneration ability in hexaploid wheat somatic tissue cultures. *Curr. Genet*. 14:387–394.

Rödin, J., and Rask, L. 1990. Characterization of matteuccin, the 2.2S storage protein of the ostrich fern. Evolutionary relationship to angiosperm seed storage proteins. *Eur. J. Biochem*. 192:101–107.

Rödin, J., Ericson, M.L., Josefsson, L.-G., and Rask, L. 1990. Characterization of a cDNA clone encoding a *Brassica napus* 12 S protein (cruciferin) subunit. Relationship between precursors and mature chains. *J. Biol. Chem*. 265:2720–2723.

Rödin, S., Sjödahl, S., Josefsson, L.-G., and Rask, L. 1992. Characterization of a *Brassica napus* gene encoding a cruciferin subunit: estimation of sizes of cruciferin gene families. *Plant Mol. Biol*. 20:559–563.

Rodkiewicz, B. 1960. Measurements of desoxyribose nucleic acid by Feulgen-photometry in nuclei of pollen grains of *Tradescantia bracteata*. *Acta Soc. Bot. Polon*. 29:211–217.

Rodkiewicz, B. 1970. Callose in cell walls during megasporogenesis in angiosperms. *Planta* 93:39–47.

Rodkiewicz, B., and Bednara, J. 1974. Distribution of organelles and starch grains during megasporogenesis in *Epilobium*. In *Fertilization in Higher Plants*, ed. H.F. Linskens, pp. 89–95. Amsterdam: North-Holland Publishing Co.

Rodkiewicz, B., and Bednara, J. 1976. Cell wall ingrowths and callose distribution in megasporogenesis in some Orchidaceae. *Phytomorphology* 26:276–281.

Rodkiewicz, B., Duda, E., and Bednara, J. 1989. Organelle aggregations during microsporogenesis in *Nymphaea*. *Flora* 183:397–404.

Rodkiewicz, B., Fyk, B., and Szczuka, E. 1994. Chlorophyll and cutin in early embryogenesis in *Capsella*, *Arabidopsis*, and *Stellaria* investigated by fluorescence microscopy. *Sex. Plant Reprod*. 7:287–292.

Rodkiewicz, B., and Górska-Brylass, A. 1967. Occurrence of callose in the walls of meiotically dividing cell in the ovule of *Orchis*. *Naturwissenschaften* 54:499–500.

Rodkiewicz, B., and Górska-Brylass, A. 1968. Callose in the walls of the developing megasporocyte and megaspores in the orchid ovule. *Acta Soc. Bot. Polon*. 37:19–28.

Rodkiewicz, B., and Mikulska, E. 1963. Electron microscope observations of cytoplasmic changes in developing megasporocyte of *Lilium candidum*. *Flora* 154:383–387.

Rodkiewicz, B., and Mikulska, E. 1965a. The development of cytoplasmic structures in the embryo sac of *Lilium candidum*, as observed with the electron microscope. *Planta* 67:297–304.

Rodkiewicz, B., and Mikulska, E. 1965b. Electron microscope observations of endoplasmic reticulum in *Lilium* megasporocyte. *Flora* 155:341–346.

Rodkiewicz, B., and Mikulska, E. 1967. The micropylar and antipodal cells of the *Lilium regale* embryo sac observed with the electron microscope. *Flora* 158:181–188.

Rodríguez-García, M.I., and Fernández, M.C. 1987. Cytoplasmic nucleoloids during microsporogenesis in *Olea europaea* L. *Biol. Cell*. 60:155–160.

Rodríguez-García, M.I., Fernández, M.C., and Alché, J.D. 1995. Immunocytochemical localization of allergenic protein (*Ole e I*) in the endoplasmic reticulum of the developing pollen grain of olive (*Olea europaea* L.). *Planta* 196:558–563.

Rodríguez-García, M.I., and García, A. 1978. Differentiation of the plastid population during microsporogenesis and the development of the pollen grain in the Liliaceae. *Biol. Cell*. 33:63–70.

Rodríguez-García, M.I., Fernández, M.C., Alché, J.D., and Olmedilla, A. 1995. Endoplasmic reticulum as a storage site for allergenic proteins in pollen grains of several Oleaceae. *Protoplasma* 187:111–116.

Rodriguez-Sotres, R., and Black, M. 1994. Osmotic potential and abscisic acid regulate triacylglycerol synthesis in developing wheat embryos. *Planta* 192:9–15.

Roeckel, P., and Dumas, C. 1993. Survival at 20°C and cryopreservation of isolated sperm cells from *Zea mays* pollen grains. *Sex. Plant Reprod*. 6:212–216.

Roeckel, P., Heizmann, P., Dubois, M., and Dumas, C. 1988. Attempts to transform *Zea mays* via pollen grains. Effect of pollen and stigma nuclease activities. *Sex. Plant Reprod*. 1:156–163.

Rogers, C.M. 1979. Distyly and pollen dimorphism in *Linum suffruticosum* (Linaceae). *Plant Syst. Evol*. 131:127–132.

Rogers, C.M., and Harris, B.D. 1969. Pollen exine deposition: a clue to its control. *Am. J. Bot*. 56:1209–1211.

Rogers, H.J., Greenland, A.J., and Hussey, P.J. 1993. Four members of the maize β-tubulin gene family are expressed in the male gametophyte. *Plant J*. 4:875–882.

Rogers, H.J., Harvey, A., and Lonsdale, D.M. 1992. Isolation and characterization of a tobacco gene with homology to pectate lyase which is specifically expressed during microsporogenesis. *Plant Mol. Biol*. 20:493–502.

Rogers, H.J., Allen, R.L., Hamilton, W.D.O., and Lonsdale, D.M. 1991. Pollen specific cDNA clones from *Zea mays*. *Biochim. Biophys. Acta* 1089:411–413.

Roggen, H.P., and Stanley, R.G. 1971. Autoradiographic studies of pear pollen tube walls. *Physiol. Plant*. 24:80–84.

Roggen, H.P.J.R. 1967. Changes in enzyme activities during the progame phase in *Petunia hybrida*. *Acta Bot. Neerl*. 16:1–31.

Roggen, H.P.J.R. 1972. Scanning electron microscopical observations on compatible and incompatible pollen–stigma interactions in *Brassica*. *Euphytica* 21:1–10.

Roggen, H.P.J.R., and Stanley, R.G. 1969. Cell-wall-hydrolysing enzymes in wall formation as measured by pollen-tube extension. *Planta* 84:295–303.

Roiz, L., Goren, R., and Shoseyov, O. 1995. Stigmatic RNase in calamondin (*Citrus reticulata* var. *austera* x *Fortunella* sp.). *Physiol. Plant*. 94:585–590.

Roiz, L., and Shoseyov, O. 1995. Stigmatic RNase in self-compatible peach (*Prunus persica*). *Int. J. Plant Sci*. 156:37–41.

Roman, H. 1948. Directed fertilization in maize. *Proc. Natl Acad. Sci. USA* 34:36–42.

Rondet, P. 1962. L'organogenèse au cours de l'embryogenèse chez l'*Alyssum maritimum* Lamk. *Compt. Rend. Acad. Sci. Paris* 255:2278–2280.

Rose, R.E., DeJesus, C.E., Moylan, S.L., Ridge, N.P., Scherer, D.E., and Knauf, V.C. 1987. The nucleotide sequence of a cDNA clone encoding acyl carrier protein (ACP) from *Brassica campestris* seeds. *Nucl. Acids Res.* 15:7197.

Rosen, W.G. 1961. Studies on pollen-tube chemotropism. *Am. J. Bot.* 48:889–895.

Rosen, W.G., and Gawlik, S.R. 1966. Fine structure of lily pollen tubes following various fixation and staining procedures. *Protoplasma* 61:181–191.

Rosen, W.G., and Thomas, H.R. 1970. Secretory cells of lily pistils. I. Fine structure and function. *Am. J. Bot.* 57:1108–1114.

Rosen, W.G., Gawlik, S.R., Dashek, W.V., and Siegesmund, K.A. 1964. Fine structure and cytochemistry of *Lilium* pollen tubes. *Am. J. Bot.* 51:61–71.

Rosenberg, L.A., and Rinne, R.W. 1988. Protein synthesis during natural and precocious soybean seed (*Glycine max* [L.] Merr.) maturation. *Plant Physiol.* 87:474–478.

Rosenfield, C.-L., Fann, C., and Loewus, F.A. 1978. Metabolic studies on intermediates in the *myo*-inositol oxidation pathway in *Lilium longiflorum* pollen. I. Conversion to hexoses. *Plant Physiol.* 61:89–95.

Rosenfield, C.-L., and Matile, P. 1979. Glycosidases in pear pollen tube development. *Plant Cell Physiol.* 20:605–613.

Ross, A.H., Manners, J.M., and Birch, R.G. 1995. Embryogenic callus production, plant regeneration and transient gene expression following particle bombardment in the pasture grass, *Cenchrus ciliaris* (Gramineae). *Aust. J. Bot.* 43:193–199.

Ross, J.H.E., and Murphy, D.J. 1992. Biosynthesis and localisation of storage proteins, oleosins and lipids during seed development in *Coriandrum sativum* and other Umbelliferae. *Plant Sci.* 86:59–70.

Ross, J.H.E., and Murphy, D.J. 1993. Differential accumulation of storage products in developing seeds and somatic cell cultures of *Daucus carota* L. *Plant Sci.* 88:1–11.

Ross, J.H.E., and Murphy, D.J. 1996. Characterization of anther-expressed genes encoding a major class of extracellular oleosin-like proteins in the pollen coat of Brassicaceae. *Plant J.* 9:625–637.

Rost, T.L., and Lersten, N.R. 1970. Transfer aleurone cells in *Setaria lutescens* (Gramineae). *Protoplasma* 71:403–408.

Roth, T.F., and Ito, M. 1967. DNA-dependent formation of the synaptinemal complex at meiotic prophase. *J. Cell Biol.* 35:247–255.

Rothenberg, M., and Hanson, M.R. 1987. Recombination between parental mitochondrial DNA following protoplast fusion can occur in a region which normally does not undergo intragenomic recombination in parental plants. *Curr. Genet.* 12:235–240.

Rothenberg, M., and Hanson, M.R. 1988. A functional mitochondrial ATP synthase proteolipid gene produced by recombination of parental genes in a *Petunia* somatic hybrid. *Genetics* 118:155–161.

Rothenberg, M., Boeshore, M.L., Hanson, M.R., and Izhar, S. 1985. Intergenomic recombination of mitochondrial genomes in a somatic hybrid plant. *Curr. Genet.* 9:615–618.

Rottmann, W.H., Brears, T., Hodge, T.P., and Lonsdale, D.M. 1987. A mitochondrial gene is lost via homologous recombination during reversion of CMS T maize to fertility. *EMBO J.* 6:1541–1546.

Rougier, M., Jnoud, N., and Dumas, C. 1988. Localization of adenylate cyclase activity in *Populus*: its relation to pollen–pistil recognition and incompatibility. *Sex. Plant Reprod.* 1:140–149.

Rougier, M., Jnoud, N., Said, C., Russell, S., and Dumas, C. 1991. Male gametophyte development and formation of the male germ unit in *Populus deltoides* following compatible pollination. *Protoplasma* 162:140–150.

Rougier, M., Jnoud, N., Saïd, C., Russell, S., and Dumas, C. 1992. Interspecific incompatibility in *Populus*: inhibition of tube growth and behaviour of the male germ unit in *P. deltoides* × *P. alba* cross. *Protoplasma* 168:107–112.

Rounsley, S.D., Ditta, G.S., and Yanofsky, M.F. 1995. Diverse roles for MADS box genes in *Arabidopsis* development. *Plant Cell* 7:1259–1269.

Roussell, D.L., Boston, R.S., Goldsbrough, P.B., and Larkins, B.A. 1988. Deletion of DNA sequences flanking an M_r 19000 zein gene reduces its transcriptional activity in heterologous plant tissues. *Mol. Gen. Genet.* 211:202–209.

Roustan, J.-P., Latché, A., and Fallot, J. 1989a. Stimulation of *Daucus carota* somatic embryogenesis by inhibitors of ethylene synthesis: cobalt and nickel. *Plant Cell Rep.* 8:182–185.

Roustan, J.-P., Latché, A., and Fallot, J. 1989b. Effet de l'acide salicylique et de l'acide acétylsalicylique sur la production d'éthylène et l'embryogenèse somatique de suspensions cellulaires de carotte (*Daucus carota* L.). *Compt. Rend. Acad. Sci. Paris* Ser. III 308:395–399.

Roustan, J.-P., Latché, A., and Fallot, J. 1990. Control of carrot somatic embryogenesis by $AgNO_3$, an inhibitor of ethylene action: effect on arginine decarboxylase activity. *Plant Sci.* 67:89–95.

Roustan, J.-P., Latché, A., and Fallot, J. 1992. Influence of ethylene on the incorporation of 3,4-[^{14}C]methionine into polyamines in *Daucus carota* cells during somatic embryogenesis. *Plant Physiol. Biochem.* 30:201–205.

Rouwendal, G.J.A., van Damme, J.M.M., and Wessels, J.G.H. 1987. Cytoplasmic male sterility in *Plantago lanceolata* L.: differences between male-sterile cytoplasms at the DNA- and RNA-level. *Theor. Appl. Genet.* 75:59–65.

Rowland, L.J., and Chourey, P.S. 1990. *In situ* hybridization analysis of sucrose synthase expression in developing kernels of maize. *Maydica* 35:373–382.

Rowley, J.R. 1963. Ubisch body development in *Poa annua*. *Grana Palynol.* 4:25–36.

Rowley, J.R. 1964. Formation of the pore in pollen of *Poa annua*. In *Pollen Physiology and Fertilization*, ed. H.F. Linskens, pp. 59–69. Amsterdam: North-Holland Publishing Co.

Rowley, J.R., and Dunbar, A. 1967. Sources of membranes for exine formation. *Svensk. Bot. Tidskr.* 61:49–64.

Rowley, J.R., and Erdtman, G. 1967. Sporoderm in *Populus* and *Salix*. *Grana Palynol.* 7:517–567.

Rowley, J.R., Mühlethaler, K., and Frey-Wyssling, A. 1959. A route for the transfer of materials through the pollen grain wall. *J. Biophys. Biochem. Cytol.* 6:537–538.

Rowley, J.R., and Skvarla, J.J. 1975. The glycocalyx and initiation of exine spinules on microspores of *Canna*. *Am. J. Bot.* 62:479–485.

Rowley, J.R., and Southworth, D. 1967. Deposition of

sporopollenin on lamellae of unit membrane dimensions. *Nature* 213:703–704.

Royo, J., Kunz, C., Kowyama, Y., Anderson, M., Clarke, A.E., and Newbigin, E. 1994. Loss of a histidine residue at the active site of *S*-locus ribonuclease is associated with self-compatibility in *Lycopersicon peruvianum*. *Proc. Natl Acad. Sci. USA* 91:6511–6514.

Rozycka, M., Khan, S., Lopez, I., Greenland, A.J., and Hussey, P.J. 1995. A *Zea mays* pollen cDNA encoding a putative actin depolymerizing factor. *Plant Physiol.* 107: 1011–1012.

Rubin, R., Levanony, H., and Galili, G. 1992. Evidence for the presence of two different types of protein bodies in wheat endosperm. *Plant Physiol.* 99:718–724.

Rubinstein, A.L., Prata, R.T.N., and Bedinger, P.A. 1995. Developmental accumulation of hydroxyproline and hydroxyproline-containing proteins in *Zea mays* pollen. *Sex. Plant Reprod.* 8:27–32.

Rubinstein, A.L., Broadwater, A.H., Lowrey, K.B., and Bedinger, P.A. 1995a. *Pex1*, a pollen-specific gene with an extensin-like domain. *Proc. Natl Acad. Sci. USA* 92:3086–3090.

Rubinstein, A.L., Márquez, J., Suárez-Cervera, M., and Bedinger, P.A. 1995b. Extensin-like glycoproteins in the maize pollen tube wall. *Plant Cell* 7:2211–2225.

Rudd, J.J., Franklin, F.C.H., Lord, J.M., and Franklin-Tong, V.E. 1996. Increased phosphorylation of a 26-kD pollen protein is induced by the self-incompaibility response in *Papaver rhoeas*. *Plant Cell* 8:713--724.

Rueda, J., and Vázquez, A.M. 1985. Effect of auxins and cytokinins upon the start of meiosis in cultured anthers of rye (*Secale cereale*). *Can. J. Genet. Cytol.* 27:759–765.

Ruiz-Avila, L., Ludevid, M.D., and Puigdomènech, P. 1991. Differential expression of a hydroxyproline-rich cell-wall protein gene in embryonic tissues of *Zea mays* L. *Planta* 184:130–136.

Ruiz-Avila, L., Burgess, S.R., Stiefel, V., Ludevid, M.D., and Puigdomènech, P. 1992. Accumulation of cell wall hydroxyproline-rich glycoprotein mRNA is an early event in maize embryo cell differentiation. *Proc. Natl Acad. Sci. USA* 89:2414–2418.

Rundle, S.J., and Nasrallah, J.B. 1992. Molecular characterization of type 1 serine/threonine phosphatases from *Brassica oleracea*. *Plant Mol. Biol.* 20:367–375.

Rundle, S.J., Nasrallah, M.E., and Nasrallah, J.B. 1993. Effects of inhibitors of protein serine/threonine phosphatases on pollination in *Brassica*. *Plant Physiol.* 103:1165–1171.

Rusche, M.L., and Mogensen, H.L. 1988. The male germ unit of *Zea mays*: quantitative ultrastructure and three-dimensional analysis. In *Sexual Reproduction in Higher Plants*, ed. M. Cresti, P. Gori and E. Pacini, pp. 221–226. Berlin: Springer-Verlag.

Rusche, M.L., Mogensen, H.L., Zhu, T., and Smith, S.E. 1995. The zygote and proembryo of alfalfa: quantitative, three-dimensional analysis and implications for biparental plastid inheritance. *Protoplasma* 189:88–100.

Russell, D.A., and Sachs, M.M. 1991. The maize cytosolic glyceraldehyde-3-phosphate dehydrogenase gene family: organ-specific expression and genetic analysis. *Mol. Gen. Genet.* 229:219–228.

Russell, D.R., Wallace, K.M., Bathe, J.H., Martinell, B.J., and McCabe, D.E. 1993. Stable transformation of *Phaseolus vulgaris* via electric-discharge mediated particle acceleration. *Plant Cell Rep.* 12:165–169.

Russell, S.D. 1979. Fine structure of megagametophyte development in *Zea mays*. *Can. J. Bot.* 57:1093–1110.

Russell, S.D. 1980. Participation of male cytoplasm during gamete fusion in an angiosperm, *Plumbago zeylanica*. *Science* 210:200–201.

Russell, S.D. 1982. Fertilization in *Plumbago zeylanica*: entry and discharge of the pollen tube in the embryo sac. *Can. J. Bot.* 60:2219–2230.

Russell, S.D. 1983. Fertilization in *Plumbago zeylanica*: gamete fusion and the fate of the male cytoplasm. *Am. J. Bot.* 70:416–434.

Russell, S.D. 1984. Ultrastructure of the sperm of *Plumbago zeylanica*. II. Quantitative cytology and three-dimensional organization. *Planta* 162:385–391.

Russell, S.D. 1985. Preferential fertilization in *Plumbago*: ultrastructural evidence for gamete-level recognition in an angiosperm. *Proc. Natl Acad. Sci. USA* 82:6129–6132.

Russell, S.D. 1986. Isolation of sperm cells from the pollen of *Plumbago zeylanica*. *Plant Physiol.* 81:317–319.

Russell, S.D. 1987. Quantitative cytology of the egg and central cell of *Plumbago zeylanica* and its impact on cytoplasmic inheritance patterns. *Theor. Appl. Genet.* 74:693–699.

Russell, S.D. 1992. Double fertilization. *Int. Rev. Cytol.* 140:357–388.

Russell, S.D., and Cass, D.D. 1981. Ultrastructure of the sperms of *Plumbago zeylanica*. 1. Cytology and association with the vegetative nucleus. *Protoplasma* 107:85–107.

Russell, S.D., and Cass, D.D. 1988. Fertilization in *Plumbagella micrantha*. *Am. J. Bot.* 75:778–781.

Russell, S.D., Cresti, M., and Dumas, C. 1990. Recent progress on sperm characterization in flowering plants. *Physiol. Plant.* 80:669–676.

Russell, S.D., and Mao, L.-J. 1990. Patterns of embryo-sac organization, synergid degeneration and cotyledon orientation in *Linum usitatissimum* L. *Planta* 182:52–57.

Russell, S.D., Rougier, M., and Dumas, C. 1990. Organization of the early post-fertilization megagametophyte of *Populus deltoides*. Ultrastructure and implications for male cytoplasmic transmission. *Protoplasma* 155:153–165.

Rutten, T.L.M., and Derksen, J. 1990. Organization of actin filaments in regenerating and outgrowing subprotoplasts from pollen tubes of *Nicotiana tabacum* L. *Planta* 180:471–479.

Rutten, T.L.M., and Derksen, J. 1992. Microtubules in pollen tube subprotoplasts: organization during protoplast formation and protoplast outgrowth. *Protoplasma* 167:231–237.

Ryan, A.J., Royal, C.L., Hutchinson, J., and Shaw, C.H. 1989. Genomic sequence of a 12S seed storage protein from oilseed rape (*Brassica napus* c.v. jet neuf). *Nucl. Acids Res.* 17:3584.

Ryczkowski, M. 1960. Changes of the osmotic value during the development of the ovule. *Planta* 55:343–356.

Ryczkowski, M. 1964. Physico-chemical properties of the central vacuolar sap in developing ovules. In *Pollen Physiology and Fertilization*, ed. H.F. Linskens, pp. 17–25. Amsterdam: North-Holland Publishing Co.

Saalbach, G., Christov, V., Jung, R., Saalbach, I., Manteuffel, R., Kunze, G., Brambarov, K., and Müntz, K. 1995. Stable expression of vicilin from *Vicia faba* with eight additional single methionine residues but failure of accumulation of legumin with an attached peptide segment in tobacco seeds. *Mol. Breed.* 1:245–258.

Saalbach, I., Pickardt, T., Machemehl, F., Saalbach, G., Schieder, O., and Müntz, K. 1994. A chimeric gene encoding the methionine-rich 2S albumin of the Brazil nut (*Bertholletia excelsa* H.B.K.) is stably expressed and inherited in transgenic grain legumes. *Mol. Gen. Genet.* 242:226–236.

Saba-El-Leil, M.K., Rivard, S., Morse, D., and Cappadocia, M. 1994. The *S11* and *S13* self incompatibility alleles in *Solanum chacoense* Bitt. are remarkably similar. *Plant Mol. Biol.* 24:571–583.

Safford, R., Windust, J.H.C., Lucas, C., de Silva, J., James, C.M., Hellyer, A., Smith, C.G., Slabas, A.R., and Hughes, S.G. 1988. Plastid-localised seed acyl-carrier protein of *Brassica napus* is encoded by a distinct, nuclear multigene family. *Eur. J. Biochem.* 174:287–295.

Sage, T.L., and Williams, E.G. 1995. Structure, ultrastructure, and histochemistry of the pollen tube pathway in the milkweed *Asclepias exaltata* L. *Sex. Plant Reprod.* 8:257–265.

Sági, L., Panis, B., Remy, S., Schoofs, H., de Smet, K., Swennen, R., and Cammue, B.P.A. 1995a. Genetic transformation of banana and plantain (*Musa* spp.) via particle bombardment. *Biotechnology* 13:481–485.

Sági, L., Remy, S., Verelst, B., Panis, B., Cammue, B.P.A., Volckaert, G., and Swennen, R. 1995b. Transient gene expression in transformed banana (*Musa* cv. Bluggoe) protoplasts and embryogenic cell suspensions. *Euphytica* 85:89–95.

Saigo, R.H., Peterson, D.M., and Holy, J. 1983. Development of protein bodies in oat starchy endosperm. *Can. J. Bot.* 61:1206–1215.

Saini, H.S., and Aspinall, D. 1981. Effect of water deficit on sporogenesis in wheat (*Triticum aestivum* L.). *Ann. Bot.* 48:623–633.

Saini, H.S., Sedgley, M., and Aspinall, D. 1984. Developmental anatomy in wheat of male sterility induced by heat stress, water deficit or abscisic acid. *Aust. J. Plant Physiol.* 11:243–253.

Saito, G.Y., Chang, Y.C., Walling, L.L., and Thomson, W.W. 1989. A correlation in plastid development and cytoplasmic ultrastructure with nuclear gene expression during seed ripening in soybean. *New Phytol.* 113:459–469.

Sakaguchi, K., Kurata, N., Takegami, M.H., and Ito, M. 1980. Inhibition of DNA and RNA syntheses and suppression of meiotic development in lily microsporocytes. *Cell Struct. Funct.* 5:367–377.

Sakai, H., Medrano, L.J., and Meyerowitz, E.M. 1995. Role of *SUPERMAN* in maintaining *Arabidopsis* floral whorl boundaries. *Nature* 378:199–203.

Salamini, F., di Fonzo, N., Fornasari, E., Gentinetta, E., Reggiani, R., and Soave, C. 1983. *Mucronate, Mc*, a dominant gene of maize which interacts with *opaque-2* to suppress zein synthesis. *Theor. Appl. Genet.* 65:123–128.

Salinas, J., Oeda, K., and Chua, N.-H. 1992. Two G-box-related sequences confer different expression patterns in transgenic tobacco. *Plant Cell* 4:1485–1493.

San, L.H., and Gelebart, P. 1986. Production of gynogenetic haploids. In *Cell Culture and Somatic Cell Genetics of Plants*, Vol. 3 , ed. I.K. Vasil, pp. 305–322. Orlando: Academic Press.

San Noeum, L.H. 1976. Haploïdes d'*Hordeum vulgare* L. par culture *in vitro* d'ovaires non fécondés. *Ann. Amélior. Plantes* 26:751–754.

Sánchez-Martínez, D., Puigdomènech, P., and Pagès, M. 1986. Regulation of gene expression in developing *Zea mays* embryos. Protein synthesis during embryogenesis and early germination of maize. *Plant Physiol.* 82:543–549.

Sánchez-Martínez, D., Gómez, J., Ludevid, M.D., Torrent, M., Puigdomènech, P., and Pagès, M. 1987. Absence of storage protein synthesis in the embryo of *Zea mays*. *Plant Sci.* 53:215–221.

Sanchez-Pina, M.A., Kieft, H., and Schel, J.H.N. 1989. Immunocytochemical detection of non-histone nuclear antigens in cryosections of developing somatic embryos from *Daucus carota* L. *J. Cell Sci.* 93:615–622.

Sand, S.A., and Christoff, G.T. 1973. Cytoplasmic-chromosomal interactions and altered differentiation in tobacco. *J. Hered.* 64:24–30.

Sanders, L.C., and Lord, E.M. 1989. Directed movement of latex particles in the gynoecia of three species of flowering plants. *Science* 243:1606–1608.

Sanders, L.C., Wang, C.-S., Walling, L.S., and Lord, E.M. 1991. A homolog of the substrate adhesion molecule vitronectin occurs in four species of flowering plants. *Plant Cell* 3:629–635.

Sanders, M.E., and Burkholder, P.R. 1948. Influence of amino acids on growth of *Datura* embryos in culture. *Proc. Natl Acad. Sci. USA* 34:516–526.

Sanford, J.C., Skubik, K.A., and Reisch, B.I. 1985. Attempted pollen-mediated plant transformation employing genomic donor DNA. *Theor. Appl. Genet.* 69:571–574.

Sangduen, N., Kreitner, G.L., and Sorensen, E.L. 1983a. Light and electron microscopy of embryo development in perennial and annual *Medicago* species. *Can. J. Bot.* 61:837–849.

Sangduen, N., Kreitner, G.L., and Sorensen, E.L. 1983b. Light and electron microscopy of embryo development in an annual × perennial *Medicago* species cross. *Can. J. Bot.* 61:1241–1257.

Sanger, J.M., and Jackson, W.T. 1971a. Fine structure study of pollen development in *Haemanthus katherinae* Baker. I. Formation of vegetative and generative cells. *J. Cell Sci.* 8:289–301.

Sanger, J.M., and Jackson, W.T. 1971b. Fine structure study of pollen development in *Haemanthus katherinae* Baker. II. Microtubules and elongation of the generative cells. *J. Cell Sci.* 8:303–315.

Sanger, J.M., and Jackson, W.T. 1971c. Fine structure study of pollen development in *Haemanthus katherinae* Baker. III. Changes in organelles during development of the vegetative cell. *J. Cell Sci.* 8:317–329.

Sangwan, R.S. 1978. Change in the amino-acid content during male gametophyte formation of *Datura metel in situ*. *Theor. Appl. Genet.* 52:221–225.

Sangwan, R.S., Bourgeois, Y., and Sangwan-Norreel, B.S. 1991. Genetic transformation of *Arabidopsis thaliana* zygotic embryos and identification of critical parameters influencing transformation efficiency. *Mol. Gen. Genet.* 230:475–485.

Sangwan, R.S., and Camefort, H. 1978. Action d'un choc thermique sur le contenu en acides aminés des anthères et des grains de pollen embryogènes du *Datura metel* L. et du *Nicotiana tabacum* L. *Compt. Rend. Acad. Sci. Paris* 287D:471–474.

Sangwan, R.S., and Camefort, H. 1982. Ribosomal bodies

specific to both pollen and zygotic embryogenesis in *Datura*. *Experientia* 38:395–397.

Sangwan, R.S., and Camefort, H. 1983. The tonoplast, a specific marker of embryogenic microspores of *Datura* cultured *in vitro*. *Histochemistry* 78:473–480.

Sangwan, R.S., and Camefort, H. 1984. Cold-treatment related structural modifications in the embryogenic anthers of *Datura*. *Cytologia* 49:473–487.

Sangwan, R.S., Ducrocq, C., and Sangwan-Norreel, B. 1993. *Agrobacterium*-mediated transformation of pollen embryos in *Datura innoxia* and *Nicotiana tabacum*: production of transgenic haploid and fertile homozygous dihaploid plants. *Plant Sci.* 95:99–115.

Sangwan, R.S., and Sangwan-Norreel, B.S. 1987a. Ultrastructural cytology of plastids in pollen grains of certain androgenic and nonandrogenic plants. *Protoplasma* 138:11–22.

Sangwan, R.S., and Sangwan-Norreel, B.S. 1987b. Biochemical cytology of pollen embryogenesis. *Int. Rev. Cytol.* 107:221–272.

Sangwan, R.S., Gauthier, D.A., Turpin, D.H., Pomeroy, M.K., and Plaxton, W.C. 1992. Pyruvate-kinase isoenzymes from zygotic and microspore-derived embryos of *Brassica napus*. *Planta* 187:198–202.

Sangwan-Norreel, B.S. 1978. Cytochemical and ultrastructural peculiarities of embryogenic pollen grains and of young androgenic embryos of *Datura innoxia*. *Can. J. Bot.* 56:805–817.

Sangwan-Norreel, B.S. 1979. Evolution du contenu en DNA nucléaire dans les gamétophytes mâles du *Datura innoxia* au cours de la période favorable à l'androgenèse. *Can. J. Bot.* 57:450–457.

Sangwan-Norreel, B.S. 1981. Evolution *in vitro* du contenu en ADN nucléaire et de la ploïdie des embryons polliniques du *Datura innoxia*. *Can. J. Bot*. 59:508–517.

Sangwan-Norreel, B.S. 1983. Male gametophyte nuclear DNA content evolution during androgenic induction in *Datura innoxia* Mill. *Z. Pflanzenphysiol.* 111:47–54.

Sano, Y. 1984. Differential regulation of waxy gene expression in rice endosperm. *Theor. Appl. Genet.* 68:467–473.

Santanen, A., and Simola, L.K. 1992. Changes in polyamine metabolism during somatic embryogenesis in *Picea abies*. *J. Plant Physiol.* 140:475–480.

Sari Gorla, M., Frova, C., Binelli, G., and Ottaviano, E. 1986. The extent of gametophytic-sporophytic gene expression in maize. *Theor. Appl. Genet.* 72:42–47.

Sarker, R.H., Elleman, C.J., and Dickinson, H.G. 1988. Control of pollen hydration in *Brassica* requires continued protein synthesis, and glycosylation is necessary for intraspecific incompatibility. *Proc. Natl Acad. Sci. USA* 85:4340–4344.

Sasaki, K., Shimomura, K., Kamada, H., and Harada, H. 1994. IAA metabolism in embryogenic and nonembryogenic carrot cells. *Plant Cell Physiol.* 35:1159–1164.

Sasaki, Y., and Harada, H. 1991. Binding form of pollen mother cell protein in the nucleosomes of lily. *Plant Physiol*. 96:1161–1166.

Sasaki, Y., Yasuda, H., Ohba, Y., and Harada, H. 1990. Isolation and characterization of a novel nuclear protein from pollen mother cells of lily. *Plant Physiol.* 94:1467–1471.

Sass, J.E., and Sprague, G.F. 1950. The embryology of "germless" maize. *Iowa State Coll. J. Sci.* 24:209–218.

Sassa, H., Hirano, H., and Ikehashi, H. 1992. Self-incompatibility-related RNases in styles of Japanese pear (*Pyrus serotina* Rehd.). *Plant Cell Physiol.* 33:811–814.

Sassa, H., Hirano, H., and Ikehashi, H. 1993. Identification and characterization of stylar glycoproteins associated with self-incompatibility genes of Japanese pear, *Pyrus serotina* Rehd. *Mol. Gen. Genet*. 241:17–25.

Sassa, H., Mase, N., Hirano, H., and Ikehashi, H. 1994. Identification of self-incompatibility-related glycoproteins in styles of apple (*Malus* × *domestica*). *Theor. Appl. Genet*. 89:201–205.

Sassa, H., Nishio, T., Kowyama, Y., Hirano, H., Koba, T., and Ikehashi, H. 1996. Self-incompatibility (*S*) alleles of the Rosaceae encode members of a distinct class of the T_2/S ribonuclease superfamily. *Mol. Gen. Genet.* 250:547–557.

Sassen, M.M.A. 1964. Fine structure of *Petunia* pollen grain and pollen tube. *Acta Bot. Neerl*. 13:175–181.

Sassen, M.M.A. 1974. The stylar transmitting tissue. *Acta Bot. Neerl*. 23:99–108.

Sastri, D.C., and Shivanna, K.R. 1976. Attempts to overcome interspecific incompatibility in *Sesamum* by using recognition pollen. *Ann. Bot*. 40:891–893.

Sastri, D.C., and Shivanna, K.R. 1979. Role of pollen-wall proteins in intraspecific incompatibility in *Saccharum bengalense*. *Phytomorphology* 29:324–330.

Satina, S. 1944. Periclinal chimeras in *Datura* in relation to development and structure (a) of the style and stigma (b) of calyx and corolla. *Am. J. Bot.* 31:493–502.

Satina, S., and Rietsema, J. 1959. Seed development. In *Blakeslee: The Genus* Datura, ed. A.G. Avery, S. Satina and J. Rietsema, pp. 181–195. New York: Ronald Press Co.

Sato, S., Willson, C., and Dickinson, H.G. 1989. The RNA content of the nucleolus and nucleolus-like inclusions in the anther of *Lilium* estimated by an improved RNase-gold labelling method. *J. Cell Sci.* 94:675–683.

Sato, S., Jones, K., de Los Dios Alche, J., and Dickinson, H.G. 1991. Cytoplasmic nucleoloids of *Lilium* male reproductive cells contain rDNA transcripts and share features of development with nucleoli. *J. Cell Sci.* 100:109–118.

Sato, S., Newell, C., Kolacz, K., Tredo, L., Finer, J., and Hinchee, M. 1993. Stable transformation via particle bombardment in two different soybean regeneration systems. *Plant Cell Rep*. 12:408–413.

Sato, S., Toya, T., Kawahara, R., Whittier, R.F., Fukuda, H., and Komamine, A. 1995. Isolation of a carrot gene expressed specifically during early-stage somatic embryogenesis. *Plant Mol. Biol.* 28:39–46.

Sato, T., Thorsness, M.K., Kandasamy, M.K., Nishio, T., Hirai, M., Nasrallah, J.B., and Nasrallah, M.E. 1991. Activity of an S-locus gene promoter in pistils and anthers of transgenic *Brassica*. *Plant Cell* 3:867–876.

Sato, Y., Hong, S.-K., Tagiri, A., Kitano, H., Yamamoto, N., Nagato, Y., and Matsuoka, M. 1996. A rice homeobox gene, *OSH1*, is expressed before organ differentiation in a specific region during early embryogenesis. *Proc. Natl Acad. Sci. USA* 93:8117–8122.

Satoh, H., and Omura, T. 1981. New endosperm mutations induced by chemical mutagens in rice, *Oryza sativa* L. *Jap. J. Breed*. 31:316–326.

Satoh, S., Kamada, H., Harada, H., and Fujii, T. 1986. Auxin-controlled glycoprotein release into the medium of embryogenic carrot cells. *Plant Physiol.* 81:931–933.

Saunders, J.W., and Bingham, E.T. 1972. Production of

alfalfa plants from callus tissue. *Crop Sci.* 12:804–808.

Sauter, J.J. 1968. Histoautoradiographische Untersuchungen zur Ribonucleinsäure-Synthese während der Meiosis bei *Paeonia tenuifolia* L. *Naturwissenschaften* 55:236.

Sauter, J.J. 1969a. Autoradiographische Untersuchungen zur RNS- und Protein-synthese in Pollenmutterzellen, jungen Pollen und Tapetumzellen während der Mikrosporogenese von *Paeonia tenuifolia*. *Z. Pflanzenphysiol.* 61:1–19.

Sauter, J.J. 1969b. Cytochemische Untersuchung der Histone in Zellen mit unterschiedlicher RNS- und Protein-Synthese. *Z. Pflanzenphysiol.* 60:434–449.

Sauter, J.J., and Marquardt, H. 1967a. Nucleohistone und Ribonukleinsäure-Synthese während der Pollenentwicklung. *Naturwissenschaften* 54:546.

Sauter, J.J., and Marquardt, H. 1967b. Die Rolle des Nukleohistons bei der RNS- und Proteinsynthese während der Mikrosporogenese von *Paeonia tenuifolia* L. *Z. Pflanzenphysiol.* 58:126–137.

Sautter, C., Leduc, N., Bilang, R., Iglesias, V.A., Gisel, A., Wen, X., and Potrykus, I. 1995. Shoot apical meristems as a target for gene transfer by microballistics. *Euphytica* 85:45–51.

Savidge, B., Rounsley, S.D., and Yanofsky, M.F. 1995. Temporal relationship between the transcription of two *Arabidopsis* MADS box genes and the floral organ identity genes. *Plant Cell* 7:721–733.

Sawhney, V.K. 1974. Morphogenesis of the stamenless 2 mutant in tomato. III. Relative levels of gibberellins in the normal and mutant plants. *J. Expt. Bot.* 25:1004–1009.

Sawhney, V.K. 1981. Abnormalities in pepper (*Capsicum annuum*) flowers induced by gibberellic acid. *Can. J. Bot.* 59:8–16.

Sawhney, V.K. 1983. Temperature control of male sterility in a tomato mutant. *J. Hered.* 74:51–54.

Sawhney, V.K., and Bhadula, S.K. 1987. Characterization and temperature regulation of soluble proteins of a male sterile tomato mutant. *Biochem. Genet.* 25:717–728.

Sawhney, V.K., and Bhadula, S.K. 1988. Microsporogenesis in the normal and male-sterile stamenless-2 mutant of tomato (*Lycopersicon esculentum*). *Can. J. Bot.* 66:2013–2021.

Sawhney, V.K., Chen, K., and Sussex, I.M. 1985. Soluble proteins of the mature floral organs of tomato (*Lycopersicon esculentum* Mill.). *J. Plant Physiol.* 121:265–271.

Sawhney, V.K., and Greyson, R.I. 1973a. Morphogenesis of the stamenless-2 mutant in tomato. I. Comparative description of the flowers and ontogeny of stamens in the normal and mutant plants. *Am. J. Bot.* 60:514–523.

Sawhney, V.K., and Greyson, R.I. 1973b. Morphogenesis of the stamenless-2 mutant in tomato. II. Modifications of sex organs in the mutant and normal flowers by plant hormones. *Can. J. Bot.* 51:2473–2479.

Sawhney, V.K., and Greyson, R.I. 1979. Interpretations of determination and canalisation of stamen development in a tomato mutant. *Can. J. Bot.* 57:2471–2477.

Sawhney, V.K., and Polowick, P.L. 1986. Temperature-induced modifications in the surface features of stamens of a tomato mutant: an SEM study. *Protoplasma* 131:75–81.

Sawhney, V.K., and Shukla, A. 1994. Male sterility in flowering plants: are plant growth substances involved? *Am. J. Bot.* 81:1640–1647.

Scalla, R., Duc, G., Rigaud, J., Lefebvre, A., and Meignoz, R. 1981. RNA containing intracellular particles in cytoplasmic male sterile faba bean (*Vicia faba* L.). *Plant Sci. Lett.* 22:269–277.

Scallon, B., Thanh, V.H., Floener, L.A., and Nielsen, N.C. 1985. Identification and characterization of DNA clones encoding group-II glycinin subunits. *Theor. Appl. Genet.* 70: 510–519.

Scallon, B.J., Dickinson, C.D., and Nielsen, N.C. 1987. Characterization of a null-allele for the Gy_4 glycinin gene from soybean. *Mol. Gen. Genet.* 208:107–113.

Scanlon, M.J., Stinard, P.S., James, M.G., Myers, A.M., and Robertson, D.S. 1994. Genetic analysis of 63 mutations affecting maize kernel development isolated from *mutator* stocks. *Genetics* 136:281–294.

Schaffner, M. 1906. The embryology of the shepherd's purse. *Ohio Naturl.* 7:1–8.

Schardl, C.L., Pring, D.R., and Lonsdale, D.M. 1985. Mitochondrial DNA rearrangements associated with fertile revertants of S-type male-sterile maize. *Cell* 43:361–368.

Schardl, C.L., Lonsdale, D.M., Pring, D.R., and Rose, K.R. 1984. Linearization of maize mitochondrial chromosomes by recombination with linear episomes. *Nature* 310:292–296.

Scharpé, A., and van Parijs, R. 1973. The formation of polyploid cells in ripening cotyledons of *Pisum sativum* L. in relation to ribosome and protein synthesis. *J. Expt. Bot.* 24:216–222.

Scheer, U., and Franke, W.W. 1972. Annulate lamellae in plant cells: formation during microsporogenesis and pollen development in *Canna generalis* Bailey. *Planta* 107:145–159.

Scheets, K., and Hedgcoth, C. 1988. Nucleotide sequence of a γ gliadin gene: comparisons with other γ gliadin sequences show the structure of γ gliadin genes and the general primary structure of γ gliadins. *Plant Sci.* 57:141–150.

Scheike, R., Gerold, E., Brennicke, A., Mehring-Lemper, M., and Wricke, G. 1992. Unique patterns of mitochondrial genes, transcripts and proteins in different male-sterile cytoplasms of *Daucus carota*. *Theor. Appl. Genet.* 83:419–427.

Schel, J.H.N., and Kieft, H. 1986. An ultrastructural study of embryo and endosperm development during *in vitro* culture of maize ovaries (*Zea mays*). *Can. J. Bot.* 64:2227–2238.

Schel, J.H.N., Kieft, H., and van Lammeren, A.A.M. 1984. Interactions between embryo and endosperm during early developmental stages of maize caryopses (*Zea mays*). *Can. J. Bot.* 62:2842–2853.

Scherer, D., Sato, A., McCarter, D.W., Radke, S.E., Kridl, J.C., and Knauf, V.C. 1992. Non-essential repeats in the promoter region of a *Brassica rapa* acyl carrier protein gene expressed in developing embryos. *Plant Mol. Biol.* 18:591–594.

Scheres, B., Wolkenfelt, H., Willemsen, V., Terlouw, M., Lawson, E., Dean, C., and Weisbeek, P. 1994. Embryonic origin of the *Arabidopsis* primary root and root meristem initials. *Development* 120:2475–2487.

Schernthaner, J.P., Matzke, M.A., and Matzke, A.J.M. 1988. Endosperm-specific activity of a zein gene promoter in transgenic tobacco plants. *EMBO J.* 7:1249–1255.

Schiavone, F.M. 1988. Microamputation of somatic embryos of the domestic carrot reveals apical control of axis elongation and root regeneration. *Development* 103:657–664.

Schiavone, F.M., and Cooke, T.J. 1987. Unusual patterns of somatic embryogenesis in the domesticated carrot: developmental effects of exogenous auxins and auxin transport inhibitors. *Cell Diffn* 21:53–62.

Schiavone, F.M., and Racusen, R.H. 1990. Microsurgery reveals regional capabilities for pattern reestablishment in somatic carrot embryos. *Devel Biol.* 141:211–219.

Schiavone, F.M., and Racusen, R.H. 1991. Regeneration of the root pole in surgically transected carrot embryos occurs by position-dependent, proximodistal replacement of missing tissues. *Development* 113:1305–1313.

Schick, R., and Stubbe, H. 1932. Die Gene von *Antirrhinum majus*. II. *Z. Indukt. Abstamm. Vererbungs.* 62:249–290.

Schickler, H., Benner, M.S., and Messing, J. 1993. Repression of the high-methionine zein gene in the maize inbred line Mo17. *Plant J.* 3:221–229.

Schiefelbein, J., Galway, M., Masucci, J., and Ford, S. 1993. Pollen tube and root-hair tip growth is disrupted in a mutant of *Arabidopsis thaliana. Plant Physiol.* 103:979–985.

Schlag, M., and Hesse, M. 1992. The formation of the generative cell in *Polystachia pubescens* (Orchidaceae). *Sex. Plant Reprod.* 5:131–137.

Schläppi, M., and Hohn, B. 1992. Competence of immature maize embryos for *Agrobacterium*-mediated gene transfer. *Plant Cell* 4:7–16.

Schlösser, K. 1961. Cytologische und cytochemische Untersuchungen über das Pollenschlauch wachstum selbststeriler Petunien. *Z. Bot.* 49:266–288.

Schmidt, H., and Schmidt, V. 1981. Untersuchungen an pollensterilen, stamenless-ähnlichen Mutanten von *Lycopersicon esculentum* Mill. II. Normalisierung von *ms-15* und *ms-33* mit Gibberellinsäure (GA_3). *Biol. Zentral.* 100:691–696.

Schmidt, R.J., Burr, F.A., and Burr, B. 1987. Transposon tagging and molecular analysis of the maize regulatory locus *opaque-2*. *Science* 238:960–963.

Schmidt, R.J., Burr, F.A., Aukerman, M.J., and Burr, B. 1990. Maize regulatory gene opaque-2 encodes a protein with a "leucine-zipper" motif that binds to zein DNA. *Proc. Natl Acad. Sci. USA* 87:46–50.

Schmidt, R.J., Ketudat, M., Aukerman, M.J., and Hoschek, G. 1992. Opaque-2 is a transcriptional activator that recognizes a specific target site in 22-kD zein genes. *Plant Cell* 4:689–700.

Schmidt, R.J., Veit, B., Mandel, M.A., Mena, M., Hake, S., and Yanofsky, M.F. 1993. Identification and molecular characterization of *ZAG1*, the maize homolog of the *Arabidopsis* floral homeotic gene *AGAMOUS*. *Plant Cell* 5:729–737.

Schmit, A.-C., and Lambert, A.-M. 1987. Characterization and dynamics of cytoplasmic F-actin in higher plant endosperm cells during interphase, mitosis, and cytokinesis. *J. Cell Biol.* 105:2157–2166.

Schmit, A.-C., and Lambert, A.-M. 1990. Microinjected fluorescent phalloidin *in vivo* reveals the F-actin dynamics and assembly in higher plant mitotic cells. *Plant Cell* 2:129–138.

Schmitz, U.K., and Kowallik, K.V. 1987. Why are plastids maternally inherited in *Epilobium*? Ultrastructural observations during microgametogenesis. *Plant Sci.* 53:139–145.

Schmitz, U.K., and Michaelis, G. 1988. Dwarfism and male sterility in interspecific hybrids of *Epilobium*. 2. Expression of mitochondrial genes and structure of the mitochondrial DNA. *Theor. Appl. Genet.* 76:565–569.

Schmülling, T., Schell, J., and Spena, A. 1988. Single genes from *Agrobacterium rhizogenes* influence plant development. *EMBO J.* 7:2621–2629.

Schmülling, T., Röhrig, H., Pilz, S., Walden, R., and Schell, J. 1993. Restoration of fertility by antisense RNA in genetically engineered male sterile tobacco plants. *Mol. Gen. Genet.* 237:385–394.

Schnall, J.A., Cooke, T.J., and Cress, D.E. 1988. Genetic analysis of somatic embryogenesis in carrot cell culture: initial characterization of six classes of temperature-sensitive variants. *Devel Genet.* 9:49–67.

Schnall, J.A., Hwang, C.H., Cooke, T.J., and Zimmerman, J.L. 1991. An evaluation of gene expression during somatic embryogenesis of two temperature-sensitive carrot variants unable to complete embryo development. *Physiol. Plant.* 82:498–504.

Schneider, T., Dinkins, R., Robinson, K., Shellhammer, J., and Meinke, D.W. 1989. An embryo-lethal mutant of *Arabidopsis thaliana* is a biotin auxotroph. *Devel Biol.* 131:161–167.

Schneitz, K., Hülskamp, M., and Pruitt, R.E. 1995. Wild-type ovule development in *Arabidopsis thaliana*: a light microscope study of cleared whole-mount tissue. *Plant J.* 7:731–749.

Schnepf, E., and Nagl, W. 1970. Über einige Strukturbesonderheiten der Suspensorzellen von *Phaseolus vulgaris*. *Protoplasma* 69:133–143.

Schöpke, C., Taylor, N., Cárcamo, R., Konan, N.K., Marmey, P., Henshaw, G.G., Beachy, R.N., and Fauquet, C. 1996. Regeneration of transgenic cassava plants (*Manihot esculenta* Crantz) from microbombarded embryogenic suspension cultures. *Nature Biotech.* 14:731–735.

Schou, O. 1984. The dry and wet stigmas of *Primula obconica*: ultrastructural and cytochemical dimorphisms. *Protoplasma* 121:99–113.

Schou, O., and Mattsson, O. 1985. Differential localization of enzymes in the stigmatic exudates of *Primula obconica*. *Protoplasma* 125:65–74.

Schrauwen, J., and Linskens, H.F. 1972. Ribonuclease in styles. *Planta* 102:277–285.

Schrauwen, J.A.M., de Groot, P.F.M., van Herpen, M.M.A., van der Lee, T., Reynen, W.H., Weterings, K.A.P., and Wullems, G.J. 1990. Stage-related expression of mRNAs during pollen development in lily and tobacco. *Planta* 182:298–304.

Schrauwen, J.A.M., Mettenmeyer, T., Croes, A.F., and Wullems, G.J. 1996. Tapetum-specific genes: what role do they play in male gametophyte development? *Acta Bot. Neerl.* 45:1–15.

Schrauwen, J.A.M., Reijnen, W.H., de Leeuw, H.C.G.M., and van Herpen, M.M.A. 1986. Response of pollen to heat stress. *Acta Bot. Neerl.* 35:321–327.

Schröder, M.-B. 1984. Ultrastructural studies on plastids of generative and vegetative cells in the family Liliaceae. 1. *Lilium martagon* L. *Biol. Zentral.* 103:547–555.

Schröder, M.-B. 1985a. Ultrastructural studies on plastids of generative and vegetative cells in the family Liliaceae. 2. *Fritillaria imperialis* and *F. meleagris*. *Biol. Zentral.* 104:21–27.

Schröder, M.-B. 1985b. Ultrastructural studies on plastids of generative and vegetative cells in Liliaceae. 3.

Plastid distribution during the pollen development in *Gasteria verrucosa* (Mill.) Duval. *Protoplasma* 124:123–129.

Schröder, M.-B. 1986a. Ultrastructural studies on plastids of generative and vegetative cells in Liliaceae. 4. Plastid degeneration during generative cell maturation in *Convallaria majalis* L. *Biol. Zentral.* 105:427–433.

Schröder, M.-B. 1986b. Ultrastructural studies on plastids of generative and vegetative cells in Liliaceae. 5. The behaviour of plastids during pollen development in *Chlorophytum comosum* (Thunb.) Jacques. *Theor. Appl. Genet.* 72:840–844.

Schröder, M.-B., and Hagemann, R. 1986. Ultrastructural studies on plastids of generative and vegetative cells in Liliaceae. 6. Patterns of plastid distribution during generative cell formation in *Aloë secundiflora* and *A. jucunda*. *Acta Bot. Neerl.* 35:243–248.

Schröder, M.-B., and Oldenburg, H. 1990. Ultrastructural studies on plastids of generative and vegetative cells in Liliaceae. 7. Plastid distribution during generative cell development in *Tulbaghia violacea* Harv. *Flora* 184:131–136.

Schroeder, H.E. 1982. Quantitative studies on the cotyledonary proteins in the genus *Pisum*. *J. Sci. Food Agr.* 33:623–633.

Schroeder, H.E. 1984. Effects of applied growth regulators on pod growth and seed protein composition in *Pisum sativum* L. *J. Expt. Bot.* 35:813–821.

Schroeder, H.E., Schotz, A.H., Wardley-Richardson, T., Spencer, D., and Higgins, T.J.V. 1993. Transformation and regeneration of two cultivars of pea (*Pisum sativum* L.). *Plant Physiol.* 101:751–757.

Schubert, R., Bäumlein, H., Czihal, A., and Wobus, U. 1990. Genomic sequence of a 12S seed storage protein gene from oat (*Avena sativa* L. cv. `Solidor'). *Nucl. Acids Res.* 18:377.

Schubert, R., Panitz, R., Manteuffel, R., Nagy, I., Wobus, U., and Bäumlein, H. 1994. Tissue-specific expression of an oat 12S seed globulin gene in developing tobacco seeds.: differential mRNA and protein accumulation. *Plant Mol. Biol.* 26:203–210.

Schuler, M.A., Schmitt, E.S., and Beachy, R.N. 1982. Closely related families of genes code for the α and α' subunits of the soybean 7S storage protein complex. *Nucl. Acids Res.* 10:8225–8244.

Schuler, M.A., Ladin, B.F., Pollacco, J.C., Freyer, G., and Beachy, R.N. 1982. Structural sequences are conserved in the genes coding for the α, α' and β-subunits of the soybean 7S seed storage protein. *Nucl. Acids Res.* 10:8245–8261.

Schultz, E.A., and Haughn, G.W. 1991. *LEAFY*, a homeotic gene that regulates inflorescence development in *Arabidopsis*. *Plant Cell* 3:771–781.

Schultz, E.A., and Haughn, G.W. 1993. Genetic analysis of the floral initiation process (FLIP) in *Arabidopsis*. *Development* 119:745–765.

Schultz, E.A., Pickett, F.B., and Haughn, G.W. 1991. The *FLO10* gene product regulates the expression domain of homeotic genes *AP3* and *PI* in *Arabidopsis* flowers. *Plant Cell* 3:1221–1237.

Schulz, P., and Jensen, W.A. 1969. *Capsella* embryogenesis: the suspensor and the basal cell. *Protoplasma* 67:139–163.

Schulz, P., and Jensen, W.A. 1971. *Capsella* embryogenesis: the chalazal proliferating tissue. *J. Cell Sci.* 8:201–227.

Schulz, P., and Jensen, W.A. 1973. *Capsella* embryogenesis: the central cell. *J. Cell Sci.* 12:741–763.

Schulz, P., and Jensen, W.A. 1974. *Capsella* embryogenesis: the development of the free nuclear endosperm. *Protoplasma* 80:183–205.

Schulz, P., and Jensen, W.A. 1977. Cotton embryogenesis: the early development of the free nuclear endosperm. *Am. J. Bot.* 64:384–394.

Schulz, P., and Jensen, W.A. 1981. Pre-fertilization ovule development in *Capsella*: ultrastructure and ultracytochemical localization of acid phosphatase in the meiocyte. *Protoplasma* 107:27–45.

Schulz, P., and Jensen, W.A. 1986. Prefertilization ovule development in *Capsella*: the dyad, tetrad, developing megaspore, and two-nucleate gametophyte. *Can. J. Bot.* 64:875–884.

Schulz, P.J., Cross, J.W., and Almeida, E. 1993. Chemical agents that inhibit pollen development: effects of the phenylcinnoline carboxylates SC-1058 and SC-1271 on the ultrastructure of developing wheat anthers (*Triticum aestivum* L. var. Yecora rojò). *Sex. Plant Reprod.* 6:108–121.

Schulz, R., and Jensen, W.A. 1968a. *Capsella* embryogenesis: the early embryo. *J. Ultrastr. Res.* 22:376–392.

Schulz, R., and Jensen, W.A. 1968b. *Capsella* embryogenesis: the synergids before and after fertilization. *Am. J. Bot.* 55:541–552.

Schulz, R., and Jensen, W.A. 1968c. *Capsella* embryogenesis: the egg, zygote, and young embryo. *Am. J. Bot.* 55:807–819.

Schumann, C.M., and Hancock, J.F. 1989. Paternal inheritance of plastids in *Medicago sativa*. *Theor. Appl. Genet.* 78:863–866.

Schuster, W. 1961. Untersuchungen über künstlich induzierte Pollensterilitât bei Sonnenblumen (*Helianthus annuus* L.). *Z. Pflanzenzüchtg* 46:389–404.

Schwab, C.A. 1971. Callose in megasporogenesis of *Diarrhena* (Gramineae). *Can. J. Bot.* 49:1523–1524.

Schwall, M., and Feix, G. 1988. Zein promoter activity in transiently transformed protoplasts from maize. *Plant Sci.* 56:161–166.

Schwartz, B.W., Yeung, E.C., and Meinke, D.W. 1994. Disruption of morphogenesis and transformation of the suspensor in abnormal suspensor mutants of *Arabidopsis*. *Development* 120:3235–3245.

Schwarz-Sommer, Z., Hue, I., Huijser, P., Flor, P.J., Hansen, R., Tetens, F., Lönnig, W.-E., Saedler, F., and Sommer, H. 1992. Characterization of the *Antirrhinum* floral homeotic MADS-box gene *deficiens*: evidence for DNA binding and autoregulation of its persistent expression throughout flower development. *EMBO J.* 11:251–263.

Schwarz-Sommer, Z., Huijser, P., Nacken, W., Saedler, H., and Sommer, H. 1990. Genetic control of flower development by homeotic genes in *Antirrhinum majus*. *Science* 250:931–936.

Scofield, S.R., and Crouch, M.L. 1987. Nucleotide sequence of a member of the napin storage protein family from *Brassica napus*. *J. Biol. Chem.* 262:12202–12208.

Scoles, G.J., and Evans, L.E. 1979. Pollen development in male-fertile and cytoplasmic male-sterile rye. *Can. J. Bot.* 57:2782–2790.

Scorza, R., Cordts, J.M., Ramming, D.W., and Emershad, R.L. 1995. Transformation of grape (*Vitis vinifera* L.) zygotic-derived somatic embryos and regeneration of transgenic plants. *Plant Cell Rep.* 14:589–592.

Scott, P., Lyne, R.L., and ap Rees, T. 1995. Metabolism of

maltose and sucrose by microspores isolated from barley (*Hordeum vulgare* L.). *Planta* 197:435–441.
Scott, R., Dagless, E., Hodge, R., Paul, W., Soufleri, I., and Draper, J. 1991. Patterns of gene expression in developing anthers of *Brassica napus*. *Plant Mol. Biol.* 17:195–207.
Scribailo, R.W., and Barrett, S.C.H. 1991a. Pollen–pistil interactions in tristylous *Pontederia sagittata* (Pontederiaceae). I. Floral heteromorphism and structural features of the pollen tube pathway. *Am. J. Bot.* 78:1643–1661.
Scribailo, R.W., and Barrett, S.C.H. 1991b. Pollen–pistil interactions in tristylous *Pontederia sagittata* (Pontederiaceae). II. Patterns of pollen tube growth. *Am. J. Bot.* 78:1662–1682.
Scutt, C.P., and Croy, R.R.D. 1992. An S5 self-incompatibility allele-specific cDNA sequence from *Brassica oleracea* shows high homology to the SLR2 gene. *Mol. Gen. Genet.* 232:240–246.
Scutt, C.P., Fordham-Skelton, A.P., and Croy, R.R.D. 1993. Okadaic acid causes breakdown of self-incompatibility in *Brassica oleracea*: evidence for the involvement of protein phosphatases in the incompatible response. *Sex. Plant Reprod.* 6:282–285.
Scutt, C.P., Gates, P.J., Gatehouse, J.A., Boulter, D., and Croy, R.R.D. 1990. A cDNA encoding an S-locus specific glycoprotein from *Brassica oleracea* plants containing the S_5 self-incompatibility allele. *Mol. Gen. Genet.* 220:409–413.
Seavey, S.R., and Bawa, K.S. 1986. Late-acting self-incompatibility in angiosperms. *Bot. Rev.* 52:195–219.
Secor, D.L., and Russell, S.D. 1988. Megagametophyte organization in a polyembryonic line of *Linum usitatissimum*. *Am. J. Bot.* 75:114–122.
Sedgley, M. 1974. The concentration of S-protein in stigmas of *Brassica oleracea* plants homozygous and heterozygous for a given S-allele. *Heredity* 33:412–416.
Sedgley, M. 1975. Flavanoids in pollen and stigma of *Brassica oleracea* and their effects on pollen germination *in vitro*. *Ann. Bot.* 39:1091–1095.
Sedgley, M. 1979. Structural changes in the pollinated and unpollinated avocado stigma and style. *J. Cell Sci.* 38:49–60.
Sedgley, M. 1981a. Ultrastructure and histochemistry of the watermelon stigma. *J. Cell Sci.* 48:137–146.
Sedgley, M. 1981b. Anatomical aspects of compatible pollen–stigma interaction. *Phytomorphology* 31:158–165.
Sedgley, M. 1981c. Early development of the *Macadamia* ovary. *Aust. J. Bot.* 29:185–193.
Sedgley, M. 1982. Anatomy of the unpollinated and pollinated watermelon stigma. *J. Cell Sci.* 54:341–355.
Sedgley, M., and Blesing, M.A. 1982. Foreign pollination of the stigma of watermelon (*Citrullus lanatus* [Thunb.] Matsum and Nakai). *Bot. Gaz.* 143:210–215.
Sedgley, M., and Blesing, M.A. 1983. Developmental anatomy of the avocado stigma papilla cells and their secretion. *Bot. Gaz.* 144:185–190.
Sedgley, M., and Buttrose, M.S. 1978. Structure of the stigma and style of the avocado. *Aust. J. Bot.* 26:663–682.
Sedgley, M., and Clarke, A.E. 1986. Immuno-gold localisation of arabinogalactan protein in the developing style of *Nicotiana alata*. *Nord. J. Bot.* 6:591–598.
Sedgley, M., and Scholefield, P.B. 1980. Stigma secretion in the watermelon before and after pollination. *Bot. Gaz.* 141:428–434.
Sedgley, M., Blesing, M.A., Bonig, I., Anderson, M.A., and Clarke, A.E. 1985. Arabinogalactan-proteins are localized extracellularly in the transmitting tissue of *Nicotiana alata* Link and Otto, an ornamental tobacco. *Micron Micros. Acta* 16:247–254.
Seffens, W.S., Almoguera, C., Wilde, H.D., Vonder Haar, R.A., and Thomas, T.L. 1990. Molecular analysis of a phylogenetically conserved carrot gene: developmental and environmental regulation. *Devel Genet.* 11:65–76.
Sehgal, C.B., and Gifford, E.M., Jr. 1979. Developmental and histochemical studies of the ovules of *Nicotiana rustica* L. *Bot. Gaz.* 140:180–188.
Sen, S.K. 1969. Chromatin-organisation during and after synapsis in cultured microsporocytes of *Lilium* in presence of mitomycin C and cycloheximide. *Expt. Cell Res.* 55:123–127.
Sen, S.K., Kundu, S.C., and Gaddipati, J.P. 1977. High speed scintillation autoradiography of DNA fibres undergoing DNA synthesis at zygotene and pachytene in the lily. *Expt. Cell Res.* 108:471–473.
Senda, M., Mikami, T., and Kinoshita, T. 1993. The sugar beet mitochondrial gene for the ATPase alpha-subunit: sequence, transcription and rearrangements in cytoplasmic male-sterile plants. *Curr. Genet.* 24:164–170.
Senda, M., Harada, T., Mikami, T., Sugiura, M., and Kinoshita, T. 1991. Genomic organization and sequence analysis of the cytochrome oxidase subunit II gene from normal and male-sterile mitochondria in sugar beet. *Curr. Genet.* 19:175–181.
Sengupta, C., and Raghavan, V. 1980a. Somatic embryogenesis in carrot cell suspension. I. Pattern of protein and nucleic acid synthesis. *J. Expt. Bot.* 31:247–258.
Sengupta, C., and Raghavan, V. 1980b. Somatic embryogenesis in carrot cell suspension. II. Synthesis of ribosomal RNA and poly $(A)^+$ RNA. *J. Expt. Bot.* 31:259–268.
Sengupta, C., Deluca, V., Bailey, D.S., and Verma, D.P.S. 1981. Post-translational processing of 7S and 11S components of soybean storage proteins. *Plant Mol. Biol.* 1:19–33.
Sengupta-Gopalan, C., Reichert, N.A., Barker, R.F., Hall, T.C., and Kemp, J.D. 1985. Developmentally regulated expression of the bean β-phaseolin gene in tobacco seed. *Proc. Natl Acad. Sci. USA* 82:3320–3324.
Serrano, L., Rochange, F., Semblat, J.P., Marque, C., Teulières, C., and Boudet, A.-M. 1996. Genetic transformation of *Eucalyptus globulus* through biolistics: complementary development of procedures for organogenesis from zygotic embryos and stable transformation of corresponding proliferating tissue. *J. Expt. Bot.* 47:285–290.
Sessa, G., and Fluhr, R. 1995. The expression of an abundant transmitting tract-specific endoglucanase (Sp41) is promoter-dependent and not essential for the reproductive physiology of tobacco. *Plant Mol. Biol.* 29:969–982.
Sessions, R.A., and Zambryski, P.C. 1995. *Arabidopsis* gynoecium structure in the wild type and in *ettin* mutants. *Development* 121:1519–1532.
Sethi, R.S., and Jensen, W.A. 1981. Ultrastructural evidence for secretory phases in the life history of stigmatic hairs of cotton: a dry type stigma. *Am. J. Bot.* 68:666–674.
Seurinck, J., Truettner, J., and Goldberg, R.B. 1990. The nucleotide sequence of an anther-specific gene. *Nucl. Acids Res.* 18:3403.

Shah, C.K., and Pandey, S.N. 1978. Histochemical studies during embryogenesis in *Limnophyton obtusifolium*. *Phytomorphology* 28:31–42.

Shannon, L.M., and Chrispeels, M.J. 1986. *Molecular Biology of Seed Storage Proteins and Lectins*. Rockville: American Society of Plant Physiologists.

Shannon, T.M., Picton, J.M., and Steer, M.W. 1984. The inhibition of dictyosome vesicle formation in higher plant cells by cytochalasin D. *Eur. J. Cell Biol.* 33:144–147.

Sharma, N., Bajaj, M., and Shivanna, K.R. 1985. Overcoming self-incompatibility through the use of lectins and sugars in *Petunia* and *Eruca*. *Ann. Bot.* 55:139–141.

Sharma, N., and Shivanna, K.R. 1982. Effects of pistil extracts on *in vitro* responses of compatible and incompatible pollen in *Petunia hybrida* Vilm. *Indian J. Expt. Biol.* 20:255–256.

Sharma, N., and Shivanna, K.R. 1983. Lectin-like components of pollen and complementary saccharide moiety of the pistil are involved in self-incompatibility recognition. *Curr. Sci.* 52:913–916.

Sharma, N., and Shivanna, K.R. 1986. Treatment of the stigma with an extract of a compatible pistil overcomes self-incompatibility in *Petunia*. *New Phytol.* 102:443–447.

Sharma, S., Singh, M.B., and Malik, C.P. 1981. Dark CO_2 fixation during germination of *Amaryllis vittata* pollen in suspension cultures. *Indian J. Expt. Biol.* 19:710–714.

Shayk, M., Kolattukudy, P.E., and Davis, R. 1977. Production of a novel extracellular cutinase by the pollen and the chemical composition and ultrastructure of the stigma cuticle of nasturtium (*Tropaeolum majus*). *Plant Physiol.* 60:907–915.

Sheffield, E., Cawood, A.H., Bell, P.R., and Dickinson, H.G. 1979. The development of nuclear vacuoles during meiosis in plants. *Planta* 146:597–601.

Sheldon, J.M., and Dickinson, H.G. 1983. Determination of patterning in the pollen wall of *Lilium henryi*. *J. Cell Sci.* 63:191–208.

Sheldon, J.M., and Dickinson, H.G. 1986. Pollen wall formation in *Lilium*: the effect of chaotropic agents, and the organisation of the microtubular cytoskeleton during pattern development. *Planta* 168:11–23.

Sheldon, J.M., and Hawes, C. 1988. The actin cytoskeleton during male meiosis in *Lilium*. *Cell Biol. Int. Rep.* 12:471–476.

Shellhammer, J., and Meinke, D. 1990. Arrested embryos from the *bio1* auxotroph of *Arabidopsis thaliana* contain reduced levels of biotin. *Plant Physiol.* 93:1162–1167.

Shen, J.B., and Hsu, F.C. 1992. *Brassica* anther-specific genes: characterization and *in situ* localization of expression. *Mol. Gen. Genet.* 234:379–389.

Shepard, J., Boothroyd, E.R., and Stern, H. 1974. The effect of colchicine on synapsis and chiasma formation in microsporocytes of *Lilium*. *Chromosoma* 44:423–437.

Sheridan, W.F. 1973. Nonaqueous isolation of nuclei from lily pollen and an examination of their histones. *Z. Pflanzenphysiol.* 68:450–459.

Sheridan, W.F., and Clark, J.K. 1987. Maize embryogeny: a promising experimental system. *Trend. Genet.* 3:3–6.

Sheridan, W.F., and Clark, J.K. 1993. Mutational analysis of morphogenesis of the maize embryo. *Plant J.* 3:347–358.

Sheridan, W.F., and Neuffer, M.G. 1980. Defective kernel mutants of maize. II. Morphological and embryo culture studies. *Genetics* 95:945–960.

Sheridan, W.F., and Neuffer, M.G. 1982. Maize developmental mutants. Embryos unable to form leaf primordia. *J. Hered.* 73:318–329.

Sheridan, W.F., and Stern, H. 1967. Histones of meiosis. *Expt. Cell Res.* 45:323–335.

Sheridan, W.F., and Thorstenson, Y.R. 1986. Developmental profiles of three embryo-lethal maize mutants lacking leaf primordia: *ptd*-1130, cp*-1418*, and *bno*-747B*. *Devel Genet.* 7:35–49.

Sheridan, W.F., Avalkina, N.A., Shamrov, I.I., Batygina, T.B., and Golubovskaya, I.N. 1996. The *mac1* gene: Controlling the commitment to the meiotic pathway in maize. *Genetics* 142:1009–1020.

Sherwood, R.T. 1995. Nuclear DNA amount during sporogenesis and gametogenesis in sexual and aposporous buffelgrass. *Sex. Plant Reprod.* 8:85–90.

Shevell, D.E., Leu, W.-M., Gillmor, C.S., Xia, G., Feldmann, K.A., and Chua, N.-H. 1994. *EMB30* is essential for normal cell division, cell expansion, and cell adhesion in *Arabidopsis* and encodes a protein that has similarity to Sec7. *Cell* 77:1051–1062.

Shewry, P.R. 1995. Plant storage proteins. *Biol. Rev.* 70:375–426.

Shewry, P.R., Kreis, M., Burgess, S.R., Parmar, S., and Miflin, B.J. 1983. The synthesis and deposition of the prolamin storage proteins (secalins) of rye. *Planta* 159:439–455.

Shewry, P.R., Pratt, H.M., Leggatt, M.M., and Miflin, B.J. 1979. Protein metabolism in developing endosperms of high-lysine and normal barley. *Cereal Chem.* 56:110–117.

Shi, L., Mogensen, H.L., and Zhu, T. 1991. Dynamics of nuclear pore density and distribution patterns within developing pollen: implications for a functional relationship between the vegetative nucleus and the generative cell. *J. Cell Sci.* 99:115–120.

Shi, L., Zhu, T., Mogensen, H.L., and Keim, P. 1996. Sperm identification in maize by fluorescence *in situ* hybridization. *Plant Cell* 8:815–821.

Shi, L., Zhu, T., Mogensen, H.L., and Smith, S.E. 1991. Paternal plastid inheritance in alfalfa: plastid nucleoid number within generative cells correlates poorly with plastid number and male plastid transmission strength. *Curr. Genet.* 19:399–401.

Shiba, H., Hinata, K., Suzuki, A., and Isogai, A. 1995. Breakdown of selfincompatibility in *Brassica* by the antisense RNA of the SLG gene. *Proc. Jap. Acad.* 71B:81–83.

Shibata, D., Steczko, J., Dixon, J.E., Andrews, P.C., Hermodson, M., and Axelrod, B. 1988. Primary structure of soybean lipoxygenase L-2. *J. Biol. Chem.* 263:6816–6821.

Shifriss, O. 1945. Male sterilities and albino seedlings in cucurbits. A study in inbreeding. *J. Hered.* 36:47–52.

Shii, C.T., Rabakoarihanta, A., Mok, M.C., and Mok, D.W.S. 1982. Embryo development in reciprocal crosses of *Phaseolus vulgaris* L. and *P. coccineus* Lam. *Theor. Appl. Genet.* 62:59–64.

Shimamoto, K., Ackermann, M., and Dierks-Ventling, C. 1983. Expression of zein in long term endosperm cultures of maize. *Plant Physiol.* 73:915–920.

Shirsat, A., Wilford, N., Croy, R., and Boulter, D. 1989. Sequences responsible for the tissue specific promoter activity of a pea legumin gene in tobacco. *Mol. Gen. Genet.* 215:326–331.

Shirsat, A.H., Meakin, P.J., and Gatehouse, J.A. 1990. Sequences 5' to the conserved 28 bp leg box element regulate the expression of pea seed storage protein gene *legA*. *Plant Mol. Biol.* 15:685–693.

Shivanna, K.R. 1965. *In vitro* fertilization and seed formation in *Petunia violacea* Lindl. *Phytomorphology* 15:183–185.

Shivanna, K.R. 1982. Pollen–pistil interaction and control of fertilization. In *Experimental Embryology of Vascular Plants*, ed. B.M. Johri, pp. 131–174. Berlin: Springer-Verlag.

Shivanna, K.R., and Cresti, M. 1989. Effects of high humidity and temperature stress on pollen membrane integrity and pollen vigour in *Nicotiana tabacum*. *Sex. Plant Reprod.* 2:137–141.

Shivanna, K.R., and Heslop-Harrison, J. 1981. Membrane state and pollen viability. *Ann. Bot.* 47:759–770.

Shivanna, K.R., Heslop-Harrison, J., and Heslop-Harrison, Y. 1981. Heterostyly in *Primula*. 2. Sites of pollen inhibition, and effects of pistil constituents on compatible and incompatible pollen-tube growth. *Protoplasma* 107:319–337.

Shivanna, K.R., Heslop-Harrison, J., and Heslop-Harrison, Y. 1983. Heterostyly in *Primula*. 3. Pollen water economy: a factor in the intramorph-incompatibility response. *Protoplasma* 117:175–184.

Shivanna, K.R., Heslop-Harrison, Y., and Heslop-Harrison, J. 1978. The pollen–stigma interaction: bud pollination in the Cruciferae. *Acta Bot. Neerl.* 27:107–119.

Shivanna, K.R., Heslop-Harrison, Y., and Heslop-Harrison, J. 1982. The pollen–stigma interaction in the grasses. 3. Features of the self-incompatibility response. *Acta Bot. Neerl.* 31:307–319.

Shivanna, K.R., Jaiswal, V.S., and Mohan Ram, H.Y. 1974a. Inhibition of gamete formation by cycloheximide in pollen tubes of *Impatiens balsamina*. *Planta* 117:173–177.

Shivanna, K.R., Jaiswal, V.S., and Mohan Ram, H.Y. 1974b. Effect of cycloheximide on cultured pollen grains of *Trigonella foenum-graecum*. *Plant Sci. Lett.* 3:335–339.

Shivanna, K.R., and Johri, B.M. 1985. *The Angiosperm Pollen. Structure and Function*. New Delhi: Wiley Eastern Ltd.

Shivanna, K.R., and Rangaswamy, N.S. 1969. Overcoming self-incompatibility in *Petunia axillaris*. I. Delayed pollination, pollination with stored pollen, and bud pollination. *Phytomorphology* 19:372–380.

Shivanna, K.R., and Rangaswamy, N.S. 1992. *Pollen Biology. A Laboratory Manual*. Berlin: Springer-Verlag.

Shivanna, K.R., and Sastri, D.C. 1981. Stigma-surface esterase activity and stigma receptivity in some taxa characterized by wet stigmas. *Ann. Bot.* 47:53–64.

Shivanna, K.R., Xu, H., Taylor, P., and Knox, R.B. 1988. Isolation of sperms from the pollen tubes of flowering plants during fertilization. *Plant Physiol.* 87:647–650.

Shoemaker, R.C., Christofferson, S.E., and Galbraith, D.W. 1987. Storage protein accumulation patterns in somatic embryos of cotton (*Gossypium hirsutum* L.). *Plant Cell Rep.* 6:12–15.

Shotwell, M.A., and Larkins, B.A. 1989. The biochemistry and molecular biology of seed storage proteins. In *The Biochemistry of Plants. A Comprehensive Treatise*, Vol. 15, ed. A. Marcus, pp. 297–345. San Diego: Academic Press.

Shotwell, M.A., Afonso, C., Davies, E., Chesnut, R.S., and Larkins, B.A. 1988. Molecular characterization of oat seed globulins. *Plant Physiol.* 87:698–704.

Shotwell, M.A., Boyer, S.K., Chesnut, R.S., and Larkins, B.A. 1990. Analysis of seed storage protein genes of oats. *J. Biol. Chem.* 265:9652–9658.

Shoup, J.R., Overton, J., and Ruddat, M. 1980. Ultrastructure and development of the sexine in the pollen wall of *Silene alba* (Caryophyllaceae). *Bot. Gaz.* 141:379–388.

Shoup, J.R., Overton, J., and Ruddat, M. 1981. Ultrastructure and development of the nexine and intine in the pollen wall of *Silene alba* (Caryophyllaceae). *Am. J. Bot.* 68:1090-1095.

Shufflebottom, D., Edwards, K., Schuch, W., and Bevan, M. 1993. Transcription of two members of a gene family encoding phenylalanine ammonia-lyase leads to remarkably different cell specificities and induction patterns. *Plant J.* 3:835–845.

Shukla, A., and Sawhney, V.K. 1992. Cytokinins in a genic male sterile line of *Brassica napus*. *Physiol. Plant.* 85:23–29.

Shukla, A., and Sawhney, V.K. 1994. Abscisic acid: one of the factors affecting male sterility in *Brassica napus*. *Physiol. Plant.* 91:522–528.

Shure, M., Wessler, S., and Fedoroff, N. 1983. Molecular identification and isolation of the *waxy* locus in maize. *Cell* 35:225–233.

Shuttuck-Eidens, D.M., and Beachy, R.M. 1985. Degradation of β-conglycinin in early stages of soybean embryogenesis. *Plant Physiol.* 78:895–898.

Siculella, L., and Palmer, J.D. 1988. Physical and gene organization of mitochondrial DNA in fertile and male sterile sunflower. CMS-associated alterations in structure and transcription of the *atpA* gene. *Nucl. Acids Res.* 16:3787–3799.

Sidorova, N.V. 1985. Characteristics of the ultrastructure of isolated embryo sacs of tobacco. *Doklady Akad. Nauk Ukr. SSR* Ser. B 12:63–66.

Siebertz, B., Logemann, J., Willmitzer, L., and Schell, J. 1989. *cis*-Analysis of the wound-inducible promoter *wun1* in transgenic tobacco plants and histochemical localization of its expression. *Plant Cell* 1:961–968.

Silcock, D.J., Francis, D., Bryant, J.A., and Hughes, S.G. 1990. Changes in nuclear DNA content, cell and nuclear size, and frequency of cell division in the cotyledons of *Brassica napus* L. during embryogenesis. *J. Expt. Bot.* 41:401–407.

Silvanovich, A., Astwood, J., Zhang, L., Olsen, E., Kisil, F., Sehon, A., Mohapatra, S., and Hill, R. 1991. Nucleotide sequence analysis of three cDNAs coding for *Poa p* IX isoallergens of Kentucky bluegrass pollen. *J. Biol. Chem.* 266:1204–1210.

Simon, A.E., Tenbarge, K.M., Scofield, S.R., Finkelstein, R.R., and Crouch, M.L. 1985. Nucleotide sequence of a cDNA clone of *Brassica napus* 12S storage protein shows homology with legumin from *Pisum sativum*. *Plant Mol. Biol.* 5:191–201.

Simon, R., Altschuler, Y., Rubin, R., and Galili, G. 1990. Two closely related wheat storage proteins follow a markedly different subcellular route in *Xenopus laevis* oocytes. *Plant Cell* 2:941–950.

Simoncioli, C. 1974. Ultrastructural characteristics of "*Diplotaxis erucoides* (L.) DC." suspensor. *Giorn. Bot. Ital.* 108:175–189.

Simpson, M.G., and Levin, G.A. 1994. Pollen ultrastructure of the biovulate Euphorbiaceae. *Int. J. Plant Sci.* 155:313–341.

Sims, T.L. 1993. Genetic regulation of self-incompatibility. *Crit. Rev. Plant Sci.* 12:129–167.

vitro germination and tube extension of *Petunia hybrida* pollen. *Proc. Koninkl. Nederl. Akad. Wetensch.* 77C:116–124.

Song, J., and Hedgcoth, C. 1994. A chimeric gene (*orf256*) is expressed as protein only in cytoplasmic male-sterile lines of wheat. *Plant Mol. Biol.* 26:535–539.

Sood, P.P., Malik, C.P., and Tewari, H.B. 1968. A histochemical study of the localization of succinicdehydrogenase in the germinating pollen grains of *Portulaca grandiflora*. *Z. Biol.* 116:215–220.

Sood, R., Prabha, K., and Gupta, S.C. 1982. Is the 'rejection reaction' inducing ability in sporophytic self-incompatibility systems restricted only to pollen and tapetum? *Theor. Appl. Genet.* 63:27–32.

Sood, R., Prabha, K., and Gupta, S.C. 1983. Overcoming self-incompatibility in *Brassica campestris* cv. Pusa Kalyani by bud and delayed pollinations. *Beitr. Biol. Pflanz.* 58:59–65.

Sopory, S.K., and Maheshwari, S.C. 1976. Development of pollen embryoids in anther cultures of *Datura innoxia*. II. Effects of growth hormones. *J. Expt. Bot.* 27:58–68.

Sørensen, M.B. 1992. Methylation of B-hordein genes in barley endosperm is inversely correlated with gene activity and affected by the regulatory gene *Lys3*. *Proc. Natl Acad. Sci. USA* 89:4119–4123.

Sørensen, M.B., Cameron-Mills, V., and Brandt, A. 1989. Transcriptional and post-transcriptional regulation of gene expression in developing barley endosperm. *Mol. Gen. Genet.* 217:195–201.

Sosnikhina, S.P., Fedotova, Y.S., Smirnov, V.G., Mikhailova, E.I., Kolomiets, O.L., and Bogdanov, Y.F. 1992. Meiotic mutants of rye *Secale cereale* L. I. Synaptic mutant *sy-1*. *Theor. Appl. Genet.* 84:979–985.

Sossountzov, L., Ruiz-Avila, L., Vignols, F., Jolliot, A., Arondel, V., Tchang, F., Grosbois, M., Guerbette, F., Miginiac, E., Delseny, M., Puigdomenèch, P., and Kader, J.-C. 1991. Spatial and temporal expression of a maize lipid transfer protein gene. *Plant Cell* 3:923–933.

Souèges, R. 1919. Les premières divisions de l'œuf et les différenciations du suspenseur chez le *Capsella bursa-pastoris* Moench (1). *Ann. Sci. Nat. Bot.* Ser. X 1:1–28.

Souèges, R. 1924. Développement de l'embryon chez le *Sagina procumbens* L. *Bull. Soc. Bot. Fr.* 71:590–614.

Souèges, R. 1936. Embryogénie des Saxifragacées. Développement de l'embryon chez le *Saxifraga granulata* L. *Compt. Rend. Acad. Sci. Paris* 202:240–242.

Souer, E., van Houwelingen, A., Kloos, D., Mol, J., and Koes, R. 1996. The *no apical meristem* gene of *Petunia* is required for pattern formation in embryos and flowers and is expressed at meristem and primordia boundaries. *Cell* 85:159–170.

Southworth, D. 1971. Incorporation of radioactive precursors into developing pollen walls. In *Pollen: Development and Physiology*, ed. J. Heslop-Harrison, pp. 115–120. London: Butterworth & Co.

Southworth, D. 1975. Lectins stimulate pollen germination. *Nature* 258:600–602.

Southworth, D. 1983. Exine development in *Gerbera jamesonii* (Asteraceae: Mutisieae). *Am. J. Bot.* 70:1038–1047.

Southworth, D. 1988. Isolation of exines from gymnosperm pollen. *Am. J. Bot.* 75:15–21.

Southworth, D. 1990. Membranes of sperm and vegetative cells in pollen of *Gerbera jamesonii*. *J. Str. Biol.* 103:97–103.

Southworth, D., and Dickinson, D.B. 1975. β-1,3-glucan synthase from *Lilium longiflorum* pollen. *Plant Physiol.* 56:83–87.

Southworth, D., and Dickinson, D.B. 1981. Ultrastructural changes in germinating lily pollen. *Grana* 20:29–35.

Southworth, D., and Jernstedt, J.A. 1995. Pollen exine development precedes microtubule rearrangement in *Vigna unguiculata* (Fabaceae): a model for pollen wall patterning. *Protoplasma* 187:79–87.

Southworth, D., and Knox, R.B. 1989. Flowering plant sperm cells: isolation from pollen of *Gerbera jamesonii* (Asteraceae). *Plant Sci.* 60:273–277.

Southworth, D., and Morningstar, P.A. 1992. Isolation of generative cells from pollen of *Phoenix dactylifera*. *Sex. Plant Reprod.* 5:270–274.

Southworth, D., Platt-Aloia, K.A., and Thomson, W.W. 1988. Freeze fracture of sperm and vegetative cells in *Zea mays* pollen. *J. Ultrastr. Mol. Str. Res.* 101:165–172.

Southworth, D., Salvatici, P., and Cresti, M. 1994. Freeze fracture of membranes at the interface between vegetative and generative cells in *Amaryllis* pollen. *Int. J. Plant Sci.* 155:538–544.

Southworth, D., Platt-Aloia, K.A., DeMason, D.A., and Thomson, W.W. 1989. Freeze-fracture of the generative cell of *Phoenix dactylifera* (Arecaceae). *Sex. Plant Reprod.* 2:270–276.

Southworth, D., Singh, M.B., Hough, T., Smart, I.J., Taylor, P., and Knox, R.B. 1988. Antibodies to pollen exines. *Planta* 176:482–487.

Souvré, A., and Albertini, L. 1982. Étude des modifications de l'ultrastructure du tapis plasmodial du *Rhoeo discolor* Hance au cours du développement de l'anthère, en relation avec les données cytochimiques et autoradiographiques. *Rev. Cytol. Biol. Végét.-Bot.* 5:151–169.

Soyfer, V.N. 1980. Hereditary variability of plants under the action of exogenous DNA. *Theor. Appl. Genet.* 58:225–235.

Spangenberg, G., Wang, Z.Y., Wu, X.L., Nagel, J., Iglesias, V.A., and Potrykus, I. 1995a. Transgenic tall fescue (*Festuca arundinacea*) and red fescue (*F. rubra*) plants from microprojectile bombardment of embryogenic suspension cells. *J. Plant Physiol.* 145:693–701.

Spangenberg, G., Wang, Z.-Y., Wu, X., Nagel, J., and Potrykus, I. 1995b. Transgenic perennial ryegrass (*Lolium perenne*) plants from microprojectile bombardment of embryogenic suspension cells. *Plant Sci.* 108:209–217.

Spanjers, A.W. 1978. Voltage variation in *Lilium longiflorum* pistils induced by pollination. *Experientia* 34:36–37.

Sparrow, A.H., Moses, M.J., and Steele, R. 1952. A cytological and cytochemical approach to an understanding of radiation damage in dividing cells. *Brit. J. Radiol.* 25:182–188.

Sparrow, A.H., Pond, V., and Kojan, S. 1955. Microsporogenesis in excised anthers of *Trillium erectum* grown on sterile media. *Am. J. Bot.* 42:384–394.

Spassova, M., Monéger, F., Leaver, C.J., Petrov, P., Atanassov, A., Nijkamp, H.J.J., and Hille, J. 1994. Characterisation and expression of the mitochondrial genome of a new type of cytoplasmic male-sterile sunflower. *Plant Mol. Biol.* 26:1819–1831.

Spena, A., and Schell, J. 1987. The expression of a heat-inducible chimeric gene in transgenic tobacco plants. *Mol. Gen. Genet.* 206:436–440.

Spena, A., Viotti, A., and Pirrotta, V. 1982. A homologous

repetitive block structure underlies the heterogeneity of heavy and light chain zein genes. *EMBO J.* 1:1589–1594.

Spena, A., Estruch, J.J., Prinsen, E., Nacken, W., van Onckelen, H., and Sommer, H. 1992. Anther-specific expression of the *rolB* gene of *Agrobacterium rhizogenes* increases IAA content in anthers and alters anther development and whole flower growth. *Theor. Appl. Genet.* 84:520–527.

Spencer, M.E., and Hodge, R. 1991. Cloning and sequencing of the cDNA encoding the major albumin of *Theobroma cacao*. Identification of the protein as a member of the Kunitz protease inhibitor family. *Planta* 183:528–535.

Spencer, M.E., and Hodge, R. 1992. Cloning and sequencing of a cDNA encoding the major storage proteins of *Theobroma cacao*. Identification of the proteins as members of the vicilin class of storage proteins. *Planta* 186:567–576.

Speranza, A., and Calzoni, G.L. 1988. *In vitro* test of self-incompatibility in *Malus domestica*. *Sex. Plant Reprod.* 1:223–227.

Speranza, A., and Calzoni, G.L. 1990. *In vitro* expression of self-incompatibility in *Malus domestica* mediated by stylar glycoproteins. *Plant Physiol. Biochem.* 28:747–754.

Speranza, A., and Calzoni, G.L. 1992. Inhibition of pollen-tube growth and protein secretion by the monovalent ionophore monensin. *Sex. Plant Reprod.* 5:232–238.

Springer, B., Werr, W., Starlinger, P., Bennett, D.C., Zakolica, M., and Freeling, M. 1986. The *shrunken* gene on chromosome 9 of *Zea mays* L. is expressed in various plant tissues and encodes an anaerobic protein. *Mol. Gen. Genet.* 205:461–468.

Springer, P.S., McCombie, W.R., Sundaresan, V., and Martienssen, R.A. 1995. Gene trap tagging of *PROLIFERA*, an essential *MCM2-3-5*-like gene in *Arabidopsis*. *Science* 268:877–880.

Sree Ramulu, K., Bredemeijer, G.M.M., Dijkhuis, P., de Nettancourt, D., and Schibilla, H. 1979. Mentor pollen effects on gametophytic incompatibility in *Nicotiana, Oenothera* and *Lycopersicum*. *Theor. Appl. Genet.* 54:215–218.

Sree Ramulu, K., Devreux, M., Ancora, G., and Laneri, U. 1976. Chimerism in *Lycopersicum peruvianum* plants regenerated from *in vitro* cultures of anthers and stem internodes. *Z. Pflanzenzüchtg* 76:299–319.

Srivastava, H.K., Sarkissian, I.V., and Shands, H.L. 1969. Mitochondrial complementation and cytoplasmic male sterility in wheat. *Genetics* 63:611–618.

Stacey, N.J., Roberts, K., and Knox, J.P. 1990. Patterns of expression of the JIM4 arabinogalactan-protein epitope in cell cultures and during somatic embryogenesis in *Daucus carota* L. *Planta* 180:285–292.

Staff, I.A., Taylor, P., Kenrick, J., and Knox, R.B. 1989. Ultrastructural analysis of plastids in angiosperm pollen tubes. *Sex. Plant Reprod.* 2:70–76.

Staff, I.A., Taylor, P.E., Smith, P., Singh, M.B., and Knox, R.B. 1990. Cellular localization of water soluble, allergenic proteins in rye-grass (*Lolium perenne*) pollen using monoclonal and specific IgE antibodies with immunogold probes. *Histochem. J.* 22:276–290.

Staiger, C.J., and Cande, W.Z. 1990. Microtubule distribution in *dv*, a maize meiotic mutant defective in the prophase to metaphase transition. *Devel. Biol.* 138:231–242.

Staiger, C.J., and Cande, W.Z. 1991. Microfilament distribution in maize meiotic mutants correlates with microtubule organization. *Plant Cell* 3:637–644.

Staiger, C.J., Goodbody, K.C., Hussey, P.J., Valenta, R., Drøbak, B.K., and Lloyd, C.W. 1993. The profilin multigene family of maize: differential expression of three isoforms. *Plant J.* 4:631–641.

Staiger, D., and Apel, K. 1993. Molecular characterization of two cDNAs from *Sinapis alba* L. expressed specifically at an early stage of tapetum development. *Plant J.* 4:697–703.

Staiger, D., Kappeler, S., Müller, M., and Apel, K. 1994. The proteins encoded by two tapetum-specific transcripts, *Sa*tap35 and *Sa*tap44, from *Sinapis alba* L. are localized in the exine cell wall layer of developing microspores. *Planta* 192:221–231.

Stålberg, K., Ellerström, M., Josefsson, L.-G., and Rask, L. 1993. Deletion analysis of a 2S seed storage protein promoter of *Brassica napus* in transgenic tobacco. *Plant Mol. Biol.* 23:671–683.

Stanley, R.G., and Linskens, H.F. 1964. Enzyme activation in germinating *Petunia* pollen. *Nature* 203:542–544.

Stanley, R.G., and Linskens, H.F. 1965. Protein diffusion from germinating pollen. *Physiol. Plant.* 18:47–53.

Stanley, R.G., and Linskens, H.F. 1967. Oxygen tension as a control mechanism in pollen tube rupture. *Science* 157:833–834.

Stanley, R.G., and Linskens, H.F. 1974. *Pollen: Biology, Biochemistry, Management*. Berlin: Springer-Verlag.

Stapleton, A.E., and Bedinger, P.A. 1992. Immature maize spikelets develop and produce pollen in culture. *Plant Cell Rep.* 11:248–252.

Staswick, P., and Chrispeels, M.J. 1984. Expression of lectin genes during seed development in normal and phytohemagglutinin-deficient cultivars of *Phaseolus vulgaris*. *J. Mol. Appl. Genet.* 2:525–535.

Staswick, P.E., Hermodson, M.A., and Nielsen, N.C. 1984. The amino acid sequence of the A_2B_{1a} subunit of glycinin. *J. Biol. Chem.* 259:13424–13430.

Stauffer, C., Benito Moreno, R.M., and Heberle-Bors, E. 1991. Seed set after pollination with *in-vitro*-matured, isolated pollen of *Triticum aestivum*. *Theor. Appl. Genet.* 81:576–580.

Staxén, I., Klimaszewska, K., and Bornman, C.H. 1994. Microtubular organization in protoplasts and cells of somatic embryo-regenerating and non-regenerating cultures of *Larix*. *Physiol. Plant.* 91:680–686.

Stead, A.D., Roberts, I.N., and Dickinson, H.G. 1979. Pollen–pistil interaction in *Brassica oleracea*. Events prior to pollen germination. *Planta* 146:211–216.

Stead, A.D., Roberts, I.N., and Dickinson, H.G. 1980. Pollen–stigma interaction in *Brassica oleracea*: the role of stigmatic proteins in pollen grain adhesion. *J. Cell Sci.* 42:417–423.

Steer, M.W. 1977a. Differentiation of the tapetum in *Avena*. I. The cell surface. *J. Cell Sci.* 25:125–138.

Steer, M.W. 1977b. Differentiation of the tapetum in *Avena*. II. The endoplasmic reticulum and Golgi apparatus. *J. Cell Sci.* 28:71–86.

Steer, M.W., and Steer, J.M. 1989. Pollen tube tip growth. *New Phytol.* 111:323–358.

Stefaniak, B. 1984. Development of rye embryo. I. Preliminary investigations on ultrastructure of 7-day-old proembryo. *Acta Soc. Bot. Polon.* 53:11–16.

Stefaniak, B. 1987. The *in vitro* development of isolated proembryos. *Acta Soc. Bot. Polon.* 56:37–42.

Steffen, K. 1963. Male gametophyte. In *Recent Advances in*

the Embryology of Angiosperms, ed. P. Maheshwari, pp. 15–40. Delhi: International Society of Plant Morphologists.

Steffensen, D.M. 1966. Synthesis of ribosomal RNA during growth and division in *Lilium*. *Expt. Cell Res.* 44:1–12.

Stein, J.C., and Nasrallah, J.B. 1993. A plant receptor-like gene, the S-locus receptor kinase of *Brassica oleracea* L., encodes a functional serine/threonine kinase. *Plant Physiol.* 101:1103–1106.

Stein, J.C., Dixit, R., Nasrallah, M.E., and Nasrallah, J.B. 1996. SRK, the stigma-specific S locus receptor kinase of *Brassica*, is targeted to the plasma membrane in transgenic tobacco. *Plant Cell* 8:429–445.

Stein, J.C., Howlett, B., Boyes, D.C., Nasrallah, M.E., and Nasrallah, J.B. 1991. Molecular cloning of a putative receptor protein kinase gene encoded at the self-incompatibility locus of *Brassica oleracea*. *Proc. Natl Acad. Sci. USA* 88:8816–8820.

Stejskal, J., and Griga, M. 1995. Comparative analysis of some isozymes and proteins in somatic and zygotic embryos of soybean (*Glycine max* [L.] Merr.). *J. Plant Physiol.* 146:497–502.

Stelly, D.M., and Palmer, R.G. 1982. Variable development in anthers of partially male-sterile soybeans. *J. Hered.* 73:101–108.

Stenram, U., Heneen, W.K., and Skerritt, J.H. 1991. Immunocytochemical localization of wheat storage proteins in endosperm cells 30 days after anthesis. *J. Expt. Bot.* 42:1347–1355.

Sterk, P., Booij, H., Schellekens, G.A., van Kammen, A., and de Vries, S.C. 1991. Cell-specific expression of the carrot EP2 lipid transfer protein gene. *Plant Cell* 3:907–921.

Sterling, C. 1955. Embryogeny in the lima bean. *Bull. Torrey Bot. Cl.* 82:325–338.

Stern H. 1956. Sulfhydryl groups and cell division. *Science* 124:1292–1293.

Stern, H. 1958. Variations in sulfhydryl concentration during microsporocyte meiosis in anthers of *Lilium* and *Trillium*. *J. Biophys. Biochem. Cytol.* 4:157–161.

Stern, H. 1961. Periodic induction of deoxyribonuclease activity in relation to the mitotic cycle. *J. Biophys. Biochem. Cytol.* 9:271–277.

Stern, H., and Hotta, Y. 1977. Biochemistry of meiosis. *Phil. Trans. Roy. Soc. Lond.* 277B:277–294.

Stern, H., and Hotta, Y. 1984. Chromosome organization in the regulation of meiotic prophase. *Symp. Soc. Expt. Biol.* 38:161–175.

Stern, H., and Kirk, P.L. 1948. The oxygen consumption of the microspores of *Trillium* in relation to the mitotic cycle. *J. Gen. Physiol.* 31:243–248.

Stern, H., and Timonen, S. 1954. The position of the cell nucleus in the pathways of hydrogen transfer: cytochrome C, flavoproteins, glutathione, and ascorbic acid. *J. Gen. Physiol.* 38:41–52.

Stern H., Westergaard, M., and von Wettstein, D. 1975. Presynaptic events in meiocytes of *Lilium longiflorum* and their relation to crossing-over: a preselection hypothesis. *Proc. Natl Acad. Sci. USA* 72:961–965.

Stettler, R.F., Koster, R., and Steenackers, V. 1980. Interspecific crossability studies in poplars. *Theor. Appl. Genet.* 58:273–282.

Stevens, V.A.M., and Murray, B.G. 1981. Studies on heteromorphic self-incompatibility systems: the cytochemistry and ultrastructure of the tapetum of *Primula obconica*. *J. Cell Sci.* 50:419–431.

Stevens, V.A.M., and Murray, B.G. 1982. Studies on heteromorphic self-incompatibility systems: physiological aspects of the incompatibility system of *Primula obconica*. *Theor. Appl. Genet.* 61:245–256.

Steward, F.C. 1963. The control of growth in plant cells. *Sci. Am.* 209 (4):104–113.

Steward, F.C., Mapes, M.O., and Mears, K. 1958. Growth and organized development of cultured cells. II. Organization in cultures grown from freely suspended cells. *Am. J. Bot.* 45:705–708.

Steward, F.C., Mapes, M.O., and Smith, J. 1958. Growth and organized development of cultured cells. I. Growth and division of freely suspended cells. *Am. J. Bot.* 45:693–703.

Steward, F.C., Mapes, M.O., Kent, A.E., and Holsten, R.D. 1964. Growth and development of cultured plant cells. *Science* 143:20–27.

Stewart, J.M., and Hsu, C.L. 1977. *In ovulo* embryo culture and seedling development of cotton (*Gossypium hirsutum* L.). *Planta* 137:113–117.

Stickland, R.G., and Wilson, K.E. 1983. Sugars and starch in developing round and wrinkled pea seeds. *Ann. Bot.* 52:919–921.

Stieglitz, H. 1977. Role of β-1,3-glucanase in post-meiotic microspore release. *Devel Biol.* 57:87–97.

Stieglitz, H., and Stern, H. 1973. Regulation of β-1,3-glucanase activity in developing anthers of *Lilium*. *Devel Biol.* 34:169–173.

Stinissen, H.M., Peumans, W.J., and Carlier, A.R. 1983. Occurrence and immunological relationships of lectins in gramineous species. *Planta* 159:105–111.

Stinissen, H.M., Peumans, W.J., and Chrispeels, M.J. 1984. Subcellular site of lectin synthesis in developing rice embryos. *EMBO J.* 3:1979–1985.

Stinissen, H.M., Peumans, W.J., and de Langhe, E. 1984. Abscisic acid promotes lectin biosynthesis in developing and germinating rice embryos. *Plant Cell Rep.* 3:55–59.

Stinson, J., and Mascarenhas, J.P. 1985. Onset of alcohol dehydrogenase synthesis during microsporogenesis in maize. *Plant Physiol.* 77:222–224.

Stinson, J.R., Eisenberg, A.J., Willing, P.R., Pe, M.E., Hanson, D.D., and Mascarenhas, J.P. 1987. Genes expressed in the male gametophyte of flowering plants and their isolation. *Plant Physiol.* 83:442–447.

Stirn, S., and Jacobsen, H.-J. 1987. Marker proteins for embryogenic differentiation patterns in pea callus. *Plant Cell Rep.* 6:50–54.

Stöger, E., Benito Moreno, R.M., Ylstra, B., Vicente, O., and Heberle-Bors, E. 1992. Comparison of different techniques for gene transfer into mature and immature tobacco pollen. *Transgen. Res.* 1:71–78.

Stöger, E., Fink, C., Pfosser, M., and Heberle-Bors, E. 1995. Plant transformation by particle bombardment of embryogenic pollen. *Plant Cell Rep.* 14:273–278.

Štorchová, H., Čapková V., and Tupý, J. 1994. A *Nicotiana tabacum* mRNA encoding a 69-kDa glycoprotein occurring abundantly in pollen tubes is transcribed but not translated during pollen development in the anthers. *Planta* 192:441–445.

Stout, A.B. 1931. Pollen-tube behavior in *Brassica pekinensis* with reference to self-incompatibility in fertilization. *Am. J. Bot.* 18:686–695.

Stout, A.B., and Chandler, C. 1933. Pollen-tube behavior in *Hemerocallis* with special reference to incompatibilities. *Bull. Torrey Bot. Cl.* 60:397–416.

Stout, A.B., and Chandler, C. 1941. Change from self-incompatibility to self-compatibility accompanying change from diploidy to tetraploidy. *Science* 94:118.

Strasburger, E. 1884. *Neue Untersuchungen über den Befruchtungsvorgang bei den Phanerogamen als Grundlage für eine Theorie der Zeugung*. Jena: Gustav Fischer.

Street, H.E., and Withers, L.A. 1974. The anatomy of embryogenesis. In *Tissue Culture and Plant Science 1974*, ed. H.E. Street, pp. 71–100. London: Academic Press.

Strokov, A.A., Bogdanov, Y.F., and Reznickova, S.A. 1973. A quantitative study of histones of meiocytes. II. Polyacrylamide gel electrophoresis of isolated histones from *Lilium* microsporocytes. *Chromosoma* 43:247–260.

Stuart, D.A., and Strickland, S.G. 1984a. Somatic embryogenesis from cell cultures of *Medicago sativa* L. I. The role of amino acid additions to the regeneration medium. *Plant Sci. Lett.* 34:165–174.

Stuart, D.A., and Strickland, S.G. 1984b. Somatic embryogenesis from cell cultures of *Medicago sativa* L. II. The interaction of amino acids with ammonium. *Plant Sci. Lett.* 34:175–181.

Sturaro, M., Vernieri, P., Castiglioni, P., Binelli, G., and Gavazzi, G. 1996. The *rea* (*r*ed *e*mbryonic *a*xis) phenotype describes a new mutation affecting the response of maize embryos to abscisic acid and osmotic stress. *J. Expt. Bot.* 47:717–728.

Sturm, A., Voelker, T.A., Herman, E.M., and Chrispeels, M.J. 1988. Correct glycosylation, Golgi-processing, and targeting to protein bodies of the vacuolar protein phytohemagglutinin in transgenic tobacco. *Planta* 175:170–183.

Sugimoto, T., Momma, M., Hashizume, K., and Saio, K. 1987. Components of storage protein in hypocotyl-radicle axis of soybean (*Glycine max* cv. Enrei) seeds. *Agr. Biol. Chem.* 51:1231–1238.

Sugiyama, T., Rafalski, A., and Söll, D. 1986. The nucleotide sequence of a wheat γ-gliadin genomic clone. *Plant Sci.* 44:205–209.

Sugiyama, T., Rafalski, A., Peterson, D., and Söll, D. 1985. A wheat HMW glutenin subunit gene reveals a highly repeated structure. *Nucl. Acids Res.* 13:8729–8737.

Sullivan, T.D., and Kaneko, Y. 1995. The maize *brittle1* gene encodes amyloplast membrane polypeptides. *Planta* 196:477–484.

Sullivan, T.D., Strelow, L.I., Illingworth, C.A., Phillips, R.L., and Nelson, O.E., Jr. 1991. Analysis of maize *brittle-1* alleles and a defective *suppressor-mutator*-induced mutable allele. *Plant Cell* 3:1337–1348.

Sumner, M.J. 1992. Embryology of *Brassica campestris*: the entrance and discharge of the pollen tube in the synergid and the formation of the zygote. *Can. J. Bot.* 70:1577–1590.

Sumner, M.J., and van Caeseele, L. 1989. The ultrastructure and cytochemistry of the egg apparatus of *Brassica campestris*. *Can. J. Bot.* 67:177–190.

Sumner, M.J., and van Caeseele, L. 1990. The development of the central cell of *Brassica campestris* prior to fertilization. *Can. J. Bot.* 68:2553–2563.

Sumner-Smith, M., Rafalski, J.A., Sugiyama, T., Stoll, M., and Söll, D. 1985. Conservation and variability of wheat α/β-gliadin genes. *Nucl. Acids Res.* 13:3905–3916.

Sun, C.-S. 1978. Androgenesis of cereal crops. In *Proceedings of Symposium on Plant Tissue Culture*, pp. 117–123. Peking: Science Press.

Sun, C.-S., Wang, C.-C., and Chu, C.-C. 1974. Cell division and differentiation of pollen grains in *Triticale* anthers cultured *in vitro*. *Sci. Sinica* 17:47–54.

Sun, M., and Ganders, F.R. 1987. Microsporogenesis in male-sterile and hermaphroditic plants of nine gynodioecious taxa of Hawaiian *Bidens* (Asteraceae). *Am. J. Bot.* 74:209–217.

Sun, S.M., Slightom, J.L., and Hall, T.C. 1981. Intervening sequences in a plant gene - comparison of the partial sequence of cDNA and genomic DNA of French bean phaseolin. *Nature* 289:37–41.

Sun, S.M., Mutschler, M.A., Bliss, F.A., and Hall, T.C. 1978. Protein synthesis and accumulation in bean cotyledons during growth. *Plant Physiol.* 61:918–923.

Sunderland, N. 1971. Anther culture: a progress report. *Sci. Prog. Oxford* 59:527–549.

Sunderland, N., Collins, G.B., and Dunwell, J.M. 1974. The role of nuclear fusion in pollen embryogenesis of *Datura innoxia* Mill. *Planta* 117:227–241.

Sunderland, N., and Dunwell, J.M. 1977. Anther and pollen culture. In *Plant Tissue and Cell Culture*, 2nd ed., ed. H.E. Street, pp. 223–265. Berkeley: University of California Press.

Sunderland, N., and Evans, L.J. 1980. Multicellular pollen formation in cultured barley anthers. II. The A, B, and C pathways. *J. Expt. Bot.* 31:501–514.

Sunderland, N., and Roberts, M. 1977. New approach to pollen culture. *Nature* 270:236–238.

Sunderland, N., and Roberts, M. 1979. Cold-pretreatment of excised flower buds in float culture of tobacco anthers. *Ann. Bot.* 43:405–414.

Sunderland, N., and Wicks, F.M. 1971. Embryoid formation in pollen grains of *Nicotiana tabacum*. *J. Expt. Bot.* 22:213–226.

Sunderland, N., and Wildon, D.C. 1979. A note on the pretreatment of excised flower buds in float culture of *Hyoscyamus* anthers. *Plant Sci. Lett.* 15:169–175.

Sunderland, N., Roberts, M., Evans, L.J., and Wildon, D.C. 1979. Multicellular pollen formation in cultured barley anthers. I. Independent division of the generative and vegetative cells. *J. Expt. Bot.* 30:1133–1144.

Sung, Z.R. 1979. Relationship of indole-3-acetic acid and tryptophan concentrations in normal and 5-methyltryptophan-resistant cell lines of wild carrots. *Planta* 145:339–345.

Sung, Z.R., Lazar, G.B., and Dudits, D. 1981. Cycloheximide resistance in carrot culture: a differentiated function. *Plant Physiol.* 68:261–264.

Sung, Z.R., and Okimoto, R. 1981. Embryonic proteins in somatic embryos of carrot. *Proc. Natl Acad. Sci. USA* 78:3683–3687.

Sung, Z.R., and Okimoto, R. 1983. Coordinate gene expression during somatic embryogenesis in carrots. *Proc. Natl Acad. Sci. USA* 80:2661–2665.

Sung, Z.R., Smith, R., and Horowitz, J. 1979. Quantitative studies of embryogenesis in normal and 5-methyltryptophan-resistant cell lines of wild carrot. *Planta* 147:236–240.

Sung, Z.R., Belachew, A., Shunong, B., and Bertrand-Garcia, R. 1992. *EMF*, an *Arabidopsis* gene required for vegetative shoot development. *Science* 258:1645–1647.

Süss, J., and Tupý, J. 1976. On the nature of RNA synthesized in pollen tubes of *Nicotiana alata*. *Biol. Plant.* 18:140–146.

Süss, J., and Tupý, J. 1978. tRNA synthesis in germinating pollen. *Biol. Plant.* 20:70–72.

Süss, J., and Tupý, J. 1979. Poly(A)+RNA synthesis in germinating pollen of *Nicotiana tabacum* L. *Biol. Plant.* 21:365–371.

Süss, J., and Tupý, J. 1982. Kinetics of uridine uptake and incorporation into RNA in tobacco pollen culture. *Biol. Plant.* 24:72–79.

Sussex, I. 1975. Growth and metabolism of the embryo and attached seedling of the viviparous mangrove, *Rhizophora mangle*. *Am. J. Bot.* 62:948–953.

Sussex, I., Clutter, M., Walbot, V., and Brady, T. 1973. Biosynthetic activity of the suspensor of *Phaseolus coccineus*. *Caryologia* 25: Suppl. 261–272.

Süssmuth, J., Dressler, K., and Hess, D. 1991. *Agrobacterium*-mediated transfer of the GUS gene into pollen of *Petunia*. *Bot. Acta* 104:72–76.

Suzuki, G., Watanabe, M., Toriyama, K., Isogai, A., and Hinata, K. 1995. Molecular cloning of members of the *S*-multigene family in self-incompatible *Brassica campestris* L. *Plant Cell Physiol.* 36:1273–1280.

Suzuki, M., Koide, Y., Hattori, T., Nakamura, K., and Asahi, T. 1995. Different sets of *cis*-elements contribute to the expression of a catalase gene from castor bean during seed formation and postembryonic development in transgenic tobacco. *Plant Cell Physiol.* 36:1067–1074.

Suzuki, T., Nakamura, C., Mori, N., and Kaneda, C. 1995. Overexpression of mitochondrial genes in alloplasmic common wheat with a cytoplasm of wheatgrass (*Agropyron trichophorum*) showing depressed vigor and male sterility. *Plant Mol. Biol.* 27:553–565.

Swain, E., Li, C.P., and Poulton, J.E. 1992a. Development of the potential for cyanogenesis in maturing black cherry (*Prunus serotina* Ehrh.) fruits. *Plant Physiol.* 98:1423–1428.

Swain, E., Li, C.P., and Poulton, J.E. 1992b. Tissue and subcellular localization of enzymes catabolizing (*R*)-amygdalin in mature *Prunus serotina* seeds. *Plant Physiol.* 100:291–300.

Swaminathan, M.S., Chopra, V.L., and Bhaskaran, S. 1962. Cytological aberrations observed in barley embryos cultured in irradiated potato mash. *Radiation Res.* 16:182–188.

Swamy, B.G.L. 1979. Embryogenesis in *Cheirostylis flabellata*. *Phytomorphology* 29:199–203.

Swamy, B.G.L. 1980. Embryogenesis in *Sagittaria sagittaefolia*. *Phytomorphology* 30:204–212.

Swamy, B.G.L., and Ganapathy, P.M. 1957. A new type of endosperm haustorium in *Nothapodytes foetida*. *Phytomorphology* 7:331–336.

Swamy, B.G.L., and Lakshmanan, K.K. 1962a. The origin of epicotylary meristem and cotyledon in *Halophila ovata* Gaudich. *Ann. Bot.* 26:243–249.

Swamy, B.G.L., and Lakshmanan, K.K. 1962b. Contributions to the embryology of the Najadaceae. *J. Indian Bot. Soc.* 41:247–267.

Swegle, M., Kramer, K.J., and Muthukrishnan, S. 1992. Properties of barley seed chitinases and release of embryo-associated isoforms during early stages of imbibition. *Plant Physiol.* 99:1009–1014.

Swift, H. 1950. The constancy of desoxyribose nucleic acid in plant nuclei. *Proc. Natl Acad. Sci. USA* 36:643–654.

Swoboda, I., Dang, T.C.H., Heberle-Bors, E., and Vicente, O. 1995a. Expression of Bet c 1, the major birch pollen allergen, during anther development. *Protoplasma* 187:103–110.

Swoboda, I., Jilek, A., Ferreira, F., Engel, E., Hoffmann-Sommergruber, K., Scheiner, O., Kraft, D., Breiteneder, H., Pittenauer, E., Schmid, E., Vicente, O., Heberle-Bors, E., Ahorn, H., and Breitenbach, M. 1995b. Isoforms of Bet v 1, the major birch pollen allergen, analyzed by liquid chromatography, mass spectrometry, and cDNA cloning. *J. Biol. Chem.* 270:2607–2613.

Syamasundar, J., and Panchaksharappa, M.G. 1975. A cytochemical study of the hypertrophied synergid in *Allium cepa* L. *Cytologia* 40:371–376.

Syamasundar, J., and Panchaksharappa, M.G. 1976. A histochemical study of some post-fertilization developmental stages in *Dipcadi montanum* Dalz. *Cytologia* 41:123–130.

Szakács, E., and Barnabás, B. 1988. Cytological aspects of *in vitro* androgenesis in wheat (*Triticum aestivum* L.) using fluorescent microscopy. *Sex. Plant Reprod.* 1:217–222.

Szakács, E., and Barnabás, B. 1995. The effect of colchicine treatment on microspore division and microspore-derived embryo differentiation in wheat (*Triticum aestivum* L.) anther culture. *Euphytica* 83:209–213.

Szymkowiak, E.J., and Sussex, I.M. 1992. The internal meristem layer (L3) determines floral meristem size and carpel number in tomato periclinal chimeras. *Plant Cell* 4:1089–1100.

Tabata, S., Sato, S., Watanabe, Y., Yamamoto, M., and Hotta, Y. 1993. Evidence of meiosis-specific regulation of gene expression in lily microsporocytes. *Plant Sci.* 89:31–41.

Tagliasacchi, A.M., Forino, L.M.C., Cionini, P.G., Cavallini, A., Durante, M., Cremonini, R., and Avanzi, S. 1984. Different structure of polytene chromosome of *Phaseolus coccineus* suspensors during early embryogenesis. 3. Chromosome pair VI. *Protoplasma* 122:98–107.

Takahashi, M. 1987. Development of omniaperturate pollen in *Trillium kamtschaticum* (Liliaceae). *Am. J. Bot.* 74:1842–1852.

Takahashi, M. 1989a. Development of the echinate pollen wall in *Farfugium japonicum* (Compositae: Senecioneae). *Bot. Mag. Tokyo* 102:219–234.

Takahashi, M. 1989b. Pattern determination of the exine in *Caesalpinia japonica* (Leguminosae: Caesalpinioideae). *Am. J. Bot.* 76:1615–1626.

Takahashi, M. 1993. Exine initiation and substructure in pollen of *Caesalpinia japonica* (Leguminosae: Caesalpinioideae). *Am. J. Bot.* 80:192–197.

Takahashi, M. 1994. Pollen development in a submerged plant, *Ottelia alismoides* (L.) Pers. (Hydrocharitaceae). *J. Plant Res.* 107:161–164.

Takahashi, M. 1995. Three-dimensional aspects of exine initiation and development in *Lilium longiflorum* (Liliaceae). *Am. J. Bot.* 82:847–854.

Takahashi, M., and Kouchi, J. 1988. Ontogenetic development of spinous exine in *Hibiscus syriacus* (Malvaceae). *Am. J. Bot.* 75:1549–1558.

Takahashi, M., and Skvarla, J.J. 1991. Exine pattern formation by plasma membrane in *Bougainvillea spectabilis* Willd. (Nyctaginaceae). *Am. J. Bot.* 78:1063–1069.

Takahashi, T., Naito, S., and Komeda, Y. 1992. The *Arabidopsis HSP18.2* promoter/*GUS* gene fusion in transgenic *Arabidopsis* plants: a powerful tool for the isolation of regulatory mutants of the heat-shock response. *Plant J.* 2:751–761.

Takaiwa, F., Kikuchi, S., and Oono, K. 1986. The structure of rice storage protein glutelin precursor

deduced from cDNA. *FEBS Lett*. 206:33–35.
Takaiwa, F., Kikuchi, S., and Oono, K. 1987. A rice glutelin gene family - a major type of glutelin mRNAs can be divided into two classes. *Mol. Gen. Genet*. 208:15–22.
Takaiwa, F., Kikuchi, S., and Oono, K. 1989. The complete nucleotide sequence of new type cDNA coding for rice storage protein glutelin. *Nucl. Acids Res*. 17:3289.
Takaiwa, F., and Oono, K. 1991. Genomic DNA sequences of two new genes for new storage protein glutelin in rice. *Jap. J. Genet*. 66:161–171.
Takaiwa, F., Oono, K., and Kato, A. 1991. Analysis of the 5' flanking region responsible for the endosperm-specific expression of a rice glutelin chimeric gene in transgenic tobacco. *Plant Mol. Biol*. 16:49–58.
Takaiwa, F., Ebinuma, H., Kikuchi, S., and Oono, K. 1987. Nucleotide sequence of a rice glutelin gene. *FEBS Lett*. 221:43–47.
Takaiwa, F., Oono, K., Wing, D., and Kato, A. 1991. Sequence of three members and expression of a new major subfamily of glutelin genes from rice. *Plant Mol. Biol*. 17:875–885.
Takats, S.T. 1959. Chromatin extrusion and DNA transfer during microsporogenesis. *Chromosoma* 10:430–453.
Takats, S.T. 1962. An attempt to detect utilization of DNA breakdown products from the tapetum for DNA synthesis in the microspores of *Lilium longiflorum*. *Am. J. Bot*. 49:748–758.
Takats, S.T. 1965. Non-random nuclear orientation during DNA synthesis in *Tradescantia* pollen grains. *J. Cell Biol*. 25:151–157.
Takats, S.T., and Wever, G.H. 1971. DNA polymerase and DNA nuclease activities in S-competent and S-incompetent nuclei from *Tradescantia* pollen grains. *Expt. Cell Res*. 69:25–28.
Takayama, S., Isogai, A., Tsukamoto, C., Ueda, Y., Hinata, K., Okazaki, K., and Suzuki, A. 1987. Sequences of *S*-glycoproteins, products of the *Brassica campestris* self-incompatibility locus. *Nature* 326:102–105.
Takegami, M.H., and Ito, M. 1975. Studies on the behavior of meiotic protoplasts. III. Features of nuclear and cell division for lily protoplasts in the cell-wall-free state. *Bot. Mag. Tokyo* 88:187–196.
Takegami, M.H., Yoshioka, M., Tanaka, I., and Ito, M. 1981. Characteristics of isolated microsporocytes from liliaceous plants for studies of the meiotic cell cycle *in vitro*. *Plant Cell Physiol*. 22:1–10.
Takei, Y., Yamauchi, D., and Minamikawa, T. 1989. Nucleotide sequence of the canavalin gene from *Canavalia gladiata* seeds. *Nucl. Acids Res*. 17:4381.
Takeno, K., Koshioka, M., Pharis, R.P., Rajasekaran, K., and Mullins, M.G. 1983. Endogenous gibberellin-like substances in somatic embryos of grape (*Vitis vinifera* x *Vitis rupestris*) in relation to embryogenesis and the chilling requirement for subsequent development of mature embryos. *Plant Physiol*. 73:803–808.
Talbot, D.R., Adang, M.J., Slightom, J.L., and Hall, T.C. 1984. Size and organization of a multigene family encoding phaseolin, the major seed storage protein of *Phaseolus vulgaris* L. *Mol. Gen. Genet*. 198:42–49.
Tamura, Y., Kitano, H., Satoh, H., and Nagato, Y. 1992. A gene profoundly affecting shoot organization in the early phase of rice development. *Plant Sci*. 82:91–99.
Tan, L.W., and Jackson, J.F. 1988. Stigma proteins of the two loci self-incompatible grass *Phalaris coerulescens*. *Sex. Plant Reprod*. 1:25–27.
Tanaka, I. 1988. Isolation of generative cells and their protoplasts from pollen of *Lilium longiflorum*. *Protoplasma* 142:68–73.
Tanaka, I. 1991. Microtubule-determined plastid distribution during microsporogenesis in *Lilium longiflorum*. *J. Cell Sci*. 99:21–31.
Tanaka, I. 1993. Development of male gametes in flowering plants. *J. Plant Res*. 106:55–63.
Tanaka, I., and Ito, M. 1980. Induction of typical cell division in isolated microspores of *Lilium longiflorum* and *Tulipa gesneriana*. *Plant Sci. Lett*. 17:279–285.
Tanaka, I., and Ito, M. 1981a. Studies on microspore development in liliaceous plants. III. Pollen tube development in lily pollens cultured from the uninucleate microspore stage. *Plant Cell Physiol*. 22:149–153.
Tanaka, I., and Ito, M. 1981b. Control of division patterns in explanted microspores of *Tulipa gesneriana*. *Protoplasma* 108:329–340.
Tanaka, I., Kitazume, C., and Ito, M. 1987. The isolation and culture of lily pollen protoplasts. *Plant Sci*. 50:205--211.
Tanaka, I., Nakamura, S., and Miki-Hirosige, H. 1989. Structural features of isolated generative cells and their protoplasts from pollen of some liliaceous plants. *Gamete Res*. 24:361–374.
Tanaka, I., Taguchi, T., and Ito, M. 1979. Studies on microspore development in liliaceous plants. I. The duration of the cell cycle and developmental aspects in lily microspores. *Bot. Mag. Tokyo* 92:291–298.
Tanaka, I., Taguchi, T., and Ito, M. 1980. Studies on microspore development in liliaceous plants. II. The behavior of explanted microspores of the lily, *Lilium longiflorum*. *Plant Cell Physiol*. 21:667–676.
Tanaka, I., and Wakabayashi, T. 1992. Organization of the actin and microtubule cytoskeleton preceding pollen germination. An analysis using cultured pollen protoplasts of *Lilium longiflorum*. *Planta* 186:473–482.
Tanaka, T., Nishihara, M., Seki, M., Sakamoto, A., Tanaka, K., Irifune, K., and Morikawa, H. 1995. Successful expression in pollen of various plant species of *in vitro* synthesized mRNA introduced by particle bombardment. *Plant Mol. Biol*. 28:337–341.
Tang, W., Diaz, P., Wilson, K.A., and Tan-Wilson, A.L. 1993. Spatial distribution of protein accumulation in soybean seed. *Phytochemistry* 33:1289–1295.
Tang, X., Hepler, P.K., and Scordilis, S.P. 1989. Immunochemical and immunocytochemical identification of a myosin heavy chain polypeptide in *Nicotiana* pollen tubes. *J. Cell Sci*. 92:569–574.
Tang, X., Lancelle, S.A., and Hepler, P.K. 1989. Fluorescence microscopic localization of actin in pollen tubes: comparison of actin antibody and phalloidin staining. *Cell Mot. Cytoskel*. 12:216–224.
Tanksley, S.D., Zamir, D., and Rick, C.M. 1981. Evidence for extensive overlap of sporophytic and gametophytic gene expression in *Lycopersicon esculentum*. *Science* 213:453–455.
Tano, S., and Takahashi, H. 1964. Nucleic acid synthesis in growing pollen tubes. *J. Biochem*. 56:578–580.
Tantikanjana, T., Nasrallah, M.E., and Nasrallah, J.B. 1996. The *Brassica* S gene family: molecular characterization of the *SLR2* gene. *Sex. Plant Reprod*. 9:107–116.
Tantikanjana, T., Nasrallah, M.E., Stein, J.C., Chen, C.-H., and Nasrallah, J.B. 1993. An alternative transcript of the *S* locus glycoprotein gene in a class II pollen-recessive self-incompatibility haplotype of *Brassica*

oleracea encodes a membrane-anchored protein. *Plant Cell* 5:657–666.

Tara, C.P., and Namboodiri, A.N. 1974. Aberrant microsporogenesis and sterility in *Impatiens sultani* (Balsaminaceae). *Am. J. Bot.* 61:585–591.

Tarasenko, L.V., and Bannikova, V.P. 1991. Ultrastructure of pollen grain development in some angiosperm species. *Phytomorphology* 41:11–20.

Tautorus, T.E., Wang, H., Fowke, L.C., and Dunstan, D.I. 1992. Microtubule pattern and the occurrence of preprophase bands in embryogenic cultures of black spruce (*Picea mariana* Mill.) and non-embryogenic cultures of jack pine (*Pinus banksiana* Lamb.). *Plant Cell Rep.* 11:419–423.

Taylor, D.C., Barton, D.L., Rioux, K.P., MacKenzie, S.L., Reed, D.W., Underhill, E.W., Pomeroy, M.K., and Weber, N. 1992. Biosynthesis of acyl lipids containing very-long chain fatty acids in microspore-derived and zygotic embryos of *Brassica napus* L. cv. Reston. *Plant Physiol.* 99:1609–1618.

Taylor, D.C., Ferrie, A.M.R., Keller, W.A., Giblin, E.M., Pass, E.W., and MacKenzie, S.L. 1993. Bioassembly of acyl lipids in microspore-derived embryos of *Brassica campestris* L. *Plant Cell Rep.* 12:375–384.

Taylor, D.C., Weber, N., Barton, D.L., Underhill, E.W., Hogge, L.R., Weselake, R.J., and Pomeroy, M.K. 1991. Triglycerol bioassembly in microspore-derived embryos of *Brassica napus* L. cv. Reston. *Plant Physiol.* 97:65–79.

Taylor, D.C., Weber, N., Underhill, E.W., Pomeroy, M.K., Keller, W.A., Scowcroft, W.R., Wilen, R.W., Moloney, M.M., and Holbrook, L.A. 1990. Storage-protein regulation and lipid accumulation in microspore embryos of *Brassica napus* L. *Planta* 181:18–26.

Taylor, J.H. 1950. The duration of differentiation in excised anthers. *Am. J. Bot.* 37:137–143.

Taylor, J.H. 1953. Autoradiographic detection of incorporation of P^{32} into chromosomes during meiosis and mitosis. *Expt. Cell Res.* 4:164–173.

Taylor, J.H. 1958. Incorporation of phosphorus-32 into nucleic acids and proteins during microgametogenesis in *Tulbaghia. Am. J. Bot.* 45:123–131.

Taylor, J.H. 1959. Autoradiographic studies of nucleic acids and proteins during meiosis in *Lilium longiflorum. Am. J. Bot.* 46:477–484.

Taylor, J.H., and McMaster, R.D. 1954. Autoradiographic and microphotometric studies of desoxyribose nucleic acid during microgametogenesis in *Lilium longiflorum. Chromosoma* 6:489–521.

Taylor, J.R.N., Schüssler, L., and Liebenberg, N.v.d.W. 1985. Protein body formation in the starchy endosperm of developing *Sorghum bicolor* (L.) Moench seeds. *S. Afr. J. Bot.* 51:35–40.

Taylor, L.P., and Jorgensen, R. 1992. Conditional male fertility in chalcone synthase-deficient *Petunia. J. Hered.* 83:11–17.

Taylor, M.G., and Vasil, I.K. 1991. Histology of, and physical factors affecting, transient GUS expression in pearl millet (*Pennisetum glaucum* (L.) R. Br.) embryos following microprojectile bombardment. *Plant Cell Rep.* 10:120–125.

Taylor, M.G., Vasil, V., and Vasil, I.K. 1993. Enhanced GUS gene expression in cereal/grass cell suspensions and immature embryos using the maize ubiquitin-based plasmid pAHC25. *Plant Cell Rep.* 12:491–495.

Taylor, P., Kenrick, J., Li, Y., Kaul, V., Gunning, B.E.S., and Knox, R.B. 1989. The male germ unit of *Rhododendron*: quantitative cytology, three-dimensional reconstruction, isolation and detection using fluorescent probes. *Sex. Plant Reprod.* 2:254–264.

Taylor, P.E., Kenrick, J., Blomstedt, C.K., and Knox, R.B. 1991. Sperm cells of the pollen tubes of *Brassica*: ultrastructure and isolation. *Sex. Plant Reprod.* 4:226–234.

Tebbutt, S.J., and Lonsdale, D.M. 1995. Deletion analysis of a tobacco pollen-specific polygalacturonase. *Sex. Plant Reprod.* 8:242–246.

Tebbutt, S.J., Rogers, H.J., and Lonsdale, D.M. 1994. Characterization of a tobacco gene encoding a pollen-specific polygalacturonase. *Plant Mol. Biol.* 25:283–297.

Telmer, C.A., Newcomb, W., and Simmonds, D.H. 1993. Microspore development in *Brassica napus* and the effect of high temperature on division *in vivo* and *in vitro*. *Protoplasma* 172:154–165.

Telmer, C.A., Newcomb, W., and Simmonds, D.H. 1995. Cellular changes during heat shock induction and embryo development of cultured microspores of *Brassica napus* cv. Topas. *Protoplasma* 185:106–112.

Temple, M., Makaroff, C.A., Mutschler, M.A., and Earle, E.D. 1992. Novel mitochondrial genomes in *Brassica napus* somatic hybrids. *Curr. Genet.* 22:243–249.

Templeman, T.S., DeMaggio, A.E., and Stetler, D.A. 1987. Biochemistry of fern spore germination: globulin storage proteins in *Matteuccia struthiopteris* L. *Plant Physiol.* 85:343–349.

Templeman, T.S., Stein, D.B., and DeMaggio, A.E. 1988. A fern spore storage protein is genetically similar to the 1.7 S seed storage protein of *Brassica napus. Biochem. Genet.* 26:595–603.

Tepfer, S.S. 1953. Floral anatomy and ontogeny in *Aquilegia formosa* var. *truncata* and *Ranunculus repens. Univ. Calif. Publ. Bot.* 25:513–647.

Tepfer, S.S., Greyson, R.I., Craig, W.R., and Hindman, J.L. 1963. *In vitro* culture of floral buds of *Aquilegia. Am. J. Bot.* 50:1035–1045.

Terada, R., Nakayama, T., Iwabuchi, M., and Shimamoto, K. 1993. A wheat histone H3 promoter confers cell division-dependent and -independent expression of the *gus* A gene in transgenic rice plants. *Plant J.* 3:241–252.

Terada, R., Nakayama, T., Iwabuchi, M., and Shimamoto, K. 1995. A type I element composed of the hexamer (ACGTCA) and octamer (CGCGGATC) motifs plays a role(s) in meristematic expression of a wheat histone H3 gene in transgenic rice plants. *Plant Mol. Biol.* 27:17–26.

Terada, S. 1928. Embryological studies in *Oryza sativa* L. *J. Coll. Agr. Hokkaido Imp. Univ.* 19:245–260.

Terasaka, O., and Niitsu, T. 1989. Peculiar spindle configuration in the pollen tube revealed by the anti-tubulin immunofluorescence method. *Bot. Mag. Tokyo* 102:143–147.

Terasaka, O., and Niitsu, T. 1990. Unequal cell division and chromatin differentiation in pollen grain cells. II. Microtubule dynamics associated with the unequal cell division. *Bot. Mag. Tokyo* 103:133–142.

Terasaka, O., and Niitsu, T. 1994. Differential roles of microtubule and actin-myosin cytoskeleton in the growth of *Pinus* pollen tubes. *Sex. Plant Reprod.* 7:264–272.

Terasaka, O., and Niitsu, T. 1995. The mitotic apparatus during unequal microspore division observed by a confocal laser scanning microscope. *Protoplasma* 189:187–193.

Terasawa, M., Shinohara, A., Hotta, Y., Ogawa, H., and Ogawa, T. 1995. Localization of RecA-like recombination proteins on chromosomes of the lily at various meiotic stages. *Genes Devel.* 9:925–934.

Tercé-Laforgue, T., and Pernollet, J.-C. 1982. Étude quantitative et qualitative de l'accumulation des gliadines au cours du développement du caryopse de blé (*Triticum aestivum* L.). *Compt. Rend. Acad. Sci. Paris* Ser. III 294:529–534.

Testillano, P.S., González-Melendi, P., Ahmadian, P., Fadón, B., and Risueño, M.C. 1995. The immunolocalization of nuclear antigens during the pollen developmental program and the induction of pollen embryogenesis. *Expt. Cell Res.* 221:41–54.

Thangavelu, M., Belostotsky, D., Bevan, M.W., Flavell, R.B., Rogers, H.J., and Lonsdale, D.M. 1993. Partial characterization of the *Nicotiana tabacum* actin gene family: evidence for pollen-specific expression of one of the gene family members. *Mol. Gen. Genet.* 240:290–295.

Thangstad, O.P., Iversen, T.-H., Slupphaug, G., and Bones, A. 1990. Immunocytochemical localization of myrosinase in *Brassica napus* L. *Planta* 180:245–248.

Thanh, V.H., and Shibasaki, K. 1977. Beta-conglycinin from soybean proteins. Isolation and immunological and physicochemical properties of the monomeric forms. *Biochim. Biophys. Acta* 490:370–384.

Theerakulpisut, P., Xu, H., Singh, M.B., Pettitt, J.M., and Knox, R.B. 1991. Isolation and developmental expression of Bcp1, an anther-specific cDNA clone in *Brassica campestris*. *Plant Cell* 3:1073–1084.

Theunis, C.H. 1990. Ultrastructural analysis of *Spinacia oleracea* sperm cells isolated from mature pollen grains. *Protoplasma* 158:176–181.

Theunis, C.H., Cresti, M., and Milanesi, C. 1991. Studies of the mature pollen of *Spinacia oleracea* after freeze substitution and observed with confocal laser scanning fluorescence microscopy. *Bot. Acta* 104:324–329.

Theunis, C.H., McConchie, C.A., and Knox, R.B. 1985. Three-dimensional reconstruction of the generative cell and its wall connection in mature bicellular pollen of *Rhododendron*. *Micron Micros. Acta* 16:225–231.

Theunis, C.H., Pierson, E.S., and Cresti, M. 1991. Isolation of male and female gametes in higher plants. *Sex. Plant Reprod.* 4:145–154.

Theunis, C.H., Pierson, E.S., and Cresti, M. 1992. The microtubule cytoskeleton and the rounding of isolated generative cells of *Nicotiana tabacum*. *Sex. Plant Reprod.* 5:64–71.

Theunis, C.H., and van Went, J.L. 1989. Isolation of sperm cells from mature pollen grains of *Spinacia oleracea*. *Sex. Plant Reprod.* 2:97–102.

Thiébaud, C.H., and Ruch, F. 1978. Cytophotometric study of nuclear differentiation during pollen development in *Tradescantia paludosa*. *Histochemistry* 57:119–128.

This, P., Goffner, D., Raynal, M., Chartier, Y., and Delseny, M. 1988. Characterization of major storage proteins of sunflower and their accumulation. *Plant Physiol. Biochem.* 26:125–132.

Thoma, S., Hecht, U., Kippers, A., Botella, J., de Vries, S., and Somerville, C. 1994. Tissue-specific expression of a gene encoding a cell wall-localized lipid transfer protein from *Arabidopsis*. *Plant Physiol.* 105:35–45.

Thomann, E.B., Sollinger, J., White, C., and Rivin, C.J. 1992. Accumulation of group 3 late embryogenesis abundant proteins in *Zea mays* embryos. Role of abscisic acid and the *viviparous-1* gene product. *Plant Physiol.* 99:607–614.

Thomas, E., Konar, R.N., and Street, H.E. 1972. The fine structure of the embryogenic callus of *Ranunculus sceleratus* L. *J. Cell Sci.* 11:95–109.

Thomas, M.K., and Dnyansagar, V.R. 1975. Carbohydrate metabolism in pollen of *Petunia nyctaginiflora* Juss. during germination and tube growth. *Indian J. Expt. Biol.* 13:268–271.

Thomas, M.S., and Flavell, R.B. 1990. Identification of an enhancer element for the endosperm-specific expression of high molecular weight glutenin. *Plant Cell* 2:1171–1180.

Thompson, A.J., Bown, D., Yaish, S., and Gatehouse, J.A. 1991. Differential expression of seed storage protein genes in the pea *legJ* subfamily; sequence of gene *legK*. *Biochem. Physiol. Pflanz.* 187:1–12.

Thompson, A.J., Evans, I.M., Boulter, D., Croy, R.R.D., and Gatehouse, J.A. 1989. Transcriptional and post-transcriptional regulation of seed storage-protein gene expression in pea (*Pisum sativum* L.). *Planta* 179:279–287.

Thompson, G.A., Boston, R.S., Lyznik, L.A., Hodges, T.K., and Larkins, B.A. 1990. Analysis of promoter activity from an α-zein gene 5' flanking sequence in transient expression assays. *Plant Mol. Biol.* 15:755–764.

Thompson, G.A., Scherer, D.E., Foxall-van Aken, S., Kenny, J.W., Young, H.L., Shintani, D.K., Kridl, J.C., and Knauf, V.C. 1991. Primary structures of the precursor and mature forms of stearoyl-acyl carrier protein desaturase from safflower embryos and requirement of ferredoxin for enzyme activity. *Proc. Natl Acad. Sci. USA* 88:2578–2582.

Thompson, K.F., and Taylor, J.P. 1971. Self-compatibility in kale. *Heredity* 27:459–471.

Thompson, R.D., and Bartels, D. 1983. Hordein messenger RNA levels in wild type and mutant barley endosperm. *Plant Sci. Lett.* 29:295–304.

Thompson, R.D., Bartels, D., and Harberd, N.P. 1985. Nucleotide sequence of a gene from chromosome 1D of wheat encoding HMW-glutenin subunit. *Nucl. Acids Res.* 13:6833–6846.

Thompson, R.D., Bartels, D., Harberd, N.P., and Flavell, R.B. 1983. Characterization of the multigene family coding for HMW glutenin subunits in wheat using cDNA clones. *Theor. Appl. Genet.* 67:87–96.

Thompson, R.D., Kemble, R.J., and Flavell, R.B. 1980. Variations in mitochondrial DNA organisation between normal and male-sterile cytoplasms of maize. *Nucl. Acids Res.* 8:1999–2008.

Thompson, R.D., and Kirch, H.-H. 1992. The *S* locus of flowering plants: when self-rejection is self-interest. *Trend. Genet.* 8:381–387.

Thompson-Coffe, C., Findlay, T.A.S., Wagner, B.A. and Orr, A.R. 1992. Changes in protein complement accompany early organogenesis in the maize ear. *Int. J. Plant Sci.* 153:31–39.

Thomson, J.A., and Schroeder, H.E. 1978. Cotyledonary storage proteins in *Pisum sativum*. II. Hereditary variation in components of the legumin and vicilin fractions. *Aust. J. Plant Physiol.* 5:281–294.

Thorsness, M.K., Kandasamy, M.K., Nasrallah, M.E., and Nasrallah, J.B. 1991. A *Brassica* S-locus gene promoter targets toxic gene expression and cell death to the pistil and pollen of transgenic *Nicotiana*. *Devel Biol.* 143:173–184.

Thorsness, M.K., Kandasamy, M.K., Nasrallah, M.E., and Nasrallah, J.B. 1993. Genetic ablation of floral cells in *Arabidopsis*. *Plant Cell* 5:253–261.

Thoyts, P.J.E., Millichip, M.I., Stobart, A.K., Griffiths, W.T., Shewry, P.R., and Napier, J.A. 1995. Expression and *in vitro* targeting of a sunflower oleosin. *Plant Mol. Biol.* 29:403–410.

Tian, H.-Q., and Yang, H.-Y. 1991. Ultrastructural observations on parthenogenesis and antipodal apogamy of *Allium tuberosum* Roxb. *Acta Bot. Sinica* 33:819–824.

Tierney, M.L., Bray, E.A., Allen, R.D., Ma, Y., Drong, R.F., Slightom, J., and Beachy, R.N. 1987. Isolation and characterization of a genomic clone encoding the β-subunit of β-conglycinin. *Planta* 172:356–363.

Tiezzi, A., Moscatelli, A., Cai, G., Bartalesi, A., and Cresti, M. 1992. An immunoreactive homolog of mammalian kinesin in *Nicotiana tabacum* pollen tubes. *Cell Mot. Cytoskel.* 21:132–137.

Tiezzi, A., Moscatelli, A., Milanesi, C., Ciampolini, C., and Cresti, M. 1987. Taxol-induced structures derived from cytoskeletal elements of the *Nicotiana* pollen tube. *J. Cell Sci.* 88:657–661.

Tilton, V.R. 1980. The nucellar epidermis and micropyle of *Ornithogalum caudatum* (Liliaceae) with a review of these structures in other taxa. *Can. J. Bot.* 58:1872–1884.

Tilton, V.R. 1981a. Ovule development in *Ornithogalum caudatum* (Liliaceae) with a review of selected papers on angiosperm reproduction. II. Megasporogenesis. *New Phytol.* 88:459–476.

Tilton, V.R. 1981b. Ovule development in *Ornithogalum caudatum* (Liliaceae) with a review of selected papers on angiosperm reproduction. IV. Egg apparatus structure and function. *New Phytol.* 88:505–531.

Tilton, V.R. 1984. Stigma, style, and pollen recognition in *Galinsoga quadriradiata* Ruiz & Pav. (Compositae). *Caryologia* 37:423–433.

Tilton, V.R., and Horner, H.T., Jr. 1980. Stigma, style, and obturator of *Ornithogalum caudatum* (Liliaceae) and their function in the reproductive process. *Am. J. Bot.* 67:1113–1131.

Tilton, V.R., and Lersten, N.R. 1981. Ovule development in *Ornithogalum caudatum* (Liliaceae) with a review of selected papers on angiosperm reproduction. III. Nucellus and megagametophyte. *New Phytol.* 88:477–504.

Tilton, V.R., and Mogensen, H.L. 1979. Ultrastructural aspects of the ovule of *Agave parryi* before fertilization. *Phytomorphology* 29:338–350.

Tilton, V.R., and Russell, S.H. 1984. Applications of *in vitro* pollination/fertilization technology. *BioScience* 34:239–242.

Tilton, V.R., Wilcox, L.W., and Palmer, R.G. 1984. Postfertilization wandlabrinthe formation and function in the central cell of soybean, *Glycine max* (L.) Merr. (Leguminosae). *Bot. Gaz.* 145:334–339.

Tilton, V.R., Wilcox, L.W., Palmer, R.G., and Albertsen, M.C. 1984. Stigma, style, and obturator of soybean, *Glycine max* (L.) Merr. (Leguminosae) and their function in the reproductive process. *Am. J. Bot.* 71:676–686.

Timmers, A.C.J., de Vries, S.C., and Schel, J.H.N. 1989. Distribution of membrane-bound calcium and activated calmodulin during somatic embryogenesis of carrot (*Daucus carota* L.). *Protoplasma* 153:24–29.

Timmers, A.C.J., Kieft, H., and Schel, J.H.N. 1995. An immunofluorescence study on calmodulin distribution during somatic and zygotic embryogenesis of carrot (*Daucus carota* L.). *Acta Bot. Neerl.* 44:19–32.

Timmers, A.C.J., Reiss, H.-D., Bohsung, J., Traxel, K., and Schel, J.H.N. 1996. Localization of calcium during somatic embryogenesis of carrot (*Daucus carota* L.). *Protoplasma* 190:107–118.

Tirlapur, U.K., and Cresti, M. 1992. Computer-assisted video image analysis of spatial variations in membrane-associated Ca^{2+} and calmodulin during pollen hydration, germination and tip growth in *Nicotiana tabacum* L. *Ann. Bot.* 69:503–508.

Tirlapur, U.K., Kranz, E., and Cresti, M. 1995. Characterisation of isolated egg cells, *in vitro* fusion products and zygotes of *Zea mays* L. using the technique of image analysis and confocal laser scanning microscopy. *Zygote* 3:57–64.

Tirlapur, U.K., and Shiggaon, S.V. 1988. Distribution of Ca^{2+} and calmodulin in the papillate cells of the stigma surface, visualized by chlorotetracycline and fluorescing calmodulin-binding phenothiazines. *Ann. Biol. India* 4:49–53.

Tirlapur, U.K., van Went, J.L., and Cresti, M. 1993. Visualization of membrane calcium and calmodulin in embryo sacs *in situ* and isolated from *Petunia hybrida* L. and *Nicotiana tabacum* L. *Ann. Bot.* 71:161–167.

Tirlapur, U.K., and Willemse, M.T.M. 1992. Changes in calcium and calmodulin levels during microsporogenesis, pollen development and germination in *Gasteria verrucosa* (Mill.) H. Duval. *Sex. Plant Reprod.* 5:214–223.

Tirlapur, U.K., Cai, G., Faleri, C., Moscatelli, A., Scali, M., Del Casino, C., Tiezzi, A., and Cresti, M. 1995. Confocal imaging and immunogold electron microscopy of changes in distribution of myosin during pollen hydration, germination and pollen tube growth in *Nicotiana tabacum* L. *Eur. J. Cell Biol.* 67:209–217.

Tirlapur, U.K., Scali, M., Moscatelli, A., Del Casino, C., Cai, G., Tiezzi, A., and Cresti, M. 1994. Confocal image analysis of spatial variations in immunocytochemically identified calmodulin during pollen hydration, germination and pollen tube tip growth in *Nicotiana tabacum* L. *Zygote* 2:63–68.

Titz, W. 1965. Vergleichende Untersuchungen über den Grad der somatischen Polyploidie an nahe verwandten diploiden und polyploiden Sippen einschließlich der Cytologie von Antipoden. *Österr. Bot. Z.* 112:101–172.

Tiwari, S.C. 1982. Callose in the walls of mature embryo sac of *Torenia fournieri*. *Protoplasma* 110:1–4.

Tiwari, S.C. 1983. The hypostase in *Torenia fournieri* Lind.: a histochemical study of the cell walls. *Ann. Bot.* 51:17–26.

Tiwari, S.C. 1989. Cytoskeleton during pollen development in *Tradescantia virginiana*: a study employing chemical fixation, freeze-substitution, immunofluorescence, and colchicine administration. *Can. J. Bot.* 67:1244–1253.

Tiwari, S.C. 1994. An intermediate-voltage electron microscopic study of freeze-substituted generative cell in pear (*Pyrus communis* L.): features with relevance to cell–cell communication between the two cells of a germinating pollen. *Sex. Plant Reprod.* 7:177–186.

Tiwari, S.C., and Gunning, B.E.S. 1986a. Cytoskeleton, cell surface and the development of invasive plasmodial tapetum in *Tradescantia virginiana* L. *Protoplasma* 133:89–99.

Tiwari, S.C., and Gunning, B.E.S. 1986b. An ultrastructural, cytochemical and immunofluorescence study of postmeiotic development of plasmodial tapetum in *Tradescantia virginiana* L. and its relevance to the pathway of sporopollenin secretion. *Protoplasma* 133:100–114.

Tiwari, S.C., and Gunning, B.E.S. 1986c. Colchicine inhibits plasmodium formation and disrupts pathways of sporopollenin secretion in the anther tapetum of *Tradescantia virginiana* L. *Protoplasma* 133:115–128.

Tiwari, S.C., and Gunning, B.E.S. 1986d. Development of tapetum and microspores in *Canna* L.: an example of an invasive but non-syncytial tapetum. *Ann. Bot.* 57: 557–563.

Tiwari, S.C., and Polito, V.S. 1988a. Spatial and temporal organization of actin during hydration, activation, and germination of pollen in *Pyrus communis* L.: a population study. *Protoplasma* 147:5–15.

Tiwari, S.C., and Polito, V.S. 1988b. Organization of the cytoskeleton in pollen tubes of *Pyrus communis*: a study employing conventional and freeze-substitution electron microscopy, immunofluorescence, and rhodamine-phalloidin. *Protoplasma* 147:100–112.

Tiwari, S.C., and Polito, V.S. 1990a. An analysis of the role of actin during pollen activation leading to germination in pear (*Pyrus communis* L.): treatment with cytochalasin D. *Sex. Plant Reprod.* 3:121–129.

Tiwari, S.C., and Polito, V.S. 1990b. The initiation and organization of microtubules in germinating pear (*Pyrus communis* L.) pollen. *Eur. J. Cell Biol.* 53:384–389.

Tiwari, S.C., Polito, V.S., and Webster, B.D. 1990. In dry pear (*Pyrus communis* L.) pollen, membranes assume a tightly packed multilamellate aspect that disappears rapidly upon hydration. *Protoplasma* 153:157–168.

Tobias, C.M., Howlett, B., and Nasrallah, J.B. 1992. An *Arabidopsis thaliana* gene with sequence similarity to the S-locus receptor kinase of *Brassica oleracea*. *Plant Physiol.* 99:284–290.

Tohda, H. 1974. Development of the embryo of *Pogonia* (Orchidaceae). *Sci. Rep. Tôhoku Univ.* Ser. IV (Biol.) 37:89–93.

Tong, Z., Wang, T., and Xu, Y. 1990. Evidence for involvement of phytochrome regulation in male sterility of a mutant of *Oryza sativa* L. *Photochem. Photobiol.* 52:161–164.

Toonen, M.A.J., Hendriks, T., Schmidt, E.D.L., Verhoeven, H.A., van Kammen, A., and de Vries, S.C. 1994. Description of somatic-embryo-forming single cells in carrot suspension cultures employing video cell tracking. *Planta* 194:565–572.

Töpfer, R., Gronenborn, B., Schell, J., and Steinbiss, H.-H. 1989. Uptake and transient expression of chimeric genes in seed-derived embryos. *Plant Cell* 1:133–139.

Töpfer, R., and Steinbiss, H.-H. 1985. Plant regeneration from cultured fertilized barley ovules. *Plant Sci.* 41:49–54.

Toriyama, K., Okada, T., Watanabe, M., Ide, T., Ashida, T., Xu, H., and Singh, M.B. 1995. A cDNA clone encoding an IgE-binding protein from *Brassica* anther has significant sequence similarity to Ca^{2+}-binding proteins. *Plant Mol. Biol.* 29:1157–1165.

Toriyama, K., Stein, J.C., Nasrallah, M.E., and Nasrallah, J.B. 1991a. Transformation of *Brassica oleracea* with an S-locus gene from *B. campestris* changes the self-incompatibility phenotype. *Theor. Appl. Genet.* 81:769–776.

Toriyama, K., Thorsness, M.K., Nasrallah, J.B., and Nasrallah, M.E. 1991b. A *Brassica* S locus gene promoter directs sporophytic expression in the anther tapetum of transgenic *Arabidopsis*. *Devel Biol.* 143:427–431.

Torrent, M., Poca, E., Campos, E., Ludevid, M.D., and Palau, J. 1986. In maize, glutelin-2 and low molecular weight zeins are synthesized by membrane-bound polyribosomes and translocated into microsomal membranes. *Plant Mol. Biol.* 7:393–403.

Torres, M.-A., Rigau, J., Puigdomènech, P., and Stiefel, V. 1995. Specific distribution of mRNAs in maize growing pollen tubes observed by whole-mount *in situ* hybridization with non-radioactive probes. *Plant J.* 8:317–321.

Torres-Ruiz, R.A., and Jürgens, G. 1994. Mutations in the *FASS* gene uncouple pattern formation and morphogenesis in *Arabidopsis* development. *Development* 120:2967–2978.

Touraev, A., Fink, C.S., Stöger, E., and Heberle-Bors, E. 1995a. Pollen selection: a transgenic reconstruction approach. *Proc. Natl Acad. Sci. USA* 92:12165–12169.

Touraev, A., Lezin, F., Heberle-Bors, E., and Vicente, O. 1995b. Maintenance of gametophytic development after symmetrical division in tobacco microspore culture. *Sex. Plant Reprod.* 8:70–76.

Traas, J., Bellini, C., Nacry, P., Kronenberger, J., Bouchez, D., and Caboche, M. 1995. Normal differentiation patterns in plants lacking microtubular preprophase bands. *Nature* 375:676–677.

Traas, J.A., Burgain, S., and de Vaulx, R.D. 1989. The organization of the cytoskeleton during meiosis in eggplant [*Solanum melongena* (L.)]: microtubules and F-actin are both necessary for coordinated meiotic division. *J. Cell Sci.* 92:541–550.

Trick, M. 1990. Genomic sequence of a *Brassica S* locus-related gene. *Plant Mol. Biol.* 15:203–205.

Trick, M., and Flavell, R.B. 1989. A homozygous *S* genotype of *Brassica oleracea* expresses two *S*-like genes. *Mol. Gen. Genet.* 218:112–117.

Trick, M., and Heizmann, P. 1992. Sporophytic self-incompatibility systems: *Brassica S* gene family. *Int. Rev. Cytol.* 140:485–524.

Triplett, B.A., and Quatrano, R.S. 1982. Timing, localization, and control of wheat germ agglutinin synthesis in developing wheat embryos. *Devel Biol.* 91:491–496.

Tröbner, W., Ramirez, L., Motte, P., Hue, I., Huijser, P., Lönnig, W.-E., Saedler, H., Sommer, H., and Schwarz-Sommer, Z. 1992. *GLOBOSA*: a homeotic gene which interacts with *DEFICIENS* in the control of *Antirrhinum* floral organogenesis. *EMBO J.* 11:4693–4704.

Trognitz, B.R., and Schmiediche, P.E. 1993. A new look at incompatibility relationships in higher plants. *Sex. Plant Reprod.* 6:183–190.

Trull, M.C., Holaway, B.L., Friedman, W.E., and Malmberg, R.L. 1991. Developmentally regulated antigen associated with calcium crystals in tobacco anthers. *Planta* 186:13–16.

Tsaftaris, A.S., and Scandalios, J.G. 1986. Spatial pattern of

Takaiwa, F., and Ishige, T. 1994. Expression and accumulation of normal and modified soybean glycinins in potato tubers. *Plant Sci.* 102:181–188.

Uwate, W.J., and Lin, J. 1980. Cytological zonation of *Prunus avium* L. pollen tubes *in vivo*. *J. Ultrastr. Res.* 71:173–184.

Uwate, W.J., and Lin, J. 1981a. Tissue development in the stigma of *Prunus avium* L. *Ann. Bot.* 47:41–51.

Uwate, W.J., and Lin, J. 1981b. Development of the stigmatic surface of *Prunus avium* L., sweet cherry. *Am. J. Bot.* 68:1165–1176.

Uwate, W.J., Lin, J., Ryugo, K., and Stallman, V. 1982. Cellular components of the midstylar transmitting tissue of *Prunus avium*. *Can. J. Bot.* 60:98–104.

Valenta, R., Duchêne, M., Pettenburger, K., Sillaber, C., Valent, P., Bettelheim, P., Breitenbach, M., Rumpold, H., Kraft, D., and Scheiner, O. 1991. Identification of profilin as a novel pollen allergen; IgE autoreactivity in sensitized individuals. *Science* 253:557–560.

Vallade, J. 1970. Développement embryonnaire chez un *Petunia hybrida* hort. *Compt. Rend. Acad. Sci. Paris* 270D:1893–1896.

Vallade, J. 1972. Structure et fonctionnement du méristème lors de la formation de la jeune racine primaire chez un *Petunia hybrida* hort. *Compt. Rend. Acad. Sci. Paris* 274D:1027–1030.

Vallade, J. 1980. Données cytologiques sur la proembryogenèse du *Petunia*; intérêt pour une interprétation morphogénétique du développement embryonnaire. *Bull. Soc. Bot. Fr. Actual. Bot.* 127:19–37.

Vallade, J., Cornu, A., Essad, S., and Alabouvette, J. 1978. Niveaux de DNA dans les noyaux zygotiques chez le *Petunia hybrida* hort. *Bull. Soc. Bot. Fr. Actual. Bot.* 125:253–258.

van Aelst, A.C., Theunis, C.H., and van Went, J.L. 1990. Freeze-fracture studies on isolated sperm cells of *Spinacia oleracea* L. *Protoplasma* 153:204–207.

van Aelst, A.C., and van Went, J.L. 1991. The ultrastructure of mature *Papaver dubium* L. pollen grains. *Acta Bot. Neerl.* 40:319–328.

van Aelst, A.C., and van Went, J.L. 1992. Ultrastructural immuno-localization of pectins and glycoproteins in *Arabidopsis thaliana* pollen grains. *Protoplasma* 168:14–19.

van Aelst, A.C., Mueller, T., Dueggelin, M., and Guggenheim, R. 1989. Three-dimensional observations on freeze-fractured frozen hydrated *Papaver dubium* pollen with cryo-scanning electron microscopy. *Acta Bot. Neerl.* 38:25–30.

van Aelst, A.C., Pierson, E.S., van Went, J.L., and Cresti, M. 1993. Ultrastructural changes of *Arabidopsis thaliana* pollen during final maturation and rehydration. *Zygote* 1:173–179.

van Boxtel, J., Berthouly, M., Carasco, C., Dufour, M., and Eskes, A. 1995. Transient expression of β-glucuronidase following biolistic delivery of foreign DNA into coffee tissues. *Plant Cell Rep.* 14:748–752.

van Driessche, E., Smets, G., Dejaegere, R., and Kanarek, L. 1981. The immuno-histochemical localization of lectin in pea seeds (*Pisum sativum* L.). *Planta* 153:287–296.

van Eldik, G.J., Vriezen, W.H., Wingens, M., Ruiter, R.K., van Herpen, M.M.A., Schrauwen, J.A.M., and Wüllems, G.J. 1995. A pistil-specific gene of *Solanum tuberosum* is predominantly expressed in the stylar cortex. *Sex. Plant Reprod.* 8:173–179.

van Eldik, G.J., Wingens, M., Ruiter, R.K., van Herpen, M.M.A., Schrauwen, J.A.M., and Wullems, G.J. 1996. Molecular analysis of a pistil-specific gene expressed in the stigma and stylar cortex of *Solanum tuberosum*. *Plant Mol. Biol.* 30:171–176.

van Engelen, F.A., de Jong, A.J., Meijer, E.A., Kuil, C.W., Meyboom, J.K., Dirkse, W.G., Booij, H., Hartog, M.V., Vandekerckhove, J., de Vries, S.C., and van Kammen, A. 1995. Purification, immunological characterization and cDNA cloning of a 47 kDa glycoprotein secreted by carrot suspension cells. *Plant Mol. Biol.* 27:901–910.

van Engelen, F.A., Hartog, M.V., Thomas, T.L., Taylor, B., Sturn, A., van Kammen, A., and de Vries, S.C. 1993. The carrot secreted glycoprotein gene EP1 is expressed in the epidermis and has sequence homology to *Brassica* S-locus glycoproteins. *Plant J.* 4:855–862.

van Engelen, F.A., Sterk, P., Booij, H., Cordewener, J.H.G., Rook, W., van Kammen, A., and de Vries, S.C. 1991. Heterogeneity and cell type-specific localization of a cell wall glycoprotein from carrot suspension cells. *Plant Physiol.* 96:705–712.

van Geyt, J., D'Halluin, K., and Jacobs, M. 1985. Induction of nuclear and cell divisions in microspores of sugarbeet (*Beta vulgaris* L.). *Z. Pflanzenzüchtg* 95:325–335.

van Heel, W.A. 1981. A S.E.M.-investigation on the development of free carpels. *Blumea* 27:499–522.

van Herpen, M.M.A., de Groot, P.F.M., Schrauwen, J.A.M., van den Heuvel, K.J.P.T., Weterings, K.A.P., and Wullems, G.J. 1992. *In-vitro* culture of tobacco pollen: gene expression and protein synthesis. *Sex. Plant Reprod.* 5:304–309.

van Herpen, M.M.A., Reijnen, W.H., Schrauwen, J.A.M., de Groot, P.F.M., Jager, J.W.H., and Wullems, G.J. 1989. Heat-shock proteins and survival of germinating pollen of *Lilium longiflorum* and *Nicotiana tabacum*. *J. Plant Physiol.* 134:345–351.

van Holst, G.-J., and Clarke, A.E. 1986. Organ-specific arabinogalactan-proteins of *Lycopersicon peruvianum* (Mill) demonstrated by crossed electrophoresis. *Plant Physiol.* 80:786–789.

van Lammeren, A.A.M. 1986. A comparative ultrastructural study of the megagametophytes in two strains of *Zea mays* L. before and after fertilization. *Agr. Univ. Wageningen Papers* 86-1:1–37.

van Lammeren, A.A.M. 1988a. Structure and function of the microtubular cytoskeleton during endosperm development in wheat: an immunofluorescence study. *Protoplasma* 146:18–27.

van Lammeren, A.A.M. 1988b. Observations on the structural development of immature maize embryos (*Zea mays* L.) during *in vitro* culture in the presence or absence of 2,4-D. *Acta Bot. Neerl.* 37:49–61.

van Lammeren, A.A.M., Bednara, J., and Willemse, M.T.M. 1989. Organization of the actin cytoskeleton during pollen development in *Gasteria verrucosa* (Mill.) H. Duval visualized with rhodamine-phalloidin. *Planta* 178:531–539.

van Lammeren, A.A.M., Keijzer, C.J., Willemse, M.T.M., and Kieft, H. 1985. Structure and function of the microtubular cytoskeleton during pollen development in *Gasteria verrucosa* (Mill.) H. Duval. *Planta* 165:1–11.

van Marrewijk, G.A.M., Bino, R.J., and Suurs, L.C.J.M. 1986. Characterization of cytoplasmic male sterility in *Petunia hybrida*. I. Localization, composition and activity of esterases. *Euphytica* 35:77–88.

van Overbeek, J., Conklin, M.E., and Blakeslee, A.F. 1942. Cultivation *in vitro* of small *Datura* embryos. *Am. J. Bot.* 29:472–477.

van Overbeek, J., Siu, R., and Haagen-Smit, A.J. 1944. Factors affecting the growth of *Datura* embryos *in vitro*. *Am. J. Bot.* 31:219–224.

van Rensburg, J.G.J., and Robbertse, P.J. 1988. Seed development of *Ornithogalum dubium*, with special reference to fertilization and the egg apparatus. *S. Afr. J. Bot.* 54:196–202.

van Rooijen, G.J.H., and Moloney, M.M. 1994. Plant seed oil bodies as carriers for foreign proteins. *Biotechnology* 13:72–77.

van Rooijen, G.J.H., Terning, L.I., and Moloney, M.M. 1992. Nucleotide sequence of an *Arabidopsis thaliana* oleosin gene. *Plant Mol. Biol.* 18:1177–1179.

van Rooijen, G.J.H., Wilen, R.W., Holbrook, L.A., and Moloney, M.M. 1992. Regulation of accumulation of mRNAs encoding a 20-kDa oil-body protein in microspore-derived embryos of *Brassica napus*. *Can. J. Bot.* 70:503–508.

van Tunen, A.J., Hartman, S.A., Mur, L.A., and Mol, J.N.M. 1989. Regulation of chalcone flavanone isomerase (CHI) gene expression in *Petunia hybrida*: the use of alternative promoters in corolla, anthers and pollen. *Plant Mol. Biol.* 12:539–551.

van Tunen, A.J., Mur, L.A., Brouns, G.S., Rienstra, J.-D., Koes, R.E., and Mol, J.N.M. 1990. Pollen- and anther-specific *chi* promoters from *Petunia*: tandem promoter regulation of the *chiA* gene. *Plant Cell* 2:393–401.

van Tunen, A.J., Mur, L.A., Recourt, K., Gerats, A.G.M., and Mol, J.N.M. 1991. Regulation and manipulation of flavonoid gene expression in anthers of *Petunia*: the molecular basis of the *Po* mutation. *Plant Cell* 3:39–48.

van Tuyl, J.M., Marcucci, M.C., and Visser, T. 1982. Pollen and pollination experiments. VII. The effect of pollen treatment and application method on incompatibility and incongruity in *Lilium*. *Euphytica* 31:613–619.

van Went, J., and Cresti, M. 1988a. Pre-fertilization degeneration of both synergids in *Brassica campestris* ovules. *Sex. Plant Reprod.* 1:208–216.

van Went, J., and Cresti, M. 1988b. Cytokinesis in microspore mother cells of *Impatiens sultani*. *Sex. Plant Reprod.* 1:228–233.

van Went, J., and Cresti, M. 1989. Cytoplasmic differentiation during tetrad formation and early microspore development in *Impatiens sultani*. *Protoplasma* 148:1–7.

van Went, J.L. 1970a. The ultrastructure of the synergids of *Petunia*. *Acta Bot. Neerl.* 19:121–132.

van Went, J.L. 1970b. The ultrastructure of the egg and central cell of *Petunia*. *Acta Bot. Neerl.* 19:313–322.

van Went, J.L. 1970c. The ultrastructure of the fertilized embryo sac of *Petunia*. *Acta Bot. Neerl.* 19:468–480.

van Went, J.L. 1981. Some cytological and ultrastructural aspects of male sterility in *Impatiens*. *Acta Soc. Bot. Polon.* 50:248–252.

van Went, J.L. 1984. Unequal distribution of plastids during generative cell formation in *Impatiens*. *Theor. Appl. Genet.* 68:305–309.

van Went, J.L., and Gori, P. 1989. The ultrastructure of *Capparis spinosa* pollen grains. *J. Submicros. Cytol. Path.* 21:149–156.

van Went, J.L., and Kwee, H.-S. 1990. Enzymatic isolation of living embryo sacs of *Petunia*. *Sex. Plant Reprod.* 3:257–262.

van Went, J.L., and Linskens, H.F. 1967. Die Entwicklung des sogenannten "Fadenapparates" im Embryosack von *Petunia hybrida*. *Züchter* 37:51–56.

van Went, J.L., and Willemse, M.T.M. 1984. Fertilization. In *Embryology of Angiosperms*, ed. B. M. Johri, pp. 273–317. Berlin: Springer-Verlag.

van der Donk, J.A.W.M. 1974a. Differential synthesis of RNA in self- and cross-pollinated styles of *Petunia hybrida* L. *Mol. Gen. Genet.* 131:1–8.

van der Donk, J.A.W.M. 1974b. Synthesis of RNA and protein as a function of time and type of pollentube-style interaction in *Petunia hybrida* L. *Mol. Gen. Genet.* 134:93–98.

van der Donk, J.A.W.M. 1975. Recognition and gene expression during the incompatibility reaction in *Petunia hybrida* L. *Mol. Gen. Genet.* 141:305–316.

van der Geest, A.H.M., Frisch, D.A., Kemp, J.D., and Hall, T.C. 1995. Cell ablation reveals that expression from the phaseolin promoter is confined to embryogenesis and microsporogenesis. *Plant Physiol.* 109:1151–1158.

van der Geest, A.H.M., Hall, G.E., Jr., Spiker, S., and Hall, T.C. 1994. The β-phaseolin gene is flanked by matrix attachment regions. *Plant J.* 6:413–423.

van der Klei, H., van Damme, J., Casteels, P., and Krebbers, E. 1993. A fifth 2S albumin isoform is present in *Arabidopsis thaliana*. *Plant Physiol.* 101:1415–1416.

van der Kley, F.K. 1954. Male sterility and its importance in breeding heterosis varieties. *Euphytica* 3:117–124.

van der Krol, A.R., Brunelle, A., Tsuchimoto, S., and Chua, N.-H. 1993. Functional analysis of *Petunia* floral homeotic MADS box gene *pMADS1*. *Genes Devel.* 7:1214–1228.

van der Leede-Plegt, L.M., van den Ven, B.C.E., Bino, R.J., van der Salm, T.P.M., and van Tunen, A.J. 1992. Introduction and differential use of various promoters in pollen grains of *Nicotiana glutinosa* and *Lilium longiflorum*. *Plant Cell Rep.* 11:20–24.

van der Maas, H.M., de Jong, E.R., van Aelst, A.C., Verhoeven, H.A., Krens, F.A., and van Went, J.L. 1994. Cytological characterization of isolated sperm cells of perennial ryegrass (*Lolium perenne* L.). *Protoplasma* 178:48–56.

van der Maas, H.M., Zaal, M.A.C.M., de Jong, E.R., Krens, F.A., and van Went, J.L. 1993a. Isolation of viable egg cells of perennial ryegrass (*Lolium perenne* L.). *Protoplasma* 173:86–89.

van der Maas, H.M., Zaal, M.A.C.M., de Jong, E.R., van Went, J.L., and Krens, F.A. 1993b. Optimization of isolation and storage of sperm cells from pollen of perennial ryegrass (*Lolium perenne* L.). *Sex. Plant Reprod.* 6:64–70.

van der Meer, I.M., Stam, M.E., van Tunen, A.J., Mol, J.N.M., and Stuitje, A.R. 1992. Antisense inhibition of flavonoid biosynthesis in *Petunia* anthers results in male sterility. *Plant Cell* 4:253–262.

van der Meer, Q.P., and van Bennekom, J.L. 1969. Effect of temperature on the occurrence of male sterility in onion (*Allium cepa* L.). *Euphytica* 18:389–394.

van der Meer, Q.P., and van Bennekom, J.L. 1973. Gibberellic acid as a gametocide for the common onion (*Allium cepa* L.). *Euphytica* 22: 239–243.

van der Meer, Q.P., and van Dam, R. 1979. Gibberellic acid as a gametocide for cole crops. *Euphytica* 28:717–722.

van der Plas, L.H.W., de Gucht, L.P.E., Bakels, R.H.A., and Otto, B. 1987. Growth and respiratory characteristics of batch and continuous cell suspension cultures derived from fertile and male sterile *Petunia hybrida*. *J. Plant Physiol.* 130:449–460.

van der Pluijm, J., and Linskens, H. F. 1966. Feinstruktur der Pollenschläuche im Griffel von *Petunia*. *Züchter* 36:220–224.

van der Pluijm, J.E. 1964. An electron microscopic investigation of the filiform apparatus in the embryo sac of *Torenia fournieri*. In *Pollen Physiology and Fertilization*, ed. H.F. Linskens, pp. 8–16. Amsterdam: North-Holland Publishing Co.

Vancanneyt, G., Sonnewald, U., Höfgen, R., and Willmitzer, L. 1989. Expression of a patatin-like protein in the anthers of potato and sweet pepper flowers. *Plant Cell* 1:533–540.

Vance, V.B., and Huang, A.H.C. 1987. The major protein from lipid bodies of maize. Characterization and structure based on cDNA cloning. *J. Biol. Chem.* 262:11275–11279.

Vance, V.B., and Huang, A.H.C. 1988. Expression of lipid body protein gene during maize seed development. Spatial, temporal, and hormonal regulation. *J. Biol. Chem.* 263:1476–1481.

VanDerWoude, W.J., and Morré, D.J. 1968. Endoplasmic reticulum-dictyosome-secretory vesicle associations in pollen tubes of *Lilium longiflorum* Thunb. *Proc. Indiana Acad. Sci.* 77:164–170.

VanDerWoude, W.J., Morré, D.J., and Bracker, C.E. 1971. Isolation and characterization of secretory vesicles in germinated pollen of *Lilium longiflorum*. *J. Cell Sci.* 8:331–351.

Varagona, M.J., Schmidt, R.J., and Raikhel, N.V. 1991. Monocot regulatory protein opaque-2 is localized in the nucleus of maize endosperm and transformed tobacco plants. *Plant Cell* 3:105–113.

Vasil, I.K. 1957. Effect of kinetin and gibberellic acid on excised anthers of *Allium cepa*. *Phytomorphology* 7:138–149.

Vasil, I.K. 1959. Cultivation of excised anthers *in vitro* - effect of nucleic acids. *J. Expt. Bot.* 10:399–408.

Vasil, I.K. 1967. Physiology and cytology of anther development. *Biol. Rev.* 42:327–373.

Vasil, I.K., and Johri, M.M. 1964. The style, stigma and pollen tube - I. *Phytomorphology* 14:352–369.

Vasil, V., and Hildebrandt, A.C. 1965. Differentiation of tobacco plants from single, isolated cells in microcultures. *Science* 150:889–892.

Vasil, V., and Vasil, I.K. 1981. Somatic embryogenesis and plant regeneration from suspension cultures of pearl millet (*Pennisetum americanum*). *Ann. Bot.* 47:669–678.

Vasil, V., and Vasil, I.K. 1982a. The ontogeny of somatic embryos of *Pennisetum americanum* (L.) K. Schum. I. In cultured immature embryos. *Bot. Gaz.* 143:454–465.

Vasil, V., and Vasil, I.K. 1982b. Characterization of an embryogenic cell suspension culture derived from cultured inflorescences of *Pennisetum americanum* (pearl millet, Gramineae). *Am. J. Bot.* 69:1441–1449.

Vasil, V., Castillo, A.M., Fromm, M.E., and Vasil, I.K. 1992. Herbicide resistant fertile transgenic wheat plants obtained by microprojectile bombardment of regenerable embryogenic callus. *Biotechnology* 10:667–674.

Vasil, V., Srivastava, V., Castillo, A.M., Fromm, M.E., and Vasil, I.K. 1993. Rapid production of transgenic wheat plants by direct bombardment of cultured immature embryos. *Biotechnology* 11:1553–1558.

Vasil'ev, A.E. 1970. Ultrastructure of the stigmatoid cells in *Lilium*. *Soviet Plant Physiol.* 17:1035–1044.

Vaughn, K.C., DeBonte, L.R., Wilson, K.G., and Schaffer, G.W. 1980. Organelle alteration as a mechanism for maternal inheritance. *Science* 208:196–198.

Vazart, B. 1970. Morphogenèse du sporoderme et participation des mitochondries à la mise en place de la primexine dans le pollen de *Linum usitatissimum* L. *Compt. Rend. Acad. Sci. Paris* 270D:3210–3212.

Vazart, B. 1971. Infrastructure de microspores de *Nicotiana tabacum* L. susceptibles de se développer en embryoïdes après excision et mise en culture des anthères. *Compt. Rend. Acad. Sci. Paris* 272D:549–552.

Vazart, B., and Vazart, J. 1965. Infrastructure de l'ovule de lin, *Linum usitatissimum* L. Les cellules du sac embryonnaire. *Compt. Rend. Acad. Sci. Paris* 261:3447–3450.

Vazart, B., and Vazart, J. 1966. Infrastructure du sac embryonnaire du lin (*Linum usitatissimum* L.). *Rev. Cytol. Biol. Végét.* 29:251–266.

Vazart, J. 1969. Organisation et ultrastructure du sac embryonnaire du lin (*Linum usitatissimum* L.). *Rev. Cytol. Biol. Végét.* 32:227–240.

Vedel, F., and Mathieu, C. 1983. Physical and gene mapping of chloroplast DNA from normal and cytoplasmic male sterile (radish cytoplasm) lines of *Brassica napus*. *Curr. Genet.* 7:13–20.

Vedel, F., Mathieu, C., Lebacq, P., Ambard-Bretteville, F., Remy, R., Pelletier, G., Renard, M., and Rousselle, P. 1982. Comparative macromolecular analysis of the cytoplasms of normal and cytoplasmic male sterile *Brassica napus*. *Theor. Appl. Genet.* 62:255–262.

Veen, H. 1963. The effect of various growth-regulators on embryos of *Capsella bursa-pastoris* growing *in vitro*. *Acta Bot. Neerl.* 12:129–171.

Veit, B., Schmidt, R.J., Hake, S., and Yanofsky, M.F. 1993. Maize floral development: new genes and old mutants. *Plant Cell* 5:1205–1215.

Vellanoweth, R.L., and Okita, T.W. 1993. Analysis of nuclear proteins interacting with a wheat α/β-gliadin seed storage protein gene. *Plant Mol. Biol.* 22:25–41.

Venema, G., and Koopmans, A. 1962. A phase-contrast microscopic study of pollengrain germination, nuclear movement and pollentube mitosis in *Tradescantia virginiana*. *Cytologia* 27:11–24.

Venkateswarlu, J., and Rao, G.R. 1958. A contribution to the life-history of *Rubia cordifolia* Linn. *J. Indian Bot. Soc.* 37:442–454.

Vennigerholz, F. 1992. The transmitting tissue in *Brugmansia suaveolens*: immunocytochemical localization of pectin in the style. *Protoplasma* 171:117–122.

Vergara, M.R., Biasini, G., Lo Schiavo, F., and Terzi, M. 1982. Isolation and characterization of carrot cell mutants resistant to α-amanitin. *Z. Pflanzenphysiol.* 107:313–319.

Vergara, R., Verde, F., Pitto, L., Lo Schiavo, F., and Terzi, M. 1990. Reversible variations in the methylation pattern of carrot DNA during somatic embryogenesis. *Plant Cell Rep.* 8:697–700.

Vergne, P., and Dumas, C. 1988. Isolation of viable wheat male gametophytes of different stages of development and variations in their protein patterns. *Plant Physiol.* 88:969–972.

Vergne, P., Riccardi, F., Beckert, M., and Dumas, C. 1993. Identification of a 32-kDa anther marker protein for androgenic response in maize, *Zea mays* L. *Theor. Appl. Genet.* 86:843–850.

Verma, D.C., and Dougall, D.K. 1978. DNA, RNA and protein content of tissue during growth and embryogenesis in wild-carrot suspension cultures. *In Vitro* 14:183–191.

Verma, S.C., Malik, R., and Dhir, I. 1977. Genetics of the incompatibility system in the crucifer *Eruca sativa* L. *Proc. Roy. Soc. Lond.* 196B:131–159.

Vernon, D.M., and Meinke, D.W. 1994. Embryogenic transformation of the suspensor in *twin*, a polyembryonic mutant of *Arabidopsis*. *Devel Biol.* 165:566–573.

Vernon, D.M., and Meinke, D.W. 1995. Late *embryo-defective* mutants of *Arabidopsis*. *Devel Genet.* 16:311–320.

Verwoert, I.I.G.S., van der Linden, K.H., Walsh, M.C., Nijkamp, H.J.J., and Stuitje, A.R. 1995. Modification of *Brassica napus* seed oil by expression of the *Escherichia coli fabH* gene, encoding 3-ketoacyl-acyl carrier protein synthase III. *Plant Mol. Biol.* 27:875–886.

Vicente, O., Benito Moreno, R.M., and Heberle-Bors, E. 1991. Pollen cultures as a tool to study plant development. *Cell Biol. Rev.* 25:295–306.

Vielle, J.P., Burson, B.L., Bashaw, E.C., and Hussey, M.A. 1995. Early fertilization events in the sexual and aposporous egg apparatus of *Pennisetum ciliare* (L.) Link. *Plant J.* 8:309–316.

Vijayaraghavan, M.R., and Bhat, U. 1983. Synergids before and after fertilization. *Phytomorphology* 33:74–84.

Vijayaraghavan, M.R., Jensen, W.A., and Ashton, M.E. 1972. Synergids of *Aquilegia formosa* - their histochemistry and ultrastructure. *Phytomorphology* 22:144–159.

Vijayaraghavan, M.R., and Prabhakar, K. 1984. The endosperm. In *Embryology of Angiosperms*, ed. B.M. Johri, pp. 319–376. Berlin: Springer-Verlag.

Vijayaraghavan, M.R., and Shukla, A.K. 1977. Absence of callose around the microspore tetrad and poorly developed exine in *Pergularia daemia*. *Ann. Bot.* 41:923–926.

Vilardell, J., Goday, A., Freire, M.A., Torrent, M., Martínez, M.C., Torné, J.M., and Pagès, M. 1990. Gene sequence, developmental expression, and protein phosphorylation of RAB-17 in maize. *Plant Mol. Biol.* 14:423–432.

Vilardell, J., Martínez-Zapater, J.M., Goday, A., Arenas, C., and Pagès, M. 1994. Regulation of the *rab*17 gene promoter in transgenic *Arabidopsis* wild-type, ABA-deficient and ABA-insensitive mutants. *Plant Mol. Biol.* 24:561–569.

Vilardell, J., Mundy, J., Stilling, B., Leroux, B., Pla, M., Freyssinet, G., and Pagès, M. 1991. Regulation of the maize *rab*17 gene promoter in transgenic heterologous systems. *Plant Mol. Biol.* 17:985–993.

Villanueva, V.R., Mathivet, V., and Sangwan, R.S. 1985. RNA, proteins and polyamines during gametophytic and androgenetic development in *Nicotiana tabacum* and *Datura innoxia*. *Plant Growth Reg.* 3:293–307.

Villar, M., Gaget, M., Rougier, M., and Dumas, C. 1993. Pollen–pistil interactions in *Populus*: β-galactosidase activity associated with pollen-tube growth in the crosses *P. nigra* × *P. nigra* and *P. nigra* × *P. alba*. *Sex. Plant Reprod.* 6:249–256.

Villar, M., Gaget, M., Said, C., Knox, R.B., and Dumas, C. 1987. Incompatibility in *Populus*: structural and cytochemical characteristics of the receptive stigmas of *Populus alba* and *P. nigra*. *J. Cell Sci.* 87:483–490.

Viotti, A., Sala, E., Alberi, P., and Soave, C. 1978. Heterogeneity of zein synthesized *in vitro*. *Plant Sci. Lett.* 13:365–375.

Vishnyakova, M.A., and Willemse, M.T.M. 1994. Pollen–pistil interaction in wheat. *Acta Bot. Neerl.* 43:51–64.

Vithanage, H.I.M.V., Gleeson, P.A., and Clarke, A.E. 1980. The nature of callose produced during self-pollination in *Secale cereale*. *Planta* 148:498–509.

Vithanage, H.I.M.V., and Heslop-Harrison, J. 1979. The pollen–stigma interaction: fate of fluorescent-labelled pollen-wall proteins on the stigma surface in rye (*Secale cereale*). *Ann. Bot.* 43:113–114.

Vithanage, H.I.M.V., and Knox, R.B. 1976. Pollen-wall proteins: quantitative cytochemistry of the origins of intine and exine enzymes in *Brassica oleracea*. *J. Cell Sci.* 21:423–435.

Vithanage, H.I.M.V., and Knox, R.B. 1977. Development and cytochemistry of stigma surface and response to self and foreign pollination in *Helianthus annuus*. *Phytomorphology* 27:168–179.

Vithanage, H.I.M.V., and Knox, R.B. 1979. Pollen development and quantitative cytochemistry of exine and intine enzymes in sunflower, *Helianthus annuus* L. *Ann. Bot.* 44:95–106.

Vithanage, H.I.M.V., and Knox, R.B. 1980. Periodicity of pollen development and quantitative cytochemistry of exine and intine enzymes in the grasses *Lolium perenne* L. and *Phalaris tuberosa* L. *Ann. Bot.* 45:131–141.

Vithanage, H.I.M.V., Howlett, B.J., Jobson, S., and Knox, R.B. 1982. Immunocytochemical localization of water-soluble glycoproteins, including group 1 allergen, in pollen of ryegrass, *Lolium perenne*, using ferritin-labelled antibody. *Histochem. J.* 14:949–966.

Vivekananda, J., Drew, M.C., and Thomas, T.L. 1992. Hormonal and environmental regulation of the carrot *lea*-class gene *Dc*3. *Plant Physiol.* 100:576–581.

Vodkin, L.O. 1981. Isolation and characterization of messenger RNAs for seed lectin and Kunitz trypsin inhibitor in soybeans. *Plant Physiol.* 68:766–771.

Vodkin, L.O., and Raikhel, N.V. 1986. Soybean lectin and related proteins in seeds and roots of Le^+ and Le^- soybean varieties. *Plant Physiol.* 81:558–565.

Vodkin, L.O., Rhodes, P.R., and Goldberg, R.B. 1983. cA lectin gene insertion has the structural features of a transposable element. *Cell* 34:1023–1031.

Voelker, T., Sturm, A., and Chrispeels, M.J. 1987. Differences in expression between two seed lectin alleles obtained from normal and lectin-deficient beans are maintained in transgenic tobacco. *EMBO J.* 6:3571–3577.

Voelker, T.A., Herman, E.M., and Chrispeels, M.J. 1989. *In vitro* mutated phytohemagglutinin genes expressed in tobacco seeds: role of glycans in protein targeting and stability. *Plant Cell* 1:95–104.

Voelker, T.A., Staswick, P., and Chrispeels, M.J. 1986. Molecular analysis of two phytohemagglutinin genes and their expression in *Phaseolus vulgaris* cv. Pinto, a lectin-deficient cultivar of the bean. *EMBO J.* 5:3075–3082.

Vogt, T., Pollak, P., Tarlyn, N., and Taylor, L.P. 1994. Pollination- or wound-induced kaempferol accumulation in *Petunia* stigmas enhances seed production. *Plant Cell* 6:11–23.

von Allmen, J.-M., Rottmann, W.H., Gengenbach, B.G.,

Harvey, A.J., and Lonsdale, D.M. 1991. Transfer of methomyl and HmT-toxin sensitivity from T-cytoplasm maize to tobacco. *Mol. Gen. Genet.* 229:405–412.

von Arnold, S., and Hakman, I. 1988. Regulation of somatic embryo development in *Picea abies* by abscisic acid (ABA). *J. Plant Physiol.* 132:164–169.

Vonder Haar, R.A., Allen, R.D., Cohen, E.A., Nessler, C.L., and Thomas, T.L. 1988. Organization of the sunflower 11S storage protein gene family. *Gene* 74:433–443.

Voss, R., Schumann, K., and Nagl, W. 1992. Phytohemagglutinin gene expression during seed development of the runner bean, *Phaseolus coccineus. Plant Mol. Biol.* 20:781–790.

Vroemen, C., Langeveld, S., Mayer, U., Ripper, G., Jürgens, G., van Kammen, A., and de Vries, S.C. 1996. Pattern formation in the *Arabidopsis* embryo revealed by position-specific lipid transfer protein gene expression. *Plant Cell* 8:783–791.

Vujičić, R., Radojević, L., and Nešković, M. 1976. Orderly arrangement of ribosomes in the embryogenic callus tissue of *Corylus avellana* L. *J. Cell Biol.* 69:686–692.

Wadsworth, G.J., and Scandalios, J.G. 1989. Differential expression of the maize catalase genes during kernel development: the role of steady-state mRNA levels. *Devel Genet.* 10:304–310.

Wagner, G., and Hess, D. 1973. *In vitro*-Befruchtungen bei *Petunia hybrida. Z. Pflanzenphysiol.* 69:262–269.

Wagner, V.T., Dumas, C., and Mogensen, H.L. 1989. Morphometric analysis of isolated *Zea mays* sperm. *J. Cell Sci.* 93:179–184.

Wagner, V.T., Kardolus, J.P., and van Went, J.L. 1989. Isolation of the lily embryo sac. *Sex. Plant Reprod.* 2:219–224.

Wagner, V.T., and Mogensen, H.L. 1988. The male germ unit in the pollen and pollen tubes of *Petunia hybrida*: ultrastructural, quantitative and three-dimensional features. *Protoplasma* 143:101–110.

Wagner, V.T., Cresti, M., Salvatici, P., and Tiezzi, A. 1990. Changes in volume, surface area, and frequency of nuclear pores on the vegetative nucleus of tobacco pollen in fresh, hydrated, and activated conditions. *Planta* 181:304–309.

Wagner, V.T., Song, Y.C., Matthys-Rochon, E., and Dumas, C. 1989. Observations on the isolated embryo sac of *Zea mays* L. *Plant Sci.* 59:127–132.

Walbot, V. 1971. RNA metabolism during embryo development and germination of *Phaseolus vulgaris. Devel Biol.* 26:369–379.

Walbot, V. 1972. Rate of RNA synthesis and tRNA end-labeling during early development of *Phaseolus. Planta* 108:161–171.

Walbot, V. 1973. RNA metabolism in developing cotyledons of *Phaseolus vulgaris. New Phytol.* 72:479–483.

Walbot, V., and Dure, L., III. 1976. Developmental biochemistry of cotton seed embryogenesis and germination. VII. Characterization of the cotton genome. *J. Mol. Biol.* 101:503–536.

Walbot, V., Brady, T., Clutter, M., and Sussex, I. 1972. Macromolecular synthesis during plant embryogeny: rates of RNA synthesis in *Phaseolus coccineus* embryos and suspensors. *Devel Biol.* 29:104–111.

Walburg, G., and Larkins, B.A. 1986. Isolation and characterization of cDNAs encoding oat 12S globulin mRNAs. *Plant Mol. Biol.* 6:161–169.

Walker, J.C., and Zhang, R. 1990. Relationship of a putative receptor protein kinase from maize to the S-locus glycoproteins of *Brassica. Nature* 345:743–746.

Walker, K.A., and Sato, S.J. 1981. Morphogenesis in callus tissue of *Medicago sativa*: the role of ammonium ion in somatic embryogenesis. *Plant Cell Tissue Organ Cult.* 1:109–121.

Wallace, J.C., Lopes, M.A., Paiva, E., and Larkins, B.A. 1990. New methods for extraction and quantitation of zeins reveal a high content of γ-zein in modified *opaque-2* maize. *Plant Physiol.* 92:191–196.

Wallace, N.H., and Kriz, A.L. 1991. Nucleotide sequence of a cDNA clone corresponding to the maize *Globulin-2* gene. *Plant Physiol.* 95:973–975.

Walling, L., Drews, G.N., and Goldberg, R.B. 1986. Transcriptional and post-translational regulation of soybean seed protein mRNA levels. *Proc. Natl Acad. Sci. USA* 83:2123–2127.

Walsh, D.J., Matthews, J.A., Denmeade, R., and Walker, M.R. 1989. Cloning of cDNA coding for an allergen of cocksfoot grass (*Dactylis glomerata*) pollen. *Int. Arch. Allergy Appl. Immunol.* 90:78–83.

Walter, C., Smith, D.R., Connett, M.B., Grace, L., and White, D.W.R. 1994. A biolistic approach for the transfer and expression of a *gus*A reporter gene in embryogenic cultures of *Pinus radiata. Plant Cell Rep.* 14:69–74.

Walters, D.A., Vetsch, C.S., Potts, D.E., and Lundquist, R.C. 1992. Transformation and inheritance of a hygromycin phosphotransferase gene in maize plants. *Plant Mol. Biol.* 18:189–200.

Walters, M.S. 1970. Evidence on the time of chromosome pairing from the preleptotene spiral stage in *Lilium longiflorum* "Croft". *Chromosoma* 29:375–418.

Walters, M.S. 1985. Meiosis readiness in *Lilium. Can. J. Genet. Cytol.* 27:33–38.

Walthall, E.D., and Brady, T. 1986. The effect of the suspensor and gibberellic acid on *Phaseolus vulgaris* embryo protein synthesis. *Cell Diffn* 18:37–44.

Wan, Y., and Lemaux, P.G. 1994. Generation of large numbers of independently transformed fertile barley plants. *Plant Physiol.* 104:37–48.

Wan, Y., Sorensen, E.L., and Liang, G.H. 1988. Genetic control of *in vitro* regeneration in alfalfa (*Medicago sativa* L.). *Euphytica* 39:3–9.

Wandelt, C., and Feix, G. 1989. Sequence of a 21 kd gene from maize containing an in-frame stop codon. *Nucl. Acids Res.* 17:2354.

Wang, C., Rathore, K.S., and Robinson, K.R. 1989. The responses of pollen to applied electrical fields. *Devel Biol.* 136:405–410.

Wang, C.-C., Chu, C.-C., Sun, C.-S., Wu, S.-H., Yin, K.-C., and Hsü, C. 1973. The androgenesis in wheat (*Triticum aestivum*) anthers cultured *in vitro. Sci. Sinica* 16:218–222.

Wang, C.-S., Walling, L.L., Eckard, K.J., and Lord, E.M. 1992a. Patterns of protein accumulation in developing anthers of *Lilium longiflorum* correlate with histological events. *Am. J. Bot.* 79:118–127.

Wang, C.-S., Walling, L.L., Eckard, K.J., and Lord, E.M. 1992b. Immunological characterization of a tapetal protein in developing anthers of *Lilium longiflorum. Plant Physiol.* 99:822–829.

Wang, C.-S., Walling, L.L., Eckard, K.J., and Lord, E.M. 1993. Characterization of an anther-specific glycoprotein in *Lilium longiflorum. Am. J. Bot.* 80:1155–1161.

Wang, F.-Y., Chen, K.-M., Wang, X.-Z., Zhou, W.-L., and

Ni, J.-F. 1995. Two-dimensional gel electrophoresis study on endosperm protein of the new line obtained by introducing DNA of sorghum into wheat. *Chinese J. Bot.* 7:2–8.

Wang, H., and Cutler, A.J. 1995. Promoters from *kin1* and *cor6.6*, two *Arabidopsis thaliana* low-temperature- and ABA-inducible genes, direct strong β-glucuronidase expression in guard cells, pollen and young developing seeds. *Plant Mol. Biol.* 28:619–634.

Wang, H., Wu, H.-M., and Cheung, A.Y. 1993. Development and pollination regulated accumulation and glycosylation of a stylar transmitting tissue-specific proline-rich protein. *Plant Cell* 5:1639–1650.

Wang, H., Wu, H.-M., and Cheung, A.Y. 1996. Pollination induces mRNA poly(A) tail-shortening and cell deterioration in flower transmitting tissue. *Plant J.* 9:715–727.

Wang, H.L., Offler, C.E., and Patrick, J.W. 1995. The cellular pathway of photosynthate transfer in the developing wheat grain. II. A structural analysis and histochemical studies of the pathway from the crease phloem to the endosperm cavity. *Plant Cell Environ.* 18:373–388.

Wang, H.L., Patrick, J.W., Offler, C.E., and Wang, X.-D. 1995. The cellular pathway of photosynthate transfer in the developing wheat grain. III. A structural analysis and physiological studies of the pathway from the endosperm cavity to the starchy endosperm. *Plant Cell Environ.* 18:389–407.

Wang, S.-Z., and Esen, A. 1986. Primary structure of a proline-rich zein and its cDNA. *Plant Physiol.* 81:70–74.

Wang, T.L., Cook, S.K., Francis, R.J., Ambrose, M.J., and Hedley, C.J. 1987a. An analysis of seed development in *Pisum sativum*. VI. Abscisic acid accumulation. *J. Expt. Bot.* 38:1921–1932.

Wang, T.L., Hadavizideh, A., Harwood, A., Welham, T.J., Harwood, W.A., Faulks, R., and Hedley, C.L. 1990. An analysis of seed development in *Pisum sativum*. XIII. The chemical induction of storage product mutants. *Plant Breed.* 105:311–320.

Wang, T.L., Smith, C.M., Cook, S.K., Ambrose, M.J., and Hedley, C.L. 1987b. An analysis of seed development in *Pisum sativum*. III. The relationship between the *r* locus, the water content and the osmotic potential of seed tissues *in vivo* and *in vitro*. *Ann. Bot.* 59:73–80.

Wang, Y.-C., and Janick, J. 1986. *In vitro* production of jojoba liquid wax by zygotic and somatic embryos. *J. Am. Soc. Hort. Sci.* 111:798–806.

Wang, Z.-Y., Wu, Z.-L., Xing, Y.-Y., Zheng, F.-G., Guo, X.-L., Zhang, W.-G., and Hong, M.-M. 1990. Nucleotide sequence of rice waxy gene. *Nucl. Acids Res.* 18:5898.

Wang, Z.-Y., Zheng, F.-Q., Shen, G.-Z., Gao, J.-P., Snustad, D.P., Li, M.-G., Zhang, J.-L., and Hong, M.-M. 1995. The amylose content in rice endosperm is related to the post-transcriptional regulation of the *waxy* gene. *Plant J.* 7:613–622.

Warmke, H.E., and Lee, S.-L.J. 1977. Mitochondrial degeneration in Texas cytoplasmic male-sterile corn anthers. *J. Hered.* 68:213–222.

Warmke, H.E., and Lee, S.-L.J. 1978. Pollen abortion in T cytoplasmic male-sterile corn (*Zea mays*): a suggested mechanism. *Science* 200:561–563.

Warmke, H.E., and Overman, M.A. 1972. Cytoplasmic male sterility in *Sorghum*. I. Callose behavior in fertile and sterile anthers. *J. Hered.* 63:102–108.

Warner, S.A.J., Scott, R., and Draper, J. 1993. Isolation of an asparagus intracellular *PR* gene (*AoPR1*) wound-responsive promoter by the inverse polymerase chain reaction and its characterization in transgenic tobacco. *Plant J.* 3:191–201.

Watanabe, M., Shiozawa, H., Isogai, A., Suzuki, A., Takeuchi, T., and Hinata, K. 1991. Existence of S-glycoprotein-like proteins in anthers of self-incompatible species of *Brassica*. *Plant Cell Physiol.* 32:1039–1047.

Watanabe, M., Takasaki, T., Toriyama, K., Yamakawa, S., Isogai, A., Suzuki, A., and Hinata, K. 1994. A high degree of homology exists between the protein encoded by *SLG* and the *S* receptor domain encoded by *SRK* in self-incompatible *Brassica campestris* L. *Plant Cell Physiol.* 35:1221–1229.

Waterkeyn, L., and Bienfait, A. 1970. On a possible function of the callosic special wall in *Ipomoea purpurea* (L.) Roth. *Grana* 10:13–20.

Watson, C.V., Nath, J., and Nanda, D. 1977. Possible mitochondrial involvement in mechanism of male sterility in maize (*Zea mays* L.). *Biochem. Genet.* 15:1113–1124.

Webb, M.C., and Gunning, B.E.S. 1990. Embryo sac development in *Arabidopsis thaliana*. I. Megasporogenesis, including the microtubular cytoskeleton. *Sex. Plant Reprod.* 3:244–256.

Webb, M.C., and Gunning, B.E.S. 1991. The microtubular cytoskeleton during development of the zygote, proembryo and free-nuclear endosperm in *Arabidopsis thaliana* (L.) Heynh. *Planta* 184:187–195.

Webb, M.C., and Gunning, B.E.S. 1994. Embryo sac development in *Arabidopsis thaliana*. II. The cytoskeleton during megagametogenesis. *Sex. Plant Reprod.* 7:153–163.

Weber, M. 1988. Formation of sperm cells in *Galium mollugo* (Rubiaceae), *Trichodiadema setuliferum* (Aizoaceae), and *Avena sativa* (Poaceae). *Plant Syst. Evol.* 161:53–64.

Weber, M. 1989. Ultrastructural changes in maturing pollen grains of *Apium nodiflorum* L. (Apiaceae), with special reference to the endoplasmic reticulum. *Protoplasma* 152:69–76.

Weber, M. 1992. The formation of pollenkitt in *Apium nodiflorum* (Apiaceae). *Ann. Bot.* 70:573–577.

Weber, M. 1994. Stigma, style, and pollen tube pathway in *Smyrnium perfoliatum* (Apiaceae). *Int. J. Plant Sci.* 155:437–444.

Weber, M., and Frosch, A. 1995. The development of the transmitting tract in the pistil of *Hacquetia epipactis* (Apiaceae). *Int. J. Plant Sci.* 156:615–621.

Wedderburn, F., and Richards, A.J. 1990. Variation in within-morph incompatibility inhibition sites in heteromorphic *Primula* L. *New Phytol.* 116:149–162.

Weeden, N.F., and Gottlieb, L.D. 1979. Distinguishing allozymes and isozymes of phosphoglucoisomerases by electrophoretic comparisons of pollen and somatic tissues. *Biochem. Genet.* 17:287–296.

Weeden, N.F., and Gottlieb, L.D. 1980. Isolation of cytoplasmic enzymes from pollen. *Plant Physiol.* 66:400–403.

Weeks, J.T., Anderson, O.D., and Blechl, A.E. 1993. Rapid production of multiple independent lines of fertile transgenic wheat (*Triticum aestivum*). *Plant Physiol.* 102:1077–1084.

Wehling, P., Hackauf, B., and Wricke, G. 1994. Phosphorylation of pollen proteins in relation to self-incompatibility in rye (*Secale cereale* L.). *Sex. Plant Reprod.* 7:67–75.

Weigel, D., and Meyerowitz, E.M. 1993. Activation of floral homeotic genes in *Arabidopsis. Science* 261:1723–1726.

Weigel, D., Alvarez, J., Smyth, D.R., Yanofsky, M.F., and Meyerowitz, E.M. 1992. *LEAFY* controls floral meristem identity in *Arabidopsis. Cell* 69:843–859.

Weinand, U., Brüschke, C., and Feix, G. 1979. Cloning of double stranded DNAs derived from polysomal mRNA of maize endosperm: isolation and characterisation of zein clones. *Nucl. Acids Res.* 6:2707–2715.

Weisenseel, M.H., and Jaffe, L.F. 1976. The major growth current through lily pollen tubes enters as K^+ and leaves as H^+. *Planta* 133:1–7.

Weisenseel, M.H., Nuccitelli, R., and Jaffe, L.F. 1975. Large electrical currents traverse growing pollen tubes. *J. Cell Biol.* 66:556–567.

Weiss, C.A., Huang, H., and Ma, H. 1993. Immunolocalization of the G protein α subunit encoded by a *GPA1* gene in *Arabidopsis. Plant Cell* 5:1513–1528.

Weissinger, A.K., Timothy, D.H., Levings, C.S., III, and Goodman, M.M. 1983. Patterns of mitochondrial DNA variation in indigenous maize races of Latin America. *Genetics* 104:365–379.

Weissinger, A.K., Timothy, D.H., Levings, C.S., III, Hu, W.W.L., and Goodman, M.M. 1982. Unique plasmid-like mitochondrial DNAs from indigenous maize races of Latin America. *Proc. Natl Acad. Sci. USA* 79:1–5.

Welk, M., Millington, W.F., and Rosen, W.G. 1965. Chemotropic activity and the pathway of pollen tube in lily. *Am. J. Bot.* 52:774–781.

Welsh, J.R., and Klatt, A.R. 1971. Effects of temperature and photoperiod on spring wheat pollen viability. *Crop Sci.* 11:864–865.

Wemmer, T., Kaufmann, H., Kirch, H.-H., Schneider, K., Lottspeich, F., and Thompson, R.D. 1994. The most abundant soluble basic protein of the stylar transmitting tract in potato (*Solanum tuberosum* L.) is an endochitinase. *Planta* 194:264–273.

Wenck, A.R., Conger, B.V., Trigiano, R.N., and Sams, C.E. 1988. Inhibition of somatic embryogenesis in orchardgrass by endogenous cytokinins. *Plant Physiol.* 88:990–992.

Weng, W.-M., Gao, X.-S., Zhuang, N.-L., Xu, M.-L., and Xue, Z.-T. 1995. The glycinin A_3B_4 mRNA from wild soybean *Glycine soja* Sieb. et. Zucc. *Plant Physiol.* 107:665–666.

Wenzel, M., Gers-Barlag, H., Schimpl, A., and Rüdiger, H. 1993. Time course of lectin and storage protein biosynthesis in developing pea (*Pisum sativum*) seeds. *Biol. Chem. Hoppe-Seyler* 374:887–894.

Werr, W., Frommer, W.-B., Maas, C., and Starlinger, P. 1985. Structure of the sucrose synthase gene on chromosome 9 of *Zea mays* L. *EMBO J.* 4:1373–1380.

Weselake, R.J., Pomeroy, M.K., Furukawa, T.L., Golden, J.L., Little, D.B., and Laroche, A. 1993. Developmental profile of diacylglycerol acyltransferase in maturing seeds of oilseed rape and safflower and microspore-derived cultures of oilseed rape. *Plant Physiol.* 102:565–571.

Weselake, R.J., Taylor, D.C., Pomeroy, M.K., Lawson, S.L., and Underhill, E.W. 1991. Properties of diacylglycerol acyltransferase from microspore-derived embryos of *Brassica napus. Phytochemistry* 30:3533–3538.

Wessler, S.R., and Varagona, M.J. 1985. Molecular basis of mutations at the waxy locus of maize: correlation with the fine structure genetic map. *Proc. Natl Acad. Sci. USA* 82:4177–4181.

West, M.A.L, Yee, K.M., Danao, J., Zimmerman, J.L., Fischer, R.L., Goldberg, R.B., and Harada, J.J. 1994. *LEAFY COTYLEDON1* is an essential regulator of late embryogenesis and cotyledon identity in *Arabidopsis. Plant Cell* 6:1731–1745.

Weterings, K., Reijnen, W., van Aarssen, R., Korstee, A., Spijkers, J., van Herpen, M., Schrauwen, J., and Wullems, G. 1992. Characterization of a pollen-specific cDNA clone from *Nicotiana tabacum* expressed during microgametogenesis and germination. *Plant Mol. Biol.* 18:1101–1111.

Weterings, K., Reijnen, W., Wijn, G., van de Heuvel, K., Appeldoorn, N., de Kort, G., van Herpen, M., Schrauwen, J., and Wullems, G. 1995. Molecular characterization of the pollen-specific genomic clone NTPg303 and *in situ* localization of expression. *Sex. Plant Reprod.* 8:11–17.

Wetherell, D.F., and Dougall, D.K. 1976. Sources of nitrogen supporting growth and embryogenesis in cultured wild carrot tissue. *Physiol. Plant.* 37:97–103.

Wetherell, D.F., and Halperin, W. 1963. Embryos derived from callus tissue cultures of the wild carrot. *Nature* 200:1336–1337.

Wetzel, C.L.R., and Jensen, W.A. 1992. Studies of pollen maturation in cotton: the storage reserve accumulation phase. *Sex. Plant Reprod.* 5:117–127.

Wever, G.H., and Takats, S.T. 1971. Isolation and separation of S-competent and S-incompetent nuclei from *Tradescantia* pollen grains. *Expt. Cell Res.* 69:29–32.

Weymann, K., Urban, K., Ellis, D.M., Novitzky, R., Dunder, E., Jayne, S., and Pace, G. 1993. Isolation of transgenic progeny of maize by embryo rescue under selective conditions. *In Vitro Cell Devel Biol.* 29P:33–37.

Wheatley, W.G., Marsolais, A.A., and Kasha, K.J. 1986. Microspore growth and anther staging in barley anther culture. *Plant Cell Rep.* 5:47–49.

Wheeler, C.T., and Boulter, D. 1967. Nucleic acids of developing seeds of *Vicia faba* L. *J. Expt. Bot.* 18:229–240.

Whipple, A.P., and Mascarenhas, J.P. 1978. Lipid synthesis in germinating *Tradescantia* pollen. *Phytochemistry* 17:1273–1274.

White, D.W.R., and Williams, E. 1976. Early seed development after crossing of *Trifolium semipilosum* and *T. repens. New Zealand J. Bot.* 14:161–168.

White, O.E. 1914. Studies of teratological phenomena in their relation to evolution and the problems of heredity. I. A study of certain floral abnormalities in *Nicotiana* and their bearing on theories of dominance. *Am. J. Bot.* 1:23–36.

Whitmore, F.W., and Kriebel, H.B. 1987. Expression of a gene in *Pinus strobus* ovules associated with fertilization and early embryo development. *Can. J. For. Res.* 17:408–412.

Wiberg, E., Råhlen, L., Hellman, M., Tillberg, E., Glimelius, K., and Stymne, S. 1991. The microspore-derived embryo of *Brassica napus* L. as a tool for studying embryo-specific lipid biogenesis and regulation of oil quality. *Theor. Appl. Genet.* 82:515–520.

Wienand, U., Langridge, P., and Feix, G. 1981. Isolation and characterization of a genomic sequence of maize coding for a zein gene. *Mol. Gen. Genet.* 182:440–444.

Wilde, H.D., Nelson, W.S., Booij, H., de Vries, S.C., and Thomas, T.L. 1988. Gene-expression programs in embryogenic and non-embryogenic carrot cultures. *Planta* 176:205–211.

Wilen, R.W., Mandel, R.M., Pharis, R.P., Holbrook, L.A., and Moloney, M.M. 1990. Effects of abscisic acid and high osmoticum on storage protein gene expression in microspore embryos of *Brassica napus. Plant Physiol.* 94:875–881.

Wilen, R.W., van Rooijen, G.J.H., Pearce, D.W., Pharis, R.P., Holbrook, L.A., and Moloney, M.M. 1991. Effects of jasmonic acid on embryo-specific processes in *Brassica* and *Linum* oilseeds. *Plant Physiol.* 95:399–405.

Wilkins, T.A., and Raikhel, N.V. 1989. Expression of rice lectin is governed by two temporally and spatially regulated mRNAs in developing embryos. *Plant Cell* 1:541–549.

Willcox, M.C., Reed, S.C., Burns, J.A., and Wynne, J.C. 1991. Effect of microspore stage and media on anther culture of peanut (*Arachis hypogaea* L.). *Plant Cell Tissue Organ Cult.* 24:25–28.

Willemse, M.T.M. 1972. Morphological and quantitative changes in the population of cell organelles during microsporogenesis of *Gasteria verrucosa. Acta Bot. Neerl.* 21:17–31.

Willemse, M.T.M. 1981. Polarity during megasporogenesis and megagametogenesis. *Phytomorphology* 31:124–134.

Willemse, M.T.M., and Bednara, J. 1979. Polarity during megasporogenesis in *Gasteria verrucosa. Phytomorphology* 29:156–165.

Willemse, M.T.M., and de Boer-de Jeu, M.J. 1981. Megasporogenesis and early megagametogenesis. *Acta Soc. Bot. Polon.* 50:111–120.

Willemse, M.T.M., and Franssen-Verheijen, M.A.W. 1986. Stylar development in the open flower of *Gasteria verrucosa* (Mill.) H. Duval. *Acta Bot. Neerl.* 35:297–309.

Willemse, M.T.M., and Kapil, R.N. 1981a. Antipodals of *Gasteria verrucosa* (Liliaceae) - An ultrastructural study. *Acta Bot. Neerl.* 30:25–32.

Willemse, M.T.M., and Kapil, R.N. 1981b. On some anomalies during megasporogenesis in *Gasteria verrucosa* (Liliaceae). *Acta Bot. Neerl.* 30:439–447.

Willemse, M.T.M., Plyushch, T.A., and Reinders, M.C. 1995. *In vitro* micropylar penetration of the pollen tube in the ovule of *Gasteria verrucosa* (Mill.) H. Duval and *Lilium longiflorum* Thunb.: conditions, attraction and application. *Plant Sci.* 108:201–208.

Willemse, M.T.M., and Reznickova, S.A. 1980. Formation of pollen in the anther of *Lilium*. I. Development of the pollen wall. *Acta Bot. Neerl.* 29:127–140.

Willemse, M.T.M., and van Lammeren, A.A.M. 1988. Structure and function of the microtubular cytoskeleton during megasporogenesis and embryo sac development in *Gasteria verrucosa* (Mill.) H. Duval. *Sex. Plant Reprod.* 1:74–82.

Willemse, M.T.M., and van Went, J.L. 1984. The female gametophyte. In *Embryology of Angiosperms*, ed. B.M. Johri, pp. 159–196. Berlin: Springer-Verlag.

Willemse, M.T.M., and Vletter, A. 1995. Appearance and interaction of pollen and pistil pathway proteins in *Gasteria verrucosa* (Mill.) H. Duval. *Sex. Plant Reprod.* 8:161–167.

Williams, B., and Tsang, A. 1991. A maize gene expressed during embryogenesis is abscisic acid-inducible and highly conserved. *Plant Mol. Biol.* 16:919–923.

Williams, E., and Heslop-Harrison, J. 1975. Membrane-bounded cisternae associated with the chromosomes at first meiotic metaphase in microsporocytes of *Rhoeo spathacea. Protoplasma* 86:285–289.

Williams, E., Heslop-Harrison, J., and Dickinson, H.G. 1973. The activity of the nucleolus organising region and the origin of cytoplasmic nucleoloids in meiocytes of *Lilium. Protoplasma* 77:79–93.

Williams, E., and White, D.W.R. 1976. Early seed development after crossing of *Trifolium ambiguum* and *T. repens. New Zealand J. Bot.* 14:307–314.

Williams, E.G. 1987. Interspecific hybridization in pasture legumes. *Plant Breed. Rev.* 5:237–305.

Williams, E.G., Clarke, A.E., and Knox, R.B., ed. 1994. *Genetic Control of Self-Incompatibility and Reproductive Development in Flowering Plants*. Dordrecht: Kluwer Academic Publishers.

Williams, E.G., and Heslop-Harrison, J. 1979. A comparison of RNA synthetic activity in the plasmodial and secretory types of tapetum during the meiotic interval. *Phytomorphology* 29:370–381.

Williams, E.G., Knox, R.B., and Rouse, J.L. 1982. Pollination sub-systems distinguished by pollen tube arrest after incompatible interspecific crosses in *Rhododendron* (Ericaceae). *J. Cell Sci.* 53:255–277.

Williams, E.G., Knox, R.B., Kaul, V., and Rouse, J.L. 1984. Post-pollination callose development in ovules of *Rhododendron* and *Ledum* (Ericaceae): zygote special wall. *J. Cell Sci.* 69:127–135.

Williams, E.G., Ramm-Anderson, S., Dumas, C., Mau, S.L., and Clarke, A.E. 1982. The effect of isolated components of *Prunus avium* L. styles on *in vitro* growth of pollen tubes. *Planta* 156:517–519.

Williamson, J.D., and Quatrano, R.S. 1988. ABA-regulation of two classes of embryo-specific sequences in mature wheat embryos. *Plant Physiol.* 86:208–215.

Williamson, J.D., Quatrano, R.S., and Cuming, A.C. 1985. E_m polypeptide and its messenger RNA levels are modulated by abscisic acid during embryogenesis in wheat. *Eur. J. Biochem.* 152:501–507.

Williamson, J.D., and Scandalios, J.G. 1992. Differential response of maize catalases to abscisic acid: Vp1 transcriptional activator is not required for abscisic acid-regulated *Cat1* expression. *Proc. Natl Acad. Sci. USA* 89:8842–8846.

Williamson, J.D., Galili, G., Larkins, B.A., and Gelvin, S.B. 1988. The synthesis of a 19 kilodalton zein protein in transgenic *Petunia* plants. *Plant Physiol.* 88:1002–1007.

Willing, R.P., Bashe, D., and Mascarenhas, J.P. 1988. An analysis of the quantity and diversity of messenger RNAs from pollen and shoots of *Zea mays. Theor. Appl. Genet.* 75:751–753.

Willing, R.P., and Mascarenhas, J.P. 1984. Analysis of the complexity and diversity of mRNAs from pollen and shoots of *Tradescantia. Plant Physiol.* 75:865–868.

Willing, R.R., and Pryor, L.D. 1976. Interspecific hybridisation in poplar. *Theor. Appl. Genet.* 47:141–151.

Wilms, H.J. 1980a. Ultrastructure of the stigma and style of spinach in relation to pollen germination and pollen tube growth. *Acta Bot. Neerl.* 29:33–47.

Wilms, H.J. 1980b. Development and composition of the spinach ovule. *Acta Bot. Neerl.* 29:243–260.

Wilms, H.J. 1981a. Ultrastructure of the developing embryo sac of spinach. *Acta Bot. Neerl.* 30:75–99.

Wilms, H.J. 1981b. Pollen tube penetration and fertilization in spinach. *Acta Bot. Neerl.* 30:101–122.

Wilms, H.J., Leferink-ten Klooster, H.B., and van Aelst, A.C. 1986. Isolation of spinach sperm cells: 1: Ultrastructure and three-dimensional construction in the mature pollen grain. In *Biotechnology and Ecology of Pollen*, ed. D.L. Mulcahy, G.B. Mulcahy and E. Ottaviano, pp. 307–312. New York: Springer-Verlag.

Wilson, D.R., and Larkins, B.A. 1984. Zein gene organization in maize and related grasses. *J. Mol. Evol.* 20:330–340.

Wilson, G.F., Rhodes, A.M., and Dickinson, D.B. 1973. Some physiological effects of viviparous genes vp_1 and vp_5 on developing maize kernels. *Plant Physiol.* 52:350–356.

Wilson, H.J., Israel, H.W., and Steward, F.C. 1974. Morphogenesis and the fine structure of cultured carrot cells. *J. Cell Sci.* 15:57–73.

Wilson, H.M. 1977. Culture of whole barley spikes stimulates high frequencies of pollen calluses in individual anthers. *Plant Sci. Lett.* 9:233–238.

Wilson, H.M., Mix, G., and Foroughi-Wehr, B. 1978. Early microspore divisions and subsequent formation of microspore calluses at high frequency in anthers of *Hordeum vulgare* L. *J. Expt. Bot.* 29:227–238.

Wing, R.A., Yamaguchi, J., Larabell, S.K., Ursin, V.M., and McCormick, S. 1989. Molecular and genetic characterization of two pollen-expressed genes that have sequence similarity to pectate lyases of the plant pathogen *Erwinia. Plant Mol. Biol.* 14:17–28.

Wintz, H., Chen, H.-C., Sutton, C.A., Conley, C.A., Cobb, A., Ruth, D., and Hanson, M.R. 1995. Expression of the CMS-associated *urfS* sequence in transgenic petunia and tobacco. *Plant Mol. Biol.* 28:83–92.

Wise, R.P., Fliss, A.E., Pring, D.R., and Gengenbach, B.G. 1987. *urf13-T* of T cytoplasm maize mitochondria encodes a 13 kD polypeptide. *Plant Mol. Biol.* 9:121–126.

Wise, R.P., Pring, D.R., and Gengenbach, B.G. 1987. Mutation to male fertility and toxin insensitivity in Texas (T)-cytoplasm maize is associated with a frameshift in a mitochondrial open reading frame. *Proc. Natl Acad. Sci. USA* 84:2858–2862.

Wobus, U., Bäumlein, H., Bassüner, R., Heim, U., Jung, R., Müntz, K., Saalbach, G., and Weschke, W. 1986. Characteristics of two types of legumin genes in the field bean (*Vicia faba* L. var. minor) genome as revealed by cDNA analysis. *FEBS Lett.* 210:74–80.

Wochok, Z.S. 1973a. Microtubules and multivesicular bodies in cultured tissues of wild carrot: changes during transition from the undifferentiated to the embryonic condition. *Cytobios* 7:87–95.

Wochok, Z.S. 1973b. DNA synthesis during development of adventive embryos of wild carrot. *Biol. Plant.* 15:107–111.

Wochok, Z.S., and Wetherell, D.F. 1972. Restoration of declining morphogenetic capacity in long term tissue cultures of *Daucus carota* by kinetin. *Experientia* 28:104–105.

Wodehouse, R.P. 1935. *Pollen Grains, Their Structure, Identification and Significance in Science and Medicine.* New York: McGraw-Hill Book Co.

Woittiez, R.D., and Willemse, M.T.M. 1979. Sticking of pollen on stigmas: the factors and a model. *Phytomorphology* 29:57–63.

Wong, K.C., Watanabe, M., and Hinata, K. 1994a. Protein profiles in pin and thrum floral organs of distylous *Averrhoa carambola* L. *Sex. Plant Reprod.* 7:107–115.

Wong, K.C., Watanabe, M., and Hinata, K. 1994b. Fluorescence and scanning electron microscopic study on self-incompatibility in distylous *Averrhoa carambola* L. *Sex. Plant Reprod.* 7:116–121.

Woodard, J.W. 1956. DNA in gametogenesis and embryogeny in *Tradescantia. J. Biophys. Biochem. Cytol.* 2:765–776.

Woodard, J.W. 1958. Intracellular amounts of nucleic acids and protein during pollen grain growth in *Tradescantia. J. Biophys. Biochem. Cytol.* 4:383–390.

Woodcock, C.L.F., and Bell, P.R. 1968a. The distribution of deoxyribonucleic acid in the female gametophyte of *Myosurus minimus. Histochemie* 12:289–301.

Woodcock, C.L.F., and Bell, P.R. 1968b. Features of the ultrastructure of the female gametophyte of *Myosurus minimus. J. Ultrastr. Res.* 22:546–563.

Woodward, J.R., Bacic, A., Jahnen, W., and Clarke, A.E. 1989. *N*-linked glycan chains on *S*-allele-associated glycoproteins from *Nicotiana alata. Plant Cell* 1:511–514.

Woodward, J.R., Craik, D., Dell, A., Khoo, K.-H., Munro, S.L.A., Clarke, A.E., and Bacic, A. 1992. Structural analysis of the N-linked glycan chains from a stylar glycoprotein associated with expression of self-incompatibility in *Nicotiana alata. Glycobiology* 2:241–250.

Worrall, D., Hird, D.L., Hodge, R., Paul, W., Draper, J., and Scott, R. 1992. Premature dissolution of the microsporocyte callose wall causes male sterility in transgenic tobacco. *Plant Cell* 4:759–771.

Wright, D.J., and Boulter, D. 1972. The characterisation of vicilin during seed development in *Vicia faba* (L.). *Planta* 105:60–65.

Wright, S.Y., Suner, M.-M., Bell, P.J., Vaudin, M., and Greenland, A.J. 1993. Isolation and characterization of male flower cDNAs from maize. *Plant J.* 3:41–49.

Wróbel, B., and Bednarska, E. 1994. Nuclear DNA content and ultrastructure of secretory cells of *Vicia faba* L. stigma. *Acta Soc. Bot. Polon.* 63:139–145.

Wu, F.S., and Murry, L.E. 1985a. Changes in protein and amino acid content during anther development in fertile and cytoplasmic male sterile *Petunia. Theor. Appl. Genet.* 71:68–73.

Wu, F.S., and Murry, L.E. 1985b. Proteolytic activity in anther extracts of fertile and cytoplasmic male sterile *Petunia. Plant Physiol.* 79:301–305.

Wu, H.-M., Wang, H., and Cheung, A.Y. 1995. A pollen tube growth stimulatory glycoprotein is deglycosylated by pollen tubes and displays a glycosylation gradient in the flower. *Cell* 82:395–403.

Wu, H.-M., Zou, J., May, B., Gu, Q., and Cheung, A.Y. 1993. A tobacco gene family for flower cell wall proteins with a proline-rich domain and a cysteine-rich domain. *Proc. Natl Acad. Sci. USA* 90:6829–6833.

Wu, S.-C., Bögre, L., Vincze, É., Kiss, G.B., and Dudits, D. 1988. Isolation of an alfalfa histone H3 gene: structure and expression. *Plant Mol. Biol.* 11:641–649.

Wu, X.-L., and Zhou, C. 1990. Nuclear divisions of isolated, *in vitro* cultured generative cells in *Hemerocallis minor* Mill. *Acta Bot. Sinica* 32:577–581.

Wu, X.-L., and Zhou, C. 1991a. A comparative study on methods for isolation of generative cells in various angiosperm species. *Acta Biol. Expt. Sinica* 24:15–23.

Wu, X.-L., and Zhou, C. 1991b. Fusion experiments of isolated generative cells in several angiosperm species. *Acta Bot. Sinica* 33:897–904.

Wu, Y., and Zhou, C. 1988. Observation on enzymatically isolated-embryo sacs in *Adenophora axilliflora*: fertilization, zygote and endosperm. *Acta Bot. Sinica* 30:210–212.

Wu, Y., and Zhou, C. 1990. An ultrastructural study on pollen protoplasts and pollen tubes germinated from them in *Gladiolus gandavensis. Acta Bot. Sinica* 32:493–498.

Wu, Y., and Zhou, C. 1992. An ultrastructural study on triggering of cell division in young pollen protoplast culture of *Hemerocallis fulva* L. *Acta Bot. Sinica* 34:20–25.

Wu, Y., Zhou, C., and Koop, H.-U. 1993. Enzymatic isolation of viable nucelli at the megaspore mother cell stage and in developing embryo sacs in *Nicotiana tabacum*. *Sex. Plant Reprod.* 6:171–175.

Wu, Y., Haberland, G., Zhou, C., and Koop, H.-U. 1992. Somatic embryogenesis, formation of morphogenetic callus and normal development in zygotic embryos of *Arabidopsis thaliana in vitro*. *Protoplasma* 169:89–96.

Wulff, H.D. 1935. Galvanotropismus bei Pollenschläuchen. *Planta* 24:602–608.

Wunderlich, R. 1954. Über das Antherentapetum mit besonderer Berücksichtigung seiner Kernzahl. *Österr. Z. Bot.* 101:1–63.

Wurtele, E.S., and Nikolau, B.J. 1992. Differential accumulation of biotin enzymes during carrot somatic embryogenesis. *Plant Physiol.* 99:1699–1703.

Wurtele, E.S., Wang, H., Durgerian, S., Nikolau, B.J., and Ulrich, T.H. 1993. Characterization of a gene that is expressed early in somatic embryogenesis of *Daucus carota*. *Plant Physiol.* 102:303–312.

Xiao, C.-M., and Mascarenhas, J.P. 1985. High temperature-induced thermotolerance in pollen tubes of *Tradescantia* and heat-shock proteins. *Plant Physiol.* 78:887–890.

Xu, B., Grun, P., Kheyr-Pour, A., and Kao, T.-H. 1990a. Identification of pistil-specific proteins associated with three self-incompatibility alleles in *Solanum chacoense*. *Sex. Plant Reprod.* 3:54–60.

Xu, B., Mu, J., Nevins, D.L., Grun, P., and Kao, T.-H. 1990b. Cloning and sequencing of cDNAs encoding two self-incompatibility associated proteins in *Solanum chacoense*. *Mol. Gen. Genet.* 224:341–346.

Xu, F., Yao, Y.-X., Fu, C.-S., and Li, Z.-L. 1993. Studies on the development of pollen wall in *Pyrus betulaefolia* Bge. (Rosaceae). *Chinese J. Bot.* 5:41–46.

Xu, H., Davies, S.P., Kwan, B.Y.H., O'Brien, A.P., Singh, M., and Knox, R.B. 1993. Haploid and diploid expression of a *Brassica campestris* anther-specific gene promoter in *Arabidopsis* and tobacco. *Mol. Gen. Genet.* 239:58–65.

Xu, H., Knox, R.B., Taylor, P.E., and Singh, M.B. 1995a. *Bcp1*, a gene required for male fertility in *Arabidopsis*. *Proc. Natl Acad. Sci. USA* 92:2106–2110.

Xu, H., Theerakulpisut, P., Taylor, P.E., Knox, R.B., Singh, M.B., and Bhalla, P.L. 1995b. Isolation of a gene preferentially expressed in mature anthers of rice (*Oryza sativa* L.). *Protoplasma* 187:127–131.

Xu, N., and Bewley, J.D. 1991. Sensitivity to abscisic acid and osmoticum changes during embryogenesis of alfalfa (*Medicago sativa*). *J. Expt. Bot.* 42:821–826.

Xu, N., and Bewley, J.D. 1995a. Embryos of alfalfa (*Medicago sativa* L.) can synthesize storage proteins after the completion of germination, in response to abscisic acid and osmoticum, if maturation drying is prevented. *Planta* 196:469–476.

Xu, N., and Bewley, J.D. 1995b. Temporal and nutritional factors modulate responses to abscisic acid and osmoticum in their regulation of storage protein synthesis in developing seeds of alfalfa (*Medicago sativa* L.). *J. Expt. Bot.* 46:675–686.

Xu, N., and Bewley, J.D. 1995c. The role of abscisic acid in germination, storage protein synthesis and desiccation tolerance in alfalfa (*Medicago sativa* L.) seeds, as shown by inhibition of its synthesis by fluridone during development. *J. Expt. Bot.* 46:687–694.

Xu, N., Coulter, K.M., and Bewley, J.D. 1990. Abscisic acid and osmoticum prevent germination of developing alfalfa embryos, but only osmoticum maintains the synthesis of developmental proteins. *Planta* 182:382–390.

Xu, N., Coulter, K.M., Krochko, J.E., and Bewley, J.D. 1991. Morphological stages and storage protein accumulation in developing alfalfa (*Medicago sativa* L.) seeds. *Seed Sci. Res.* 1:119–125.

Xu, S.-X. (S.Y. Zee). 1991. Changes in microtubule cytoskeleton of generative cells during mitosis and cytokinesis. *Chinese J. Bot.* 3:77–80.

Xu, S.-X. (S.Y. Zee), Zhu, C., and Hu, S.-Y. 1990. Changes in the organization of microtubules during generative cell division and sperm formation in *Lilium*. *Acta Bot. Sinica* 32:821–826.

Xu, S.-X. (S.Y. Zee), Zhu, C., and Hu, S.-Y. 1993. Microtubule-cytoskeleton changes during generative cell division and sperm formation in the pollen tubes of *Zantedeschia aethiopica*. *Chinese J. Bot.* 5:18–22.

Xu, X., and Li, B. 1994. Fertile transgenic indica rice plants obtained by electroporation of the seed embryo cells. *Plant Cell Rep.* 13:237–242.

Xu, Z.H., Huang, B., and Sunderland, N. 1981. Culture of barley anthers in conditioned media. *J. Expt. Bot.* 32:767–778.

Xue, J., Pihlgren, U., and Rask, L. 1993. Temporal, cell-specific, and tissue-preferential expression of myrosinase genes during embryo and seedling development in *Sinapis alba*. *Planta* 191:95–101.

Xue, Y., Carpenter, R., Dickinson, H.G., and Coen, E.S. 1996. Origin of allelic diversity in *Antirrhinum* S locus RNases. *Plant Cell* 8:805–814.

Xue, Y., Collin, S., Davies, D.R., and Thomas, C.M. 1994. Differential screening of mitochondrial cDNA libraries from male-fertile and cytoplasmic male-sterile sugar-beet reveals genome rearrangements at *atp6* and *atpA* loci. *Plant Mol. Biol.* 25:91–103.

Xue, Z.-T., Xu, M.-L., Shen, W., Zhuang, N.L., Hu, W.-M., and Shen, S.C. 1992. Characterization of a *Gy4* glycinin gene from soybean *Glycine max* cv. Forrest. *Plant Mol. Biol.* 18:897–908.

XuHan, X., and van Lammeren, A.A.M. 1993. Microtubular configurations during the cellularization of coenocytic endosperm in *Ranunculus sceleratus* L. *Sex. Plant Reprod.* 6:127–132.

XuHan, X., and van Lammeren, A.A.M. 1994. Microtubular configurations during endosperm development in *Phaseolus vulgaris*. *Can. J. Bot.* 72:1489–1495.

Yabe, N., Takahashi, T., and Komeda, Y. 1994. Analysis of tissue-specific expression of *Arabidopsis thaliana* HSP90-family gene *HSP81*. *Plant Cell Physiol.* 35:1207–1219.

Yadegari, R., de Paiva, G.R., Laux, T., Koltunow, A.M., Apuya, N., Zimmerman, J.L., Fischer, R.L., Harada, J.J., and Goldberg, R.B. 1994. Cell differentiation and morphogenesis are uncoupled in *Arabidopsis raspberry* embryos. *Plant Cell* 6:1713–1729.

Yaklich, R.W., and Herman, E.M. 1995. Protein storage vacuoles of soybean aleurone cells accumulate a unique glycoprotein. *Plant Sci.* 107:57–67.

Yaklich, R.W., Vigil, E.L., Erbe, E.F., and Wergin, W.P. 1992. The fine structure of aleurone cells in the soybean seed coat. *Protoplasma* 167:108–119.

Yakovlev, M.S., and Yoffe, M.D. 1957. On some peculiarities in the embryogeny of *Paeonia* L. *Phytomorphology* 7:74–82.

Yamada, Y. 1965. Studies on the histological and cytological changes in the tissues of pistil after pollination. *Jap. J. Bot.* 19:69–82.

Yamagata, H., and Tanaka, K. 1986. The site of synthesis and accumulation of rice storage proteins. *Plant Cell Physiol.* 27:135–145.

Yamagata, H., Sugimoto, T., Tanaka, K., and Kasai, Z. 1982. Biosynthesis of storage proteins in developing rice seeds. *Plant Physiol.* 70:1094–1100.

Yamagata, H., Tamura, K., Tanaka, K., and Kasai, Z. 1986. Cell-free synthesis of rice prolamin. *Plant Cell Physiol.* 27:1419–1422.

Yamaguchi, H., and Kakiuchi, H. 1983. Electrophoretic analysis of mitochondrial DNA from normal and male sterile cytoplasms in rice. *Jap. J. Genet.* 58:607–611.

Yamaguchi-Shinozaki, K., Mundy, J., and Chua, N.-H. 1989. Four tightly linked *rab* genes are differentially expressed in rice. *Plant Mol. Biol.* 14:29–39.

Yamaguchi-Shinozaki, K., Mino, M., Mundy, J., and Chua, N.-H. 1990. Analysis of an ABA-responsive rice gene promoter in transgenic tobacco. *Plant Mol. Biol.* 15:905–912.

Yamakawa, S., Shiba, H., Watanabe, M., Shiozawa, H., Takayama, S., Hinata, K., Isogai, A., and Suzuki, A. 1994. The sequences of S-glycoproteins involved in self-incompatibility of *Brassica campestris* and their distribution among Brassicaceae. *Biosci. Biotech. Biochem. Japan* 58:921–925.

Yamakawa, S., Watanabe, S., Hinata, K., Suzuki, A., and Isogai, A. 1995. The sequences of *S*-receptor kinases (SRK) involved in self-incompatibility and their homologies to S-locus glycoproteins of *Brassica campestris*. *Biosci. Biotech. Biochem. Japan* 59:161–162.

Yamakawa, S., Watanabe, M., Isogai, A., Takayama, S., Satoh, S., Hinata, K., and Suzuki, A. 1993. The cDNA sequence of NS_3-glycoprotein from *Brassica campestris* and its homology to related proteins. *Plant Cell Physiol.* 34:173–175.

Yamamoto, S., Nishihara, M., Morikawa, H., Yamauchi, D., and Minamikawa, T. 1995. Promoter analysis of seed storage protein genes from *Canavalia gladiata* D.C. *Plant Mol. Biol.* 27:729–741.

Yamauchi, D., and Minamikawa, T. 1986. *In vivo* studies on protein synthesis in developing seeds of *Canavalia gladiata* D.C. *Plant Cell Physiol.* 27:1033–1041.

Yamauchi, D., and Minamikawa, T. 1987. Synthesis of canavalin and concanavalin A in maturing *Canavalia gladiata* seeds. *Plant Cell Physiol.* 28:421–430.

Yamauchi, D., and Minamikawa, T. 1990. Structure of the gene encoding concanavalin A from *Canavalia gladiata* and its expression in *Escherichia coli* cells. *FEBS Lett.* 260:127–130.

Yamauchi, D., Nakamura, K., Asahi, T., and Minamikawa, T. 1988. cDNAs for canavalin and concanavalin A from *Canavalia gladiata* seeds. Nucleotide sequence of cDNA for canavalin and RNA blot analysis of canavalin and concanavalin A mRNAs in developing seeds. *Eur. J. Biochem.* 170:515–520.

Yamaura, A. 1933. Karyologische und embryologische Studien über einige *Bambus*-Arten. *Bot. Mag. Tokyo* 47:551–555.

Yan, H., Yang, H.-Y., and Jensen, W.A. 1989. An electron microscopic study on *in vitro* parthenogenesis in sunflower. *Sex. Plant Reprod.* 2:154–166.

Yan, H., Yang, H.-Y., and Jensen, W.A. 1991a. Ultrastructure of the micropyle and its relationship to pollen tube growth and synergid degeneration in sunflower. *Sex. Plant Reprod.* 4:166–175.

Yan, H., Yang, H.-Y., and Jensen, W.A. 1991b. Ultrastructure of the developing embryo sac of sunflower (*Helianthus annuus*) before and after fertilization. *Can. J. Bot.* 69:191-202.

Yan, L., Wang, X., Teng, X., Ma, Y., and Liu, G. 1986. Actin and myosin in pollens and their role in the growth of pollen tubes. *Chinese Sci. Bull.* 31:267–272.

Yang, C., Xing, L., and Zhai, Z. 1992. Intermediate filaments in higher plant cells and their assembly in a cell-free system. *Protoplasma* 171:44–54.

Yang, H., Zhou, C., Cai, D., Yan, H., Wu, Y., and Chen, X. 1986. *In vitro* culture of unfertilized ovules in *Helianthus annuus* L. In *Haploids of Higher Plants* In Vitro, ed. H. Hu and H. Yang, pp. 182–191. Beijing: China Academic Publishers.

Yang, H.-Y., and Zhou, C. 1979. Experimental researches on the two pathways of pollen development in *Oryza sativa* L. *Acta Bot. Sinica* 21:345–351.

Yang, H.Y., and Zhou, C. 1982. *In vitro* induction of haploid plants from unpollinated ovaries and ovules. *Theor. Appl. Genet.* 63:97–104.

Yang, H.-Y., and Zhou, C. 1984. Observations on megasporogenesis and megagametophyte development in *Paulownia* sp. and *Sesamum indicum* by enzymatic maceration technique. *Acta Bot. Sinica* 26:355–358.

Yang, H.-Y., and Zhou, C. 1989. Isolation of viable sperms from pollen of *Brassica napus, Zea mays* and *Secale cereale*. *Chinese J. Bot.* 1:80–84.

Yang, L.J., Barratt, D.H.P., Domoney, C., Hedley, C.L., and Wang, T.L. 1990. An analysis of seed development in *Pisum sativum*. X. Expression of storage protein genes in cultured embryos. *J. Expt. Bot.* 41:283–288.

Yano, M., Isono, Y., Satoh, H., and Omura, T. 1984. Gene analysis of sugary and shrunken mutants of rice, *Oryza sativa* L. *Jap. J. Breed.* 34:43–49.

Yano, M., Okuno, K., Kawakami, J., Satoh, H., and Omura, T. 1985. High amylose mutants of rice, *Oryza sativa* L. *Theor. Appl. Genet.* 69:253–257.

Yanofsky, M.F., Ma, H., Bowman, J.L., Drews, G.N., Feldmann, K.A., and Meyerowitz, E.M. 1990. The protein encoded by the *Arabidopsis* homeotic gene *agamous* resembles transcription factors. *Nature* 346:35–39.

Yao, J.-L., Wu, J.-H., Gleave, A.P., and Morris, B.A.M. 1996. Transformation of *Citrus* embryogenic cells using particle bombardment and production of transgenic embryos. *Plant Sci.* 113:175–183.

Yenofsky, R.L., Fine, M., and Liu, C. 1988. Isolation and characterization of a soybean (*Glycine max*) lipoxygenase-3 gene. *Mol. Gen. Genet.* 211:215–222.

Yeung, E.C. 1980. Embryogeny of *Phaseolus*: the role of the suspensor. *Z. Pflanzenphysiol.* 96:17–28.

Yeung, E.C., and Brown, D.C.W. 1982. The osmotic environment of developing embryos of *Phaseolus vulgaris*. *Z. Pflanzenphysiol.* 106:149–156.

Yeung, E.C., and Cavey, M.J. 1988. Cellular endosperm formation in *Phaseolus vulgaris*. I. Light and scanning electron microscopy. *Can. J. Bot.* 66:1209–1216.

Yeung, E.C., and Clutter, M.E. 1978. Embryogeny of *Phaseolus coccineus*: growth and microanatomy. *Protoplasma* 94:19–40.

Yeung, E.C., and Clutter, M.E. 1979. Embryogeny of *Phaseolus coccineus*: the ultrastructure and development of the suspensor. *Can. J. Bot.* 57:120–136.

Yeung, E.C., and Meinke, D.W. 1993. Embryogenesis in angiosperms: development of the suspensor. *Plant Cell* 5:1371–1381.

Yeung, E.C., and Sussex, I.M. 1979. Embryogeny of *Phaseolus coccineus*: the suspensor and the growth of the embryo-proper *in vitro*. *Z. Pflanzenphysiol.* 91:423–433.

Yeung, E.C., Thorpe, T.A., and Jensen, C.J. 1981. *In vitro* fertilization and embryo culture. In *Plant Tissue Culture. Methods and Applications in Agriculture*, ed. T.A. Thorpe, pp. 253–271. New York: Academic Press.

Yibrah, H.S., Manders, G., Clapham, D.H., and von Arnold, S. 1994. Biological factors affecting transient transformation in embryogenic suspension cultures of *Picea abies*. *J. Plant Physiol.* 144:472–478.

Yin, Y., and Beachy, R.N. 1995. The regulatory regions of the rice tungro bacilliform virus promoter and interacting nuclear factors in rice (*Oryza sativa* L.). *Plant J.* 7:969–980.

Ylstra, B., Busscher, J., Franken, J., Hollman, P.C.H., Mol, J.N.M., and van Tunen, A.J. 1994. Flavonols and fertilization in *Petunia hybrida*: localization and mode of action during pollen tube growth. *Plant J.* 6:201–212.

Ylstra, B., Touraev, A., Benito Moreno, R.M., Stöger, E., van Tunen, A.J., Vicente, O., Mol, J.N.M., and Heberle-Bors, E. 1992. Flavonols stimulate development, germination, and tube growth of tobacco pollen. *Plant Physiol.* 100:902–907.

Yokota, E., McDonald, A.R., Liu, B., Shimmen, T., and Palevitz, B.A. 1995. Localization of a 170 kDa myosin heavy chain in plant cells. *Protoplasma* 185:178–187.

Yoo, J., and Jung, G. 1995. DNA uptake by imbibition and expression of a foreign gene in rice. *Physiol. Plant.* 94:453–459.

You, R., and Jensen, W.A. 1985. Ultrastructural observations of the mature megagametophyte and the fertilization in wheat (*Triticum aestivum*). *Can. J. Bot.* 63:163–178.

Youle, R.J., and Huang, A.H.C. 1981. Occurrence of low molecular weight and high cysteine containing albumin storage proteins in oilseeds of diverse species. *Am. J. Bot.* 68:44–48.

Young, E.G., and Hanson, M.R. 1987. A fused mitochondrial gene associated with cytoplasmic male sterility is developmentally regulated. *Cell* 41:41–49.

Young, L.C.T., and Stanley, R.G. 1963. Incorporation of tritiated nucleosides thymidine, uridine and cytidine in nuclei of germinating pine pollen. *Nucleus* 6:83–90.

Yu, H.-S., Hu, S.-Y., and Russell, S.D. 1992. Sperm cells in pollen tubes of *Nicotiana tabacum* L.: three-dimensional reconstruction, cytoplasmic diminution, and quantitative cytology. *Protoplasma* 168:172–183.

Yu, H.-S., Hu, S.-Y., and Zhu, C. 1989. Ultrastructure of sperm cells and the male germ unit in pollen tubes of *Nicotiana tabacum*. *Protoplasma* 152:29–36.

Yu, H.-S., Huang, B.-Q., and Russell, S.D. 1994. Transmission of male cytoplasm during fertilization in *Nicotiana tabacum*. *Sex. Plant Reprod.* 7:313–323.

Yu, H.-S., and Russell, S.D. 1992. Male cytoplasmic diminution and male germ unit in young and mature pollen of *Cymbidium goeringii*: a 3-dimensional and quantitative study. *Sex. Plant Reprod.* 5:169–181.

Yu, H.-S., and Russell, S.D. 1993. Three-dimensional ultrastructure of generative cell mitosis in the pollen tube of *Nicotiana tabacum*. *Eur. J. Cell Biol.* 61:338–348.

Yu, H.-S., and Russell, S.D. 1994a. Populations of plastids and mitochondria during male reproductive cell maturation in *Nicotiana tabacum* L.: a cytological basis for occasional biparental inheritance. *Planta* 193:115–122.

Yu, H.-S., and Russell, S.D. 1994b. Male reproductive cell development in *Nicotiana tabacum*: male germ unit associations and quantitative cytology during sperm maturation. *Sex. Plant Reprod.* 7:324–332.

Yuffá, A.M., de García, E.G., and Nieto, M.S. 1994. Comparative study of protein electrophoretic patterns during embryogenesis in *Coffea arabica* cv Catimor. *Plant Cell Rep.* 13:197–202.

Yunes, J.A., Neto, G.C., da Silva, M.J., Leite, A., Ottoboni, L.M.M., and Arruda, P. 1994. The transcriptional activator opaque2 recognizes two different target sequences in the 22-kD-like α-prolamin gene. *Plant Cell* 6:237–249.

Zachgo, S., de Andrade Silva, E., Motte, P., Tröbner, W., Saedler, H., and Schwarz-Sommer, Z. 1995. Functional analysis of the *Antirrhinum* floral homeotic *DEFICIENS* gene *in vivo* and *in vitro* by using a temperature-sensitive mutant. *Development* 121:2861–2875.

Zaki, M., and Dickinson, H. 1995. Modification of cell development *in vitro*: the effect of colchicine on anther and isolated microspore culture in *Brassica napus*. *Plant Cell Tissue Organ Cult.* 40:255–270.

Zaki, M., and Kuijt, J. 1994. Ultrastructural studies on the embryo sac of *Viscum minimum*. II. Megagametogenesis. *Can. J. Bot.* 72:1613–1628.

Zaki, M., and Kuijt, J. 1995. Ultrastructural studies on the embryo sac of *Viscum minimum*. I. Megasporogenesis. *Protoplasma* 185:93–105.

Zaki, M.A.M. and Dickinson, H.G. 1990. Structural changes during the first divisions of embryos resulting from anther and free microspore culture in *Brassica napus*. *Protoplasma* 156:149–162.

Zaki, M.A.M., and Dickinson, H.G. 1991. Microspore-derived embryos in *Brassica*: the significance of division symmetry in pollen mitosis I to embryogenic development. *Sex. Plant Reprod.* 4:48–55.

Žárský, V., Říhová, L., and Tupý, J. 1987. Interference of pollen diffusible substances with peroxidase catalyzed reaction. *Plant Sci.* 52:29–32.

Žárský, V., Čapková, V., Hrabětová, E., and Tupý J. 1985. Protein changes during pollen development in *Nicotiana tabacum* L. *Biol. Plant.* 27:438–444.

Žárský, V., Garrido, D., Eller, N., Tupý, J., Vicente, O., Schöffl, F., and Heberle-Bors, E. 1995. The expression of a small heat shock gene is activated during induction of tobacco pollen embryogenesis by starvation. *Plant Cell Environ.* 18:139–147.

Žárský, V., Garrido, D., Říhová, L., Tupý, J., Vicente, O., and Heberle-Bors. E. 1992. Derepression of the cell cycle by starvation is involved in the induction of tobacco pollen embryogenesis. *Sex. Plant Reprod.* 5:189–194.

Zee, S.Y. 1992. Confocal laser scanning microscopy of microtubule organizational changes in isolated generative cells of *Allemanda neriifolia* during mitosis. *Sex. Plant Reprod.* 5:182–188.

Zee, S.Y., and Aziz-Un-Nisa. 1991. Mitosis and microtubule organizational changes in isolated generative cells of *Allemanda neriifolia*. *Sex. Plant Reprod.* 4:132–137.

Zee, S.-Y., and O'Brien, T.P. 1971. Aleurone transfer cells and other structural features of the spikelet of millet. *Aust. J. Biol. Sci.* 24:391–395.

Zee, S.-Y., and Siu, I.H.P. 1990. Studies on the ontogeny of the pollinium of a massulate orchid (*Peristylus spiranthes*). *Rev. Palaeobot. Palynol.* 64:159–164.

Zee, S.-Y., Wu, S.C., and Yue, S.B. 1979. Morphological and SDS-polyacrylamide gel electrophoretic studies of pro-embryoid formation in the petiole explants of Chinese celery. *Z. Pflanzenphysiol.* 95:397–403.

Zee, S.-Y., and Ye, X.L. 1995. Changes in the pattern of organization of the microtubular cytoskeleton during megasporogenesis in *Cymbidium sinense*. *Protoplasma* 185:170–177.

Zeijlemaker, F.C.J. 1956. Growth of pollen tubes *in vitro* and their reaction on potential differences. *Acta Bot. Neerl.* 5:179–186.

Zenkteler, M. 1962. Microsporogenesis and tapetal development in normal and male-sterile carrots (*Daucus carota*). *Am. J. Bot.* 49:341–348.

Zenkteler, M. 1967. Test-tube fertilization of ovules in *Melandrium album* Mill. with pollen grains of several species of the Caryophyllaceae family. *Experientia* 23:775.

Zenkteler, M. 1990. *In vitro* fertilization and wide hybridization in higher plants. *Crit. Rev. Plant Sci.* 9:267–297.

Zenkteler, M., Maheswaran, G., and Williams, E.G. 1987. *In vitro* placental pollination in *Brassica campestris* and *B. napus*. *J. Plant Physiol.* 128:245–250.

Zenkteler, M., Misiura, E., and Guzowska, I. 1975. Studies on obtaining hybrid embryos in test tubes. In *Form, Structure and Function in Plants*, ed. H.Y. Mohan Ram, J.J. Shaw and C.K. Shaw, pp. 180–187. Meerut: Sarita Prakashan.

Zenkteler, M., and Nitzsche, W. 1984. Wide hybridization experiments in cereals. *Theor. Appl. Genet.* 68:311–315.

Zenkteler, M., and Nitzsche, W. 1985. *In vitro* culture of ovules of *Triticum aestivum* at early stages of embryogenesis. *Plant Cell Rep.* 4:168–171.

Zhan, X.-Y., Wu, H.-M., and Cheung, A.Y. 1996. Nuclear male sterility induced by pollen-specific expression of a ribonuclease. *Sex. Plant Reprod.* 9:35–43.

Zhang, F., and Boston, R.S. 1992. Increases in binding protein (BiP) accompany changes in protein body morphology in three high-lysine mutants of maize. *Protoplasma* 171:142–152.

Zhang, G., Gifford, D.J., and Cass, D.D. 1993. RNA and protein synthesis in sperm cells isolated from *Zea mays* L. pollen. *Sex. Plant Reprod.* 6:239–243.

Zhang, G., Campenot, M.K., McGann, L.E., and Cass, D.D. 1992a. Flow cytometric characteristics of sperm cells isolated from pollen of *Zea mays* L. *Plant Physiol.* 99:54–59.

Zhang, G., Williams, C.M., Campenot, M.K., McGann, L.E., and Cass, D.D. 1992b. Improvement of longevity and viability of sperm cells isolated from pollen of *Zea mays* L. *Plant Physiol.* 100:47–53.

Zhang, G., Williams, C.M., Campenot, M.K., McGann, L.E., Cutler, A.J., and Cass, D.D. 1995. Effects of calcium, magnesium, potassium and boron on sperm cells isolated from pollen of *Zea mays* L. *Sex. Plant Reprod.* 8:113–122.

Zhang, H.-Q., Croes, A.F., and Linskens, H.F. 1982. Protein synthesis in germinating pollen of *Petunia*: role of proline. *Planta* 154:199–203.

Zhang, H.-Q., Bohdanowicz, J., Pierson, E.S., Li, Y.-Q., Tiezzi, A., and Cresti, M. 1995. Microtubular organization during asymmetrical division of the generative cell in *Gagea lutea*. *J. Plant Res.* 108:269–276.

Zhang, J.Z., Gomez-Pedrozo, M., Baden, C.S., and Harada, J.J. 1993. Two classes of isocitrate lyase genes are expressed during late embryogeny and postgermination in *Brassica napus* L. *Mol. Gen. Genet.* 238:177–184.

Zhang, J.Z., Laudencia-Chingcuanco, D.L., Comai, L., Li, M., and Harada, J.J. 1994. Isocitrase lyase and malate synthase genes from *Brassica napus* L. are active in pollen. *Plant Physiol.* 104:857–864.

Zhang, J.Z., Santes, C.M., Engel, M.L., Gasser, C.S., and Harada, J.J. 1996. DNA sequences that activate isocitrate lyase gene expression during late embryogenesis and during postgerminative growth. *Plant Physiol.* 110:1069–1079.

Zhang, S., Chen, L., Qu, R., Marmey, P., Beachy, R., and Fauquet, C. 1996. Regeneration of fertile transgenic indica (group 1) rice plants following microprojectile transformation of embryogenic suspension culture cells. *Plant Cell Rep.* 15:465–469.

Zhang, W., McElroy, D., and Wu, R. 1991. Analysis of rice *Act1* 5′ region activity in transgenic rice plants. *Plant Cell* 3:1155–1165.

Zhang, X.S., and O'Neill, S.D. 1993. Ovary and gametophyte development are coordinately regulated by auxin and ethylene following pollination. *Plant Cell* 5:403–418.

Zhao, J.-P., Simmonds, D.H., and Newcomb, W. 1996. Induction of embryogenesis with colchicine instead of heat in microspores of *Brassica napus* L. cv. Topas. *Planta* 198:433–439.

Zhao, Y., Leisy, D.J., and Okita, T.W. 1994. Tissue-specific expression and temporal regulation of the rice glutelin Gt3 gene are conferred by at least two spatially separated *cis*-regulatory elements. *Plant Mol. Biol.* 25:429–436.

Zheng, Y., He, M., Hao, S., and Huang, B. 1992. The ultrastructural evidence on the origin of protein bodies in the rough endoplasmic reticulum of developing cotyledons of soybean. *Ann. Bot.* 69:377–383.

Zheng, Y.-Z., He, M.-Y., Hu, A.-L., and Hao, S. 1990. Patterns of the transition of vacuoles into protein bodies in developing cotyledon cells of soybean. *Acta Bot. Sinica* 32:97–102.

Zheng, Z., Kawagoe, Y., Xiao, S., Li, Z., Okita, T., Hau, T.L., Lin, A., and Murai, N. 1993. 5′ distal and proximal *cis*-acting regulator elements are required for developmental control of a rice seed storage protein *glutelin* gene. *Plant J.* 4:357–366.

Zheng, Z., Sumi, K., Tanaka, K., and Murai, N. 1995. The bean seed storage protein β-phaseolin is synthesized, processed, and accumulated in the vacuolar type-II protein bodies of transgenic rice endosperm. *Plant Physiol.* 109:777–786.

Zhong, H., Bolyard, M.G., Srinivasan, C., and Sticklen, M.B. 1993. Transgenic plants of turfgrass (*Agrostis palustris* Huds.) from microprojectile bombardment of embryogenic callus. *Plant Cell Rep.* 13:1–6.

Zhong, H., Zhang, S., Warkentin, D., Sun, B., Wu, T., Wu, R., and Sticklen, M.B. 1996. Analysis of the functional activity of the 1.4-kb 5′-region of the rice actin

1 gene in stable transgenic plants of maize (*Zea mays* L.). *Plant Sci.* 116:73–84.

Zhou, C. 1987a. A study of fertilization events in living embryo sacs isolated from sunflower ovules. *Plant Sci.* 52:147–151.

Zhou, C. 1987b. Cell-biological studies on artificially isolated generative cells from angiosperm pollen. *Acta Bot. Sinica* 29:117–122.

Zhou, C. 1988. Isolation and purification of generative cells from fresh pollen of *Vicia faba* L. *Plant Cell Rep.* 7:107–110.

Zhou, C. 1989a. A study on isolation and culture of pollen protoplasts. *Plant Sci.* 59:101–108.

Zhou, C. 1989b. Cell divisions in pollen protoplast culture of *Hemerocallis fulva* L. *Plant Sci.* 62:229–235.

Zhou, C., and Wu, X.-L. 1990. Mass isolation and purification of generative cells from pollen grains. *Acta Bot. Sinica* 32:404–406.

Zhou, C., and Yang, H.-Y. 1980. Anther culture and androgenesis of *Hordeum vulgare* L. *Acta Bot. Sinica* 22:211–215.

Zhou, C., and Yang, H.-Y. 1982. Enzymatic isolation of embryo sacs in angiosperms: isolation and microscopical observation on fixed materials. *Acta Bot. Sinica* 24:403–407.

Zhou, C., and Yang, H.Y. 1985. Observations on enzymatically isolated, living and fixed embryo sacs in several angiosperm species. *Planta* 165:225–231.

Zhou, C., and Yang, H.Y. 1991. Microtubule changes during the development of generative cells in *Hippeastrum vittatum* pollen. *Sex. Plant Reprod.* 4:293–297.

Zhou, C., Yang, H.-Y., and Xu, S.-X. 1990. Fluorescence microscopic observations on actin filament distribution in corn pollen and *Gladiolus* pollen protoplasts. *Acta Bot. Sinica* 32:657–662.

Zhou, C., Zee, S.Y., and Yang, H.Y. 1990. Microtubule organization of *in situ* and isolated generative cells in *Zephyranthes grandiflora* Lindl. *Sex. Plant Reprod.* 3:213–218.

Zhou, C., Orndorff, K., Allen, R.D., and DeMaggio, A.E. 1986a. Direct observations on generative cells isolated from pollen grains of *Haemanthus katherinae* Baker. *Plant Cell Rep.* 5:306–309.

Zhou, C., Orndorff, K., Daghlian, C.P., and DeMaggio, A.E. 1988. Isolated generative cells in some angiosperms: a further study. *Sex. Plant Reprod.* 1:97–102.

Zhou, C., Yang, H., Tian, H., Liu, Z., and Yan, H. 1986b. *In vitro* culture of unpollinated ovaries in *Oryza sativa* L. In *Haploids of Higher Plants* In Vitro, ed. H. Hu and H. Yang, pp. 165–181. Beijing: China Academic Publishers.

Zhou, G.Y., Weng, J., Gong, Z., Zhen, Y., Yang, W., Shen, W., Wang, Z., Tao, Q., Huang, J., Qian, S., Lin, G., Ying, M., Xue, D., Hong, A., Xu, Y., Chen, S., and Duan, X. 1988. Molecular breeding of agriculture. A technique for introducing exogenous DNA into plants after self pollination. *Sci. Agr. Sinica* 21:1–6.

Zhou, J.-Y. 1980. Pollen dimorphism and its relation to the formation of pollen embryos in anther culture of wheat (*Triticum aestivum*). *Acta Bot. Sinica* 22:117–121.

Zhou, X., and Fan, Y.-L. 1993. The endosperm-specific expression of a rice prolamin chimeric gene in transgenic tobacco plants. *Transgen. Res.* 2:141–146.

Zhou, X., Han, Y., Yang, W., and Xi, T. 1992. Somatic embryogenesis and analysis of peroxidase in cultured lettuce (*Lactuca sativa* L.) cotyledons. *Ann. Bot.* 69:97–100.

Zhu, C., and Liu, B. 1990. Microtubules in generative and sperm cells of *Amaryllis* pollen tubes. *Chinese J. Bot.* 2:1–6.

Zhu, C., Hu, S., Xu, L., Li, X., and Shen, J. 1980. Ultrastructure of sperm cell in mature pollen grain of wheat. *Sci. Sinica* 23:371–379.

Zhu, J.-K., Damsz, B., Kononowicz, A.K., Bressan, R.A., and Hasegawa, P.M. 1994. A higher plant extracellular vitronectin-like adhesion protein is related to the translational elongation factor-1α. *Plant Cell* 6:393–404.

Zhu, Q., Doerner, P.W., and Lamb, C.J. 1993. Stress induction and developmental regulation of a rice chitinase promoter in transgenic tobacco. *Plant J.* 3:203–212.

Zhu, T., Ma, Z.-L., and Li, W.-T. 1990. Ultrastructural localization of adenosine triphosphatase activity in stigma cells of *Populus lasiocarpa* and its changes during development. *Acta Bot. Sinica* 32:91–96.

Zhu, T., Mogensen, H.L., and Smith, S.E. 1990. Generative cell composition and its relation to male plastid inheritance patterns in *Medicago sativa. Protoplasma* 158:66–72.

Zhu, T., Mogensen, H.L., and Smith, S.E. 1991. Quantitative cytology of the alfalfa generative cell and its relation to male plastid inheritance patterns in three genotypes. *Theor. Appl. Genet.* 81:21–26.

Zhu, T., Mogensen, H.L., and Smith, S.E. 1992. Heritable paternal cytoplasmic organelles in alfalfa sperm cells: ultrastructural reconstruction and quantitative cytology. *Eur. J. Cell Biol.* 59:211–218.

Zhu, T., Mogensen, H.L., and Smith, S.E. 1993. Quantitative, three-dimensional analysis of alfalfa egg cells in two genotypes: implications for biparental plastid inheritance. *Planta* 190:143–150.

Zhu, Z.-P., Shen, R.-J., and Tang, X.-H. 1980a. Studies on the developmental biology of embryogenesis in higher plants. II. Biochemical changes during embryogeny in *Oryza sativa* L. *Acta Phytophysiol. Sinica* 6:141–148.

Zhu, Z.-P., Shen, R.-J., and Tang, X.-H. 1980b. Studies on the developmental biology of embryogenesis in higher plants. III. Kinetic changes of nucleic acids and protein during embryogenesis of wheat (*Triticum vulgaris* L.). *Acta Bot. Sinica* 22:122–126.

Zhu, Z.-P., Shen, R.-J., and Tang, X.-H. 1989. The biosynthesis of storage protein and changes of free amino acids content during seed development of rice. *Chinese J. Bot.* 1:123–130.

Zhu, Z.-Q., Sun, J.-S., and Wang, J.-J. 1978. Cytological investigation on androgenesis of *Triticum aestivum. Acta Bot. Sinica* 20:6–12.

Zhukova, G.Y., and Sokolovskaya, T.B. 1977. Ultrastructure of antipodals of *Aconitum napellus* L. (Ranunculaceae) embryo sac before fertilization. *Bot. Zh.* 62:1600–1611.

Ziebur, N.K., Brink, R.A., Graf, L.H., and Stahmann, M.A. 1950. The effect of casein hydrolysate on the growth *in vitro* of immature *Hordeum* embryos. *Am. J. Bot.* 37:144–148.

Zimmerman, J.L. 1993. Somatic embryogenesis: a model for early development of higher plants. *Plant Cell* 5:1411–1423.

Zimmerman, J.L., Apuya, N., Darwish, K., and O'Carroll, C. 1989. Novel regulation of heat shock genes during carrot somatic embryo development. *Plant Cell* 1:1137–1146.

Zonia, L.E., and Tupý, J. 1995. Lithium treatment of *Nicotiana tabacum* microspores blocks polar nuclear migration, disrupts the partitioning of membrane-associated Ca^{2+}, and induces symmetrical mitosis. *Sex. Plant Reprod.* 8:152–160.

Zou, J., Abrams, G.D., Barton, D.L., Taylor, D.C., Pomeroy, M.K., and Abrams, S.R. 1995. Induction of lipid and oleosin biosynthesis by (+)-abscisic acid and its metabolites in microspore-derived embryos of *Brassica napus* L. cv. Reston. *Plant Physiol.* 108:563–571.

Zou, J.-T., Zhan, X.-Y., Wu, H.-M., Wang, H., and Cheung, A.Y. 1994. Characterization of a rice pollen-specific gene and its expression. *Am. J. Bot.* 81:552–561.

Zuberi, M.I., and Dickinson, H.G. 1985. Pollen–stigma interaction in *Brassica*. III. Hydration of the pollen grains. *J. Cell Sci.* 76:321–336.

Zúbková, M., and Sladký, Z. 1975. The possibility of obtaining seeds following placental pollination *in vitro*. *Biol. Plant.* 17:276–280.

zur Nieden, U., Manteuffel, R., Weber, E., and Neumann, D. 1984. Dictyosomes participate in the intracellular pathway of storage proteins in developing *Vicia faba* cotyledons. *Eur. J. Cell Biol.* 34:9–17.

zur Nieden, U., Neumann, D., Bucka, A., and Nover, L. 1995. Tissue-specific localization of heat-stress proteins during embryo development. *Planta* 196:530–538.

zur Nieden, U., Neumann, D., Manteuffel, R., and Weber, E. 1982. Electron microscopic immunocytochemical localization of storage proteins in *Vicia faba* seeds. *Eur. J. Cell Biol.* 26:228–233.

Index